									17 7A	18 0
			13 3A	14 4A	15 5A	16 6A		1 **H** 1.0079	2 **He** 4.00260
				5 **B** 10.81	6 **C** 12.011	7 **N** 14.0067	8 **O** 15.9994	9 **F** 18.9984	10 **Ne** 20.179
10	11 1B	12 2B	13 **Al** 26.9815	14 **Si** 28.0855	15 **P** 30.9738	16 **S** 32.06	17 **Cl** 35.453	18 **Ar** 39.948
28 **Ni** 58.69	29 **Cu** 63.546	30 **Zn** 65.39	31 **Ga** 69.72	32 **Ge** 72.59	33 **As** 74.9216	34 **Se** 78.96	35 **Br** 79.904	36 **Kr** 83.80
46 **Pd** 106.42	47 **Ag** 107.868	48 **Cd** 112.41	49 **In** 114.82	50 **Sn** 118.71	51 **Sb** 121.75	52 **Te** 127.60	53 **I** 126.905	54 **Xe** 131.29
78 **Pt** 195.08	79 **Au** 196.967	80 **Hg** 200.59	81 **Tl** 204.383	82 **Pb** 207.2	83 **Bi** 208.980	84 **Po** (209)	85 **At** (210)	86 **Rn** (222)
110 **Uun** (271)	111 **Uuu** (272)							

63 **Eu** 151.96	64 **Gd** 157.25	65 **Tb** 158.925	66 **Dy** 162.50	67 **Ho** 164.930	68 **Er** 167.26	69 **Tm** 168.934	70 **Yb** 173.04	71 **Lu** 174.967
95 **Am** (243)	96 **Cm** (247)	97 **Bk** (247)	98 **Cf** (251)	99 **Es** (254)	100 **Fm** (257)	101 **Md** (260)	102 **No** (259)	103 **Lr** (262)

See inside back cover for the names of the elements and their atomic masses.

Note: Atomic masses shown are the 1983 IUPAC values (maximum of six significant figures). Group numbers 1–18 in blue are the recommended notations.

Introduction to
General, Organic, and
Biological Chemistry

Introduction to General, Organic, and Biological Chemistry

Michael S. Matta
Antony C. Wilbraham
Dennis D. Staley

Southern Illinois University at Edwardsville

D. C. Heath and Company
Lexington, Massachusetts   Toronto

Address editorial correspondence to:
D. C. Heath and Company
125 Spring Street
Lexington, MA 02173

Acquisitions: Richard Stratton
Development: June Goldstein
Editorial Production: Ron Hampton
Design: Henry J. Rachlin
Photo Research: Toni Michaels
Art Editing: Penny Peters
Production Coordination: Charles Dutton

DEDICATION:

To Our Parents
M. S. M.
A. C. W.
D. D. S.

Published simultaneously in Canada.

Printed in the United States of America.

International Standard Book Number: 0-669-33309-3

Library of Congress Catalog Number: 95-68944

10 9 8 7 6 5 4 3 2

Preface

Introduction to General, Organic, and Biological Chemistry is intended as a survey of chemistry primarily for nursing and allied health students. In writing this book we were guided by two convictions: first, that by incorporating chemical applications, we could sustain student interest and demonstrate that chemistry is an absolute prerequisite to understanding how living organisms work; and second, that even though composed of diverse topics, chemistry can be rendered logical and coherent when presented against a background of simple principles.

Applications

Every chapter describes applications of chemical principles to the life sciences. The connection between chemistry and the health sciences is highlighted in a chapter-opening feature called *Case in Point*. The *Case in Point* aims to provide students with an incentive to learn the chapter material and gain an understanding of the reasons that the chemistry covered in the chapter is important in the health-related sciences. Each *Case in Point* presents a vignette related to the chapter's content. The chemical background needed to understand the chemistry of the *Case in Point* is provided in the chapter, then applied in a *Follow-up to the Case in Point*. Additional health-related applications and other subjects of interest are highlighted in 74 brief essays entitled *A Closer Look*. Many other health-related applications are embedded in text.

Coherent Presentation of Principles

Our primary aim in teaching general and organic chemistry is to prepare students to understand the chemistry of living cells. We have therefore paid particular attention in the general and organic sections of the book to the fundamental physical phenomena and chemical reactions needed to establish a firm foundation for the study of biochemistry. In the general chemistry section, we introduce scientific measurement, atomic and molecular structure, chemical bonding, stoichiometry, reaction equilibria and dynamics, states of matter, aqueous systems, acid-base chemistry, and nuclear chemistry. The organic chemistry section begins with the structures of hydrocarbons, extends to the reactions of important functional groups in simple molecules, and then branches out to describe the complex, naturally occurring molecules at the boundary between organic chemistry and biochemistry. We devote considerable space to the role played by weak forces in determining the physical properties of organic molecules. An understanding of this role in turn gives insight into the factors underlying the biologically active conformation of proteins, nucleic acids, and cell membranes.

In keeping with our pedagogical aims, the book is designed so that chemical themes particularly relevant to biochemistry recur strategically throughout the text. Chapter 6, for example, presents a general introduction to oxidation reactions that is further developed in an organic context in Chapter 13 and reexamined in a biological context in Chapter 23. A section introducing esters and anhydrides of phosphoric acid in Chapter 14 helps demystify discussions of nucleic acid structure in Chapter 20, the biological importance of ATP in Chapter 23, and the energy yields of metabolic processes in subsequent chapters. Similarly, buffers are introduced in Chapter 9, and their importance in maintaining the body's acid-base balance is stressed in Chapter 22, and they are treated further in discussions of lactic acidosis in Chapter 24 and diabetic ketosis in Chapter 25.

Problem Solving

Many sections contain worked *Examples* that show detailed solutions to problems. The text features approximately 150 worked examples. The chapters also contain nearly 300 *Practice Exercises* that give students on-the-spot feedback so that they can see whether they have grasped an important concept, learned an important skill, or should review the material. The answers to all *Practice Exercises* are given at end of the book. Approximately 1900 *Exercises, Additional Exercises,* and *Self-Test (Review)* questions at the end of the chapters provide more practice in mastering the material. The *Exercises* are grouped and keyed to the sections of the text, again providing an opportunity for students to identify problem areas. The first exercise in each of these sections is similar to one presented as an *Example* or *Practice Exercise* in the chapter. The *Additional Exercises* that follow the *Exercises* take a global view of the chapter, incorporating material from more than one section. Answers to selected *Exercises* and *Additional Exercises* are provided at the end of the book. Finally, each chapter contains a *Self-Test* that students can use to review personal progress in meeting learning goals.

Study Aids

A *Chapter Outline* at the beginning of each chapter itemizes the topics to be covered in the chapter. Major topics in each chapter are introduced by *numbered section headings.* Each section contains one or more learning objectives labeled as *Aims.* A *Focus* statement in the margin summarizes the contents of the section. These statements as a group constitute both a good chapter outline and an excellent checklist for review. Throughout the text, *Key Terms* are printed in bold type and are defined where they first appear. Key terms are grouped at the end of the chapter and are included (with definitions) in a *Glossary* at the back of the text, for convenient reference. Each chapter also features a *Summary,* which briefly integrates the chapter's major ideas. In addition, the text contains margin notes that include interesting applications of environmental chemistry, biological chemistry, and health-related chemistry, as well as problem-solving tips supplementary to the many found in the main text.

Supplementary Materials

Supplementary materials for *Introduction to General, Organic, and Biological Chemistry* include the following:

Study Guide   By Danny V. White of American River College and Joanne White. The chapters in the *Study Guide* provide additional explanations of important topics, abundant exercises for more practice in learning new concepts and skills, and self-tests.

Student Solutions Manual   By Michael S. Matta, Antony C. Wilbraham, and Dennis D. Staley. The *Student Solutions Manual* provides detailed solutions to all *Practice Exercises,* end-of-chapter odd-numbered *Exercises* and *Additional Exercises,* and annotated answers to the *Self-Tests.*

Flash Cards   A set of 200 cards including hundreds of key terms and important equations, reactions, and chemical structures.

Experiments for Introduction to General, Organic, and Biological Chemistry   By Michael S. Matta, Antony C. Wilbraham, and Dennis D. Staley. The laboratory manual contains 45 experiments so that instructors can choose those most relevant to their course. The order of experimental topics follows the topics in the text. Each experiment has an introduction, hypotheses, objectives, a list of required materials and equipment, detailed descriptions of procedures, prelaboratory exercises, and laboratory report sheets with postlaboratory exercises. Suggestions for demonstrations and structural studies are also included.

Complete Solutions Manual   By Michael S. Matta, Antony C. Wilbraham, and Dennis D. Staley. The *Complete Solutions Manual* supplies detailed solutions to all *Practice Exercises,* all end-of-chapter *Exercises* and *Additional Exercises,* as well as annotated answers to the *Self-Tests.*

Test Item File   By Michael S. Matta, Antony C. Wilbraham, and Dennis D. Staley. The print version of the *Test Item File* includes approximately 1500 multiple-choice questions arranged by chapter.

Computerized Test Item File   The computerized *Test Item File* is available for IBM or Macintosh computers.

Instructor's Guide   By Michael S. Matta, Antony C. Wilbraham, and Dennis D. Staley. The *Instructor's Guide* offers suggested lecture schedules for one-semester, two-semester, two-quarter, and three-quarter courses. Sequences of laboratory experiments keyed to the text are also provided, as well as references to audiovisual and multimedia materials.

Instructor's Guide to Accompany Experiments for Introduction to General, Organic, and Biological Chemistry.   By Michael S. Matta, Antony C. Wilbraham, and Dennis D. Staley. The instructor's laboratory guide provides procedures for the preparation of laboratory materials and solution and answers to the prelaboratory and postlaboratory questions. Material disposal guidelines and time estimates for completing each experiment are also included.

Transparencies A set of 76 full-color transparencies of selected illustrations and tables from the text is available for classroom use.

Acknowledgments

We are grateful for the enthusiasm and unflagging energy of the staff at D. C. Heath and Company. We thank June Goldstein, Development Editor, for the unstinting gift of her editorial talents and especially for her kindness and patience towards sometimes irascible authors; Ron Hampton, Production Editor, for his careful supervision of this project; Henry Rachlin, Designer, for his functional and attractive design; and Richard Stratton, Acquisitions Editor, for his support and faith in the project.

Others have made significant contributions to this book. Dick Morel made many suggestions that have been incorporated into the manuscript, figures, and photographs. Copy for several of the supplements was expertly prepared by John Matta of Moon Over Maine Graphics. Elizabeth and Leigh Anne Staley helped with many details of manuscript preparation.

We'd also like to thank the accuracy reviewers of the manuscript, galleys, pages, and supplements: Beverly Foote, Leland Harris, University of Arizona; Ruiess Van Fossen Bravo, Indiana University of Pennsylvania; John Goodenow, Lawrence Technological University; Robert Howell, University of Cincinnati; Catherine Keenan, Chaffey College; Thomas Nycz, Broward Community College; Roger Penn, Sinclair Community College; David B. Shaw, Madison Area Technical College; and Peggy Zitek.

Finally, we want to acknowledge the fine work of the reviewers of the manuscript: Beatrice Arnowich, Queensborough Community College; John Barbas, Valdosta State College; Kenneth F. Cerny, University of Massachusetts; Jack L. Dalton, Boise State University; John E. Davidson, Eastern Kentucky University; David V. Frank, Ferris State University; Leland Harris, University of Arizona; Robert G. Howell, University of Cincinnati–Raymond Walters College; David Hunter, Kaskaskia College; Philip M. Jaffe, Oakton Community College; Catherine A. Keenan, Chaffey College; Frank R. Milio, Towson State University; Kenneth E. Miller, Milwaukee Area Technical College; Frazier W. Nyasulu, University of Washington; Thomas J. Nycz, Broward Community College; Richard Peterson, Memphis State University; Edith M. Rand, East Carolina University; Theresa A. Salerno, Mankato State University; David Saltzman, Santa Fe Community College; James Schooler, Jr., North Carolina Central University; David B. Shaw, Madison Area Technical College; Danny V. White, American River College; Karen Wiechelman, University of Southwestern Louisiana; and James E. Wiedman, Kaskaskia College.

M.S.M.
A.C.W.
D.D.S.

Chapter Opener

The connection between chemistry and the health sciences is highlighted in a chapter-opening feature called Case in Point. The chemical principles needed to understand the chemistry of the Case in Point are presented in the chapter, then applied to a Follow-up to the Case in Point.

CHAPTER OUTLINE

CASE IN POINT: Magnetic resonance imaging

3.1 Atoms

A CLOSER LOOK: Images of Atoms

3.2 Subatomic particles

3.3 Atomic number and mass number

3.4 Isotopes and atomic mass

FOLLOW-UP TO THE CASE IN POINT: Magnetic resonance imaging

3.5 Electronic structure of atoms

A CLOSER LOOK: Electrons, Flames, and Fireworks

3.6 Electron configurations

We are already familiar with some of the properties of many forms of matter. Coal is a black solid that burns in air; water is a colorless, odorless, tasteless liquid; table salt is a crystalline solid with a characteristic taste; aspirin is a white solid with a sour taste. However, being able to identify some properties of matter is not the same as knowing what matter is made of. Much of nursing and the allied health fields is concerned with exerting control over matter. For example, should a patient receive more or less medication? What medication should be given? Because of the strong relationship between matter and almost everything that happens, anyone can benefit from knowing what matter is made of. This chapter briefly outlines some of what is known about the makeup of matter. We will see that matter is composed of only about 100 kinds of basic building blocks, called *atoms*. We also will learn that atoms are composed of only three kinds of particles that are even more fundamental than atoms. The arrangement of these particles in atoms is responsible for the infinite variety of matter in the world. This arrangement also leads to certain properties of matter that are valuable to the health sciences, as illustrated by the following Case in Point.

CASE IN POINT: Magnetic resonance imaging

 Jamal is a 19-year-old college sophomore. He is also a good student and a gifted athlete. Although he weighs only 165 pounds, Jamal is generally considered to be the best defensive player on his football team, where he anchors the backfield. Late in one of Jamal's games, the 230-pound fullback of the opposing team stormed toward the line of scrimmage. As Jamal rushed to close the hole in the line, he and the fullback collided head-on. Jamal lay motionless on the turf. When Jamal remained unconscious for a few minutes, the team physician decided he should be taken to the hospital. During the trip, Jamal began to regain consciousness, but he had no recollection of the game or the circumstances of his injury. Jamal's speech was slurred, and he complained of a headache and double vision. The emergency room physician found that Jamal showed no signs of paralysis or numbness in his arms or legs. He suspected that Jamal had suffered a concussion and ordered a scan of Jamal's brain by magnetic resonance imaging (MRI) (see figure). What is an MRI?

What property of atoms does the MRI exploit? What did Jamal's physician find? We will learn the answers to these questions in Section 3.4.

Magnetic resonance imaging (MRI) allows physicians to "see" inside the human body.

FOLLOW-UP TO THE CASE IN POINT: Magnetic resonance imaging

Jamal, the student-athlete introduced in the Case in Point, was knocked unconscious in a football game. From careful observation of Jamal's behavior upon regaining consciousness, the attending physician hypothesized that Jamal suffered a concussion, which in severe cases can be accompanied by a noticeable swelling of the brain. Although a concussion is a serious matter, patients usually recover completely. However, the doctor was concerned that a blood clot might have formed in Jamal's brain. This would be an even more serious matter than a concussion. In order to examine this possibility, the doctor ordered a magnetic resonance image (MRI) of Jamal's brain (see figure). MRI is a method that allows physicians to see inside the human body without surgery. MRI involves measurement of a phenomenon called *nuclear magnetic resonance*, a behavior exhibited by the nuclei of certain isotopes. Hydrogen-1 (^{1}H), phosphorus-31 (^{31}P), and carbon-13 (^{13}C) are among the stable isotopes that can be detected by MRI and that are reasonably abundant in the body. To conduct MRI, the subject's body is placed in a strong magnetic field and bombarded with radiowaves similar to those of an ordinary FM radio. As these radiowaves emerge from the body, they provide information about the location and environment of the isotopic nuclei. This information can be used by a computer to construct a three-dimensional image of the body's interior. The computer may assign different colors to the results of its calculations. These colors are helpful in making clear distinctions between tissues in the final images. In Jamal's case, the MRI was completely normal, with no evidence of a blood clot on the brain. After a few days' recovery, he returned to school. Jamal was advised by his physician to quit football.

MRI of a single patient would not be very useful. Diagnosis usually requires comparison of MRIs for normal subjects with those for patients who are ill. The difference between these MRIs is very important in diagnosing the disease. For example, comparison of an MRI of the brain of a normal subject with that of an ill one could reveal a small tumor that would not have been apparent if the physician had no idea of the appearance of a normal image.

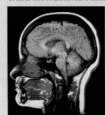

MRI of the brain. MRI is often used instead of X rays as a diagnostic tool because it does not involve ionizing radiation.

Closer Look Essays

Health-related and other interesting applications are explored in essays entitled A Closer Look. These special-interest boxes cover topics such as biomedical implants, cisplatin, food irradiation, the greenhouse effect, lactose intolerance, octane ratings of gasoline, DNA fingerprinting, and drug strategies for reducing serum cholesterol.

A Closer Look

Chlorofluorocarbons and the Ozone Layer

Halocarbon molecules that contain chlorine as well as fluorine are known as *chlorofluorocarbons*, or *CFCs*, or *Freons*.

Freon 11   Freon 12   Freon 13

CFCs are gases or low boiling liquids that are chemically inert, nontoxic, nonflammable, and insoluble in water. These properties made them good candidates for refrigerants in air conditioners and as propellants for aerosol cans of hair sprays, deodorants, and inhalation medications (see figure, part a).

Ozone is an important natural component of the stratosphere, the layer of the atmosphere that ranges from 11 to 48 km above the Earth. The ozone molecules shield the Earth's plants and animals from life-destroying ultraviolet radiation. When an ozone molecule in the stratosphere absorbs ultraviolet radiation, it is converted to an oxygen molecule (O_2) and an oxygen atom ($O\cdot$).

$$O_3(g) \xrightarrow[\text{radiation}]{\text{Ultraviolet}} O_2(g) + O\cdot(g)$$

The chemical inertness of CFCs allows them to remain in the environment for a long time. Eventually, they find their way into the stratosphere, where the carbon-chlorine bond in CFCs is broken by energy from ultraviolet light. The chlorine atom ($Cl\cdot$) that is produced combines with an ozone molecule in the stratosphere, to give a chlorine oxide radical ($ClO\cdot$) and an oxygen molecule.

$$Cl\cdot(g) + O_3(g) \longrightarrow ClO\cdot(g) + O_2(g)$$

The $ClO\cdot$ radical then reacts with an oxygen atom ($O\cdot$), formed when ozone absorbs ultraviolet light, to produce another chlorine atom and an oxygen molecule.

$$ClO\cdot(g) + O\cdot(g) \longrightarrow Cl\cdot(g) + O_2(g)$$

This process is repeated many times. It has been estimated that the breaking of a single C—Cl bond of a CFC molecule results in the destruction of 4000 or more ozone molecules in the stratosphere. The destruction of the ozone layer permits larger amounts of harmful ultraviolet radiation to reach the Earth. The effect of ozone depletion is manifested in an increased incidence of skin cancers and crop damage. In 1985, scientists discovered a "hole" in the ozone layer over Antarctica (see figure, part b). Following the concern that this discovery generated, the Montreal Protocol on Substances that Deplete the Ozone Layer went into effect in January 1989. It was signed by 24 nations. The protocol calls for CFCs to be phased out by the year 1996. In the meantime, alternatives are being sought.

One class of compounds that show promise as alternatives are the hydrofluorocarbons, or HFCs. They are possible substitutes for ozone-depleting CFCs because they contain no chlorine and therefore cannot catalyze ozone destruction. However, an air-conditioning system designed for CFC refrigerant will not operate on HFC refrigerants. Most new cars and trucks sold in the United States are now equipped with air-conditioning systems that use HFCs.

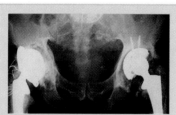

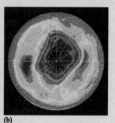

CFCs are used to deliver precise doses of medication directly into the lungs of asthmatics (a). The hole in the ozone layer over Antarctica is clearly visible in this NASA photograph (b).

(a)   (b)

361

A Closer Look

Biomedical Implants

People have been replacing various body parts since at least 300 B.C. Modern medicine has made great strides in developing materials for surgical implantation of artificial knees, hips, and other body parts. One of the first surgically implanted materials was surgical steel. Made of iron (67%), chromium (18%), nickel (12%), and molybdenum (3%), this alloy was used to make plates and screws for joining broken bones. Surgical steel has high mechanical strength and resists corrosion in the body. More recently, titanium- and cobalt-based alloys are being used as implant materials (see figure). These new alloys are less dense than surgical steel yet stronger. They have another advantage over surgical steel: They do not react chemically with living tissue. Interestingly, however, this lack of reactivity is sometimes a problem.

Hip joint replacement allows the patient to regain full movement of the leg.

Bone or tissue will not bond to the implant, making it difficult for the implant to make a strong attachment. We will see in A Closer Look: Bioactive Materials how it is possible to get implants to bond with living tissue and bone, heal rapidly, and form a mechanically strong structure.

Worked Examples
The text contains approximately 150 worked examples, each titled for easy reference. The examples include detailed solutions.

Atoms in reactants = Atoms in products

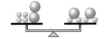

2 molecules of hydrogen   2 molecules of water
1 molecule of oxygen

$$2H_2 + O_2 \longrightarrow 2H_2O$$

The equation is balanced. There are 4 hydrogen atoms and 2 oxygen atoms on each side of the equation.

(a)

$$CH_4 + 2O_2 \longrightarrow CO_2 + 2H_2O$$

The equation is balanced. There are 4 oxygen atoms, 4 hydrogen atoms, and 1 carbon atom on each side of the equation.

(b)

Figure 6.3
When a chemical equation is balanced, the number and kinds of atoms in the reactants are the same as in the products.

EXAMPLE 6.7 **Writing a balanced equation**

Balance the equation for the reaction that appears in Figure 6.4:

$$AgNO_3(aq) + Cu(s) \longrightarrow Cu(NO_3)_2(aq) + Ag(s)$$

SOLUTION

Since this reaction involves the polyatomic nitrate ion, we can save time balancing the equation if we consider the nitrate ion as a unit. Place a coefficient of 2 in front of $AgNO_3$ to balance the nitrate ion:

$$2AgNO_3(aq) + Cu(s) \longrightarrow Cu(NO_3)_2(aq) + Ag(s)$$

But now the silver (Ag) is not balanced; there are two atoms in the reactants but only one atom in the products. Place a coefficient of 2 in front of Ag; now the equation is balanced and the coefficients are correct:

$$2AgNO_3(aq) + Cu(s) \longrightarrow Cu(NO_3)_2(aq) + 2Ag(s)$$

Figure 6.4
Copper metal reacts with a solution of silver nitrate to produce silver-metal and a solution of copper(II) nitrate.

---

402 CHAPTER 13 Aldehydes and Ketones

EXAMPLE 13.4 **Identifying the relative degree of oxidation**

List the compounds 2-propanol, propane, and propanone in order from most reduced to most oxidized.

SOLUTION

Write the structure of each compound:

$$\underset{\text{2-Propanol}}{\overset{\overset{\displaystyle OH}{|}}{CH_3CHCH_3}} \qquad \underset{\text{Propane}}{CH_3CH_2CH_3} \qquad \underset{\text{Propanone}}{\overset{\overset{\displaystyle O}{\|}}{CH_3CCH_3}}$$

Propane is most reduced; it has the maximum number of hydrogens. Propanone is the most oxidized. It has the same number of oxygens as 2-propanol but fewer hydrogens. The order is propane, 2-propanol, and propanone.

PRACTICE EXERCISE 13.5
Indicate the most oxidized compound in each pair.
(a) 1-butyne and 1-butene
(b) propanal and propane
(c) cyclohexane and cyclohexanol
(d) 3-pentanol and 3-pentanone

Practice Exercises
Nearly 300 practice exercises found throughout the book provide on-the-spot feedback so students can check their understanding of an important concept or skill. Answers are provided at the back of the book.

Aim Statements
Aim statements accompany each numbered section and provide learning objectives for the student.

Focus Statements
Focus statements summarize the content of each numbered section.

Marginal Notes
Marginal notes provide interesting applications of environmental chemistry, biological chemistry, health-related chemistry, and problem-solving tips.

7.3 Pressure

AIMS: To define pressure and explain how it can be measured. To convert pressure measurements between units of kilopascals, millimeters of mercury, and atmospheres.

Focus

Collisions of gas particles with objects in their paths generate gas pressure.

Mercury, the liquid metal that is used in barometers and some thermometers, is toxic, and mercury-containing devices should be handled with care. Mercury spills should be cleaned up immediately. Prolonged inhalation of mercury vapor can lead to nervous disorders.

Moving bodies exert forces when they collide with other bodies. Although a gas particle is a moving body, the force exerted by a single tiny gas particle is very small. However, it is not hard to imagine that many simultaneous collisions would produce a measurable force on an object. **Gas pressure** *is the result of simultaneous collisions of billions upon billions of gas particles on an object.*

Atmospheric pressure

Air exerts pressure because molecules in the atmosphere collide with objects in their paths. Atmospheric pressure can be measured with a mercury barometer (Fig. 7.2). Daily measurements with mercury barometers show that air particles at sea level exert enough push or pressure to support a column of mercury about 760 mm high. The *atmosphere*, a unit of pressure, is derived from these observations. *One* **atmosphere** *is defined as the pressure required to support 760 mm of mercury.* Atmospheres (atm) and millimeters of mercury (mm Hg) are the most commonly used units of pressure measurement in chemistry. Another pressure unit is the *torr,* named for the Italian scientist Evangelista Torricelli (1608–1647). *One* **torr** *is 1/760 atm or 1 mm Hg. The SI unit of pressure is the* **pascal (Pa),** named for the French mathematician and physician Blaise Pascal (1623–1662). The

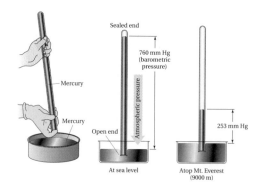

Figure 7.2
A mercury barometer is made by filling a tube, sealed at one end, with mercury. The tube is then inverted into a reservoir of mercury. Normal atmospheric pressure pushing on the mercury in the reservoir supports a column of mercury about 760 mm high. At 9000 m altitude, the *air* exerts enough push to support a column of mercury only 253 mm high; the atmospheric pressure is therefore 253 mm Hg at 9000 m elevation.

24.1 Catabolism

AIMS: To write an equation for the catabolism of one mole of glucose. To name the three stages of aerobic catabolism of glucose.

Focus

Aerobic catabolism of glucose is accomplished in three stages.

One tomato and one potato each contain about 3.5 g of sugar; an orange contains about 12.5 g, and a banana, about 18 g.

Glycolysis
glykos (Greek): sugar or sweet
lysis (Greek): splitting

The glucose produced by digestion of the carbohydrates a person eats enters the bloodstream. In a normal individual, this glucose is extracted from the bloodstream by body cells. Once inside the cells, the glucose is completely oxidized to carbon dioxide and water.

$$C_6H_{12}O_6 + O_2 \longrightarrow 6CO_2 + 6H_2O$$
Glucose   Oxygen   Carbon   Water
dioxide

The paths that cells use to oxidize glucose completely to carbon dioxide involve many individual chemical reactions. But all aerobic cells have adopted essentially the same three-stage master plan to do the job. Briefly these three stages, shown in Figure 24.1, consist of initial breakdown in *glycolysis*, further degradation to *acetyl coenzyme A,* and, finally, complete oxidation in the *citric acid cycle.*

Illustrations and Photographs

The illustrations and photographs provide visual representations to help students understand chemistry and make it more inviting to them.

---

564 CHAPTER 18 Amino Acids, Peptides, and Proteins

Alpha helix

In some proteins, regions of the backbone of the peptide chain are coiled into a spiral shape called an **alpha helix,** *similar to a corkscrew.* As Figure 18.3 shows, a corkscrew must be turned in a right-handed, or clockwise, direction to penetrate a cork. The alpha helixes of proteins are always right-handed. The helixes are held together by hydrogen bonds, shown in Figure 18.4, formed between the hydrogen of an N—H of a peptide bond and the carbonyl oxygen of another peptide bond group four residues away in the same peptide chain.

The tightness of coiling is such that 3.6 amino acid residues of the peptide backbone make each full turn of the alpha helix. There are no amino acid residue side chains inside the alpha helix; they are located on the outside. The cyclic amino acid proline does not fit well into the peptide backbone of alpha helixes. Alpha helixes in long protein chains often end at a place where proline residues occur in the primary structure. Proline is

Figure 18.3
A corkscrew must be turned in a right-handed, or clockwise, direction to penetrate a cork.

Figure 18.4
A peptide chain twisted into a right-handed alpha helix constitutes a protein's secondary structure. The N-terminal to C-terminal direction is from top to bottom. The dotted lines show the hydrogen bonds between the carbonyl oxygen of one amino acid residue and the N—H hydrogen of another, four amino acid residues further down the chain.

To N-terminal

Hydrogen bond

To C-terminal

8.6 Solubility **225**

(a)   (b)   (c)

Figure 8.10
When a seed crystal is added to a supersaturated solution (a), the excess solute crystallizes (b), (c).

original temperature, the excess solute does not always immediately crystallize. *Such a solution, which contains more solute than it can theoretically hold at the given temperature, is* **supersaturated.** In a supersaturated solution, a dynamic equilibrium cannot exist between the dissolved solute and the undissolved solid because there is no undissolved solid. To initiate crystallization in a supersaturated solution, we could scratch the inside surface of the container or simply add a very small crystal—a *seed crystal*—of the solute (Fig. 8.10).

Chapter Summaries
Each chapter has a summary section to reinforce and integrate key concepts in the chapter.

SUMMARY

Aldehydes and ketones contain the carbonyl group ($-C=O$). The carbonyl carbon in an aldehyde has at least one hydrogen attached (R—CHO), but the carbonyl carbon in a ketone has no hydrogens (R—CO—R). Formaldehyde (H_2CO) is the simplest aldehyde; acetone (CH_3COCH_3) is the simplest ketone. The physical and chemical properties of aldehydes and ketones are influenced by the very polar carbonyl group. Molecules of aldehydes and ketones can attract each other through polar-polar interactions. These compounds have higher boiling points than the corresponding alkanes but lower boiling points than the corresponding alcohols. Aldehydes and ketones can accept hydrogen bonds, and those with low molar mass are completely soluble in water.

Aldehydes and ketones are produced by the oxidation of primary alcohols and secondary alcohols, respectively. Aldehydes are usually more reactive than ketones and are good reducing agents. An aldehyde can be oxidized to the corresponding carboxylic acid, but ketones resist further oxidation. Addition reactions are characteristic of both aldehydes and ketones. Addition of water to the carbon-oxygen bond of the carbonyl group forms hydrates. Addition of an alcohol produces hemiacetals and hemiketals. The reaction of alcohols with hemiketals and hemiacetals produces acetals and ketals, respectively.

The most important industrial aldehydes and ketones are formaldehyde, acetaldehyde, acetone, and methyl ethyl ketone. A 40% aqueous solution of formaldehyde, called formalin, is commonly used to preserve biological specimens. Many aldehydes and ketones have fragrant aromas.

Summary of Reactions   **415**

SUMMARY OF REACTIONS

Here are the reactions of aldehydes and ketones presented in this chapter.

Aldehydes

1. Preparation of aldehydes:

Primary alcohol → Aldehyde

2. Oxidation of aldehydes:

Carboxylic acid

3. Reaction of aldehydes with Tollens' reagent:

$$R-\overset{O}{\overset{\|}{C}}-H + 2Ag^+ + 2OH^- \longrightarrow$$

Silver ions

$$R-\overset{O}{\overset{\|}{C}}-OH + 2Ag(s) + H_2O$$

Carboxylic acid   Metallic silver

4. Reaction of aldehydes with Benedict's reagent:

$$R-\overset{O}{\overset{\|}{C}}-H + 2Cu^{2+} + 2OH^- \longrightarrow$$

Copper(II) ion (blue solution)

$$R-\overset{O}{\overset{\|}{C}}-O^- + 2Cu^+ + H_2O$$

Copper(I) ion (red precipitate of Cu_2O)

5. Addition of water to aldehydes:

$$R-\overset{O}{\overset{\|}{C}}-H + H-OH \rightleftharpoons R-\overset{OH}{\underset{OH}{C}}-H$$

Aldehyde hydrate

6. Addition of alcohol to an aldehyde followed by a second reaction with an alcohol:

$$R-\overset{O}{\overset{\|}{C}}-H \xrightarrow{RO-H} R-\overset{OH}{\underset{OR}{C}}-H \xrightarrow{RO-H}$$

Hemiacetal

$$R-\overset{OR}{\underset{OR}{C}}-H + H_2O$$

Acetal

Ketones

1. Preparation of ketones:

$$R-\overset{OH}{\underset{H}{C}}-R \xrightarrow{Oxidation} R-\overset{O}{\overset{\|}{C}}-R$$

Secondary alcohol   Ketone

2. Addition of alcohol to a ketone followed by a second reaction with an alcohol:

$$R-\overset{O}{\overset{\|}{C}}-R \xrightarrow{RO-H} R-\overset{OH}{\underset{OR}{C}}-R \xrightarrow{RO-H} R-\overset{OR}{\underset{OR}{C}}-R + H_2O$$

Hemiketal   Ketal

Key Terms

Key terms are printed in bold type and are defined where they first appear. They are also grouped at the end of the chapter and in a glossary at the back of the text, for convenient reference.

End-of-Chapter Exercises

The end-of-chapter exercises are keyed to chapter sections. Additional exercises are not keyed to any section or topic and incorporate material from more than one section. The odd-numbered exercises are answered in the back of the text.

---

4. Hydrolysis of amides. (The amide may be simple, monosubstituted, or disubstituted.)

$$R-\overset{\overset{\displaystyle O}{\|}}{C}-\ddot{N}H_2 + HO-H \xrightarrow[\text{H}^+ \text{ or OH}^-]{\text{Heat}} R-\overset{\overset{\displaystyle O}{\|}}{C}-OH + H-\ddot{N}H_2$$

KEY TERMS

Alkaloid (15.8)	Anilide (15.4)	Free amine (15.2)	Polyamide (15.5)
Amide (15.4)	Antihistamine (15.7)	Hallucinogen (15.7)	Protonated amine (15.2)
Amine (15.1)	Arylammonium ion (15.2)	Hypnotic (15.9)	Quaternary ammonium
Alkylammonium ion (15.2)	Barbiturate (15.9)	Neurotransmitter (15.7)	salt (15.3)
Ammonium salt (15.2)	Decongestant (15.7)	Opiate (15.8)	Sedative (15.9)

EXERCISES

Amines (Sections 15.1, 15.2, 15.3)

15.15 Name or write a structural formula for the following amines. Classify them as primary, secondary, or tertiary amines.
(a) $(CH_3)_2NH$ (b) p-chloro-N-methylaniline
(c) NH_2 (on benzene ring with Cl)
(d) diethylmethylamine

15.16 Write structural formulas and name the following amines. Classify each amine as primary, secondary, or tertiary.
(a) diethylamine (b) $(CH_3CH_2)_2NCH_3$
(c) butylamine (d) $CH_3\overset{\displaystyle H}{N}$ (attached to benzene ring)

15.17 What is the meaning of the term *heterocyclic amine*?

15.18 Draw the structure and give the name of (a) an aromatic heterocyclic amine and (b) an aliphatic heterocyclic amine.

15.19 Draw structural formulas for (a) pyrimidine and (b) purine. Derivatives of these two compounds are found in what biologically important molecules?

15.20 Draw the structure of pyrrole. List some of the naturally occurring molecules that contain the pyrrole ring system.

15.21 Why are amines weak bases?

15.22 Draw the general formulas for (a) an unprotonated (free) amine and (b) a protonated amine.

15.24 Draw the structure and name the organic product for each of the following reactions.
(a) $CH_3NH_3^+I^- + NaOH$
(b) $(CH_3)_2NH + CH_3Cl$
(c) $CH_3I + NH_3$
(d) $NH_3^+Cl^- + NaOH$ (on benzene ring)
(e) $(CH_3CH_2)_3N + CH_3CH_2Cl$

15.25 Write an equation for the dissociation of the dimethylammonium ion. Why does an aqueous solution of dimethylammonium chloride test acidic?

15.26 Draw the structure of tetramethylammonium iodide. What happens if this compound is treated with sodium hydroxide?

Amides (Sections 15.4, 15.5, 15.6)

15.27 Name or write structural formulas for the following amides.
(a) $CH_3\overset{\overset{\displaystyle O}{\|}}{C}NH_2$ (b) $CH_3CH_2\overset{\overset{\displaystyle O}{\|}}{C}NHCH_3$
(c) $\overset{\overset{\displaystyle O}{\|}}{C}NHCH_3$ (on benzene ring)
(d) N-ethyl-N-methylpropanamide
(e) acetanilide

---

246   CHAPTER 8 Water and Aqueous Systems

8.58 Calculate the volume of 0.50 M $MgSO_4$ you must dilute to make 2.5 L of 0.15 M $MgSO_4$.

Colloids and Suspensions (Section 8.9)

8.59 Distinguish between a solution and a colloid.

8.60 Distinguish between a suspension and a solution.

8.61 Describe the Tyndall effect.

8.62 Explain Brownian motion.

Colligative Properties (Section 8.10)

8.63 What effect does dissolving a solute have on the boiling and freezing points of a liquid?

8.64 How does the vapor pressure of a solution compare with the vapor pressure of the solvent of that solution?

8.65 Why does seawater evaporate more slowly than fresh water at the same temperature?

8.66 How does the boiling point of seawater compare with the boiling point of distilled water?

8.67 What is the boiling point of (a) a solution containing 0.50 mol glucose in 1000 g H_2O and (b) a solution containing 1.5 mol NaCl in 1000 g H_2O?

8.68 What effect does the addition of 1.00 mol $CaCl_2$ to 1000 g H_2O have on the freezing point of water?

8.76 The solubility of oxygen in water at 20 °C is 5.4 mg O_2 per 100 g H_2O (assuming standard pressure). If all the oxygen above the water were removed, how would the solubility of the oxygen change?

8.77 What happens in each of the following situations?
(a) A dialyzing bag containing a solution of sodium chloride is immersed in pure water.
(b) A dialyzing bag containing sodium chloride and a dispersion of a large protein in water is immersed in pure water.
(c) A dialyzing bag containing a dispersion of a protein in water is immersed in a concentrated solution of sodium chloride.

8.78 How many grams of water are in 166 g of the hydrate $CoCl_2 \cdot 6H_2O$?

8.79 On an equal molar basis, why does an electrolyte depress the freezing point of water to a greater extent than a nonelectrolyte?

8.80 Explain each of the following facts.
(a) Finely ground sugar dissolves more quickly in a glass of water than does a large sugar cube.
(b) Fish need oxygen to survive. Goldfish in a crowded bowl have a better chance of survival in cold water than in warm water.
(c) Electronic equipment that can be damaged by

xvi

Self-Tests
Each chapter's review material ends with a Self-Test that students can use to gauge their progress in meeting learning goals.

SELF-TEST (REVIEW)

True/False

1. A dehydrogenation reaction is a reduction reaction.
2. Hydrogen bonding accounts for the relatively high boiling point of acetaldehyde.
3. The reaction of equal moles of an alcohol and an aldehyde gives an acetal.
4. Oxidation of a tertiary alcohol gives a ketone.
5. You would expect propanal to have a higher boiling point than propanol.
6. Propanal should have higher water solubility than hexanal.
7. One mole of methanol would release more energy upon complete oxidation than one mole of methane.
8. Both aldehydes and ketones have a carbonyl group.
9. All aldehydes and ketones give a positive Tollens' test.
10. Four pairs of electrons are shared in the carbonyl bond formed between an oxygen and a carbon atom.

Multiple Choice

11. Which of the following statements about the carbon-oxygen double bond of the carbonyl group is *false*?
    (a) The bond is polar.
    (b) The carbon has a partial negative charge.
    (c) The bonding electrons are unequally shared between the carbon and oxygen.
    (d) The oxygen has two unshared pairs of electrons.
12. Acetaldehyde would be likely to form hydrogen bonds with
    (a) formaldehyde.   (b) octane.   (c) water.
    (d) acetone.
13. On the basis of your knowledge of intermolecular forces, which of the following would you expect to have the highest boiling point?
    (a) propanal   (b) propane   (c) acetone
    (d) 1-propanol

14. Which of the following compounds contains a di-ether linkage?
    (a) a hemiacetal   (b) chloral hydrate
    (c) a ketal   (d) camphor
15. Which of the following substances is used as a preservative of biological specimens?
    (a) paraldehyde   (b) formalin
    (c) methyl ethyl ketone   (d) cinnamaldehyde
16. Which of the following substances can undergo an addition reaction with methanol?
    (a) propane   (b) methyl ethyl ether
    (c) propanol   (d) propanal
17. A structural isomer of 2-butanone is
    (a) diethyl ether.   (b) *tert*-butyl alcohol.
    (c) diethyl ketone.   (d) butanal.
18. Which of the following compounds would release the most energy upon oxidation to carbon dioxide?
    (a) ethanol   (b) acetic acid, CH_3COOH
    (c) ethane   (d) acetaldehyde
19. The oxidation of 2-methyl-2-butanol with $K_2Cr_2O_7$ and H_2SO_4 would give
    (a) 2-methyl-2-butanone.
    (b) isopropyl alcohol and ethane.
    (c) 2-methyl-2-butanal.
    (d) none of the above.
20. In a positive Tollens' test,
    (a) silver ions are oxidized to silver atoms.
    (b) the aldehyde is an oxidizing agent.
    (c) a silver mirror is formed.
    (d) more than one are correct.
21. In the reaction of substance *A* with substance *B*, substance *A* loses oxygen. Which of the following is true?
    (a) Substance *B* is an oxidizing agent.
    (b) Substance *A* is reduced.
    (c) Substance *B* is reduced.
    (d) Substance *A* is a reducing agent.

Brief Contents

1 **Matter, Change, and Energy**
Ideas About the Material World

2 **Scientific Measurement**
A Fundamental Skill

3 **Matter and the Structure of Atoms**
Nature's Building Blocks

4 **The Periodic Table**
Organizing the Elements

5 **Chemical Bonds**
Holding Atoms Together

6 **Chemical Reactions**
Equations, Equilibria, and Reaction Rates

7 **States of Matter**
Molecules in Motion

8 **Water and Aqueous Systems**
The Story of a Unique Compound

9 **Acids, Bases, Salts, and Buffers**
Hydrogen Ions in Chemistry

10 **Nuclear Chemistry**
Radioactivity and Ionizing Radiations

11 **Carbon Chains and Rings**
The Foundations of Organic Chemistry

12 **Halocarbons, Alcohols, and Ethers**
The Polar Bond in Organic Molecules

13 **Aldehydes and Ketones**
Introduction to the Carbonyl Group

14 **Acids and Their Derivatives**
Reactions of the Carboxyl Group

15 **Amines and Amides**
Organic Nitrogen Compounds

16 **Carbohydrates**
The Structure and Chemistry of Sugars

17 **Lipids**
A Potpourri of Fatty Molecules

18 **Amino Acids, Peptides, and Proteins**
Molecular Structures and Biological Roles

19 **Enzymes**
Catalysis of the Reactions of Life

20 **Nucleic Acids**
The Molecular Basis of Heredity

21 **Digestion and Nutrition**
Materials for Living

22 **Body Fluids**
Maintaining the Body's Internal Environment

23 **Energy and Life**
Sources and Uses of Energy in Living Organisms

24 **Carbohydrates in Living Organisms**
At the Core of Metabolism

25 **Lipid Metabolism**
Fat Chemistry in Cells

26 **Metabolism of Nitrogen Compounds**
Nitrogen and Life

Contents

Matter, Change, and Energy: 1

Ideas About the Material World
Case in Point: **Sickle cell anemia**   2

1.1   Chemistry   3
1.2   The scientific method   3
A Closer Look: **Chemicals and Chemophobia**   5
Follow-up to the Case in Point: **Sickle cell anemia**   7
1.3   Temperature   8
A Closer Look: **Clinical Thermometers**   9
1.4   Physical states of matter   10
1.5   Mixtures   11
1.6   Physical properties and physical changes   14
1.7   Chemical reactions and chemical properties   15
1.8   Elements and compounds   17
A Closer Look: **Essential Elements for Life**   19
1.9   Mass conservation   20
1.10   Energy interconversion and conservation   21
1.11   Exothermic and endothermic processes   22

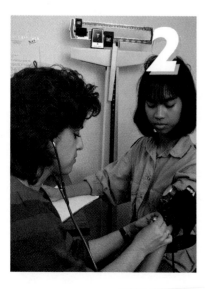

Scientific Measurement: 28

A Fundamental Skill
Case in Point: **Maturity-onset diabetes**   29

2.1   Measurements and units   30
2.2   Systems of measurement   31
2.3   Scientific notation   32
2.4   Significant figures in measurements   35
2.5   Significant figures in calculations   36
2.6   Length and volume   38
A Closer Look: **Sizes of Organisms Affecting Health**   40
2.7   Solving problems and converting units   43
2.8   Units of mass   46

Follow-up to the Case in Point: **Maturity-onset diabetes** 49

2.9 Density and specific gravity 49

A Closer Look: **Urinometers** 52

2.10 Units of heat and the heat capacity of matter 53

A Closer Look: **Counting Food Calories** 55

Matter and the Structure of Atoms: 61

Nature's Building Blocks

Case in Point: **Magnetic resonance imaging** 62

3.1 Atoms 63

A Closer Look: **Images of Atoms** 64

3.2 Subatomic particles 64

3.3 Atomic number and mass number 65

3.4 Isotopes and atomic mass 67

Follow-up to the Case in Point: **Magnetic resonance imaging** 71

3.5 Electronic structure of atoms 72

A Closer Look: **Electrons, Flames, and Fireworks** 73

3.6 Electron configurations 75

The Periodic Table: 83

Organizing the Elements

Case in Point: **Two faces of selenium** 84

4.1 Development of the periodic table 85

4.2 Metals and nonmetals 85

A Closer Look: **Diamond Tools for Surgery** 86

4.3 Organization of the modern periodic table 87

A Closer Look: **Biomedical Implants** 89

4.4 Trends in atomic size 91

A Closer Look: **The Origin of the Elements** 93

4.5 Trends in ionization energy 93

4.6 Representative elements 94

A Closer Look: **Semiconductors** 98

A Closer Look: **Bioactive Materials** 100

Follow-up to the Case in Point: **Two faces of selenium** 101

4.7 Transition elements and inner transition elements 103

Chemical Bonds: 108

Holding Atoms Together

Case in Point: **A new era in medicinal drug discovery** 109

5.1 Valence electrons 110

5.2 Electron dot structures 110

5.3 Molecules and ions 111

5.4 Formation of ions 113

A Closer Look: **A Healthy Diet of Ions** 116

A Closer Look: **Superconductors** 121

5.6 Covalent bonds and molecular compounds 121

5.7 Coordinate covalent bonds 125

A Closer Look: **Cisplatin: Coordinate Covalent Bonds and Chemotherapy** 126

5.8 Bond polarity 127

5.9 Attractions between molecules 129

5.10 Shapes of molecules 131

Follow-up to the Case in Point: **A new era of medicinal drug discovery** 132

A Closer Look: **Buckyball: A Third Form of Carbon** 134

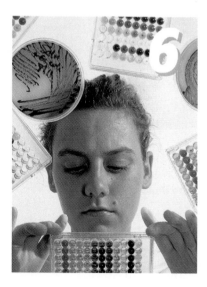

Chemical Reactions: 141

Equations, Equilibria, and Reaction Rates

Case in Point: **Effect of cold on the body** 142

6.1 The mole 143

6.2 Chemical equations 146

6.3 Chemical calculations 150

6.4 Oxidation and reduction reactions 153

6.5 Heat and entropy changes 156

6.6 Free energy 160

6.7 Reaction rate 162

Follow-up to the Case in Point: **Effect of cold on the body** 164

A Closer Look: **Catalase** 166

6.8 Reversible reactions 167

A Closer Look: **The Rain Forest, Equilibrium, and Medicine** 169

States of Matter: 179

Molecules in Motion

Case in Point: **Heimlich maneuver**   180

7.1   Kinetic-molecular theory   181
7.2   Kinetic energy   181
7.3   Pressure   183
7.4   Avogadro's hypothesis   184
7.5   Diffusion   186
7.6   Behavior of gases   187
A Closer Look: **Measuring Blood Pressure**   189
A Closer Look: **Breathing**   191
7.7   The gas laws   193
A Closer Look: **Exchange of Physiologic Gases**   194
Follow-up to the Case in Point: **Heimlich maneuver**   196
7.8   The ideal gas law   201
7.9   Liquids   203
7.10   Solids   206

Water and Aqueous Systems: 212

The Story of a Unique Compound

Case in Point: **Drinking water and health**   213

8.1   Water and hydrogen bonds   214
8.2   Ice   217
8.3   Aqueous solutions   218
Follow-up to the Case in Point: **Drinking water and health**   219
8.4   Solvation   219
8.5   Solution formation   223
8.6   Solubility   224
A Closer Look: **Kidney Stones**   226
A Closer Look: **What Do the Foam of Soda and the Bends Have in Common?**   227
8.7   Units of concentration   228
8.8   Molarity   231
8.9   Colloids and suspensions   235
8.10   Colligative properties   237
8.11   Osmosis and dialysis   239
A Closer Look: **Hemodialysis**   243

Acids, Bases, Salts, and Buffers: 249
Hydrogen Ions in Chemistry

Case in Point: **Antacids**   250

9.1   Hydrogen ions from water   251

9.2   Acids and bases   253

A Closer Look: **Acids and Dental Health**   255

9.3   The pH scale   260

A Closer Look: **pH of Body Fluids**   262

9.4   Measurement of pH   267

9.5   Neutralization   269

Follow-up to the Case in Point: **Antacids**   271

9.6   Equivalents   271

9.7   Normality   274

9.8   Titration   276

9.9   Salts of weak acids and bases   278

9.10   Buffers   279

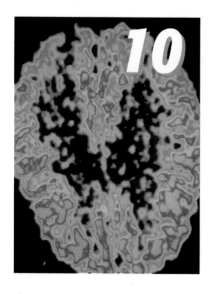

Nuclear Chemistry: 286
Radioactivity and Ionizing Radiations

Case in Point: **Hyperthyroidism**   287

10.1   Radiation   288

A Closer Look: **Food Irradiation**   292

10.2   Half-life   292

10.3   Carbon-14   295

A Closer Look: **Radon in Homes**   297

10.4   Nuclear reactions   298

10.5   Nuclear fission   298

A Closer Look: **The Nuclear Reactor Debate**

10.6   Nuclear fusion   300

10.7   Radiation detection   302

10.8   The biological effects of radiation   303

10.9   Units of radiation   304

10.10   Radiation in medicine   306

Follow-up to Case in Point: **Hyperthyroidism**   307

10.11   Background radiation   310

Carbon Chains and Rings: 315
The Foundations of Organic Chemistry
Case in Point: **Carbon monoxide poisoning**   316

11.1   Carbon compounds   317
11.2   Hydrocarbons   317
11.3   Carbon-carbon bonds   320
11.4   Straight-chain alkanes   321
11.5   Branched-chain alkanes   325
11.6   Structural isomers   330
11.7   Cycloalkanes   331
11.8   Multiple bonds   334
11.9   Aromatic compounds   338
A Closer Look: **Hydrocarbons and Health**   343

11.10   Sources of hydrocarbons   344
A Closer Look: **Octane Ratings of Gasoline**   345

11.11   Properties of hydrocarbons   346
Follow-up to the Case in Point: **Carbon monoxide poisoning**   347
A Closer Look: **The Greenhouse Effect**   348

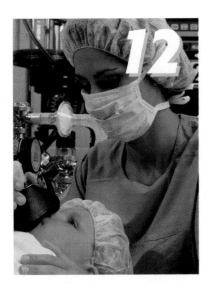

Halocarbons, Alcohols, and Ethers: 354
The Polar Bond in Organic Molecules
Case in Point: **A mistaken conviction**   355

12.1   Halocarbons   356
A Closer Look: **Chlorofluorocarbons and the Ozone Layer**   361

12.2   Halocarbons from alkenes   362
12.3   Alcohols   367
A Closer Look: **Phenolic Antiseptics and Disinfectants**   371

12.4   Making alcohols   373
12.5   Elimination reactions   373
12.6   Alkoxides   375
12.7   Ethers   376
A Closer Look: **Halocarbon and Ether Anesthetics**   377
Follow-up to the Case in Point: **A mistaken conviction**   378

12.8   Physical properties   379
12.9   Sulfur compounds   381
12.10   Polyfunctional compounds   383

Aldehydes and Ketones: 391

Introduction to the Carbonyl Group

Case in Point: **Fetal alcohol syndrome** 392

13.1 Aldehydes and ketones 393

13.2 The carbonyl group 397

A Closer Look: **Flavors and Fragrances** 399

13.3 Redox reactions in organic chemistry 400

13.4 Redox reactions of aldehydes and ketones 402

Follow-up to the Case in Point: **Fetal alcohol syndrome** 405

13.5 Aldehyde detection 406

13.6 Additions to the carbonyl group 408

13.7 Uses of aldehydes and ketones 410

A Closer Look: **The Chemistry of Vision** 412

Acids and Their Derivatives: 420

Reactions of the Carboxyl Group

Case in Point: **Acne and tretinoin** 421

14.1 Carboxylic acids 422

Follow-up to the Case in Point: **Acne and tretinoin** 425

14.2 The carboxyl group 427

14.3 Acidity of carboxylic acids 429

A Closer Look: **Hard Water and Water Softening** 435

14.4 Synthesis of carboxylic acids 435

14.5 Carboxylic acid anhydrides 437

14.6 Carboxylic esters 439

A Closer Look: **Ester Local Anesthetics** 442

A Closer Look: **Aspirin** 444

14.7 Thioesters 446

14.8 Ester hydrolysis 446

14.9 Phosphoric acids, anhydrides, and esters 448

14.10 Esters of nitric and nitrous acids 451

Amines and Amides: 459

Organic Nitrogen Compounds

Case in Point: **Antidepressants** 460

15.1 Amines   461
15.2 Amine basicity   465
15.3 Preparation of amines   467
15.4 Amides   468

A Closer Look: **The Sulfonamide Antibiotics**   469
A Closer Look: **Amide Local Anesthetics**   471

15.5 Preparation of amides   472
15.6 Amide hydrolysis   476
15.7 Important amines   476

Follow-up to the Case in Point: **Antidepressants**   479

15.8 Alkaloids   481
15.9 Barbiturates   484

Carbohydrates: 491

The Structure and Chemistry of Sugars

Case in Point: **Lactose intolerance** 492

16.1 Carbohydrate structure and stereochemistry   493
16.2 Monosaccharides   497
16.3 Cyclic structures   501
16.4 Haworth projections   502
16.5 Glycosides   505
16.6 Polysaccharides   507

A Closer Look: **Dietary Fiber**   510

16.7 Disaccharides   511
16.8 Sucrose and lactose   511

Follow-up to the Case in Point: **Lactose intolerance**   513

16.9 Reducing and nonreducing sugars   514

A Closer Look: **Tests for Blood Sugar in Diabetes**   515

16.10 Chitin, heparin, and acid mucopolysaccharides   516

Lipids: 522

A Potpourri of Fatty Molecules

Case in Point: **Obesity** 523

17.1 Waxes 524

17.2 Triglycerides 524

Follow-up to the Case in Point: **Obesity** 525

A Closer Look: **Cosmetic Creams** 526

17.3 Lipid composition of cell membranes 529

A Closer Look: **Cells** 530

17.4 Structure of liposomes and cell membranes 534

17.5 Steroids 536

17.6 Steroid hormones 539

17.7 Male sex hormones 540

17.8 Female sex hormones 542

17.9 The pill 543

17.10 Plant steroids 545

17.11 Prostaglandins and leukotrienes 545

Amino Acids, Peptides, and Proteins: 551

Molecular Structures and Biological Roles

Case in Point: **Runner's high** 552

18.1 The amino acids 553

18.2 Stereoisomers of amino acids 556

18.3 Zwitterions 557

18.4 Peptides 559

A Closer Look: **Aspartame** 561

Follow-up to the Case in Point: **Runner's high** 563

18.5 Primary structure of proteins 563

18.6 Secondary structure of proteins 563

18.7 Tertiary structure of proteins 567

A Closer Look: **X-ray Crystallography** 569

18.8 Quaternary structure of proteins 570

18.9 Hemoglobin function 571

18.10 Sickle cell anemia 572

A Closer Look: **Electrophoresis** 573

18.11 Glycoproteins 574

18.12 Denaturation 574

A Closer Look: **Glycoproteins: Control of Blood Glucose in Diabetes** 575

Enzymes: 581
Catalysis of the Reactions of Life
Case in Point: **Enzyme therapy for heart attacks** 582

19.1 Enzymes 583
19.2 Names of enzymes 584
19.3 Enzyme specificity 584
19.4 Enzyme-substrate complexes 584
19.5 Active sites 586
19.6 Cofactors 587
19.7 Enzyme assay 587
19.8 Effects of pH, temperature, and heavy metals 589
19.9 Induction and degradation of enzymes 590
A Closer Look: **Lead Poisoning** 591

19.10 Control of enzyme activity 592
A Closer Look: **HIV Protease and Its Inhibition** 593

19.11 Zymogens 595
19.12 Blood clotting 597
A Closer Look: **Trypsin: Anatomy of an Enzyme** 598

Follow-up to the Case in Point: **Enzyme therapy for heart attacks** 600

19.13 Antibiotics and other therapeutic drugs 601
A Closer Look: **The Toxicity of Pesticides** 603

Nucleic Acids: 607
The Molecular Basis of Heredity
Case in Point: **Treatment for an inherited disease** 608

20.1 Nucleic acids, nucleotides, and nucleosides 609
20.2 The DNA double helix 614
20.3 The central dogma 616
20.4 Replication 617
A Closer Look: **The HIV Reverse Transcriptase** 619

20.5 Genes 621
A Closer Look: **The Human Genome Project** 622
A Closer Look: **DNA Fingerprinting** 623

20.6 Classes of RNA 624
20.7 Transcription 625
20.8 Translation 627
20.9 The genetic code 631
20.10 Gene mutations and molecular diseases 633

A Closer Look: **Oncogenes and Tumor Suppressor Genes** 634

20.11 Recombinant DNA and gene therapy 636

Follow-up to the Case in Point: **Treatment for an inherited disease** 637

Digestion and Nutrition: 643
Materials for Living

Case in Point: **Osteoporosis** 644

21.1 Essential needs 645

21.2 Digestion 645

21.3 The stomach 646

21.4 The small intestine 647

A Closer Look: **Blood Lipoproteins and Heart Disease** 651

A Closer Look: **Therapies for Cystic Fibrosis** 652

21.5 Balanced diet 653

21.6 Proteins and amino acids in the diet 654

A Closer Look: **The Nutrition Pyramid** 655

21.7 Fat-soluble vitamins 656

21.8 Water-soluble vitamins 659

21.9 Minerals 664

Follow-up to the Case in Point: **Osteoporosis** 666

21.10 Trace elements 670

Body Fluids: 675
Maintaining the Body's Internal Environment

Case in Point: **A diabetic imbalance** 676

22.1 Body water 677

A Closer Look: **Artificial Skin** 678

22.2 Blood 679

A Closer Look: **Blood: Risks and Replacements** 682

22.3 Antibodies and interferons 682

22.4 Blood buffers 686

22.5 Oxygen and carbon dioxide transport 687

22.6 The urinary system 690

22.7 Acid-base balance 691

22.8 Acidosis and alkalosis 693

Follow-up to the Case in Point: **A diabetic imbalance** 696

22.9 Water and salt balance 696

Energy and Life: 702

Sources and Uses of Energy in Living Organisms

Case in Point: **A mysterious fatigue**   703

23.1 Metabolism   704
23.2 Photosynthesis   704

A Closer Look: **Phototherapies**   706

23.3 The energy and carbon cycle   707
23.4 Adenosine triphosphate   708
23.5 Cellular energetics   710
23.6 Oxidative phosphorylation   711

A Closer Look: **Mitochondria**   715

A Closer Look: **Oxygen, Disease, and Aging**   718

Follow-up to the Case in Point: **A mysterious fatigue**   720

23.7 Cellular work   720

Carbohydrates in Living Organisms: 729

At the Core of Metabolism

Case in Point: **Carbohydrate loading**   730

24.1   Catabolism   731
24.2   Glucose oxidation   732
24.3   Glycolysis   733
24.4   Acetyl coenzyme A   737
24.5   The citric acid cycle   738
24.6   ATP yield   742
24.7   Lactic fermentation   743

A Closer Look: **Monitoring Heart Attacks with Serum LDH Tests**   744

24.8   Oxygen debt   745
24.9   Glucose storage   748
24.10   Glycogen breakdown   749

Follow-up to the Case in Point: **Carbohydrate loading**   751

24.11   Metabolic regulation   752

A Closer Look: **Nitric Oxide and Carbon Monoxide as Second Messengers**   754

24.12   Control of glycogenolysis   755
24.13   Glucose absorption   756

Lipid Metabolism: 762

Fat Chemistry in Cells

Case in Point: **Carnitine deficiency** 763

25.1 Body lipids 764
25.2 Fat mobilization 764
25.3 Fatty acid oxidation 765
A Closer Look: **Lipid Storage Diseases** 766
Follow-up to the Case in Point: **Carnitine deficiency** 768
25.4 ATP yield 771
25.5 Glycerol metabolism 772
25.6 The key intermediate—acetyl CoA 773
25.7 Fatty acid synthesis 774
25.8 Ketone bodies 776
25.9 Ketosis 777
25.10 Cholesterol synthesis 778
25.11 Blood cholesterol 779
A Closer Look: **Drug Strategies for Reducing Serum Cholesterol** 781

Metabolism of Nitrogen Compounds: 785

Nitrogen and Life

Case in Point: **A painful episode** 786

26.1 Nitrogen fixation 787
26.2 Protein turnover 788
26.3 Transamination reactions 788
26.4 The urea cycle 790
A Closer Look: **Hyperammonemia** 792
26.5 Catabolism of amino acids 795
26.6 Synthesis of amino acids 797
26.7 Defects of amino acid metabolism 799
26.8 Hemoglobin and bile pigments 799
A Closer Look: **The Amino Acidurias** 800
A Closer Look: **Hyperbilirubinemia** 802
26.9 Purines and pyrimidines 803
Follow-up to the Case in Point: **A painful episode** 804

Glossary   A1

Answers to Selected Exercises   A23

Photograph Credits   A58

Index   A59

Health-Related Topics

Acidosis   Sec. 22.4

Acids and dental health   Sec. 9.2   Closer Look, page 255

Acne and tretinoin   Chap. 14   Case in Point and Sec. 14.1

Aging   Sec. 23.6   Closer Look, page 718

AIDS   Sec. 19.10   Closer Look, page 593

Alkaloids   Sec. 15.8

Alkalosis   Sec. 22.4

Amide local anesthetics   Sec. 15.4   Closer Look, page 471

Amino acidurias   Sec. 26.7   Closer Look, page 800

Amphetamines   Sec. 15.7

Anabolic steroids   Sec. 17.7

Anesthetics: amides   Sec. 15.4   Closer Look, page 471

Anesthetics: esters   Sec. 14.6   Closer Look, page 442

Anesthetics: ethers   Sec. 12.7   Closer Look, page 377

Anesthetics: halocarbon   Sec. 12.7   Closer Look, page 377

Angiotensin converting enzyme   Sec. 19.13

Antacids   Chap. 9   Case in Point and Sec. 9.5

Antibiotics   Sec. 19.13

Antibodies   Sec. 22.3

Antidepressants   Chap. 15   Case in Point and Sec. 15.7

Antidiuretic hormone (ADH)   Sec. 22.9

Antihistamines   Sec. 15.7

Antiseptics   Sec. 12.3   Closer Look, page 371

Artificial skin   Sec. 22.1   Closer Look, page 678

Aspartame   Sec. 18.4   Closer Look, page 561

Aspirin   Sec. 14.6   Closer Look, page 444

Atherosclerosis   Sec. 25.11

Barbiturates   Sec. 15.9

Bends   Sec. 8.7   Closer Look, page 227

Beriberi   Sec. 21.8

Bile salts   Sec. 21.4

Bile pigments   Sec. 26.8

Bioactive materials   Sec. 4.6   Closer Look, page 100

Biological effects of radiation   Sec. 10.8

Biomedical implants   Sec. 4.3   Closer Look, page 89

Blood   Sec. 22.2

Blood-brain barrier   Sec. 15.7

Blood buffers   Sec. 22.4

Blood cholesterol   Sec. 25.11

Blood clotting   Sec. 19.12

Blood glucose in diabetes   Sec. 18.11   Closer Look, page 575

Blood lipoproteins   Sec. 21.4   Closer Look, page 651

Blood pressure   Sec. 7.6   Closer Look, page 189

Blood: risks and replacements   Sec. 22.2   Closer Look, page 682

Blood sugar in diabetes   Sec. 16.9   Closer Look, page 515

Body fluids, pH   Sec. 9.3   Closer Look, page 262

Breathing   Sec. 7.6   Closer Look, page 191

Buckyball   Sec. 5.10   Closer Look, page 134

Calories in food   Sec. 2.10   Closer Look, page 55

Carbohydrate loading   Chap. 24   Case in Point and Sec. 24.9

Carbon monoxide, second messenger   Sec. 24.11   Closer Look, page 754

Carbon monoxide poisoning   Chap. 11   Case in Point and Sec. 11.11

Carboxyhemoglobin   Sec. 18.9

Carnitine deficiency   Chap. 25   Case in Point and Sec. 25.3

CAT scan   Sec. 10.10

Catalase   Sec. 6.7   Closer Look, page 166

Cells   Sec. 17.3   Closer Look, page 530

Chemophobia   Sec. 1.1   Closer Look, page 5

Chemotherapy: cisplatin   Sec. 5.7   Closer Look, page 126

Chlorofluorocarbons   Sec. 12.1   Closer Look, page 361

Cholesterol   Sec. 17.5

Chronic fatigue syndrome   Chap. 23   Case in Point and Sec. 23.6

Clinical thermometers   Sec. 1.2   Closer Look, page 9

Cocaine   Sec. 15.8

Collagen   Sec. 18.6

Cosmetic creams   Sec. 17.2   Closer Look, page 526

Crenation   Sec. 8.11

Cystic fibrosis   Sec. 21.4   Closer Look, page 652

Decongestants   Sec. 15.7

Dental health   Sec. 9.2   Closer Look, page 255

Diabetes: maturity-onset   Chap. 2   Case in Point and Sec. 2.8

Diabetic imbalance   Chap. 22   Case in Point and Sec. 22.8

Dialysis   Sec. 8.11

Diamond tools for surgery   Sec. 4.2   Closer Look, page 86

Dietary fiber   Sec. 16.6   Closer Look, page 510

Digestion   Sec. 21.2

Disinfectants   Sec. 12.3   Closer Look, page 371

DNA fingerprinting   Sec. 20.5   Closer Look, page 623

Drinking water and health   Chap. 8   Case in Point and Sec. 8.3

Drug strategies   Sec. 25.11   Closer Look, page 781

Electrophoresis   Sec. 18.10   Closer Look, page 573

Enkephalins   Chap. 18   Case in Point and Sec. 18.4

Enzyme assay   Sec. 19.7

Enzyme inhibitors   Sec. 19.10

Enzyme therapy, heart attacks   Chap. 19   Case in Point and Sec. 19.12

Essential amino acids   Sec. 21.6

Essential elements for life   Sec. 1.8   Closer Look, page 19

Ester anesthetics   Sec. 14.6   Closer Look, page 442

Ether anesthetics   Sec. 12.7   Closer Look, page 377

Fat-soluble vitamins   Sec. 21.7

Female sex hormones   Sec. 17.8

Fetal alcohol syndrome   Chap. 13   Case in Point and Sec. 13.4

Flavors   Sec. 13.2   Closer Look, page 399

Food irradiation   Sec. 10.1   Closer Look, page 292

Fragrances   Sec. 13.2   Closer Look, page 399

Gene mutations   Sec. 20.10

Gene therapy   Sec. 20.11

Genes   Sec. 20.5

Goiter   Sec. 21.9

Gout   Chap. 26   Case in Point and Sec. 26.9

Greenhouse effect   Sec. 11.11   Closer Look, page 348

Halocarbon anesthetics   Sec. 12.7   Closer Look, page 377

Hard water   Sec. 14.3   Closer Look, page 435

Heart disease   Sec. 21.4   Closer Look, page 651

Heimlich maneuver   Chap. 7   Case in Point and Sec. 7.7

Hemodialysis   Sec. 8.11   Closer Look, page 243

Hemoglobin   Sec. 18.8

Hemolysis   Sec. 8.11

Hepatitis   Sec. 26.8

HIV protease   Sec. 19.10   Closer Look, page 593

HIV reverse transcriptase   Sec. 20.3   Closer Look, page 619

Hormones   Sec. 17.5

Human genome project   Sec. 20.5   Closer Look, page 622

Hydrocarbons and health   Sec. 11.9   Closer Look, page 343

Hyperammonemia   Sec. 26.8   Closer Look, page 792

Hyperbaric oxygenation   Sec. 18.10

Hyperbilirubinemia   Sec. 26.4   Closer Look, page 802

Hyperthyroidism   Chap. 10   Case in Point and Sec. 10.10

Hypothermia   Chap. 6   Case in Point and Sec. 6.7

Hypnotic   Sec. 15.9

Immunoglobulins   Sec. 22.3

Inherited disease, treatment   Chap. 20   Case in Point and Sec. 20.11

Insulin   Sec. 24.13

Interferons   Sec. 22.3

Ions in diet   Sec. 5.5   Closer Look, page 116

Jaundice   Sec. 26.8

Ketoacidosis   Sec. 25.9
Ketone bodies   Sec. 25.8
Ketosis   Sec. 25.9
Kidney stones   Sec. 8.6   Closer Look, page 226
Kwashiorkor   Sec. 21.6

Lactose intolerance   Chap. 16   Case in Point and Sec. 16.8
Lead poisoning   Sec. 19.8   Closer Look, page 591
Leukotrienes and prostaglandins   Sec. 17.11
Lipid storage diseases   Sec. 25.3   Closer Look, page 768
Local anesthetics   Sec. 15.4   Closer Look, page 471
Lou Gehrig's disease   Sec. 23.6   Closer Look, page 718

Magnetic resonance imaging (MRI)   Chap. 3   Case in Point and Sec. 3.4
Male sex hormones   Sec. 17.7
Marasmus   Sec. 21.6
Medicinal drug discovery   Chap. 5   Case in Point and Sec. 5.10
Medicine and the rain forest   Sec. 6.8   Closer Look, page 169
Methemoglobinemia   Sec. 18.8
Mitochrondria   Sec. 23.6   Closer Look, page 715
Molecular diseases   Sec. 20.10
Monitoring heart attacks   Sec. 24.7   Closer Look, page 744
Myoglobin   Sec. 18.11

Neurotransmitters   Sec. 15.7
Night blindness (nyctalopia)   Sec. 21.7
Nitric oxide, second messenger   Sec. 24.11   Closer Look, page 754
Nitroglycerin   Sec. 14.10
Nuclear reactor debate   Sec. 10.5   Closer Look, page 298
Nutrition pyramid   Sec. 21.5   Closer Look, page 655

Obesity   Chap. 17   Case in Point and Sec. 17.2
Oncogenes   Sec. 20.10
Opiates   Sec. 15.8
Oral contraceptives   Sec. 17.9

Organisms affecting health   Sec. 2.6   Closer Look, page 40
Origin of the elements   Sec. 4.4   Closer Look, page 93
Osteoporosis (osteomalacia)   Chap. 21   Case in Point and Sec. 21.9
Oxygen, disease, and aging   Sec. 23.6   Closer Look, page 718
Ozone layer   Sec. 12.1   Closer Look, page 361

Pancreatitis   Sec. 19.11
Parkinson's disease   Sec. 15.7
Pellagra   Sec. 21.8
Penicillin   Sec. 19.13
Pesticide toxicity   Sec. 19.13   Closer Look, page 603
Phenols as antiseptics   Sec. 12.3   Closer Look, page 371
Phenylketonuria (PKU)   Sec. 26.7
Phototherapies   Sec. 23.2   Closer Look, page 706
Physiological gas exchange   Sec. 7.6   Closer Look, page 194
Pill   Sec. 17.9
Prostaglandins and leukotrienes   Sec. 17.11

Radiation in medicine   Sec. 10.9
Radiation sickness   Sec. 10.8
Radioactive fallout   Sec. 10.5
Radioisotopes in medicine   Sec. 10.10
Radon in homes   Sec. 10.11   Closer Look, page 297
Recombinant DNA   Sec. 20.11
Rickets   Sec. 21.7
Runner's high   Chap. 18   Case in Point and Sec. 18.4

Sedative   Sec. 15.9
Selenium toxicity   Chap. 4   Case in Point and Sec. 4.6
Semiconductors   Sec. 4.6   Closer Look, page 98
Serum cholesterol reduction   Sec. 25.11   Closer Look, page 781
Serum LDH tests   sec. 24.7   Closer Look, page 744
Scurvy   Sec. 21.8
Sickle cell anemia   Chap. 1   Case in Point, Sec. 1.2, and Sec. 18.10
Steroid hormones   Sec. 17.6
Steroids   Sec. 17.5

Sulfonamide antibiotics   Sec. 15.4   Closer Look, page 469

Superconductors   Sec. 5.5   Closer Look, page 121

Trypsin   Sec. 19.11   Closer Look, page 598

Tumor suppressor genes   Sec. 20.10   Closer Look, page 634

Urinometers   Sec. 2.9   Closer Look, page 52

Vasopressin   Sec. 22.9

Viruses   Sec. 22.3

Vision chemistry   Sec. 13.7   Closer Look, page 412

Vitamins, fat soluble   Sec. 21.7

Vitamins, water soluble   Sec. 21.8

Water softening   Sec. 14.3   Closer Look, page 435

Water-soluble vitamins   Sec. 21.8

X-ray crystallography   Sec. 18.7   Closer Look, page 569

Introduction to General, Organic, and Biological Chemistry

Matter, Change, and Energy

Ideas about the Material World

Chemistry deals with everything in the natural world: our bodies and the world around us.

CHAPTER OUTLINE

CASE IN POINT: Sickle cell anemia

1.1   Chemistry

A CLOSER LOOK: Chemicals and Chemophobia

1.2   The scientific method

FOLLOW-UP TO THE CASE IN POINT: Sickle cell anemia

1.3   Temperature

A CLOSER LOOK: Clinical Thermometers

1.4   Physical states of matter

1.5   Mixtures

1.6   Physical properties and physical changes

1.7   Chemical reactions and chemical properties

1.8   Elements and compounds

A CLOSER LOOK: Essential Elements for Life

1.9   Mass conservation

1.10 Energy interconversion and conservation

1.11 Exothermic and endothermic processes

Students often ask, "Why study chemistry?" One answer to this question is that chemistry has produced many practical things, making it a vital, dynamic force in our culture. The food we eat, the clothes we wear, the house we live in, and the television we watch are all made possible to some degree by chemistry. Sometimes the question is more specific. "I want to be a health care [or other allied field] professional. Why should *I* study chemistry?" The answer is that nowhere is chemistry's practicality more evident than in medicine and its allied health sciences: nursing, nutrition and dietetics, clinical laboratory science or medical technology, inhalation therapy, X-ray technology, and the like. An understanding of chemistry is also necessary to the practice of any life science, such as agriculture or forestry.

The reason for chemistry's importance to these fields is simple. The more we know about how something is made and how it works, the more likely we will be able to keep it in good working order. And the more likely we will be able to repair it when it breaks down. Chemistry is concerned with how things are made and how they interact with one another. The human body is a complex chemical factory, so its composition and workings are the concern of chemistry. Many of today's ideas about diet and exercise stem from applications of chemistry to medicine and the life sciences. Applications of chemical principles to the design of medical drugs have produced medicines for such diseases as *a*cquired *i*mmune *d*eficiency *sy*ndrome (AIDS), high blood pressure, and depression. In order to emphasize the connection between chemistry and the allied health sciences, our first Case in Point addresses the root cause of sickle cell anemia, a disease of the blood.

CASE IN POINT: Sickle cell anemia

 In 1904, James Harrick, a Chicago physician, observed that the red blood cells of people afflicted with a particular inherited blood disease are shaped rather like the blade of a common garden sickle (see figure). The disease, sickle cell anemia, got its name from this observation. Harrick documented the painful physical symptoms of the disease but was unable to find its cause. Was there a single cause for this chronic and often fatal disease? Would identifying the cause lead to an effective treatment or cure? It was not until 1949 that American chemist Linus Pauling showed that sickle cell anemia has a chemical basis. We will see in Section 1.2 what reasoning Pauling used and what he learned.

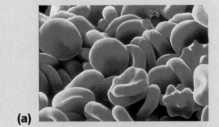

(a)

(b)

Compared with normal red blood cells (a), sickled cells (b) have a clearly different shape.

1.1 Chemistry

AIMS: *To define the terms* **matter** *and* **chemistry.** *To explain why knowledge of chemistry is important to health care workers.*

Our study of the natural sciences—chemistry, biology, and physics—aims to improve our understanding of nature. **Chemistry** *deals with the structure of* **matter**—*the stuff that things are made of—and the changes that matter undergoes.* All matter consists of chemicals, a fact that is sometimes ignored, as discussed in A Closer Look: Chemicals and Chemophobia, on page 5.

Health professions share many common interests with chemistry and the other natural sciences. For example, the structure of matter interests chemists, physicists, and biologists, and it also is of interest in medicine and its allied health fields. Health professionals certainly would be interested in a chemist's new material for bone replacement. Similarly, they would welcome a physicist's new method for "seeing" inside the human body without the need for surgery or a biologist's new insight about the way that AIDS is transmitted. In many parts of this book you will not be able to tell where chemistry ends and biology, physics, or health science begins. You should not try. The more you know about science, the less meaningful you will find such distinctions.

The remainder of this chapter introduces some fundamental aspects of the principles and language of chemistry. By the time you complete it, you will be on your way to applying some of these principles and language to challenges in the allied health fields.

1.2 The scientific method

AIM: *To describe the scientific method and show how the terms* **observation, hypothesis, experiment, theory,** *and* **law** *relate to this method.*

Chemists study many interesting materials and processes for a variety of reasons. They may see the potential for a practical application—a chemical might be examined as the treatment for a human disease—or they might simply be curious about something they have seen. Two chemists seldom approach a problem in exactly the same way because of differences in their personalities, outlook on science, training, resources, and so forth. Despite these differences, both chemists will probably study the problem using a systematic approach that is responsible for most of the advances in modern chemistry. *This systematic approach to the solution of chemical and other scientific problems is called the* **scientific method.** Penicillin and the polio vaccine were discovered and the secrets of heredity were unveiled through application of the scientific method (Fig. 1.1). You may find it helpful to refer to Figure 1.1 as we discuss the scientific method in the following paragraphs.

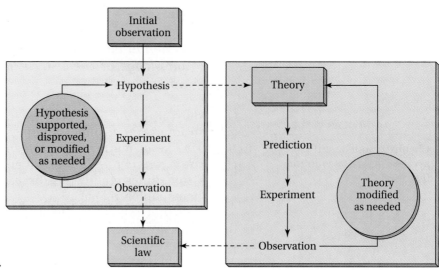

Figure 1.1
Components of the scientific method.

Health care workers routinely use the scientific method to diagnose patients. They generally begin a diagnosis by observing symptoms, then formulate a hypothesis about what is wrong, and do tests (experiments) to confirm or reject the hypothesis.

Although there is no unique way to solve a scientific problem, generally, scientists collect facts about the behavior of the material or phenomenon being studied. These facts are usually obtained from **observations**—*recognizing when something happens and clearly stating what happens.* Imagine that we accidentally drop a small amount of baking soda on some lemon juice while we are cooking. We would *observe* that the baking soda foams when it contacts the lemon juice (Fig. 1.2), and we might be interested enough to study the process by performing some **experiments**—*carefully designed and controlled procedures for obtaining useful information.* In order to decide what experiments will give us useful information, we must develop a **hypothesis**—*a tentative explanation for an observation.* Our first

Figure 1.2
The initial observation:
Baking soda foams when it is sprinkled on lemon juice.

hypothesis about the observation of the foaming of baking soda in lemon juice might be: *Citrus fruit juices cause baking soda to foam.* Our hypothesis may be correct or incorrect, since it is only a tentative explanation of our observations.

Now that we have a hypothesis, we can perform experiments to test the hypothesis. We might conduct an experiment in which we add baking soda to orange juice instead of lemon juice. This new experiment will help us support or disprove our hypothesis. When we add baking soda to orange juice, the soda foams, so our hypothesis—citrus fruit juices cause baking soda to foam—is supported. If the baking soda did not foam—if the experiment did not give the expected results—the hypothesis would have been disproved. We would have to scrap the hypothesis or adjust it to take into account the new findings. We could perform another experiment with baking soda, this time adding it to grapefruit juice. We would observe that the baking soda foams, further supporting our hypothesis. This cycle of observation, formulation of a hypothesis, and experimentation to test the hypothesis is typical of a scientific study. After a hypothesis has been formulated, a scientific study often progresses through many rounds of testing

A Closer Look

Chemicals and Chemophobia

The ink in a ballpoint pen is made from chemicals, and so is the pen itself. Laundry detergent is chemical, as is its container. Rose petals are chemical, and so is the bee that pollinates flowers. Every animal in the entire world is chemical. Even your favorite hamburger or pizza is *all* chemical. And so are *all* fruits and vegetables. We cannot think of any material thing that *is not chemical.*

Unfortunately, the word *chemical* is synonymous with *hazardous substance* or *toxic material* or *poison* to a large segment of the general public. This misperception elicits the unreasonable fear of chemicals called *chemophobia.* The reasons for widespread chemophobia are understandable. Miracle drugs or newborn babies are seldom referred to as chemicals. Rather, we are bombarded with reports that our health is threatened by chemicals: fluoride in drinking water, radon gas in buildings, and asbestos in schools. Drug addiction is sometimes referred to as "chemical dependency." Media reports, isolated incidents, emotions, and misconceptions have helped to fuel chemophobia.

Everything is chemical. We eat, drink, breathe, wear, ride in, sit on, and work with chemicals.

In our thinking about chemicals, we must recognize that chemicals can be used wisely or unwisely. Problems that exist because of chemicals are usually caused by the unwise uses of chemicals and not by the chemicals themselves. A better knowledge of chemistry among the general public would help people evaluate reports of the risks posed by chemicals. Knowledge of the possible uses and abuses of chemicals would certainly help alleviate chemophobia in the general public.

a hypothesis by further experimentation. There is an important relationship between a hypothesis and experiments. The hypothesis can be used to design new experiments, and the results of new experiments indicate whether the hypothesis needs to be rejected or adjusted. In our experiments, the hypothesis that juices of citrus fruits cause baking soda to foam led us to test the hypothesis with other citrus fruit juices.

Although all our experiments support our hypothesis, we are still uncertain as to whether the hypothesis will stand up to further testing. Later experiments might disprove the hypothesis. For example, we do not know for certain what results lime juice or kumquat juice will give. However, if we are reasonably confident that our hypothesis is valid, we might expand our inquiry to ask *why* baking soda foams when it is added to citrus fruit juices. These juices contain mostly water, so we might formulate a second hypothesis: *The water in citrus juices makes the baking soda foam.* If we test the hypothesis with another experiment by adding baking soda to tap water, we observe that the baking soda does *not* foam (Fig. 1.3a). Our hypothesis is *not* supported. We will have to reject it and formulate another hypothesis to explain the foaming of baking soda when it comes into contact with citrus fruit juices. What else do we know about the juices? All the citrus juices taste more or less sour. Could it be that sourness has something to do with making baking soda foam? Water is not sour, and it does not make baking soda foam. We can try a third hypothesis: *Citrus fruit juices make baking soda foam because they are sour.* Vinegar is sour, and we can use it to test this hypothesis by an experiment—adding a small amount of baking soda to a teaspoon of vinegar. We observe that the baking soda foams vigorously, and our hypothesis is supported (Fig. 1.3b). We have developed a reasonable hypothesis about the action of citrus fruit juices on baking soda: Citrus fruit juices make baking soda foam because they are sour. We have not proven this hypothesis because we have not tested all sour liquids. In general, a hypothesis cannot be proved because additional experiments might disprove it. Nevertheless, our work has given us some good information to use if we want to pursue a detailed study of what makes baking soda foam when it is added to certain liquids.

Sometimes the results of many observations or experiments can be organized into a general statement called a *law.* The law of gravity—what goes up must come down—is this kind of general statement. *A **scientific law** expresses a principle that is true for every experiment done so far,* and in this sense, a law is usually considered to be proved. The principles expressed as scientific laws are so general and so universally applicable that it is very unlikely that any future experiment will disprove them. *Similarly, a **theory** is a hypothesis or a set of hypotheses that are supported by the results of many experiments.* Like laws and hypotheses, theories cannot be proved absolutely, because they provide only tentative—although experimentally well-supported—explanations for experimental results. For this reason, theories can change if the outcomes of new experiments demand it. Theories are very valuable, since a well-developed theory allows scientists to *predict* the results of experiments that have not yet been done. More important, the total of all scientific theories constitutes scientists' best explanation for how the physical universe functions.

(a)

(b)

Figure 1.3
Further observations:
(a) Baking soda does not foam in tap water.
(b) Baking soda does foam when added to vinegar.

FOLLOW-UP TO THE CASE IN POINT: Sickle cell anemia

Recall from the opening to this chapter that Linus Pauling (see figure) discovered the root cause of sickle cell anemia. Pauling made this discovery by applying the scientific method—observations, experiments, and hypotheses—to his work. In the late 1940s, scientists already knew a fair amount about red blood cells. They knew that red blood cells carry oxygen in the blood and that red blood cells contain mostly hemoglobin, which is the actual carrier of oxygen. They also knew that the sickle-shaped red blood cells of patients with sickle cell anemia are less effective in carrying oxygen than the red blood cells of healthy people. Nobody knew why the hemoglobin of patients with sickle cell anemia was a less effective oxygen carrier. From these *observations,* Pauling formulated a *hypothesis:* Since the oxygen-carrying ability of red blood cells depends on hemoglobin, the hemoglobin of patients with sickle cell anemia is somehow defective; the oxygen-carrying ability of hemoglobin and the shapes of the red blood cells might be affected by this defect. He then performed *experiments* to test his hypothesis. The experiments consisted of extracting and analyzing the hemoglobin from the red blood cells of a healthy person and from a patient with sickle cell anemia. These experiments supported his hypothesis. The hemoglobin of the patient with sickle cell anemia was slightly different from the hemoglobin of the healthy person. Pauling tested his hypothesis further with the hemoglobin of additional healthy people and more patients with sickle cell anemia. The results were always the same: The hemoglobin of patients with sickle cell anemia is slightly different from the hemoglobin of healthy subjects. With a good working hypothesis in hand, Pauling and others were eventually able to pinpoint the exact chemical difference between normal hemoglobin and sickle cell hemoglobin.

Because of Pauling's application of the scientific method to the root cause of sickle cell anemia, other scientists were able to expand our understanding of this disease. Using the scientific method, these scientists have learned that the slight difference in sickle cell and normal hemoglobin causes a vast difference in the biological function of sickled and normal red blood cells. The hemoglobin of patients with

Linus Pauling (1901–1994) won the 1954 Nobel Prize in chemistry and the 1963 Nobel Peace Prize. His ideas and discoveries embraced all aspects of chemistry.

sickle cell anemia carries less oxygen than the hemoglobin of healthy people. Sickle cell hemoglobin that is not carrying oxygen tends to clump in red blood cells. This clumping deforms the cells, causing them to assume their odd sickle shapes that we saw in the figure on page 2. The sickled cells block blood vessels, thereby cutting off supplies of vital oxygen to tissues. Although there is as yet no cure for sickle cell anemia, further research has suggested ways to alleviate the painful episodes of the disease. It suggested, for example, that the administration of oxygen might be helpful, since sickle cell hemoglobin is less likely to clump in red blood cells when it is carrying oxygen than when it is not. Today, new methods are being tried to prevent the clumping of sickle cell hemoglobin even when it does not carry oxygen.

The sickle cell story is a fairly typical application of the scientific method, and many other similar stories could be told. The story also shows that solving problems in the modern health sciences often demands an understanding of the underlying chemistry.

1.3 Temperature

AIM: To convert between Fahrenheit, Celsius, and Kelvin temperature scales.

The temperature of warm-blooded animals, such as humans, is regulated at a constant value by body systems, but the temperature of cold-blooded animals, such as reptiles and amphibians, changes to match that of the surrounding air or water. Certain microorganisms, the thermophilic bacteria, live in hot springs at temperatures as high as 90 °C.

The practice of chemistry almost always involves the recognition of differences between two observations. The difference in the way that air and water behave when placed in containers is a simple example of an observed difference between two substances. One of the easiest differences to observe is the difference in the temperatures of two objects. It is important to understand how temperature is determined, since the effects of temperature on matter play an important role in the sciences.

Almost all substances expand with an increase in temperature and contract as the temperature decreases. This is the basis for the common mercury-in-glass thermometer. The mercury in the thermometer rises up the stem when the temperature increases because the mercury expands more than the volume of the bulb that holds it.

Two readily determined temperatures serve as reference temperatures for the *Celsius scale*, a temperature scale named after Anders Celsius (1701–1744). This scale uses the temperatures at which water freezes to ice and boils to make steam as its reference values. *In the* **Celsius scale,** *the freezing temperature of water is 0 °C and the boiling temperature of water is 100 °C.* The space between these two fixed temperatures is divided into 100 equal intervals, or degrees.

On the **Fahrenheit scale,** devised by Daniel Fahrenheit (1686–1736), *the freezing point of water is 32 °F and the boiling point of water is 212 °F.* The space between these two fixed points is divided into 180 equal divisions, or degrees. A degree Fahrenheit is 100/180, or 5/9, of a degree Celsius (Figs. 1.4a and b). The relationship between the Celsius and Fahrenheit scales is given by the equation

$$\frac{(°F - 32)}{180} = \frac{°C}{100}$$

If we rearrange this equation, we get

$$°C = \tfrac{5}{9}(°F - 32) \qquad \text{and} \qquad °F = (\tfrac{9}{5}°C) + 32$$

A third temperature scale used in the physical sciences is the *Kelvin scale* (Fig. 1.4c), named after Lord Kelvin (1824–1907), a British physicist and mathematician. *On the* **Kelvin scale,** *the freezing point of water is 273 K (the degree sign is not used) and the boiling point of water is 373 K.* A change of one degree on the Kelvin scale is the same as that on the Celsius scale. *The zero point on the Kelvin scale,* **absolute zero,** *is −273 °C.* The relationship between a temperature on the Celsius scale and one on the Kelvin scale is

$$K = °C + 273$$

A Closer Look: Clinical Thermometers describes some of the different kinds of thermometers used to determine body temperature in homes and clinical settings.

Dental and medical instruments are usually sterilized by heating them in chambers called *autoclaves* at 121 °C and 15 lb/in^2 of pressure for 15 minutes. Sterilization is also achieved with dry heat at 170 °C for 90 minutes. Either combination of temperature, pressure, and time is sufficient to kill dangerous bacteria and viruses.

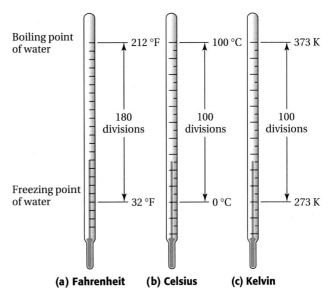

Figure 1.4
A comparison of the Fahrenheit, Celsius, and Kelvin temperature scales. Since there are 100 degrees Celsius for every 180 degrees Fahrenheit, a change of 1 degree on the Fahrenheit scale is the same as 100/180 or 5/9 of a degree Celsius. A change of 1 degree on the Celsius scale, however, is the same a change of 1 degree on the Kelvin scale.

A Closer Look

Clinical Thermometers

"Do you have a fever?" This is usually one of the first questions you are asked if you don't feel well. The average oral temperature of a healthy adult is about 37 °C or 98.6 °F. A normal rectal temperature is 0.6 °C or 1°F higher than the temperature taken orally. If you have a fever, your body temperature may be as high as 106 °F.

Until a few years ago, the mercury-in-glass thermometer was the only type of clinical thermometer in general use. Mercury thermometers are easily broken, which could expose people to toxic mercury metal. Today, electronic thermometers are commonly used in doctors' offices and in hospitals (see figure). A disposable plastic sleeve covers the temperature probe to prevent the spread of infection. Electronic thermometers are very fast and accurate.

Another kind of commonly found thermometer is a temperature-sensitive tape that changes

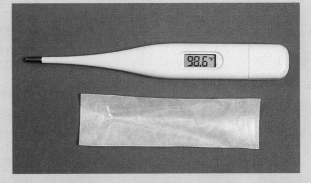

An electronic thermometer.

color as it changes temperature. The tape contains liquid crystals encased in a narrow, flexible plastic strip. As the temperature of the strip changes, the color of the liquid crystals changes, and the temperature reading becomes visible on the scale. Liquid-crystal thermometers are not as accurate as mercury or electronic thermometers, but they are inexpensive, easy to use, and virtually unbreakable.

EXAMPLE 1.1	**Converting Fahrenheit temperature**

If a patient has a temperature of 101 °F, what is the temperature in (a) °C and (b) K?

SOLUTION

(a) $°C = \frac{5}{9}(°F - 32) = \frac{5}{9}(101 - 32) = \frac{5}{9} \times 69 = 38.3 °C$

(b) To convert from Fahrenheit to Kelvin, we must first convert to Celsius. Using the answer from part (a):

$$K = °C + 273 = 38.3 °C + 273 = 311 \text{ K}$$

EXAMPLE 1.2	**Converting Celsius temperature**

Convert 10 °C to (a) °F and (b) K.

SOLUTION

(a) $°F = \left(\frac{9}{5} °C\right) + 32 = \left(\frac{9}{5} \times 10\right) + 32 = 18 + 32 = 50 °F$

(b) $K = °C + 273 = 10 + 273 = 283 \text{ K}$

EXAMPLE 1.3	**Converting Kelvin temperature**

What is 194 K in (a)°C and (b)°F?

SOLUTION

(a) $°C = K - 273 = 194 - 273 = -79 °C$

(b) To convert from Kelvin to Fahrenheit, we must first convert to Celsius. Using the answer from part (a):

$$°F = \left(\frac{9}{5} °C\right) + 32 = \left(\frac{9}{5} \times -79\right) + 32 = -142 + 32 = -110 °F$$

PRACTICE EXERCISE 1.1

Surgical instruments are sterilized by heating at 170 °C for 1.5 hours. Convert 170 °C to (a) °F and (b) K.

PRACTICE EXERCISE 1.2

Lithium metal melts at 453 K. What is this temperature in (a) °C and (b) °F?

1.4 Physical states of matter

AIM: To name, characterize, and list examples of the three states of matter.

Focus

There are three physical states of matter.

We have already observed that matter takes different forms. Water certainly looks different from air, and water and air look different from iron. Matter exists in three different physical states: *solid*, *liquid*, or *gas*. Each state of matter has certain unique characteristics (Fig. 1.5).

(a) **(b)** **(c)**

Figure 1.5
The three physical states of matter. (a) A solid has its own shape and volume, no matter what container it is in. (b) A liquid conforms to the shape but not to the volume of its container. (c) A gas will occupy all of the volume of its container.

A **solid** *has a definite shape and volume.* The shape of a solid does not depend on the shape of the container. Bone, ice, coal, and iron are examples of solids.

A **liquid** *is a form of matter that flows, has a fixed volume, and takes the shape of its container.* Examples of liquids are water, oil, and milk.

A **gas** *takes the shape and completely occupies the volume of its container.* Air is a gas, and it fills the empty space in a vessel that contains it. The term *gas* is commonly used for substances that are gases at room temperature. **Vapor** *is a substance that, although a gas, is generally a liquid or solid at room temperature.* Steam, the gaseous form of water, is a vapor. Humid air is air that contains a small amount of water vapor.

PRACTICE EXERCISE 1.3

What is the physical state of each of the following at room temperature?
(a) air   (b) syrup   (c) copper   (d) gasoline

1.5 Mixtures

AIMS: *To classify a sample of matter as a pure substance or a mixture. To classify a mixture as homogeneous or heterogeneous.*

Focus

Most matter consists of mixtures.

Very few of the chemicals that we encounter are pure in the way that chemists think of purity. Ordinary refined cane sugar is a substance that is essentially chemically pure—table sugar contains only sugar. However, mixtures of substances are much more common. **Mixtures** *consist of a blend of two or more substances.* Milk, blood, brine (salty water), wood, and gasoline are mixtures.

The most important characteristic of mixtures is that they may have *variable compositions*. Homogenized milk is a mixture of milk and butterfat, but it might contain 1%, 2%, or 4% butterfat. Blood is a mixture of water, various chemicals, and cells. Blood composition varies somewhat from one individual to another, and each person's blood composition varies to a certain extent with health, nutrition, and activity.

Brine (a solution of table salt and water) and wood are examples of two classifications of mixtures. Brine is a *homogeneous mixture. A **homogeneous mixture** is completely uniform—its components are evenly distributed throughout the sample. Homogeneous mixtures are also called* **solutions.** As Table 1.1 shows, solutions may be gases, liquids, or solids. If we examine any portion of a jar of brine, we find that it is uniform—one portion looks just like any other portion. A second jar of brine might contain different proportions of salt and water, since mixtures can have varying compositions, but this brine also would be uniform and therefore a homogeneous mixture. Wood, on the other hand, is a *heterogeneous mixture. A **heterogeneous mixture** is not uniform—its components are not evenly distributed.* If we examine a block of grainy wood, we can see that the substances that compose the wood are not evenly distributed throughout the block.

Some mixtures are easy to separate into their various components. Consider the gray-colored mixture produced by stirring together powdered yellow sulfur and black iron filings. The individual particles of sulfur and iron can be readily distinguished from one another under a magnifying glass. The mixture is easy to separate. Figure 1.6 shows how the iron filings can be removed from the mixture with a magnet, leaving the sulfur behind.

Tap water is a homogeneous mixture of water plus a variety of other substances that are dissolved in it. One method used to purify water is distillation. In the distillation of water, the water is heated to form steam, which is then condensed by cooling to give water (Fig. 1.7). Water collects in a receiver, and the solid substances originally dissolved in the water remain in the distillation flask. Distilled water is a pure substance except for any dissolved gases it may contain.

Homogeneous
homo- (Greek): the same
genes (Greek): kind

Heterogeneous
heteros (Greek): different

Figure 1.6
Physical change. Sulfur and iron filings can be mixed and separated without changing their composition.

Table 1.1 Some Common Types of Solutions

System	Example
gas–gas	Air (carbon dioxide and oxygen in nitrogen)
liquid–gas	Moist air (water vapor in air)
gas–liquid	Soda water (carbon dioxide in water)
liquid–liquid	Vinegar (acetic acid in water)
	Vodka (ethanol in water)
solid–liquid	Brine (table salt in water)
	Syrup (sugar in water)
solid–solid	Gold alloy (copper in gold)

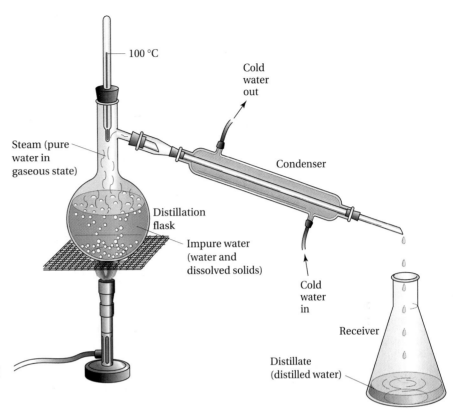

Figure 1.7
The distillation of water. Water is boiled in a flask; the resulting steam is passed through a water-cooled condenser, and the steam condenses to water. The condensed steam is called the *distillate,* or *distilled water.*

EXAMPLE 1.4	Devising the separation of mixtures

How would we separate the following mixtures?

(a) sawdust from small rocks   (b) sand from sugar

(c) iron filings from zinc filings

SOLUTION

(a) Add water to the mixture. The sawdust floats, and the rocks sink.

(b) Add water to dissolve the sugar. Pour the mixture onto a piece of cloth. The sand remains on the cloth, and the sugar solution goes through. Evaporate the water to recover the sugar.

(c) Use a magnet to attract the iron filings.

PRACTICE EXERCISE 1.4

Classify each of the following as homogeneous or heterogeneous mixtures:

(a) solder (lead and tin)   (b) tea   (c) a slice of bread

(d) a chocolate-chip cookie

1.6 Physical properties and physical changes

AIM: To identify physical changes and physical properties of matter.

Sublimation is an effective method for purifying chemical compounds. A mixture is heated, and the compounds that sublime are collected on a cold condenser placed above the vessel. The ancient Chinese separated several hormones, compounds with dramatic effects on the body in minute amounts (Chapter 17), from urine by sublimation.

Freeze-drying involves the sublimation of ice. A water solution of the desired material is frozen, and the ice is removed by sublimation at low pressures. Freeze-drying extends the shelf life of many valuable biological compounds such as proteins and nucleic acids. Instant coffee crystals are produced by freeze-drying brewed coffee.

The physical state of a substance—gas, liquid, or solid—is one of the physical properties of that substance. *A **physical property** is a quality or condition of a substance that can be observed without changing the substance's composition.* Other physical properties of matter include color, solubility, odor, hardness, mass, density, electrical conductivity, magnetism, melting temperature, and boiling temperature.

Physical properties can help us identify substances (Table 1.2). Color and the absence of color are examples of physical properties. Gold can be distinguished from silver by the difference in the colors of gold and silver. Gold and silver can be distinguished from diamond because diamond is colorless.

The physical properties of a substance can be quite different at different temperatures. For example, water is a liquid at 25 °C and freezes to solid ice at 0 °C. *A conversion without causing a change in the composition of the substance is called a **physical change.*** The melting of ice, the freezing of water, the vaporization of water into steam, the condensation of steam to water—all are examples of physical changes (Fig. 1.8). These physical changes do not change the *composition* of the water. We can tell that the composition of the water has not changed because water that has been frozen and melted has the same properties as water that has been converted into steam and then condensed back to water.

Table salt, a white solid at room temperature, also can exist in each of the three physical states. However, the temperatures at which the changes of state occur in salt are much higher than for the corresponding physical changes in water. Salt melts at 801 °C and boils at 1413 °C.

Many substances undergo physical change from solid to liquid or from liquid to gas as the temperature is raised. One exception is iodine. This purple solid changes into a purple vapor without passing through a liquid state (Fig. 1.9). *The change of a substance from a solid to a gas or vapor without passing through the liquid state is called **sublimation.*** A purple solid unchanged in appearance and composition from the original solid is

Table 1.2 Physical Properties of Some Common Substances

Substance	State	Color	Melting point (°C)	Boiling point (°C)
oxygen	gas	colorless	−218	−183
bromine	liquid	red-brown	−7	59
water	liquid	colorless	0	100
sulfur	solid	yellow	113	445
sodium chloride (table salt)	solid	white	801	1413
iron	solid	silver-white (freshly cut)	1535	2750

Figure 1.9
The physical change in which a solid is converted directly into a gas or vapor is called *sublimation.* Iodine crystals sublime when gently heated; iodine vapor forms iodine crystals when it is cooled.

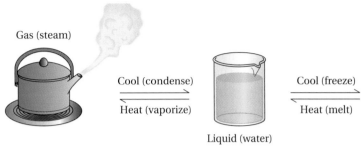

Gas (steam)

Cool (condense) / Heat (vaporize)

Cool (freeze) / Heat (melt)

Liquid (water)

Solid (ice)

Figure 1.8
The three physical states of matter are solid, liquid, and gas. For water, these states are ice, water, and steam. Physical changes can take place among these states of matter when temperature changes occur.

obtained when iodine vapor is cooled. Ice also sublimes under certain conditions. Have you ever seen the ice cubes stored in a freezer get smaller over time? At the low temperature of the freezer, the ice cubes sublime to water vapor without becoming liquid water. A freshly washed sheet hung on a clothesline outdoors in cold weather freezes solid. Although the temperature does not climb above 32 °F, the sheet still dries because the ice on the sheet sublimes.

1.7 Chemical reactions and chemical properties

AIM: To identify chemical changes and chemical properties of matter.

Focus

Chemical reactions cause changes in the composition of matter.

When we heat pure cane sugar in a pan, a physical change occurs as the sugar melts to a colorless liquid. However, the liquid soon turns brown and emits the smell of caramel as the temperature is raised. If we strongly heat the contents of the pan, they eventually char into a black mass (Fig. 1.10). If the pan is fitted with a lid, we also might see some liquid water forming on its underside. When we allow the black material to cool, it does not turn back into sugar. Since the material did not change back to sugar when we

Figure 1.10
When sugar (a) is heated, it melts in a physical change (b), and then it decomposes to carbon and water in a chemical change (c).

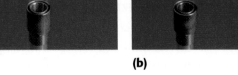

(a)   **(b)**   **(c)**

Cooking food involves many chemical reactions. When food is cooked, many complex biological compounds are broken down into simpler compounds. The partial breakdown of complex compounds makes the food more tender and easier to digest.

(a)

(b)

Figure 1.11
Chemical change.
(a) A chemical reaction occurs when a mixture of sulfur and iron filings is heated.
(b) The reactants are converted into a product that has different chemical and physical properties than the reactants.

stopped heating, the sugar must have been changed to one or more new substances. That is, the composition of the sugar changed when it was heated. If we were to analyze the black material, we would learn that it is mostly carbon, the same material that is the major component of charcoal The heating of the sugar has produced two new substances—water and carbon. This behavior does not conform to the definition of a physical change—a physical change does not change the composition of a substance. The change in composition of the sugar upon strong heating is an indication that a *chemical change* has taken place. The decomposition of the sugar to different substances is therefore the result of a **chemical reaction**—*a change in the composition of matter. In a chemical reaction, the starting substances,* **reactants,** *are changed into new substances,* **products.** In the example given, the sugar is the reactant. Water and carbon and any other substances produced by the heating are the products of the reaction. Chemists use an arrow as a shorthand form of the phrase "are changed into":

$$\text{Reactants} \longrightarrow \text{Products}$$

As another example of the difference between physical changes and chemical reactions, recall how sulfur and iron filings can be separated unchanged from a mixture. This separation is an example of a physical change. If the mixture is heated, however, a chemical change occurs, and the sulfur and iron are changed into a nonmagnetic substance, iron sulfide (Fig. 1.11). Using the shorthand expression, we could write this change as

$$\text{Iron + sulfur} \xrightarrow{\text{Heat}} \text{Iron sulfide}$$
$$\text{Reactants} \qquad\qquad \text{Product}$$

Just as every substance has physical properties, it also has chemical properties. *A substance's behavior in chemical reactions—its chemical reactivity—constitutes its* **chemical properties.** Chemical properties are observed only when a substance is undergoing a chemical change.

Is there any easy way to tell whether a change is physical or chemical? It is not always obvious, but here are a few guidelines. A color change, such as the leaves turning in the fall, or an odor change, such as rotting meat, often accompanies a chemical change. Thus, a change in color and a change in odor can be chemical properties. Physical changes, especially those involving a change of state, usually can be reversed. Ice can be melted and then the water refrozen. The components of mixtures can be separated and then mixed back together. Most chemical changes are not so easily reversed. Once iron has reacted with oxygen to form rust, as it often does on a car, we cannot easily reverse the process. Once meat has begun to spoil, we cannot restore it to freshness, and autumn leaves cannot be restored to spring green.

PRACTICE EXERCISE 1.5

Classify each of the following as a physical or chemical change:

(a) gasoline evaporates from a spill   (b) a diamond is cut
(c) a candle burns   (d) a snowflake melts

1.8 Elements and compounds

AIM: To distinguish between an element and a compound.

Your early education consisted of learning the elements, or fundamentals, of reading, writing, and arithmetic. Later in school you used these educational elements within more complex subjects. Similarly, in chemistry, some substances—which are chemically fundamental—are called *elements*. Other substances are *compounds*, which contain elements but which are chemically more complex. **Elements** *are the simplest forms of matter.* They cannot be separated into simpler substances by physical methods or chemical reactions. Elements are the building blocks for all other substances, and they can combine with one another in chemical reactions to form compounds. **Compounds** *are substances that can be broken down to simpler substances only by chemical reactions.* Every element and every compound has its own unique set of properties.

Distinguishing elements and compounds

A common experiment—heating pure sugar—will help show the difference between elements and compounds. Recall from the preceding section that the products of the strong heating of sugar are carbon and water. This result proves that sugar is a compound, not an element, because the sugar is broken down into two distinct substances, carbon and water, in a chemical reaction. And the possible decompositions are not finished. The water that comes from the breakdown of the sugar could be broken down into hydrogen and oxygen by another chemical reaction; it too is a compound. Carbon, hydrogen, and oxygen cannot be broken down into simpler substances; they are elements.

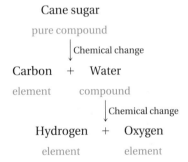

Each element has properties in some way different from those of every other element. For example, helium is a gas that is lighter than air; iron is a silver-gray solid that conducts electric current and rusts in moist air. Sulfur is a brittle yellow solid that burns with a suffocating odor.

In general, the chemical and physical properties of compounds are quite different from those of their component elements. For example, sugar is a white, sweet solid that dissolves in water. Carbon is a black, tasteless solid that does not dissolve in water. Water is a tasteless, odorless, colorless liquid, but oxygen and hydrogen are tasteless, odorless, colorless gases. Table salt (or sodium chloride, to use its chemical name) is a nontoxic compound, but it is composed of the elements sodium, a soft metal that reacts

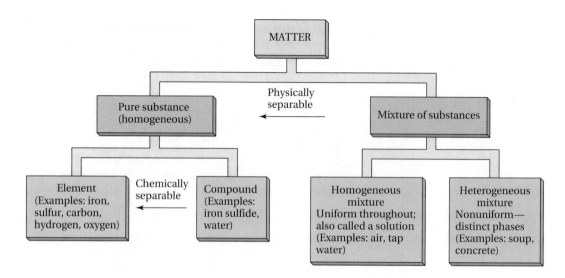

Figure 1.12
The classification
of matter.

explosively with water, and chlorine, a pale green poisonous gas. Elements previously mentioned in this chapter include oxygen, hydrogen, iron, sulfur, and iodine. Iron sulfide and water, also mentioned previously, are compounds. Figure 1.12 is a flowchart that will help you distinguish among mixtures, elements, and compounds.

Chemical symbols

Carbon, hydrogen, oxygen, sodium, and chlorine are only a few of the 90 natural elements. There are 110 elements, but some must be made artificially. All matter in the universe is composed of elements. Each element is represented by a symbol. The symbols for most elements consist of the first one or two letters of the name of the element, as shown in the table on the inside back cover. Since some elemental symbols are derived from older Latin names, however, the symbol is not always consistent with the common name. Table 1.3 lists these exceptions. Table 1.4 gives the origins of the names of some of the elements. A Closer Look: Essential Elements for Life discusses some of the elements that are needed for human life to exist.

Table 1.3 Symbols and Latin Names for Some Elements

Element	Symbol	Latin name
sodium	Na	*natrium*
potassium	K	*kalium*
antimony	Sb	*stibium*
copper	Cu	*cuprum*
gold	Au	*aurum*
silver	Ag	*argentum*
iron	Fe	*ferrum*
lead	Pb	*plumbum*
mercury	Hg	*hydrargyrum*
tin	Sn	*stannum*
tungsten	W	from *wolfram* (German)

Table 1.4 Origin of the Names of Some of the Elements

Element	Symbol	Comments
berkelium	Bk	First isolated at the University of California, Berkeley
californium	Cf	Discovered at the University of California
curium	Cm	After Pierre and Marie Curie
nobelium	No	After the Nobel Institute in Stockholm, Sweden, where it was first discovered
phosphorus	P	Greek: *phosphoros,* bringer of light; certain forms of this element give off light
polonium	Po	Named by its codiscoverer, Marie Curie, after her native land, Poland
radium	Ra	Latin: *radius,* ray; this element emits penetrating rays
technecium	Tc	Greek: *technetos,* artificial; this element does not occur in nature
uranium	U	Latin: *uranus,* heaven

A Closer Look

Essential Elements for Life

The most abundant elements in the human body and other living things are oxygen (O), carbon (C), hydrogen (H), and nitrogen (N). Together these four elements account for approximately 96% of our body weight. Seventeen other elements are essential for maintaining a healthy body. They make up the remaining 4% (see figure). The compound water is made from the elements oxygen and hydrogen and accounts for about 60% of our body weight.

The phrase "We are what we eat" is true. The foods we eat and the liquids we drink provide our bodies with the elements we need to live. For example, dairy products provide the elements calcium (Ca), magnesium (Mg), and phosphorus (P). These elements are required for building strong bones and teeth. Some elements required for life are present in our bodies in very small amounts. Some of these required trace elements, such as copper (Cu), are extremely toxic in larger doses.

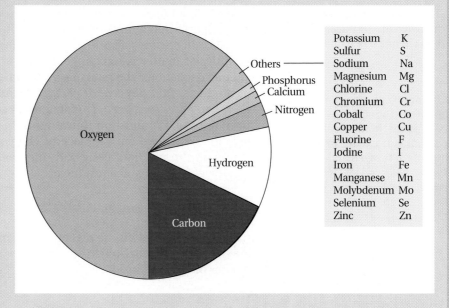

The elements and their relative abundance by weight in the human body.

| EXAMPLE 1.5 | Writing symbols for elements |

Write the chemical symbol for each of the following trace elements:
(a) iron, a component of hemoglobin
(b) fluorine, helps prevent dental caries
(c) iodine, needed for proper thyroid function
(d) manganese, aids in fat digestion

SOLUTION

(a) Fe   (b) F   (c) I   (d) Mn

PRACTICE EXERCISE 1.6

What chemical elements are represented by the following symbols?
(a) S, found in proteins
(b) Cl, important component of body fluids
(c) Mg, necessary for proper nerve function
(d) Co, found in vitamin B_{12}

1.9 Mass conservation

AIM: To state the law of conservation of mass.

Focus

The mass of substances involved in a physical change or a chemical reaction is conserved.

The law of conservation of mass applies to the action of an ordinary photographic flashbulb. The components needed to make the bulb flash (oxygen and magnesium) are sealed inside the bulb. When a flashbulb goes off, we can see that a chemical reaction has occurred. If we weigh the bulb before and after firing, we find that the mass is the same.

Mass *is the quantity of matter an object contains.* The gram, abbreviated *g*, is a measure of mass. When we melt 10 g of ice, for example, we get 10 g of water. (To put these masses in perspective, a penny has a mass of about 3 g, and a quarter has a mass of about 6 g.) If we convert this water into steam, the mass of steam is also 10 g. During these changes of physical state, the quantity of matter is unchanged; mass is conserved. This is also true of chemical reactions; the total mass of reactants always equals the total mass of products. *In a physical or chemical change, mass is neither created nor destroyed—it is conserved.* This statement is known as the **law of conservation of mass.** Combustion (burning) is an example of a familiar chemical reaction. When we burn a piece of charcoal, atmospheric oxygen is consumed in the reaction along with the charcoal. The products are carbon dioxide gas, water, and a residue of ash:

$$\text{Charcoal} + \text{Oxygen} \xrightarrow{\text{Combustion}} \text{Carbon dioxide} + \text{Water} + \text{Ash}$$

If we accounted for all the reactants and products, we would find that the mass of the charcoal plus the oxygen consumed equals the sum of the masses of carbon dioxide, water, and ash.

1.10 Energy interconversion and conservation

AIMS: To define energy. To state and explain the law of conservation of energy.

Focus

Forms of energy can be interconverted, but energy is neither created nor destroyed.

All chemical and physical changes involve the absorption or emission of energy. **Energy** *is the capacity for overcoming resistance and doing work.* When we walk or run, we use energy to overcome resistance caused by the force of gravity that otherwise would hold us in one place.

Heat

Heat *is energy that is being transferred between two objects.* Heat is not the same as temperature. The heat that is transferred to an object may or may not raise the temperature of the object. However, in many instances, the transfer of heat to an object results in an increase in its temperature. When you cup your hands around a mug of hot tea, your hands feel hot because heat has transferred from the hot mug to your hands. The heat has raised the temperature of your hands. If you hold a snowball, heat transfers from your hands to the snowball, and the loss of heat lowers the temperature of your hands. The transferred heat causes the snowball to melt. When two objects of different temperatures are placed in contact, heat always transfers from the object at higher temperature to the object at lower temperature until the temperature of both objects is the same.

Interconvertibility of forms of energy

Other forms of energy are radiant, chemical, electrical, mechanical, and nuclear. Each form of energy can be converted into other forms of energy. Green plants illustrate how one form of energy may be converted to another. Sunlight is radiant energy. Green plants use the radiant energy of sunlight to do the work of making carbon dioxide and water into complex chemicals. These chemicals store chemical energy that can be used by plants for life processes. Green plants therefore survive by acquiring energy in one form, radiant energy, and using it in another, chemical energy.

As another example of the interconvertibility of forms of energy, consider what happens when natural gas is burned. This chemical reaction releases the chemical energy stored in the gas. The chemical energy could be given off as heat and used to warm a house.

Our bodies perform functions similar to combustion. Food contains stored chemical energy that is released when our bodies break it down to simpler substances. Some of the energy is used to keep us alive and well; the remainder is used as heat to keep our temperature at 37 °C.

Conservation of energy

It is not difficult to see that various forms of energy may be interconverted and that various forms of energy may be used to do work. If we were to measure the work done or the energy stored in a process, we would find that it

The immense heat generated by power plants, industries, and automobiles contributes to heating of the air and bodies of water, a form of pollution called *thermal pollution*. The design and use of machines that run as efficiently as possible—doing a maximum amount of work with minimum heat loss—would help reduce thermal pollution.

is always less than expected on the basis of the energy input. It appears that some energy is destroyed. This is not so, however. In every process involving work, some energy is unavoidably lost as heat. When we pedal a bicycle, for example, some of the mechanical energy we supply by pedaling is lost as heat generated by the friction of the tires rubbing against the road, by the rubbing of the chain and sprocket, and so forth. However, in pedaling a bicycle, or in any other chemical or physical process, all the energy involved can be accounted for as work, stored energy, or heat. This principle is embodied in the **law of conservation of energy**—*in any chemical or physical process, energy may be converted from one form to another, but it is neither created nor destroyed.*

1.11 Exothermic and endothermic processes

AIM: To define and give examples of exothermic and endothermic reactions.

Focus

Physical and chemical processes absorb or release heat.

As we have seen, chemical and physical changes involve heat. *The term* **exothermic** *refers to processes that release heat to the surroundings.* **Endothermic** *processes absorb heat from the surroundings.* Both exothermic and endothermic processes are illustrated in Figure 1.13. The condensation of steam from a kettle of boiling water, a physical change, is an

Exothermic
exo- (Greek): out of
thermos (Greek): heat

Endothermic
endo- (Greek): within

Cold packs are available for the first-aid treatment of sprains and other injuries. These packs, which do not need refrigeration, contain chemicals in separate compartments. When the barrier between the compartments is broken, the chemicals mix in an endothermic process, lowering the temperature of the pack. Hot packs function similarly, except the chemicals that are mixed produce heat in an exothermic process.

Figure 1.13
The red-hot poker cools in an exothermic process; the ice is converted to water and steam in endothermic processes.

exothermic process. If we accidentally allow the steam to condense to water on our skin, we can be scalded as heat transfers from the steam to our skin. The burning of coal or any other fuel in oxygen is an exothermic process—heat is produced in the reaction:

$$\text{Coal} + \text{Oxygen} \longrightarrow \text{Carbon dioxide} + \text{Heat}$$

The melting of ice in a glass of water is an endothermic physical process.Since heat always flows from a warmer object to a cooler one, ice absorbs heat from the water. As the ice melts, the surrounding water loses heat and becomes colder.

Many chemical reactions require heat to make them occur. The production of lime (calcium oxide), used in making mortar from limestone (calcium carbonate), is an example of an endothermic chemical reaction:

$$\text{Limestone} + \text{Heat} \longrightarrow \text{Lime}$$

or,

$$\text{Calcium carbonate} + \text{Heat} \longrightarrow \text{Calcium oxide} + \text{Carbon dioxide}$$

The reactants must be heated to make the reaction occur. If a source of heat is supplied, the reaction proceeds to products as shown. If the source of heat is removed before all the reactants are converted to products, not enough heat is available to keep the reaction going, and it soon stops.

SUMMARY

Every substance consists of matter, has mass, and occupies space. Chemistry is a science that deals with the structure of matter and the changes that matter undergoes. The results of observations and experiments are summarized in laws. Hypotheses and theories attempt to explain these results.

Temperature differences are among the easiest measurements to make. Convenient reference temperatures used to set up temperature scales are the freezing temperature and the boiling temperature of water. Three temperature scales are Celsius, Fahrenheit, and Kelvin.

Matter can exist in three different physical states: solids, liquids, and gases. Mixtures are composed of two or more substances. Mixtures whose compositions are not uniform are heterogeneous mixtures; those whose compositions are uniform are homogeneous mixtures. Solutions are homogeneous mixtures of gases, liquids, or solids.

The physical properties of a substance can be used to identify a substance. Physical properties can be measured without changing the substance's composition. During a physical change, such as freezing, there is no change in the composition of the substance.

Chemical changes in matter are called chemical reactions. When a chemical reaction occurs, the starting substances (reactants) are changed into new substances (products) of different composition. The chemical properties of a substance are observed when a chemical change takes place.

A substance that cannot be broken down into two or more simple substances is an element. A substance that can be broken down into two or more simple substances by chemical reactions is called a compound. Most elements can be represented by a one- or two-letter symbol derived from a current name or from an older Latin name.

For a physical change or chemical reaction, the law of conservation of mass states that mass is conserved; mass is neither created nor destroyed.

All changes in nature are accompanied by changes in energy. The law of conservation of energy states that energy cannot be created or destroyed, but it can be changed from one form into another.

KEY TERMS

Absolute zero (1.3)
Celsius temperature scale (1.3)
Chemical property (1.7)
Chemical reaction (1.7)
Chemistry (1.1)
Compound (1.8)
Element (1.8)
Endothermic (1.11)
Energy (1.10)
Exothermic (1.11)

Experiment (1.2)
Fahrenheit temperature scale (1.3)
Gas (1.4)
Heat (1.10)
Heterogeneous mixture (1.5)
Homogeneous mixture (1.5)
Hypothesis (1.2)

Kelvin temperature scale (1.3)
Law (1.2)
Law of conservation of energy (1.10)
Law of conservation of mass (1.9)
Liquid (1.4)
Mass (1.9)
Matter (1.1)
Mixture (1.5)

Observation (1.2)
Physical change (1.6)
Physical property (1.6)
Product (1.7)
Reactant (1.7)
Scientific method (1.2)
Solid (1.4)
Solution (1.5)
Sublimation (1.6)
Theory (1.2)
Vapor (1.4)

EXERCISES

Science and the Scientific Method (Sections 1.1, 1.2)

1.7 What is the difference between a theory and a law?

1.8 Distinguish between a theory and a hypothesis.

1.9 Classify these statements as true or false.
  (a) Scientific laws are the only useful result of the scientific method.
  (b) A good *theory* to live by is that lightning never strikes twice in the same place.
  (c) You go to the doctor with a sore big toe. After the examination, the doctor tells you that she thinks you have gout and that she would like to run some blood tests. At this point, the doctor has a hypothesis about your sore toe.

1.10 Classify these statements as true or false.
  (a) A scientific hypothesis is tested by experiments.
  (b) A scientific law should be able to predict future behavior.
  (c) Once a scientific theory has been accepted, it will never change.

Temperature (Section 1.3)

1.11 Convert 80 °C to (a) °F and (b) K.

1.12 What is −20.2 °F in (a) °C and (b) K?

1.13 Grain alcohol boils at 351.5 K. What is the boiling point in (a) °C and (b) °F?

1.14 A patient has a temperature of 103 °F. What is this in (a) °C and (b) K?

1.15 The highest recorded temperature in the United States, 56.7 °C, was recorded in Death Valley, California, on July 10, 1913. Express this temperature in (a) °F and (b) K.

1.16 The element lead melts at 327.5 °C and boils at 1740 °C. Express these temperatures in (a) K and (b) °F.

1.17 Liquid oxygen boils at 90.2 K. What is this temperature in (a) °C and (b) °F?

1.18 The directions for baking a cake are to set the oven at 170 °C. If the oven dial is calibrated in degrees Fahrenheit, where would you set the control?

Physical States of Matter (Section 1.4)

1.19 What are the physical states of matter? Give two examples, different from those in the text, for each physical state of matter.

1.20 Match each term with the state of matter on the left. More than one state can match each term.
  (a) solid   1. flows
  (b) liquid   2. indefinite shape
  (c) gas   3. definite volume

1.21 At room temperature, nitrogen and oxygen are gases. The gaseous forms of rubbing alcohol and gasoline are vapors. Use these substances to explain the difference between a gas and a vapor.

1.22 In which state of matter do the following exist at room temperature?
  (a) diamond   (b) oxygen   (c) cooking oil
  (d) mercury   (e) clay   (f) neon

Mixtures (Section 1.5)

1.23 Distinguish between a heterogeneous mixture and a homogeneous mixture.

1.24 Identify each of the following samples of matter as homogeneous or heterogeneous:
  (a) soil   (b iron   (c) oxygen   (d) glass
  (e) table sugar   (f) sulfur   (g) ocean water
  (h) cough syrup

1.25 Which of the following samples of matter are mixtures?
  (a) beer   (b) pure table salt   (c) clouds
  (d) orange juice   (e) an egg   (f) ice
  (g) gasoline   (h) candle wax

1.26 Classify each of the substances in Exercise 1.25 as homogeneous or heterogeneous.

Properties of Matter (Sections 1.6, 1.7)

1.27 What is the difference between a physical change and a chemical change?

1.28 Classify each of the following as a physical or chemical change:
(a) breaking a piece of glass (b) burning oil
(c) cooking a steak (d) dissolving sugar in water
(e) souring of milk (f) stretching an elastic band

1.29 What physical properties would you use when attempting to separate
(a) sugar from iron filings or (b) salt from water?

1.30 Devise a way to separate sand from a mixture of cork, sand, salt, and water.

1.31 Which of the following are physical properties of water?
(a) colorless
(b) changes from a liquid to a solid at 0 °C
(c) decomposed by electricity into the elements hydrogen and oxygen
(d) condenses at 212 °F
(e) liquid at room temperature

1.32 Name at least two physical properties that are used to distinguish between
(a) water and gasoline, (b) copper and aluminum,
(c) nitrogen gas and oxygen gas.

1.33 Distinguish between physical and chemical properties. Give two examples of each for iron.

1.34 State several properties (physical or chemical) that could be used to distinguish between
(a) copper and silver, (b) water and gasoline,
(c) water and a solution of table salt in water.

1.35 Give one or more reasons why each of the following is a chemical change.
(a) Milk sours.
(b) An Alka-Seltzer tablet fizzes when dropped in water.
(c) A bicycle frame rusts.
(d) Dynamite explodes.

1.36 Classify the following properties of the element silicon as chemical or physical properties:
(a) blue-gray color (b) brittle
(c) insoluble in water (d) melts at 1410 °C
(e) reacts vigorously with fluorine (f) shiny

Elements and Compounds (Section 1.8)

1.37 How can we distinguish between an element and a compound? Between a compound and a mixture?

1.38 Classify each of the samples of matter in Exercise 1.24 as an element, a compound, or a mixture.

1.39 When a small amount of a red powder is heated in the absence of air, it darkens and then changes into a shiny, silvery liquid.
(a) Is the red powder an element or a compound? Explain.
(b) Can we classify the shiny liquid with certainty? Why or why not?

1.40 Passing electricity through a liquid causes the formation of gas bubbles even though the temperature stays below the boiling point of the liquid. Is the liquid an element or a compound? How do you know?

1.41 Write the chemical symbol for each of the following elements:
(a) calcium (b) oxygen (c) potassium
(d) gold (e) iron (f) hydrogen

1.42 Name the chemical elements that are represented by these symbols:
(a) Pb (b) S (c) P (d) Na (e) Ag (f) Ne

Mass and Energy Conservation (Sections 1.9, 1.10, 1.11)

1.43 What is the law of conservation of mass?

1.44 When an iron nail rusts, it actually increases in weight. Explain how this can be true in terms of the law of conservation of mass.

1.45 State the law of conservation of energy.

1.46 Discuss the statement, "Energy may be converted from one form to another, but it is never lost."

Additional Exercises

1.47 Arrange the following temperatures from the coldest to the hottest: (a) 70 °C (b) 170 °F (c) 340 K

1.48 An autoclave, used to sterilize items, operates at 245 °F. What does this temperature correspond to on (a) the Celsius scale and (b) the Kelvin scale?

1.49 Convert each of the following to degrees Celsius:
(a) 120 K (b) 300 °F (c) 293 K (d) 60 °F

1.50 Identify each of the following as either a physical or chemical property.
(a) Sulfur burns in air.
(b) Iron melts at 1534 °C.
(c) Copper tarnishes in air.
(d) Iodine sublimes when heated.

1.51 Convert each of the following to Kelvin:
(a) −40 °C (b) −40 °F (c) 25 °C (d) 212 °F

1.52 What is one requirement of all physical and chemical changes?

1.53 Give an example, other than those in the text, of the following conversions between forms of energy:
(a) light to electrical (b) chemical to mechanical
(c) electrical to chemical (d) chemical to radiant

1.54 Every substance has a unique set of physical and chemical properties that sets it apart from other substances. Identify each of the following as a physical or chemical property, and identify the substance.
(a) Colorless
(b) Does not conduct electricity
(c) Tends to form round droplets
(d) Produces a gas when sodium metal is dropped on it
(e) Changes from a liquid to a solid at 273 K
(f) Density of 0.998 g/cm^3 at 20 °C
(g) Condenses at 100 °C
(h) Liquid at room temperature

1.55 Even though we may sometimes feel that it is true, criticize the scientific soundness of Murphy's law: "Nothing is as easy as it looks. Everything takes longer than you expect. And if anything can go wrong, it will. At the worst possible time."

1.56 The following two tables document the five most abundant elements in the human body and the earth's atmosphere, oceans, and crust.

Human body		Earth's atmosphere, oceans, and crust	
Element	Mass percent	Element	Mass percent
Oxygen	64.8%	Oxygen	49.8%
Carbon	18.1%	Silicon	26.5%
Hydrogen	10.0%	Aluminum	7.6%
Nitrogen	3.1%	Iron	4.8%
Calcium	1.9%	Calcium	3.4%

Compare the two tables and propose explanations to account for the similarities and differences.

SELF-TEST (REVIEW)

True/False

1. The symbol of every element is the first one or two letters in its English name.
2. A compound can be chemically broken down into two or more elements.
3. Homogeneous mixtures have a definite composition.
4. In every physical and chemical change both mass and energy are conserved.
5. When you buy gasoline, you are buying chemical potential energy.
6. When a solution is heated, a gas is formed; therefore, a chemical reaction most certainly occurred.
7. Pure water is both a homogeneous substance and a compound.
8. Theories are proven by experiments.
9. A match weighs less after it is burned; therefore, some of the mass of the match has been *destroyed* when the match burns.
10. A change in color is a good indicator of chemical change.

Multiple Choice

11. Which of the following is *not* part of the scientific method?
    (a) hypothesis
    (b) principle
    (c) theory
    (d) experiment

12. A substance melts at −5 °C and boils at 19 °C. At room temperature, about 23 °C, which of these statements must be true?
    (a) The substance is an element.
    (b) The substance has an indefinite volume.
    (c) The substance is a liquid.
    (d) The substance is a compound.

13. Which of the following is an *incorrect* chemical symbol?
    (a) potassium, K   (b) sulfur, S
    (c) aluminum, Al   (d) iron, FE

14. An example of an exothermic process is
    (a) cooking potatoes.
    (b) breaking glass.
    (c) melting iron.
    (d) exploding a firecracker.

15. The state of matter characterized by a definite volume but an indefinite shape is
    (a) gas.   (b) liquid.   (c) mixture.   (d) solid.

16. Which of the following is heterogeneous?
    (a) a dollar bill   (b) tincture of iodine
    (c) a gold ring   (d) a stick of margarine

17. Which of the following is *not* a form of energy?
    (a) light   (b) pressure
    (c) heat   (d) electricity

18. Which of the following is *not* a physical change?
    (a) melting wax   (b) freezing alcohol
    (c) cutting a tree limb   (d) souring of milk

19. The temperature 120 K is the same as _____ °C.
    (a) 393   (b) −153   (c) −393   (d) 153

20. One hundred grams of charcoal is burned in an experiment. The ashes from this reaction weigh 12 g. This experiment appears to violate the
    (a) law of unequal weights.
    (b) law of conservation of mass.
    (c) law of chemical reactions.
    (d) law of conservation of energy.

21. Every mixture must
    (a) have a variable composition.
    (b) be homogeneous.
    (c) contain at least two compounds.
    (d) be a solution.

22. Dry ice (solid carbon dioxide) characteristically changes directly from the solid state to a vapor. This physical change is called
    (a) evaporation.   (b) condensation.
    (c) gasification.   (d) sublimation.

23. Which of the following is a chemical change?
    (a) melting wax   (b) burning a candle
    (c) boiling water   (d) dissolving sugar in water

24. Which of the following is the highest temperature?
    (a) 450 K   (b) 450 °C
    (c) 450 °F   (d) All represent the same temperature.

25. A physical property of water is that
    (a) it is a liquid at room temperature.
    (b) it boils at 100 °C.
    (c) it is tasteless.
    (d) all of the above

26. Homogeneous matter that has a variable composition is called
    (a) a liquid.   (b) an element.
    (c) a compound.   (d) a solution.

27. Which of the following is a chemical property of butter?
    (a) It has a greasy feel.
    (b) It turns rancid if left unrefrigerated.
    (c) It is light yellow in color.
    (d) It melts at 40 °C.

28. The temperature of 14 °F is equivalent to the temperature _____ °C.
    (a) 14   (b) 26   (c) −10   (d) 32

29. Which of the following is generally *not* an indication of a chemical change?
    (a) energy change   (b) color change
    (c) easily reversed   (d) odor change

30. When heated, a piece of wire changes from yellow-orange to black, and its mass increases. From this we can conclude that
    (a) a physical change has taken place.
    (b) mass is not conserved in this reaction.
    (c) the original wire was a compound.
    (d) the black color is due to the formation of a new compound.

2

Scientific Measurement

A Fundamental Skill

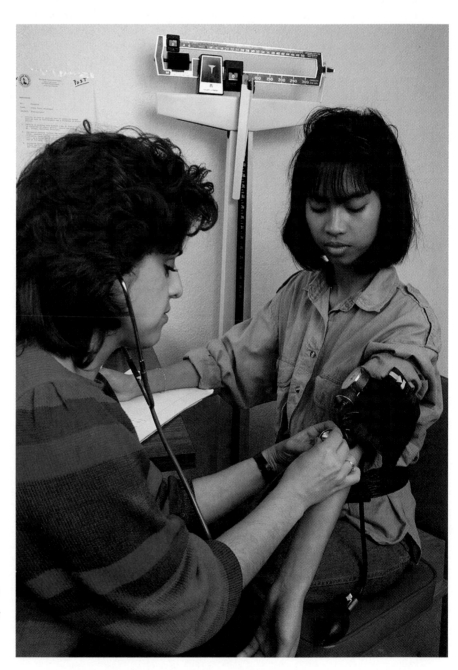

Accurate measurements play an important role in the daily work of an allied health professional. Here, a student experiences firsthand how a blood pressure measurement is made.

CHAPTER OUTLINE

CASE IN POINT: Maturity-onset diabetes

2.1   Measurements and units

2.2   Systems of measurement

2.3   Scientific notation

2.4   Significant figures in measurements

2.5   Significant figures in calculations

2.6   Length and volume

A CLOSER LOOK: Sizes of Organisms Affecting Health

2.7   Solving problems and converting units

2.8   Units of mass

FOLLOW-UP TO THE CASE IN POINT: Maturity-onset diabetes

2.9   Density and specific gravity

A CLOSER LOOK: Urinometers

2.10 Units of heat and the heat capacity of matter

A CLOSER LOOK: Counting Food Calories

Measurements are a part of daily life. We may weigh ourselves in the morning, put a cup of flour in our brownie recipe, or put 5 gallons of gasoline in the car. We use a clock to measure time. We judge from the size marked on a garment whether the piece of clothing will fit. If we are buying a cabinet to house our stereo, we will want to measure the cabinet to be sure that our stereo will fit on the shelves.

Allied health professionals are very concerned with measurements. In hospital laboratories, skilled clinical chemists and medical technologists try to answer questions related to the diagnosis, treatment, and prevention of disease of patients. What is the hemoglobin content of the blood of an anemic patient? How much cholesterol is in the blood serum of a patient at risk of heart disease? What is the composition of gases in the blood of a patient suspected of having emphysema? Only measurements can answer these questions. The following Case in Point, which considers a problem experienced by a woman named Marjorie, illustrates a typical example of the importance of measurements in health applications.

CASE IN POINT: Maturity-onset diabetes

Marjorie, a 45-year-old woman, has had generally good health. Lately, however, she has felt tired and has begun to lose weight. Also, she has noticed that she urinates more frequently than usual. After listening to her symptoms and performing a thorough physical examination, her physician suspects that Marjorie might be suffering from a disease called mature-onset diabetes, also called *type II diabetes*. Diabetes is a family of diseases in which the uptake of glucose (blood sugar) by tissues is impaired. As a result of this impairment, the amount of glucose in the blood and urine of untreated diabetics reaches abnormally high levels. In order to confirm the diagnosis, the doctor has Marjorie admitted to the hospital for tests. One of these tests is the glucose tolerance test. What is measured in the glucose tolerance test, and how are these measurements reported? What were the results of Marjorie's glucose tolerance test? These questions are answered in Section 2.8.

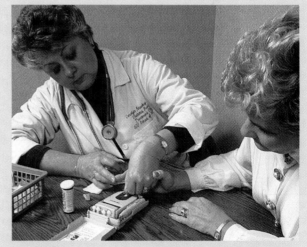

Abnormally high levels of glucose are present in both blood and urine samples of an untreated diabetic.

2.1 Measurements and units

AIM: To distinguish between the accuracy and precision of a measurement.

Focus

Measuring is a fundamental skill.

The diagnosis of disease depends on both qualitative and quantitative measurements. The fact that a certain bacteria stains a particular color is a qualitative measurement and can help identify the bacteria. A quantitative measurement of the amount of the compound uric acid in the blood can aid in the diagnosis of gout.

Measurements can be of two kinds, quantitative and qualitative. **Quantitative** *measurements are evaluations that give results as numbers, usually with units.* All quantitative measurements use units as reference standards for comparison. For example, when we fill up our gas tanks, the delivery pump is calibrated in gallons or liters. We are delivering a volume of gasoline in units of gallons or liters. The gallon or the liter is the reference standard for the volume of the gasoline. **Qualitative** *measurements are evaluations that do not lend themselves to expression as numbers.* Some measurements can be made either qualitatively or quantitatively. Consider, for example, a sick person with a fever. If we touch the person's forehead, we might decide, "Yes, this person feels feverish." This is a qualitative or subjective evaluation. It comes through a combination of our sense of touch and our previous knowledge of what "feverish" feels like, as opposed to "normal," for a person. If we need to be more objective, we can use a thermometer. Calibrated by the manufacturer against a standard temperature source, a thermometer eliminates guesswork and gives a quantitative measurement. The temperature consists of a specific number and a unit, the degree, that can be recorded and used for future reference.

Measurements must be accurate and precise to be useful. The words *accuracy* and *precision* mean the same thing to many people, but to scientists their meanings are very different. **Accuracy** *is concerned with how close a measurement comes to the actual dimension or true value of whatever is measured.* **Precision** *is concerned with the reproducibility of the measurement.*

We can use darts stuck in a dart board to illustrate accuracy and precision in making measurements (Fig. 2.1). If we let the bull's-eye of the dart board represent the true value of what we are measuring, the closeness of a dart to the bull's-eye is a measure of our accuracy—how close we have come to the true value. Consider three possible outcomes of tossing three darts at the board:

1. *All the darts stick in the bull's-eye.* Each dart in the bull's-eye represents an accurate measurement of the true value, so our measurement is very accurate. Since all the measurements are closely grouped, our measurement is also very reproducible and therefore very precise. High accuracy with high precision is the ideal in measurement, just as it is in darts.

2. *All the darts are closely grouped, but they are not in the bull's-eye.* In this instance, our precision is excellent because our tosses are very reproducible, but they are all inaccurate because they do not reflect the true value. To make a more accurate measurement, we must improve the accuracy of each toss.

3. *The darts are spread randomly around the target.* In this instance, our shots are neither accurate nor precise—we have missed the target and have failed to toss the darts reproducibly. Inaccurate and imprecise measurements are useless for obtaining scientific information.

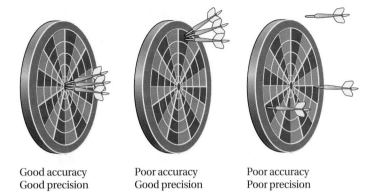

| Good accuracy | Poor accuracy | Poor accuracy |
| Good precision | Good precision | Poor precision |

Figure 2.1
A game of darts can be used to demonstrate the differences between accuracy and precision.

Note that the precision of a measurement with a particular device depends on more than one measurement, but an individual measurement may be accurate or inaccurate.

The accuracy of real measurements often depends on the quality of the measuring device. Any instrument whose calibration is in doubt must be checked or discarded, for a measurement can be no more accurate than the measuring instrument. Precision often depends on the skill of the person making the measurements. The operator must take care to obtain the best precision that is possible with the instrument being used.

2.2 Systems of measurement

AIM: To cite the advantages of the metric system.

Focus

The metric system, or the SI, is the preferred system of scientific measurement.

Makeshift systems of measurement through the ages caused confusion and disputes. The rise of commerce and science in the late 1700s dictated the need for a standard system of measurement. One such system that enabled people to communicate their measurements without ambiguity was the **metric system**—*a system of measurement units based on multiples of 10.*

The metric system was originally established in France in 1790. *However, a revised version, the* **International System of Units** *(abbreviated* **SI,** *after the French name* Le Système International d'Unités*) was adopted by international agreement in 1960.* The SI system has seven base units (Table 2.1). Other SI units of measurement, such as volume, density, and pressure,

Table 2.1 SI Base Units of Measurement

Quantity measured	Unit	SI symbol
length	meter	m
mass	kilogram	kg
time	second	s (or sec)
electric current	ampere	A
temperature	kelvin	K
amount of substance	mole	mol
luminous intensity	candela	cd

Table 2.2 Some Units of Measurement Used in This Text

Quantity	SI unit	Symbol	Non-SI unit	Symbol
length	meter	m	centimeter	cm
volume	——	——	liter	L
mass	kilogram	kg	gram	gm
density	——	——	grams per cubic centimeter	g/cm^3
temperature	kelvin	K	degree celsius	°C
time	second	s or sec	——	——
pressure	pascal	Pa	atmosphere	atm
			millimeters of mercury	mm Hg
energy	joule	J	calorie	cal
amount of substance	mole	mol	——	——

are derived from these base units. The metric system is important because of its simplicity and convenience. Since all the units are based on 10 or multiples of 10, conversions between units are simple. Although we could report all measured quantities in SI units, it is not always convenient to do so. Sometimes non-SI units are preferred for convenience or practical reasons. Table 2.2 lists some SI and non-SI units of measurement that are used in this text.

2.3 *Scientific notation*

AIM: To write numbers in scientific notation.

Focus

Numbers can be conveniently expressed in scientific notation.

Many scientific measurements involve very small or very large numbers. For example, the diameter of a typical bacterium is about 0.000001 m, and the distance from the Earth to the Sun is about 93,000,000 miles. In order to express large and small numbers in a convenient form and to simplify calculations, scientists often express numbers in a form called *exponential notation.* For example, the number 7800 can be written in exponential form as 7.8×10^3. The superscript 3—the *exponent*—tells us how many times the coefficient 7.8 must be *multiplied* by 10 to equal the number 7800. To give another example, the number 0.00043 can be written in exponential form as 4.3×10^{-4}. The negative exponent -4 tells us that the coefficient 4.3 must be *divided* four times by 10 to equal the number 0.00043. That is,

$$4.3 \times 10^{-4} = 4.3 \div 10 \div 10 \div 10 \div 10 = 0.00043.$$

Scientific notation is a type of exponential notation. *In* **scientific notation,** *a number is written as the product of two numbers: a coefficient between 1 and 10 and a power of 10.* For example, 1.06×10^5 is already in scientific notation because the coefficient 1.06 is between 1 and 10. The number 0.106×10^6 (which has the same value as 1.06×10^5) is in exponential notation but is not in scientific notation because the coefficient, 0.106, is not between 1 and 10.

To convert a decimal number such as 10,300.00 to scientific notation, we move the decimal point so that there is one nonzero digit to the left of it. If the original number is greater than 10, we move the decimal point to the left. By moving the decimal point 4 places to the left,

$$1\ 0\ 3\ 0\ 0\ .\ 0\ 0$$

becomes

$$1\ .\ 0\ 3\ 0\ 0\ 0\ 0$$

The exponent equals the number of places the decimal point is moved—in this example, 4 places—and it has a positive sign. The original number in scientific notation is 1.030000×10^4.

If a number is less than 1, such as 0.0000050, we move the decimal point to the right until there is one nonzero digit to its left. By moving the decimal point to the right 6 places,

$$0\ .\ 0\ 0\ 0\ 0\ 0\ 5\ 0$$

becomes

$$0\ 0\ 0\ 0\ 0\ 5\ .\ 0$$

The exponent equals the number of places the decimal has moved—6 places—but in this case the sign is negative. The original number in scientific notation is 5.0×10^{-6}.

EXAMPLE 2.1   **Writing numbers in scientific notation**

Represent each number in scientific notation.
(a) 0.000000356   (b) 346,000

SOLUTION

(a) Move the decimal to the right to give a number between 1 and 10.

$$0\ .\ 0\ 0\ 0\ 0\ 0\ 0\ 3\ 5\ 6$$

Because the decimal was moved 7 places to the right, the exponent is a negative 7. The number in scientific notation is 3.56×10^{-7}.

(b) Move the decimal 5 places to the left.

$$3\ 4\ 6\ 0\ 0\ 0\ .\ 0$$

The exponent is a positive 5; 3.46×10^5

PRACTICE EXERCISE 2.1

Represent the following numbers in scientific notation.

(a) 368.8   (b) 0.000607   (c) 67,000,000,000   (d) 0.0102

Addition and subtraction of numbers in exponential form

When numbers in exponential form are added or subtracted, all the exponents must be the same.

EXAMPLE 2.2 | **Addition of numbers written in exponential notation**

Calculate the sum of 3.04×10^3 and 2.3×10^2.

SOLUTION

Rewrite the second number so that the exponents of both numbers are identical, and then add.

$$
\begin{aligned}
&3.04 \times 10^3 \\
&\underline{+0.23 \times 10^3} \\
&3.27 \times 10^3
\end{aligned}
$$

PRACTICE EXERCISE 2.2

Add or subtract the following numbers. Express the answers in scientific notation.
(a) $0.73 \times 10^5 + 6.6 \times 10^4$
(b) $1.39 \times 10^{-2} + 9.5 \times 10^{-3}$
(c) $9.823 \times 10^{-5} - 7.34 \times 10^{-6}$

Multiplication and division of numbers in exponential form

When we multiply numbers in exponential form, we multiply the coefficients and add the exponents. When we divide numbers in exponential form, we divide the coefficients and subtract the exponent of the denominator from the exponent of the numerator.

EXAMPLE 2.3 | **Multiplying and dividing numbers expressed in exponential form**

Simplify this expression:

$$
\frac{(8.6 \times 10^5) \times (2.5 \times 10^{-2})}{5.0 \times 10^4}
$$

SOLUTION

Multiply and divide the coefficients:

$$
\frac{8.6 \times 2.5}{5.0} = 4.3
$$

Multiply and divide the exponents by adding the two exponents that are being multiplied and then subtracting the exponent in the denominator:

$$
5 + (-2) - 4 = -1
$$

The value of the exponent is -1, and the answer is 4.3×10^{-1}.

PRACTICE EXERCISE 2.3

Carry out the following operations. Express the answers in scientific notation.

(a) $(5.2 \times 10^{-3}) \times (12.5 \times 10^{-4})$

(b) $(2.21 \times 10^3) \div (2.6 \times 10^{-6})$

(c) $\dfrac{6.84 \times 10^{-3}}{(1.60 \times 10^{-7}) \times (2.50 \times 10^2)}$

2.4 Significant figures in measurements

AIM: To identify the number of significant figures in a measurement.

Focus

The significant figures in a measurement include one estimated digit.

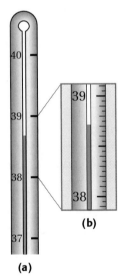

Figure 2.2
The thermometer reading in (a) can be reported to three significant figures, 38.7; the 7 is an educated guess. In (b) the reading can be reported to four significant figures, 38.72; the 2 is estimated.

Rulers, thermometers, and measuring cups are a few of the devices we can use to make measurements. No matter how careful we are in making our measurements, however, all of them have a degree of uncertainty. The amount of uncertainty depends on the measuring device we use. Most people would agree that they can measure the length of a needle more accurately with a ruler that is calibrated in hundredths than with a ruler calibrated in tenths. But can we convey to someone that our measurement is known with more or less certainty? We can, if we use *significant figures.*

The **significant figures** *in a measurement include all the digits that we know with certainty plus one digit that we estimate.* You may have taken a person's oral temperature using a mercury-in-glass thermometer. If the thermometer you use is marked in intervals of 1 degree, you can read the temperature to three significant figures. Figure 2.2(a) shows a thermometer indicating that a person's temperature is greater than 38 °C but less than 39 °C. The mercury level appears to be about two-thirds the distance between 38 °C and 39 °C. You might report the temperature as 38.7 °C. Someone else might look at the same thermometer and report the temperature as 38.8 °C. Both measurements have three significant figures. The first two digits are known with certainty, and the last digit is estimated by the individual making the measurement. If you use a thermometer with marks indicating one-twentieth of a degree (Fig. 2.2b), you could measure a temperature to four significant figures, 38.72 °C. The first three digits, 38.7, are known with certainty. You must estimate the digit in the hundredths place.

Finding the number of significant figures in a measurement is important, especially when doing calculations using measured values. In the following discussion, the significant figures are printed in color. The SI unit of length, the meter (abbreviated m), is used as the unit throughout the discussion. (The longest dimension of a page in this book is about one-fifth of a meter.)

We can determine the number of significant figures in a measurement by four simple rules.

1. Every nonzero digit is significant. Each of the following measurements has three significant figures: 154 m, 2.34 m, 1.96×10^4 m.

2. Zeros between nonzero digits are significant: 5706 m, 4.002 m, and 1.102 $\times 10^{-2}$ m all have four significant figures.

3. Leading zeros, those located in front of a number, are never significant: 0.0034 m, 0.000067 m, and 0.86 m all have two significant figures.

4. Zeros located after a decimal point at the end of a number are significant: 6.00 m, 0.0450 m, and 15.0 $\times 10^3$ m all have three significant figures. A measurement of 3000 m has one significant figure. If this number is an exact measurement rather than an approximation, it should be expressed to four significant figures by writing it in scientific notation, 3.000×10^3 m.

EXAMPLE 2.4

Counting significant figures

Give the number of significant figures in each measurement.

(a) 7.402 m   (b) 0.03040 m   (c) 4.700×10^6 m   (d) 0.0003 m
(e) 1.080 m   (f) 8.20×10^{-4} m

SOLUTION

The significant figures are printed in color.

(a) 7.402 m, 4   (b) 0.03040 m, 4   (c) 4.700×10^6 m, 4
(d) 0.0003 m, 1   (e) 1.080 m, 4   (f) 8.20×10^{-4} m, 3

PRACTICE EXERCISE 2.4

How many significant figures are in each of the following measurements?

(a) 16.030 m   (b) 0.00450 m   (c) 4.5000 m   (d) 9.80×10^{-4} m
(e) 7864 m   (f) 0.543 m   (g) 4.00×10^3 m   (h) 0.0502 m

2.5 Significant figures in calculations

AIM: To use the rules for significant figures in calculations to correctly round numbers.

Focus

Calculated answers must be rounded to the correct number of significant figures.

When we use measurements in calculations, we may obtain an answer that has more digits than are justified by the significant figures in the measurements. When this happens, we must round the calculated answer to the correct number of significant figures. To round an answer, we first determine how many significant figures the answer should have based on the arithmetic operation and the rules outlined below. If the digit immediately to the right of the last digit that is kept is less than 5, then that digit and all the digits to the right are dropped. If the digit immediately to the right of the last digit that is kept is 5 or greater, the value of the digit in the last significant place is increased by 1.

EXAMPLE 2.5 **Rounding numbers**

Round each measurement to the number of significant figures given in brackets.

(a) 5.678 m [3]   (b) 0.008234 m [2]   (c) 65890 m [2]

(d) 0.82394 m [1]   (e) 4.06×10^4 m [2]   (f) 78.006 m [3]

SOLUTION

Look at the digit in color, and drop it and subsequent digits if it is less than 5. When the digit in color is 5 or greater, the digit immediately to the left is increased by 1.

(a) 5.678 m; 5.68 m

(b) 0.008234 m; 0.0082 m or 8.2×10^{-3} m

(c) 65890 m; 6.6×10^4 m

(d) 0.82394 m; 0.8 m

(e) 4.06×10^4 m; 4.1×10^4 m

(f) 78.006 m; 78.0 m

PRACTICE EXERCISE 2.5

Round each of these measurements to two significant figures.

(a) 9.364×10^{-4} m   (b) 0.04605 m   (c) 8642 m

(d) 0.0317 m   (e) 7.038 m   (f) 3.861×10^7 m

Multiplication and division

Sometimes it is necessary to round numbers that are obtained by multiplication or division of measurements. A simple rule applies to finding the correct number of significant figures in the answer. *When multiplying and dividing measurements, the answer can have no more significant figures than the measurement with the smaller number of significant figures.* For example, what is the area of a room that measures 5.3 m by 4.74 m? Multiplying the length times the width on a calculator gives an answer of 25.122 m². The correct answer is 25 m². The answer must be rounded to two significant figures because the measurement 5.3 m has only two significant figures.

Addition and subtraction

There is also a simple rule for finding the number of significant figures when measurements are added or subtracted. *When adding and subtracting measurements, the answer can have no more digits to the right of the decimal point than the measurement with the least number of decimal places.* For example, what is the perimeter of the room that measures 5.3 m by 4.74 m? The *perimeter* is the distance around the edge of the room. It is the sum of the length plus the width plus the length plus the width:

$$
\begin{array}{r}
5.3\ \text{m} \\
4.74\ \text{m} \\
5.3\ \text{m} \\
\underline{4.74\ \text{m}} \\
20.08\ \text{m}
\end{array}
$$

The correct answer, 20.1 m, is rounded to one decimal place because the measurement 5.3 has only one decimal place.

> **PRACTICE EXERCISE 2.6**
> Find the area and perimeter of each room with the following dimensions.
> (a) 2.34 m by 4.2 m   (b) 12.0 m by 6.52 m
> (c) 8 m by 2.6 m   (d) 8.0 m by 2.6 m

Exact numbers

There are some numbers, called *exact numbers,* that do not affect how the answer to a calculation is rounded. Some exact numbers are definitions, for example: 60 seconds = 1 minute and 100 centimeters = 1 meter. We obtain another type of exact number by counting, for example: 26 students in a class and 36 pills in a bottle. Every exact number is considered to have an unlimited number of significant figures (Fig. 2.3). When we use exact numbers in a calculation, they are not used as a basis for rounding off an answer. For example, if the 36 pills in the bottle weigh 92.9 grams, we can express the average weight of each pill to three significant figures (92.9 grams/36 pills = 2.58 grams/pill).

Figure 2.3
A pharmacist dispenses pills in exact numbers. Here, a prescription is being filled with exactly 36 pills.

2.6 Length and volume

AIMS: To name the metric units of length and volume. To list and define the commonly used metric prefixes.

When we make a measurement, we must assign the units to the numerical value, expressed to the correct number of significant figures. Without the units, you cannot communicate the resulting measurement. Imagine your confusion if you were instructed to "Run 20." Your immediate response probably would be "Run 20 feet, laps, miles, or what?" We'll discuss two important measurements and units—those of length and volume—in this section.

Units of length

Length is an important property of matter. *The basic metric unit of length, or linear measure, is the* **meter.** The familiar yardstick is 36 inches long. A meter (39.4 inches when measured to three significant figures) is slightly longer than a yard. Most meter rules are divided into 100 equal parts. These

Meter
metrum (Latin):   to measure

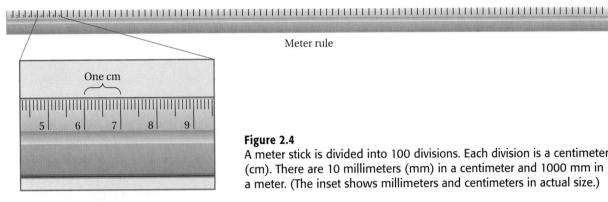

Figure 2.4
A meter stick is divided into 100 divisions. Each division is a centimeter (cm). There are 10 millimeters (mm) in a centimeter and 1000 mm in a meter. (The inset shows millimeters and centimeters in actual size.)

Clinical laboratory technicians (med techs) work with other health professionals to provide physicians with diagnostic information. Quantitative measurements include the counts of various cells in the blood and the amounts of various substances dissolved in the blood, urine, and spinal fluid.

smaller units are also given a name by adding a prefix to the name of the base unit, the meter. Thus, for a meter rule divided into 100 equal parts, each division is called a *centimeter* (Fig. 2.4). The prefix *centi-* means 1/100 or one-hundredth. Table 2.3 lists some of the prefixes in common use in the metric system. Table 2.4 summarizes the relationship among units of length in the metric system. A Closer Look: Sizes of Organisms Affecting Health uses some of these metric units.

Table 2.3 Prefixes in Common Use in the Metric System

Prefix	Symbol	Meaning
kilo	k	1000 times larger than the unit it precedes (1000 or 10^3)
centi	c	100 times smaller than the unit it precedes (1/100 or 10^{-2})
milli	m	1000 times smaller than the unit it precedes (1/1000 or 10^{-3})
micro	μ	1 million times smaller than the unit it precedes (1/1,000,000 or 10^{-6}) (μ is the Greek letter mu)
nano	n	1000 million (1 billion) times smaller than the unit it precedes (1/1,000,000,000 or 10^{-9})

Table 2.4 Metric Units for Length

Unit	Symbol	Relationship	Examples (approximate values)	
kilometer	km	$1 \text{ km} = 10^3 \text{ m}$	diameter of earth	13,000 km
meter (base unit)	m		length of football field	90 m
decimeter	dm	$10^1 \text{ dm} = 1 \text{ m}$	height of door	22 dm
centimeter	cm	$10^2 \text{ cm} = 1 \text{ m}$	height of dining table	75 cm
			length of paper clip	3 cm
millimeter	mm	$10^3 \text{ mm} = 1 \text{ m}$	thickness of 12 pages of this book	1 mm
micrometer	μm	$10^6 \text{ } \mu\text{m} = 1 \text{ m}$	diameter of human hair	80 μm
			diameter of bacterial cell	1 μm
nanometer	nm	$10^9 \text{ nm} = 1 \text{ m}$	thickness of cell membrane	7 nm
			length of carbon-hydrogen bond	0.1 nm

Sizes of Organisms Affecting Health

Infections by bacteria and viruses are responsible for such health problems as pneumonia and the common cold. Trying to imagine just how small these organisms are may help you grasp the sizes of some metric units. We can begin with a printed period (.) as a reference dimension. The diameter of the period is about 0.5 mm (500 μm), which is as small an object as a person with normal vision can comfortably see. Bacteria that cause scarlet fever are about 0.75 μm in diameter. This is about 700 times smaller than the diameter of the period. Although they cannot be seen with the unaided eye, these and many other bacteria can be seen with an ordinary light microscope. A microscope can detect circular objects about 0.5 μm in diameter. Viruses are often much smaller than bacteria. Poliovirus, a small virus, has a diameter of 0.025 μm (25 nm), making it 30 times smaller than scarlet fever bacteria and undetectable by the light microscope. The poliovirus and many other viruses can be visualized with the help of more powerful instruments called *electron microscopes*.

The dust mite is an organism that affects many people's health. Larger than bacteria and viruses, the average length of a dust mite (see figure) is 300 μm (0.3 mm). Dust mites have eight legs and are related to ticks and spiders. "House dust allergies," which cause wheezing, sneezing, and runny or stuffy noses, are often caused by these mighty mites or, more precisely, by their excrement. A single gram of house dust can contain as many as

(a)

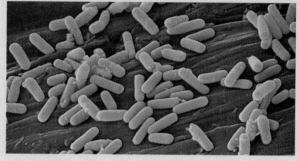
(b)

Just because you cannot see an object with the naked eye doesn't mean that the object doesn't exist. Some microscopic and submicroscopic objects are essential to life; others cause illness and even death. (a) Scanning electron micrograph of the head of a dust mite (360 ×), (b) bacteria on a pin point (2050 ×).

250,000 bits of dust mite excreta. These particles continue to cause allergic symptoms long after the mites that produced them have died.

Units of volume

The space occupied by matter is called **volume.** The volume of a cube 10 centimeters (cm) on each edge is 1000 cubic centimeters (10 cm × 10 cm × 10 cm = 1000 cm^3). *This volume is a* **liter** (Fig. 2.5). The liter (L) is the basic metric unit of volume. In terms of a more familiar volume, 1 liter of milk is slightly larger than 1 quart (qt) of milk (1 L = 1.057 qt). The milliliter (mL) is a smaller unit of volume than the liter. A milliliter is one-thousandth part of a liter (Fig. 2.5). Since a liter is defined as 1000 cm^3, the milliliter and the cubic centimeter have the same volume. These two units are used interchangeably. Table 2.5 summarizes the relationships among units of volume.

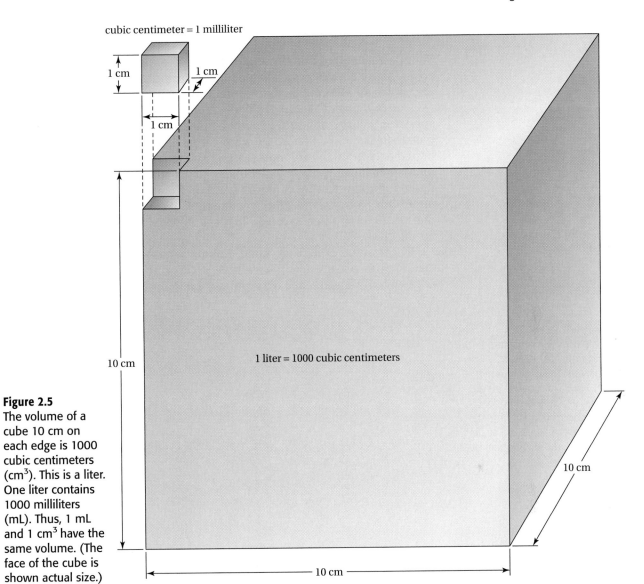

Figure 2.5
The volume of a cube 10 cm on each edge is 1000 cubic centimeters (cm^3). This is a liter. One liter contains 1000 milliliters (mL). Thus, 1 mL and 1 cm^3 have the same volume. (The face of the cube is shown actual size.)

Table 2.5 Metric Units for Volume

Unit	Symbol	Relationship	Examples (approximate values)	
liter (base unit)	L		blood in body	5 L
milliliter	mL	10^3 mL = 1 L	cup of water	200 mL
cubic centimeter	cm^3	1 cm^3 = 1 mL	chicken egg	50 mL
			drop of water	0.05 cm^3
microliter	μL	10^6 μL = 1 L	human egg	5×10^{-4} μL
			hemoglobin molecule	6×10^{-17} μL

Measuring volumes

Volume measurements are important in chemistry and the health sciences. Many devices are available for measuring liquid volumes, and some are shown in Figure 2.6. Volumetric devices are often labeled *to contain* (TC) or *to deliver* (TD) a specified volume. A volumetric flask contains a specified volume of liquid when it is filled to the calibration mark. Volumetric flasks are available in many sizes. Graduated cylinders are calibrated to contain the volumes marked on the cylinder. They are also useful for dispensing approximate volumes of a liquid. Pipets and burets are calibrated to deliver the volumes marked on them. They are used when it is important to transfer a more accurately known volume of a liquid from one vessel to another. Syringes are used to measure small liquid volumes for injection.

The volume of matter—solid, liquid, or gas—changes with temperature. For this reason, accurate volume-measuring devices are calibrated at a stated temperature, usually 20 °C, which is about room temperature.

Figure 2.6
Apparatus used for measuring liquid volumes: (a) IV bag, (b) volumetric flask, graduated cylinder, buret, graduated pipet, volumetric pipet, and syringe. (*TC* means *to contain* and *TD* means *to deliver.*) These devices are available in many sizes and have many uses.

(a)

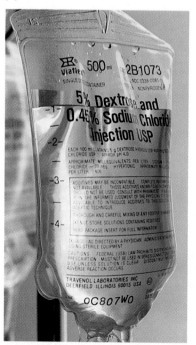

(b)

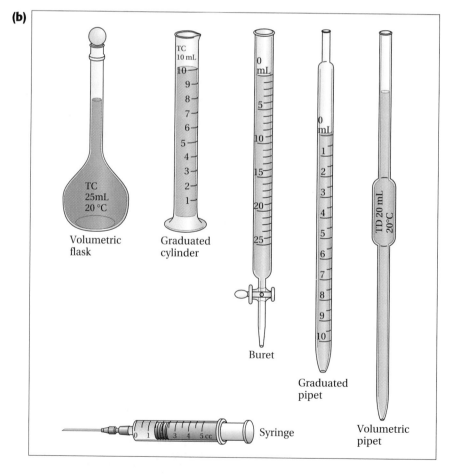

2.7 Solving problems and converting units

AIMS: *To apply good problem solving techniques to the solution of various problems. To convert measurements within the metric system.*

We can view chemistry as a series of problems that come in many different forms. In this course we will need to find the answers to many problems that require math skills. Working these problems is similar to playing tennis. The more we practice tennis, the more proficient we become. And just like a tennis player who improves with practice, we become better problem solvers by doing problems. After you have read a problem, the first thought that is likely to cross your mind is "Where do I begin?" There is no standard, correct answer to this question, because there is no one correct way to solve every problem. In fact, there are generally a number of approaches available to solve any single problem.

Despite the absence of an exclusive approach to solve every problem, there are some useful guidelines for problem solving that apply to almost every problem:

1. *Read the problem carefully to be certain that you know what the problem is asking.* If the problem is long, you may need to read it a number of times. If the problem has a numerical answer, be aware of the unit of the answer.

2. *Write down the fact(s) given in the problem.* Facts may include a number, a measurement, or a relationship.

3. *Supply any other necessary information.* You may need to remember or look up other facts or relationships to solve the problem.

4. *Ignore extra information.* There may be facts given in a problem that are not needed to solve the problem. Do not be misled.

5. *Sketch a picture of the problem.* A drawing often helps you see the relationship between what is given and what you are asked for in a problem.

6. *Look for a way to break down complex problems into two or more simpler problems.* The solutions to these simpler problems can then be combined to form the solution to the more complex problem.

7. *Be organized.* You or anyone else should be able to go back and follow the steps you went through in solving a problem.

8. *Be productively persistent.* Working at a problem for 2 hours is a sign of persistence, but it is not necessarily good. If you are stuck, *get help.* You can learn from other good problem solvers, including your tutor, classmates, and instructor.

9. *Check your work.* Reread the problem. Have you found what is asked for? Is your answer reasonable? You can often estimate an approximate answer for a quick check. Have you copied down the facts correctly? Check your math, and check the units. Recheck if you have time.

Many problems in chemistry can be solved using a method called *dimensional analysis* (sometimes called the *factor-label,* or *unit-factor,*

Some medicines are prescribed on the basis of the body weight of the patient. A particular drug is prescribed at a dosage of 2.4 mg of drug per 1 kg of body weight. Dimensional analysis is a convenient way to calculate the amount of medicine needed for a 55-kg patient.

$$\frac{55 \, \cancel{kg}}{1} \times \frac{2.4 \, mg}{1 \, \cancel{kg}} = 132 \, mg$$
$$= 1.3 \times 10^2 \, mg$$

method). As the name implies, this approach uses the units (dimensions) that are part of the measurement to help solve (analyze) the problem. We will first demonstrate the use of dimensional analysis by converting a length given in one unit to a length in another unit. In later examples we will show that dimensional analysis can be used to convert a volume given in one unit to a volume in another unit.

Length and volume conversions

Dimensional analysis uses **conversion factors**—*ratios that show relationships between two different units of measure*—to solve problems. Conversion factors allow us to convert one unit of measure to another. Suppose we want to know how many meters there are in 36 cm. We can write a conversion factor if we know the relationship between meters and centimeters. A meter and a centimeter are related by the expression

$$100 \, cm = 1 \, m$$

(we also could express 100 cm as 1×10^2 cm or as 10^2 cm). We can read the equation two ways: 100 cm equals 1 m or 1 m equals 100 cm. Hence there are two conversion factors:

$$\frac{100 \, cm}{1 \, m} \quad or \quad \frac{1 \, m}{100 \, cm}$$

Remember that we are converting from centimeters to meters. We will use the second conversion factor because centimeters will cancel in multiplication. In this problem we have

$$36 \, \cancel{cm} \times \frac{1 \, m}{100 \, \cancel{cm}} = 36 \times 10^{-2} \, m$$

In scientific notation, the answer is 3.6×10^{-1} m. Had the problem been to convert 36 m into centimeters, we would have used the first conversion factor:

$$36 \, \cancel{m} \times \frac{100 \, cm}{1 \, \cancel{m}} = 36 \times 100 \, cm = 3.6 \times 10^3 \, cm$$

In every conversion problem we want to cancel the unit to be converted (the given) and end up with the unit we are asked to find (the unknown).

The important things to remember when working conversion problems are

1. Start with the unit to be converted.

2. Find a relationship between the unit to be converted and the unit of the unknown.

3. Use this relationship to write a conversion factor such that the units to be converted cancel and you are left with the units called for in the problem.

4. Once the problem is set up so that the given units cancel, do the arithmetic.

5. Report the answer to the correct number of significant figures.

Sometimes it is necessary to use more than one conversion factor to make the desired conversion. The general procedure for working the problem is still the same (see Example 2.7).

| **EXAMPLE 2.6** | **Converting meters to kilometers** |

Express 24.5 m in kilometers.

SOLUTION

The quantity to be converted is in meters (24.5 m). The expression relating the units is 10^3 m = 1 km. The conversion factors are 1 km/10^3 m and 10^3 m/1 km. (Since we want to cancel the meter unit and be left with the kilometer unit, we will use the first conversion factor.) Set up the relationship and do the arithmetic:

$$24.5 \, \cancel{m} \times \frac{1 \, km}{10^3 \, \cancel{m}} = 24.5 \times 10^{-3} \, km$$

The answer, in scientific notation to three significant figures, is 2.45×10^{-2} km. (The answer should have three significant figures because there are three significant figures in the measured quantity 24.5 m. The 1 km and 10^3 m in the calculation are defined quantities—there are exactly 10^3 m in 1 km—so these numbers do not affect the number of significant figures in the answer.)

| **EXAMPLE 2.7** | **Converting kilometers to centimeters** |

How many centimeters is 0.054 km?

SOLUTION

The quantity to be converted is in kilometers (0.054 km). The problem requires two conversion steps: kilometers to meters and then meters to centimeters. We use the following relationships: 1 km = 10^3 m and 1 m = 10^2 cm. The conversion factors are 10^3 m/1 km and 10^2 cm/1 m. Set up the relationship and do the arithmetic:

$$0.054 \, \cancel{km} \times \frac{10^3 \, \cancel{m}}{1 \, \cancel{km}} \times \frac{10^2 \, cm}{1 \, \cancel{m}} = 0.054 \times 10^5 \, cm$$

The answer, in scientific notation to two significant figures, is 5.4×10^3 cm

| **EXAMPLE 2.8** | **Converting milliliters to liters** |

Convert 392 mL to liters.

SOLUTION

The quantity to be converted is in milliliters (392 mL). The expression relating the units (see Table 2.5) is 10^3 mL = 1 L. The conversion factor is 1 L/10^3

mL. Set up the relationship and do the arithmetic:

$$392 \text{ mL} \times \frac{1 \text{ L}}{10^3 \text{ mL}} = 392 \times 10^{-3} \text{ L}$$

The answer, in scientific notation to three significant figures, is 3.92×10^{-1} L.

PRACTICE EXERCISE 2.7

A physician requests 54 cm of silk to suture (stitch up) a laceration. Convert 54 cm to (a) meters, (b) millimeters, and (c) micrometers.

PRACTICE EXERCISE 2.8

A patient receives an injection of 0.015 L of a medication. What is 0.015 L in (a) milliliters and (b) microliters?

PRACTICE EXERCISE 2.9

The recommended dosage of a certain drug is 5.0 mg of the drug for each kilogram of body weight. (a) How many milligrams of this drug should an 85-kg person receive? Express this dosage in (b) grams and (c) micrograms.

2.8 Units of mass

AIM: To distinguish between the mass and weight of an object.

Focus

A balance is used to determine mass.

Mass *is the quantity of matter an object contains.* The mass of an object is measured by comparing it to a standard mass of 1 kilogram, the basic unit of mass in the metric system. *A **kilogram** (kg) is defined as the mass of 1 L of water at 4 °C.* (Water freezes at 0 °C.) Imagine a cube of water at 4 °C measuring 10 cm on each edge. This cube would have a volume of 1 L and a mass of 1000 grams (g), or 1 kilogram. *A **gram** is defined as the mass of 1 cm³ of water at 4 °C.*

Table 2.6 shows the relationships among units of mass. The mass of an object can be measured with a platform balance—a delicate seesaw. An

Table 2.6 Metric Units for Mass

Unit	Symbol	Relationship	Examples (approximate values)	
kilogram (base unit)	kg	$1 \text{ kg} = 10^3 \text{ g}$	human adult (male)	70 kg
			human adult (female)	50 kg
			human baby	3 kg
			roasting chicken	1.5 kg
gram	g	$1 \text{ g} = 10^{-3} \text{ kg}$	chicken egg	50 g
			a nickel	5 g
			a page of this book	3 g
			a 1-cm cube of water	1 g
milligram	mg	$1 \text{ mg} = 10^{-3} \text{ g}$	eyelash	0.1 mg
microgram	μg	$1 \text{ } \mu\text{g} = 10^{-6} \text{ g}$	grain of salt	100 μg

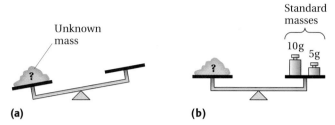

Figure 2.7
Balance operation: (a) Object of unknown mass is placed on the left-hand side. Balance is disturbed. (b) Standard masses are placed on the right-hand side to restore balance. Mass of the unknown object is the same as the sum of the standard masses added to the right-hand side; in this case, it is equal to 15 g.

Clinical laboratory technicians analyze blood plasma and serum for substances such as glucose and protein. The values are reported in units of mass per volume, such as mg (glucose)/100 mL and g (protein)/100 mL.

object of unknown mass is placed on one side, and standard masses are added to the other until the beam is in a position of balance (Fig. 2.7). When the beam is balanced, the unknown mass is equal to the sum of the standard masses. Two laboratory balances are shown in Figure 2.8.

Mass is not the same as weight. **Weight** *is a force—a measure of the pull on a given mass by the Earth's gravity.* Although the weight of an object can change with its location, its mass remains constant regardless of its location. A block of wood on Earth weighs about 6 times its weight on the Moon, since the force of gravity on Earth is about 6 times that on

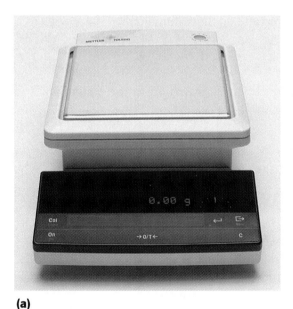

(a)

Figure 2.8
Balances in common use in the laboratory:
(a) top-loading electronic balance
(b) single-pan analytical balance

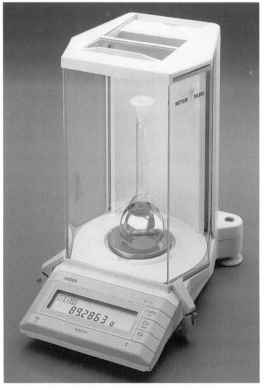

(b)

the Moon. Whether on Earth or on the Moon, the mass of the wood does not change.

EXAMPLE 2.9   **Converting grams to kilograms**

The pharmacist gives you a bottle of pills with a mass of 3.5×10^2 g. What is this mass in kilograms?

SOLUTION

The quantity to be converted is in grams (3.5×10^2 g). The expression relating the units is 10^3 g = 1 kg. Since we want the answer in kg, the conversion factor is 1 kg/10^3 g. Set up the relationship and do the arithmetic:

$$3.5 \times 10^2 \, g \times \frac{1 \, \text{kg}}{10^3 \, g} = 3.5 \times 10^{-1} \, \text{kg}$$

The answer, in scientific notation to two significant figures, is 3.5×10^{-1} kg.

EXAMPLE 2.10   **Converting milligrams to grams**

An aspirin tablet contains 138 mg of aspirin. Convert 138 mg to grams.

SOLUTION

The quantity to be converted is in milligrams (138 mg). The expression relating the units is 10^3 mg = 1 g. The conversion factor is 1 g/10^3 mg. Set up the relationship and do the arithmetic:

$$138 \, mg \times \frac{1 \, \text{g}}{10^3 \, mg} = 138 \times 10^{-3} \, g$$

The answer, in scientific notation to three significant figures, is 1.38×10^{-1} g.

PRACTICE EXERCISE 2.10

Convert a mass of 8.4 g to (a) kilograms, (b) milligrams, and (c) micrograms.

PRACTICE EXERCISE 2.11

Which is the largest mass?
(a) $5 \times 10^4 \, \mu g$ (b) 0.5×10^4 mg   (c) 6×10^{-1} g
(d) 0.5×10^{-4} kg   (e) 0.5 g

Now that we are familiar with such units as mass and volume, it is time to see how measurements that involve these units are used to solve the health questions posed in the Case in Point for this chapter.

FOLLOW-UP TO THE CASE IN POINT: Maturity-onset diabetes

Marjorie, the woman discussed in the Case in Point, has the symptoms of maturity-onset (type II) diabetes. This disease can be diagnosed by the glucose tolerance test, which measures the ability of tissue to absorb glucose from blood. Marjorie was told to fast before the test and was then given a dose of about 75 g of glucose. Her blood glucose level was then measured at regular intervals over a period of several hours. In the accompanying figure, the results of Marjorie's glucose tolerance test are compared with those of a normal subject. The amount of blood glucose of normal people rises from 60 to 110 mg in 1 dL (1 dL 5 1 deciliter or 0.1 L) of blood to about 120 to 140 mg/dL of blood in 1 hour after receiving glucose. Blood glucose levels drop to normal in 2 or 3 hours.

In Marjorie's case, the blood glucose level at the start of the test was already higher than normal and rose to about 240 mg/dL of blood in 1 hour before declining to its original level. Fasting glucose levels consistently above 120 mg/dL of blood are usually considered diagnostic of diabetes.

The glucose tolerance test therefore confirmed the doctor's suspicion that Marjorie has diabetes. The successful diagnosis of diabetes through use of the glucose tolerance test, which involves three measurements—mass of glucose, volume of blood, and time—vividly demonstrates the importance of measurement in the health sciences. Notice that the doctor applied the scientific method to diagnose Marjorie's illness. Based on the observation of symptoms, the doctor tentatively identified Marjorie's problem as diabetes and then tested the hypothesis. The glucose tolerance test helped confirm the hypothesis.

The results of a glucose tolerance test.

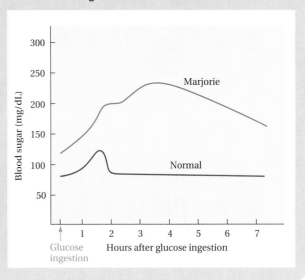

2.9 Density and specific gravity

***AIMS**: To find the mass, volume, or density of a substance when any two of these values are given. To state some useful applications of the measurement of specific gravity.*

Focus

The relationship between mass and volume is called *density*.

Suppose that we place a 25-g lead cylinder on one pan of a balance and a 25-g aluminum cylinder on the other pan, as shown in Figure 2.9. We can see from the figure that it takes a much larger volume of aluminum than of lead to make equal masses of the two metals. This experiment suggests that the relationship between an object's mass and its volume is a characteristic physical property of the object. This property is called *density*.

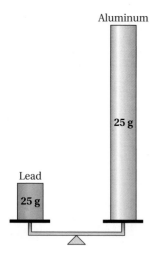

Figure 2.9
A 25-g sample of lead is compared with a 25-g sample of aluminum. The volume of the lead sample is much smaller than that of the aluminum because lead has a much higher density than aluminum.

Density

Density *is the ratio of the mass of an object to its volume:*

$$\text{Density} = \frac{\text{Mass}}{\text{Volume}}$$

When mass is measured in grams and volume in cubic centimeters, density has units of grams per cubic centimeter (g/cm^3). We can see from Table 2.7 that as the density of a substance increases, the volume of a given mass of that substance decreases. The densities of materials are extremely important. If bones were as dense as lead, for example, you would collapse under your own weight. Table 2.8 gives the densities of some common substances.

The density of a substance usually decreases as its temperature increases because its volume usually increases as the temperature is raised. Water is an important exception. Ice is less dense than liquid water; therefore, ice floats on water.

Table 2.7 Relationship Between Volume and Density for Identical Masses of Common Substances

Substance	Cube of substance (face shown actual size)	Mass (g)	Volume (cm³)	Density (g/cm³)
lithium		10.0	18.8	0.532
water		10.0	10.0	1.00
aluminum		10.0	3.70	2.70
lead		10.0	0.880	11.4

Table 2.8 Densities of Some Common Substances

Substance	Density at 25 °C (g/cm³)
gold	19.2
mercury	13.5
lead	11.3
iron	7.2
aluminum	2.7
bone	1.7–2.0
milk	1.03
water	0.997
water	1.000 (4 °C)
ice	0.92 (0 °C)
gasoline	0.70
wood	0.5
air	0.0013

| EXAMPLE 2.11 | Relating mass and volume to density |

A copper penny has a mass of 3.1 g and a volume of 0.35 cm³. What is the density of copper?

SOLUTION

Insert the values for mass and volume into the equation for density and do the arithmetic:

$$\text{Density} = \frac{\text{Mass}}{\text{Volume}} = \frac{3.1\ \text{g}}{0.35\ \text{cm}^3} = 8.9\ \text{g/cm}^3$$

| EXAMPLE 2.12 | Determining volume from mass and density |

What is the volume of a sample of cough syrup if it has a mass of 50.0 g and a density of 0.950 g/cm³?

SOLUTION

Rearrange the equation in Example 2.11 to find volume, substitute the values for mass and density, and do the arithmetic:

$$\text{Volume} = \frac{\text{Mass}}{\text{Density}} = \frac{50.0\ \cancel{\text{g}}}{0.950\ \cancel{\text{g}}/\text{cm}^3} = 52.6\ \text{cm}^3$$

| EXAMPLE 2.13 | Determining mass from volume and density |

What is the mass of 5.0 L of air if the density of air is 1.3×10^{-3} g/cm³?

SOLUTION

Rearrange the equation of Example 2.11 to find mass, substitute values for volume and density, and then do the arithmetic:

$$\text{Mass} = \text{Volume} \times \text{Density} = 5.0\ \cancel{L} \times \frac{10^3\ \cancel{\text{cm}^3}}{1\ \cancel{L}} \times \frac{1.3 \times 10^{-3}\ \text{g}}{1\ \cancel{\text{cm}^3}} = 6.5\ \text{g}$$

PRACTICE EXERCISE 2.12

Imagine you have a small piece of gold with a volume of 0.50 cm³.
(a) What is its mass if the density of gold is 19.2 g/cm³?
(b) What is the piece of gold worth if the market value is $12 per gram?

PRACTICE EXERCISE 2.13

A stone removed from the gallbladder of a patient has a volume of 2.20 cm³ and a mass of 1.89 g. What is the density of the gallstone? Will the gallstone float on water?

Specific gravity

Specific gravity *is the ratio of the density of a substance to the density of a reference substance, usually at the same temperature.* Water at 4 °C, which has a density of 1.00 g/cm³, is a convenient reference and is commonly used for this measurement:

$$\text{Specific gravity} = \frac{\text{Density of substance (g/cm}^3\text{)}}{\text{Density of water (g/cm}^3\text{)}}$$

Since the units in the equation cancel, specific gravity is a unitless number.

A **hydrometer** *is an instrument used to measure specific gravity of a liquid.* The depth to which the hydrometer sinks depends on the specific gravity of the liquid. The calibration mark on the hydrometer stem at the surface of the liquid indicates the specific gravity. A Closer Look: Urinometers discusses how the specific gravity of urine can be helpful in diagnosing diseases.

A Closer Look

Urinometers

Urine is a valuable source of clinical information. Urine analysis has long been used in the diagnosis of infections, diseases, and other abnormalities. Normal urine is a clear, pale-yellow liquid. It does not contain sugar, bacteria, blood cells, or significant amounts of protein. The presence of these substances in urine can indicate a disease state. The urine of a person who is dehydrated, for example, is much darker in color, and its specific gravity is higher than that of normal urine.

Clinical laboratories sometimes use special hydrometers, called *urinometers,* to measure the specific gravity of urine samples. Pure water has a specific gravity of 1.000. The specific gravity of a normal urine sample is in the range 1.010 to 1.030, depending on the amount of dissolved material it contains (see figure). The urine of a person with untreated diabetes mellitus may contain larger than normal amounts of dissolved material, chiefly the sugar glucose, and its specific gravity would be greater than 1.030. The urine of a person with diabetes insipidus (Bright's disease) is almost colorless with an abnormally low specific gravity because it contains very little dissolved material.

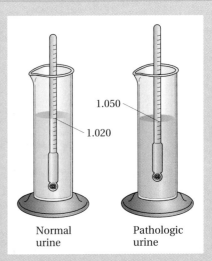

The specific gravity of a urine sample can help a physician in the diagnosis of disease.

In recent years, other methods for finding the specific gravity of urine have largely supplanted urinometers. One alternative method uses an instrument called a *refractometer* to measure the degree to which light is bent, or refracted, as it passes through a thin film of urine. The degree of refraction of light, called the *refractive index,* is related to the specific gravity of the urine.

2.10 Units of heat and the heat capacity of matter

***AIMS:** To name and define the units of heat. To find the mass, heat change, temperature change, or specific heat when any three of these values are given.*

Focus

Different substances require different amounts of heat to raise their temperatures by the same amount.

One measure of heat is the calorie. *One* **calorie** *(cal) is the quantity of heat that raises the temperature of 1 g of pure water 1 °C.* The **kilocalorie** (kcal) is used to measure large amounts of heat; 1 kcal equals 1000 cal. *The* **joule** *(J) is the SI unit of energy, including heat.* We can carry out conversions between calories and joules using the relationships

$$1 \text{ cal} = 4.18 \text{ J} \quad \text{and} \quad 1 \text{ J} = 0.239 \text{ cal}$$

The term *Calorie,* used in nutrition, is actually the kilocalorie. Notice that the food Calorie is written with a capital C. The statement "10 g of sugar has 41 Calories" means that the sugar will release 41 kcal of heat if it is burned completely to carbon dioxide and water by fire. The same amount of heat is also released when 10 g of sugar is "burned" to carbon dioxide and water by your body. Table 2.9 gives the caloric value of various foods. A Closer Look: Counting Food Calories describes the relationship between body weight and the caloric value of food eaten.

The quantity of heat required to change an object's temperature by exactly 1 °C is the **heat capacity** *of that object.* Heat capacity depends partly on mass. The greater the mass of an object, the greater is its heat capacity. A massive steel girder has a much larger heat capacity than a small steel nail, for example. Likewise, the heat capacity of a cup of water is much greater than that of a drop of water. Besides mass, however, the heat capacity of an object also depends on the kind of substance from which it is made. For this reason, different substances of the same mass may have different heat capacities. *The* **specific heat capacity,** *or simply the* **specific heat,** *of a substance is the quantity of heat required to raise the*

Table 2.9 Caloric Values of Some Foods

Food	Calories*	Food	Calories*
tomato, 5 oz	30	orange juice, 8 oz	100
carrot, raw, 8 oz	45	swiss cheese, 1 oz	105
bread, rye, 1 slice	55	cookie, 3-in.	110
bread, white, 1 slice	60	cornflakes, 1 oz	110
honey, 1 tablespoon	60	shrimp, 3 oz	110
corn, 5-in. ear	65	beef, lean, 2 oz	115
apple, 5 oz	70	haddock, fried, 3 oz	135
egg, boiled, large	80	milk, 8 oz	165
banana, 5 oz	85	noodles, 8 oz	200
potato, 5 oz	90	peanuts, roasted, 2 oz	210
butter, 1 tablespoon	100	hamburger patty, 3 oz	245

*One Calorie equals 1 kcal.

Because of its high specific heat, hot water can cause serious burns. Exercise care when heating water for any purpose.

The water content of the body is about 70% by mass. This large percentage of water and the high specific heat of water help the body maintain a relatively constant temperature.

Table 2.10 Specific Heat of Some Common Substances

Substance	Specific heat capacity (cal/(g × °C))
water	1.00
ethanol	0.58
ice	0.50
wood	0.42
steam	0.40
chloroform	0.23
aluminum	0.21
glass	0.12
iron	0.11
silver	0.06
mercury	0.33

temperature of 1 g of the substance 1 °C. As Table 2.10 shows, water has a high specific heat and metals have low specific heats. One calorie of heat raises the temperature of 1 g of water by 1 °C, but the same amount of heat raises the temperature of 1 g of iron by 9 °C. Therefore, water has a specific heat capacity 9 times that of iron. Specific heat is needed to calculate how much the temperature of a substance increases when it absorbs a known number of calories. Specific heat is also used to calculate how many calories are required to heat a known mass of a substance from one temperature to another. The following equation relates these quantities:

$$\text{Specific heat} \frac{\text{(cal)}}{\text{(g} \times \text{°C)}} = \frac{\text{Heat (cal)}}{\text{Mass (g)} \times \text{Change in temperature (°C)}}$$

EXAMPLE 2.14   Calculating the heat involved in a temperature change

How many (a) calories and (b) kilocalories would you need to raise the temperature of 11.0 g of water from 10.0 to 25.0 °C? [Specific heat of water = 1.00 cal/(g × °C).]

SOLUTION

We can rearrange the equation for specific heat to solve for calories:

$$\text{Heat (cal)} = \text{Specific heat} \frac{\text{(cal)}}{\text{(g} \times \text{°C)}} \times \text{Mass (g)} \times \text{Change in temperature (°C)}$$

(a) Substituting values for mass, temperature, and specific heat, we get:

$$\text{Heat (cal)} = 1.00 \ \frac{\text{cal}}{g \times °C} \times 11.0 \ g \times (25.0 - 10.0)°C = 165 \text{ cal}$$

(b) Since 1 kcal = 10^3 cal and we want the answer in kilocalories, the conversion factor is 1 kcal/10^3 cal:

$$165 \ cal \times \frac{1 \text{ kcal}}{10^3 \ cal} = 0.165 \text{ kcal} \quad \text{or} \quad 1.65 \times 10^{-1} \text{ kcal}$$

EXAMPLE 2.15   Calculating a mass from a temperature change and a specific heat

To raise the temperature of a mass of water from 25.0 to 50.0 °C requires 0.85 kcal. What is the mass of the water in (a) grams and (b) kilograms?

SOLUTION

(a) Rearrange the equation for specific heat to solve for mass:

$$\text{Mass (g)} = \frac{\text{Heat (cal)}}{\text{Specific heat} \frac{\text{(cal)}}{\text{(g} \times \text{°C)}} \times \text{Change in temperature (°C)}}$$

Counting Food Calories

You may have a friend who eats copious amounts of food but never seems to gain an ounce, while another acquaintance claims to gain weight just by looking at food. The weight-control problem that many people experience is physically and emotionally complex. Nevertheless, the relationship between the caloric value of food and body fat is fairly simple. The chemical energy contained in food is generally either expended through activity or stored as fat. A pound of fat stores an amount of chemical energy that is equivalent to approximately 3500 Calories. Even while sleeping, the average adult expends about 0.45 Calorie per hour for every pound of body weight. A person weighing 150 pounds would burn 540 Calories during 8 hours of sleep. This same person would have to sleep for more than 50 hours to lose 1 pound of fat. Clearly, sleeping is not a very good way to lose weight. The same person walking briskly would expend 325 Calories per hour. At this rate, the pound of fat would be lost in 11 hours. If this person is a very energetic swimmer, he or she could expend 640 Calories per hour and lose the pound of fat in 5.5 hours. A typical student expends between 2300 and 3100 Calories per day. As long as the energy output equals the caloric intake, the

You would have to walk about 30 miles (50 kilometers) to burn off 1 pound of fat!

student will neither gain nor lose weight. If you have a hearty appetite but are inactive, chances are good that you will gain weight. The combination of regular exercise and nutritious foods that contain the proper number of Calories will permit most people to achieve and maintain the proper weight.

Since the equation calls for heat in calories and it is given in kilocalories, we must convert kilocalories to calories. The conversion factor is 10^3 cal/1 kcal, so

$$0.85 \ \cancel{kcal} \times \frac{10^3 \ cal}{1 \ \cancel{kcal}} = 8.5 \times 10^2 \ cal$$

Now, using this value, the specific heat of water [1.00 cal/(g × °C)], and the given temperatures, we can calculate the mass of water:

$$\text{Mass (g)} = \frac{8.5 \times 10^2 \ cal}{1.00 \ cal/(g \times °C) \times (50.0 - 25.0)°C}$$

$$= \frac{8.5 \times 10^2 \ \cancel{cal}}{1.00 \ \cancel{cal}/(g \times \cancel{°C}) \times (25.0)\cancel{°C}}$$

$$= 34 \ g$$

(b) Since $1 \text{ kg} = 10^3 \text{ g}$ and we want the answer in kilograms, the conversion factor is $1 \text{ kg}/10^3 \text{ g}$ and

$$34 \text{ g} \times \frac{1 \text{ kg}}{10^3 \text{ g}} = 0.034 \text{ kg} \qquad \text{or} \qquad 3.4 \times 10^{-2} \text{ kg}$$

PRACTICE EXERCISE 2.14

The specific heat of iron is $0.11 \text{ cal}/(\text{g} \times °\text{C})$. How many calories are required to heat 125 g of iron from 25 °C to its melting point, 1535 °C?

PRACTICE EXERCISE 2.15

A sample of aluminum weighs 67 g. If 854 calories are required to raise the temperature of this sample from 25 to 85 °C, calculate the specific heat of aluminum.

SUMMARY

Measurements can be qualitative or quantitative. The accuracy of a measurement is concerned with how close a measurement comes to the true value; the precision of a measurement is concerned with its reproducibility. Measurements are reported with the proper number of significant figures. The significant figures in a measurement contain the number of digits that are known with certainty plus one digit that is estimated.

The metric system, based on the number 10, is the system of measurement used by scientists. In the metric system, the basic unit of length is the meter. The kilometer is 1000 m, and the centimeter is 0.01 m. The space occupied by matter is called volume. The basic unit of volume is the liter. A liter has a volume of 1000 cm^3. One cubic centimeter has the same volume as 1 mL. The quantity of matter an object contains is its mass. The mass of an object is constant and independent of gravity. The basic unit of mass is the kilogram; smaller units are the gram and the milligram. Weight is not the same as mass. Weight is a measure of the pull of gravity on an object of a given mass. The ratio of the mass of an object to its volume is its density. The most common unit of density is grams per cubic centimeter. Specific gravity, the ratio of the density of a substance to the density of water, has no units.

Heat is measured quantitatively in calories. One calorie is the quantity of heat required to raise the temperature of 1 g of pure water by 1 °C. A larger unit, the kilocalorie, is equal to 1000 cal. The SI unit of heat is the joule; 1 cal is equivalent to 4.18 J. An object's ability to store or release heat is its heat capacity. For any given substance, the greater the mass of an object, the larger is its heat capacity. The heat capacity of 1 g of a substance is the specific heat capacity, or specific heat, of that substance.

KEY TERMS

Accuracy (2.1)	International System of	Meter (2.6)	Significant figure (2.4)
Calorie (2.10)	Units (2.2)	Metric system (2.2)	Specific gravity (2.9)
Conversion factor (2.7)	Joule (2.10)	Precision (2.1)	Specific heat (2.10)
Density (2.9)	Kilocalorie (2.10)	Qualitative (2.1)	Specific heat capacity
Gram (2.8)	Kilogram (2.8)	Quantitative (2.1)	(2.10)
Heat capacity (2.10)	Liter (2.6)	SI (2.2)	Volume (2.6)
Hydrometer (2.9)	Mass (2.8)	Scientific notation (2.3)	Weight (2.8)

EXERCISES

SI Measurements (Sections 2.1, 2.2)

2.16 Give the basic SI units for (a) length, (b) volume, (c) temperature, and (d) mass.

2.17 State whether the following units are used to measure length, volume, or mass.
(a) kg   (b) cm^3 (c) mm   (d) μL

2.18 What do the following prefixes mean?
(a) centi-   (b) kilo-   (c) milli-   (d) nano-

2.19 What are the prefixes for the following multipliers of a gram?
(a) 10^{-6} (b) 1/1000   (c) 10^3 (d) 0.01

Scientific Notation (Section 2.3)

2.20 Write the following numbers in scientific notation.
(a) The distance from the Earth to the Sun is 93,000,000 miles.
(b) A carbon atom has a diameter of 0.00000000015 meters.
(c) The number of atoms of gold in 1 gram of gold is 3,060,000,000,000,000,000,000.

2.21 Express the following numbers in decimal notation.
(a) The number of seconds in a day is 8.64×10^4.
(b) The diameter of a human hair is 7.5×10^{-6} meters.
(c) There are 5×10^8 red blood cells in a drop of blood.

2.22 Express the answers to the following problems using scientific notation.
(a) $(8.6 \times 10^3) \times (2.5 \times 10^{-6})$
(b) $\dfrac{(9 \times 10^{-6}) + (6.1 \times 10^{-5})}{3.5 \times 10^{-3}}$
(c) $\dfrac{9.12 \times 10^4}{(7.84 \times 10^6) - (2.3 \times 10^5)}$

2.23 Complete the following operations. Express the answers in scientific notation.
(a) $(10^4) \times (10^{-6}) \times (10^{-3})$
(b) $(5.60 \times 10^{-2}) \times (9.8 \times 10^{-7})$
(c) $\dfrac{(4.6 \times 10^5) - (2 \times 10^4)}{(3.9 \times 10^{-2}) + (5 \times 10^{-3})}$

Significant Figures (Sections 2.4, 2.5)

2.24 Round off each of these measurements to three significant figures.
(a) 98.473 L   (b) 0.000 763 21 cg   (c) 57.048 m
(d) 9500   (e) 12.17 °C   (f) 0.0074983×10^4 mm

2.25 Round off each of these measurements to two significant figures.
(a) 0.0642 μL (b) 945.78 K   (c) 0.6793 g
(d) 3000 km   (e) 528×10^7 m   (f) 3.768 mm

2.26 Write each of the rounded-off measurements in Exercise 2.24 in scientific notation.

2.27 Write each of the rounded-off measurements in Exercise 2.25 in scientific notation.

2.28 Comment on this statement: "When two measurements are added together, the answer can have no more significant figures than the measurement with the least number of significant figures."

2.29 Comment on this statement: "When a number is rounded off, the last significant figure is dropped if it is less than 5."

2.30 Round off each calculator answer correctly.
(a) 12.4 mg + 3.54 mg + 21 mg = 36.94 mg
(b) 2.36 m $\times$ 8.4 m = 19.824 m^2
(c) 87.2 μL − 35.56 μL = 51.64 μL
(d) 6.466 g $\div$ 0.225 = 28.737778 g

2.31 Round off each calculator answer correctly.
(a) 254 K − 57.8 K = 196.2 K
(b) 25.64 cm^2 $\div$ 1.4 cm = 18.314286 cm
(c) 5.40 m $\times$ 3.21 m $\times$ 1.871 m = 32.431914 m^3
(d) 5.47 cm^3 + 11 cm^3 + 87.300 cm^3 = 103.770 cm^3.

2.32 Express each of the rounded-off answers in Exercise 2.30 in scientific notation.

2.33 Write each rounded-off answer in Exercise 2.31 in scientific notation.

Units of Length, Volume, and Mass (Sections 2.6, 2.7, 2.8)

2.34 Make the following conversions:
(a) 79 g to kg   (b) 2.62 cm to m
(c) 2.00 L to mL   (d) 560 mcal to cal

2.35 Convert these measurements:
(a) 45.6 μm to m   (b) 0.0234 kg to g
(c) 17 mL to L   (d) 988 g to cg

2.36 Express 0.26 kg in (a) grams, (b) milligrams, and (c) centigrams.

2.37 Express 42.6 mm in (a) centimeters, (b) meters, and (c) micrometers.

2.38 Which of the following linear measures is the longest?
(a) 5×10^2 mm   (b) 5×10^{-2} m
(c) 60 cm   (d) 5×10^2 cm

2.39 Which is the largest mass?
(a) 5×10^{-3} kg   (b) 2×10^{-1} g
(c) 5×10^2 mg   (d) 0.006 kg

2.40 A rectangular metal tank measuring 40.0 cm long and 0.500 m wide is filled with water to a depth of 30.0 mm. What is the volume of water in
(a) cubic centimeters and (b) liters?

2.41 A glass aquarium is 25.0 cm wide and 80.0 cm long. How many liters of water must be added to the aquarium to achieve a depth of 34.0 cm?

2.42 List these in order from smallest to largest:
(a) a centimeter   (b) a kilometer   (c) a micrometer
(d) a millimeter   (e) a decimeter   (f) a meter

2.43 Which is larger?
(a) a centigram or a milligram
(b) a liter or a centiliter
(c) a millisecond or a centisecond
(d) a calorie or a kilocalorie
(e) a microliter or a milliliter
(f) a nanogram or a microgram

2.44 How is the mass of an object measured?

2.45 Astronauts in space are said to be weightless. Explain why it is incorrect to say that they are massless.

Density and Specific Gravity (Section 2.9)

2.46 A chunk of lead has a mass of 1.34 kg and a volume of 119 mL. What is its density?

2.47 What is the density of a solution of sulfuric acid if 312 mL has a mass of 375 g?

2.48 A solid has a density of 0.920 g/cm^3 and a mass of 20.0 g. What is its volume?

2.49 What is the volume of 60.0 g of mercury? The density of mercury is 13.5 g/cm^3.

2.50 What is the mass, in grams, of a cube of aluminum, 4.0 cm on each edge? The density of aluminum is 2.7 g/cm^3.

2.51 A block of wood measures 20.0 cm × 30.0 cm × 40.0 cm. If the density of wood is 0.55 g/cm^3, what is the mass of the piece of wood in (a) grams, (b) kilograms, and (c) milligrams?

2.52 A shiny, gold-colored bar of metal weighing 42.1 g has a volume of 4.5 cm^3. Is the metal bar pure gold?

2.53 A small coin with a volume of 0.66 cm^3 weighs 5.78 g. Is the coin made of pure silver? (The density of silver is 10.5 g/cm^3.)

2.54 The density of air at 25 °C and 1 atm of pressure is 1.3 × 10^{-3} g/cm^3. What is the volume of 20.0 g of air?

2.55 How large a container, in liters, would you need to hold 7.2 kg of gasoline? (The density of gasoline is 0.70 g/cm^3.)

2.56 What is the mass, in kilograms, of a 1.00-L block of copper? (The density of copper is 8.92 g/cm^3.)

2.57 What is the mass, in kilograms, of 1.75 L of mercury?

2.58 How tall is a rectangular block of wood measuring 4.2 cm wide and 8.5 cm deep that has a mass of 246.0 g? (The specific gravity of the wood is 0.42.)

2.59 Ice has a specific gravity of 0.92 at 0 °C. A 1450.0-g block of ice is 12.0 cm wide and 12.0 cm deep. How tall is the block of ice?

Heat and Heat Capacity (Section 2.10)

2.60 What is the definition of calorie? How does a calorie compare with a Calorie?

2.61 What is a joule? How does a joule compare with a calorie?

2.62 How many (a) calories and (b) kilocalories are required to raise the temperature of 25.0 g of water from 20.0 to 70.0 °C?

2.63 How many (a) joules and (b) kilojoules are required to raise the temperature of 155 g of water from 27.0 to 87.0 °C?

2.64 A quantity of heat equal to 850 cal is added to 45 g of water at 15 °C. What will be (a) the change in temperature and (b) the final temperature of the water?

2.65 A 75.0-g sample of water at 85.0 °C loses 1250 cal of heat. What is the final temperature of the water?

2.66 How many (a) kilocalories and (b) kilojoules are required to raise the temperature of 170.0 g of ethanol (grain alcohol) from 10.0 to 45.0 °C?

2.67 How many (a) kilocalories and (b) kilojoules are required to raise the temperature of 68.0 g of aluminum from 20.0 to 125.0 °C?

2.68 What is the mass of a silver necklace if its temperature changes from 30.0 to 230 °C when it absorbs 3.50 × 10^2 cal of heat?

2.69 The temperature of an iron nail increases 68 °C when it absorbs 87 cal of heat. What is the mass of the nail?

2.70 A 66.0-g sample of chloroform at 77 °C loses 222 cal of heat. What is (a) the change in temperature and (b) the final temperature of the chloroform?

2.71 A quantity of heat equal to 4.44 cal is added to 5.40 g of mercury at 25.0 °C. What will be (a) the change in temperature and (b) the final temperature of the mercury?

Additional Exercises

2.72 How many significant figures are in each of these numbers?
(a) 527   (b) 0.022   (c) 1.630 × 10^{-2} (d) 9.034
(e) 75,900   (f) 9.00   (g) 1.5000 × 10^8 (h) 55.330

2.73 Which of these numbers have more than two significant figures?
(a) 0.00072   (b) 8.07 × 10^4 (c) 155
(d) 6.00 × 10^{-3} (e) 8700   (f) 1.0003

2.74 List at least two advantages to using SI units of measure.

2.75 An average American consumes food that provides 2.4×10^3 kcal of energy a day. How many calories per second is this?

2.76 Comment on the precision and accuracy of these basketball free-throw shooters: (a) 95 of 100 shots are made; (b) 99 of 100 shots hit the front of the rim and bounce off; (c) 25 of 100 shots are made; the rest rebound in various directions after hitting the rim or backboard.

2.77 The average human heart pumps blood at the rate of 2.0×10^2 mL/s. How many gallons of blood does an average heart pump in 1 day?

2.78 Make the following conversions, expressing your answers in scientific notation.
(a) 7.0×10^{-2} km to millimeters
(b) 0.045 cm to micrometers
(c) 8.9×10^3 ng to milligrams
(d) 22 cg to grams
(e) 7.84×10^2 km to meters
(f) 1560 cm^3 to liters

2.79 Which of the following linear measures is the longest?
(a) 4×10^3 cm   (b) 4×10^5 mm   (c) 4 m
(d) 0.04 km   (e) 4×10^8 nm   (f) 4×10^7 μm

2.80 Water weighing 35.4 g is added to an empty flask weighing 87.432 g. The flask with the water weighs 146.72 g after a rubber stopper is added. Express the weight of the stopper to the correct number of significant figures.

2.81 What is the SI base unit of (a) time, (b) temperature, (c) amount of substance, (d) length, and (e) mass?

2.82 A rectangular fish tank measures 87 cm long, 45.3 cm wide, and 41 cm high. What is the volume of the tank in (a) cubic centimeters and (b) liters?

2.83 A flask that can hold 71.8 g of water at 4 °C can hold only 57.7 g of ethyl alcohol. What is the specific gravity of ethyl alcohol?

2.84 A cheetah has been clocked at 7.0×10^1 mi/h over a 100-yd distance. What is this speed in meters per second?

2.85 The density of acetic acid is 1.05 g/mL. What volume, in milliliters, must be measured to provide 50.0 g of acetic acid?

2.86 Arrange these length measurements from largest to smallest.
(a) 7.2×10^{-2} km   (b) 8.8 ft   (c) 5.7×10^3 cm
(d) 2.1 m   (e) 0.03 miles

2.87 Four students determined the mass of the same silver bracelet using different balances. The correct mass of the bracelet had been determined previously by an experienced analyst as 39.564 g. Describe the accuracy and precision of each student's measurements based on the information given in the following table.

	Mass of bracelet (g)			
	Lissa	**Manuel**	**Yukari**	**Leigh Anne**
Weighing 1	39.69	39.60	37.97	38.55
Weighing 2	39.21	39.55	37.99	35.61
Weighing 3	40.12	39.58	37.96	44.32

2.88 Mount Everest, the tallest mountain in the world, has an elevation of 29,028 ft. (a) How many meters is this? (b) How many kilometers?

2.89 An atom of zinc weighs 1.086×10^{-22} g. How many atoms of zinc are in 5.00 g of zinc?

2.90 Many cars now average 30.0 miles per gallon of gasoline. Express this mileage in units of kilometers per liter.

2.91 A candy bar can provide 220 kcal of energy. What mass of water could this same amount of energy raise from the freezing point to the boiling point?

2.92 Why is there a temperature inscribed on the volumetric flask and pipet in Figure 2.6?

2.93 A balloon is inflated to a volume of 842 L with 14.3 g of helium. (a) What is the density of helium in grams per liter? (b) What is this density in grams per cubic centimeter?

SELF-TEST (REVIEW)

True/False

1. Equal masses of two different substances have the same heat capacity.
2. The calorie is a unit of heat energy.
3. A substance with a relatively high specific heat would undergo a relatively large temperature change upon heating.
4. A cm^3 is the same volume as a mL.
5. The volume of a cube of platinum 1.25 cm on a side is 1.953125 cm^3.
6. On a windless day a balloon rises when released. The gas inside the balloon is more dense than the air into which it is released.
7. When a substance expands on heating, its density decreases.

8. When the temperature of equal masses of two different substances increases 5 °C, the substance with the lower specific heat absorbs the most energy.

9. The measurement 6.543 cm has four significant figures.

10. Each of these is the same volume: 16 mL, 0.016 L, 16 cm³.

Multiple Choice

11. The basic unit of length in the metric system is the
    (a) cubic centimeter.   (b) yard.   (c) meter.
    (d) liter.

12. Calculate the density of silver if a 2.8-g sample of the metal occupies 0.25 cm³.
    (a) 0.70 g/cm³   (b) 0.089 cm³/g   (c) 11 g/cm³
    (d) 7.0 g/cm³

13. Which of these equalities is correct?
    (a) 10^3 cg = 1 g   (b) $10^6 \mu$L = 1.0 L
    (c) 10^3 km = 1 m   (d) 10^3 mL = 1 cm³

14. Three students using the same thermometer find the boiling point of benzene at atmospheric pressure to be 82.3 °C, 82.4 °C, and 82.3 °C, respectively. Comment on the accuracy and precision of these measurements if the boiling point of benzene is 80.0 °C.
    (a) good accuracy and good precision
    (b) good accuracy and poor precision
    (c) poor accuracy and good precision
    (d) poor accuracy and poor precision

15. How many of these measurements have three significant figures? 14.03 g   0.017 km 5.30×10^2 mL   6.780 cg   0.172 L
    (a) 1   (b) 2   (c) 3   (d) 4

16. How many grams are there in 540 cg?
    (a) 5.4   (b) 54   (c) 5.4×10^4 (d) 5.4×10^{-2}

17. To 50 g of water at 20 °C is added 0.50 kcal of heat energy. What is the final temperature of the water?
    (a) 20 °C   (b) 25 °C   (c) 30 °C   (d) 40 °C

18. What is the total mass of three beakers weighing 143.7 g, 87.68 g, and 101.00 g, respectively?
    (a) 332.38 g   (b) 332.4 g
    (c) 332 g   (d) 3.3×10^2 g

19. The density of air is 1.29 g/L. What volume would 5.20 kg of air occupy?
    (a) 6.71×10^3 L   (b) 4.03×10^3 L   (c) 0.248 L
    (d) 3.14×10^3 L

20. A substance can be described as dense if it has
    (a) a large mass.
    (b) a small mass in a large volume.
    (c) a small volume.
    (d) a large mass in a small volume.

21. Express 0.042 mL in μL.
    (a) 4.2×10^{-2} (b) 4.2   (c) 4.2×10^1
    (d) 4.2×10^2

22. An astronaut takes a piece of gold to the Moon. The density of this piece of gold on the Moon compared with its density on Earth is
    (a) less because it weighs less.
    (b) more because there is less gravitational attraction.
    (c) the same because the mass has not changed.
    (d) less because the ratio of mass to volume is smaller.

23. What is the mass of a cube of platinum 1.40 cm on a side? The density of platinum is 21.4 g/cm³.
    (a) 30.0 g   (b) 7.81 g   (c) 58.6 g   (d) 15.3 g

24. How much energy, in calories, is required to heat a cup of water (225 g) from room temperature to boiling (25 °C → 100 °C)?
    (a) 3.0 cal   (b) 5.6×10^3 cal   (c) 1.7×10^4 cal
    (d) 0.33 cal

25. A glass of milk is poured from a full one-gallon container of milk. The milk in the glass when compared with the milk in the container has
    (a) a different specific heat.
    (b) the same heat capacity.
    (c) a lower heat capacity.
    (d) a higher heat capacity.

26. What is the mass of 245 mL of ether if the density of ether is 0.725 g/mL?
    (a) 338 g   (b) 2.96×10^{-3} g   (c) 425 g   (d) 178 g

27. If a substance contracts when it freezes, its specific gravity will
    (a) increase.   (b) decrease.   (c) stay the same.
    (d) vary depending on the substance.

28. A 1.00×10^2 mL flask is filled with 77.0 g of an organic solvent. The density of this solvent is
    (a) 0.770 g/mL.   (b) 1.30 g/mL.   (c) 77.0 g.
    (d) 0.130 g/mL.

29. If the following masses were arranged from smallest to largest values, which one would be in the middle, 1.0 cg, 1.0 mg, 1.0 ng, 1.0 kg, or 1.0 g?
    (a) 1.0 g   (b) 1.0 ng   (c) 1.0 mg   (d) 1.0 cg

30. Heat is
    (a) measured in K.   (b) measured in calories.
    (c) measured in °C.   (d) not a form of energy.

Matter and the Structure of Atoms

3

Nature's Building Blocks

If the nucleus of an atom could be enlarged to the size of a walnut and placed at the center of a football stadium, the closest electrons would be at the outer perimeter of the stadium.

CHAPTER OUTLINE

CASE IN POINT: Magnetic resonance imaging

3.1 Atoms

A CLOSER LOOK: Images of Atoms

3.2 Subatomic particles

3.3 Atomic number and mass number

3.4 Isotopes and atomic mass

FOLLOW-UP TO THE CASE IN POINT: Magnetic resonance imaging

3.5 Electronic structure of atoms

A CLOSER LOOK: Electrons, Flames, and Fireworks

3.6 Electron configurations

We are already familiar with some of the properties of many forms of matter. Coal is a black solid that burns in air; water is a colorless, odorless, tasteless liquid; table salt is a crystalline solid with a characteristic taste; aspirin is a white solid with a sour taste. However, being able to identify some properties of matter is not the same as knowing what matter is made of. Much of nursing and the allied health fields is concerned with exerting control over matter. For example, should a patient receive more or less medication? What medication should be given? Because of the strong relationship between matter and almost everything that happens, anyone can benefit from knowing what matter is made of. This chapter briefly outlines some of what is known about the makeup of matter. We will see that matter is composed of only about 100 kinds of basic building blocks, called *atoms*. We also will learn that atoms are composed of only three kinds of particles that are even more fundamental than atoms. The arrangement of these particles in atoms is responsible for the infinite variety of matter in the world. This arrangement also leads to certain properties of matter that are valuable to the health sciences, as illustrated by the following Case in Point.

CASE IN POINT: Magnetic resonance imaging

Jamal is a 19-year-old college sophomore. He is also a good student and a gifted athlete. Although he weighs only 165 pounds, Jamal is generally considered to be the best defensive player on his football team, where he anchors the backfield. Late in one of Jamal's games, the 230-pound fullback of the opposing team stormed toward the line of scrimmage. As Jamal rushed to close the hole in the line, he and the fullback collided head-on. Jamal lay motionless on the turf. When Jamal remained unconscious for a few minutes, the team physician decided he should be taken to the hospital. During the trip, Jamal began to regain consciousness, but he had no recollection of the game or the circumstances of his injury. Jamal's speech was slurred, and he complained of a headache and double vision. The emergency room physician found that Jamal showed no signs of paralysis or numbness in his arms or legs. He suspected that Jamal had suffered a concussion and ordered a scan of Jamal's brain by magnetic resonance imaging (MRI) (see figure). What is an MRI?

What property of atoms does the MRI exploit? What did Jamal's physician find? We will learn the answers to these questions in Section 3.4.

Magnetic resonance imaging (MRI) allows physicians to "see" inside the human body.

3.1 Atoms

AIM: To summarize modern atomic theory.

John Dalton had more than one good idea. In 1807 he discovered a relationship among gases that led to Dalton's law of partial pressures (Sec. 7.7).

Atom
atomos (Greek):   indivisible

The concept of fundamental forms of matter began with the ancient Greeks. In the fourth century B.C., Democritus of Abdera, one of the most famous thinkers of antiquity, proposed that matter is composed of tiny indivisible particles called *atoms*. However, Democritus' ideas about atoms were not much use in explaining chemical phenomena. This situation did not change until the English chemist John Dalton (1766–1844) stated his own atomic theory. Dalton's atomic theory differed from that of Democritus in two important ways: It was based on solid experimental evidence for the existence of atoms, and it helped explain chemical reactions at the level of individual atoms. Modern atomic theory, only slightly changed from that proposed by Dalton, can be summarized in five statements:

1. All elements are composed of small particles. *An* **atom** *is the smallest particle into which an element can be divided and still retain its identity as the element.*

2. Atoms of the same element are identical in most ways, and the atoms of one element are different from those of any other element.

3. Atoms of different elements can combine only in certain ratios in forming compounds. For example, the ratio of hydrogen (H) atoms to oxygen (O) atoms in the compound water is always 2 to 1.

4. Chemical reactions involve the union, separation, or rearrangement of atoms.

5. In all chemical reactions, the atoms keep their identities; atoms of one element are not changed into atoms of another by chemical reaction.

How small is an atom? Suppose we could take a small cube of the soft, gray, metallic element lead (Pb) and cut it into the smallest possible pieces without destroying the "leadness" of each particle. Since this is a purely theoretical problem, we can assume that we have a tool to do the cutting. The final products of our efforts would be lead atoms, the smallest representative particles of the element lead. Fortunately, the problem is only theoretical, because lead atoms are very small. One gram of lead contains 2,900,000,000,000,000,000,000 (2.9×10^{21}) atoms. By comparison, the Earth's human population is only about 5,000,000,000 (5×10^9).

Lead is one example of the small size of atoms. The human body contains about 3.0×10^{27} atoms of hydrogen chemically combined with oxygen to form water. Even a tiny sample of any element contains an immense number of atoms. A Closer Look: Images of Atoms describes how scientists have recently been able to obtain images of individual atoms.

Images of Atoms

Many generations of scientists lived out their professional lives without hope of seeing individual atoms. This situation changed in the 1970s with the development of an instrument called the *scanning-tunneling microscope* (STM). An STM can generate images of atoms found on the surface of a material, as shown in the accompanying figure. The STM does not actually "see" atoms. Atomic images are generated from differences in electric current between the surface and a very thin, electrified metal probe. The closer the tip of the probe is to the surface, the greater is the current. When the probe is passed over the material, differences in the current indicate hills and valleys on its surface. These differences can be recorded and used by a computer to generate a kind of topographic map of the surface. The map constitutes the images of the atoms on the surface.

The scanning-tunneling microscope (STM) provides us with images of atoms.

STM images of atoms have helped confirm the evidence for the existence of atoms that was obtained from many experiments performed over a period of nearly 200 years. The evidence for atoms from previous experiments was indirect but so powerful that the existence of atoms has been firmly accepted for more than a century.

3.2 Subatomic particles

AIM: *To distinguish between protons, electrons, and neutrons in terms of their relative masses, relative charges, and location in the atom.*

Focus

Atoms are composed of electrons, protons, and neutrons.

Electron
elektron (Greek):   shining beam

Proton
protos (Greek):   first

One revision of early atomic theories concerns the idea that atoms are indivisible, a belief held until about a century ago. Today, atoms can be broken down into more fundamental particles. Chemistry is mainly concerned with only three subatomic particles: *electrons, protons,* and *neutrons.*

Electrons, protons, and neutrons

Electrons are a part of the atoms of all the elements. **Electrons** *are very small, light subatomic particles.* The negative electrical charge on the electron is defined as one unit of negative charge. The mass of the electron is 1840 times lighter than that of a hydrogen atom, which is the lightest atom that exists. *The* **proton,** *a positively charged subatomic particle, carries a single unit of positive charge and is 1840 times heavier than an electron.* A proton is what remains when a hydrogen atom is stripped of its only electron. As we will see later, protons are found in every living organism. The number of protons in the blood has an important bearing on the correct functioning of the body, as we will see in Chapters 7 and 22. **Neutrons** *are subatomic particles with no charge, but their mass nearly equals that of the proton.*

Putting the atom together

The fundamental building blocks of all atoms are the electron, the proton, and the neutron. (More details of the structure of the atom are discussed in Section 3.5.) *The* **nucleus** *(plural: nuclei) of the atom, composed of the protons and neutrons, has a positive charge and contains most of the mass of the atom in a very small volume.* The electrons are outside the nucleus. Although electrons contribute almost no mass to the atom, they occupy most of the volume. Unlike electrical charges attract, so negatively charged electrons are attracted to the positively charged nucleus. Objects with equal numbers of positive and negative charges are electrically neutral. Atoms are electrically neutral because the number of electrons equals the number of protons.

3.3 Atomic number and mass number

AIM: To use the atomic number and mass number of an element to find the number of protons, electrons, and neutrons.

Focus

The number of protons in an atomic nucleus identifies the element.

Now that we have a general description of the makeup of atoms, you may be wondering why one element is different from any other element. The answer to this question, which was discovered by the English physicist Henry Moseley in the early 1900s, is simple: *Different elements have different numbers of protons in their atoms.* Later it was shown that *the various elements behave differently in chemical reactions because of differences in the number of electrons in their atoms.* The chemistry of a hydrogen atom, with one proton and one electron, is different from the chemistry of a lithium atom, with three protons and three electrons.

Atomic number

If you look at the table inside the back cover, you will see an atomic number listed for each element. *The* **atomic number** *of an element is the number of protons in the nucleus of an atom of that element.* Hydrogen (atomic number 1) contains one proton in its nucleus, carbon (atomic number 6) contains six protons in its nucleus, and so forth. Atoms are electrically neutral, so the number of protons in the nucleus of an atom must equal the number of electrons around its nucleus. If we know the atomic number of an element, we also know the number of electrons in an atom of that element because the two numbers are the same.

PRACTICE EXERCISE 3.1

How many protons are there in an oxygen atom? How many electrons are there in an oxygen atom?

PRACTICE EXERCISE 3.2

What is the element with 11 electrons in each of its atoms?

Atomic mass units

The actual mass of a proton or a neutron is a very small number: 1.67×10^{-24} g. Even compared with this small mass, the mass of an electron is negligible: only 9.11×10^{-28} g. These mass values are so small they are inconvenient to work with. Fortunately, chemistry usually requires *relative* comparisons of the masses of atoms. For example, it is more useful to know that the mass of an oxygen atom is 16 times the mass of a hydrogen atom than to know the actual masses of oxygen and hydrogen atoms in grams. The unit of relative comparison of the masses of atoms is the **atomic mass unit** *(amu)—which is defined as one-twelfth the mass of a carbon atom that contains six protons and six neutrons.* Thus, for all practical purposes, we can consider the mass of a single proton or a single neutron as 1 amu. Table 3.1 summarizes the properties of the subatomic particles.

Mass number

Virtually the entire mass of an atom is concentrated in its nucleus. The total mass of all the electrons of the heaviest atom is not even one-tenth the mass of a single proton. *The total number of protons and neutrons in the nucleus is the* **mass number** *of an atom.* Table 3.2 gives the composition of some atoms of the first 10 elements. Hydrogen (atomic number 1) has a mass number of 1, and helium (atomic number 2) has a mass number of 4.

Table 3.1 Properties of Subatomic Particles

Particle	Symbol	Electrical charge	Relative mass (amu)	Actual mass (g)
electron	e^-	$1-$	1/1840	9.11×10^{-28}
proton	p or H^+	$1+$	1	1.67×10^{-24}
neutron	n	0	1	1.67×10^{-24}

Table 3.2 Some Atoms of the First 10 Elements

Name	Symbol	Atomic number	Composition of the nucleus — Protons	Composition of the nucleus — Neutrons	Mass number	Number of electrons
hydrogen	H	1	1	0	1	1
helium	He	2	2	2	4	2
lithium	Li	3	3	4	7	3
beryllium	Be	4	4	5	9	4
boron	B	5	5	6	11	5
carbon	C	6	6	6	12	6
nitrogen	N	7	7	7	14	7
oxygen	O	8	8	8	16	8
fluorine	F	9	9	10	19	9
neon	Ne	10	10	10	20	10

From these data we can conclude that an atom of helium is 4 times heavier than an atom of hydrogen. Likewise, we can see that an atom of neon (mass number 20) is 5 times heavier than an atom of helium.

EXAMPLE 3.1

Finding the composition of atoms

How many protons, electrons, and neutrons are in the following atoms?

	Atomic number	Mass number
(a) Lithium (Li)	3	7
(b) Aluminum (Al)	13	27
(c) Neon (Ne)	10	20

SOLUTION

(a) For an atom, the number of electrons equals the number of protons and is given by the atomic number. Lithium has three protons and three electrons. We get the number of neutrons by subtracting the atomic number from the mass number. Li has $7 - 3 = 4$ neutrons.

(b) Al has 13 protons, 13 electrons, and $27 - 13 = 14$ neutrons

(c) Ne has 10 protons, 10 electrons, and $20 - 10 = 10$ neutrons

PRACTICE EXERCISE 3.3

Complete the following table.

Atomic number	Mass number	Number of protons	Number of neutrons	Number of electrons	Symbol of element
4	——	——	5	——	——
——	——	11	12	——	——
——	52	——	——	24	——
——	40	18	——	——	——

3.4 Isotopes and atomic mass

AIMS: *To define an atomic mass unit. To use the concept of isotopes to explain why the atomic masses of elements are not whole numbers. To calculate the average atomic mass of an element from isotope data.*

Focus

Atomic masses of elements are not always whole numbers.

Not all atoms of the same element have identical masses. This is so because the number of neutrons may vary even though the nuclei all contain the same number of protons. *Atoms that have the same number of protons but different numbers of neutrons are called* **isotopes.** Since isotopes of an element have different numbers of neutrons, they also have different mass numbers and different atomic masses. Despite these differences, isotopes are chemically alike because they have identical numbers of protons and electrons. It is important to remember that electrons are responsible for the chemical behavior of an element.

Isotope
isotopos (Greek):   same place

Representing the composition of isotopes

To symbolize the composition of an isotope of a particular atom, it is customary to write its mass number as a superscript (a number slightly above) and its atomic number as a subscript (a number slightly below) to the left of the chemical symbol. Atoms of hydrogen with a mass number of 1 (designated hydrogen-1) and helium with a mass number of 4 (designated helium-4) are represented as

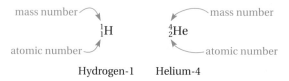

mass number $\rightarrow$ $_1^1$H $_2^4$He $\leftarrow$ mass number

atomic number $\rightarrow$ $\leftarrow$ atomic number

Hydrogen-1   Helium-4

EXAMPLE 3.2

Writing compositions of isotopes

Two isotopes of carbon are designated by mass number as carbon-12 and carbon-13. Give the chemical symbol for each.

SOLUTION

Carbon is atomic number 6, so all atoms of carbon have six protons. The atomic number is written as a subscript and the mass number is written as a superscript to the left of the chemical symbol for carbon. The chemical symbol for carbon-12 is $_6^{12}$C. The symbol for carbon-13 is $_6^{13}$C.

PRACTICE EXERCISE 3.4

Two isotopes of potassium are potassium-39 and potassium-41. Write the chemical symbol for each.

EXAMPLE 3.3

Finding the number of neutrons in atoms

How many neutrons are in the following atoms?
(a) $_8^{16}$O   (b) $_{21}^{45}$Sc   (c) $_{33}^{75}$As   (d) $_{38}^{88}$Sr   (e) $_{47}^{108}$Ag

SOLUTION

Recall that the superscript is the mass number and the subscript is the atomic number. Mass number minus atomic number equals the number of neutrons. (a) 8   (b) 24   (c) 42   (d) 50   (e) 61

PRACTICE EXERCISE 3.5

Determine the number of neutrons in the following atoms:

(a) ^{44}Ca   (b) ^{84}Kr   (c) ^{195}Pt

Atomic mass

Since the mass of an atom is concentrated in the nucleus, and since the mass of both a proton and a neutron is 1 amu, the mass of a single atom expressed in atomic mass units is a whole number. However, chemists work with large numbers of atoms, and the atomic masses they use are not whole numbers. The periodic table inside the front cover shows the atomic masses that chemists use. In the periodic table, the number above the symbol for each element is the atomic number. The number below each symbol is the atomic mass. We can see that the atomic mass of hydrogen (H) is 1.0079 amu. We would expect the atomic mass to be 1.0000 amu if all hydrogen atoms contain only 1 proton. The reason that the atomic masses of the elements—large collections of atoms—are not whole numbers is because of the existence of isotopes. *The* **atomic mass** *of an element in the periodic table is an average mass of the mixture of isotopes that reflects the masses and relative abundance of the elements as they occur in nature.*

The percent abundances of the isotopes of some familiar elements, along with the atomic masses from the periodic table, are listed in Table 3.3.

About 250 isotopes occur in nature. More than four times this number have been made using nuclear reactors. Many of the synthetic isotopes are useful in the medical fields.

Table 3.3 Natural Isotopes of Some Familiar Elements

Name	Symbol	Mass (amu)	Natural percent abundance	Atomic mass
hydrogen	$^{1}_{1}$H	1.0078	99.985	
	$^{2}_{1}$H	2.0141	0.015	1.0079
	$^{3}_{1}$H	3.0160	negligible	
helium	$^{3}_{2}$He	3.0160	0.0001	4.0026
	$^{4}_{2}$He	4.0026	99.9999	
carbon	$^{12}_{6}$C	12.000	98.89	12.011
	$^{13}_{6}$C	13.003	1.11	
nitrogen	$^{14}_{7}$N	14.003	99.63	14.007
	$^{15}_{7}$N	15.000	0.37	
oxygen	$^{16}_{8}$O	15.995	99.759	
	$^{17}_{8}$O	16.995	0.037	15.999
	$^{18}_{8}$O	17.999	0.204	
sulfur	$^{32}_{16}$S	31.972	95.00	
	$^{33}_{16}$S	32.971	0.76	32.064
	$^{34}_{16}$S	33.967	4.22	
	$^{36}_{16}$S	35.967	0.014	
zinc	$^{64}_{30}$Zn	63.929	48.89	
	$^{66}_{30}$Zn	65.926	27.81	
	$^{67}_{30}$Zn	66.927	4.11	65.37
	$^{68}_{30}$Zn	67.925	18.57	
	$^{70}_{30}$Zn	69.925	0.62	

Ratio of chlorine atoms in natural abundance: three $^{35}_{17}Cl$ to one $^{37}_{17}Cl$

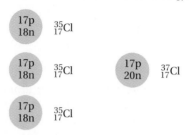

Figure 3.1
The average atomic mass
of chlorine is 35.5 amu.

Total number of protons in
three $^{35}_{17}Cl$ atoms and one $^{37}_{17}Cl$ atom $= 4 \times 17 = 68$

Total number of neutrons in
three $^{35}_{17}Cl$ atoms and one $^{37}_{17}Cl$ atom $= (3 \times 18) + 20 = 74$

Total mass (amu) of all the protons and neutrons $= 68 + 74 = 142$ amu

Average mass (amu) of one atom $= \frac{142}{4} = 35.5$ amu

Most elements occur as two or more isotopes. For example, chlorine has two isotopes, both of which have 17 protons in their atomic nuclei. One isotope has 18 neutrons and an atomic mass of 35 amu (designated as chlorine-35 or $^{35}_{17}Cl$). The other isotope is chlorine-37. In nature, the isotopes of chlorine occur in a nearly 3:1 ratio. Based on this information, we would expect the average atomic mass of chlorine to be closer to 35 amu than to 37 amu. How would we *calculate* the average atomic mass for chlorine? Figure 3.1 shows how the calculation is made. The average atomic mass calculated by this method is 35.5 amu. In the periodic table of the elements (inside the front cover) the atomic mass for chlorine (Cl) is 35.453 amu. This is because the actual ratio is not exactly 3:1 and because the masses of protons and neutrons are not exactly 1 amu. In calculating average atomic masses, the relative abundance of each isotope must be taken into account.

EXAMPLE 3.4

Calculating an average atomic mass

Element X has two natural isotopes. One, with a relative abundance of 20%, has a mass number of 10; the other, with a relative abundance of 80%, has a mass number of 11. Estimate the average atomic mass for the element from these figures. What is the true identity and atomic number of element X?

SOLUTION

The mass that each isotope contributes toward the average atomic mass is equal to the product of its mass multiplied by its relative abundance expressed as a fraction. To express percent as a fraction, divide by 100 or express as parts per 100 parts. Thus, 20% = 20/100 = 0.20.

$$^{10}X: \quad 10 \text{ amu} \times 0.20 = \quad 2.0 \text{ amu}$$
$$^{11}X: \quad 11 \text{ amu} \times 0.80 = \quad 8.8 \text{ amu}$$
$$\text{Total} \quad \overline{10.8 \text{ amu}}$$

The average mass of element X is 10.8. Therefore, element X is boron, B, with atomic mass 10.81 and atomic number 5.

PRACTICE EXERCISE 3.6

The element gallium has two naturally occurring isotopes: gallium-69 and gallium-71. The relative abundances are 60.1% and 39.9%, respectively. Calculate the average atomic mass of gallium.

Water made from deuterium is known as *heavy water* and often written as D_2O. It is also used to trace the role of water in many biological reactions.

There are three known isotopes of hydrogen. Each of these isotopes has one proton. The most common hydrogen isotope has no neutrons. It has an atomic mass of 1 amu and is called *hydrogen-1* (^1H) or *hydrogen*. The second isotope has one neutron and an atomic mass of 2 amu; it is called either *hydrogen-2* (^2_1H) or *deuterium*. The third isotope has two neutrons in the nucleus and an atomic mass of 3 amu. This isotope is called *hydrogen-3* (^3_1H) or *tritium*. Tritium nuclei are unstable and slowly disintegrate, as we will discuss in Chapter 10. Almost all hydrogen in nature is hydrogen-1. The other two isotopes are present in only trace amounts, and the average atomic mass for hydrogen in the periodic table is 1.0079 amu. We will see in Chapter 10 that many isotopes with unstable nuclei are useful in the diagnosis and treatment of disease. Isotopes with stable nuclei also have important uses in medicine. One of these uses is relevant to the Case in Point for this chapter.

FOLLOW-UP TO THE CASE IN POINT: Magnetic resonance imaging

Jamal, the student-athlete introduced in the Case in Point, was knocked unconscious in a football game. From careful observation of Jamal's behavior upon regaining consciousness, the attending physician hypothesized that Jamal suffered a concussion, which in severe cases can be accompanied by a noticeable swelling of the brain. Although a concussion is a serious matter, patients usually recover completely. However, the doctor was concerned that a blood clot might have formed in Jamal's brain. This would be an even more serious matter than a concussion. In order to examine this possibility, the doctor ordered a magnetic resonance image (MRI) of Jamal's brain (see figure). MRI is a method that allows physicians to see inside the human body without surgery. MRI involves measurement of a phenomenon called *nuclear magnetic resonance,* a behavior exhibited by the nuclei of certain isotopes. Hydrogen-1 (^1H), phosphorus-31 (^{31}P), and carbon-13 (^{13}C) are among the stable isotopes that can be detected by MRI and that are reasonably abundant in the body. To conduct MRI, the subject's body is placed in a strong magnetic field and bombarded with radiowaves similar to those of an ordinary FM radio. As these radiowaves emerge from the body, they provide information about the location and environment of the isotopic nuclei. This information can be used by a computer to construct a three-dimensional image of the body's interior. The computer may assign different colors to the results of its calculations. These colors are helpful in making clear distinctions between tissues in the final images. In Jamal's case, the MRI was completely normal, with no evidence of a blood clot on the brain. After a few days' recovery, he returned to school. Jamal was advised by his physician to quit football.

MRI of a single patient would not be very useful. Diagnosis usually requires comparison of MRIs for normal subjects with those for patients who are ill. The difference between these MRIs is very important in diagnosing the disease. For example, comparison of an MRI of the brain of a normal subject with that of an ill one could reveal a small tumor that would not have been apparent if the physician had no idea of the appearance of a normal image.

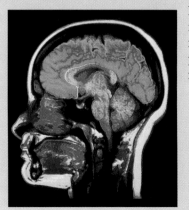

MRI of the brain. MRI is often used instead of X rays as a diagnostic tool because it does not involve ionizing radiation.

3.5 Electronic structure of atoms

AIMS: *To distinguish among principal energy level, energy sublevel, and atomic orbital. To describe the shapes of s, p, and d orbitals.*

Focus

Energies of electrons in atoms are confined to certain values.

After the discovery of electrons and atomic nuclei, scientists were eager to learn how these particles are put together to form atoms. An early model of the structure of the atom envisioned tiny electrons embedded in a large nucleus, rather like blueberries in a muffin. In 1913, Niels Bohr (1885–1962), a young Danish physicist, made a revolutionary proposal.

Energies of electrons

Bohr suggested that the electrons in atoms have certain fixed energies. Bohr's idea can be illustrated by a ladder. The rungs of a ladder are analogous to the permissible energies of electrons. The lowest rung corresponds to the lowest energy level (Fig. 3.2). Just as a person climbs up or down a ladder by going from rung to rung, an electron can jump from one energy level to another. However, the regions between the energy levels are forbidden. A person on a ladder cannot stand between the rungs, and electrons in an atom cannot have an energy between the energy levels. To jump from one energy level to another, an electron must gain or lose just the right amount of energy. *A* **quantum** *of energy is the amount of energy required to move an electron from its present energy level to the next higher one.* Thus the energies of electrons are said to be *quantized.* The term *quantum leap,* used to describe an abrupt change, comes from this concept. The amount of energy gained or lost by every electron as it jumps between energy levels is not the same. Unlike the evenly spaced rungs of a ladder, the energy levels in an atom are *not equally spaced.* The energy levels of electrons become closer together as the electrons move farther from the nucleus. The outermost electrons are the easiest to move from one energy level to a higher level, as described in A Closer Look: Electrons, Flames, and Fireworks.

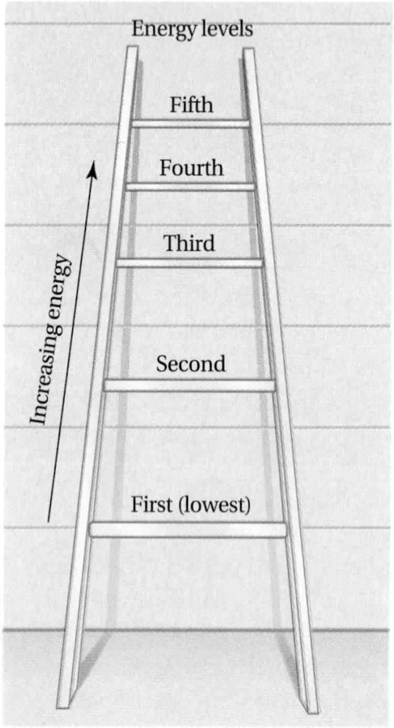

Location of electrons

Where are an atom's electrons in relation to the nucleus? Perhaps surprisingly, there is no exact answer to this question. Consider a thrown baseball, which has a fairly large mass and moves at most about 100 miles per hour. The position of the baseball at any instant can be determined with great accuracy. An electron, however, has a very low mass and moves much faster than a baseball. It is impossible to pinpoint the location of such an object at any instant. All anyone can do is describe the chances that an electron will be in a certain place at any instant. As a result of the uncertainty inherent in electron movements, the locations of electrons in atoms are generally described by *atomic orbitals. An* **atomic orbital** *is a region in space where there is a good chance of finding an electron.* An atomic orbital is usually portrayed as a blurry cloud that is most dense where there is a good chance of finding the electron and less dense where the chance of finding the elec-

Figure 3.2
The energy levels in an atom are analogous to the rungs on a ladder. Electrons have only certain allowed energies. The higher the energy level occupied by the electron, the more energetic the electron.

A Closer Look

Electrons, Flames, and Fireworks

The spectacular colors of fireworks always draw oohs and aahs from the crowd watching them. Did you know that the different colors produced by flames and fireworks result from burning various elements and compounds? Burning iron filings and carbon particles produces gold light. Burning sodium compounds produces yellow light, strontium compounds give red light, barium compounds give green light, and copper compounds give blue light (see figure).

These vivid colors are the result of the electronic structures of atoms. As you may recall, the energy of an electron in an atom depends on the atomic orbital that the electron occupies. The electrons in higher energy levels are held less tightly to the nucleus than those in lower energy levels. Less tightly held electrons can jump to a higher energy level or even be kicked out of the atoms if enough energy is supplied. Heating small particles of an element, or a compound containing the element, supplies enough energy to cause the changes. Because of the law of conservation of energy, the energy absorbed by the atom is emitted as the electrons drop back to lower energy levels or when separated atoms and electrons recombine. Some of this energy is emitted as heat, but some of the energy also can be emitted as visible light. We see this light as the burning particles from fireworks streak across the sky. The color of the light that we see depends on its energy. The energy of light decreases in the order purple, blue, green, yellow, orange, and red.

The tendency of elements and compounds to lose and regain electrons in a flame is used in a

The spectacular colors of fireworks are due to light emission by excited atoms; compounds of barium give green light, and compounds of strontium produce red light.

method for chemical analysis called *atomic absorption*. This method measures the color and amount of light emitted by a hot substance. The color of light emitted by the substance is characteristic of the elements in the substance, and the amount of light is proportional to the mass of the substance. Atomic absorption permits the identification and quantification of very small amounts of elements and compounds. Amounts of elements and compounds in mixtures also can be determined.

tron is small. We can imagine the negative charge spread out over the entire atomic orbital. Figure 3.3 illustrates a cloudy volume of the kind typically used to show an atomic orbital. There is at least a small chance of finding the electron a considerable distance from the nucleus. Atomic orbitals are usually drawn so that there is a 90% chance of finding the electron somewhere within its boundaries. Different shapes for atomic orbitals are denoted by letters: *s orbitals* are spherical, and *p orbitals* are dumbbell-

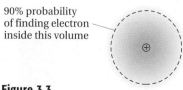

90% probability
of finding electron
inside this volume

Figure 3.3
In an atomic orbital, the probability
of finding an electron in a certain
volume of space is shown as a
blurry cloud of negative charge.
The cloud is most dense where the
probability of finding the electron is
large and is least dense where the
probability is small.

shaped; the shapes of *d orbitals* and *f orbitals* are far more complex. Figure 3.4 shows the shapes of *s*, *p*, and *d* orbitals. The shapes of the *f* orbitals are not shown.

The energies of electrons in atoms and the atomic orbitals that they occupy are related. We can think of the energies of the electrons that occupy atomic orbitals in terms of *principal energy levels* and *sublevels*. The principal energy levels are designated in order of increasing energy as 1, 2, 3, 4, 5, 6, and so forth. Within each principal energy level, electrons have slightly different energies corresponding to *energy sublevels*. The number of energy sublevels within a principal energy level is equal to the number of the prin-

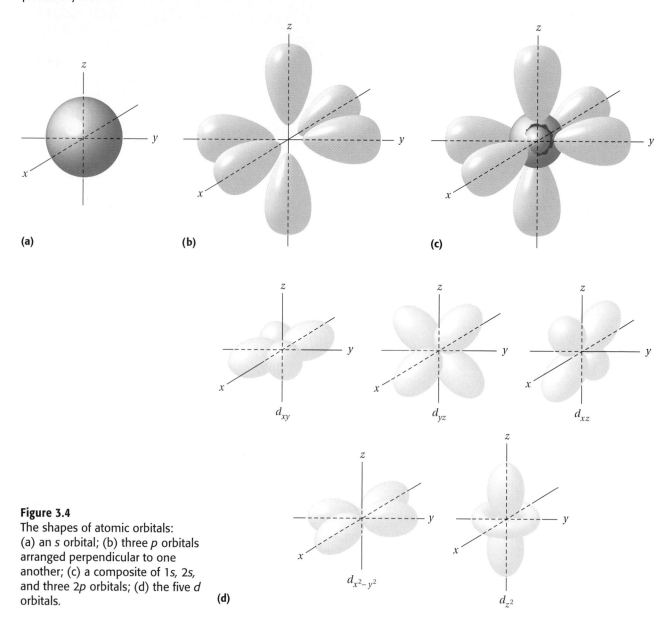

(a)   (b)   (c)

d_{xy} d_{yz} d_{xz}

$d_{x^2-y^2}$ d_{z^2}

(d)

Figure 3.4
The shapes of atomic orbitals:
(a) an *s* orbital; (b) three *p* orbitals
arranged perpendicular to one
another; (c) a composite of 1*s*, 2*s*,
and three 2*p* orbitals; (d) the five *d*
orbitals.

Table 3.4 Summary of Principal Energy Levels, Sublevels, and Orbitals

Principal energy level	Number of sublevels	Type of sublevel
1	1	$1s$ (1 orbital)
2	2	$2s$ (1 orbital), $2p$ (3 orbitals)
3	3	$3s$ (1 orbital), $3p$ (3 orbitals), $3d$ (5 orbitals)
4	4	$4s$ (1 orbital), $4p$ (3 orbitals), $4d$ (5 orbitals), $4f$ (7 orbitals)

cipal level. The first principal energy level has only one sublevel, designated $1s$. The second principal energy level has two sublevels, designated $2s$ and $2p$. The $2p$ sublevel is of higher energy than the $2s$ sublevel and consists of three p orbitals of equal energy, with the long axis of each p orbital perpendicular to the other two. It is convenient to label the axes $2p_x$, $2p_y$, and $2p_z$. The second principal energy level has a total of four orbitals, designated $2s$, $2p_x$, $2p_y$, and $2p_z$. The third principal energy level has three sublevels, designated $3s$, $3p$, and $3d$. The $3d$ sublevel consists of five d orbitals of equal energy. Thus the third principal energy level has nine orbitals. The fourth principal energy level has four sublevels, designated $4s$, $4p$, $4d$, and $4f$. The $4f$ sublevel consists of seven f orbitals of equal energy. The fourth principal energy level, then, has 16 orbitals. The sublevels and orbitals for each principal energy level are summarized in Table 3.4.

PRACTICE EXERCISE 3.7

How many orbitals are in (a) the $3p$ sublevel, (b) the $2s$ sublevel, (c) the $4f$ sublevel, (d) the $4p$ sublevel, (e) the $3d$ sublevel, and (f) the third principal energy level?

3.6 Electron configurations

AIM: To use the Aufbau principle, the Pauli exclusion principle, and Hund's rule to write the electron configuration of representative elements.

Focus

The electron configurations of most elements can be written using three rules.

It is the nature of things to seek the lowest possible energy. High-energy systems are unstable, and unstable systems undergo changes that lose energy and become more stable. In the world of the atom, electrons and the nucleus interact to make the most stable arrangement, or configuration, possible. *The ways in which electrons are arranged around the nuclei of atoms are called* **electron configurations.**

Three rules govern the filling of atomic orbitals by electrons within the principal energy levels: the **Aufbau principle,** the **Pauli exclusion principle,** and **Hund's rule.** It is important to realize that these rules provide a way to obtain correct electron configurations for atoms. It is not the way that atoms are formed.

Aufbau

Aufbau (German):   building up

1. *The Aufbau principle:* Electrons enter orbitals of lowest energy first. We can use a box (☐) to represent an atomic orbital. The Aufbau diagram is a useful device for remembering the sequence of energy levels (Fig. 3.5). As we can see from this diagram, the filling of atomic orbitals does not follow a simple pattern beyond the second energy level. For example, the 4s orbital is lower in energy than the 3d orbital, and the 4f orbital is lower in energy than the 5d orbital.

It is helpful to remember that the number of sublevels equals the number of the principal energy level and that the orbitals in a given sublevel in a principal energy level are always of equal energy.

The maximum number of electrons that can occupy a given energy level is given by the formula $2n^2$, where n is the principal energy level. The number of electrons allowed in each of the first four energy levels is as follows:

		Increasing energy (increasing distance from nucleus) →			
Energy level n		1	2	3	4
Maximum number of electrons allowed		2	8	18	32

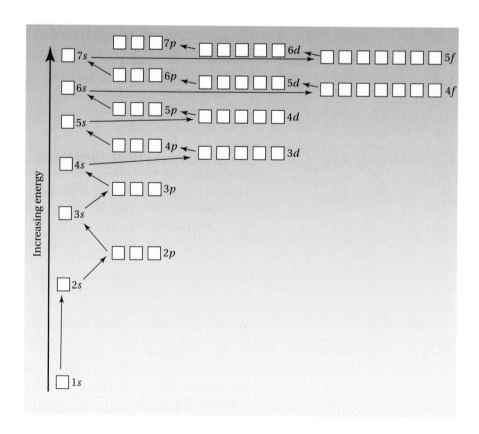

Figure 3.5
The Aufbau diagram. To determine the order of filling energy levels in an atom with many electrons, begin with the 1s orbital, follow the arrows, and fill in order of increasing energy.

Table 3.5 Electron Configurations for Some Selected Elements

Element	1s	2s	2p_x	2p_y	2p_z	3s	Electron configuration
H	↑						$1s^1$
He	↑↓						$1s^2$
Li	↑↓	↑					$1s^2 2s^1$
C	↑↓	↑↓	↑	↑			$1s^2 2s^2 2p^2$
N	↑↓	↑↓	↑	↑	↑		$1s^2 2s^2 2p^3$
O	↑↓	↑↓	↑↓	↑	↑		$1s^2 2s^2 2p^4$
F	↑↓	↑↓	↑↓	↑↓	↑		$1s^2 2s^2 2p^5$
Ne	↑↓	↑↓	↑↓	↑↓	↑↓		$1s^2 2s^2 2p^6$
Na	↑↓	↑↓	↑↓	↑↓	↑↓	↑	$1s^2 2s^2 2p^6 3s^1$

2. *The Pauli exclusion principle:* An atomic orbital may describe at most two electrons. We can think of electrons as tiny objects that spin in clockwise or counterclockwise directions. To occupy the same orbital, two electrons must have opposite spins. When two electrons occupy the same orbital and have opposite spins, we say that the spins are *paired*. A vertical arrow (↑) indicates an electron and its direction of spin (↑ or ↓); we can write an orbital containing paired electrons as ↑↓ .

3. *Hund's rule:* When electrons occupy orbitals of equal energy, one electron enters each orbital until all the orbitals contain one electron with spins parallel (in the same direction, as in ↑ ↑ ↑). Second electrons then add to each orbital so that their spins are paired with the first electron in the orbital.

Consider the electron configurations of atoms of the nine elements in Table 3.5. An oxygen atom, for example, contains eight electrons. The orbital of lowest energy, $1s$, gets one electron and then a second of opposite spin. The next orbital to fill is $2s$. Three electrons then go, one each, into the three $2p$ orbitals, which have equal energy. The remaining electron now pairs with an electron occupying one of the $2p$ orbitals.

PRACTICE EXERCISE 3.8

Arrange the following sublevels in order of decreasing energy: $2p$, $4s$, $3s$, $3d$, and $3p$.

Writing electron configurations

A convenient shorthand method for showing the electron configuration of an atom involves writing the energy level and the symbol for every sublevel occupied by an electron. A superscript indicates the number of electrons occupying that sublevel. As shown in Table 3.6, for hydrogen, with one electron in a $1s$ orbital, the electron configuration is written $1s^1$; for helium, with two electrons in a $1s$ orbital, it is $1s^2$; for oxygen, with two electrons in a $1s$ orbital, two electrons in a $2s$ orbital, and four electrons in $2p$ orbitals, it is $1s^2 2s^2 2p^4$.

Table 3.6 The Electron Configurations of the First 36 Elements

Element	Atomic number	Electron structure*	Element	Atomic number	Electron structure*
H	1	$1s^1$	K	19	[Ar] $4s^1$
He	2	$1s^2$	Ca	20	[Ar] $4s^2$
Li	3	$1s^2 2s^1$	Sc	21	[Ar] $3d^1 4s^2$
Be	4	$1s^2 2s^2$	Ti	22	[Ar] $3d^2 4s^2$
B	5	$1s^2 2s^2 2p^1$	V	23	[Ar] $3d^3 4s^2$
C	6	$1s^2 2s^2 2p^2$	Cr	24	[Ar] $3d^5 4s^1$
N	7	$1s^2 2s^2 2p^3$	Mn	25	[Ar] $3d^5 4s^2$
O	8	$1s^2 2s^2 2p^4$	Fe	26	[Ar] $3d^6 4s^2$
F	9	$1s^2 2s^2 2p^5$	Co	27	[Ar] $3d^7 4s^2$
Ne	10	$1s^2 2s^2 2p^6$	Ni	28	[Ar] $3d^8 4s^2$
Na	11	[Ne] $3s^1$	Cu	29	[Ar] $3d^{10} 4s^1$
Mg	12	[Ne] $3s^2$	Zn	30	[Ar] $3d^{10} 4s^2$
Al	13	[Ne] $3s^2 3p^1$	Ga	31	[Ar] $3d^{10} 4s^2 4p^1$
Si	14	[Ne] $3s^2 3p^2$	Ge	32	[Ar] $3d^{10} 4s^2 4p^2$
P	15	[Ne] $3s^2 3p^3$	As	33	[Ar] $3d^{10} 4s^2 4p^3$
S	16	[Ne] $3s^2 3p^4$	Se	34	[Ar] $3d^{10} 4s^2 4p^4$
Cl	17	[Ne] $3s^2 3p^5$	Br	35	[Ar] $3d^{10} 4s^2 4p^5$
Ar	18	[Ne] $3s^2 3p^6$	Kr	36	[Ar] $3d^{10} 4s^2 4p^6$

*For elements beyond neon, the configurations are simplified by using the symbol for the preceding noble gas rather than its complete electron configuration.

EXAMPLE 3.5

Writing electron configurations of atoms

Use Figure 3.5 to write the electron configuration of (a) silicon and (b) cobalt.

SOLUTION

Silicon has 14 electrons; cobalt has 27 electrons. Using Figure 3.5, start placing electrons in the orbitals with the lowest energy ($1s$). Remember that there is a maximum of two electrons in each orbital and that electrons do not pair up in orbitals of equal energy until necessary.

(a) Silicon

(b) Cobalt

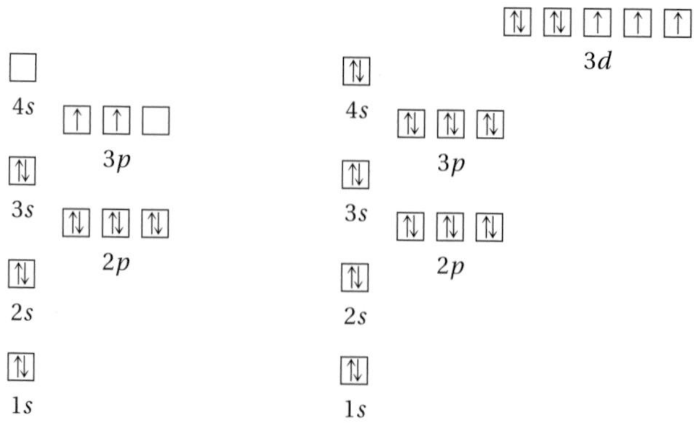

Si: $1s^2 2s^2 2p^6 3s^2 3p^2$

Co: $1s^2 2s^2 2p^6 3s^2 3p^6 3d^7 4s^2$

PRACTICE EXERCISE 3.9

Write electron configurations for atoms of the elements boron and fluorine. How many unpaired electrons does (a) an atom of boron and (b) an atom of fluorine have?

PRACTICE EXERCISE 3.10

How many electrons are in the highest occupied energy level of the following atoms?

(a) arsenic   (b) sulfur   (c) phosphorus   (d) bromine

SUMMARY

More than 100 elements exist, and each is composed of atoms. The atoms of a given element are different from the atoms of all other elements. Atoms are exceedingly small.

Atoms contain negatively charged electrons, positively charged protons, and electrically neutral neutrons. The proton has a mass 1840 times the mass of the electron. A proton and a neutron are nearly identical in mass. The nucleus of an atom is composed of protons and neutrons and contains most of the mass of an atom in a very small volume. The electrons surround the nucleus and occupy most of the volume of the atom.

The number of protons in the nucleus of an element's atoms is the atomic number of that element; the total number of protons and neutrons is the mass number. The atoms of a given element all contain the same number of protons, but the number of neutrons may vary. Atoms with the same number of protons but a different number of neutrons are isotopes. Because most elements occur as two or more isotopes, atomic masses are not whole numbers.

In the Bohr model of the atom, the energies of electrons are quantized. The modern description of the electrons in atoms does not define the exact path of an electron but shows the chance of finding an electron in a region of space called an atomic orbital. The ways in which electrons are arranged around the nuclei of atoms are called electron configurations. Correct electron configurations for atoms may be written by using the Aufbau principle, the Pauli exclusion principle, and Hund's rule. The Aufbau principle gives the sequence in which the orbitals are filled; the Pauli exclusion principle states that a maximum of only two electrons can occupy each orbital; and Hund's rule states that the electrons pair up only after each orbital in a sublevel is occupied by a single electron.

KEY TERMS

Atom (3.1)
Atomic mass (3.4)
Atomic mass unit (3.3)
Atomic number (3.3)
Atomic orbital (3.5)

Atomic theory (3.1)
Aufbau principle (3.6)
Electron (3.2)
Electron configuration (3.6)

Hund's rule (3.6)
Isotope (3.4)
Mass number (3.3)
Neutron (3.2)
Nucleus (3.2)

Pauli exclusion principle (3.6)
Proton (3.2)
Quantum (3.5)

EXERCISES

Atoms and Their Components (Sections 3.1, 3.2)

3.11 Describe Dalton's contributions to atomic theory.

3.12 Both Democritus and Dalton arrived at the concept of an atom, but by different means. Explain.

3.13 Explain a chemical reaction using modern atomic theory.

3.14 Does atomic theory support the law of conservation of matter (Sec. 1.11)? Explain.

3.15 What is the charge and the relative mass of (a) a proton, (b) a neutron, and (c) an electron?

3.16 Describe the composition of the nucleus of the atom and its position relative to the atom as a whole.

Atomic Number and Mass Number (Section 3.3)

3.17 What did Moseley discover, and why was it significant?

3.18 What makes an atom of carbon uniquely different from atoms of all other elements?

3.19 What are the atomic numbers of (a) fluorine, (b) nitrogen, (c) carbon, and (d) strontium?

3.20 How many protons are in the nucleus of the following atoms?
(a) calcium   (b) helium   (c) silicon   (d) silver

3.21 What is an atomic mass unit?

3.22 What is the total mass, in amu, of 10 protons, 10 electrons, and 10 neutrons?

3.23 An atom has 22 neutrons and 18 protons in its nucleus. What is the atomic number, mass number, and symbol of this element?

3.24 Complete the following table.

Element	Atomic number	Number of neutrons	Mass number	Number of protons
Au	___	___	197	___
___	26	___	56	___
___	___	___	136	56
___	___	1	2	___

Isotopes and Atomic Mass (Section 3.4)

You may use the periodic table for these exercises.

3.25 Name three ways that isotopes of an element differ.

3.26 Explain why the atomic masses for most of the elements are not whole numbers.

3.27 The mass number of an isotope of oxygen is 17. State (a) the name, (b) the symbol, and (c) the number of protons and neutrons in its nucleus.

3.28 A natural isotope of carbon has the symbol $^{14}_{6}C$. State (a) the atomic number, (b) the mass number, (c) the number of protons, and (d) the number of neutrons for this isotope.

3.29 An isotope of cobalt used in the treatment of cancer has 33 neutrons. Write the symbol for this isotope.

3.30 Complete the following table.

Symbol	Atomic number	Mass number	Number of electrons	Number of neutrons
$^{81}_{35}Br$	___	___	___	___
___	___	11	5	___
___	40	94	___	___
___	___	___	36	46

3.31 List the number of protons, neutrons, and electrons in each of the following atoms.
(a) $^{27}_{13}Al$ (b) $^{65}_{29}Cu$ (c) $^{37}_{17}Cl$ (d) $^{7}_{3}Li$

3.32 Give the number of protons, electrons, and neutrons in each of the following atoms.
(a) $^{79}_{35}Br$ (b) $^{52}_{24}Cr$ (c) $^{22}_{10}Ne$ (d) $^{127}_{53}I$

3.33 A sample of silver as it occurs in nature is 51.8% of isotope $^{107}_{47}Ag$ and 48.2% of isotope $^{109}_{47}Ag$. What is the average atomic mass of silver? (Compare your result with the value given in the periodic table.)

3.34 Ninety-two percent of the atoms of an element have a mass of 28.0 amu, 5.0% of the atoms have a mass of 29.0 amu, and the remaining atoms have a mass of 30.0 amu. Calculate the average atomic mass, and identify the element.

Atomic Structure and Electron Configuration (Sections 3.5, 3.6)

3.35 The energies of electrons are said to be *quantized.* Explain.

3.36 What is an *atomic orbital*?

3.37 What are the three rules that govern the filling of atomic orbitals by electrons?

3.38 Which rule or principle is linked to the following?
(a) the maximum number of electrons in an orbital
(b) the filling of orbitals within a sublevel
(c) the order in which energy levels are filled

3.39 What is the maximum number of electrons allowed in each of the following sublevels?
(a) $3d$ (b) $2p$ (c) $4s$ (d) $5f$

3.40 What is the maximum number of electrons allowed in each of the following sublevels?
(a) $4p$ (b) $1s$ (c) $4f$ (d) $4d$

3.41 What is the meaning of each number and letter in the symbol $4d^3$?

3.42 What is meant by $3s^2$?

3.43 An atom of an element has two electrons in the first energy level and four electrons in the second energy level.
(a) Write the electron configuration, and name the element.
(b) How many unpaired electrons does an atom of this element have?

3.44 An atom of an element has two electrons in the first energy level and seven electrons in the second energy level.
(a) Write the electron configuration, and name the element.
(b) How many unpaired electrons does an atom of this element have?

3.45 How many electrons are in the third energy level of the following atoms?
(a) chlorine (b) oxygen (c) potassium

3.46 How many electrons are in the second energy level of the following atoms?
(a) fluorine (b) phosphorus (c) beryllium

3.47 Give the symbol and name of the elements whose atoms have the following configurations.
(a) $1s^2 2s^2 2p^6 3s^1$ (b) $1s^2 2s^2 2p^3$
(c) $1s^2 2s^2 2p^6 3s^2 3p^2$ (d) $1s^2 2s^2 2p^6 3s^2 3p^6 4s^1$

3.48 Give the name and symbol of the elements whose atoms have the following configurations.
(a) $1s^2 2s^2 2p^6 3s^2 3p^4$ (b) $1s^2 2s^2 2p^1$
(c) $1s^2 2s^2 2p^6 3s^2 3p^6 3d^2 4s^2$ (d) $1s^2 2s^2 2p^5$

3.49 Explain why one of the electrons in an atom of lithium goes into the second energy level.

3.50 Why does one electron in an atom of potassium go into the fourth energy level instead of squeezing into the third energy level along with the eight already there?

3.51 Write electron configurations for the three elements that are identified only by the following atomic numbers: (a) 15, (b) 9, and (c) 18.

3.52 Write electron configurations for atoms of the elements (a) selenium, (b) vanadium, and (c) nickel. (Use Figure 3.5 on page 76 as a guide.)

Additional Exercises

3.53 Sketch the shapes of the s and p orbitals. How are the p orbitals oriented in relation to one another?

3.54 What is the Pauli exclusion principle, and how does it affect the filling of atomic orbitals by electrons?

3.55 In your own words, state Hund's rule. Draw electron configurations (see Table 3.5) for the elements Si, P, and S, and use them as examples to explain how Hund's rule is applied.

3.56 Explain why each of the following electron configurations is incorrect.
(a) $1s^2 2s^1 2p^1$ (b) $1s^2 2s^2 2p^7$
(c) $1s^2 2s^2 2p^6 3s^2 3p^6 3d^{10}$ (d) $1s^2 2s^2 2p^6 3s^2 3p^5 4s^2$

3.57 How many electrons can occupy (a) an s orbital, (b) a p orbital, and (c) a d orbital?

3.58 Explain why the following electron configurations are not possible for atoms of elements in their most stable state.
(a) $1s^2 2s^2 2p^6 3s^3$ (b) $1s^2 2s^2 2p^4 3s^1$
(c) $1s^2 2s^3 2p^1$ (d) $1s^2 2s^2 2p^6 3s^2 3p^1 4s^2$

3.59 Convert each of these tabular electron configurations into the shorthand form, and name the element.

3.60 Complete the following table.

	Element	Symbol	Atomic number	Mass number	Number of protons	Number of neutrons
(a)	___	^3_1H	___			___
(b)	cobalt	___	___	60	___	___
(c)	___	___	17	37	___	___
(d)	___	___		90	___	52
(e)	___	$^{235}_{92}U$	___	___	___	___
(f)	chlorine	___	___	___	17	18
(g)	___	___	___	234	94	___
(h)	___	$^{32}_{16}S$	___	___	___	___
(i)	iron	___	___	56	___	___
(j)	calcium	___	___	___	___	28

3.61 Answer the following questions about these isotopes of elements. Q is a symbol for any element.

$^{10}_5Q$ $^{34}_{16}Q$ $^{17}_8Q$ $^{18}_8Q$ $^{36}_{18}Q$ $^{36}_{16}Q$ ^9_4Q

(a) List any pairs of isotopes of the same element.
(b) List any pairs of isotopes with the same mass number.
(c) List any pairs of isotopes with the same number of neutrons.
(d) List any pairs of isotopes with the same number of electrons.

SELF-TEST (REVIEW)

True/False

1. Atoms of elements are electrically neutral.
2. The mass of an electron is equal to the mass of a neutron.
3. The charge on all protons is identical.
4. Two electrons repel one another.
5. The atomic number of an element is the sum of the number of protons and neutrons in an atom.
6. Atomic masses of atoms are expressed in the unit amu.

7. The orbital with the lowest energy is $1s$.
8. In the electron configuration of any atom, the superscripts add up to the atomic number of the element.
9. The mass numbers of all gold atoms are identical.
10. Orbitals of an atom are quantized.

Multiple Choice

11. Which of these statements is true?
    (a) For all atoms except the most abundant isotope of hydrogen the atomic number is less than the mass number.
    (b) Atoms are the smallest particles of matter.
    (c) The atomic number of aluminum is 27.
    (d) Every atom of the same element has the same mass.
12. An atom of phosphorus-31 has
    (a) 15 protons, 15 neutrons, and 16 electrons.
    (b) 15 protons, 16 neutrons, and 15 electrons.
    (c) 16 protons, 16 neutrons, and 15 electrons.
    (d) 16 protons, 15 neutrons, and 16 electrons.
13. Which of these particles has the smallest mass?
    (a) alpha particle   (b) neutron   (c) proton
    (d) electron
14. Bohr is more closely linked to the idea of
    (a) the discovery of the nucleus of the atom.
    (b) quantized energy of electrons.
    (c) discovery of the neutron.
    (d) protons in orbitals.
15. The maximum number of electrons in the third principal energy level is
    (a) 3.   (b) 8.   (c) 18.   (d) 32.
16. We would expect the electron configuration of oxygen to show two unpaired electrons according to
    (a) the Pauli exclusion principle.
    (b) Bohr's rule.
    (c) the Aufbau principle.
    (d) Hund's rule.
17. Which of the following statements is true?
    (a) The mass of an electron is equal to the mass of a neutron.
    (b) The mass number is equal to the number of protons in the nucleus of the atom.
    (c) The charge of all protons is the same.
    (d) Isotopes contain different numbers of electrons.
18. How many electrons are in the fourth energy level of a calcium atom?
    (a) 20   (b) 18   (c) 8   (d) 2
19. Isotopes of an element
    (a) contain the same number of neutrons.
    (b) have different nuclear charges.
    (c) must have the same atomic number.
    (d) must have the same mass.

20. How many electrons are in the third energy level of an atom of sulfur?
    (a) 4   (b) 6   (c) 8   (d) 16
21. Which of the following is *not* part of modern atomic theory?
    (a) Atoms of different elements can combine only in certain ratios in forming compounds.
    (b) Atoms of the same element are identical.
    (c) Atoms of one element are changed into atoms of another by chemical reaction.
    (d) All elements are composed of atoms.
22. An orbital is best described as
    (a) an electron's path.
    (b) a probability region.
    (c) an energy level.
    (d) an energy sublevel.
23. The maximum numbers of electrons and orbitals, respectively, in the $4d$ energy sublevel are
    (a) 6 and 3.   (b) 10 and 5.   (c) 2 and 1.
    (d) 14 and 7.
24. Which of these statements is true?
    (a) Atoms are mostly empty space.
    (b) Protons and electrons make up the nucleus.
    (c) Atoms are changed as a result of a chemical change.
    (d) All of the above.
25. An atom composed of 14 protons, 14 electrons, and 14 neutrons is symbolized as
    (a) $^{28}_{14}Si$.   (b) $^{14}_{7}N$.   (c) $^{42}_{28}Ni$.   (d) $^{28}_{14}Ni$.
26. The smallest part of an element that can take part in a chemical reaction is a(n)
    (a) nucleus.   (b) proton.   (c) atom.
    (d) gas.
27. The number 37 in the isotope chlorine-37 represents
    (a) the atomic number.
    (b) the mass number.
    (c) the sum of the protons and electrons.
    (d) none of these.
28. The nucleus of an atom is
    (a) positively charged and has a high density.
    (b) positively charged and has a low density.
    (c) negatively charged and has a high density.
    (d) negatively charged and has a low density.
29. The two isotopes of some element Z have a mass of 280 and 283 amu, respectively. If these isotopes existed in a 50:50 mixture, the average atomic mass of the element would be
    (a) 110.   (b) 140.
    (c) 281.5.   (d) 563.
30. The electron configuration of an atom is $1s^2 2s^2 2p^3$. The element is
    (a) boron.   (b) nitrogen.   (c) neon.
    (d) magnesium.

The Periodic Table

Organizing the Elements

Some biologically and medically important elements. Clockwise from left: red phosphorus, magnesium ribbon, zinc, calcium, and iodine.

CHAPTER OUTLINE

CASE IN POINT: Two faces of selenium

4.1 Development of the periodic table

4.2 Metals and nonmetals

A CLOSER LOOK: Diamond Tools for Surgery

4.3 Organization of the modern periodic table

A CLOSER LOOK: Biomedical Implants

4.4 Trends in atomic size

A CLOSER LOOK: The Origin of the Elements

4.5 Trends in ionization energy

4.6 Representative elements

A CLOSER LOOK: Semiconductors

A CLOSER LOOK: Bioactive Materials

FOLLOW-UP TO THE CASE IN POINT: Two faces of selenium

4.7 Transition elements and inner transition elements

Near the end of the nineteenth century, science was in turmoil. Physicists found that ideas about matter and energy that had been relied on for over a century were inadequate when they began to probe the submicroscopic world of atoms. Chemists were confused about the reasons for the similarity in the chemistry of certain elements. Within about 25 years, however, the modern model of the atom had emerged (Sec. 3.5). The new atomic model helped to clarify the relationship between atomic structure and the chemical and physical properties of the elements. The modern periodic table summarizes much of what we know about this relationship. We will see in this chapter that the information in the periodic table is useful for understanding the similarities and differences in the chemical and physical properties of the elements. We also will see how this knowledge has been put to work in the design of materials that are useful in the allied health sciences.

Ninety of the elements of the periodic table occur naturally. Of these 90 elements, only about 25 appear to be essential for life. Half these essential elements are required in such small amounts that, until recent times, their necessity to good nutrition was overlooked. The following Case in Point tells the story of one essential trace element that has caused health problems in several regions of the world.

CASE IN POINT: Two faces of selenium

Over the years, many people in certain regions of China suffered the loss of hair and fingernails and showed signs of nervous system disorders. Chinese in another part of the country, Keshan Province, also suffered, but in a different way. Large numbers of these people had a mysterious affliction, called *Keshan disease,* that damages heart muscle and leads to congestive heart failure. In the United States, cowboys who drove their herds across the Great Plains noticed that their cattle and horses sometimes developed muscle weakness, poor vision, and hair loss. Often, the afflicted animals would wander, stumble, and eventually die from respiratory failure. The cowboys called the condition "blind staggers." Scientists in China and the United States have identified selenium as the root cause of these human and animal health problems (see figure). Where does the selenium come from? What can be done to alleviate its harmful effects?

We will learn more about the vital role of selenium in health in Section 4.6.

Selenium is an essential trace element; insufficient selenium can cause congestive heart failure, and too much selenium can cause nervous disorders.

4.1 Development of the periodic table

AIMS: To explain the origin of the periodic table. To distinguish between a period and a group in the periodic table.

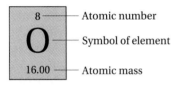

Figure 4.1
The symbolism used for each element in the periodic table.

The International Union of Pure and Applied Chemistry (IUPAC) recommends that the groups in the periodic table be numbered 1 to 18 from left to right. This system is also included in the periodic table on the inside front cover.

About 70 elements had been described by the mid-1800s, but no common feature that would relate the elements with similar properties had been found. Dmitri Mendeleev (1834–1907), a Russian chemist, had more success than most. Mendeleev listed the elements in several vertical columns in order of increasing atomic weight. Noticing a regular (periodic) recurrence of their physical and chemical properties, he arranged the columns so that elements with the most similar properties were side by side. Mendeleev constructed the first **periodic table**—*an arrangement of the elements according to similarities in their properties.* Numerous blank spaces had to be left in the table because there were no known elements with the appropriate properties. By studying the properties of the elements adjacent to blank spaces, Mendeleev and others were able to predict the physical and chemical properties of the missing elements. When these missing elements were discovered and found to have properties similar to those predicted, scientists rapidly accepted Mendeleev's table. In 1913, Henry Moseley, a young English physicist, determined the nuclear charge—also called the *atomic number*—of the atoms of the elements. Recall that the atomic number of an element is the number of protons in the nucleus of an atom, and it also is the number of electrons that surround the nucleus (Sec. 3.3). Moseley arranged the elements in a table by ascending order of atomic number, which is the way that periodic tables are arranged today.

The most commonly used modern periodic table is shown on the inside front cover of this text. In this table the elements are arranged in seven horizontal rows in order of increasing atomic number. Each element is identified by its symbol placed in a block, as shown in Figure 4.1. The atomic number of the element is shown above the symbol, and the atomic mass is shown below the symbol. In some periodic tables, the name of the element is also included below the symbol. *The horizontal rows of the periodic table are called* **periods;** *there are seven periods. The vertical columns are called* **groups,** *or families.* Each group is identified by an Arabic numeral and the letters *A* or *B*.

4.2 Metals and nonmetals

AIMS: To distinguish among a metal, a nonmetal, and a metalloid. To compare the physical properties of metallic and nonmetallic elements.

The heavy diagonal zigzag line on the periodic table shown in Figure 4.2 separates the metallic elements from the nonmetallic elements. *The elements to the left of the line are* **metals** *and those to the right are* **nonmetals.** Most of the elements in the periodic table are metals. You are probably familiar with some of the physical properties of metals: They have a high

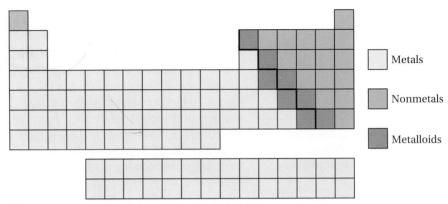

Figure 4.2
Distribution of metals, nonmetals, and metalloids in the periodic table.

Metals

Nonmetals

Metalloids

Figure 4.3
The metals gold and mercury, the nonmetals sulfur and bromine, and the metalloids boron and silicon.

Two metals, gallium and cesium, melt below body temperature. Gallium has a melting point of 30 °C and so will melt when held in the hand. Gallium is used in high-temperature thermometers because it stays in the liquid state over a wide temperature range.

luster when clean, are ductile (can be drawn into wires) and malleable (can be beaten into thin sheets), and have high electrical conductivity. Except for mercury, which is liquid, all the metals are solids at room temperature. Although the names of many metals end in *-ium* (lithium, sodium), there are a few exceptions to this rule, which you may recognize: copper, gold, iron, lead, mercury, nickel, platinum, silver, tin, and zinc.

In sharp contrast, some of the nonmetallic elements are gases, others are brittle solids, and one, bromine, is an orange-brown liquid. Most nonmetals are poor conductors of electricity. *The elements that border on the zigzag line exhibit properties of both metals and nonmetals and are called* **metalloids.** Boron and silicon are examples of metalloids. All the metalloids are solids at room temperature. Examples of some metals, nonmetals, and metalloids are shown in Figure 4.3. A diamond is pure carbon, a nonmetallic element. Some unique applications of diamond, the hardest substance known, are described in A Closer Look: Diamond Tools for Surgery.

A Closer Look

Diamond Tools for Surgery

Diamond's hardness and resistance to chemical reactions make this form of ordinary carbon an ideal material for the cutting edges of surgical scalpels and such instruments as dental drills and wheels for grinding bone. Natural diamonds are very expensive, so scientists have developed methods to deposit diamonds on metals in the form of thin films. To produce a diamond film, a mixture of hydrogen and methane (CH_4) is passed over a very hot filament of tungsten wire at 2400 °C above the surface to be coated. Microscopic diamond crystals form (see figure) and combine to produce a continuous thin diamond film on the

(a) Microscopic view of a diamond film.
(b) A diamond-edged scalpel.

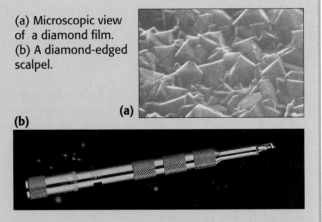

(a)

(b)

surface. Diamond-coated cutting instruments are long-wearing, supersharp, and superresistant to chemical attack.

4.3 Organization of the modern periodic table

AIMS: To state the periodic law. To classify the elements into four categories according to the configuration of their outermost electrons. To write the electron configuration of elements by using the periodic table.

Focus

There is a relationship between an element's location in the periodic table and its electron configuration.

The elements in any group of the periodic table have similar physical and chemical properties. The properties of the elements in the periods change from group to group. Because of this situation, we have the **periodic law:** *When the elements are arranged in order of increasing atomic number, there is a periodic repetition of their physical and chemical properties.* If we consider the relative sizes of atoms of different elements, there is a periodic decrease in atomic size with increasing atomic number (Fig. 4.4). Starting with Li (lithium), atomic size decreases until we reach F (fluorine). Then with Na (sodium) the atomic size increases, again gradually decreasing until we reach Cl (chlorine). Then there is an increase to K (potassium), followed by a decrease to Br (bromine). The decrease in atomic size is a *trend,* and the fact that the trend repeats makes it a *periodic trend.*

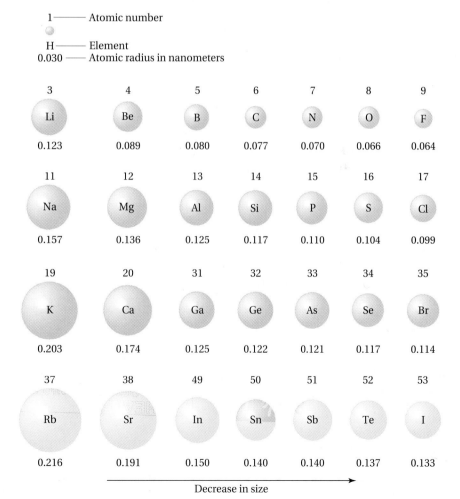

Figure 4.4
Relative atomic sizes of most of the representative elements. When the elements are arranged in order of increasing atomic number, we see a decrease in atomic size as we move across a period from left to right. (The atomic radii are given in nanometers.)

The periodic repetition of elemental properties such as atomic size occurs because the electron configuration of an element plays the greatest part in determining the element's physical and chemical properties. Therefore, to better understand the periodic law, we'll explore how the arrangement of elements in the periodic table is related to their electron configuration. In the following section we will see how this relationship leads to periodic behavior.

Categories of elements

Chemists traditionally have divided the elements into three categories according to their chemical and physical properties. These categories are *representative elements, transition metals,* and *inner transition metals.* Figure 4.5 shows the location of each category on the periodic table. Examination of the electron configurations of the elements reveals striking similarities within each category, as we will see in what follows:

1. Representative elements. *Groups 1A through 7A and Group 0 make up the* **representative elements,** so named because they exhibit a wide variety of both physical and chemical properties. Note that this category includes some metals, metalloids, and nonmetals. In terms of electron configuration, the elements of Groups 1A through 7A (sometimes called the *A-group elements*) have outermost *s* or *p* sublevels that are only partially filled.

The **noble gases** *belong to Group 0.* The elements in this group were in the past sometimes referred to as *inert gases* because of their reluctance to undergo chemical reactions. When we examine the electron configuration

Incandescent light bulbs are filled with the noble gas argon to prolong the life of the bulb. An air-filled bulb would burn out relatively quickly because of the reaction of the metal filament with oxygen in the air.

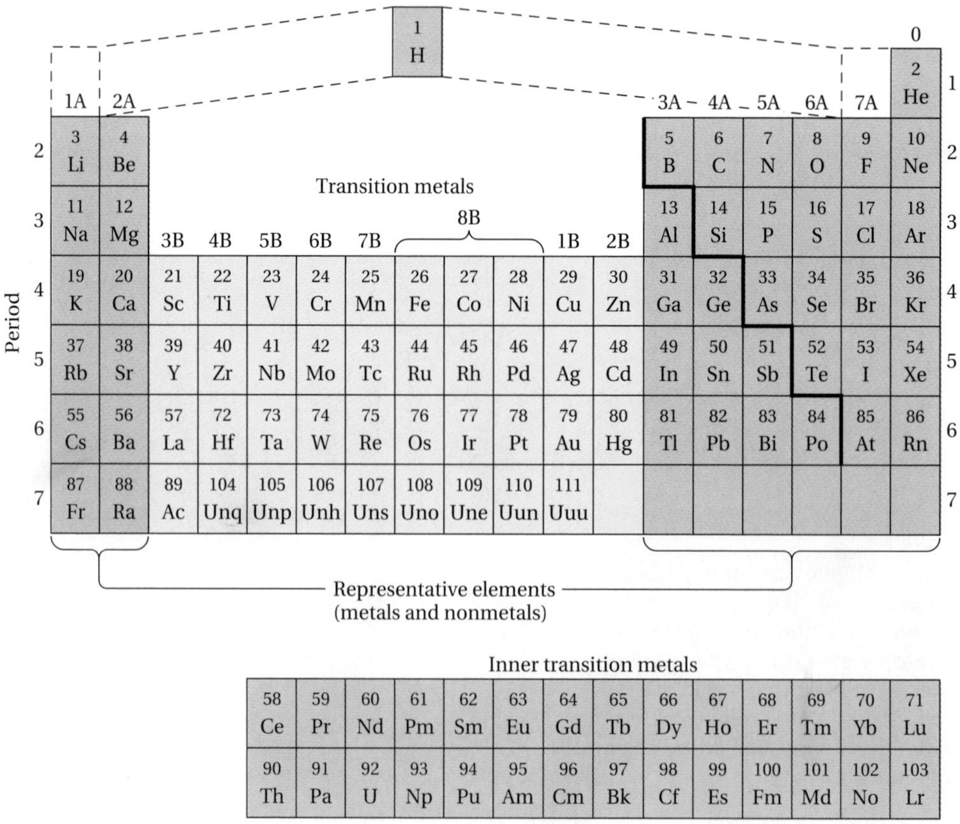

Figure 4.5
Besides being divided into metals and non-metals, the elements are commonly divided into representative elements, transition metals, and inner transition metals. This figure shows the positions of these categories of elements in the periodic table. Hydrogen is positioned centrally above the table because it can be included in both Groups 1A and 7A.

of the noble gases, we find that with the exception of helium, all have outermost s and p sublevels that are filled. Helium is an exception. It contains a $1s$ outermost energy level that is filled.

For any representative element, with the exception of the noble gases, the group number is equal to the number of electrons in the outermost occupied energy level. The electron configurations for the elements in Group 1A (lithium, sodium, potassium, rubidium, and cesium), for example, all have one electron in the outermost energy level.

Lithium: $1s^2\ 2s^1$

Sodium: $1s^2\ 2s^2\ 2p^6\ 3s^1$

Potassium: $1s^2\ 2s^2\ 2p^6\ 3s^2\ 3p^6\ 4s^1$

Carbon and silicon, in Group 4A, have four electrons in the outermost energy level.

Carbon: $1s^2\ 2s^2\ 2p^2$

Silicon: $1s^2\ 2s^2\ 2p^6\ 3s^2\ 3p^2$

Oxygen and sulfur, in Group 6A, have six electrons in the outermost energy level.

2. Transition metals. Transition metals *are the B-group elements.* Their outermost s sublevel and the nearby d sublevel contain electrons. Transition metals are characterized by having electrons present in the d orbitals. The importance of some transition metals in health-related matters appears in A Closer Look: Biomedical Implants.

3. Inner transition metals. *The* **inner transition metals** *are elements whose outermost* s *sublevel and the nearby* f *sublevel contain electrons.* The inner transition metals are characterized by filling of f orbitals.

A Closer Look

Biomedical Implants

People have been replacing various body parts since at least 300 B.C. Modern medicine has made great strides in developing materials for surgical implantation of artificial knees, hips, and other body parts. One of the first surgically implanted materials was surgical steel. Made of iron (67%), chromium (18%), nickel (12%), and molybdenum (3%), this alloy was used to make plates and screws for joining broken bones. Surgical steel has high mechanical strength and resists corrosion in the body. More recently, titanium- and cobalt-based alloys are being used as implant materials (see figure). These new alloys are less dense than surgical steel yet stronger. They have another advantage over surgical steel: They do not react chemically with living tissue. Interestingly, however, this lack of reactivity is sometimes a problem.

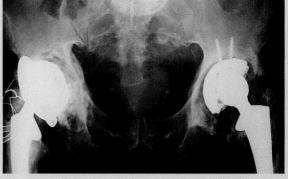

Hip joint replacement allows the patient to regain full movement of the leg.

Bone or tissue will not bond to the implant, making it difficult for the implant to make a strong attachment. We will see in A Closer Look: Bioactive Materials how it is possible to get implants to bond with living tissue and bone, heal rapidly, and form a mechanically strong structure.

Blocks of elements

If we consider both the electron configurations and the positions of elements in the periodic table, another pattern emerges. The periodic table can be divided into sections, or blocks, that correspond to the levels that are filled with electrons (see Fig. 4.6).

The s block is the part of the periodic table that contains the elements with s^1 and s^2 electron configurations. It is composed of the elements in Groups 1A and 2A and the noble gas helium.

The p block is composed of elements in Groups 3A, 4A, 5A, 6A, 7A, and 0, with the exception of helium.

The transition metals belong to the d block, and the inner transition metals fall in the f block.

Writing electron configurations by using the periodic table

Because of the relationship between an element's electron configuration and its position in the periodic table, you can find the electron configuration of any element by using the table shown in Figure 4.6. Read the table left to right and top to bottom until the element of interest is reached. The following example shows how to read the table to determine an element's electron configuration.

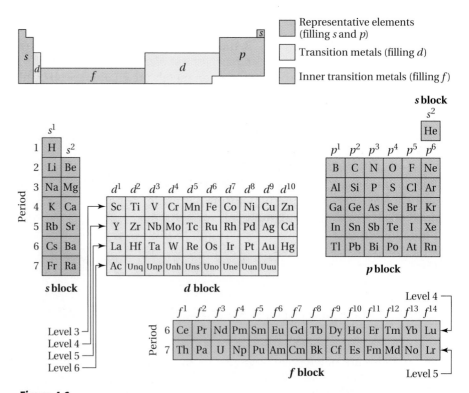

Figure 4.6
A block diagram identifies groups of elements according to the type of orbital being filled.

EXAMPLE 4.1   **Determining electron configurations**

Use the periodic table in Figure 4.6 to write the electron configuration of (a) fluorine and (b) vanadium.

SOLUTION

(a) Fluorine, with atomic number 9, has nine electrons. Reading the periodic table in Figure 4.6, we complete the first period, $1s^2$. The second period contains the remaining seven electrons. In reading the second period, start on the left at $2s^1$ and move right through $2s^2$ to $2p^5$ at fluorine.

   The complete electron configuration for fluorine is $1s^2 2s^2 2p^5$. For all representative elements, the period number corresponds to the principal energy level for the elements in that level.

(b) Vanadium has 23 electrons. Using Figure 4.6, we find that the first three periods are $1s^2 2s^2 2p^6 3s^2 3p^6$. Next, we have $4s^2$ and, finally, $3d^3$. Vanadium is the third element in the d block.

   The complete configuration is $1s^2 2s^2 2p^6 3s^2 3p^6 4s^2 3d^3$. For all transition metals, the outermost electrons are in a d sublevel with a principal energy level that is one less than the period number. Inner transition metals have their outermost electrons in a principal energy level two less than the period number.

PRACTICE EXERCISE 4.1

Write the electron configuration for (a) nitrogen, (b) sodium, (c) nickel, and (d) strontium.

PRACTICE EXERCISE 4.2

Write the electron configuration of the

(a) noble gas in period 4, (b) the element in Group 3A, period 4, and

(c) the element in Group 1A, period 5.

4.4 Trends in atomic size

AIM: To describe how the atomic radii of atoms vary within a group and within a period in the periodic table.

Focus

Atomic radii decrease across a period and increase down a group.

In Section 4.3 we used atomic size as an example of periodic characteristic. Now we will explore the periodic trend in atomic size that was introduced in Figure 4.4.

Recall from Section 3.5 that an atom does not have a sharply defined boundary to set the limit of its size. Therefore, the size of an atom cannot be measured directly. There are, however, several ways to estimate the relative

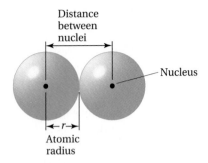

Distance between nuclei

Nucleus

←r→
Atomic radius

Figure 4.7
The radius *r* of an atom is half the distance between the nuclei in identical atoms attached to each other.

sizes of atoms. An estimate that can be used is based on the **atomic radius**—*half the distance between the nuclei of identical atoms attached to each other* (Fig. 4.7). Atomic radii for most of the representative elements are presented in Figure 4.4.

Period trends in atomic radius

Atomic size generally decreases from left to right across a period. As we go across a period, we remain in the same principal energy level. Each element has one proton and one electron more than the preceding element. Because electrons are being added to the same principal energy level, the effect of the increasing nuclear charge on the outermost electrons is to pull them closer to the nucleus. Atomic size therefore decreases.

Group trends in atomic radius

Atomic size generally increases as we move down a group of the periodic table. As we descend, electrons are added to successively higher principal energy levels, and the nuclear charge increases. With each step down a group, the electrons in the outermost orbital are shielded from the nucleus by an additional layer of electrons. These inner electrons are called the *core electrons*. The presence of the core electrons decreases the pull of the nucleus on the outermost electrons. Therefore, atomic size increases as we go down a group.

EXAMPLE 4.2

Comparing sizes of atoms

Identify the larger atom of each pair of elements.
(a) fluorine and chlorine
(b) sodium and sulfur

SOLUTION

(a) The size of atoms increases moving down a group. Since chlorine is below fluorine in Group 7A, chlorine is the larger of the two atoms.
(b) The size of atoms decreases moving from left to right across a period. Since sodium is to the left of sulfur, sodium is the larger of the two atoms.

PRACTICE EXERCISE 4.3
Explain why fluorine has a smaller atomic radius than both oxygen and chlorine.

Our study of the elements tells us about the atomic structure and properties of the elements as they exist, but we have not discussed how the elements were formed. For a discussion of this interesting topic, see A Closer Look: The Origin of the Elements.

The Origin of the Elements

Where did the elements—and the protons, neutrons, and electrons they consist of—originate? A widely accepted theory proposes that the material universe began with an indescribably violent explosion of a dense, homogeneous collection of matter and energy—the "big bang." At the instant of the big bang, the temperature was many billions of degrees, hot enough to decompose neutrons into protons and electrons. Within 3 minutes, the temperature cooled down to about 400,000 K—70 times the average temperature of the Sun. Further cooling was required before the electrons and protons combined to form atoms. This cooling, which may have taken as long as 500,000 years, produced many large clouds of hydrogen atoms. Over eons, gravity caused the clouds of hydrogen atoms to contract, causing them to heat up to the point at which the electrons were stripped away and proton fusion was possible. The fusion of two protons created helium nuclei and a large amount of energy; proton fusion is still occurring in the Sun and in other stars today. After the supply of protons was used up, the helium clouds contracted and heated up to temperatures allowing the fusion of helium nuclei. The fusion of helium nuclei produced the nuclei of heavier elements such as oxygen and carbon. This cycle of contraction and heating resulted in the fusion of heavier and heav-

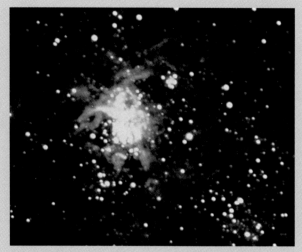

A supernova.

ier nuclei until iron nuclei were formed. Iron nuclei are the most stable nuclei. Unlike the fusion of lighter nuclei, the fusion of iron nuclei with other nuclei requires adding energy instead of producing it. This energy probably came from explosions of large, unstable stars called *supernovas* (see figure). Because supernova explosions provided the energy for the fusion of heavy nuclei, astronomers propose that the debris of these explosions contains many elements, heavy and light. According to the big bang theory, our planet Earth, with its wealth of chemical elements, is the debris of a supernova explosion. It is this "star dust" that contains all the elements essential for life.

4.5 Trends in ionization energy

AIM: To explain period and group trends of ionization energies in the periodic table.

As we will further explain in Chapter 5, an ion is formed when an atom gains or loses an electron. *The energy required to remove an electron from an atom in the gaseous state is the* **ionization energy.** Removing one electron from an atom results in the formation of a positive ion with a 1+ charge. For example,

$$Na(gas) \longrightarrow Na^+(gas) + e^-$$

The energy required to remove this first outermost electron is called the *first ionization energy.* To remove the outermost electron from the 1+ gaseous ion requires an amount of energy called the *second ionization energy,* and so forth.

Table 4.1 First Ionization Energies of Representative Elements in Periods 1, 2, 3, and 4*

Period	1A	2A	3A	4A	5A	6A	7A	0
				Group				
1	H 310							He 566
2	Li 124	Be 215	B 191	C 260	N 335	O 314	F 402	Ne 497
3	Na 118	Mg 175	Al 138	Si 188	P 252	S 239	Cl 299	Ar 363
4	K 100	Ca 141	Ga 138	Ge 182	As 226	Se 225	Br 272	Kr 323

*Values are given in kilocalories per mole, a unit of energy.

Period trends in ionization energy

For the representative elements, the first ionization energy generally increases as we move from left to right—from metals to nonmetals—across a period. The nuclear charge is increasing, but the shielding effect of the core electrons is constant. A greater attraction of the nucleus for the electron therefore leads to the increase in ionization energy. The period trends of ionization energies show that metals give up electrons more readily than nonmetals (Table 4.1).

Group trends in ionization energy

In general, the ionization energy decreases as we move down a group of the periodic table, as shown in Table 4.1. Since the sizes of the atoms are increasing as we descend, the outermost electrons are farther from the nucleus and more easily removed.

Now that we have investigated the relationships between electron configurations and periodic trends, we can examine a few of the elements' well-known and interesting properties. We will be able to appreciate more fully why properties are characteristic of groups and in some instances be able to tell why a certain group of elements has a certain characteristic property.

4.6 Representative elements

AIM: To identify an element as a representative element, noble gas, alkali metal, alkaline earth metal, or halogen.

Group 0 elements: Noble gases

Focus

Elements in Group A and Group 0 represent diverse physical and chemical properties.

Neon, argon, xenon, krypton, and radon are Group 0 elements. The name *inert gas,* or *rare gas,* was used to describe these elements by early chemists who were disconcerted by their chemical aloofness, or refusal to combine with other elements. In 1962, however, Canadian chemist Neil Bartlett combined xenon and fluorine to produce xenon tetrafluoride (XeF_4). Since that

Figure 4.8
Helium is used to fill balloons and blimps.

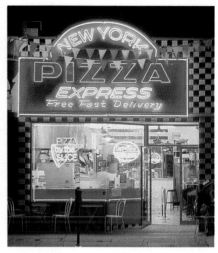

Figure 4.9
Neon lights.

Group 0

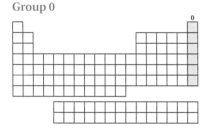

Voices in a helium-oxygen atmosphere sound like the voice of Donald Duck.

time, compounds of argon, krypton, and radon also have been produced. Nevertheless, compared with all other elements, Group 0 elements are extremely unreactive. For this reason, they are now called *noble gases,* a name that again emphasizes their tendency to exist as separate atoms rather than in combination with other atoms.

Due to their relative lack of reactivity, the noble gases have many uses. Helium is used to fill weather balloons and blimps (Fig. 4.8) because, although it is more dense than hydrogen, it is not explosive. Both helium and neon are used in artificial atmospheres such as required in deep-sea diving. Neon, argon, krypton, and xenon are used to produce the inert atmosphere needed for photographic bulbs and aluminum welding. High voltage through glass tubes containing one of the noble gases produces the familiar neon lights (Fig. 4.9).

Group 1A

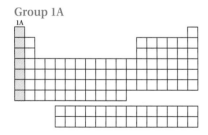

Alkali
al aqali (Arabic):   the ashes

The alkali metal lithium is a component of the compound lithium carbonate, used to treat manic-depressive illness, a major psychiatric disorder.

Group 1A elements: Alkali metals

Group 1A elements are metals with low densities, low melting points, and good electrical conductivity. *Many compounds of sodium and potassium were isolated from aqueous extracts of wood ash (alkali) by early chemists, so the Group 1A elements are called* **alkali metals.** Alkali metals are not found in nature in the uncombined state. In their pure form these metals are soft enough to be cut with a knife (Fig. 4.10)—the freshly cut surface is shiny, but it dulls on exposure to air due to rapid reaction with oxygen and moisture. *The reaction of a substance with oxygen is an* **oxidation reaction;** *the product of the reaction of a substance with oxygen is called an* **oxide** (we will expand our coverage of oxidation reactions in Chapters 6 and 13). A sample of sodium (Na) that stands in air quickly acquires a coating of sodium oxide (Na_2O). Each of the alkali metals reacts violently with cold water, producing hydrogen gas (Fig. 4.11). Because alkali metals react vigorously with water, they should not come into contact with skin;

Figure 4.10
The alkali metals are soft. Here we see sodium metal being cut with a knife.

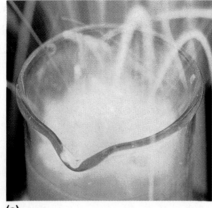 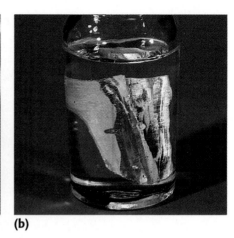

(a)   (b)

Figure 4.11
(a) Sodium metal, as well as the other alkali metals, reacts violently with water.
(b) These metals must be stored under kerosene to prevent reaction with the moisture and oxygen in air.

in the laboratory they are usually stored under kerosene to protect the metal from oxygen or moisture in the air.

Group 2A

Group 2A elements: Alkaline earth metals

Group 2A elements, called **alkaline earth metals,** *are extracted from mineral ores that since early times have been called "earths."* Alkaline earth metals are not found uncombined in nature, but they are less chemically reactive than the Group 1A metals and need not be stored under kerosene. The alkaline earth metals are harder than the alkali metals and have a gray-white luster when freshly cut, but they quickly tarnish in air. Magnesium oxide, the compound produced when magnesium metal is burned in oxygen (oxidized), is the active ingredient of milk of magnesia (Fig. 4.12).

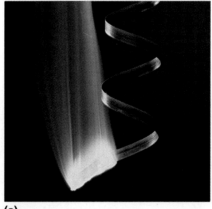 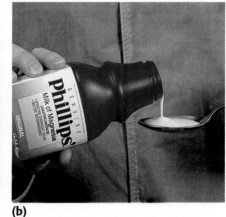

(a)   (b)

Figure 4.12
(a) When magnesium metal is burned in air, the product is magnesium oxide.
(b) A suspension of magnesium oxide in water, known as milk of magnesia, has medicinal properties.

Group 3A

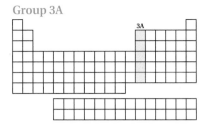

Group 3A elements: Aluminum group

There is a significant difference among the elements in this group because they are on the borderline between metals and nonmetals. The nonmetallic solid boron at the top of the group is followed by the four metallic elements aluminum, gallium, indium, and thallium.

Aluminum is a strong metal of low density that is especially corrosion resistant because, like magnesium in Group 2A, it reacts rapidly with the oxygen in air to form a thin but tough protective coating of aluminum oxide. Like many metals, aluminum is more useful for many applications if it is alloyed with other metals. *An **alloy** is a solid solution of a metal dissolved in one or more other metals.* Aluminum-magnesium alloys are used widely as lightweight structural material in aircraft production and the manufacture of cooking ware. Aluminum does not exist in the uncombined state in nature but is a major component of many rocks and minerals. Aluminum is commonly found as corundum (impure aluminum oxide), a very hard material used as an industrial abrasive. Rubies are aluminum oxide in which a few of the aluminums are replaced by chromium. The last three metals—gallium, indium, and thallium—are rare and have few practical uses. Gallium has an interesting industrial use in thermometers because of its extraordinary wide liquid range; it melts at 30 °C and boils at 2403 °C.

Group 4A

Group 4A elements: Carbon group

Group 4A elements continue the trend we saw in Group 3A. Carbon, at the top of the group, is a nonmetal; silicon and germanium are metalloids—borderline between metals and nonmetals—and both tin and lead are metals.

Diamond and graphite are two forms of carbon (Fig. 4.13). Diamond behaves like a typical nonmetal and is not an electrical conductor, but graphite has some properties of a metal and is a good conductor of electricity. A third form of carbon was discovered as recently as 1985. This new form of carbon will be discussed in Chapter 5, in A Closer Look: Buckyball on page 134. Carbon-containing compounds, of which there are millions, are called *organic compounds*. Organic compounds are essential components of all living things.

Silicon, the second most abundant element of earth, occurs in nature in the combined state as sand and in rocks, soils, and clays. Both silicon and germanium are insulators at low temperatures but conduct electricity at high temperatures. These two elements form the foundation of transistor technology, as described in A Closer Look: Semiconductors.

Tin and lead have properties typical of metals. Tin is important in the manufacture of tinplate, in which a thin coating of tin is applied to an iron can to prevent it from rusting. Prior to 1986, plumber's solder was an alloy of tin and lead. Tetraethyl lead, an organic lead-containing compound, was once used as an antiknock additive in gasoline. Leaded gasoline has been phased out of use in recent years because of lead's high toxicity to humans.

Figure 4.13
A diamond, the hardest known substance, is pure carbon. Graphite, another form of carbon, is a good conductor of electricity and is used in pencil leads.

A Closer Look

Semiconductors

Semiconductors are materials that conduct electricity better than insulators such as diamond but less well than conductors such as copper. Thanks to semiconductors, consumer electronic goods such as radios, stereo equipment, home computers, and hand calculators are relatively inexpensive and readily available. Electronic goods built with semiconductors require much less electricity and are much more reliable than yesterday's vacuum tube counterparts. Modern cardiac pacemakers and sensors that monitor pulse, respiration rate, heartbeat, blood pressure, and temperature all use semiconductor technology.

Pure silicon and germanium are semiconductors and conduct electricity to a limited extent when a voltage is applied. However, the conductivity of silicon and germanium markedly increases when they are *doped* (small amounts of certain other atoms are added). This property makes semiconductors useful in electronic devices such as transistors (amplifiers of electronic signals) and rectifiers (converters of alternating current to direct current).

The manufacture of a silicon semiconductor begins with an extremely pure silicon crystal (see figure, part a)—less than one foreign atom in 10 billion. An electronic device might contain two kinds of doped silicon. For the first type, silicon might be doped with about one part per million (1 ppm) of arsenic. Silicon doped with arsenic has arsenic atoms incorporated into the silicon crystal (see figure, part b). When arsenic is incorporated into the silicon structure, one of its negatively charged electrons is free to move to silicon when a voltage is applied to the crystal. A crystal of arsenic-doped silicon with its mobile electrons is called a *donor* or *n*-type semiconductor (the *n* stands for negative). The second type of doping produces an *acceptor* or *p*-type semiconductor (the *p* stands for positive). Silicon doped with 1 ppm of boron is typical (see figure, part c). When boron atoms are incorporated into the silicon structure, there is a deficiency of one electron around each boron atom, leaving *positive holes* that can accept an electron from silicon.

When a *p*-type and an *n*-type semiconductor are joined, there is a strong tendency for the mobile electrons in the donor to move into the positive holes in the acceptor. The *p*-type and *n*-type semiconductors are joined in highly creative ways to produce the many useful electronic devices we enjoy today. Joined *p*-type and *n*-type materials can be created by selectively doping sections of a semiconductor surface. Incredibly complex miniature circuits (integrated circuits) can be fabricated in this way. Integrated circuits containing millions of components can be placed on semiconductor chips smaller than a dime.

Group 5A elements: Nitrogen group

Group 5A

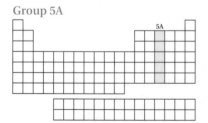

Another periodic change, a change from a gaseous nonmetal to a solid metal, is introduced with Group 5A. The first element, nitrogen, is a gas at room temperature. In descending order, the next elements are phosphorus (a solid nonmetal), arsenic, and antimony. Arsenic and antimony are at the borderline between metals and nonmetals. The last element, bismuth, is a metal.

Nitrogen is an element essential to living organisms. It is present in DNA and proteins. Although 80% of air is nitrogen, it cannot be used to make these essential substances. Fortunately, bacteria that live in the root nodules of legumes (peas and beans) can "fix" atmospheric nitrogen, which

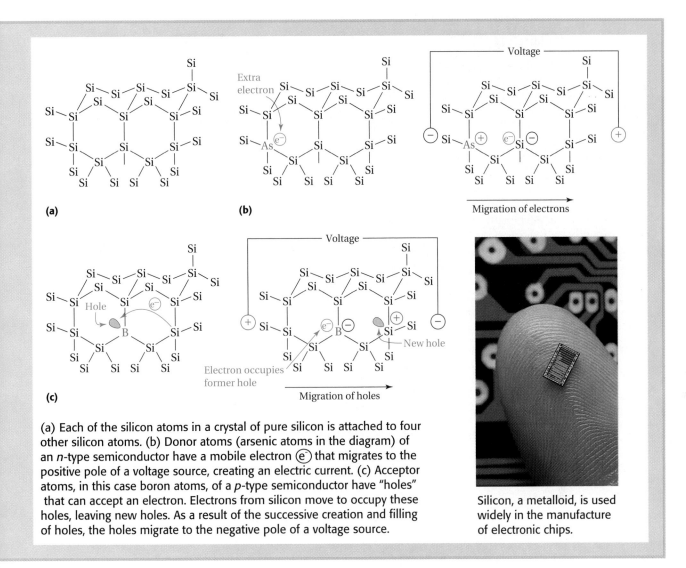

(a) Each of the silicon atoms in a crystal of pure silicon is attached to four other silicon atoms. (b) Donor atoms (arsenic atoms in the diagram) of an *n*-type semiconductor have a mobile electron (e⁻) that migrates to the positive pole of a voltage source, creating an electric current. (c) Acceptor atoms, in this case boron atoms, of a *p*-type semiconductor have "holes" that can accept an electron. Electrons from silicon move to occupy these holes, leaving new holes. As a result of the successive creation and filling of holes, the holes migrate to the negative pole of a voltage source.

Silicon, a metalloid, is used widely in the manufacture of electronic chips.

can then be used by plants to synthesize proteins and other biologically important nitrogen-containing compounds.

Phosphorus is also vital to living organisms. It is present in DNA, bones and teeth, and ATP, the principal energy-storage compound in living systems. Phosphorus compounds are found in phosphate rock. Elemental phosphorus exists in a white form and a red form. The white form is very reactive and is usually stored under water to prevent reaction with oxygen from the air. Red phosphorus is a less active form used in the manufacture of matches. Hydroxyapatite, a phosphorus-containing substance of teeth and bone, is the focus of attention in A Closer Look: Bioactive Materials.

Arsenic, antimony, and bismuth occur in nature in chemical combination with sulfur. They are not essential to living organisms. Alloys contain-

Bioactive Materials

The replacement of bones by artificial implants is an accepted medical practice. However, there is still a need for new materials for implants and for finding creative ways to integrate these materials into the body. Artificial materials developed for use in the body must tolerate the effects of chemical and physical contact between living tissue and the materials. Recently, biomedical research has turned toward the development of bioactive materials. Bioactive substances are so much like the natural materials they replace that they become incorporated into the living tissue. A synthetic version of hydroxyapatite, the crystalline substance that makes up 65% of the dry weight of bones and teeth, is one example. When powdered hydroxyapatite is heated, it becomes very much like human teeth and bones. New bone growth can permeate artificial joints made from this material (see figure). A continuous lattice is formed between the bone and the implant. The end result is an implant with the strength and resilience of natural bone. The resulting graft is often so strong that bone

Implants are coated with bioactive materials to ensure that they form a strong bond with surrounding tissue and bone.

implants that have accidentally received hard blows have caused failure of the bone and not the implant.

ing antimony and bismuth are useful for making very accurate metal castings because they expand as they solidify. This ensures that they will fill the mold completely, making a sharp, clear casting.

Group 6A elements: Oxygen group

Group 6A

The Group 6A elements are oxygen, sulfur, selenium, tellurium, and polonium. Oxygen is a gas. Sulfur is a nonmetal that occurs free in nature as a brittle yellow solid. Selenium is a nonmetal, and tellurium is a metalloid. Both elements are solids. Polonium, the last element in the group, is a radioactive metal that occurs only in trace quantities in radium-containing ores.

Oxygen is the most abundant element. It accounts for 20% by volume of the air we breathe, 60% by mass of the human body, and 50% by mass of the Earth's crust. Most oxygen is combined in the silicate rocks of the Earth's crust. It is produced by plants in photosynthesis. Oxygen is used in the practice of medicine, in the manufacture of steel (to remove the impurities), and along with acetylene in oxyacetylene welding.

Sulfur is essential to living organisms, where it is found in proteins. Sulfur occurs in the elemental state (Fig. 4.14) in large underground deposits

Figure 4.14
Crystals of sulfur.

from which it can be extracted by the Frasch process. Sulfur is also a minor component of coal and petroleum. When sulfur-containing fuel is burned, sulfur dioxide, a serious air pollutant, is invariably a product. The major use of sulfur is in the manufacture of sulfuric acid and in the vulcanization of rubber. In rubber vulcanization, soft latex rubber is heated with sulfur to produce the elastic, tough rubber of automobile tires and many other products. Sulfuric acid is the most widely used industrial chemical. It is used in petroleum refining and in the manufacture of pharmaceuticals, for example.

Selenium is a semiconductor. It is a poor conductor of electricity in the dark, but its conductivity increases greatly upon exposure to light. Because of this property, selenium is used in photoelectric cells in exposure meters for cameras and light-sensitive switches. The xerographic process of photocopying machines also depends on the photoconductivity of selenium. Selenium is used in glass to create the ruby-red color of traffic lights and other warning signals. Selenium is an element that is essential for life. However, as the Follow-up to the Case in Point will explain, it is also extremely toxic. Tellurium is one of the rarest elements. Its compounds are toxic, and the element itself plays no known natural role in living organisms.

FOLLOW-UP TO THE CASE IN POINT: Two faces of selenium

Earlier in this chapter we mentioned people in China who lost their hair and fingernails. We also mentioned "blind staggers" affecting cattle and horses. Both the people in certain regions of China and the animals with blind staggers suffered from the same affliction: selenium poisoning. The people and the animals were poisoned by the food they ate. Certain regions of China are rich in selenium, and grains and vegetables grown in these soils are sufficiently toxic to cause selenium poisoning. Cattle and horses living on the selenium-rich soils in a belt from North Dakota to New Mexico can get blind staggers from the grass they eat. Today, the threat of selenium poisoning has been largely removed, and the remedy is the same for people and animals: They are provided with food that is grown in soils without excessive amounts of selenium.

We might conclude that selenium is harmful for people and animals. However, selenium has two faces, one threatening and the other kind and caring. Too much selenium is harmful, but selenium is also absolutely essential to good health. The many cases of congestive heart failure in the Keshan Province of China were the result of severe selenium deficiencies. People who developed Keshan disease ate crops that were grown in selenium-deficient soils. When these people's diets were supplemented with selenium salts, heart disease abated. Selenium deficiencies also affect people in the United States. Residents of the Pacific Northwest, New England, and Florida, where selenium levels in soil are very low, have three times the chance of dying from a heart attack or stroke than people who live on the Great Plains, where soils are selenium-rich. Cancer studies also portray the two faces of selenium. Selenium appears to reduce the occurrence of certain cancers in animals. However, as shown in animal testing, large doses cause cancer.

Exactly how much selenium we need is not known, but it is probably in the range of 0.05 to 0.2 mg/day. This range is very narrow but crucial; as little selenium as 1 mg/day can produce toxic effects. In the food we eat, most vegetables, especially spinach, are good sources of selenium. Many other foods, such as wheat, also contain sufficient amounts of selenium.

Group 7A

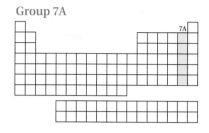

Halogen

hals (Greek): salt
genes (Greek): born

Halogen-containing compounds called *chlorofluorocarbons* (CFCs) are used as refrigerants in air conditioners. Release of CFCs into the atmosphere contributes to the destruction of the protective ozone layer in the upper atmosphere.

The irritating odor of household bleach is due to the release of chlorine gas, a halogen. Contact with and inhalation of chlorine gas should be avoided at all times. The gas can irritate and damage the mucous membranes of the eyes, nose, throat, and respiratory tract.

Group 7A elements: Halogens

The **halogens** *are fluorine, chlorine, bromine, iodine, and astatine.* In contrast with Groups 3A through 6A, which contain both metallic and nonmetallic elements, the halogens form a homogeneous family—they are all nonmetals. The first two elements, fluorine and chlorine, are yellowish green gases. Bromine is a dark red liquid, and iodine is a purple-black crystalline solid with a metallic sheen. Bromine is the only nonmetallic element that is a liquid at room temperature. Tincture of iodine, a dilute solution of iodine in alcohol, was used formerly as an antiseptic. The last element, astatine, is a rare radioactive solid that has not been well investigated.

The halogens do not exist in nature in the uncombined state, but their compounds are fairly abundant. The name *halogen* means "salt producer," and these elements are so named because they are usually found as a class of compounds, called *salts,* of Groups 1A or 2A metals. For example, the salts sodium chloride, sodium bromide, and sodium iodide are found in seawater and salt beds, and calcium fluoride is the mineral fluorspar. The free halogens react vigorously with many substances and must be handled with extreme caution, but compounds of fluorine, chlorine, and iodine are essential to our well-being and must be included in our diet. Fluorine, as fluoride ion, is beneficial in the formation and maintenance of healthy teeth; chlorine, as chloride ion, is an important component of the blood and other body fluids; and iodine, as the iodide ion, is necessary to prevent goiter, an enlargement of the thyroid gland.

The halogens have many other uses in the home and industry. A dilute solution of chlorine is used as a bleaching and disinfecting agent. Silver salts of sodium hypochlorite, a compound of chlorine (silver chloride) and bromine (silver bromide) are light-sensitive and are used to make photographic film. Fluorine is used in the manufacture of Teflon, the familiar nonstick coating applied to frying pans and other cookware.

Hydrogen

Hydrogen: A group by itself

Hydrogen, the lightest of the elements, is the most abundant element in the galaxy; more than 99% of the galaxy's atoms are hydrogen. However, only about 0.9% of the mass of the Earth's crust, oceans, and atmosphere is hydrogen; essentially all this hydrogen is in compounds with other elements. It forms an explosive mixture with oxygen and reacts violently with many other elements. Hydrogen is usually put at the top of Group 1A in the periodic table. It is not, however, a metal; nor is it a good conductor of heat or electricity at room temperature, like the alkali metals. In some versions of the periodic table hydrogen also appears at the top of Group 7A. This position has some validity because, like the halogens, hydrogen has one electron less than helium, the noble gas it precedes.

4.7 Transition elements and inner transition elements

AIM: To identify some uses of common transition metals and inner transition metals.

The transition and inner transition elements are typical metals (Fig. 4.15); they have a metallic luster and are very good conductors of electricity and heat. They have a wide range of uses. Tungsten, a hard, brittle solid with a melting point of 3400 °C, is used in the manufacture of light bulb filaments. At the other end of the scale is mercury, the only metal that is liquid at room temperature. Mercury has a freezing point of −38 °C and is used in thermometers. The excellent reflective qualities of silver (the high luster) make it the ideal coating for mirrors. Copper wire production in enormous quantities attests to the high electrical conductivity of copper. In addition, an alloy of copper and beryllium is used as a substitute for steel in the manufacture of nonsparking tools, and steels with widely different characteristics are made by adding small amounts of cobalt, chromium, nickel, or vanadium to iron. We need transition metals to function normally: Iron is a component of hemoglobin, cobalt is an integral part of vitamin B_{12}, and both zinc and copper are necessary components of many vital compounds in living organisms.

The transition and inner transition metals vary greatly in their chemical reactivity. The elements scandium, yttrium, and lanthanum are similar to Group 1A and 2A metals—they react with oxygen on exposure to air and react with water to liberate hydrogen. In contrast, the metals platinum and gold are extremely unreactive and do not react with oxygen.

PRACTICE EXERCISE 4.4

Give the symbol and name for the element that occupies each of the designated positions in the periodic table.

(a) Period 2, Group 4A (b) Period 3, Group 7A
(c) Period 5, Group 6B (d) Period 4, Group 5B

Figure 4.15
Some transition metals:
(*l* to *r*) Ti, V, Cr, Mn, Fe, Co, Ni, and Cu.

SUMMARY

The periodic table organizes the elements into groups (vertical columns) and periods (horizontal rows) in order of increasing atomic number. Most of the elements in the periodic table are metals. The nonmetals are confined to a triangular area on the right-hand side of the table and are separated from the metals by a diagonal zigzag line. The table is constructed so that elements that have similar chemical properties are in the same group. The properties of any element are generally intermediate between those of its neighbors on either side in the same period and similar to those elements above and below it in the same group. The elements in Groups 1A through 7A are called the representative elements because they exhibit a wide range of both physical and chemical properties. The noble gases make up Group 0. The elements in Groups 2A and 3A are interrupted in periods 4 and 5 by the transition metals and in periods 6 and 7 by the inner transition metals.

Elements that have similar properties also have similar electron configurations and are members of the same group. The atoms of the noble gas elements have their outermost s and p sublevels filled; the outermost s and p sublevels of the representative elements are only partially filled; the outermost s and nearby d sublevels of transition metals contain electrons; and the outermost s and nearby f sublevels of inner transition metals contain electrons.

The regular changes in electron configuration of the elements cause gradual changes in the physical and the chemical properties of the elements within a group and within a period. Atomic radii generally decrease from left to right in a given period and increase within a given group. The ionization energy—the energy required to remove an electron from an atom—generally increases from left to right across a period and decreases down a group.

KEY TERMS

Alkali metal (4.6)
Alkaline earth metal (4.6)
Alloy (4.6)
Atomic radius (4.4)
Group (4.1)
Halogen (4.6)

Inner transition metal (4.3)
Ionization energy (4.5)
Metal (4.2)
Metalloid (4.2)
Noble gas (4.3)

Nonmetal (4.2)
Oxidation reaction (4.6)
Oxide (4.6)
Period (4.1)
Periodic law (4.3)

Periodic table (4.1)
Representative element (4.3)
Transition metal (4.3)

EXERCISES

Development of the Periodic Table (Section 4.1)

4.5 What criterion did Mendeleev use in arranging his periodic table?

4.6 What criterion was used in constructing the modern periodic table?

4.7 (a) What is a *group* on the periodic table? (b) How is a group identified?

4.8 (a) What is a *period* on the periodic table? (b) How is a period identified?

4.9 What are the group numbers of (a) chlorine, (b) nitrogen, (c) aluminum, and (d) cesium?

4.10 Name the element at each of the following locations in the periodic table.
(a) Group 1A, period 3   (b) Group 3A, period 2
(c) Group 7A, period 4   (d) Group 6A, period 2

Metals and Nonmetals (Section 4.2)

4.11 What are properties of metals?

4.12 What are properties of nonmetals?

4.13 From the elements Ba, C, Si, Ar, Hg, Au, P, and Br, select the following
(a) a solid nonmetal   (b) a liquid metal
(c) a gaseous nonmetal   (d) a metalloid
(e) a nonmetal that is a good electrical conductor.

4.14 List the symbols of all the nonmetals in the first five periods, excluding the metalloids and the noble gases.

4.15 Identify the following elements as metals or nonmetals.
(a) P   (b) K   (c) S   (d) Fe   (e) Ar   (f) Ca

4.16 Which of the following metallic elements are in the same group?
Na   Mg   Al   Ba   Fe   Ca

Organization of the Modern Periodic Table (Section 4.3)

4.17 What is the periodic law?

4.18 What is a periodic trend?

4.19 Where are the representative elements and the inner transition elements located in the periodic table?

4.20 Where are the transition elements and the noble gas elements in the periodic table?

4.21 How is an element's outer electron configuration related to its position in the periodic table?

4.22 Which sections of the periodic table correspond to the various energy sublevels?

4.23 Use Figure 4.6 to write the electron configuration of (a) fluorine, (b) aluminum, (c) zinc, and (d) tin.

4.24 What are the symbols for all the elements that have the following outer configuration?
(a) s^2 (b) s^2p^2 (c) s^2d^2

4.25 Describe the electron configuration of the outermost energy level for (a) the noble gas elements and (b) the representative elements.

4.26 Describe the electron configuration of the outermost energy level for (a) the transition metals and (b) the inner transition metals.

4.27 Write the electron configuration of (a) calcium, (b) bromine, (c) sodium, and (d) manganese.

4.28 What elements have the following electron configurations?
(a) $1s^2 2s^2 2p^6 3s^2 3p^6 4s^2 3d^7$
(b) $1s^2 2s^2 2p^6 3s^2 3p^6 4s^1$
(c) $1s^2 2s^2 2p^6 3s^2 3p^6 4s^2 3d^{10} 4p^5$
(d) $1s^2 2s^2 2p^6 3s^2 3p^6 4s^2 3d^{10} 4p^6 5s^2 4d^6$

Trends in the Periodic Table (Sections 4.4, 4.5)

4.29 (a) Arrange the following elements in order of increasing atomic size: boron, fluorine, oxygen, and lithium.
(b) Does your arrangement demonstrate a periodic or a group trend?

4.30 Indicate which element in each of the following pairs has the greatest atomic radius.
(a) sodium and lithium
(b) strontium and magnesium
(c) carbon and germanium
(d) selenium and oxygen

4.31 What does the term *ionization energy* mean?

4.32 (a) In general, would you expect metals or nonmetals to have higher ionization energies?
(b) Why?

4.33 Indicate which element in each of the following pairs has the greatest ionization energy.
(a) lithium and boron
(b) magnesium and strontium
(c) cesium and aluminum

4.34 How do we determine the atomic radius?

Representative Elements (Sections 4.6, 4.7)

4.35 Give the name and symbol for the element that is at each of the following locations in the periodic table:
(a) alkali metal in period 4, (b) noble gas in period 3,
(c) halogen in period 2, and (d) alkaline earth metal in period 6.

4.36 Give the name and symbol for elements that are
(a) halogens, (b) alkali metals, and (c) alkaline earth metals.

4.37 Elements are divided into three major categories.
(a) Name the categories. (b) What is the approximate percentage of elements in each category?

4.38 Are the following elements representative metals, transition metals, or inner transition metals?
(a) uranium (b) mercury (c) barium (d) molybdenum

4.39 Which of the following is a noble gas:
H, Te, He, Zr, Ce?

4.40 Which of the following is a halogen: Cd, Ba, I, Al, Fe?

4.41 Which of the following are alkali metals: Mg, P, Na, Li, Ra?

4.42 Why is helium used to fill weather balloons?

4.43 State two properties (chemical or physical) we can use to distinguish between alkali metals and alkaline earth metals.

4.44 Why are magnesium and aluminum used in light-weight structural materials?

4.45 The elements gallium and mercury are used in thermometers. Explain.

4.46 Why is gold used to crown teeth and as an electrical contact in sensitive electronic equipment?

Additional Exercises

4.47 Do metallic properties increase or decrease for the series Si, Ge, Sn, Pb? Explain.

4.48 Identify the correct choice(s) for each statement.
(a) A noble gas is N, F, Kr, Br, B.
(b) The representative elements include Cu, Ca, He, Cl, Zn.
(c) An alkali metal is Cd, Rb, Mg, Al, Be.
(d) A halogen is Ba, Cr, F, Se, N.
(e) Alkaline earth metals are Na, Sr, Ca, Li, Rb.
(f) An inner transition metal is Ag, Pt, Po, Pu, Cd.
(g) The transition metals include Ca, Cu, Cs, Co, Cr.

4.49 Describe four physical properties that are usually characteristic of metals.

4.50 Give four physical properties that we usually observe for nonmetals.

4.51 Sketch the periodic table, and identify areas where we would find
(a) the inner transition metals,
(b) the representative elements, and
(c) the transition metals.

4.52 Sketch the periodic table, and identify areas where we would find
(a) the metallic elements,
(b) the nonmetallic elements, and
(c) the metalloids.

4.53 Give the name and symbol of the following elements.
(a) an alkaline earth metal in the fourth period
(b) a noble gas with only two electrons in its outer shell
(c) an element in the second period with four valence electrons
(d) a halogen with a partially filled third energy level
(e) a third-period element with one electron in its outer energy level
(f) a second-period element with a completely filled energy level
(g) an element with only two electrons in its fourth energy level
(h) the alkali metal in period 2

SELF-TEST (REVIEW)

True/False

1. In his periodic table, Mendeleev arranged the elements in ascending order of atomic number.

2. The outermost s or d sublevels are only partially filled for the representative elements.

3. The element in Group 4A, period 3, is gallium, Ga.

4. Atomic size decreases as we move from left to right across a period.

5. The energy change that accompanies the addition of an electron to a gaseous atom is the ionization energy.

6. Solid nonmetals are ductile.

7. Boron and silicon are metalloids.

8. Alkali metals are stored under water to protect them from the oxygen in air.

9. Diamond, a form of carbon, is an excellent conductor of electricity.

10. Eighty percent of the air we breathe is oxygen.

Multiple Choice

11. The electron configuration for the element whose atomic number is 21 is
(a) $1s^2 2s^2 2p^6 3p^6 3d^3$.
(b) $1s^2 2s^2 2p^6 3s^2 3p^6 4s^2 4p^1$.
(c) $1s^2 2s^2 2p^6 3s^2 3p^7 4s^2$.
(d) $1s^2 2s^2 2p^6 3s^2 3p^6 3d^1 4s^2$.

12. The representative elements include
(a) Cu, Co, and Cd.   (b) Ni, Fe, and Zn.
(c) Al, Mg, and Li.   (d) Hg, Cr, and Ag.

13. In the modern periodic table, the elements are arranged
(a) according to their similarities in properties.
(b) by ascending order of atomic number.
(c) by ascending order of atomic mass.
(d) both (a) and (b) are correct.

14. The horizontal rows of the periodic table are called
(a) groups.   (b) families.   (c) periods.
(d) transitions.

15. When writing electron configurations of transition metals, the sublevel that is being filled with electrons last is (a) s.   (b) p.   (c) d.   (d) f.

16. How many electrons are in an atom of the element in the third period of Group 4A?
(a) 3   (b) 4   (c) 14   (d) 31

17. Atomic size generally
(a) decreases as we move from right to left across a period.
(b) decreases as we move down a group.
(c) remains constant within a period.
(d) decreases as we move from left to right across a period.

18. The energy required to remove an electron from a gaseous atom is the
(a) electron attraction.   (b) ionization affinity.
(c) electron affinity.   (d) ionization energy.

19. An element that is a liquid nonmetal at room temperature is
(a) chlorine.   (b) phosphorus.   (c) bromine.
(d) arsenic.

20. An example of a noble gas compound is
(a) RaF_2.   (b) NaH.   (c) XeF_4.   (d) KKF_4.

21. The alkali metals do not include
(a) K.   (b) Ca.   (c) Na.   (d) Cs.

22. An element that is a semiconductor is
(a) carbon.   (b) silicon.   (c) sulfur.
(d) iodine.

23. Which atom has the smallest ionization energy?
(a) boron   (b) carbon   (c) aluminum
(d) silicon

24. The alkaline earth metals
(a) are less reactive than the alkali metals.
(b) must be stored under oil.
(c) are sometimes called the alkali metals.
(d) do not react with water, even when heated.

25. An element that occurs in nature in the uncombined state is
(a) potassium.   (b) sodium.   (c) sulfur.
(d) aluminum.

26. The most abundant element on earth is
(a) nitrogen.   (b) carbon.   (c) oxygen.
(d) aluminum.

27. The first ionization energy of sodium is 118.5 kcal/mol. The second ionization energy is most probably
(a) 95.4 kcal/mol.   (b) 130.2 kcal/mol.
(c) 118.5 kcal/mol.   (d) 1067.8 kcal/mol.

28. $1s^2 2s^2 2p^6 3s^2 3p^3$ is the electron configuration of an element. What element is it?
(a) silver   (b) phosphorus   (c) nitrogen
(d) selenium

29. Which of these is a transition metal?
(a) cesium   (b) copper   (c) tellurium
(d) tin

30. Which elements are ductile and malleable?
(a) metalloids   (b) nonmetals   (c) metals
(d) both (b) and (c)

5

Chemical Bonds

Holding Atoms Together

The precise shapes of crystals result from the organization of their atomic building blocks in regular repeating patterns. This photograph is a kidney stone magnified 240 times.

CHAPTER OUTLINE

CASE IN POINT: A new era of medicinal drug discovery

5.1   Valence electrons

5.2   Electron dot structures

5.3   Molecules and ions

5.4   Formation of ions

5.5   Ionic bonds and ionic compounds

A CLOSER LOOK: A Healthy Diet of Ions

A CLOSER LOOK: Superconductors

5.6   Covalent bonds and molecular compounds

5.7   Coordinate covalent bonds

A CLOSER LOOK: Cisplatin: Coordinate Covalent Bonds and Chemotherapy

5.8   Bond polarity

5.9   Attractions between molecules

5.10 Shapes of molecules

FOLLOW-UP TO THE CASE IN POINT: A new era of medicinal drug discovery

A CLOSER LOOK: Buckyball: A Third Form of Carbon

Although there are only 111 elements, the potential chemical combinations of atoms makes the formation of an enormous number and variety of substances possible. And with this diversity of substances comes the diversity of chemical and physical properties exhibited by nature's creations: the tanginess of salt, the resilience of skin, the brilliance of diamond, the fragrance of new-mown hay.

If the properties of different substances result from the different ways in which nature arranges atoms, what binds the atoms together in these arrangements? In this chapter we will study chemical bonds and how they are formed.

Thousands of compounds are used in health-related applications. How have these compounds been identified? In the pharmaceutical industry, for example, how does a medical researcher know what compound may have a medicinal use? This is the subject of the Case in Point.

CASE IN POINT: A new era of medicinal drug discovery

Traditionally, promising compounds for use as medicinal drugs have been isolated from plants or microorganisms or prepared by chemists in the laboratory and then screened for medical effectiveness by tests in animals and finally in humans. All these procedures are costly and time-consuming. History shows that for every 10,000 substances tested, about 20 prove promising enough to enter animal trials, about 10 proceed to human trials, and only 1 gains approval as a new drug. Can this process be made more efficient and more cost-effective so that new pharmaceuticals for the treatment of deadly diseases such as AIDS can be brought to the marketplace more quickly and at a more reasonable cost? We will examine the possibility in Section 5.10.

Taxol, a promising antitumor agent, is obtained from the bark of the Pacific yew tree. Taxol also has been synthesized in the laboratory.

5.1 Valence electrons

AIM: To use the periodic table to find the number of valence electrons in an atom.

Focus

The number of valence electrons in an element determines the chemistry of the element.

Valence
valere (Latin): to be strong

The electron configuration of an atom reveals important information. The number of **valence electrons**—*the electrons in the outermost principal energy level of an element's atoms*—largely determines the chemical properties of that element. Recall from our discussion of the periodic table (Sec. 4.3) that the representative (Group A) elements in a particular group generally have the same number of electrons in the outermost principal energy level. In other words, they have the same number of valence electrons. The number of valence electrons in an atom of a Group A element corresponds to the group number of that element. For example, the elements in Group 1A, hydrogen, lithium, sodium, and so forth, all have one valence electron. Carbon and silicon, in Group 4A, have four valence electrons; nitrogen and phosphorus, in Group 5A, have five; and oxygen and sulfur, in Group 6A, have six. The exceptions to this rule are the noble gases that make up Group 0. Of these, helium has two valence electrons, and all the others have eight. Refer to Table 3.6 to find the number of electrons in the outermost principal energy levels for atoms of the first 36 elements.

PRACTICE EXERCISE 5.1

How many valence electrons do each of the following biologically important atoms have?

(a) potassium (b) carbon
(c) magnesium (d) oxygen

5.2 Electron dot structures

AIM: To draw electron dot structures for the representative elements.

Focus

Valence electrons can be depicted by electron dot structures.

Since valence electrons are usually the only electrons involved in changes in electronic structure, it is customary to show only the valence electrons in **electron dot structures.** *Electron dot structures depict valence electrons as dots; the inner electrons and the atomic nuclei are represented by the symbol for the element being considered.* Electron dot structures are also called Lewis dot structures after G. N. Lewis (1875–1946), whose ideas concerning the nature of chemical bonds are very much a part of modern chemistry. Table 5.1 shows the electron dot structures for the atoms of selected elements. As we saw in the preceding section, the number of valence electrons of Group A elements corresponds to the group number of the element in the periodic table.

Table 5.1 Electron Dot Structures of Selected Elements

Period	Group							
	1A	2A	3A	4A	5A	6A	7A	0
1	H·							He:
2	Li·	·Be·	·B·	·C·	·N·	:O·	:F·	:Ne:
3	Na·	·Mg·	·Al·	·Si·	·P·	:S·	:Cl·	:Ar:
4	K·	·Ca·	·Ga·	·Ge·	·As·	:Se·	:Br·	:Kr:

EXAMPLE 5.1

Writing electron dot structures

Write the electron dot structure for (a) barium and (b) tellurium.

SOLUTION

(a) Barium is in Group 2A; it has two valence electrons. Write the symbol, Ba, with two dots. · Ba ·

(b) Tellurium is in Group 6A; it has six valence electrons. :Te·

PRACTICE EXERCISE 5.2

Give the electron dot structures of the atoms of the following elements that are found in living organisms.

(a) carbon   (b) sulfur   (c) phosphorus   (d) magnesium

5.3 Molecules and ions

AIM: To distinguish between a molecule and an ion.

Focus

Most elements are found in a combined state.

Except for the noble gases, most elements exist in some form of chemical combination. Elements and compounds can be symbolized by chemical formulas. **Chemical formulas** *are shorthand notations that show the kinds and numbers of atoms in the smallest representative unit of a substance.* For many substances, the smallest representative unit is a **molecule**—*an electrically neutral group of tightly bound atoms that acts as a unit.* For example, the molecule formed from two hydrogen atoms is symbolized as H_2. The atomic symbol, H, shows what kind of atom is in the molecule; the subscript, 2, indicates how many atoms are in the molecule. The formula for the compound glucose, a simple sugar, is $C_6H_{12}O_6$. A molecule of glucose is composed of 6 atoms of carbon, 12 atoms of hydrogen, and 6 atoms of oxygen. Similarly, a molecule of glycine, $C_2H_5NO_2$, is composed of 2 carbon atoms, 5 hydrogen atoms, 2 oxygen atoms, and 1 nitrogen atom (the subscript 1, as in N_1, is not used in writing formulas). There are three ways to classify the substances formed when atoms combine chemically: *homonuclear molecules, heteronuclear molecules,* and *ionic compounds.*

Homonuclear molecules

Molecules composed of identical atoms are called **homonuclear molecules.** When two identical atoms are combined, the resulting molecule is called a *homonuclear diatomic molecule.* The atoms of seven nonmetallic elements—hydrogen, oxygen, nitrogen, and the halogens: fluorine, chlorine, bromine, and iodine—can form molecules of this type. The chemical formulas for the homonuclear diatomic molecules are H_2, O_2, N_2, F_2, Cl_2, Br_2, and I_2. Oxygen and nitrogen are the major components of air, and both are essential for life. The halogens are highly reactive and toxic. Many municipalities add chlorine to their drinking water to kill dangerous bacteria.

Heteronuclear molecules

Molecules formed from different elements are called **heteronuclear molecules.** The elements in heteronuclear molecules are usually nonmetals. *A collection of a large number of identical heteronuclear molecules is called a* **molecular compound.** Heteronuclear molecules may be composed of only two atoms or of many atoms. A large heteronuclear molecule may contain hundreds or even thousands of atoms. Examples of three small heteronuclear molecules are carbon monoxide, with the formula CO; carbon dioxide, with the formula CO_2; and water, with the formula H_2O. Carbon monoxide is a deadly gas when breathed in sufficient amount. In normal breathing, carbon dioxide and water vapor are exhaled. An example of a larger heteronuclear molecule is lidocaine, $C_{14}H_{22}ON_2$, a local anesthetic.

Ionic compounds

Ion
ienai (Greek): to go or move

All atoms and molecules are electrically neutral, but some atoms may gain or lose electrons. When this happens, the atoms are said to be *ionized. Ionized atoms have one or more units of positive or negative charge and are called* **ions.** *Compounds that contain ions are called* **ionic compounds.** Ionic compounds often consist of positively charged metal ions and negatively charged nonmetal ions. The numbers of negative and positive charges are equal, so ionic compounds are electrically neutral. Table salt (sodium chloride, NaCl) is a familiar ionic compound. Sodium chloride is composed of equal numbers of positively charged sodium ions (Na^+) and negatively charged chloride ions (Cl^-). The superscript plus sign ($^+$) indicates one unit of positive charge on the sodium ion; the minus sign indicates one unit of negative charge on the chloride ion. Sodium ions and chloride ions play an important role in biological systems. Not only must these ions be present, but they also must exist in proportions that maintain healthy systems in the organism.

5.4 Formation of ions

AIMS: To describe the formation of a cation and an anion. To determine the charge of Group A element cations and anions. To learn the names and formulas of common polyatomic ions.

Except for the noble gases, the isolated atoms of all the elements are quite energetic, which makes them quite unstable. Atoms lose energy and gain stability when they form compounds. In compounds, unstable atoms attain stable electron configurations like those of noble gas atoms. In ionic compounds, atoms have lost or gained valence electrons and have attained stable noble gas electron configurations. *The loss of one or more valence electrons from atoms produces positively charged ions called* **cations.** A sodium ion is a cation. *The gain of one or more valence electrons produces negatively charged ions called* **anions.** A chloride ion is an anion.

Cations

The most common cations are produced by the loss of electrons from metal atoms, which usually have up to three valence electrons that are easily removed. Sodium, in Group 1A, is typical. Sodium atoms have a total of 11 electrons, including one valence electron. By losing their single valence electron, sodium atoms attain the same electron configuration as neon, a noble gas (Fig. 5.1). Since the number of protons in the sodium nucleus is still 11, the lack of one unit of negative charge produces a sodium ion with a charge of 1+. We can show the electron loss, or ionization, of the sodium atom by drawing the complete electron configuration of the atom—or, more simply, by using electron dot structures as follows:

$$\text{Na} \cdot \xrightarrow[\text{(ionization)}]{\text{Loss of valence electron}} \text{Na}^+ \quad + \quad e^-$$

| Sodium atom (electrically neutral: charge = 0) | | Sodium ion (plus sign indicates 1 unit of positive charge) | Electron (minus sign indicates 1 unit of negative charge) |

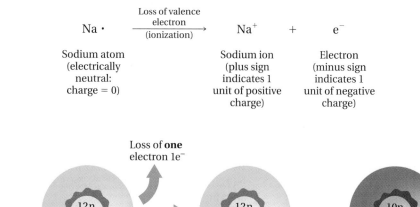

Figure 5.1
A sodium atom loses an electron to become a positively charged sodium ion. The sodium ion has an electron configuration like that of the noble gas neon.

Magnesium belongs to Group 2A and therefore has two valence electrons. Magnesium atoms attain the electron configuration of neon by losing both valence electrons. The loss of the valence electrons produces a magnesium ion, a cation with twice the positive charge of a sodium ion:

$$\cdot \text{Mg} \cdot \longrightarrow \text{Mg}^{2+} + 2e^-$$

Magnesium	Magnesium
atom	ion

The cations of Group 1A elements always have a charge of 1+; Group 2A elements always have a charge of 2+. The charge on transition metal cations may vary, however. For example, iron cations exist as Fe^{2+} or Fe^{3+}; copper cations exist as Cu^+ or Cu^{2+}.

Cations are named by using the name of the element followed by the word *ion*. If a metal atom forms more than one ion, Roman numerals are usually used to distinguish which ion is being discussed. For example, the cation Fe^{2+} is the iron(II) ion (read as "iron two ion"); Fe^{3+} is named iron(III) ion ("iron three ion"); and so forth. An older way of naming cations referred to Fe^{2+} as the *ferrous ion* and Fe^{3+} as the *ferric ion*. These *-ous* (for the lower charge) and *-ic* (for the higher charge) endings are still used occasionally to name transition metal ions.

There are important chemical differences between metals and their ions. Sodium metal, for example, reacts explosively with water, but sodium ions are essential components of biological systems. Likewise, magnesium and iron, important structural materials, are found in biological systems only as their ions.

Table 5.2 lists the formulas and names of some common cations. The table includes one polyatomic cation: the ammonium ion (NH_4^+). **Polyatomic ions** *are tightly bound groups of atoms that behave as a unit and*

Polyatomic
poly-(Greek): many
atomos (Greek): atoms

Table 5.2 Some Common Cations

Element(s)	Ion	Ion name
lithium	Li^+	lithium ion
sodium	Na^+	sodium ion
potassium	K^+	potassium ion
beryllium	Be^{2+}	beryllium ion
magnesium	Mg^{2+}	magnesium ion
calcium	Ca^{2+}	calcium ion
strontium	Sr^{2+}	strontium ion
barium	Ba^{2+}	barium ion
aluminum	Al^{3+}	aluminum ion
silver	Ag^+	silver ion
iron	Fe^{2+}	iron(II) ion or ferrous ion
	Fe^{3+}	iron(III) ion or ferric ion
copper	Cu^+	copper(I) ion or cuprous ion
	Cu^{2+}	copper(II) ion or cupric ion
nitrogen and hydrogen	NH_4^+	ammonium ion

carry a charge. Ammonium ions are a toxic waste product of body cells and must be removed to maintain good health.

Anions

Atoms of nonmetallic elements often attain noble gas electron configurations by gaining electrons. For example, chlorine belongs to Group 7A. Chlorine atoms need one more valence electron to achieve the electron configuration of the nearest noble gas, argon (Fig. 5.2). A gain of one electron converts a chlorine atom into a chloride ion, an anion with a single negative charge:

$$: \overset{\cdot \cdot}{\underset{\cdot \cdot}{Cl}} \cdot \quad + \ e^- \quad \xrightarrow[\text{(ionization)}]{\text{Gain of one valence electron}} \quad : \overset{\cdot \cdot}{\underset{\cdot \cdot}{Cl}} :^-$$

Chlorine atom   Chloride ion (Cl^-)

Anions of the other halogens, *halide ions,* are formed similarly, and all halide ions have a charge of $1-$. An oxygen atom, with six valence electrons, attains the electron configuration of neon by gaining two electrons. The resulting anion, an *oxide ion,* has a charge of $2-$ and is written as O^{2-}:

$$: \overset{\cdot \cdot}{\underset{\cdot}{O}} \cdot \quad + \ 2e^- \quad \longrightarrow \quad : \overset{\cdot \cdot}{\underset{\cdot \cdot}{O}} :^{2-}$$

Oxygen atom   Oxide ion

Table 5.3 shows the formulas of some common anions. The names of all these anions have an *-ide* ending. As with cations, there are also polyatomic anions. Table 5.4 lists the most important polyatomic anions. Observe that the names of most polyatomic anions end in *-ite* or *-ate.* Two important exceptions are the cyanide ion (CN^-) and the hydroxide ion (OH^-). Notice also the use of the prefixes *bi-, hypo-,* and *per-* in naming polyatomic anions. Of the polyatomic ions listed in Table 5.4, the bicarbonate (HCO_3^-), monohydrogen phosphate (HPO_4^{2-}), and dihydrogen phosphate ($H_2PO_4^-$) ions are essential components of biological systems, where they help remove excess quantities of protons in body fluids by reacting with them. The cyanide ion (CN^-) is extremely poisonous to living creatures because it blocks a process called *cellular respiration,* the cell's means of producing energy. The role of ions in health is discussed further in A Closer Look: A Healthy Diet of Ions.

Table 5.3 Some Common Anions

Element	Ion	Ion name
fluorine	F^-	fluoride ion
chlorine	Cl^-	chloride ion
bromine	Br^-	bromide ion
iodine	I^-	iodide ion
oxygen	O^{2-}	oxide ion
sulfur	S^{2-}	sulfide ion

The principal ions of the blood are Na^+, Cl^-, and HCO_3^-. Within tissue cells, the principal ions are K^+, Mg^{2+}, HPO_4^{2-}, and SO_4^{2-}.

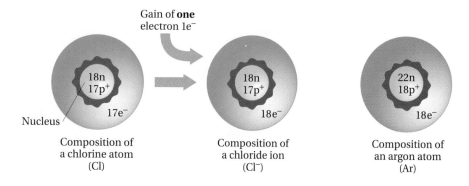

Figure 5.2
A chlorine atom gains an electron to become a negatively charged chloride ion. The chloride ion has an electron configuration like that of the noble gas argon.

Gain of **one**
electron 1e⁻

Nucleus

Composition of a chlorine atom (Cl)

Composition of a chloride ion (Cl^-)

Composition of an argon atom (Ar)

A Healthy Diet of Ions

A diet suitable for normal health and growth must include the appropriate kinds and amounts of proteins, carbohydrates, fats, vitamins, and minerals (see figure). Minerals supply a number of different ions that the human body uses in a variety of ways. Iron(II) ions (Fe^{2+}) are an essential part of hemoglobin, the protein molecule in red blood cells involved in oxygen transport. Iodide ions (I^-) are needed to make hormones in the thyroid gland that help regulate the body's chemical reactions. Calcium ions (Ca^{2+}) and phosphate ions (PO_4^{3-}) help build strong teeth and bones. Magnesium ions (Mg^{2+}) are needed for the proper functioning of many *enzymes*—biological molecules that speed up chemical reactions—and play a role in muscle contraction. Sodium ions (Na^+) play a role in regulating the amount of fluid that bathes body cells and in maintaining a healthy nervous system. Bicarbonate ions (HCO_3^-) and monohydrogen phosphate ions (HPO_4^{2-}) contribute to keeping body fluids such as blood in healthy condition.

Although present in relatively minute amounts, the ions of elements such as copper (Cu^{2+}), cobalt (Co^{2+}), manganese (Mn^{2+}), zinc (Zn^{2+}), and fluorine (F^-) are essential for good

The proper intake of all nutrients—proteins, carbohydrates, fats, vitamins, and minerals—is essential for a healthy body.

health. For example, a deficiency in zinc impairs normal growth, in some severe cases causing dwarfism. Cobalt deficiency can cause anemia. Dietary sources and uses of some biologically important ions are summarized in the accompanying table.

Some Biologically Important Ions

Ion name	Ion formula	Dietary sources	Use
iron(II)	Fe^{2+}	red meat, whole grains, green leafy vegetables, dried beans, apricots	oxygen transport in blood, present in enzymes necessary for respiration
calcium	Ca^{2+}	milk products, whole grains, green leafy vegetables	part of bones and teeth, blood clotting, nerve contraction
magnesium	Mg^{2+}	potatoes, red meat, green leafy vegetables, corn, whole grains	present in enzymes, present in nerve and muscle cells and bone
sodium	Na^+	seafood, table salt	extracellular fluid regulation, nerve impulses, muscle contraction
potassium	K^+	bananas, milk, oranges, meats	regulation of heartbeat, maintenance of intracellular ion concentrations
chloride	Cl^-	seafood, table salt	maintenance of intracellular fluid balance

Table 5.4 Some Common Polyatomic Anions

Ion	Name
OH^-	hydroxide ion
CN^-	cyanide ion
CO_3^{2-}	carbonate ion
HCO_3^-	hydrogen carbonate ion (bicarbonate ion)
CH_3COO^-	acetate ion (ethanoate ion)
SO_3^{2-}	sulfite ion
HSO_3^-	hydrogen sulfite ion
SO_4^{2-}	sulfate ion
HSO_4^-	hydrogen sulfate ion
PO_4^{3-}	phosphate ion
HPO_4^{2-}	monohydrogen phosphate ion
$H_2PO_4^-$	dihydrogen phosphate ion
NO_2^-	nitrite ion
NO_3^-	nitrate ion
ClO^-	hypochlorite ion
ClO_2^-	chlorite ion
ClO_3^-	chlorate ion
ClO_4^-	perchlorate ion
MnO_4^-	permanganate ion

EXAMPLE 5.2   Illustrating the formation of an ion

Write an equation for the formation of the (a) selenide ion and (b) strontium ion.

SOLUTION

(a) A selenium atom gains two electrons to form a selenide ion, attaining an electron configuration like that of the noble gas krypton:

$$\cdot \ddot{Se} \cdot \; + \; 2e^- \; \longrightarrow \; :\ddot{Se}:^{2-}$$

(b) A strontium atom loses two electrons to form a strontium ion, attaining an electron configuration like that of the noble gas krypton:

$$\cdot Sr \cdot \; \longrightarrow \; Sr^{2+} + 2e^-$$

PRACTICE EXERCISE 5.3

What charge will the cation have when the following elements lose their valence electrons?

(a) calcium (Ca)   (b) potassium (K)   (c) aluminum (Al)

PRACTICE EXERCISE 5.4

What charge will the anion have when the following elements gain valence electrons and attain noble gas configurations?

(a) sulfur (S)   (b) nitrogen (N)   (c) fluorine (F)

5.5 Ionic bonds and ionic compounds

AIMS: To define an ionic bond and give examples of ionic compounds. To name an ionic compound when given its chemical formula. To write the formula for an ionic compound when given its name.

Focus

Ionic compounds are composed of oppositely charged ions in fixed ratios.

A knowledge of ions allows us to explore ionic compounds—the electrically neutral groups of ions mentioned in Section 5.3. Since oppositely charged particles attract each other, it is not surprising that oppositely charged cations and anions are held tightly together in ionic compounds. *The attractions between oppositely charged ions constitute* **ionic bonds.**

Formulas of ionic compounds

Ionic compounds are compounds in which ions are joined by ionic bonds. Since ionic compounds are electrically neutral, the positive charges of the cations in any sample of an ionic compound must equal the negative charges of the anions. This is easily illustrated by using a few of the ions in Tables 5.2, 5.3, and 5.4. Sodium ions (charge $1+$) can combine only in a 1:1 ratio with chloride ions (charge $1-$). The charges cancel to give an electrically neutral compound with the formula NaCl:

$$Na^+: \quad \text{Charge of } 1+$$
$$Cl^-: \quad \underline{\text{Charge of } 1-}$$
$$\text{Net charge } 0$$

In the formula for an ionic compound, the cation is always written first. The substance gets its name, sodium chloride, from the sodium ion and chloride ion. The ions are named in the order in which they appear in the formula.

In writing the formula for any ionic compound, the positive charge of the cation must be balanced exactly by the negative charge of the anion. Stated another way, the net ionic charge of the formula must be zero. In the compound magnesium chloride, composed of magnesium ions (Mg^{2+}) and chloride ions (Cl^-), the ions must combine in a 1:2 ratio. Each magnesium ion with its $2+$ charge must combine with (or be balanced by) two chloride ions, each with a $1-$ charge. The formula is $MgCl_2$.

Now let's use aluminum ions (Al^{3+}) and oxide ions (O^{2-}) to get the formula of aluminum oxide. In this instance, we need two Al^{3+} ions (a total of $6+$) and three O^{2-} ions (a total of $6-$) for the charges to cancel. The formula of aluminum oxide is Al_2O_3.

What is the formula of copper(II) oxide? The copper(II) ion is Cu^{2+}. The oxide ion is O^{2-}. The simplest (smallest) ratio and the correct formula of copper(II) oxide is CuO, *not* Cu_2O_2. The formula of an ionic compound must contain the ions in the simplest ratio of whole numbers. Polyatomic ions are treated as units in writing formulas. For example, one sodium ion (Na^+) is combined with one hydroxide ion (OH^-) in sodium hydroxide, NaOH. When more than one polyatomic ion is involved, the formula is written with the polyatomic ion enclosed in parentheses, and a subscript indicating the number of polyatomic ions is placed outside and to the right of

There are many ionic compounds that have medicinal uses. For example, ammonium carbonate, $(NH_4)_2CO_3$, is an expectorant; potassium permanganate, $KMnO_4$, is a topical anti-infective agent; magnesium sulfate (Epsom salts), $MgSO_4$, is a cathartic; barium sulfate, $BaSO_4$, is a radiopaque substance used in X-ray work.

the second parenthesis. Aluminum sulfate provides an example. We need two aluminum ions, Al^{3+}, (total of $6+$) to balance three sulfate ions, (total of $6-$). The formula of aluminum sulfate is $Al_2(SO_4)_3$. The compound is designated by naming the cation followed by the name of the anion.

EXAMPLE 5.3 | **Inferring formulas from ionic charges**

What is the formula of potassium sulfide?

SOLUTION

This compound is composed of potassium ions (K^+) and sulfide ions (S^{2-}). The formula must be composed of two potassium ions (total of $2+$) and one sulfide ion (total of $2-$). The formula is K_2S.

> **PRACTICE EXERCISE 5.5**
> Write correct formulas for the compounds formed by the following pairs of ions.
> (a) Mg^{2+}, Br^- (b) Al^{3+}, Cl^- (c) Ca^{2+}, HCO_3^- (d) Mg^{2+}, PO_4^{3-}

EXAMPLE 5.4 | **Naming ionic compounds**

Write the name of (a) $AlCl_3$ and (b) $CuSO_4$.

SOLUTION

Name the cation followed by the name of the anion.
(a) Aluminum chloride

(b) The copper ion can have one of two different charges. In this compound, the copper ion has a $2+$ charge [copper(II)] because one copper ion is balanced by one sulfate ion with a $2-$ charge. The name is copper(II) sulfate.

> **PRACTICE EXERCISE 5.6**
> Name the following compounds.
> (a) MgO (b) Na_2S (c) $Mg(OH)_2$ (d) NH_4Cl

EXAMPLE 5.5 | **Writing formulas of ionic compounds**

Give the formula of (a) ammonium carbonate and (b) iron(II) nitrate.

SOLUTION

(a) The carbonate ion is CO_3^{2-}. It will take two ammonium ions, NH_4^+, to balance the $2-$ charge. The ammonium ions are put in parentheses, and a subscript 2 is placed to the right of the second parenthesis to show that there are two NH_4^+ polyatomic ions. The formula of ammonium carbonate is $(NH_4)_2CO_3$.

(b) The iron(II) ion is Fe^{2+}. It takes two nitrate ions, NO_3^-, to balance. The nitrate ions are put in parentheses, and a subscript 2 is placed to the right of the second parenthesis to show that there are two of these polyatomic ions. The formula of iron(II) nitrate is $Fe(NO_3)_2$.

PRACTICE EXERCISE 5.7

Give the formulas for the following compounds.

(a) potassium iodide   (b) magnesium fluoride
(c) barium sulfate   (d) copper(II) phosphate

Structure of ionic compounds

Most ionic compounds are crystalline solids at room temperature. **Crystals** *consist of matter in which the component molecules, atoms, or ions are arranged in repeating three-dimensional patterns.* The composition of a crystal of sodium chloride is typical. In a sodium chloride crystal, each sodium ion is surrounded by six chloride ions and each chloride ion is surrounded by six sodium ions to form a three-dimensional chessboard pattern, as shown in Figure 5.3. As a whole, each sodium chloride crystal, no matter what its size, is electrically neutral. A number of complex materials containing both metals and nonmetals have very interesting and useful properties, as seen in A Closer Look: Superconductors.

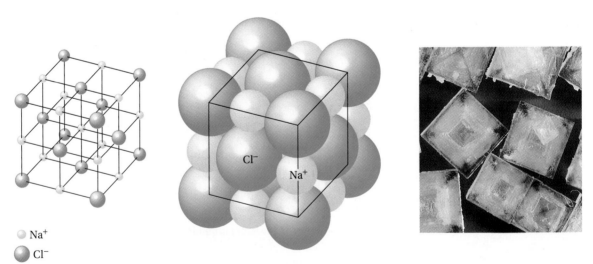

○ Na$^+$
● Cl$^-$

Figure 5.3
Sodium ions and chloride ions form a three-dimensional array in a crystal of sodium chloride.

Superconductors

Superconductivity is the absence of electrical resistance in certain metals, alloys, and compounds. The potential applications of superconducting materials are boundless. Imagine frictionless high-speed trains that could run suspended on a pad of air above magnetic tracks! Or miniature supercomputers many times more powerful than any computer built today. These inventions could become reality through superconductor technology. Superconductors are already being used in a few medical applications. For example, magnetic resonance imaging (see Chapter 2) uses magnets wound from superconducting wire.

An intrinsic property of superconductors, diamagnetism, will be important for such applications as suspended trains. A substance in which all the electrons are spin-paired is *diamagnetic*. Superconducting substances become diamagnetic below the superconductor's critical tempera-

A magnet hovers in the air above a superconducting ceramic immersed in liquid nitrogen.

ture, the temperature at which a material becomes superconducting. Diamagnets are repelled by magnetic fields. In fact, a diamagnet repels a magnet with a force equal to the strength of its magnetic field. A magnet placed on top of a superconducting ceramic hangs suspended in the air above (see figure). This phenomenon, *levitation*, occurs because the repulsive forces between the superconductor and the magnet are perfectly balanced.

5.6 Covalent bonds and molecular compounds

AIMS: *To define a covalent bond and give examples of covalent compounds. To draw electron dot structures for simple covalent molecules containing single, double, or triple bonds.*

Unlike atoms that gain or lose electrons in forming ionic compounds, some atoms *share* electrons to attain noble gas electron configurations—particularly atoms of hydrogen and the nonmetallic elements in Groups 4A, 5A, 6A, and 7A. **Covalent bonds** *are the result of electron sharing between atoms. The atoms in molecules are joined by covalent bonds, and large collections of heteronuclear molecules are called* **covalent (or molecular) compounds.** All living organisms contain many covalent compounds as well as ions.

Single covalent bonds

The hydrogen molecule, H_2, is the simplest molecule. One hydrogen atom has a single valence electron. By sharing electrons to form a homonuclear diatomic molecule, each hydrogen attains the electron configuration of helium, which has two valence electrons. *A **single covalent bond** is formed as a result of sharing a pair of electrons:*

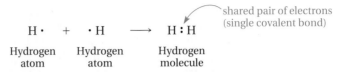

$$H \cdot \quad + \quad \cdot H \quad \longrightarrow \quad H : H$$

Hydrogen atom   Hydrogen atom   Hydrogen molecule

shared pair of electrons (single covalent bond)

The chemical formula for the hydrogen molecule is H_2. Another way to show the pair of electrons in a covalent bond is to use a dash, as in H—H for hydrogen. Notations of this type are called *structural formulas.* **Structural formulas** *are chemical formulas that show the arrangement of atoms in molecules and polyatomic ions.* There are two things to remember about the dashes between the atoms in structural formulas: They *always* indicate a *pair* of shared electrons, and they are *never* used to show ionic bonds.

An examination of the formula of hydrogen, H_2, brings up a difference between the chemical formulas of ionic and covalent compounds. The chemical formulas of ionic compounds show the lowest whole-number ratio of ions in a formula. The chemical formulas of covalent compounds are correctly described only as *molecular formulas. A* **molecular formula** *shows the actual number of atoms of each element in a molecule.* Figure 5.4 illustrates the essential difference between ionic and covalent compounds.

The halogens also form single covalent bonds in their diatomic molecules. Fluorine can serve as a general example. Since a fluorine atom has seven valence electrons, it needs one more electron to attain the electron configuration of neon. By electron sharing, two fluorine atoms attain the

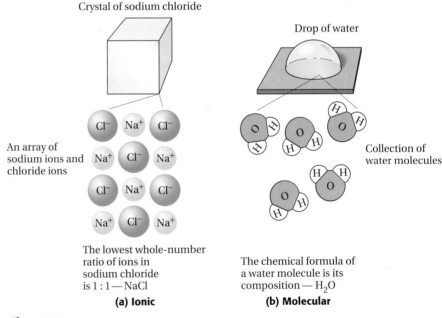

Crystal of sodium chloride

Drop of water

An array of sodium ions and chloride ions

Collection of water molecules

The lowest whole-number ratio of ions in sodium chloride is 1 : 1 — NaCl

(a) Ionic

The chemical formula of a water molecule is its composition — H_2O

(b) Molecular

Figure 5.4
Ionic and molecular compounds compared: (a) ionic; (b) molecular.

electron configuration of neon and in the process form a single covalent bond:

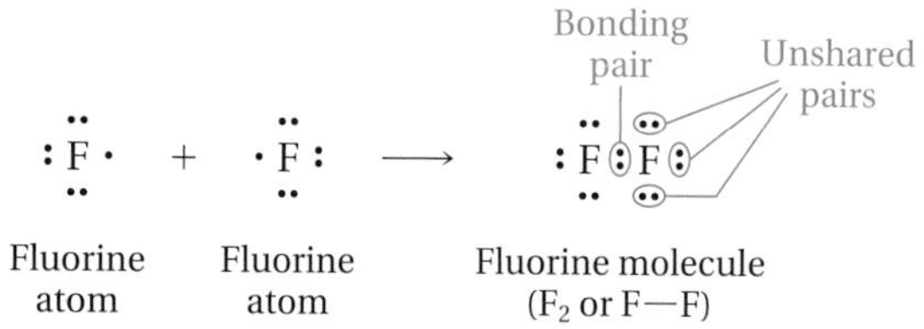

$$: \overset{\displaystyle ..}{\underset{\displaystyle ..}{F}} \cdot \ + \ \cdot \overset{\displaystyle ..}{\underset{\displaystyle ..}{F}} : \ \longrightarrow \ : \overset{\displaystyle ..}{\underset{\displaystyle ..}{F}} \overset{..}{\underset{..}{\odot}} F \overset{..}{\underset{..}{\odot}}$$

Bonding pair

Unshared pairs

Fluorine atom   Fluorine atom   Fluorine molecule (F_2 or F—F)

Notice that each fluorine atom has one pair of bonding electrons and three pairs of unshared valence electrons. *We call these pairs of unshared valence electrons* **unshared pairs of electrons** *or simply* **unshared pairs.**

PRACTICE EXERCISE 5.8

Draw electron dot structures for the diatomic halogen molecules (a) chlorine, (b) bromine, and (c) iodine.

Double and triple covalent bonds

Atoms can sometimes share more than one pair of electrons to attain noble gas electron configurations. **Double covalent bonds** *have two shared pairs of electrons, and* **triple covalent bonds** *have three shared pairs.* We can write the oxygen molecule, O_2, as if oxygen atoms, with six valence electrons, share two of these electrons with another oxygen atom to form a double covalent bond:

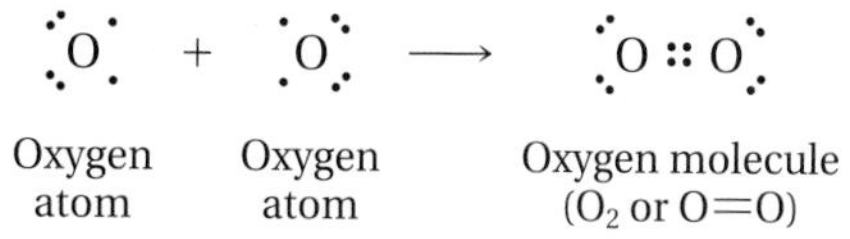

$$\overset{\displaystyle .}{\underset{\displaystyle .}{\cdot O \cdot}} \ + \ \overset{\displaystyle .}{\underset{\displaystyle .}{\cdot O \cdot}} \ \longrightarrow \ \overset{\displaystyle .}{\underset{\displaystyle .}{\cdot O}} :: \overset{\displaystyle .}{\underset{\displaystyle .}{O \cdot}}$$

Oxygen atom   Oxygen atom   Oxygen molecule (O_2 or O=O)

We can write the oxygen-oxygen double covalent bond in oxygen as two dashes (O=O). However, research studies of the oxygen molecule show that the actual bonding is less than a double covalent bond but more than a single covalent bond. Each oxygen in the oxygen molecule has two unshared pairs of electrons.

The nitrogen molecule contains a triple covalent bond. Each nitrogen atom has five valence electrons and needs three more to attain the electron configuration of neon:

$$: \overset{\displaystyle .}{N} \cdot \ + \ \cdot \overset{\displaystyle .}{N} : \ \longrightarrow \ : N \overset{..}{\underset{..}{:}} N :$$

Nitrogen atom   Nitrogen atom   Nitrogen molecule (N_2 or N≡N)

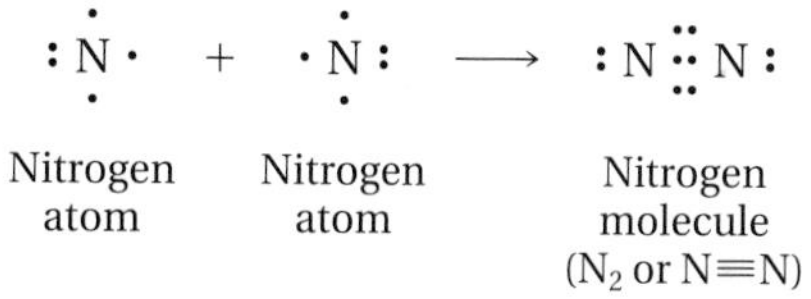

Each nitrogen in the nitrogen molecule has one unshared pair of electrons.

The examples of covalent bonds considered so far have all been in homonuclear diatomic molecules.

Heteronuclear molecules

Heteronuclear molecules are assembled in much the same way as homonuclear molecules. We will discuss three simple examples: water, ammonia, and carbon dioxide. Water is a triatomic molecule in which two hydrogen atoms share one electron each with one oxygen atom:

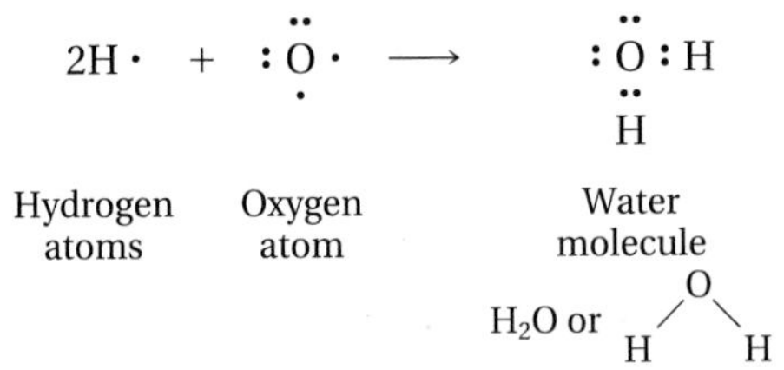

$$2H\cdot \ + \ :\overset{\cdot\cdot}{\underset{\cdot}{O}}\cdot \ \longrightarrow \ :\overset{\cdot\cdot}{\underset{\cdot\cdot}{O}}:\underset{|}{\overset{}{H}}$$

| Hydrogen atoms | Oxygen atom | Water molecule |

$$H_2O \text{ or } \quad \overset{O}{\underset{H \qquad H}{\diagup \ \diagdown}}$$

The hydrogen and oxygen atoms attain noble gas configurations by electron sharing. The water molecule has two unshared pairs of electrons.

Ammonia, an important component of fertilizers, is formed in a similar way:

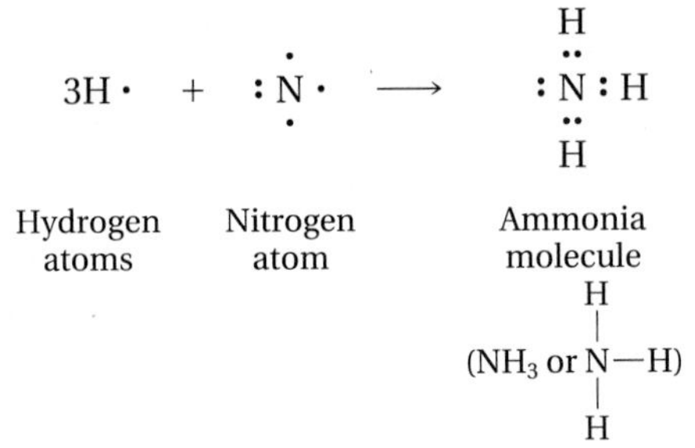

$$3H\cdot \ + \ :\overset{\cdot}{\underset{\cdot}{N}}\cdot \ \longrightarrow \ \overset{H}{:\overset{\cdot\cdot}{N}:H}$$

| Hydrogen atoms | Nitrogen atom | Ammonia molecule |

$$(NH_3 \text{ or } \overset{H}{\underset{H}{\overset{|}{N}-H}})$$

The ammonia molecule has one unshared pair of electrons.

A carbon atom has four valence electrons and needs four more to attain a noble gas configuration. The carbon dioxide molecule contains two oxygen atoms that each share two electrons with carbon to form two carbon-oxygen double bonds:

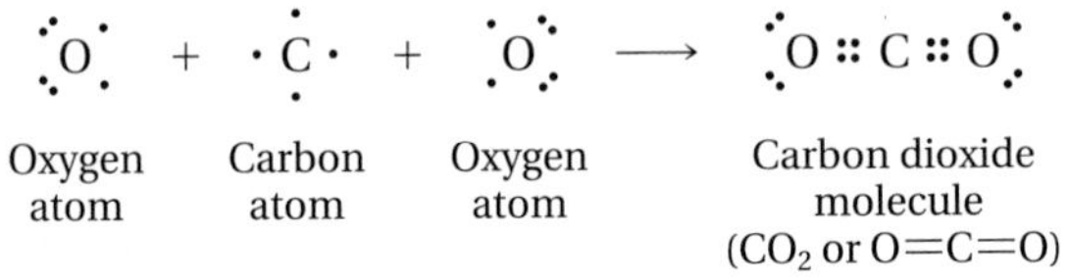

An increased amount of carbon dioxide in the atmosphere is a major contributor to an increase in the greenhouse effect. This change may increase the Earth's average temperature and result in significant climatic changes.

$$:\overset{\cdot}{\underset{\cdot\cdot}{O}}: \ + \ \cdot\overset{\cdot}{C}\cdot \ + \ :\overset{\cdot}{\underset{\cdot\cdot}{O}}: \ \longrightarrow \ :\overset{\cdot\cdot}{O}::C::\overset{\cdot\cdot}{O}:$$

| Oxygen atom | Carbon atom | Oxygen atom | Carbon dioxide molecule |

$$(CO_2 \text{ or } O{=}C{=}O)$$

EXAMPLE 5.6 **Drawing electron dot structures**

Hydrogen chloride (HCl) is a diatomic molecule with a single covalent bond. Draw the electron dot structure for HCl.

SOLUTION

To form a single covalent bond, a hydrogen atom and a chlorine atom must share a pair of electrons. First, write electron dot structures for the two atoms; then show the electron sharing:

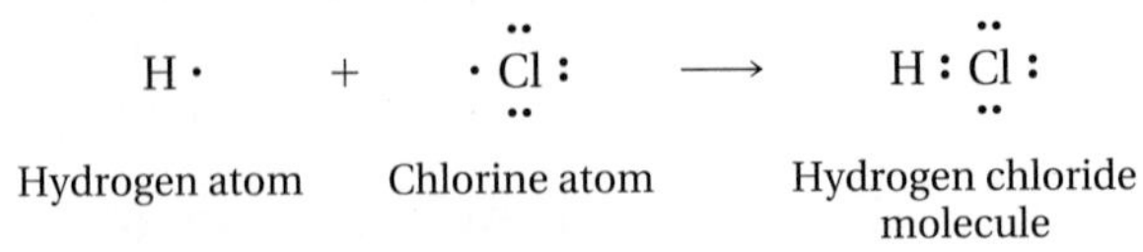

$$H\cdot \quad + \quad \cdot\overset{\cdot\cdot}{\underset{\cdot\cdot}{Cl}}: \quad \longrightarrow \quad H:\overset{\cdot\cdot}{\underset{\cdot\cdot}{Cl}}:$$

| Hydrogen atom | Chlorine atom | Hydrogen chloride molecule |

Through electron sharing, hydrogen and chlorine atoms attain the electron configurations of the noble gases helium and argon, respectively.

PRACTICE EXERCISE 5.9

Draw electron dot structures for the following covalent molecules that have only single covalent bonds:

(a) H_2S (b) PH_3 (c) ClF   (d) CH_4

5.7 Coordinate covalent bonds

AIM: To describe the formation of a coordinate covalent bond.

Focus

One atom supplies the pair of bonding electrons in a coordinate covalent bond.

Sometimes it is necessary for one atom to contribute *both* the bonding electrons in a covalent bond. *A covalent bond in which one atom contributes both bonding electrons is a* **coordinate covalent bond.** An example of coordinate covalent bonding is found in carbon monoxide. In the last section we saw that a carbon atom is four electrons short of the electronic configuration of neon and an oxygen atom is two electrons short. Yet it is possible for both atoms to achieve noble gas electron configurations by coordinate covalent bonding. To see how, let's examine a double covalent bond between carbon and oxygen:

$$\overset{\cdot\cdot}{\underset{\cdot}{C}}\cdot \quad + \quad \cdot\overset{\cdot\cdot}{\underset{\cdot\cdot}{O}}\cdot \quad \longrightarrow \quad \cdot\overset{\cdot\cdot}{C}::\overset{\cdot\cdot}{O}\cdot$$

Carbon atom   Oxygen atom

With eight electrons, the oxygen atom now has the electron configuration of neon. The carbon atom does not have the neon electron configuration; carbon has only six electrons. The dilemma is solved if the oxygen also donates one of its unshared pairs of electrons to make a coordinate covalent bond:

$$\overset{\cdot\cdot}{\underset{\cdot\cdot}{C}}::\overset{\curvearrowleft}{\underset{\cdot\cdot}{O}}\cdot \quad \longrightarrow \quad \overset{\cdot\cdot}{\underset{\cdot\cdot}{C}}::\overset{\cdot\cdot}{\underset{\cdot\cdot}{O}}:$$

Carbon monoxide
molecule

Coordinate covalent bonds are shown in structural formulas as arrows pointing from the atom donating the pair of electrons to the atom receiving them. The structural formula of carbon monoxide, with two covalent bonds and one coordinate covalent bond, is C≡O. Once formed, a coordinate covalent bond is no different from any other covalent bond because electrons are indistinguishable from each other. The only difference between a coordinate covalent bond and other covalent bonds is the source of the bonding electrons. The polyatomic ammonium ion (NH_4^+) has a coordinate covalent bond. It is formed when a hydrogen ion, a proton, is attracted to the unshared electron pair of an ammonia molecule. Hydrogen becomes a cation when a hydrogen atom loses its valence electron:

Unshared electron pair

$$H^+ \quad + \quad :\overset{\overset{\textstyle H}{\cdot\cdot}}{\underset{\underset{\textstyle H}{\cdot\cdot}}{N}}:H \quad \longrightarrow \quad \left[\overset{\overset{\textstyle H}{\cdot\cdot}}{\underset{\underset{\textstyle H}{\cdot\cdot}}{H:N:H}}\right]^+$$

Proton
(hydrogen cation)

Ammonia
molecule
(NH_3)

Ammonium ion
(NH_4^+)

Cisplatin: Coordinate Covalent Bonds and Chemotherapy

Cisplatin is an important cancer chemotherapy drug used in the treatment of cancer. Cisplatin has the formula $Pt(NH_3)_2Cl_2$. The cisplatin molecule has two coordinate covalent bonds. Each coordinate covalent bond is formed between the platinum ion, Pt^{2+}, and an unshared pair of electrons of each ammonia molecule:

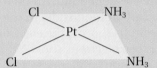

Cancers of the ovaries, testes, lungs, and other organs are treated using cisplatin. The anticancer activity of the compound arises from its ability to bind to the genetic material (DNA) to prevent the division of tumor cells (see figure). Although this compound was first isolated in the mid-1800s, its use as a potential cancer chemotherapy agent was not recognized until the mid-1960s. The ability of cisplatin to prevent the division of cells was discovered accidently during an experiment studying the effect of electricity on bacterial growth. The compound was formed during the experiment by a reaction between the platinum ions from platinum electrodes and ammonia and chloride ions in the bacteria culture medium.

Cisplatin bonds to two nitrogen atoms in DNA, thereby preventing the division of tumor cells.

A Closer Look: Cisplatin: Coordinate Covalent Bonds and Chemotherapy describes an important anticancer compound that contains coordinate covalent bonds.

EXAMPLE 5.7   **Drawing the electron dot structure of a polyatomic ion**

The polyatomic hydronium ion (H_3O^+) contains a coordinate covalent bond. It forms when a proton is attracted to an unshared electron pair of a water molecule. Write the electron dot structure for the hydronium ion.

SOLUTION

$$H^+ \quad + \quad \overset{\textstyle H}{\underset{\displaystyle \cdot\cdot}{:\overset{\cdot\cdot}{O}:H}} \quad \longrightarrow \quad \left[\overset{\textstyle H}{\underset{\displaystyle \cdot\cdot}{H:\overset{\cdot\cdot}{O}:H}} \right]^+$$

Proton   Water molecule   Hydronium ion
              (H_2O)   (H_3O^+)

5.8 Bond polarity

AIMS: **To distinguish between nonpolar and polar covalent bonds. To use electronegativity values to determine whether a bond is ionic, polar covalent, or nonpolar covalent.**

Focus

Covalent bonds joining unlike atoms are usually polar.

The bonding pairs of electrons in covalent bonds are pulled, as in a tug of war, between the nuclei of the atoms sharing the electrons. In the homonuclear diatomic molecules of hydrogen, oxygen, nitrogen, and the halogens, the atoms involved in the struggle are of equal strength, and the contest is a standoff. The covalent bonds in these molecules are said to be *nonpolar. Covalent bonds in which the bonding electrons are shared equally by both atoms are* **nonpolar covalent bonds.**

Polar bonds

The situation changes when two different atoms are joined by a covalent bond. Since one atom is always able to attract electrons more strongly than the other, it gains a greater share of the bonding electrons. *When the bonding electron pair is unequally distributed between two atoms, the bond becomes a* **polar covalent bond,** *or simply a* **polar bond.** It follows that the atom with the stronger electron attraction acquires a slight negative charge. The other atom acquires a slight positive charge. *The ability of an atom to attract electrons to itself in a covalent bond is called the atom's* **electronegativity.**

Table 5.5 presents the electronegativities of some common elements. Let's consider the covalent bond in the heteronuclear diatomic molecule hydrogen chloride (HCl). Table 5.5 shows that hydrogen has an electronegativity of 2.1. The electronegativity of chlorine is higher with a value of 3.0. In the tug of war between hydrogen and chlorine atoms for the shared pair, chlorine has the edge, but it never wins completely. Thus the covalent bond in hydrogen chloride is polar, with chlorine acquiring a slight negative charge and hydrogen acquiring a slight positive charge. The polarity of the bond can be represented as

$$\overset{\delta+}{H}\text{---}\overset{\delta-}{Cl}$$

Table 5.5 Electronegativity Values for Some Elements

H 2.1							He 0.0
Li 1.0	Be 1.5	B 2.0	C 2.5	N 3.0	O 3.5	F 4.0	Ne 0.0
Na 0.9	Mg 1.2	Al 1.5	Si 1.8	P 2.1	S 2.5	Cl 3.0	Ar 0.0
K 0.8	Ca 1.0	Ga 1.6	Ge 1.8	As 2.0	Se 2.4	Br 2.8	Kr 0.0

Table 5.6 Electronegativity Differences and Bond Types

Bond system	Electronegativity difference	Type of bond
H—H	0	covalent (nonpolar)
$\overset{\delta+}{H}—\overset{\delta-}{Cl}$	0.9	covalent (polar)
Na$^+$ Cl$^-$	2.1	ionic

In this symbolism, the Greek letter lowercase delta (δ) is used to show that atoms involved in the covalent bond acquire only partial charges less than 1+ or 1−. The minus and plus signs show that chlorine has acquired a partial negative charge and that hydrogen has acquired a partial positive charge. We also can represent the polarity of the bond as

$$\overset{\longrightarrow}{H—Cl}$$

In this notation, the arrow shows that the bonding electrons are pulled toward chlorine and away from hydrogen.

Electronegativities can help us predict whether two atoms will form an ionic bond, a polar bond, or a nonpolar covalent bond. If the electronegativity difference between two atoms is greater than about 2.0, a bond between the two atoms is usually ionic; if it is less than about 2.0 but greater than 0.0, the bond is usually polar covalent; if it is 0.0, the bond is nonpolar. Table 5.6 shows this relationship between electronegativity values and the type of bond formed for a few substances.

EXAMPLE 5.8

Predicting the polarity of chemical bonds

What type of bond—polar covalent, nonpolar covalent, or ionic—will form between atoms of the following pairs of elements?

(a) Na and Cl   (b) N and H
(c) F and F   (d) C and Cl

SOLUTION

(a) The electronegativity difference between Na (0.9) and Cl (3.0) is 2.1. The difference is greater than 2; the bond is ionic.

(b) The electronegativity difference between N (3.0) and H (2.1) is 0.9. The difference is less than 2 but greater than zero; the bond is polar covalent.

(c) When identical atoms share electrons, the electronegativity difference is zero, and the bond is nonpolar covalent.

(d) The electronegativity difference between C (2.5) and Cl (3.0) is 0.5; the bond is polar covalent.

PRACTICE EXERCISE 5.10

Identify the bonds between atoms of the following pairs of elements as ionic, nonpolar covalent, or polar covalent.

(a) H and Br   (b) K and Cl   (c) C and O   (d) Cl and F

PRACTICE EXERCISE 5.11

Which covalent bond is the most polar?

(a) H—Cl   (b) H—Br   (c) H—S   (d) H—C

Polar molecules

The presence of a polar bond in a molecule often makes the entire molecule polar. In the hydrogen chloride molecule, the partial charges on the hydrogen and chlorine atoms are electrically charged regions, or poles, one slightly positive and one slightly negative. *A molecule that has oppositely charged electrical poles is a* **polar molecule.** Hydrogen chloride is a polar molecule. Since it has *two* poles, hydrogen chloride also could be called a *dipolar molecule.*

The effect of individual polar bonds on the polarity of an entire molecule depends on the shape of the molecule and the orientation of the polar bonds. A carbon dioxide molecule, for example, has two polar bonds and is linear:

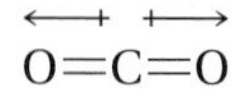

$$\overset{\longleftarrow\;+\;\longrightarrow}{O=C=O}$$

The carbon and oxygens lie along the same axis; the bond polarities cancel because the poles at both ends of the molecule are negative. Carbon dioxide is a nonpolar molecule.

The water molecule also has two polar bonds, but the situation is different from that of carbon dioxide because the molecule is bent:

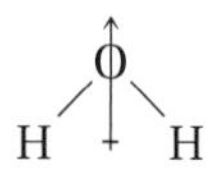

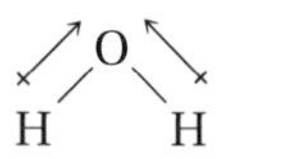

The O—H bonds   The polarity of
are polar   the H_2O molecule

The bond polarities do not cancel, so a water molecule is polar.

PRACTICE EXERCISE 5.12

Is the hydrogen fluoride molecule polar? Explain.

5.9 Attractions between molecules

AIM: To name and describe the weak attractive forces that hold molecules to each other.

The type of bonding in a compound dramatically affects its physical properties. For example, most ionic compounds are crystalline solids at room temperature. However, some covalent compounds are gases, some are liquids, and some are solids at room temperature. Table 5.7 summarizes the differences between ionic and covalent substances. The greater variety of

Table 5.7 Characteristics of Ionic and Covalent Compounds

Characteristic	Ionic compound	Covalent compound
representative unit	oppositely charged ions	molecule
bond formation	transfer of one or more electrons between atoms	sharing of electron pairs between atoms
type of elements	metallic and nonmetallic	nonmetallic
physical state	solid	solid, liquid, or gas
melting point	high (usually above 300 °C)	low (usually below 300 °C)
solubility in water	usually high	high to low
electrical conductivity of aqueous solution	good conductor	poor conductor or nonconducting

physical properties among covalent compounds occurs because of many weak attractions between covalent molecules. There are several types of attractions between molecules, each of which is much weaker than either an ionic or a covalent bond. We should not, however, underestimate the power of these forces simply because they are weak. Among other things, they play an important role in the organization of molecules in living systems. Although covalent and ionic bonds hold the atoms of biological molecules together, the weak forces hold the molecules to each other. Without the weak forces, there would be no life.

Weak bonding forces contribute to maintenance of the three-dimensional structures of proteins and other biologically important molecules.

Van der Waals forces

The weakest attractions between molecules are collectively called **van der Waals forces** after the Dutch chemist Johannes van der Waals (1837–1923). Two major van der Waals forces are dispersion forces and dipole interactions.

Dispersion forces, *the weakest of all molecular interactions, vary in strength depending on a molecule's size and shape.* The origin of dispersion forces is obscure, but they are thought to be caused by temporary dipoles created by the motion of electrons. They are sufficiently strong to cause nonpolar bromine to exist as a liquid and iodine to exist as a solid at room temperature.

Dipole interactions *result from attractions between polar molecules.* Considerably stronger than dispersion forces, dipole interactions occur when polar molecules are attracted to one another (Fig. 5.5). Both dipole interactions between polar molecules and attractions between oppositely charged ions result from electrostatic forces, but the charges on dipolar molecules are only a fraction of those on ions. Dipole interactions are therefore much weaker than ionic bonds. Dipolar interactions in water, for example, result in a weak attraction of water molecules for one another. Each O—H bond in the water molecule is highly polar; the oxygen, because of its greater electronegativity, acquires a slightly negative charge; the hydrogens acquire a slightly positive charge. Polar molecules attract one another, so the positive region of one water molecule attracts the negative region of another. We will explore the attractions between water molecules in more detail in Chapter 8.

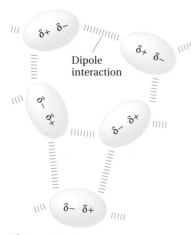

Figure 5.5
Polar molecules are attracted to one another by weak dipole interactions.

5.10 Shapes of molecules

AIM: To explain the shapes of simple covalently bonded molecules by using electron pair repulsion theory.

Just as a photograph or sketch may fail to do justice to a person's appearance, electron dot structures and structural formulas cannot reflect the three-dimensional shapes of molecules. Examples of space-filling and ball-and-stick molecular models for hydrogen chloride (HCl), water (H_2O), ammonia (NH_3), and methane (CH_4) are shown in Figure 5.6. The shapes of molecules are of interest to researchers in the health fields because complex molecules in biological systems must fit together correctly for important chemical reactions to occur. The fit of biological molecules is also important in the discovery of medicinal drugs. Historically, the discovery of therapeutic drugs has been an expensive, time-consuming process. In the Follow-up to the Case in Point we will now learn about another method that could facilitate the discovery of new pharmaceuticals.

We can get some basic ideas about molecular shapes by studying simple molecules such as methane (CH_4). A carbon atom shares its four valence electrons with four hydrogen atoms in the methane molecule, which also may be written as its electron dot structure or its structural formula:

$$H : \overset{\displaystyle H}{\underset{\displaystyle H}{C}} : H \qquad H - \overset{\displaystyle H}{\underset{\displaystyle H}{C}} - H$$

Methane   Methane

Neither formula really describes the shape of the methane molecule.

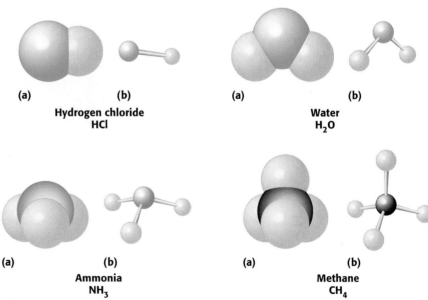

(a)   (b)
Hydrogen chloride
HCl

(a)   (b)
Water
H_2O

(a)   (b)
Ammonia
NH_3

(a)   (b)
Methane
CH_4

Figure 5.6
Space-filling (a) and ball-and-stick (b) models of some molecular compounds.

FOLLOW-UP TO THE CASE IN POINT: A new era of medicinal drug discovery

Pharmaceutical chemists hope to eliminate much of the tedious screening and preliminary testing of potential pharmaceutical drugs by using a method called *molecular modeling*. Molecular modeling blends experimental facts and theories about molecules. With the use of molecular modeling, molecular properties such as shapes and charges can be predicted. The method also can provide information about the ways that molecules interact with each other in chemical reactions. Molecular modeling has been embraced by many branches of chemistry, but it has been greeted with special enthusiasm in the pharmaceutical industry. Because a drug requires 10 to 12 years of testing to gain approval for therapeutic use, no currently available drug was discovered by molecular modeling. However, pharmaceutical chemists are using the method to identify promising variations of compounds already in drug use. They are also searching for new candidate compounds for medicinal use. To see how this might be done, suppose we want to find an ion that will prevent the passage of calcium ions into heart muscle through the passages called *ion channels*. Such substances, called *calcium ion-channel blockers*, can relieve angina, or heart pain. If we know the molecular structure of the ion channel, we can use computerized molecular modeling to test a large number of ions as calcium ion-channel blockers. Our modeling could show which ions will enter the ion channel and make a tight fit, much as a cork blocks the opening of a bottle. If our modeling studies could help us identify the top 20 potential channel blockers out of, say, 10,000 compounds, we could prepare these compounds for experimental testing.

The software programs for molecular modeling studies contain fundamental information about chemical bonding and the shapes of ions and molecules. When an ionic or molecular structure is introduced into the program, essentially as a rough drawing, the computer calculates such properties as the bond lengths and the overall three-dimensional shape of the ion or molecule (see figure). In most instances, the three-dimensional structures produced by the computer are in excellent agreement with experimentally determined structures. Two or more structures can then be brought together to study how they interact, as in the case of a calcium ion channel and an ion that could be used as a blocker.

Today, thanks to relatively inexpensive, powerful computers, the average chemist can explore many detailed aspects of molecular structure. The savings in time and money gained through the use of molecular modeling can be enormous. Today's pharmaceutical chemists have only begun to tap the potential of molecular modeling, but this powerful tool for the study of molecular structures and interactions has a bright future.

A computer-generated version of the structure of Taxol, a potential antitumor agent.

Tetrahedron
tetra (Greek): four
hedra (Greek): face

Hydrogens in the methane molecule are at the four corners of a geometric solid, the regular **tetrahedron,** *as shown in Figure 5.7. In this arrangement, all the H—C—H angles are 109.5 degrees—the tetrahedral angle. One concept that explains this shape says that since electron pairs repel, molecules adjust their shapes so that the valence electron pairs are as far apart as possible.* In a molecule such as methane, with four bonding electron pairs and

Regular tetrahedron

Figure 5.7
The hydrogens in methane are at the corners of a regular tetrahedron.

no unshared pairs, the bond pairs are farthest apart when the angle between the central carbon and its attached hydrogens is 109.5 degrees, the H—C—H bond angle found by experiment. Any other arrangement tends to bring two bonding pairs of electrons closer together.

The structure of a diamond is a three-dimensional framework in which all the atoms are carbons joined by single covalent bonds (Fig. 5.8). The incredible symmetry inherent in the diamond structure is a result of bonding electron pair repulsions. As in methane, all the bond angles (in this instance C—C—C) are 109.5 degrees. Diamond cutting requires breaking a multitude of these bonds. A Closer Look: Buckyball: A Third Form of Carbon describes another form of carbon that has recently been discovered.

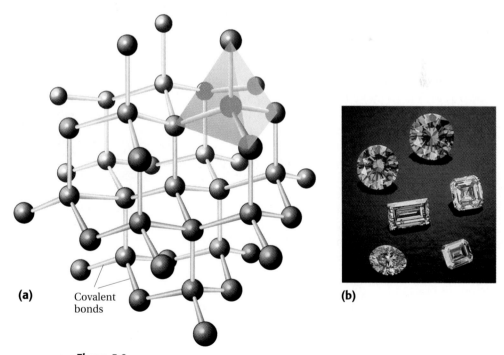

(a) Covalent bonds

(b)

Figure 5.8
The arrangement of carbons in diamond, (a), is manifested in the shape of diamonds, (b).

Buckyball: A Third Form of Carbon

For many years scientists believed that carbon came in two basic forms: hard, shiny diamond and soft, dull graphite. Both these forms of carbon have their uses. Diamond, besides being used for jewelry, is important in making metal-cutting tools because it is so hard. Graphite, in which carbon atoms are bonded together in flat plates that slide easily over each other, has many applications as a lubricant.

Recently, chemists have discovered a fascinating third form of carbon. This substance has the molecular formula C_{60}. Although C_{60} had been detected previously in interstellar space, it was unknown on Earth. The carbons of the C_{60} molecule are arranged to form a spherical molecule (see figure, part a). Because of its structure, chemists have nicknamed the C_{60} compound "buckyball." The name comes from R. Buckminster Fuller (1895–1983), inventor of the geodesic dome, which the buckyball closely resembles (see figure, part b). You also might recognize the arrangement of the carbons of buckyball in a soccer ball. The positions of the carbon atoms in buckyball are fixed at the corners of pentagons and hexagons, placed exactly in the arrangement of these shapes that is sewn together to make a soccer ball. Despite its late discovery on Earth, buckyball is a minor component in soot obtained by burning such ordinary objects as a plastic milk bottle.

Scientists are now working to discover practical applications for buckyball. Buckyball molecules would seem to be naturals as lubricants; because they are spherical, they could act as molecule-sized ball bearings. Buckyball with metal atoms trapped in the carbon cage could prove useful as electrical semiconductors or superconductors.

There are also exciting potential uses in medicine. Buckyball has the ability to "tie up" energetic molecules in the body that are linked to the formation of cancer cells. It may someday, therefore, be part of a preventative medicine for cancer. It has been proposed that radioactive atoms enclosed inside buckyball molecules could be used for antitumor therapy in cancer patients. In a test-tube experiment, a compound made by modifying buckyball has been shown to block the action of an enzyme needed for reproduction of the HIV virus, the causative agent in AIDS. Only time will tell whether someday buckyball or similar molecules will be used to deliver medicines attached to or enclosed in the carbon cage of specific cells of the body.

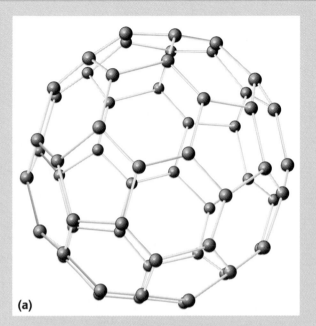

(a)

(b)

(a) Buckyball, the latest isotope of carbon and (b) the geodesic dome of the religious center at Southern Illinois University at Edwardsville, Illinois.

Effect of unshared pairs on molecular shape

Unshared pairs of electrons are important when we are trying to predict the shapes of molecules. Like the carbon in methane (CH_4), the nitrogen in ammonia (NH_3) is surrounded by four pairs of valence electrons that take a tetrahedral orientation (Fig. 5.9a). Note, however, that one of the pairs in ammonia is an unshared pair. The unshared pair is held closer to the nitrogen than the bonding pairs and strongly repels them, pushing them closer together so that each H—N—H bond angle is only 107 degrees.

In a water molecule, oxygen forms single covalent bonds with two hydrogen atoms, and there are two unshared pairs. The bonding and unshared pairs form a tetrahedral arrangement around the central oxygen (Fig. 5.9b). Thus the water molecule is flat but bent. With two unshared pairs repelling the bonding pairs, the H—O—H bond angle is compressed to about 105 degrees.

A tetrahedral arrangement does not always minimize repulsions between unshared pairs. Carbon dioxide, for example, is a linear molecule (Fig. 5.9c). The carbon in CO_2 has no unshared pairs, and the double bonds joining the oxygens to the carbon are farthest apart when the O—C—O bond angle is 180 degrees (when the atoms are in a straight line).

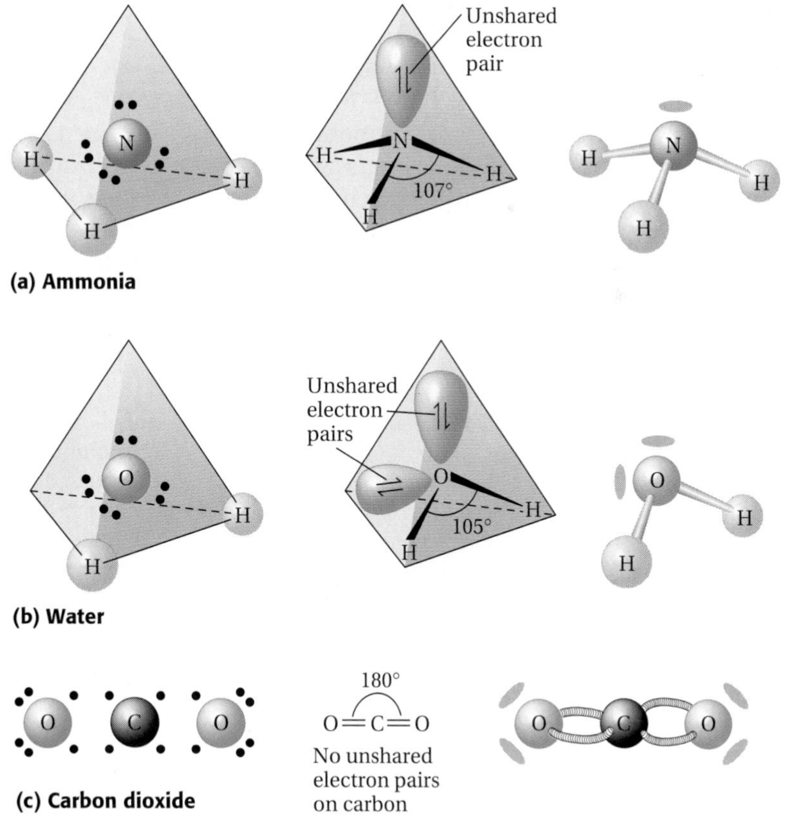

The shapes of molecules are very important in biological systems. Even a minor change in the shape of a molecule can result in a significant change in the ability of the molecule to function as needed in the organism.

(a) Ammonia

(b) Water

(c) Carbon dioxide

Figure 5.9
The unshared pairs of electrons and the hydrogens in ammonia (a) and water (b) molecules are at the corners of a tetrahedron; the carbon dioxide molecule (c) is linear.

PRACTICE EXERCISE 5.13

Just as carbon and hydrogen form methane (CH_4), carbon and chlorine form carbon tetrachloride (CCl_4). Draw the electron dot structure for CCl_4, and discuss the shape of the molecule.

PRACTICE EXERCISE 5.14

Would you predict a molecule of PH_3 to be planer (flat)? Explain.

SUMMARY

Atoms are held together in compounds by chemical bonds. Chemical bonds result from the sharing or transfer of valence electrons between pairs of atoms. The transfer of one or more valence electrons between atoms produces positively and negatively charged ions—cations and anions.

The attraction between an anion and a cation is an ionic bond. A substance with an ionic bond is an ionic compound; nearly all ionic compounds are crystalline solids at room temperature. These solids consist of positive and negative ions packed in an orderly arrangement with the total positive charge balanced by the total negative charge. Ionic solids are electrically neutral. Many ionic compounds contain polyatomic ions. A polyatomic ion is a group of covalently bonded atoms that carries a charge.

When atoms share electrons to gain the stable electron configuration of a noble gas, they form covalent bonds. A shared pair of valence electrons constitutes a single covalent bond. Sometimes two or three pairs of electrons may be shared to give double or triple covalent bonds, respectively. When like atoms are joined by a covalent bond, the bonding electrons are shared equally, and the bond is nonpolar. When the atoms in a bond are not the same, the bonding electrons are shared unequally, and the bond is polar. Some molecules are polar because they contain polar covalent bonds.

The attractions between opposite poles of polar molecules constitute dipole interactions. The dipole interaction is one of several weak attractions between molecules that determine whether a covalent compound will be a solid, a liquid, or a gas.

As a general rule, molecules adjust their three-dimensional shapes so that the valence electron pairs around a central atom are as far apart as possible.

KEY TERMS

Anion (5.4)
Cation (5.4)
Chemical formula (5.3)
Coordinate covalent bond (5.7)
Covalent bond (5.6)
Covalent compound (5.6)
Crystal (5.5)
Dipole interaction (5.9)
Dispersion forces (5.9)

Double covalent bond (5.6)
Electron dot structures (5.2)
Electronegativity (5.8)
Heteronuclear molecule (5.3)
Homonuclear molecule (5.3)
Ion (5.3)

Ionic bond (5.5)
Ionic compound (5.3)
Molecular compound (5.3)
Molecular formula (5.6)
Molecule (5.3)
Nonpolar covalent bond (5.8)
Polar bond (5.8)
Polar covalent bond (5.8)
Polar molecule (5.8)

Polyatomic ion (5.4)
Single covalent bond (5.6)
Structural formula (5.6)
Tetrahedron (5.10)
Triple covalent bond (5.6)
Unshared pair (5.6)
Valence electrons (5.1)
van der Waals forces (5.9)

EXERCISES

Valence Electrons (Sections 5.1, 5.2)

5.15 Define the term *valence electron*.

5.16 What do the dots in an electron dot structure represent?

5.17 Write the number of valence electrons for each of the following atoms.
(a) nitrogen   (b) sulfur   (c) fluorine
(d) lithium

5.18 Indicate the number of valence electrons of an element in each group.
(a) alkaline earth metal   (b) halogen
(c) Group 4A   (d) alkali metal

5.19 In which of the following sets do all the elements have the same number of valence electrons?
(a) O, Fe, He   (b) O, S, Cl   (c) Mg, Ca, Ba
(d) F, I, Br   (e) Na, Mg, Al

5.20 Write the electron dot structure for the period 3 element with the same number of valence electrons as
(a) neon   (b) potassium   (c) oxygen   (d) boron

Ions (Sections 5.3, 5.4)

5.21 (a) What are cations? (b) How and why are cations produced?

5.22 (a) What are anions? (b) How and why are anions produced?

5.23 How many electrons must be lost by each of the following atoms to attain a noble gas structure?
(a) Ba   (b) Mg   (c) Li   (d) Al   (e) Cs

5.24 Write the formula of the anion of the following nonmetallic elements.
(a) fluorine   (b) nitrogen   (c) sulfur   (d) bromine

5.25 Write the formula for the cation of the following metallic elements.
(a) calcium   (b) sodium   (c) potassium
(d) strontium

5.26 Name each of the following ions, and identify each as an anion or a cation.
(a) F^- (b) Cu^{2+} (c) P^{3-}
(d) H^+ (e) S^{2-} (f) Ca^{2+}

Ionic Bonds and Ionic Compounds (Section 5.5)

5.27 Which of the following pairs of elements are most likely to form ionic compounds?
(a) magnesium and sodium
(b) lithium and bromine
(c) neon and potassium
(d) nitrogen and oxygen

5.28 Which pairs of elements will not form ionic compounds?
(a) sulfur and chlorine   (b) calcium and oxygen
(c) potassium and sulfur   (d) strontium and chlorine

5.29 What is an ionic bond?

5.30 Why are ionic compounds electrically neutral?

5.31 Write the formula of the compound formed from each of the following pairs of ions.
(a) K^+, S^{2-} (b) Ca^{2+}, O^{2-} (c) Na^+, SO_4^{2-}
(d) Al^{3+}, PO_4^{3-}

5.32 Write the formula of the ionic compounds formed from each of the following pairs of ions.
(a) Fe^{2+}, Cl^- (b) NH_4^+, NO_3^- (c) Al^{3+}, S^{2-}
(d) Mg^{2+}, OH^-

5.33 Write the names and formulas of the ions in the following compounds.
(a) KCl   (b) $KMnO_4$ (c) $CaSO_4$ (d) $MgBr_2$

5.34 Identify the formulas and names of the ions in the following ionic compounds.
(a) KOH   (b) Li_2CO_3 (c) $AgNO_3$ (d) LiBr

5.35 Name these compounds.
(a) NaBr   (b) Ag_2O
(c) $Ca(HSO_4)_2$ (d) FeO
(e) CuCN   (f) $NaHCO_3$

5.36 Write formulas for the following compounds.
(a) iron(III) bromide   (b) beryllium oxide
(c) ammonium sulfate   (d) magnesium phosphate
(e) lithium oxide   (f) ammonium carbonate

5.37 Which of the following substances are ionic?
(a) H_2O_2 (b) Na_2O
(c) SO_2 (d) CuS
(e) N_2H_4 (f) CO

5.38 Which of the following species would *not* be present in a crystal of sodium chloride?
(a) Na^{2+} (b) Cl^+ (c) Cl^- (d) Na^+ (e) Cl_2
(f) Na_2Cl_2

5.39 What is the charge of the cation in each of the following compounds?
(a) Cu_2CO_3 (b) FeS   (c) $HgSO_3$ (d) $Fe_2(SO_4)_3$

5.40 Name each of the compounds in Exercise 5.39.

Covalent Bonds and Molecular Compounds (Section 5.6)

5.41 (a) Draw the electron dot structure for sulfur.
(b) Why is the structure the same as that for oxygen?

5.42 Draw the electron dot structure for fluorine, chlorine, and bromine. Explain any similarities.

5.43 Draw the electron dot structure for a sulfide ion. Is there a relationship between this structure and the electron dot structure of a chloride ion?

5.44 Draw electron dot structures for nitride, oxide, and fluoride ions. Explain any similarities.

5.45 Explain why atoms form chemical bonds, and describe the difference between an ionic bond and a covalent bond.

5.46 Why do atoms of sodium and chlorine combine to form the ionic compound sodium chloride, but two atoms of chlorine combine to form the diatomic covalent chlorine molecule?

5.47 Explain why helium is monatomic but hydrogen exists as diatomic molecules.

5.48 Briefly explain why the noble gases are unreactive.

5.49 The following covalent molecules have only single covalent bonds. Draw an electron dot structure for each.
(a) H_2O (b) PCl_3 (c) H_2O_2 (d) CH_3Cl

5.50 Draw electron dot structures for each of the following covalent molecules.
(a) NH_3 (b) $CHCl_3$ (c) HF (d) C_2H_6

5.51 Classify the following compounds as ionic or covalent.
(a) $BeCl_2$ (b) HBr (c) Na_2S (d) H_2S
(e) SrO (f) H_2O

5.52 Which of the following is an ionic compound?
(a) SO_2 (b) Al_2O_3 (c) $CuCO_3$ (d) CH_4
(e) NH_4NO_3 (f) SiO_2

Polar Bonds and Polar Molecules (Sections 5.7, 5.8, 5.9)

5.53 What is meant by the term *electronegativity?*

5.54 What must be true about the electronegativities of the two atoms joined by a polar covalent bond?

5.55 Examine Table 5.5 on page 127, and discuss group and period trends in electronegativity values.

5.56 Examine Table 5.5, and link metallic and nonmetallic elements with high and low electronegativity values.

5.57 Determine what type of bond—ionic, polar covalent, or nonpolar covalent—will form between atoms of the following elements, and show the polarity of the bond if it is polar covalent.
(a) Ca and Cl (b) C and S (c) Mg and F
(d) N and O (e) H and O (f) S and O

5.58 The bonds between the following pairs of elements are covalent. Arrange them according to polarity, naming the most polar first.
(a) H—F (b) H—Cl (c) H—O (d) H—H
(e) H—C (f) S—Cl

Molecular Shapes (Section 5.10)

5.59 Explain the shapes of the methane and water molecules.

5.60 If one of the hydrogens in methane is replaced by chlorine, explain how the shape of the molecule will change (if at all).

Additional Exercises

5.61 How many atoms of each element are represented by each formula?
(a) NaH_2PO_4 (b) $(NH_4)_2SO_4$ (c) $KMnO_4$ (d) $Al(OH)_3$

5.62 What effect does the unshared pair of electrons have on the shape of an ammonia molecule?

5.63 In which group on the periodic table would an element with the following characteristics be found?
(a) The element gains two electrons to form an anion.
(b) The element has four valence electrons.
(c) The element loses one electron to form a cation.
(d) The element has a stable electron configuration.

5.64 Complete the following table by writing in the correct formula for the compound formed from each pair of ions.

	S^{2-}	ClO_3^-	Cl^-	SO_3^{2-}
Cu^{2+}	____	____	____	____
Li^+	____	____	____	____
Al^{3+}	____	____	____	____
NH_4^+	____	____	____	____

5.65 Draw the electron dot structure for each of the following molecules.
(a) BrCl (one single bond)
(b) HCN (one single and one triple bond)
(c) SO_2 (one single and one double bond)
(d) HCCH (two single bonds and one triple bond)

5.66 Which of the molecules in Exercise 5.65 has a coordinate covalent bond?

5.67 A hydrogen atom occasionally gains an electron to form an anion. What is the charge and name of this ion?

5.68 Explain why compounds containing C—N and C—O single bonds can form coordinate covalent bonds with H^+, but compounds containing only C—H and C—C single bonds cannot.

5.69 What relationship exists between the electron dot structure of an element and the location of the element in the periodic table?

5.70 Use the periodic table to complete the following table.

Element	Group	Electron dot structure	Ion	Name of Ion	Cation or anion
Ca	2A	· Ca ·	Ca^{2+}	calcium ion	cation
Br	____	____	____	____	____
P	____	____	____	____	____
____	____	____	____	sodium ion	____
____	____	____	____	sulfide ion	____
____	N/A	N/A	____	iron(III) or ferric ion	____

5.71 Complete the following table by giving the formula and name for each compound formed from each pair of ions.

	Oxide	Nitrate	Iodide	Phosphate	Hydroxide
Potassium	__ __	__ __	__ __	__ __	__ __
Iron(III)	__ __	__ __	__ __	__ __	__ __
Magnesium	__ __	__ __	__ __	__ __	__ __
Copper(I)	__ __	__ __	__ __	__ __	__ __

SELF-TEST (REVIEW)

True/False

1. A strontium atom gains two electrons in forming a strontium ion, Sr^{2+}.

2. In general, the electronegativity values of nonmetallic elements are greater than the electronegativity values of metallic elements.

3. To attain a noble gas electron structure, a nitrogen atom must lose its five valence electrons.

4. The electron dot formula for each metal in Group 2A has two dots.

5. A molecule with polar bonds must itself always be polar.

6. In a polar covalent bond, the more electronegative atom has a slight positive charge.

7. A triple covalent bond involves a total of six electrons.

8. The net charge of an ionic compound is zero or neutral.

9. The greater the difference in electronegativities of two atoms covalently bonded, the greater is the polarity.

10. The bonds in diatomic elements such as H_2, Cl_2, and I_2 are nonpolar covalent.

Multiple Choice

11. The chemical formula for potassium sulfide is
    (a) PS   (b) K_2S
    (c) K_2SO_3 (d) KS_2

12. Which of the following compounds has polar covalent bonds? (Use Table 5.5 on page 127 if necessary.)
    (a) NaCl   (b) I_2
    (c) CaF_2 (d) H_2S

13. What is the correct electron dot structure for the element selenium (Se)?

    (a) $\cdot$ Se $\cdot$ (b) $\overset{\cdot\cdot}{\underset{\cdot\cdot}{:}}$ Se $:$ (c) $\cdot$ Se $\cdot$ (d) $\cdot$ Se $\cdot$

14. When H^+ forms a bond with H_2O to form the hydronium ion H_3O^+, this bond is called a coordinate covalent bond because
    (a) both electrons come from the oxygen atom.
    (b) it forms an especially strong bond.
    (c) the electrons are equally shared.
    (d) the oxygen no longer has eight valence electrons.

15. Which of the following statements is *not* true about the compound strontium chloride $SrCl_2$?
    (a) It is a gas at room temperature.
    (b) It dissolves easily in water.
    (c) It is made up of oppositely charged ions.
    (d) It has a high melting point.

16. Which of these polyatomic ions has a charge of 2–?
    (a) perchlorate
    (b) monohydrogen phosphate
    (c) cyanide   (d) nitrate

17. In forming chemical bonds, atoms attain
    (a) lower energy.
    (b) the electron configuration of noble gas atoms.
    (c) higher stability.
    (d) all of the above.

18. If a bonding pair of electrons is unequally distributed between two atoms, the bond is called
    (a) an ionic bond.
    (b) a nonpolar covalent bond.
    (c) an electrovalent bond.
    (d) a polar covalent bond.

19. The correct name of the compound $Al_2(SO_3)_3$ is
    (a) aluminum sulfide.   (b) aluminum(III) sulfite.
    (c) dialuminum sulfate.   (d) aluminum sulfite.

20. The ions of each of the following elements have an ionic charge of 1+ except
    (a) hydrogen.   (b) rubidium.   (c) fluorine.
    (d) sodium.

21. The electron dot structure for the hydroxide polyatomic ion OH^- is

    (a) $\left[O:H\right]^-$ (b) $\left[:\overset{\cdot\cdot}{\underset{\cdot\cdot}{O}}:H\right]^-$ (c) $\left[:\overset{\cdot\cdot}{H}:\overset{\cdot\cdot}{\underset{\cdot\cdot}{O}}:\right]^-$

    (d) $\left[\cdot\overset{\cdot}{\underset{\cdot}{O}}:H\right]^-$

22. Ionic bonds are characterized by all the following except
    (a) a transfer of electrons.
    (b) unequal sharing of electrons.
    (c) formation between atoms of a metal and a nonmetal.
    (d) forces of attraction between oppositely charged ions.

23. Which of the following molecules would you expect to have the same general shape as an NH_3 molecule?
    (a) SO_3 (b) PCl_3 (c) CH_3Cl (d) BH_3

24. The electron dot structure for hydrogen sulfide, H_2S is

    (a) $H:\overset{\cdot\cdot}{\underset{\cdot\cdot}{S}}:$ (b) $H:\overset{\cdot\cdot}{\underset{\cdot\cdot}{S}}:H$ (c) $H\cdot\overset{\cdot}{\underset{\cdot}{S}}\cdot H$

    (d) $H:H:\overset{\cdot\cdot}{\underset{\cdot\cdot}{S}}:$

25. The formula for calcium cyanide is
    (a) $CaCN_2$.   (b) CaCN.   (c) $Ca(CN)_2$.
    (d) Ca_2CN.

26. Which of the following compounds would not have covalent bonds?
    (a) NO_2 (b) Cs_2O (c) N_2O (d) H_2O_2

27. A particle has 9 protons and 10 electrons. Its charge is
(a) zero.　　(b) 1−.　　(c) 9+.　　(d) 1+.

28. The compound HgO has the name
(a) mercury oxide.　　(b) mercurous oxide.
(c) mercury(II) oxide.　　(d) mercury(IV) oxide.

29. In which of the following compounds are the bonds most polar?
(a) CO　　(b) I_2　　(c) NH_3　　(d) CBr_4

30. A polyatomic cation
(a) is composed of two or more different atoms.
(b) has a positive charge.
(c) has a Roman numeral as part of its name.
(d) (a) and (b) are correct.

Chemical Reactions

Equations, Equilibria, and Reaction Rates

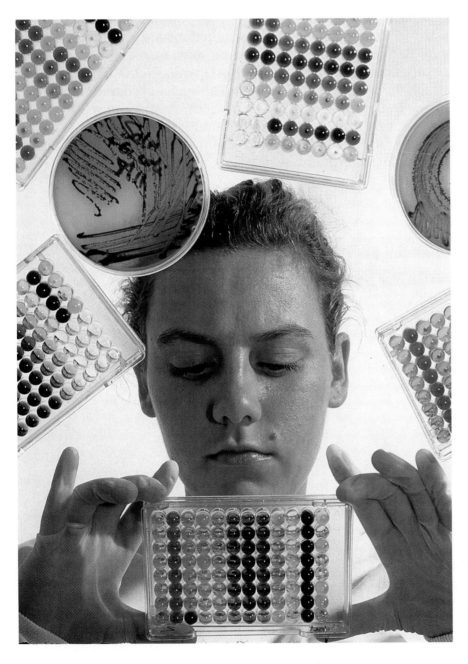

Visible changes that are critical to medical diagnoses are the result of chemical reactions. Here, the technician uses characteristic color changes to distinguish among types of bacteria.

CHAPTER OUTLINE

CASE IN POINT: Effect of cold on the body

6.1 The mole

6.2 Chemical equations

6.3 Chemical calculations

6.4 Oxidation and reduction reactions

6.5 Heat and entropy changes

6.6 Free energy

6.7 Reaction rate

FOLLOW-UP TO THE CASE IN POINT: Effect of cold on the body

A CLOSER LOOK: Catalase

6.8 Reversible reactions

A CLOSER LOOK: The Rain Forest, Equilibrium, and Medicine

Previous chapters have been concerned with the composition of atoms, ions, and molecules. Now we turn to chemical reactions, the means by which elements and compounds form different substances. Examples of several kinds of chemical reactions in which the change that occurs is visible are shown in the chapter-opening figure. Our understanding of these and other chemical reactions is important to our study of the health sciences. For example, we can comprehend why biological reactions needed for energy production by cells give products only in the presence of cellular molecules called *enzymes*, why our body temperature is a warm 37°C, and why exhaling carbon dioxide keeps us alive. As we will see in this chapter, the speed at which reactants are converted to products in chemical reactions depends on the temperature at which the reaction is carried out. The speed of chemical reactions in the body also depends on the body's temperature. When the temperature drops too low, the chemical reactions needed to sustain life are affected. This is exemplified in the following Case in Point, which considers the effect of temperature on chemical reactions in the body.

CASE IN POINT: Effect of cold on the body

During the winter of 1993–1994, a cold wave swept over the entire eastern United States. Heavy snows and temperatures as low as −40 °F assaulted regions of the country unaccustomed to such frigid conditions. Many people were unprepared or unable to keep warm enough to avoid the threat to their health caused by the low temperatures. Bill, a 60-year-old homeless male, was one of the cold's victims (see figure). On a night when the temperature registered −22 °F, a police officer in a large metropolitan area found him in a make-shift cardboard tent built over a ventilation grate in the sidewalk outside an office building. Bill was unresponsive to the officer's attempts to rouse him. He was taken to the emergency room of the city hospital, where his body temperature was found to be 34 °C (93 °F). What is Bill's condition called? What does his condition have to do with the chemical reactions in Bill's body? We will see in Section 6.7.

Numbed by the cold and unable to feel pain, your body's vital signs slow. Finally, your heart stops beating.

6.1 The mole

***AIMS:** To define formula mass, molar mass, and mole and show the relationship among these terms. To calculate the mass of a given number of moles of a substance and the number of moles in a given mass of a substance. To recognize Avogadro's number and state its relationship to the mole.*

Focus

Formula masses lead to the mole concept.

Figure 6.1
Limestone cliffs are composed of calcium carbonate (formula mass: 100.1 amu).

Recall from Chapter 3 that elements can exist as isotopes. Because the isotopes of an element have different masses, the atomic masses given in the periodic table are the *average* masses for atoms of the element. These average atomic masses can be used to calculate the average masses of representative units of matter. A *representative unit* is the smallest unit of a substance that has the formula of the substance: one atom of a monatomic element such as sodium (Na), one homonuclear molecule such as hydrogen (H_2) or chlorine (Cl_2), one heteronuclear molecule such as carbon monoxide (CO) or glucose ($C_6H_{12}O_6$), or one ion such as chloride ion (Cl^-) or oxide ion (O^{2-}). The representative unit also could be the formula of an ionic compound such as sodium sulfate (Na_2SO_4). We can add the average atomic masses of the atoms in any chemical formula to obtain the *formula mass*. The **formula mass** *is the average mass in atomic mass units (amu) of a representative unit of any pure substance, no matter what its chemical nature: an atom, an ion, or an ionic or molecular compound.* The formula masses for a metallic element, a diatomic molecule, a molecular compound, and an ionic compound are illustrated in Figure 6.1.

EXAMPLE 6.1

Calculating a formula mass

What is the formula mass of sodium chloride?

SOLUTION

Sodium chloride is an ionic compound with the chemical formula NaCl. We start by looking up the average atomic masses of sodium (23.0 amu) and chlorine (35.5 amu) in the periodic table. (For all our calculations, the average atomic masses will be rounded off to one digit after the decimal point.) The mass of the electron lost by a sodium atom and gained by a chlorine atom in ion formation is negligible. By adding the average mass of one atom of sodium and the average mass of one atom of chlorine, we obtain a formula mass of 58.5 amu:

$$
\begin{aligned}
& 1 \text{ atomic mass of Na} = 23.0 \text{ amu} \\
&\underline{+\ \ 1 \text{ atomic mass of Cl} = 35.5 \text{ amu}} \\
& \text{Formula mass of NaCl} = 58.5 \text{ amu}
\end{aligned}
$$

PRACTICE EXERCISE 6.1

Find the formula mass of (a) potassium bromide (KBr), (b) calcium sulfate ($CaSO_4$), (c) sodium bicarbonate ($NaHCO_3$), and (d) glucose ($C_6H_{12}O_6$).

Table 6.1 Comparison of the Mass of One Mole of Water and One Mole of Glucose

Substance	Molecular formula	Formula mass (amu)
water	H_2O	18.0
glucose	$C_6H_{12}O_6$	180.0

Formula masses enable us to compare the relative masses of representative units of different pure substances. We can use the molecular compounds water (H_2O) and glucose ($C_6H_{12}O_6$) to demonstrate this. The formula mass of one molecule of water is 18.0 amu, and the formula mass of glucose is 180 amu. Therefore, the mass of one molecule of glucose is 10 times greater than the mass of one molecule of water (Table 6.1). Since a molecule of glucose is 10 times heavier than a molecule of water, it follows that if we want equal numbers of molecules of glucose and water, all we need to do is to take a mass of glucose 10 times greater than a given mass of water. We can say with confidence that 18 g of water and 180 g of glucose contain the same number of molecules (representative units). To go further with this idea, imagine that we could weigh a mass of each element equal to that element's atomic mass in grams. The samples all would contain the same number of atoms.

Mole

moles (Latin): heap or pile

The number of representative units in the formula mass in grams of a substance is important in chemistry. *One* **mole** *(1 mol) can be defined as the number of carbon atoms contained in 12.01 grams of carbon.* The number of carbon atoms in 1 mol of carbon is 6.02×10^{23}, or 602,000,000,000,000,000,000,00. Similarly, 1 mol is equal to 6.02×10^{23} representative units of any element, ion, molecular compound, or ionic compound. *The number 6.02×10^{23} is called* **Avogadro's number** in honor of Amedeo Avogadro de Quarenga, the nineteenth-century Italian scientist and lawyer whose work made its calculation possible. Avogadro's number is incomprehensibly large. If we started counting peas at the rate of one pea per second, it would take 1.7×10^{16} years to count 6.02×10^{23} peas. On the other hand, with 5.6 billion helpers (the entire population of the world), we could do the job in 3.4×10^6 years—3.4 million years!

The hemoglobin molecules in red blood cells carry oxygen from your lungs to your tissue cells. Even though your body produces a million million (10^{12}) hemoglobin molecules each second, this is still only 0.00005 mol of hemoglobin molecules produced in a year.

Notice mass of carbon in the definition of the mole is the average atomic mass in grams. The mass of 1 mol of any substance must be its formula mass in grams because 1 mol of carbon atoms and 1 mol of any other substance each contain Avogadro's number of representative units. *One* **molar mass** *of a substance is the mass in grams of 1 mol of the substance* (Table 6.2). One

Table 6.2 Chemical Formulas, Molar Mass, and Moles

Substance	Representative particle	Chemical formula	Formula mass (amu)	Molar mass (mass of 1 mol, g)
atomic hydrogen	atom	H	1.0	1.0
hydrogen gas	molecule	H_2	2.0	2.0
water	molecule	H_2O	18.0	18.0
sodium ion	ion	Na^+	23.0	23.0
sodium metal	atom	Na	23.0	23.0
sodium chloride	formula unit	NaCl	58.5	58.5
glucose	molecule	$C_6H_{12}O_6$	180.0	180.0

Figure 6.2
A comparison of the molar masses of carbon (left), copper (top), sulfur (right), and mercury (bottom).

molar mass of any pure substance contains the same number of representative units as 1 molar mass of any other substance, whether the units are atoms, ions, molecules, or the formulas of ionic compounds. For example, both 1 molar mass of water and 1 molar mass of glucose contain 1 mol of molecules. Figure 6.2 compares molar masses of carbon, copper, sulfur and mercury.

EXAMPLE 6.2   **Calculating the molar mass of carbon dioxide**

What is the molar mass of carbon dioxide?

SOLUTION

The chemical formula of carbon dioxide is CO_2. The molar mass of carbon (C) is 12.0 g, and the molar mass of oxygen (O) is 16.0 g. By adding the mass of 1 mol of carbon and 2 mols of oxygen, we get a molar mass of 44.0 g:

$$1 \text{ mol C} = 1 \times 12.0 \text{ g} = 12.0 \text{ g}$$
$$\underline{2 \text{ mol O} = 2 \times 16.0 \text{ g} = 32.0 \text{ g}}$$
$$\text{Molar mass of } CO_2 = 44.0 \text{ g}$$

PRACTICE EXERCISE 6.2

Find the molar mass of each of the following substances.

(a) methane, CH_4 (b) sucrose, $C_{12}H_{22}O_{11}$

(c) calcium nitrate, $Ca(NO_3)_2$ (d) chlorine, Cl_2

Chemistry often requires that we find the number of moles of a substance when the mass is known. Conversely, it may be necessary to convert the number of moles of a substance to mass. The following examples show how to make these conversions.

EXAMPLE 6.3   **Making a mole-mass conversion**

Calculate the mass in grams of 0.200 mol CO_2.

SOLUTION

From Example 6.2 we know that the molar mass of CO_2 (or the mass of 1 mol of CO_2) is 44.0 g. The conversion factor for moles to grams is

$$\frac{44.0 \text{ g } CO_2}{1 \text{ mol } CO_2}$$

Therefore,

$$0.200 \cancel{\text{mol } CO_2} \times \frac{44.0 \text{ g } CO_2}{1 \cancel{\text{mol } CO_2}} = 8.80 \text{ g } CO_2$$

PRACTICE EXERCISE 6.3

Find the mass in grams of each of the following.

(a) 1.00 mol H_2O_2 (b) 0.200 mol Cl_2 (c) 0.250 mol K_2SO_4
(d) 3.00×10^{-2} mol $CaCO_3$

EXAMPLE 6.4

Converting mass to moles

How many moles is 4.0 g of methane (CH_4)?

SOLUTION

We can calculate that 1 mol CH_4 has a molar mass of 16.0 g. To convert grams to moles, the conversion factor is

$$\frac{1 \text{ mol } CH_4}{16.0 \text{ g } CH_4}$$

Therefore,

$$4.0 \cancel{\text{g } CH_4} \times \frac{1 \text{ mol } CH_4}{16.0 \cancel{\text{g } CH_4}} = 0.25 \text{ mol } CH_4$$

PRACTICE EXERCISE 6.4

How many moles is (a) 2.00 g NaOH, (b) 115 g Na, (c) 4.90 g H_2SO_4, and (d) 4.10 g Na_3PO_4?

6.2 Chemical equations

AIM: To write balanced chemical equations.

Focus

Chemical equations describe chemical reactions.

In chemical reactions, bonds are broken and new bonds are formed as reactants are converted into products. For example, chemical bonds in starch molecules (reactants) are broken as food is degraded into glucose molecules (products) in the chemical reactions of digestion. There are several ways to describe a chemical reaction. For example, we could say, "Carbon burns in the presence of oxygen to produce carbon dioxide." Or we could write a word equation:

Carbon + Oxygen $\longrightarrow$ Carbon dioxide

More often, chemists write equations using chemical formulas. The physical state of a substance in a reaction is sometimes important and is identified by putting (*s*) for a solid, (*l*) for a liquid, and (*g*) for a gas. The symbol (*aq*) after a formula describes a substance dissolved in water; it is an aqueous solution of the substance. Using symbols, we can write the chemical equation for the combustion of solid carbon and gaseous oxygen to produce gaseous carbon dioxide as

$$C(s) \quad + \quad O_2(g) \quad \longrightarrow \quad CO_2(g)$$

Carbon   Oxygen   Carbon dioxide

Balancing equations

Chemical equations must be balanced to represent chemical reactions correctly. The law of conservation of matter always holds true in a chemical reaction. *In a* **balanced chemical equation,** *each side of the equation must represent the same number of atoms of each element.* The equation for the burning of carbon is balanced; there is one carbon atom on each side of the equation, and there are two oxygen atoms on each side. Balanced equations are a convenient way of keeping track of atoms in a chemical reaction and of providing an accurate description of a chemical process.

Chemical equations are balanced by trial and error, but there are rules we can follow:

1. Use the correct formulas for all the reactants and products in the reaction.

2. Write the formulas for the reactants on the left and the formulas for the products on the right with an arrow in between. If there are two or more reactants or products, their formulas are separated by plus signs.

3. Count the number of atoms of each element in the reactants and products. A polyatomic ion appearing unchanged on both sides of the equation is counted as a single unit.

4. Balance the elements (or polyatomic ions) one at a time by putting small whole numbers, called *coefficients,* in front of the appropriate formula. When no coefficient is written, it is assumed to be 1. It is best to begin with elements that appear only once on each side of the equation. *Do not attempt to balance an equation by changing the subscripts in the chemical formula of a substance.*

5. Check each atom or polyatomic ion to be sure the equation is balanced.

6. Finally, make sure that all the coefficients are whole numbers and that they have the lowest possible ratio.

EXAMPLE 6.5	Balancing a chemical equation

Hydrogen gas and oxygen gas can react to give liquid water. Write a balanced equation for this reaction.

SOLUTION

Since the chemical formulas and physical states of the reactants and products are known, we can write

$$H_2(g) + O_2(g) \longrightarrow H_2O(l)$$

Hydrogen is balanced, but oxygen is not. If we put a coefficient of 2 in front of H_2O, the oxygen becomes balanced:

$$H_2(g) + O_2(g) \longrightarrow 2H_2O(l)$$

Now, however, there are twice as many hydrogen atoms in the product as there are in the reactants. Put a coefficient of 2 in front of H_2 and the equation is balanced (Fig. 6.3a):

$$2H_2(g) + O_2(g) \longrightarrow 2H_2O(l)$$

Check the coefficients. They are small whole numbers and in their lowest possible ratio—$2(H_2)$, $1(O_2)$, and $2(H_2O)$—so they are correct. Suppose we had written the equation as

$$4H_2(g) + 2O_2(g) \longrightarrow 4H_2O(l)$$

This equation is balanced but incorrect because the coefficients are not in their lowest possible ratio. Each of them can be divided by 2.

EXAMPLE 6.6	Writing and balancing a chemical equation

Write a balanced equation for the reaction of methane gas with oxygen gas. The products are carbon dioxide gas and steam.

SOLUTION

The unbalanced reaction is

$$CH_4(g) + O_2(g) \longrightarrow CO_2(g) + H_2O(g)$$

Since carbon is balanced, we can proceed to balance hydrogen by putting a coefficient of 2 in front of H_2O:

$$CH_4(g) + O_2(g) \longrightarrow CO_2(g) + 2H_2O(g)$$

The equation is still unbalanced because there are only two atoms of oxygen in the reactants but four atoms of oxygen in the products. Place a coefficient of 2 in front of O_2; this gives four oxygen atoms on each side of the equation. Now the equation will balance with the coefficients in their lowest possible ratio of 1:2:1:2 (see Fig. 6.3b).

$$CH_4(g) + 2O_2(g) \longrightarrow CO_2(g) + 2H_2O(g)$$

Atoms in reactants = Atoms in products

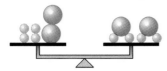

2 molecules of hydrogen   2 molecules of water
1 molecule of oxygen

$$2H_2 + O_2 \longrightarrow 2H_2O$$

The equation is balanced. There are 4 hydrogen atoms and 2 oxygen atoms on each side of the equation.

(a)

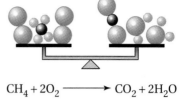

$$CH_4 + 2O_2 \longrightarrow CO_2 + 2H_2O$$

The equation is balanced. There are 4 oxygen atoms, 4 hydrogen atoms, and 1 carbon atom on each side of the equation.

(b)

Figure 6.3
When a chemical equation is balanced, the number and kinds of atoms in the reactants are the same as in the products.

EXAMPLE 6.7   **Writing a balanced equation**

Figure 6.4
Copper metal reacts with a solution of silver nitrate to produce silver-metal and a solution of copper(II) nitrate.

Balance the equation for the reaction that appears in Figure 6.4:

$$AgNO_3(aq) + Cu(s) \longrightarrow Cu(NO_3)_2(aq) + Ag(s)$$

SOLUTION

Since this reaction involves the polyatomic nitrate ion, we can save time balancing the equation if we consider the nitrate ion as a unit. Place a coefficient of 2 in front of $AgNO_3$ to balance the nitrate ion:

$$2AgNO_3(aq) + Cu(s) \longrightarrow Cu(NO_3)_2(aq) + Ag(s)$$

But now the silver (Ag) is not balanced; there are two atoms in the reactants but only one atom in the products. Place a coefficient of 2 in front of Ag; now the equation is balanced and the coefficients are correct:

$$2AgNO_3(aq) + Cu(s) \longrightarrow Cu(NO_3)_2(aq) + 2Ag(s)$$

EXAMPLE 6.8   **Writing a balanced equation**

Balance the following reaction:

$$NaOH(aq) + H_2SO_4(aq) \longrightarrow Na_2SO_4(aq) + H_2O(l)$$

| Sodium hydroxide | Sulfuric acid | Sodium sulfate | Water |

SOLUTION

The polyatomic sulfate ion and oxygen are balanced, but both sodium and hydrogen require balancing. Begin by placing a coefficient of 2 in front of NaOH to balance sodium:

$$2NaOH(aq) + H_2SO_4(aq) \longrightarrow Na_2SO_4(aq) + H_2O(l)$$

The hydrogen and oxygen are now unbalanced. A coefficient of 2 in front of H_2O balances the equation:

$$2NaOH(aq) + H_2SO_4(aq) \longrightarrow Na_2SO_4(aq) + 2H_2O(l)$$

PRACTICE EXERCISE 6.5

Balance the following equations:
(a) $C(s) + O_2(g) \longrightarrow CO(g)$ (b) $Al(s) + S(s) \longrightarrow Al_2S_3(s)$
(c) $BaCl_2(aq) + Na_2SO_4(aq) \longrightarrow BaSO_4(s) + NaCl(aq)$
(d) $Na(s) + Br_2(l) \longrightarrow NaBr(s)$

PRACTICE EXERCISE 6.6

Rewrite these word equations as balanced chemical equations.
(a) nitrogen(g) + hydrogen$(g) \longrightarrow$ ammonia(g)
(b) sodium(s) + oxygen$(g) \longrightarrow$ sodium oxide(s)
(c) hydrogen(g) + sulfur$(s) \longrightarrow$ hydrogen sulfide(g)

6.3 Chemical calculations

AIMS: To interpret a balanced chemical equation in terms of particles, moles, or mass. To perform the following types of stoichiometric calculations: mole-mole, mole-mass, and mass-mass.

Focus

Chemists need balanced equations to do calculations.

Almost everything we buy—soap, bread, medicines, paint—contains chemicals manufactured by some industrial process. These processes must be done efficiently and with the least waste possible because waste costs money and needlessly consumes resources. This is where balanced chemical equations are valuable. Balanced equations are recipes, telling us what amounts of reactants to mix and what amounts of products to expect. Consider the equation for making carbon dioxide:

$$C(s) \quad + \quad O_2(g) \quad \longrightarrow \quad CO_2(g)$$
Carbon   Oxygen   Carbon dioxide

We could read this equation in several ways. For example, the equation tells us that if one atom of carbon and one molecule of oxygen react, we get one molecule of carbon dioxide. As you may recall, however, we use large numbers of units such as moles, not single atoms and molecules. Thus we also can translate the equation to read:

$$1 \text{ mol C atoms} + 1 \text{ mol } O_2 \text{ molecules} \longrightarrow 1 \text{ mol } CO_2 \text{ molecules}$$

Since we can calculate the molar mass of a substance, we also have the relationship

$$12.0 \text{ g C} + 32.0 \text{ g } O_2 \longrightarrow 44.0 \text{ g } CO_2$$

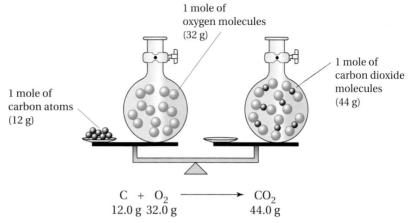

$$C + O_2 \longrightarrow CO_2$$
12.0 g 32.0 g   44.0 g

Figure 6.5
Mass of reactants must always equal mass of products.

Because the equation was balanced at the start, the total mass of all the reactants (44.0 g) is equal to the total mass of the products (44.0 g), and mass is conserved (Fig. 6.5).

EXAMPLE 6.9

Interpreting a balanced equation

Explain the meaning of this balanced equation:

$$2H_2(g) + O_2(g) \longrightarrow 2H_2O(l)$$

SOLUTION

The equation means:

2 molecules H_2 + 1 molecule O_2 $\longrightarrow$ 2 molecules H_2O
2 mol H_2 + 1 mol O_2 $\longrightarrow$ 2 mol H_2O
4.0 g H_2 + 32.0 g O_2 $\longrightarrow$ 36.0 g H_2O

When we know the quantity of one substance in a reaction, we can calculate the quantity of any other substance provided we have the balanced equation. *Quantity* usually means the amount of a substance expressed in grams or moles. Calculations using balanced equations are called *stoichiometric* calculations. **Stoichiometry**—*the calculation of quantities in balanced equations*—is a form of chemical bookkeeping. Stoichiometry often involves the following calculations:

Stoichiometry
stoicheion (Greek): element
metron (Greek): measure

1. Find the moles of products in a reaction when given the moles of reactants.

2. Convert moles of a substance into grams or grams of a substance into moles.

3. Find the grams of reactants that will give a specific mass of product or the grams of product formed from a specific mass of reactant.

Now let's look at some worked examples that show how we can do these calculations.

EXAMPLE 6.10   **Making a mole-mole conversion**

How many moles of water are produced when 0.42 mol oxygen reacts with hydrogen? The balanced equation is

$$2H_2(g) + O_2(g) \longrightarrow 2H_2O(l)$$

SOLUTION

From the equation, 1 mol O_2 produces 2 mol H_2O. The mole ratio relating H_2O and O_2 is

$$\frac{2 \text{ mol } H_2O}{1 \text{ mol } O_2}$$

Use this factor to calculate moles of H_2O from 0.42 mol O_2:

$$0.42 \text{ mol } O_2 \times \frac{2 \text{ mol } H_2O}{1 \text{ mol } O_2} = 0.84 \text{ mol } H_2O$$

EXAMPLE 6.11   **Making a mass-mole conversion**

Calculate the number of moles of O_2 required to produce 3.6 g H_2O according to the balanced equation

$$2H_2(g) + O_2(g) \longrightarrow 2H_2O(l)$$

SOLUTION

First, convert 3.6 g H_2O to moles of H_2O. The molar mass of H_2O is 18.0 g/mol. Therefore,

$$3.6 \text{ g } H_2O \times \frac{1 \text{ mol } H_2O}{18.0 \text{ g } H_2O} = 0.20 \text{ mol } H_2O$$

From the equation, 1 mol O_2 is required to produce 2 mol H_2O. The mole ratio relating O_2 and H_2O in the equation is

$$\frac{1 \text{ mol } O_2}{2 \text{ mol } H_2O}$$

Thus, 0.20 mol H_2O requires

$$0.20 \text{ mol } H_2O \times \frac{1 \text{ mol } O_2}{2 \text{ mol } H_2O} = 0.10 \text{ mol } O_2$$

EXAMPLE 6.12   **Making a mass-mass conversion**

Calculate the number of grams of liquid H_2O produced by the reaction of 8.0 g gaseous O_2 with hydrogen gas. The balanced equation is

$$2H_2(g) + O_2(g) \longrightarrow 2H_2O(l)$$

SOLUTION

First, change 8.0 g O_2 to moles. The molar mass of O_2 is 32.0 g/mol. Thus,

$$8.0 \text{ g } O_2 \times \frac{1 \text{ mol } O_2}{32.0 \text{ g } O_2} = 0.25 \text{ mol } O_2$$

Calculate moles of H_2O produced by 0.25 mol O_2. From the equation, the mole ratio is

$$\frac{2 \text{ mol } H_2O}{1 \text{ mol } O_2}$$

Thus, 0.25 mol O_2 produces

$$0.25 \text{ mol } O_2 \times \frac{2 \text{ mol } H_2O}{1 \text{ mol } O_2} = 0.50 \text{ mol } H_2O$$

Finally, convert 0.50 mol H_2O to grams. The molar mass of H_2O is 18.0 g/mol. Therefore,

$$0.50 \text{ mol } H_2O \times \frac{18.0 \text{ g } H_2O}{1 \text{ mol } H_2O} = 9.0 \text{ g } H_2O$$

PRACTICE EXERCISE 6.7

This is the balanced equation for the reaction of methane with oxygen:

$$CH_4(g) + 2O_2(g) \longrightarrow CO_2(g) + 2H_2O(l)$$

Now calculate the following quantities.

(a) The number of moles of O_2 to produce 0.50 mol CO_2

(b) The number of moles of H_2O produced from 0.20 mol CH_4

(c) The number of grams of CH_4 needed to react with 25.6 g O_2

6.4 Oxidation and reduction reactions

AIMS: To identify the substances being oxidized and reduced in a redox reaction. To identify the oxidizing agent and the reducing agent in a redox reaction.

Focus

Oxidation reactions are important sources of energy.

Chemists put chemical reactions with similarities into groups. Among the most important categories of reactions to the health sciences are *oxidations* and *reductions*. Oxidation reactions, for example, are the principal source of energy on Earth. The combustion of gasoline in an automobile engine and the burning of coal in a furnace are oxidation reactions. So is the "burning" of food by our bodies. Reduction reactions are also important to the function of living cells. The reduction of carbon dioxide to glucose via photosynthesis is essential for sustaining life on Earth.

In Chapter 4 we stated the original definition of *oxidation:* the combination of a substance with oxygen to give oxides. Today, the definition of

Oxidation reactions are important in the production of energy in living organisms.

oxidation is much broader, having been expanded to include hydrogen and electrons: **Oxidation** *is a gain of oxygen, a loss of hydrogen, or a loss of electrons by a substance.* The advantage of the new definition is that it has much wider application than the old one. Reduction is similarly defined: **Reduction** *is a loss of oxygen, a gain of hydrogen, or a gain of electrons by a substance.* All oxidation reactions are accompanied by reduction reactions; there is no oxidation without reduction and no reduction without oxidation. *Since oxidation always accompanies reduction and vice versa, oxidation-reduction reactions are often called* **redox reactions.** We can use the definitions of oxidation and reduction to identify what is oxidized and what is reduced in redox reactions. Let's consider the reaction of hydrogen and oxygen gases to form liquid water:

$$2H_2(g) + O_2(g) \longrightarrow 2H_2O(l)$$

In this redox reaction, hydrogen is oxidized because it gains oxygen. Oxygen is reduced because it gains hydrogen. Hydrogen is a **reducing agent**—*a substance that causes reduction*—because it reduces oxygen. Oxygen is an **oxidizing agent**—*a substance that causes oxidation*—because it oxidizes hydrogen.

When carbon burns in air, it oxidizes to carbon dioxide by combining with oxygen:

$$C(s) \quad + \quad O_2(g) \quad \longrightarrow \quad CO_2(g)$$

Carbon   Oxygen   Carbon dioxide

Oxygen is the oxidizing agent in this reaction. Because oxidation is always accompanied by reduction, oxygen must be reduced by carbon. Carbon is the reducing agent.

In the reaction between a metal and a nonmetal to produce an ionic compound, electrons are transferred from atoms of the metal to atoms of the nonmetal. For making magnesium sulfide, the reaction is

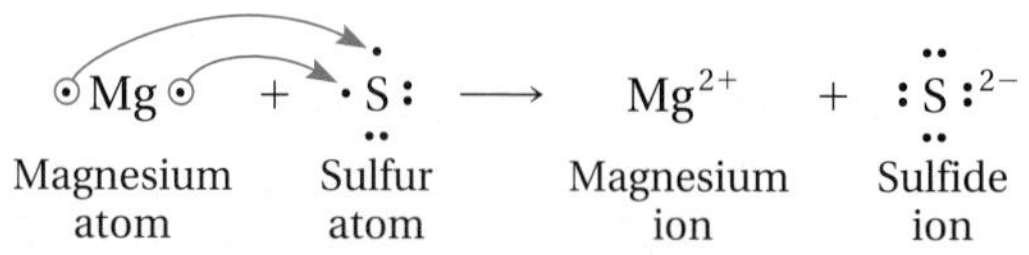

Magnesium atom   Sulfur atom   Magnesium ion   Sulfide ion

The net result of this reaction is a transfer of two electrons from a magnesium atom to a sulfur atom. The magnesium atom is oxidized (loses electrons) to form a magnesium ion, and simultaneously, the sulfur atom is reduced (gains electrons) to form a sulfide ion:

Oxidation: $\cdot Mg \cdot \longrightarrow Mg^{2+} + 2e^-$ (loss of electrons)

Magnesium atom   Magnesium ion

Reduction: $\cdot \ddot{S} : + 2e^- \longrightarrow : \ddot{\ddot{S}} :^{2-}$ (gain of electrons)

Sulfur atom   Sulfide ion

Magnesium is the reducing agent; sulfur is the oxidizing agent. To summarize:

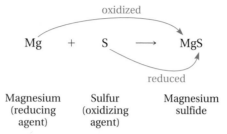

| Magnesium (reducing agent) | Sulfur (oxidizing agent) | Magnesium sulfide |

When iron turns to rust, it oxidizes to iron(III) oxide:

$$4Fe(s) + 3O_2(g) \longrightarrow 2Fe_2O_3(s)$$

Iron   Oxygen   Iron(III) oxide

The iron atoms are oxidized by losing three electrons to form iron(III) ions. Oxygen atoms gain two electrons from iron to form oxide ions (O^{2-}). Iron is the reducing agent, and oxygen is the oxidizing agent.

EXAMPLE 6.13 **Identifying the elements oxidized and reduced**

Consider another oxidation-reduction reaction, one that we balanced earlier (Example 6.7):

$$2AgNO_3(aq) + Cu(s) \longrightarrow Cu(NO_3)_2(aq) + 2Ag(s)$$

Which metal is oxidized, and which metal is reduced?

SOLUTION

Begin by rewriting the equation and showing the ions:

$$2Ag^+ + 2NO_3^- + Cu \longrightarrow Cu^{2+} + 2NO_3^- + 2Ag$$

There has been a transfer of two electrons from a single copper atom (Cu) to two silver ions (Ag^+):

Oxidation: $Cu \longrightarrow Cu^{2+} + 2e^-$ (loss of electrons)

Reduction: $2Ag^+ + 2e^- \longrightarrow 2Ag$ (gain of electrons)

The Ag^+ gains electrons and is reduced; Cu loses electrons and is oxidized.

EXAMPLE 6.14 **Identifying oxidizing and reducing agents**

Methane gas, a compound of carbon and hydrogen, burns in oxygen to form oxides of carbon and hydrogen:

$$CH_4(g) + 2O_2(g) \longrightarrow CO_2(g) + 2H_2O(g)$$

What are the oxidizing and reducing agents in this reaction?

SOLUTION

Methane is oxidized to carbon dioxide by the gain of oxygen (or loss of hydrogen); oxygen is reduced to water by the gain of hydrogen. Oxygen is the oxidizing agent for the oxidation of methane. Methane is the reducing agent for the reduction of oxygen.

PRACTICE EXERCISE 6.8

Determine what is oxidized and what is reduced in the following reactions. Identify the oxidizing agent and reducing agent in each case.

(a) $2Na(s) + S(s) \longrightarrow Na_2S(s)$ (b) $2K(s) + Cl_2(g) \longrightarrow 2KCl(s)$

(c) $4Al(s) + 3O_2(g) \longrightarrow 2Al_2O_3(s)$ (d) $N_2(g) + O_2(g) \longrightarrow NO_2(g)$

PRACTICE EXERCISE 6.9

In each of the following reactions, decide which reactant is oxidized and which reactant is reduced. Which reactant is the reducing agent, and which is the oxidizing agent?

(a) $H_2(g) + Cl_2(g) \longrightarrow 2HCl(g)$ (b) $N_2(g) + 3H_2(g) \longrightarrow 2NH_3(g)$

(c) $S(s) + Cl_2(g) \longrightarrow SCl_2(l)$ (d) $2Li(s) + F_2(g) \longrightarrow 2LiF(s)$

6.5 Heat and entropy changes

AIMS: *To distinguish between exothermic and endothermic reactions. To explain how changes in energy and entropy affect the favorability of a reaction occurring.*

Focus

Heat and entropy changes determine whether reactants will give products.

It is the nature of physical and chemical systems to attain the lowest possible energy. Because of its position, water at the top of a hill has more energy than water at the bottom of a hill. This energy of position is released as water flows downhill. A wound spring is more energetic than an unwound spring. When the spring unwinds, energy is released. Similarly, chemical reactions proceed when reactants can form products of lower energy than the reactants. In this section we will explore the two factors that together determine whether a reaction will occur. The first of these factors is *heat changes*.

Heat changes

Recall from Section 1.11 that chemical reactions that release heat, a form of energy, are referred to as *exothermic* reactions. Many reactions release heat when they occur. Body temperature is 37 °C because of heat released by exothermic reactions. There are many other examples of exothermic reactions, such as the combustion of carbon. Charcoal, which is almost

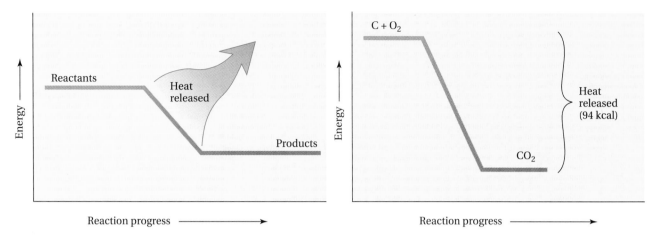

Figure 6.6 (a)
Energy diagram for an exothermic reaction. The energy in the bonds of both carbon, C, and oxygen, O_2, is higher than the energy in the bonds of carbon dioxide, CO_2. Therefore, heat is released when C burns in O_2 to form CO_2.

pure carbon, burns. Each mole of carbon (12.0 g) burned in the combustion of charcoal releases 94 kcal of heat:

$$C(s) + O_2(g) \longrightarrow CO_2(g) + 94 \text{ kcal}$$

The heat released in the combustion of carbon must have a source, since the law of conservation of energy states that energy is neither created nor destroyed in an ordinary physical or chemical process. The heat source is the energy stored in oxygen and carbon—or, more specifically, the energy stored in oxygen-oxygen bonds that hold oxygen molecules together and carbon-carbon bonds that hold atoms of carbon together. In other words, the energy stored in the carbon-oxygen bonds of carbon dioxide (the product of the reaction) is less than the energy stored in the oxygen-oxygen bonds of oxygen and the carbon-carbon bonds of carbon (the reactants). When carbon is burned, some of the energy that was stored in the reactants is released as heat. Figure 6.6(a) displays this concept in an *energy diagram.* Energy diagrams show the relative energies of the reactants and products and the amount of heat that is absorbed or released when the reactants are transformed to products.

PRACTICE EXERCISE 6.10

When hydrogen burns in oxygen, energy is released as heat. Draw an energy diagram for this reaction if the balanced equation is

$$2H_2(g) + O_2(g) \longrightarrow 2H_2O(l) + 116 \text{ kcal}$$

Although the preceding discussion might seem to suggest that the release of heat is necessary for reactants to be converted to products, many chemical reactions are *endothermic,* or heat-absorbing. An example of an

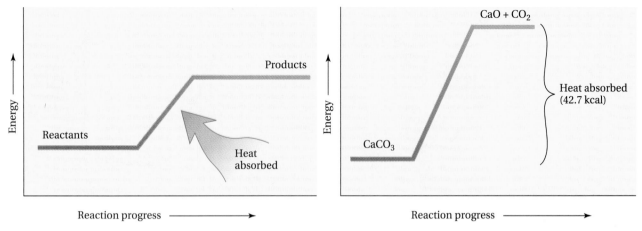

Figure 6.6 (b)
Energy diagram for an endothermic reaction. The energy in the bonds of calcium carbonate, $CaCO_3$, is lower than the energy in the bonds of calcium oxide, CaO, and carbon dioxide, CO_2. Therefore, heat is absorbed when $CaCO_3$ is decomposed to CaO and CO_2.

endothermic chemical reaction is the decomposition of calcium carbonate to calcium oxide and carbon dioxide at 850 °C; 42.7 kcal of heat is absorbed for each mole of calcium carbonate decomposed (Fig. 6.6b):

$$CaCO_3(s) \ + \ 42.7 \text{ kcal} \ \xrightarrow{850\,°C} \ CaO(s) \ + \ CO_2(g)$$

| Calcium carbonate | Calcium oxide | Carbon dioxide |

An example of a *physical* process that is endothermic is the melting of ice. One mole of ice (18.0 g) at 25 °C absorbs 1.44 kcal of heat from its surroundings as it melts from a solid to a liquid:

$$H_2O(s) + 1.44 \text{ kcal} \longrightarrow H_2O(l)$$

Since many processes release heat and many others absorb heat, heat changes alone cannot govern whether a chemical reaction or physical process will occur. Besides heat changes, we must consider *entropy*.

Entropy

Entropy
en- (Greek): in
tropos (Greek): direction

Physical and chemical systems tend to move in the direction of maximum chaos or disorder. **Entropy** *is a measure of the disorder of a system.* A handful of marbles is ordered in the sense that all the marbles are collected in one place, in the hand. It is highly improbable that when permitted to fall, the marbles will land in the same arrangement. Instead, the marbles scatter in all directions—their entropy will increase. Entropy operates in a similar way at the level of atoms and molecules. On an atomic and molecular scale:

1. The entropy of a gas is greater than the entropy of a liquid or a solid at the same temperature, and the entropy of a liquid is usually greater than that of a solid. Entropy therefore increases in reactions in which solid

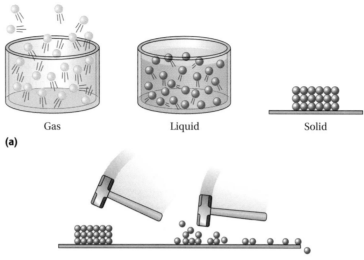

Figure 6.7
Entropy is a measure of the disorder of a system.
(a) The entropy of a gas is greater than the entropy of a liquid or a solid.
(b) Entropy increases when a substance is divided into parts.
(c) Entropy increases with an increase in temperature.

reactants give liquid or gaseous products or in which liquid reactants give gaseous products.

2. Entropy increases when a substance is divided into parts. Entropy therefore increases in chemical reactions in which the total number of products is greater than the total number of reactants.

3. Entropy increases with a temperature increase because the motions of molecules become more chaotic. Molecules are therefore more disordered at high temperatures than at low temperatures.

Figure 6.7 illustrates these concepts.

Living cells, which are highly organized structures, constantly expend energy to overcome the disorganizing effects of entropy. When they lose this capacity, they die. From this point of view, the difference between life and death is one of organization.

PRACTICE EXERCISE 6.11

Which of the following systems has the higher entropy?

(a) a new pack of playing cards or playing cards in use

(b) salt dissolved in water or a crystal of salt

(c) a cup of water at room temperature or a cup of water at 75 °C

(d) a 1-g sugar crystal or 1 g of powdered sugar

Table 6.3 How Changes in Energy and Entropy Affect a Reaction

Heat	Entropy	Will reaction occur?
decreases (exothermic)	increases (more disorder in products than in reactants)	always
increases (endothermic)	increases	depends on whether unfavorable energy change is offset by favorable entropy change
decreases (exothermic)	decreases (less disorder in products than in reactants)	depends on whether unfavorable entropy change is offset by favorable energy change
increases (endothermic)	decreases	never

Chemical reactions tend to occur whenever one or more of the factors just described make the entropy of the products greater than the entropy of the reactants. In a given reaction, therefore, both the heat and entropy changes may act in a favorable direction, act in an unfavorable direction, or act in opposition to each other. Table 6.3 summarizes these ideas on how heat and entropy affect chemical reactions.

Our first example of a reaction, the combustion of carbon, has both the heat change and the entropy change in its favor; the reaction is exothermic, *and* the entropy increases as one of the reactants, *solid* carbon, is converted to *gaseous* carbon dioxide. Hence the reactants go to products.

The other chemical reaction mentioned, the decomposition of solid calcium carbonate to give calcium oxide and carbon dioxide, shows the effect of temperature on a reaction. In this reaction there is an entropy increase, since two products, one of which is a *gas,* are formed from one reactant that is a *solid.* However, the entropy increase is not great enough to offset the highly unfavorable heat change at ordinary temperatures, so the reaction does not occur below 850 °C. At 850 °C (and higher temperatures), however, the entropy of the products relative to the reactants increases to the point where it offsets the unfavorable heat change, and so the reaction occurs.

Heat and entropy changes work in opposition in the melting of ice. The process is endothermic but still occurs because an unfavorable change, the absorption of heat, is more than offset by a favorable entropy change, the increase in entropy as *solid* water (ice) goes to *liquid* water.

6.6 Free energy

AIM: To describe the change in free energy in an exergonic reaction.

Focus

Free energy is a measure of energy available for work.

In many processes, some energy is released that becomes available to do work. *The energy made available for work by a process is called* **free energy.** Free energy released by water that is flowing downhill can be used to drive a water wheel, and free energy released by a spring that is winding down can be used to turn the hands of a clock. Free energy is also released in many

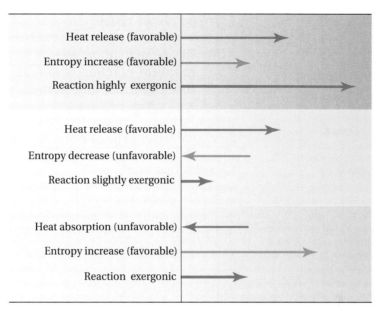

Figure 6.8
The relationship among heat, entropy, and free energy

chemical reactions. *Reactions that release free energy are called* **exergonic** *reactions*. The free energy changes that occur in chemical reactions include both the heat and entropy changes that occur when reactants are converted to products. The use of free energy allows us to discuss energy changes in chemical reactions without knowing the details of the heat and entropy changes that occur. However, the size and direction of these changes determine whether a reaction occurs, as described previously, and they also determine the *amount* of free energy released in an exergonic reaction. For example, a reaction that proceeds with a large release of heat and a large increase in entropy is more exergonic (Fig. 6.8) than one in which, say, a large release of heat is nearly offset by a large decrease in entropy (Fig. 6.8). Free energy is the most important energy quantity in living cells because it is produced and harnessed by cells for the work they must do to stay alive. Table 6.4 gives the free energy values for some exergonic reactions.

Table 6.4 Free Energy Values for Some Exergonic Reactions

Reaction	Free energy (kcal/mol)
$H_2(g)$ + $Cl_2(g)$ $\longrightarrow$ $2HCl(g)$ Hydrogen Chlorine Hydrogen chloride	45.6
$S(s)$ + $O_2(g)$ $\longrightarrow$ $SO_2(g)$ Sulfur Oxygen Sulfur dioxide	71.7
$C_6H_{12}O_6(s)$ + $6O_2(g)$ $\longrightarrow$ $6CO_2(g)$ + $6H_2O(l)$ Glucose Oxygen Carbon dioxide Water	686.0

6.7 Reaction rate

**AIM: *To use activation energy to explain how the rate of a
chemical reaction is influenced by temperature, catalysts,
concentration, and particle size of reactants.***

We now have some idea of what causes a reaction to occur. If it does occur,
how fast will it go? The **rate of reaction**— *the number of atoms, ions, or molecules that react in a given time to form products*—depends on the reaction.
We know that iron is oxidized to iron(III) oxide (rust) and that carbon (or
coal) is oxidized to carbon dioxide, but iron is converted to rust much faster
than coal is converted to carbon dioxide at room temperature. Iron exposed
to moist air gets a thin coat of rust in a few days, but a chunk of coal remains
unchanged even after many centuries. To see why reactions occur at different rates, we must consider the factors that influence rates.

Activation energy

Most chemical reactions involve the transfer of atoms from one molecule to
another, so contact between the reactants is very important. For any reaction, the reacting particles must collide. We might assume that the more
often they collide, the faster the reaction should go. However, chemical
reactions also involve chemical bond breaking and bond making—
processes that involve energy changes. The colliding particles must have
sufficient energy to enable these processes to occur. Otherwise, the particles collide without reacting. *The minimum energy the colliding particles
must have in order to react is the* **activation energy.** Activation energy is a
barrier that the reactants must cross to be converted to products (Fig. 6.9).

The concept of activation energy explains why carbon does not change
but iron rusts in air at room temperature. With carbon, the necessary activation energy is so high, compared with the energy available at room tem-

Figure 6.9
The activation energy barrier must
be crossed before reactants are
converted to products.

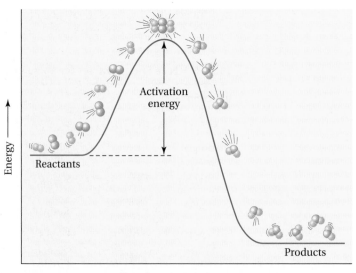

perature, that collisions of oxygen with carbon are insufficiently energetic to break oxygen-oxygen and carbon-carbon bonds. Virtually none of the collisions is energetic enough to surmount the energy barrier; the reaction rate is essentially zero. Iron reacts at room temperature because the activation energy barrier for its oxidation is lower than the barrier for the oxidation of carbon. Even at a low temperature, some collisions between iron and oxygen are energetic enough for the reactants to give products.

Without outside intervention, some chemical reactions are too slow to be practical, and others are too fast to be convenient. Imagine how useless coal would be if people had to wait for it to release its energy of combustion at room temperature, or how we could benefit if our cars rusted 10 times more slowly. Fortunately, there are several ways to affect the rates of chemical reactions.

Effect of temperature on reaction rates

Raising the temperature normally speeds up a reaction; lowering it usually slows it down (Fig. 6.10). At higher temperatures, the motions of the reactants are more chaotic and more energetic than at lower temperatures. Thus the collision frequency increases, and more and more colliding molecules become energetic enough to overcome the activation energy barrier to form products.

As mentioned, coal does not burn at a measurable rate at room temperature. If we supply an "energy prod" in the form of heat, however, the result is dramatic. When we hold a flame to coal, the reactants (carbon and oxygen) collide with greater frequency; some of these collisions are at a high enough energy that the product (carbon dioxide) forms. The heat released by the reaction then supplies enough energy to get more carbon and oxygen over the activation energy barrier. When the flame is removed, the reaction continues—the coal burns—as long as there is a supply of coal and oxygen.

PRACTICE EXERCISE 6.12
Refrigerated food stays fresh for long periods of time, but the same food stored at room temperature spoils quickly. Why?

Figure 6.10
The effect of temperature on reaction rate. The reaction at the higher temperature (at right) is the faster of the two.

FOLLOW-UP TO THE CASE IN POINT: Effect of cold on the body

At the beginning of this chapter we learned about Bill, a homeless person exposed to the cold. Bill's exposure to cold weather lowered his body temperature from a normal 37 °C (98.6 °F) to 34 °C (93 °F). Bill was suffering from general hypothermia, depression of the inner body temperature. Hypothermia is often fatal, especially when the body's internal temperature drops below 34 °C. Bill recovered from his hypothermia when he warmed up at the hospital.

Hypothermia is an example of the effect of temperature on chemical reactions. Our bodies can be thought of as massive collections of chemical reactions. When the air temperature is mild, these reactions produce enough heat to keep the body at 37 °C. In cold weather, more of this body heat is transferred to the air. If the heat produced by chemical reactions cannot keep up with the heat loss, a vicious cycle can set in. As the body's temperature begins to drop, the heat-producing reactions slow down. Since less heat is produced in a given time, the body's temperature drops still further. The result of this cycle can lead to death by hypothermia (see figure).

A person's susceptibility to hypothermia varies with such factors as physical condition and age. Senior citizens are more susceptible than young adults. Seniors generally produce less heat than younger people. The heat that is generated may be less efficiently transferred to the extremities because of impaired circulation. The risk of hypothermia in people of all ages can be reduced by dressing to keep body heat from escaping in cold weather. Layers of dry clothing should be worn with mittens covering the hands. Air spaces between the layers help to insulate the body from cold. Since about 30% of the heat that escapes the body is lost through the head, hoods and face masks are helpful as well.

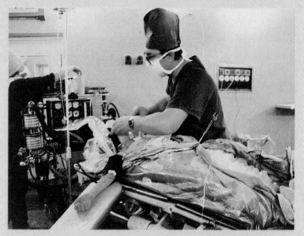

In some medical procedures, such as open heart surgery, a patient's body temperature is deliberately lowered 10 °C or more. Normally, the brain is damaged if its oxygen supply is stopped for 5 minutes; at the lower temperature, however, the heart can be stopped for an hour or more without brain damage due to oxygen starvation.

Effect of catalysts

An increase in temperature is not always the best solution for increasing the rate of a reaction. A catalyst, if one can be found, is often better. **Catalysts** *are substances that speed up reactions by providing reactants with a reaction path of lower activation energy than they could normally take* (Fig. 6.11). *Catalysts neither are used up in a reaction nor appear as reactants or products in the equation for the reaction.* Presented with a lower-energy pathway, a larger fraction of reactants at a given temperature can cross the activation energy barrier to products. The reaction of hydrogen and oxygen at room temperature is negligible, for instance, but with a trace of catalyst (finely divided platinum metal is often used) the reaction is rapid:

$$2H_2(g) + O_2(g) \xrightarrow{Pt} 2H_2O(l)$$

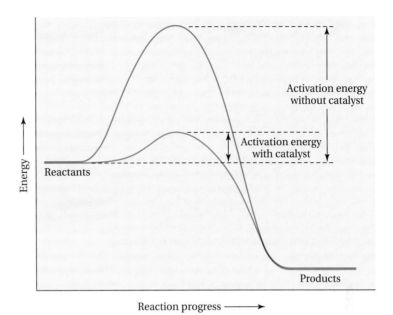

Figure 6.11
A catalyst increases the rate of a reaction by lowering the activation energy.

The concentration of clinically significant enzymes found in the blood can be determined by measuring the disappearance of reactants or the appearance of products of reactions these enzymes catalyze. When doing these measurements, the temperature must be regulated closely because the reaction rate will vary with the temperature.

Air enriched with oxygen, or sometimes pure oxygen, is used in hospitals for patients who have respiratory problems. Materials that burn in air will burn faster in oxygen-enriched air or pure oxygen. Where oxygen is in use, precautions must be taken against the generation of sparks, and of course, smoking is banned.

Although catalysts are very useful for speeding up reactions, they *cannot* make a reaction proceed if it would not do so in the absence of a catalyst. Catalysts called *enzymes* are required for increasing the rates of biological reactions. The temperature of our bodies is held at 37 °C and cannot be raised or lowered significantly without imperiling our lives. Yet without catalysts, few reactions in the body would proceed fast enough to be useful at this temperature. We could not live without enzymes to catalyze these reactions. A Closer Look: Catalase describes an enzyme that plays an important role in keeping body cells healthy.

Effect of concentration

The number of reacting particles in a given volume affects the rate at which reactions occur. Increasing the concentration of the reactants—cramming more particles into a fixed volume—increases the collision frequency and therefore the reaction rate. An ember in a campfire illustrates the effect. An ember glows dully in air, which is 20% oxygen, but if plunged into pure oxygen, it would immediately burst into flame. A lit cigarette would behave much like an ember. This is why smoking is forbidden wherever oxygen is being used.

Particle size and surface area

The total surface area of a solid or liquid reactant has an important effect on reaction rate. The larger the surface area for a given mass, the greater is the collision frequency, and the faster is the reaction. A whole log burns more slowly than it would burn if it were chopped into many small pieces, because the pieces have a larger total surface area. The best way to increase

A Closer Look

Catalase

Catalase is a large molecule of the family of biological compounds called *enzymes*. Like all enzymes, it is present in the body in very small quantities. Catalase is one of several enzymes that catalyze the decomposition of hydrogen peroxide (H_2O_2). Hydrogen peroxide is formed in body cells as a natural waste product of reactions of oxygen. However, even small amounts of hydrogen peroxide in the cells can inflict severe damage or even cause death. Catalase molecules prevent the damage by catalyzing the breakdown of hydrogen peroxide to water and oxygen:

$$2H_2O_2(aq) \xrightarrow{\text{Catalase}} 2H_2O(l) + O_2(g)$$

The catalytic action of catalase is readily visible in the use of hydrogen peroxide for the treatment of minor cuts and abrasions. A 3% solution of hydrogen peroxide is commonly used for this purpose. The solution is stable until it is applied to the wound. The peroxide preparation begins to foam the instant it is applied to the wound as the catalase in the blood in the wound catalyzes the decomposition of the hydrogen peroxide (see figure). The foaming is caused by the release of oxygen gas. The antiseptic (bacteria-killing) properties of hydrogen peroxide are thought to be due to the oxygen, which is initially generated in a particularly reactive form called *active oxygen.*

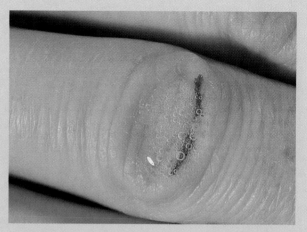

When hydrogen peroxide solution is poured onto an open wound (or a piece of fresh ground beef is dropped into a solution of hydrogen peroxide), the enzyme catalase in the blood rapidly decomposes the hydrogen peroxide to water and active oxygen.

Catalase molecules contain iron(III) ions. These transition metal ions are responsible for the rapid catalysis of the breakdown of hydrogen peroxide. When the iron(III) ions are removed from the protein part of catalase, the protein molecules lose their catalytic activity. The decomposition of hydrogen peroxide is also catalyzed by iron(III) ions in the absence of an enzyme. For example, iron(III) chloride ($FeCl_3$) and iron(III) nitrate ($Fe(NO_3)_3$) catalyze the decomposition of hydrogen peroxide in the laboratory.

the surface area of solid reactants is usually by dissolving them and thus separating the particles. For example, a marble on the floor is more accessible than a marble in the middle of a box of marbles. Homogeneous mixtures of reactants usually react more rapidly than heterogeneous mixtures. When forced to use heterogeneous mixtures of reactants, we can often grind solids to a fine powder first. Chemical reactions can be slowed down by reducing the surface area of the reacting particles. Automobile manufacturers paint and undercoat cars to reduce the expanse of metal surface exposed to the elements and thereby slow down their products' inevitable corrosion.

PRACTICE EXERCISE 6.13

Name at least two factors that influence the rate of a chemical reaction, and explain these effects using collisions between reactants.

6.8 Reversible reactions

AIMS: *To explain chemical equilibrium in terms of a reversible reaction. To write the expression for the equilibrium constant K_{eq} of a reaction. To predict the equilibrium position of a reaction from a given K_{eq} value. To state Le Châtelier's principle and use it to predict changes in the equilibrium position due to changes in concentration and temperature.*

Focus

Some chemical reactions are reversible.

So far it seems reasonable that all the reactants in a reaction are used up and become products. For all practical purposes, this is true of very exergonic reactions. However, some reactions are only slightly exergonic. In these reactions, not all the reactants are converted to products. One example of this kind of reaction is the reaction of sulfur dioxide with oxygen to give sulfur trioxide, an industrial chemical used to make sulfuric acid (H_2SO_4):

$$2SO_2(g) \ + \ O_2(g) \ \xrightarrow{\text{Pt}} \ 2SO_3(g)$$

Sulfur   Oxygen   Sulfur
dioxide   trioxide

This reaction proceeds very slowly below 500 °C but is quite fast in the presence of a platinum catalyst at 725 °C. When 2 mol SO_2 and 1 mol O_2 are mixed at 725 °C, the amount of SO_3 formed is always less than the expected 2 mol. A mixture of SO_2, O_2, and SO_3 is obtained. In a second experiment, a mixture of SO_2, O_2, and SO_3 of composition identical to that of the first experiment is obtained when 2 mol SO_3 is kept at 725 °C. The first experiment shows that SO_2 and O_2 react to form a certain amount of SO_3. The second experiment indicates that SO_3 decomposes to SO_2 and O_2 at the same temperature as the first experiment. Both the formation of SO_3 and its decomposition occur simultaneously. The formation of SO_3 from SO_2 and O_2 is an example of a reversible reaction. *In* **reversible reactions** *the conversion of reactants into products (the forward reaction) and the conversion of products into reactants (the reverse reaction) occur simultaneously.* Two arrows are used to show that a reaction is reversible.

$$2SO_2(g) \ + \ O_2(g) \ \underset{}{\overset{\text{Pt}}{\rightleftharpoons}} \ 2SO_3(g)$$

Sulfur   Oxygen   Sulfur
dioxide   trioxide

Reversibility and chemical equilibrium

Let's consider what happens when SO_2 and O_2 are mixed at 725 °C. The SO_2 and O_2 molecules collide, and SO_3 molecules are formed. The rate of the reverse process, the decomposition of SO_3 to SO_2 and O_2 is zero at first, since there is no SO_3 to decompose to SO_2 and O_2. As the amount of SO_3 builds up by the forward reaction, however, some of the SO_3 molecules decompose to SO_2 and O_2 by the reverse reaction. As the SO_2 and O_2 are used up, their amount decreases. Fewer collisions occur between them, and

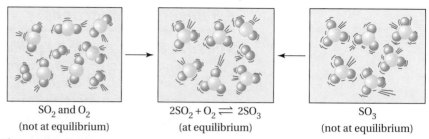

SO$_2$ and O$_2$ 2SO$_2$ + O$_2$ ⇌ 2SO$_3$ SO$_3$
(not at equilibrium)   (at equilibrium)   (not at equilibrium)

Figure 6.12
Molecules of SO$_2$ and O$_2$ react to give SO$_3$; molecules of SO$_3$ decompose to give SO$_2$ and O$_2$. When the rates of the forward and reverse reactions are the same, molecules of SO$_2$, O$_2$, and SO$_3$ are present in the mixture and a state of dynamic chemical equilibrium exists.

Equilibrium
aequilibrium (Latin): in balance

so the forward reaction slows down. As the amount of SO$_3$ builds up, the reverse reaction speeds up. Eventually, the forward and reverse rates are the same; as much of the products are going to reactants as reactants are going to products. At this point, the reaction has reached **chemical equilibrium**—*a dynamic state in which the forward and reverse reactions are taking place at the same rate*—but there is *no change* in the actual amounts of reactants and products in the system (Fig. 6.12). Although our example of equilibrium involves a fairly simple chemical reaction, other equilibria have an impact on the entire world, as discussed in A Closer Look: The Rain Forest, Equilibrium, and Medicine.

Equilibrium position

We usually say that an equilibrium reaction *favors* the products when the products are more abundant than the reactants. In the following equation, a reactant, A, is converted to a product, B, in a reversible reaction:

A ⇌ B
1%   99%

At equilibrium, 0.01 mol of A and 0.99 mol of B are present. The product, B, is favored because it is more abundant than the reactant, A, at equilibrium. The longer arrow indicates the direction of the predominant reaction.

If at equilibrium the mixture contains 0.99 mol of A and 0.01 of B, however, the *reactants* are favored, as indicated in the following equation:

A ⇌ B
99%   1%

*The **equilibrium position** of a reaction—the relative amounts of reactants and products present at equilibrium—*indicates whether the reactants or the products are favored in a reversible reaction.

In principle, all reactions are reversible. That is, reactants go to products in the forward direction and products go to reactants in the reverse direction. In practice, however, the products of highly exergonic reactions are so favored over reactants that the reactions are essentially irreversible (only products can be detected at equilibrium). Some reactions may give such small amounts of products that the products are not detectable.

A Closer Look

The Rain Forest, Equilibrium, and Medicine

What connection can there be between equilibrium processes, the tropical rain forest, and medicines used to treat diseases? In brief, an equilibrium process helps keep atmospheric carbon dioxide tolerably low and the rain forests—a potential source of useful medicines—healthy.

Life on Earth depends on a careful balance that has evolved over eons. This balance results from equilibria involving many chemical and physical processes. For example, an equilibrium between atmospheric carbon dioxide and the carbon compounds in plants helps promote the healthy growth of the massive tropical rain forests while it simultaneously reduces the amount of atmospheric carbon dioxide (see figure). Some scientists believe that the uptake of atmospheric carbon dioxide by plants is crucial to keeping the Earth's surface from becoming warmer.

Healthy tropical rain forests have a significance beyond the equilibrium between atmospheric carbon dioxide and carbon compounds in plants. Vast numbers of plants and insects that live exclusively in these regions could be sources of new medicines. Some estimates indicate that only about 2% of all plants and insects that live in the tropical rain forests have been discovered. Few of the known organisms have been tested to see if they contain substances that might be of medical benefit. The importance of plants in medicine is indicated by the fact that fewer than 3% of all flowering plants on Earth have been examined for medicinally useful substances, yet about 40% of drugs currently in use are derived from living organisms. The salicylates, from which aspirin derives, were originally found in the willow and

The rain forest, a source of current and future medicinals.

meadowsweet plants. Morphine, an important painkiller, is derived from the opium poppy. Digitalis, an important drug in the treatment of heart disease, comes from the foxglove plant. Vincristine and vinblastine, two anticancer drugs effective against Hodgkin's disease and a form of leukemia, come from a plant found in Madagascar known as the rosy periwinkle.

Today, rain forests are being destroyed at an alarming rate. At the same time, the burning of fuels in cars and for electricity releases record amounts of carbon dioxide into the atmosphere. Ecologists and other scientists are concerned that deforestation of the tropical rain forests diminishes the forests' ability to handle the additional carbon dioxide load. They are also concerned that the deforestation leads to the irretrievable loss of untested organisms every day. These are compelling reasons to halt the destruction of rain forests for agriculture, timbering, and other development.

Chemists usually say there is "no reaction" for these reactions. Slightly exergonic reactions are reversible, and products go to reactants only to a certain extent.

Catalysts speed up the forward and reverse reactions equally, since the activation barrier is lowered by the same amount in both the forward and reverse directions. Therefore, catalysts do not affect the amounts of reactants and products present at equilibrium, but they do decrease the time it takes for equilibrium to be established.

Shifting the equilibrium position

There is a delicate balance between reactants and products in a system at equilibrium. Any change in the conditions of the system constitutes a *stress* on the system and disrupts this balance. Adding or removing reactants and products, increasing or lowering the temperature, or changing pressure are examples of stresses that will disrupt the equilibrium position. A system's response to a stress is predicted by **Le Châtelier's principle:** *If a system in dynamic equilibrium experiences a change in conditions, the system changes to relieve the stress.* The ways that systems respond to two types of stress that are easily applied to systems in chemical equilibrium—changes in the concentration of reactants and products and changes in temperature—are described as follows:

1. *Changes in concentration of reactants and products.* Changing the amount of any reactant or product in a system at equilibrium disturbs the equilibrium position. The system responds to this stress by trying to restore the original equilibrium. When carbon dioxide dissolves in water, for example, it reacts with the water to a slight extent to form a substance known as *carbonic acid:*

$$CO_2(aq) + H_2O(l) \rightleftharpoons H_2CO_3(aq)$$

| Carbon dioxide | Water | Carbonic acid |

Mostly reactants (CO_2 and H_2O) are present at equilibrium; the amount of carbonic acid is only about 1% of the total number of carbon-containing molecules CO_2 and H_2CO_3:

$$CO_2 + H_2O \rightleftharpoons H_2CO_3$$

99%   1%

We could disturb the equilibrium by adding more carbon dioxide. Suppose we add enough carbon dioxide so that the amount of carbon dioxide increases to 99.5% and the carbonic acid is only 0.5%. As soon as we do, the carbon dioxide reacts with water to form more carbonic acid and restore it to the original 1%. In other words, *adding a reactant always pushes a reversible reaction in the direction of products.* If we remove carbon dioxide, the amount of carbon dioxide decreases to less than 99% and carbonic acid is greater than 1%. Carbonic acid rapidly decomposes to carbon dioxide and water as the system readjusts itself to 1% CO_2 and 99% H_2CO_3. *Removing a reactant always pulls a reversible reaction in the direction of reactants.*

Our bodies use the removal of reactants to keep carbonic acid at low, safe levels in the blood. Blood contains dissolved carbonic acid in equilibrium with carbon dioxide and water. When we exhale carbon dioxide, the equilibrium shifts toward carbon dioxide and water, reducing the amount of carbonic acid. The same principle applies to adding or removing products: *By adding a product to a system at equilibrium, we push a reaction in the direction of formation of reactants; by removing a product, we pull a reaction in the direction of formation of products.*

Chemists often remove products to increase the yield of a desired product. As products are removed from a reaction mixture, the system continually attempts to readjust itself to restore equilibrium with the products. Because they are being removed as fast as they are formed, however, the reaction never builds up enough products to establish an equilibrium. The reactants continue to give products until they are completely used up.

2. *Changes in temperature.* Increasing the temperature causes the equilibrium position of a reaction to shift in the direction that absorbs heat. The production of SO_3 is an exothermic reaction as written:

$$2SO_2(g) + O_2(g) \rightleftharpoons 2SO_3(g) + \text{heat}$$

Heat can be considered as a product of this reaction. The addition of heat to the reaction causes the decomposition of the SO_3 into SO_2 and O_2. The equilibrium position is pushed to the left in favor of the reactants, and the yield of the product, SO_3, decreases. The removal of heat by cooling the reaction pulls the equilibrium to the right, and the product yield increases.

PRACTICE EXERCISE 6.14

What effect does lowering the temperature have on the equilibrium position in the decomposition of calcium carbonate? The reaction is

$$CaCO_3(s) + \text{heat} \rightleftharpoons CO_2(g) + CaO(s)$$

Equilibrium constants

Consider a hypothetical reaction in which a mol of reactant A and b mol of reactant B react to give c mol of product C and d mol of product D at equilibrium:

$$aA + bB \rightleftharpoons cC + dD$$

Since the reaction is at equilibrium, there is no *change* in the amounts of A, B, C, or D at any given instant.

An expression relating the amounts of reactants and products is

$$K_{eq} = \frac{[C]^c[D]^d}{[A]^a[B]^b}$$

The symbol K_{eq} is an **equilibrium constant**—*a number that relates the relative amounts of reactants and products in a chemical reaction at equilibrium.* The *square brackets* indicate that the amounts of substances are in *moles per liter.* As the equation shows, K_{eq} is the ratio of product concentrations to reactant concentrations, with each concentration raised to a power given by the number of moles of that substance appearing in the balanced chemical equation. The value of K_{eq} for a reaction depends on the conditions. If the conditions change, the value of K_{eq} also changes.

EXAMPLE 6.15

Writing an equilibrium constant expression

Dinitrogen tetroxide (N_2O_4), a colorless gas, and nitrogen dioxide (NO_2), a reddish brown gas, exist in equilibrium with each other. The chemical reaction is

$$N_2O_4(g) \rightleftharpoons 2NO_2(g)$$

Write the expression for the equilibrium constant.

SOLUTION

Since the reaction is at equilibrium, there is no net change in the amounts of N_2O_4 or NO_2 at any given instant:

$$K_{eq} = \frac{[NO_2]^2}{[N_2O_4]}$$

EXAMPLE 6.16   **Calculating an equilibrium constant**

A liter of a gas mixture at 100 °C at equilibrium contains 0.0045 mol dinitrogen tetroxide and 0.030 mol nitrogen dioxide. Calculate the equilibrium constant K_{eq} for the reaction

$$N_2O_4(g) \rightleftharpoons 2NO_2(g)$$

SOLUTION

Substitute the given data in the equilibrium constant expression of Example 6.15:

$$K_{eq} = \frac{[NO_2]^2}{[N_2O_4]} = \frac{0.030 \text{ mol/L} \times 0.030 \text{ mol/L}}{0.0045 \text{ mol/L}}$$

$$= 0.20 \text{ mol/L}$$

EXAMPLE 6.17   **Calculating an equilibrium constant**

One mole of hydrogen gas and one mole of iodine are sealed in a 1-L flask and allowed to react at 450 °C. At equilibrium, 1.56 mol hydrogen iodide is present. Calculate K_{eq} for the reaction

$$H_2(g) + I_2(g) \rightleftharpoons 2HI(g)$$

SOLUTION

We know from the balanced equation that 1 mol hydrogen and 1 mol iodine form 2 mol hydrogen iodide. To make 1.56 mol hydrogen iodide therefore requires 0.78 mol hydrogen and 0.78 mol iodine. Since we started with 1 mol each of hydrogen and iodine, at equilibrium we have

$$\text{mol } H_2 = 1.00 - 0.78 = 0.22 = \text{mol } I_2$$

Substitute these concentrations in the equation for K_{eq}:

$$K_{eq} = \frac{[HI]^2}{[H_2][I_2]}$$

$$= \frac{1.56 \text{ mol/L} \times 1.56 \text{ mol/L}}{0.22 \text{ mol/L} \times 0.22 \text{ mol/L}}$$

$$= 50.3 = 5.0 \times 10^1 \quad \text{(to two significant figures)}$$

PRACTICE EXERCISE 6.15

Give the equilibrium constant expression for the formation of ammonia from hydrogen and nitrogen. The reaction is

$$3H_2(g) + N_2(g) \rightleftharpoons 2NH_3(g)$$

PRACTICE EXERCISE 6.16

Analysis of an equilibrium mixture of nitrogen, hydrogen, and ammonia contained in a 1-L flask at 300 °C gives the following results: hydrogen, 0.15 mol; nitrogen, 0.25 mol; and ammonia, 0.10 mol. Calculate K_{eq} for the reaction

$$3H_2(g) + N_2(g) \rightleftharpoons 2NH_3(g)$$

PRACTICE EXERCISE 6.17

The decomposition of hydrogen iodide at 450 °C in a 1-L container produces an equilibrium mixture that contains 0.500 mol hydrogen. The equilibrium constant is 0.0200 for the reaction

$$2HI(g) \rightleftharpoons H_2(g) + I_2(g)$$

How many moles of iodine and hydrogen iodide are present in the equilibrium mixture?

Equilibrium constants are valuable chemical information. Among other things, we can use them to tell whether products or reactants are more favored at equilibrium. Since an equilibrium constant is always written as a ratio of products to reactants, a value of K_{eq} greater than 1 means that products are favored over reactants. Conversely, a value of K_{eq} less than 1 means that reactants are favored over products.

PRACTICE EXERCISE 6.18

A chemist has determined the equilibrium constants for several reactions. The results are (a) $K_{eq} = 1 \times 10^4$, (b) $K_{eq} = 5.4$, (c) $K_{eq} = 0.002$, and (d) $K_{eq} = 7 \times 10^{-6}$. In which reactions are the products favored over the reactants?

SUMMARY

The mass of a representative unit of an atom, ion, molecule, or ionic compound expressed in atomic mass units is its formula mass. The formula mass of any substance in grams is 1 mol of the substance. A mole contains 6.02×10^{23} (Avogadro's number) representative units of the substance. One mole of any substance contains the same number of representative units as 1 mol of any other substance. The formula mass in grams of 1 mol of a substance is called the molar mass. Moles are used to do chemical arithmetic, or stoichiometry. Stoichiometric calculations involving chemical reactions begin with a balanced equation. Balanced equations are necessary because mass must be conserved: There must be as many of the same kinds of atoms in the products of a reaction as in the reactants.

Oxidation-reduction, or redox, reactions are an important category of chemical reactions. Oxidation is the gain of oxygen, loss of hydrogen, or loss of electrons by a substance. Reduction involves a loss of oxygen, gain of hydrogen, or gain of electrons. An oxidation reaction is always accompanied by a reduction reaction. The substance that does the oxidizing (is reduced) is called an oxidizing agent; the substance that does the reducing (is oxidized) is called a reducing agent.

The natural tendency is for all things to go to lower energy. Chemical reactions can release energy that becomes available to do work. This energy is called free energy.

Reaction rates are determined by an activation energy, the minimum energy that reactants must have to go to products (or, in the reverse direction, that products must have to go to reactants). The higher the activation energy barrier, the slower is the reaction. Chemists help reactants overcome the activation barrier in a number of ways, but increasing the temperature at which the reaction is done

and using a catalyst are the two most effective methods.

In principle, all reactions are reversible. That is, reactants go to products in the forward direction and products go to reactants in the reverse direction. The point at which the rates of conversion of reactants to products and vice versa are equal is the position of equilibrium. In practice, highly exergonic reactions are irreversible (go completely to products). Slightly exergonic reactions go to products only to a certain extent. Changes in the position of equilibrium may be predicted by applying Le Châtelier's principle. The equilibrium constant K_{eq} of a reversible reaction is essentially a measure of the ratio of products to reactants at equilibrium.

KEY TERMS

Activation energy (6.7)
Avogadro's number (6.1)
Balanced chemical equation (6.2)
Catalyst (6.7)
Chemical equilibrium (6.8)

Entropy (6.5)
Equilibrium constant (6.8)
Equilibrium position (6.8)
Exergonic (6.6)
Formula mass (6.1)
Free energy (6.6)
Le Châtelier's principle (6.8)

Molar mass (6.1)
Mole (6.1)
Oxidation (6.4)
Oxidizing agent (6.4)
Rate of reaction (6.7)
Redox reaction (6.4)
Reducing agent (6.4)

Reduction (6.4)
Reversible reaction (6.8)
Stoichiometry (6.3)

EXERCISES

Chemical Quantities (Section 6.1)

6.19 What is the formula mass (in atomic mass units) of the following substances?
(a) nitrogen, N_2
(b) sodium nitrate, $NaNO_3$
(c) methane, CH_4
(d) magnesium, Mg
(e) nitrogen atom, N
(f) ammonium sulfate, $(NH_4)_2SO_4$

6.20 Find the molar mass of
(a) bromine, Br_2,
(b) neon, Ne,
(c) magnesium chloride, $MgCl_2$,
(d) cane sugar, $C_{12}H_{22}O_{11}$,
(e) calcium phosphate, $Ca_3(PO_4)_2$, and
(f) iron, Fe.

6.21 What is the mass in grams of
(a) 1.00 mol O_2,
(b) 2.50 mol NaCl,
(c) 0.400 mol $Mg(NO_3)_2$,
(d) 5.45 mol Al,
(e) 0.250 mol Na_2SO_4, and
(f) 1.50 mol NH_3?

6.22 Find the mass, in grams, of
(a) 2.48 mol CO_2,
(b) 0.750 mol Au,
(c) 5.40×10^{-2} mol CH_4,
(d) 0.0680 mol SO_2,
(e) 1.64 mol $Ca(OH)_2$, and
(f) 3.25 mol C_3H_8O.

6.23 How many moles is
(a) 141 g $CaCl_2$,
(b) 1.80 g H_2O,
(c) 1.00 g NaOH,
(d) 3.55 g C,
(e) 50.0 g $Ba(NO_3)_2$, and
(f) 2.20 g CO_2?

6.24 Calculate the moles in
(a) 5.32 g S,
(b) 1.45 g K_2O,
(c) 0.0500 g C_2H_6,
(d) 864 g $(NH_4)_2SO_4$,
(e) 5.50 g SiO_2, and
(f) 0.750 g KOH.

Chemical Equations (Section 6.2)

6.25 What is the law of conservation of matter?

6.26 Why must chemical equations be balanced?

6.27 Balance the following equations.
(a) $H_2(g) + Cl_2(g) \longrightarrow HCl(g)$

(b) $Al(s) + HCl(aq) \longrightarrow AlCl_3(aq) + H_2(g)$

(c) $C_2H_6(g) + O_2(g) \longrightarrow CO_2(g) + H_2O(g)$

(d) $CaCO_3(s) + HCl(aq) \longrightarrow CaCl_2(aq) + H_2O(l) + CO_2(g)$

(e) $C_3H_6(g) + O_2(g) \longrightarrow CO_2(g) + H_2O(g)$

(f) $Mg(s) + Cl_2(g) \longrightarrow MgCl_2(s)$

6.28 Rewrite these word equations as balanced chemical equations.
(a) sodium + chlorine $\longrightarrow$ sodium chloride
(b) sulfur + oxygen $\longrightarrow$ sulfur dioxide
(c) calcium chloride + sodium sulfate $\longrightarrow$
      sodium chloride + calcium sulfate
(d) calcium carbonate $\longrightarrow$
      calcium oxide + carbon dioxide

Stoichiometry (Section 6.3)

6.29 Answer the following questions using this balanced equation:

$$3H_2(g) + N_2(g) \longrightarrow 2NH_3(g)$$

(a) How many moles of NH_3 are produced when 5.0 mol H_2 reacts?
(b) How many moles of H_2 are needed to react with 14.4 mol N_2?
(c) How many grams of N_2 are needed to produce 3.4 g NH_3?
(d) How many grams of NH_3 are produced when 2.40 g H_2 reacts?

6.30 The balanced equation for the burning of acetylene gas, C_2H_2, is

$$2C_2H_2(g) + 5O_2(g) \longrightarrow 4CO_2(g) + 2H_2O(l)$$

(a) Calculate the number of moles of CO_2 produced by the reaction of 15.0 mol O_2 with C_2H_2.
(b) How many moles of O_2 are needed to react with 3.0 mol C_2H_2?
(c) How many grams of water are produced when 22 g CO_2 is produced?
(d) Calculate the number of grams of C_2H_2 needed to react with oxygen to produce 6.0 g water.

Oxidation-Reduction (Section 6.4)

6.31 Define the term *redox*.

6.32 Define *oxidation* and *reduction*.

6.33 Identify the reactant that is oxidized and the reactant that is reduced in the following reactions. Name the oxidizing agent and reducing agent in each case.
(a) $Mg(s) + Br_2(l) \longrightarrow MgBr_2(s)$
(b) $2Fe(s) + 3Cl_2(g) \longrightarrow 2FeCl_3(s)$
(c) $Cu(s) + S(s) \longrightarrow CuS(s)$
(d) $2C(s) + O_2(g) \longrightarrow 2CO_2(g)$
(e) $2P(s) + 3H_2(g) \longrightarrow 2PH_3(g)$

6.34 Which of the following would most likely be oxidizing agents, and which would most likely be reducing agents?
(a) Mn^{4+} (b) Cl_2 (c) K   (d) K^+

Heat and Entropy Changes (Sections 6.5, 6.6)

6.35 What is an exothermic reaction? Name an exothermic reaction you have observed today.

6.36 Draw the energy diagram for the combustion of methane in oxygen. The reaction is

$$CH_4(g) + 2O_2(g) \longrightarrow CO_2(g) + 2H_2O(l) + 192\,kcal$$

6.37 Draw an energy diagram for the formation of NO from its elements:

$$N_2(g) + O_2(g) + 43.3\,kcal \longrightarrow 2NO(g)$$

6.38 What is the meaning of the term *entropy*?

6.39 The products in a chemical reaction are more ordered than the reactants. Is this entropy change favorable or unfavorable?

6.40 What is *free energy*?

6.41 Which system has the higher entropy?
(a) 100 g of ice or 100 g of water
(b) a teaspoon of table salt or a teaspoon of table salt dissolved in water
(c) iced tea or hot tea

6.42 Predict the entropy change in each of the following reactions.
(a) $2H_2O_2(l) \longrightarrow 2H_2O(l) + O_2(g)$
(b) $CaCO_3(s) \longrightarrow CaO(s) + CO_2(g)$
(c) $NH_3(g) + HCl(g) \longrightarrow NH_4Cl(s)$
(d) $2NaHCO_3(s) \longrightarrow Na_2CO_3(s) + H_2O(g) + CO_2(g)$

Reaction Rate (Section 6.7)

6.43 What is meant by *rate of reaction?*

6.44 Are chemical reactions always fast? Why? Give examples that prove your point.

6.45 Explain the collision theory of reactions.

6.46 Does every collision between reacting particles lead to products? What other factor is involved?

6.47 How does each of the following factors affect the rate of a chemical reaction?
(a) temperature   (b) particle size

6.48 How is the rate of a reaction influenced by (a) a catalyst and (b) an increase in reactant concentration?

Reversible Reactions and Equilibria (Section 6.8)

6.49 In your own words, define a reversible reaction.

6.50 A reversible reaction has reached a state of dynamic chemical equilibrium. What does this information tell you?

6.51 Predict what will happen if a catalyst is added to a slow reversible reaction. What happens to the equilibrium position?

6.52 What is Le Châtelier's principle? Use it to explain why carbonated drinks go flat when they are left open.

6.53 The industrial production of ammonia is described by the reversible reaction

$$N_2(g) + 3H_2(g) \rightleftharpoons 2NH_3(g) + 22 \text{ kcal}$$

What effect do the following changes have on the equilibrium position?
(a) addition of heat   (b) addition of catalyst
(c) removal of heat

6.54 The equilibrium constants for three reactions are (a) $K_{eq} = 4 \times 10^3$, (b) $K_{eq} = 0.5$, and (c) $K_{eq} = 1 \times 10^{-5}$. Comment on the favorability of product formation in each reaction.

Additional Exercises

6.55 How many molecules are in 2.50 mol hydrogen peroxide, H_2O_2?

6.56 What effect do the following changes have on the position of equilibrium for this reversible reaction?

$$PCl_5(g) + \text{heat} \rightleftharpoons PCl_3(g) + Cl_2(g)$$

(a) addition of Cl_2 (b) removal of heat
(c) removal of PCl_3 as it is formed

6.57 Balance the following equations.
(a) $BaO_2(s) \longrightarrow BaO(s) + O_2(g)$
(b) $CS_2(l) + Cl_2(g) \longrightarrow CCl_4(l) + S_2Cl_2(g)$
(c) $(NH_4)_2Cr_2O_7(s) \longrightarrow Cr_2O_3(s) + H_2O(g) + N_2(g)$
(d) $MnO_2(s) + HCl(aq) \longrightarrow MnCl_2(aq) + H_2O(l) + Cl_2(g)$
(e) $CO(g) + Fe_2O_3(s) \longrightarrow Fe(s) + CO_2(g)$
(f) $Ba(CN)_2(aq) + H_2SO_4(aq) \longrightarrow BaSO_4(s) + HCN(aq)$
(g) $Li(s) + O_2(g) \longrightarrow Li_2O(s)$

6.58 Calculate the mass of (a) 2.40 mol NaOH, (b) 0.780 mol $Ca(CN)_2$, (c) 1.93 mol $BaSO_4$, and (d) 0.045 mol O_2.

6.59 Bromine chloride, BrCl, decomposes to form chlorine and bromine:

$$2BrCl(g) \rightleftharpoons Cl_2(g) + Br_2(g)$$

At a certain temperature the equilibrium constant for the reaction is 11.1, and the equilibrium mixture contains 4.00 mol Cl_2. How many moles of Br_2 and BrCl are present in the equilibrium mixture?

6.60 Explain each of the following based on factors that determine the rate of a reaction. (a) A campfire is "fanned" to help get it going. (b) An explosion in a grain elevator is blamed on dust. (c) A pinch of powdered manganese dioxide causes hydrogen peroxide to decompose rapidly to water and oxygen even though the manganese dioxide is not itself changed.

6.61 How many moles is each of the following?
(a) 937 g $Ca(C_2H_3O_2)_2$ (b) 79.3 g Cl_2
(c) 5.96 g KOH   (d) 37.3 g C_6H_6O

6.62 Would the formation of reactants or products be favored in each of the following reactions?
(a) $H_2(g) + F_2(g) \rightleftharpoons 2HF(g)$ $K_{eq} = 1 \times 10^{13}$
(b) $2H_2O(g) \rightleftharpoons 2H_2(g) + O_2(g)$ $K_{eq} = 6 \times 10^{-28}$

6.63 Carbon disulfide, which is important industrially as a solvent, is prepared by the reaction of coke with sulfur dioxide:

$$5C(s) + 2SO_2(g) \longrightarrow CS_2(l) + 4CO(g)$$

(a) How many moles of CS_2 form when 6.30 mol carbon reacts?
(b) How many moles of carbon are needed to react with 7.24 mol SO_2?
(c) How many moles of SO_2 are required to make 182 mol CS_2?

6.64 The reactant Y in a hypothetical reversible chemical reaction is a bright yellow color, and the product B is a bright blue. The equation for the reaction is Y $\rightleftharpoons$ B; 1 mol of Y reacts to give 1 mol of B. In each instance you will mix an equal number of moles of Y and B. What will be the color of the reaction mixture at equilibrium assuming the following?
(a) $K_{eq} < 1 \times 10^{-3}$
(b) $K_{eq} = 1$
(c) $K_{eq} > 1 \times 10^{3}$

6.65 A yellow gas Y reacts with a colorless gas C to produce a blue gas B according to this equation:

$$C(g) + Y(g) \rightleftharpoons 3B(g) + \text{heat}$$

The reaction is exothermic and slightly exergonic. The equilibrium constant is slightly larger than 1. The system is initially at equilibrium and is green in color. What, if anything, happens to the equilibrium and the color of the system when each of the following *stresses* is applied to the system?
(a) adding a large quantity of C
(b) cooling the system
(c) adding a catalyst
(d) removing B from the system

SELF-TEST (REVIEW)

True/False

1. An atom of neon is heavier than an atom of argon.
2. Equations are balanced by changing the subscripts of the formulas in the equation.
3. Atoms are conserved in a chemical reaction.
4. The condensation of steam to liquid water is an exothermic process.
5. When a sandwich is eaten and digested, its entropy increases.
6. An endothermic reaction absorbs heat.
7. The speed of a reaction can be increased by increasing reactant concentration or decreasing particle size.
8. In an exothermic reaction, the energy of the reactants is lower than the energy of the products.
9. The K_{eq} for a certain reaction is 2×10^{-7}. For this reaction at equilibrium, the concentration of the reactants is greater than the concentration of the products.
10. Oxygen is a reactant in many oxidation reactions.

Multiple Choice

11. The following equation

$$C_3H_8(g) + 5O_2(g) \longrightarrow 3CO_2(g) + 4H_2O(g) + \text{heat}$$

tells us all the following except
(a) 1 mol C_3H_8 reacts with oxygen to give 4 mol H_2O.
(b) when 3 g CO_2 is produced, 4 g H_2O is produced.
(c) 44 g C_3H_8 reacts with 160 g O_2.
(d) heat is given off in this reaction.

12. The activation energy of a reaction
(a) is not affected by a catalyst.
(b) can be considered an energy barrier.
(c) is always higher in an endothermic reaction than in an exothermic reaction.
(d) is usually the same for a forward and a reverse reaction.

13. In the balanced equation for the formation of sodium chloride from its elements

sodium + chlorine $\longrightarrow$ sodium chloride + heat

(a) the reaction is endothermic.
(b) the formula for chlorine is Cl.
(c) the coefficient 3 is used in front of Na.
(d) none of the above.

14. One mole of magnesium hydroxide $Mg(OH)_2$
(a) has a mass of 41.3 g.
(b) contains 6.02×10^{23} oxygen atoms.
(c) contains $16 \times (6.02 \times 10^{23})$ oxygen atoms.
(d) has a mass of 58.3 g.

15. The K_{eq} of a reaction is 4×10^7. At equilibrium,
(a) the reactants are favored.
(b) the products are favored.
(c) the reactants and products are present in equal amounts.
(d) the rate of the forward reaction is much greater than the rate of the reverse reaction.

16. Which of these is *not* a diatomic element?
(a) hydrogen   (b) helium
(c) bromine   (d) oxygen

17. The molar mass of ozone, O_3, is
(a) 16.0 g.   (b) 32.0 g.
(c) 48.0 g.   (d) 96.0 g.

18. When this equation is balanced

$$CS_2(l) + O_2(g) \longrightarrow CO_2(g) + SO_2(g)$$

the coefficients are
(a) 1, 1, 1, and 2.
(b) 2, 6, 2, and 4.
(c) 1, 3, 1, and 2.
(d) 1, 2, 1, and 1.

19. In which of these systems is entropy (disorder) decreasing?
    (a) air escaping from a tire
    (b) melting snow
    (c) dissolving salt in water
    (d) cooling a liquid

20. The chemical formula of aspirin is $C_9H_8O_4$. What is the mass of 0.40 mol aspirin?
    (a) 45 g   (b) 11 g   (c) 72 g   (d) 160 g

21. For a reaction with a K_{eq} = 1.0, which of the following statements is true?
    (a) The mixture contains equal amounts of reactants and products at equilibrium.
    (b) The activation energies for the forward and reverse reactions are equal.
    (c) The K_{eq} does not change if more reactant is added to the system.
    (d) All of the above are true.

22. How many grams of laughing gas, N_2O, are produced by the decomposition of 20.0 g NH_4NO_3?

$$NH_4NO_3(s) + \text{heat} \longrightarrow N_2O(g) + 2H_2O(g)$$

   (a) 11.0 g   (b) 7.50 g   (c) 44.0 g   (d) 2.20 g

23. Chemical equilibrium is reached when
    (a) the concentration of reactants equals the concentration of products.
    (b) the reaction stops.
    (c) the rate of the forward reaction equals the rate of the reverse reaction.
    (d) All of the above are true.

24. Which of the following terms appears when this equation is balanced?

$$\_\_Fe_2O_3(s) + \_\_C(s) \longrightarrow \_\_Fe(s) + \_\_CO(g)$$

   (a) $6CO(g)$ (b) $4C(s)$ (c) $2Fe(s)$ (d) $2Fe_2O_3(s)$

25. Which of the following does not affect the rate of a chemical reaction?
    (a) the equilibrium position
    (b) the temperature
    (c) the concentration of reactants
    (d) the presence of a catalyst

26. Which of the following changes will cause the equilibrium position to move to the left?

$$4NH_3(g) + 3O_2(g) \rightleftharpoons 2N_2(g) + 6H_2O(g) + \text{heat}$$

   (a) adding more O_2
   (b) heating the reaction
   (c) removing H_2O
   (d) adding a catalyst

27. How many moles of oxygen molecules are in 8.0 g oxygen gas?
    (a) 0.25   (b) 0.50   (c) 4.0   (d) 128

28. An equation for the formation of ice from water is

$$H_2O(l) \longrightarrow H_2O(s) + \text{heat}$$

   Which of the following statements is true?
   (a) This reaction is a phase change.
   (b) This reaction is exothermic.
   (c) This equation represents a decrease in entropy.
   (d) All of the above are true.

States of Matter

7

Molecules in Motion

Hot air is less dense than cold air; a balloon and its cargo float gently upward as the air inside it is heated.

CHAPTER OUTLINE

CASE IN POINT: Heimlich maneuver

7.1   Kinetic-molecular theory

7.2   Kinetic energy

7.3   Pressure

7.4   Avogadro's hypothesis

7.5   Diffusion

7.6   Behavior of gases

A CLOSER LOOK: Measuring Blood Pressure

A Closer Look: Breathing

A CLOSER LOOK: Exchange of Physiologic Gases

7.7   The gas laws

FOLLOW-UP TO THE CASE IN POINT: Heimlich maneuver

7.8   The ideal gas law

7.9   Liquids

7.10   Solids

We know that matter exists as solids, liquids, and gases. Moreover, matter can change from one state to another as it absorbs or releases heat. Solids melt and liquids vaporize when heated. Liquids solidify and gases liquefy when cooled. Why do changes in the state of matter occur with changes in temperature? We will find out in this chapter with the help of a model called the *kinetic-molecular theory.*

The chapter begins with the physical behavior of gases. The exchange of oxygen for waste gases such as carbon dioxide in the blood is important to our health. A knowledge of gas behavior is also important in anesthesia and respiration therapy. An understanding of how gas pressure changes when a gas is compressed into a smaller volume has helped to save many lives, as illustrated by the following Case in Point.

CASE IN POINT: Heimlich maneuver

Marla, a first-year college student, joined a few classmates for dinner at a local restaurant. The students were talking and laughing when her friend Clive, a nursing student, noticed that Marla was choking. Clive quickly rose from his chair and stood behind Marla. He placed his arms around her waist and clasped his hands just under Marla's breastbone (see figure). When he made a sharp, upward thrust, a piece of food lodged in Marla's windpipe was expelled, and she gasped for air. What procedure did Clive apply to help Marla? What behavior of gases does the procedure involve? We will find out in Section 7.8.

When Clive realized that Marla was choking and couldn't breathe, he clasped his hands together under Marla's breastbone and gave a sharp, upward thrust.

7.1 Kinetic-molecular theory

AIM: To state the kinetic-molecular theory.

The word *kinetic* means "motion." *The* **kinetic-molecular theory** *states that the atoms, ions, or molecules of gases, liquids, and solids are in constant motion and that there is no attraction between gas particles.* Consider the kinetic-molecular theory as it applies to gases. Gas particles move independently and randomly in space, traveling in straight lines and changing direction by rebounding from collisions with one another or with other objects (Fig. 7.1).

Speed of particles in a gas

The average speed of particles in a gas at 20 °C is about 1600 km/h. The use of an average speed is appropriate because some particles are moving faster and some slower. At the average speed some molecules in a bottle of perfume opened in New York should reach San Francisco in about 3 h. This does not happen, though, because their unimpeded path of travel in one direction is very short. The gas particles are said to have a short *mean free path.* That is, the perfume molecules are constantly striking gas molecules in air and rebounding in other directions. The aimless path they take is called a *random walk.*

The physical state of a gas concerns its volume, temperature, and pressure. Gases in sealed containers are called *contained gases.* The volume of a contained gas is the volume of the container, since gases adjust their shape to fit the contours of their containers. Gases that are not in sealed containers are called *uncontained gases.* Air is usually an uncontained gas, although it may be contained by sealing it in a closed vessel.

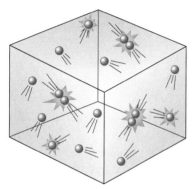

Figure 7.1
Gas particles have random and chaotic movements. They are constantly colliding with one another and with the walls of the container.

7.2 Kinetic energy

AIMS: To calculate how the average kinetic energy of particles varies with the temperature. To discuss the significance of absolute zero, giving its value in degrees Kelvin and degrees Celsius.

Energy is required to make a stationary object move. A tennis ball moves when it acquires energy by being struck by a tennis racket. *The energy an object has because of its motion is* **kinetic energy.** Since gas particles are in motion, it follows that they have kinetic energy.

Average kinetic energy and temperature

When a gas is heated, the particles in the gas absorb thermal energy. Some of it goes into increasing the energy *within* the particles, and the rest goes into increasing the *motion* of the particles—increasing their average kinetic energy.

Particles of all substances at the same temperature have the same average kinetic energy. It is the *average* kinetic energy that is of concern because in any sample of a gas some particles are nearly at rest, some are moving slowly, and others are moving very rapidly. *Temperature is a measure of the* **average kinetic energy** *of the particles in a substance*. Thus particles of all substances at the same temperature have the same average kinetic energy. An increase in the temperature of a substance signifies an increase in the average energy of the particles; a decrease in temperature signifies a decrease in the average kinetic energy.

PRACTICE EXERCISE 7.1

Compare the kinetic energy of oxygen gas molecules in a tank of oxygen with the kinetic energy of oxygen molecules in air at the same temperature as the tank of oxygen.

PRACTICE EXERCISE 7.2

Does the average kinetic energy of the gas particles in an inflated life raft increase or decrease if the sun heats the life raft from 20 to 35 °C?

PRACTICE EXERCISE 7.3

How does the average kinetic energy of the particles in a block of ice at 0 °C compare with the average kinetic energy of the particles in a gas-filled dirigible at 0 °C?

Absolute zero

There is no theoretical limit to the temperature to which a substance can be raised. But there is a lower limit. There should be some low temperature at which the particles of all substances stop moving. Particles would have no kinetic energy below this temperature because they would have no motion. Since temperature measures the average kinetic energy of the smallest components of matter, taking a substance to a temperature *below* this lower limit—absolute zero—is impossible. Absolute zero does exist, though, and it falls at −273 ° on the Celsius temperature scale. The Kelvin temperature scale (Sec. 1.4) is based on the idea of an absolute zero that occurs at this temperature. Zero degrees Kelvin (0 K) is −273 ° on the Celsius scale. The Kelvin scale is extremely useful because it gives a direct measure of the average kinetic energies of the particles of a substance—the particles in a substance at 200 K have twice the average kinetic energy of the particles in a substance at 100 K.

PRACTICE EXERCISE 7.4

By what factor does the average kinetic energy of the molecules of a gas in an aerosol container increase when the temperature is raised from 300 K to 900 K?

7.3 Pressure

AIMS: *To define pressure and explain how it can be measured. To convert pressure measurements between units of kilopascals, millimeters of mercury, and atmospheres.*

Moving bodies exert forces when they collide with other bodies. Although a gas particle is a moving body, the force exerted by a single tiny gas particle is very small. However, it is not hard to imagine that many simultaneous collisions would produce a measurable force on an object. **Gas pressure** *is the result of simultaneous collisions of billions upon billions of gas particles on an object.*

Atmospheric pressure

Air exerts pressure because molecules in the atmosphere collide with objects in their paths. Atmospheric pressure can be measured with a mercury barometer (Fig. 7.2). Daily measurements with mercury barometers show that air particles at sea level exert enough push or pressure to support a column of mercury about 760 mm high. The *atmosphere,* a unit of pressure, is derived from these observations. *One* **atmosphere** *is defined as the pressure required to support 760 mm of mercury.* Atmospheres (atm) and millimeters of mercury (mm Hg) are the most commonly used units of pressure measurement in chemistry. Another pressure unit is the *torr,* named for the Italian scientist Evangelista Torricelli (1608–1647). *One* **torr** *is 1/760 atm or 1 mm Hg. The SI unit of pressure is the* **pascal (Pa),** named for the French mathematician and physician Blaise Pascal (1623–1662). The

Mercury, the liquid metal that is used in barometers and some thermometers, is toxic, and mercury-containing devices should be handled with care. Mercury spills should be cleaned up immediately. Prolonged inhalation of mercury vapor can lead to nervous disorders.

Figure 7.2
A mercury barometer is made by filling a tube, sealed at one end, with mercury. The tube is then inverted into a reservoir of mercury. Normal atmospheric pressure pushing on the mercury in the reservoir supports a column of mercury about 760 mm high. At 9000 m altitude, the *air* exerts enough push to support a column of mercury only 253 mm high; the atmospheric pressure is therefore 253 mm Hg at 9000 m elevation.

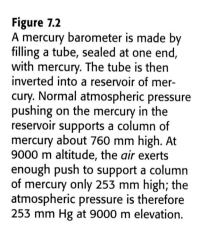
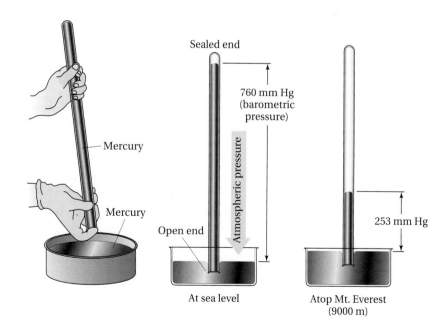

In cases of carbon monoxide poisoning, in which oxygen transport by hemoglobin in red blood cells is impaired, a patient is placed in an oxygen-enriched chamber at up to 3 atm of pressure. This hyperbaric oxygen treatment increases the amount of oxygen that dissolves in the blood, increasing the amount of oxygen that reaches tissue cells.

pascal is a small unit, only about 1/100,000 the size of an atmosphere; 1 atm = 101.3 kPa. Local atmospheric pressure varies slightly from day to day, so meteorologists use barometers to record changes in atmospheric pressure because such changes ("highs" and "lows") can be used to forecast weather. Special barometers (altimeters) are used to measure altitude. Since the air layer around the Earth thins out as altitude increases, atmospheric pressure decreases with altitude. The atmospheric pressure at the top of Mount Everest, at an elevation of 9000 m, is only 253 mm Hg.

Standard temperature and pressure

In working with gases, scientists often find it useful to compare their results with a standard set of conditions. *A temperature of 0 °C or 273 K and a pressure of 1 atm (760 mm Hg or 101.3 kPa) are called the* **standard temperature and pressure (STP)** *in gas work.*

EXAMPLE 7.1   **Converting mm Hg to atmospheres**

What pressure in atmospheres does a gas exert at 1520 mm Hg?

SOLUTION $1520 \text{ mm Hg} \times \dfrac{1 \text{ atm}}{760 \text{ mm Hg}} = 2.00 \text{ atm}$

EXAMPLE 7.2   **Converting atmospheres to mm Hg**

How many millimeters of mercury does a gas exert at 2.54 atm of pressure?

SOLUTION $2.54 \text{ atm} \times \dfrac{760 \text{ mm Hg}}{1 \text{ atm}} = 1.93 \times 10^3 \text{ mm Hg}$

PRACTICE EXERCISE 7.5

Convert 325 mm Hg to atmospheres of pressure.

PRACTICE EXERCISE 7.6

The pressure at the top of Mount Everest is 253 mm Hg. Is this greater than or less than 0.30 atm?

7.4 Avogadro's hypothesis

AIMS: To state and use Avogadro's hypothesis. To define molar gas volume and give the values of STP. To calculate the mass, volume, or number of molecules of a gas sample at STP, given the other two quantities.

Focus

A gas is mostly empty space.

The particles that make up different gases have different sizes. For example, chlorine molecules (Cl_2), with their large numbers of electrons, protons, and neutrons, are bigger and occupy more volume than hydrogen molecules (H_2), with only two protons and two electrons. Early scientists recog-

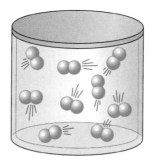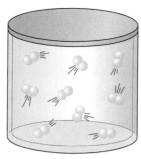

Figure 7.3
The volume of a container easily accommodates an equal number of large or small gas particles as long as the particles are not tightly packed. When the particles are tightly packed, large particles take up more space than small particles.

nized this, and many were stunned in 1811 when they heard of **Avogadro's hypothesis:** *Equal volumes of gases at the same temperature and pressure contain equal numbers of particles.* To these scientists, it was as if Avogadro were suggesting that two rooms of the same size could be filled by the same number of objects, no matter whether the objects were marbles or oranges.

What Avogadro had in mind is not so mysterious; it makes sense if the particles in a gas are very far apart with nothing but space in between (Fig. 7.3). We can suspend 10 marbles or 10 oranges in a ballroom without decreasing the volume of empty space in the room (the volume of the room minus the total volume of the marbles or oranges) very much. The same concept can be applied to gases. This was Avogadro's great insight. At STP, 1 mol (6.02×10^{23} particles) of any gas occupies 22.4 L. *The* **molar volume of a gas** *at STP is 22.4 L.*

We also can understand Avogadro's hypothesis by thinking of the current explanation for gas pressure. The same number of particles in a given volume at the same temperature should exert the same pressure because they have the same average kinetic energy and are contained within an equal amount of space. The same pressure will be caused by the impact of the molecules of two gases at identical temperature and volume on the walls of their containers. Thus, whenever we have equal volumes of gases at the same temperature and pressure, the volumes must contain equal numbers of particles.

EXAMPLE 7.3

Calculating the number of molecules in a gas at STP

How many nitrogen molecules are in 5.82 L nitrogen gas at standard temperature and pressure (STP)?

SOLUTION

Since 22.4 L nitrogen at STP contains 1 mol or 6.02×10^{23} nitrogen molecules, 5.82 L N_2 contains

$$5.82 \, \cancel{L} \times \frac{6.02 \times 10^{23} \text{ nitrogen molecules}}{22.4 \, \cancel{L}} = 1.56 \times 10^{23} \text{ nitrogen molecules}$$

186 CHAPTER 7 States of Matter

EXAMPLE 7.4   Finding the volume of a gas at STP

Determine the volume, in liters, occupied by 0.228 mol of a gas at STP.

SOLUTION

Since 1 mol of a gas has a volume of 22.4 L at STP, the volume of 0.228 mol is

$$0.228 \text{ mol} \times \frac{22.4 \text{ L}}{1 \text{ mol}} = 5.11 \text{ L}$$

PRACTICE EXERCISE 7.7
(a) How many moles is 3.25 L oxygen at STP?
(b) How many oxygen molecules does it contain?

PRACTICE EXERCISE 7.8
What is the volume occupied by 2.40 mol of any gas at STP?

PRACTICE EXERCISE 7.9
Hydrogen reacts with oxygen to give water according to the reaction

$$2H_2(g) + O_2(g) \longrightarrow 2H_2O(g)$$

What volume of hydrogen (liters) at STP is required to combine with 3.2 mol oxygen?

EXAMPLE 7.5   Determining the volume of a known mass of a gas at STP

Calculate the volume in liters occupied by 4.80 g hydrogen gas at STP.

SOLUTION

Hydrogen is a diatomic molecule (H_2). One mole of hydrogen gas has a mass of 2.00 g and occupies a volume of 22.4 L at STP. Therefore, the volume occupied by 4.80 g is

$$4.80 \text{ g } H_2 \times \frac{1 \text{ mol } H_2}{2.00 \text{ g } H_2} \times \frac{22.4 \text{ L}}{1 \text{ mol } H_2} = 53.8 \text{ L}$$

PRACTICE EXERCISE 7.10
What is the volume of a container if it holds 6.54 g carbon monoxide at STP?

7.5 Diffusion

AIMS: To predict which of two gases will diffuse at the faster rate.

Focus

Light gases diffuse faster than heavy gases.

The strength of the force exerted when a moving body strikes a stationary object depends on the mass and the speed, or velocity, of the body. The formula that relates kinetic energy KE to the mass m and velocity v of a body is

$$KE = \tfrac{1}{2}mv^2$$

Suppose that a ball bearing with a mass of 2 g traveling at 5 m/s has just enough kinetic energy to shatter a pane of glass. A ball bearing with a mass of 1 g would need to travel at slightly more than 7 m/s to have the same kinetic energy. The lighter ball bearing would therefore have to move faster to shatter the same pane of glass.

There is an important principle here: *When two bodies of different mass have the same kinetic energy, the lighter body is moving faster.* We can apply this principle to the **diffusion** of gases—*that is, the spread of a gas through space or through another gas.* The particles in two gases at the same temperature have the same average kinetic energy. If gas particles act anything like ball bearings, a gas of low molar mass should diffuse faster than a gas of high molar mass when the temperatures of the two gases are the same. We can see this is true by comparing two balloons, one filled with helium and the other filled with air. Many youngsters have gone to bed with helium-filled balloons floating against the ceilings of their rooms and awakened to the disappointment of much smaller balloons. Fast-moving helium atoms, with atomic masses of only 4 amu, diffuse through small pores in the balloon in a relatively short time. Balloons filled with air stay inflated longer because the main components of air, oxygen molecules (molar mass 32 amu) and nitrogen molecules (molar mass 28 amu), move slower and therefore diffuse more slowly than helium atoms. The rate of diffusion is related only to the speed of the particles, since the pores in a balloon are large enough for helium atoms and air molecules to pass through freely.

Diffusion
diffusus (Latin):   spread out

A respiratory therapist helps patients regain and maintain normal breathing functions after surgery or when recovering from respiratory illness or injury. A respiratory therapist must pass a registry examination and work under the direction of a physician.

PRACTICE EXERCISE 7.11

Which gas, hydrogen (molar mass 2 amu) or chlorine (molar mass 71 amu), diffuses faster at (a) 20 °C and (b) 70 °C?

7.6 Behavior of gases

AIMS: To use the kinetic-molecular theory to predict the change in pressure of a gas after the following change: amount of gas, container size, and temperature. To state Dalton's law of partial pressures. To find the partial pressure for a gas in a mixture of gases.

Focus

The kinetic-molecular theory explains the physical behavior of gases.

The atmosphere is too vast for anyone to have any control over atmospheric pressure, but contained gases are a different matter. We can add gas or remove some of it. And we can squeeze or expand the container or heat or cool the gas. When we do any of these things, we change a condition of the gas. How can we find out what happens to the gas? At one time, there was only one way—by trying it. That is what such eminent scientists as Robert Boyle (1627–1691) and Jacques Charles (1746–1823) did. These scientists studied the effects of changes in the pressure, volume, and temperature of contained gases. From their results they proposed a set of rules collectively known as the *gas laws* (discussed in Sec. 7.8).

The gas laws apply exactly to *ideal* gases only. Ideal gases would not liquefy and then solidify as they are cooled; they would have no volume at absolute zero; and they would obey the gas laws at all conditions of temperature and pressure. There are no ideal gases, however, only *real* ones.

Real gases are made of real atoms; they liquefy and solidify at low temperatures, and since they are made of real atoms, they occupy real volumes. At low pressures and within certain ranges of temperature, however, most real gases obey the gas laws. In science, development of a theory usually follows experiment, and the gas laws were instrumental in the evolution of the kinetic-molecular theory. The theory is now available, and with it we can often explain how gases behave without resorting to formal mathematical expressions. In this section we will examine some simple examples of how the kinetic-molecular theory is used to explain gas behavior. The emphasis in these examples will be on gas pressure.

Effect of adding gas

When we pump up a tire, the pressure inside it increases. The pressure exerted by an enclosed gas is caused by collisions of gas particles with the walls of the container. By adding gas, you increase the number of gas particles, which increases the number of collisions and therefore the gas pressure. Doubling the number of gas particles doubles the pressure if the temperature and volume of the gas do not change (Fig. 7.4). Tripling the number of gas particles triples the pressure, and so forth. With a powerful pump and a strong container you could generate very high gas pressures, but once the pressure exceeds the strength of the container, the container will rupture. Overinflated balloons burst for this reason.

The pressure exerted by the impact of air particles on the interior walls of automobile tires is enough to lift a car's entire mass. Gas pressure is used to advantage for many other purposes, one of which is described in A Closer Look: Measuring Blood Pressure. The heavy metal cylinders of oxygen and other gases used in hospitals illustrate the practicality of compressing gases. These sturdy cylinders can hold gases at very high pressures, which is convenient because a large quantity of gas stored in a small volume can be readily carted from place to place.

Figure 7.4
(a) When gas is pumped into a closed rigid container, the pressure increases in proportion to the number of gas particles added, as shown on a pressure gauge.
(b) If the number of particles doubles, the pressure doubles.
(c) If the pressure exceeds the strength of the container, the container bursts.

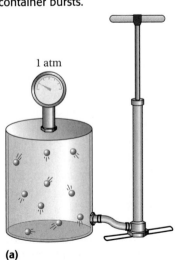

1 atm

(a)

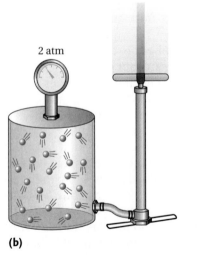

2 atm

(b)

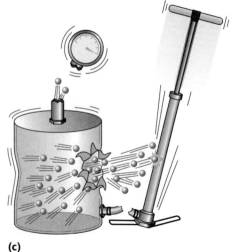

(c)

A Closer Look

Measuring Blood Pressure

Blood pressure is expressed as two numbers. These numbers represent the systolic pressure and the diastolic pressure. Systolic pressure is the maximum pressure in the arteries when the heart pumps; diastolic pressure is the pressure in the arteries when the heart relaxes between beats. Blood pressure varies with age, physical exercise, emotional stress, and health. Its measurement allows the diagnosis of high blood pressure (hypertension). A blood pressure of 120/80 means a systolic pressure of 120 mm Hg above atmos-

Pressure	Systolic (mm Hg)	Diastolic (mm Hg)
normal	139 or less	89 or less
borderline	140 to 159	90 to 94
high	160 or more	95 or more

pheric pressure and a diastolic pressure of 80 mm Hg above atmospheric pressure.

The pressure needed to stop the circulation of blood in an artery is easily attained by pumping air into an inflatable cuff wrapped around the upper arm. The accompanying figure shows the procedure for measuring blood pressure.

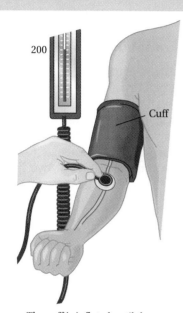

The cuff is inflated until the artery collapses and blood stops circulating, no sounds are heard through the stethoscope. A cuff pressure of 200 mm Hg is shown.

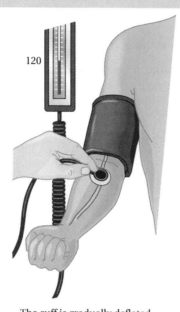

The cuff is gradually deflated. The pressure in the cuff when the first spurt of blood is heard as a sound through the stethoscope is the maximum pressure in the artery when the heart pumps. A systolic pressure of 120 mm Hg is shown.

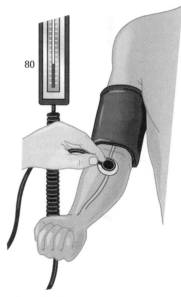

The cuff is further deflated. The pressure in the cuff when artery sounds are no longer heard is the pressure in the arteries when the heart relaxes between beats. A diastolic pressure of 80 mm Hg is shown.

Measuring blood pressure.

PRACTICE EXERCISE 7.12

A manufacturer of an aerosol deodorant wishes to produce a new, large, economy-sized package that will hold twice as much product as the present 150-mL regular size, but the pressure in both containers must be kept the same. What will the volume of the larger package have to be?

Effect of removing gas

Letting the air out of tires, the propellant out of an aerosol can, or the gas out of a storage cylinder decreases the pressure in the container because there are fewer particles left. As you have probably already guessed, halving the number of particles in a given volume of gas decreases the pressure by one-half. When a sealed container of gas under pressure is opened, the gas always expands from the higher pressure to the lower pressure because there is more empty space for the gas particles to occupy. The outward flow continues until the gas pressures inside and outside the container are equal.

Real gases cool when they expand and heat when they are compressed. When the gas is rapidly released from an aerosol can, the can becomes cooler because the expanding gas absorbs some thermal energy from the container and its contents as it escapes. Compression of a real gas always increases its temperature.

Effect of changing the size of the container

We also can increase or decrease the pressure exerted by a contained gas when we reduce or expand the size of the container. The more a gas is compressed, the greater the pressure it exerts on its container; reducing the volume of the container by half has the same effect on pressure as doubling the quantity of gas. Increasing the volume of the container has just the opposite effect; by doubling the volume, we halve the gas pressure because there are now half as many gas particles in a fixed volume. Indeed, we live and breathe because of pressure differences caused by expansion and contraction of our diaphragm muscles. A Closer Look: Breathing describes the breathing process in more detail.

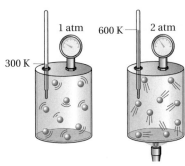

Figure 7.5
The effect of temperature on pressure of a gas. When the gas in the container is heated from 300 K to 600 K, the kinetic energy of the gas particles doubles, and they strike the sides of the container with twice the force at 600 K as they do at 300 K; the pressure exerted by the gas therefore doubles.

Effect of heating or cooling a gas

Raising the temperature of an enclosed gas increases the gas pressure because the kinetic energy of gas particles increases as they absorb thermal energy. Naturally, fast-moving particles bombard the walls of their container harder than slow-moving particles. Since the kinetic energy of gas particles doubles with a doubling of the Kelvin temperature, doubling the Kelvin temperature of an enclosed gas doubles the gas pressure (Fig. 7.5).

A heated gas sealed in a container can generate enormous pressure. Highway blowouts almost always occur after continuous traveling at high speeds has overheated worn or defective tires, and an aerosol can thrown in a fire is an explosion hazard. As the temperature of an enclosed gas decreases, the particles move more slowly and strike the container walls with less force. By halving the Kelvin temperature of a gas in a rigid container, we decrease the gas pressure by one-half.

PRACTICE EXERCISE 7.13

What happens to the volume of a balloon when it is taken outside on a cold winter day? Why?

Breathing

The thoracic cavity is a sealed chamber in which the lungs are suspended like two inflatable bags open to the atmosphere. At the bottom of the thoracic chamber is the diaphragm muscle, which increases the size of the chamber when it contracts and decreases the size of the cavity when it relaxes. When you are ready to inhale, your diaphragm contracts and moves down, your rib cage expands, and the volume of your thoracic cavity increases. Since the cavity is sealed, the pressure in the cavity is momentarily less than atmospheric pressure. As soon as this happens, however, some air moves into your lungs, expanding them and decreasing the volume of the thoracic cavity just enough to equalize the internal and external pressures. When you are ready to exhale, your diaphragm relaxes and moves up, your rib cage contracts, and the volume of your thoracic cavity decreases. The pressure in the cavity momentarily becomes greater than atmospheric, so some air is pushed out of the lungs, contracting them and increasing the volume of the thoracic cavity.

An artificial respirator can be used to keep a person alive when he or she is unable to breathe. The respirator encloses the chest of the patient and uses rhythmic changes in pressure to force air into and out of the lungs.

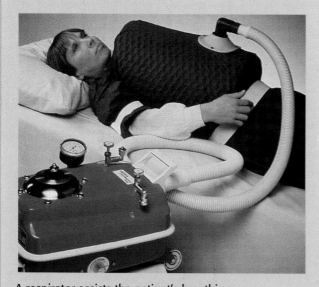

A respirator assists the patient's breathing.

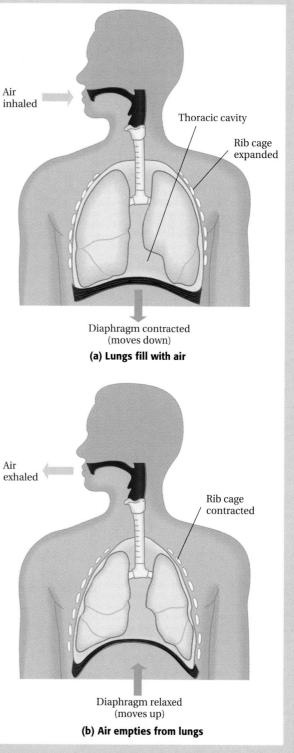

Air inhaled

Thoracic cavity

Rib cage expanded

Diaphragm contracted (moves down)

(a) Lungs fill with air

Air exhaled

Rib cage contracted

Diaphragm relaxed (moves up)

(b) Air empties from lungs

Breathing. (a) Volume of the thoracic cavity increases; pressure in lungs decreases; lungs fill with air. (b) Volume of the thoracic cavity decreases; pressure in lungs increases; air empties from lungs.

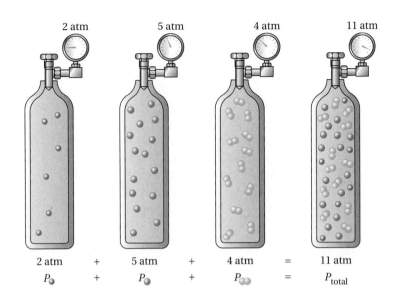

Figure 7.6
Dalton's law of partial pressures. The sum of the pressures exerted by the gas in each container is the same as the total pressure exerted by a mixture of the gases in the same volume as long as the temperature stays the same. Dalton's law holds because each gas exerts its own pressure independent of the pressure exerted by the other gases.

Dalton's law of partial pressures

Many gases, including air, are mixtures. Since the particles in a gas at the same temperature all have the same kinetic energy, it follows that if we know the pressure exerted by each gas in a mixture we can add the individual pressures and get the total gas pressure. *The contribution each gas makes to the total is the* **partial pressure** *exerted by that gas.* **Dalton's law of partial pressures** *states that at constant volume and temperature, the total pressure exerted by a mixture of gases is equal to the sum of the partial pressures* (Fig. 7.6). In a mixture of three gases, for example, the total pressure is the sum of the partial pressures of the three gases:

$$P_{total} = P_1 + P_2 + P_3$$

P_{total} is the total gas pressure, and P_1, P_2, and P_3 are the partial pressures of the individual gases.

EXAMPLE 7.6   **Calculating the partial pressure of a gas**

Air contains oxygen, nitrogen, carbon dioxide, and trace amounts of other gases (Table 7.1). What is the partial pressure of oxygen (P_{O_2}) at 1 atm of pressure if $P_{N_2} = 593.4$ mm Hg, $P_{CO_2} = 0.3$ mm Hg, and $P_{others} = 7.1$ mm Hg?

SOLUTION

$$1 \text{ atm} = 760 \text{ mm Hg} = P_{total}$$
$$P_{total} = P_{O_2} + P_{N_2} + P_{CO_2} + P_{others}$$
$$P_{O_2} = P_{total} - (P_{N_2} + P_{CO_2} + P_{others})$$
$$= 760 \text{ mm} - (593.4 \text{ mm} + 0.3 \text{ mm} + 7.1 \text{ mm}) = 159.2 \text{ mm Hg}$$

PRACTICE EXERCISE 7.14

Determine the total pressure of a gas mixture that contains oxygen, nitrogen, and helium if the partial pressures of the gases are

$$P_{O_2} = 230 \text{ mm Hg} \qquad P_{N_2} = 440 \text{ mm Hg} \qquad P_{He} = 150 \text{ mm Hg}$$

Give your answer in millimeters of mercury and atmospheres.

The proportionate pressure exerted by each gas in a mixture does not change as the temperature, pressure, or volume changes. This fact has important implications for aviators and mountain climbers. When the total atmospheric pressure is reduced to 253 mm Hg (one-third atmosphere), as it is atop Mount Everest, the partial pressure of oxygen is reduced to 53 mm Hg (one-third the partial pressure of oxygen at 1 atm). This oxygen pressure is insufficient for respiration (breathing air with a P_{O_2} of about 80 mm Hg supplies the minimum amount of oxygen), and mountaineering expeditions therefore must carry a supply of oxygen. The interiors of airplanes are pressurized with air for the same reason. The difference between the partial pressures of oxygen and carbon dioxide in the atmosphere and in our bodies is also extremely important, as we will learn in A Closer Look: Exchange of Physiologic Gases.

Breathing air that is too high or too low in oxygen can be hazardous to our health. Premature babies often have trouble breathing and must be given oxygen-enriched (about 40% O_2) air to survive. Pure oxygen is not used because of its link to retinal damage.

7.7 The gas laws

AIM: To make calculations involving pressure-volume (Boyle's law), temperature-volume (Charles's law), temperature-pressure, and the general gas law.

Focus

The gas laws describe the effects of pressure, volume, and temperature on a constant quantity of a contained gas.

A good understanding of the kinetic-molecular theory often allows us to solve a gas problem; the doubling of the pressure of a gas in a rigid container by doubling the Kelvin temperature is one example. With other changes, however, it is sometimes easier to put the values into formulas called *gas laws*. Although formulas are used in the following examples, try to think each problem through beforehand, using the kinetic-molecular theory, to see whether the answer makes sense.

Pressure-volume relationship at constant temperature (Boyle's law)

As a scuba diver ascends after a dive, the surrounding water pressure decreases. A scuba diver must continue exhaling during the ascension or the increasing air volume in the lungs due to the decreasing external pressure will damage the lungs.

The British chemist Robert Boyle investigated the effect of pressure on the volume of a contained gas while the temperature was held constant. In 1662, he stated **Boyle's law:** *For a given mass of gas at constant temperature, the volume of the gas varies inversely with pressure.* Or when one goes up, the other goes down. In Figure 7.7 on page 195, a volume of 1 L (V_1) is at a pressure of 1 atm (P_1). When the volume is increased to 2 L (V_2), the pressure decreases to 0.5 atm (P_2). Observe that the product $P_1 \times V_1$ (1 atm × 1 L = 1 L × atm) is the same as the product of $P_2 \times V_2$ (0.5 atm × 2 L = 1 L × atm). When the volume is decreased to 0.5 L, the pressure of the gas increases to 2 atm. Once again, the product of pressure times volume

A Closer Look

Exchange of Physiologic Gases

Oxygen and carbon dioxide are the physiologic gases. We breathe to deliver oxygen to the body and to carry carbon dioxide away. The exchange of oxygen and carbon dioxide between the body and the atmosphere is caused by differences in their partial pressures in the air we breathe and in body tissues. The different gases diffuse from regions of higher partial pressure to regions of lower partial pressure (see figure).

Venous blood that circulates from the tissues to the lungs is depleted of oxygen (low partial pressure of O_2), but it is loaded with carbon dioxide (high partial pressure of CO_2). In the alveoli of the lungs (small, grapelike clusters of spongy tissue where gases are exchanged), the partial pressure of oxygen is higher than in the blood, and oxygen diffuses into the venous blood that passes by. The partial pressure of alveolar carbon dioxide is low compared with the partial pressure of carbon dioxide in the blood. Thus carbon dioxide diffuses from the blood to the lungs to be exhaled.

The arterial blood that circulates *from* the lungs is now depleted of carbon dioxide (low partial pressure) and loaded with oxygen (high partial pressure). Body tissues use oxygen and produce carbon dioxide, so the partial pressure of oxygen at the tissues is low and that of carbon dioxide is high. Oxygen diffuses from the blood into the tissues, and carbon dioxide diffuses into the blood. The blood, now depleted of oxygen and rich in carbon dioxide, returns to the lungs, where oxygen is picked up and carbon dioxide is exhaled as before.

The partial pressures of oxygen and carbon dioxide in the blood become unbalanced in such respiratory conditions as emphysema. Normal blood has a P_{O_2} of 80 to 100 mm Hg and a P_{CO_2} of 40 mm Hg, but the P_{O_2} may be 40 mm Hg and the P_{CO_2} 70 mm Hg in a patient with emphysema.

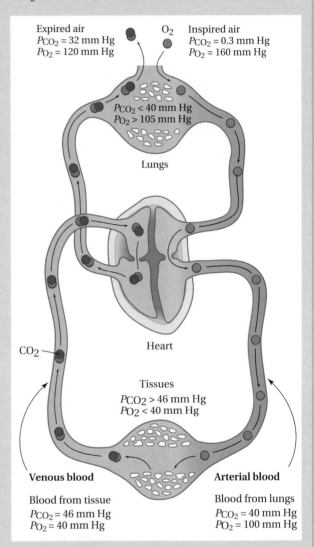

Blood circulation showing the partial pressure of oxygen and carbon dioxide in various parts of the body.

equals 1.0 L × atm. Since the product of pressure and volume at any two sets of conditions is always constant at a given temperature, Boyle's law can be written as

$$P_1 \times V_1 = P_2 \times V_2$$

provided the temperature is held constant.

Figure 7.7
Boyle's law. Pressure-volume relationships are studied by using a cylindrical container fitted with a piston. When the pressure of a gas at constant temperature decreases (P_2), the volume increases (V_2); when the pressure increases (P_3), the volume decreases (V_3).

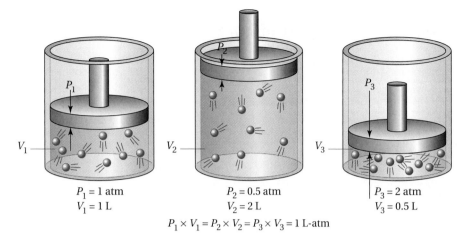

$P_1 = 1$ atm
$V_1 = 1$ L

$P_2 = 0.5$ atm
$V_2 = 2$ L

$P_3 = 2$ atm
$V_3 = 0.5$ L

$P_1 \times V_1 = P_2 \times V_2 = P_3 \times V_3 = 1$ L-atm

EXAMPLE 7.7   **Applying Boyle's law**

A balloon is filled with 14 L of helium gas at exactly 1.0 atm. What is the volume when the balloon rises to an altitude where the pressure is only 0.25 atm? (Assume that the temperature remains constant.)

SOLUTION

We know that $P_1 = 1.0$ atm, $V_1 = 14$ L, and $P_2 = 0.25$ atm. We need to solve for V_2. Divide both sides of the expression for Boyle's law by P_2 to obtain V_2:

$$P_1 \times V_1 = P_2 \times V_2$$

$$V_2 = \frac{P_1 \times V_1}{P_2}$$

Substitute values for P_1, V_1, and P_2:

$$V_2 = \frac{1.0 \ \cancel{atm} \times 14 \ L}{0.25 \ \cancel{atm}} = 56 \ L$$

The balloon will expand to a volume of 56 L at 0.25 atm. We can check the answer by calculating $P_1 \times V_1$ and $P_2 \times V_2$. If our answer is correct, both products should be identical:

$$P_1 \times V_1 = 1 \text{ atm} \times 14 \text{ L} = 14 \text{ L} \times \text{atm}$$
$$P_2 \times V_2 = 0.25 \text{ atm} \times 56 \text{ L} = 14 \text{ L} \times \text{atm}$$

PRACTICE EXERCISE 7.15

The pressure on 3.60 L of an anesthetic gas is changed from 780 to 376 mm Hg. What will the new volume be if the temperature remains constant?

The Case in Point presented at the beginning of this chapter illustrates a practical application of Boyle's law, as we will now see.

FOLLOW-UP TO THE CASE IN POINT: Heimlich maneuver

You may recall that Clive was able to dislodge food stuck in Marla's windpipe. The procedure she used is called the *Heimlich maneuver*. By applying the Heimlich maneuver, Clive may have saved Marla from choking to death. The Heimlich maneuver takes advantage of the fact that gas pressure increases when the volume of the gas is decreased. When she choked, Marla's lungs were essentially stoppered by food, much as a cork could be used to stopper a plastic bottle. When the stoppered bottle is squeezed, the volume of air inside it decreases, and the air pressure increases. If the increase in pressure is sufficient, the cork will be expelled. When Clive performed the Heimlich maneuver, the upward thrust of his arms rapidly reduced the volume of the gases in Marla's lungs. The decrease in volume of the gases created a surge of pressure that was sufficient to expel the food from Marla's windpipe.

Temperature-volume relationship at constant pressure (Charles's law)

In the early 1800s, a French physicist, Jacques Charles, investigated the effect of temperature on the volume of a gas at constant pressure. In every experiment he observed an increase in the gas's volume with an increase in temperature. His observations are summarized in **Charles's law:** *The volume of a fixed mass of gas is directly proportional to its Kelvin temperature if the pressure is kept constant.* Figure 7.8 shows the effect of increasing the temperature of 1 L (V_1) of a gas from 300 K (T_1) to 600 K. (Temperature in gas law problems is always in degrees Kelvin.) When the temperature is increased to 600 K (T_2), the volume increases to 2 L (V_2). Observe that the ratio $V_1 \div T_1$ (1 L ÷ 300 K = 0.0033 L/K) is equal to the ratio $V_2 \div T_2$ (2 L ÷ 600 K = 0.0033 L/K). Since the ratio of volume to Kelvin temperature at any two sets of conditions is constant, Charles's law can be written as

$$\frac{V_1}{T_1} = \frac{V_2}{T_2}$$

provided the pressure is held constant.

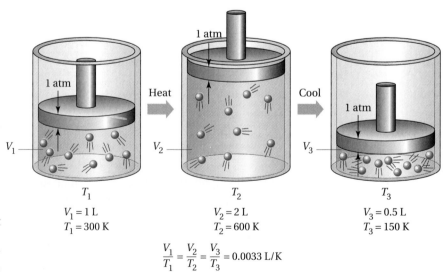

Figure 7.8
Charles's law. When a gas at constant pressure is heated, the volume increases; when a gas at constant pressure is cooled, the volume decreases.

T_1
$V_1 = 1$ L
$T_1 = 300$ K

T_2
$V_2 = 2$ L
$T_2 = 600$ K

T_3
$V_3 = 0.5$ L
$T_3 = 150$ K

$$\frac{V_1}{T_1} = \frac{V_2}{T_2} = \frac{V_3}{T_3} = 0.0033 \text{ L/K}$$

EXAMPLE 7.8   Applying Charles's law

A toy balloon, inflated in an air-conditioned room at 21 °C, has a volume of 3.5 L. It is heated to a temperature of 46 °C. What is the new volume of the balloon if the pressure remains constant?

SOLUTION

Temperature in all gas problems is expressed as degrees Kelvin. Convert degrees Celsius to degrees Kelvin (Sec. 1.4):

$$T_1 = 21 \text{ °C} + 273 = 294 \text{ K}$$
$$T_2 = 46 \text{ °C} + 273 = 319 \text{ K}$$

We know that $V_1 = 3.5$ L, $T_1 = 294$ K, and $T_2 = 319$ K. We can use Charles's law to calculate V_2.

Multiply both sides of the expression for Charles's law by T_2 to obtain V_2:

$$\frac{V_1}{T_1} = \frac{V_2}{T_2}$$

$$V_2 = \frac{V_1 \times T_2}{T_1}$$

Substituting values for T_1, V_1, and T_2, we get

$$V_2 = \frac{3.5 \text{ L} \times 319 \text{ } K}{294 \text{ } K} = 3.8 \text{ L}$$

The balloon will expand from a volume of 3.5 L at 21 °C to a volume of 3.8 L at 46 °C.

Check the answer by calculating $V_1 \div T_1$ and $V_2 \div T_2$. If the answer is correct, both ratios should be identical:

$$\frac{V_1}{T_1} = \frac{3.5 \text{ L}}{294 \text{ K}} = 0.012 \text{ L/K}$$

$$\frac{V_2}{T_2} = \frac{3.8 \text{ L}}{319 \text{ K}} = 0.012 \text{ L/K}$$

PRACTICE EXERCISE 7.16

A sample of gas occupies 14.2 L at 145 °C. What will be its volume at 27 °C if the pressure does not change?

Temperature-pressure relationship at constant volume

The pressure-temperature relationship explains why you are warned not to incinerate aerosol containers. What happens to the pressure when a contained gas is heated? What might eventually happen to the container?

As we have mentioned, an aerosol can thrown on a fire builds up a very high pressure. This illustrates the **temperature-pressure relationship at constant volume:** *The pressure of a gas is directly proportional to the Kelvin temperature if the volume is kept constant.* If the Kelvin temperature is doubled, for example, the gas pressure is doubled (Fig. 7.9). Like Charles's law, the temperature-pressure relationship involves direct proportions. The ratios $P_1 \div T_1$ and $P_2 \div T_2$ should be constant at constant volume. (Work out these

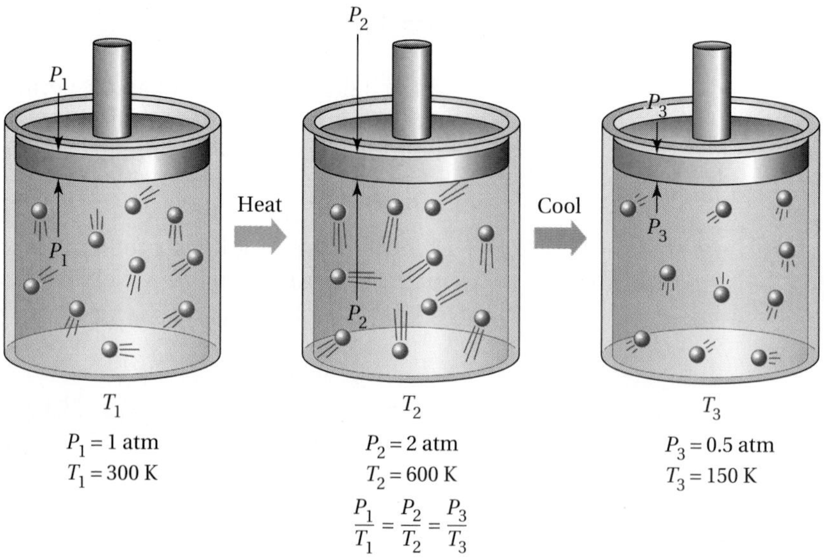

Figure 7.9
Temperature-pressure relationship at constant volume. When a gas at constant volume is heated, the pressure increases; when a gas at constant volume is cooled, the pressure decreases.

ratios for the quantities shown in Figure 7.9 to assure yourself that this is true.) Therefore, we can write the temperature-pressure relationship as

$$\frac{P_1}{T_1} = \frac{P_2}{T_2}$$

provided the volume is held constant.

EXAMPLE 7.9   **Applying the temperature-pressure relationship**

The gas left in a used aerosol can is at a pressure of exactly 1.0 atm at 27 °C (room temperature). If this can is thrown onto a fire, what is the internal pressure of the gas when its temperature reaches 927 °C?

SOLUTION

Convert degrees Celsius to degrees Kelvin:

$$T_1 = 27\ °C + 273 = 300\ K$$
$$T_2 = 927\ °C + 273 = 1200\ K$$

We know that $P_1 = 1.0$ atm, $T_1 = 300$ K, and $T_2 = 1200$ K. We can solve the equation for P_2. Rearranging the expression to obtain P_2, we get

$$\frac{P_1}{T_1} = \frac{P_2}{T_2}$$

$$P_2 = \frac{P_1 \times T_2}{T_1}$$

By inserting values for P_1, T_2, and T_1 into the equation and working it out, we obtain

$$P_2 = \frac{1.0\ \text{atm} \times 1200\ \cancel{K}}{300\ \cancel{K}} = 4.0\ \text{atm}$$

This is the pressure of a gas at constant volume and exactly 1 atm of pressure at 927 °C. We can check our answer by calculating $P_1 \div T_1$ and $P_2 \div T_2$ to see if the ratios are identical:

$$\frac{P_1}{T_1} = \frac{1.0 \text{ atm}}{300 \text{ K}} = 0.0033 \text{ atm/K}$$

$$\frac{P_2}{T_2} = \frac{4.0 \text{ atm}}{1200 \text{ K}} = 0.0033 \text{ atm/K}$$

PRACTICE EXERCISE 7.17

A gas has a pressure of 36 mm Hg at 355 °C. What will be the pressure at -125 °C if the volume does not change?

The combined gas law

The three gas laws just discussed can be combined into a single expression called the **combined gas law:**

$$\frac{P_1 \times V_1}{T_1} = \frac{P_2 \times V_2}{T_2}$$

If you have been wondering how to remember the expressions for the other gas laws, there is really no need; the other laws can be obtained from the combined gas law by holding one quantity—pressure, volume, or temperature—constant.

To illustrate, suppose we hold temperature constant ($T_1 = T_2$):

$$\frac{P_1 \times V_1}{T_1} = \frac{P_2 \times V_2}{T_2}$$

Therefore,

$$P_1 \times V_1 = \frac{P_2 \times V_2 \times T_1}{T_2}$$

We see that T_1 and T_2 cancel, and $P_1 \times V_1 = P_2 \times V_2$ (Boyle's law).

We can rearrange the combined gas law and cancel P_1 and P_2 if pressure is constant:

$$\frac{P_1 \times V_1}{T_1} = \frac{P_2 \times V_2}{T_2}$$

Therefore,

$$\frac{V_1}{T_1} = \frac{P_2 \times V_2}{P_1 \times T_2}$$

$$\frac{V_1}{T_1} = \frac{V_2}{T_2} \quad \text{(Charles's law)}$$

Likewise, we can obtain the temperature-pressure relationship law from the combined gas law by holding the *volume* of a gas constant:

$$\frac{P_1 \times V_1}{T_1} = \frac{P_2 \times V_2}{T_2}$$

Therefore,

$$\frac{P_1 \times V_1}{T_1 \times V_2} = \frac{P_2}{T_2}$$

$$\frac{P_1}{T_1} = \frac{P_2}{T_2}$$

The combined gas law has other uses as well. We can calculate changes in pressure, volume, or temperature when none of these quantities is held constant.

EXAMPLE 7.10   **Applying the combined gas law**

A cylinder of compressed oxygen gas has a volume of 30 L and 65 atm of pressure at 27 °C. The cylinder of gas is cooled until the pressure is 15 atm. What is the new temperature of the gas in the cylinder?

SOLUTION

Convert degrees Celsius to degrees Kelvin:

27 °C + 273 = 300 K

We know that $T_1 = 300$ K, $P_1 = 65$ atm, and $P_2 = 15$ atm; both V_1 and V_2 are equal to 30 L. We can rearrange the combined gas law to obtain T_2:

$$\frac{P_1 \times V_1}{T_1} = \frac{P_2 \times V_2}{T_2}$$

$$T_2 = \frac{P_2 \times V_2 \times T_1}{P_1 \times V_1}$$

By substituting the known quantities into the expression for the combined gas law, we obtain

$$T_2 = \frac{15 \ \cancel{atm} \times 30 \ \cancel{L} \times 300 \ \text{K}}{65 \ \cancel{atm} \times 30 \ \cancel{L}} = 69 \ \text{K}$$

We convert 69 K back to degrees Celsius:

K = °C + 273

°C = K − 273

69 K − 273 = −204 °C

PRACTICE EXERCISE 7.18

A container with a volume of 2.0 L is occupied by a gas at a pressure of 2.85 atm at 15 °C. If the pressure of the gas increases to 8.45 atm as the temperature is raised to 125 °C, what is the new volume?

7.8 The ideal gas law

AIM: To state and use the ideal gas law.

Focus

The ideal gas law relates the pressure, volume, temperature, and number of moles of a gas.

Sometimes we may wish to calculate the number of moles of gas in a fixed volume at a known temperature and pressure. Such a calculation is possible if the combined gas law is modified. The modification may be understood by recognizing that the volume occupied by a gas at a specified temperature and pressure is directly proportional to the number of particles in the gas. Since the number of moles n of gas is also directly proportional to the number of particles, moles must be directly proportional to volume as well. Moles may be introduced into the combined gas law by placing n in the denominator on each side of the equation:

$$\frac{P_1 \times V_1}{T_1 \times n_1} = \frac{P_2 \times V_2}{T_2 \times n_2}$$

This equation says that $(P \times V)/(T \times n)$ is a constant for any ideal gas because both sides of the equation are equal. If we could evaluate this constant, we could then calculate the number of moles of gas at any specified conditions of P, V, and T.

We have sufficient information to evaluate the constant because we know that 1 mol of every ideal gas occupies 22.4 L at STP. Inserting the appropriate values of P, V, T, and n into the right side of the equation, we obtain

$$\frac{P_1 \times V_1}{T_1 \times n_1} = \frac{1 \text{ atm} \times 22.4 \text{ L}}{273 \text{ K} \times 1 \text{ mol}} = 0.0821 \frac{\text{L} \times \text{atm}}{\text{K} \times \text{mol}}$$

The constant obtained, 0.0821(L × atm)/(K × mol), is usually designated by the letter R *and is called the* **ideal gas constant.** The numerical value of R depends on the units used. For example, if 760 torr is used for standard pressure, R is 62.4(L × torr)/(K × mol). By rearranging the equation to

$$\frac{P_1 \times V_1}{T_1 \times n_1} = R$$

and dropping the subscript, we obtain the usual form of the **ideal gas law**

$$P \times V = n \times R \times T$$

One advantage of the ideal gas law over the combined gas law is that it permits us to solve for the number of moles of a contained gas when P, V, and T are known.

The ideal gas law may be applied to most real gases at ordinary temperatures and pressures. Real gases deviate from the ideal gas law at high pressures and low temperatures because of the tendency of gas particles to adhere to one another at these conditions and because the volume of the gas particles becomes a significant part of the total volume of the gas. The following examples and problems give some applications of the ideal gas law.

EXAMPLE 7.11

Applying the ideal gas law

A rigid steel cylinder with a volume of 30.0 L is filled with nitrogen gas to a final pressure of 185 atm at 27 °C. How many moles of N_2 gas does the cylinder contain?

SOLUTION

We convert degrees Celsius to degrees Kelvin:

$$27 \text{ °C} + 273 = 300 \text{ K}$$

The conditions specified are $P = 185$ atm, $V = 30.0$ L, and $T = 300$ K. Since we want to find the number of moles of N_2 gas, we can rearrange the ideal gas law to obtain n:

$$n = \frac{P \times V}{R \times T}$$

Substituting the known quantities into the equation, we obtain

$$n = \frac{185 \text{ atm} \times 30.0 \text{ L}}{0.0821 \dfrac{\text{L} \times \text{atm}}{\text{K} \times \text{mol}} \times 300 \text{ K}} = 225 \text{ mol}$$

PRACTICE EXERCISE 7.19

When a rigid hollow sphere containing 420 L of helium gas is heated from 350 K to 620 K, the pressure of the gas increases to 18 atm. How many moles of helium does the sphere contain?

EXAMPLE 7.12

Calculating a mass from the ideal gas law

A deep underground cavern contains 2.24×10^6 L of methane gas at a pressure of 15.0 atm and a temperature of 42 °C. How many grams of methane does this natural gas deposit contain?

SOLUTION

The problem calls for an answer in grams, but the ideal gas law permits us to obtain the solution only in moles. We must first find the number of moles and then convert to grams.

We know that $P = 15.0$ atm, $V = 2.24 \times 10^6$ L, and $T = 315$ K (42 °C + 273). As in the preceding problem,

$$n = \frac{P \times V}{R \times T}$$

Substituting the known quantities into the equation, we obtain

$$n = \frac{15.0 \text{ atm} \times 2.24 \times 10^6 \text{ L}}{0.0821 \dfrac{\text{L} \times \text{atm}}{\text{K} \times \text{mol}} \times 315 \text{ K}} = 1.30 \times 10^6 \text{ mol}$$

The cavern contains 1.30×10^6 mol of methane.

Since 1 mol of methane has a mass of 16.0 g, the number of grams of methane is calculated as

$$1.30 \times 10^6 \; \text{mol CH}_4 \times \frac{16.0 \text{ g CH}_4}{1 \text{ mol CH}_4} = 20.8 \times 10^6 = 2.08 \times 10^7 \text{ g CH}_4$$

The cavern contains nearly 21 million grams of methane gas.

The amount of air inhaled and exhaled during breathing, called the *tidal volume,* can vary over an almost tenfold range. At rest, the tidal volume is 0.5 L. During heavy exercise, volume can be up to 4.5 L.

PRACTICE EXERCISE 7.20

A child has a lung capacity of 2.2 L. How many grams of air do her lungs hold at a pressure of 1.0 atm and a normal body temperature of 37 °C? Air is a mixture, but we may assume a "molar mass" of 29 for air because air is about 20% oxygen (molar mass of $O_2 = 32$) and 80% nitrogen (molar mass of $N_2 = 28$). Thus 0.20×32 g/mol + 0.80×28 g/mol = 29 g/mol.

7.9 Liquids

AIMS: *To define vapor pressure and explain how a change in temperature can cause a change in vapor pressure. To define boiling point, normal boiling point, and melting point and explain how these vary with the strength of intermolecular forces.*

Focus

Attractive forces cause some substances to be liquids.

We have seen that many properties of gases can be explained by assuming that gas particles have motion and there is no attraction between them. The particles that constitute liquids have motion, but unlike gas particles, the particles of a liquid are held together by weak attractions such as van der Waals forces and dipolar attractions (Sec. 5.9). As a result of their motion, the particles of a liquid are free to slide past one another, but because of their attraction, most do not have enough kinetic energy to escape the liquid. Since there is little space between the particles of a liquid, liquids are much less compressible and much denser than gases. However, liquids flow and take the shapes of their containers because of the movement of their particles. The extent to which forces that tend to hold a liquid together are offset by motions that tend to force it apart determines many important properties of liquids.

Vaporization

As a person sweats, the water in the perspiration *vaporizes. In* **vaporiza- tion**—*the conversion to a vapor of a liquid at a temperature below its boiling temperature*—the water in the perspiration is turned from liquid water into water vapor, and the skin dries. *When vaporization of a liquid in an open container occurs, it is commonly called* **evaporation.** The attractive forces between particles in a liquid are stronger within the liquid than at the surface. Evaporation occurs when molecules at the surface of the liquid that

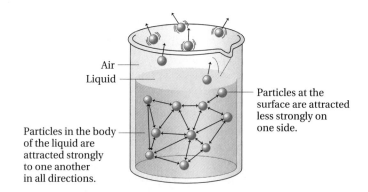

Air
Liquid

Particles at the surface are attracted less strongly on one side.

Particles in the body of the liquid are attracted strongly to one another in all directions.

Figure 7.10
Evaporation takes place when particles escape from the surface of the liquid.

have a certain minimum kinetic energy break away and go into the vapor state. Some escaping particles collide with air molecules and return to the liquid, but others escape completely (Fig. 7.10). A liquid evaporates faster when heated because the kinetic energy of its particles increases, enabling them to overcome the attractive forces keeping them in the liquid. Although it requires heat, evaporation itself is a cooling process because the particles with the highest kinetic energy (highest temperature) escape first. The particles left in the liquid have a lower average temperature than the particles that have escaped. When water molecules in perspiration evaporate from your skin's surface, the remaining perspiration is cooler and therefore cools you by absorbing body heat.

Vapor pressure

Vaporization of a liquid in a closed container is somewhat different from evaporation because no particles can escape into the atmosphere. When a partially filled container of liquid is sealed, some of the particles in the liquid vaporize and collide with the walls of the container. *The pressure created in the container by the vapor above the liquid is the* **vapor pressure** *of the liquid.* As the container stands, the number of particles entering the vapor increases, but some are also returning to the liquid state. These returning particles are said to be *condensing* to the liquid state. Eventually, the number of liquid particles vaporizing equals the number of vapor particles condensing. The container is now saturated with vapor, and a *dynamic equilibrium* exists between gas and liquid. At equilibrium, the particles in the system are still vaporizing and condensing, but there is no *net change* in either number of particles. One sign that equilibrium is established is that the inner walls of the container "sweat" because liquid that once vaporized is now condensing. An increase in the temperature of a contained liquid also increases the vapor pressure, since the particles in the warmed liquid have an increased kinetic energy.

The heat of vaporization and boiling point of a liquid

As you may recall, the evaporation of liquid in an open container increases with heat, and ever larger numbers of particles at the liquid's surface break the attractive forces keeping them in the liquid. The remain-

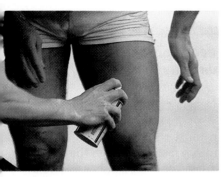

Figure 7.11
The pain of some sports injuries can be relieved with a spray of ethyl chloride. As the ethyl chloride rapidly evaporates from the skin, it produces a freezing effect that deadens nerves and reduces pain.

Figure 7.12
An autoclave uses high pressure to achieve the high temperatures needed to sterilize surgical equipment.

ing particles in the liquid grow more and more agitated as they absorb thermal energy and their average kinetic energy increases. Eventually, a temperature is reached where bubbles of vapor form throughout the liquid. The vapor pressure in these bubbles is equal to the atmospheric pressure, so vapor escapes to the atmosphere. This stage marks the onset of boiling. *The **boiling point** is the temperature at which the vapor pressure of the liquid is equal to the external pressure. The boiling point of a liquid at a pressure of 1 atm is the **normal boiling point.*** Odd as it may seem, boiling is a cooling process like evaporation. As in evaporation, the particles with the highest kinetic energy escape first when the liquid is at the boiling point. If no more heat is supplied, the temperature of the liquid will drop below its boiling point. If more heat is supplied, however, more particles acquire enough kinetic energy to escape. The result is the usual cooling effect on the remaining liquid; therefore, the temperature of a boiling liquid never rises above its boiling point. No matter how much heat is supplied, the liquid only boils faster. Eventually, all the liquid boils away.

The use of ethyl chloride (C_2H_5Cl) as a local anesthetic demonstrates the cooling effect of boiling. Ethyl chloride boils at 12.3 °C, but it may be kept as a liquid in a pressurized container. When squirted on warm skin, it boils, rapidly removing a large quantity of heat from the skin. The result of this rapid boiling is a numbing or anesthetic effect (Fig 7.11).

The boiling point of a liquid changes as the pressure changes. At lower external pressures, the boiling point decreases because particles in the liquid need less kinetic energy to escape. The normal boiling point of water is 100 °C. Denver, which is 1600 m above sea level, has an average atmospheric pressure of only 640 mm Hg. Water in Denver boils at 95 °C. Hard-boiling an egg takes longer in Denver than it does in Boston, which is at sea level. Cooking an egg is even more of a challenge on top of the 9000-meter peak of Mount Everest, with its atmospheric pressure of 253 mm Hg. On Mount Everest, water boils at 70 °C. At higher external pressure, a liquid's boiling point increases; particles in such a liquid need more kinetic energy to escape. Pressure cookers reduce cooking time because at high pressure water boils at well above 100 °C. Autoclaves are pressure cookers used to sterilize medical instruments (Fig. 7.12). Most harmful bacteria are killed by autoclaving them at 120 °C for 15 minutes, although especially stubborn bacteria may require longer times.

*The number of calories required to change 1 g of a liquid to gas at the boiling point at 1 atm is the liquid's **heat of vaporization.*** When a vapor condenses to a liquid, the opposite transfer of heat occurs. The heat of vaporization is passed to anything on which the vapor condenses. The temperature of the condensed liquid does not decrease until the heat of vaporization is lost. *The heat released when 1 g of a gas at 1 atm condenses to a liquid at the boiling point is the **heat of condensation.*** Heats of vaporization and condensation have the same numerical value because the amount of heat absorbed during vaporization at a liquid's boiling point is the same as the amount of heat released when the vapor condenses to the liquid. As you might expect, various liquids have different heats of vaporization and boiling points because the intermolecular attractive forces vary.

7.10 Solids

AIM: To define heat of vaporization, heat of condensation, heat of fusion, and heat of solidification and explain how these quantities vary with the strength of intermolecular forces.

The motions in liquids are chaotic, but in most solids the particles are packed against one another in a highly organized fashion. Rather than sliding from place to place, they vibrate about fixed points. Solids are dense and incompressible; because of the fixed positions of their particles, they do not flow to fit the shape of their container.

When a solid is heated, its particles vibrate more rapidly—their average kinetic energy increases. At the **melting point**—*the temperature at which a solid changes to a liquid*—disruptive vibrations of some of the particles are strong enough to overcome the interactions that hold them in fixed positions; the solid begins to melt to a liquid (Fig. 7.13). Just as an extra amount of heat is necessary to transform a liquid to a gas, addition of an extra amount of heat is necessary to transform a solid at its melting point temperature into a liquid. *The* **heat of fusion** *is the number of calories needed to completely transform 1 g of a solid at its melting point to 1 g of a liquid at the same temperature.* A mixture of ice cubes and water stays at 0 °C until all the ice cubes have melted; when the ice has completely melted, the temperature of the liquid rises if we add more heat. When a melted liquid resolidifies, it gives up heat to whatever it touches before the temperature of the solid goes below the melting point. *The heat given off by a solidifying liquid is called the* **heat of solidification.** Since solidification is the reverse of fusion, the heat of fusion and the heat of solidification have the same numerical value.

In general, ionic solids (held together by ionic bonds) have higher melting points than molecular solids (held together by weak forces). Sodium chloride, an ionic compound, has a melting point of 801 °C, but hydrogen chloride, a molecular compound, melts at −112 °C. Not all solids melt. Wood, for example, decomposes when heated.

Figure 7.13
The particles in a solid do not change places but vibrate about fixed points. A solid melts when its temperature is raised to a point at which the vibrations of the particles become so intense that they disrupt the ordered structure.

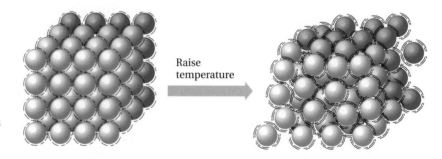

Raise temperature

SUMMARY

There are three states of matter: gas, liquid, and solid. The kinetic-molecular theory concerns the motion of particles in matter and the forces of attraction between them. The theory assumes that the volume occupied by a gas is mostly empty space and that the particles of a gas are far apart, move rapidly, and have chaotic motion.

Gases expand to fill any volume available to them, and they diffuse from a region of high gas concentration to one of lower concentration. The collision of the particles in a gas with the walls of the container is gas pressure. Gas pressure is measured in atmospheres or millimeters of mercury; 1 atm of pressure, at sea level, is 760 mm Hg. The pressure and volume of a fixed mass of gas are inversely related—if one decreases, the other increases (Boyle's law). The volume of a gas at con-stant pressure is directly related to its Kelvin temperature (Charles's law); the pressure of a fixed volume of gas is directly related to its Kelvin temperature. At a pressure of 1 atm and a temperature of 273 K, the volume of 1 mol of any gas is 22.4 L—this is called the molar volume. The total pressure in a mixture of gases is equal to the sum of the partial pressures of each gas present (Dalton's law).

When the temperature of a gas is lowered, the particles slow down. As the temperature is lowered further, the gas condenses to a liquid. Liquids have a fixed volume and a characteristic vapor pressure. A liquid boils when its vapor pressure equals the external pressure. Solids are rigid and have fixed volumes. In a solid, the movement of particles is restricted to vibrations about fixed points.

KEY TERMS

Atmosphere (7.3)
Average kinetic energy (7.2)
Avogadro's hypothesis (7.4)
Boiling point (7.9)
Boyle's law (7.7)
Charles's law (7.7)
Combined gas law (7.7)

Dalton's law of partial pressures (7.6)
Diffusion (7.5)
Evaporation (7.9)
Gas pressure (7.3)
Heat of condensation (7.9)
Heat of fusion (7.10)
Heat of solidification (7.10)

Heat of vaporization (7.9)
Ideal gas constant (7.8)
Ideal gas law (7.8)
Kinetic energy (7.2)
Kinetic-molecular theory (7.1)
Melting point (7.10)
Molar volume of a gas (7.4)
Normal boiling point (7.9)

Partial pressure (7.6)
Pascal (7.3)
Standard temperature and pressure (STP) (7.3)
Temperature-pressure relationship (7.7)
Torr (7.3)
Vaporization (7.9)
Vapor pressure (7.9)

EXERCISES

Kinetic-Molecular Theory (Section 7.1)

7.21 Use the kinetic-molecular theory to explain each of the following properties of gases.
(a) Gases have a relatively low density. The density of water is 1.00 g/cm^3, the density of oxygen is 0.00143 g/cm^3, and that of hydrogen is 0.000090 g/cm^3.
(b) Gases take the shape of any closed container.

7.22 Explain each of the following observations based on the kinetic-molecular theory.
(a) Gases mix easily. The carbon dioxide that you exhale mixes rapidly with the air, which is mostly nitrogen and oxygen.
(b) It takes much longer for the odor of a gas to travel across a room than you expect based on the average speed of gas particles.

Temperature and Average Kinetic Energy (Section 7.2)

7.23 How can you decrease the average kinetic energy of helium atoms in a helium-filled balloon?

7.24 What happens to the average kinetic energy of water molecules in your body when you get a fever?

7.25 What is the significance of absolute zero?

7.26 A student reports a temperature as -24 K. You know there is an error in the measurement. Explain.

7.27 A cylinder of nitrogen gas is cooled from 320 K (47 °C) to 160 K (-113 °C). By what factor does the average kinetic energy of the nitrogen molecules decrease?

7.28 The temperature of the gas in an aerosol container is 0 °C (273 K). To what new temperature must the gas be raised to increase the average kinetic energy of the gas molecules by a factor of 3?

Pressure (Section 7.3)

7.29 What is gas pressure, and in what units is it measured?

7.30 What causes atmospheric pressure, and why is it higher at sea level than on the top of a mountain?

7.31 What pressure in (a) atmospheres and (b) kilopascals does a gas exert at 190 mm Hg?

7.32 How many (a) kilopascals and (b) millimeters of mercury pressure does hydrogen gas exert at 1.25 atm?

7.33 A patient's blood pressure is 114/84 (mm Hg). What is her blood pressure expressed in (a) atmospheres and (b) kilopascals?

7.34 The pressure inside a kitchen pressure cooker is 225 kPa. Express this pressure in (a) atmospheres and (b) millimeters of mercury.

Avogadro's Hypothesis (Section 7.4)

7.35 What does STP represent?

7.36 What is the importance of STP?

7.37 What is the value of standard temperature on the (a) Celsius and (b) Kelvin temperature scales?

7.38 Express standard pressure in units of (a) atmospheres, (b) kilopascals, and (c) millimeters of mercury.

7.39 A nitrogen-filled balloon has a volume of 2.80 L at STP. How many moles of nitrogen is this?

7.40 What is the volume occupied by 0.15 mol argon gas at STP?

7.41 Hydrogen and oxygen react to give water according to the equation

$$2H_2(g) + O_2(g) \longrightarrow 2H_2O(g)$$

How many liters of hydrogen at STP are needed to produce 0.25 mol water?

7.42 Hydrogen and nitrogen combine to form ammonia:

$$3H_2(g) + N_2(g) \longrightarrow 2NH_3(g)$$

How many liters of ammonia are formed at STP when 0.75 mol hydrogen reacts with nitrogen?

7.43 What is the volume in liters occupied by 14.8 g oxygen gas at STP?

7.44 What is the volume in liters of a balloon filled with 62.6 g carbon dioxide, CO_2, gas at STP?

Characteristics of Gases (Sections 7.5, 7.6)

7.45 A gas is compressed from 7.5 to 2.5 L. What happens to the pressure assuming the temperature remains constant?

7.46 A metal cylinder contains 2.50 mol helium gas at STP. What happens to the pressure in the cylinder if an additional 2.50 mol of gas is added to the cylinder but the temperature and volume do not change?

7.47 The gas in an aerosol container has a pressure of 651 mm Hg at 27 °C (300 K). What will be the new pressure if the temperature is lowered to −173 °C (100 K)?

7.48 A gas with a volume of 6.4 L is expanded into a container with a volume of 25.6 L. What happens to the pressure if the temperature remains constant?

7.49 A gas mixture containing carbon dioxide, oxygen, and nitrogen has a pressure of 480 mm Hg. If P_{CO_2} = 150 mm Hg and P_{O_2} = 220 mm Hg, what is P_{N_2}?

7.50 Two liters each of nitrogen, argon, and oxygen, all at 1 atm pressure, are forced into a single container with a volume of 1 L. Assume that the temperature does not change. What will be the new total pressure and the partial pressure of each component gas?

The Gas Laws (Section 7.7)

7.51 A gas with a volume of 8.4 L and a pressure of 2.8 atm is allowed to expand until the pressure drops to 0.70 atm. Assuming the temperature remains constant, find the new volume.

7.52 A gas is compressed at constant temperature from 27 to 3.0 L. If the initial pressure of the gas is 0.5 atm, what is the final pressure?

7.53 A sample of carbon monoxide gas has a volume of 3.6 L at 2.0 atm. What volume will the gas occupy at 225 mm Hg if the temperature remains constant?

7.54 Calculate the volume of neon gas in liters at a pressure of 0.45 atm if the pressure of a 1250-mL filled balloon is originally 810 mm Hg. Assume a constant temperature.

7.55 A gas with a volume of 4.60 L at 250 °C is heated until its volume is 11.5 L. What is the new temperature of the gas if the pressure is constant?

7.56 Seven liters of nitrogen at −120 °C are warmed to 250 °C. What is the new volume if the pressure is unaltered?

7.57 At 35 °C the gas in an oxygen-filled gas cylinder has a pressure of 12 atm. The cylinder is left in the sun, and the temperature of the gas increases to 60 °C. What is the pressure in the cylinder?

7.58 The pressure in an automobile tire is 2.2 atm at 22 °C. After an hour-long trip, the pressure increases to 2.5 atm. Assuming the volume has not changed, find the temperature of the air in the tire.

7.59 At STP a gas has a volume of 4.2 L. What must the temperature be changed to in order to double the pressure and double the volume of the gas?

7.60 A 6.0-L sample of oxygen at a temperature of $-25\ °C$ has a pressure of 780 mm Hg. What will be the new volume if the temperature is increased to 150 °C and the pressure decreased to 660 mm Hg?

Ideal Gas Law (Section 7.8)

7.61 Use the gas laws and the kinetic-molecular theory to complete these statements. Unless otherwise stated, assume a constant amount of gas.
   (a) As the volume of a gas increases at constant temperature, its pressure _____ .
   (b) As the temperature of a gas increases and its pressure decreases, its volume _____ .
   (c) At constant pressure, a decrease in the volume of a gas is caused by a/an _____ in its temperature.
   (d) An increase in the volume and pressure of a gas is caused by a/an _____ in its temperature.
   (e) At constant volume, a decrease in the temperature of a gas causes a/an _____ in the pressure.

7.62 Find the number of liters occupied, at STP, by (a) 6.20 mol H_2, (b) 6.8 g N_2, and (c) 2.35 mol CO_2.

7.63 What volume will 12.0 g oxygen gas occupy at 25 °C and a pressure of 0.520 atm?

7.64 What is the pressure, in atmospheres, of a 5.50-L container at 65 °C that contains 18.5 g nitrogen gas?

7.65 A 25.0-L flask, at 25.0 °C, contains 4.50 mol CO_2. What is the pressure in the flask?

7.66 A balloon is filled with helium to a volume of 66 L at 1.00 atm pressure and 27 °C. When the balloon is released, it rises to an altitude where the pressure is 380 mm Hg and the temperature is -23 °C. What is the volume of the balloon at this new altitude?

7.67 Use the kinetic-molecular theory to explain (a) the expansion of a gas when it is heated at constant pressure, (b) the diffusion of gases, (c) the increase in pressure of a gas heated at constant volume, and (d) the decrease in volume of a gas when the pressure is increased at constant temperature.

Liquids and Solids (Sections 7.9, 7.10)

7.68 Describe, on a particle basis, what happens when a liquid boils.

7.69 Distinguish among vaporization, vapor pressure, and boiling point.

7.70 Explain what happens at the melting point of a substance.

7.71 Comment on the fact that ionic solids melt at a higher temperature than molecular substances.

Additional Exercises

7.72 The pressure of a 25.0-L sample of H_2 gas at 20 °C is increased from 2.00 to 10.0 atm at constant temperature. What is the final volume of the gas in liters?

7.73 The equation for the combustion of acetylene, C_2H_2, is

$$2C_2H_2(g) + 5O_2(g) \longrightarrow 4CO_2(g) + 2H_2O(g)$$

   (a) At STP, how many liters of oxygen are required to react with 16.4 L of acetylene?
   (b) How many liters of carbon dioxide (at STP) are produced when 12.8 g acetylene is burned?

7.74 Why do some liquids evaporate more rapidly at room temperature than others?

7.75 Sketch a simple graph that shows how the pressure of a gas varies with a change in temperature (at constant volume).

7.76 Molecules of two gases at the same temperature have the same kinetic energy. Why would they not diffuse at the same rate?

7.77 The density of carbon monoxide at STP is 1.25 g/L. Assuming no change in the pressure, what is the mass of 1.00 L carbon monoxide at 55 °C?

7.78 Explain each of the following observations on the basis of your knowledge of the kinetic-molecular theory and the forces of attraction that exist between the particles in matter.
   (a) Water evaporates faster at 40 °C than 20 °C.
   (b) A burn from steam at 100 °C is worse than a burn from water at 100 °C.
   (c) A cap "pops" off a bottle of root beer that has been kept in the trunk of a car on a hot day.
   (d) Diethyl ether, $C_4H_{10}O$, has a vapor pressure of 442 mm Hg at 20 °C, whereas the vapor pressure of water at this temperature is only 17.5 mm Hg.
   (e) A melting ice cube cools a glass of tea.
   (f) Foods are dehydrated (water is removed) by using low pressures rather than high heat.

7.79 A balloon contains an equal number of moles of three different gases. How do the partial pressures of the three gases compare?

7.80 A gas has a volume of 55.8 mL at 125 °C and 560 mm Hg. What is the volume of this gas at STP?

7.81 A mixture of ice and water kept at 0 °C in an insulated container is in a state of dynamic equilibrium. Explain what this means.

7.82 What is the mass of 4.50 mol ethane gas, C_2H_6, at STP?

7.83 Sketch a simple graph that shows how the volume of a gas varies with a change in pressure (at constant temperature).

7.84 The heat of fusion of calcium fluoride is much higher than the heat of fusion of carbon tetrachloride. Explain.

7.85 A balloon is filled with 7.0 g carbon dioxide, CO_2. What is the volume (at STP) of the balloon in liters?

SELF-TEST (REVIEW)

True/False

1. When a gas at constant volume is cooled, the pressure of the gas decreases.
2. Particles in a solid are motionless.
3. The pressure of any one gas in a mixture of gases is independent of the pressure exerted by the other gases.
4. Increasing the number of gas particles in a closed, rigid container causes the pressure in the container to increase.
5. Assuming constant pressure, if the absolute temperature of a gas is doubled, we would expect the volume to double.
6. The average kinetic energy of the molecules in liquid water at 80 °C is the same as the average kinetic energy of the molecules in oxygen gas at 80 °C.
7. At STP, 88 L carbon dioxide contains the same number of molecules as 88 L hydrogen.
8. As a liquid evaporates, it cools.
9. The molar volume of a gas at STP depends on the size of the molecules of the gas.
10. A pressure of 1 mm Hg is 1/760 of an atmosphere.

Multiple Choice

11. The assumptions of the kinetic-molecular theory for gases include all the following *except*
    (a) gas particles move at high speeds.
    (b) there are only slight attractive forces between the gas particles.
    (c) the gas particles are far apart.
    (d) gas particles move in straight lines between collisions.
12. Standard temperature and pressure (STP) is defined as
    (a) 0 K and 1 mm Hg.   (b) 0 °F and 1 atm.
    (c) 273 K and 1 atm.   (d) 273 °C and 760 mm Hg.
13. A real gas behaves most like an ideal gas at relatively
    (a) low temperature and high pressure.
    (b) low temperature and low pressure.
    (c) high temperature and high pressure.
    (d) high temperature and low pressure.

14. The height of the column of mercury in a barometer would decrease as the barometer was carried up a mountain because
    (a) the pressure increases.
    (b) there are fewer particles colliding with the mercury in the dish.
    (c) the force of gravity changes.
    (d) all of the above are true.
15. The volume of a gas is doubled while the temperature is held constant. The pressure of the gas
    (a) is reduced by one-half.
    (b) remains unchanged.   (c) is doubled.
    (d) varies depending on the nature of the gas.
16. Absolute zero
    (a) is −273 °C.
    (b) is the temperature at which the motion of particles stops.
    (c) is zero degrees Kelvin.
    (d) is all of the above.
17. At a certain temperature and pressure, 0.20 mol carbon dioxide has a volume of 3.1 L. A 3.1-L sample of hydrogen at the same temperature and pressure
    (a) has the same mass.
    (b) contains the same number of atoms.
    (c) has a higher density.
    (d) contains the same number of molecules.
18. Which of these gases, Ar, F_2, CO, and N_2O, diffuses the fastest?
    (a) Ar   (b) F_2 (c) CO   (d) N_2O
19. A gas cylinder has a pressure of 20.0 atm. What is this pressure in millimeters of mercury?
    (a) 1.52×10^4 mm Hg   (b) 38.0 mm Hg
    (c) 0.026 mm Hg   (d) 1520 mm Hg
20. A box with a volume of 22.4 L contains 1.0 mol nitrogen and 2.0 mol hydrogen at 0 °C. Which of the following statements is true?
    (a) The total pressure in the box is 1.0 atm.
    (b) The partial pressures of N_2 and H_2 are equal.
    (c) The total pressure is 1520 mm Hg.
    (d) The partial pressure of N_2 is 1.0 atm.
21. As the temperature of the gas in a balloon decreases,
    (a) the volume increases.
    (b) the average kinetic energy of the gas particles decreases.
    (c) the pressure increases.
    (d) all of the above are true.
22. Equal masses of ice and water are placed in a container in a refrigerator that maintains the temperature at exactly 0.0 °C. You would expect
    (a) all the ice to melt.
    (b) all the water to freeze.
    (c) the ice cube to alternately freeze and melt.
    (d) the ice to melt and water to freeze at the same time, and the amounts of each to remain equal.

23. A sample containing 3.00 mol hydrogen at STP
  (a) has a volume of 67.2 L.
  (b) has a mass of 6.0 g.
  (c) contains 1.81×10^{24} molecules.
  (d) all of the above.

24. Water could be made to boil at 92 °C by
  (a) heating the water faster.
  (b) decreasing the pressure above the water.
  (c) putting the water in a pressure cooker.
  (d) increasing the pressure above the water.

25. How large a balloon would be needed to hold 4.80×10^2 g O_2 at STP?
  (a) 672 L   (b) 15 L   (c) 336 L   (d) 480 L

26. Which of the following compounds has the lowest melting point?
  (a) N_2O (b) K_2O (c) CaI_2 (d) Li_2O

27. Which of the following changes would *not* cause an increase in the pressure of a gaseous system?
  (a) The container is made larger.
  (b) Additional amounts of the same gas are added to the container.
  (c) The temperature is increased.
  (d) A different gas is added to the container.

28. When the number of molecules that escape from a liquid exactly equals the number that return to it,
  (a) the liquid is evaporating.
  (b) equilibrium has been achieved.
  (c) the heat of vaporization of the liquid is low.
  (d) the liquid is boiling.

29. A liquid with a high vapor pressure at room temperature would be expected to
  (a) have a relatively low boiling point.
  (b) evaporate rapidly if left in an open container.
  (c) make your skin feel cool when spilled on it.
  (d) all of the above

30. A gas occupies 40.0 mL at −123 °C. What volume does it occupy at 27.0 °C?
  (a) 182 mL   (b) 8.80 mL   (c) 80.0 mL
  (d) 20.0 mL

31. A gas sample occupies 225 mL at 765 mm Hg. What volume does the gas occupy at 475 mm Hg?
  (a) 140 mL   (b) 362 mL   (c) 1.60×10^3 mL
  (d) 7.16×10^{-3} mL

8

Water and Aqueous Systems

The Story of a Unique Compound

Yellowstone Park in winter. Life on planet Earth depends on water in all its forms: liquid water, ice, and water vapor.

CHAPTER OUTLINE

CASE IN POINT: Drinking water and health

8.1 Water and hydrogen bonds

8.2 Ice

8.3 Aqueous solutions

FOLLOW-UP TO THE CASE IN POINT: Drinking water and health

8.4 Solvation

8.5 Solution formation

8.6 Solubility

A CLOSER LOOK: Kidney Stones

A CLOSER LOOK: What Do the Foam of Soda and the Bends Have in Common?

8.7 Units of concentration

8.8 Molarity

8.9 Colloids and suspensions

8.10 Colligative properties

8.11 Osmosis and dialysis

A CLOSER LOOK: Hemodialysis

Liquid water covers about three-quarters of the surface of the Earth in oceans, lakes, and rivers. Ice dominates the vast polar regions of the globe, appears as icebergs in the oceans, and whitens the temperate zones as snow in winter. Water vapor from the evaporation of surface water and from steam spouted from geysers and volcanoes is ever-present in the Earth's atmosphere. Water is the foundation of everything that lives; without it neither plant nor animal life could exist.

The human body is over 60% water. Body water is inside cells (intracellular) and outside cells (extracellular). Intracellular water is the medium in which much of the chemical activity of the body takes place. Extracellular water is the principal component of body fluids and secretions. These fluids are essential for digestion, circulation, temperature regulation, and waste elimination. This chapter tells the chemical story of water. It is a unique story, because water is a unique compound. Although we may take water for granted, making water available to drink and for other purposes is of great concern to public health officials and government leaders. Some events that threatened the water supplies of millions of people are described in the following Case in Point.

CASE IN POINT: Drinking water and health

We usually do not have to worry that we are risking our health when we drink water. However, do not take for granted that an ample supply of drinking water will always be available. Early in 1993, an outbreak of water-borne disease in Milwaukee, Wisconsin, demonstrated that constant vigilance is needed to maintain supplies of safe drinking water. Because a typical person drinks about 2 L of water every day, the maintenance of safe supplies of drinking water is a major concern to public health authorities. This concern was intensified in the spring and summer of 1993 by the flood that covered large areas of the midwestern United States. The flood knocked out more than 50 water treatment plants along the Mississippi River and its tributaries. Engulfed in flood water, the 250,000 residents of Des Moines, Iowa, endured the hardship of being without safe water for 12 days (see figure). Public health professionals need to be knowledgeable about the health risks from impure water and the methods used to make it safe. These topics are discussed in Section 8.3.

"Water, water everywhere and not a drop to drink." These words rang true when flooding caused by the rain-swollen Mississippi River disabled the water treatment facility in Des Moines, Iowa, on July 12, 1993.

8.1 *Water and hydrogen bonds*

AIM: To explain the unusual properties of water, focusing on the hydrogen-bonding between water molecules.

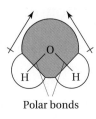

Polar bonds

Molecule has net polarity

Figure 8.1
The water molecule is bent. The bond polarities are equal, but the two dipoles do not cancel each other. There is a net polarity for the molecule; water is polar.

Water is a simple triatomic molecule, with chemical formula H_2O and structural formula H—O—H. Each O—H covalent bond in the water molecule is highly polar. Because of its greater electronegativity, oxygen wins a rather lopsided victory in its tug of war for the electron pair of the covalent O—H bond and acquires a partial negative charge. The losing hydrogens acquire a partial positive charge.

Although the water molecule's O—H bonds are polar, water would not be a polar molecule if the structure of water were linear, since the polarities of the bonds would cancel. The water molecule is not linear, however; the H—O—H bond angle is 105 degrees. Because of the water molecule's bent shape, the O—H bond polarities do not cancel and the water molecule as a whole is polar (Fig. 8.1)—the region around oxygen is slightly negatively charged, and the region around the hydrogens is slightly positively charged.

One property of polar molecules is that they attract one another. There is a dipolar attraction between the positive region of one water molecule and the negative region of another (Sec. 5.9). This attraction between water molecules is an example of *hydrogen bonding*. **Hydrogen bonds** *are attractive forces in which hydrogen covalently bonded to a very electronegative atom is strongly attracted and therefore weakly bonded to an unshared electron pair of an electronegative atom in the same molecule or in a nearby molecule.* The hydrogen bond is the strongest of the weak attractions between molecules, with about 5% of the strength of an average covalent bond. Hydrogen bonding always involves hydrogen because it is the only chemically reactive element whose valence electrons are not shielded from the nucleus by a layer of underlying electrons. The unequal sharing of electrons in a very polar covalent bond between hydrogen and an electronegative atom such as oxygen, nitrogen, or fluorine leaves the hydrogen quite

Figure 8.2
Hydrogen bonding in water. Hydrogen bonds are formed when the electron-deficient hydrogen in a polar covalent bond interacts with the unshared electron pairs on an electronegative atom. Since oxygen has two unshared electron pairs, each water molecule may have as many as four water molecules hydrogen-bonded to it, as shown here.

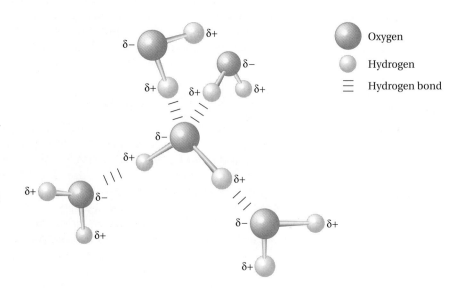

electron deficient. The hydrogen makes up for its deficiency by interacting with an unshared electron pair on a nearby electronegative atom. Figure 8.2 shows hydrogen bonding between water molecules. In water, the individual molecules change positions by sliding over one another, but because of hydrogen bonding, most do not have enough kinetic energy to escape the liquid. Hydrogen bonds are the cause of many of the unique and important properties of water.

High surface tension of water

You have probably seen a glass so filled with water that the water surface bulges above the rim. Or you may have noticed that water forms nearly spherical droplets at the end of a medicine dropper or when sprayed on a greasy surface. A steel needle floats if placed carefully on the surface of water, but it sinks immediately if it breaks through the surface. Apparently, the surface of water acts like a skin. *This property, called* **surface tension,** *is explained by the ability of water to form hydrogen bonds.* The molecules in the bulk of the liquid are completely surrounded by, and hydrogen bonded to, adjacent water molecules. At the surface of the liquid, water molecules experience an uneven attraction because they are hydrogen bonded on only one side. They are not attracted to the air because they cannot form hydrogen bonds with air molecules (Fig. 8.3). As a result, the surface molecules tend to be drawn into the body of the liquid, producing the surface tension.

All liquids have a surface tension, but water's surface tension is higher than most. The surface tension of water may be decreased by adding a *wetting agent* such as soap or detergent. *Soaps and detergents are* **surfactants** *(surface active agents).* When a detergent is added to beads of water on a greasy surface, surface tension is reduced, the beads collapse, and the water spreads out.

Low vapor pressure of water

The vapor pressure of a liquid is explained by the tendency of the molecules at the surface to escape. Hydrogen bonding among water molecules accounts for water's unusually low vapor pressure. Since hydrogen bonds cause relatively strong attractions between water molecules, the tendency of these molecules to escape is low.

High specific heat of water

You may recall that it takes 1 cal (4.18 J) of heat to raise the temperature of 1 g of water 1 °C; the specific heat capacity of water is nearly the same at 1 cal/g (4.18 J/g) at all temperatures between 0 and 100 °C. The specific heat of water is higher than that of many other substances. Iron has a specific heat of only 0.107 cal/g. Imagine we have equal masses of iron and water: It takes about 10 times less heat to raise the temperature of a gram of iron by 1 °C than to raise the temperature of a gram of water by 1 °C.

Water's high heat capacity helps to moderate daily air temperatures around large bodies of water. On a warm day, water absorbs heat from its warmer environment, lowering the air temperature. On a cool night, the

It is hydrogen bonding that holds proteins, DNA, and many other biologically important molecules in their unique shapes.

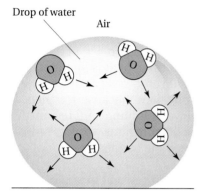

Drop of water

Air

Figure 8.3
Water molecules at the surface experience an uneven attraction. Because they cannot hydrogen bond with air molecules, they tend to be drawn into the body of the liquid, producing surface tension.

transfer of heat is from the water to its cooler environment, raising the air temperature.

High heat of vaporization and high boiling point of water

Water, because of hydrogen bonding, has the highest heat of vaporization of any known liquid. It takes 540 cal to convert 1 g of water at 100 °C into 1 g of steam at 100 °C. This is so because an extensive network of hydrogen bonds holds the molecules in liquid water tightly together. The heat of vaporization of liquid ammonia, which is less hydrogen bonded than water, is 327 cal/g. Liquid methane, which has no hydrogen bonding, has a heat of vaporization of only 122 cal/g. Temperatures in the tropics would be much higher than they are if water did not absorb heat while evaporating from the surface of the surrounding oceans.

The reverse of vaporization is condensation. When 1 g of steam at 100 °C condenses to 1 g of water at 100 °C, 540 cal of heat is liberated; the heat of condensation of water is therefore 540 cal/g. You can get a severe burn if you allow steam to condense on your hand because this heat is absorbed by your skin. The condensation of water is important in controlling local temperatures. Temperatures in the polar regions would be much colder if water vapor did not release its heat while condensing out of the air.

Compounds with low molar masses are often gases at normal atmospheric pressure (Table 8.1). Ammonia (NH_3) is a typical example. Ammonia, with a molar mass of 17 g, is a gas above −33 °C. Yet water, with a molar mass of 18 g, is a liquid and has a boiling point of 100 °C. The reason is hydrogen bonding. Hydrogen bonding is more extensive in water than in ammonia, and it takes much more heat to disrupt the attractions among water molecules than among ammonia molecules. This is one reason why water has become the medium in which living cells conduct their chemical reactions. If the hydrogen bonding in water were as weak as it is in ammonia, water would be a gas at the temperatures found in our biosphere.

PRACTICE EXERCISE 8.1

Why does water have a much higher boiling point (100 °C) than methane (−164 °C) when their molar masses are almost the same?

Table 8.1 Melting and Boiling Points of Some Substances with Low Molar Mass

Name of substance	Formula	Molar mass (g)	Melting point (°C)	Boiling point (°C)
methane	CH_4	16	−183	−164
ammonia	NH_3	17	−77.7	−33.3
water	H_2O	18	0	100
neon	Ne	20	−249	−246
methanol	CH_3OH	32	−93.9	64.9
hydrogen sulfide	H_2S	34	−85.5	−60.7

8.2 Ice

AIM: To explain the low density and high heat of fusion of ice.

As a typical liquid is cooled, it contracts slightly. Its density increases because its volume decreases, but its mass remains constant. If the cooling continues, the liquid eventually solidifies. Because the density of most solids is greater than that of its melted form, a solid sinks in its own liquid. Figure 8.4 shows solid paraffin wax at the bottom of a container of melted paraffin wax but ice cubes floating on water. As water is cooled, it first behaves like a typical liquid; its density gradually increases until the temperature reaches 4 °C. Below 4 °C, however, the density of water starts to decrease (Table 8.2). At 0 °C ice begins to form. Ice has about 10% greater volume and, therefore, a lower density than water, so it floats on water. This is unusual behavior for a solid. Ice is one of very few solids that floats on its own liquid.

Why does ice behave so differently? Because there is a good deal of space in its structure, more than in the structure of water. The structure of ice is an ordered open framework of water molecules arranged like a honeycomb (Fig. 8.5). Hydrogen bonding holds the water molecules in place; at low temperatures they do not have sufficient energy to break out of the rigid framework. When ice melts, the framework collapses, the water molecules pack together, and the water becomes more dense.

The fact that ice floats and water is most dense at 4 °C has important consequences for living organisms. The ice on a pond acts as an insulator for the water beneath, preventing it from freezing except under extreme conditions. Since the water at the bottom of a frozen pond is warmer than 0 °C, aquatic life is able to survive. If ice were more dense than water, all bodies of water would tend to freeze solid during the winter months, and aquatic life would be destroyed.

Figure 8.4
Solid paraffin wax sinks in melted wax (on right); ice floats on water (on left).

Table 8.2 Density of Water and Ice

Physical state	Temperature (°C)	Density (g/cm³)
water	100	0.9584
water	50	0.9881
water	25	0.9971
water	10	0.9997
water	4	1.0000*
water	0	0.9998
ice	0	0.9168

*Most dense.

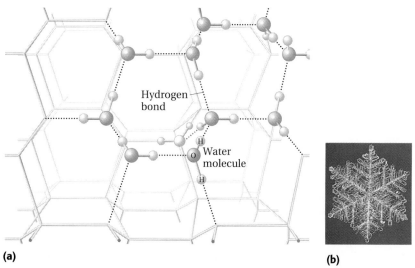

(a)

(b)

Figure 8.5
(a) Ice is an ordered open framework of water molecules held together by hydrogen bonds. Compare this structure with that of the diamond in Figure 5.8.
(b) The exquisite symmetry of a snowflake results from this orderly arrangement.

The increase in the volume of ice as water freezes is also important. Creatures that live in lakes and ponds would soon deplete the available oxygen supply if the water were covered with a solid sheet of ice. However, as surface water freezes, it expands to the point where its surface area is greater than that of the water below it. The resulting strains cause the ice to crack and relieve the stress. Enough oxygen to support life usually reaches the underlying water through the cracks.

Injury from frostbite results from the expansion of water on the outside and inside of tissue cells as it turns into crystals of ice. The cells rupture, and the damage becomes apparent when the tissue thaws.

Water molecules require a considerable amount of kinetic energy to return to a liquid state from a frozen state. Ice melts at 0 °C, a high temperature for such a small molecule (see Table 8.1). The heat of fusion of ice is 80 cal/g. We can get an idea of how much energy this is by realizing that while 80 cal only changes 1 g of ice at 0 °C to 1 g of water at 0 °C, it also will raise the temperature of 1 g of water from 0 to 80 °C. We take advantage of water's high heat of fusion when we put an ice pack on a sprained ankle. The ice pack absorbs large amounts of body heat as it melts.

8.3 Aqueous solutions

AIM: To define solution, aqueous solution, solute, and solvent and give an example of each.

Focus

Water is an excellent solvent.

Chemically pure water never exists in nature because water dissolves many of the substances in and around it. Tap water contains varying amounts of dissolved minerals and gases, many of which are essential constituents of our diets. Untreated water can contain contaminants that make people sick. You may recall that the Case in Point in the introduction to this chapter concerned the safety of drinking water. In our follow-up, we will examine some potentially dangerous contaminants of water supplies and how these contaminants can be removed.

Aqueous
aqua (Latin): water

Water samples containing dissolved substances are **aqueous solutions.** *The dissolving medium is the* **solvent,** *and the dissolved particles are the* **solute.** If we dissolve sodium chloride in water, water is the solvent, and sodium chloride is the solute.

Solutions are homogeneous mixtures. They are also stable; sodium chloride does not settle out when its solutions are allowed to stand. Solute particles can be either ionic or molecular, but their average diameters are usually less than 1.0 nm (10^{-9} m). If a solution is filtered, both the solute and the solvent pass through the filter unimpeded. (Solvents and solutes may be gases, liquids, or solids, as was previously shown in Table 1.1.)

Blood plasma contains sodium (Na^+), potassium (K^+), magnesium (Mg^{2+}), and calcium (Ca^{2+}) cations and chloride (Cl^-) and bicarbonate (HCO_3^-) anions.

Substances that dissolve most readily in water include ionic compounds, such as sodium sulfate and magnesium chloride, and polar covalent molecules, such as sugar and hydrogen chloride. Nonpolar covalent molecules, such as methane and those in grease and gasoline, do not dissolve in water. Grease will, however, dissolve in gasoline. Why is there a difference? To answer this question, we will consider the structures of the sol-

FOLLOW-UP TO THE CASE IN POINT: Drinking water and health

The major concern of health officials in the midwestern floods of 1993 was the possibility of contamination of public water supplies by microorganisms such as bacteria, viruses, and protozoa. These microbes are the most rapidly acting agents of waterborne illness. A protozoa contaminated the municipal water in Milwaukee in 1993. A person infected by this protozoa could experience weeks of diarrhea. Bacteria in water supplies has been responsible for past epidemics of cholera and typhus, both deadly diseases. Certain strains of the common bacterium *Escherischia coli (E. coli)* sometimes appear in water supplies. These *E. coli* strains and coliform bacteria can cause acute gastrointestinal distress, diarrhea, and other flulike symptoms. The viruses that carry hepatitis A and poliomyelitis, among other serious diseases, are also carried by water. Slower-acting but still harmful contaminants in drinking water include such heavy metal ions as barium, cadmium, chromium, lead, mercury, and silver. In addition, more than 300 organic compounds, some known to be carcinogenic (cancer-causing), have been identified in drinking water supplies.

Although it is possible to remove all the microorganisms, inorganic and organic contaminants, and suspended solids from water for drinking, the expense of doing so would be prohibitive. Safe drinking water can be produced in large volumes by some fairly simple procedures in water treatment plants. Much of the particulate matter in water is removed by coagulation and settling. Coagulating agents may be added to the water to aid in removing contaminants. Aluminum sulfate ($Al_2(SO_4)_3$) and iron(III) chloride ($FeCl_3$) are good coagulating agents. These compounds produce fluffy solids that slowly settle out of the water. Most of the suspended particles of organic matter also settle out. Under carefully controlled conditions, more than 99% of potentially harmful microorganisms and heavy metal ions may be adsorbed on these solid deposits. The deposits are removed by filtering the water through sand. Large populations of oxygen-using bacteria (aerobic bacteria) are then permitted to oxidize the remaining dissolved organic matter into solids that separate as a sludge. Air is bubbled into the water to provide oxygen for the aerobic bacteria as they do their work. The purified water is often disinfected by chlorinating it. A low level of dissolved chlorine generally kills any remaining harmful bacteria and protects the water supply as it is distributed through the water supply system.

vent and the solute and the forces of attraction between them in the following section.

8.4 Solvation

AIMS: *To describe the process of solvation. To explain the statement that "like dissolves like" and use it to predict the solubility of one substance in another. To calculate the percent of water in a given hydrate. To distinguish between electrolytes and nonelectrolytes, giving examples of each.*

Focus
Substances dissolve in water because they are polar or ionic.

Water molecules are in continuous motion because of their kinetic energy. When a crystal of sodium chloride is placed in water, the water molecules collide with, and exert an attractive force on, the ions on the surface of the crystal. The attractive forces between the bombarding solvent molecules (H_2O) and the solute (NaCl) will overcome the attractive forces within the

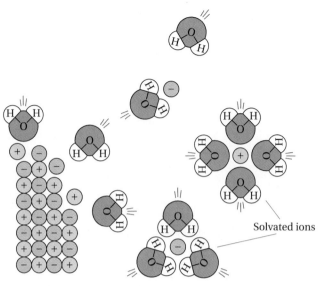

Solvated ions

Figure 8.6
When an ionic solid dissolves in water, the ions become solvated.

Surface of ionic solid

solute, and the sodium chloride crystal will dissolve (Fig. 8.6). When an ionic solid dissolves in water, the negatively and positively charged ions become surrounded by solvent molecules. **Solvation** *is the interaction of a solute particle with solvent molecules.* A solute ion is stabilized by solvation relative to the stability of the ion in the undissolved solid. Solvation of ions is important because ionic compounds that do not become solvated do not dissolve. Water dissolves polar nonionic compounds as well as ionic compounds. Many of these compounds are solvated by water molecules because they form hydrogen bonds with water.

In some ionic compounds, the attractive forces within the crystal are stronger than the attractive forces exerted by water. These compounds cannot be solvated and are therefore insoluble in water. Barium sulfate ($BaSO_4$) and calcium carbonate ($CaCO_3$) are examples of insoluble ionic compounds. Because of its insolubility and opacity to X rays, barium sulfate is given to patients before X-ray examinations of their gastrointestinal (GI) tract. With a lining of barium sulfate, the outline of the GI tract is visible on an X ray (see Fig. 10.10 on page 306).

What about dissolving grease in gasoline? Both grease and gasoline are composed of nonpolar molecules. They mix to form a solution, but not because the solute and solvent are favorably attracted. Rather, it is because there are no particularly unfavorable repulsive forces between the two. As a rule, polar solvents dissolve ionic and polar compounds; nonpolar solvents dissolve nonpolar compounds. This relationship can be summed up in the expression "like dissolves like."

PRACTICE EXERCISE 8.2

Which of the following substances dissolve in water? Give reasons for your choice.

(a) HCl   (b) NaI   (c) NH_3 (d) $MgSO_4$ (e) CH_4

(f) $CaCO_3$ (g) gasoline

Water of hydration

Water molecules are an integral part of the crystal structure of many substances. *The water in the crystal is called* **water of hydration** *or* **water of crystallization.** Calcium sulfate dihydrate ($CaSO_4 \cdot 2H_2O$) is commonly known as gypsum. (It is conventional to use a dot between the formula of the compound and the number of water molecules in the compound in writing the formula of a hydrate.) Gypsum is an important hydrate in making casts to hold fractured bones in place. When gypsum is heated, it loses some of its water of hydration and crumbles to a white powder—plaster of paris:

$$2CaSO_4 \cdot 2H_2O(s) \xrightarrow[125\,^\circ C]{\text{Heat}} (CaSO_4)_2 \cdot H_2O(s) \ + \ 3H_2O(g)$$

Gypsum   Plaster of paris
(calcium sulfate   (calcium sulfate
dihydrate)   hemihydrate)

When water is added to plaster of paris, the reverse process occurs, and crystals of the dihydrate salt "set" to a hard solid. Casts that hold fractured bones in place are made by covering the injured limb with a layer of cotton wrapped with gauze soaked in wet plaster of paris. The cast gets warm as the plaster sets because the rehydration process is quite exothermic.

When an aqueous solution of copper(II) sulfate is allowed to evaporate, deep blue crystals of copper(II) sulfate pentahydrate are deposited. The chemical formula for this compound is $CuSO_4 \cdot 5H_2O$. These blue crystals always contain five molecules of water for every pair of ions, one copper(II) and one sulfate. Copper(II) sulfate pentahydrate crystals are dehydrated by heating them above 250 °C. Water of hydration is driven off as steam and the blue crystals crumble to an off-white powder (Fig. 8.7):

$$CuSO_4 \cdot 5H_2O(s) \xrightarrow[250\,^\circ C]{\text{Heat}} \quad CuSO_4(s) \quad + \quad 5H_2O(g)$$

Copper(II) sulfate   Copper(II) sulfate   Steam
pentahydrate   (anhydrous)

Figure 8.7
Crystals of copper(II) sulfate pentahydrate (a) lose their blue color and crumble to an off-white powder when heated (b).

(a)   **(b)**

EXAMPLE 8.1   **Determining percent by mass of water in a hydrate**

Calculate the percent by mass of water in washing soda ($Na_2CO_3 \cdot 10H_2O$).

SOLUTION

The molar mass of $Na_2CO_3 \cdot 10H_2O$ is 286 g. The mass of water within the molar mass of $Na_2CO_3 \cdot 10H_2O$ is 180 g (10×18 g). And, finally,

$$\text{Water (percent by mass)} = \frac{180 \cancel{g}}{286 \cancel{g}} \times 100 = 62.9\%$$

PRACTICE EXERCISE 8.3

What is the percent by mass of water in $CuSO_4 \cdot 5H_2O$?

Electrolytes and nonelectrolytes

Most water contains some dissolved minerals and is therefore an electrolytic solution. Do not handle or use electrical equipment near water or if you have wet hands.

Compounds that conduct an electric current in aqueous solution or the molten state are **electrolytes** (Fig. 8.8). All ionic compounds are electrolytes. Sodium chloride, magnesium sulfate, copper sulfate, and sodium hydroxide are typical examples of water-soluble electrolytes. They conduct electricity in solution and in the molten state. Barium sulfate is an example of an ionic compound that conducts electricity in the molten state but not in aqueous solution because it is insoluble.

Compounds that do not conduct an electric current in aqueous solution or the molten state are **nonelectrolytes.** Since they are nonionic, many mol-

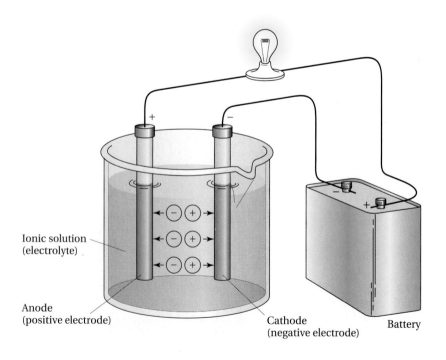

Figure 8.8
Diagram of a simple apparatus for testing conductivity. If the test solution is an electrolyte—that is, conducts an electric current—the bulb lights. Nonelectrolytes do not conduct a current, and the bulb does not light.

Ionic solution (electrolyte)

Anode (positive electrode)

Cathode (negative electrode)

Battery

ecular compounds are nonelectrolytes. Most compounds of carbon such as cane sugar and rubbing alcohol are nonelectrolytes.

Some very polar molecular compounds are electrolytes because, although they are not ionic, they ionize when they dissolve in water or melt. Neither ammonia (NH_3) nor hydrogen chloride (HCl) is an electrolyte; both are polar covalent compounds. However, an aqueous solution of ammonia conducts electricity because the ammonia reacts with water to form some ammonium ions (NH_4^+) and hydroxide ions (OH^-) in the solution (Sec. 9.2). Similarly, hydrogen chloride produces hydronium ions (H_3O^+) and chloride ions (Cl^-) (Sec. 9.2); a hydrogen chloride solution is therefore an electrolyte.

Electrolytes are important in the body, where they help to maintain acid-base balance and water balance (Chap. 22). The cation electrolytes in the body are largely sodium (Na^+), potassium (K^+), calcium (Ca^{2+}), and magnesium (Mg^{2+}) ions. The major anion electrolytes are hydrogen carbonate (HCO_3^-), chloride (Cl^-), sulfate (SO_4^{2-}), monohydrogen phosphate (HPO_4^{2-}), and dihydrogen phosphate ($H_2PO_4^-$) ions.

8.5 Solution formation

AIM: To list three factors that determine how fast a soluble substance dissolves.

The nature of the solvent and solute affects *whether* a substance will dissolve, but several other factors determine *how fast* a soluble substance dissolves.

Agitation

If we place sodium chloride crystals in a flask containing water and we shake the flask and its contents, the crystal disappears more quickly than if we do not shake the flask. *Agitation* makes solutes dissolve more rapidly because it brings more solvent molecules into contact with the solute. This is why it helps to stir lemonade after adding sugar.

Temperature

Temperature influences the rate at which solutes dissolve. A crystal of salt dissolves much more rapidly in hot water than in cold water because the kinetic energy of the water molecules—and hence the frequency and force of their collisions with the crystal surface—is greater at the higher temperature.

Particle size

Another factor that determines the rate of dissolution of a solute is its *particle size*. Dissolving is a surface phenomenon. A crystal crushed to a powder dissolves more rapidly than a single crystal because a greater surface area of ions is exposed to the colliding water molecules.

8.6 Solubility

AIMS: *To distinguish among saturated, unsaturated, and super-saturated solutions. To use Henry's law to solve gas solubility problems.*

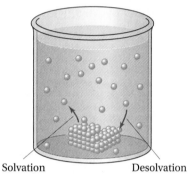

Solvation   Desolvation

Figure 8.9
When the rate of solvation (dissolving) equals the rate of desolvation (crystallization), a state of dynamic equilibrium exists.

When substances in body fluids exceed their solubility levels, they come out of solution and crystallize. These crystals produce painful and crippling medical conditions when they form in the gallbladder (gallstones), the kidneys (kidney stones), or the joints (gout).

If we add 36 g sodium chloride to 100 mL water at 25 °C, all the salt dissolves. However, if we add one more gram of salt and stir, no matter how vigorously or how long, only 0.2 g of the last portion goes into solution. This may seem surprising, since the kinetic-molecular theory says that water molecules are in a state of continuous motion and must continue to bombard the excess solid, removing and solvating ions. This suggests that all the sodium chloride should eventually disappear, but this does not happen. What does take place is an exchange. As new particles from the solid are solvated and enter into solution (dissolve), an equal number of particles already in solution become desolvated and deposit as a solid (crystallize). If we were to examine the undissolved crystals of sodium chloride over a period of time, we might notice a change in their shape or location, but their mass would remain constant.

Since particles move from the solid to the solvated state and back to the solid again, there is no net change in the overall system; this is another example of a *dynamic equilibrium* (Fig. 8.9). A state of dynamic equilibrium exists between the solution and the undissolved solute provided the temperature remains constant. The sodium chloride solution is now **saturated**—*the solution contains the maximum amount of solute* (36.2 g) that can be dissolved in 100 mL of water at 25 °C. *The* **solubility** *of a substance is the amount of substance that dissolves in a given quantity of a solvent at a given temperature to produce a saturated solution.* Solubility is usually expressed in grams of solute per 100 mL of solvent. A solution that contains less solute than a saturated solution is **unsaturated.**

PRACTICE EXERCISE 8.4
What mass of NaCl can be dissolved in 7.50×10^2 mL of water at 25 °C?

Effects of temperature on solubility

The solubility of most substances increases as the temperature of the solvent is increased. At 25 °C, the solubility of sodium chloride in water is 36.2 g per 100 mL. When the temperature is raised to 100 °C, the solubility increases to 39.1 g per 100 mL. There are some solutes that decrease in solubility as the temperature is increased. The solubility of sodium sulfate in water drops from 50 g per 100 mL at 40 °C to 43 g per 100 mL at 80 °C. Gases such as chlorine, hydrogen chloride, sulfur dioxide, ammonia, and the components of air become less soluble in water as the temperature of the solution is raised.

Supersaturated solutions

If we raise the temperature of a saturated solution until all the excess solid dissolves and then allow the system to cool slowly and undisturbed to its

(a)

(b)

(c)

Figure 8.10
When a seed crystal is added to a supersaturated solution (a), the excess solute crystallizes (b), (c).

original temperature, the excess solute does not always immediately crystallize. *Such a solution, which contains more solute than it can theoretically hold at the given temperature, is* **supersaturated.** In a supersaturated solution, a dynamic equilibrium cannot exist between the dissolved solute and the undissolved solid because there is no undissolved solid. To initiate crystallization in a supersaturated solution, we could scratch the inside surface of the container or simply add a very small crystal—a *seed crystal*—of the solute (Fig. 8.10).

The rate at which excess solute deposits on the surface of the seed is often impressively rapid. Rainmaking based on scientific principles is done by seeding clouds, which are made of air supersaturated with water vapor. When tiny silver iodide (AgI) crystals are dusted on a cloud, water molecules attracted to the ionic particles come together and form droplets that act as seeds for other water molecules. The water droplets grow and eventually fall as rain when they are heavy enough.

The amount of solutes in the various fluids of the body must remain relatively constant. Sometimes, however, owing to disease or poor diet, the fluid and electrolyte balance is disturbed, and the substances in the body fluids exceed their solubility levels and start to crystallize. If crystallization occurs in the kidneys, kidney stones form (see A Closer Look: Kidney Stones). The nature of kidney stones can vary, but many are composed of calcium phosphate, $Ca_3(PO_4)_2$. Sometimes cholesterol crystals start to precipitate in the gallbladder, aggregating with other matter to become gallstones. Gallstones may then block the bile duct, stopping the flow of bile to the digestive tract and causing severe discomfort.

Solubility of gases in liquids

Henry's law *states that at a given temperature the solubility* S *of a gas in a liquid is directly proportional to the pressure* P *of the gas above the liquid.* In other words, as the pressure of the gas above the liquid increases, the solubility of the gas increases; as the pressure of the gas decreases, the solubility

Kidney Stones

The movement of a kidney stone through the urinary tract is one of the most agonizing episodes experienced by a human being. Urine is a solution that is supersaturated with many substances. Small amounts of materials in the urine help keep these substances dissolved. For example, it would take about 2.5 L of water to dissolve the uric acid normally present in 100 mL of urine. Under certain conditions, however, such as an increase in concentration or a change in the acidity of the urine, some of the substances are deposited as solids. These deposits might occur in the kidney or in the bladder. Eventually, these deposits may form jagged kidney stones, called *renal calculi* (see figure). The stones can be very small or sometimes as large as a peach. Acidic urine often produces stones composed of calcium oxalate (CaC_2O_4) or uric acid and its salts, products of the body's breakdown of nitrogen compounds (Sec. 26.11). Alkaline urine is more likely to produce stones composed of calcium phosphate ($Ca_3(PO_4)_2$), calcium carbonate ($CaCO_3$), or magnesium ammonium phosphate ($MgNH_4PO_4$). Physicians have

These kidney stones are composed of magnesium ammonium hydrogen phosphate.

some success in dissolving some types of kidney stones, and attention to diet can help prevent recurrences. If the stones are predominantly calcium oxalate, removing foods containing oxalates such as spinach and rhubarb from the diet is recommended.

A device called a *lithotripter* pulverizes kidney stones by bombarding them with highly focused shock waves. A single instrument can treat about 1500 patients a year, eliminating kidney stones without surgery.

of the gas decreases. Or, in much the same way as we write the gas laws:

$$\frac{S_1}{P_1} = \frac{S_2}{P_2}$$

The relationship between gas pressure and gas solubility is very important, as we see in A Closer Look: What Do the Foam of Soda and the Bends Have in Common?

EXAMPLE 8.2

Calculating the solubility of a gas

If the solubility of a gas in water is 0.77 g/L at 3.5 atm of pressure, what is its solubility, in grams per liter, at 1.0 atm of pressure? The temperature is held constant at 25 °C.

SOLUTION

Use the same general method used for solving gas law problems in the preceding chapter. We know that $P_1 = 3.5$ atm, $S_1 = 0.77$ g/L, and $P_2 = 1.0$ atm. We need to find S_2. We can begin by rearranging Henry's law to give S_2:

$$S_2 = \frac{S_1 \times P_2}{P_1}$$

What Do the Foam of Soda and the Bends Have in Common?

The foaming of soda drinks and the painful, often fatal "bends" that are an occupational hazard of deep-sea divers who surface too rapidly illustrate the effect of pressure on the solubilities of gases in liquids. A capped bottle of soda is a solution of carbon dioxide in water (see figure, part a). The carbon dioxide in solution is in equilibrium with a higher than atmospheric pressure of carbon dioxide above the liquid. When the bottle is uncapped, the equilibrium is disturbed, and the pressurized gas in the bottle escapes into the atmosphere (see figure, part b). With a lower carbon dioxide pressure now above the liquid, the solubility of dissolved carbon dioxide is reduced. The carbon dioxide that leaves solution forms bubbles that cause the soda to froth as they make their way to the atmosphere (see figure, part c). Eventually, the system reestablishes equilibrium at the lower pressure (see figure, part d).

Like air, water exerts pressure on anything immersed in it; the deeper divers go, the higher the pressure exerted by the mass of water above them. The pressure of the air that divers breathe must be the same or slightly greater than the water pressure; otherwise, the air supply hoses may collapse.

Nitrogen, not very soluble in blood at normal atmospheric pressure, has a higher solubility at higher pressures. As a result, the blood of a diver who surfaces too rapidly behaves like the liquid in a rapidly opened bottle of soda. At the lower pressure of the atmosphere, the excess nitrogen leaves solution and forms bubbles. Nitrogen diffuses from the blood vessels slowly. The bubbles of

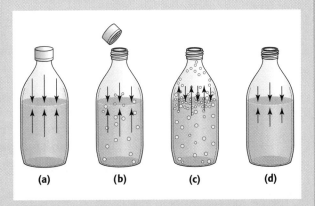

(a) Bottle sealed. High CO_2 pressure above liquid; concentration of CO_2 in the liquid is high. System is at equilibrium. (b) Seal removed. Equilibrium disturbed. Low CO_2 pressure above liquid; concentration of CO_2 in solution is high. (c) System changes to reestablish equilibrium. Low CO_2 pressure above liquid; CO_2 starts to leave solution. (d) System at equilibrium again. Low CO_2 pressure above the liquid; concentration of CO_2 in solution is low.

nitrogen are trapped in the blood, causing the blood to froth. The diver has the bends—or, in a more descriptive but inaccurate phrase, the diver's blood "boils." Bends may be prevented or its effects lessened by placing the diver in a decompression chamber at high air pressure. Pressure is then slowly lowered so that undissolved nitrogen can diffuse out of the blood vessels before bubbles have a chance to form.

The incidence of the bends has been greatly reduced by replacing the nitrogen in breathing mixtures with helium. Helium is less soluble and diffuses out of blood vessels faster than nitrogen. Less helium dissolves in the blood in the first place. The amount that does dissolve diffuses out of the blood vessels before bubbles can form.

Substitute values in this equation:

$$S_2 = \frac{0.77 \text{ g/L} \times 1.0 \text{ atm}}{3.5 \text{ atm}} = 0.22 \text{ g/L}$$

When the pressure is reduced to 1.0 atm, the solubility of the gas decreases to 0.22 g/L.

PRACTICE EXERCISE 8.5

A gas has a solubility in water at 0 °C of 3.6 g/L at a pressure of 1 atm. What pressure is required to produce an aqueous solution containing 9.5 g/L of the same gas at 0 °C?

8.7 Units of concentration

AIM: To define solution concentration and work problems involving percent (volume/volume), percent (mass/volume), parts per million, and parts per billion.

Focus

The concentration of a solution must be expressed in quantitative units.

To be useful, concentrations of solutions must be expressed in quantitative and consistent ways. *The* **concentration** *of a solution is a measure of the amount of solute that is dissolved in a given quantity of the solution.* Since most chemical reactions, including those in your body, take place in solution, and since chemical compounds react in definite proportions, it is important to be able to express solution concentrations quantitatively.

Concentrated or dilute?

These two terms—*concentrated* and *dilute*—are imprecise and are used only in a relative sense. We would probably describe an aqueous solution of sodium chloride containing a teaspoon of salt in a cup of water as dilute when comparing it with another sodium chloride solution containing a handful of salt in a cup of water. The former solution might be described as concentrated if it were compared with another solution containing only a pinch of salt in a cup of water. By describing a solution as concentrated or dilute, we are expressing the concentrations of the solutions only in a very general way.

Volume/volume percent

The concentration of a liquid solute can be expressed as a percent of the solution by volume. *Percent* means "parts per hundred." If 20 mL of rubbing alcohol is diluted to a total volume of 100 mL with water, the final solution is 20% alcohol by volume. The volume/volume percent can be written as 20 percent (volume/volume), 20 percent (v/v), or 20% (v/v).

$$\text{Percent by volume} = \frac{\text{Volume of solute}}{\text{Volume of solution}} \times 100$$

The denominator in the expression is equal to the total volume of the final solution.

Aqueous solutions of hydrogen peroxide used as antiseptics contain 3% (v/v) H_2O_2.

EXAMPLE 8.3   Determining a percent by volume

What is the percent by volume of ethanol (ethyl alcohol, C_2H_5OH) in a solution when 75 mL of ethanol is diluted to a volume of 250 mL with water?

SOLUTION

The volume of ethanol is 75 mL in a total volume of 250 mL. Therefore,

$$\text{Percent by volume} = \frac{75 \text{ mL}}{250 \text{ mL}} \times 100 = 30\% \text{ ethanol (v/v)}$$

PRACTICE EXERCISE 8.6

If 10 mL of acetic acid (CH_3COOH) is diluted with water to a total solution volume of 200 mL, what is the percent by volume of acetic acid in the solution?

Mass/volume percent

A commonly used relationship to express the concentrations of solids dissolved in liquids is mass/volume percent. It is usually convenient to weigh the solute in grams and to measure the volume of the total solution in milliliters. Percent (mass/volume) is the number of grams of solute per 100 mL of solution:

$$\text{Percent (mass/volume)} = \frac{\text{Mass of solute (g)}}{\text{Solution volume (mL)}} \times 100$$

A solution containing 7 g sodium chloride in 100 mL solution is 7 percent (mass/volume) or 7% (m/v).

| EXAMPLE 8.4 | Calculating a mass/volume percent |

How many grams of glucose ($C_6H_{12}O_6$) would you need to prepare 2.0 L of 2.0% glucose (m/v) solution?

SOLUTION

A 2.0% glucose (m/v) solution means that 100 mL of the solution contains 2.0 g glucose. Therefore, 1 L (1000 mL) of the solution contains 20 g glucose. The calculation is

$$\frac{2.0 \text{ g}}{100 \text{ mL}} \times \frac{1000 \text{ mL}}{1 \text{ L}} = 20 \text{ g/L}$$

To make 2.0 L of 2.0% glucose (m/v) requires

$$2.0 \text{ L} \times \frac{1000 \text{ mL}}{1 \text{ L}} \times \frac{2.0 \text{ g}}{100 \text{ mL}} = 40 \text{ g glucose}$$

The required mass of glucose is therefore 40 g.

| EXAMPLE 8.5 | Determining a mass/volume percent |

The physiologic saline solution used in hospitals is 0.90% NaCl (m/v). How many grams of sodium chloride are needed to make 500 mL of physiologic saline?

SOLUTION

A 0.90% NaCl (m/v) solution contains 0.90 g sodium chloride in 100 mL. Therefore, to make 500 mL requires 5 times the amount of sodium chloride, or 4.5 g. The calculation is done as follows:

$$500 \text{ mL} \times \frac{0.90 \text{ g}}{100 \text{ mL}} = 4.5 \text{ g NaCl}$$

PRACTICE EXERCISE 8.7

How many grams of magnesium sulfate do you need to make 250 mL of a 1.6% $MgSO_4$ (m/v) solution?

PRACTICE EXERCISE 8.8

A solution contains 2.7 g $CuSO_4$ in 75 mL of solution. What is the percent (mass/volume) of the solution?

Percentage composition often can be misleading. When we read a label that says a product contains 5% glucose, what does it mean? It is probably percent (mass/volume), but we cannot be certain unless the units are specified. When using percentages to express concentration, be sure to state the units.

Parts per million, parts per billion

Parts per million (ppm) and parts per billion (ppb) are convenient units for expressing concentrations of extremely dilute solutions. They are particularly useful for expressing the concentrations of trace pollutants and toxic chemicals in air and water. A concentration of 1 ppm means 1 mg of solute per kilogram of solution; if water is the solvent, 1 ppm means 1 mg of solute per liter of solution. (Recall that 1 L of water has a mass of 1 kg.) You can see that 1 mg is one-millionth part of a kilogram (10^{-6} kg) by the unit conversion:

$$\frac{1 \text{ mg solute}}{1 \text{ kg solution}} \times \frac{1 \text{ g solute}}{10^3 \text{ mg solute}} \times \frac{1 \text{ kg solution}}{10^3 \text{ g solution}} = \frac{1 \text{ g solute}}{10^6 \text{ g solution}} = 1 \text{ ppm}$$

We can calculate parts per million by dividing the grams of solute by the grams of solution and multiplying the result by 10^6:

$$ppm = \frac{\text{g solute}}{\text{g solution}} \times 10^6$$

Similarly, 1 microgram (μg) of solute per kilogram of solution is 1 ppb because 1 μg is one-billionth of a kilogram (10^{-9} kg). The expression for parts per billion is

$$ppb = \frac{\text{g solute}}{\text{g solution}} \times 10^9$$

A solution containing 0.1 mg lead chloride ($PbCl_2$) per liter is 0.1 ppm $PbCl_2$

or 100 ppb $PbCl_2$. These low concentrations are difficult to visualize, but 1 ppm could be represented by a pebble with a volume of 1 cm^3 in a cube of concrete measuring 1 m along each edge; 1 m^3 is equal to 1 million cubic centimeters. A concentration of 1 ppb could be represented by a pebble with a volume of 1 cm^3 in a cube of concrete measuring 10 m along each edge; 1000 m^3 is equal to 1 billion cm^3.

EXAMPLE 8.6

Calculating parts per million

A 2.0-kg sample of water contains 3.0 mg sodium fluoride. What is the concentration of NaF in parts per million?

SOLUTION

Convert masses to grams: 2.0 kg = 2.0×10^3 g and 3.0 mg = 3.0×10^{-3} g; thus,

$$\text{ppm} = \frac{3.0 \times 10^{-3} \text{ g NaF}}{2.0 \times 10^3 \text{ g solution}} \times 10^6 = 1.5 \text{ ppm NaF}$$

PRACTICE EXERCISE 8.9

The blood of a child suspected of having lead poisoning is analyzed for lead(II) ions (Pb^{2+}). The child's blood is found to contain 0.003 mg of lead ions per 100 g of blood. What is the concentration of lead(II) ions in parts per billion?

8.8 Molarity

***AIMS:** To define and work problems involving the molarity of a solution. To explain how to prepare a dilute solution from a more concentrated solution of known molarity.*

Focus

Molarity is an important unit of concentration in chemistry.

The concentrations of the different electrolytes in blood plasma vary widely. A normal sodium concentration is about 0.14 *M*, while chloride ion concentration is about 0.10 *M*. The concentration of potassium ion is much lower, about 0.004 *M*.

Just as the mole is an important measure of the quantity of matter in chemistry, *molarity* is an important unit of concentration. **Molarity** *(abbreviated* M) *is defined as the number of moles of a solute dissolved in 1 L of solution.* (Note that the volume is the volume of the *total solution*, not the volume of the solvent.)

Determination of solution molarities

If we dissolve 0.5 mol of a solute in water and dilute with water to a volume of exactly 1 L, we get a 0.5 molar (0.5 mol/L) or 0.5 *M* solution. To determine the molarity of a solution, we therefore need to know the number of moles in 1 L of the solution. We can find the molarity of the solution by this relationship:

$$\text{Molarity } M = \frac{\text{Number of moles of solute}}{\text{Number of liters of solution}}$$

EXAMPLE 8.7   **Finding molarity from moles of solute**

What is the molarity of a solution made by adding 2 mol glucose to sufficient water to give 5 L of solution?

SOLUTION

Divide the number of moles by the volume in liters to obtain molarity
(mol/L):

$$\frac{2 \text{ mol glucose}}{5 \text{ L solution}} = 0.4 \text{ mol/L}$$

The solution is 0.4 molar (0.4 M) in glucose.

EXAMPLE 8.8   **Finding molarity from mass of solute**

Physiologic saline is 0.90% NaCl (w/v). What is its molarity?

SOLUTION

Since the saline solution is 0.90% NaCl (w/v), it contains 0.90 g NaCl per 100
mL or 9.0 g NaCl per 1000 mL (9.0 g NaCl per liter). However, molarity is
moles per liter. Therefore, we must convert from grams per liter to moles
per liter. To do this, we need the molar mass of sodium chloride:

$$\text{Molar mass NaCl} = 23.0 \text{ g} + 35.5 \text{ g} = 58.5 \text{ g}$$

$$\text{Moles} = \frac{\text{Mass (g)}}{\text{Molar mass NaCl}} = \frac{9.0 \text{ g}}{58.5 \text{ g/mol}} = 0.15 \text{ mol}$$

$$\text{Molarity} = \frac{\text{Moles}}{\text{Liter}} = \frac{0.15 \text{ mol}}{1 \text{ L}} = 0.15 \text{ } M \text{ solution}$$

PRACTICE EXERCISE 8.10

A solution contains 0.70 mol NaCl in 250 mL of solution. What is the
molarity of the solution?

PRACTICE EXERCISE 8.11

An aqueous solution has a volume of 2.0 L and contains 36.0 g glucose.
If the molar mass of glucose is 180 g, what is the molarity of the solution?

EXAMPLE 8.9   **Determining moles and mass of solute**

How many moles of solute are present in 1.5 L of 0.20 M Na_2SO_4? What is the
mass of solute?

SOLUTION

Rearrange the expression for molarity to give:

Number of moles of solute = molarity M × number of liters of solution

Multiply the molarity of the solution by the volume in liters:

$$\frac{0.20 \text{ mol}}{1 \cancel{L}} \times 1.5 \cancel{L} = 0.30 \text{ mol}$$

To determine the mass of solute, multiply the molar mass of Na_2SO_4 by the number of moles:

$$\frac{142 \text{ g}}{1 \cancel{mol}} \times 0.30 \cancel{mol} = 43 \text{ g}$$

EXAMPLE 8.10   **Calculating moles and mass of solute**

Calculate the moles of solute and grams of solute in 5.0×10^2 mL of 0.040 M KNO_3.

SOLUTION

Use the same relationship as Example 8.9 but first convert milliliters to liters:

$$5.0 \times 10^2 \cancel{mL} \times \frac{1 \cancel{L}}{1000 \cancel{mL}} \times \frac{0.040 \text{ mol}}{1 \cancel{L}} = 0.020 \text{ mol}$$

To find grams of solute, multiply the number of moles by the molar mass of KNO_3:

$$0.020 \cancel{mol} \times \frac{101 \text{ g}}{1 \cancel{mol}} = 2.0 \text{ g}$$

PRACTICE EXERCISE 8.12

How many moles of solute are in 250 mL of 2.0 M $CaCl_2$? How many grams of $CaCl_2$ is this?

Dilution of solutions of known molarity

Solutions of known molarity are usually available in hospital and clinical laboratories. Often, more dilute solutions of known concentration are required. We can make an aqueous solution less concentrated by diluting with water. The number of moles of solute does not change when a solution is diluted. Therefore,

Number of moles of solute = number of moles of solute

Before dilution   *After* dilution

Rearranging the equation

$$\text{Molarity } M = \frac{\text{Number of moles of solute}}{\text{Number of liters of solution}}$$

we get

Number of moles of solute = molarity $M \times$ number of liters of solution

Since the total moles remains constant upon dilution, we can write

$$M_1 \times V_1 = M_2 \times V_2$$

In this equation, M_1 and V_1 are the initial solution's molarity and volume, and M_2 and V_2 are the final solution's molarity and volume. Although the volume in liters was used to get the equation, volume can be in liters or milliliters when we use the equation. This is so because the units of volume cancel:

$$M_1 \times V_1 \text{ (L)} = M_2 \times V_2 \text{ (L)}$$

or

$$M_1 \times V_1 \text{ (mL)} = M_2 \times V_2 \text{ (mL)}$$

We will usually have a variety of volume-measuring devices available to us, and our choice of what to use depends on how accurate the solution concentration must be. The following example shows how a dilute solution is prepared.

EXAMPLE 8.11

Diluting a solution

How would you prepare 100 mL of 0.40 M $MgSO_4$ from a stock solution of 2.0 M $MgSO_4$?

SOLUTION

We can first use the equation to calculate the required volume of 2.0 M $MgSO_4$. We know that $M_1 = 2.0\ M$, $M_2 = 0.40\ M$, and $V_2 = 100$ mL. We need to find V_1.

Rearrange the equation to give V_1:

$$M_1 \times V_1 = M_2 \times V_2$$

Therefore,

$$V_1 = \frac{M_2 \times V_2}{M_1}$$

Now substitute values in this equation:

$$V_1 = \frac{0.40\ \cancel{M} \times 100 \text{ mL}}{2.0\ \cancel{M}} = 20 \text{ mL}$$

Suppose the dilute solution we are making must be very accurate. We can measure 20 mL of the 0.40 M $MgSO_4$ with a pipet and carefully transfer the solution to a 100-mL volumetric flask. (A buret also can be used to measure volumes accurately.) Add distilled water to the flask—up to the etched line—to dilute the contents to exactly 100 mL and shake. We now have 100 mL of 0.40 M $MgSO_4$.

PRACTICE EXERCISE 8.13

You need 250 mL of 0.20 M NaCl, but the only supply of sodium chloride you have available is a solution of 1.0 M NaCl. How do you prepare the required solution? Assume that you have the appropriate volume-measuring devices on hand.

8.9 Colloids and suspensions

AIM: To distinguish among solutions, suspensions, and colloidal dispersions.

Focus

Substances may be dispersed or suspended in water.

We have been dealing with water and aqueous solutions for most of this chapter. We will now shift our emphasis slightly to mixtures of water with substances that do not form true solutions. Two such mixtures are suspensions and colloids.

Suspensions

A piece of clay shaken with water forms a muddy mixture. This type of mixture is a *suspension*. The clay particles become suspended in the water, but they quickly settle when shaking stops. *In a* **suspension,** *the component particles are much larger than in a solution.* The particles in a typical suspension have an average diameter greater than 100 nanometers (nm), but in a solution, the particle size is usually about 1 nm. Suspensions are heterogeneous mixtures—at least two substances, clay and water in the example, can be clearly identified. If the muddy water is filtered, the suspended clay particles are trapped by the filter, and clear water passes through.

Colloids

Colloid
kolla (Greek):   glue
eides (Greek):   like

Particles of a size between those in true solution and those in suspension (between 1 and 100 nm) are *colloidal. Water that contains colloidal particles is a* **colloid** *or* **colloidal dispersion;** the particles are the *dispersed phase,* and the water is the *dispersion medium.* The word *colloidal* comes from two Greek words meaning "looks like glue." Animal glues are colloids and consist of protein particles dispersed in water, but many colloids are not glues and do not resemble them in any way. Blood, paint, aerosol sprays, and smoke are also colloids (Table 8.3).

Table 8.3 Some Colloidal Systems

Dispersed phase	Dispersion medium	Type	Example
	System		
gas	liquid	foam	whipped cream
gas	solid	foam	marshmallow
liquid	liquid	emulsion	milk, mayonnaise
liquid	gas	aerosol	fog, aerosol sprays
solid	gas	smoke	dust in air
solid	liquid	sols and gels	egg white, jellies, paint, colloidal gold, blood, starch in water

Table 8.4 Properties of Solutions, Colloids, and Suspensions

	System		
Property	*Solution*	*Colloid*	*Suspension*
particle type	ions, atoms, small molecules	large molecules or particles	large particles or aggregates
particle size (approximate)	0.1–1 nm	1–100 nm	100 nm and larger
effect of light	no scattering	gives Tyndall effect	gives Tyndall effect
effect of gravity	stable, does not separate	stable, does not separate	unstable, sediment forms
filtration	particles not retained on filter	particles not retained on filter	particles retained on filter
uniformity	homogeneous	borderline	heterogeneous

Figure 8.11
The Tyndall effect. The path of a beam of light is made visible as light is reflected in all directions from particles in colloidal dispersion or suspension. The Tyndall effect is not observed with true solutions.

The properties of colloids differ from those of solutions and suspensions. Many colloids are cloudy or milky in appearance but look clear when they are very dilute. The particles in a colloid cannot be retained by a filter paper and do not settle out with time.

Colloidal particles exhibit the **Tyndall effect**—*scattering visible light (the light we see by) in all directions.* You can see a beam of light passed through a colloid just as you see a sunbeam in a dusty room. Suspensions also exhibit the Tyndall effect, but solutions never do (Fig. 8.11).

Flashes of light (scintillations) are seen when colloids are studied under a microscope. Colloids scintillate because the particles reflecting and scattering the light move erratically. The chaotic movement of colloidal particles is called *Brownian motion* after the English botanist Robert Brown (1773–1858). Brownian motion is caused by the water molecules of the medium colliding with the small, dispersed colloidal particles. The buffeting action exerts such a force on the particles that they cannot settle.

Colloidal particles also may absorb charged particles from the surrounding medium onto their surface. Depending on their characteristics, the colloidal particles will absorb positively charged particles and become positively charged or absorb negatively charged particles and become negatively charged. All the particles in a system have the same charge. The repulsion of the like-charged particles prevents them from forming aggregates and keeps them dispersed throughout the medium. Charged colloidal particles form aggregates and precipitate from the dispersion when ions of opposite charge are added. Table 8.4 summarizes the properties of solutions, colloids, and suspensions.

Emulsions

Emulsions *are colloidal dispersions of liquids in liquids.* An emulsifying agent is essential for the formation of an emulsion and for maintaining its stability. Although oils and greases are not soluble in water, they readily form a colloidal dispersion if soap or detergent is added to the water. Soaps and detergents are emulsifying agents.

Milk is an emulsion of fat droplets in water. The fat droplets are stabilized by casein, a milk protein. Mayonnaise is an emulsion of water-insoluble vegetable oils and water. It is stabilized by egg yolk, which is rich in lecithins, fatty molecules that are good emulsifying agents.

The fats we eat must be emulsified before they can be digested. The emulsifying agents in the body are the bile salts that are stored in the gallbladder and secreted into the small intestine as bile. Unemulsified fats cannot be broken down by water-soluble enzymes in the digestive juices, so people who have their gallbladders removed must restrict their intake of fats and oils.

8.10 Colligative properties

AIMS: *To describe on a molecular level how a solution can have a lower vapor pressure than the pure solvent of the solution. To calculate the freezing point depression and the boiling point elevation of aqueous solutions.*

Focus

The properties of solutions are different from those of their solvents.

Colligative
colligatus (Latin):   grouped together

When a solute is dissolved in a solvent, the physical properties of the solution are different in certain respects from those of the pure solvent. Some of these differences are due to changes in the colligative properties of the solution. **Colligative properties** *are properties that depend on the number of particles dissolved in a given mass of solvent.* There are four common colligative properties of solutions: vapor pressure lowering, boiling point elevation, freezing point depression, and osmotic pressure. The first three are discussed here, and the next section considers osmotic pressure.

Vapor pressure lowering

A solution that contains a nonvolatile solute—one that does not vaporize—always has a lower vapor pressure than the pure solvent. An aqueous sodium chloride solution is an example. Some sodium ions and some chloride ions are at the surface of the liquid, where they are surrounded by shells of water of solvation. The formation of these shells ties up much of the surface water and reduces the number of solvent molecules that have enough kinetic energy to escape as vapor. Thus, when equilibrium between the solution and its vapor is reached, there are fewer solvent molecules in the vapor; the solution has a lower vapor pressure than the pure solvent. The decrease in the vapor pressure is directly proportional to the number of particles the solute makes in solution. For example, the vapor pressure lowering caused by dissolving 0.1 mol sodium chloride in 1000 g water is twice that caused by dissolving 0.1 mol glucose in 1000 g water. One molar mass of the ionic compound sodium chloride produces 2 mol of particles, 1 mol of sodium ions and 1 mol of chloride ions, whereas 1 mol of glucose, a molecular compound, produces only 1 mol of glucose particles.

Ethylene glycol, $HOCH_2CH_2OH$, is added to water in the cooling system of your car to give a solution with a freezing point below the freezing point of water and a boiling point above the boiling point of water.

Boiling point elevation

When the vapor pressure of a solvent is decreased by the addition of a nonvolatile solute, the solution must be raised to a higher temperature than the pure solvent to reach the boiling point. The boiling point of water increases by 0.52 °C for every mole of particles of solute dissolved in 1000 g water.

EXAMPLE 8.12

Calculating boiling point elevation

We have two solutions: Solution (a) contains 1 mol NaCl in 1000 g H_2O, and solution (b) contains 1 mol glucose in 1000 g H_2O. What is the boiling point of each solution?

SOLUTION

(a) One mole of NaCl produces 2 mol of particles in solution (1 mol Na^+ and 1 mol Cl^-). Since the boiling point of H_2O increases by 0.52 °C for every mole of particles, 2 mol of particles produces an increase of

$$2 \text{ mol} \times \frac{0.52 \text{ °C}}{1 \text{ mol}} = 1.04 \text{ °C}$$

The solution therefore boils at 101.04 °C.

(b) One mole of glucose produces 1 mol of particles. Since the boiling point of H_2O increases by 0.52 °C for every mole of particles, the solution boils at 100.52 °C.

PRACTICE EXERCISE 8.14

What is the boiling point of a solution that contains 1 mol $CaCl_2$ dissolved in 1000 g water?

Freezing point depression

Addition of 1 mol of particles of any solute to 1000 g of water lowers the freezing point by 1.86 °C. When 1 mol (58.5 g) of sodium chloride is added to 1000 g of water, for example, the solution freezes at −3.72 °C instead of 0 °C. People take advantage of the freezing point depression of aqueous solutions to melt ice on sidewalks; ice sprinkled with salt melts and forms a solution with a lower freezing point than the air temperature. Salt will not melt ice when the air temperature is extremely low because the solubility of salt in water—and consequently the freezing point depression that can be obtained—is limited.

PRACTICE EXERCISE 8.15

We have two solutions: Solution (a) contains 1 mol Na_2SO_4 dissolved in 1000 g H_2O, and solution (b) contains 1 mol $MgSO_4$ dissolved in 1000 g H_2O. What is the freezing point of each solution?

8.11 Osmosis and dialysis

AIMS: To distinguish among semipermeable, osmotic, and dialyzing membranes. To describe what happens to a red blood cell when it is placed into a hypertonic solution, a hypotonic solution, or an isotonic solution. To explain how dialysis can be used to separate dissolved ions from large biologic molecules.

Permeate
permeate (Latin):   to go through

Before refrigeration was available, meat was preserved by salting. When bacteria are placed in a very concentrated sodium chloride solution, water flows from the bacteria cell to the surroundings. The bacteria "dry up" and die.

From the biological point of view, the most important colligative property is **osmosis**—*the spontaneous movement of water across a semipermeable membrane from a more dilute solution to a more concentrated solution.*

Osmosis

Materials such as filter paper are permeable to both solvent and solute particles. Other materials, such as the walls of metal containers, are impermeable. Between these two are semipermeable membranes—natural materials like cell membranes and synthetics such as cellophane. **Semipermeable membranes** *have extremely small pores that allow molecules and ions below a certain size to pass through but exclude larger particles.* **Osmotic membranes** *are very selective semipermeable membranes that permit only water molecules to pass through.*

Suppose that two compartments containing pure water are separated by an osmotic membrane. Equal numbers of water molecules pass through the membrane from either side; the water is in dynamic equilibrium. As a result, the water levels in both compartments stay the same, as shown in Figure 8.12(a). The equilibrium can be disturbed by dissolving a solute in one of the compartments to form a solution. Now some of the water molecules in the compartment containing the dissolved solute are tied up as water of solvation, making them unavailable for passage through the membrane (Fig. 8.12b). The number of mobile water molecules in the compartment containing the pure water is not reduced, however, and an imbalance exists. More water molecules pass from the compartment containing pure water than in the opposite direction. The result is a net flow of water from the compartment containing pure water to the compartment containing the dissolved solute—osmosis. Osmosis also

Figure 8.12
Apparatus to illustrate osmosis. (a) System in a state of dynamic equilibrium; levels stay the same. (b) Equilibrium disturbed by addition of solute to one compartment; net flow of water molecules into the solution.

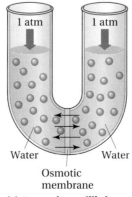
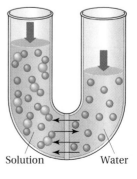

Water   Water   Solution   Water
Osmotic
membrane
(a) Dynamic equilibrium   **(b) Equilibrium disturbed**

occurs with solutions of different solute concentrations. There is always a net flow of water from the more dilute to the more concentrated solution. It is useful to remember that in osmosis *water always seeks solute.* That is, water moves through a selective, semipermeable membrane in such a direction as to dilute the more concentrated solution.

EXAMPLE 8.13

Inferring the direction of flow of water in osmosis

Two solutions, 1% NaCl (m/v) and 4% NaCl (m/v), are separated by an osmotic membrane. Initially, the levels of the two solutions are the same.

(a) In which direction is there a net flow of water?

(b) Which solution becomes more concentrated?

(c) Which solution increases in volume?

SOLUTION

The process that occurs is osmosis. In osmosis there is always a net flow of water from the more dilute to the more concentrated solution. Therefore, (a) net flow of water is from 1% NaCl to 4% NaCl, (b) 1% NaCl becomes more concentrated, and (c) 4% NaCl increases in volume.

PRACTICE EXERCISE 8.16

Two sugar solutions are separated by an osmotic membrane. Solution A is twice the concentration of solution B. If both solution levels are initially the same, (a) which solution increases in volume and (b) which solution becomes more dilute?

Osmotic pressure

The net movement of water from the compartment of lower solute concentration to the compartment of higher solute concentration continues until the number of mobile water molecules in both compartments is equal. At this point, the flow of water molecules through the membrane is the same from either side. The system is again at equilibrium, as shown in Figure 8.12(c). The difference in the levels of liquid in the two compartments at

Figure 8.12 con't.
(c) Equilibrium restored; no net flow of water. The difference in the levels of the liquids in the two compartments at equilibrium is the osmotic pressure. (d) By applying an external pressure equal to the osmotic pressure, the liquid levels can be made the same.

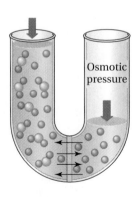

(c) Equilibrium restored

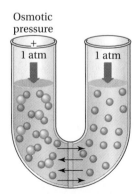

(d) External pressure equals osmotic pressure

A person lost at sea cannot survive by drinking seawater. The concentration of salt in seawater, 3.0% (m/v), is much higher than the concentration of salt in the blood, 0.9% (m/v). Drinking seawater causes the salt concentration in the extracellular fluid of the body to increase. Water leaves the tissue cells, and the cells become dehydrated.

A patient in a hospital is often given an intravenous injection of an aqueous solution. The composition of the solution must be monitored carefully. The introduction of a solution of the wrong concentration into the bloodstream could be harmful.

Plasmolysis
plasma (Greek): formed material
lysis (Greek): splitting

equilibrium equals the osmotic pressure of the system. As Figure 8.12(d) shows, the original system can be restored by applying **osmotic pressure—** *an opposing external pressure sufficient to make the levels of differing solution strength in the two compartments the same.*

Osmosis and osmotic pressure play an important role in maintaining a correct balance between water and dissolved substances in cells. Among other things, body cells contain solutions of salts and biological molecules (intracellular fluid) enclosed in a semipermeable membrane. Every cell has an internal solute concentration that is greater than, less than, or equal to the solute concentration of its external fluid environment. Depending on the relationship between the internal and external concentrations, the external solutions are called *hypertonic, hypotonic,* or *isotonic:*

1. **Hypertonic solutions** *contain larger numbers of solute particles than the intercellular fluid.* A cell placed in a hypertonic solution shrinks because there is a net flow of intracellular water to the extracellular fluid. As the cell shrivels, it takes a *crenated (scalloped)* appearance (Fig. 8.13a).

2. **Hypotonic solutions** *contain fewer numbers of solute particles than the intracellular fluid.* A cell placed in a hypotonic solution swells because the net flow of water is from the extracellular fluid to the intracellular fluid. If the cell swells too much, the membrane ruptures (see Fig. 8.13b), just as an overinflated balloon bursts. *When cells in general rupture, the event is called* **plasmolysis;** *when red blood cells rupture, it is called* **hemolysis.** Crenated or plasmolyzed cells are useless.

3. **Isotonic solutions** *contain the same numbers of solute particles as the intracellular fluid.* A cell placed in an isotonic solution retains its original shape (see Fig. 8.13c). When body fluids are replaced intravenously, the injected fluid must be isotonic with the contents of red blood cells; otherwise, crenation or hemolysis will occur. An aqueous solution containing 0.90% NaCl (m/v) (physiologic saline) and one containing 5.5% glucose (m/v) (called *dextrose* or *blood sugar solution*) are often used for intravenous injections because they are isotonic with red blood cells.

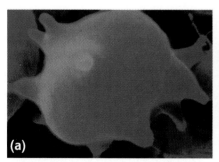 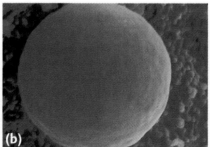 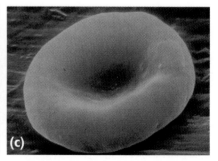

Figure 8.13
Cell membranes are semipermeable membranes.
(a) Red blood cells placed in a hypertonic solution shrink and become crenated because there is a net movement of water out of the cells by osmosis.
(b) Red blood cells placed in a hypotonic solution swell and eventually rupture (undergo hemolysis) because there is a net movement of water into the cells by osmosis. **(c)** Red blood cells placed in an isotonic solution retain their normal size and shape; there is no net movement of water.

Dialysis

Cell membranes are *dialyzing membranes*—semipermeable membranes that are considerably more permeable than osmotic membranes, permitting the passage of ions and other small molecules; large molecules are retained. **Dialysis** *is the process in which ions and small molecules other than water pass through a semipermeable membrane* (often cellophane).

In dialysis, *solute always seeks water.* When an aqueous solution of sodium chloride contained in a cellophane dialyzing bag is surrounded by water, there is a net movement of sodium ions and chloride ions from the high-concentration region to the low-concentration region—from the salt solution to the water—until both solutions are of equal strength (Fig. 8.14).

Now suppose we dialyze a solution of sodium chloride containing protein or other molecules too large to get through the pores of the dialyzing membrane. There is a net movement of sodium ions and chloride ions through the membrane into the water, but the protein molecules are too large to escape. When equilibrium is reached, the sodium chloride concentrations on either side of the membrane are about the same.

Dialysis is a useful procedure for removing salts from solutions containing very large molecules (macromolecules). By replacing the water surrounding a dialysis bag, we can eventually remove most of the sodium chloride from a dialysis bag and obtain a solution containing only protein (Fig. 8.15). The artificial kidney machine is a sophisticated dialyzer. This device removes waste products from the blood by **hemodialysis**—*dialysis of the blood.* The process is described in A Closer Look: Hemodialysis.

A person who experiences a physical injury may go into shock because of a sudden increase in the permeability of the membranes of the capillaries. Large, colloidally dispersed molecules, such as proteins normally retained in the blood, can then leave the bloodstream and leak into the spaces between cells. This damage disrupts the nutrient- and oxygen-carrying function of the blood. If a patient in shock is left untreated, death may ensue.

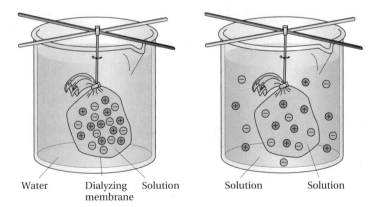

Water   Dialyzing   Solution   Solution   Solution
               membrane

Figure 8.14
In dialysis, there is a net movement of ions from the region of high concentration to the region of low concentration.

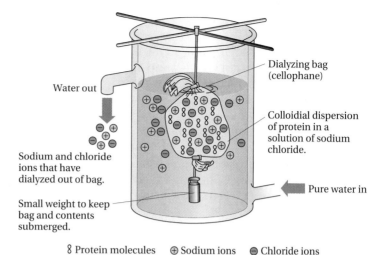

Figure 8.15
Salt-free protein solutions can be obtained by dialysis. Small molecules and ions pass through the membrane. The protein molecules are too large to pass through, and thus they remain in the dialyzing bag.

A Closer Look

Hemodialysis

The kidneys form a complex dialyzing system that removes waste products of metabolism from the blood. When the kidneys are not functioning properly, these waste products are not excreted in the urine and can build up to toxic levels in the blood. This life-threatening condition can be corrected by using a kidney machine to dialyze the blood—hemodialysis (see figure).

In hemodialysis, the patient's blood is circulated through a long, coiled, cellophane dialyzing tube that has a carefully selected pore size. This tube is immersed in a bath containing a solution that is isotonic with the solutes that must be retained but is lacking the toxic waste products. Wastes therefore dialyze out of the blood. The isotonic solution consists of 0.6% NaCl (m/v), 0.04% KCl (m/v), 0.2% NaHCO$_3$ (m/v), and 1.5% glucose (m/v). A typical treatment might last 6 hours, and the solution in the bath would be changed every 2 hours. This procedure ensures that waste products that have already been removed from the blood do not move into the blood again.

Kidney machines help keep people with kidney failure alive, and many lead a fairly normal life as long as they have regular dialysis. Kidney machines are also used as temporary life supports for people awaiting kidney transplantation.

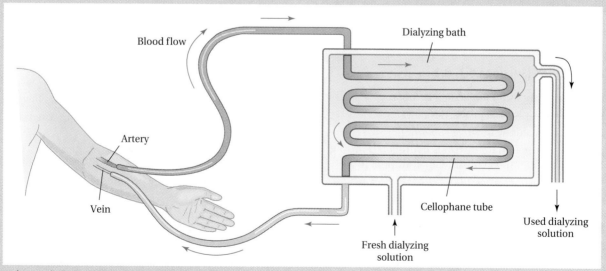

In hemodialysis, the patient's blood is cleansed of life-threatening toxic waste products.

SUMMARY

Hydrogen bonding between polar water molecules accounts for the high boiling point (100 °C) and high freezing point (0° C) of water, and it explains why ice floats. Water is a polar liquid and an excellent solvent for many substances. Aqueous solutions are homogeneous mixtures of ions or molecules in water. The relative amount of the dissolved substance in water can be expressed imprecisely as dilute or concentrated or expressed quantitatively as moles per liter (molarity), percent, or parts per million.

The solubility of a solute depends on solute-solvent interactions; a good rule to remember is that "like dissolves like." Changes in temperature and pressure of a system change the solubility of a solute. The rate at which a solute dissolves is influenced by a number of factors, including temperature of the solvent and particle size of the solute. True solutions are distinguished from colloidal dispersions and suspensions on the basis of particle size.

Substances that are in solution as ions are electrolytes. A solution of an electrolyte will conduct an electric current; a solution of a nonelectrolyte is nonconducting.

The addition of a solute to a solvent lowers the freezing point and vapor pressure and raises the solution's boiling point. These effects, together with osmotic pressure, are known as colligative properties. Colligative properties depend only on the number of particles in solution and not on the particle type. In osmosis, water flows spontaneously across a semipermeable membrane from a region of low concentration of solute to a region of higher concentration of solute. Cell membranes are semipermeable membranes. Osmosis and osmotic pressure are important in maintaining the correct balance between water and electrolytes in cells. Cells are properly maintained in an isotonic environment, but they shrink in a hypertonic solution and they swell and may burst in a hypotonic solution.

KEY TERMS

Aqueous solution (8.3)
Colligative property (8.10)
Colloid (8.9)
Colloidal dispersion (8.9)
Concentration (8.7)
Dialysis (8.11)
Electrolyte (8.4)
Emulsion (8.9)
Hemodialysis (8.11)
Hemolysis (8.11)

Henry's law (8.6)
Hydrogen bond (8.1)
Hypertonic solution (8.11)
Hypotonic solution (8.11)
Isotonic solution (8.11)
Molarity (8.8)
Nonelectrolyte (8.4)
Osmosis (8.11)
Osmotic membrane (8.11)
Osmotic pressure (8.11)

Plasmolysis (8.11)
Saturated solution (8.6)
Semipermeable membrane (8.11)
Solubility (8.6)
Solute (8.3)
Solvation (8.4)
Solvent (8.3)
Supersaturated solution (8.6)

Surface tension (8.1)
Surfactant (8.1)
Suspension (8.9)
Tyndall effect (8.9)
Unsaturated solution (8.6)
Water of crystallization (8.4)
Water of hydration (8.4)

EXERCISES

Water and Ice (Sections 8.1, 8.2)

8.17 Explain why water molecules are polar.

8.18 Name four physical properties of water.

8.19 What is meant by *hydrogen bonding?*

8.20 Draw a sketch that illustrates hydrogen bonding between a water molecule and an ammonia molecule.

8.21 Explain how hydrogen bonding is responsible for the high surface tension of water.

8.22 Why does water bead up on a newly waxed car?

8.23 How are vapor pressure and boiling point related?

8.24 Explain why water has a relatively high boiling point.

8.25 Distinguish between the structures of water and ice, and explain why ice floats on water.

8.26 How does freezing water contribute to the weathering (breakdown) on rocks?

8.27 Why does your body produce sweat when you exercise strenuously?

8.28 Temperatures below -2 °C are harmful to grape vines. Why do vineyard owners spray a mist of water on their vines when they expect a severe frost?

Aqueous Solutions (Sections 8.3, 8.4, 8.5, 8.6)

8.29 Why is water an excellent solvent for most ionic compounds and polar covalent molecules but not for nonpolar compounds?

8.30 Which of the following substances dissolve in water?
(a) CH_4 (b) KCl
(c) He   (d) $MgSO_4$
(e) cane sugar   (f) $NaHCO_3$

8.31 Distinguish between a solution and an aqueous solution.

8.32 Describe how an ionic compound dissolves in water.

8.33 What is the percent by mass of water in gypsum (calcium sulfate dihydrate)?

8.34 Calculate the percent by mass of water in plaster of paris.

8.35 Distinguish between an electrolyte and a nonelectrolyte.

8.36 Must a compound be soluble in water to be an electrolyte?

8.37 Name and give the formula for some of the principal cation electrolytes in the body.

8.38 Give the names and formulas of the important anion electrolytes in the fluids of the human body.

8.39 Name three factors that influence the rate at which a solute dissolves in a solvent.

8.40 Explain how the factors in Exercise 8.39 agree with kinetic-molecular theory.

8.41 The solubility of KCl in water is 34.0 g KCl/100 mL H_2O at 20 °C. A warm solution containing 50.0 g KCl in 1.30×10^2 mL H_2O is cooled to 20 °C.
(a) How many grams of KCl remain dissolved?
(b) How many grams come out of solution?

8.42 A solution contains 26.5 g NaCl in 75.0 mL H_2O at 20 °C. Determine whether the solution is unsaturated, saturated, or supersaturated. (The solubility of NaCl at 20 °C is 36.0 g/100 mL H_2O.)

8.43 What is the effect of pressure on the solubility of gases in liquids?

8.44 Why does a bottled carbonated beverage "bubble" when the cap is removed?

8.45 The solubility of a gas in water at room temperature and 1 atm pressure is 0.5 g/L. What is the concentration of the gas in solution if the pressure is raised to 6 atm and the temperature stays the same?

8.46 How does the solubility of a gas in a liquid change when the pressure of the gas over the liquid is cut in half?

Solution Concentrations (Sections 8.7, 8.8)

8.47 A solution contains 16 mL methanol, CH_3OH, in 150 mL of solution. What is the percent (volume/volume) of this solution?

8.48 What is the percent (v/v) of a solution that is 5.0 mL H_2O_2 in 230 mL of solution?

8.49 Calculate the number of grams of solute required to make
(a) 2.5 L of physiologic saline solution, 0.90% NaCl (w/v).
(b) 75 mL of 4.0% $MgCl_2$ (w/v).
(c) 350 mL of 2.0% glucose (w/v).
(d) 250 mL of 0.10% $MgSO_4$ (w/v).

8.50 What is the concentration, in percent (w/v), of the following solutions?
(a) 28 g KCl in 670 mL of solution
(b) 32 g $NaNO_3$ in 2.0 L of solution
(c) 75 g K_2SO_4 in 1500 mL of solution
(d) 4.5 g HCOOH (formic acid) in 380 mL of solution

8.51 A body of polluted water is analyzed for mercury compounds. A sample has a mass of 50 g and contains 0.02 mg mercury. What is the concentration of mercury in (a) parts per million and (b) parts per billion?

8.52 Barium sulfate is used in X-raying the gastrointestinal tract. At 20 °C the solubility of $BaSO_4$ is 0.00024 g/100 mL of water.
(a) What is the concentration of $BaSO_4$ in ppm?
(b) What is the molarity of the solution?

8.53 Calculate the molarity of the following solutions.
(a) 1.0 mol KCl in 750 mL of solution
(b) 0.50 mol $MgCl_2$ in 1.5 L of solution
(c) 5.85 g NaCl in 2.0 L of solution
(d) 2.65 g ethyl alcohol, C_2H_5OH, in 145 mL of solution

8.54 Calculate the number of moles of solute in each of the following solutions.
(a) 1.0 L of 0.50 M NaCl
(b) 4.5 L of 0.036 M KI
(c) 530 mL of 2.0 M KNO_3

8.55 How many moles of K_2SO_4 are in 55 mL of solutions of the following concentrations?
(a) 1.2 M K_2SO_4
(b) 0.10 M K_2SO_4
(c) 3.0 M K_2SO_4

8.56 Calculate the grams of solute in each of the following solutions.
(a) 25 mL of 0.10 M NaCl
(b) 1.7 L of 0.85 M $KMnO_4$
(c) 0.3 L of 2.0 M $CaCl_2$
(d) 725 mL of 0.36 M $AgNO_3$

8.57 How many milliliters of 2.0 M NaCl must you dilute to make 250 mL of 0.50 M NaCl?

8.58 Calculate the volume of 0.50 M MgSO$_4$ you must dilute to make 2.5 L of 0.15 M MgSO$_4$.

Colloids and Suspensions (Section 8.9)

8.59 Distinguish between a solution and a colloid.

8.60 Distinguish between a suspension and a solution.

8.61 Describe the Tyndall effect.

8.62 Explain Brownian motion.

Colligative Properties (Section 8.10)

8.63 What effect does dissolving a solute have on the boiling and freezing points of a liquid?

8.64 How does the vapor pressure of a solution compare with the vapor pressure of the solvent of that solution?

8.65 Why does seawater evaporate more slowly than fresh water at the same temperature?

8.66 How does the boiling point of seawater compare with the boiling point of distilled water?

8.67 What is the boiling point of (a) a solution containing 0.50 mol glucose in 1000 g H$_2$O and (b) a solution containing 1.5 mol NaCl in 1000 g H$_2$O?

8.68 What effect does the addition of 1.00 mol CaCl$_2$ to 1000 g H$_2$O have on the freezing point of water?

Osmosis and Dialysis (Section 8.11)

8.69 Describe the process of osmosis.

8.70 Describe the process of dialysis.

8.71 A 2% sucrose solution is separated from a 4% sucrose solution by an osmotic membrane. Initially, the two solution levels are the same, but after a short period of time the levels are different.
(a) What process is taking place?
(b) In which direction is there a net flow of water?
(c) Which solution level will rise and which will fall?

8.72 An osmotic membrane separates a 3% NaBr solution from a 1% NaBr solution. Over time, the once-even solution levels change.
(a) Which solution level drops?
(b) Which solution becomes more concentrated?

8.73 What happens to red blood cells that are placed in the following solutions?
(a) 2% NaCl   (b) pure water   (c) 0.9% NaCl

8.74 What happens to red blood cells that are placed in the following solutions?
(a) 1% glucose   (b) 7% glucose   (c) 5.5% glucose

Additional Exercises

8.75 The volume of a 0.06 M KCl solution is reduced from 340 to 200 mL by boiling off some of the water. What is the molarity of the new solution?

8.76 The solubility of oxygen in water at 20 °C is 5.4 mg O$_2$ per 100 g H$_2$O (assuming standard pressure). If all the oxygen above the water were removed, how would the solubility of the oxygen change?

8.77 What happens in each of the following situations?
(a) A dialyzing bag containing a solution of sodium chloride is immersed in pure water.
(b) A dialyzing bag containing sodium chloride and a dispersion of a large protein in water is immersed in pure water.
(c) A dialyzing bag containing a dispersion of a protein in water is immersed in a concentrated solution of sodium chloride.

8.78 How many grams of water are in 166 g of the hydrate CoCl$_2 \cdot$6H$_2$O?

8.79 On an equal molar basis, why does an electrolyte depress the freezing point of water to a greater extent than a nonelectrolyte?

8.80 Explain each of the following facts.
(a) Finely ground sugar dissolves more quickly in a glass of water than does a large sugar cube.
(b) Fish need oxygen to survive. Goldfish in a crowded bowl have a better chance of survival in cold water than in warm water.
(c) Electronic equipment that can be damaged by moisture is often packed for shipment with small packets of anhydrous CaSO$_4$.
(d) Adding more sugar to a glass of tea that has crystals of sugar at the bottom doesn't make the tea taste any sweeter.
(e) Vinegar and oil salad dressing separates into two layers. (Vinegar is a mixture of acetic acid, CH$_3$COOH, and water.)

8.81 Which of these statements describes (a) a suspension, (b) a solution, and (c) a colloidal dispersion?
1. particles can be filtered
2. heterogeneous mixture
3. gives Tyndall effect
4. particle size less than 1.0 nm
5. particles are invisible to the naked eye
6. does not settle out on standing

8.82 You have 6.00 mL of 8.4 M NH$_4$Cl. How much water must be added to give a solution that is 1.0 M?

8.83 Complete the following table:

	Substance		
	C$_6$H$_{12}$O$_6$	NH$_4$NO$_3$	LiF
Mass (g)	56.8	_____	14.6
Volume (mL)	450	8.00 $\times$ 10^2	_____
Percent (w/v)	_____	2.80%	_____
Molarity (M)	_____	_____	1.50

SELF-TEST (REVIEW)

True/False

1. The rate at which a solute dissolves can generally be increased by stirring.
2. Carbon tetrachloride, CCl_4, is an electrolyte.
3. The particles in a suspension can be removed by filtration.
4. One hundred milliliters of a 5.0 M sodium chloride solution is more concentrated than 1.0 L of a 1.0M sodium chloride solution.
5. Water beads up on a waxed surface because of the hydrogen bonding between water molecules.
6. Solutions are homogeneous.
7. The amount of sodium chloride in 100 mL of a 5.0 M NaCl solution is greater than that in 1.0 L of a 1.0 M NaCl solution.
8. Grinding a solute should increase the rate of its dissolving.
9. As an open bottle of a carbonated beverage warms, the concentration of dissolved carbon dioxide decreases.
10. Fifty milliliters of a 16% (v/v) solution of ethanol in water would contain 42 mL of water.

Multiple Choice

11. If a crystal added to an aqueous solution dissolves, then we know that the original solution
    (a) is saturated.   (b) is unsaturated.
    (c) is supersaturated.   (d) is really an emulsion.
12. A solute in an aqueous solution has all the following properties *except* that
    (a) it can be removed by filtering.
    (b) it does not settle out on standing.
    (c) it is molecular in size and cannot be seen.
    (d) it is uniformly distributed throughout the solvent.
13. A membrane that permits the passage of water and particles in solution but not of particles of colloidal size is
    (a) an osmotic membrane.
    (b) an impervious membrane.
    (c) a dialyzing membrane.
    (d) a permeable membrane.
14. The density of ice is less than the density of water because
    (a) ice is lighter than water.
    (b) of water's high specific heat.
    (c) of hydrogen bonding between the H_2O molecules.
    (d) of solvation effects.

15. If 1 mol of each of the following solutes is added to the same amount of water, which solution has the highest boiling point?
    (a) ethanol, C_2H_5OH
    (b) calcium phosphate, $Ca_3(PO_4)_2$
    (c) magnesium acetate, $Mg(C_2H_3O_2)_2$
    (d) copper(I) chloride, CuCl
16. How many grams of potassium iodide would you need to make 400 mL of 0.200 M KI?
    (a) 13.3 g   (b) 83.8 g   (c) 26.6 g   (d) 332 g
17. Red blood cells are hemolyzed (ruptured) when placed in an aqueous solution. With respect to the red blood cells, this solution is
    (a) hypotonic.   (b) isotonic.
    (c) hypertonic.   (d) hygroscopic.
18. Which of these substances is *not* an electrolyte?
    (a) $AlCl_3$ (b) CCl_4 (c) Na_2SO_4 (d) LiCl
19. Homogenized milk would be best described as a(n)
    (a) suspension.   (b) true solution.
    (c) isotonic solution.   (d) colloidal dispersion.
20. Suppose that you would like to prepare 200 mL of a 0.50 M NaBr solution using a 2.0 M NaBr stock solution. How many milliliters of the stock solution should you use?
    (a) 25 mL   (b) 100 mL   (c) 150 mL   (d) 50 mL
21. Which of these substances would you expect to dissolve most readily in the nonpolar solvent hexane C_6H_{14}?
    (a) $MgCl_2$ (b) CCl_4 (c) HCl   (d) Na_2CO_3
22. A certain wine is 14.0% ethanol (m/v). What mass of ethanol, C_2H_5OH, is in 1 L of wine?
    (a) 6.4 g   (b) 71.4 g   (c) 140 g   (d) 64.0 g
23. Which of the following is *not* a colligative property of a solution?
    (a) osmotic pressure
    (b) supersaturation
    (c) vapor pressure lowering
    (d) freezing point depression
24. Anhydrous cobalt(II) chloride can be used to detect moisture in the air because it changes color as it undergoes hydration.

    $$CoCl_2(s) + 6H_2O(l) \longrightarrow CoCl_2{\cdot}6H_2O(s)$$

    Blue   Red

    What is the percentage of water in the hydrate $CoCl_2{\cdot}6H_2O$?
    (a) 45.4%   (b) 7.6%
    (c) 20.7%   (d) 22.7%

25. What kind of bond exists between water molecules in the liquid state?
    (a) nonpolar bond   (b) hydrogen bond
    (c) ionic bond   (d) covalent bond

26. If the pressure of a gas over a liquid is doubled, you would expect the solubility of the gas to
    (a) double.
    (b) halve.
    (c) remain the same.
    (d) vary depending on the gas.

27. When two solutions, a 2% $CaCl_2$ solution and a 5% $CaCl_2$ solution, are separated by an osmotic membrane,
    (a) the height of the 2% solution rises.
    (b) water molecules flow into the 5% solution.
    (c) the 2% solution becomes more dilute.
    (d) the 5% solution has a lower osmotic pressure.

28. Both water and ammonia have higher boiling points than we would expect based on their molecular weights because of
    (a) surface tension.   (b) boiling point elevation.
    (c) hydrogen bonding.   (d) osmotic pressure.

29. A mixture of protein molecules, potassium chloride, glucose, and colloidal-size starch is placed inside a dialyzing bag. If this bag is placed in a flowing stream of distilled water, which of these substances would remain in the bag?
    (a) glucose and starch   (b) starch and protein
    (c) protein and KCl   (d) glucose and protein

30. How many grams of glucose must be dissolved in 1.0 L of water to make a solution that has a glucose concentration of 1 ppm?
    (a) 1.0 g   (b) 10^{-3} g   (c) 10^3 g   (d) 10^2 g

Acids, Bases, Salts, and Buffers

Hydrogen Ions in Chemistry

A commercial remedy for acid indigestion is Alka-Seltzer. The bubbles of carbon dioxide result from the reaction of an acid (citric acid) with a base (sodium bicarbonate).

CHAPTER OUTLINE

CASE IN POINT: Antacids

9.1 Hydrogen ions from water

9.2 Acids and bases

A CLOSER LOOK: Acids and Dental Health

9.3 The pH scale

A CLOSER LOOK: pH of Body Fluids

9.4 Measurement of pH

9.5 Neutralization

FOLLOW-UP TO THE CASE IN POINT: Antacids

9.6 Equivalents

9.7 Normality

9.8 Titration

9.9 Salts of weak acids and bases

9.10 Buffers

You may have heard it said that someone who speaks sharply "has an acid tongue" or that the cafeteria's coffee "tastes like battery acid." Underlying these descriptions is the fact that acids and bases play a central role in much of the chemistry that affects our daily lives, including the correct functioning of our bodies. This chapter is about acids and bases and why they are biologically important.

Acid indigestion is a simple example of an effect of acids on the body. Affecting millions each year, acid indigestion is not usually a serious health problem. However, the acid produced in the stomach can cause pain and discomfort, as described in the following Case in Point.

CASE IN POINT: Antacids

Robert, a man of 52 years, liked to raid the refrigerator before bedtime. One night he snacked on a sandwich he made from some leftover spicy meatballs. He was awakened at about 3 A.M. with a strong burning sensation in his chest and his throat. Robert's discomfort was quite intense; however, he could relieve it somewhat by sitting up. Eventually, he went back to sleep in his sitting position. Although he felt fine the next day, Robert saw his physician, just in case. After listening to Robert's symptoms and performing a thorough examination, the doctor told Robert that he had probably suffered from heartburn and acid reflux. The doctor made a few suggestions and sent Robert home. What is the major acid involved in heartburn and acid reflux? What did the doctor suggest as a remedy for Robert's symptoms, should they recur? What chemistry of acids and bases is involved in the remedy? We will find out in Section 9.5.

The consequences of this midnight snack may not be so enjoyable!

9.1 Hydrogen ions from water

AIMS: To write the equation for the self-dissociation of water. To write the equation for K_w and use it to calculate a value of [H⁺] given a value of [OH⁻], or vice versa. To classify a solution as neutral, acidic, or basic.

Water molecules are highly polar. Moreover, water molecules are in continuous motion, even at room temperature. Occasionally, the collisions between water molecules are energetic enough that a hydrogen ion (H⁺) is transferred from one water molecule to another. (You may recall that a hydrogen ion is a proton—a hydrogen atom that has lost its electron.) *A water molecule that loses a hydrogen ion becomes a negatively charged* **hydroxide ion (OH⁻),** *and the water molecule that accepts the hydrogen ion becomes a positively charged* **hydronium ion (H₃O⁺):**

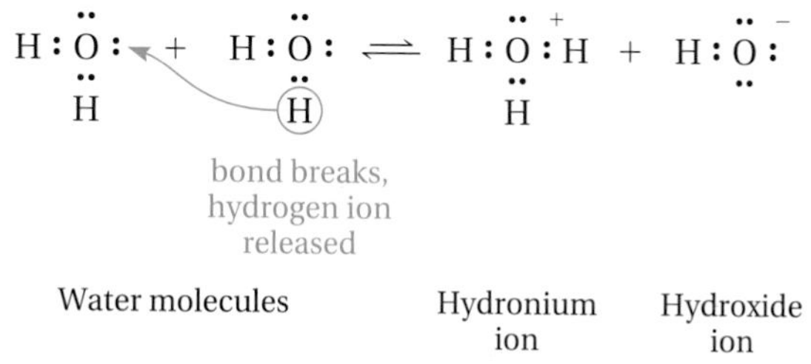

bond breaks,
hydrogen ion
released

Water molecules   Hydronium   Hydroxide
                              ion   ion

The dissociation of water into hydronium ions and hydroxide ions is called the **self-dissociation of water.** We also can write this dissociation as

$$H_2O \rightleftharpoons H^+ + OH^-$$

Hydrogen ion   Hydroxide ion

Hydrogen ions in aqueous solution have several aliases—some chemists call them protons; others prefer to call them hydrogen ions or hydronium ions. In this book we use either H^+ or H_3O^+ to represent hydrogen ions in aqueous solution, depending on the topic being discussed.

The ion-product constant of water

In pure water, only one water molecule in about six hundred million self-dissociates to form hydrogen ions and hydroxide ions. At 25 °C, the [H⁺] and [OH⁻] are only 1×10^{-7} mol/L (1×10^{-7} M). Here [H⁺] is read as "the hydrogen ion concentration" and [OH⁻] is read as "the hydroxide ion concentration." The concentration units within brackets are understood to be molar. *The H^+ concentration can be multiplied by the OH^- concentration to give a constant called* K_w, *the* **ion-product constant of water:**

$$K_w = [H^+][OH^-] = [1 \times 10^{-7}][1 \times 10^{-7}] = 1 \times 10^{-14}$$

The value of K_w is constant at 25 °C; it is always 1×10^{-14}. If [H⁺] goes up, [OH⁻] must go down. When [H⁺] is 1×10^{-3} M, [OH⁻] must be 1×10^{-11} M.

Or when $[H^+]$ is 1×10^{-9} M, $[OH^-]$ must be 1×10^{-5} M. The calculation of $[H^+]$ or $[OH^-]$ when one of these quantities is known is shown in the following example.

EXAMPLE 9.1

Calculating a hydroxide ion concentration

What is the hydroxide concentration of a solution containing a hydrogen ion concentration of 2.0×10^{-6} M?

SOLUTION

The hydroxide ion and hydrogen ion concentrations are related to each other by the ion-product constant of water K_w:

$$K_w = [H^+][OH^-]$$

The hydroxide ion concentration in the preceding expression can be obtained by dividing both sides of the equation by $[H^+]$:

$$[OH^-] = \frac{K_w}{[H^+]}$$

Insert the values for $[H^+]$ and K_w into the equation:

$$[OH^-] = \frac{1 \times 10^{-14}}{2.0 \times 10^{-6}}$$

Solve the problem by dividing 1×10^{-14} by 2.0×10^{-6}. We will consider K_w to be an exact number, and 2.0×10^{-6} has two significant figures. We can keep two significant figures in the answer:

$$[OH^-] = \frac{1 \times 10^{-14}}{2.0 \times 10^{-6}} = 0.50 \times 10^{-8}$$

$$= 5.0 \times 10^{-9} \, M$$

Check the answer by multiplying $[H^+]$ times $[OH^-]$ to be certain that the product of these two quantities is 1×10^{-14}.

PRACTICE EXERCISE 9.1

What is the hydrogen ion concentration in the following aqueous solutions?

(a) $[OH^-] = 2.6 \times 10^{-3}$ M (b) $[OH^-] = 7.2 \times 10^{-7}$ M

(c) $[OH^-] = 9.3 \times 10^{-8}$ M

Neutrality, acidity, and basicity

Chemists describe pure water as neutral. In fact, they describe *any aqueous solution in which [H⁺] and [OH⁻] are 1 × 10⁻⁷ M as a* **neutral solution.** **Acidic solutions** *are those in which [H⁺] is greater than 1 × 10⁻⁷ M.* Solutions containing H^+ concentrations of 0.01 M, 0.1 M, 10^{-5} M, or 10^{-6} M

would all be acidic, since all these concentrations are greater than 1×10^{-7} M. **Basic (alkaline) solutions** *are those in which [OH⁻] is greater than 1×10^{-7} M.* Solutions containing OH⁻ concentrations of 0.001 M or 10^{-4} M would be basic, since these concentrations exceed 1×10^{-7} M.

| EXAMPLE 9.2 | **Determining acidity and [OH⁻] in an aqueous solution** |

Is an aqueous solution in which $[H^+] = 1 \times 10^{-5}$ M acidic, basic, or neutral? What is the [OH⁻] in this solution?

SOLUTION

The $[H^+] = 1 \times 10^{-5}$ M, which is greater than 1×10^{-7} M. Therefore, the solution is acidic. The [OH⁻] can be calculated from the expression

$$K_w = [H^+][OH^-]$$

Dividing both sides of the equation by $[H^+]$ gives

$$[OH^-] = \frac{K_w}{[H^+]}$$

Substitute numerical values for $[H^+]$ and K_w, and perform the division:

$$[OH^-] = \frac{1 \times 10^{-14}}{1 \times 10^{-5}} = 1 \times 10^{-9} \, M$$

PRACTICE EXERCISE 9.2

If the hydroxide ion concentration of an aqueous solution is 1×10^{-3} M, what is $[H^+]$? Is this solution acidic, basic, or neutral?

9.2 Acids and bases

AIMS: *To define and give examples of Arrhenius acids and bases. To use the Brønsted-Lowry theory to classify substances as acids or bases or as hydrogen ion donors or hydrogen ion acceptors. To use the extent of dissociation and the acid dissociation constant K$_a$ to distinguish between strong and weak acids and strong and weak bases.*

Focus

Acids produce hydronium ions and bases produce hydroxide ions in aqueous solution.

In 1887, the Swedish chemist Svante Arrhenius (1859–1927) proposed a revolutionary concept: **Acids** *are compounds that contain hydrogens that dissociate to yield hydrogen ions (H⁺) in aqueous solution;* **bases** *are compounds that dissociate to yield hydroxide ions (OH⁻) in aqueous solution.* A selection of acids and bases found in the home is pictured in Figure 9.1.

Acid
acidus (Latin):   sour

(a)

(b)

Figure 9.1
Many acids (a) and bases (b) are found in the home.

Table 9.1 Some Common Acids

Name	Formula
hydrochloric acid	HCl
nitric acid	HNO_3
sulfuric acid	H_2SO_4
phosphoric acid	H_3PO_4
acetic acid	CH_3COOH
carbonic acid	H_2CO_3

Acids

Table 9.1 lists some common acids. Hydrochloric acid is a typical acid. The hydrogen chloride molecule is a polar covalent molecule that dissociates to form hydrogen ions and chloride ions in aqueous solution:

$$\overset{\delta+ \quad \delta-}{H-Cl}(g) \xrightarrow{\text{Water}} \overbrace{H^+(aq) \quad + \quad Cl^-(aq)}^{\text{Hydrochloric acid, HCl}(aq)}$$

Hydrogen chloride   Hydrogen ion   Chloride ion

Hydrochloric acid, HCl(aq), contains one dissociable hydrogen and is therefore a *monoprotic* ("one proton") acid. Sulfuric acid, $H_2SO_4(aq)$, and phosphoric acid, $H_3PO_4(aq)$, contain two and three dissociable protons, respectively. Sulfuric acid is a diprotic ("two proton") acid, and phosphoric acid is a triprotic ("three proton") acid.

Acids have a sour taste. Some acids are very corrosive to the skin and can cause serious or even fatal burns. All acids should be handled with proper safety precautions. The corrosive power of acids has important implications in dental hygiene, as described in A Closer Look: Acids and Dental Health.

Not all compounds containing hydrogen are acids. Only the hydrogen in very polar bonds—those bonds in which hydrogen is joined to a very electronegative element—dissociates. Hydrogen ions are released when compounds that contain such bonds dissolve in water because the hydrogen ions are stabilized by solvation. Hydrogens in nonpolar bonds do not dissociate. The C—H bonds in methane, CH_4, are nonpolar, so methane has no dissociable hydrogens and is not an acid. Acetic acid, CH_3COOH, is different. Although each molecule contains four hydrogens, acetic acid is a monoprotic acid. We can see why from the structural formula:

$$H-\overset{\overset{\displaystyle H}{|}}{\underset{\underset{\displaystyle H}{|}}{C}}-\overset{\overset{\displaystyle O}{||}}{C}-O-H$$

Acetic acid
(CH_3COOH)

The three hydrogens attached to carbon are in nonpolar bonds and do not dissociate; only the hydrogen attached to the very electronegative oxygen is dissociable. As you see more written formulas, you will be able to recognize dissociable hydrogens.

PRACTICE EXERCISE 9.3

Identify the following acids as monoprotic, diprotic, or triprotic.
(a) H_2CO_3 (b) H_3PO_4 (c) HBr

Bases

Table 9.2 lists some bases. The most common base is sodium hydroxide, NaOH. Lye is an impure form of sodium hydroxide and is a major con-

Acids and Dental Health

The chief cause of dental decay is lactic acid ($C_3H_6O_3$). Lactic acid is formed in the mouth by the action of specific bacteria, such as *Streptococcus mutans*, on sugars present in sticky plaque on tooth surfaces. If the concentration of hydrogen ions becomes high enough, these ions can speed erosion (decay) of the minerals of the teeth, much as the erosion of surfaces of stone statues and buildings has been hastened by acid rain in recent years. The enamel of teeth is in direct contact with plaque, food particles, bacteria, and saliva so that it is the first part of the tooth to decay. Once the enamel is penetrated, dental caries or cavities result. As the damage progresses, the underlying material, called *dentin*, is also destroyed. Decay may eventually reach the pulp of a tooth, which contains blood vessels, fibrous tissue, and nerves, and cause a toothache. The anatomy of a tooth is shown in the accompanying figure.

Applying substances containing fluoride ions directly to the teeth or adding fluoride ions to drinking water makes teeth more resistant to acids. Enamel and dentin are hard because they are made mainly of crystalline forms of two mineral-like compounds of calcium. One is calcium carbonate, $CaCO_3$, a substance made up of limestone and chalk. The other is hydroxyapatite,

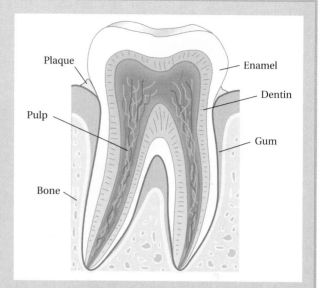

The anatomy of a human tooth.

$[Ca_3(PO_4)_2]_3 \cdot Ca(OH)_2$. Fluoride ions replace hydroxide ions in hydroxyapatite to form fluoroapatite, $[Ca_3(PO_4)_2]_3 \cdot CaF_2$, which forms a more acid-resistant enamel. Fluoride ions, usually as sodium fluoride, NaF, are added to drinking water in some cities. As a result, there has been a drastic reduction in tooth decay. Fluoride ions are toxic in high concentrations, and excessive amounts can discolor (mottle) or damage teeth. Vigorous use of a toothbrush, especially with a fluoride toothpaste, is an important aid in keeping teeth cavity-free.

Solutions of phosphoric acid are often used in dentistry to prepare the surface of teeth when doing fillings.

stituent of some drain cleaners. Sodium metal reacts vigorously with water to form sodium hydroxide with the evolution of hydrogen gas:

$$2Na(s) + 2H_2O(l) \longrightarrow 2NaOH(aq) + H_2(g)$$

Sodium | Water | Sodium hydroxide | Hydrogen

Potassium also reacts vigorously with water to produce potassium hydroxide (KOH):

$$2K(s) + 2H_2O(l) \longrightarrow 2KOH(aq) + H_2(g)$$

Potassium | Water | Potassium hydroxide | Hydrogen

Both sodium hydroxide and potassium hydroxide are ionic solids. They dis-

Table 9.2 Some Common Bases

Name	Formula
potassium hydroxide	KOH
sodium hydroxide	NaOH
calcium hydroxide	$Ca(OH)_2$
magnesium hydroxide	$Mg(OH)_2$

sociate completely into the metal ions and hydroxide ions when dissolved in water:

$$NaOH(s) \xrightarrow{\text{H}_2\text{O}} Na^+(aq) + OH^-(aq)$$

Sodium   Sodium   Hydroxide
hydroxide   ion   ion

$$KOH(s) \xrightarrow{\text{H}_2\text{O}} K^+(aq) + OH^-(aq)$$

Potassium   Potassium   Hydroxide
hydroxide   ion   ion

PRACTICE EXERCISE 9.4

Write balanced equations for the reaction of (a) calcium metal with water and (b) magnesium metal with water.

Alkali
al aqali (Arabic): the ashes

The Group 1A elements, which include sodium and potassium, were originally named the alkali metals because their reaction with water produces alkaline solutions.

Both sodium hydroxide and potassium hydroxide are very soluble in water, and concentrated solutions are readily prepared. Such solutions have a bitter taste and feel slippery. They are extremely corrosive to the skin and can cause painful, deep, slow-healing wounds if not immediately washed off. Calcium hydroxide, $Ca(OH)_2$, and magnesium hydroxide, $Mg(OH)_2$, both hydroxides of Group 2A metals, are not very soluble in water. Consequently, their solutions are always very dilute, even when saturated, and the concentration of hydroxide ions is correspondingly low. A saturated solution of calcium hydroxide contains 0.165 g $Ca(OH)_2$ per 100 g water. Limewater is an aqueous solution of calcium hydroxide. Magnesium hydroxide is much less soluble than calcium hydroxide; a saturated solution contains only 0.0009 g $Mg(OH)_2$ per 100 g water.

PRACTICE EXERCISE 9.5

Write the reaction for the dissociation of magnesium hydroxide in water.

Brønsted-Lowry acids and bases

One problem with the Arrhenius theory is that it is not comprehensive enough. For example, aqueous solutions of ammonia (NH_3), sodium carbonate (Na_2CO_3), and sodium bicarbonate (sodium hydrogen carbonate, $NaHCO_3$) are basic, but none of these compounds is a hydroxide. In 1923, Johannes Brønsted (Danish; 1897–1947) and Thomas Lowry (English; 1874–1936) proposed a new theory. *The **Brønsted-Lowry theory** defines an acid as a **hydrogen ion donor** and a base as a **hydrogen ion acceptor.*** All the acids and bases included in Arrhenius' theory are also acids and bases according to the Brønsted-Lowry theory. The only difference is that some compounds that were not included in Arrhenius' theory can be classified as acids and bases.

The Brønsted-Lowry theory explains why solutions of ammonia are basic. Ammonia gas is very soluble in water and dissolves to give aqueous ammonia, $NH_3(aq)$. According to the Brønsted-Lowry theory, ammonia molecules in aqueous ammonia act as a base because they accept hydrogen ions from water:

$$NH_3(aq) \quad + \quad H_2O(l) \quad \rightleftharpoons \quad NH_4^+(aq) \quad + \quad OH^-(aq)$$

Ammonia	Water	Ammonium ion	Hydroxide ion
(H^+ acceptor, Brønsted-Lowry base)	(H^+ donor, Brønsted-Lowry acid)	(H^+ donor, Brønsted-Lowry acid)	(H^+ acceptor, Brønsted-Lowry base)

In this reaction, ammonia is the hydrogen ion acceptor and therefore a Brønsted-Lowry base. Water, the hydrogen ion donor, is a Brønsted-Lowry acid. As a result of the transfer of a hydrogen ion from water to ammonia, the hydroxide ion concentration is greater in the solution than it is in pure water, and aqueous solutions of ammonia are basic.

The basicity of sodium carbonate and sodium bicarbonate can be explained in a similar way. When these two compounds dissolve in water, they dissociate into sodium ions and carbonate (CO_3^{2-}) or sodium ions and bicarbonate (HCO_3^-) ions:

$$Na_2CO_3(aq) \longrightarrow 2Na^+(aq) + CO_3^{2-}(aq)$$

Sodium carbonate	Sodium ions	Carbonate ion

$$NaHCO_3(aq) \longrightarrow Na^+(aq) + HCO_3^-(aq)$$

Sodium bicarbonate	Sodium ion	Bicarbonate ion

Both the carbonate and the bicarbonate ions are strong enough Brønsted-Lowry bases to pull a hydrogen ion from a water molecule:

$$CO_3^{2-}(aq) \quad + \quad H_2O(l) \quad \rightleftharpoons \quad HCO_3^-(aq) \quad + \quad OH^-(aq)$$

Carbonate ion	Water molecule	Bicarbonate ion	Hydroxide ion
(H^+ acceptor, Brønsted-Lowry base)	(H^+ donor, Brønsted-Lowry acid)	(H^+ donor, Brønsted-Lowry acid)	(H^+ acceptor, Brønsted-Lowry base)

$$HCO_3^-(aq) \quad + \quad H_2O(l) \quad \rightleftharpoons \quad H_2CO_3(aq) \quad + \quad OH^-(aq)$$

Bicarbonate ion	Water molecule	Carbonic acid	Hydroxide ion
(H^+ acceptor, Brønsted-Lowry base)	(H^+ donor, Brønsted-Lowry acid)	(H^+ donor, Brønsted-Lowry acid)	(H^+ acceptor, Brønsted-Lowry base)

Once again, $[OH^-]$ is greater than in pure water, so solutions of these compounds are basic.

The Brønsted-Lowry theory is also applicable to acids. As we saw in Section 9.2, hydrogen chloride gas dissolves in water to give an aqueous solution, HCl(aq). Consider the dissociation that occurs in this solution:

$$HCl(aq) \quad + \quad H_2O(l) \quad \longrightarrow \quad H_3O^+(aq) \quad + \quad Cl^-(aq)$$

Hydrogen chloride	Water molecule	Hydronium ion	Chloride ion
(H^+ donor, Brønsted-Lowry acid)	(H^+ acceptor, Brønsted-Lowry base)	(H^+ donor, Brønsted-Lowry acid)	(H^+ acceptor, Brønsted-Lowry base)

In this reaction, hydrogen chloride and hydronium ion are hydrogen ion donors and are therefore Brønsted-Lowry acids; water and chloride ion are hydrogen ion acceptors and are Brønsted-Lowry bases. However, the solvated chloride ion is such a weak base that it cannot regain a hydrogen ion, and the dissociation of HCl in aqueous solution is essentially irreversible. The concentration of hydronium ions is much higher than the hydroxide ion concentration, and aqueous solutions of HCl are acidic. Aqueous solutions of any Brønsted-Lowry acid always contain an excess of hydronium ions over hydroxide ions and are therefore acidic.

You may have noticed that water has a split personality, sometimes receiving and sometimes donating a hydrogen ion. In aqueous solutions, Brønsted-Lowry bases are substances capable of pulling a hydrogen ion from water to form a hydroxide ion, and Brønsted-Lowry acids are substances capable of donating a hydrogen ion to water to form a hydronium ion.

PRACTICE EXERCISE 9.6

Write equations for the dissociation of HNO_3 and H_2CO_3 in water. Identify the hydrogen ion donor and hydrogen ion acceptor in each case.

Strong and weak acids and bases

Chemists classify acids as strong or weak depending on whether all the molecules dissociate in water. For practical purposes, **strong acids** *are 100% dissociated in aqueous solution.* Hydrochloric, nitric, and sulfuric acids are all strong acids in aqueous solution because the equilibrium for their dissociation lies far to the right:

$$HCl(aq) \quad \longrightarrow \quad H^+(aq) \quad + \quad Cl^-(aq)$$

Hydrochloric acid	Hydrogen ion	Chloride ion

$$HNO_3(aq) \longrightarrow H^+(aq) \quad + \quad NO_3^-(aq)$$

Nitric acid	Hydrogen ion	Nitrate ion

$$H_2SO_4(aq) \longrightarrow 2H^+(aq) \quad + \quad SO_4^{2-}(aq)$$

Sulfuric acid	Hydrogen ion	Sulfate ion

Weak acids *are less than 100% dissociated in aqueous solution.* The dissociation of acetic acid, a weak acid, is

$$H_3COOH(aq) \rightleftharpoons H^+(aq) + CH_3COO^-(aq)$$

| Acetic acid | Hydrogen ion | Acetate ion |

The equilibrium in this reaction lies far to the left. Over 99% of acetic acid molecules exist in solution as undissociated covalent molecules, CH_3COOH. Less than 1% are dissociated at any instant. The equilibrium constant expression for the dissociation of acetic acid is

$$K_a = \frac{[H^+][CH_3COO^-]}{[CH_3COOH]}$$

The equilibrium constant K_a *for a weak acid is called the* **acid dissociation constant** *or* **ionization constant.** The dissociation constant for acetic acid at 25 °C is 1.8×10^{-5}. The size of the K_a of a weak acid depends on what percentage of the acid molecules are in the dissociated form. Therefore, the dissociation constants can be used to compare the relative strengths of weak acids. Small K_a values indicate that a small percentage of the acid molecules are dissociated. Large K_a values indicate a large percentage of dissociation of the acid. An acid with a K_a of 1×10^{-3} is a stronger acid than an acid with a K_a of 1×10^{-5}, even though both acids are weak. Table 9.3 lists the dissociation constants for some common acids at 25 °C. For diprotic and triprotic acids, the dissociation constant in the table is the value for the dissociation of only one hydrogen ion of the acid molecule.

PRACTICE EXERCISE 9.7

Which acid in Table 9.3 is (a) the strongest and (b) the weakest?

Just as there are strong acids and weak acids, there are also strong bases and weak bases. All the bases in Table 9.2 are classified as **strong bases** *because they are 100% dissociated into metal ions and hydroxide ions in aqueous solution.* Unlike strong bases, **weak bases** *produce less than 100% of the possible number of hydroxide ions in aqueous solution.* Aqueous ammonia is an example of a weak base:

$$NH_3(aq) + H_2O(l) \rightleftharpoons NH_4^+(aq) + OH^-(aq)$$

| Ammonia | Water | Ammonium ion | Hydroxide ion |

Table 9.3 Dissociation Constants for Some Acids

Acid	Equilibrium reaction	Dissociation constant (K_a)
hydrochloric acid	$HCl \rightleftharpoons H^+ + Cl^-$	approx. 10^3
phosphoric acid	$H_3PO_4 \rightleftharpoons H^+ + H_2PO_4^-$	7.5×10^{-3}
carbonic acid	$H_2CO_3 \rightleftharpoons H^+ + HCO_3^-$	4.3×10^{-7}
acetic acid	$CH_3COOH \rightleftharpoons H^+ + CH_3COO^-$	1.8×10^{-5}

The equilibrium greatly favors NH_3 and H_2O; approximately 99% of the ammonia is unprotonated, and only 1% is in the form of NH_4^+ and OH^-. There is never very much NH_4^+ and OH^- in aqueous ammonia solutions.

Strong and weak, concentrated and dilute

Concentrated solutions of strong acids and strong bases are extremely corrosive and can produce severe and disfiguring chemical burns if splashed onto the skin or into the eyes. Use the recommended protective clothing and in particular your safety goggles when handling these substances.

The words *strong* and *weak* refer to the extent of dissociation of an acid or base; the words *concentrated* and *dilute* tell how much of an acid or a base is dissolved in solution. Gastric juice in the stomach is a dilute solution of hydrochloric acid, a strong acid. A drop of 1 *M* hydrochloric acid added to 1 L of water becomes more dilute, but hydrochloric acid is still a strong acid; the acid that is present in both solutions is 100% dissociated. Acetic acid is a weak acid even in a concentrated solution because it is less than 100% dissociated. Phosphoric and carbonic acids are two biochemically important weak acids. In addition to being a weak acid, carbonic acid is also unstable and readily decomposes into carbon dioxide and water:

$$H_2CO_3(aq) \rightleftharpoons H_2O(l) + CO_2(aq)$$

| Carbonic acid | Water | Carbon dioxide |

Carbonic acid helps the body dispose of water and carbon dioxide formed as waste products in tissue cells. In tissue cells, water and carbon dioxide combine to form carbonic acid. In the lungs, the carbonic acid decomposes to carbon dioxide gas and water vapor, which are expelled by breathing.

9.3 The pH scale

AIMS: *To calculate the pH of a solution, given the [H^+] or [OH^-]. To calculate the [H^+] or [OH^-], given the pH of a solution.*

Focus

The pH scale is used to express hydrogen ion concentrations.

The expression of hydrogen ion concentration in molarity can be rather cumbersome. Imagine the difficulty in reading a table in which the hydrogen ion concentrations of blood are reported as numbers such as 4.00×10^{-8} *M*. We can simplify a hydrogen ion concentration by expressing it as a *pH*. The **pH** *of a solution is the negative of the logarithm of the hydrogen ion concentration:*

$$\text{pH} = -\log[\text{H}^+]$$

Many shampoos are advertised as being "pH balanced." Normally this means that the shampoo has been formulated to give a pH in the range of 5 to 8.

The usual range of the pH scale for expressing hydrogen ion concentrations is 0 to 14. You may recall from our discussion of K_w that in a neutral solution the hydrogen ion concentration and the hydroxide ion concentration are 1×10^{-7} *M*. Neutral solutions always have a pH of 7. Any solution with a pH below 7 is acidic. For example, a solution of pH 6.9 is very slightly acidic. A solution of pH 0 is strongly acidic. Any solution with a pH above 7 is basic. A solution of pH 7.1 is very slightly basic. A solution of pH 14 is strongly basic. Figure 9.2 gives pH values for a variety of liquids found in the home.

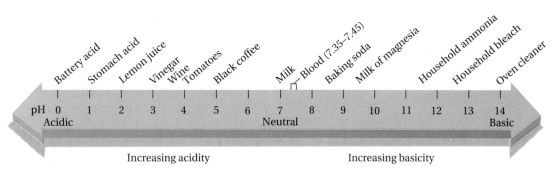

Figure 9.2
The approximate pH values of various substances.

Recall:
$1\ M = 1 \times 10^0\ M$

Within the normal range of pH, 0 to 14, the hydrogen ion concentration may have any value lower than $1\ M\ H^+$. The hydroxide ion concentration may be any value lower than $1\ M\ OH^-$. The relationship among $[H^+]$, $[OH^-]$, and pH is presented in Table 9.4. Most fluids in the human body are about pH 7, but there are some important exceptions. See A Closer Look: pH of Body Fluids.

PRACTICE EXERCISE 9.8

Do the following pH values represent acidic or basic solutions? Which solution is the most acidic, and which is the most basic?

(a) pH 9.9   (b) pH 7.3   (c) pH 6.6   (d) pH 10.9?

Table 9.4 Relationship Among [H⁺], [OH⁻], and pH

	$[H^+]$ (mol/L)	$[OH^-]$ (mol/L)	pH
Increasing acidity ↑	1×10^{0}	1×10^{-14}	0
	1×10^{-1}	1×10^{-13}	1
	1×10^{-2}	1×10^{-12}	2
	1×10^{-3}	1×10^{-11}	3
	1×10^{-4}	1×10^{-10}	4
	1×10^{-5}	1×10^{-9}	5
	1×10^{-6}	1×10^{-8}	6
Neutral	1×10^{-7}	1×10^{-7}	7
Increasing basicity ↓	1×10^{-8}	1×10^{-6}	8
	1×10^{-9}	1×10^{-5}	9
	1×10^{-10}	1×10^{-4}	10
	1×10^{-11}	1×10^{-3}	11
	1×10^{-12}	1×10^{-2}	12
	1×10^{-13}	1×10^{-1}	13
	1×10^{-14}	1×10^{0}	14

A Closer Look

pH of Body Fluids

The human body can be thought of as compartments containing many substances in solution or in suspension. The pH of fluids in each of these compartments is important to the proper functioning of the body. The accompanying table shows that plasma, the fluid of the blood, and the interstitial fluid, the fluid that surrounds cells, is slightly basic. So is the mildly basic cerebrospinal fluid, the fluid that surrounds the brain and spinal cord. The intracellular fluid of liver cells, the fluid inside the cells, has pH 6.9

and is very mildly acidic. In contrast, gastric juice, the fluid of the stomach, has a low pH. The stomach's acidity is important in digestion of food. Whereas the enzymes that catalyze most chemical reactions in the body operate most efficiently near pH 7, the digestive enzyme of the stomach is most effective at low pH. A relatively low pH environment is also found in lysosomes. Lysosomes are a class of subcellular particles, called *organelles*, inside cells. Lysosomal enzymes are responsible for a kind of digestion, recycling old cell parts so that they can be rebuilt. Like the stomach enzyme, the enzymes of lysosomes are efficient catalysts at the lower pH. The pH of urine and perspiration, both waste products of the body, can vary in pH. Both urine and perspiration help to rid the body of excess hydrogen ions formed by chemical reactions in the body. The pH of urine decreases with heavy exercise or in some diseases such as untreated diabetes. The body produces large amounts of hydrogen ions at these conditions.

pH of Some Body Fluids

Fluid	pH*
Gastric juice	1.5–3.0
Perspiration	4.0–6.8
Urine	5.0–7.0
Lysosomal fluid	5.5–6.5
Saliva	6.4–7.4
Milk	6.8
Intracellular fluid	6.9
Cerebrospinal fluid (fluid of brain and spinal cord)	7.3
Blood	7.35–7.45
Interstitial fluid (fluid surrounding cells)	7.4
Pancreatic juice	7.8–8.0

* ▨ = acidic, ▨ = basic.

Finding the pH when the hydrogen ion concentration is known

The pH of a solution is easy to calculate when $[H^+]$ is a number such as 1×10^{-x} M. Notice that when $[H^+]$ is exactly 1×10^{-x} M, the pH is equal to the exponent x with the sign changed. A solution in which $[H^+]$ is exactly 1×10^{-7} M has a pH of 7. A concentration of 1.0 M hydrogen ions can be expressed as 1×10^0 M, so the solution has pH 0. Sometimes the problem is more complicated. For example, what is the pH of a solution in which $[H^+]$ is 6.0×10^{-9} M? The pH of a solution of *any* hydrogen ion concentration can be found on an electronic calculator by using the log function key, as shown in the following examples.

EXAMPLE 9.3   **Calculating the pH from the hydrogen ion concentration**

The hydrogen ion concentration of a solution is 6.00×10^{-9} M. What is the pH of the solution?

SOLUTION

The expression that relates pH and $[H^+]$ is

$$pH = -\log [H^+]$$

Insert the concentration of H^+ in the expression:

$$pH = -\log (6.00 \times 10^{-9})$$

(a)

(b)

Figure 9.3
Use your calculator to find the pH of a solution if the hydrogen ion concentration is 6.00×10^{-9} M. The pH = 8.2.

Evaluate the logarithm of 6.00×10^{-9}. The number could be entered in a calculator directly as 0.00000000600, but counting the large number of zeros could lead to an error. It would be better to enter 6.00×10^{-9} in the calculator by using the exponential function key (EE or exp). Enter the 6.00 on the calculator, press the exponential function key (Fig. 9.3), and then enter the power of ten (−9). On some calculators you may need to enter the 9 first and then change the sign. The calculator screen should display 6.00 and −09 or something similar (Fig. 9.3a). Now obtain the log of 6.00×10^{-9} by pressing the log key. The calculator display will show −8.22184875 (Fig. 9.3b). This is more numbers than is justified on the basis of how well the original hydrogen ion concentration is known. The rules presented in Chapter 2 for determining significant figures in an answer do not apply when logarithms are involved in calculations. In the pH calculations in this book, we will keep *one digit* in the answer past the decimal point. Therefore, round the calculator value to −8.2. This value, after reversing its sign, is the pH:

$$pH = -(-8.2)$$
$$= 8.2$$

A solution containing 6.00×10^{-9} M hydrogen ions has pH 8.2.

EXAMPLE 9.4   **Calculating the pH from the hydrogen ion concentration**

What is the pH of a solution 0.00300 M in hydrogen ions?

SOLUTION

The relationship that relates pH and $[H^+]$ is

$$pH = -\log [H^+]$$

Insert the numerical value of $[H^+]$:

$$pH = -\log [0.00300]$$

Find the logarithm of 0.00300. A hydrogen ion concentration expressed as 0.00300 M can be entered in the calculator directly as 0.00300 or as 3.00×10^{-3} using the exponential function key. Pressing the log key gives −2.52287745. Rounding the answer to one decimal place gives −2.5. After reversing the sign, we obtain 2.5. The solution has pH 2.5.

PRACTICE EXERCISE 9.9

Determine the pH of the following solutions.

(a) $[H^+] = 2.00 \times 10^{-10}\ M$ (b) $[H^+] = 0.0015\ M$

(c) $[H^+] = 1.00 \times 10^{-2}\ M$ (d) $[H^+] = 7.71 \times 10^{-11}\ M$

PRACTICE EXERCISE 9.10

What is the pH of a solution if the hydrogen ion concentration is $2.0 \times 10^{-5}\ M$?

Finding the hydrogen ion concentration when the pH is known

Sometimes the hydrogen ion concentration is needed when only the pH of a solution is known. When we want to find the hydrogen ion concentration of a solution of known pH, we can work from the pH. We have already seen that a solution containing $1 \times 10^{-7}\ M$ hydrogen ions has a pH of 7. Therefore, the hydrogen ion concentration and pH of an aqueous solution are related by the equation

$$[H^+] = 10^{-pH}$$

This relationship works for finding the hydrogen ion concentration of any solution when the pH is known. Use of the 10^x and y^x keys on an electronic calculator to find the numerical value of 10^{-pH} is shown in Example 9.5.

EXAMPLE 9.5

Calculating the hydrogen ion concentration from pH

What is the hydrogen ion concentration of a solution of pH 5.3?

SOLUTION

The pH is known, so find hydrogen ion concentration with the equation

$$[H^+] = 10^{-pH}$$

Insert the numerical value of the pH into the equation:

$$[H^+] = 10^{-5.3}$$

(a)

(b)

Figure 9.4
Use your calculator to find the hydrogen ion concentration of a solution if the pH is 5.3. The $[H^+] = 5.0 \times 10^{-6}\ M$.

To solve for $[H^+]$, use either the 10^x or the y^x key on a calculator. To use the 10^x key, enter -5.3, the negative value of the pH, and then press the 10^x key. To use the y^x key, enter 10, press the y^x key, enter -5.3, and then press the equal (=) key. The use of either the 10^x or the y^x key gives 5.012×10^{-6} (Fig. 9.4a) or possibly 0.000005012 (Fig. 9.4b). Rounding to one decimal place gives 5.0×10^{-6}. The answer to the problem is

$$[H^+] = 5.0 \times 10^{-6}\ M$$

Check the answer by converting the $[H^+]$ back to pH on the calculator. The log of 5.0×10^{-6} is -5.3. A change of sign gives 5.3. Our calculation is correct, because this is the starting pH. We also can check answers of problems in which we convert $[H^+]$ to pH. All we must do is convert the pH back to $[H^+]$ by the method shown in this example.

PRACTICE EXERCISE 9.11

Calculate [H$^+$] for each of the following solutions.

(a) pH 5.0   (b) pH 5.8   (c) pH 12.2   (d) pH 2.6

Finding the pH when hydroxide ion concentration is known

Solutions of bases are often made by dissolving metal hydroxides in water. The hydroxide ion concentration of such a solution is known. For example, the hydroxide ion concentration of a 0.1 M NaOH solution is 0.1 M OH$^-$. We might need to know the pH of such a solution, but we do not have the hydrogen ion concentration. We know, however, that the hydrogen ion and hydroxide ion concentrations are related by K_w, the ion-product constant of water. The pH of a solution containing a known concentration of a base can be calculated using K_w. The calculation is shown in the following two examples.

EXAMPLE 9.6 **Calculating pH from the hydroxide ion concentration**

The hydroxide ion concentration of a solution is 2.3×10^{-4} M. What is the pH of the solution?

SOLUTION

The [OH$^-$] = 2.3×10^{-4} M, but we need [H$^+$] in order to obtain pH. Once we know [H$^+$], we can find the pH of the solution. Set up the problem to find [H$^+$] by recalling that

$$K_w = [H^+][OH^-] = 1 \times 10^{-14}$$

Dividing both sides of the equation by [OH$^-$] gives

$$[H^+] = \frac{K_w}{[OH^-]}$$

Insert the values of K_w and [OH$^-$] into the equation to obtain [H$^+$]:

$$[H^+] = \frac{1 \times 10^{-14}}{2.3 \times 10^{-4}} = 4.3 \times 10^{-11} \, M$$

The hydrogen ion concentration is 4.3×10^{-11} M.

Now calculate the pH of a solution in which [H$^+$] is 4.3×10^{-11} M (refer to Example 9.3 if you need help).

$$
\begin{aligned}
pH &= -\log [H^+] \\
&= -\log (4.3 \times 10^{-11}) \\
&= -(-10.4) \\
&= 10.4
\end{aligned}
$$

A solution in which [OH$^-$] is 2.3×10^{-4} M is pH 10.4.

EXAMPLE 9.7 **Calculating pH from the hydroxide ion concentration**

What is the pH of a solution of [OH$^-$] = 4.0×10^{-11} M?

SOLUTION

We are given [OH$^-$] and need the pH. To calculate pH, we must first get [H$^+$]. Calculate [H$^+$], recalling from the preceding example that

$$[\mathrm{H}^+] = \frac{K_w}{[\mathrm{OH}^-]}$$

Insert the values of K_w and [OH$^-$] into the equation:

$$[\mathrm{H}^+] = \frac{1 \times 10^{-14}}{4.0 \times 10^{-11}} = 2.5 \times 10^{-4}\, M$$

The hydrogen ion concentration is $2.5 \times 10^{-4}\, M$.

Use the [H$^+$] to calculate the pH, taking the log on a calculator:

$$
\begin{aligned}
\mathrm{pH} &= -\log\,[\mathrm{H}^+]\\
&= -\log\,(2.5 \times 10^{-4})\\
&= -(-3.6)\\
&= 3.6
\end{aligned}
$$

The pH of a solution in which [OH$^-$] = $4 \times 10^{-11}\, M$ is 3.6.

Finding hydroxide ion concentrations when pH is known

The relationship among [H$^+$], [OH$^-$], and K_w also can be used to find the hydroxide ion concentration of a solution when only the pH is known. The calculation is shown in the following example.

EXAMPLE 9.8 **Calculating the hydroxide ion concentration from the pH**

What is the hydroxide ion concentration of a solution of pH 9.1?

SOLUTION

The pH is known, and the hydroxide ion concentration is called for. We do not have an equation that relates the pH directly to [OH$^-$], so we will need to use two relationships. We can find the hydrogen ion concentration from the pH using the relationship

$$[\mathrm{H}^+] = 10^{-\mathrm{pH}}$$

Once we have [H$^+$], we can use the relationship

$$K_w = [\mathrm{H}^+][\mathrm{OH}^-]$$

to find the hydrogen ion concentration.

Find the hydrogen ion concentration:

$$
\begin{aligned}
[\mathrm{H}^+] &= 10^{-\mathrm{pH}}\\
&= 10^{-9.1}\\
&= 7.9 \times 10^{-10}\, M
\end{aligned}
$$

Now the desired [OH$^-$] can be calculated using the known values of K_w and [H$^+$]:

$$K_w = [\mathrm{H}^+][\mathrm{OH}^-]$$

or

$$[OH^-] = \frac{K_w}{[H^+]}$$

$$= \frac{1 \times 10^{-14}}{7.9 \times 10^{-10}} = 1.3 \times 10^{-5}\, M$$

The hydroxide ion concentration of a solution of pH 9.1 is $1.3 \times 10^{-5}\, M$.

PRACTICE EXERCISE 9.12

What are the hydroxide ion concentrations for aqueous solutions with the following pH values?

(a) pH 6   (b) pH 8.3   (c) pH 12.7

Table 9.4 summarizes the relationship among $[H^+]$, $[OH^-]$, and pH. As the previous examples have shown, problems involving pH can be solved with three formulas:

$$K_w = [H^+][OH^-]$$
$$pH = -\log [H^+]$$
$$[H^+] = 10^{-pH}$$

When you are faced with a pH problem, identify what is given and what is wanted in the answer. Think about the relationships involved in the problem. Then apply the formulas in the correct order to obtain the correct answer.

PRACTICE EXERCISE 9.13

Calculate the pH for each of the following solutions.

(a) $[H^+] = 5.0 \times 10^{-6}$ (b) $[H^+] = 8.3 \times 10^{-10}$
(c) $[OH^-] = 2.0 \times 10^{-5}$ (d) $[OH^-] = 4.5 \times 10^{-11}$

PRACTICE EXERCISE 9.14

Calculate $[H^+]$ and $[OH^-]$ for solutions of pH (a) 3.8, (b) 5.7, and (c) 10.3.

9.4 Measurement of pH

AIM: To describe methods of measuring the pH of a solution.

Focus

pH is measured with acid-base indicators or a pH meter.

Maintaining pH within certain bounds is of great importance in many areas of interest to scientists and nonscientists. For example, the body maintains the pH of the blood between 7.35 and 7.45. A change in the blood of only one pH unit higher or lower can be fatal. The chlorinated water in swimming pools is kept near pH 7.0 by the addition of chemicals to neutralize hydrochloric acid produced by the sunlight-catalyzed reaction of the chlo-

Figure 9.5
Trees and other plants are damaged by acid rain.

The juices from many common fruits and vegetables are natural acid-base indicators. Examples include carrot stems, beets, rhubarb, purple cabbage leaves, and cherries.

rine with water. Conservationists are very concerned about the destruction of plant and animal life in lakes and rivers by acid rain (Fig. 9.5), which sometimes has a pH lower than 3. To adjust pH or to study problems involving acidity or basicity, it is first necessary to measure pH. The pH is usually measured using acid-base indicators or pH meters.

The simplest way to tell whether the pH of a solution lies above or below a certain value is to use acid-base indicators—dyes that are weak acids whose dissociated and undissociated forms are different colors. The dissociation of an indicator, HIn, in water is expressed as

$$\text{HIn}(aq) \rightleftharpoons \text{H}^+(aq) + \text{In}^-(aq)$$

With phenolphthalein, a common indicator of complex molecular structure, the undissociated form (HIn) is colorless and the dissociated form (In$^-$) is pink. Alkaline solutions turn pink when a small amount of phenolphthalein is added, since at pH values greater than 10.0 nearly all the indicator is in its dissociated form due to the presence of a substantial concentration of hydroxide ions:

$$\text{HIn}(aq) + \text{OH}^-(aq) \rightleftharpoons \text{In}^-(aq) + \text{H}_2\text{O}(l)$$

Acidic solutions remain colorless when phenolphthalein is added, since at pH values less than 8.0 nearly all the phenolphthalein is in its undissociated form due to the presence of a substantial concentration of hydrogen ions:

$$\text{In}^-(aq) + \text{H}^+(aq) \rightleftharpoons \text{HIn}(aq)$$

We can say whether a solution has a pH of less than 8.0 or greater than 10.0 simply by looking at the color of a solution to which phenolphthalein has been added. The pH ranges in which color changes occur for some other common indicators are shown in Figure 9.6.

Acid-base indicators are also sold as pH test papers (Fig. 9.7). One especially convenient test paper is impregnated with several indicators yielding a rainbow of colors. When dipped into a solution of unknown pH, the test paper changes to a specific color; the pH of the solution may be determined to within about 0.1 pH unit of the true value by comparing the color of the wet test paper with a color reference chart.

Figure 9.6
The color changes of some acid-base indicators.

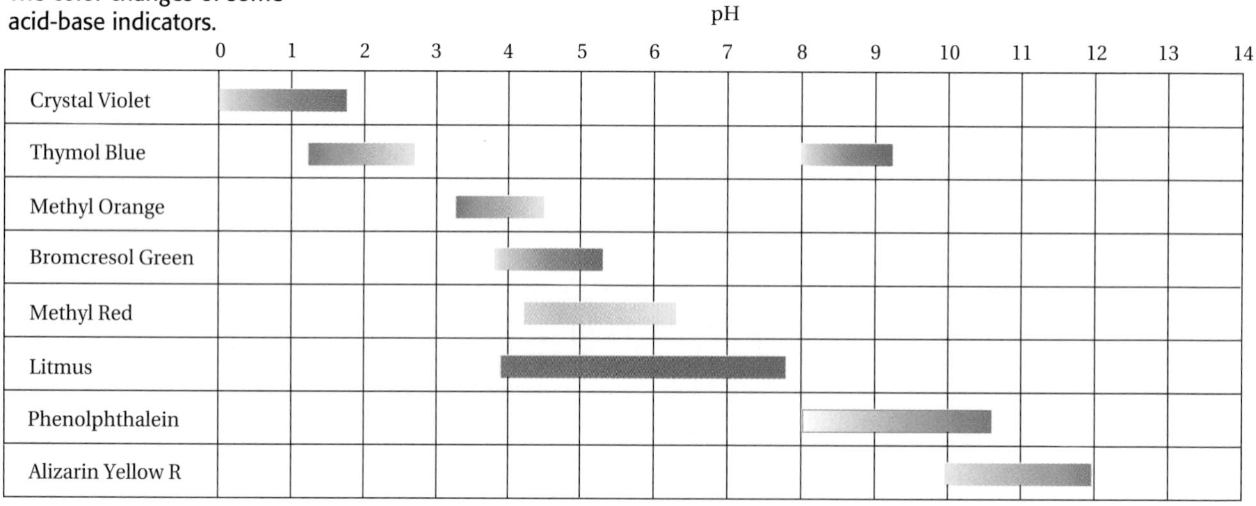

Indicator	pH range (0–14)
Crystal Violet	~0–2
Thymol Blue	~1–2.5 and ~8–9.5
Methyl Orange	~3–4.5
Bromcresol Green	~3.5–5.5
Methyl Red	~4–6
Litmus	~4–8.5
Phenolphthalein	~8–10
Alizarin Yellow R	~10–12

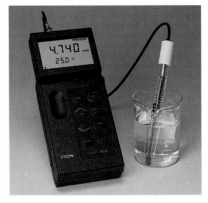

Figure 9.7
The approximate pH of an aqueous solution can be determined by using pH test papers. After a drop of the sample is applied to the paper, the color is compared with a standard chart. What is the pH of the urine sample?

Figure 9.8
A pH meter is used when an accurate determination of the pH of a solution is required.

Measurement of pH is rapid and accurate with a pH meter. A pH meter consists of a pair of special electrodes connected to a millivoltmeter that has a readout in pH units (Fig. 9.8). The meter is calibrated by immersing the electrodes in a solution of known pH. The electrodes are then rinsed with distilled water and dipped into a solution of unknown pH; the pH of the unknown is read from the meter scale. Values of pH obtained with a sensitive meter are typically accurate to within 0.01 pH unit of the true pH. The color or turbidity of a solution has little effect on the reading obtained. This, along with its accuracy, makes the pH meter of particular value in measuring the pH of blood, urine, and other body fluids, as well as for monitoring sewage and industrial effluents for environmental contamination.

9.5 Neutralization

AIM: *To write and balance an equation for a neutralization reaction.*

Focus

Acids and bases react to produce salts and water.

Salt
sal (Latin): salt

The product of the reaction of an acid with a base is given the general name **salt.** *Reactions in which acids and bases react in aqueous solution to produce salts and water are called* **neutralization reactions.** Here are three examples of neutralization reactions:

$$HCl(aq) \ + \ NaOH(aq) \longrightarrow NaCl(aq) \ + \ H_2O(l)$$

| Hydrochloric acid | Sodium hydroxide | Sodium chloride | Water |

$$H_2SO_4(aq) \ + \ 2NaOH(aq) \longrightarrow Na_2SO_4(aq) \ + \ 2H_2O(l)$$

| Sulfuric acid | Sodium hydroxide | Sodium sulfate | Water |

$$H_2SO_4(aq) \ + \ Ca(OH)_2(aq) \longrightarrow CaSO_4(aq) \ + \ 2H_2O(l)$$

| Sulfuric acid | Calcium hydroxide | Calcium sulfate | Water |

Weak bases such as aqueous ammonia, $NH_3(aq)$, can be neutralized by acids, but in the balanced equation it does not appear that water is produced:

$$NH_3(aq) \ + \ HCl(aq) \ \longrightarrow \ NH_4Cl(aq)$$

Ammonia   Hydrochloric   Ammonium
acid   chloride

$$2NH_3(aq) \ + \ H_2CO_3(aq) \ \longrightarrow \ (NH_4)_2CO_3(aq)$$

Ammonia   Carbonic   Ammonium
acid   carbonate

To understand why water does not appear in these equations, we must recognize that aqueous ammonia contains ammonium ions and hydroxide ions (Sec. 9.2):

$$NH_3(aq) \ \longrightarrow \ NH_4^+(aq) + OH^-(aq)$$

When we substitute NH_4^+ and OH^- for NH_3 in the neutralization equations, it becomes more apparent that a salt and water are produced by neutralizing the base with the acid:

$$NH_4^+(aq) \ + \ OH^-(aq) \ + \ HCl(aq) \ \longrightarrow \ NH_4Cl(aq) \ + \ H_2O(l)$$

Ammonium   Hydroxide   Hydrochloric   Ammonium   Water
ion   ion   acid   chloride

$$2NH_4^+(aq) \ + \ 2OH^-(aq) \ + \ H_2CO_3(aq) \ \longrightarrow \ (NH_4)_2CO_3(aq) \ + \ 2H_2O(l)$$

Ammonium   Hydroxide   Carbonic   Ammonium   Water
ion   ion   acid   carbonate

Many salts have medical applications. Table 9.5 lists some of these. Over-the-counter antacid preparations are among the most common medical uses of salts and bases. Antacids are formulated to neutralize stomach acids. You may recall that when we left Robert, the subject of the Case in

Table 9.5 Some Salts and Their Medical Applications

Name	Formula	Application
barium sulfate	$BaSO_4$	gastrointestinal studies
calcium sulfate (plaster of paris)	$(CaSO_4)_2 \cdot H_2O$	plaster casts
magnesium sulfate (Epsom salts)	$MgSO_4 \cdot 7H_2O$	cathartic
potassium permanganate	$KMnO_4$	disinfectant and fungicide
silver nitrate	$AgNO_3$	cauterizing agent; photographic emulsions
sodium bicarbonate (baking soda)	$NaHCO_3$	antacid; intravenous injection
sodium chloride (table salt)	$NaCl$	intravenous injection
sodium sulfate (Glauber's salt)	$Na_2SO_4 \cdot 10H_2O$	cathartic

FOLLOW-UP TO THE CASE IN POINT: Antacids

Robert experienced discomfort from heartburn and acid reflux. Hydrochloric acid is normally present in the stomach to aid digestion, but an excess can cause heartburn and a feeling of nausea. Acid reflux occurs when stomach acid backs up ("refluxes") into the esophagus, the tube that food passes through on the way to the stomach. Chronic acid reflux can cause scarring of this passage. Acid indigestion and heartburn often are caused by overeating or eating spicy foods. These actions trigger the excess production of hydrochloric acid in the stomach. The physician suggested that Robert stop eating before bedtime. The doctor also suggested that Robert should have an antacid preparation handy. Antacids are sold to millions of people every year. Some of the antacids Robert found at his neighborhood pharmacy are shown in the accompanying figure. All these antacids are formulated to neutralize stomach acid. You may wonder why Alka-Seltzer can act as an antacid when it contains citric acid. When Alka-

Seltzer is added to water, the reaction between citric acid and sodium bicarbonate produces sodium citrate, an antacid.

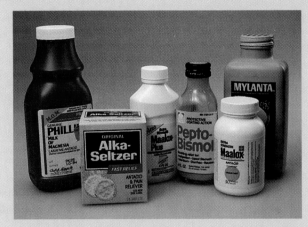

Some popular formulations that provide relief for acid indigestion.

The intake of too much antacid can lead to metabolic alkalosis, a condition in which the pH of the blood is higher (more alkaline) than normal.

Point for this chapter, he was suffering loss of sleep from the discomfort of heartburn and acid reflux. Above, we learn of some suggested remedies.

PRACTICE EXERCISE 9.15

Identify the products and write balanced equations for the following neutralization reactions. (If you need help, go back to Sec. 6.2.)

(a) $HNO_3(aq) + KOH(aq) \longrightarrow$

(b) $HCl(aq) + Mg(OH)_2(aq) \longrightarrow$

(c) $NH_4^+(aq) + OH^-(aq) + H_3PO_4(aq) \longrightarrow$

9.6 Equivalents

AIM: To calculate the equivalent mass and number of equivalents of any acid or base.

Focus

Neutralization reactions require equivalent amounts of acids and bases.

The numbers of hydrogen ions and hydroxide ions must be equal in a neutralization reaction. Some acids and bases are neutralized by reacting in a 1:1 mole ratio:

$$HCl(aq) + NaOH(aq) \longrightarrow NaCl(aq) + H_2O(aq)$$

1 mol 1 mol 1 mol 1 mol

When sodium hydroxide reacts with sulfuric acid, however, the ratio is 2:1. Two moles of base (hydroxide ion) are required to neutralize 1 mol of H_2SO_4:

$$H_2SO_4(aq) \ + \ 2NaOH(aq) \ \longrightarrow \ Na_2SO_4(aq) \ + \ 2H_2O(l)$$

1 mol	2 mol	1 mol	2 mol

Three moles of base (hydroxide ion) are required to neutralize 1 mol of H_3PO_4:

$$H_3PO_4(aq) \ + \ 3NaOH(aq) \ \longrightarrow \ Na_3PO_4(aq) \ + \ 3H_2O(l)$$

1 mol	3 mol	1 mol	3 mol

EXAMPLE 9.9   **Finding the amount of acid needed to neutralize a base**

How many moles of sulfuric acid are required to neutralize 0.50 mol sodium hydroxide?

SOLUTION

Set up the relationship between moles of sodium hydroxide and sulfuric acid by writing a balanced equation for the reaction:

$$H_2SO_4(aq) \ + \ 2NaOH(aq) \ \longrightarrow \ Na_2SO_4(aq) \ + \ 2H_2O(aq)$$

1 mol	2 mol	1 mol	2 mol

The equation shows that 2 mol NaOH is required to neutralize 1 mol H_2SO_4. This can be expressed as the ratio

$$\frac{1 \text{ mol } H_2SO_4}{2 \text{ mol NaOH}}$$

Take the moles of NaOH times this ratio to obtain the amount of sulfuric acid:

$$0.50 \text{ mol NaOH} \times \frac{1 \text{ mol } H_2SO_4}{2 \text{ mol NaOH}} = 0.25 \text{ mol } H_2SO_4$$

It takes 0.25 mol H_2SO_4 to neutralize 0.50 mol NaOH.

PRACTICE EXERCISE 9.16

How many moles of potassium hydroxide are required to neutralize 0.20 mol phosphoric acid?

To attain neutralization of an acidic or basic solution, an equal number of moles of hydrogen ions and hydroxide ions must react to form water. The way to get equal numbers of moles of hydrogen ions and hydroxide ions is to take *equivalent* amounts of the acids and bases. *In acid-base chemistry, an* **equivalent** *is the mass of acid or base that can furnish or accept exactly 1 mol of hydrogen ions (H^+).* One equivalent (1 equiv) of an acid is the amount in grams that will furnish 1 mol of hydrogen ions. For example, 1 molar mass of hydrochloric acid equals 1 equiv of HCl; 1 molar mass of sulfuric acid equals 2 equiv of H_2SO_4; and 1 mol of phosphoric acid equals 3 equiv of H_3PO_4. One equivalent (1 equiv) of a base is the amount in grams needed to accept 1 mol of hydrogen ions. For example, 1 molar mass of potassium hydroxide equals 1 equiv of KOH, and 1 molar mass of calcium hydroxide equals 2 equiv of $Ca(OH)_2$. We can see from the definition of an equivalent that 1 equiv of an acid will neutralize exactly 1 equiv of a base.

EXAMPLE 9.10   **Calculating the mass of 1 equivalent of an acid**

What is the mass of 1 equiv of nitric acid?

SOLUTION

We must find the molar mass and number of dissociable hydrogens to find the mass of 1 equiv of nitric acid. The molar mass of nitric acid (HNO_3) is 63 g (review Sec. 6.1 if you need help); HNO_3 can furnish 1 mol of hydrogen ions. Therefore,

$$1 \text{ equiv } HNO_3 = \frac{63 \text{ g } HNO_3}{1} = 63 \text{ g } HNO_3$$

EXAMPLE 9.11   **Calculating the mass of 1 equivalent of a base**

What is the mass of 1 equiv of calcium hydroxide ($Ca(OH)_2$)?

SOLUTION

To find the mass of 1 equiv of $Ca(OH)_2$, we need its molar mass and the number of moles of hydrogen ions that it can accept in a neutralization reaction. The molar mass of $Ca(OH)_2$ is 74 g, and 1 mol of the compound contains two hydroxide ions that can accept 2 mol of hydrogen ions in neutralization:

$$Ca(OH)_2(aq) + 2H^+(aq) \longrightarrow Ca^{2+}(aq) + 2H_2O(l)$$

Therefore,
$$1 \text{ equiv } Ca(OH)_2 = \frac{74 \text{ g}}{2} = 37 \text{ g } Ca(OH)_2$$

PRACTICE EXERCISE 9.17
Determine the mass of 1 equiv for (a) KOH, (b) HCl, and (c) HBr.

EXAMPLE 9.12   **Determining the equivalents in a diprotic acid**

How many equivalents is 4.9 g sulfuric acid?

SOLUTION

A new twist to this problem is that we are given a certain mass of sulfuric acid and asked to find the number of equivalents it contains. We can find the molar mass and the number of dissociable hydrogens for sulfuric acid. This will give us the mass of 1 equiv of the acid. When we know the mass of 1 equiv, we can find the number of equivalents in any fraction or multiple of 1 equiv.

The molar mass of H_2SO_4 is 98 g. Sulfuric acid has two dissociable hydrogens:

$$H_2SO_4(aq) \longrightarrow 2H^+(aq) + SO_4{}^{2-}(aq)$$

Therefore,

$$1 \text{ equiv } H_2SO_4 = \frac{98 \text{ g } H_2SO_4}{2} = 49 \text{ g } H_2SO_4$$

Calculate the number of equivalents in 4.9 g H_2SO_4 by the following relationship:

$$\text{Equivalents } H_2SO_4 = 4.9 \text{ g } H_2SO_4 \times \frac{1 \text{ equiv}}{49 \text{ g } H_2SO_4} = 0.10 \text{ equiv}$$

PRACTICE EXERCISE 9.18

How many equivalents is each of the following?

(a) 3.7 g $Ca(OH)_2$ (b) 98 g H_2SO_4 (c) 9.8 g H_3PO_4

(d) 10 g NaOH

9.7 Normality

AIMS: To define and calculate the normality of a solution. To describe the procedure for preparing a dilute solution of known normality from a solution of more concentrated known normality.

Focus

Concentrations of acids and bases are sometimes expressed in normalities.

You may recall from the discussion of solution concentrations in Section 8.8 that molarity is the number of moles of a compound dissolved in 1 L of solution. In dealing with solutions of acids and bases, however, we are usually more interested in how many equivalents of acid or base a solution contains. The equivalents contained in solutions of acids and bases can be expressed as *normalities. A solution containing 1 equiv of an acid or base per liter has a* **normality** *of 1.0.* That is, the solution is 1 normal (1 *N*).

Relationship between molarity and normality

The numerical values of normality and molarity are equal for acids and bases that give 1 equiv of H^+ or OH^- per mole. For example, a solution containing 1 mol NaOH per liter is 1 *M* and also 1 *N*. A solution containing 1 mol H_2SO_4 per liter is 1 *M*, but it is 2 *N*. By the same token, a 1 *M* solution of H_3PO_4 is 3 *N*.

PRACTICE EXERCISE 9.19

What is the normality of the following solutions?

(a) 2 M HCl (b) 1 M NaOH (c) 0.1 M CH$_3$COOH
(d) 0.2 M H$_2$SO$_4$ (e) 0.3 M H$_3$PO$_4$

Equivalents from volume and normality

The number of equivalents of an acid or base in a given volume of solution of known normality can be found from the relationship

$$\text{Equivalent of acid or base} = \text{volume (liters)} \times \text{normality}$$
$$= V\text{(liters)} \times N$$

A 1 N H$_2$SO$_4$ solution contains 1 equiv/L of H$_2$SO$_4$. Therefore, 2 L of 1 N H$_2$SO$_4$ contains 2 equiv of H$_2$SO$_4$:

$$V \times N = 2\, \cancel{L} \times \frac{1 \text{ equiv H}_2\text{SO}_4}{1\, \cancel{L}} = 2 \text{ equiv H}_2\text{SO}_4$$

equivalent = equiv
milliequivalent = mEq

In most clinical settings, volume is measured in milliliters. Because 1 mL equals 1/1000 L, 1 mL of a 1 N solution of an acid or base contains 1 milliequivalent (1/1000 equiv) of the acid or base. The product of the volume of a solution in milliliters multiplied by the normality of the solution gives the number of milliequivalents (mEq) of the acid or base present:

$$\text{mEq of acid or base} = \text{volume (milliliters)} \times \text{normality}$$
$$= V\text{(mL)} \times N$$

For example, 50 mL of 1 N H$_2$SO$_4$ contains 50 mEq of H$_2$SO$_4$:

$$V \times N = 50\, \cancel{\text{mL}} \times \frac{1 \text{ mEq}}{1\, \cancel{\text{mL}}} = 50 \text{ mEq H}_2\text{SO}_4$$

In clinical work, the concentrations of important ions in the blood and other body fluids are often reported in mEq/L. For example, the bicarbonate ion (HCO$_3^-$) concentration of normal blood plasma is 27 mEq/L, that of the monohydrogen phosphate ion (HPO$_4^{2-}$) is 2 mEq/L, and that of the chloride ion is 103 mEq/L.

PRACTICE EXERCISE 9.20

How many equivalents are there in each of the following:
(a) 0.5 L of 1 N NaOH
(b) 1 L of 0.5 N HCl
(c) 0.2 L of 0.1 N H$_2$SO$_4$

PRACTICE EXERCISE 9.21

How many milliequivalents are there in each of the following:
(a) 25 mL of 0.1 N HCl
(b) 50 mL of 0.2 N H$_3$PO$_4$
(c) 10 mL of 0.5 N HNO$_3$?

Dilution of solutions of known normality

Solutions of known normality can be made less concentrated by diluting with water, using the relationship given by the equation

$$N_1 \times V_1 = N_2 \times V_2$$

Here N_1 and V_1 are the initial solution's normality and volume, and N_2 and V_2 are the final solution's normality and volume. This relationship is similar to the one used for the dilution of solutions of known molarity (Sec. 8.8).

EXAMPLE 9.13	**Preparing a solution of known normality by dilution**

We need to make 250 mL of 0.100 N sodium hydroxide from a stock solution containing 2.00 N sodium hydroxide. How many milliliters of the stock solution must we dilute to 250 mL to get the required solution?

SOLUTION

The normality (N_2 = 0.100 N) and volume (V_2 = 250 mL) of the final solution are known. The only missing piece of information is the volume of the initial solution (V_1). Use the following relationship to find V_1:

$$N_1 \times V_1 = N_2 \times V_2$$

Divide both sides of the equation by N_1 to give V_1, then insert the values N_1 = 2.00 N, N_2 = 0.100 N, and V_2 = 250 mL:

$$V_1 = \frac{N_2 \times V_2}{N_1} = \frac{0.100\,\cancel{N} \times 250\text{ mL}}{2.00\,\cancel{N}} = 12.5 \text{ mL}$$

We must dilute 12.5 mL of 2.0 N NaOH to 250 mL to make a solution of 0.100 N NaOH.

PRACTICE EXERCISE 9.22

How would you prepare 5.0×10^2 mL of 0.20 N sulfuric acid from a stock solution of 4.0 N sulfuric acid?

9.8 Titration

AIM: To calculate the normality of an acid or base at the equivalence point of a titration.

Focus

Acids and bases are neutralized by titrations.

It is often necessary to determine the normality of an acid or base in a sample. The method most commonly used is called **titration,** *which is a neutralization reaction in which the concentration of one of the reactants, either the acid or the base, is unknown.* In a typical titration, a known volume of acid, of unknown concentration, is transferred to a flask. Several drops of an acid-base indicator solution (such as phenolphthalein) are then added. A solution of known concentration of sodium hydroxide is then slowly added

Figure 9.9
An acid-base titration. (a) A known volume of acid of unknown concentration is added to the flask. A few drops of the acid-base indicator phenolphthalein are added to the acidic solution. (b) Standardized sodium hydroxide is added slowly from a buret. In regions where the hydroxide is high, the solution turns pink. The pink color disappears when the contents of the flask are mixed. (c) The equivalence point is reached when the solution in the flask is just neutralized: the indicator turns from colorless to pink and remains pink when the contents of the flask are mixed.

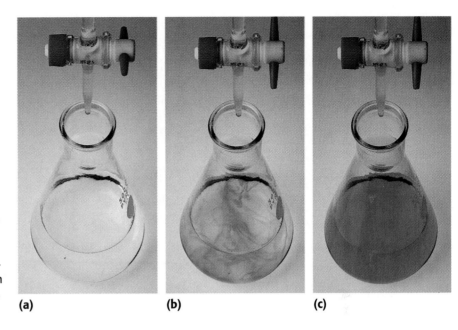

(a)   (b)   (c)

Equivalence
aequi (Latin):   equal
valere (Latin):   to be strong

from a buret until the solution in the flask is just neutralized (Fig. 9.9). *The point of neutralization in a titration is called the* **equivalence point.** The equivalence point usually can be closely approximated by the use of an acid-base indicator or a pH meter. If the indicator is phenolphthalein, the equivalence point is reached when the indicator changes from colorless to pink. The volume of base required to reach the equivalence point is obtained from the difference between the initial and final buret readings.

At the equivalence point, the numbers of equivalents of acid and base are equal. It is now possible to calculate the number of equivalents of acid or base in an unknown sample. If N_A and N_B are the normalities of the acid and base solutions and V_A and V_B are the volumes of the acid and base solutions required to give a neutral solution, then the equivalents of acid and base are given by

$$\text{Equivalents of acid } = N_A \times V_A$$
$$\text{Equivalents of base} = N_B \times V_B$$

Since the numbers of equivalents of acid and base are equal at the equivalence point,

$$N_A \times V_A = N_B \times V_B$$

The volumes V_A and V_B may be expressed in liters or milliliters, provided the same unit is used for both. For practical reasons, milliliters are usually used for reporting solution volumes in analyses.

EXAMPLE 9.14   **Finding the normality of a base from a neutralization reaction**

If 35.0 mL of 0.200 N hydrochloric acid is required to neutralize 25.0 mL of an unknown base, what is the normality of the base?

SOLUTION

The unknown and known quantities are related by the expression

$$N_A \times V_A = N_B \times V_B$$

Divide both sides of the preceding equation by V_B to obtain N_B, and then insert numerical values for V_A, N_A, and V_B:

$$N_B = \frac{V_A \times N_A}{V_B} = \frac{35.0 \text{ mL} \times 0.200 \text{ N}}{25.0 \text{ mL}} = 0.280 \text{ N}$$

EXAMPLE 9.15

Calculating the amount of acid needed to neutralize a base

How many milliliters of 0.500 N sulfuric acid are required to neutralize 50.0 mL of 0.200 N potassium hydroxide?

SOLUTION

We can use the relationship

$$N_A \times V_A = N_B \times V_B$$

to find the volume of acid V_A. Dividing both sides of the preceding equation by N_A gives V_A. Insert the numerical values and units of V_B, N_B, and N_A, and then calculate the answer:

$$V_A = \frac{V_B \times N_B}{N_A} = \frac{50.0 \text{ mL} \times 0.200 \text{ N}}{0.500 \text{ N}} = 20.0 \text{ mL}$$

PRACTICE EXERCISE 9.23

How many milliliters of 0.200 N sodium hydroxide must be added to 75.0 mL of 0.0500 N hydrochloric acid to make a neutral solution?

PRACTICE EXERCISE 9.24

What is the normality of a solution of a base if 25.0 mL is neutralized by 75.0 mL of 0.400 N acid?

9.9 Salts of weak acids and bases

AIM: To explain why some salts give acidic or basic aqueous solutions.

Focus

Solutions of a salt are not neutral if the salt undergoes hydrolysis.

Since salts are made by neutralizing an acid with a base, we might assume that solutions of salts should be neutral, and many are. Nevertheless, solutions of some salts are acidic, and others are basic. Solutions of sodium chloride and potassium sulfate are neutral, a solution of ammonium chloride is acidic, and a solution of sodium acetate is basic.

Salt of a weak acid and strong base

Sodium acetate is the salt of acetic acid, a weak acid, and sodium hydroxide, a strong base. In solution, the salt is completely dissociated:

$$CH_3COONa(aq) \longrightarrow CH_3COO^-(aq) + Na^+(aq)$$

Sodium acetate   Acetate ion   Sodium ion

But now the acetate ion, an H^+ acceptor, establishes an equilibrium with water, forming some undissociated acetic acid and hydroxide ions:

$$CH_3COO^-(aq) + H_2O(l) \rightleftharpoons CH_3COOH(aq) + OH^-(aq)$$

Acetate ion	Water	Acetic acid	Hydroxide ion
(H^+ acceptor, Brønsted-Lowry base)	(H^+ donor, Brønsted-Lowry acid)		(Makes the solution basic)

Hydrolysis
hydro (Greek):   water
lysis (Greek):   splitting

This process is called *hydrolysis,* since it splits (lyses) the bond between hydrogen and oxygen in a water molecule. The solution contains a hydroxide ion concentration greater than the hydrogen ion concentration, and the solution is basic.

Salt of a strong acid and weak base

Ammonium chloride (NH_4Cl) is the salt of hydrochloric acid, a strong acid, and ammonia, a weak base. It is completely dissociated in solution:

$$NH_4Cl(s) \longrightarrow NH_4^+(aq) + Cl^-(aq)$$

Ammonium chloride	Ammonium ion	Chloride ion

The ammonium ion is a strong enough acid to donate a hydrogen ion to a water molecule, although the equilibrium is strongly to the left:

$$NH_4^+(aq) + H_2O(l) \rightleftharpoons NH_3(aq) + H_3O^+(aq)$$

Ammonium ion	Water	Ammonia	Hydronium ion
(H^+ donor, Brønsted-Lowry acid)	(H^+ acceptor, Brønsted-Lowry base)		(Makes the solution acidic)

This process, also hydrolysis, results in the formation of some unprotonated ammonia and hydronium ions. The hydronium ion concentration is therefore greater than the hydroxide ion concentration, and a solution of ammonium chloride is acidic. To summarize, salts formed from weak acids and strong bases give basic solutions. Salts formed from strong acids and weak bases give acidic solutions.

9.10 Buffers

AIMS: To define a buffer and show with equations how a buffer works. To explain the importance of buffers in biologic systems.

Focus

Buffers are solutions that resist changes in pH.

The addition of 10 mL of 0.1 M sodium hydroxide to 1 L of pure water increases the pH by 4.0 pH units—from 7.0 to 11.0. A solution that is 0.1 M in both acetic acid and sodium acetate has pH 4.74. When moderate amounts of either acid or base are added to this solution, the pH changes little. For example, the addition of 10 mL of 0.1 M sodium hydroxide to 1 L of the solution increases the pH by only 0.01 pH unit—from 4.74 to 4.75.

The solution containing acetic acid and acetate anions is a typical example of a *buffer*. **Buffers** *are solutions in which the pH remains relatively constant when small amounts of acid or base are added.* A buffer can be a solution of a weak acid and one of its salts or a solution of a weak base and one of its salts.

The acetic acid–acetate buffer is able to resist drastic changes in pH because the acetic acid and its anion act as reservoirs of neutralizing power for the hydrogen ions and hydroxide ions added to the solution. Let's see how this works. The sodium acetate in the buffer solution is completely dissociated:

$$CH_3COONa(aq) \longrightarrow CH_3COO^-(aq) + Na^+(aq)$$

Sodium acetate   Acetate ion   Sodium ion

When an acid, a source of hydrogen ions, is added to the solution, the acetate ions act as a "proton sponge" to mop up these hydrogen ions. Recall that acetic acid does not dissociate extensively in water. Therefore, most of the acetate ion becomes acetic acid, and the pH does not appreciably change:

$$CH_3COO^-(aq) + H^+(aq) \rightleftharpoons CH_3COOH(aq)$$

Acetate   Hydrogen   Acetic acid
ion   ion

When a base, a source of hydroxide ions, is added to the solution, the acetic acid and the base react to produce neutral water:

$$CH_3COOH(aq) + OH^-(aq) \rightleftharpoons CH_3COO^-(aq) + H_2O(l)$$

Acetic acid   Hydroxide   Acetate   Water
            ion   ion

The acetate ion is not a strong enough base to accept hydrogen ions from water extensively, and again, the pH changes very little.

An acetic acid–acetate buffer becomes ineffective when there are no more acetate ions available to accept hydrogen ions or acetic acid molecules to donate hydrogen ions. When an acid is added to a buffer, the pH holds steady until most of the acetate ions have been used up. At this point the solution is no longer a buffer. If only small amounts of additional acid are then added, the pH drops sharply. When a base is added to acetic acid–acetate buffer, the pH holds steady until most of the acetic acid has been used up. If more base is then added, the pH sharply increases. When excessive amounts of either acid or base have been added, the *buffer capacity* of the solution is said to have been exceeded. Table 9.6 lists some buffer

Table 9.6 Important Buffer Systems

Buffer name	Buffer components
acetic acid–acetate ion	CH_3COOH/CH_3COO^-
dihydrogen phosphate ion–monohydrogen phosphate ion	$H_2PO_4^-/HPO_4^{2-}$
carbonic acid–bicarbonate ion	H_2CO_3/HCO_3^-
ammonium ion–ammonia	NH_4^+/NH_3

systems; the carbonic acid–bicarbonate ion and the dihydrogen phosphate–monohydrogen phosphate buffer systems are especially important in controlling pH in the body. We will discuss them again with body fluids in Chapter 22.

EXAMPLE 9.16   **Demonstrating how a buffer works**

Show how the carbonic acid–bicarbonate (H_2CO_3/HCO_3^-) buffer can sponge up hydrogen ions and hydroxide ions.

SOLUTION

The carbonic acid–bicarbonate buffer is a solution of carbonic acid (H_2CO_3) and bicarbonate ions (HCO_3^-). When a base is added to this buffer, it reacts with H_2CO_3 to produce neutral water. The pH changes very little:

$$H_2CO_3(aq) \; + \; OH^-(aq) \; \rightleftharpoons \; HCO_3^-(aq) \; + \; H_2O(l)$$

Carbonic acid   Hydroxide ion   Bicarbonate ion   Water

When an acid is added to the buffer, it reacts with HCO_3^- to produce undissociated carbonic acid, and again, the pH changes very little:

$$HCO_3^-(aq) \; + \; H^+(aq) \; \rightleftharpoons \; H_2CO_3(aq)$$

Bicarbonate ion   Hydrogen ion   Carbonic acid

PRACTICE EXERCISE 9.25

Write reactions to show what happens when (a) acid is added to a solution of HPO_4^{2-} and (b) base is added to a solution of $H_2PO_4^-$.

PRACTICE EXERCISE 9.26

Give the reactions for the addition of base to the ammonium ion–ammonia (NH_4^+/NH_3) buffer.

SUMMARY

Water molecules dissociate into hydrogen ions (H^+) and hydroxide ions (OH^-). Both the concentrations of these ions in pure water are equal to $1 \times 10^{-7}\, M$. A substance that raises the hydrogen ion concentration in solution is an acid. A substance that lowers the hydrogen ion concentration is a base. Hydrochloric acid (HCl) and sulfuric acid (H_2SO_4) are completely dissociated in solution and are strong acids, but acetic acid (CH_3COOH) is only partially dissociated and is a weak acid. Sodium hydroxide (NaOH) and calcium hydroxide ($Ca(OH)_2$) are strong bases, while ammonia (NH_3) is a weak base.

The pH scale, which has a normal range from 0 to 14, is used to denote the hydrogen ion concentration of a solution. On this scale 0 is strongly acidic, 14 is strongly basic, and 7 is neutral; pure water has a pH of 7.

In the reaction of an acid with a base, hydrogen ions and hydroxide ions react to produce water. This reaction, called neutralization, is usually carried out by titration. The equivalence point in a titration occurs when the numbers of reacting hydrogen ions and hydroxide ions are equal; the number of equivalents of acid is the same as the number of equivalents of base. An equivalent of an acid is the mass of the acid that provides 1 mol of hydrogen ions. A solution that contains one equivalent of an acid or a base in a liter of solution (1 equiv per liter) is a one normal (1 N) solution.

A salt forms when an acid is neutralized by a base. Salts consist of an anion from the acid and a cation from the base. Salts of strong acid–strong base reactions produce neutral solutions with water;

salts formed from weak acids or weak bases hydrolyze in water to produce solutions that are basic or acidic. Body fluids contain many salts, or electrolytes, that are essential for maintaining fluid balance and acid-base balance.

Solutions that resist changes in pH are called buffers. The pH of body fluids is kept within its normal range by buffers.

KEY TERMS

Acid (Arrhenius) (9.2)
Acid dissociation constant, K_a (9.2)
Acidic solution (9.1)
Base (Arrhenius) (9.2)
Basic (alkaline) solution (9.1)
Brønsted-Lowry theory (9.2)

Buffer (9.10)
Equivalence point (9.8)
Equivalent (equiv) of acid or base (9.6)
Hydronium ion (9.1)
Hydrogen ion acceptor (9.2)
Hydrogen ion donor (9.2)

Hydroxide ion (9.1)
Ion-product constant for water, K_w (9.1)
Neutral solution (9.1)
Neutralization reaction (9.5)
Normality (9.7)
pH (9.3)

Salt (9.5)
Self-dissociation of water (9.1)
Strong acid (9.2)
Strong base (9.2)
Titration (9.8)
Weak acid (9.2)
Weak base (9.2)

EXERCISES

Self-Dissociation of Water (Section 9.1)

9.27 Write two equations for the self-dissociation of water.

9.28 Write the expression for the ion-product constant for water.

9.29 What is the value of K_w at 25 °C?

9.30 What are the concentrations of H^+ and OH^- in pure water at 25 °C?

9.31 Calculate the hydrogen ion concentration $[H^+]$ for an aqueous solution in which $[OH^-]$ is 1×10^{-5} M. Is the solution acidic, basic, or neutral?

9.32 Calculate the hydroxide ion concentration $[OH^-]$ for an aqueous solution in which $[H^+]$ is 1×10^{-4} M. Is the solution acidic, basic, or neutral?

Acids and Bases (Section 9.2)

9.33 How did Arrhenius describe (a) an acid and (b) a base?

9.34 Name the following common acids and bases. Identify any acids as monoprotic, diprotic, or triprotic.
(a) H_3PO_4 (b) $Ca(OH)_2$ (c) HNO_3
(d) CH_3COOH (e) KOH (f) H_2CO_3

9.35 Write a balanced equation for the reaction of lithium metal with water.

9.36 How are acids and bases defined by the Brønsted-Lowry theory?

9.37 What advantage does the Brønsted-Lowry theory have over the theory proposed by Arrhenius?

9.38 Identify each of the reactants in the following equations as a proton donor (acid) or a proton acceptor (base).
(a) $HNO_3(aq) + H_2O(l) \longrightarrow H_3O^+(aq) + NO_3^-(aq)$
(b) $CH_3COOH(aq) + H_2O(l) \rightleftharpoons$
$$H_3O^+(aq) + CH_3COO^-(aq)$$
(c) $H_2O(l) + NH_3(aq) \rightleftharpoons NH_4^+(aq) + OH^-(aq)$
(d) $H_2O(l) + CH_3COO^-(aq) \rightleftharpoons$
$$CH_3COOH(aq) + OH^-(aq)$$
(e) $NH_4^+(aq) + H_2O(l) \rightleftharpoons H_3O^+(aq) + NH_3(aq)$
(f) $HCO_3^-(aq) + H_2O(l) \rightleftharpoons H_2CO_3(aq) + OH^-(aq)$

9.39 Distinguish between a strong and a weak acid.

9.40 Why are $Mg(OH)_2$ and $Ca(OH)_2$ called strong bases when their saturated solutions are only mildly basic?

9.41 Identify each of the following compounds as strong or weak, acid or base.
(a) $NaOH$ (b) HCl (c) $Ca(OH)_2$
(d) H_2SO_4 (e) NH_3

9.42 For each of the compounds given in Exercise 9.41, write an equation for its dissociation in water.

pH (Sections 9.3, 9.4)

9.43 Write the expression for the pH of an aqueous solution.

9.44 (a) Show by calculation that the pH of pure water is equal to 7.0.
(b) What is the normal range for the pH of blood?

9.45 Calculate the pH for the following aqueous solutions, and indicate whether the solution is acidic or basic.
(a) $[OH^-] = 1 \times 10^{-9} \, M$
(b) $[H^+] = 1 \times 10^{-3} \, M$
(c) $[OH^-] = 1 \times 10^{-5} \, M$
(d) $[H^+] = 1 \times 10^{-11} \, M$

9.46 Determine the hydrogen ion concentrations for aqueous solutions that have the following pH values.
(a) pH 7   (b) pH 13   (c) pH 2   (d) pH 11

Acid-Base Reactions (Sections 9.5, 9.6, 9.7)

9.47 What is a neutralization reaction?

9.48 What are two products of any acid-base neutralization reaction?

9.49 Write balanced equations for the following acid-base reactions and name the salts produced.
(a) $HCl(aq) + KOH(aq) \longrightarrow$
(b) $H_2SO_4(aq) + KOH(aq) \longrightarrow$
(c) $H_3PO_4(aq) + Ca(OH)_2(aq) \longrightarrow$
(d) $CH_3COOH(aq) + NaOH(aq) \longrightarrow$

9.50 How many moles of hydrochloric acid are required to neutralize (a) 3.2 mol NH_4OH, (b) 0.82 mol $Mg(OH)_2$, and (c) 1.6 mol KOH?

9.51 How many moles of NaOH are needed to neutralize (a) 2.0 mol H_2SO_4, (b) 12 mol H_3PO_4, and (c) 0.56 mol HCl?

9.52 Determine the mass of 1 equivalent for (a) LiOH, (b) H_2SO_4, (c) $Mg(OH)_2$, and (d) H_3PO_4.

9.53 How many equivalents is (a) 0.88 g KOH, (b) 148 g H_3PO_4, (c) 45.6 g HCl, (d) 18.8 g $Mg(OH)_2$, (e) 0.96 g NaOH, and (f) 75.0 g $Ca(OH)_2$?

9.54 What is the molarity of each of the following solutions?
(a) 4 N KOH   (b) 0.80 N H_2SO_4
(c) 2.6 N $Ca(OH)_2$ (d) 1.3 N HCl

9.55 What is the normality of each of the following solutions?
(a) 1.8 M NaOH   (b) 2.1 M H_3PO_4
(c) 5.0 M KOH   (d) 0.18 M HNO_3

9.56 How many milliliters of 2.0 N NaOH would you need to dilute with water to make 250 mL of 0.10 N NaOH?

9.57 A 25-mL sample of 0.50 N H_2SO_4 is diluted to 250 mL with distilled water. What is the normality of this solution?

Titration (Section 9.8)

9.58 A colorless solution of unknown pH turns blue when tested with the acid-base indicator thymol blue, and it remains colorless when tested with phenolphthalein. What is the approximate pH of the solution? How could you determine the pH more accurately?

9.59 A colorless solution of unknown pH turns yellow when tested with the acid-base indicators methyl red and thymol blue. What is the approximate pH of the solution?

9.60 A student carried out several titrations. The volume of each solution of unknown concentration and the volume and normality of the standard solution used are given. Calculate the normality for each unknown.
(a) 25.0 mL NaOH required 15.0 mL of 0.100 N HCl.
(b) 20.0 mL H_3PO_4 required 20.0 mL of 0.400 N NaOH.
(c) 25.0 mL NaOH required 18.4 mL of 0.150 N HNO_3.
(d) 50.0 mL CH_3COOH required 24.6 mL of 0.135 N NaOH.
(e) 40.0 mL H_2SO_4 required 24.5 mL of 0.180 N KOH.

9.61 Calculate the missing value at the equivalence point for the following acid-base reactions.

Acid		Base	
(a) 25 mL 0.10 N		_____ mL 0.050 N	
(b) 38.6 mL 0.25 N		_____ mL 0.14 N	
(c) 12.2 mL _____ N		20.0 mL 0.092 N	
(d) _____ mL 0.88 N		38.0 mL 1.02 N	
(e) 45.5 mL 0.98 N		25 mL _____ N	

9.62 Determine the normality of the following solutions.
(a) 460 mL of solution contains 16.2 g KOH.
(b) 150 mL of solution contains 9.8 g H_2SO_4.
(c) 1.54 g NaOH in 1.50 L of solution.
(d) 1.00 L of solution contains 12.6 g HNO_3.

Salts of Weak Acids and Bases (Section 9.9)

9.63 Characterize the salts that hydrolyze water.

9.64 Why are solutions of salts that hydrolyze water not pH 7?

9.65 Predict whether an aqueous solution of the following salts will be acidic, basic, or neutral.
(a) Na_2SO_4 (b) NH_4Cl (c) $NaHCO_3$
(d) NH_4NO_3

9.66 Classify aqueous solutions of the following salts as acidic, basic, or neutral.
(a) CH_3COONa (b) NaCl   (c) KCl
(d) Na_2CO_3

9.67 Arrange the following solutions in order of increasing acidity:
(a) 0.5 N HNO$_3$ (b) 0.5 M ammonium chloride
(c) 0.5 M sodium acetate   (d) 0.5 N KOH
(e) 0.5 M KCl

Buffers (Section 9.10)

9.68 Use the phosphate buffer (H$_2$PO$_4^-$/HPO$_4^{2-}$) to illustrate how a buffer system works. Show by means of equations how the pH of a solution can be kept almost constant when small amounts of acid or base are added.

9.69 Write the reactions that occur when acid is added to (a) the acetic acid–acetate buffer and (b) the carbonic acid–bicarbonate buffer.

Additional Exercises

9.70 How many moles of H$_3$PO$_4$ are in 3.0 L of 0.40 M H$_3$PO$_4$?

9.71 The pH of a swimming pool should be constantly monitored.
(a) How could you determine whether pool water was acidic or basic?
(b) If the water was too acidic, what could you do to correct the problem?

9.72 How many grams of potassium hydroxide would be needed to completely neutralize 7.82 g sulfuric acid in water solution?

9.73 Would you expect an aqueous solution of a weak or a strong acid to be the best conductor of electricity? Explain.

9.74 What is the molarity and the normality of a 1.0-L solution of sulfurous acid that contains 75.0 g H$_2$SO$_3$?

9.75 Vinegar is an aqueous solution of acetic acid. In a titration, 28.0 mL of 0.30 M NaOH was necessary to neutralize 20.0 mL of vinegar. What is the molarity of the vinegar?

9.76 Complete the following table for each solution.

Solution	[H$^+$]	[OH$^-$]	pH	Acidic or basic
X		1×10^{-4}		
Y			9.0	
Z	1×10^{-12}			

9.77 Use the Brønsted-Lowry theory to classify each reactant as an acid or a base.
(a) CH$_3$COOH + H$_2$O $\rightleftharpoons$ CH$_3$COO$^-$ + H$_3$O$^+$
(b) CN$^-$ + H$_2$O $\rightleftharpoons$ HCN + OH$^-$
(c) H$_2$PO$_4^-$ + OH$^-$ $\rightleftharpoons$ HPO$_4^{2-}$ + H$_2$O

SELF-TEST (REVIEW)

True/False

1. In an acidic solution, [H$^+$] is greater than [OH$^-$].
2. Every compound that contains hydrogen atoms will be an acid when dissolved in water.
3. Hydrochloric acid is a strong acid that is diprotic.
4. The pH of a solution that is 1.0 M HCl is zero.
5. If the [H$^+$] in a solution increases, the [OH$^-$] must decrease.
6. Formic acid, HCOOH, which adds the sting to ant bites, is a weak acid because its K_a is 1.8×10^{-4}.
7. The product of a neutralization reaction between a strong acid and a strong base is a salt plus water.
8. Ammonium ion, NH$_4^+$, is a Brønsted-Lowry base.
9. Strong acids are those that are highly dissociated.
10. An Arrhenius base is a hydrogen ion acceptor.

Multiple Choice

11. A 12.0 M solution of an acid that dissociates completely in solution would be termed
    (a) concentrated and weak.
    (b) strong and dilute.
    (c) dilute and weak.
    (d) concentrated and strong.
12. What is the equivalent mass of chromic acid, H$_2$CrO$_4$?
    (a) 59 g   (b) 236 g
    (c) 2 g   (d) 118 g
13. If the [H$^+$] in a solution is 1×10^{-1} M, then the [OH$^-$]
    (a) is 1×10^{-1} M.   (b) is 1×10^{-15} M.
    (c) is 1×10^{-13} M.   (d) cannot be determined.
14. Which of the following acids is monoprotic?
    (a) CH$_3$COOH   (b) H$_2$PO$_4^-$ (c) H$_2$SO$_4$
    (d) H$_3$PO$_4$

15. The following equation is representative of what type of reaction?

$$H_2SO_4 + Mg(OH)_2 \longrightarrow MgSO_4 + 2H_2O$$

   (a) hydrolysis   (b) combination
   (c) neutralization   (d) buffering

16. A 50.0-mL sample of hydrobromic acid, HBr, is titrated to the equivalence point with 24.0 mL of 1.5 N NaOH. What is the concentration of HBr?
   (a) 1.4 N (b) 0.72 N
   (c) 3.1 N (d) 0.38 N

17. The hydrolysis of the salt of a weak base and a strong acid should give a solution that is
   (a) weakly basic.   (b) neutral.
   (c) weakly acidic.   (d) strongly basic.

18. Which of the following solutions is the most basic?
   (a) $[H^+] = 1 \times 10^{-2}$
   (b) $[OH^-] = 1 \times 10^{-4}$
   (c) $[H^+] = 1 \times 10^{-11}$
   (d) $[OH^-] = 1 \times 10^{-13}$

19. What is the normality of a solution containing 30.0 g of acetic acid in 1.0 L of solution?
   (a) 0.50 N (b) 1.0 N
   (c) 1.3 N (d) 2.0 N

20. According to the Brønsted-Lowry theory, water
   (a) can act as an acid by accepting protons.
   (b) can be neither an acid nor a base.
   (c) acts as a base when it accepts a proton.
   (d) can accept but not donate protons.

21. A solution with a pH of 5.0
   (a) is basic.
   (b) has a hydrogen ion concentration of 5.0 M.
   (c) is neutral.
   (d) has a hydroxide concentration of 1×10^{-9} M.

22. Which of the following is an Arrhenius base?
   (a) LiOH   (b) NH_3 (c) $H_2PO_4^-$
   (d) CH_3COOH

23. An acid has a K_a value of 1.6×10^{-10}. Which of the following statements is true?
   (a) An aqueous solution of this acid would be basic.
   (b) This is a strong acid.
   (c) This acid dissociates slightly in aqueous solution.
   (d) This is a dilute acid.

24. Which of the following would not make a good buffering system?
   (a) bicarbonate and carbonic acid
   (b) ammonia and ammonium ion
   (c) nitrate and nitric acid
   (d) acetate and acetic acid

25. How much of a 5.0 N H_2SO_4 stock solution would you need to make 400 mL of 0.40 N H_2SO_4?
   (a) 32 mL   (b) 50 mL   (c) 10 mL   (d) 80 mL

26. The pH of a solution with a concentration of 0.01 M hydrochloric acid is
   (a) 10^{-2}.   (b) 12.   (c) 2.   (d) 10^{-12}.

27. A solution labeled 0.2 N H_2SO_4 contains
   (a) 0.2 mol H_2SO_4 per liter.
   (b) 0.4 equivalents of H_2SO_4 per liter.
   (c) 0.1 mol H_2SO_4 per liter.
   (d) 0.4 mol H_2SO_4 per liter.

10

Nuclear Chemistry

Radioactivity and Ionizing Radiations

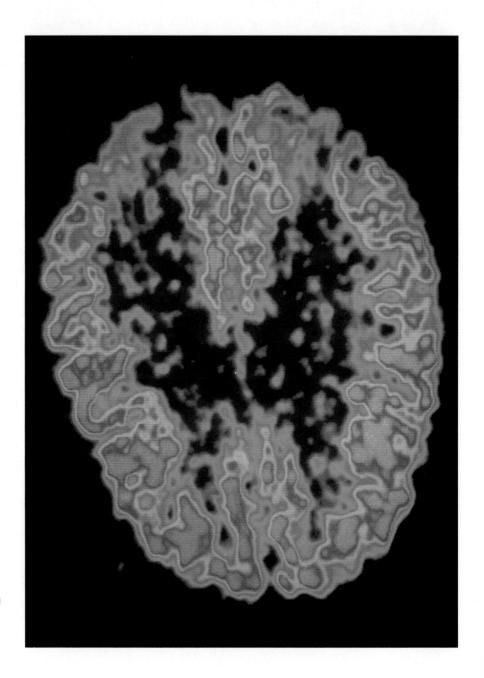

Nuclear chemistry can be used to probe the body's interior without surgery. This scan of the brain of a patient with Alzheimer's disease shows regions of high (red) and low (blue) brain activity.

CHAPTER OUTLINE

CASE IN POINT: Hyperthyroidism

10.1 Radiation

A CLOSER LOOK: Food Irradiation

10.2 Half-life

10.3 Carbon-14

A CLOSER LOOK: Radon in Homes

10.4 Nuclear reactions

10.5 Nuclear fission

A CLOSER LOOK: The Nuclear Reactor Debate

10.6 Nuclear fusion

10.7 Radiation detection

10.8 The biological effects of radiation

10.9 Units of radiation

10.10 Radiation in medicine

FOLLOW-UP TO THE CASE IN POINT: Hyperthyroidism

10.11 Background radiation

In the preceding chapters the focus has been on chemical reactions in which atoms generally attain the electron configurations of the noble gases. This chapter deals with nuclear reactions. In nuclear reactions, certain isotopes called *radioisotopes* gain stability by making changes within their nuclei. These changes are accompanied by the emission of large amounts of energy. Unlike a chemical reaction, a nuclear reaction is not affected by changes in temperature or pressure, and it cannot be stopped. In addition to studying nuclear reactions, we will examine radioisotopes and the use of radiation and radioisotopes in medical diagnosis and therapy. One of these uses is illustrated in the Case in Point.

CASE IN POINT: Hyperthyroidism

Annabelle was beginning to look and feel much older than her 60 years. By the time she saw her physician, she was noticeably nervous, had lost weight, and lacked the energy to perform even simple physical tasks such as light housekeeping. Annabelle's doctor noted that she had a slight fever and that her heart rate was higher than normal. The doctor suspected that Annabelle's symptoms might be due to hyperthyroidism—an overactive thyroid gland. The thyroid, a gland located at the base of the throat, is very efficient in extracting iodide ions from the bloodstream and using them in the synthesis of an iodine-containing compound called *thyroxine* (see figure). Thyroxine enters the bloodstream from the thyroid and is transferred to body tissues, where it helps to control energy production. (The structure of thyroxine is shown in Sec. 21.9.) If the thyroid gland becomes overactive, it produces abnormally large amounts of thyroxine. The doctor reasoned that Annabelle's symptoms might have resulted from an excessive amount of thyroxine due to hyperthyroidism. As part of the search for Annabelle's problem, the doctor ordered that she be tested for hyperthyroidism. How is this test performed and how are the results interpreted? We will find out in Section 10.10.

Seafood is an important source of iodine in our diet.

10.1 Radiation

AIMS: To define radioisotope, radioactive decay, and transmutation. To characterize alpha, beta, and gamma radiation by composition and penetrating power.

Focus

Isotopes with unstable nuclei are radioactive.

Transmutation
transmutare (Latin): to change

In 1896, the French chemist Antoine H. Becquerel discovered that uranium ores emit invisible rays that fog photographic plates wrapped in thick, black paper. This meant that the rays emitted by the ore were more energetic than visible light rays because they could pass through thick paper. One of his research students, a young Polish woman named Marie Sklodowska (1867–1934), working with Pierre Curie, showed that these penetrating rays come from uranium atoms. *The ability of certain elements to emit penetrating rays that fog photographic plates is called* **radioactivity,** *and the penetrating rays are called* **radiation.** In the course of working together, Marie and Pierre married, and together they won a Nobel Prize in physics in 1903. After Pierre's death in 1906, Madame Curie (Fig. 10.1) went on to win another Nobel Prize, this time in chemistry, in 1911.

Many studies have since shown that *certain isotopes—***radioisotopes—** *are radioactive because they have unstable nuclei.* The stability of the nucleus depends on its ratio of protons to neutrons; too many or too few neutrons leads to an unstable nucleus. The unstable nucleus loses energy by emitting radiation, eventually achieving a more stable state. *The emission of radiation by unstable nuclei, called* **radioactive decay,** *results in the production of atoms of a different element. The conversion of an atom of one element into an atom of another by the emission of radiation is called* **transmutation.**

Several types of radiation can be emitted during radioactive decay. They include alpha particles (α), beta particles (β), and gamma rays (γ). Table 10.1 summarizes the properties of different types of radiation.

Table 10.1 Characteristics of Some Ionizing Radiations

Radiation	Composition	Symbol	Charge	Mass (amu)	Common source	Penetrating power	Shielding
alpha particle	helium nucleus	$\alpha, {}^4_2\text{He}$	2+	4	radium-226	low (0.05 mm body tissue)	paper, clothing
beta particle	electron	$\beta, {}^0_{-1}\text{e}$	1−	$\frac{1}{1837}$	carbon-14	moderate (4 mm body tissue)	metal foil
gamma ray	high-energy electromagnetic radiation	γ	0	0	cobalt-60	very high (penetrates body easily)	lead, concrete
X ray	high-energy electromagnetic radiation	X ray	0	0	X-ray machine	very high (penetrates body easily)	lead, concrete

Alpha emission

Alpha emission *consists of alpha particles. An* **alpha particle** *is a helium nucleus*—a particle containing two protons and two neutrons. It has a double positive charge ($_2^4\text{He}^{2+}$). In writing nuclear reactions, the charge on the helium nucleus is omitted and the alpha particle is written as $_2^4\text{He}$ or α.

The disintegration of the radioisotope uranium-238 produces the more stable (but still radioactive) isotope thorium-234 and an alpha particle:

$$_{92}^{238}\text{U} \xrightarrow[\text{decay}]{\text{Radioactive}} {}_{90}^{234}\text{Th} + {}_2^4\text{He}$$

Uranium-238   Thorium-234   Alpha particle

When an atom loses an alpha particle, the atomic number of the product atom is lowered by 2, and its mass number is lowered by 4. Because of their large mass and charge, alpha particles do not travel very far and are not very penetrating. They are easily stopped by a sheet of paper and are unable to penetrate even the dead cells on the surface of the skin. They cannot reach internal organs from outside the body. However, alpha-particle emitters can damage cells of internal organs if they are ingested or inhaled.

Figure 10.1
Marie Sklodowska Curie in her laboratory.

EXAMPLE 10.1   **Writing a balanced nuclear equation for alpha emission**

The radioisotope americium-241 is often used in the smoke detectors seen in homes, apartments, and offices. As it decays, this isotope emits alpha particles. In the absence of smoke, the alpha particles cause the surrounding molecules of gas in air to become ions and produce a constant electric current. Smoke interrupts the current and causes the alarm to sound. When americium-241 emits an alpha particle, it decays to an isotope of neptunium. Write a balanced equation for this nuclear reaction.

SOLUTION

Americium-241 disintegrates to neptunium and an alpha particle, so we can write

$$_{95}^{241}\text{Am} \longrightarrow \text{Np} + {}_2^4\text{He}$$

The total number of neutrons and protons must be conserved. The neptunium isotope must have a mass number of $241 - 4 = 237$ and an atomic number of $95 - 2 = 93$. The balanced nuclear equation is

$$_{95}^{241}\text{Am} \longrightarrow {}_{93}^{237}\text{Np} + {}_2^4\text{He}$$

PRACTICE EXERCISE 10.1

The disintegration of the radioisotope radon-222 produces an isotope of the element polonium and an alpha particle. Write a balanced nuclear equation for this transmutation.

Beta emission

Beta emission *consists of beta particles.* **Beta particles** *are fast-moving electrons formed by the decomposition of a neutron into a proton and an electron:*

$$\ce{^{1}_{0}n} \longrightarrow \ce{^{1}_{1}H} + \ce{^{0}_{-1}e}$$

Neutron   Proton   Electron
(beta particle)

The proton stays in the nucleus, and the electron is ejected from the atom. Carbon-14 emits a beta particle as it undergoes radioactive decay to nitrogen-14:

$$\ce{^{14}_{6}C} \longrightarrow \ce{^{14}_{7}N} + \ce{^{0}_{-1}e}$$

Carbon-14   Nitrogen-14   Beta
(radioactive)   (stable)   particle

The mass number of nitrogen-14 is the same as that of carbon-14, but the atomic number increases by 1 because the nucleus now contains an additional proton.

Beta particles have a mass about 8000 times smaller than alpha particles, and they have a smaller charge. Consequently, they are much more penetrating. They can penetrate the living cells of the skin to a depth of about 4 mm. If exposure is prolonged, beta particles can burn the skin. The cells of the cornea of the eye are particularly vulnerable to damage by beta particles. Beta particles are stopped by aluminum foil or a thin piece of wood. Inhaled or eaten, beta emitters can cause severe internal damage such as destruction of the blood-forming tissues of the bone marrow.

EXAMPLE 10.2   **Writing a nuclear equation for beta decay**

The radioisotope phosphorus-32 is used in the treatment of a form of leukemia. What different element is formed when phosphorus-32 decays by beta emission? Write a balanced equation for the process.

SOLUTION

Since we are given that phosphorus-32 decays to an unknown element and a beta particle, we can write

$$\ce{^{32}_{15}P} \longrightarrow \, ? + \ce{^{0}_{-1}e}$$

The new element has a mass number of 32, the same as phosphorus-32. The atomic number (number of protons) increases by 1, so $15 + 1 = 16$; the different element is sulfur, atomic number 16. The equation for this nuclear reaction is

$$\ce{^{32}_{15}P} \longrightarrow \ce{^{32}_{16}S} + \ce{^{0}_{-1}e}$$

A radioisotope of the element iodine is used to monitor thyroid activity. The isotope, iodine-131, decays to an isotope of the element xenon (Xe) by emission of a beta particle. Complete the equation for this nuclear decay process.

$$^{131}_{53}\text{I} \longrightarrow \text{Xe} + {}^{0}_{-1}\text{e}$$

Gamma radiation

Visible light (the light we see by) is electromagnetic radiation. **Gamma radiation** *also is electromagnetic radiation but of much higher energy.* Gamma rays are often emitted by the nuclei of disintegrating radioactive atoms along with alpha or beta particles:

$$^{226}_{88}\text{Ra} \longrightarrow {}^{222}_{86}\text{Rn} + {}^{4}_{2}\text{He} + \gamma$$

Radium-226   Radon-222   Alpha particle   Gamma ray

$$^{234}_{90}\text{Th} \longrightarrow {}^{234}_{91}\text{Pa} + {}^{0}_{-1}\text{e} + \gamma$$

Thorium-234   Protactinium-234   Beta particle   Gamma ray

Since gamma rays have no mass and no electrical charge, the emission of gamma radiation does not alter the atomic number or mass number of an atom.

The radioisotope cobalt-60 is used in the treatment of cancer. Cobalt-60 emits both beta particles and gamma radiation. Write the equation for its nuclear decay process.

Another form of radiation, *X radiation,* is produced by specially built machines. *Except for their origins, gamma rays and* **X rays** *are essentially the same;* they are extremely penetrating and potentially very dangerous. Both gamma rays and X rays pass easily through paper, wood, and the human body. They can be stopped (though not completely) by several feet of concrete or several inches of lead (Fig. 10.2). Since gamma rays and X rays penetrate the body easily, they can reach internal organs and kill cells

Figure 10.2
The relative penetrating power of alpha, beta, gamma, and X radiation. Note that X rays and gamma rays are not stopped completely, even by lead.

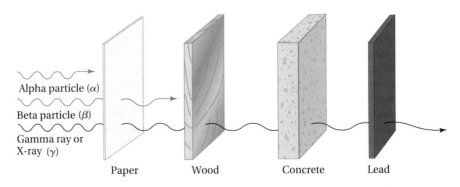

Alpha particle (α)

Beta particle (β)

Gamma ray or X-ray (γ)

Paper   Wood   Concrete   Lead

Food Irradiation

Exposing food to radiation can destroy the fungi, bacteria, parasites, and insects that infest food products. The process, known as *irradiation,* is simple: Expose food products to a dose of radiation from an X-ray beam, or to gamma rays from a radioactive cobalt-60 source, or to a beam of high-energy electrons. Food does not become radioactive when it is irradiated.

Food irradiation is not a new idea. The irradiation of wheat and flour to control insects is permitted in the United States (see table). Spices and herbs, irradiated to reduce insects and molds, have been marketed since 1983. In 1985, the Food and Drug Administration (FDA) approved the use of low radiation doses to slow spoilage of fruit and vegetables and to control *Trichinella spiralis* in pork, the parasite that causes trichinosis. Irradiated foods also have been used by the military and for space missions.

A low radiation dose in food irradiation is up to 100 kilorads—the equivalent of 2 million chest X rays. This dose of radiation inhibits the sprouting of onions and potatoes during storage and extends the shelf life of fruits and vegetables by 1 to 2 weeks. A higher radiation dose, up to 1000 kilorads, significantly reduces *Salmonella* bacteria in fish, poultry, and other meats. This higher radiation dose also greatly extends the shelf life of strawberries and other fruits that mold quickly (see figure). High radiation doses are used to sterilize foods, giving them shelf lives of years.

Irradiated foods show little change in appearance or taste from nonirradiated foods. There is, however, concern among some people that chemical substances produced during irradiation, called *radiolytic products,* might be harmful. At this time there is no evidence to support this concern from tests done with laboratory animals fed irradiated food. Further, there have been no known adverse effects to humans in countries where irradiated foods have been consumed for years.

FDA (Food and Drug Administration) Approvals for Food Irradiation

Food	Purpose	Dose (krad)	Date of approval
wheat, flour	insect disinfestation	20–50	1963
spices	insect disinfestation	3000	1983
pork	control of *Trichinella spiralis*	30–100	1985
fresh fruit, vegetables	delay of ripening	100	1986
poultry	bacterial decontamination/ extend shelf life	150–300	1990

Irradiation greatly extends the shelf life of foods that mold quickly, like strawberries.

in healthy tissue. Damage to embryonic cells can cause birth defects. Radiation also can be put to beneficial uses, as described in A Closer Look: Food Irradiation.

10.2 Half-life

AIM: To use the half-life of a radioisotope to calculate the amount of isotope remaining at a given time.

Every radioisotope has a characteristic rate of decay measured by its half-life. *A* **half-life** $(t_{1/2})$ *is the time required for one-half of the atoms of a radioisotope to emit alpha, beta, or gamma radiation as it decays to products.* After one half-life, one-half the original radioactive atoms remain unchanged, and the other half have decayed into atoms of a new element.

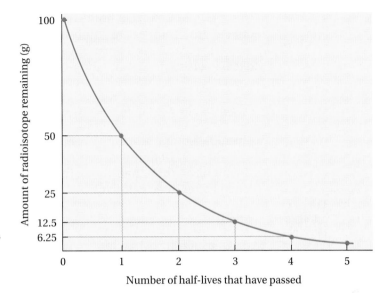

Figure 10.3
Decay curve for a radioisotope. Each half-life reduces the quantity of the radioisotope by a factor of 2. Shown here is the decay curve for 100 g of iodine-131, which has a half-life of 8 days.

After a second half-life, only one-quarter of the original radioactive atoms remain, and three-quarters have decayed to atoms of a new element. This pattern continues until all the radioisotope has decayed. The decay curve for the radioisotope iodine-131, which has a half-life of 8 days, is shown in Figure 10.3. If, for example, we begin with 100 g of iodine-131, after four half-lives or 32 days, only 6.25 g of iodine-131 will remain.

The stability of a radioisotope is indicated by its half-life: The longer the half-life, the more stable is the isotope. Table 10.2 shows half-lives of

Table 10.2 Half-lives and Radiation of Some Naturally Occurring and Artificially Produced Radioisotopes

Isotope	Half-life	Radiation emitted
Natural		
carbon-14	5.73×10^3 years	beta
potassium-40	1.28×10^9 years	beta, gamma
radon-222	3.8 days	alpha
radium-226	1.6×10^3 years	alpha, gamma
thorium-234	25 days	beta, gamma
uranium-238	4.5×10^9 years	alpha
Artificial		
tritium-3	12.3 years	beta
phosphorus-32	14.3 days	beta
cobalt-60	5.3 years	beta, gamma
strontium-90	28.1 years	beta
technetium-99m*	6 hours	gamma
iodine-131	8 days	beta
cesium-137	30 years	beta
americium-243	7.37×10^3 years	alpha

*The m associated with the mass number means that the isotope is metastable, that is, highly energetic.

isotopes that vary from hours to thousands of years. Some artificially produced radioisotopes are highly unstable and have half-lives of a few minutes. This feature is a great advantage in nuclear medicine, since rapidly decaying isotopes are not long-term biological radiation hazards.

EXAMPLE 10.3 **Estimating the amount of radioactive decay**

The radioisotope nitrogen-13 emits beta radiation and decays to carbon-13 with a $t_{1/2} = 10$ minutes. Suppose we start with 12.0 g of nitrogen-13. (a) How long is three half-lives? (b) How many grams of that isotope will remain after three half-lives?

Number of half-lives	Grams of nitrogen-13 remaining
0	12.0
1	6.0
2	3.0
3	1.5

SOLUTION

(a) One half-life is 10 minutes. Three half-lives is 3×10 minutes $= 30$ minutes.

(b) One-half the mass of the isotopes decays each half-life as shown in the table to the left.

After three half-lives, 1.5 g nitrogen-13 will remain.

PRACTICE EXERCISE 10.4

A patient with a thyroid disorder is given a solution containing 10.0 mg of iodine-131 to drink.

(a) How much iodine-131 is left after 24 days?

(b) How much iodine-131 has decayed?

EXAMPLE 10.4 **Calculating the percent of an isotope remaining**

Number of half-lives	Technetium-99m (percent)
0	100.0
1	50.0
2	25.0
3	12.5
4	6.25

Technetium-99m, with a half-life of 6.0 hours, is used in the detection of brain tumors. What percentage of this isotope remains after 1 day?

SOLUTION

One day (24 hours) is four half-lives. The percent decreases by one-half each half-life. The mass of technetium-99m remaining is shown in the table to the left.

After 1 day (24 hours), 6.25% of the technetium-99m will remain.

PRACTICE EXERCISE 10.5

Phosphorus-32 is a radioisotope used in the detection of tumors. It is a beta emitter with a half-life of 14.3 days.

(a) What is the mass of phosphorus-32 in a 1.00-mg sample of the isotope after 57.2 days?

(b) What percent of the original sample remains?

10.3 Carbon-14

AIM: To explain how radioisotopes can be used to date objects.

Archaeologists use the half-lives of naturally occurring radioisotopes to establish the age of archaeological finds. Carbon-14, with a half-life of 5730 years, has been used most extensively. The production of carbon-14 takes place in the upper atmosphere, where atoms of nitrogen-14 are bombarded with neutrons. The reaction is written

$$^{14}_{7}\text{N} \quad + \quad ^{1}_{0}\text{n} \quad \longrightarrow \quad ^{14}_{6}\text{C} \quad + \quad ^{1}_{1}\text{H}$$

$$\text{Nitrogen-14} \qquad \text{Neutron} \qquad \text{Carbon-14} \qquad \text{Proton}$$

Radioactive carbon-14 is mixed uniformly with stable carbon-12, and both isotopes react with oxygen to form the carbon dioxide in our atmosphere:

$$^{14}\text{C} \quad + \quad \text{O}_2 \quad \longrightarrow \quad ^{14}\text{CO}_2$$

$$\text{Carbon-14} \qquad \text{Oxygen} \qquad \begin{array}{c}\text{Carbon dioxide}\\\text{(radioactive)}\end{array}$$

$$^{12}\text{C} \quad + \quad \text{O}_2 \quad \longrightarrow \quad ^{12}\text{CO}_2$$

$$\text{Carbon-12} \qquad \text{Oxygen} \qquad \begin{array}{c}\text{Carbon dioxide}\\\text{(nonradioactive)}\end{array}$$

Plants use atmospheric carbon dioxide containing both isotopes to make many carbon-containing compounds. Some of these plants are eaten by animals, so all living organisms contain the isotopes carbon-14 and carbon-12. The ratio of these isotopes remains at a constant value for as long as the organism lives. Since no new carbon is incorporated after the organism dies, the carbon-14 to carbon-12 ratio starts changing because the carbon-14 decays. Every 5730 years (the half-life for carbon-14), half the carbon-14 atoms originally present in the organism when it was living decay into nitrogen-14:

$$^{14}_{6}\text{C} \longrightarrow ^{14}_{7}\text{N} + ^{0}_{-1}\text{e}$$

By making the reasonable assumption that the carbon-14 to carbon-12 ratio has been the same in living systems for tens of thousands of years, and by comparing the carbon-14 to carbon-12 ratio obtained for living organisms with that of their ancient remains, scientists are able to date artifacts up to 40,000 years old.

EXAMPLE 10.5   Dating an object using radioisotopes

A sample of bone from a saber-toothed tiger recovered from the tar pits at La Brea, Los Angeles, California, was subjected to carbon-14 dating. The test results showed that the carbon-14 content in the tiger bone was only one-quarter of that in a sample of fresh bone. When did the tiger die?

SOLUTION

Because the carbon-14 content of the tiger bone was only one quarter of that of fresh bone, two carbon-14 half-lives had passed ($1/2 \times 1/2 = 1/4$)

after the animal had died. Since carbon-14 has a half-life of 5730 years, we can estimate that this saber-toothed tiger was roaming around southern California about 11,000 years ago, or about 9000 B.C.

PRACTICE EXERCISE 10.6

A piece of charcoal unearthed during an archaeological dig is analyzed. It contains an amount of carbon-14 that is only 12.5% the amount detected in a newly made piece of charcoal. How old is the unearthed piece of charcoal?

Although carbon-14 is useful for dating some human artifacts, much longer-lived radioisotopes are used to establish the dates of the Earth's geologic periods. One such isotope, uranium-238, decays through a complex series of radioactive intermediates to the stable isotope lead-206 (Fig. 10.4). The age of certain rocks is determined by measuring the ratio of uranium-238 to lead-206. Since the half-life of uranium-238 is 4.5×10^9 years, it is possible to use this method to date rocks as old as the solar system (4.5×10^9 years).

One of the radioisotopes produced in the decay series of uranium-238 is the radioactive gas radon-222. We will read about this gas in A Closer

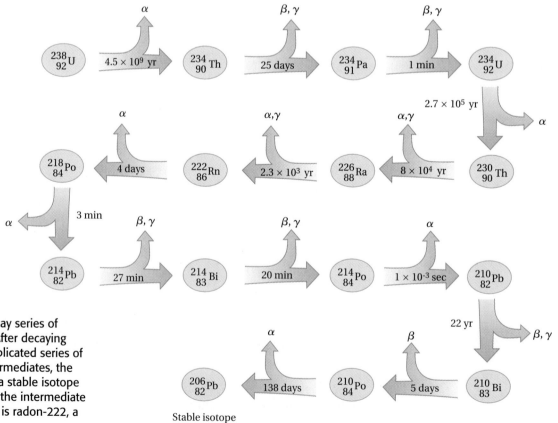

Figure 10.4
The natural decay series of uranium-238. After decaying through a complicated series of radioactive intermediates, the end product is a stable isotope of lead. One of the intermediate decay products is radon-222, a radioactive gas.

Look: Radon in Homes and learn why the accumulation of radon in basements is a potential health hazard.

Radon in Homes

Radon-222 is a product of the decay series of uranium-238. Radon-222, a radioactive gas, emits alpha particles as it decays with a half-life of about 4 days. Chemically, it is as inert as the other noble gases, but because it is a radioactive gas, it is a dangerous air pollutant. The story of the effects of environmental radon on health is still being written, but it begins with uranium and silver miners, who in the past contracted lung cancer at a much higher rate than other people. The higher risk of lung cancer among miners was traced to radon. Today, the accumulation of radon gas in homes in some areas of the United States is cause for concern and is a potential public health hazard.

After the health threat of radon was recognized and testing begun, the air in houses constructed on the tailings of uranium and phosphate mines often tested positive for the gas. Additional testing turned up measurable quantities of radon in the air inside some homes in almost every part of the United States. The radon-222 found in homes comes mainly from the decay of radioactive rocks that contain uranium-238. As the radon forms, it diffuses from the earth to the atmosphere. In the atmosphere, the gas is diluted enough by mixing with air to keep its concentration at low, relatively safe levels. Trouble starts when radon gas seeps from the earth through cracks in the floors and walls of basements that lack adequate ventilation. Radon is denser than air and collects in basements and the lower levels of a house.

As a result of concerns about radiation from radon, the Environmental Protection Agency (EPA) has set an acceptable upper limit for the amount of radioactivity from radon in air: 200 picocuries (a picocurie is 10^{-12} curie) of radioactivity per liter of air. We can compare this limit with the highest level of radioactivity from radon ever recorded. A home in Pennsylvania that was built on granite that was rich in uranium had 2700 picocuries per liter.

Radon test kits are available commercially to determine the presence of radon gas in homes and office buildings.

It is quite easy to test a home for radon, and testing is recommended in parts of the country where the gas can be a problem (see figure). When unacceptable levels of radon are present, the solution may be simple or complex. Often, improvement of the ventilation in a basement and other areas of a house where air stagnates will reduce the concentration of radon. Cracks in basement floors and walls can be sealed to keep radon from seeping into the house. Sometimes covering dirt floors with concrete helps.

How concerned should people be about radon in their homes? This question is difficult to answer. The EPA recommendations have been questioned by some critics who are experts in the field. These critics say that the mineral dust and radiation levels found in mines are usually much higher than in homes; therefore, we cannot use the mine studies to draw conclusions about radon in homes. New studies may clarify the risks stemming from the presence of environmental radon.

10.4 Nuclear reactions

AIM: To write balanced equations for nuclear transformation reactions.

Nuclear transformation reactions *occur when high-energy particles bombard the nucleus of an atom to produce a different isotope.* Some nuclear transformations occur in nature; as we noted earlier, the production of carbon-14 from nitrogen-14 occurs in the upper atmosphere, where nitrogen atoms are bombarded with neutrons. Many other nuclear transformations are done in laboratories or in nuclear reactors. The earliest artificial nuclear transformation was performed as long ago as 1919 by Ernest Rutherford. Rutherford bombarded nitrogen gas with alpha particles to produce an unstable isotope of fluorine:

$$\underset{\text{Nitrogen-14}}{^{14}_{7}\text{N}} + \underset{\substack{\text{Alpha}\\\text{particle}}}{^{4}_{2}\text{He}} \longrightarrow \underset{\text{Fluorine-18}}{^{18}_{9}\text{F}}$$

The fluorine isotope quickly decomposes to an isotope of oxygen and a proton:

$$\underset{\text{Fluorine-18}}{^{18}_{9}\text{F}} \longrightarrow \underset{\text{Oxygen-17}}{^{17}_{8}\text{O}} + \underset{\text{Proton}}{^{1}_{1}\text{H}}$$

This experiment eventually led to discovery of the proton.

James Chadwick's 1932 discovery of the neutron also involved a nuclear transformation experiment. The neutrons were produced when beryllium-9 was bombarded with alpha particles:

$$\underset{\text{Beryllium-9}}{^{9}_{4}\text{Be}} + \underset{\substack{\text{Alpha}\\\text{particle}}}{^{4}_{2}\text{He}} \longrightarrow \underset{\text{Carbon-12}}{^{12}_{6}\text{C}} + \underset{\text{Neutron}}{^{1}_{0}\text{n}}$$

When new elements are created, the amounts are often exceedingly small. When element 104 was synthesized, only 3000 atoms (with a half-life of 4.5 seconds) were made.

None of the elements with atomic numbers above 92 (uranium) occur in nature. All of them are radioactive. These elements have been synthesized in nuclear reactors and nuclear accelerators, which accelerate the bombarding particles to very high speeds.

10.5 Nuclear fission

AIM: To describe how a nuclear fission chain reaction is controlled in a nuclear reactor.

When the nuclei of certain isotopes are bombarded with neutrons, they split into smaller fragments. *The splitting of nuclei is called* **nuclear fission.** Uranium-235 and plutonium-239 are fissionable materials. A fissionable atom such as uranium-235 breaks into two fragments of roughly the same size when struck by a neutron. At the same time, more neutrons are released (Fig. 10.5). The released neutrons react with the nuclei of other uranium-235 atoms, continuing fission by a chain reaction. *In a* **chain reaction,** *some of the neutrons produced react with other atoms, producing more neutrons that react with still more atoms.*

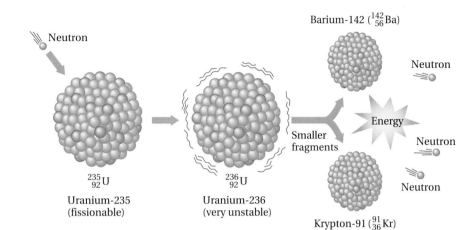

Figure 10.5
In nuclear fission, uranium-235 breaks into two fragments; energy and more neutrons are released. The released neutrons cause other uranium-235 atoms to split.

Fission
fissus (Latin): to split

Fission unleashes enormous amounts of energy. The fission of 1 kg of uranium-235 releases an amount of energy equal to that generated in the explosion of 20,000 tons of dynamite. If a nuclear chain reaction is uncontrolled, the energy release is nearly instantaneous. Atomic bombs are devices that start uncontrolled nuclear chain reactions.

It is possible to control fission so that energy is released more slowly. Controlled fission takes place in the nuclear reactor at the heart of every nuclear power plant (Fig. 10.6). Carbon is used to moderate (slow down) the neutrons so that they are more easily captured by the reactor fuel. Cadmium control rods are used to absorb the neutrons to slow the chain reaction. In the controlled fission reaction within the nuclear reactor, much of the energy generated appears as heat. The heat is removed from the reactor

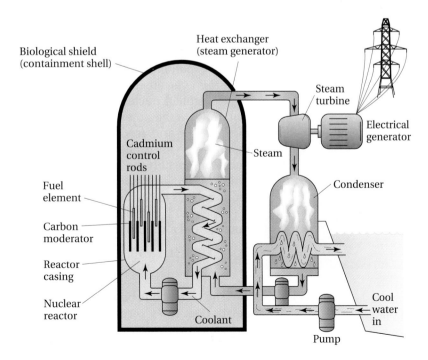

Figure 10.6
A nuclear reactor generates electrical energy by a controlled fission process.

core by a suitable coolant fluid (liquid sodium, water, or a gas), and it is used to generate steam to drive a turbine. The spinning turbine generates electricity.

Once a nuclear reactor is started, it will be highly radioactive for many generations. Shields are used to protect the reactor structure from radiation damage, and walls of high-density concrete are designed to protect the operating personnel. After a period of operation, the fuel elements in a nuclear reactor lose their efficiency. This occurs because the amount of fissionable material left in the fuel element has dropped too low and because the concentration of fission by-products is too high. When this happens, the spent fuel elements are removed from the reactor core and replaced with new ones. Some of the radioisotopes in the spent fuel elements are in demand; iodine-131 and cesium-137, for example, have direct medical applications. So some of the "waste" is reprocessed, and the desired radioisotopes are isolated along with any remaining fissionable isotopes such as uranium-235. A Closer Look: The Nuclear Reactor Debate describes some of the concerns associated with nuclear reactors.

10.6 Nuclear fusion

AIM: To describe a nuclear fusion reaction.

Every day about 1.5×10^{18} kcal of energy from the sun reaches the surface of the earth. Less than 1% of the sun's energy is used for photosynthesis. If an economical way of converting the excess solar energy into more useful forms could be devised, many environmental problems could be alleviated.

Focus

Fusion unites the nuclei of small atoms.

Fusion
fusus (Latin): melted together

The sun is an extraordinary energy source. The energy released from the sun is the result of **nuclear fusion**—*the combination of atomic nuclei.* Fusion reactions occur continuously in the sun, providing us with an endless source of energy. In the sun, hydrogen nuclei (protons) fuse to make helium nuclei. The reaction requires two beta particles:

$$4{}^{1}_{1}\text{H} + 2{}^{0}_{-1}\text{e} \longrightarrow {}^{4}_{2}\text{He} + \text{energy}$$

Fusion takes place only at exceedingly high temperatures, which is why fusion reactions are known as *thermonuclear reactions.* This large initial energy expenditure is a sound investment; the energy that fusion releases as heat is even greater.

Controlled nuclear fusion is appealing as an energy source on earth. Not only are the potential fuels, such as deuterium (${}^{2}_{1}\text{H}$) and tritium (${}^{3}_{1}\text{H}$), inexpensive and readily available, but fusion products are usually not radioactive. One problem with fusion is in achieving the high temperatures necessary to start the reaction. Another problem is in containing the reaction. No known structural material can withstand the exceedingly high temperatures necessary for fusion.

One reaction scientists are studying is the combination of a deuterium nucleus and a tritium nucleus to form a helium nucleus:

$$\underset{\text{Deuterium}}{{}^{2}_{1}\text{H}} + \underset{\text{Tritium}}{{}^{3}_{1}\text{H}} \longrightarrow \underset{\text{Helium}}{{}^{4}_{2}\text{He}} + \underset{\text{Neutron}}{{}^{1}_{0}\text{n}} + \text{energy}$$

The research is being carried out in a device called a *tokamak.* In this device (Fig. 10.7), very strong magnetic fields are used to contain and heat

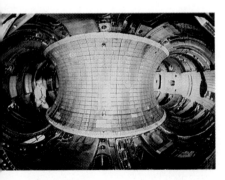

Figure 10.7
The tokamak fusion test reactor at Princeton University.

A Closer Look

The Nuclear Reactor Debate

About 10% of the electrical energy in the United States is supplied by nuclear power plants. Despite this benefit, many people are opposed to nuclear plants because they feel that they are unsafe. Some of these fears are legitimate. Incidents at Three-Mile Island and Chernobyl reinforce public concerns and dampen public support for nuclear power. In 1979, the Three-Mile Island nuclear power station near Harrisburg, Pennsylvania, suffered a sudden coolant loss around the reactor core. Although the incident released little radioactive material into the environment, it took seven years for specialists to clean up the site and get the adjacent undamaged reactor working again. The damaged reactor will be highly radioactive for generations to come.

The incident that occurred on April 26, 1986, in Chernobyl, Ukraine (see figure), was far more devastating than Three-Mile Island. An explosion blew the reactor apart, and an estimated 10 tons of burning radioactive fuel was hurled up to 3000 feet into the atmosphere and distributed around the world by the prevailing winds. More than 30 workers at the plant and rescue workers died in the explosion or shortly after the incident from acute radiation exposure, and more than 100,000 people were evacuated from the immediate surrounding area because of excessive doses of radiation in the contaminated air, water, and soil. Because of the slow-acting effects of low levels of radiation, the final human toll of the Chernobyl incident has yet to be tallied.

A week after the Chernobyl incident the radioactive cloud reached the British Isles, about 1300 miles away. Tens of thousands of sheep from the Welsh hills were so contaminated with radio-

The Chernobyl nuclear power plant after the accident in 1986.

activity from the fallout that they could not be slaughtered for meat for many months. In the state of Washington, some 4000 miles from Chernobyl, increased levels of iodine-131 and cesium-137 were reported in milk, vegetables, and rain water. The levels of contamination, however, were not considered harmful.

Today, the past promise of nuclear power as a clean, efficient source of energy is in considerable doubt. One of the lessons of Three-Mile Island and Chernobyl is that very stringent safeguards must be enforced by the nuclear power industry in order to protect public health and safety.

the reaction. Temperatures of 3,000,000 °C have been achieved, and fusion power has been generated for very short periods of time. Scientists predict, however, that it will be several decades before continuous fusion will be a reality on earth.

10.7 Radiation detection

AIM: To describe three methods of detecting radiation and state each method's limitations.

The best protection against radiation exposure is distance. If you double your distance from a radiation source, your exposure will decrease by a factor of 4.

The monitoring, cleanup, and disposal of radioactive materials in the workplace are the responsibility of personnel called *health physicists.*

The radiation emitted by radioisotopes (and X rays) is ionizing radiation. **Ionizing radiation** *knocks electrons off some atoms of the bombarded substance to produce ions.* This radiation is potentially harmful, and people are unable to detect ionizing radiation with any of the five senses. Therefore, they must rely on various instruments and devices to do the job.

Geiger counters *One of the most widely used radiation detectors is the Geiger-Müller counter, more commonly called the* **Geiger counter** (Fig. 10.8). The detector in a Geiger counter is a small, metal, gas-filled (usually argon) tube at low pressure. The tube is connected to a power supply and has a central wire electrode. When ionizing radiation penetrates the thin window at one end of the tube, the gas inside becomes ionized. Since ions are produced, the gas becomes an electrical conductor. Each time a Geiger tube is exposed to radiation, a current flows. The bursts of current drive electronic counters or cause audible clicks from a built-in speaker. Geiger counters are used primarily to detect beta radiation; alpha particles cannot pass through the thin window, and most gamma rays and X rays pass through the gas, causing few ionizations.

Scintillation counters *A* **scintillation counter** *is a detector that uses a specially coated surface, or phosphor, to detect radiation.* Ionizing radiation striking the phosphor of a scintillation detector produces bright flashes of light (scintillations). This is similar to a TV, in which the electrons from the TV "gun" striking the phosphor of the screen produce a pattern of scintillations as the TV picture. With radiation detection, the number of flashes and their respective energies are detected electronically and converted into

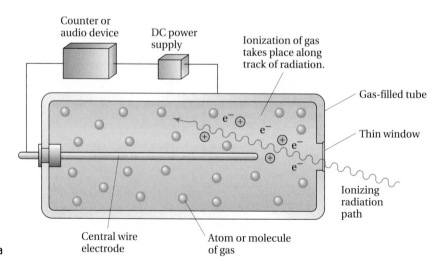

Figure 10.8
A Geiger-Müller radiation detector tube. The Geiger-Müller tube is used chiefly in the detection of beta radiation.

electronic pulses that are measured and recorded. Scintillation counters have many applications in medicine. A scintillation counter is at the heart of an apparatus called a *scanner*. A scanner is a device that is used to detect radiation from radioisotopes in the body. After a patient receives a radioisotope, the scanner moves slowly across the patient's body above the region where the organ containing the radioisotope is located and produces a scan or image of the organ. From the scan, the radiologist can determine such conditions as disease of an organ, a blood clot, or a tumor.

Film badges   Film badges are the most important radiation detectors for persons near radiation sources. All personnel who work in X-ray departments in hospitals wear film badges. A film badge consists of several layers of photographic film covered with black lightproof paper encased in a plastic or metal holder. It is worn all the time a person is at work (Fig. 10.9). At frequent specific intervals, depending on the type of work involved, the film is removed and developed. The strength and type of radiation exposure are determined from any darkening of the encased film. Records are kept of the results. Film badges do not protect an individual from radiation exposure; they merely serve as precautionary devices. Our only protections against radiation are to keep a safe distance away and use adequate shielding.

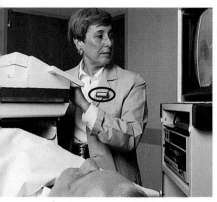

Figure 10.9
The film badge radiation detector (encircled above) is worn by people who work with radioisotopes and other sources of ionizing radiation. The film darkens in proportion to the amount of radiation it receives.

10.8 *The biological effects of radiation*

AIM: To distinguish between radiation-induced genetic damage and somatic damage.

Focus

Radiation causes chemical changes in the body.

What happens when a biological system is exposed to radiation? The energies of alpha, beta, and gamma radiation as well as X rays far exceed the bond energies that hold atoms together. Consequently, as they pass through a sample of matter, these radiations cause molecules to fragment and ionize. Since biological systems are more than 60% water, water molecules suffer most of the initial effects when a person is exposed to radiation. Gamma radiation, for example, can knock an electron off a water molecule to form an unstable ion.

$$H_2O \xrightarrow[\text{radiation}]{\text{Gamma}} H_2O^+ + e^-$$

This H_2O^+ ion immediately reacts with a neighboring water molecule to form a stable hydronium ion, H_3O^+, and a hydroxyl radical, $\cdot OH$:

$$H_2O^+ + H_2O \longrightarrow \underset{\substack{\text{Hydronium} \\ \text{ion}}}{H_3O^+} + \underset{\substack{\text{Hydroxyl} \\ \text{radical}}}{\cdot OH}$$

The hydroxyl radical is an example of a **free radical,** *a species with one unpaired electron.* This free radical has the electron dot structure

$$\cdot \overset{\cdot\cdot}{O} \!:\! H,$$

which is usually written as $\cdot OH$. The electrically neutral hydroxyl radical has one less electron than a negatively charged hydroxide ion,

$$: \overset{\cdot\cdot}{\underset{\cdot\cdot}{O}} \!:\! H^- \, .$$

Hydroxyl radicals are unstable and highly reactive. They initiate most of the biological effects of radiation by taking an electron from any nearby molecule to become a stable hydroxide ion. In the body, the electron can be taken from a host of molecules: enzymes, nerve fibers, and even DNA. The molecule that loses its electron to the hydroxyl radical, however, now becomes a free radical. This free radical in turn takes an electron from another molecule. A single hydroxyl radical can initiate a very large number of free radicals. Each free radical in turn initiates a chemical change. The radiation-induced chemical changes result in biological damage.

Tissues and organs whose cells grow rapidly and are constantly being replaced are affected the most by radiation. These include bone marrow, lymph nodes, intestinal mucosa, and the body's reproductive system. Radiation damage can be *genetic* or *somatic*. **Genetic damage** *affects the body's reproductive system and genetic material; it harms the offspring.* Genetic effects may not become apparent for several generations. **Somatic damage** *affects the irradiated organism during its own lifetime.* Radiation burns and cancer are examples of somatic effects. In cancer, cells reproduce in an uncontrolled manner because their growth regulation is damaged by the radiation. The major cancer problem associated with radiation exposure is leukemia. In leukemia there is excessive growth of white blood cells.

An acute exposure to a large dose of radiation produces fatigue, nausea, diarrhea, and a decrease in white blood cells. These are the symptoms of radiation sickness. With a higher exposure, death will result from damage to the gastrointestinal tract and to the central nervous system. What level of radiation exposure is safe? How much radiation can a radiologist, for example, be permitted to receive?

In setting health standards for exposure to radiation, scientists generally assume that any exposure to radiation has the potential to cause injury, no matter how low the dose. However, there is some evidence to suggest that biological systems have repair mechanisms that can counter the effects of very low radiation levels. While the controversy exists, it is prudent to assume that even low levels of radiation represent some danger and that we should avoid any unnecessary exposure.

Tobacco leaves are usually contaminated with the environmental pollutant lead-210 ($t_{1/2}$ = 20.4 years). This radioisotope is inhaled when a cigarette is smoked. Lead-210 and its decay products are eventually distributed in the spleen, liver, and bone marrow. As a result, these tissues are continually subjected to harmful alpha- and beta-particle emissions.

10.9 Units of radiation

AIM: To characterize the following units of measurement of radiation: curie, roentgen, rad, and rem.

Focus

Radiation is measured in several units.

It is often necessary to measure radiation accurately. Several units of measurement that describe specific features about a radiation source and its radiation are discussed in this section. We will see that the *curie* is the unit that describes the activity of a radiation source. In studying the effect of radiation on the body, however, a knowledge of the activity of the source is not of much help. Of much more importance are the effects that the radiation produces in the body. The biological effects of radiation are described by three units, the *roentgen,* the *rad,* and the *rem.*

The curie You may recall that radioisotopes decay at specific rates called half-lives. *Intensity of radiation depends on the number of atoms that decay in a given period of time; this breakdown is called the* **rate of decay.** The *curie* (named after Marie Curie) is the unit used to measure the rate of radioactive decay. *One* **curie** *(Ci) is* 3.7×10^{10} *disintegrations per second (dps);* this number is the disintegrations per second from 1 g of pure radium. The curie measures only the activity of a radiation source; it does not describe the source or type of radiation emitted. Any source that gives 3.7×10^{10} dps is, by definition, 1 Ci.

Cobalt-60 sources are gamma emitters used in many hospitals for radiation therapy. These sources are rated in curies. One curie is a very intense radiation source, and so, for many applications in the health sciences, fractions of a curie are used: The *millicurie,* mCi, is 10^{-3} Ci or 3.7×10^{7} dps; the *microcurie,* μCi, is 10^{-6} Ci or 3.7×10^{4} dps.

The roentgen Like the curie, the *roentgen* (named for Wilhelm Roentgen, the discoverer of X rays) is a measure of the output from a radiation source. However, the roentgen measures the capacity of the source to cause ionization, not the rate at which the source disintegrates. *One* **roentgen** *(R) is the amount of X rays or gamma radiation that produces ions carrying* 2.1×10^{9} *units of electrical charge in 1 cm^3 of dry air at 0 °C and 1 atm pressure.* We can see that the roentgen unit is a measure of a radiation beam's capacity for causing ionization in air.

The roentgen is used to measure the radiation output from a radiation source. For example, if a person is placed for 1 minute in front of a radiation source, such as an X-ray beam, emitting 1 R/min, that person will receive an exposure (an *exposed dose*) of 1 R. Ionization of body tissue is proportional to radiation exposure measured in roentgens. The greater the number of ionizations in body tissue, the greater is the damage. A person receiving a single dose of about 600 R to the entire body (a *total-body dose*) will die within 3 weeks.

Ionizing radiation is indiscriminate and kills normal cells as well as cancer cells. Because of this, a side effect of radiation therapy is radiation sickness.

The rad The *rad* (*r*adiation *a*bsorbed *d*ose) is a measure of the energy absorbed by an object exposed to a radiation source. *One* **rad** *is the quantity of ionizing radiation that delivers 100 erg of energy to 1 g of a substance.* (The erg is a very small unit of energy; 4.2×10^{7} erg is equivalent to only 1 cal.) With soft tissue, such as muscle and internal organs, the rad and the roentgen are equivalent; the delivery of 1 R to soft tissue results in the absorption of 1 rad, so a single total-body dose of 600 rad also will kill a human being.

The rem *The* **rem** *(***r***oentgen* **e***quivalent* **m***edical) is a measure of a quantity of radiation.* The quantity differs with the type of radiation exposure. The absorption of 1 rem of *any* ionizing radiation produces the same biological effects as the absorption of either 1 R or 1 rad of gamma rays or X rays. Since biological damage caused by 1 rad of alpha particles differs from that caused by 1 rad of gamma rays or X rays, the rem measure is used to compensate. For example, 1 rad of alpha radiation causes about 15 times more radiation damage than 1 rad of X rays; a 1-rem dose of alpha radiation is therefore 15 times less radiation than a 1-rem dose of X rays.

10.10 Radiation in medicine

AIMS: To describe how X rays are used to take pictures of parts of the body. To explain how radioactive tracers are used in diagnosis of disease. To explain how radiation therapy is used to treat cancer.

Focus

Radiation is widely used for medical diagnosis and therapy.

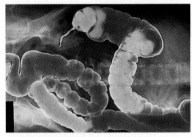

Figure 10.10
The colon is clearly visible in this radiograph. To produce such an image, the patient is first given an enema with a suspension of barium sulfate.

Figure 10.11
The CAT scanner produces detailed images of the tissues and organs inside the body.

X-ray technologists take and develop X-ray pictures and assist with other imaging techniques. They must carefully explain each procedure to the patient. X-ray technologists must be aware of the harmful effects of X radiation.

Slowly the pioneers of radiochemistry became painfully aware of the biological hazards of ionizing radiation. Both Marie Curie and her daughter Irene Joliot-Curie, who spent their lives studying radioactive materials, died of leukemia, probably as the result of exposure to excessive doses of radiation. Madame Curie's husband, Pierre, deliberately exposed a region of his skin to a radioactive source and received a serious, slow-healing burn. (Pierre died in 1906 at the age of 47, a traffic fatality in an age of horse-drawn vehicles.) Many other researchers in this field died of cancer or suffered from radiation sickness before they realized the necessity for care. Despite these hazards, when used properly, radioisotopes and other radiation sources are indispensable tools in modern medical diagnosis and therapy.

Radiation as a diagnostic tool

The most common diagnostic tools involving radiation are X rays and radioactive tracers.

X rays   X rays are useful in the diagnosis of many medical conditions. Because of their penetrating power, X rays pass through many substances, including the human body, that are opaque to visible light. When ionizing radiation passes through matter, energy is absorbed. The amount of energy absorbed from a single source depends on the density of the matter. In the body, hard tissue (bones and teeth) absorbs more energy than soft tissue. *Radiographs*—X-ray photographs—take advantage of these properties.

To make a radiograph, a photographic film encased in a lightproof metal container is placed on one side of the body. X rays are passed through the body from the other side. The radiation exposes the film in proportion to the amount passing through the body. Since bone absorbs more radiation than soft tissue, those regions of the film exposed to X rays passing through bone are lighter. In this way, broken bones and dental cavities can be located.

For gastrointestinal studies, the patient is frequently given a "barium cocktail" (an aqueous suspension of barium sulfate) to drink. Barium sulfate is a dense compound that absorbs radiation much like bone and teeth, so the stomach and intestines are clearly outlined in the radiograph (Fig. 10.10). Radiographs also reveal gallstones. The CAT (*c*omputer-*a*ssisted *t*omography) scanner is a device that produces sophisticated three-dimensional X-ray pictures of internal structures of the body (Fig. 10.11).

Radioactive tracers   When introduced into the body, a radioactive isotope of an element behaves chemically like its nonradioactive counterparts. The only difference is that it emits radiation. Therefore, its location in the body

Table 10.3 Typical Radiation Doses Used for Diagnostic Procedures

Organ	Dose (rem)
liver	0.3
lung	2.0
thyroid	50.0

can be traced by using a radiation detector. Radioactive tracers offer several advantages over other diagnostic procedures: Exceedingly small amounts are needed because modern detection devices are very sensitive; they can be selected so that they travel to or accumulate in certain parts and organs of the body; and those with short half-lives can be chosen so that the radioactivity soon decays. Typical radiation doses used in some diagnostic procedures are given in Table 10.3. Since the 1940s, over a hundred different radioisotopes have been used as tracers, but only a few are used routinely. Some of the radioisotopes commonly used in diagnostic procedures are given in Table 10.4. The test for hyperthyroidism performed on Annabelle, the 60-year-old woman in the Case in Point, involved an application of iodine-123 in the diagnosis of hyperthyroidism. After our Follow-up to the Case in Point we will examine applications of phosphorus-32 and technetium-99m in more detail.

FOLLOW-UP TO THE CASE IN POINT: Hyperthyroidism

A common procedure for diagnosing thyroid malfunction is to measure the uptake of iodide ions by using iodine-123 as the tracer element. This isotope is radioactive and emits gamma radiation as it decays with a half-life of 13.3 hours. Since iodine-123 behaves chemically in the body just like nonradioactive iodine, it is similarly concentrated by the thyroid.

Annabelle was given a fruit-flavored drink containing a very small, carefully measured dose of sodium iodide in which some of the iodide ions were iodine-123. After about 2 hours, the amount of iodide ion taken up by the thyroid was measured by scanning her throat with a scintillation detector. The amount of iodine taken up was then compared with the uptake by normal individuals (see figure). Annabelle's iodine uptake was more than 30% greater than normal. This amount of iodine uptake is considered to indicate hyperthyroidism. Further testing showed that Annabelle did not have thyroid cancer, which has been found to cause hyperthyroidism. Annabelle's physician had a number of options for treatment, ranging from drug therapy to removal of the overactive gland. The doctor recommended that treatment begin with a drug intended to reduce the ability of the thyroid to produce thyroxine (an antithyroid drug). After a few weeks of the antithyroid drug therapy, Annabelle's fever had disappeared and her heart rate had slowed to normal. She gained weight and felt energetic enough to live a normal daily life.

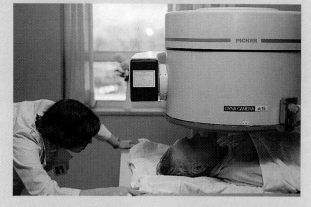

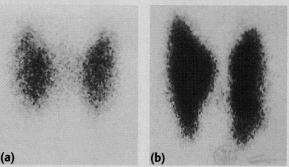

(a) (b)

A scanner (top) employs a scintillation counter as the radiation detector to determine the distribution and concentration of radioisotopes in the body of a patient. The scanner produces an image of the thyroid of the patient after administration of iodine-123. Shown in these scans are (a) a normal thyroid and (b) an overactive thyroid.

Table 10.4 Some Radioisotopes and Their Applications in Medical Diagnoses

Isotope	Emission	Half-life	Application/diagnosis
phosphorus-32	beta	14.3 days	detection of tumors
chromium-51	gamma	27.8 days	blood volume, red blood cell life-time, spleen scan
iron-59	beta	45.1 days	iron turnover (anemia), bone marrow function
strontium-85	gamma	64 days	bone scan
technetium-99m	gamma	6 hours	brain, kidney, liver, thyroid, heart muscle, and bone scan
iodine-123	gamma	13.3 hours	thyroid function, thyroid cancer
iodine-131	gamma, beta	8.0 days	thyroid function, thyroid cancer
gold-198	beta	2.7 days	liver scan
thallium-200	gamma, beta	1.1 days	blood flow in coronary artery

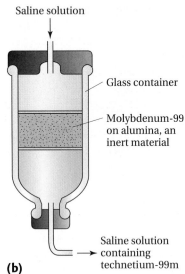

(a)

Saline solution

Glass container

Molybdenum-99 on alumina, an inert material

Saline solution containing technetium-99m

(b)

Figure 10.12
The technetium-99m generator, or the *molybdenum cow:* (a) a typical commercial generator package; (b) inside the generator.

Phosphorus-32   Phosphate ions, a normal component of blood, are often present in higher than normal concentrations in certain tumors. This characteristic provides a way to distinguish cancer cells from normal cells by using phosphate ions labeled with phosphorus-32 (half-life 14.3 days) as a tracer. Since phosphorus-32 is a weak beta emitter, the scintillation counter must be placed close to the radiation source. This requirement limits the use of the isotope to cancers of the skin or cancers exposed through surgical procedures. In the latter case, tiny beta-particle radiation detectors assist the surgeon by defining the limits of the cancerous area.

Technetium-99m   All the isotopes of the element technetium are radioactive, and all are artificially produced. Because it emits gamma radiation and has a short half-life (6 hours), one isotope, technetium-99m, is an ideal radioisotope for use in nuclear medicine. It is important in the study of the brain, kidneys, liver, thyroid, heart muscle, and bone. The m associated with the mass number means that the isotope produced is metastable (that is, highly energetic). This metastable isotope loses its excess energy as gamma radiation to yield a more stable radioisotope of technetium:

$$^{99m}_{43}\text{Tc} \longrightarrow {}^{99}_{43}\text{Tc} + \gamma$$

The source of technetium-99m is the decay of the radioisotope molybdenum-99.

$$^{99}_{42}\text{Mo} \longrightarrow {}^{99m}_{43}\text{Tc} + {}^{0}_{-1}\text{e}$$

A minor disadvantage of technetium-99m is its short half-life. When technetium-99m is required, the radiologist "milks" a *molybdenum cow.* This generating device (Fig. 10.12) contains molybdenum-99 and is used to obtain solutions containing technetium-99m. After about 2 days, the cow has to be sent to be recharged with molybdenum-99.

For brain scans, technetium-99m is usually administered intravenously as a solution of sodium pertechnetate (NaTcO_4) in normal saline. Tumor cells absorb pertechnetate ions (TcO_4^-) differently from normal cells, and this differentiation is detected by scanning the head about 1 hour after administration of the tracer element.

Radiation therapy

Radiation therapy *uses ionizing radiation to kill cancer cells.* Together with surgery, and in some cases chemotherapy, it is the major method of treatment for cancer. Ionizing radiation destroys the cells' ability to reproduce. Cancerous cells usually divide rapidly, making them more sensitive to radiation damage than normal cells. Although it would seem that radiation therapy is the answer to treating cancer, the problem is much more complex. The presence of oxygen increases radiation damage in cells, but the cells at the center of a tumor are poorly supplied with both blood and oxygen. These cells divide slowly and are therefore more radiation resistant than normal cells. Some radioisotopes that are used for radiation therapy, both internally and externally, are given in Table 10.5. Typical doses of radiation administered during radiation therapy are shown in Table 10.6.

A person who receives radiation therapy is at some risk from the radiation itself. The benefits and the risks have to be considered by patient and doctor before the treatment is given. There are three ways in which radiation for radiation therapy is delivered: teletherapy, radioisotope seeds, and radiopharmaceuticals.

Teletherapy   *Teletherapy*, a term that literally means "to cure at a distance," is the use of a narrow beam of high-intensity ionizing radiation to destroy cancerous tissue. The gamma radiation emitted by teletherapy units usually comes from the radioisotopes cobalt-60 and cesium-137, although in some cases X rays are used. Cobalt-60 units with a source intensity of over 1000 Ci are quite common (Fig. 10.13). Since cobalt-60 is a powerful beta and gamma emitter, only 5 g of cobalt-60 is required to produce this high radiation intensity. Since the half-life of cobalt-60 is 5.3 years, the radiation output must be determined periodically so that the radiation dose is known accurately. When the output becomes too low for practical use, the source is changed.

To destroy cancerous tissues without damaging the healthy tissue surrounding a tumor requires a narrow, intense beam of radiation. Ionizing radiation cannot be focused, so blocks of lead or tungsten metal are used to absorb the radiation outside the field of treatment. One technique to minimize damage to healthy tissue, particularly in the treatment of tumors deep inside the body, is to use a rotating radiation unit. A very narrow beam of radiation is aimed at the tumor while the radiation source revolves completely around the patient. The patient is carefully positioned so that the cancerous region is in the radiation beam all the time. The cancerous

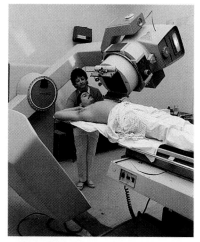

Figure 10.13
A cobalt-60 radiation therapy unit is used in the treatment of cancer.

Table 10.6 Typical Radiation Therapy Doses

Condition	Dose (rem)
brain tumor	6000–7000
skin cancer	5000–6000
lung cancer	6000
lymphoma	4500

Table 10.5 Some Radioisotopes Used in Radiation Therapy

Isotope	Half-life	Application
phosphorus-32	14.3 days	treatment of lymphomas and a form of leukemia
cobalt-60	5.3 years	cancer treatment
iodine-131	8.0 days	thyroid cancer treatment
cesium-137	30 years	cancer treatment
iridium-192	74.2 days	implanted in tumor
radium-226	1600 years	implanted in tumor

region receives a high dose of radiation, but the radiation to the skin and surrounding tissue is spread in a belt all the way around the patient. A person undergoing prolonged cobalt-60 treatment or deep X-ray therapy may receive up to 6000 rem or more to a small part of the body. Most patients undergoing cancer treatment experience unpleasant side effects such as nausea, hair loss, a weakened immune system, and fatigue.

Radioisotope seeds Salts of radioisotopes are powdered and carefully sealed in gold or platinum tubes to make needles or seeds. The encapsulated isotopes are then implanted directly into tumors and left there for predetermined periods of time. Cobalt-60 encapsulated in gold needles is often used to treat cervical cancer. The gamma emitters strontium-90 and radium-226 are also often used in this way. Since the radioactive isotope is in a sealed container, the excretions of patients undergoing treatment are not radioactive.

Radiopharmaceuticals Radiopharmaceuticals are unsealed radioactive substances that are given orally or intravenously. They are usually radioisotopes of gold, iodine, or phosphorus. We mentioned earlier that certain isotopes concentrate in specific regions of the body. The administration of low doses of iodine-131 is used to map the thyroid region in diagnostic work, but if larger doses are given, the emitted radiation (gamma and beta) irradiates the thyroid and provides radiation therapy. Since the radioisotope is unsealed, the excretions of the patient being treated are radioactive and must be handled with appropriate precautions.

10.11 Background radiation

AIM: To define background radiation *and give some of its common sources.*

No matter who we are or where we live, we cannot escape exposure to ionizing radiation. Radioisotopes are present in soil and rocks, in the air we breathe, in the water we drink, and in the food we eat. Of the 350 or so naturally occurring isotopes, about 50 are radioactive. *The radiation we receive from natural sources*—**background radiation**—consists almost entirely of cosmic rays and gamma rays from potassium-40, thorium-232, and the decay of uranium-238 (see A Closer Look: Radon in Homes, on page 297).

Background radiation exposes us to about 300 millirem (mrem) per year—or about 25 rem in a lifetime. There are a few places on earth (the monazite sand areas of India and Brazil) with soils so rich in thorium that people living there receive background radiation of about 2 rem per year.

Besides naturally occurring background radiation, amounts occur through the environmental pollutants from nuclear reactor incidents and from nuclear power station wastes. These isotopes, which can be spread worldwide, include cesium-137, strontium-90, and iodine-131. People living close to a nuclear power plant do not get any significant additional amounts of radiation. However, if there is an accident, the situation could be very different. It is estimated that people living in a town near the Chernobyl nuclear power plant several days after the accident, were receiving as much as 1 rem per hour.

Because of the widespread use of X rays and radioisotopes in medicine, we are also subjected to what might be called *medical background radiation.* The amount of radiation used for medical purposes varies widely. Some people never have a dental or chest X ray, but at the other end of the scale, a cancer patient might receive a dose of 6000 rem concentrated at a small region where a tumor is located. The average person receives about 60 mrem per year from dental and medical X rays.

The human body incorporates radioisotopes that produce a measurable internal background radiation. The predominant isotope is potassium-40, a gamma emitter; there are lesser amounts of carbon-14, thorium-232, and uranium-238. Radium-226, which is an alpha and gamma emitter, is a Group 2 element and chemically resembles calcium. Since it tends to accumulate in the skeleton, radium-226 is called a *bone-seeker.* The major danger from these isotopes is that they are incorporated into the tissues and skeleton after ingestion. Strontium-90, which resembles calcium, is also a bone-seeker and is incorporated into bones; iodine-131, like the non-radioactive isotope iodine-127, concentrates in the thyroid. Young children are at a far greater risk from the accumulation of the artificially produced background radioisotopes than adults. Adults would use strontium-90 only for bone repair, but young children would use it to build bones. Thus the body burden of strontium-90 in young children would be significantly higher than in adults, and because the half-life of strontium-90 is 28 years, they would carry most of the radioisotope for the rest of their lives. Table 10.7 lists the average radiation dose received by each person in the United States, from a variety of radiation sources, each year.

Any exposure to ionizing radiation is potentially harmful. Having an occasional diagnostic X ray is not likely to cause harm, but any exposure to ionizing radiation should be avoided if at all possible. No dose of ionizing radiation, however small, is completely safe.

The more complex an organism, the more likely it will suffer some effects from exposure to radiation. Bacteria and viruses are relatively simple organisms and are less affected by exposure to radiation than more complex organisms such as mammals.

Table 10.7 Average Amount of Radiation Received Annually by Each Person in the United States

Source	Average dose per year (mrem)
Natural background	
air (including radon), food, water	35
cosmic rays	40
radioisotopes naturally inside the human body	40
rocks, soil, bricks, and buildings	65
Artificial: medical and other sources	
air travel	1
consumer products (cigarette smoking)	20
diagnostic X rays (dental, chest X ray, etc.)	70
nuclear medicine (radiotherapy)	15
nuclear power industry and global fallout	2
television	2
TOTAL	290*

*Maximum permissible dose for occupational workers—5000 mrem/year.

SUMMARY

Isotopes with unstable nuclei are radioactive and are called radioisotopes. The nuclei of radioisotopes decay to stable nuclei plus alpha, beta, or gamma emissions. Alpha particles (positively charged helium nuclei) are easily stopped by a sheet of paper. Fast-moving beta particles are more penetrating than alpha particles; they are stopped by aluminum foil. Gamma radiation is electromagnetic radiation similar to visible light, but it is much more energetic. Gamma radiation has no mass or charge. It is extremely penetrating; lead bricks and concrete reduce the intensity of gamma radiation but do not completely stop it.

The decay of radioisotopes is represented by nuclear equations. When a radioactive nucleus emits an alpha particle, its atomic number decreases by 2 and its mass number decreases by 4. When a beta particle is emitted, the atomic number increases by 1; the mass number stays the same. When a gamma ray is emitted, the atomic number and the mass number stay the same.

Every radioisotope decays at a characteristic rate. A half-life is the time required for one-half the nuclei in a radioisotope to decay. The product nuclei may or may not be radioactive. Half-lives vary from fractions of a second to millions of years.

Nuclear fission occurs when fissionable isotopes are bombarded with neutrons and split into smaller species. Fission releases enormous amounts of energy. Controlled fission is the energy source in every nuclear power plant. In nuclear fusion, small nuclei fuse to make heavier nuclei. The sun's energy is released when hydrogen nuclei fuse to make helium nuclei. Fusion releases even more energy than fission.

Radiation may be detected with a Geiger-Müller counter or a scintillation counter. The film badge is the most important personnel radiation detector. The radiation its wearer receives is detected by darkening of the encased film.

The units of radiation measurement are the curie (radiation intensity), the roentgen (radiation exposure), and the rad (radiation absorbed dose). Another unit, the rem (roentgen equivalent medical), compensates for the differences in biological damage caused by 1 rad of alpha radiation, 1 rad of beta radiation, or 1 rad of gamma radiation or X rays.

Radioisotopes and X rays are important tools in the diagnosis of organ malfunctions and diseases. All radiation damages cells, especially dividing cells. Many of the radioisotopes selected for medical use have short half-lives, which helps to minimize biological damage.

KEY TERMS

Alpha emission (10.1)
Alpha particle (10.1)
Background radiation (10.11)
Beta emission (10.1)
Beta particle (10.1)
Chain reaction (10.5)
Curie (10.9)

Fission (10.5)
Free radical (10.8)
Fusion (10.6)
Gamma radiation (10.1)
Geiger counter (10.7)
Genetic damage (10.8)
Half-life (10.2)
Ionizing radiation (10.7)

Nuclear transformation reaction (10.4)
Rad (10.9)
Radiation (10.1)
Radiation therapy (10.10)
Radioactive decay (10.1)
Radioactivity (10.1)
Radioisotope (10.1)

Rate of decay (10.9)
Rem (10.9)
Roentgen (10.9)
Scintillation counter (10.7)
Somatic damage (10.8)
Transmutation (10.1)
X ray (10.1)

EXERCISES

Radioisotopes (Sections 10.1, 10.2, 10.3, 10.4)

10.7 Write the symbol, including the charge, for (a) an alpha particle, (b) a beta particle, and (c) a gamma ray.

10.8 Describe the type of shielding required for protection against (a) X rays, (b) beta-particle emission, (c) gamma radiation, and (d) alpha-particle emission.

10.9 Distinguish between an *isotope* and a *radioisotope*.

10.10 In what way does a nuclear reaction differ from a chemical reaction?

10.11 Each of the following radioisotopes emits alpha particles on decay. Write balanced nuclear equations for their decay processes.
(a) radium-226 ($^{226}_{88}$Ra)   (b) thorium-230 ($^{230}_{90}$Th)
(c) uranium-233 ($^{233}_{92}$U)   (d) radon-222 ($^{222}_{86}$Rn)

10.12 Write a balanced nuclear equation for the beta decay of each of the following radioisotopes.
(a) strontium-90 ($^{90}_{38}$Sr)   (b) nitrogen-13 ($^{13}_{7}$N)
(c) lithium-8 ($^{8}_{3}$Li)   (d) sodium-24 ($^{24}_{11}$Na)

10.13 Iron-59 has a half-life of 45 days. How much iron-59 will remain in a 0.2-g sample after 135 days?

10.14 Radium-226 has a half-life of 1600 years. How much radium-226 will remain in a 16.0-mg sample after 8000 years?

10.15 The mass of strontium-90 in a sample is found to have decreased from 16.8 to 2.1 g in a period of 84.3 years. Calculate the half-life of strontium-90.

10.16 In 21 years the mass of a sample of cobalt-60 decreased from 88.0 to 5.50 mg. What is the half-life of cobalt-60?

Fission and Fusion (Sections 10.5, 10.6)

10.17 Distinguish between *nuclear fission* and *nuclear fusion*.

10.18 What is the difference between the nuclear reactions taking place in the sun and the nuclear reactions taking place in a nuclear reactor?

10.19 What is a nuclear chain reaction?

10.20 How is a nuclear chain reaction moderated in a nuclear reactor?

Detection and Measurement (Sections 10.7, 10.8)

10.21 Why are X rays called *ionizing* radiation?

10.22 Why must a film badge be worn when working with radiation sources?

10.23 Define the units of radiation (a) the curie and (b) the roentgen.

10.24 Distinguish between the radiation units the rad and the rem.

Radiation and Health (Sections 10.9, 10.10, 10.11)

10.25 Describe how some radioactive isotopes—called *tracers*—are useful in the diagnosis of certain medical disorders.

10.26 Explain why ionizing radiation is used in the treatment of cancer.

10.27 How can iodine-131 be used in radiation therapy of the thyroid gland?

10.28 List the symptoms of radiation sickness that might accompany radiation therapy.

Additional Exercises

10.29 The following radioactive nuclei decay by emitting alpha particles. Write the product of the decay process for each.
(a) $^{238}_{94}$Pu   (b) $^{210}_{83}$Bi   (c) $^{174}_{72}$Hf   (d) $^{210}_{84}$Po   (e) $^{230}_{90}$Th

10.30 A patient is administered 20 mg of iodine-131. How much of this isotope will remain in the body after 40 days if half-life for iodine-131 is 8 days?

10.31 Complete the following nuclear reactions.
(a) $^{32}_{15}$P $\longrightarrow$ _____ $+ _{-1}^{0}$e
(b) _____ $\longrightarrow$ $^{14}_{7}$N $+ _{-1}^{0}$e
(c) $^{238}_{92}$U $\longrightarrow$ $^{234}_{90}$Th $+$ _____
(d) $^{141}_{56}$Ba $\longrightarrow$ _____ $+ _{-1}^{0}$e

10.32 What is the significant difference between radiation exposed dose and radiation absorbed dose?

10.33 Identify the most stable isotope in each of the following pairs.
(a) $^{14}_{6}$C, $^{12}_{6}$C   (b) $^{3}_{1}$H, $^{1}_{1}$H   (c) $^{16}_{8}$O, $^{18}_{8}$O   d) $^{14}_{7}$N, $^{15}_{7}$N

10.34 What part of an atom undergoes change during radioactive decay? Explain.

10.35 The half-life of the radioisotope xenon-133 is 5 days. If you start today with 2.00×10^6 atoms of this isotope, how many would remain after 5 weeks?

10.36 Describe three kinds of radiation that are products of radioactive decay.

10.37 Explain the process of transmutation. Write at least three nuclear equations to illustrate your answer.

10.38 Complete the following nuclear reactions.
(a) $^{6}_{3}$Li $+ ^{1}_{0}$n $\longrightarrow$ $^{4}_{2}$He $+$ _____ $+$ energy
(b) $^{14}_{7}$N $+ ^{4}_{2}$He $\longrightarrow$ $^{1}_{1}$H $+$ _____ $+$ energy
(c) $^{39}_{18}$Ar $\longrightarrow$ $^{39}_{19}$K $+$ _____
(d) $^{235}_{92}$Cl $+$ _____ $\longrightarrow$ 2^{118}_{46}Pd

10.39 What type(s) of radiation can be measured using a Geiger-Müller detector?

10.40 Write a nuclear equation for each of the following word equations.
(a) Radon-222 emits an alpha particle to form polonium-218.
(b) Strontium-90 decays to yttrium-90 as it emits a beta particle.
(c) Phosphorus-32 produces sulfur-32 plus a beta particle.
(d) Carbon-14 emits a beta particle as it decays to nitrogen-14.
(e) Radium-230 is produced when thorium-234 emits an alpha particle.
(f) When polonium-210 emits an alpha particle the product is lead-206.

SELF-TEST (REVIEW)

True/False

1. Beta particles are emitted when every radioisotope decays.
2. Gamma radiation has a negative charge.
3. The film badge is a personal radiation monitor.
4. Background radiation is zero inside a brick building.
5. After 10 half-lives, a radioisotope will have completely decayed.
6. If clothes contaminated with a radioisotope are burned, the radioactivity is destroyed.
7. Gamma radiation and X radiation are high-energy electromagnetic radiation.
8. A Geiger-Müller counter is the best alpha particle detector.
9. A curie is the unit used to measure the rate of radioactive decay.
10. A radioisotope has a half-life of 12 minutes. After 36 minutes, only one-third of the radioactive atoms initially present will remain unchanged.

Multiple Choice

11. Which of the following has the most penetrating power?
    (a) gamma radiation
    (b) beta-particle emission
    (c) background radiation
    (d) alpha-particle emission
12. Which of the following processes results in a "splitting" of atoms?
    (a) decomposition reaction   (b) fusion reaction
    (c) fission reaction   (d) ionizing reaction
13. Which of the following units is the best measure of the biological damage caused by radiation?
    (a) rad   (b) roentgen   (c) rem   (d) curie
14. Background radiation comes from
    (a) the air.   (b) the ground.   (c) water.
    (d) all of these.

15. Polonium-214 undergoes radioactive decay to give an alpha particle and
    (a) $^{210}_{82}Pb$.   (b) $^{218}_{86}Rn$.   (c) $^{214}_{82}Pb$.   (d) $^{210}_{82}Po$.
16. The radioisotope radon-222 has a half-life of 3.8 days. How much of a 10.0 g sample of radon-222 would be left after approximately 15 days?
    (a) 2.50 g   (b) 6.25 g   (c) 0.625 g   (d) 1.25 g
17. Uncontrolled nuclear chain reactions
    (a) take place in nuclear reactors.
    (b) are always fusion reactions.
    (c) never produce radioactive by-products.
    (d) are characteristic of atomic bombs.
18. Which of the following could stop the penetration of an alpha particle?
    (a) 2 feet of lead   (b) aluminum foil
    (c) a piece of paper   (d) all of these
19. Which of the following naturally occurring radioisotopes would be most useful in dating objects thought to be millions of years old?
    (a) carbon-14, $t_{1/2} = 5.73 \times 10^3$ years
    (b) potassium-40, $t_{1/2} = 1.28 \times 10^9$ years
    (c) thorium-234, $t_{1/2} = 25$ days
    (d) radon-222, $t_{1/2} = 3.8$ days
20. The conversion of an atom of one element into an atom of another element by the emission of radiation is called
    (a) *ionization*.   (b) *transmutation*.
    (c) *roentgen*.   (d) none of these.
21. If an isotope undergoes beta-particle emission,
    (a) the mass number changes.
    (b) the atomic number changes.
    (c) protons are given off.
    (d) the number of neutrons remains the same.
22. Barium sulfate is used when the gastrointestinal tract is X rayed because
    (a) it is a beta-particle emitter.
    (b) it concentrates in the small intestine.
    (c) it absorbs radiation.
    (d) it is an alpha-particle emitter.

Carbon Chains and Rings

The Foundations of Organic Chemistry

Organic chemistry is the chemistry of the compounds of carbon. Daily, we encounter numerous carbon-based compounds. Paraffin wax, one of the many organic products obtained from petroleum, is used to make candles.

11

CHAPTER OUTLINE

CASE IN POINT: Carbon monoxide poisoning

11.1   Carbon compounds

11.2   Hydrocarbons

11.3   Carbon-carbon bonds

11.4   Straight-chain alkanes

11.5   Branched-chain alkanes

11.6   Structural isomers

11.7   Cycloalkanes

11.8   Multiple bonds

11.9   Aromatic compounds

A CLOSER LOOK: Hydrocarbons and Health

11.10 Sources of hydrocarbons

A CLOSER LOOK: Octane Ratings of Gasoline

11.11 Properties of hydrocarbons

FOLLOW-UP TO THE CASE IN POINT: Carbon monoxide poisoning

A CLOSER LOOK: The Greenhouse Effect

The element carbon constitutes only two-tenths of 1% of the mass of the elements found in the Earth's crust. Oxygen, silicon, aluminum, and iron are much more abundant, but carbon is the basis of life on Earth. Nearly one out of every ten atoms in our bodies is a carbon atom. Most of the rest are hydrogen and oxygen, mainly combined as water. Since *all* organic material is carbon based, so is the food our bodies burn as fuel.

Over 10 million organic compounds have been isolated or synthesized. Many have proved useful as medicines, plastics, fibers, fertilizers, insecticides, and a host of other products. In this and most of the following chapters we will explore the immense structural and chemical diversity of carbon compounds.

The combustion of carbon compounds is an important source of energy. Carbon compounds in coal and natural gas are burned for heat throughout the world. Like other chemical reactions, this common and widespread chemical reaction of carbon compounds must be handled with respect, as illustrated in the Case in Point.

CASE IN POINT: Carbon monoxide poisoning

 Claudine lives in the upper midwestern United States. A friend arrived early one morning to take Claudine on a shopping trip. Her friend rang Claudine's doorbell, but oddly, no one answered. The friend looked through a garage window and saw Claudine's car which suggested that she was at home. Concerned, the friend went to the nearest public phone and called the police. When they entered the house, the police found Claudine near the front door. She was conscious but disoriented. Her breathing was shallow, and her pulse was weak. Her face appeared unusually flushed. There was no strong odor in the house, and there was no evidence of foul play. Claudine was immediately taken to the hospital, where she was treated for carbon monoxide poisoning. We will find out how Claudine was poisoned and how she was treated in Section 11.11.

Carbon monoxide is a colorless, odorless, tasteless, and nonirritating gas. Although it is difficult to detect with the senses, people can be warned of carbon monoxide's presence by installing a carbon monoxide detector.

11.1 Carbon compounds

AIM: To describe the ways that carbon atoms are bonded in organic compounds.

Focus

Carbon atoms combine with other carbon atoms and with atoms of other elements.

Organic compounds exhibit dramatically different chemical and physical properties because they have different structures. At standard conditions, some are solids, some are liquids, and some are gases. Some taste sweet, and others taste sour. Some are poisons, but others are essential for life. In order to understand these properties of organic molecules, it is necessary to explore and understand their structures. Three simple principles can provide a basic understanding of the structure and chemistry of organic molecules:

1. Carbon atoms can form covalent bonds with hydrogen atoms.
2. Carbon atoms can form covalent bonds with other carbon atoms to build carbon chains.
3. Carbon atoms can form covalent bonds with other elements, especially oxygen, nitrogen, phosphorus, sulfur, and the halogens.

We begin our study of organic chemistry with bonding between carbon and hydrogen atoms, and then we will examine bonding between carbon atoms. We will discuss carbon's bonding to other elements in later chapters.

11.2 Hydrocarbons

AIMS: To draw the electron dot and a structural formula of methane. To use orbital hybridization to describe the shape of the methane molecule.

Focus

Methane is the simplest organic molecule.

Hydrocarbons *are compounds whose molecular structures contain only hydrogen and carbon.* The simplest hydrocarbons are the **alkanes**—*hydrocarbons that contain only single covalent bonds.* The smallest organic molecule is the alkane called *methane* (CH_4). Methane, a gas at standard temperature and pressure, is the major component of natural gas. It is sometimes called "marsh gas" because it is formed by the action of bacteria on decaying vegetation in swamps and other marshy areas.

The carbon-hydrogen bond in methane

A carbon atom contains four valence electrons. Four hydrogen atoms, each with one valence electron, can form four covalent carbon-hydrogen bonds. This combination is a molecule of methane:

$$\cdot\dot{\underset{\cdot}{C}}\cdot \quad + \quad 4H\cdot \quad \longrightarrow \quad H:\overset{H}{\underset{H}{\overset{\cdot\cdot}{\underset{\cdot\cdot}{C}}}}:H$$

Carbon atom | Hydrogen atoms | Methane molecule

There is an important principle here: *Because a carbon atom contains four valence electrons, in making organic compounds it nearly always forms four*

Anaerobic (in the absence of oxygen) digestion of sewage sludge produces methane gas. It is estimated that each ton of solid waste can be digested to produce 85 m³ of methane.

covalent bonds. Remembering this principle will help you to write complete and correct structures for organic molecules.

As you can see, the molecular formula of methane is CH_4. However, molecular formulas are not very useful in organic chemistry. Molecular formulas cannot, by themselves, contain enough information about a molecule, such as the arrangement of the atoms and the bonding between the atoms. For this reason, organic chemists prefer to write **structural formulas**—*formulas that show covalent bonds and the arrangement of atoms in each molecule.* Organic chemists usually abbreviate the bonding electron pairs in carbon-containing molecules as short lines. It will help to remember that the line between the atomic symbols represents *two electrons.* For example:

$$ H-\overset{\displaystyle H}{\underset{\displaystyle H}{C}}-H \qquad \text{Line represents shared electron pair} $$

Methane
molecule

The shape of methane

You may recall from Section 5.10 that electron-pair repulsion theory predicted a tetrahedral shape for the methane molecule. This prediction agrees with experimental work that shows that methane is a completely symmetrical tetrahedral molecule; all the C—H bonds are identical, and all the H—C—H bond angles are 109.5 degrees (Fig. 11.1).

As we discussed in Section 3.5, electron configurations in atoms can be described by atomic orbitals. The bonding in methane and other molecules also can be described by atomic orbitals. When one atomic orbital overlaps with another atomic orbital, each with one electron, the pair of electrons is shared, and a covalent bond is formed. We can illustrate this concept with the hydrogen molecule, H_2. The 1s atomic orbitals of two hydrogen atoms overlap in the formation of a hydrogen molecule. Two electrons, one from each hydrogen atom, are available for sharing and a covalent bond is formed (Fig. 11.2).

However, if we attempt to generate a picture of methane by simple atomic orbital overlap, the method fails. A carbon atom has only two unpaired electrons, one in each of two 2p orbitals:

$$ \boxed{\uparrow\downarrow} \quad \boxed{\uparrow\downarrow} \quad \boxed{\uparrow} \quad \boxed{\uparrow} \quad \boxed{} $$
$$ 1s \qquad 2s \qquad\qquad 2p $$

If the two electrons of the 2p orbitals were each shared with an electron of the 1s orbital of hydrogen atoms, we would get a molecule with the formula

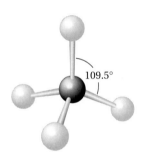

Figure 11.1
Ball-and-stick model of the methane molecule.

109.5°

Figure 11.2
The overlap of the 1s orbitals of two hydrogen atoms, each with one electron, produces a covalent electron pair bond. The product is a hydrogen molecule, H_2.

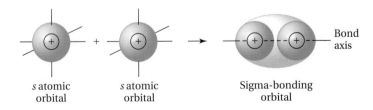

s atomic
orbital

s atomic
orbital

Sigma-bonding
orbital

Bond
axis

CH_2 instead of CH_4. One solution to this dilemma is **orbital hybridization**—*the mixing of two or more different atomic orbitals.* In methane, a carbon atom's 2s orbital and three 2p orbitals mix to form four identical sp^3 (read "s-p-three") hybrid atomic orbitals, each with one electron. Note that the number of hybrid orbitals that are formed is equal to the number of orbitals that are mixed. Because each sp^3 hybrid orbital is composed of 75% of dumbbell-shaped p orbitals and only 25% of a spherical s orbital, its appearance is similar to a fattened p atomic orbital. Figure 11.3 shows how the balloon-shaped sp^3 orbitals point toward the corners of a

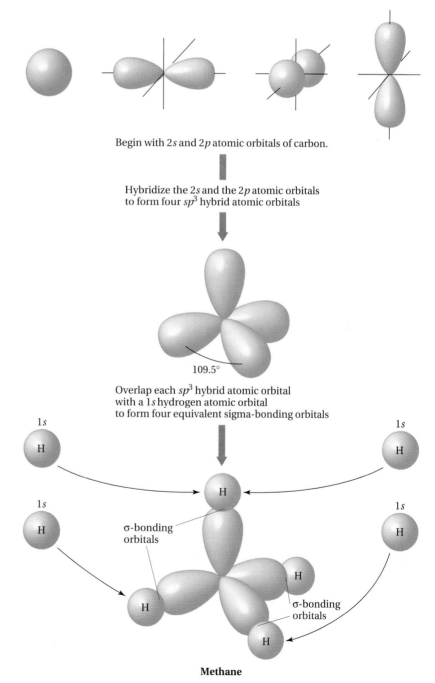

Begin with 2s and 2p atomic orbitals of carbon.

Hybridize the 2s and the 2p atomic orbitals
to form four sp^3 hybrid atomic orbitals

109.5°

Overlap each sp^3 hybrid atomic orbital
with a 1s hydrogen atomic orbital
to form four equivalent sigma-bonding orbitals

1s

1s

1s

1s

σ-bonding
orbitals

σ-bonding
orbitals

Methane

Figure 11.3
Steps leading to an orbital description of the methane molecule. Hybridization of the 2s and 2p orbitals of a carbon atom produces four sp^3 hybrid atomic orbitals that point to the corners of a tetrahedron. Each of these hybrid orbitals overlaps with a 1s orbital of hydrogen to produce a methane molecule.

tetrahedron. The tetrahedral angle of 109.5 degrees between any two hybrid orbitals minimizes unfavorable interactions among the orbitals. Now the four sp^3 orbitals of carbon, each of which has one electron, can overlap with the $1s$ orbitals of four hydrogen atoms, each with one electron. The product is methane, CH_4. The carbon-hydrogen bonds in methane are all of a type called *sigma bonds*. Imagine slowly rotating the bond between a carbon nucleus and a hydrogen nucleus. The bond looks exactly the same throughout the rotation because it is symmetrical along the bond axis. *A covalent bond that is completely symmetrical along the axis connecting two atomic nuclei is a* **sigma bond** (the Greek letter sigma is σ). The covalent bond in the hydrogen molecule is a sigma bond. Carbon-hydrogen single bonds, carbon-carbon single bonds, and single bonds between carbon and other elements are also sigma bonds.

The structure of the resulting methane molecule is in agreement with experiment; it has four identical C—H bonds and is tetrahedral. Since sp^3 orbitals extend farther in space than either s or p orbitals, the overlap of a carbon sp^3 orbital with a hydrogen $1s$ orbital is greater than is possible with either a $2s$ or a $2p$ orbital of carbon. This greater overlap results in an unusually strong covalent bond.

11.3 Carbon-carbon bonds

AIM: To show with an electron dot and a structural formula how a carbon-carbon bond is formed in ethane, C_2H_6.

The unique ability of carbon to make stable carbon-carbon bonds and form chains is the major reason for the vast number of organic molecules. A few other elements, notably silicon, form short chains, but most silicon chains are unstable in an oxygen environment.

Like methane, ethane is a gas at standard temperature and pressure. Let us see how ethane could be formed from carbon and hydrogen. Two carbon atoms could share a pair of electrons to form a carbon-carbon covalent bond:

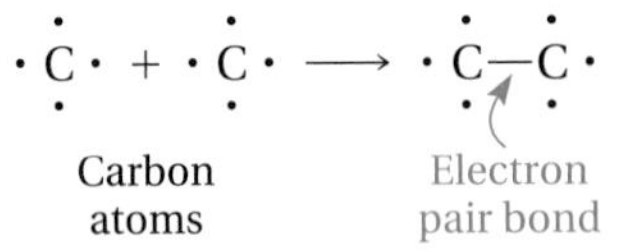

$$\cdot \overset{\cdot}{\underset{\cdot}{C}} \cdot \ + \ \cdot \overset{\cdot}{\underset{\cdot}{C}} \cdot \ \longrightarrow \ \cdot \overset{\cdot}{C} \!-\! \overset{\cdot}{C} \cdot$$

Carbon atoms   Electron pair bond

Now the remainder of the valence shell electrons of each carbon could be paired with the electrons from six hydrogen atoms to make a molecule of ethane:

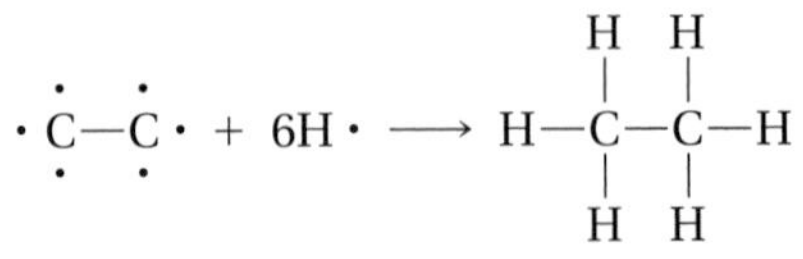

$$\cdot \overset{\cdot}{\underset{\cdot}{C}} \!-\! \overset{\cdot}{\underset{\cdot}{C}} \cdot \ + \ 6H \cdot \ \longrightarrow \ H \!-\! \overset{\overset{\displaystyle H}{|}}{\underset{\underset{\displaystyle H}{|}}{C}} \!-\! \overset{\overset{\displaystyle H}{|}}{\underset{\underset{\displaystyle H}{|}}{C}} \!-\! H$$

Ethane

11.4 Straight-chain alkanes

AIM: To name and recognize structural, condensed, and molecular formulas of the straight-chain hydrocarbons containing up to 10 carbon atoms.

Focus

Straight-chain alkanes can be constructed using more than two carbon atoms.

We can build alkanes with more than two carbons by the same method used to build ethane. *Alkanes with more than two carbons strung together like the links of a chain are called* **straight-chain alkanes.** There is no need to resort to electron dot structures for each chain. We can just write the symbol for carbon as many times as we need in order to get the chain length and then fill in with hydrogens and lines representing covalent bonds. Remember that each carbon has four covalent bonds.

EXAMPLE 11.1

Writing structural formulas for straight-chain alkanes

Draw complete structural formulas for the straight-chain alkanes that have three and four carbons.

SOLUTION

Draw three carbons and four carbons in a row, and then put in lines to show the carbon-carbon bonds.

$$C{-}C{-}C \qquad C{-}C{-}C{-}C$$

Next, add enough covalent bonds to give each carbon a total of four bonds.

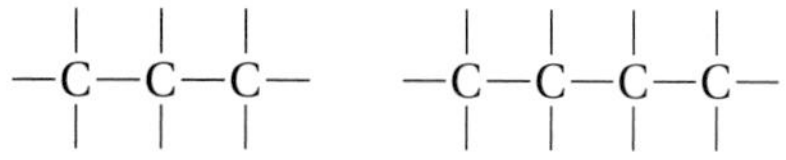

$$\begin{array}{ccc} | & | & | \\ -C{-}C{-}C{-} \\ | & | & | \end{array} \qquad \begin{array}{cccc} | & | & | & | \\ -C{-}C{-}C{-}C{-} \\ | & | & | & | \end{array}$$

Finally, complete the structures by adding hydrogens to fill out the carbon-hydrogen bonds.

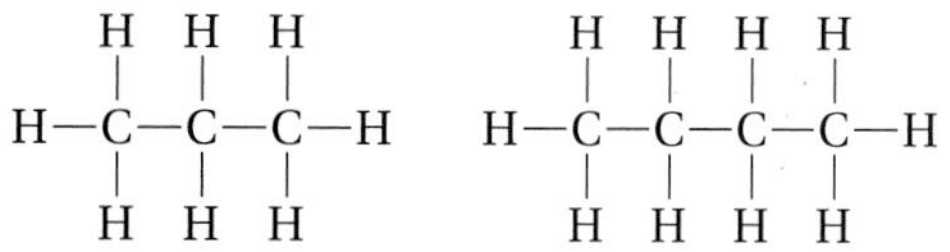

$$\begin{array}{ccc} H & H & H \\ | & | & | \\ H{-}C{-}C{-}C{-}H \\ | & | & | \\ H & H & H \end{array} \qquad \begin{array}{cccc} H & H & H & H \\ | & | & | & | \\ H{-}C{-}C{-}C{-}C{-}H \\ | & | & | & | \\ H & H & H & H \end{array}$$

PRACTICE EXERCISE 11.1

Draw complete structural formulas for the straight-chain alkanes with five and six carbons.

Drawing structural formulas

Sometimes we will find it convenient to draw complete structural formulas—that is, to show *all* the atoms and bonds in a molecule. There are, however, many ways to draw shorthand or *condensed structural formulas.*

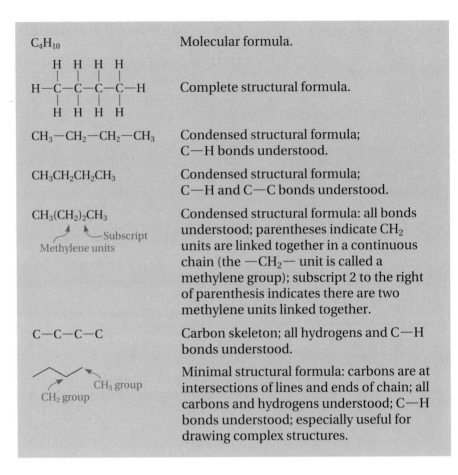

C_4H_{10}	Molecular formula.
	Complete structural formula.
$CH_3\!-\!CH_2\!-\!CH_2\!-\!CH_3$	Condensed structural formula; C—H bonds understood.
$CH_3CH_2CH_2CH_3$	Condensed structural formula; C—H and C—C bonds understood.
$CH_3(CH_2)_2CH_3$	Condensed structural formula: all bonds understood; parentheses indicate CH_2 units are linked together in a continuous chain (the —CH_2— unit is called a methylene group); subscript 2 to the right of parenthesis indicates there are two methylene units linked together.
C—C—C—C	Carbon skeleton; all hydrogens and C—H bonds understood.
	Minimal structural formula: carbons are at intersections of lines and ends of chain; all carbons and hydrogens understood; C—H bonds understood; especially useful for drawing complex structures.

Figure 11.4
Ways to represent structural formulas—in this case for a straight-chain alkane containing four carbons.

Condensed structural formulas *leave out part of the structure of a molecule.* What has been left out has to be understood by the person looking at the structure. Figure 11.4 shows several ways to draw condensed structural formulas for a straight-chain alkane containing four carbons.

> **PRACTICE EXERCISE 11.2**
> Draw all versions of the condensed structural formulas for straight-chain alkanes containing five and six carbons.

Origins of names of organic compounds

Table 11.1 shows the names and condensed structures for straight-chain alkanes containing up to 10 carbons. These names are recommended by the International Union of Pure and Applied Chemistry (IUPAC). *The* **IUPAC system** *is the basis for a precise, internationally accepted system of naming compounds.* We will use the IUPAC system to name many compounds.

Table 11.1 Structural Formulas of the First 10 Straight-chain Alkanes

Name	Molecular formula	Structural formula	Boiling point (°C)
methane	CH_4	CH_4	−161.0
ethane	C_2H_6	CH_3CH_3	−88.5
propane	C_3H_8	$CH_3CH_2CH_3$	−42.0
butane	C_4H_{10}	$CH_3CH_2CH_2CH_3$	0.5
pentane	C_5H_{12}	$CH_3CH_2CH_2CH_2CH_3$	36.0
hexane	C_6H_{14}	$CH_3CH_2CH_2CH_2CH_2CH_3$	68.7
heptane	C_7H_{16}	$CH_3CH_2CH_2CH_2CH_2CH_2CH_3$	98.5
octane	C_8H_{18}	$CH_3CH_2CH_2CH_2CH_2CH_2CH_2CH_3$	125.6
nonane	C_9H_{20}	$CH_3CH_2CH_2CH_2CH_2CH_2CH_2CH_2CH_3$	150.7
decane	$C_{10}H_{22}$	$CH_3CH_2CH_2CH_2CH_2CH_2CH_2CH_2CH_2CH_3$	174.1

Propane, often used for heating in rural areas where natural gas is not available, is more dense than air. If a propane leak develops in a building, propane can collect in a basement and is an explosion hazard. Because of its density, it is difficult to air out a propane-filled room.

All organic compounds can be named by the IUPAC system, but many are still called by their *common names,* which bear no relation to the IUPAC name or to the molecular structures of the compounds themselves. Common names for organic compounds isolated from nature often reflect the source of the compounds—for example, penicillin from the mold *Penicillium notatum.* Like the IUPAC names for many complex organic compounds, the IUPAC name for penicillin is too long and unwieldy for everyday use. In such instances, we will use the common names exclusively. The common names for even some simple organic compounds are so deeply ingrained in organic chemistry and biochemistry that we will mention them along with the IUPAC names.

PRACTICE EXERCISE 11.3

Refer to Table 11.1 to name the following alkanes:

(a) 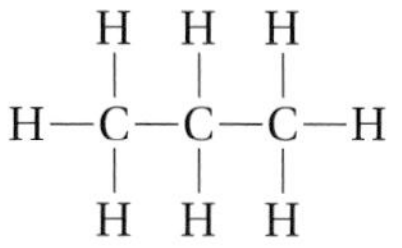
```
    H  H  H
    |  |  |
H—C—C—C—H
    |  |  |
    H  H  H
```

(b)
```
    H  H  H  H  H  H
    |  |  |  |  |  |
H—C—C—C—C—C—C—H
    |  |  |  |  |  |
    H  H  H  H  H  H
```

(c)
```
    H  H  H  H  H
    |  |  |  |  |
H—C—C—C—C—C—H
    |  |  |  |  |
    H  H  H  H  H
```

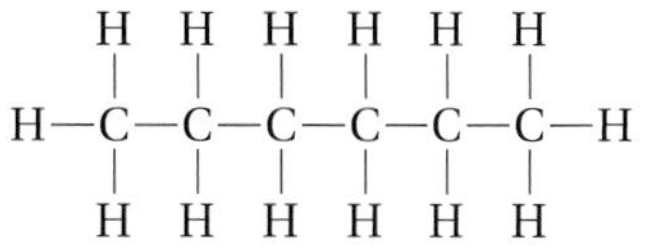

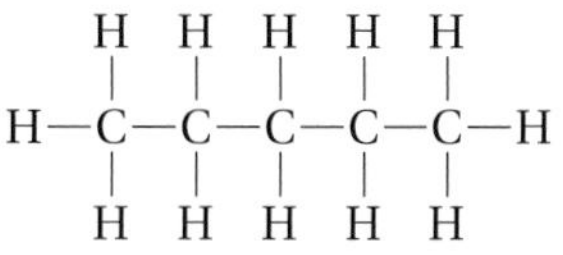

True shapes of straight-chain alkanes

Although they are usually drawn in a straight line, the carbon chains of straight-chain alkanes are actually zigzag shaped, as shown for hexane in Figure 11.5(a). The zigzag shape occurs because the bonds extending from

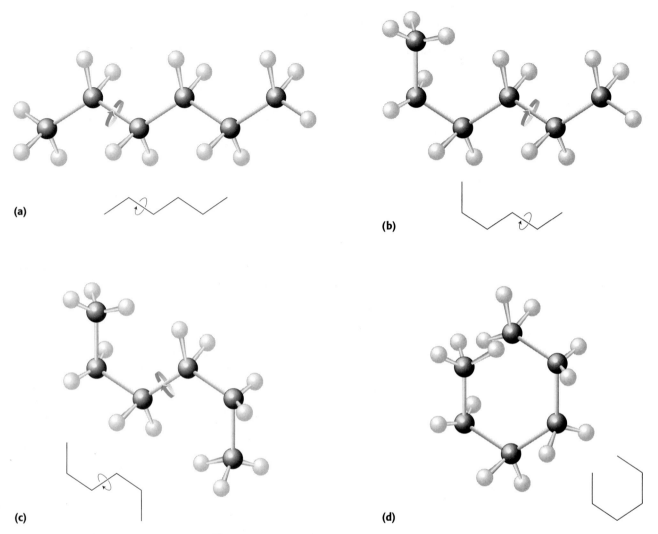

(a)

(b)

(c)

(d)

Figure 11.5
Ball-and-stick model of the extended form of hexane (a). Structures (b), (c), and (d) show a few of the possible coilings of the chain by rotations about carbon-carbon bonds.

The female mushroom fly produces a mixture of straight-chain alkanes from C_{15} to C_{30} that attracts the male fly. Such insect attractants are called *pheromones*.

the carbons in the chain are in a tetrahedral arrangement, and the C—C—C bond angles are 109.5 degrees, the normal tetrahedral bond angle. The zigzag form of a straight-chain alkane is the most stable shape the molecule can achieve because it puts the constituent atoms in the chain as far apart as they can be. However, the carbons in alkanes can rotate about carbon-carbon single bonds, as shown in Figure 11.5(b–d). At room temperature the freedom of rotation is nearly complete. Thus a beaker of hexane molecules is dynamic—it resembles the activity of a can of live worms much more than the inactivity on a plate of spaghetti.

11.5 Branched-chain alkanes

AIMS: To name an alkane according to IUPAC rules, given the structural formula of the alkane. To draw the structural formula of an alkane, given its IUPAC name.

Focus

The carbon chains of alkanes can be branched.

Hydrocarbon groups can substitute for hydrogens on the straight-chain alkanes. For example, the middle hydrogen of propane could be replaced by the hydrocarbon group CH_3—, as shown in the following structures:

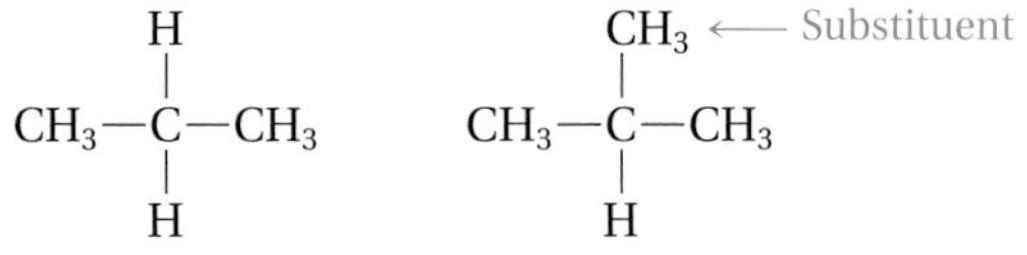

$$
\begin{array}{cc}
\begin{array}{c}
\text{H} \\
| \\
CH_3-\underset{\underset{\text{H}}{|}}{\overset{}{C}}-CH_3 \\
\end{array}
&
\begin{array}{c}
CH_3 \ \longleftarrow \ \text{Substituent}\\
| \\
CH_3-\underset{\underset{\text{H}}{|}}{\overset{}{C}}-CH_3 \\
\end{array}
\\
\text{Propane} &
\end{array}
$$

Groups that replace hydrogen on carbon are called **substituents.** *When the substituents are hydrocarbon groups, they are called* **alkyl groups.**

Naming branched-chain alkanes

Alkyl groups are named by removing the *-ane* ending from the alkane name and adding *-yl*. For example, the group CH_3— is derived from the hydrocarbon methane (the parent hydrocarbon) and is called a *methyl* group. The group CH_3CH_2—, derived from the parent hydrocarbon ethane, is called an *ethyl* group, and the group $CH_3CH_2CH_2$—, derived from the parent hydrocarbon propane, is a *propyl* group. The straight-chain alkyl group derived from butane is called a *butyl* group ($CH_3CH_2CH_2CH_2$—). The presence of alkyl groups as substituents in a hydrocarbon molecule turns straight-chain alkanes into *branched-chain alkanes. In a* **branched-chain alkane,** *hydrocarbon substituents (alkyl groups) are attached to a parent straight-chain alkane.* The following structures are examples of branched-chain alkanes. Hydrogens have been omitted in these structures.

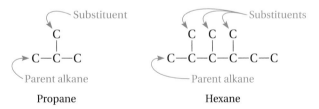

We will use five rules to help us determine the IUPAC names of branched-chain hydrocarbons:

1. Find the longest chain of carbons in the molecule. This chain is used as the parent name of the compound.

2. Number the carbons in the main chain in sequence, starting at the end that will give the alkyl substituents attached to the chain the *smallest numbers.*

3. Give the appropriate position number and name for each alkyl group.

4. Attach an appropriate prefix to groups that appear more than once in the structure: *di-* (twice), *tri-* (three times), *tetra-* (four), and *penta-* (five).

5. List the names of alkyl substituents in alphabetical order. For purposes of alphabetizing, the prefixes *di-, tri-,* and so on are ignored.

Examples 11.2 to 11.4 show how to apply these rules to naming branched-chain hydrocarbons. If we know the IUPAC name of a compound, the structure is easy to reconstruct, as shown in Examples 11.5 and 11.6.

EXAMPLE 11.2

Naming a branched-chain alkane

What is the IUPAC name for the following branched-chain alkane?

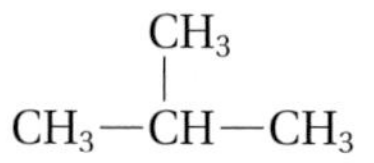

$$CH_3-\underset{\underset{CH_3}{|}}{CH}-CH_3$$

SOLUTION

Number the longest chain of carbons in the molecule (*rule 1*). Since the longest chain contains three carbons, the parent name of the compound will be *propane.*

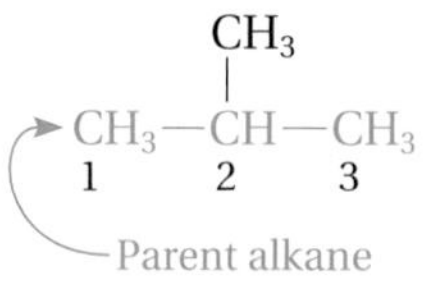

$$\underset{1\qquad 2\qquad 3}{CH_3-\underset{\underset{CH_3}{|}}{CH}-CH_3}$$
Parent alkane

Propane

The chain could be numbered in other ways as follows, but notice that in this molecule, a methyl substituent always ends up in the 2 position.

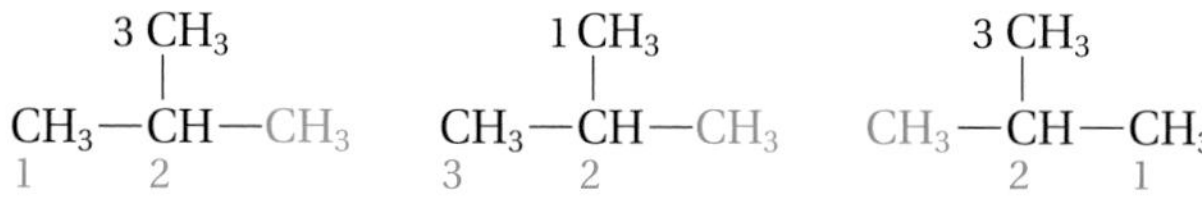

$$\underset{1\qquad 2}{CH_3-\underset{\underset{3\ CH_3}{|}}{CH}-CH_3} \qquad \underset{3\qquad 2}{CH_3-\underset{\underset{1\ CH_3}{|}}{CH}-CH_3} \qquad \underset{2\qquad 1}{CH_3-\underset{\underset{3\ CH_3}{|}}{CH}-CH_3}$$

The 2 position is the lowest number the methyl substituent can have (*rule 2*). Hyphens are used to separate position numbers and substituent names. The substituent is a 2-methyl group, no matter how the numbering is done (*rule 3*).

Rules 4 and 5 do not apply, since we have only one substituent.

Name the compound by combining the substituent number and name, *2-methyl*, with the name of the parent chain, *propane*. The name of this compound appears to be *2-methylpropane*. The number 2 is not necessary here, however, because there is no other place a methyl substituent can be attached to one propane. The compound is *methylpropane*. Notice that we write the name of the alkane as one word, *not* as two words, as in methyl propane.

EXAMPLE 11.3   **Naming a branched-chain hydrocarbon**

Use the IUPAC rules to name this branched-chain alkane:

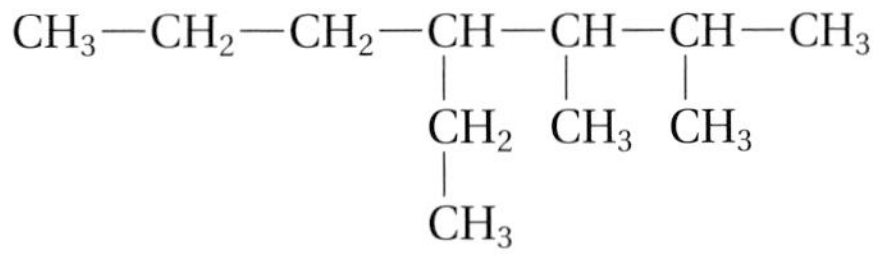

$$CH_3-CH_2-CH_2-CH-CH-CH-CH_3$$
$$\begin{array}{ccc} | & | & | \\ CH_2 & CH_3 & CH_3 \end{array}$$
$$\begin{array}{c} | \\ CH_3 \end{array}$$

SOLUTION

The longest chain in our example contains seven carbons, so the parent straight-chain alkane is heptane (*rule 1*).

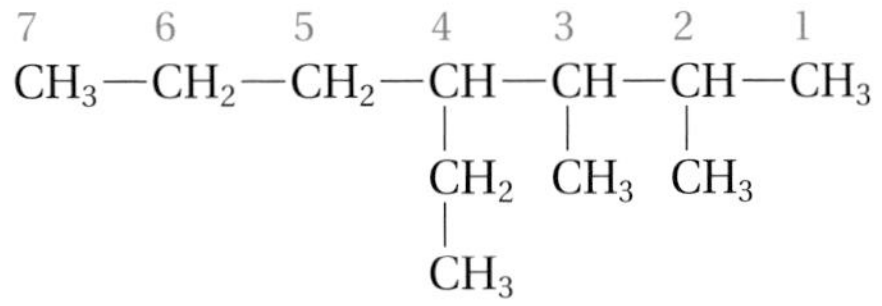

$$\overset{7}{C}H_3-\overset{6}{C}H_2-\overset{5}{C}H_2-\overset{4}{C}H-\overset{3}{C}H-\overset{2}{C}H-\overset{1}{C}H_3$$
$$\begin{array}{ccc} | & | & | \\ CH_2 & CH_3 & CH_3 \end{array}$$
$$\begin{array}{c} | \\ CH_3 \end{array}$$

Number the chain carbons to give the lowest possible numbers to the substituents (*rule 2*). This has already been done in the preceding structure. In this instance, the numbers go from right to left, which places the substituents at carbon atoms 2, 3, and 4. If the chain were numbered from left to right, the substituents would be at positions 4, 5, and 6—these are higher numbers and would violate the rule.

The example contains two methyl groups and one ethyl group. In this example the substituents and positions are 2-methyl, 3-methyl, and 4-ethyl (*rule 3*).

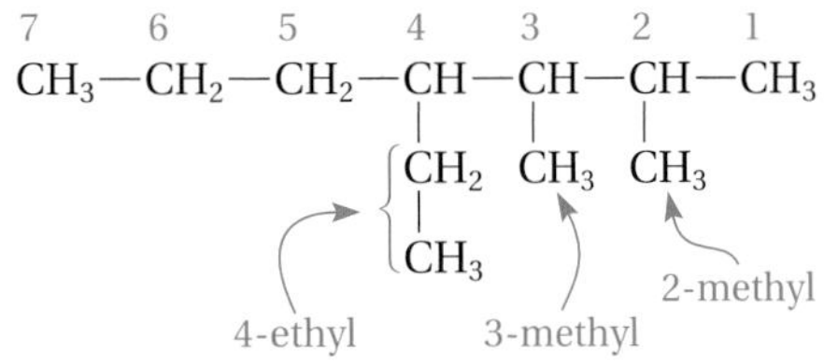

$$\overset{7}{C}H_3-\overset{6}{C}H_2-\overset{5}{C}H_2-\overset{4}{C}H-\overset{3}{C}H-\overset{2}{C}H-\overset{1}{C}H_3$$
$$\begin{array}{ccc} | & | & | \\ CH_2 & CH_3 & CH_3 \end{array}$$
$$\begin{array}{c} | \\ CH_3 \end{array}$$

4-ethyl   3-methyl   2-methyl

This example has two methyl substituents. To name these, group them together by using the prefix *di-* to make the word *dimethyl* (*rule 4*). Also designate the positions of the methyl groups. Indicate the name and numbers as *2,3-dimethyl*. Commas are used to separate numbers in naming organic compounds.

Now list the substituent names in alphabetical order, ignoring the prefixes in alphabetizing (*rule 5*). The substituents in the example are written *4-ethyl-2,3-dimethyl*.

Recalling that the parent name is heptane, combine the substituent names with the name of the parent to obtain the name of the compound. According to the IUPAC rules, the name of the compound is *4-ethyl-2,3-dimethylheptane*.

EXAMPLE 11.4 **Naming alkanes by the IUPAC system**

Name the following compound according to the IUPAC system:

$$
\begin{array}{c}
\mathrm{CH_3} \\
| \\
\mathrm{CH_3-CH_2-C-CH_3} \\
| \\
\mathrm{CH_2} \\
| \\
\mathrm{CH_2} \\
| \\
\mathrm{CH_3}
\end{array}
$$

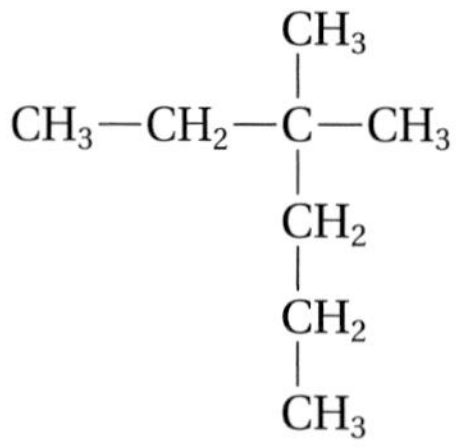

SOLUTION

Notice that the longest chain of carbons is *not* written in a straight line in the molecule.

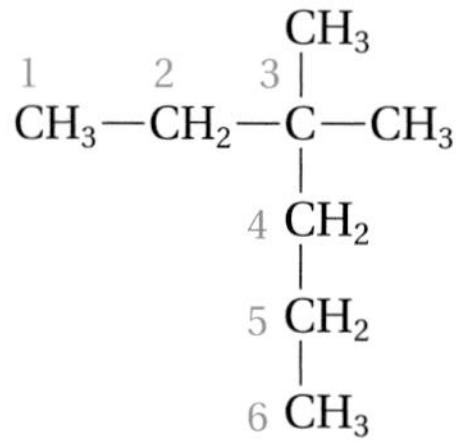

$$
\begin{array}{c}
\quad\quad\quad\quad \mathrm{CH_3} \\
{\scriptstyle 1 \quad\quad 2 \quad\quad 3}| \\
\mathrm{CH_3-CH_2-C-CH_3} \\
| \\
{\scriptstyle 4}\ \mathrm{CH_2} \\
| \\
{\scriptstyle 5}\ \mathrm{CH_2} \\
| \\
{\scriptstyle 6}\ \mathrm{CH_3}
\end{array}
$$

The longest chain has six carbon atoms, so the parent straight-chain alkane name is hexane. The chain is numbered so that the methyl substituents have the lowest possible numbers. When a carbon atom carries two identical substituents, the number is repeated. The compound is 3,3-dimethylhexane.

PRACTICE EXERCISE 11.4

Name the following compounds according to the IUPAC system.

(a) 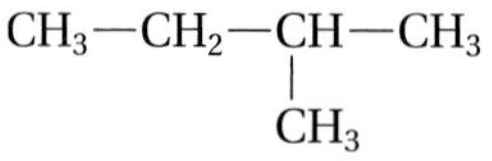 $\mathrm{CH_3-CH_2-CH-CH_3}$
$\quad\quad\quad\quad\quad\quad\ |$
$\quad\quad\quad\quad\quad\ \mathrm{CH_3}$

(b) 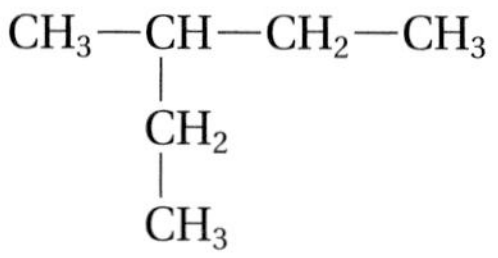 $\mathrm{CH_3-CH-CH_2-CH_3}$
$\quad\quad\quad\quad\ |$
$\quad\quad\quad\ \mathrm{CH_2}$
$\quad\quad\quad\quad\ |$
$\quad\quad\quad\ \mathrm{CH_3}$

(c) 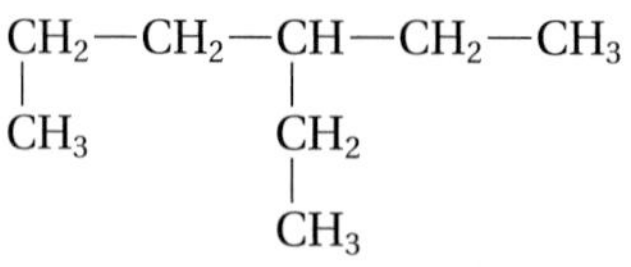 $\mathrm{CH_2-CH_2-CH-CH_2-CH_3}$
$\quad\ |\quad\quad\quad\quad\ |$
$\ \mathrm{CH_3}\quad\quad\ \mathrm{CH_2}$
$\quad\quad\quad\quad\quad\ |$
$\quad\quad\quad\quad\ \mathrm{CH_3}$

EXAMPLE 11.5 **Constructing a structural formula from a name**

Draw a condensed structural formula for 3,3-dimethylpentane.

SOLUTION

Draw five carbons and connect them with a line to show the single carbon-carbon bonds of pentane, the parent hydrocarbon. Number the carbons.

$$\underset{\substack{\\1\quad 2\quad 3\quad 4\quad 5}}{C-C-C-C-C}$$

Attach two methyl groups to the 3 position of the structure.

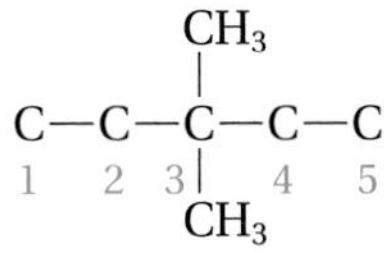

$$\begin{array}{c} CH_3 \\ | \\ C-C-C-C-C \\ \scriptstyle 1\ \ \ 2\ \ \ 3|\ \ \ 4\ \ \ 5 \\ CH_3 \end{array}$$

Complete the structure by filling in hydrogens so that each carbon has four single covalent bonds. This gives the structure of 3,3-dimethylpentane.

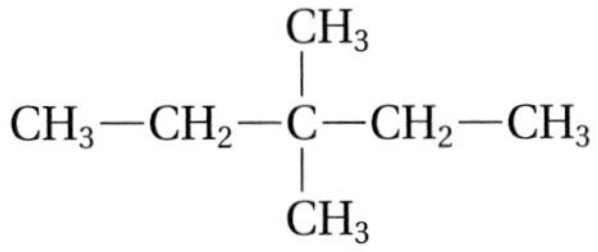

$$\begin{array}{c} CH_3 \\ | \\ CH_3-CH_2-C-CH_2-CH_3 \\ | \\ CH_3 \end{array}$$

3,3-Dimethylpentane

EXAMPLE 11.6　　Constructing a structural formula from a name

Draw a condensed structural formula for 2,2,4-trimethylpentane.

SOLUTION

As in the preceding example, draw and number the parent chain, which is pentane.

$$\underset{\substack{\\1\quad 2\quad 3\quad 4\quad 5}}{C-C-C-C-C}$$

According to the name, the structure has two methyl groups in the 2 position and one methyl group in the 4 position. It will be convenient to add only the carbons of the methyl groups and fill out the structure with hydrogens later.

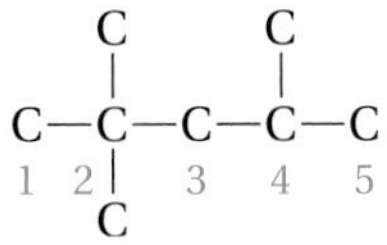

$$\begin{array}{c} C\qquad\quad C \\ |\qquad\quad | \\ C-C-C-C-C \\ \scriptstyle 1\ \ 2|\ \ \ 3\ \ \ 4\ \ \ 5 \\ C \end{array}$$

Now fill in the hydrogens to produce the completed structure.

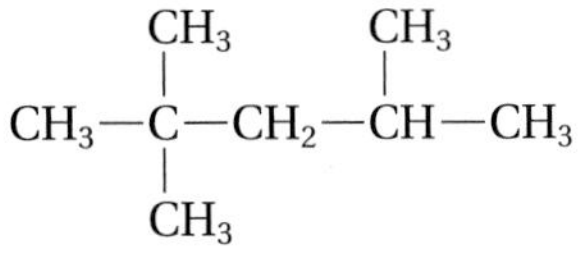

$$\begin{array}{c} CH_3\qquad\quad CH_3 \\ |\qquad\qquad\quad | \\ CH_3-C-CH_2-CH-CH_3 \\ | \\ CH_3 \end{array}$$

2,2,4-Trimethylpentane

PRACTICE EXERCISE 11.5

Draw a condensed structural formula for each of the following compounds:

(a) 2,3-dimethylhexane

(b) 3-ethylheptane

(c) 3-ethyl-2,4-dimethyloctane

11.6 Structural isomers

AIM: To name and draw structural isomers of hydrocarbons.

Focus

Structural isomers have the same molecular formula but different molecular structures and properties.

We can draw the structures of two or more alkanes that have the same molecular formula but different molecular structures. *Compounds that have the same molecular formula but different molecular structures are called* **structural isomers.** For example, there are two different molecules with the formula C_4H_{10}: butane and methylpropane. They are structural isomers of each other.

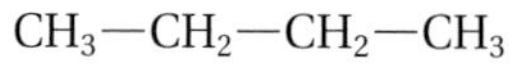

$$CH_3-CH_2-CH_2-CH_3 \qquad CH_3-\overset{\overset{\textstyle CH_3}{|}}{CH}-CH_3$$

Butane   Methylpropane
(C_4H_{10}) (C_4H_{10})

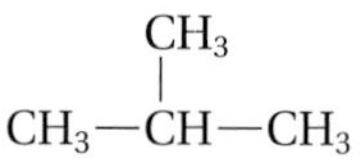

Structural isomerism possible with carbon compounds helps explain the immense number of organic compounds. Octane, C_8H_{18}, has 18 structural isomers. Decane, $C_{10}H_{22}$, has 75 isomers. The alkane with 20 carbon atoms, $C_{20}H_{42}$, has 336,319 isomers. All these isomers do not occur in nature, but any one of them could theoretically be made in the laboratory.

The existence of structural isomers is the reason organic chemists do not often use molecular formulas. To be clear, they need to use IUPAC names or structural formulas. Structural isomers have different physical properties, such as boiling points or melting points, and different chemical reactivities. For example, butane boils at 0 °C, and its structural isomer, methylpropane, boils at −10.2 °C. In general, the more highly branched the hydrocarbon structure, the lower is its boiling point compared with its other structural isomers. Pentane, C_5H_{12}, provides another example.

$$CH_3-CH_2-CH_2-CH_2-CH_3 \qquad CH_3-\overset{\overset{\textstyle CH_3}{|}}{CH}-CH_2-CH_3 \qquad CH_3-\overset{\overset{\textstyle CH_3}{|}}{\underset{\underset{\textstyle CH_3}{|}}{C}}-CH_3$$

Pentane   Methylbutane   Dimethylpropane
(C_5H_{12}) (C_5H_{12}) (C_5H_{12})
(bp 36 °C)   (bp 28 °C)   (bp 9 °C)

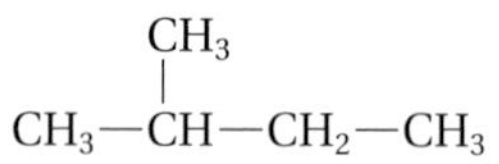

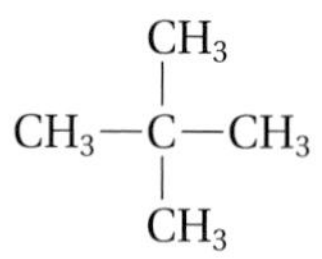

PRACTICE EXERCISE 11.6

There are five structural isomers with the molecular formula C_6H_{14}. Draw and name each one. (For convenience, you may wish to draw only the carbon skeleton for each structure.)

11.7 Cycloalkanes

AIM: To name and draw structural isomers for cycloalkanes.

Focus

Ends of carbon chains can join to form rings.

The ends of the carbon chains of some alkanes are joined. *Alkanes with joined ends are called* **cycloalkanes,** and the structures formed are called *rings* (Table 11.2). Many biologically important molecules such as cholesterol and the steroid hormones contain carbon rings. The parent cyclo-

Table 11.2 Names and Structures for Some Cycloalkanes

Name	Structure	Shorthand structure
cyclopropane	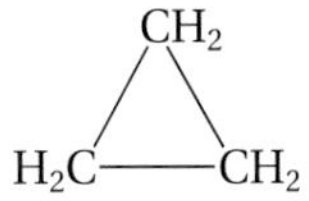 CH_2 / H_2C——CH_2	 △
cyclobutane	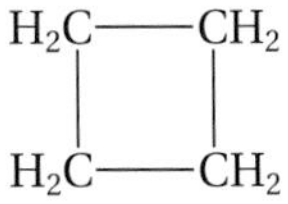 H_2C——CH_2 / H_2C——CH_2	□
cyclopentane	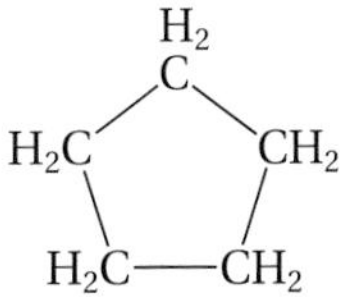 H_2C / H_2C CH_2 / H_2C——CH_2	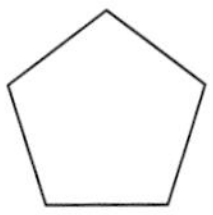 ⬠
cyclohexane	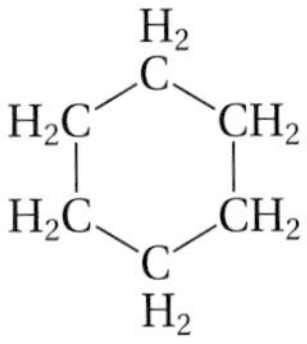 H_2C / H_2C CH_2 / H_2C CH_2 / H_2C	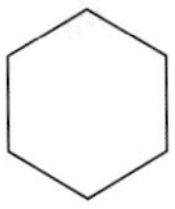 ⬡
cycloheptane	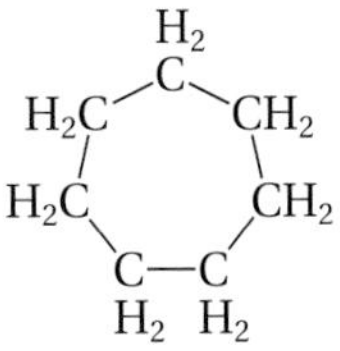 H_2C / H_2C CH_2 / H_2C CH_2 / C—C / H_2 H_2	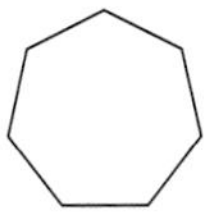 ⬡
cyclooctane	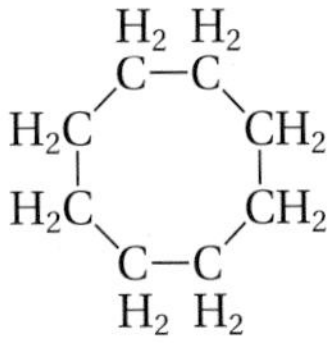 H_2 H_2 / C—C / H_2C CH_2 / H_2C CH_2 / C—C / H_2 H_2	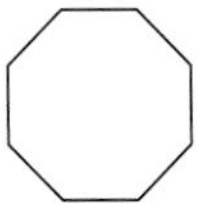 ⯃

alkanes are named somewhat like the parent straight-chain alkanes. In this case, however, the prefix *cyclo-* is attached to the name of the parent hydrocarbon.

Substituted cycloalkanes

Substituted cycloalkanes are numbered and named according to the same IUPAC rules as straight-chain alkanes. Here are a few examples:

Ethylcyclohexane
(a number is not
needed if there
is only one
substituent)

1,3-Dimethylcyclobutane

1,3,6-Trimethylcyclooctane

The simplest cycloalkane, cyclopropane, C_3H_6, is used as an inhalation anesthetic. Since cyclopropane is nonpolar, it is not very soluble in blood. However, sufficient cyclopropane dissolves and is transported by the blood to the brain, where it is transferred to brain cells. Cyclopropane is both flammable and explosive and must be handled carefully.

Rings containing from 3 to more than 30 carbons are found in nature. Five- and six-membered rings, the cyclopentanes and cyclohexanes, are the most abundant. Of the common unsubstituted cycloalkanes, cyclopropane and cyclobutane are gases at standard temperature and pressure. Cyclopropane is sometimes used as a surgical anesthetic. The remainder of the cycloalkanes are oily liquids.

True shapes of cycloalkanes

Carbon-carbon bonds are easier to break in smaller rings than in larger ones. In cyclopropane and cyclobutane rings, the C—C—C bond angles are 60 and 90 degrees, respectively. These rings are severely strained by this compression of the normal tetrahedral bond angle, which is 109.5 degrees. As a result, cyclopropane and cyclobutane rings are much less stable than cyclopentane or cyclohexane rings.

Placing carbon in a cycloalkane ring locks the carbon so that its freedom to rotate is severely restricted. The highly strained cyclopropane ring is locked into a flat, rigid shape (Fig. 11.6). Cyclobutane rings are also highly strained. In fact, it is impossible to make ball-and-stick models of cyclopropane and cyclobutane without using flexible springs instead of rigid sticks to join the ring carbons. The C—C—C bond angles in cyclopentane rings are nearly normal—about 108 degrees—and cyclopentane rings are not very strained. Cyclopentane and larger rings are also rather flexible; they readily "pucker" or bend. The puckered form of cyclopentane resembles an open envelope. If cyclohexane rings were flat, their C—C—C bond angles would be 120 degrees, and the ring would be strained, since 120 degrees is much larger than 109.5 degrees. In cyclohexane and larger rings, however, puckering reduces ring strain and produces normal tetrahedral bond angles of 109.5 degrees. One puckered shape that a cyclohexane molecule takes resembles a reclining chair. The cyclohexane ring can flip from one chair form to another through an intermediate shaped like a boat. The

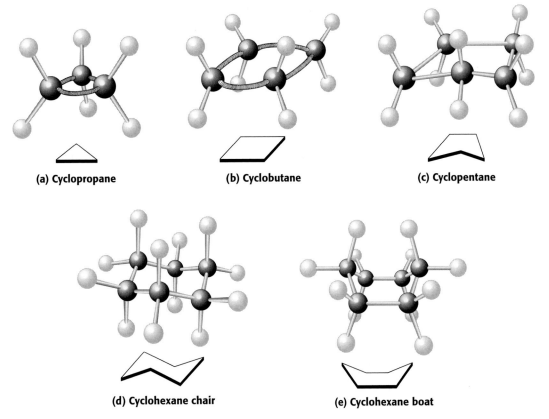

(a) Cyclopropane   (b) Cyclobutane   (c) Cyclopentane

(d) Cyclohexane chair   (e) Cyclohexane boat

Figure 11.6

Ball-and-stick models showing the true shapes of cycloalkane rings. Flexible connectors (springs) must be used to connect the ring carbon atoms in strained rings such as cyclopropane (a) and cyclobutane (b), since the holes in the carbon atoms of the models are drilled at angles of 109.5 degrees and the angles in the three-membered and four-membered rings are only 60 and 90 degrees, respectively. The cyclopentane ring (c) is puckered and shaped like an open envelope. Cyclohexane rings are also puckered and can take the shape of a chair (d) or a boat (e).

chair and boat forms of cyclohexane are in dynamic equilibrium, but the chair forms are more stable and are therefore highly favored over the boat.

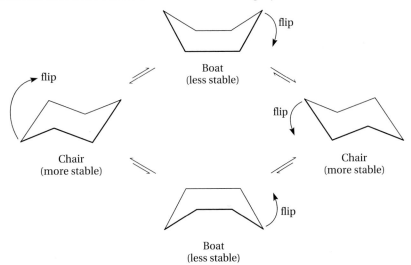

flip

Boat
(less stable)

flip

flip

Chair
(more stable)

Chair
(more stable)

flip

Boat
(less stable)

The reason the boat form of cyclohexane is less stable than the chair form is that ring hydrogens in the boat form infringe on each other's space, which is an unfavorable situation. These interfering hydrogens are called *flagpole hydrogens* because they resemble flagpoles on the bow and stern of the cyclohexane boat. In the chair form of cyclohexane, the ring hydrogens are in positions where they do not interfere with each other.

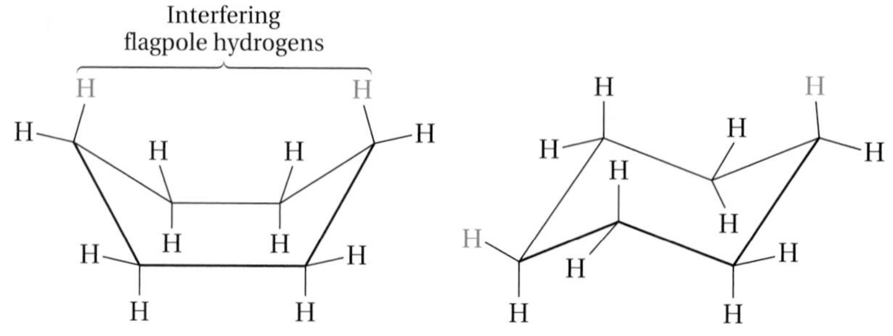

(a) Boat   (b) Chair

PRACTICE EXERCISE 11.7

Draw a carbon skeleton for each of the following:

(a) 1,2,3-trimethylcyclobutane

(b) methylcyclohexane   (c) 1,3-dimethylcyclopentane

11.8 Multiple bonds

AIM: To name and draw structural isomers of alkenes, cycloalkenes, and alkynes, identifying cis and trans geometric isomers.

The straight-chain alkanes, branched-chain alkanes, and cycloalkanes have—are *saturated* with—the largest number of hydrogens that making four bonds to carbon will permit. *For this reason, the straight-chain alkanes, branched-chain alkanes, and cycloalkanes are called* **saturated hydrocarbons.** In saturated hydrocarbons, all the carbon-carbon bonds in alkanes and cycloalkanes are single bonds. But multiple bonds between carbons also exist. *Organic compounds containing carbon-carbon double bonds are called* **alkenes.** *Alkenes contain fewer than the largest possible number of hydrogens in their molecular structures; for this reason, they are sometimes called* **unsaturated compounds.** This is the carbon-carbon double bond found in alkenes:

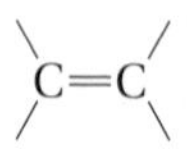

$$\backslash\!\!\!\diagup C = C \diagup\!\!\!\backslash$$

Naming alkenes

To name an alkene by the IUPAC system, we find the longest straight chain in the *molecule that contains the double bond*. This is the parent alkene. It gets the root name of the alkane with the same number of carbons but the ending -*ene* rather than -*ane*. For example, the alkene with two carbons is called *ethene;* the alkene with three carbons is called propene.

$$ \underset{\substack{\text{Ethene} \\ \text{(ethylene)}}}{\overset{\displaystyle \underset{H}{\overset{H}{\diagdown}} C = C \underset{H}{\overset{H}{\diagup}}}{}} \qquad \underset{\substack{\text{Propene} \\ \text{(propylene)}}}{CH_3 - \overset{H}{\underset{|}{C}} = \overset{H}{\underset{|}{C}} - H} $$

Ethene and propene, the smallest alkenes, are often called by the common names *ethylene* and *propylene.*

In naming alkenes, we number the longest chain containing the double bond so that the positions of the carbon atoms of the double bond get the lowest possible numbers. Substituents on the chain are named and numbered the same way as for the alkanes. Here are some examples of the structures and IUPAC names of simple alkenes.

$$ \underset{\text{1-Butene}}{H - \overset{H}{\underset{|}{C}} = \overset{H}{\underset{|}{C}} - CH_2 - CH_3} \qquad \underset{\text{2-Butene}}{CH_3 - \overset{H}{\underset{|}{C}} = \overset{H}{\underset{|}{C}} - CH_3} \qquad \underset{\text{4-Methyl-2-pentene}}{CH_3 - \overset{CH_3}{\underset{|}{CH}} - \overset{H}{\underset{|}{C}} = \overset{H}{\underset{|}{C}} - CH_3} $$

(1-butene and 2-butene are structural isomers of C_4H_8)

Double bonds are also found in ring compounds. The same change in the parent cycloalkane name from -*ane* to -*ene* applies in naming the cycloalkenes. Some examples of the cycloalkenes are given here.

Cyclobutene Cyclopentene Cyclohexene

Shape of the double bond

There is no rotation about carbon-carbon double bonds. Experimental evidence shows that the four bonds that project from a pair of double-bonded carbons lie in a plane and are 120 degrees apart (Fig. 11.7). This is the maximum separation that can be obtained without breaking bonds, and it is what would be predicted by the electron pair repulsion theory.

Recall that we used orbital hybridization to obtain a description of the shape of the methane molecule. The orbital description of the carbon-carbon double bond in ethene relies on hybridization as well, but the hybridization of carbon in ethene is different from that in methane. Use Figure 11.8 to visualize the steps leading to the structure of ethene. Remember that each orbital represents one electron.

In ethene, the $2s$ orbital of carbon mixes with *two* $2p$ orbitals to give three hybrid atomic orbitals called sp^2. Note that the number of hybrid orbitals is always the same as the number of orbitals mixed. Each hybrid orbital is separated from the other two by 120 degrees. Overlap of one of the sp^2 hybrid orbitals of one carbon with that of another results in formation

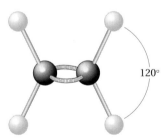

Figure 11.7
Ball-and-stick model of ethene (common name ethylene). Rotation about the double bond is prohibited. All the atoms are in the plane of the page.

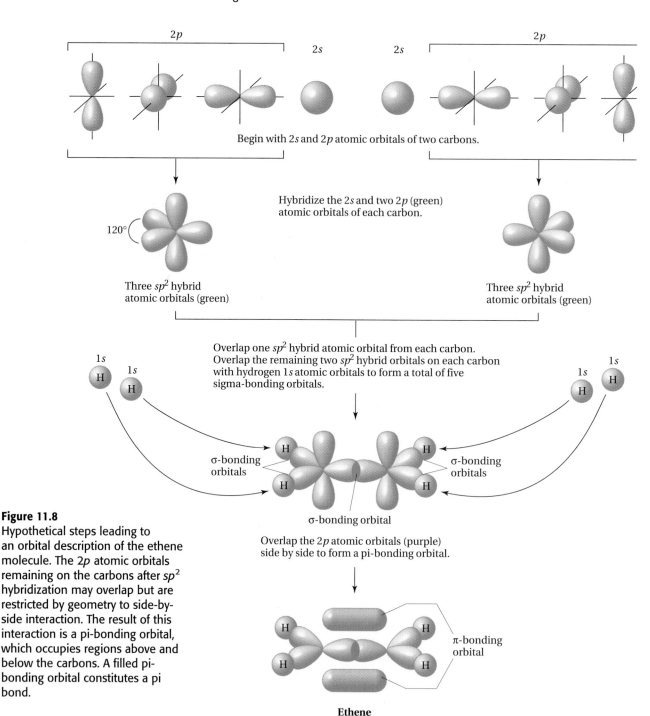

Figure 11.8
Hypothetical steps leading to an orbital description of the ethene molecule. The 2p atomic orbitals remaining on the carbons after sp^2 hybridization may overlap but are restricted by geometry to side-by-side interaction. The result of this interaction is a pi-bonding orbital, which occupies regions above and below the carbons. A filled pi-bonding orbital constitutes a pi bond.

of a carbon-carbon sigma bond. Similar overlap of 1s orbitals of hydrogen with the remaining sp^2 orbitals on the carbons gives four carbon-hydrogen sigma bonds. Each carbon still has a 2p orbital. These 2p orbitals overlap side by side to form a type of bond that we have not seen before. *A covalent bond formed by side-by-side overlap of p atomic orbitals is called a* **pi bond** (the Greek letter pi is π).

The completed bonds of the ethene molecule consist of five sigma bonds and one pi bond. Ethene is a molecule held together by five sigma bonds and one pi bond.

Like a sigma bond, the pi bond is a two-electron covalent bond. In writing structural formulas, we draw the pi bond as a line like any other covalent bond. Pi bonds are weaker than sigma bonds, however; so in chemical reactions of carbon-carbon double bonds the pi bond is broken in preference to the sigma bond.

Geometric isomerism

Geometric isomerism arises from a difference in orientation of atoms that leads to a difference in molecular shapes. Many biological reactions are shape-sensitive. For example, the odor of a compound depends on both the size and the shape of its molecule. A change from a *cis* to a *trans* form changes the shape of a molecule and may very well change its odor.

Trans
trans (Latin):   across
cis
cis (Latin):   on this side

The lack of rotation around carbon-carbon double bonds has an important structural implication. Looking at the structure of 2-butene in Figure 11.9, we see that there are two possible arrangements of the methyl groups with respect to the rigid double bond. *One arrangement has the substituents on opposite sides of the double bond; this is called the* **trans configuration.** *The other arrangement has the methyl groups on the same side of the double bond; this is called the* **cis configuration.** Trans-*2-butene and* cis-*2-butene are* **geometric isomers**—*they have the same molecular formulas but differ in the geometry of their substituents.* Like other structural isomers, isomeric 2-butenes are distinguishable by their different physical and chemical properties. The groups on the carbons of the double bond need not be the same. Geometric isomerism is possible whenever there is one hydrogen and one substituent on each carbon of the double bond. For example:

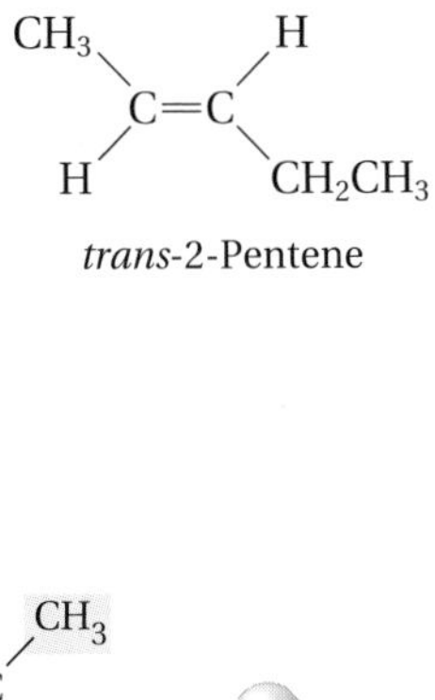

trans-2-Pentene

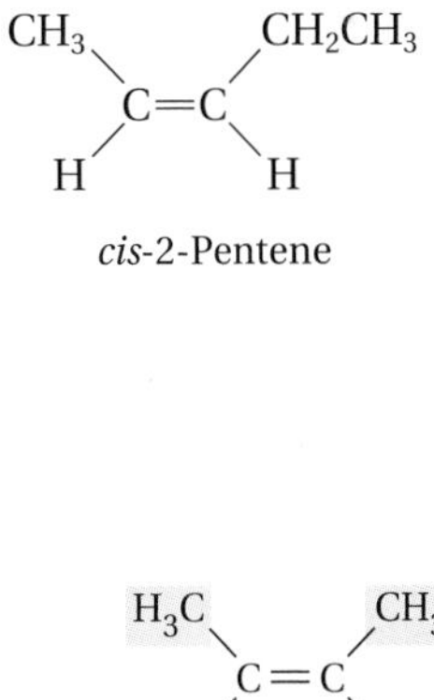

cis-2-Pentene

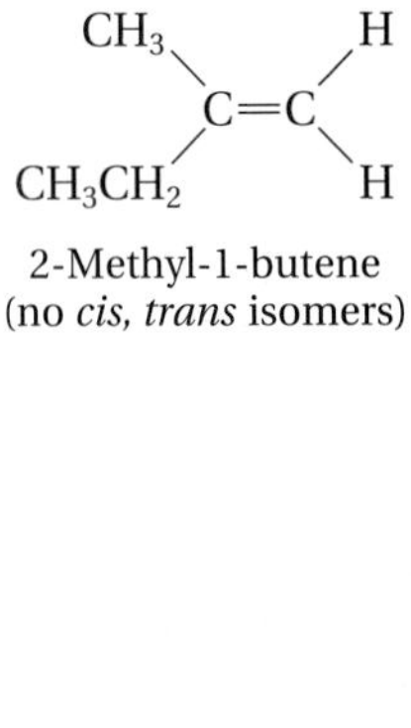

2-Methyl-1-butene
(no *cis, trans* isomers)

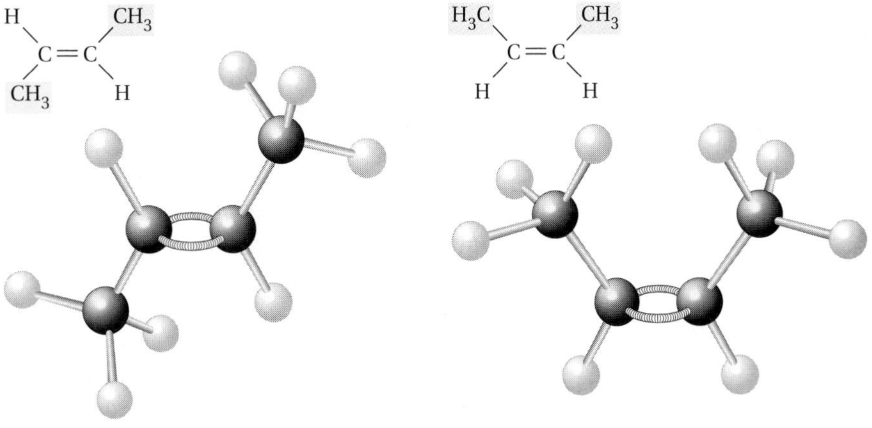

Figure 11.9
Structural formulas and ball-and-stick models of the geometric isomers of 2-butene.

(a) *trans*-**2-Butene (bp 1 °C)**

(b) *cis*-**2-Butene (bp 4 °C)**

PRACTICE EXERCISE 11.8

Draw structural formulas for the following alkenes. If a compound has geometric isomers, draw both the *cis* and the *trans* forms.

(a) 1-pentene   (b) 2-hexene   (c) 2-methyl-2-hexene

(d) 2,3-dimethyl-2-butene

Alkynes

Organic compounds containing carbon-carbon triple bonds are called **alkynes.** Like alkenes, alkynes are unsaturated compounds. This is the carbon-carbon triple bond found in alkynes:

$$-C\equiv C-$$

Alkynes are quite rare in nature. The simplest alkyne is the gas ethyne, C_2H_2. The common name for ethyne is *acetylene*—the fuel burned in oxyacetylene torches used in welding. The single bonds that extend from the carbons involved in the carbon-carbon triple bond of ethyne are separated by the maximum bond angle, 180 degrees (Fig. 11.10). Ethyne is a linear molecule.

The development of an orbital picture for ethyne requires us to go through the same steps that we used for methane and ethene: hybridization of carbon atomic orbitals and orbital overlap. As with methane and ethene, remember that each orbital represents one electron. The description of ethyne that best fits the experimental data is obtained if the $2s$ atomic orbital of carbon is mixed with *only one* of the three available $2p$ atomic orbitals (Fig. 11.11). The result of this mixing is two sp atomic orbitals; two $2p$ orbitals remain unused. Overlap of one of the sp hybrid orbitals of one carbon with that of another carbon produces a carbon-carbon sigma bond. The remaining sp orbital on each carbon overlaps with the $1s$ orbital of a hydrogen to form two more sigma-bonding orbitals. The remaining pair of p atomic orbitals on each carbon now overlaps side by side to form *two* pi bonds. The bonding in ethyne consists of three sigma bonds and two pi bonds. Each pi bond has two regions in which the probability of finding the bonding electrons is high, so the pi bonds in ethyne resemble four sausages tightly packed around the carbon-carbon sigma bond.

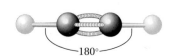

Figure 11.10
Ball-and-stick model of ethyne (common name acetylene).

11.9 Aromatic compounds

AIM: To describe the bonding, structure, and chemical properties of benzene and other simple aromatic and fused-ring aromatic compounds.

Focus

Benzene is the simplest arene.

The organic compounds we have seen thus far are all *aliphatic.* **Aliphatic** *compounds contain only carbon-carbon single, double, or triple bonds.* Now we will turn to the group of carbon compounds known as *arenes.* **Arenes** *are a class of carbon compounds that have structures based on the parent molecule benzene,* C_6H_6. All arenes contain a benzene ring or a

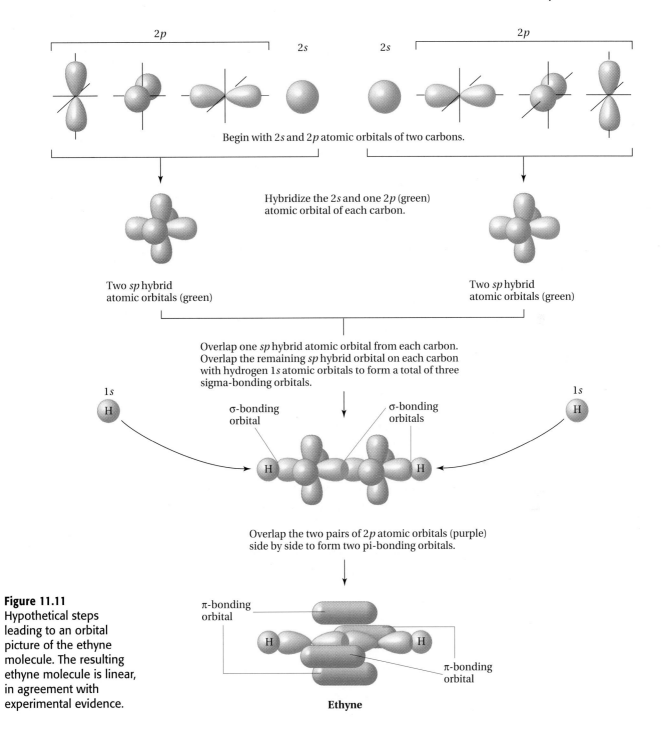

Figure 11.11
Hypothetical steps leading to an orbital picture of the ethyne molecule. The resulting ethyne molecule is linear, in agreement with experimental evidence.

$2p$

$2s$

$2s$

$2p$

Begin with $2s$ and $2p$ atomic orbitals of two carbons.

Hybridize the $2s$ and one $2p$ (green) atomic orbital of each carbon.

Two sp hybrid atomic orbitals (green)

Two sp hybrid atomic orbitals (green)

Overlap one sp hybrid atomic orbital from each carbon. Overlap the remaining sp hybrid orbital on each carbon with hydrogen $1s$ atomic orbitals to form a total of three sigma-bonding orbitals.

$1s$

H

σ-bonding orbital

σ-bonding orbitals

$1s$

H

H

H

Overlap the two pairs of $2p$ atomic orbitals (purple) side by side to form two pi-bonding orbitals.

π-bonding orbital

H

H

π-bonding orbital

Ethyne

related system of rings. **Arenes** *are also called* **aromatic compounds** *because the earliest known examples of these compounds had pleasant odors.* Benzene, C_6H_6, is the simplest arene. The benzene molecule has a six-membered carbon ring with a hydrogen attached to each carbon. This leaves one electron on each carbon free to participate in a double bond.

"Connecting the dots" to complete the structural formula of benzene, we see that there are *two different ways* to form three double bonds:

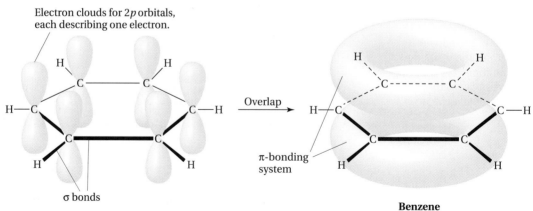

One extreme (single bond)

Another extreme (double bond)

or

These two structural formulas only describe the extremes of electron sharing between any two adjacent carbons in benzene. One extreme is a normal single bond; the other is a normal double bond. *These two extremes of bonding—single and double bonds—are called* **resonance structures.** When resonance structures can be drawn for a certain molecule or ion, that molecule or ion is more stable than others for which only one structural formula can be drawn. As a result, benzene resists chemical reactions to which any ordinary alkene easily succumbs.

A more accurate description of bonding in benzene is an average picture or **resonance hybrid** *of these forms.* (The offspring of a jackass and a mare, two extreme forms of the horse family, is a hybrid, a neuter mule. The hybrid resembles each of its parents but also has its own unique characteristics.) When organic chemists talk about benzene and related arenes, they usually call this hybrid bonding *aromatic character* or *aromaticity.* The benzene ring is a perfectly flat molecule; ring puckering or bending would make it less stable.

The orbital description of benzene is helpful in furthering our understanding of the low chemical reactivity of benzene and related arenes compared with alkenes such as ethene. As mentioned previously, all orbitals represent one electron. The carbons of benzene are sp^2 hybridized, and the formation of the carbon-hydrogen and carbon-carbon sigma bonds of benzene is similar to their formation in ethene. Once again, one $2p$ atomic orbital is available on each carbon. In benzene, however, each of these p orbitals overlaps side by side with *both* of its neighbors (Fig. 11.12). The

Electron clouds for $2p$ orbitals, each describing one electron.

Overlap

π-bonding system

σ bonds

Benzene

Figure 11.12
Side-by-side overlap of $2p$ atomic orbitals results in an extensive pi-bonding orbital in the benzene molecule.

result is a pi bond in which there is a high probability of finding the electrons in doughnut-shaped regions above and below the carbon ring. The orbital picture shows that the pi electrons in benzene are no longer identifiable with any particular carbon. Thus the electrons in the pi-bonding orbitals of benzene are said to be *delocalized*. This bonding is quite different from that in the carbon-carbon double bond of an alkene. In the double bond of an alkene, the pi electrons are *localized* above and below the two carbons in the bonding orbitals.

The low chemical reactivity of benzene and related compounds compared with that of alkenes results from their extensive pi-bonding systems. The more the electrons in pi bonds are smeared over a molecule, the lower the energy of the bonding electrons and the more stable the molecule. Therefore, the pi bonds of benzene are less chemically reactive than those of ordinary alkenes.

When drawing an abbreviated structure of benzene, chemists often inscribe either a complete or dashed circle in the ring to indicate the delocalized pi-bonding electrons. The inscribed circle is useful for showing the aromatic character of benzene and related compounds, but it does not give any indication of *how many* electrons are involved. For this reason, we will stick with the time-honored benzene structure shown at the right.

The benzene ring is found in many compounds used medicinally: the analgesic aspirin, the decongestant ephedrine, the painkiller and cough suppressant codeine, and the antibiotic sulfa drugs.

Alternative representations of benzene

Naming arenes

Most compounds containing alkyl substituents attached to the benzene ring are named as derivatives of benzene. Here are some examples of benzene derivatives with one substituent:

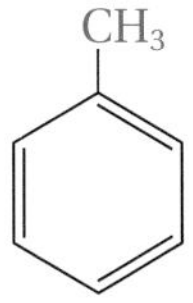

Methylbenzene
(toluene)

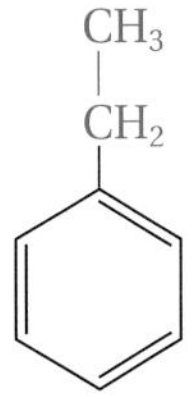

Ethylbenzene

Another name for methylbenzene is *toluene.* Sometimes the benzene ring is named as a substituent on an alkane or cycloalkane. In such instances, the benzene ring is called a *phenyl group,* as in 3-phenylhexane or phenyl-cyclohexane.

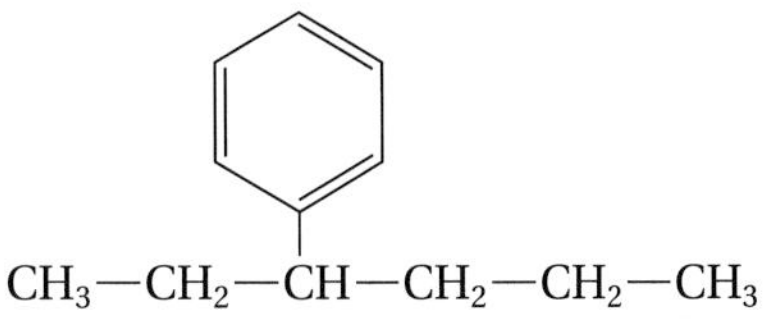

$CH_3-CH_2-CH-CH_2-CH_2-CH_3$

3-Phenylhexane

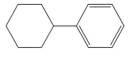

Phenylcyclohexane

PRACTICE EXERCISE 11.9

Name the following compounds:

(a) [benzene ring]—CH₂CH₂CH₃   (b) CH₃—CH—CH₃ [attached to benzene ring]

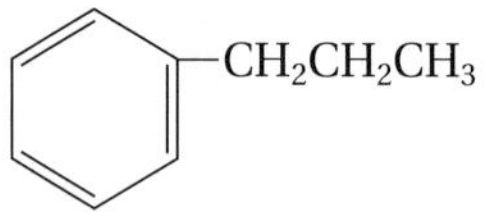

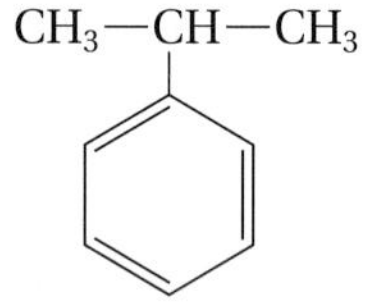

(c) CH₃—CH—CH₂—CH—CH—CH₃ [with CH₃ groups above the first and third CH, and benzene ring below]

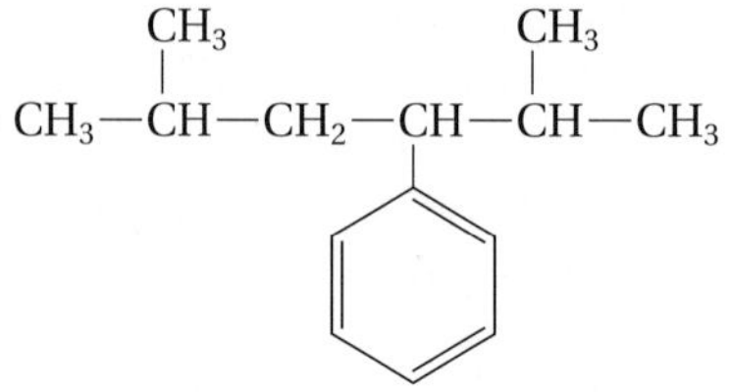

Structural isomers of benzene derivatives

There are derivatives of benzene with two substituents of the same kind. Such derivatives are called *disubstituted benzenes.* There are three different structural isomers of the liquid aromatic compound dimethylbenzene, $C_6H_4(CH_3)_2$. The dimethylbenzenes are also called *xylenes.* The difference between the structural isomers of dimethylbenzene is the relative positions of the methyl groups on the benzene ring. We can distinguish among the possible positions of the methyl groups by assigning the methyl groups the numbers 1,2; 1,3; or 1,4. It is also permissible in naming disubstituted benzenes to use the prefixes *ortho-, meta-,* and *para-* (abbreviated *o, m,* and *p*) in place of numbers. *Ortho-* means that the substituents are on adjacent carbons on the benzene ring. *Meta-* substituents are separated by one ring carbon, and *para-* substituents are separated by two ring carbons. The physical properties of the structural isomers of dimethylbenzene are different, as indicated by their boiling points.

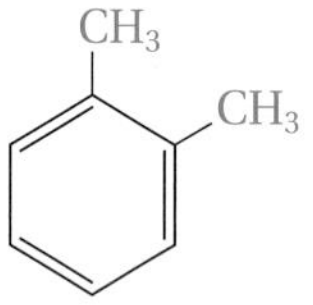

1,2-Dimethylbenzene
(*o*-dimethylbenzene,
o-xylene)
(bp 144 °C)

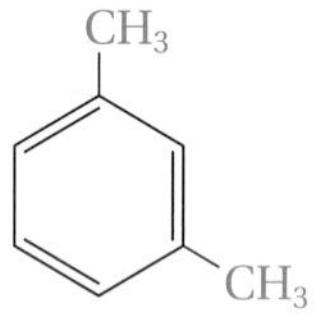

1,3-Dimethylbenzene
(*m*-dimethylbenzene,
m-xylene)
(bp 139 °C)

1,4-Dimethylbenzene
(*p*-dimethylbenzene,
p-xylene)
(bp 138 °C)

Fused-ring aromatics

Fused-ring (polycyclic) aromatic compounds *are derivatives of benzene in which carbons are shared between benzene rings.* Naphthalene, used in mothballs, is the simplest fused-ring aromatic. Anthracene is found in anthracite coal, and the carbon skeleton of phenanthrene forms the basic structure of the steroids, among which are the sex hormones. Benzpyrene is

Hydrocarbons and Health

Some hydrocarbons have medical applications. Cyclopropane is an anesthetic. Mineral oil is a mixture of hydrocarbons of high molar mass. Petroleum jelly (Vaseline is one brand name) is a semisolid mixture of solid and liquid aliphatic hydrocarbons of high molar mass. Mineral oil and petroleum jelly (see figure) are used to soften skin and protect it from exposure to water. The hydrocarbon components of mineral oil and petroleum jelly have low vapor pressures, and these products are essentially odorless. Mineral oil is indigestible and is sometimes used as a mild laxative because of its lubricating properties.

Petroleum jelly, gasoline, paint thinners, and mineral oil are all examples of hydrocarbons found in the home.

Many polycyclic aromatic compounds are carcinogenic (cancer-causing). Smoking is a chief source of these carcinogenic chemicals. More than 300 carcinogens have been identified in cigarette smoke. One such compound is the hydrocarbon benzpyrene. This carcinogen is converted by the body into highly reactive compounds that react with DNA molecules in cell nuclei to cause the transformation of normal cells into cancerous ones. Benzene is implicated as a causative agent in leukopenia (low white cell count) and in certain types of leukemia. Toluene poses less of a hazard than benzene, although prolonged exposure to its vapors can cause headaches, nausea, and vomiting. Breathing toluene vapors also can produce a narcotic effect. The permissible exposure of chemical workers to benzene and other aromatic hydrocarbons is regulated by law.

Every year hospitals admit patients who have swallowed gasoline. The saturated aliphatic hydrocarbons in gasoline are not particularly toxic, since they are chemically unreactive in our bodies. The main threat to health from the ingestion of gasoline is from toxic additives. Gasoline in the lungs can damage lung tissue, causing the symptoms of pneumonia. Anyone who swallows gasoline or other liquid hydrocarbons should not be made to vomit. That would increase the chance of the hydrocarbon getting from the stomach into the lungs. The greatest threat to health from gasoline is its high flammability. Gasoline should never be used near a source of flames or sparks.

an aromatic hydrocarbon that poses a threat to health. A Closer Look: Hydrocarbons and Health describes some applications and hazards of aliphatic and aromatic hydrocarbons to health.

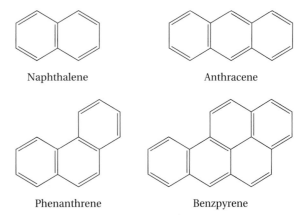

Naphthalene   Anthracene

Phenanthrene   Benzpyrene

11.10 Sources of hydrocarbons

AIM: To characterize hydrocarbons in terms of their source and use.

Petroleum and coal are the sources of most hydrocarbons used in the home, in industry, and in the laboratory. Major carbon compounds for organic chemicals and energy needs are supplied by petroleum, natural gas, and coal. Petroleum, or crude oil, originated in marine life buried in the sediments of the oceans millions of years ago. A combination of heat and pressure, and perhaps the action of bacteria, changed this residue of marine life into petroleum and gas. Vast deposits of crude oil were discovered in the United States in 1859 and in the Middle East in 1908. Crude oil is a complex mixture of hydrocarbons, mostly alkanes and cycloalkanes; it also includes small amounts of other organic compounds that contain sulfur, oxygen, and nitrogen. The components of crude oil can be separated by *refining*. **Petroleum refining** *consists of distilling crude oil to divide it into fractions of hydrocarbons according to boiling point.* Fractional distillation separates petroleum into "fractions" of hydrocarbons: butane (gas), octane (gasoline), dodecane (kerosene), fuel oil, lubricating oil, paraffin wax, and tar—a thick black residue. The distillation fractions, each of which contains several different hydrocarbons, are shown in Table 11.3. Modern petroleum refining also employs **cracking**—*a controlled process by which catalysts are used to break down or rearrange natural hydrocarbons into more useful molecules.* The hydrocarbon composition of gasoline affects its performance as a fuel, as discussed in A Closer Look: Octane Ratings of Gasoline.

Natural gas is found overlying deposits of oil or in separate pockets in rock. Natural gas is an important source of alkanes of low molecular mass. A typical natural gas composition is methane 80%, ethane 10%, propane 4%, and butane 2%. The balance is nitrogen and higher-molecular-mass hydrocarbons. Most natural gas contains a small percentage of helium and is an important source of this noble gas. Natural gas is used as a fuel and is distributed through pipes; the individual gases may be liquefied and sold as fuel for heating, portable cookers, and welding equipment.

Table 11.3 Fractions Obtained from Crude Oils*

Fraction	Composition of carbon chains	Boiling range (°C)
natural gas	C_1 to C_4	below 20
petroleum ether (solvent)	C_5 to C_6	30 to 60
naphtha (solvent)	C_7	60 to 90
gasoline	C_6 to C_{12}	75 to 200
kerosene	C_{12} to C_{15}	200 to 300
fuel oils, mineral oil	C_{15} to C_{18}	300 to 400
lubricating oil, petroleum jelly, greases, paraffin wax	C_{16} to C_{24}	over 400

*The undistilled residue is asphalt.

A Closer Look

Octane Ratings of Gasoline

When we fill the tank of an automobile, we must decide what octane of gasoline to buy. Today's automobile manufacturers generally recommend the octane rating that will give good engine performance. If our engine is out of tune, however, or if we choose an octane rating that is lower than recommended, we may experience engine knock. The engine may ping if we accelerate too rapidly or knock and lose power when the automobile goes up a steep hill. The knock characteristics of a gasoline depend on the structure of the hydrocarbons in the fuel. Straight-chain hydrocarbons cause engine knock and are relatively poor fuels for today's automobiles. Branched-chain hydrocarbons have better antiknock characteristics. The octane or antiknock rating is an arbitrary scale defined by the petroleum industry. The knock created by burning heptane in a standard engine is assigned a value of zero octane; the knock from 2,2,4-trimethylpentane (isooctane) is assigned a value of 100 octane. The octane rating of a gasoline is equal to the percentage of isooctane in a heptane-isooctane mixture that performs the same way in the standard engine (see figure). For example, a gasoline giving the same knock as a mixture of 90% isooctane and 10% heptane has an octane

Gasoline is a mixture of many different hydrocarbons. The octane rating of a gasoline, shown on the delivery pump, gives an indication of the engine performance you may expect using that grade of gasoline.

rating of 90. The percentage of branched-chain hydrocarbons in gasoline can be increased by catalytic cracking in petroleum refining (see Sec. 11.11). High-octane gasoline is produced by reforming the structure of the C_4–C_{18} straight-chain hydrocarbons to give a highly branched C_6–C_{10} fraction with octane ratings of 90 to 110. Gasoline available today is usually 87 octane, 89 octane, and 92 octane.

Geologists think that coal originated about 300 million years ago. When huge tree ferns and mosses growing in swampy tropical regions died, they formed thick layers of decaying vegetation, which were eventually covered by soil and rock. The pressure of soil, rocks, and earth with the heat from the Earth's interior eventually turned these remains into coal (Fig. 11.13a). Coal consists largely of condensed ring compounds of very high molecular mass. These compounds have a high proportion of carbon compared with hydrogen. Coal is obtained from underground and surface mines, where it is found in seams (Fig. 11.13b) from 1 to 3 m thick. In the United States, coal mines are usually less than 100 m deep, and much of the coal is so close to the surface that it is strip-mined. Many coal mines in Europe and other parts of the world go down 1000 to 1500 m. The hydrocarbons in coal are mostly aromatic—benzene, naphthalene, anthracene, and a host of others. Distillation of coal yields coal gas, coke, ammonia, and coal tar. Coal gas consists mainly of hydrogen, methane, and carbon monoxide, all of which are flammable and good fuels. Coke, the solid residue of almost pure car-

(a)

(b)

(c)

Figure 11.13
Shown is (a) fossil tree-fern leaf imprint in coal, (b) an exposed coal seam, and (c) a sample of coal tar.

bon, is also used as a fuel in many industrial processes. Ammonia is often converted to ammonium sulfate for use as a fertilizer. Coal tar (Fig. 11.13c) can be further distilled to yield many useful chemicals, including benzene, toluene, naphthalene, phenol, and pitch.

Over the past few years, growth in the world's demand for petroleum has outstripped the discovery of new deposits. Since petroleum is a nonrenewable resource, there eventually will be insufficient quantities to support the world's energy and other demands. Scientists and engineers are working to develop replacement sources of energy such as the sun, winds, tides, and geothermal springs. As we mentioned in Section 10.6, nuclear fusion could provide a clean and abundant source of energy.

Some of the clothes you wear and the music you play (CDs and tapes) are made from the same finite supply of hydrocarbons that you burn in your car and use to heat your home. Does this give you cause for concern?

11.11 Properties of hydrocarbons

AIM: To characterize the physical and chemical properties of hydrocarbons.

Focus

Hydrocarbons float on water and burn in air.

The public transit system of the city of Santa Fe, New Mexico, operates entirely on compressed natural gas.

Anyone who has seen footage of the oil slick from a wrecked oil tanker can observe two physical properties common to all hydrocarbons: They are insoluble in water and less dense than water (they float). If the tanker has been involved in a collision, a chemical property common to all hydrocarbons often is evident: combustion. Under controlled conditions, however, the burning of hydrocarbons enormously benefits people. Of the more than 500,000 tons of crude oil and 560 million tons of coal produced yearly in the United States, most is burned to produce energy. Methane, the major constituent of natural gas, burns with a hot, clean flame:

$$CH_4(g) + 2O_2(g) \rightarrow CO_2(g) + 2H_2O(g) + heat$$

Propane and butane are also good heating fuels and are often sold in liquid

Figure 11.14
Propane is the fuel that is also known as *bottled gas*.

form under pressure in tanks as liquid petroleum gas (LPG), as shown in Figure 11.14.

A sufficient supply of oxygen is necessary to oxidize a hydrocarbon fuel completely to carbon dioxide and obtain the greatest amount of heat. Complete combustion of a hydrocarbon gives a blue flame. Incomplete combustion gives a yellow flame, due to the formation of small, glowing carbon particles that are deposited as soot when they cool. Carbon monoxide, an odorless toxic gas, is also formed along with carbon dioxide and water during incomplete combustion. Generally, unsaturated compounds, such as the aromatic compounds found in coal, are sootier fuels than the alkanes.

The burning of hydrocarbons found in coal and petroleum are not without environmental hazards, as described in A Closer Look: The Greenhouse Effect, on the following page. For the effect of the products of incomplete combustion of hydrocarbons on people's health, recall Claudine from the Case in Point for this chapter. Claudine was the victim of carbon monoxide poisoning. In the Follow-up, below, we will see how she was poisoned and how she was treated for her condition.

FOLLOW-UP TO THE CASE IN POINT: Carbon monoxide poisoning

Claudine's house has a furnace that burns natural gas piped in from the local gas company. The incomplete combustion of the gas produces highly toxic but odorless carbon monoxide. Ordinarily, the carbon monoxide and other combustion products are carried out of the house by a flue pipe. Unknown to Claudine, the flue pipe worked loose from the furnace, permitting carbon monoxide to escape into the house. Before the incident in which she was hospitalized, Claudine had experienced increasingly severe headaches caused by the increasing level of carbon monoxide in her house. The paramedics who transported Claudine to the hospital recognized the characteristic flushed skin of carbon monoxide poisoning, a result of interference of the gas with the body's use of oxygen. Carbon monoxide combines with the hemoglobin of the blood, making it useless as an oxygen carrier. Fortunately, the carbon monoxide can be replaced by oxygen. The paramedics administered oxygen, and by the time Claudine arrived at the hospital, her condition had significantly improved. Claudine suffered no permanent effects from her ordeal. Fortunately, she was discovered before she suffered permanent damage to her nervous system or brain or

even death. Others are not so lucky. Every year hundreds of people lose their lives as a result of carbon monoxide poisoning from faulty heating devices.

Less dramatic incidents of carbon monoxide poisoning are experienced by millions of people every day. Cigarette smoke is a source of low-level carbon monoxide poisoning. A burning cigarette produces a great deal of carbon monoxide. It can take several hours to replace the carbon monoxide in a smoker's blood after only one cigarette. Low levels of carbon monoxide poisoning can cause headaches, dizziness, nausea, and sluggishness. The reduced amount of oxygen in the blood of smokers makes the heart pump harder to supply oxygen to tissues. As a result, smoking contributes to the incidence of heart disease and heart attacks. Even nonsmokers can suffer from low-level carbon monoxide poisoning. The air around traffic jams often reaches dangerously high concentrations of carbon monoxide. Automobiles recycle gases from the combustion of gasoline in order to burn more of the carbon monoxide to carbon dioxide. This has helped to lower the level of carbon monoxide in automobile exhaust emissions.

A Closer Look

The Greenhouse Effect

Since the onset of the industrial revolution in the 19th century, human activities have been adding vast amounts of carbon dioxide into the Earth's atmosphere. People burn fossil fuels, petroleum and coal, for energy. Grasslands and forests are being burned to clear land for crops. Each year, more than 8 billion (8×10^9) metric *tons* of carbon dioxide pour into the Earth's atmosphere (1 metric ton is 1000 kg). The amount of atmospheric carbon dioxide is increasing at about 0.5% per year and is already 25% higher than preindustrial levels.

The increased concentration of atmospheric carbon dioxide due to burning of fossil fuels and plants has led to a problem called the *greenhouse effect*. Ordinarily, much of the Sun's energy would be lost by the radiation of heat from the Earth's surface back into space. However, high concentrations of carbon dioxide in the atmosphere reflect the heat back to Earth, making the surface warmer (see figure). Other gases such as methane are also thought to contribute to the warming effect. The warming of the Earth's surface by reflected heat is called the *greenhouse effect* because it operates in much the same way as the panes of glass in a greenhouse. The glass in a greenhouse is transparent to visible light but absorbs infrared radiation, which is absorbed as heat. The heat reflects from the glass, raising the temperature inside the building. Scientists have predicted that a rise of only a few degrees in the average temperature of the Earth is likely to disrupt weather patterns. These weather changes would affect the growing of food crops. Polar ice caps also could melt, inundating coastal cities with water. Because of concerns about the greenhouse effect, scientists are investigating the possibility of converting carbon dioxide into solid compounds that could be readily disposed of without adding to the atmosphere's carbon dioxide load.

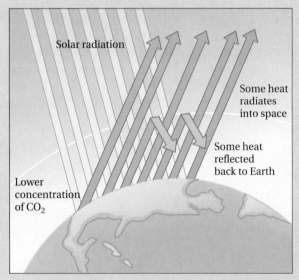

(a)

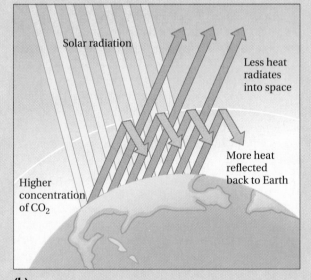

(b)

The Earth is warmed by the Sun's radiant energy. (a) The warming effect is controlled because some heat radiates from the Earth's surface back into space. (b) The *greenhouse effect* results when higher concentrations of atmospheric carbon dioxide reflect more heat back to Earth, making its surface warmer.

SUMMARY

Organic chemistry deals with carbon compounds. Carbon makes covalent bonds with other carbons to form chain and ring compounds. Hydrocarbons are compounds containing only carbon and hydrogen. Many hydrocarbons exhibit structural isomerism; structural isomers have the same molecular formula but different molecular structures. Alkanes, alkenes, and alkynes are the principal classes of aliphatic hydrocarbons.

Alkanes contain only carbon-carbon single bonds. The four bonds extending from each carbon in an alkane are arranged in a regular tetrahedron; the angle between any two bonds is 109.5 degrees. The groups attached to single bonds in straight-chain alkanes rotate freely about the bonds at room temperature.

Alkenes contain one or more carbon-carbon double bonds. The two bonds that extend from each carbon of a carbon-carbon double bond are 120 degrees apart. All four bonds extending from the pair of carbons in the carbon-carbon double bond lie in the same plane. The carbons in the carbon-carbon double bond do not rotate. Alkenes that have one group besides hydrogen attached to each carbon of the double bond exhibit geometric isom-erism, a special case of structural isomerism. Geometric isomers are *cis* or *trans* according to whether substituent groups—one on each carbon of the double bond—are on the same side (*cis*) or on opposite sides (*trans*) of the double bond. Like structural isomers, geometric isomers differ in their physical and chemical properties.

Alkynes contain one or more carbon-carbon triple bonds. The bonds that extend from the triple bond are 180 degrees apart. Ethyne (C_2H_2), the simplest alkyne, is a linear molecule.

Aromatic hydrocarbons are related to the hydrocarbon benzene. Benzene is rather unusual among hydrocarbons. It appears to be a six-membered carbon ring containing three alternating double bonds. However, the interior bonds of the benzene ring are hybrid or aromatic bonds—somewhere in between double and single bonds. Benzene is less reactive than alkenes. Many aromatic compounds, such as naphthalene, are fused-ring compounds.

Aliphatic hydrocarbons come from petroleum. Aromatic hydrocarbons come from coal. All hydrocarbons, aliphatic and aromatic, have three properties in common: They are insoluble in water, they are less dense than water, and they burn in air.

KEY TERMS

Aliphatic compound (11.9)
Alkane (11.2)
Alkene (11.8)
Alkyl group (11.5)
Alkyne (11.8)
Arene (11.9)
Aromatic compound (11.9)
Branched-chain alkane (11.5)

Cis configuration (11.8)
Condensed structural formula (11.4)
Cracking (11.10)
Cycloalkane (11.7)
Fused-ring aromatic compound (11.9)
Geometric isomer (11.8)
Hydrocarbon (11.2)
IUPAC system (11.4)

Orbital hybridization (11.2)
Petroleum refining (11.10)
Pi bond (11.8)
Polycyclic aromatic compound (11.9)
Resonance hybrid (11.9)
Resonance structure (11.9)
Saturated compound (11.8)

Sigma bond (11.2)
Straight-chain alkane (11.4)
Structural formula (11.2)
Structural isomer (11.6)
Substituent (11.5)
Trans configuration (11.8)
Unsaturated compound (11.8)

EXERCISES

Saturated Hydrocarbons (Sections 11.2–11.7)

11.10 Draw complete structural formulas and condensed structural formulas and give the IUPAC names for the first five straight-chain alkanes with an odd number of carbon atoms.

11.11 Name the alkanes or alkyl groups that have the following formulas.

(a) CH_3CH_2- (b) $CH_3(CH_2)_6CH_3$ (c) C_3H_8

(d)
$$H-\overset{\overset{\displaystyle H}{|}}{\underset{\underset{\displaystyle H}{|}}{C}}-\overset{\overset{\displaystyle H}{|}}{\underset{\underset{\displaystyle H}{|}}{C}}-\overset{\overset{\displaystyle H}{|}}{\underset{\underset{\displaystyle H}{|}}{C}}-\overset{\overset{\displaystyle H}{|}}{\underset{\underset{\displaystyle H}{|}}{C}}-\overset{\overset{\displaystyle H}{|}}{\underset{\underset{\displaystyle H}{|}}{C}}-H$$

(e) $CH_3CH_2CH_2CH_2CH_3$ (f) $CH_3CH_2CH_2-$

11.12 What system is used for naming branched-chain alkanes? Briefly state the rules.

11.13 What is a substituent? What is the name of a branched-chain alkane substituent?

11.14 Draw structural formulas for the following branched-chain alkanes.
   (a) 2,3-dimethylbutane
   (b) 2,2-dimethylbutane
   (c) 2,2,3-trimethylbutane
   (d) 3,4-diethyl-4,5-dimethyloctane
   (e) 4-ethyl-2,3,4-trimethylnonane
   (f) 3,5-diethyl-2,3-dimethyl-5-propyldecane

11.15 Give the IUPAC name for each of the following structural formulas.

(a)
$$CH_3-\overset{\overset{\displaystyle CH_3}{|}}{\underset{\underset{\displaystyle CH_3}{|}}{CH}}$$

(b)
$$CH_3-CH-CH_2-CH-CH-CH_3$$
$$\overset{}{\underset{\displaystyle CH_3}{|}}\overset{}{\underset{\displaystyle CH_2}{|}}\ \overset{}{\underset{\displaystyle CH_3}{|}}$$
$$\overset{}{\underset{\displaystyle CH_3}{|}}$$

(c)
$$CH_3-\overset{\overset{\displaystyle CH_3}{|}}{CH}-CH_2$$
$$\overset{}{\underset{\displaystyle CH_2}{|}}$$
$$CH_3-CH_2$$

(d)
$$CH_3-\overset{\overset{\displaystyle CH_3}{|}}{CH}-CH_2$$
$$CH_2-CH_2-CH_3$$

11.16 What are structural isomers?

11.17 Draw one structural isomer of each of the following.

(a)
$$CH_3-CH-CH_2$$
$$\overset{}{\underset{\displaystyle CH_3}{|}}\ \overset{}{\underset{\displaystyle CH_3}{|}}$$

(b)
$$CH_3-CH-CH-CH_3$$
$$\overset{}{\underset{\displaystyle CH_3}{|}}\ \overset{}{\underset{\displaystyle CH_3}{|}}$$

(c)
$$CH_3-CH-CH_2-CH_2$$
$$\overset{}{\underset{\displaystyle CH_2}{|}}\ \overset{}{\underset{\displaystyle CH_3}{|}}$$
$$\overset{}{\underset{\displaystyle CH_3}{|}}$$

(d)
$$CH_3-CH-CH_3$$
$$\overset{}{\underset{\displaystyle CH_3}{|}}$$

(e)
$$CH_3-\overset{\overset{\displaystyle CH_3}{|}}{CH}-CH_2-CH $$
$$\overset{}{\underset{\displaystyle CH_2}{|}}$$

(e)
$$CH_3-\overset{\overset{\displaystyle CH_3}{|}}{\underset{\underset{\displaystyle CH_2}{|}}{CH}}-CH_2-\overset{\overset{\displaystyle CH_3}{|}}{\underset{\underset{\displaystyle CH_2}{|}}{CH}}$$
$$CH_3\ \ \ CH_3$$

(f)
$$CH_3\ \ \ CH_3$$
$$\overset{}{\underset{\displaystyle CH_2}{|}}\ \ \overset{}{\underset{\displaystyle CH_2}{|}}$$
$$CH_2-CH-CH_2$$
$$\overset{}{\underset{\displaystyle CH_2}{|}}$$
$$CH_3$$

11.18 Which of the following are structural isomers?

(a)
$$CH_3\overset{\overset{\displaystyle CH_3}{|}}{\underset{\underset{\displaystyle CH_3}{|}}{CH}}$$

(b)
$$CH_3-CH-CH_3$$
$$CH_3-CH_2$$

(c)
$$CH_3-CH_2$$
$$CH_2-CH_3$$

(d)
$$CH_3-\overset{\overset{\displaystyle CH_3}{|}}{\underset{\underset{\displaystyle CH_3}{|}}{C}}-CH_3$$

(e)
$$CH_3-\overset{\overset{\displaystyle CH_3}{|}}{\underset{\underset{\displaystyle CH_3}{|}}{C}}-CH_2-CH_3$$

(f)
$$CH_3-CH_2$$
$$CH_3$$

11.19 A student incorrectly names a compound 1,3-dimethylpropane. Draw a structural formula and correctly name this compound.

11.20 Explain why cyclopropane and cyclobutane are strained molecules but cyclopentane and cyclohexane are not.

11.21 Name the following substituted cycloalkanes.

(a) 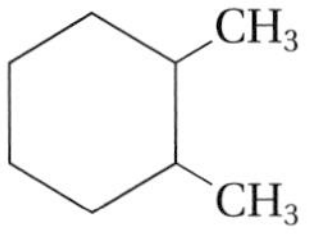 [structure with CH₃ groups on cyclohexane]   (b) 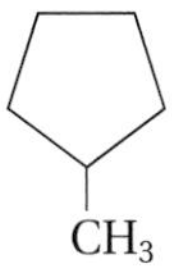 [cyclopentane with CH₃]   (c) 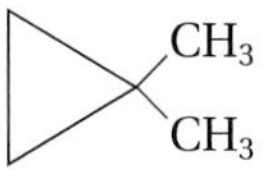 [cyclopropane with two CH₃ groups]

Unsaturated Hydrocarbons (Section 11.8)

11.22. Explain the difference between saturated and unsaturated hydrocarbons.

11.23 How are alkenes named by the IUPAC system?

11.24 Show how the lack of rotation about a carbon-carbon double bond leads to geometric isomerism. Use the geometric isomers of 2-pentene to illustrate your answer.

11.25 Does 2,4-dimethyl-2-hexene have geometric isomerism? Explain.

11.26 Draw a structural formula for each of the following alkenes. Include both *cis* and *trans* forms if the compound has geometric isomerism.
(a) 2-pentene   (b) 2-methyl-2-pentene

11.27 Draw a carbon skeleton for each of the following alkenes. Include both *cis* and *trans* forms if the compound has geometric isomerism.
(a) 2-pentene   (b) 3-hexene

11.28 Give a systematic name for each of the following alkenes.

(a) $CH_3CHCH_2CH{=}CH_2$ (b)
     |
    CH_3

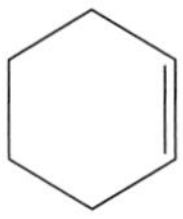

(c) $CH_3CH{=}CH_2$ (d) 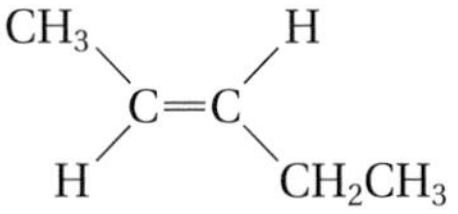 [structure: CH₃ and H on one carbon, H and CH₂CH₃ on the other, C=C]

11.29 Name each of these unsaturated compounds.

(a) $CH_3CH_2CH{=}CH_2$ (b) $CH_3(CH_2)_5CH{=}CCH_3$
                                     |
                                   CH_3

(c) [structure: CH₃ and H on one carbon, CH₃ and CH₃ on the other, C=C] 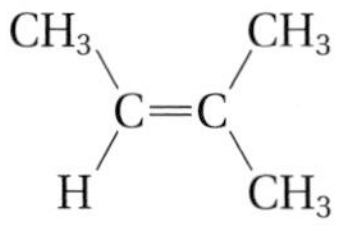   (d) $CH_2{=}CHCH_2CH_2$
                                          |
                                         CH_2CH_3

11.30 Draw a structural formula for each alkene with the molecular formula C_5H_{10}. Name each compound.

Aromatic Compounds (Section 11.9)

11.31 What is meant by the term *aromatic character?*

11.32 Describe the shape of a benzene molecule.

11.33 Explain why both the following structures represent 1,2-diethylbenzene.

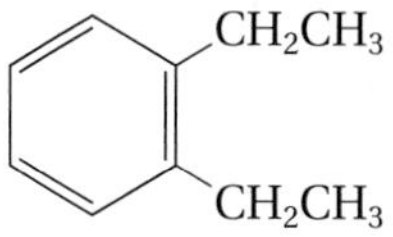

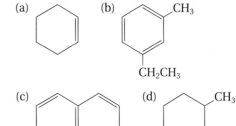

11.34 How many different compounds can be formed when two ethyl groups are attached to a benzene ring?

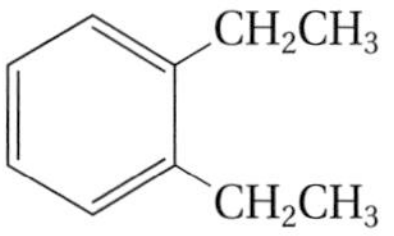

11.35 Name or draw a structural formula for the following compounds, and say which are aromatic.

(a) [cyclohexene]   (b) [benzene ring with CH₃ and CH₂CH₃]

(c) 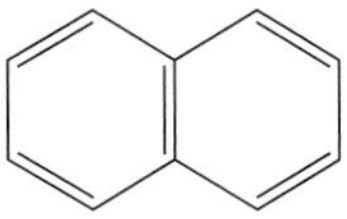 [naphthalene]   (d) [cyclohexane with CH₃]

(e) toluene   (f) ethylbenzene

11.36 Draw a structural formula or name the following compounds. Classify each compound as an aliphatic or aromatic compound.
(a) *p*-diethylbenzene
(b) anthracene
(c) 2-methyl-3-phenylpentane
(d) *p*-xylene

(e) 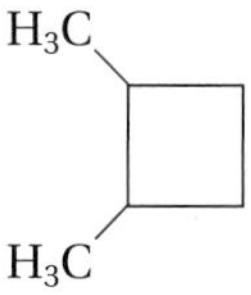 [cyclobutane with two H₃C groups]

(f) 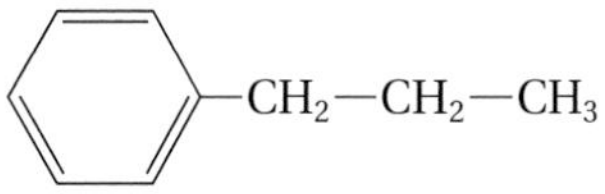 [benzene ring with $-CH_2-CH_2-CH_3$]

Petroleum and Coal (Sections 11.10, 11.11)

11.37 What is the major difference between crude oil and coal as a source of hydrocarbons?

11.38 Define the terms (a) *petroleum refining* and (b) *cracking.*

Additional Exercises

11.39 Briefly describe (a) hydrocarbon, (b) alkane, (c) alkene, (d) alkyne, (e) cycloalkane, and (f) arene.

11.40 Write a balanced equation for the complete combustion of pentane.

11.41 Match the following:

(a) no carbon-carbon bond rotation
(b) structural isomer
(c) cyclopropane
(d) cracking
(e) ethene
(f) contains a carbon-carbon triple bond
(g) methane
(h) butane
(i) benzene
(j) *trans*-2-butene

(1) ethyne
(2) straight-chain alkane
(3) carbon-carbon double and triple bonds
(4) geometric isomer
(5) simplest aromatic triple bond compound
(6) strained ring
(7) petroleum refining
(8) contains carbon-carbon double bond
(9) marsh gas
(10) same formula, different structure

11.42 Draw the structural formula for each of the following:
(a) 3,3-dimethylheptane   (b) 4,4-dipropylnonane
(c) methylpropane   (d) 3-ethylhexane

11.43 Give the IUPAC name for

(a) [pentagon structure]   (b) CH_3—CH=C—CH_3 with CH_3 below the C

(c) [cyclohexane structure with CH₃ groups]   (d) [cyclobutane structure with CH₃ groups]

11.44 The following names are incorrect. Explain. (*Hint:* Write a structural formula for the incorrectly named compound and then rename it correctly.)
(a) 2-ethylbutane
(b) 3-methylcyclohexane
(c) 1,2-dimethylpentane
(d) 4-ethylhexane
(e) 3-dimethylheptane
(f) 2-ethyl-3,3-dimethylbutane

11.45 How many *cycloalkanes* have the molecular formula C_5H_{10}? Draw each of their structures and give the correct IUPAC names.

11.46 Draw condensed structural formulas for the following compounds; show all carbon-carbon bonds.
(a) 4-methyl-2-pentene
(b) 2-phenylheptane
(c) *trans*-2-heptene
(d) 3-methylcyclohexene
(e) 3,3-dimethyl-1-butene
(f) 4,5-diethyl-3,6,7-trimethyldecane
(g) 5,5-diethyl-2,6,7-trimethyl-3-octene

SELF-TEST (REVIEW)

True/False

1. All unsaturated compounds have double bonds.
2. Hydrocarbons readily dissolve in water.
3. The designations *cis* and *trans* are used to distinguish structural isomers.
4. An alkyne has a triple covalent bond between two carbon atoms.
5. Any hydrocarbon molecule containing seven carbon atoms will be called a *heptene*.
6. Most alkenes tend to be more chemically reactive than benzene.
7. The hybridization of the bonding orbitals in the carbon-carbon double bond of an alkene is sp^3.
8. The boiling points of hydrocarbons tend to increase with increasing chain length.
9. Structural isomers and geometric isomers have identical molecular formulas.
10. Cycloalkanes with more than four carbon atoms in the ring are planar (flat) molecules.

Multiple Choice

11. All the following are unsaturated hydrocarbons except
    (a) 3-octene.   (b) benzene.   (c) cyclobutane.
    (d) ethyne.
12. *Cis-trans* geometric isomerism is possible in
    (a) 1-butane.   (b) 1-butene.   (c) 2-pentene.
    (d) pentane.

13. Hydrocarbons in general are
    (a) soluble in water.   (b) chemically very reactive.
    (c) less dense than water.   (d) all of the above.

14. Rotation about a carbon-carbon bond is possible in
    (a) benzene.   (b) ethyne.   (c) propane.
    (d) ethene.

15. The name of an alkyl group that contains two carbon atoms is
    (a) diphenyl.   (b) ethyl.   (c) dimethyl.
    (d) propyl.

16. Name this compound: $CH_3CH(CH_3)C(CH_3)_3$.
    (a) 2,2,3-trimethylbutane
    (b) tetramethylpropane
    (c) 1,1,1,2-tetramethylpropane
    (d) 2,3,3,3-tetramethylpropane

17. In the electron dot structure of butane, C_4H_{10}, what is the total number of electrons shared between all carbon atoms?
    (a) 4   (b) 6   (c) 8   (d) 10

18. A hydrocarbon with nine carbon atoms joined by single bonds in a straight chain is called
    (a) heptane.   (b) nonane.   (c) hexene.
    (d) trinane.

19. Structural isomers have
    (a) the same melting points.
    (b) the same molecular formulas.
    (c) the same densities.
    (d) all of the above.

20. A structural isomer of hexane is
    (a) 2,2-dimethylbutane.   (b) cyclohexane.
    (c) benzene.   (d) 2-methylpentene.

21. Which of the following words or phrases do not describe benzene?
    (a) aromatic   (b) puckered ring
    (c) hybrid bonding   (d) arene

22. A correct name for benzene with one CH_3— substituent attached to the ring is
    (a) benzylmethane.   (b) methene benzene.
    (c) methylbenzene.   (d) methylene benzene.

23. The number of fused rings in naphthalene is
    (a) 2.   (b) 3.   (c) 4.   (d) 5.

24. The major products of combustion of an alkane are
    (a) carbon and water.
    (b) water and carbon monoxide.
    (c) carbon dioxide and oxygen.
    (d) water and carbon dioxide.

25. The name of $(CH_3)_2-C=CH_2$ is
    (a) methylpropene.   (b) dimethylethene.
    (c) trimethylmethene.   (d) 1-butene.

26. The most stable form of cyclohexane
    (a) is planar.
    (b) is called the chair form.
    (c) is called the inert form.
    (d) is called the boat form.

27. How many isomers, including geometric isomers, are there for the molecular formula C_4H_8?
    (a) three   (b) four   (c) five   (d) six

12 Halocarbons, Alcohols, and Ethers

The Polar Bond in Organic Molecules

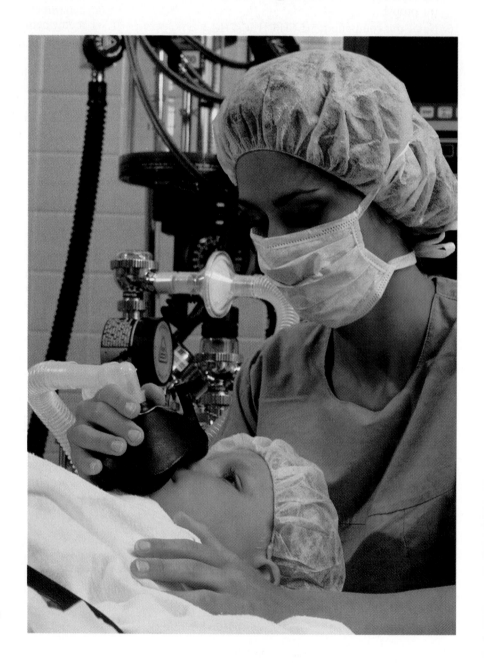

A general anesthetic produces unconsciousness and insensitivity to pain.

CHAPTER OUTLINE

CASE IN POINT: A mistaken conviction

12.1 Halocarbons

A CLOSER LOOK: Chlorofluorocarbons and the Ozone Layer

12.2 Halocarbons from alkenes

12.3 Alcohols

A CLOSER LOOK: Phenolic Antiseptics and Disinfectants

12.4 Making alcohols

12.5 Elimination reactions

12.6 Alkoxides

12.7 Ethers

A CLOSER LOOK: Halocarbon and Ether Anesthetics

FOLLOW-UP TO THE CASE IN POINT: A mistaken conviction

12.8 Physical properties

12.9 Sulfur compounds

12.10 Polyfunctional compounds

In this chapter we will begin our study of organic compounds containing carbon bonded to other elements. Some of these compounds are important biological molecules. Others are used in the health sciences as antiseptics and anesthetics. The compounds considered here are those with carbon-halogen bonds, carbon-oxygen single bonds, and carbon-sulfur single bonds.

Among the compounds we will study in this chapter are alcohols, which contain carbon-oxygen bonds. Ethylene glycol, an alcohol, played a prominent role in a guilty verdict in a murder trial in St. Louis, Missouri. The case is worth noting not only because it supposedly involved an alcohol but also because it shows how scientists subsequently established the innocence of the accused murderer.

CASE IN POINT: A mistaken conviction

 In 1989, a woman brought her ill 3-month-old son to the emergency room of a St. Louis hospital, where the attending physician diagnosed the symptoms of ethylene glycol poisoning. After the child recovered, he was taken from his parents and placed in a foster home. The unfortunate child died soon after a visit from his mother, who was charged with first-degree murder for poisoning her son. A commercial laboratory and a hospital laboratory claimed to have found ethylene glycol in the boy's blood and in a bottle of milk he was fed by his mother shortly before he died. The mother was convicted and sentenced to life in prison for the murder of her oldest son by ethylene glycol poisoning. However, this murder case was not as open-and-shut as it seemed, as we will learn in Section 12.7.

Ethylene glycol is the principal ingredient in antifreeze solutions. Although it has a sweet taste, ethylene glycol is quite toxic.

12.1 Halocarbons

AIMS: To name and draw structures of simple halocarbons. To name and draw the structures of the common alkyl groups containing up to four carbon atoms.

Focus

Compounds containing carbon-halogen bonds are called halocarbons.

Table 12.1 Names of Halogens as Substituent Groups

Halogen	Substituent name
fluorine	fluoro-
chlorine	chloro-
bromine	bromo-
iodine	iodo-

Halocarbons *are a class of organic compounds containing covalently bonded halogens: fluorine, chlorine, bromine, or iodine.* For example, the compounds CH_3Cl and CH_3Br are halocarbons. Although very few halocarbons are found in nature, they are nevertheless readily prepared and used for many purposes, such as anesthetics and insecticides. *Halocarbons in which a halogen is attached to a carbon of an aliphatic chain are* **alkyl halides;** *halocarbons in which a halogen is attached to a carbon of an arene ring are* **aryl halides.**

The IUPAC rules for naming halocarbons are very similar to those we cited in Chapter 11 for naming substituted alkanes, except that halogen groups must now be added to our repertoire of substituents. Table 12.1 shows the names used for halogen groups when they are substituents on carbon. Common names of a few simple halocarbons are still used. These common names consist of two parts. The first part names the hydrocarbon entity of the molecule as an alkyl group, such as methyl or ethyl. The second part names the halogen as if it were an ion. Methyl chloride, CH_3Cl, is an example. Remember, however, that the bonding in a halocarbon is covalent, not ionic. Alkyl groups besides those of methyl, ethyl, and propyl covered in Chapter 11 have been given names. Table 12.2 shows some of these alkyl groups and the vinyl group ($CH_2=CH-$), which is derived from ethene. The table also describes the use of the prefix *iso-* and the terms *secondary* and *tertiary* in alkyl group names. Some structures and IUPAC and common names (in parentheses) for some simple halocarbons follow. Studying them and Examples 12.1 and 12.2 will help you understand the naming of halocarbons. Example 12.3 shows how to write a structure when the compound name is given.

$$CH_3-Cl$$

Chloromethane
(methyl chloride)

$$CH_3-\overset{\overset{\displaystyle CH_3}{|}}{\underset{\underset{\displaystyle Br}{|}}{C}}-CH_3$$

2-Bromo-2-methylpropane
(*tert*-butyl bromide)

Chloroethene
(vinyl chloride)

$$CH_3-\overset{\overset{\displaystyle CH_3}{|}}{CH}-I$$

2-Iodopropane
(isopropyl iodide)

$$CH_3-CH_2-\overset{\overset{\displaystyle F}{|}}{CH}-CH_3$$

2-Fluorobutane
(*sec*-butyl fluoride)

Chlorobenzene
(phenyl chloride)

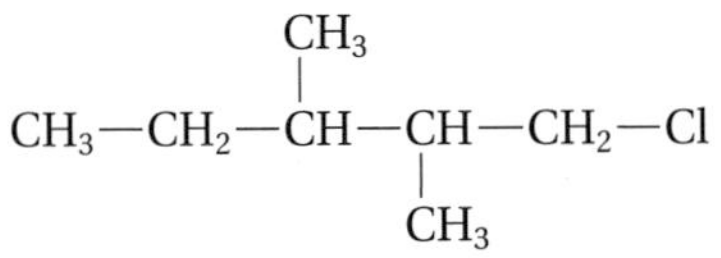

$$CH_3-CH_2-CH-CH-CH_2-Cl$$

with CH_3 above the third carbon and CH_3 below the fourth carbon

1-Chloro-2,3-dimethylpentane
(no common name)

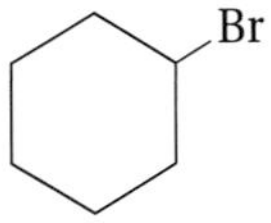

Br (on cyclohexane ring)

Bromocyclohexane
(cyclohexyl bromide)

Table 12.2 Names of Some Common Alkyl Groups

Name	Alkyl group	Ball-and-stick models	Remarks
isopropyl	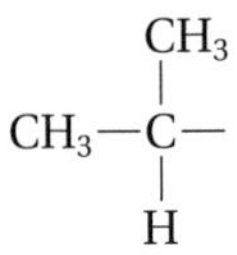 $CH_3-\underset{H}{\overset{CH_3}{C}}-$	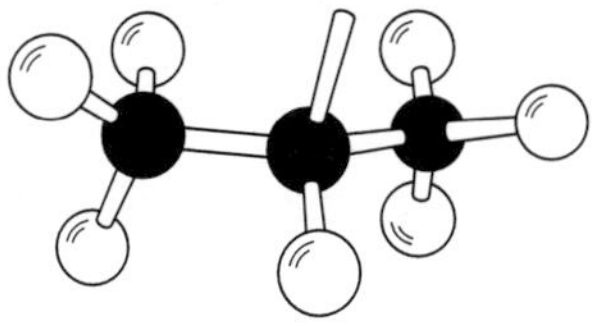	The prefix *iso-* is reserved for carbon chains that are straight except for the presence of a methyl group on the carbon second from the unsubstituted end of the longest chain.
isobutyl	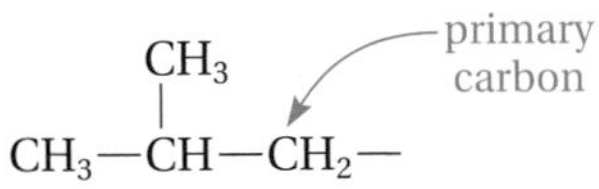 $CH_3-CH-CH_2-$ with CH_3 above; primary carbon	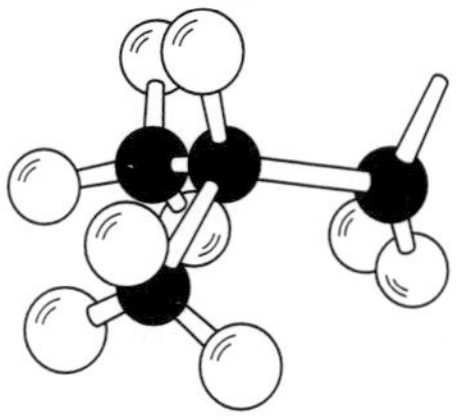	Note the use of the prefix *iso-*. The carbon joining this alkyl group to another group is bonded to one other carbon; it is a *primary carbon*.
secondary butyl (*sec*-butyl)	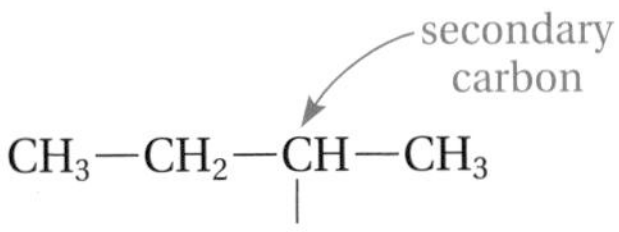 $CH_3-CH_2-CH-CH_3$; secondary carbon	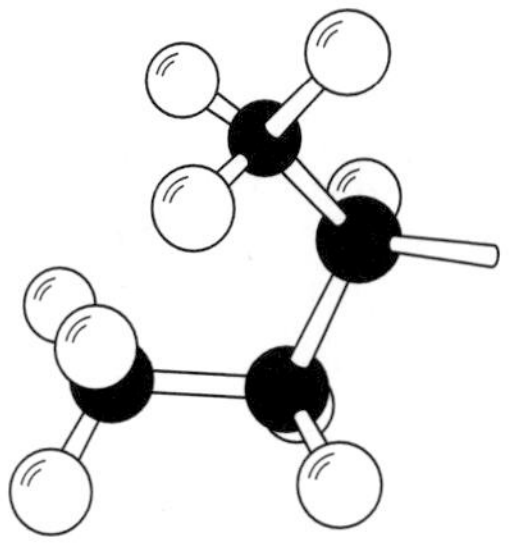	The carbon joining this alkyl group to another group is bonded to two other carbons; it is a *secondary carbon*.
tertiary butyl (*tert*-butyl)	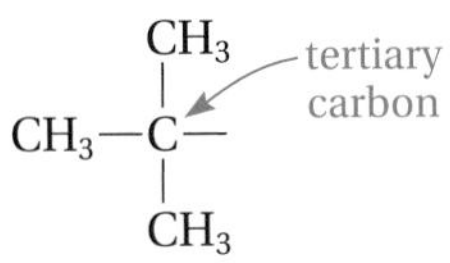 $CH_3-\underset{CH_3}{\overset{CH_3}{C}}-$; tertiary carbon	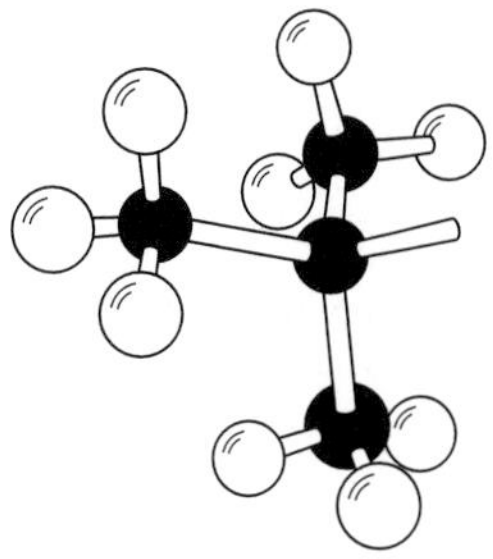	The carbon joining this alkyl group to another group is bonded to three carbons; it is a *tertiary carbon*.
vinyl	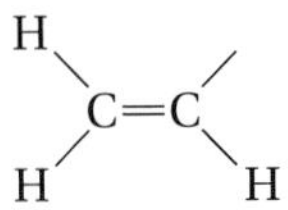 $\underset{H}{\overset{H}{C}}=\underset{H}{C}$	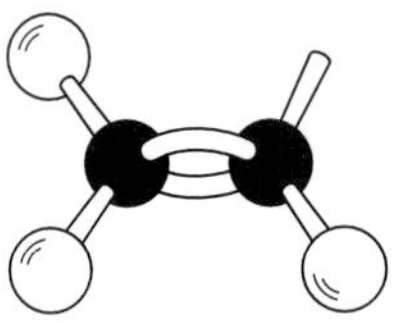	When used as an alkyl group in giving compounds common names, this group, derived from ethene, is called *vinyl*.

EXAMPLE 12.1	Naming an alkyl halide

What are the IUPAC and common names for $CH_3CH_2CH_2Cl$?

SOLUTION

You may wish to review the rules given in Chapter 11 for the IUPAC naming of alkanes and alkyl substituents. The rules will be the same for halocarbons, except that the halogen must be named as a substituent. The longest carbon chain in our molecule contains three carbons, making propane the parent alkane.

$$CH_3CH_2CH_2Cl$$
$$321$$

A chlorine atom substituent is called *chloro-*; this substituent is given the lowest possible number on the parent chain—in this case position 1. The IUPAC name for the compound is *1-chloropropane.* We write the common name of the compound by naming the $CH_3CH_2CH_2$— portion of the molecule as the alkyl group, *propyl,* and adding the word *chloride.* The common name for this compound is *propyl chloride.*

EXAMPLE 12.2	Naming an alkyl fluoride

Give the IUPAC and common names for the following compound.

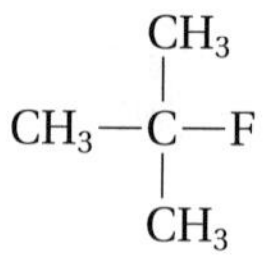

$$\begin{array}{c} CH_3 \\ | \\ CH_3-C-F \\ | \\ CH_3 \end{array}$$

SOLUTION

The longest chain is three carbons long, making propane the parent alkane.

$$\begin{array}{c} CH_3 \\ | \\ CH_3-C-F \\ 3 \quad 2 \;| \\ 1\; CH_3 \end{array}$$

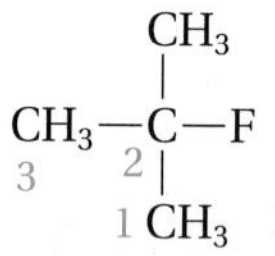

Fluoro and *methyl* groups are substituents on the propane chain, both in the 2 position. Listing the substituent names in alphabetical order and attaching the name of the parent alkane gives the IUPAC name for the compound: *2-fluoro-2-methylpropane.* The common name for the alkyl group with three methyl groups bonded to the same carbon is *tertiary butyl,* which we abbreviate as *tert*-butyl in writing the name (see Table 12.2). Since the halogen substituent is fluorine, the common name for the compound is *tertiary butyl fluoride,* written as *tert*-butyl fluoride.

PRACTICE EXERCISE 12.1

Write the IUPAC name for each of the following halocarbons.

(a) 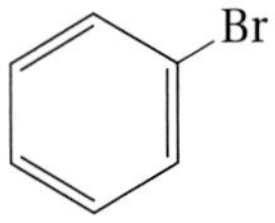 (b) CH_3CH_2Cl (c) 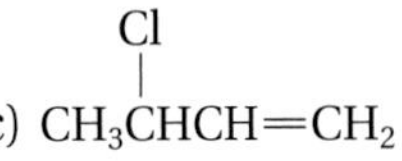 $CH_3\overset{Cl}{\underset{|}{C}}HCH{=}CH_2$

EXAMPLE 12.3

Writing a condensed structural formula for a halocarbon

Write a condensed structural formula for 2-bromo-3,3-dimethylpentane.

SOLUTION

Use the information in the name to construct this halocarbon. The parent alkane is pentane, so write a chain of five carbons.

$$\underset{1}{C}-\underset{2}{C}-\underset{3}{C}-\underset{4}{C}-\underset{5}{C}$$

Add the substituents at the proper positions. Substituents are *bromo-* at the 2 position and two *methyl* groups on the 3 position of the parent chain.

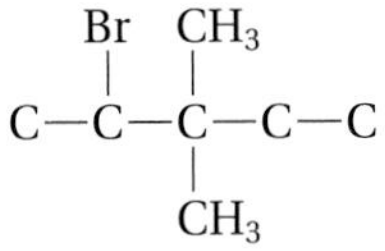

$$C-\overset{Br}{\underset{}{C}}-\overset{CH_3}{\underset{\underset{CH_3}{|}}{C}}-C-C$$

Fill in the structure with the number of hydrogens needed to give each carbon four covalent bonds. The structure of 2-bromo-3,3-dimethylpentane must be

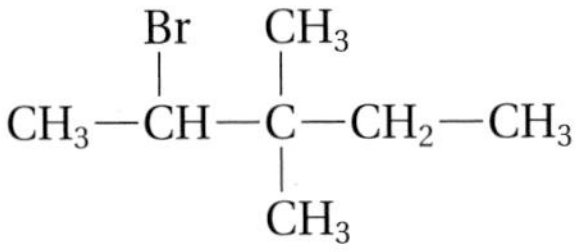

$$CH_3-\overset{Br}{\underset{}{C}}H-\overset{CH_3}{\underset{\underset{CH_3}{|}}{C}}-CH_2-CH_3$$

2-Bromo-3,3-dimethylpentane

PRACTICE EXERCISE 12.2

Give the structural formula for each of the following.

(a) isopropyl chloride
(b) 1-iodo-2,2-dimethylpentane
(c) 1-bromo-4-ethylbenzene

Some alkyl and aryl halocarbons currently in use for a variety of purposes are shown in Table 12.3. Some halocarbons that were widely used in the 1980s, however, are no longer made or used, for reasons discussed in A Closer Look: Chlorofluorocarbons and the Ozone Layer.

Table 12.3 Some Halocarbons and Their Uses

Halocarbon	Use
CH_3—CH_2—Cl Chloroethane (ethyl chloride)	A local anesthetic, its rapid evaporation on the skin (bp 13 °C) cools nerve endings and cuts down transmission of pain.
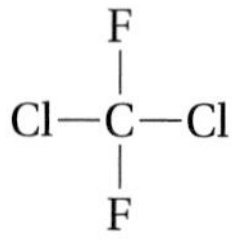 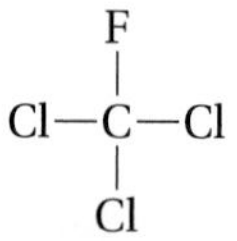 Dichlorodifluoromethane (Freon 12) and Trichloromonofluoromethane (Freon 11)	Although they were once widely used as refrigerants, Freons are now banned. They are permitted, however, as propellants to deliver inhalation aerosols of medication to asthmatics. (Freons are also known as *chlorofluorocarbons*, or CFCs. CFCs cause depletion of ozone in the stratosphere.)
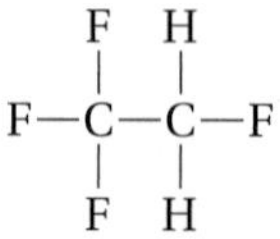 1,1,1,2-tetrafluoroethane (HFC-134a)	Hydrofluorocarbons (HFCs) have been developed as replacements for CFCs (Freons) in car air-conditioning units. Most cars are equipped with systems that use HFC-134a. HFCs do not cause ozone depletion.
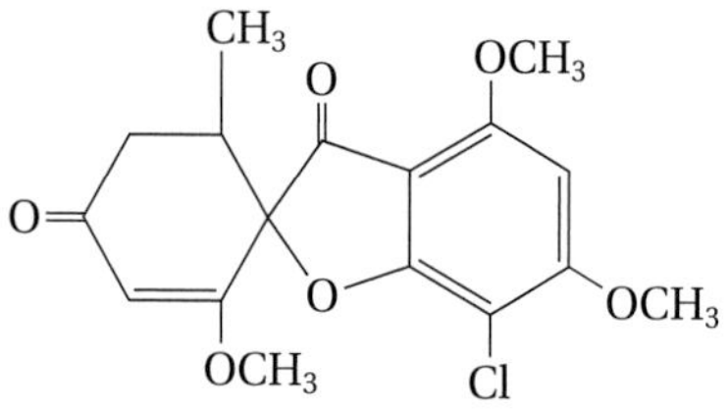 Griseofulvin	Used in treatment of fungal infections. Obtained from *Penicillium*.
 *p*-Dichlorobenzene (1,4-dichlorobenzene)	Used as a moth repellent.
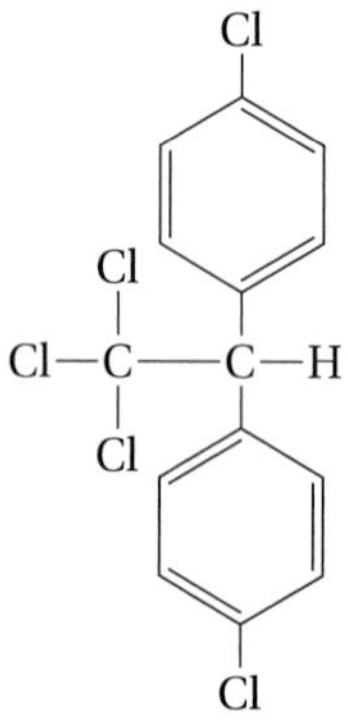 Dichlorodiphenyltrichloroethane (DDT)	Synthesized in 1874, the insecticidal properties of DDT were not recognized until the 1940s. Its use has been banned in the United States because of its persistence in the environment; it is nonbiodegradable.

Chlorofluorocarbons and the Ozone Layer

Halocarbon molecules that contain chlorine as well as fluorine are known as *chlorofluorocarbons,* or *CFCs,* or *Freons.*

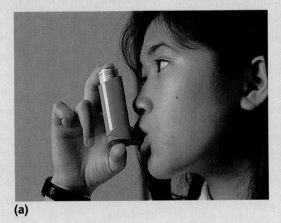

$$
\begin{array}{ccc}
\text{Cl} & \text{Cl} & \text{Cl Cl} \\
| & | & |\ \ | \\
\text{Cl}-\text{C}-\text{F} & \text{Cl}-\text{C}-\text{F} & \text{F}-\text{C}-\text{C}-\text{F} \\
| & | & |\ \ | \\
\text{Cl} & \text{F} & \text{F F}
\end{array}
$$

Freon 11   Freon 12   Freon 13

CFCs are gases or low boiling liquids that are chemically inert, nontoxic, nonflammable, and insoluble in water. These properties made them good candidates for refrigerants in air conditioners and as propellants for aerosol cans of hair sprays, deodorants, and inhalation medications (see figure, part a).

Ozone is an important natural component of the stratosphere, the layer of the atmosphere that ranges from 11 to 48 km above the Earth. The ozone molecules shield the Earth's plants and animals from life-destroying ultraviolet radiation. When an ozone molecule in the stratosphere absorbs ultraviolet radiation, it is converted to an oxygen molecule (O_2) and an oxygen atom ($O\cdot$).

$$ O_3(g) \xrightarrow[\text{radiation}]{\text{Ultraviolet}} O_2(g) + O\cdot(g) $$

The chemical inertness of CFCs allows them to remain in the environment for a long time. Eventually, they find their way into the stratosphere, where the carbon-chlorine bond in CFCs is broken by energy from ultraviolet light. The chlorine atom ($Cl\cdot$) that is produced combines with an ozone molecule in the stratosphere, to give a chlorine oxide radical ($ClO\cdot$) and an oxygen molecule.

$$ Cl\cdot(g) + O_3(g) \longrightarrow ClO\cdot(g) + O_2(g) $$

The $ClO\cdot$ radical then reacts with an oxygen atom ($O\cdot$), formed when ozone absorbs ultraviolet light, to produce another chlorine atom and an oxygen molecule.

$$ ClO\cdot(g) + O\cdot(g) \longrightarrow Cl\cdot(g) + O_2(g) $$

This process is repeated many times. It has been estimated that the breaking of a single C—Cl bond of a CFC molecule results in the destruction of 4000 or more ozone molecules in the stratosphere. The destruction of the ozone layer permits larger amounts of harmful ultraviolet radiation to reach the Earth. The effect of ozone depletion is manifested in an increased incidence of skin cancers and crop damage. In 1985, scientists discovered a "hole" in the ozone layer over Antarctica (see figure, part b). Following the concern that this discovery generated, the Montreal Protocol on Substances that Deplete the Ozone Layer went into effect in January 1989. It was signed by 24 nations. The protocol calls for CFCs to be phased out by the year 1996. In the meantime, alternatives are being sought.

One class of compounds that show promise as alternatives are the hydrofluorocarbons, or HFCs. They are possible substitutes for ozone-depleting CFCs because they contain no chlorine and therefore cannot catalyze ozone destruction. However, an air-conditioning system designed for CFC refrigerant will not operate on HFC refrigerants. Most new cars and trucks sold in the United States are now equipped with air-conditioning systems that use HFCs.

CFCs are used to deliver precise doses of medication directly into the lungs of asthmatics (a). The hole in the ozone layer over Antarctica is clearly visible in this NASA photograph (b).

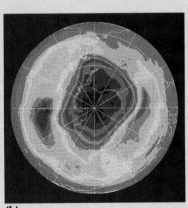

(a)   (b)

12.2 Halocarbons from alkenes

***AIMS:** To distinguish among halogenation, hydrohalogenation, and hydrogenation reactions. To contrast an addition reaction of an alkene with a substitution reaction with benzene. To use Markovnikov's rule to predict the major products in an addition reaction. To describe a polymerization reaction.*

Focus

Halogens and hydrogen halides add to multiple carbon-carbon bonds.

Up to this point we have focused on the structures of hydrocarbon chains and rings, which are essential components of every organic compound. Yet in most chemical reactions involving organic molecules, the hydrocarbon skeletons of the molecules are chemically inert. The chemistry of the alkanes is relatively limited. Most organic chemistry involves substituents attached to hydrocarbon chains. These groups act as a unit and often contain oxygen, nitrogen, sulfur, phosphorus, or halogens. Such groups are called **functional groups**—*the chemically functional parts of the molecule.*

The double and triple bonds of alkenes and alkynes are chemically reactive and are considered functional groups. So are the carbon-halogen groups of halocarbons. Carbon-carbon single bonds are not easy to break. However, recall from Section 11.8 that the pi bond of the double bond in alkenes is somewhat weaker than a carbon-carbon single bond. It is possible for a compound of general structure X—Y to react with a double bond under appropriate conditions. *A reaction in which two molecules of reactant combine to form a single product is called an* **addition reaction.** Here is the general equation for addition reactions of alkenes:

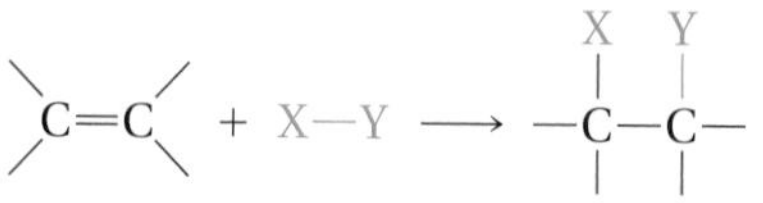

$$\diagup C=C \diagdown + X-Y \longrightarrow -\overset{X}{\underset{|}{C}}-\overset{Y}{\underset{|}{C}}-$$

Addition reactions are an important method of introducing new functional groups into organic molecules. Halocarbons can be produced by addition reactions in which *the adding reactant or reagent X—Y is a halogen molecule, such as Cl—Cl* (**halogenation**) *or a hydrogen halide, such as H—Cl* (**hydrohalogenation**). Although *hydrogenation* is a method for producing alkanes rather than halocarbons, we include the *addition of hydrogen, H—H, to double bonds* (**hydrogenation**) in this section on additions to alkenes.

Halogenation

When the reagent X—Y is a halogen molecule such as chlorine or bromine, the product of the reaction is a disubstituted halocarbon:

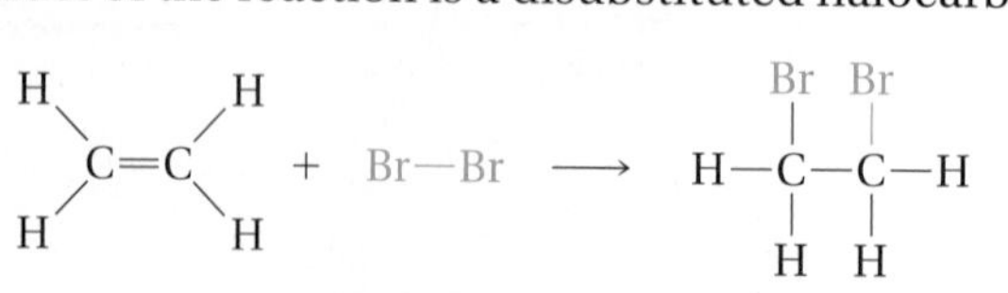

Ethene (colorless) + Bromine (brownish orange) → 1,2-Dibromoethane (colorless)

(a)

(b)

Figure 12.1
(a) The beaker on the left contains an alkene; the beaker on the right is a solution of bromine in dichloromethane.
(b) The bromine solution loses its brownish orange color when it is added to the alkene.

The addition of bromine to carbon-carbon multiple bonds is often used as a chemical test for unsaturation in an organic molecule. Bromine has a brownish orange color, but most organic compounds of bromine are colorless. The test for unsaturation is done by adding a few drops of a 1% solution of bromine in dichloromethane (CH_2Cl_2) to the suspected alkene. Loss of the brownish orange color is a positive test for unsaturation (Fig. 12.1).

PRACTICE EXERCISE 12.3

Write the structure for the expected product from each of the following reactions.

(a) $CH_2{=}CHCH_2CH_3 + Br_2 \longrightarrow$ (b) ⬡ $+ Cl_2 \longrightarrow$

(c) $CH_3CH{=}CHCH_3 + I_2 \longrightarrow$

Hydrohalogenation

Hydrogen halides such as HBr or HCl also can add to the double bond. Since the product contains only one substituent, it is called a *monosubstituted* halocarbon. The addition of hydrogen chloride to ethene is an example of hydrohalogenation.

$$
\underset{\substack{\text{Ethene}\\\text{(ethylene)}}}{\overset{\displaystyle H\!\!\diagdown\!\!\!\underset{H\diagup}{}C{=}C\overset{H}{\underset{H}{}}}{}} \;+\; \underset{\substack{\text{Hydrogen}\\\text{chloride}}}{H{-}Cl} \longrightarrow \underset{\substack{\text{Chloroethane}\\\text{(ethyl chloride)}}}{H{-}\overset{H}{\underset{H}{C}}{-}\overset{Cl}{\underset{H}{C}}{-}H}
$$

The addition of hydrogen halides to alkenes more complex than ethene can, in principle, give equal amounts of two different structural isomers, as shown for propene:

$$
\underset{\substack{\text{Propene}\\\text{(propylene)}}}{H_3C\!\!\diagup\!\!\overset{H}{}C{=}C\overset{H}{\underset{H}{}}} + H{-}Cl \longrightarrow \underset{\substack{\text{2-Chloropropane}\\\text{(isopropyl chloride)}\\\text{(major product)}}}{CH_3{-}\overset{Cl}{\underset{H}{C}}{-}\overset{H}{\underset{H}{C}}{-}H} + \underset{\substack{\text{1-Chloropropane}\\\text{(propyl chloride)}\\\text{(minor product)}}}{CH_3{-}\overset{H}{\underset{H}{C}}{-}\overset{Cl}{\underset{H}{C}}{-}H}
$$

In reality, more of one structural isomer is formed. This follows **Markovnikov's rule:** *The major product of hydrohalogenation is the one in which the hydrogen of the hydrogen halide ends up on the carbon of the double bond that already has more hydrogens.* This general rule was devised by Vladimir Markovnikov (1838–1904), a Russian chemist. You can usually be certain of writing the correct major product in reactions of this type by thinking of where the hydrogen of the adding reagent is going and remembering that "birds of a feather flock together."

The benzene ring resists addition reactions, since addition would destroy the stable aromatic electron system. If benzene is treated with a halogen in the presence of a catalyst, however, *substitution* of a ring hydrogen by a halogen group occurs. *In* **aromatic substitution,** *the hydrogen on a* *benzene ring is replaced by another group.* Iron compounds are often used as catalysts for aromatic substitution reactions—a rusty nail dropped in the reaction flask works fine.

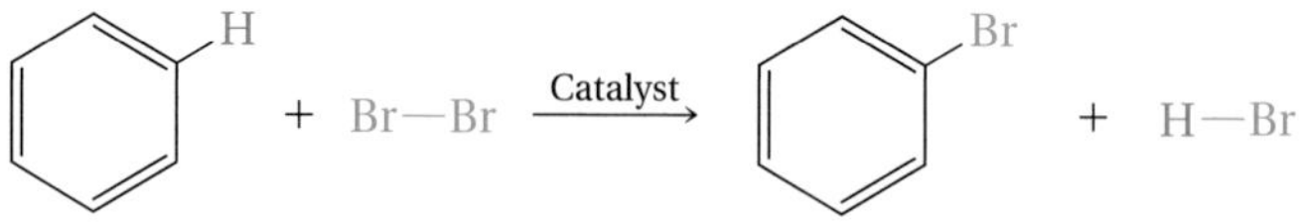

Benzene   Bromine   Bromobenzene   Hydrogen
                         (phenyl bromide) bromide

EXAMPLE 12.4

Predicting the major product of hydrohalogenation

Which is the major addition product in the following reaction?

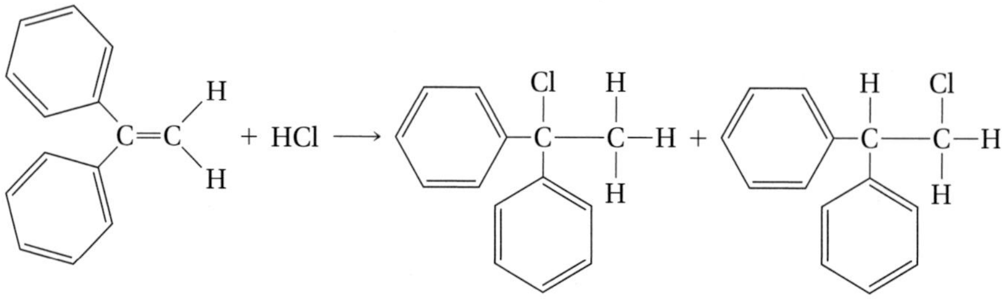

SOLUTION

The reaction is a hydrohalogenation in which hydrogen chloride adds to the double bond of the alkene. Markovnikov's rule predicts that the hydrogen of HCl will end up on the carbon of the reacting double bond that has more hydrogens; the chlorine of HCl will end up on the carbon that has fewer attached hydrogens. In the reactant, one of the carbons of the double bond has two hydrogens; the other carbon of the double bond has none. The major product of the reaction should be

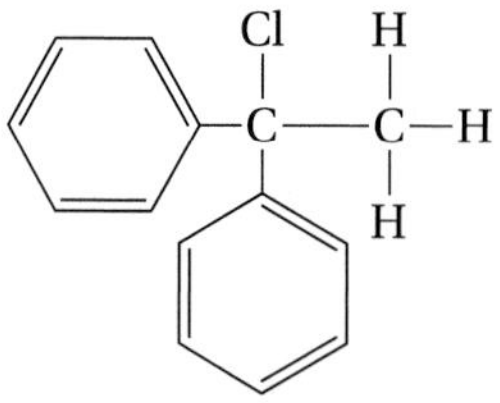

Major product

PRACTICE EXERCISE 12.4

Give the structure(s) for the expected product(s) from each of the following reactions.

(a)

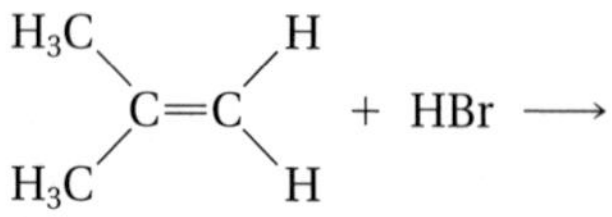

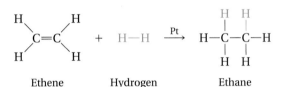

Hydrogenation

Besides halogens and hydrogen halides, hydrogen also adds to carbon-carbon double bonds to give alkanes. These hydrogenation reactions usually require a catalyst. Finely divided platinum (Pt) or palladium (Pd) are often used. Figure 12.2 shows an industrial chemist working with an assortment of catalysts. Here are two examples:

Figure 12.2
A collection of catalysts. Catalysts are used to speed up a wide variety of chemical reactions.

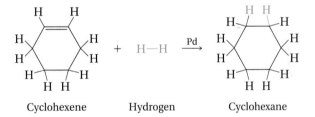

Hydrogenation of a double bond is a *reduction reaction* (see Sec. 6.4)—ethene is reduced to ethane, for example, and cyclohexene is reduced to cyclohexane. We will examine reduction reactions of organic compounds in more detail in Chapter 13.

Benzene resists hydrogenation, just as it resists halogenation and hydrohalogenation. At high temperatures and high pressures of hydrogen and in the presence of a catalyst, however, three molecules of hydrogen reduce one molecule of benzene to cyclohexane:

Polymerization reactions

The flammability of hydrocarbons decreases as the halogen content increases. Because they cannot be burned at ordinary combustion temperatures, the destruction of polyhalogenated hydrocarbons (such as PVC) is a serious waste-disposal problem.

Many useful **polymers**—*long molecules formed from smaller molecular units*—are made from alkene *monomers*. **Monomers** *are molecules that add to each other to form the repeating units of the polymer.* Many halocarbon polymers have useful properties. Two examples are polyvinyl chloride and polytetrafluoroethene. Vinyl chloride ($CH_2{=}CHCl$) is the monomer of polyvinyl chloride.

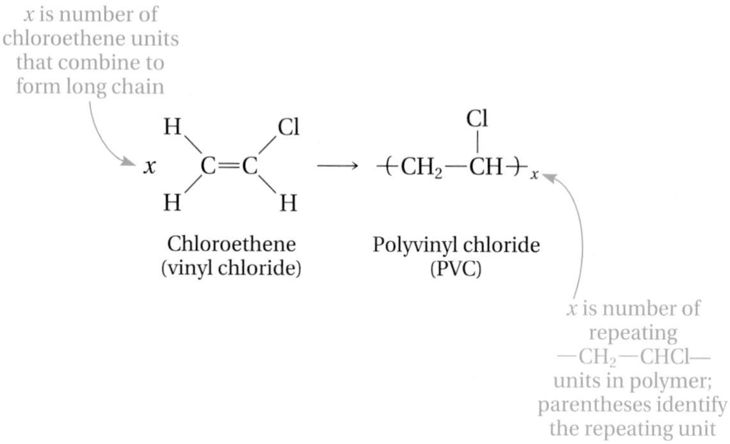

x is number of chloroethene units that combine to form long chain

$$x \quad \begin{array}{c} H \\ \diagdown \\ H \end{array} C{=}C \begin{array}{c} Cl \\ \diagup \\ \diagdown \\ H \end{array} \quad \longrightarrow \quad \left(CH_2{-}\overset{\overset{\displaystyle Cl}{|}}{CH}\right)_x$$

Chloroethene
(vinyl chloride)

Polyvinyl chloride
(PVC)

x is number of repeating $-CH_2-CHCl-$ units in polymer; parentheses identify the repeating unit

Polyvinyl chloride (PVC) is used for pipes in plumbing and is also produced in sheets for use as tough plastic covering on upholstery. Polytetrafluoroethene (Teflon or PTFE) is the product of the polymerization of tetrafluoroethene ($F_2C{=}CF_2$) monomers.

$$xCF_2{=}CF_2 \quad \longrightarrow \quad \left(CF_2{-}CF_2\right)_x$$

Tetrafluoroethene

Teflon (PTFE)

Teflon is resistant to most chemicals. A coating of Teflon on the inside surfaces of pots and pans makes them nonstick. Teflon is used in the manufacture of medical instruments and is also produced as a tape for sealing pipe fittings.

Other alkene molecules can react to form polymers, as shown here for ethene:

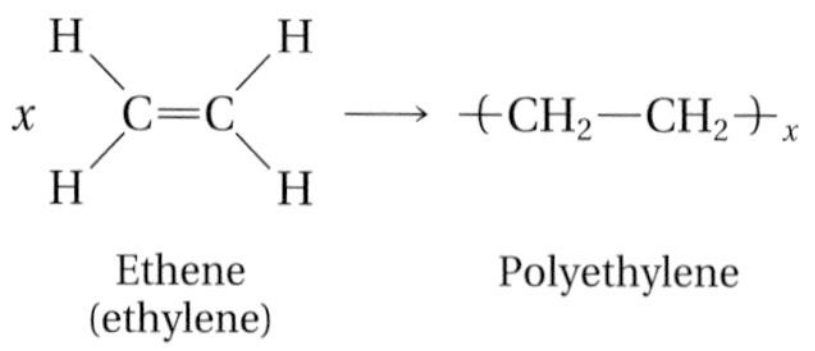

$$x \quad \begin{array}{c} H \\ \diagdown \\ H \end{array} C{=}C \begin{array}{c} H \\ \diagup \\ \diagdown \\ H \end{array} \quad \longrightarrow \quad \left(CH_2{-}CH_2\right)_x$$

Ethene
(ethylene)

Polyethylene

Ethene is the monomer of polyethylene. Polyethylene is an important industrial product. Chemically resistant and easy to clean, it is used to make refrigerator dishes, plastic milk bottles, laboratory wash bottles, and many other familiar items. By shortening or lengthening the carbon chains, chemists can control the physical properties of polyethylene. Polyethylene containing relatively short chains ($x = 100$) has the consistency of paraffin wax; that with long chains ($x = 1000$) is harder and more rigid, like plastic milk containers. Polypropylene is prepared by polymerization of propene.

$$x\overset{\overset{\displaystyle CH_3}{|}}{CH}{=}CH_2 \longrightarrow \overset{\overset{\displaystyle CH_3}{|}}{+CH}{-}CH_2+_x$$

Propene   Polypropylene
(propylene)

Polypropylene makes a stiffer polymer than polyethylene, and it is used extensively in utensils and containers.

Polystyrene is prepared by the polymerization of styrene (vinyl benzene):

$$x CH_2{=}CH \longrightarrow +CH_2{-}CH+_x$$

Styrene   Polystyrene
(vinyl benzene)

Polystyrene is a poor heat conductor when produced as a foam. It is used to insulate homes and in the manufacture of molded items such as coffee cups and coolers.

Polyisoprene is the polymer that constitutes natural rubber. It is used in the manufacture of boots, tires, and rubber tubing. The polyisoprene molecule contains *cis* double bonds:

$$x CH_2{=}\overset{\overset{\displaystyle }{|}}{C}CH{=}CH_2 \longrightarrow \overset{+CH_2 \qquad CH_2+_x}{\underset{H_3C \qquad\quad H}{C{=}C}}$$
$$\quad\;\; \overset{|}{CH_3}$$

Isoprene   Polyisoprene

Figure 12.3 shows a small selection of the wide variety of articles made from polymers.

Figure 12.3
Synthetic polymers play an important role in our daily lives.

12.3 Alcohols

AIMS: To name and draw structures of alcohols, glycols, and phenols. To identify an alcohol as being primary, secondary, or tertiary. To identify the uses of some common alcohols.

Alcohols *are compounds in which one hydrogen of the water molecule is replaced by a hydrocarbon chain or ring. The functional group —OH in alcohols is called a* **hydroxy function** *or* **hydroxyl group.** In organic chemistry, the symbol **R** is often used to represent any carbon chains or carbon rings. Therefore, the general formula of an alcohol can be written as ROH.

hydroxyl group

$$\overset{\displaystyle ..}{\underset{H \qquad H}{O}}$$ $$\overset{\displaystyle ..}{\underset{R \qquad H}{O}}$$

Water molecule   Alcohol molecule

Chemists often arrange aliphatic alcohols into structural categories according to the number of R groups attached to the *carbon* that is bonded to the hydroxyl group. The R groups in a molecule may be the same or may be different. Only one R group and two hydrogens are attached to the C—OH of a primary (abbreviated 1°) alcohol. Two R groups and one hydrogen are attached to the C—OH of a secondary (2°) alcohol. A tertiary (3°) alcohol has three R groups and no hydrogens attached to the C—OH.

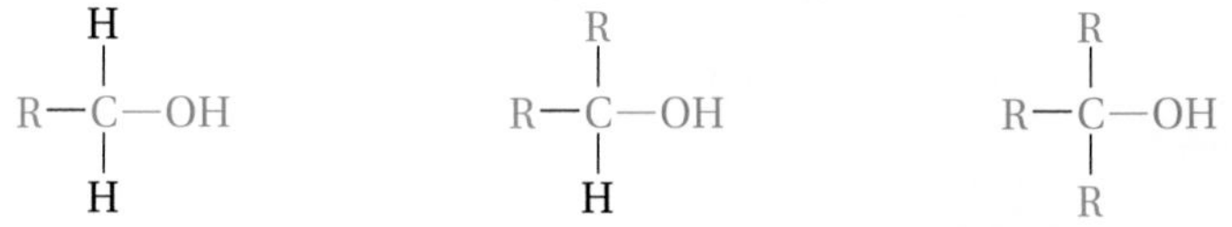

$$
\begin{array}{ccc}
\text{H} & \text{R} & \text{R} \\
| & | & | \\
\text{R—C—OH} & \text{R—C—OH} & \text{R—C—OH} \\
| & | & | \\
\text{H} & \text{H} & \text{R}
\end{array}
$$

Primary (1°) alcohol   Secondary (2°) alcohol   Tertiary (3°) alcohol

---

EXAMPLE 12.5   **Determining whether an alcohol is primary, secondary, or tertiary**

Identify each of the following compounds as a primary, secondary, or tertiary alcohol.

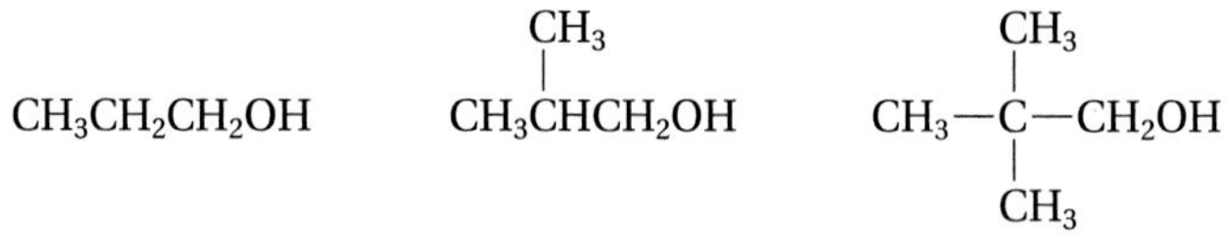

$$
\begin{array}{ccc}
 & \text{CH}_3 & \text{CH}_3 \\
 & | & | \\
\text{CH}_3\text{CH}_2\text{CH}_2\text{OH} & \text{CH}_3\text{CHCH}_2\text{OH} & \text{CH}_3\text{—C—CH}_2\text{OH} \\
 & & | \\
 & & \text{CH}_3
\end{array}
$$

SOLUTION

The alcohols in this example have increasingly complex hydrocarbon chains. However, it is the number of alkyl groups bonded to the carbon in C—OH that determines whether an alcohol is primary, secondary, or tertiary: A primary alcohol has one R group and two hydrogens attached, a secondary alcohol has two R groups and one hydrogen attached, and a tertiary alcohol has three R groups and no hydrogens attached. In each of the examples, the carbon bonded to the hydroxyl group has only one R group and two hydrogens attached. Therefore, each of these alcohols is a primary alcohol.

Naming alcohols

To name straight-chain and substituted alcohols by the IUPAC system, we drop the *-e* ending of the name of the parent alkane and add the ending *-ol.* The parent alkane is the longest continuous chain of carbons that includes the carbon attached to the hydroxyl group. In numbering the longest continuous chain, we give the position of the hydroxyl group the lowest possible number. Alcohols containing two, three, and four —OH substituents are named *diols, triols,* and *tetrols,* respectively. Common names of aliphatic alcohols are written in the same way as those for the halocarbons. The alkyl group—methyl, for example—is named and followed by the word *alcohol,* as in methyl alcohol. Compounds with more than one —OH substituent are called *glycols.* Here are some simple aliphatic alcohols along with their IUPAC and common names:

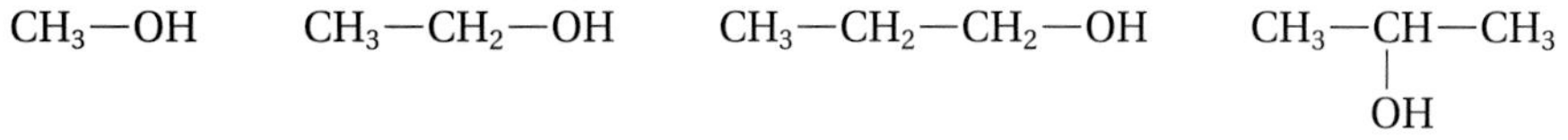

$$CH_3-OH \qquad CH_3-CH_2-OH \qquad CH_3-CH_2-CH_2-OH$$

$$CH_3-\overset{\displaystyle |}{\underset{\displaystyle OH}{CH}}-CH_3$$

| Methanol (methyl alcohol) | Ethanol (ethyl alcohol) | 1-Propanol (propyl alcohol) | 2-Propanol (isopropyl alcohol) |

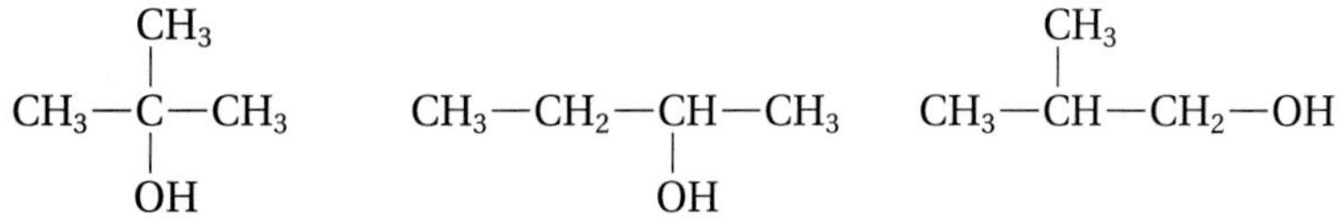

$$CH_3-\overset{\displaystyle CH_3}{\underset{\displaystyle OH}{\overset{\displaystyle |}{\underset{\displaystyle |}{C}}}}-CH_3 \qquad CH_3-CH_2-\overset{\displaystyle |}{\underset{\displaystyle OH}{CH}}-CH_3 \qquad CH_3-\overset{\displaystyle CH_3}{\overset{\displaystyle |}{CH}}-CH_2-OH$$

| 2-Methyl-2-propanol (*tert*-butyl alcohol) | 2-Butanol (*sec*-butyl alcohol) | 2-Methyl-1-propanol (isobutyl alcohol) |

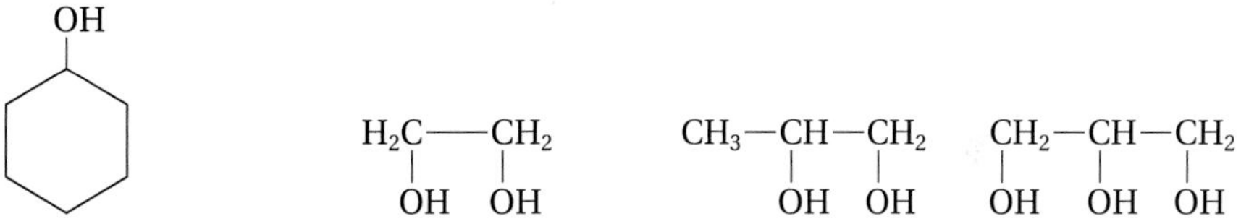

$$H_2C\overset{\displaystyle |}{\underset{\displaystyle OH}{}}-\overset{\displaystyle |}{\underset{\displaystyle OH}{CH_2}} \qquad CH_3-\overset{\displaystyle |}{\underset{\displaystyle OH}{CH}}-\overset{\displaystyle |}{\underset{\displaystyle OH}{CH_2}} \qquad CH_2-\overset{\displaystyle |}{\underset{\displaystyle OH}{CH}}-\overset{\displaystyle |}{\underset{\displaystyle OH}{CH_2}}$$

| Cyclohexanol (cyclohexyl alcohol) | 1,2-Ethanediol (ethylene glycol) (the common name has an *-ene* ending, but the molecule contains no double bond) | 1,2-Propanediol | 1,2,3-Propanetriol (glycerol) |

PRACTICE EXERCISE 12.5

Name the following alcohols.

(a) $CH_3CH_2CH_2CH_2OH$

(b) 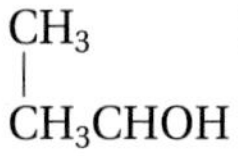 CH_3
    $\quad |$
    CH_3CHOH

(c) 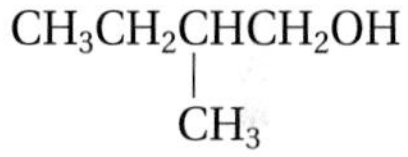 $CH_3CH_2CHCH_2OH$
    $\qquad\qquad |$
    $\qquad\qquad CH_3$

(d) 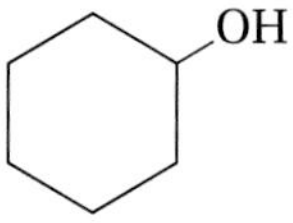 [cyclohexane ring]—OH

(e) 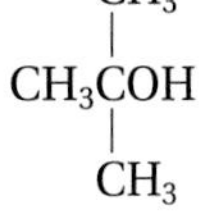 $\overset{\displaystyle CH_3}{\underset{\displaystyle CH_3}{CH_3\overset{\displaystyle |}{\underset{\displaystyle |}{C}}OH}}$

PRACTICE EXERCISE 12.6

Classify each of the alcohols in Practice Exercise 12.5 as primary, secondary, or tertiary.

Uses of alcohols

Many aliphatic alcohols are used in the laboratory, in the health sciences, and in industry. Isopropyl alcohol, a colorless, nearly odorless liquid (bp 82 °C), is the rubbing alcohol used for massages. It is also used as a base for perfumes, creams, lotions, and other cosmetics. Ethylene glycol (bp 197 °C) is the principal ingredient of antifreeze. Its advantages over other high-boiling liquids are its solubility in water and a freezing point of −17.4 °C. If

Drinking during pregnancy increases the chance of miscarriage, may cause low birth weight, and is linked to fetal alcohol syndrome. Characteristics of this disorder include abnormal limb development, facial abnormalities such as cleft palate, heart defects, and lower-than-average intelligence.

water is added to ethylene glycol, the mixture freezes at an even lower temperature—a 50% (v/v) aqueous solution of ethylene glycol freezes at $-36\ °C$. Glycerol is a viscous, sweet-tasting, water-soluble liquid. It is used as a lubricant in suppositories and as a moistening agent in cosmetics, foods, and drugs. Glycerol is also an important component of fats and oils.

Ethanol (bp 78.5 °C) is the most important alcohol. Also called *grain alcohol*, much ethanol is produced by yeast fermentation of sugar. Archaeological evidence indicates that cave dwellers had produced alcoholic beverages by using fermentation. Since those early times, people have pursued and perfected that art. Virtually every grain, fruit, or vegetable has been fermented: corn, wheat, rye, rice, grapes, tomatoes, dandelions, elderberries, cherries, potatoes, and even cactus pulp. The *proof* of an alcoholic beverage is twice the alcohol content. For example, 100 proof spirits contain 50% ethanol. Ethanol itself is tasteless and odorless. The differences in taste among alcoholic beverages are the result of other products formed during the fermentation process. Among these products are the fusel oils, which include propyl, isopropyl, and other straight- and branched-chain aliphatic alcohols. Fusel oils are often toxic and are one cause of the hangover effects of excessive drinking.

Most ethanol is produced by the fermentation of corn starch and cane sugar. Ethanol is an alternative fuel for use in automobiles. It is mixed with gasoline to form *gasohol*. At a concentration of 10% ethanol, gasohol can be used in a standard automobile engine.

Ethanol is a hypnotic or sleep-inducing drug. Its abuse is a serious public health problem. Some experts estimate that one in nine adult Americans have a drinking problem of some kind. There is no evidence that moderate drinking is harmful.

Ethanol for industrial use is often *denatured*. **Denatured alcohol** *is ethanol that contains an added substance to make it toxic.* Methanol, sometimes called *wood alcohol* because before 1925 it was prepared by the distillation of wood, is often the denaturant. The inability to remember the difference between grain alcohol and wood alcohol has spelled tragedy for many people. Wood alcohol is extremely toxic: 10 mL has been reported to cause permanent blindness—and as little as 30 mL, death. Some undenatured ethanol is used in laboratories under careful supervision. Pure ethanol is labeled *absolute alcohol*. A preparation containing 5% water is also available. Nobody should ever attempt to drink laboratory alcohol.

Phenols

Phenols *are compounds in which a hydroxyl group is attached directly to an aromatic ring.* Many phenolic compounds are found in nature. Shown below are phenol, the parent compound, and some natural and synthetic derivatives.

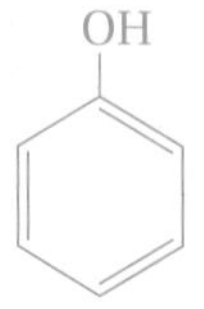

Phenol 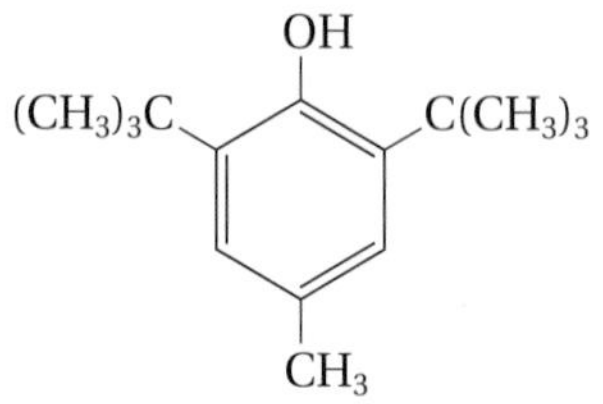   BHT 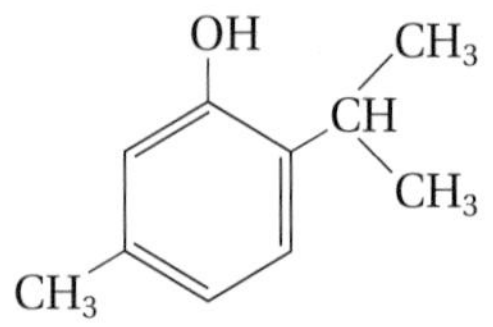   Thymol

Phenol was formerly used as an antiseptic in hospitals (see A Closer Look: Phenolic Antiseptics and Disinfectants). A synthetic derivative of phenol, butylated hydroxy toluene (BHT), is widely used as a food preservative. BHT is often put in food wrappers rather than in the food itself to keep it from imparting a rather antiseptic flavor to the product.

A Closer Look

Phenolic Antiseptics and Disinfectants

Phenol, also called *carbolic acid,* was first used as a medical antiseptic in 1867 by Joseph Lister. He demonstrated that the incidence of infections after surgical procedures was dramatically reduced when solutions of phenol were used to clean the surgical instruments, the operating room, and the patient's skin. Today's use of phenol for medicinal purposes is more limited. When phenol is absorbed through the skin, ingested, or inhaled, it is toxic. It also can cause severe chemical burns. However, it is available as an antiseptic in aqueous or alcoholic solutions that contain a maximum of 1.5% phenol. Phenol is an ingredient in a variety of throat lozenges that can contain up to 50 mg phenol per lozenge. It is a useful topical drug for treating sore throats because it numbs the inflamed area.

A number of derivatives of the parent phenol are used today as antiseptics. Many mouthwashes and throat lozenges include alkyl-substituted phenols as their active ingredients for pain relief (see figure). The compound 4-hexylresorcinol has a superior antibacterial action to that of phenol. It is also much less toxic. Therefore, it is an ingredient of choice in mouthwashes and throat lozenges.

4-hexylresorcinol

Phenols and methyl phenols (cresols) are commonly used as disinfectants in hospitals and around the home. Disinfectants are formulated for use on inanimate objects, not living tissues. Phenol's germicidal properties result because these chemicals have the ability to disrupt the microorganisms' cell wall permeability.

The naturally occurring phenol eugenol is found in cloves and is used in dentistry to relieve toothaches. Methyl salicylate, another phenol derivative commonly known as oil of wintergreen, is a flavoring agent and is used in liniments to relieve sore muscles. Another derivative of phenol, thymol, is used as an antiseptic in mouthwash preparations.

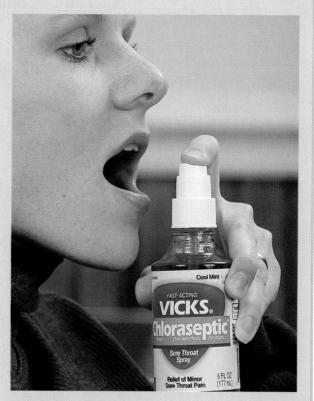

Phenol and its derivatives are the active ingredients in many of the preparations used for oral hygiene.

12.4 Making alcohols

AIM: To illustrate the synthesis of alcohols by addition and displacement reactions.

An organic chemist who wants to make an alcohol may find over a dozen ways to do the job. We will discuss two ways: addition of water to carbon-carbon double bonds of alkenes and reactions of halocarbons with hydroxide ions.

Addition of water to alkenes

The general reaction for adding water to the double bond in an alkene is

$$\begin{array}{c}\diagdown \\ C = C \\ \diagup \\\end{array} + \text{H—OH} \xrightarrow{\text{H}^+} \begin{array}{c} \text{H} \quad \text{OH} \\ | \quad\quad | \\ -C - C - \\ | \quad\quad | \end{array}$$

Alkene   Water   Alcohol

The addition of water to an alkene is called a **hydration reaction.** Hydration reactions usually require heating of the alkene and water at about 100 °C in the presence of a trace of strong acid. The acid serves as a catalyst for the reaction; hydrochloric or sulfuric acid is generally used. For alkenes with alkyl substituents on the double bond, Markovnikov's rule applies to the preparation of alcohols, just as it does to the preparation of halocarbons. Hydrogen goes to the carbon of the double bond with more hydrogens; —OH goes to the other carbon. This is shown for the addition of water to propene:

$$\begin{array}{c} \text{H} \quad\quad \text{H} \\ \diagdown \quad\quad \diagup \\ C = C \\ \diagup \quad\quad \diagdown \\ \text{H}_3\text{C} \quad\quad \text{H} \end{array} + \text{H—OH} \xrightarrow[100\,°\text{C}]{\text{HCl}} \begin{array}{c} \text{H} \quad\quad \text{H} \\ | \quad\quad\quad | \\ \text{CH}_3 - C\!-\!\!-\!\!-\!C - \text{H} \\ | \quad\quad\quad | \\ \text{OH} \quad\quad \text{H} \end{array} + \begin{array}{c} \text{H} \quad \text{H} \\ | \quad\quad | \\ \text{CH}_3 - C - C - \text{H} \\ | \quad\quad | \\ \text{H} \quad \text{OH} \end{array}$$

Propene   2-Propanol   1-Propanol
(propylene)   (isopropyl alcohol)   (minor product)
                 (major product)

Reactions of halocarbons with hydroxide ions

The reactions of halocarbons with hydroxide ions to give alcohols are examples of *displacement reactions. In a* **displacement reaction,** *a reactant replaces a substituent group (—X) on another reactant.* The substituent group —X that is displaced is usually a halogen, most often a chloro-, bromo-, or iodo- substituent. The general displacement reaction for the preparation of alcohols from halocarbons is

$$\text{R—X} \quad + \quad \text{OH}^- \xrightarrow[100\,°\text{C}]{\text{H}_2\text{O}} \text{R—OH} + \text{X}^-$$

Halocarbon   Hydroxide   Alcohol   Halide
               ion   ion

Chemists usually use aqueous solutions of sodium or potassium hydroxide as the source of hydroxide ions. Fluoro- groups are not easily displaced, and fluorocarbons are seldom, if ever, used to make alcohols. Here are a few specific examples of displacement reactions:

$$CH_3-I \quad + \quad KOH \quad \xrightarrow[100\,°C]{H_2O} \quad CH_3-OH \quad + \quad KI$$

Iodomethane (methyl iodide)	Potassium hydroxide	Methanol	Potassium iodide

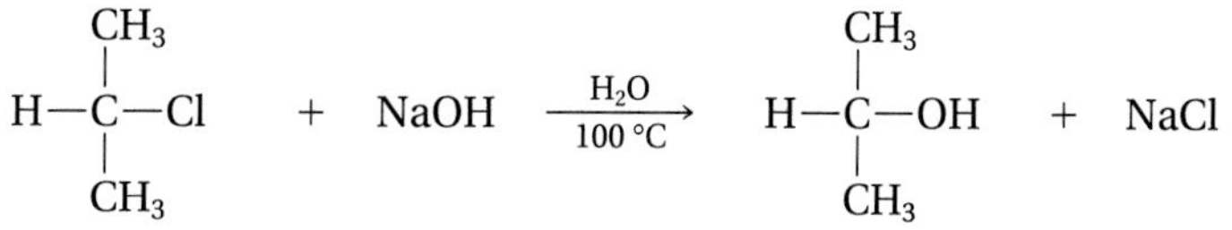

2-Chloropropane (isopropyl chloride) + Sodium hydroxide → 2-Propanol (isopropyl alcohol) + Sodium chloride

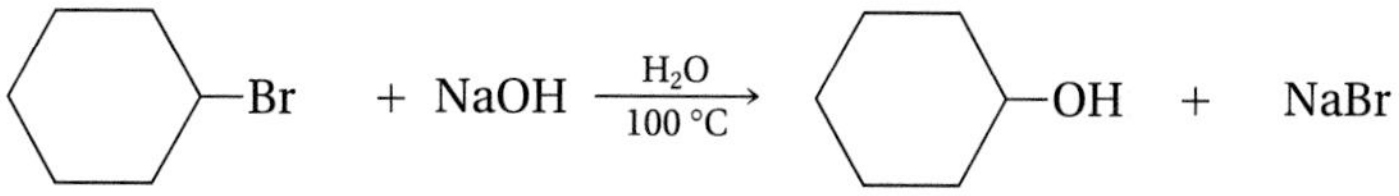

Bromocyclohexane (cyclohexyl bromide) + NaOH → Cyclohexanol (cyclohexyl alcohol) + Sodium bromide

Phenols can be prepared from aromatic halides. Because of the stability of the benzene ring, the reaction must be performed at high pressure and a temperature of 350 °C.

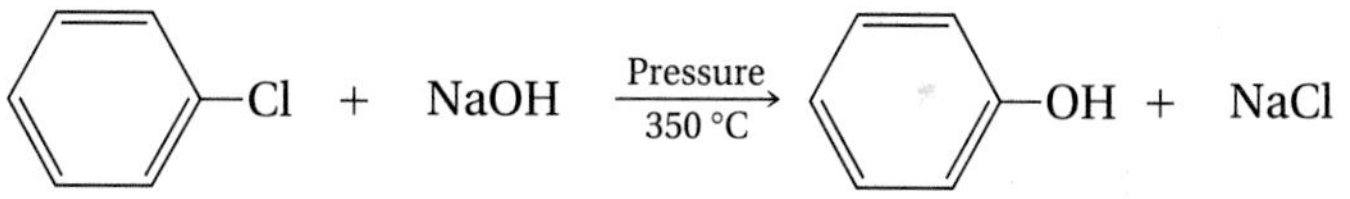

Chlorobenzene (phenyl chloride) + Sodium hydroxide $\xrightarrow[350\,°C]{Pressure}$ Phenol + Sodium chloride

12.5 Elimination reactions

AIM: To use Saytzeff's rule to predict the major product of an elimination reaction.

You may recall that hydrogen halides and water add to double bonds to yield halocarbons and alcohols, respectively. However, the addition process also can be reversed—water can be removed from alcohols and hydrogen halides or halogens can be removed from halocarbons to produce alkenes. *Reversals of addition reactions are* **elimination reactions.** The general reaction for elimination is

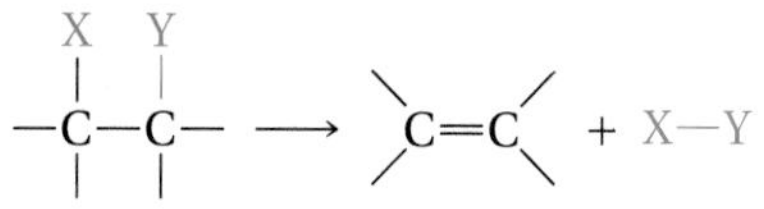

Dehydrohalogenation

Dehydrohalogenation *is the elimination of hydrogen halides such as HCl or HBr from a carbon-carbon bond.* The elimination of a hydrogen halide usually requires heating the halocarbon in a concentrated solution of a strong base such as sodium hydroxide or potassium hydroxide. As the acidic hydrogen halide is formed, it is immediately neutralized by the strong base to form a salt:

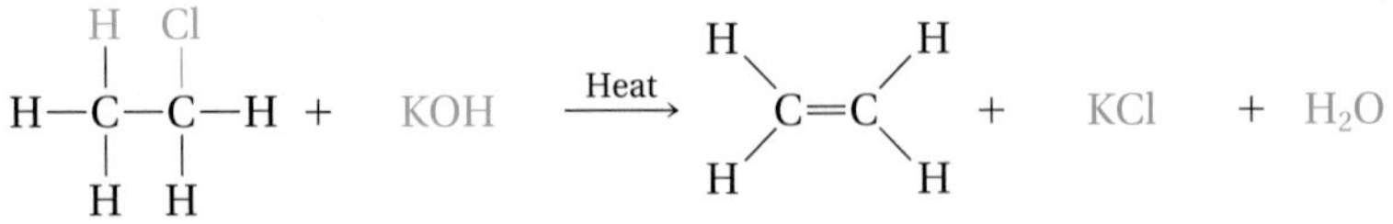

$$
\underset{\substack{\text{Chloroethane}\\ \text{(ethyl chloride)}}}{H-\overset{\displaystyle H}{\underset{\displaystyle H}{C}}-\overset{\displaystyle Cl}{\underset{\displaystyle H}{C}}-H} + \underset{\substack{\text{Potassium}\\ \text{hydroxide}}}{KOH} \xrightarrow{\text{Heat}} \underset{\substack{\text{Ethene}\\ \text{(ethylene)}}}{\overset{H}{\underset{H}{}}C=C\overset{H}{\underset{H}{}}} + \underset{\substack{\text{Potassium}\\ \text{chloride}}}{KCl} + \underset{\text{Water}}{H_2O}
$$

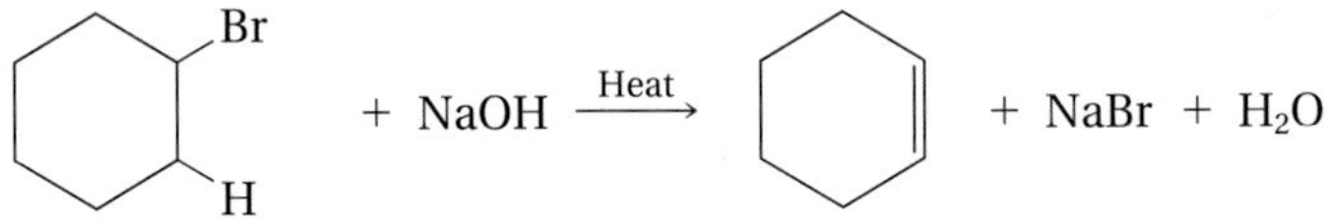

Bromocyclohexane (cyclohexyl bromide) + NaOH $\xrightarrow{\text{Heat}}$ Cyclohexene + NaBr + H₂O

Dehydration

Concentrated sulfuric acid is such a strong dehydrating agent that it can remove the elements of water from organic compounds that themselves contain no free water. For example, concentrated sulfuric acid removes water from table sugar, $C_{12}H_{22}O_{11}$, leaving a residue of carbon. In a similar reaction, concentrated sulfuric acid can char paper, cotton, and wool and destroy skin tissue.

Dehydration *is the elimination of water.* To prepare an alkene from an alcohol, a solution of the alcohol in concentrated sulfuric acid is usually heated to 180 °C. The alkene is collected by distillation as it is formed.

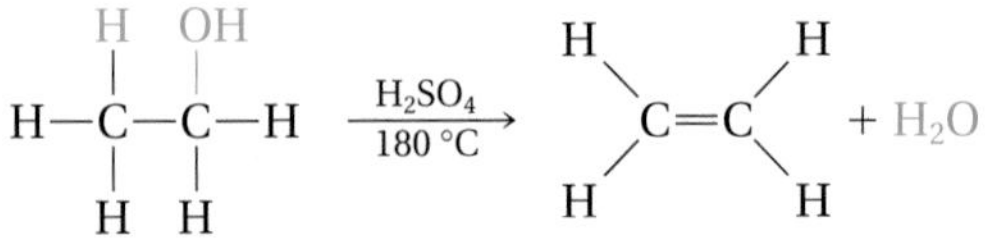

$$
H-\overset{\displaystyle H}{\underset{\displaystyle H}{C}}-\overset{\displaystyle OH}{\underset{\displaystyle H}{C}}-H \xrightarrow[180\,°C]{H_2SO_4} \overset{H}{\underset{H}{}}C=C\overset{H}{\underset{H}{}} + H_2O
$$

bp 78.5 °C   bp −104 °C

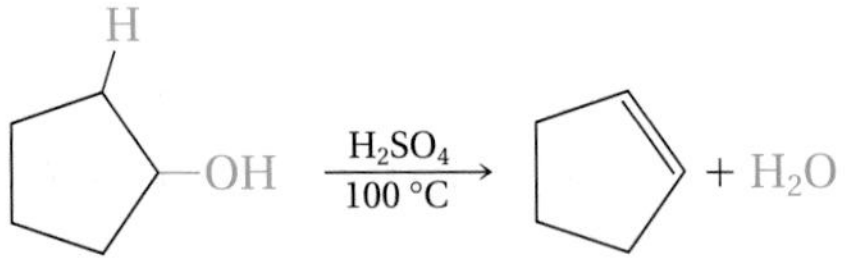

cyclopentanol (bp 140 °C) $\xrightarrow[100\,°C]{H_2SO_4}$ cyclopentene (bp 44 °C) + H₂O

Dehalogenation

Dehalogenation *is the elimination of halogen from a carbon-carbon bond of a halocarbon.* Alkenes also can be prepared by elimination of halogen from an alkyl dichloride or dibromide using finely divided zinc metal as a reagent.

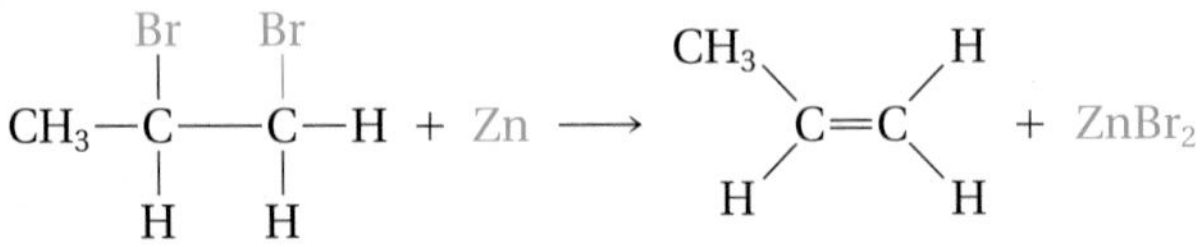

$$
CH_3-\overset{\displaystyle Br}{\underset{\displaystyle H}{C}}-\overset{\displaystyle Br}{\underset{\displaystyle H}{C}}-H + Zn \longrightarrow \overset{CH_3}{\underset{H}{}}C=C\overset{H}{\underset{H}{}} + ZnBr_2
$$

Direction of elimination

Sometimes more than one alkene can be produced during an elimination reaction. In such instances, we can predict which alkene will be the major product by **Saytzeff's rule:** *The major product of elimination is the alkene with the largest number of carbon groups on the carbons of the double bond.* For example:

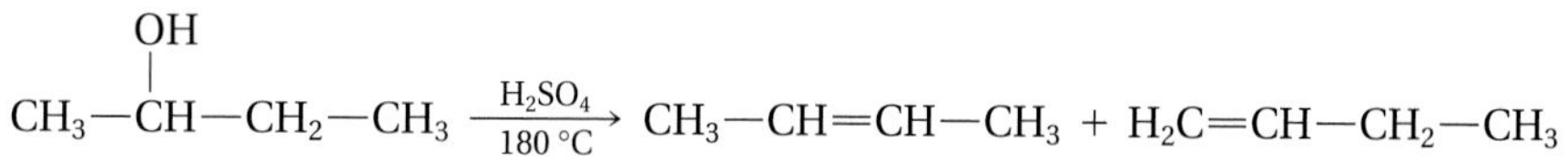

$$CH_3-\overset{\overset{\displaystyle OH}{|}}{CH}-CH_2-CH_3 \xrightarrow[180\,°C]{H_2SO_4} CH_3-CH=CH-CH_3 + H_2C=CH-CH_2-CH_3$$

2-Butanol (*sec*-butyl alcohol)	2-Butene Major product (90%) (2 carbon groups on the carbons of the double bond)	1-Butene Minor product (10%) (1 carbon group on the carbons of the double bond)

PRACTICE EXERCISE 12.7

Write the product or products you would expect to get from each of the following elimination reactions.

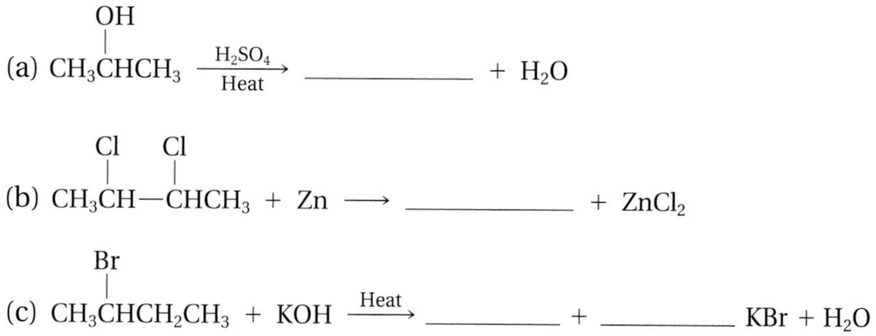

(a) $CH_3\overset{\overset{\displaystyle OH}{|}}{C}HCH_3 \xrightarrow[\text{Heat}]{H_2SO_4}$ _____ $+ H_2O$

(b) $CH_3\overset{\overset{\displaystyle Cl}{|}}{C}H-\overset{\overset{\displaystyle Cl}{|}}{C}HCH_3 + Zn \longrightarrow$ _____ $+ ZnCl_2$

(c) $CH_3\overset{\overset{\displaystyle Br}{|}}{C}HCH_2CH_3 + KOH \xrightarrow{\text{Heat}}$ _____ $+$ _____ $KBr + H_2O$

12.6 Alkoxides

AIM: To explain the behavior of alcohols and phenols as weak acids.

Water dissociates to a slight extent to hydronium ions and hydroxide ions. Alcohols are very similar to water in this respect:

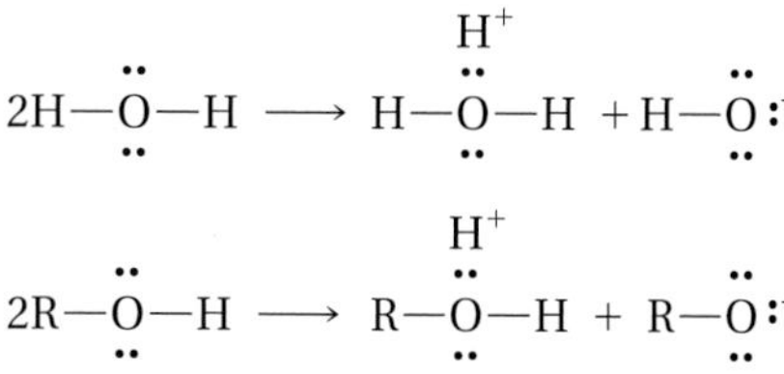

$$2H-\overset{\cdot\cdot}{\underset{\cdot\cdot}{O}}-H \longrightarrow H-\overset{H^+}{\overset{\cdot\cdot}{\underset{\cdot\cdot}{O}}}-H + H-\overset{\cdot\cdot}{\underset{\cdot\cdot}{O}}\!:^-$$

$$2R-\overset{\cdot\cdot}{\underset{\cdot\cdot}{O}}-H \longrightarrow R-\overset{H^+}{\overset{\cdot\cdot}{\underset{\cdot\cdot}{O}}}-H + R-\overset{\cdot\cdot}{\underset{\cdot\cdot}{O}}\!:^-$$

And just as reactive metals such as sodium react vigorously with water to form the strong bases called *hydroxides, reactive metals react vigorously with alcohols to form strong bases called* **alkoxides:**

$$2H\text{—}OH + 2Na \longrightarrow 2H\text{—}O^- + 2Na^+ + H_2$$

Hydroxide
ion

$$2R\text{—}OH + 2Na \longrightarrow 2R\text{—}O^- + 2Na^+ + H_2$$

Alkoxide
ion

Because of their weak acidic character, aqueous solutions of alcohols are essentially neutral. Phenols are much stronger acids than alcohols, but they are still very weak acids. Phenols produce slightly acidic solutions.

PRACTICE EXERCISE 12.8

Write the equation for the reaction of sodium with each of the following:
(a) methanol   (b) phenol.

12.7 Ethers

AIMS: To name and draw structures of ethers. To illustrate the synthesis of an ether from a halocarbon and an alkoxide ion.

Focus

Ethers are disubstituted derivatives of water.

Ethers *are compounds in which both hydrogens of water are replaced by carbon chains or rings.* The general formula for ethers is R—O—R. The R stands for any alkyl or aryl group.

Ö
H   H

Ö
R   R

Water molecule   Ether molecule

The alkyl or aryl groups joined by the ether linkage are named in alphabetical order and are followed by the word *ether.* For example:

$CH_3CH_2\text{—}O\text{—}CH_3$

Ethylmethyl ether

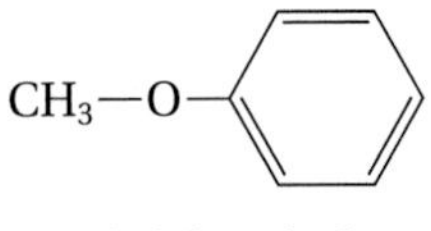

$CH_3\text{—}O\text{—}$⬡

Methylphenyl ether
(anisole)

Ethylmethyl ether and methylphenyl ether are asymmetrical ethers because the R groups attached to the oxygen are different. When both R groups are the same, the ether is symmetrical. Symmetrical ethers are named by using the prefix *di-.* For example:

$CH_3CH_2\text{—}O\text{—}CH_2CH_3$

Diethyl ether

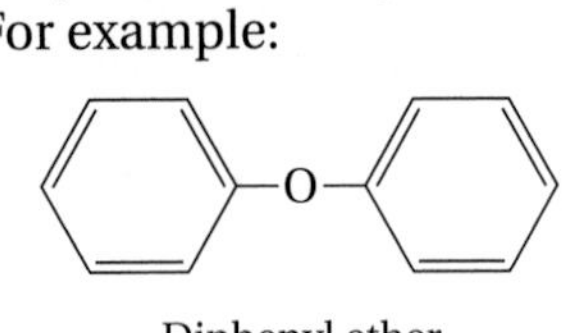

⬡—O—⬡

Diphenyl ether
(phenyl ether)

Many modern anesthetics contain halogen and ether functional groups, as described in A Closer Look: Halocarbon and Ether Anesthetics.

Halocarbon and Ether Anesthetics

Anesthetics have alleviated a great deal of pain and suffering during surgery. *Local anesthetics* make one part of the body insensitive to pain but leave the patient conscious. *General anesthetics* act on the brain to produce unconsciousness and insensitivity to pain. Many general anesthetics are halocarbons or ethers or contain both kinds of functional groups in their molecular structures.

Diethyl ether (C_2H_5—O—C_2H_5), the first general anesthetic, was introduced into surgery in 1846 by William Morton, a Boston dentist (see figure). In 1847, chloroform ($CHCl_3$) was introduced as a general anesthetic. Both diethyl ether and chloroform can cause undesirable side effects. Today, neither compound is used as an anesthetic in the Western Hemisphere.

Why are compounds such as diethyl ether and chloroform general anesthetics? The potency of an anesthetic is related to its solubility in fatty tissue. One theory for the action of general anesthetics is that they dissolve in the fatlike membranes of nerve cells of the brain (neurons). This changes the properties of the membranes. As a consequence, the activity of the neurons is depressed, leading to anesthesia.

The fat-solubility theory carries over into modern anesthetics. Today's anesthetics include relatively nonpolar fluorine-containing organic compounds such as halothane (Fluothane, $CF_3CHBrCl$). Enflurane (Ethrane, Efrane, $CHFClCF_2$—O—CHF_2) and isoflurane (Forane, CF_3CHCl—O—CHF_2) are also used. All these compounds are inhalant anesthetics. These compounds are nonflammable and relatively safe, and the patient recovers rapidly from their effects.

Halothane is nonexplosive. The start of anesthesia is rapid, but slower than for anesthetics of greater solubility in membranes such as enflurane and isoflurane.

Enflurane is a stable liquid that is somewhat less volatile than halothane. Enflurane provides rapid anesthesia and rapid recovery for the patient. Although enflurane is broken down in the liver to produce fluoride ions, elevated fluoride levels in the blood are not considered a problem.

Isoflurane has physical, pharmacologic, and clinical properties that are similar to those of halothane and enflurane. This anesthetic is a more potent muscle relaxant than halothane, and induction of anesthesia is relatively rapid.

General anesthesia is usually induced with the administration of an intravenous anesthetic, regardless of the inhalant anesthetic subsequently used for anesthesia maintenance. The most commonly used induction agent is the barbiturate thiopental, also known as Pentothal (see Sec. 15.10). In many instances, low concentrations of halothane, enflurane, or isoflurane are used in conjunction with nitrous oxide (N_2O).

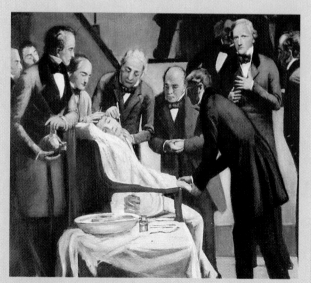

The first use of ether as an anesthetic in 1846 is depicted in this painting.

Just as alcohols can be prepared by using hydroxide ions to displace halogen from a halocarbon, ethers can be prepared by using alkoxide ions. For example:

$$CH_3-I \quad + \quad CH_3CH_2O^-Na^+ \quad \longrightarrow \quad CH_3CH_2-O-CH_3 \quad + \quad NaI$$

| Iodomethane | Sodium ethoxide | Ethylmethyl ether | Sodium iodide |

The ether linkage is often found in rings. *Rings that contain elements other than carbon are called* **heterocyclic rings** *or heterocycles.* Here are some common oxygen-containing heterocycles:

| Furan | Tetrahydrofuran | Pyran | Tetrahydropyran |

The fundamental ring structures of furan and pyran are found in many natural sugars. Compounds containing an oxygen atom in a three-membered ring are called *epoxides.* Epoxyethane is the simplest example.

Epoxyethane
(ethylene oxide)

The sex attractant of the female gypsy moth is the following epoxide. A very small quantity of the synthetic material is placed in traps to attract the male moth. Trapping helps to control the pest, which damages deciduous forests and fruit trees in eastern North America.

Other epoxides are used to make cements and adhesives. With the exception of the epoxides, the ether linkage is very resistant to chemical modification.

Since they are three-membered rings, epoxide rings are highly strained. Therefore, epoxides are much more reactive than other ethers. For example,

FOLLOW-UP TO THE CASE IN POINT: A mistaken conviction

You may recall from the Case in Point that a mother was accused in 1989 of murdering her oldest son by ethylene glycol poisoning. She was convicted and sentenced to life in prison. The case might have ended there, except for a quirk of fate and the efforts of Dr. William Sly and Dr. James Shoemaker of the St. Louis University School of Medicine. The quirk of fate was that while the accused mother was in custody in 1990, she gave birth to a second son, who soon began to exhibit the same symptoms as his late older brother. It was impossible that she had poisoned her youngest son. The St. Louis University scientists, who had followed the case on television, recognized that a rare inherited disease, methylmalonic acidemia, has symptoms very similar to those of ethylene glycol poisoning. Contrary to findings reported in previous blood tests, new tests undertaken by Dr. Shoemaker revealed no evidence of ethylene glycol in the blood of either child. Dr. Piero Rinaldo of the Yale University School of Medicine verified from blood samples that both sons had been born with methylmalonic acidemia. In September 1991, the mother's conviction was reversed and all charges were dismissed. Thanks to science and caring scientists, she has resumed her life with her husband and remaining son.

the epoxide ring of epoxyethane is easily opened. In aqueous solution containing a trace of strong acid, the product is ethylene glycol:

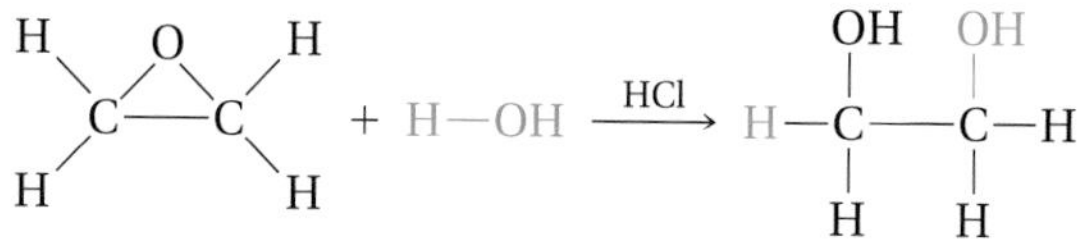

Although very useful as antifreeze, ethylene glycol is toxic and can be fatal if ingested, a fact that led to the accusation of murder described in the Case in Point early in this chapter.

12.8 Physical properties

AIM: To relate differences in boiling point and solubility to the molecular structures of hydrocarbons, halocarbons, alcohols, and ethers.

Focus

The physical properties of organic molecules depend on their molecular structure.

Thus far we have seen the aliphatic and aromatic hydrocarbons, halocarbons, alcohols, and ethers. Except for an occasional comment, not much has been said about their physical properties. This is not an oversight. Discussion of the physical properties of all four classes of compounds at one time will help us understand why the properties are what they are.

Boiling points

Hydrocarbons and halocarbons of low molar mass tend to be gases or low-boiling liquids. Hydrocarbon molecules such as the alkanes are nonpolar. The electron pair in a carbon-hydrogen or carbon-carbon bond is about equally shared by the nuclei of the elements involved. The carbon-halogen bond is only slightly polar. You may recall from Section 8.1 that attractions between molecules, because of hydrogen bonding, require that the molecules contain hydrogen attached to very electronegative atoms such as oxygen. There is no hydrogen bonding in hydrocarbons and halocarbons. Consequently, the forces that hold hydrocarbon or halocarbon molecules together in the liquid state are very weak. Table 11.1 shows that all alkanes containing fewer than five carbons are gases at room temperature.

Boiling points of closely related organic compounds usually increase as molar mass increases. The data in Table 12.4 show this principle. Remember that a pure liquid boils when enough heat energy has been supplied to let molecules in the liquid escape. The sum of weak forces holding heavy nonpolar molecules together in a liquid is greater than the sum of weak forces holding light nonpolar molecules together.

Like water, alcohols are capable of intermolecular hydrogen bonding. Alcohols therefore boil at higher temperatures than alkanes and halocarbons containing comparable numbers of atoms (Table 12.5).

Ethers usually have lower boiling points than alcohols of comparable molar mass, but they have higher boiling points than comparable hydrocarbons and halocarbons.

Table 12.4 Molar Masses and Boiling Points of the Chloromethanes Compared with Those of Methane

Molecular structure	Name	Molar mass (g)	Boiling point (°C)
CH_4	methane	16	−161
CH_3Cl	chloromethane (methyl chloride)	50.5	−24
CH_2Cl_2	dichloromethane (methylene chloride)	85.0	40
$CHCl_3$	trichloromethane (chloroform)	129.5	61
CCl_4	tetrachloromethane (carbon tetrachloride)	154.0	74

Table 12.5 Boiling Points of Alcohols and Comparable Alkyl and Aryl Chlorides

Alcohol	Boiling point (°C)	Alkyl chloride	Boiling point (°C)
CH_3OH	65	CH_3Cl	−64
CH_3CH_2OH	78	CH_3CH_2Cl	13
$CH_3CH_2CH_2OH$	97	$CH_3CH_2CH_2Cl$	47
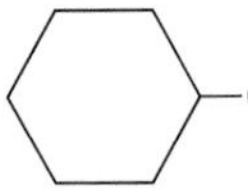 ⬡—OH	162	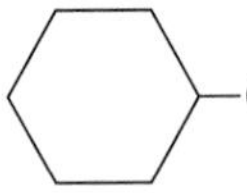 ⬡—Cl	143
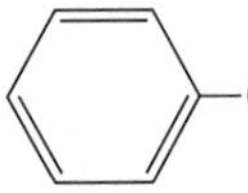 ⬡—OH	182	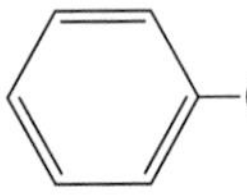 ⬡—Cl	132

Solubility in water

The hydrocarbon parts of chains and rings of organic molecules are **hydrophobic** *("water-hating")—repelled by water.* Oil and water don't mix. If we mix two nonpolar liquids, however, they form a solution. A good rule of thumb is that "like dissolves like."

With the principle that "like dissolves like" in mind, how would we expect alcohols to behave with respect to their solubilities in water? Since alcohols are derivatives of water, we might expect them to have similar properties. And to a point, this is correct. Alcohols of up to four carbons are soluble in water in all proportions. The solubility of alcohols with four or more carbons in the chain is usually much less. For example, the solubility of 1-butanol is only 7.9 g/100 mL of water. The reason is that alcohols consist of two parts: the carbon chain and the hydroxyl group. These parts are in opposition to each other. The carbon chain is nonpolar and hydrophobic, but the hydroxyl group forms hydrogen bonds with water. *Groups such*

Hydrophilic
hudor (Greek): water
philos (Greek): loving

as —OH that interact strongly with water, usually by hydrogen bonding, are called **hydrophilic** *("water-loving") groups.* Alcohols with short carbon chains are soluble in water. Those with longer carbon chains will not dissolve. And some alcohols whose carbon chains are not too long are only slightly soluble.

Ethers are more soluble in water than hydrocarbons and halocarbons but less soluble than alcohols of approximately the same molar mass. The reason is that the oxygens in ethers are hydrogen-bond acceptors, but ethers have no hydroxyl hydrogens to donate in hydrogen bonding. This lower solubility compared with alcohols is overcome in molecules with more than one ether linkage. Dioxane, a cyclic compound with two ether linkages, is soluble in water in all proportions; diethyl ether, with the same number of carbons but only one ether linkage, is not.

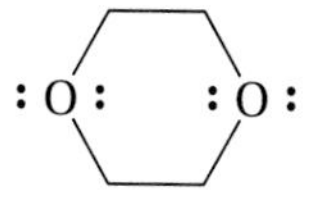

$$:O: \qquad :O: \qquad\qquad CH_3CH_2-\overset{..}{\underset{..}{O}}-CH_2CH_3$$

Dioxane
(soluble in water
in all proportions)

Diethyl ether
(solubility in water:
8 g/100 mL)

PRACTICE EXERCISE 12.9

Name and classify by functional groups the following compounds, and identify the one that is most polar.

(a) CH_3OCH_3 (b) CH_3CH_2Cl (c) CH_3CH_2OH
(d) $CH_3CH_2CH_3$

PRACTICE EXERCISE 12.10

Arrange the compounds in Practice Exercise 12.9 in order of increasing polarity then comment on their relative boiling points and water solubilities.

12.9 Sulfur compounds

AIM: To identify names, structures, and uses of some common thiols, thioethers, and disulfides.

Focus

Thiols and thioethers are sulfur analogues of alcohols and ethers.

Just as alcohols and ethers are organic derivatives of water, thioalcohols and thioethers are organic derivatives of hydrogen sulfide, H_2S.

Thioalcohols and thioethers

Thioalcohols, *compounds with the general formula R—S—H, are also called* **thiols.** *Thiols are also called* **mercaptans**—*a name coined because of their ready ability to "capture" or react with the element mercury.* The odor of the fluid that skunks eject to protect themselves from predators comes partially from two thiols.

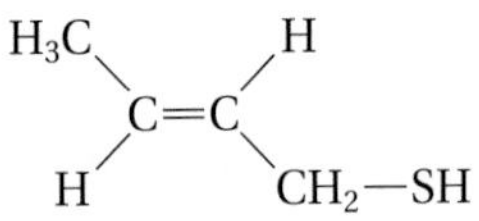

trans-2-Butenethiol

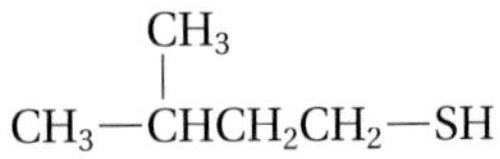

3-Methyl-1-butanethiol

Thiol

theion (Greek): sulfur

The odor from a skunk can be removed from a contaminated object by treatment with a solution of the following composition: 1 quart 3% H_2O_2, 1/4 cup baking soda, and 1 teaspoon liquid soap. After treatment, the object should be rinsed with tap water.

Sulfur compounds are well known among chemists as the most foul smelling of all organic compounds. Sometimes, however, even an odor can be put to good use. Natural gas is odorless, but the gas piped to a gas range has a characteristic smell. The gas company has added a trace of methanethiol or ethanethiol to the natural gas so that leaks can be detected.

$$CH_3\!-\!SH \qquad CH_3CH_2\!-\!SH$$

Methanethiol
(methyl mercaptan)

Ethanethiol
(ethyl mercaptan)

In very low concentrations, the odors of some thiols and sulfides are even desirable in cooking. The pungent smell of onions comes from propanethiol.

$$CH_3CH_2CH_2\!-\!SH$$

Propanethiol
(propyl mercaptan)

Thioethers, *compounds with the general formula R—S—R, are also called sulfides.* The characteristic aroma and flavor of garlic comes from divinyl sulfide.

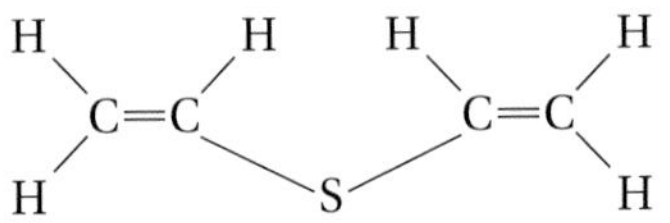

Divinyl sulfide

Disulfides

Sulfur atoms can form relatively stable bonds with other sulfur atoms. The sulfur-sulfur bond is not as stable as the carbon-carbon bond, and sulfur does not form long chains. But **disulfides,** *organic compounds of the general structural formula R—S—S—R,* are found in many proteins, especially those of hair, hooves, and nails (see Sec. 18.6). A third compound in the defense fluid of skunks is a disulfide:

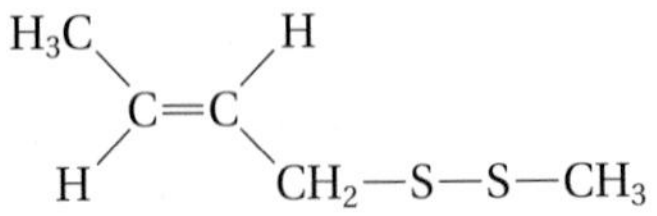

Disulfides are prepared by the mild oxidation of thiols. Thiols permitted to stand in air spontaneously oxidize to disulfides:

$$2CH_3S\!-\!H \xrightarrow{\ O_2\ } CH_3S\!-\!SCH_3$$

Methanethiol   Methyl disulfide

Once formed, disulfides can be further oxidized to sulfonic acids by hydrogen peroxide:

$$CH_3S\!-\!SCH_3 \xrightarrow{\ H_2O_2\ } 2CH_3SO_3H$$

Methyl disulfide   Methyl sulfonic acid

Disulfides are easily reduced to thiols. Although many reducing agents are suitable, hydrogen works well:

$$CH_3S\!-\!SCH_3 \xrightarrow{\ H_2\ } 2CH_3\!-\!SH$$

Methyl disulfide   Methanethiol

12.10 Polyfunctional compounds

AIM: To recognize the functional groups of a given polyfunctional molecule.

Focus

Many organic molecules contain more than one functional group.

So far our introduction to functional group chemistry has been focused on molecules with only one kind of functional group. **Polyfunctional** *organic compounds contain two or more functional groups.* Tetrahydrocannabinol is an example of a polyfunctional molecule, as shown in the following example.

EXAMPLE 12.6   **Identifying functional groups**

Identify the functional groups in the tetrahydrocannabinol molecule.

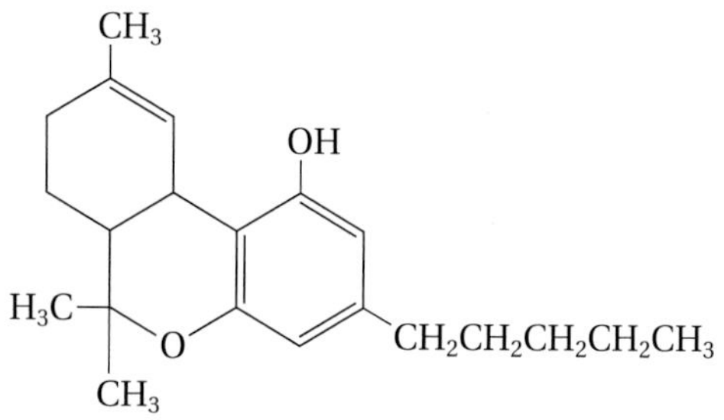

Tetrahydrocannabinol

SOLUTION

Inspect the molecule to find the functional groups, such as multiple carbon-carbon bonds, hydroxyl groups, ether linkages, and so forth. The tetrahydrocannabinol molecule contains three functional groups: a cyclic ether linkage, a phenolic hydroxyl group, and a carbon-carbon double bond. These functional groups are in color in the following structure.

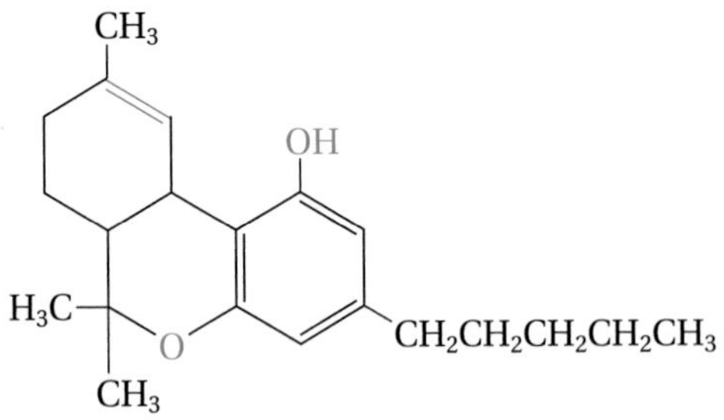

Tetrahydrocannabinol is the active ingredient of the *Cannabis sativa* (marijuana) plant. The effects of marijuana use on health and society are still being debated. One interesting finding is that smoking marijuana reduces the pressure of the optic fluid and may therefore be helpful in relieving the symptoms of glaucoma, a serious eye disease.

Another polyfunctional compound is hexachlorophene, which is both an aromatic halocarbon and a phenol:

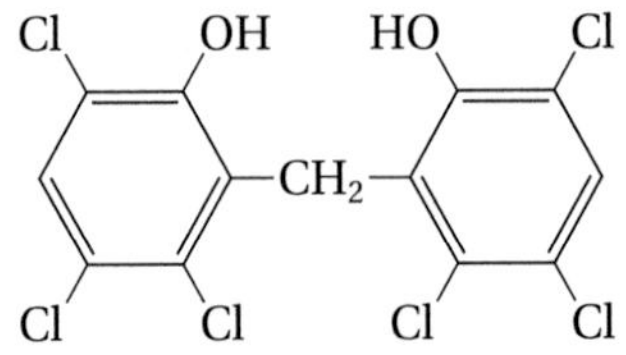

Hexachlorophene

Hexachlorophene is an antiseptic. Until recently, it was an important ingredient in germicides and soaps used in hospitals. Its use was curbed, however, when it was found that the babies of female hospital workers who frequently washed their hands with hexachlorophene soap had a higher incidence of birth defects than babies born to women in the general population.

PRACTICE EXERCISE 12.11

The structure that follows is a urushiol, a family of compounds that are the irritants in poison ivy. It is a polyfunctional compound with the following structure. How many functional groups can you identify?

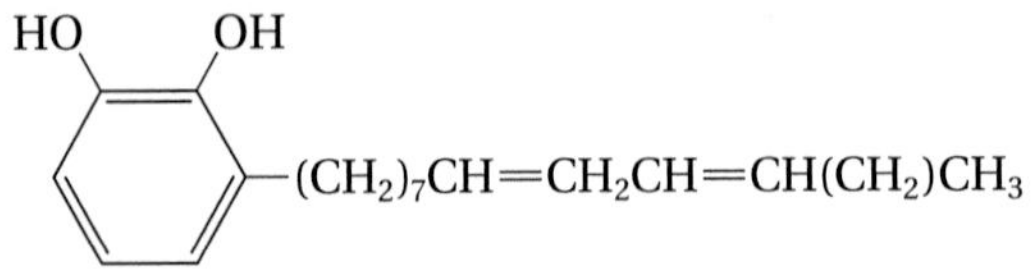

PRACTICE EXERCISE 12.12

Estradiol is an important female sex hormone. Describe the structural features and identify the functional groups in the estradiol molecule.

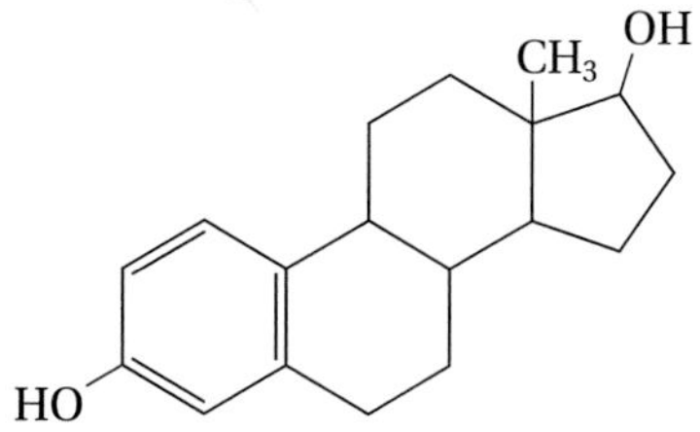

SUMMARY

Halocarbons are like hydrocarbons except that one or more hydrogens is replaced by a halogen—fluorine, chlorine, bromine, or iodine. Halocarbons may be saturated or unsaturated aliphatic compounds. They also may be aromatic. Many useful polymers are made from haloalkenes.

Replacement of a hydrocarbon hydrogen by a hydroxyl group, —OH, gives alcohols. Alcohols are organic derivatives of water, R—OH. They may be primary, secondary, or tertiary. Methanol, ethanol, isopropyl alcohol, ethylene glycol, and glycerol are alcohols of commercial importance. Glycerol is also a major constituent of plant and animal fats and oils. Aromatic alcohols are called phenols. Ethers have the general formula R—O—R.

The chemistry of the alkenes, halocarbons, alcohols, and ethers is related. Halocarbons are synthesized from alkenes by halogenation and hydrohalogenation. Alcohols are prepared by hydration of alkenes or from halocarbons by displacement of a halide ion by a hydroxide ion. In additions of hydrogen halides to double bonds, Markovnikov's rule tells us: If more than one halo-

carbon can be formed, then the hydrogen of the adding reagent usually ends up on the carbon of the double bond that has the most hydrogens. Ethers are made by displacement reactions, except that halocarbons react with alkoxide ions rather than hydroxide ions.

Halocarbons and alcohols undergo elimination reactions to form alkenes. When two or more alkenes can be formed by elimination of a hydrogen halide, Saytzeff's rule tells us: The alkene with the most carbon groups attached to the double bond is the major product.

Alcohol molecules hydrogen bond to each other and to water. Alcohols therefore have higher boiling points and greater water solubility than hydrocarbons, halocarbons, or ethers.

Thioalcohols (thiols) and thioethers (sulfides) are sulfur derivatives of alcohols and ethers; they have the general structures R—SH and R—S—R, respectively. These sulfur-containing groups and disulfide groups (R—S—S—R) are found in many biological molecules.

REACTION SUMMARY

Here is a summary of the reactions covered in this chapter.

Halogenation (the halogen is usually Cl_2 or Br_2):

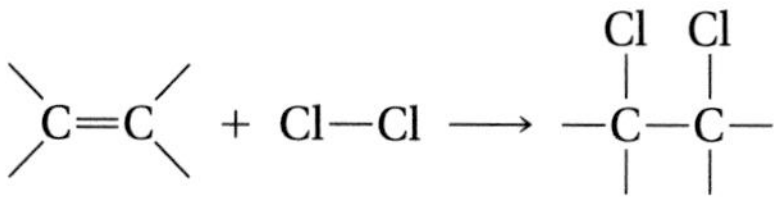

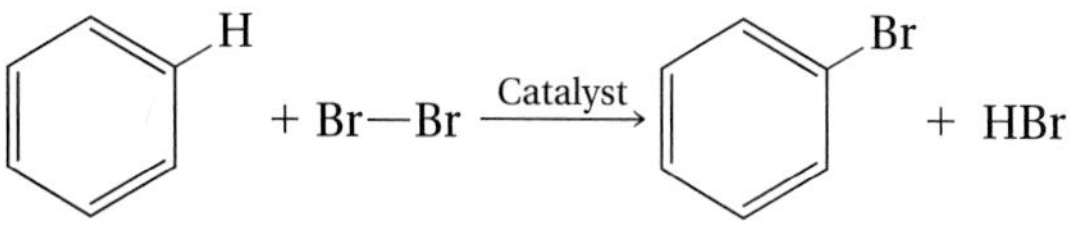

Hydrohalogenation (the hydrohalogen is usually HCl, HBr, or HI):

$$\text{C=C} + \text{H—Br} \longrightarrow -\overset{\overset{\displaystyle H}{|}}{C}-\overset{\overset{\displaystyle Br}{|}}{C}-$$

Hydration:

$$\text{C=C} + \text{H—OH} \longrightarrow -\overset{\overset{\displaystyle H}{|}}{C}-\overset{\overset{\displaystyle OH}{|}}{C}-$$

Polymerization:

$$x\text{C=C} \xrightarrow{\text{Catalyst}} +\overset{|}{\underset{|}{C}}-\overset{|}{\underset{|}{C}}+_x$$

Displacement (Br⁻, Cl⁻, or I⁻ is usually displaced):

$$\text{R—Cl} + {}^{-}\text{OH} \longrightarrow \text{R—OH} + \text{Cl}^{-}$$

$$\text{R—I} + {}^{-}\text{OR} \longrightarrow \text{R—OR} + \text{I}^{-}$$

Dehydrohalogenation (Br⁻, Cl⁻, or I⁻ is usually displaced):

$$-\overset{\overset{\displaystyle H}{|}}{C}-\overset{\overset{\displaystyle Cl}{|}}{C}- + \text{OH}^{-} \longrightarrow \text{C=C} + \text{Cl}^{-} + \text{H}_2\text{O}$$

Dehydration:

$$-\overset{\overset{\displaystyle H}{|}}{C}-\overset{\overset{\displaystyle OH}{|}}{C}- \xrightarrow{\text{H}_2\text{SO}_4} \text{C=C} + \text{H}_2\text{O}$$

Dehalogenation:

$$-\overset{\overset{\displaystyle Br}{|}}{C}-\overset{\overset{\displaystyle Br}{|}}{C}- + \text{Zn} \longrightarrow \text{C=C} + \text{ZnBr}_2$$

KEY TERMS

Addition reaction (12.2)	Denatured alcohol (12.3)	Hydration reaction (12.4)	Phenol (12.3)
Alcohol (12.3)	Displacement reaction	Hydrogenation (12.2)	Polyfunctional molecule
Alkoxide (12.6)	(12.4)	Hydrohalogenation (12.2)	(12.10)
Alkyl halide (12.1)	Disulfide (12.9)	Hydrophilic (12.8)	Polymer (12.2)
Aromatic substitution	Elimination reaction (12.5)	Hydrophobic (12.8)	Saytzeff's rule (12.5)
(12.2)	Ether (12.7)	Hydroxy function (12.3)	Thioalcohol (12.9)
Aryl halide (12.1)	Functional group (12.2)	Hydroxyl group (12.3)	Thioether (12.9)
Dehalogenation (12.5)	Halocarbon (12.1)	Markovnikov's rule (12.2)	Thiol (12.9)
Dehydration (12.5)	Halogenation (12.2)	Mercaptan (12.9)	
Dehydrohalogenation	Heterocyclic ring (12.7)	Monomers (12.2)	
(12.5)			

EXERCISES

Halocarbons (Sections 12.1, 12.2)

12.13 Name or write a structural formula for the following halocarbons.
(a) *m*-dichlorobenzene   (b) CH₂=CHCH₂Cl

(c) $\text{CH}_3\overset{\overset{\displaystyle CH_3}{|}}{\text{CH}}\text{CH}_2\overset{\overset{\displaystyle Cl}{|}}{\text{CH}}\text{CH}_2\text{Cl}$

(d) 1,2-dichlorocyclohexane

12.14 Write a structural formula for or name each of the following.
(a) 1,2,2-trichlorobutane   (b) 1,3,5-tribromobenzene

(c) $\overset{\overset{\displaystyle Cl}{|}}{}$
$\text{CH}_3\text{CHCH}_2\text{CH}_3$

(d) CH₂=CHBr

12.15 Write the structural formula(s) for the product(s) from the following reactions.
(a) CH₃CH₂CH=CH₂ + Cl₂ ⟶
(b) CH₃CH₂CH=CH₂ + HBr ⟶
(c) 3-methyl-2-pentene + hydrogen (with platinum catalyst)

12.16 Write structural formula(s) and name(s) of product(s) of each reaction.

(a) CH₃CH₂CH=CH₂ + H₂ $\xrightarrow{\text{Pt}}$
(b) cyclohexene + bromine ⟶
(c) CH₃CH₂CH=C(CH₃)₂ + HBr ⟶

12.17 Write structural formulas and give IUPAC names for all the isomers of C₃H₆Cl₂.

12.18 Write structural formulas and give IUPAC names for all the isomers of C_4H_9Br.

Polymers and Polymerization (Section 12.2)

12.19 What is the structure of the repeating units in a polymer in which the monomer is (a) 1-butene and (b) 1,2-dichloroethene?

12.20 The *trans* isomer of polyisoprene is a hard material known as *gutta percha*. What is the structure of the repeating unit in this polymer?

Alcohols (Sections 12.3, 12.4, 12.5, 12.6)

12.21 Write the IUPAC name or give the structural formula for each of the following alcohols.

(a) $CH_3CH_2CHCH_3$
      $\qquad\quad$ |
      $\qquad\quad$ OH

(b) CH_3CHCH_2OH
      $\qquad\quad$ |
      $\qquad\quad$ CH_3

(c) 1,2-ethanediol (d) *p*-bromophenol

12.22 Give the structural formula or IUPAC name for each of the following alcohols.

(a) 3-methyl-2-butanol (b) cyclopentanol

$\qquad\qquad\quad CH_3$
$\qquad\qquad\quad$ |
(c) $CH_3CH_2CCH_2CH_3$
$\qquad\qquad\quad$ |
$\qquad\qquad\quad$ OH

(d) $CH_3CHCH_2CH_2OH$
$\qquad\qquad\quad$ |
$\qquad\qquad\quad CH_3$

12.23 Identify each of the alcohols in Exercise 12.22 as primary, secondary, or tertiary.

12.24 Which of the alcohols in Exercise 12.22 are not secondary alcohols?

12.25 Describe two ways of synthesizing alcohols. Use specific examples to illustrate the reactions involved.

12.26 Name three types of elimination reactions that produce alkenes from halocarbons and alcohols.

12.27 Write the structures and give the names of the alcohols produced in the following reactions.

(a) $CH_2{=}CH_2 + H_2O \xrightarrow[100\,°C]{H^+}$

(b) $CH_3CH{=}CHCH_3 + H_2O \xrightarrow[100\,°C]{H^+}$

(c) 2-methyl-1-butene $+ H_2O \xrightarrow[100\,°C]{H^+}$

12.28 What are the structures and IUPAC names of the alcohols formed in these reactions?

(a) 3-hexene $+ H_2O \xrightarrow[100\,°C]{H^+}$

$\qquad\quad CH_3$
$\qquad\quad$ |
(b) $CH_3CH{-}Br + NaOH \xrightarrow[100\,°C]{H_2O}$

(c) CH_3CH_2 $\qquad$ H
$\qquad\qquad$ \ $\qquad$ /
$\qquad\qquad$ $C{=}C$ $\qquad + H_2O \xrightarrow[100\,°C]{H^+}$
$\qquad\qquad$ / $\qquad$ \
$\qquad$ H_3C $\qquad$ CH_3

12.29 What are the structures and names of the alkenes that result when the following compounds undergo elimination?

(a) cyclohexanol (b) 1,2-dibromopropane

$\qquad\qquad CH_3 \qquad\qquad\qquad\qquad$ Br
$\qquad\qquad$ | $\qquad\qquad\qquad\qquad\qquad$ |
(c) $CH_3{-}C{-}OH$ (d) $CH_3CH_2CHCH_2CH_3$
$\qquad\qquad$ |
$\qquad\qquad CH_3$

12.30 Write the structures and give the names of the alkenes produced in the following elimination reactions.

$\qquad$ Br $\quad$ Br $\qquad\qquad\qquad\qquad CH_3$
$\qquad$ | $\qquad$ | $\qquad\qquad\qquad\qquad$ |
(a) $CH_3CH{-}CCH_2CH_3$ (b) $CH_3CH_2C{-}OH$
$\qquad\qquad$ | $\qquad\qquad\qquad\qquad\qquad\qquad$ |
$\qquad\qquad CH_3 \qquad\qquad\qquad\qquad\qquad CH_3$

(c) 2-methyl-2-butanol (d) bromocyclohexane

Ethers (Section 12.7)

12.31 Name or write structural formulas for the following ethers.

(a) ethylphenyl ether
(b) furan
(c) $CH_3CH_2OCH_2CH_2CH_2CH_3$
(d) $CH_3CH_2CH_2OCH_2CH_2CH_3$

12.32 Write structural formulas or name each of the following ethers.

(a) $CH_3OCH_2CH_3$ (b) $CH_2{=}CHOCH{=}CH_2$
(c) tetrahydropyran (d) ethylpropyl ether

12.33 What are the structures of the ethers produced by the following reactions:

(a) ⬡$-$ONa $+ CH_3Br \longrightarrow$ _____ $+$ NaBr

(b) $CH_3OK +$ ⬡$-$Cl $\longrightarrow$ _____ $+$ KCl

(c) $CH_3CHONa + CH_3CH_2Br \longrightarrow$ _____ $+$ NaBr
$\qquad$ |
$\qquad CH_3$

12.34 Classify each of these compounds as an alcohol, a phenol, or an ether.

(a)

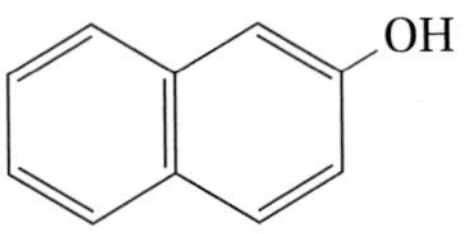

(b)

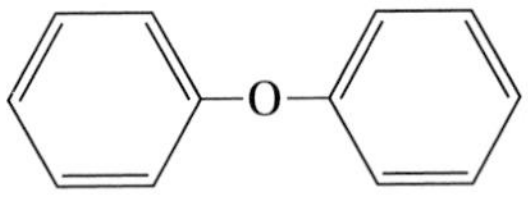

(c)

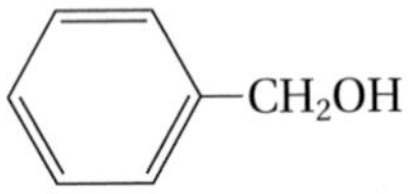

(d)

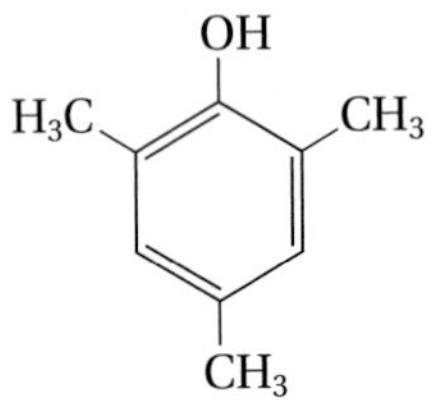

(e) 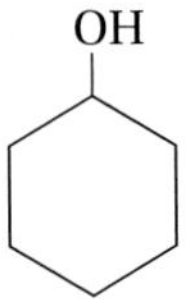 OH

(f) $CH_3CH_2\overset{\displaystyle}{\underset{\displaystyle CH_3}{C}HOH}$

Physical Properties of Organic Molecules (Section 12.8)

12.35 Explain why ethanol, CH_3CH_2OH, is soluble in water in all proportions but decanol, $CH_3(CH_2)_9OH$, is almost insoluble in water.

12.36 Explain why diethyl ether is more soluble in water than dihexyl ether. Would you expect propane to be more soluble than diethyl ether in water? Why?

12.37 Show how hydrogen bonds form between molecules of the following pairs of compounds: (a) water-water, (b) water-methanol, and (c) methanol-methanol.

12.38 Describe why hydrogen bonds can form between molecules of water and diethyl ether but not between molecules of diethyl ether.

Polyfunctional Molecules (Section 12.10)

12.39 The compound whose structure is drawn here has no common name, but it has a number of functional groups and other structural features. See how many you can identify.

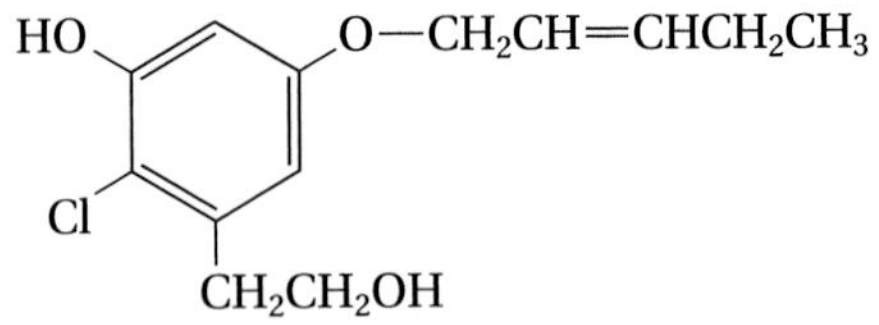

12.40 Cholesterol is a compound that is in our diet and also synthesized in the liver. Sometimes it is deposited on the inner walls of blood vessels, causing hardening of the arteries. Describe the structural features and functional groups of this important molecule.

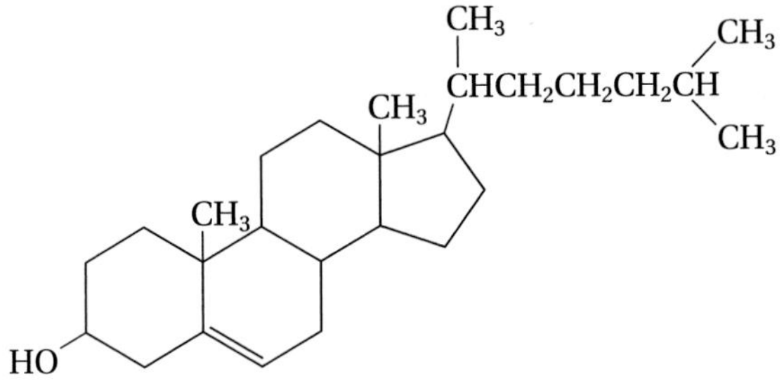

Additional Exercises

12.41 Which member of each of the following pairs of compounds would you predict to have the higher boiling point? Explain your answers.

(a)

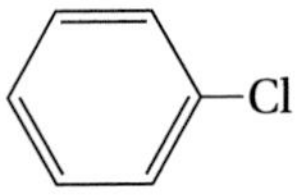

or

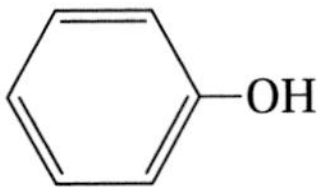

(b) $CH_3OCH_2CH_3$ or $CH_3CH_2CH_2OH$

(c) $CH_3-\overset{\displaystyle CH_3}{\underset{\displaystyle CH_3}{C}}-OH$ or $CH_3CH_2CH_2CH_2OH$

(d) CH_3Cl or CH_3OH

(e)

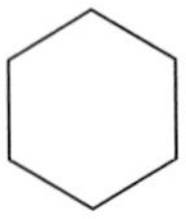

or

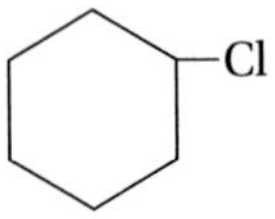

12.42 Both compounds A and B have the molecular formula C_2H_6O. When compound A reacts with sodium metal, hydrogen gas is produced. Compound B does not react with sodium metal. The boiling points of compounds A and B are 78.5 and $-23.7\,°C$, respectively. Write the structural formulas and names of the two compounds.

12.43 Write the structural formulas for the major and minor products from the following reactions.

(a) 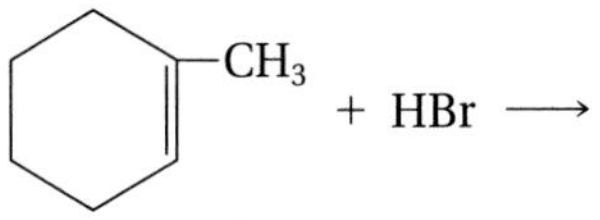 $+$ HBr $\longrightarrow$

(b) 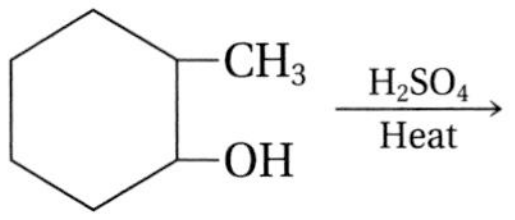 $\xrightarrow[\text{Heat}]{H_2SO_4}$

(c) 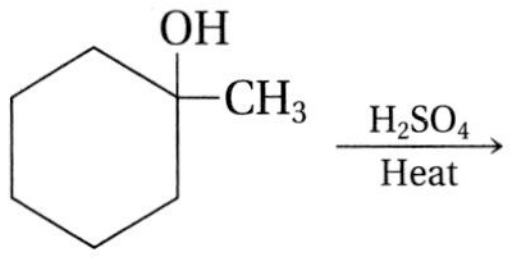 $\xrightarrow[\text{Heat}]{H_2SO_4}$

(d) 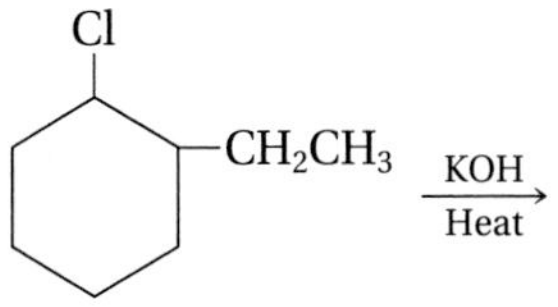 $\xrightarrow[\text{Heat}]{\text{KOH}}$

12.44 Complete the following reactions by writing the structural formulas of the products.

(a) 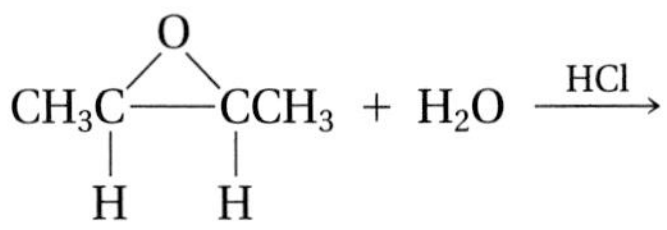 $+$ H_2O $\xrightarrow{\text{HCl}}$

(b) 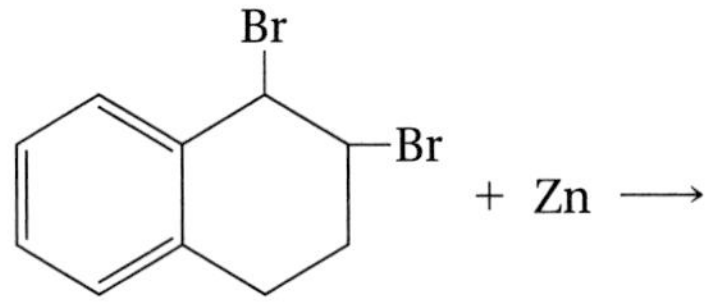 $+$ Zn $\longrightarrow$

(c) 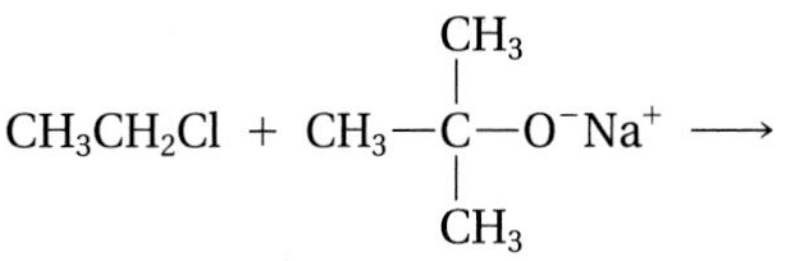 CH_3CH_2Cl $+$ $\longrightarrow$

(d) 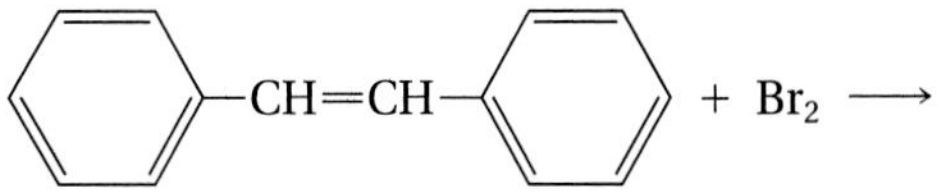 $+$ Br_2 $\longrightarrow$

12.45 List the following compounds in the expected order of increasing solubility in water.

(a) 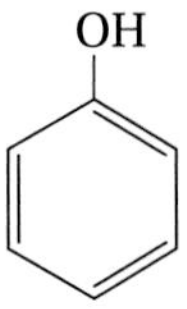   (b)

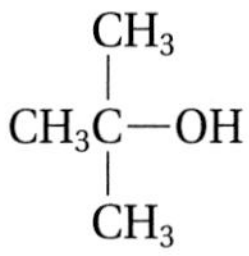

(c) $CH_3CH_2OCH_2CH_3$ (d)

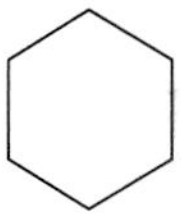

12.46 Give the reagents necessary to make the following transformations.

(a) 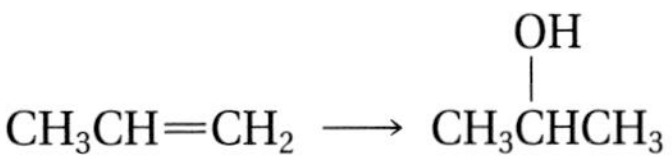 $CH_3CH{=}CH_2$ $\longrightarrow$

(b) 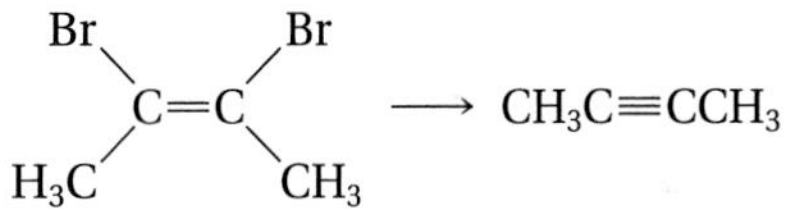 $\longrightarrow$ $CH_3C{\equiv}CCH_3$

(c) 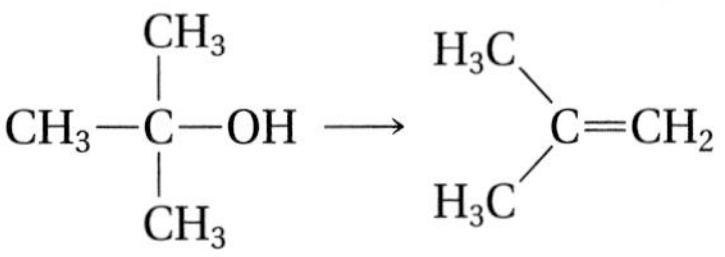 $\longrightarrow$

(d) CH_3CH_2OH $\longrightarrow$ $CH_3CH_2O^-Na^+$

SELF-TEST (REVIEW)

True/False

1. As a general rule, the hydrocarbon portion of halocarbons is chemically inert.

2. A disulfide molecule contains a single covalent bond between two sulfur atoms.

3. Although similar in molar mass, ethanol has a higher boiling point than dimethyl ether.

4. The hydrocarbon chain end of a decanol molecule is hydrophilic.

5. The reaction of an alcohol and sodium hydroxide gives an alkoxide.

6. Both hydrogenation and halogenation are examples of addition reactions.

7. Pyran is an example of a cyclic ether.

8. Although similar in molar mass, chloroethane is more soluble in water than 1-propanol.

9. A propylene glycol molecule has two hydroxyl groups.

10. Both aromatic and aliphatic alcohols are weakly acidic.

Multiple Choice

11. A hydrocarbon added to an orangish solution of bromine in carbon tetrachloride turns the solution colorless. The hydrocarbon is probably
   (a) 2-butene.   (b) isopropyl alcohol.
   (c) ethyl ether.   (d) benzene.

12. A low-molar-mass alcohol dissolved in water gives
   (a) an acidic solution.   (b) a neutral solution.
   (c) a basic solution.   (d) none of the above.

13. An ether can be formed by the reaction of a halocarbon with
   (a) an alkene.   (b) an alkoxide ion.
   (c) an alcohol.   (d) water.

14. Which of the following reactions is most typical of alkenes?
   (a) substitution   (b) addition
   (c) replacement   (d) elimination

15. Which of the following compounds, all of similar molar mass, would you expect to be most soluble in water?
   (a) $CH_3CH_2—Cl$ (b) $CH_3CH_2CH_2—F$
   (c) $CH_3CH_2CH_2CH_3$ (d) $CH_3CH_2CH_2—OH$

16. The reaction of water with ethene in the presence of an acid catalyst
   (a) results in the formation of ethanol.
   (b) is an example of a replacement reaction.
   (c) produces a polymer.
   (d) More than one answer is correct.

17. A compound once widely used as an anesthetic is
   (a) $CH_3CH_2CH_2OH$.   (b) $CH_3CH_2—O—CH_2CH_3$.
   (c) $CH_2—CH_2$ (d) $CH_3CH_2CH_2SH$
       $\diagdown O \diagup$

18. Which of the following compounds, all of similar molar mass, has the highest boiling point?
   (a) $CH_3CH_2—F$ (b) $CH_3CH_2CH_3$
   (c) $CH_3—O—CH_3$ (d) $CH_3CH_2—OH$

19. Which of the following compounds is found in antifreeze?
   (a) glycerol   (b) ethylene glycol   (c) propanol
   (d) propanethiol

20. Another name for isobutyl chloride is
   (a) 2-chloro-2-methylpropane.
   (b) chlorobutane.
   (c) 2-chloro-3-methylbutane.
   (d) 1-chloro-2-methylpropane.

21. Which of the following could be used as starting materials to produce 1-butene as the major product? (*Hint:* Use Saytzeff's rule.)
   (a) 2-butanol (with H_2SO_4)
   (b) 1,2-dibromobutane (with zinc)
   (c) 2-chlorobutane (with base)
   (d) all of the above

22. Using Markovnikov's rule, what would you predict the major organic product of the following reaction to be?

$$CH_3—CH_2—\underset{\underset{CH_3}{|}}{C}=CH—CH_3 + HBr \longrightarrow$$

   (a) 3-methylpentane
   (b) 3-bromo-3-methylpentane
   (c) 2-bromo-3-methylpentane
   (d) 2,3-dibromo-3-methylpentane

23. Phenols are characterized by
   (a) their behavior as bases.   (b) ether linkages.
   (c) an —OH group attached to a benzene ring.
   (d) their use as flavoring agents.

24. Which of the following compounds *does not* contain the element sulfur?
   (a) methyl ethyl sulfide   (b) propyl mercaptan
   (c) 2-propanethiol   (d) epoxyethane

25. A correct name for the secondary alcohol containing four carbon atoms is
   (a) 2-butanol.   (b) 3-butanol.
   (c) 2-methyl-2-propanol.   (d) ethanediol.

26. Benzene typically undergoes which of the following types of reactions?
   (a) elimination   (b) addition   (c) substitution
   (d) polymerization

Aldehydes and Ketones

Introduction to the Carbonyl Group

The aroma and flavor of vanilla, nutmeg, cloves, and many other spices are due to aldehydes and ketones.

CHAPTER OUTLINE

CASE IN POINT: Fetal alcohol syndrome

13.1 Aldehydes and ketones

13.2 The carbonyl group

A CLOSER LOOK: Flavors and Fragrances

13.3 Redox reactions of organic compounds

13.4 Redox reactions involving aldehydes and ketones

FOLLOW-UP TO THE CASE IN POINT: Fetal alcohol syndrome

13.5 Aldehyde detection

13.6 Additions to the carbonyl group

13.7 Uses of aldehydes and ketones

A CLOSER LOOK: The Chemistry of Vision

The carbonyl group consists of a carbon atom and an oxygen atom joined by a double bond ($C{=}O$). Present in carbohydrates, fats, proteins, and steroids, the carbonyl group is one of the most common functional groups in nature. In this chapter we will learn about the chemistry of the carbonyl group in two classes of compounds, aldehydes and ketones. This is only a brief introduction to an important functional group that will appear again and again in following chapters.

Although aldehydes and ketones are important components of cells, some of the harmful effects of drinking ethanol (CH_3CH_2OH) result from the effect of an aldehyde, acetaldehyde (CH_3CHO), on a variety of bodily functions. These effects are magnified in a developing fetus, as described in the Case in Point.

CASE IN POINT: Fetal alcohol syndrome

 Marla is an alcoholic and has been in and out of treatment for her drinking problems since she was a teenager. Her physician is aware of her alcoholism and at one point in the past helped place Marla in an alcohol abuse center. Her treatment had gone well, but when Marla became pregnant, her physician became concerned. The doctor told Marla that it was especially important that she abstain from alcohol during her pregnancy. The physician was concerned that the child might be born with *fetal alcohol syndrome,* a condition that can result when pregnant women consume alcohol. We will learn more about the relationships among ethanol consumption, acetaldehyde, and fetal alcohol syndrome in Section 13.4.

The surgeon general warns that pregnant women should not drink alcoholic beverages because of the risk of birth defects.

13.1 Aldehydes and ketones

AIM: To describe the carbon-oxygen bond of the carbonyl group of aldehydes and ketones.

The functional group known as the **carbonyl group** $(\,{>}C{=}O)$—*a carbon atom and an oxygen atom joined by a double bond*—is found in compounds called *aldehydes* and *ketones*.

Structures of aldehydes and ketones

Aldehydes *are organic compounds in which the carbonyl carbon—the carbon to which the oxygen is bonded—is always joined to at least one hydrogen.* The general formula for an aldehyde is

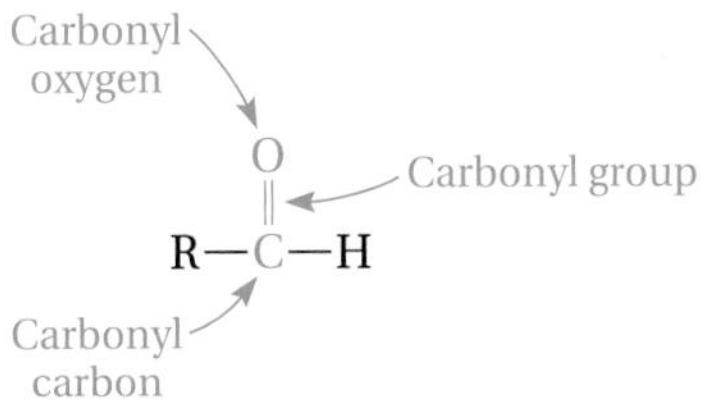

Carbonyl oxygen

Carbonyl group

$$R-\underset{\underset{\text{Carbonyl carbon}}{}}{\overset{\overset{O}{\parallel}}{C}}-H$$

This structural formula is often abbreviated to RCHO.

Ketones *are organic compounds in which the carbonyl carbon is joined to two other carbons:*

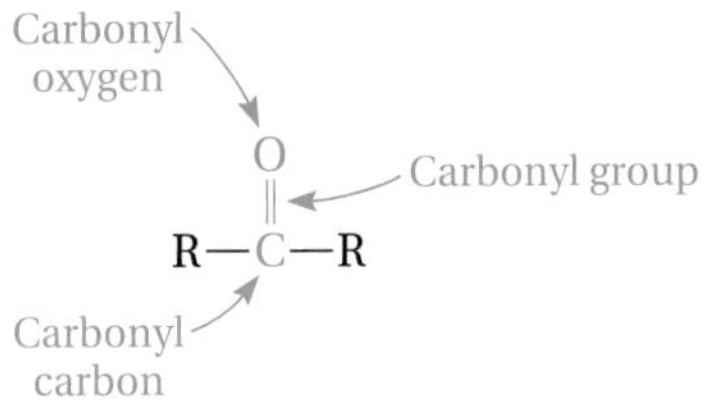

Carbonyl oxygen

Carbonyl group

$$R-\underset{\underset{\text{Carbonyl carbon}}{}}{\overset{\overset{O}{\parallel}}{C}}-R$$

The abbreviated form for a ketone is RCOR.

Note the similarity in structure of aldehydes and ketones. Because they both contain the carbonyl group, the chemistry of aldehydes and ketones is similar. Both aldehydes and ketones are highly reactive, but aldehydes are generally the more reactive of the two classes.

Naming aldehydes and ketones

The IUPAC system may be used for naming aldehydes. We must first identify the longest hydrocarbon chain that contains the carbonyl carbon. The *-e* ending of the hydrocarbon is replaced by *-al* to designate an aldehyde. Using the IUPAC system, we name the aldehydes methanal, ethanal, propanal, butanal, and so forth. In naming substituted aldehydes, the longest chain is counted starting from the carbon of the aldehyde group.

EXAMPLE 13.1

Naming a substituted aldehyde by the IUPAC system

What is the IUPAC name for the following compound?

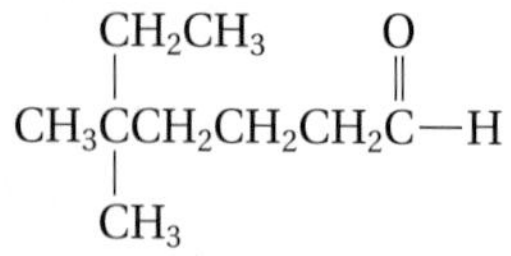

$$
\begin{array}{c}
\quad\ \ \text{CH}_2\text{CH}_3 \qquad\ \ \text{O} \\
\quad\ \ | \qquad\qquad\quad\ \| \\
\text{CH}_3\text{CCH}_2\text{CH}_2\text{CH}_2\text{C}-\text{H} \\
\quad\ \ | \\
\quad\ \ \text{CH}_3
\end{array}
$$

SOLUTION

The longest continuous chain contains seven carbons, making heptane the parent alkane. In counting the carbons in the chain, we number the chain so that the carbon of the aldehyde group is at position 1. The carbonyl carbon is understood to be at position 1 and will not appear in the final name of the aldehyde.

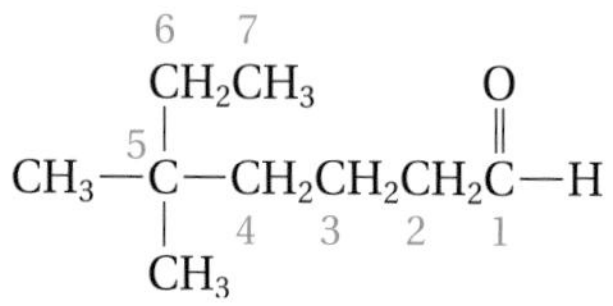

$$
\begin{array}{c}
\quad\ \ ^6\ \ ^7 \\
\quad\ \ \text{CH}_2\text{CH}_3 \qquad\ \ \text{O} \\
\quad\ ^5| \qquad\qquad\quad\ \| \\
\text{CH}_3-\overset{}{\text{C}}-\text{CH}_2\text{CH}_2\text{CH}_2\text{C}-\text{H} \\
\quad\ | \ \ _4 \ \ _3 \ \ _2 \ \ _1 \\
\quad\ \text{CH}_3
\end{array}
$$

Obtain the aldehyde name by dropping the -*e* ending from the alkane name and adding the ending -*al;* our aldehyde is a *heptanal.* The substituents are named in the same way as for hydrocarbons. The molecule has two methyl groups in the 5 position of the parent chain, which are designated *5,5-dimethyl.* Combine this with the parent name to obtain the complete name of the aldehyde: *5,5-dimethylheptanal.*

PRACTICE EXERCISE 13.1

Give the IUPAC names for each of the following aldehydes.

(a) $\text{CH}_3\text{CH}_2\text{CHO}$

(b) $\qquad\qquad\ \ \text{CH}_3$
$\qquad\qquad\qquad\ \ |$
$\qquad\ \text{CH}_3\text{CH}_2\text{CHCH}_2\text{CHO}$

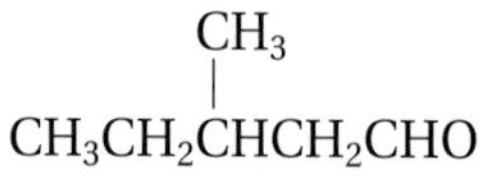

(c) $\text{CH}_3\text{CH}_2\text{CH}_2\text{CH}_2\text{CHO}$

(d) $\qquad\ \text{Cl}$
$\qquad\qquad\ |$
$\qquad\ \text{CH}_3\text{CHCH}_2\text{CHO}$

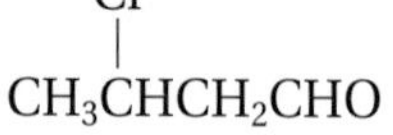

Ketones also can be named by the IUPAC system. We show that a compound is a ketone by changing the ending of the longest carbon chain that contains the carbonyl group from -*e* to -*one,* as demonstrated in Table 13.1. If there are several locations in the chain where the carbonyl group could be placed, its position is designated by the lower number.

EXAMPLE 13.2

Naming ketones by the IUPAC system

Using the IUPAC system, name the following two structural isomers of $\text{C}_5\text{H}_{10}\text{O}$:

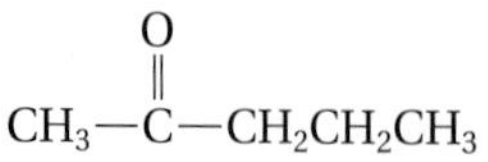

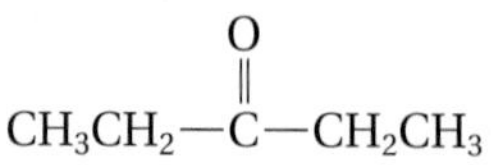

$$
\begin{array}{cc}
\quad\ \ \text{O} & \qquad\qquad\ \ \text{O} \\
\quad\ \ \| & \qquad\qquad\ \ \| \\
\text{CH}_3-\text{C}-\text{CH}_2\text{CH}_2\text{CH}_3 & \quad \text{CH}_3\text{CH}_2-\text{C}-\text{CH}_2\text{CH}_3
\end{array}
$$

Table 13.1 Some Common Aldehydes and Ketones

Condensed formula	Structural formula	IUPAC name	Common name
Aldehydes			
HCHO	$$H-\overset{\overset{\displaystyle O}{\|\|}}{C}-H$$	methanal	formaldehyde
CH_3CHO	$$CH_3-\overset{\overset{\displaystyle O}{\|\|}}{C}-H$$	ethanal	acetaldehyde
CH_3CH_2CHO	$$CH_3-CH_2-\overset{\overset{\displaystyle O}{\|\|}}{C}-H$$	propanal	propionaldehyde
$CH_3CH_2CH_2CHO$	$$CH_3-CH_2-CH_2-\overset{\overset{\displaystyle O}{\|\|}}{C}-H$$	butanal	butyraldehyde
C_6H_5CHO	$$\text{(benzene ring)}-\overset{\overset{\displaystyle O}{\|\|}}{C}-H$$	benzaldehyde	benzaldehyde
$C_6H_5CH=CHCHO$	$$\text{(benzene ring)}-CH=CH-\overset{\overset{\displaystyle O}{\|\|}}{C}-H$$	3-phenyl-2-propenal	cinnamaldehyde
Ketones			
CH_3COCH_3	$$CH_3-\overset{\overset{\displaystyle O}{\|\|}}{C}-CH_3$$	propanone	acetone (dimethyl ketone)
$CH_3COC_2H_5$	$$CH_3-\overset{\overset{\displaystyle O}{\|\|}}{C}-CH_2-CH_3$$	butanone	methyl ethyl ketone
$C_6H_5COC_6H_5$	$$\text{(two benzene rings)}-\overset{\overset{\displaystyle O}{\|\|}}{C}-$$	diphenylmethanone	benzophenone (diphenyl ketone)
$C_6H_{10}O$	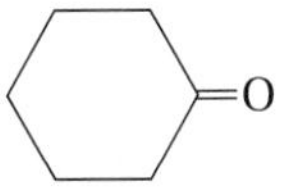 $$\text{(cyclohexane ring)}=O$$	cyclohexanone	cyclohexanone

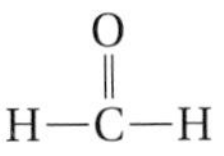

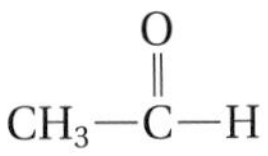

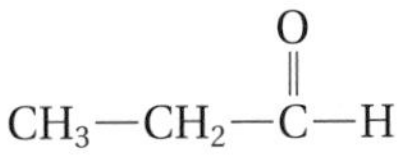

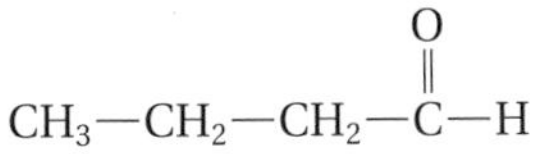

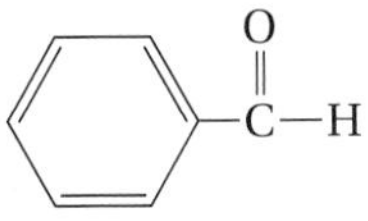

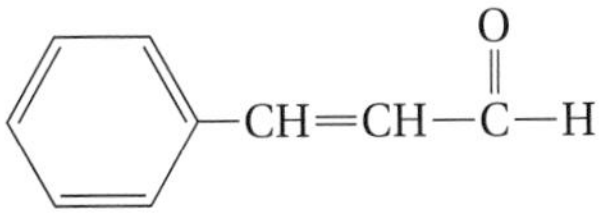

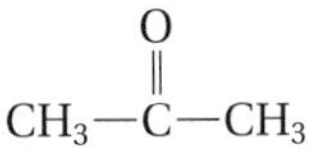

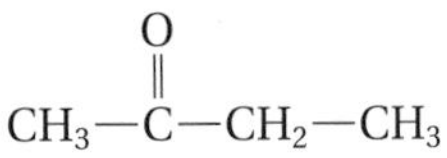

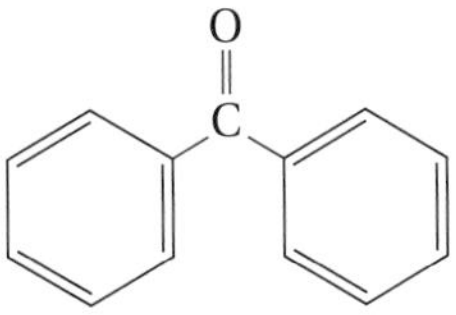

SOLUTION

In both compounds the longest chains that contain the carbonyl group have five carbons, making pentane the parent alkane. Indicate that the compounds are ketones by dropping the *-e* ending from pentane and adding the ending *-one*. The compounds are *pentanones*. Give the position of the carbonyl groups in the carbon chain. The numbering that gives the lowest position number to the isomer on the left is 2 and on the right, 3.

These compounds are 2-pentanone and 3-pentanone, respectively.

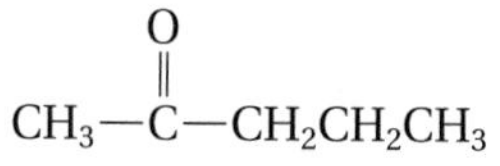

$$CH_3-\overset{\overset{\textstyle O}{\|}}{C}-CH_2CH_2CH_3 \qquad CH_3CH_2-\overset{\overset{\textstyle O}{\|}}{C}-CH_2CH_3$$

2-Pentanone 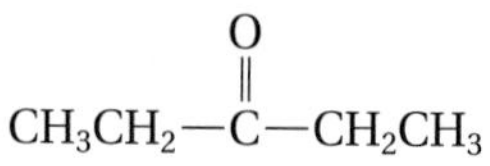   3-Pentanone

PRACTICE EXERCISE 13.2

Give the IUPAC name for each of the following ketones.

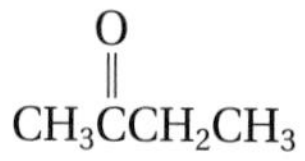

(a)
$$CH_3CH_2CH_2\overset{\overset{\textstyle O}{\|}}{C}CH_2CH_3$$

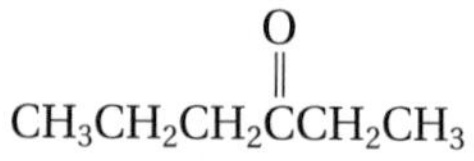

(b)
$$CH_3\overset{\overset{\textstyle O}{\|}}{C}CH_2CH_3$$

(c)
$$CH_3\overset{\overset{\textstyle CH_3}{|}}{C}HCH_2\overset{\overset{\textstyle O}{\|}}{C}CH_3$$

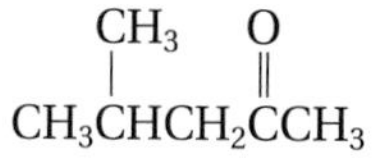

Common names for aldehydes and ketones are frequently used. The common names for methanal and ethanal are *formaldehyde* and *acetaldehyde*, respectively. The common names of the ketones are obtained by naming each of the alkyl groups attached to the carbonyl carbon and adding the word *ketone*. The sole exception is dimethyl ketone, which almost everyone calls *acetone*, a versatile solvent.

EXAMPLE 13.3   **Writing a structure for a ketone**

Write the structure for ethylisopropyl ketone.

SOLUTION

The common name of a ketone indicates the alkyl or aryl groups bonded to either side of the carbonyl group. Write a carbonyl group and attach ethyl and isopropyl groups to obtain the structure of the ketone.

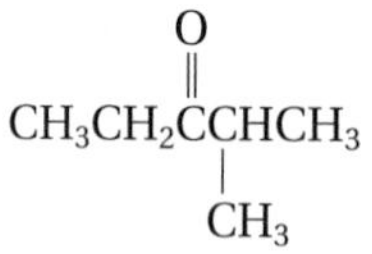

$$CH_3CH_2\overset{\overset{\textstyle O}{\|}}{C}\underset{\underset{\textstyle CH_3}{|}}{C}HCH_3$$

PRACTICE EXERCISE 13.3

Write the structure of each of the following compounds.

(a) acetaldehyde   (b) acetone   (c) diethyl ketone

(d) 3-methylbutanal

13.2 The carbonyl group

AIMS: *To name and draw structures of simple aldehydes and ketones. To explain how intermolecular interactions of the carbonyl group affect the boiling point and water solubility of aldehydes and ketones.*

Focus

The polar carbonyl group influences the physical properties of aldehydes and ketones.

Electron sharing in the carbon-oxygen double bond of the carbonyl group is similar to that of the carbon-carbon double bond in alkenes. Compare ethylene, $H_2C{=}CH_2$, with formaldehyde, $H_2C{=}O$ (Fig. 13.1). A carbon atom has only four unshared electrons in its valence shell, whereas oxygen has six. This means that carbon can make four covalent bonds, as in ethylene, but oxygen can make only two, as in the carbonyl group. The oxygen in a carbonyl group has two unshared pairs of electrons, but the carbon in ethylene has none.

In $C{=}C$ bonds, the bonding electrons are shared equally. But oxygen is much more electronegative than carbon. The $C{=}O$ bond is very polar because oxygen draws the bonding electrons away from carbon. The oxygen in the carbonyl group carries a partial negative charge, and the carbon carries a partial positive charge:

Hydrogen bonding cannot take place between molecules of aldehydes or ketones because they lack O—H bonds. However, polar-polar interactions are possible:

Polar-polar interaction

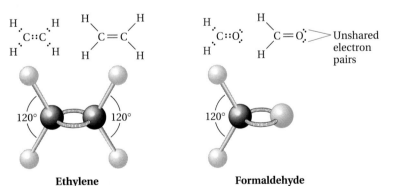

Ethylene   **Formaldehyde**

Figure 13.1
Comparison of the structural formulas and ball-and-stick models of ethylene and formaldehyde. The oxygen of the carbonyl group of formaldehyde has two unshared pairs of electrons.

Boiling points

Aldehydes and ketones with low molar masses are very volatile and highly flammable.

Aldehydes and ketones cannot form intermolecular hydrogen bonds because they lack hydroxyl (—OH) groups. Consequently, they have boiling points lower than those of the corresponding alcohols. The aldehydes and ketones can attract one another through polar-polar interactions of their carbonyl groups, however, and their boiling points are higher than those of the corresponding alkanes. Table 13.2 shows how the boiling points increase in the sequence alkane, aldehyde, and alcohol for compounds containing one and two carbons. Except for formaldehyde, which is an irritating, pungent gas, all aldehydes and ketones are either liquids or solids at 20 °C (Table 13.3). The vapors of aldehydes and ketones contribute to the pleasant odors of many natural products, as described in A Closer Look: Flavors and Fragrances.

Table 13.2 Some Characteristics of One- and Two-Carbon Alkanes, Aldehydes, and Alcohols

Compound	Formula	Boiling point (°C)	Comments
One Carbon			
methane	CH_4	−161	no hydrogen bonding or polar-polar interactions
formaldehyde	HCHO	−21	polar-polar interactions
methanol	CH_3OH	65	hydrogen bonding
Two Carbons			
ethane	C_2H_6	−89	no hydrogen bonding or polar-polar interactions
acetaldehyde	CH_3CHO	20	polar-polar interactions
ethanol	CH_3CH_2OH	78	hydrogen bonding

Table 13.3 Physical Constants of Some Aldehydes and Ketones

Compound	Melting point (°C)	Boiling point (°C)	Solubility in water (g/100 mL)
Aldehydes			
formaldehyde	−92	−21	completely miscible
acetaldehyde	−123	20	completely miscible
butyraldehyde	−99	76	4
benzaldehyde	−26	179	0.3
Ketones			
acetone	−95	56	completely miscible
methyl ethyl ketone	−86	80	25
diethyl ketone	−42	101	5
benzophenone	48	306	insoluble

Flavors and Fragrances

All pleasant odors and tastes come from chemicals. Aldehydes are responsible for the delightful odors of vanilla, cinnamon, and almonds. Other odors such as that of camphor, which has been used in medicine for thousands of years, are due to ketones.

Perfumes are made by blending certain odoriferous substances. Fine perfumes may contain more than 100 ingredients. Perfumes are classified by the dominant odor. Odors of rose, gardenia, lily of the valley, and jasmine are the floral group. Aromas of clove, nutmeg, cinnamon, and carnation provide the spicy blends. Cedar and sandalwood make up the woody group. The Orientals group combines the spicy and woody odors with the sweet odors of balsam and vanilla and accentuates the mixture with the odors of musk and civet. The odors of aldehydes dominate the aldehydic group, which have fruity character.

Natural products of vegetable and animal origin traditionally have been used as starting materials for perfume manufacture. Mixtures of essential oils (which give perfumes their odor or *essence*) and waxes are obtained from these natural products by extraction into organic solvents. Removal of the solvent gives a residue called a *concrete*. If the concrete is treated with ethanol, the wax remains behind and the essential oil dissolves in the alcohol. Essential oils are removed from citrus peel by pressing. Individual compounds used in perfumes may be isolated from the essential oils by distillation.

Some animal secretions contain odorous compounds that improve the lasting quality of perfumes. These substances and their components act as fixatives, preventing the more volatile ingredients in the perfume from evaporating too rapidly. Such materials are usually used as solutions in ethanol and give high-quality perfume its strength, character, and tenacity. The traditional animal products include castor or castoreum from the beaver, civet from the civet cat (see figure), ambergris from the sperm whale, and musk from the musk deer. Today, because of ecological concerns, chemists synthesize the compounds that produce the musk odor. Musk, the most commonly known musky substance, has a penetrating, persistent odor. Natural musk is produced by a gland under the skin of the abdomen of the male musk deer. The odorous substance in musk is a ketone, muscone, which has the chemical name 3-methylcyclopentadecanone. The civet cat, a native of Africa, southern Europe, and Asia, marks its territory with another ketone, civetone. Civetone, like the musk of the deer, has a musky odor that is excellent for use in perfumes.

The civet cat and musk deer produce secretions that are used as fixatives in perfumes—they give the perfume a lasting odor.

Solubility in water

Aldehydes and ketones can form hydrogen bonds with polar water molecules (Fig. 13.2). Formaldehyde, acetaldehyde, and acetone are soluble in water in all proportions. As the length of the hydrocarbon chain increases, water solubility decreases; when the carbon chain exceeds five

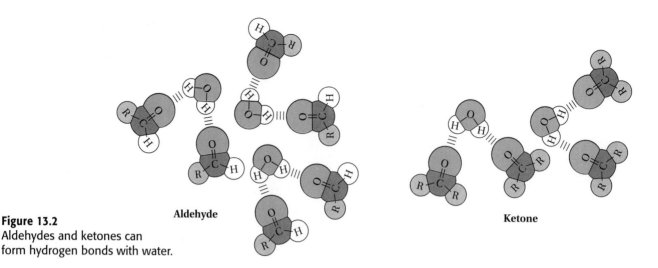

Figure 13.2
Aldehydes and ketones can form hydrogen bonds with water.

or six carbons, solubility of both aldehydes and ketones is very low. All aldehydes and ketones are soluble in nonpolar solvents.

> **PRACTICE EXERCISE 13.4**
>
> Arrange the following substances in order of increasing solubility in water:
>
> (a) acetone   (b) butanal   (c) pentanol
>
> (d) benzophenone   (e) benzaldehyde

13.3 *Redox reactions of organic compounds*

AIMS: *To describe the processes of oxidation and reduction in organic chemistry in terms of the loss or gain of oxygen, hydrogen, or electrons. To relate the energy content of a molecule to its degree of oxidation or reduction.*

Focus

Many reactions of organic compounds involve oxidation or reduction.

Now that we are familiar with functional groups in organic chemistry, we can examine oxidation-reduction reactions of organic molecules. We are interested in redox reactions of organic molecules because they are important in energy production in living organisms. Oxidation reactions are energy-releasing, and the more reduced a carbon compound is, the more energy the compound can release upon its complete oxidation. You may recall the principles of redox reactions from Section 6.4: *Oxidation reactions* involve a gain of oxygen, a loss of hydrogen, or a loss of electrons; *reduction reactions* involve a loss of oxygen, a gain of hydrogen, or a gain of electrons. Oxidation and reduction reactions must be coupled; if a compound is oxidized in a reaction, some other compound in the reaction must be reduced. We will discuss electron transfers in redox reactions in living organisms in Chapter 23. Our concern with redox reactions in this chapter will focus on those reactions that involve oxygen and hydrogen. For example, methane, a saturated hydrocarbon, can be oxidized in steps to carbon dioxide by alternately gaining oxygens and losing hydrogens. Methane can be oxidized to methanol, then to formaldehyde, then to formic acid, and finally to carbon dioxide.

$$
\underset{\substack{\text{Methane}\\ \text{Most energetic}\\ \text{molecule}}}{\text{H}-\overset{\displaystyle H}{\underset{\displaystyle H}{\text{C}}}-\text{H}}
\xrightarrow[\text{Oxidation}]{\substack{\text{Gain of}\\ \text{oxygen}}}
\underset{\substack{\text{Methanol}\\ \text{(methyl}\\ \text{alcohol)}}}{\text{H}-\overset{\displaystyle OH}{\underset{\displaystyle H}{\text{C}}}-\text{H}}
\xrightarrow[\text{Oxidation}]{\substack{\text{Loss of}\\ \text{hydrogen}}}
\underset{\substack{\text{Methanal}\\ \text{(formaldehyde)}}}{\text{H}-\overset{\displaystyle O}{\text{C}}-\text{H}}
\xrightarrow[\text{Oxidation}]{\substack{\text{Gain of}\\ \text{oxygen}}}
\underset{\substack{\text{Methanoic}\\ \text{acid}\\ \text{(formic acid)}}}{\text{H}-\overset{\displaystyle O}{\text{C}}-\text{OH}}
\xrightarrow[\text{Oxidation}]{\substack{\text{Loss of}\\ \text{hydrogen}}}
\underset{\substack{\text{Carbon dioxide}\\ \text{Least energetic}\\ \text{molecule}}}{\text{O}=\text{C}=\text{O}}
$$

In any series consisting of an alkane, alcohol, aldehyde (or ketone), carboxylic acid, and carbon dioxide, the alkane is the least oxidized (most reduced) compound and the carbon dioxide is the most oxidized (least reduced) compound. In biological systems, fats are similar in molecular structure to alkanes, and sugars are similar to alcohols. The more reduced a compound is, the more energy it can release upon its complete oxidation to carbon dioxide. Therefore, the complete oxidation of a carbon atom in a fat can produce more energy than the complete oxidation of a carbon atom in a sugar. The oxidation of carbon compounds to carbon dioxide can be reversed. For example, in photosynthesis, green plants can extract carbon dioxide from the atmosphere and produce sugars by a series of reduction reactions (see Sec. 23.2).

Many fat molecules have long hydrocarbon chains as part of their structure. As such, they are highly reduced molecules and have a relatively high energy content compared with other nutrient molecules.

We can use the loss and gain of hydrogen to find the relative degree of oxidation of organic molecules that contain carbon-carbon double bonds and carbon-carbon triple bonds. For example, ethane (an alkane) can lose hydrogen and go to ethene (an alkene) and then to ethyne (an alkyne).

Least oxidized (most reduced)

$$
\underset{\text{Ethane}}{\text{H}-\overset{\displaystyle H}{\underset{\displaystyle H}{\text{C}}}-\overset{\displaystyle H}{\underset{\displaystyle H}{\text{C}}}-\text{H}}
\xrightarrow[\text{Oxidation}]{\substack{\text{Loss of hydrogen}\\ \text{(dehydrogenation)}}}
\underset{\text{Ethene}}{\overset{\displaystyle H}{\underset{\displaystyle H}{\text{C}}}=\overset{\displaystyle H}{\underset{\displaystyle H}{\text{C}}}}
\xrightarrow[\text{Oxidation}]{\substack{\text{Loss of hydrogen}\\ \text{(dehydrogenation)}}}
\underset{\text{Ethyne}}{\text{H}-\text{C}\equiv\text{C}-\text{H}}
$$

Most oxidized (least reduced)

Each of these losses of hydrogen represents an oxidation of the compound that loses the hydrogen. Ethane is the least oxidized compound, and ethyne is the most oxidized. The fewer the number of hydrogens on a carbon-carbon bond, the more oxidized is the bond. *A redox reaction involving the loss of hydrogen from an organic molecule is called a* **dehydrogenation reaction.** The loss of hydrogen from ethane to give ethene and the loss of hydrogen from methanol to give methanal are examples of dehydrogenation reactions. Strong heating and a catalyst are usually necessary to make dehydrogenation reactions occur in the laboratory. In living organisms, dehydrogenation reactions are catalyzed by enzymes called *dehydrogenases* and occur at very mild conditions. Like other oxidation reactions, dehydrogenation reactions can be reversed. Alkynes can be reduced to alkenes and alkenes can be reduced to alkanes by addition of hydrogen to the double bond.

EXAMPLE 13.4	Identifying the relative degree of oxidation

List the compounds 2-propanol, propane, and propanone in order from most reduced to most oxidized.

SOLUTION

Write the structure of each compound:

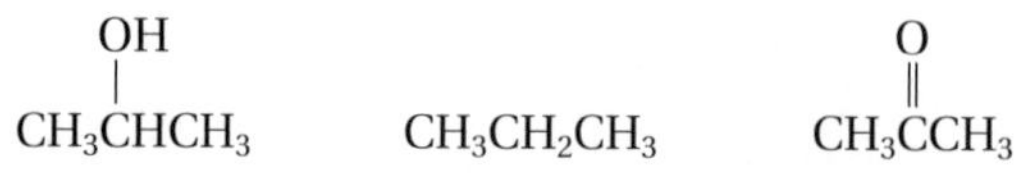

$$\underset{\text{2-Propanol}}{\overset{\overset{\displaystyle OH}{|}}{CH_3CHCH_3}} \qquad \underset{\text{Propane}}{CH_3CH_2CH_3} \qquad \underset{\text{Propanone}}{\overset{\overset{\displaystyle O}{\|}}{CH_3CCH_3}}$$

Propane is most reduced; it has the maximum number of hydrogens. Propanone is the most oxidized. It has the same number of oxygens as 2-propanol but fewer hydrogens. The order is propane, 2-propanol, and propanone.

> **PRACTICE EXERCISE 13.5**
>
> Indicate the most oxidized compound in each pair.
>
> (a) 1-butyne and 1-butene
>
> (b) propanal and propane
>
> (c) cyclohexane and cyclohexanol
>
> (d) 3-pentanol and 3-pentanone

13.4 Redox reactions involving aldehydes and ketones

AIMS: To write structures for the aldehyde and ketone products (if any) of the oxidation of primary, secondary, and tertiary alcohols. To write structures for the products of the reduction of aldehydes and ketones.

The transformation of an alcohol to an aldehyde or ketone can be accomplished by an oxidation reaction. Conversely, aldehydes and ketones can be reduced to alcohols.

Oxidation of alcohols

Primary alcohols can be oxidized to aldehydes, and secondary alcohols can be oxidized to ketones. These oxidations are represented as follows:

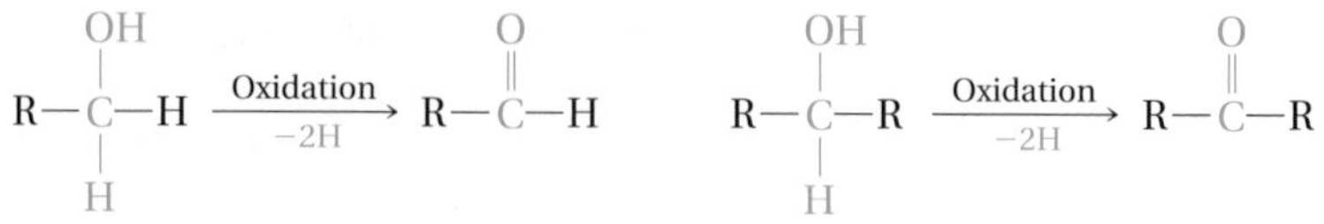

$$\underset{\substack{\text{Primary} \\ \text{alcohol}}}{\overset{\overset{\displaystyle OH}{|}}{\underset{\underset{\displaystyle H}{|}}{R-C-H}}} \xrightarrow[-2H]{\text{Oxidation}} \underset{\text{Aldehyde}}{\overset{\overset{\displaystyle O}{\|}}{R-C-H}} \qquad \underset{\substack{\text{Secondary} \\ \text{alcohol}}}{\overset{\overset{\displaystyle OH}{|}}{\underset{\underset{\displaystyle H}{|}}{R-C-R}}} \xrightarrow[-2H]{\text{Oxidation}} \underset{\text{Ketone}}{\overset{\overset{\displaystyle O}{\|}}{R-C-R}}$$

Tertiary alcohols are not oxidized because there is no hydrogen to remove from the carbon bearing the hydroxyl group.

Oxidation of the primary alcohols methanol and ethanol by warming them at about 50 °C with acidified potassium dichromate ($K_2Cr_2O_7$) produces formaldehyde and acetaldehyde, respectively:

$$
\underset{\substack{\text{Methanol} \\ \text{(methyl alcohol)} \\ \text{(bp 65 °C)}}}{H-\overset{\displaystyle OH}{\underset{\displaystyle H}{\overset{|}{\underset{|}{C}}}}-H}
\xrightarrow[\text{H}_2\text{SO}_4]{\text{K}_2\text{Cr}_2\text{O}_7}
\underset{\substack{\text{Methanal} \\ \text{(formaldehyde)} \\ \text{(bp } -21 \text{ °C)}}}{H-\overset{\displaystyle O}{\overset{\|}{C}}-H}
\qquad
\underset{\substack{\text{Ethanol} \\ \text{(ethyl alcohol)} \\ \text{(bp 78 °C)}}}{CH_3-\overset{\displaystyle OH}{\underset{\displaystyle H}{\overset{|}{\underset{|}{C}}}}-H}
\xrightarrow[\text{H}_2\text{SO}_4]{\text{K}_2\text{Cr}_2\text{O}_7}
\underset{\substack{\text{Ethanal} \\ \text{(acetaldehyde)} \\ \text{(bp 21 °C)}}}{CH_3-\overset{\displaystyle O}{\overset{\|}{C}}-H}
$$

The preparation of an aldehyde by this method is often a problem because aldehydes are easily oxidized to carboxylic acids:

$$
\underset{\text{Aldehyde}}{R-\overset{\displaystyle O}{\overset{\|}{C}}-H}
\xrightarrow[\text{H}_2\text{SO}_4]{\text{K}_2\text{Cr}_2\text{O}_7}
\underset{\text{Carboxylic acid}}{R-\overset{\displaystyle O}{\overset{\|}{C}}-OH}
$$

Further oxidation is not a problem with aldehydes that have low boiling points, such as acetaldehyde, because the product can be distilled from the reaction mixture as it is formed.

Oxidation of the secondary alcohol 2-propanol by warming with acidified potassium dichromate produces acetone:

$$
\underset{\substack{\text{2-Propanol} \\ \text{(isopropyl alcohol)}}}{CH_3-\overset{\displaystyle OH}{\underset{\displaystyle H}{\overset{|}{\underset{|}{C}}}}-CH_3}
\xrightarrow[\text{H}_2\text{SO}_4]{\text{K}_2\text{Cr}_2\text{O}_7}
\underset{\substack{\text{Propanone} \\ \text{(acetone)}}}{CH_3-\overset{\displaystyle O}{\overset{\|}{C}}-CH_3}
$$

Ketones are resistant to further oxidation. There is no need to remove them from the reaction mixture during the course of the reaction.

| **EXAMPLE 13.5** | **Writing an equation for oxidation of an alcohol** |

Write an equation for the oxidation of 4-methyl-2-hexanol.

SOLUTION

A secondary alcohol is oxidized to a ketone. The equation is

$$
\underset{}{CH_3\overset{\displaystyle OH}{\overset{|}{C}}HCH_2\overset{\displaystyle CH_3}{\overset{|}{C}}HCH_2CH_3}
\xrightarrow[\text{H}_2\text{SO}_4]{\text{K}_2\text{Cr}_2\text{O}_7}
CH_3\overset{\displaystyle O}{\overset{\|}{C}}CH_2\overset{\displaystyle CH_3}{\overset{|}{C}}HCH_2CH_3
$$

Since the reactant is a secondary alcohol, oxidation to a carboxylic acid will not occur.

PRACTICE EXERCISE 13.6

What products are expected when the following compounds are oxidized?

(a) $CH_3CH_2CH_2CH_2OH$

(b)
$$OH$$
$$CH_3CH_2CHCH_3$$

(c)
$$OH$$
$$CH_3CH_2CCH_3$$
$$CH_3$$

(d)
$$OH$$

PRACTICE EXERCISE 13.7

Give the name and structure of the alcohol you must oxidize to make the following compounds.

(a) CH_3CH_2CHO

(b)
$$O$$
$$CH_3CH_2CCH_3$$

(c) CH_3CH_2CHCHO
$$CH_3$$

Oxidation of alcohols in the liver

Ethanol is oxidized in the liver to acetaldehyde. Acetaldehyde is then oxidized to acetic acid and finally to carbon dioxide and water:

$$CH_3CH_2-OH \longrightarrow CH_3-\overset{O}{\overset{\|}{C}}-H \longrightarrow CH_3-\overset{O}{\overset{\|}{C}}-OH \longrightarrow CO_2 + H_2O$$

Ethanol   Acetaldehyde   Acetic acid   Carbon   Water
                                                            dioxide

Consumption of large quantities of ethanol causes the buildup of high concentrations of acetaldehyde in the blood. This can lead to a sharp decrease in blood pressure, a more rapid heartbeat, and a generally uncomfortable feeling—a hangover. Continued overindulgence in ethanol eventually leads to yellowing and hardening of the liver, called *cirrhosis* (Fig. 13.3), because of the sustained high levels of acetaldehyde. Cirrhosis is an irreversible degeneration of functioning liver cells.

Methanol, sometimes called *wood alcohol,* is extremely toxic. When methanol enters the body, it is quickly absorbed into the bloodstream and passes to the liver, where it is oxidized to formaldehyde. Formaldehyde is a very reactive compound. It destroys the catalytic power of enzymes and causes liver tissue to become hard. This is why formaldehyde solutions are

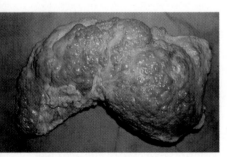

Figure 13.3
Chronic alcohol consumption can cause cirrhosis of the liver.

FOLLOW-UP TO THE CASE IN POINT: Fetal alcohol syndrome

Fortunately, Marla was able to follow her physician's advice, and at the end of a full-term pregnancy, she gave birth to a healthy baby girl. The doctor's concern about Marla's history of alcoholism resulted from the knowledge that alcohol consumption can cause great damage to the fetus. The fetus processes ethanol more slowly than the mother, so the injurious effects of alcohol and its oxidation product, acetaldehyde, affect the fetus for a longer time. A pregnant woman who drinks alcohol passes these effects in a magnified form to her unborn child. The damaging effects of alcohol on the unborn are called *fetal alcohol effects* (FAE). The range of FAE can be mild to severe. Some FAE are so broad and long-lasting that they can cripple a child's physical, social, and academic development. *Fetal alcohol syndrome* (FAS) is the name given to extremely severe FAE. FAS children can be underweight or below normal height and have abnormally small heads or other abnormal facial features, crossed eyes, underdeveloped jawbones,

cleft palates, and dysfunction of the central nervous system. FAS is the leading cause of mental retardation in children, and the retardation may be profound. The symptoms of FAE/FAS last a lifetime, and new symptoms often appear as the child grows up. Some children who do not have visible signs of FAE/FAS at birth develop them over time. For example, an infant who appears normal at birth may display slowed development in walking or talking as a toddler.

When a woman drinks heavily while she's pregnant, the chances are 30% to 40% that her baby will have FAE/FAS. If the mother also smokes, the chances of FAE/FAS increase, and the severity is intensified. FAE/FAS researchers recommend that women abstain from alcohol consumption and smoking during pregnancy. Fathers are not relieved of responsibility in preventing FAE/FAS. One study shows the birth weight of babies born to fathers who drink regularly averages 181 g less than children born to fathers who drink only occasionally.

used to preserve biological specimens. When methanol is ingested, temporary or permanent blindness may occur.

The consumption of even moderate amounts of ethanol by pregnant women can have devastating effects on the unborn fetus. In the Case in Point for this chapter, you may recall that Marla, a pregnant woman with a history of alcoholism, was strongly advised by her doctor to avoid alcohol during her pregnancy. In the Follow-up to the Case in Point, above, we will learn more about the consequences of using alcohol during pregnancy.

Reduction of aldehydes and ketones

Aldehydes and ketones can be reduced to alcohols by the addition of hydrogen, H—H, to the —C=O double bond. Reductions of an aldehyde and a ketone can be represented as follows:

As we can see from these reactions, the reduction of aldehydes produces primary alcohols, and the reduction of ketones produces secondary alcohols. A variety of reagents are available for the reduction of —C=O groups

An important biological example of reduction of an aldehyde occurs in fermentation. At the last step in the metabolism of glucose, yeast and other organisms reduce acetaldehyde to ethanol.

to —CH—OH groups. For example, hydrogen gas can be used with a platinum or palladium catalyst.

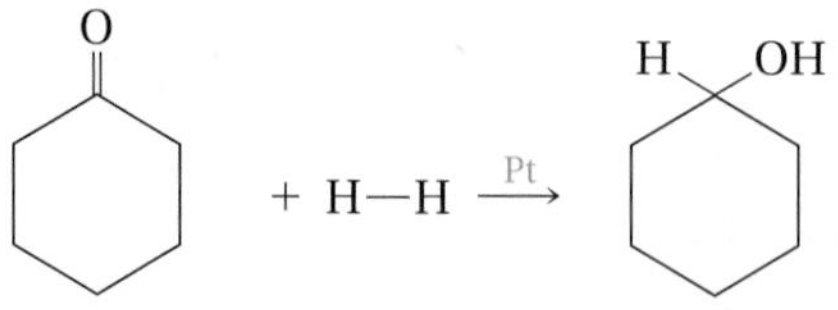

Cyclohexanone   Cyclohexanol

The hydrogenation of C=O bonds is similar to the hydrogenation of C=C bonds discussed previously. Other reducing reagents for C=O bonds include compounds called *hydrides*. Lithium aluminum hydride ($LiAlH_4$) and sodium tetrahydroborate ($NaBH_4$) are often used.

Propanal   1-Propanol (propyl alcohol)

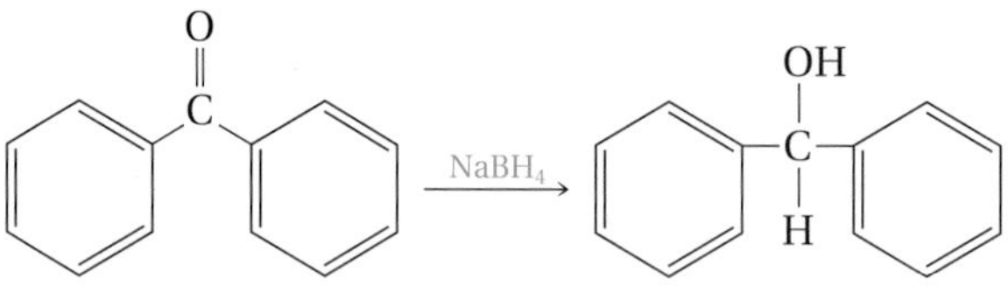

Diphenylmethanone (benzophenone)   Diphenylmethanol

Reductions of aldehydes and ketones in living organisms are catalyzed by dehydrogenase enzymes. Enzymes work reversibly, and dehydrogenases catalyze both the oxidation of alcohols and the reduction of aldehydes and ketones.

13.5 Aldehyde detection

AIM: To describe the results of a Tollens', a Benedict's, or a Fehling's test on an aldehyde, a ketone, and an alpha-hydroxy ketone.

Focus

Aldehydes can be distinguished from ketones by using mild oxidizing agents.

Aldehydes and ketones react with a wide variety of compounds. In general, however, aldehydes are more reactive than ketones. Chemists have taken advantage of the ease with which an aldehyde can be oxidized to develop several visual tests for their detection. The most widely used tests for aldehyde detection are Tollens', Benedict's, and Fehling's.

Tollens' test

The mild oxidizing agent used in this test, **Tollens' reagent,** *is an alkaline solution of silver nitrate.* It is clear and colorless. To prevent the silver ions from precipitating as silver oxide (Ag_2O) at the high pH, a few drops of an ammonia solution are added. The ammonia forms a water-soluble complex with silver ions:

$$Ag^+(aq) + 2NH_3(aq) \rightarrow [Ag(NH_3)_2]^+(aq)$$

When an aldehyde is oxidized with Tollens' reagent, the corresponding carboxylic acid is formed, and simultaneously, silver ions are reduced to metallic silver. For example, acetaldehyde goes to acetic acid. The silver is usually deposited as a mirror on the inside surface of the reaction vessel. The appearance of a silver mirror is a positive test for an aldehyde. If acetaldehyde is treated with Tollens' reagent, the reaction is

$$
\underset{\substack{\text{Acetaldehyde}}}{CH_3-\overset{\displaystyle O}{\overset{\displaystyle \|}{C}}-H} + \underset{\substack{\text{Tollens'}\\\text{reagent}}}{2[Ag(NH_3)_2]^+} + 2OH^- \longrightarrow \underset{\substack{\text{Acetic acid}\\\text{(as ammonium salt)}}}{CH_3-\overset{\displaystyle O}{\overset{\displaystyle \|}{C}}-O^-NH_4^+} + \underset{\substack{\text{Silver}\\\text{(mirror)}}}{2Ag(s)} + 3NH_3 + H_2O
$$

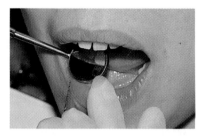

Figure 13.4
The reflective coating on a dentist's mirror is produced on the back of a sheet of glass when formaldehyde reduces silver ions in solution to silver metal. The formaldehyde is oxidized to formic acid in the process.

The aldehyde, acetaldehyde, is oxidized to a carboxylic acid, acetic acid; it is a reducing agent. The silver ions are reduced to metallic silver; they are oxidizing agents. Mirrors (Fig. 13.4) are often silvered by using Tollens' reagent. The commercial process uses glucose or formaldehyde as the reducing agent.

PRACTICE EXERCISE 13.8

You have two unlabeled test tubes, one containing pentanal and the other containing 2-pentanone. What simple test could you do to find out which tube contains pentanal and which tube contains 2-pentanone?

Benedict's and Fehling's tests

Benedict's *and* **Fehling's reagents** *are deep blue alkaline solutions of copper sulfate* of slightly differing compositions. When an aldehyde is oxidized with Benedict's or Fehling's reagents, a brick red precipitate of copper(I) oxide (Cu_2O) is obtained. The reaction with acetaldehyde is

$$
\underset{\substack{\text{Acetaldehyde}}}{CH_3-\overset{\displaystyle O}{\overset{\displaystyle \|}{C}}-H} + \underset{\substack{\text{Copper(II)}\\\text{ion complex}\\\text{(blue solution)}}}{2Cu^{+2}} + 5OH^- \longrightarrow \underset{\substack{\text{Acetic acid}\\\text{(as acetate}\\\text{ion)}}}{CH_3-\overset{\displaystyle O}{\overset{\displaystyle \|}{C}}-O^-} + \underset{\substack{\text{Copper(I)}\\\text{oxide}\\\text{(red precipitate)}}}{Cu_2O(s)} + 3H_2O
$$

The acetaldehyde is oxidized to acetic acid; copper(II) ions (Cu^{2+}) are reduced to copper(I) ions (Cu^+).

Alpha-hydroxy ketones

Ketones are not usually oxidized by mild oxidizing agents such as Tollens' and Benedict's solutions. However, ketones that contain a carbonyl group attached to a carbon that bears a hydroxyl group give positive tests with Tol-

Persons in good health do not have sugar in their urine. Clinitest tablets are used to screen for sugar in urine. The reaction of the tablet with the urine gives colored products corresponding to differing levels of sugar. The reaction chemistry is that of a Benedict's test.

lens', Benedict's, and Fehling's reagents. *These compounds are called* **alpha-hydroxy ketones,** *which have this general formula:*

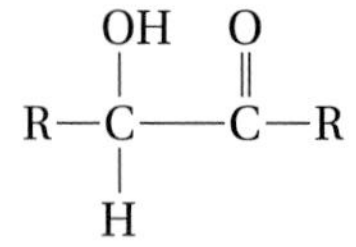

$$
\begin{array}{c}
\text{OH}\quad\text{O} \\
|\qquad\parallel \\
\text{R}-\text{C}-\!-\text{C}-\text{R} \\
| \\
\text{H}
\end{array}
$$

Alpha-hydroxy ketone

PRACTICE EXERCISE 13.9

Determine which of the following substances give a positive test (red precipitate) with Benedict's reagent.

(a)
$$
\begin{array}{c}
\text{O} \\
\parallel \\
\text{CH}_3\text{CCH}_3
\end{array}
$$

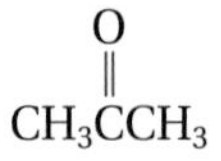

(b) CH_3CH_2CHO

(c)
$$
\begin{array}{c}
\text{OH} \\
| \\
\text{CH}_3\text{CHCHO}
\end{array}
$$

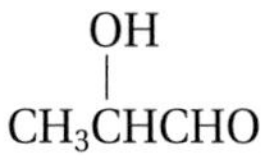

(d)
$$
\begin{array}{c}
\text{O} \\
\parallel \\
\text{Ph}-\!\overset{}{\text{C}}-\text{Ph}
\end{array}
$$

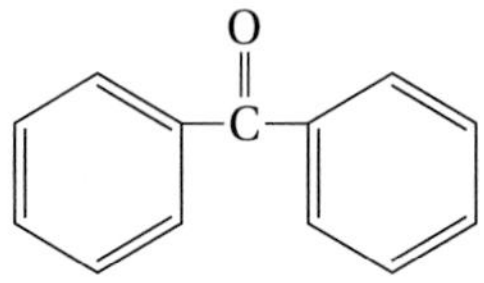

(e)
$$
\begin{array}{c}
\text{HO}\ \ \text{O} \\
|\quad\ \parallel \\
\text{CH}_3\text{CHCCH}_2\text{CH}_3
\end{array}
$$

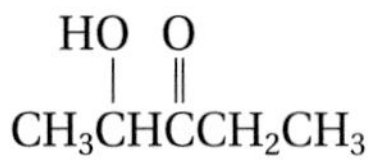

13.6 Additions to the carbonyl group

AIM: To illustrate with equations the formation of a hydrate, a hemiacetal and an acetal, and a hemiketal and a ketal.

The most characteristic reactions of both aldehydes and ketones are *addition* reactions. A wide variety of compounds add to the carbon-oxygen double bond of the carbonyl group. We have already seen additions to carbon-carbon double bonds in Sections 12.2 and 12.4, and additions to carbon-oxygen double bonds are quite similar. Our main concern here is with the addition of water and alcohols to the carbonyl group.

Addition of water

Water will add to most aldehydes and ketones to form **hydrates**—*compounds that have two hydroxyl groups on the same carbon.* The reaction is reversible. With few exceptions, the equilibrium lies to the left, in favor of the starting materials.

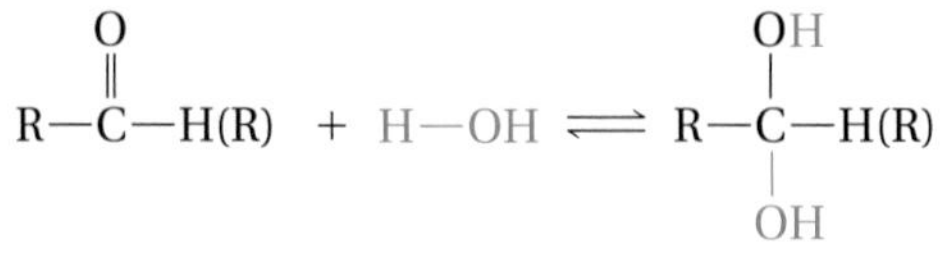

$$
\begin{array}{c}
\text{O} \\
\parallel \\
\text{R}-\text{C}-\text{H(R)} \ +\ \text{H}-\text{OH} \ \rightleftharpoons \ \text{R}-\overset{\displaystyle\text{OH}}{\underset{\displaystyle\text{OH}}{\text{C}}}-\text{H(R)}
\end{array}
$$

Carbonyl group of aldehyde or ketone · · · Water · · · Hydrate

One carbonyl compound that forms a stable hydrate is chloral. The reaction is

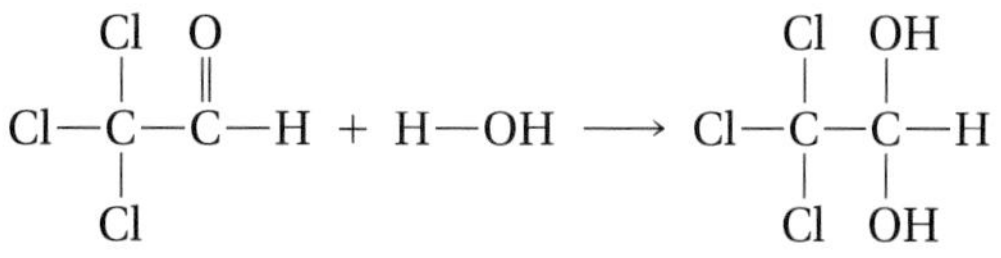

$$Cl-\underset{\underset{Cl}{|}}{\overset{\overset{Cl}{|}}{C}}-\overset{\overset{O}{||}}{C}-H \ + \ H-OH \longrightarrow Cl-\underset{\underset{Cl}{|}}{\overset{\overset{Cl}{|}}{C}}-\underset{\underset{OH}{|}}{\overset{\overset{OH}{|}}{C}}-H$$

Chloral · · · · · · · · · · · Water · · · · · · · · · Chloral
hydrate

Chloral hydrate is a colorless crystalline solid that is very soluble in water. It is one of the few stable organic molecules that has two hydroxyl groups on the same carbon. Chloral hydrate is a powerful sedative (relaxant) and hypnotic (sleep inducer).

Addition of alcohols

The addition of alcohols to aldehydes and ketones is very similar to the addition of water. *The product of the reaction between an aldehyde and an alcohol is called a* **hemiacetal.** A hemiacetal is a compound with the general structure

$$R-\underset{\underset{OR}{|}}{\overset{\overset{OH}{|}}{C}}-H$$
Hemiacetal carbon

Hemiacetal

With acetaldehyde and methanol, the reaction for formation of a hemiacetal is

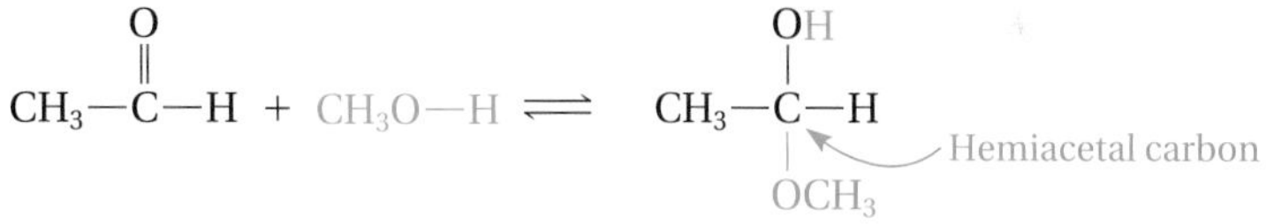

$$CH_3-\overset{\overset{O}{||}}{C}-H \ + \ CH_3O-H \ \rightleftharpoons \ CH_3-\underset{\underset{OCH_3}{|}}{\overset{\overset{OH}{|}}{C}}-H$$
Hemiacetal carbon

Acetaldehyde · · · · · Methanol · · · · · Methyl hemiacetal
of acetaldehyde

The addition of the alcohol is readily reversible, so many hemiacetals are quite unstable. In the presence of an excess of alcohol, a hemiacetal rapidly reacts with another molecule of alcohol to produce a relatively stable *acetal. An* **acetal** *is a compound with the general structure*

$$R-\underset{\underset{OR}{|}}{\overset{\overset{OR}{|}}{C}}-H$$
Acetal carbon

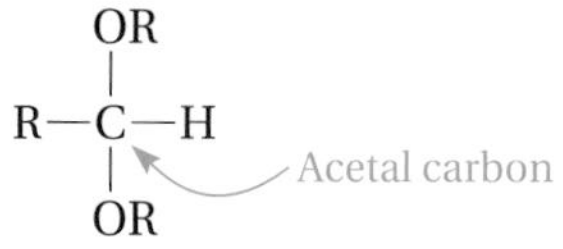

Acetal

Acetals contain a carbon that has a hydrogen and two OR groups attached and therefore contain a diether linkage. The following reaction shows the formation of an acetal from a hemiacetal:

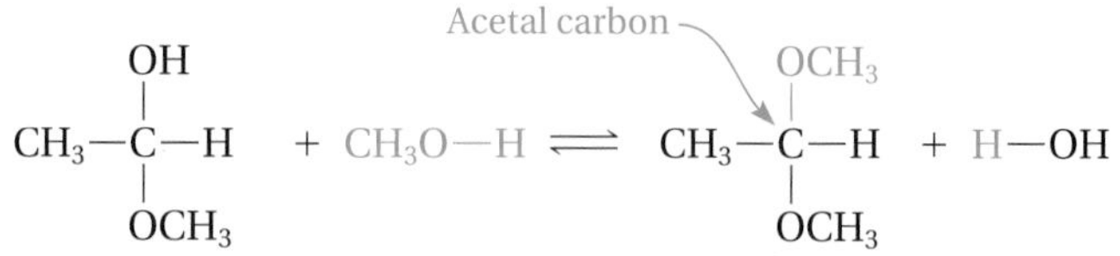

$$
\begin{array}{c}
\text{OH} \\
|\\
\text{CH}_3\text{—C—H} \\
|\\
\text{OCH}_3
\end{array}
\; + \; \text{CH}_3\text{O—H} \; \rightleftharpoons \;
\begin{array}{c}
\text{Acetal carbon} \\
\text{OCH}_3 \\
|\\
\text{CH}_3\text{—C—H} \\
|\\
\text{OCH}_3
\end{array}
\; + \; \text{H—OH}
$$

| Methyl hemiacetal of acetaldehyde | Methanol | Dimethyl acetal of acetaldehyde | Water |

Ketones undergo similar additions with alcohols to form *hemiketals* and *ketals*. **Hemiketals** *are compounds with the general structure*

$$
\begin{array}{c}
\text{OH} \\
|\\
\text{R—C—R} \\
|\\
\text{OR}
\end{array}
\quad \text{Hemiketal carbon}
$$

Hemiketal

Ketals *are compounds with the general structure*

$$
\begin{array}{c}
\text{OR} \\
|\\
\text{R—C—R} \\
|\\
\text{OR}
\end{array}
\quad \text{Ketal carbon}
$$

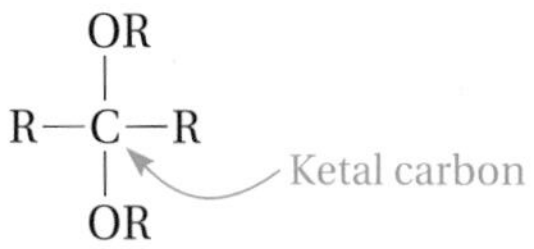

Ketal

Hemiketals and ketals have two carbons attached to their central carbons; hemiacetals and acetals have only one carbon attached. Hemiketals are often unstable, but the ketals are relatively stable. The following equation shows the formation of a hemiketal and a ketal from a ketone and an alcohol.

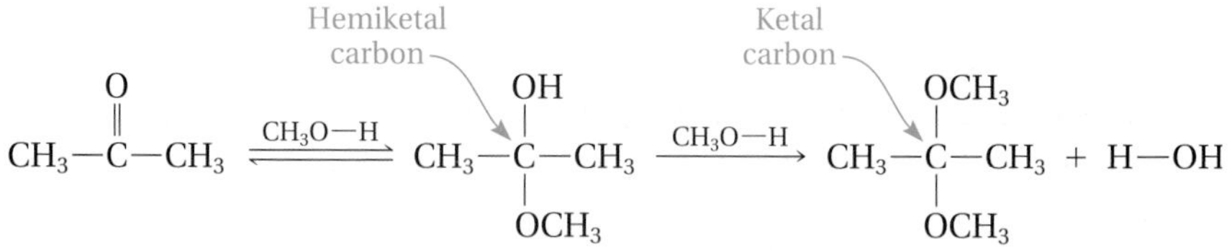

$$
\begin{array}{c}
\text{O} \\
\|\\
\text{CH}_3\text{—C—CH}_3
\end{array}
\;\xrightleftharpoons{\text{CH}_3\text{O—H}}\;
\begin{array}{c}
\text{Hemiketal carbon} \\
\text{OH} \\
|\\
\text{CH}_3\text{—C—CH}_3 \\
|\\
\text{OCH}_3
\end{array}
\;\xrightarrow{\text{CH}_3\text{O—H}}\;
\begin{array}{c}
\text{Ketal carbon} \\
\text{OCH}_3 \\
|\\
\text{CH}_3\text{—C—CH}_3 \\
|\\
\text{OCH}_3
\end{array}
\; + \; \text{H—OH}
$$

| Ketone (acetone) | Hemiketal | Ketal | Water |

The hemiacetal, hemiketal, acetal, and ketal groups are important in the chemistry of sugars.

13.7 Uses of aldehydes and ketones

AIM: To state the names and uses of some important aldehydes and ketones.

Focus

Aldehydes and ketones have many important uses.

A wide variety of aldehydes and ketones have been isolated from plants and animals. Many of them, particularly those with high molar masses, have fragrant or penetrating odors. They are usually known by their common names, which indicate their natural source or perhaps a characteristic property. Aromatic aldehydes are often used as flavoring agents. Benzaldehyde, also known as *oil of bitter almonds,* is a constituent of the almond. It is a colorless liquid with a pleasant almond odor. Cinnamaldehyde imparts the characteristic odor to oil of cinnamon. Vanillin, which is responsible for the popular vanilla flavor, was once obtainable only from the podlike capsules of certain climbing orchids. Today most vanillin is synthetically produced.

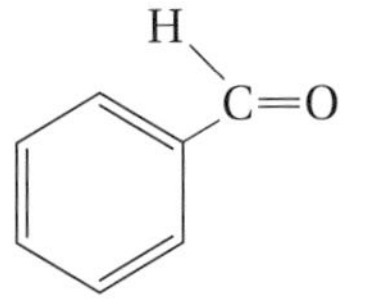

Benzaldehyde 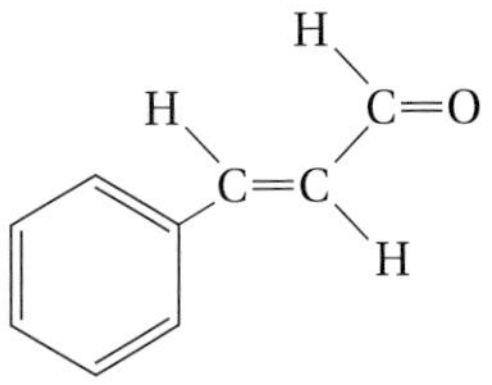   Cinnamaldehyde 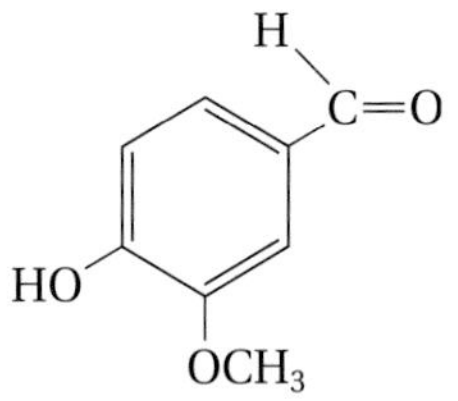   Vanillin

Vanillin is an interesting molecule because it has a number of different functional groups. It possesses an aldehyde group and an aromatic ring, making it an aromatic aldehyde. It also contains an ether linkage and a phenol functional group.

Camphor is a naturally occurring ketone obtained from the bark of the camphor tree. It has a penetrating and fragrant odor. Long known for its medicinal properties, it is an analgesic often found in liniments. Two other natural ketones, beta-ionine and muscone, are both used in perfumes. Beta-ionine is the scent of violets. Muscone, obtained from the scent gland of the male musk deer, has a ring structure containing 15 carbons.

The sex hormones are included in the long list of naturally occurring aldehydes and ketones. The molecule of the male sex hormone testosterone has a cyclic ketone group. One of the primary female sex hormones, progesterone, has two ketone functional groups.

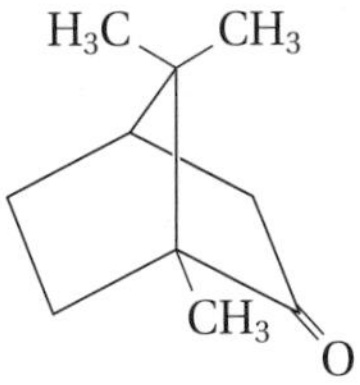

Camphor 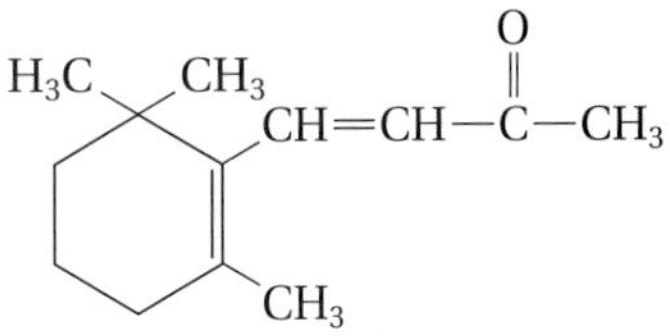   Beta-ionine 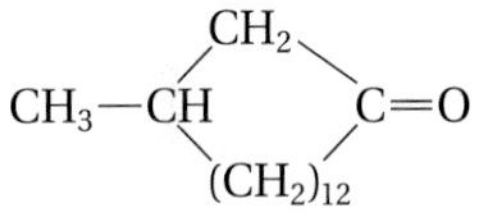   Muscone

EXAMPLE 13.6

Recognizing the structural features of an organic molecule

What structural features can you identify in the cinnamaldehyde molecule?

SOLUTION

Structural features include functional groups, carbon rings, and multiple bonds. The cinnamaldehyde molecule has a carbonyl (aldehyde) group and a phenyl group arranged in a *trans* configuration about a carbon-carbon double bond.

PRACTICE EXERCISE 13.10

Cortisone is secreted by the adrenal gland and has been used to treat arthritic conditions. How many functional groups can you identify?

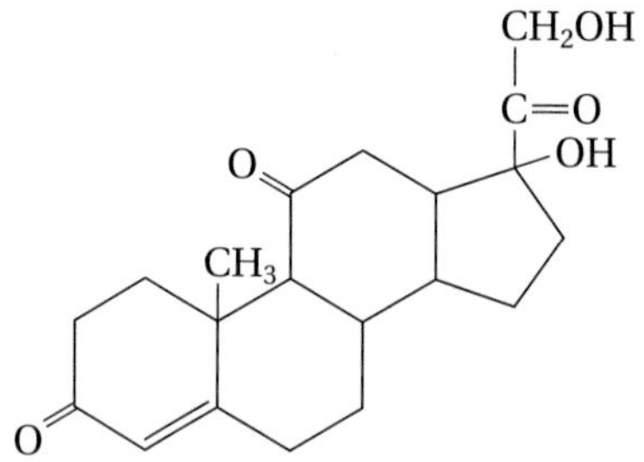

The naturally occurring aldehyde 11-*cis*-retinal is essential to the production of light-sensitive molecules within the retina of the eye. How this aldehyde permits us to see is described in A Closer Look: The Chemistry of Vision.

A Closer Look

The Chemistry of Vision

The chemistry of vision is very complex. How does the action of light on the retina of the eye produce a visual image in the brain? Not all the chemical mechanisms involved in the visual cycle are known. However, it is known that a *cis-trans* isomerization plays a central role. This isomerization involves the change from a *cis* double bond in the aldehyde 11-*cis*-retinal to a *trans* double bond in its isomer all-*trans*-retinal, as seen below.

Cones and rods, so named because of their distinctive shapes, are the two kinds of photoreceptor cells in the retina (see figure). Cones function in bright light and are responsible for color perception; rods function in dim light and are responsible for black and white vision. Rods and

Scanning electron micrograph (×1755) of the retina showing photoreceptors (rods and cones).

cones contain *chromophores*, molecules that absorb light. The chromophore in the rod cells is 11-*cis*-retinal. When the 11-*cis*-retinal is bound to a protein called *opsin*, it forms the light-sensi-

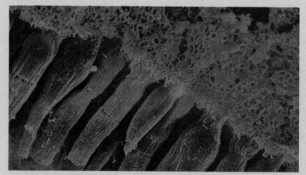

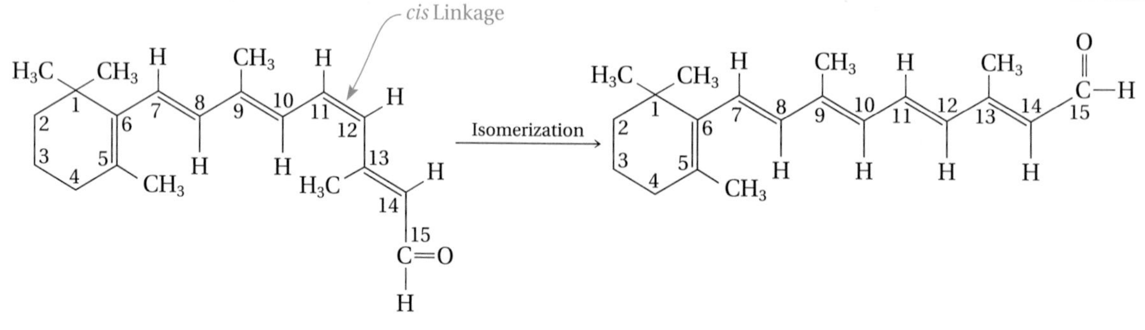

11-*cis*-Retinal   all-*trans*-Retinal

Phenol   Formaldehyde

Loss of water

Bakelite

Figure 13.5
Bakelite is a polymer of formaldehyde and phenol.

The simplest aldehyde, formaldehyde, is a colorless gas with an irritating odor. It is very important industrially but inconvenient to handle in the gaseous state. Formaldehyde is usually available as a 40% aqueous solution, known as *formalin,* or as a white, solid polymer, known as *paraformaldehyde.* When paraformaldehyde is gently heated, it decomposes and gives off formaldehyde:

$$HO\text{-}(CH_2O)_x\text{-}H \xrightarrow{\text{Heat}} x H\text{-}\overset{\displaystyle O}{\overset{\displaystyle \|}{C}}\text{-}H$$

Paraformaldehyde   Formaldehyde

Formalin is used to preserve biological specimens. The formaldehyde in solution combines with protein in tissues to make them hard and insoluble in water. This prevents the specimen from decaying. Formalin also can be used as a general antiseptic. The greatest use of formaldehyde is in the manufacture of synthetic resins. When it is polymerized with phenol, a phenol-formaldehyde resin known as *Bakelite* is formed (Fig. 13.5). Bakelite is an

tive molecule *rhodopsin.* As you can see, 11-*cis*-retinal has a molecular shape that permits it to fit exactly on the surface of the opsin molecule:

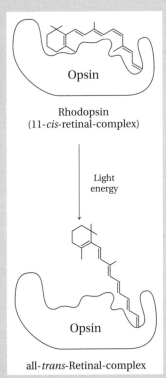

Rhodopsin
(11-*cis*-retinal-complex)

Light energy

Opsin

all-*trans*-Retinal-complex

When rhodopsin absorbs light energy, the *cis* double bond of the chromophore is broken, and the molecule re-forms as the *trans* isomer, all-*trans*-retinal. This change in shape of the chro-

mophore makes it impossible for the *trans* isomer to remain bound to opsin in the same way as the *cis* isomer. As a consequence, the shape of the entire rhodopsin molecule changes. The altered rhodopsin then undergoes further structural changes during which a nerve impulse is generated at the optic nerve and transmitted to the brain. It is this event that permits sight. Subsequently, the rhodopsin complex breaks apart, all-*trans*-retinal is liberated from opsin, and the cycle repeats.

The photosensitivity of rod cells depends on a continuous supply of the chromophore 11-*cis*-retinal. Where does this chromophore come from? It turns out that the precursor for 11-*cis*-retinal is all-*trans*-retinol, an alcohol, better known as *vitamin A.* Liver, eggs, butter, fish, and carrots are rich dietary sources of vitamin A. In carrots it is obtained indirectly from beta-carotene, the pigment that makes carrots orange. When one molecule of beta-carotene is cleaved during digestion, it produces two molecules of vitamin A (see Sec. 21.7):

all-*trans*-Retinol
(Vitamin A)

excellent electrical insulator. At one time it also was used for making billiard balls.

Acetaldehyde is a colorless, volatile liquid with an irritating odor. It is a versatile starting material and is used for the manufacture of many compounds. When acetaldehyde is heated with an acid catalyst, it polymerizes to a liquid called *paraldehyde*:

Acetaldehyde   Paraldehyde

Paraldehyde was once used as a sedative and hypnotic. Its use has diminished because of its unpleasant odor and the discovery of more effective substitutes.

The most important industrial ketone is acetone, a colorless, volatile liquid that boils at 56 °C. It is used as a solvent for resins, plastics, and varnishes. Moreover, it is miscible with water in all proportions. Acetone is produced in the body as a by-product of fat metabolism. Its concentration is usually less than 1 mg/100 mL of blood. However, in uncontrolled diabetes mellitus, acetone is produced in larger quantities, causing its level in the body to increase dramatically. It is expelled by the body as a waste product in urine and in severe cases may even be detected in the breath.

Methyl ethyl ketone is used industrially to remove wax from lubricating oils when they are refined. It is also a common solvent in nail polish remover.

SUMMARY

Aldehydes and ketones contain the carbonyl group (—C=O). The carbonyl carbon in an aldehyde has at least one hydrogen attached (R—CHO), but the carbonyl carbon in a ketone has no hydrogens (R—CO—R). Formaldehyde (H_2CO) is the simplest aldehyde; acetone (CH_3COCH_3) is the simplest ketone. The physical and chemical properties of aldehydes and ketones are influenced by the very polar carbonyl group. Molecules of aldehydes and ketones can attract each other through polar-polar interactions. These compounds have higher boiling points than the corresponding alkanes but lower boiling points than the corresponding alcohols. Aldehydes and ketones can accept hydrogen bonds, and those with low molar mass are completely soluble in water.

Aldehydes and ketones are produced by the oxidation of primary alcohols and secondary alcohols, respectively. Aldehydes are usually more reactive than ketones and are good reducing agents. An aldehyde can be oxidized to the corresponding carboxylic acid, but ketones resist further oxidation. Addition reactions are characteristic of both aldehydes and ketones. Addition of water to the carbon-oxygen bond of the carbonyl group forms hydrates. Addition of an alcohol produces hemiacetals and hemiketals. The reaction of alcohols with hemiketals and hemiacetals produces acetals and ketals, respectively.

The most important industrial aldehydes and ketones are formaldehyde, acetaldehyde, acetone, and methyl ethyl ketone. A 40% aqueous solution of formaldehyde, called formalin, is commonly used to preserve biological specimens. Many aldehydes and ketones have fragrant aromas.

SUMMARY OF REACTIONS

Here are the reactions of aldehydes and ketones presented in this chapter.

Aldehydes

1. Preparation of aldehydes:

$$
\underset{\text{Primary alcohol}}{R-\overset{\displaystyle OH}{\underset{\displaystyle H}{\vert\,\vert}}{\!C}-H} \xrightarrow{\text{Oxidation}} \underset{\text{Aldehyde}}{R-\overset{\displaystyle O}{\overset{\|}{C}}-H}
$$

2. Oxidation of aldehydes:

$$
R-\overset{\displaystyle O}{\overset{\|}{C}}-H \xrightarrow{\text{Oxidation}} \underset{\text{Carboxylic acid}}{R-\overset{\displaystyle O}{\overset{\|}{C}}-OH}
$$

3. Reaction of aldehydes with Tollens' reagent:

$$
R-\overset{\displaystyle O}{\overset{\|}{C}}-H + \underset{\substack{\text{Silver}\\\text{ions}}}{2Ag^+} + 2OH^- \longrightarrow
$$

$$
\underset{\substack{\text{Carboxylic}\\\text{acid}}}{R-\overset{\displaystyle O}{\overset{\|}{C}}-OH} + \underset{\substack{\text{Metallic}\\\text{silver}}}{2Ag(s)} + H_2O
$$

4. Reaction of aldehydes with Benedict's reagent:

$$
R-\overset{\displaystyle O}{\overset{\|}{C}}-H + \underset{\substack{\text{Copper(II) ion}\\\text{(blue solution)}}}{2Cu^{2+}} + 2OH^- \longrightarrow
$$

$$
R-\overset{\displaystyle O}{\overset{\|}{C}}-O^- + \underset{\substack{\text{Copper(I) ion}\\\text{(red precipitate}\\\text{of }Cu_2O)}}{2Cu^+} + H_2O
$$

5. Addition of water to aldehydes:

$$
R-\overset{\displaystyle O}{\overset{\|}{C}}-H + H-OH \rightleftharpoons \underset{\substack{\text{Aldehyde}\\\text{hydrate}}}{R-\overset{\displaystyle OH}{\underset{\displaystyle OH}{\vert\,\vert}}{\!C}-H}
$$

6. Addition of alcohol to an aldehyde followed by a second reaction with an alcohol:

$$
R-\overset{\displaystyle O}{\overset{\|}{C}}-H \underset{}{\overset{RO-H}{\rightleftharpoons}} \underset{\text{Hemiacetal}}{R-\overset{\displaystyle OH}{\underset{\displaystyle OR}{\vert\,\vert}}{\!C}-H} \xrightarrow{RO-H}
$$

$$
\underset{\text{Acetal}}{R-\overset{\displaystyle OR}{\underset{\displaystyle OR}{\vert\,\vert}}{\!C}-H} + H_2O
$$

Ketones

1. Preparation of ketones:

$$
\underset{\substack{\text{Secondary}\\\text{alcohol}}}{R-\overset{\displaystyle OH}{\underset{\displaystyle H}{\vert\,\vert}}{\!C}-R} \xrightarrow{\text{Oxidation}} \underset{\text{Ketone}}{R-\overset{\displaystyle O}{\overset{\|}{C}}-R}
$$

2. Addition of alcohol to a ketone followed by a second reaction with an alcohol:

$$
R-\overset{\displaystyle O}{\overset{\|}{C}}-R \overset{RO-H}{\rightleftharpoons} \underset{\text{Hemiketal}}{R-\overset{\displaystyle OH}{\underset{\displaystyle OR}{\vert\,\vert}}{\!C}-R} \xrightarrow{RO-H} \underset{\text{Ketal}}{R-\overset{\displaystyle OR}{\underset{\displaystyle OR}{\vert\,\vert}}{\!C}-R} + H_2O
$$

KEY TERMS

Acetal (13.6)
Aldehyde (13.1)
Alpha-hydroxy ketone (13.5)

Benedict's reagent (13.5)
Carbonyl group (13.1)
Dehydrogenation reaction (13.3)

Fehling's reagent (13.5)
Hemiacetal (13.6)
Hemiketal (13.6)
Hydrate (13.6)

Ketal (13.6)
Ketone (13.1)
Tollens' reagent (13.5)

EXERCISES

The Carbonyl Group (Sections 13.1, 13.2)

13.11 What is the carbonyl group? Explain why a carbon-carbon double bond is nonpolar, but a carbon-oxygen double bond is very polar.

13.12 Aldehydes and ketones are carbonyl compounds. Draw their general formulas.

13.13 Name the following aldehydes and ketones.
(a) CH_3CHO
(b) 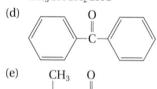 ⬡—CHO
(c) CH_3
         |
    CH_3CHCH_2CHO
(d)   O
        ||
    ⬡—C—⬡
(e) CH_3 O
        | ||
    $CH_3CHCH_2CCH_2CH_3$

13.14 Name the following aldehydes and ketones.
(a)   O
       ||
    CH_3CCH_3
(b) (cyclopentanone with =O)
(c)   O
          ||
    $CH_3CH_2CCH_2CH_2CH_3$
(d) ⬡—CH_2CHO
(e) CH_3 Cl
        | |
    $CH_3CHCH_2CHCH_2CHO$

13.15 Propane ($CH_3CH_2CH_3$) and acetaldehyde (CH_3CHO) have the same molar mass, but propane boils at $-42\ °C$, whereas acetaldehyde boils at $20\ °C$. Why are the boiling points so different?

13.16 Explain why aldehydes and ketones cannot form intermolecular hydrogen bonds but can form hydrogen bonds with water molecules.

Redox Reactions of Organic Compounds (Section 13.3)

13.17 Classify each reaction as an oxidation or a reduction.
(a) ethene $\longrightarrow$ ethyne
(b) decanol $\longrightarrow$ decanal
(c) cyclopentanone $\longrightarrow$ cyclopentane
(d) 3-hexanol $\longrightarrow$ 3-hexanone

13.18 Justify each classification in Exercise 13.17 in terms of loss or gain of oxygen and/or hydrogen.

Preparation of Aldehydes and Ketones (Section 13.4)

13.19 Give the names and structures of the expected oxidation products from
(a) 1-propanol
(b)   OH
           |
    $CH_3CH_2CCH_2CH_3$
           |
          CH_3
(c)   OH
          |
    $CH_3CH_2CHCH_2CH_3$
(d) cyclohexanol

13.20 Write the structure of the product of the oxidation of each compound. Name each product.
(a) CH_3
        |
    CH_3CHOH
(b) $CH_3(CH_2)_6CH_2OH$
(c) 2-methyl-1-butanol
(d) acetaldehyde

13.21 Write the name and structure for the aldehyde or ketone that must be reduced to make each of the following alcohols.
(a) methanol
(b)   OH
           |
    CH_3CHCH_3
(c) CH_3
        |
    CH_3CHCH_2OH

13.22 Write the name and structure for the aldehyde or ketone that must be reduced to make each of the following alcohols.
(a) 2-methylcyclopentanol   (b) CH_3CH_2OH
(c) 2-hexanol

Detection of Aldehydes (Section 13.5)

13.23 Which of the oxidation products in Exercise 13.20 will give a positive Benedict's test?

13.24 Which of the carbonyl compounds that were reduced in Exercise 13.22 will give a positive Tollens' test?

Addition Reactions (Section 13.6)

13.25 Write the structure for the hemiacetal that is formed when ethanol is added to propanal.

13.26 Write the structure for the acetal that is formed when an excess of methanol is added to ethanal.

13.27 Which of the following structures are acetals, hemiacetals, or neither?

(a) 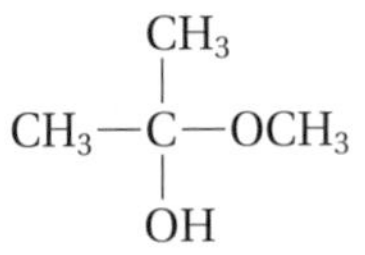
$$CH_3-\underset{\underset{OH}{|}}{\overset{\overset{CH_3}{|}}{C}}-OCH_3$$

(b)
$$CH_3-\underset{\underset{OH}{|}}{\overset{\overset{OH}{|}}{C}}-CH_3$$

(c)
$$CH_3CH_2\underset{\underset{OCH_2CH_3}{|}}{CH}OCH_2CH_3$$

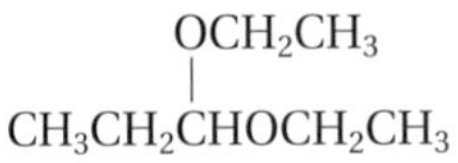

(d)

$$\text{(cyclohexyl)}-O-\underset{}{CHCH_2CH_3}$$ with OCH_3 on the CH

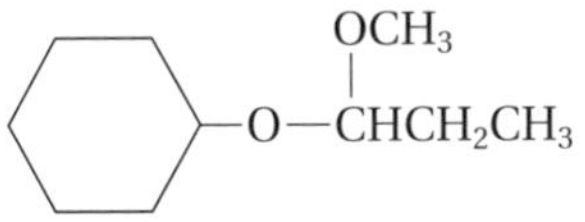

13.28. Which of the following are acetals, hemiacetals, or neither?

(a) 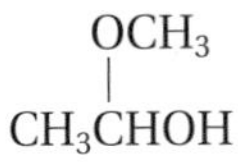
$$CH_3\underset{\underset{OCH_3}{|}}{CH}OH$$

(b) 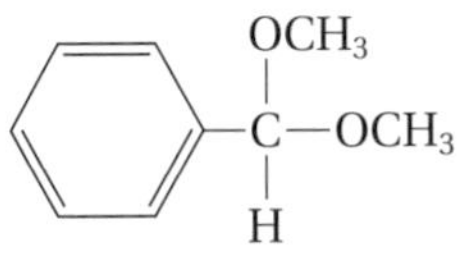
$$\text{(phenyl)}-\underset{\underset{H}{|}}{\overset{\overset{OCH_3}{|}}{C}}-OCH_3$$

(c)
$$CH_3CH_2-\underset{\underset{OCH_3}{|}}{CH}OH$$

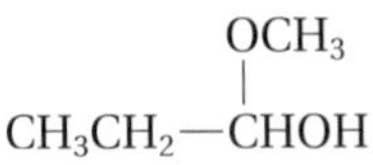

(d) 
$$CH_3\underset{\underset{OH}{|}}{CH}OCH_2CH_3$$

Additional Exercises

13.29 Give the product of each of the following reactions. If there is none, write "no reaction."

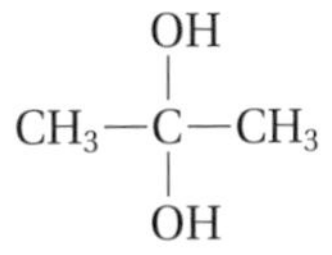

(a) 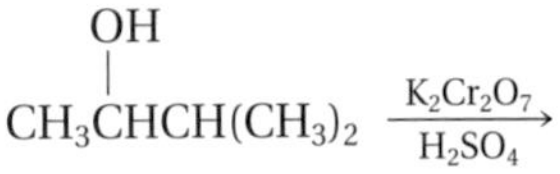
$$CH_3\underset{\underset{OH}{|}}{CH}CH(CH_3)_2 \xrightarrow[H_2SO_4]{K_2Cr_2O_7}$$

(b) 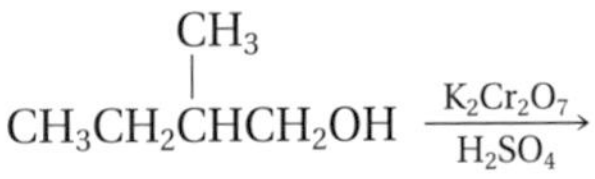
$$CH_3CH_2\underset{\underset{CH_3}{|}}{CH}CH_2OH \xrightarrow[H_2SO_4]{K_2Cr_2O_7}$$

(c) 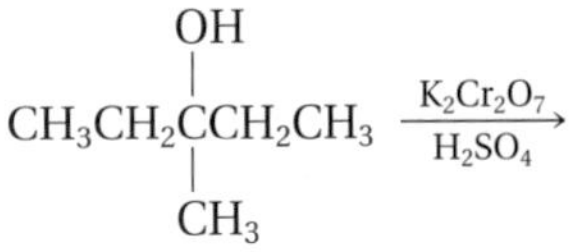
$$CH_3CH_2\underset{\underset{CH_3}{|}}{\overset{\overset{OH}{|}}{C}}CH_2CH_3 \xrightarrow[H_2SO_4]{K_2Cr_2O_7}$$

(d) 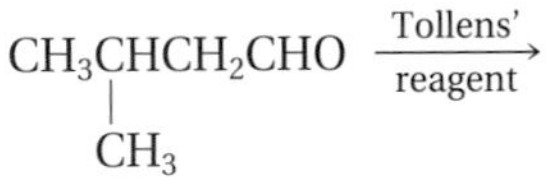
$$CH_3\underset{\underset{CH_3}{|}}{CH}CH_2CHO \xrightarrow[\text{reagent}]{\text{Tollens'}}$$

(e) 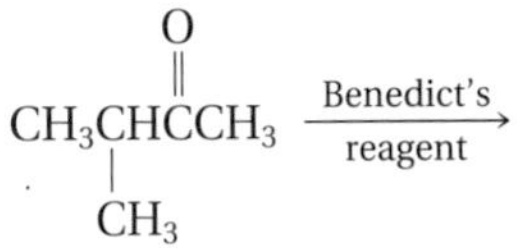
$$CH_3\underset{\underset{CH_3}{|}}{\overset{\overset{O}{\|}}{CH}}CCH_3 \xrightarrow[\text{reagent}]{\text{Benedict's}}$$

(f) $CH_3CHO + H_2 \xrightarrow[\text{Pressure}]{\text{Ni + Heat}}$

(g) $CH_3CHO + CH_3OH \longrightarrow$

(h) 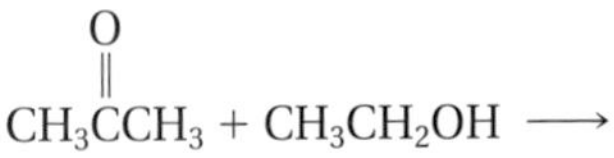
$$CH_3\overset{\overset{O}{\|}}{C}CH_3 + CH_3CH_2OH \longrightarrow$$

13.30 Give the major organic product of each of the following reactions.

(a) 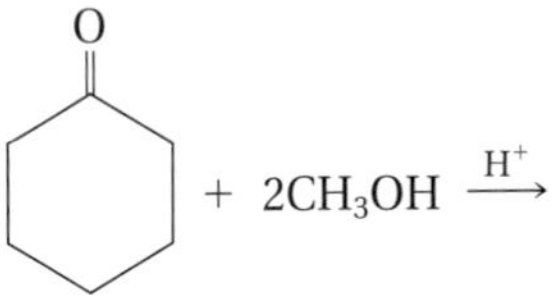
$$\text{(cyclohexanone)} + 2CH_3OH \xrightarrow{H^+}$$

(b) 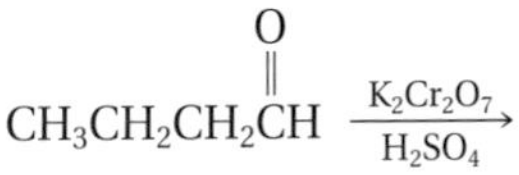
$$CH_3CH_2CH_2\overset{\overset{O}{\|}}{CH} \xrightarrow[H_2SO_4]{K_2Cr_2O_7}$$

(c) 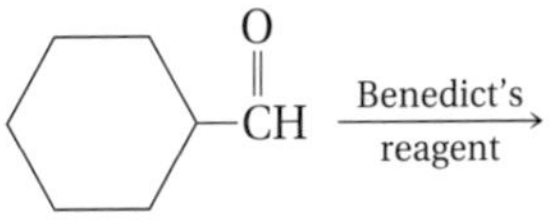
$$\text{(cyclohexyl)}-\overset{\overset{O}{\|}}{CH} \xrightarrow[\text{reagent}]{\text{Benedict's}}$$

(d) 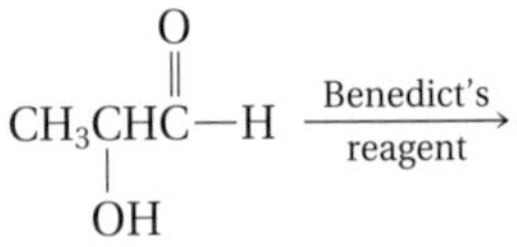
$$CH_3\underset{\underset{OH}{|}}{CH}\overset{\overset{O}{\|}}{C}-H \xrightarrow[\text{reagent}]{\text{Benedict's}}$$

13.31 Identify the following structures as acetals, hemi-acetals, ketals, or hemiketals.

(a) 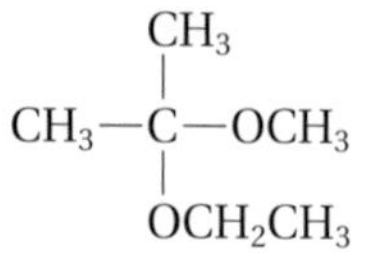
$$CH_3-\underset{\underset{\displaystyle OCH_2CH_3}{\overset{\displaystyle CH_3}{|}}}{\overset{}{C}}-OCH_3$$

(b)
$$CH_3-\underset{\overset{\displaystyle |}{OCH_3}}{CH}-OH$$

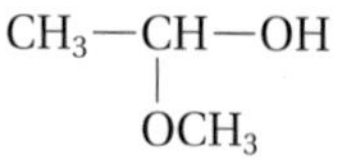

(c)
$$\begin{array}{c} H_2 \\ C \\ H_2C \qquad O \\ | \qquad | \\ H_2C \qquad CH-OCH_3 \\ C \\ H_2 \end{array}$$

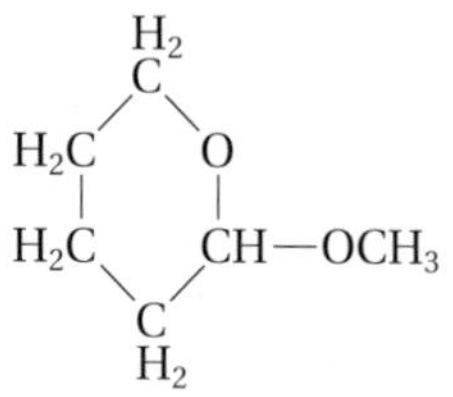

(d) 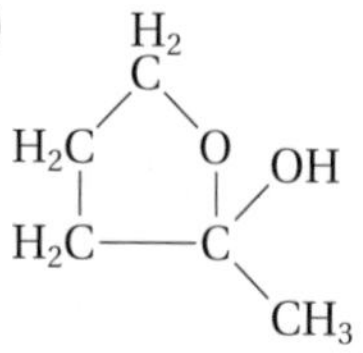
$$\begin{array}{c} H_2 \\ C \\ H_2C \qquad O \quad OH \\ | \qquad | \diagup \\ H_2C-\underset{\underset{\displaystyle CH_3}{}}{C} \end{array}$$

(e) 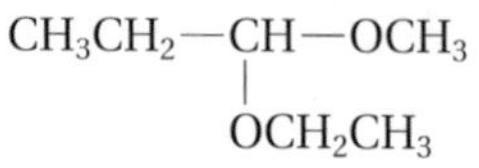
$$CH_3CH_2-\underset{\overset{\displaystyle |}{OCH_2CH_3}}{CH}-OCH_3$$

(f) 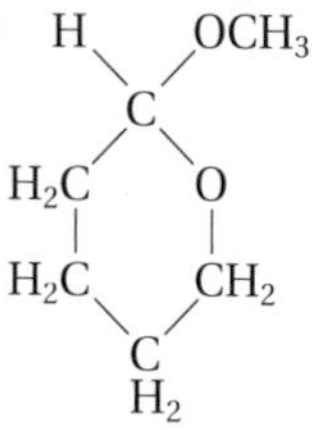
$$\begin{array}{c} H \qquad OCH_3 \\ \diagdown \quad \diagup \\ C \\ H_2C \qquad O \\ | \qquad | \\ H_2C \qquad CH_2 \\ C \\ H_2 \end{array}$$

13.32 Name the following compounds by the IUPAC system.

(a) $CH_3CH_2CH_2CHO$

(b) 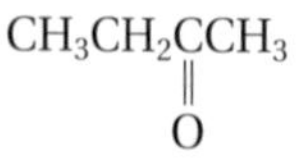 $CH_3CH_2\underset{\underset{\displaystyle O}{\|}}{C}CH_3$

(c) $ClCH_2CH_2CHO$

(d) 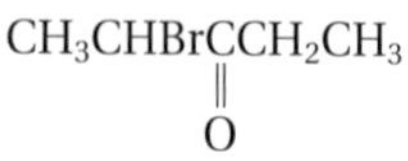 $CH_3CHBr\underset{\underset{\displaystyle O}{\|}}{C}CH_2CH_3$

(e) 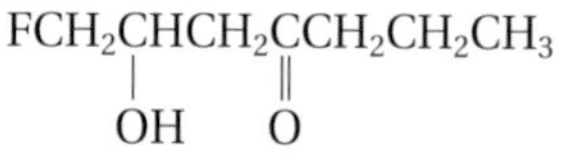 $FCH_2\underset{\underset{\displaystyle OH}{|}}{CH}CH_2\underset{\underset{\displaystyle O}{\|}}{C}CH_2CH_2CH_3$

(f) 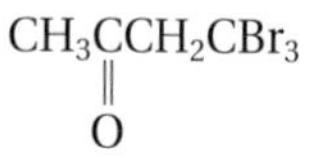 $CH_3\underset{\underset{\displaystyle O}{\|}}{C}CH_2CBr_3$

13.33 The boiling points of ethyl methyl ether and propanone are 11 and 56 °C, respectively, but their molecular weights are very similar. Explain.

13.34 Which pairs of the following molecules form hydrogen bonds? Draw structures to show the hydrogen bond, if formed.
(a) acetone and diethyl ether
(b) acetone and acetone
(c) acetone and water
(d) acetone and acetaldehyde
(e) acetaldehyde and water

13.35 You are given a compound and told that it is either $CH_3CH_2COCH_3$ or $CH_3CH_2CH_2CHO$. What laboratory test(s) would you perform to identify the compound?

13.36 Draw structural formulas for the following.
(a) 3-hexanone   (b) butyraldehyde   (c) propanal
(d) diisobutyl ketone   (e) methanal
(f) methyl phenyl ketone

13.37 Name the following structures. (Where appropriate, give both the common name and the IUPAC name.)

(a) HCHO

(b) 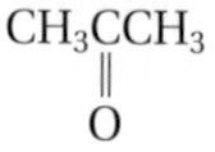 $CH_3\underset{\underset{\displaystyle O}{\|}}{C}CH_3$

(c) $CH_3CH_2CH_2CHO$

(d) 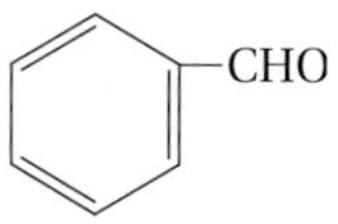 a benzene ring with —CHO

(e) 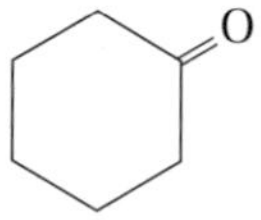 a cyclohexane ring with =O

(f) 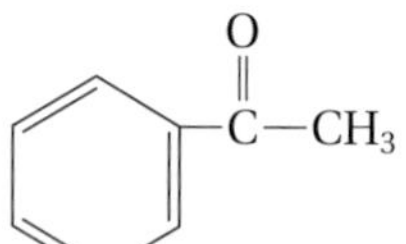 a benzene ring with $-\underset{\underset{\displaystyle O}{\|}}{C}-CH_3$

(g) 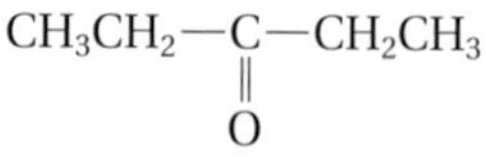 $CH_3CH_2-\underset{\underset{\displaystyle O}{\|}}{C}-CH_2CH_3$

13.38 Write equations for the following reactions. Identify the class of compound(s) formed.
(a) the oxidation of 2-propanol
(b) the addition of water to acetone
(c) the reduction of formaldehyde
(d) the stepwise addition of two moles of ethanol to one mole of acetaldehyde

13.39 Which member of each of these pairs of compounds would release more energy upon complete oxidation?
(a) butanol, butanal
(b) ethene, ethane
(c) carbon dioxide, formaldehyde
(d) isopropyl alcohol, acetone

SELF-TEST (REVIEW)

True/False

1. A dehydrogenation reaction is a reduction reaction.
2. Hydrogen bonding accounts for the relatively high boiling point of acetaldehyde.
3. The reaction of equal moles of an alcohol and an aldehyde gives an acetal.
4. Oxidation of a tertiary alcohol gives a ketone.
5. You would expect propanal to have a higher boiling point than propanol.
6. Propanal should have higher water solubility than hexanal.
7. One mole of methanol would release more energy upon complete oxidation than one mole of methane.
8. Both aldehydes and ketones have a carbonyl group.
9. All aldehydes and ketones give a positive Tollens' test.
10. Four pairs of electrons are shared in the carbonyl bond formed between an oxygen and a carbon atom.

Multiple Choice

11. Which of the following statements about the carbon-oxygen double bond of the carbonyl group is *false?*
    (a) The bond is polar.
    (b) The carbon has a partial negative charge.
    (c) The bonding electrons are unequally shared between the carbon and oxygen.
    (d) The oxygen has two unshared pairs of electrons.
12. Acetaldehyde would be likely to form hydrogen bonds with
    (a) formaldehyde.   (b) octane.   (c) water.
    (d) acetone.
13. On the basis of your knowledge of intermolecular forces, which of the following would you expect to have the highest boiling point?
    (a) propanal   (b) propane   (c) acetone
    (d) 1-propanol

14. Which of the following compounds contains a di-ether linkage?
    (a) a hemiacetal   (b) chloral hydrate
    (c) a ketal   (d) camphor
15. Which of the following substances is used as a preservative of biological specimens?
    (a) paraldehyde   (b) formalin
    (c) methyl ethyl ketone   (d) cinnamaldehyde
16. Which of the following substances can undergo an addition reaction with methanol?
    (a) propane   (b) methyl ethyl ether
    (c) propanol   (d) propanal
17. A structural isomer of 2-butanone is
    (a) diethyl ether.   (b) *tert*-butyl alcohol.
    (c) diethyl ketone.   (d) butanal.
18. Which of the following compounds would release the most energy upon oxidation to carbon dioxide?
    (a) ethanol   (b) acetic acid, CH_3COOH
    (c) ethane   (d) acetaldehyde
19. The oxidation of 2-methyl-2-butanol with $K_2Cr_2O_7$ and H_2SO_4 would give
    (a) 2-methyl-2-butanone.
    (b) isopropyl alcohol and ethane.
    (c) 2-methyl-2-butanal.
    (d) none of the above.
20. In a positive Tollens' test,
    (a) silver ions are oxidized to silver atoms.
    (b) the aldehyde is an oxidizing agent.
    (c) a silver mirror is formed.
    (d) more than one are correct.
21. In the reaction of substance A with substance B, substance A loses oxygen. Which of the following is true?
    (a) Substance B is an oxidizing agent.
    (b) Substance A is reduced.
    (c) Substance B is reduced.
    (d) Substance A is a reducing agent.

14

Acids and Their Derivatives

Reactions of the Carboxyl Group

Citric acid, a carboxylic acid, gives oranges and other citrus fruits their sour taste and low pH.

CHAPTER OUTLINE

CASE IN POINT: Acne and tretinoin

14.1   Carboxylic acids

FOLLOW-UP TO THE CASE IN POINT: Acne and tretinoin

14.2   The carboxyl group

14.3   Acidity of carboxylic acids

A CLOSER LOOK: Hard Water and Water Softening

14.4   Synthesis of carboxylic acids

14.5   Carboxylic acid anhydrides

14.6   Carboxylic esters

A CLOSER LOOK: Ester Local Anesthetics

A CLOSER LOOK: Aspirin

14.7   Thioesters

14.8   Ester hydrolysis

14.9   Phosphoric acids, anhydrides, and esters

14.10   Esters of nitric and nitrous acids

What causes the sting of ant venom, the sour taste of unripe fruit, and the tart taste of vinegar? All are caused by carboxylic acids (RCO_2H), which along with their derivatives are the major topic of this chapter. We also will examine some organic derivatives of inorganic acids that are important to biology and health. One carboxylic acid that has proven to be valuable in the treatment of *acne vulgaris,* or common acne, is the subject of the following Case in Point.

CASE IN POINT: Acne and tretinoin

Lana passed her first 12 years of life without a blemish on her skin. When she was 13 years old, Lana noticed her first pimple. Soon her face was covered with red blotches, some of which formed white caps of pus. She also began to find blackheads around her nose and on her cheeks. Although most adolescents experience common acne, Lana's acne seemed worse than most. When Lana developed a large boil on her nose, her parents decided to seek medical advice. The doctor agreed that Lana's acne was worse than usual and gave Lana a cream containing a carboxylic acid called *tretinoin* to apply to her face (see figure). Within a few weeks, Lana's skin had cleared up considerably. What is the chemical structure of tretinoin? How does this carboxylic acid work to relieve acne? We will learn about this remarkable compound in Section 14.1.

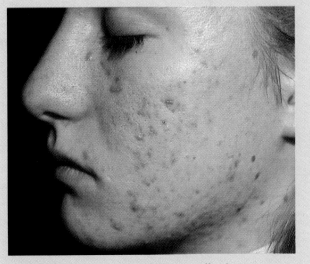

Preparations containing tretinoin are effective treatments for acne.

14.1 Carboxylic acids

AIMS: *To name and draw structures of common carboxylic acids and organic acid salts. To distinguish among dicarboxylic acids, polyfunctional carboxylic acids, and fatty acids.*

Focus

Carboxylic acids are widely distributed in nature.

Carboxylic acids *are a class of organic compounds characterized by the* **carboxyl group,** *whose name comes from the words* carbonyl *and* hydroxyl, *its two functional groups.*

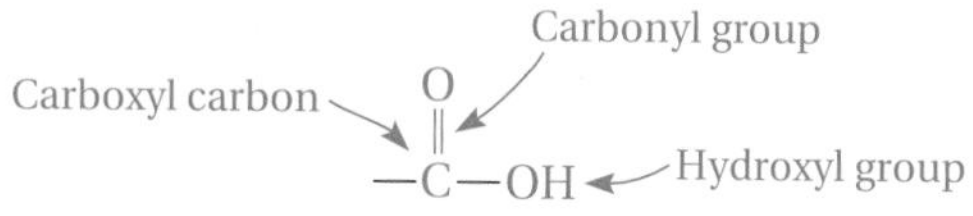

Carbonyl group

Carboxyl carbon

$$-\overset{\overset{\displaystyle O}{\|}}{C}-OH$$ ← Hydroxyl group

Carboxyl group
(also written —CO_2H or —COOH)

The general formula of a carboxylic acid is RCOOH. Carboxylic acids are weak acids because they dissociate to a slight extent in aqueous solution to give a carboxylate ion and a hydrogen ion:

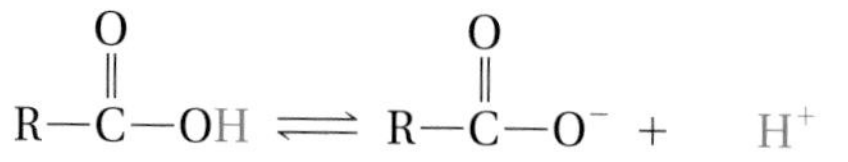

$$R-\overset{\overset{\displaystyle O}{\|}}{C}-OH \rightleftharpoons R-\overset{\overset{\displaystyle O}{\|}}{C}-O^- + H^+$$

Carboxylate ion   Hydrogen ion

Naming carboxylic acids

In the IUPAC system for naming carboxylic acids, the *-e* ending of the parent alkane is replaced by the ending *-oic acid*. The parent alkane is the hydrocarbon with the longest continuous carbon chain containing the carboxyl group. The first four straight-chain aliphatic acids (common names in parenthesis) are

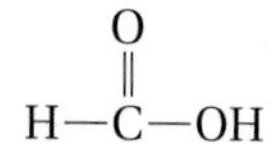

$$H-\overset{\overset{\displaystyle O}{\|}}{C}-OH \qquad CH_3-\overset{\overset{\displaystyle O}{\|}}{C}-OH \qquad CH_3CH_2-\overset{\overset{\displaystyle O}{\|}}{C}-OH \qquad CH_3CH_2CH_2-\overset{\overset{\displaystyle O}{\|}}{C}-OH$$

Methanoic acid (formic acid) 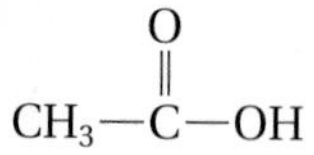   Ethanoic acid (acetic acid) 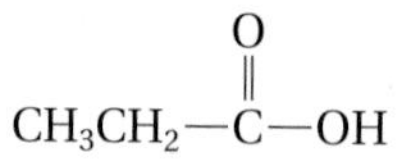   Propanoic acid (propionic acid) 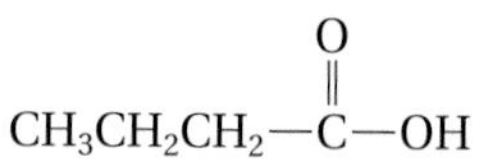   Butanoic acid (butyric acid)

In naming carboxylic acids by the IUPAC system, we number the carboxyl carbon as carbon 1. Each substituent on the chain is identified by its name and a number indicating its position on the chain. Here are the structures and IUPAC names of two substituted carboxylic acids:

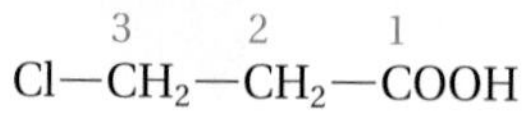

$$\overset{3}{C}l-\overset{2}{C}H_2-\overset{1}{C}H_2-COOH \qquad\qquad \overset{4}{C}H_3-\overset{3}{C}H-\overset{2}{C}H-\overset{1}{C}OOH$$
$$\qquad\qquad\qquad\qquad\qquad\qquad\qquad\qquad | \quad\; |$$
$$\qquad\qquad\qquad\qquad\qquad\qquad\qquad\qquad OH \; Br$$

3-Chloropropanoic acid 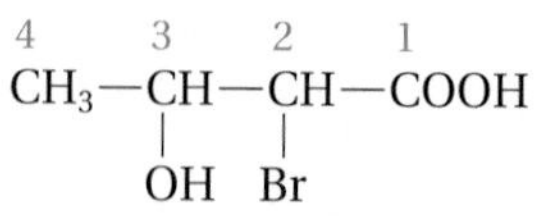   2-Bromo-3-hydroxybutanoic acid

The simplest aromatic acid is benzoic acid, a name derived from the parent hydrocarbon benzene:

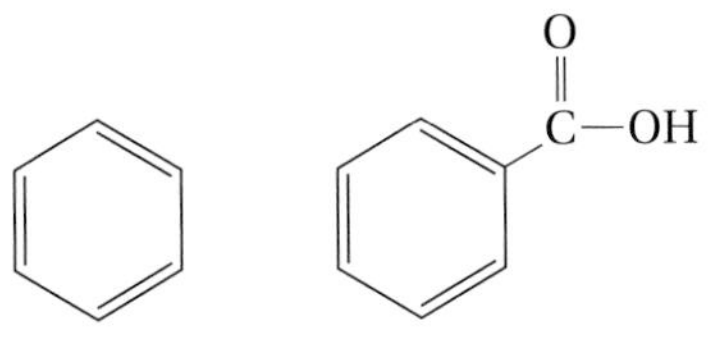

Benzene Benzoic acid

EXAMPLE 14.1 **Naming a carboxylic acid**

Give the IUPAC name for the following carboxylic acid.

$$\begin{array}{c} Br \\ | \\ HOCH_2CH_2CH_2CHCO_2H \end{array}$$

SOLUTION

We number the longest carbon chain so that the carboxyl group is in position 1.

$$\begin{array}{c} Br \\ | \\ HOCH_2CH_2CH_2CHCO_2H \\ 5 \quad 4 \quad 3 \quad 2 \quad 1 \end{array}$$

The longest carbon chain contains five carbons, making *pentane* the parent alkane. We remove the *-e* ending from the word *pentane* and add *-oic acid* to the root name *pentan-*. Our compound is a *pentanoic acid.* Now we name and number the substituents in the carbon chain; the name for a —Br substituent is *bromo,* and the name for an —OH substituent is *hydroxy.* The substituents in our compound are *2-bromo* and *5-hydroxy.* We combine the substituent names and numbers in alphabetical order with the parent acid name to obtain the complete name of the carboxylic acid: *2-bromo-5-hydroxypentanoic acid.*

Carboxylic acids are abundant and widely distributed in nature. Many have common names derived from a Greek or Latin word describing their natural sources. For example, methanoic acid is usually called by its common name, *formic acid.* Formic acid is produced by ants. The common name for ethanoic acid is *acetic acid.* Acetic acid is produced when wine turns sour and becomes vinegar. The pungent aroma of vinegar comes from its acetic acid. Common names for many carboxylic acids are used more often than IUPAC names.

Concentrated acetic acid is known as *glacial* acetic acid because it freezes to an icelike solid in cold stockrooms or laboratories that are below 17 °C.

Fatty acids

The straight-chain carboxylic acids were first isolated from fats and are called **fatty acids.** Propionic acid, the three-carbon acid, literally means

Lactobacillic acid (shown below), a fatty acid from a bacterial source, contains the highly strained cyclopropyl ring.

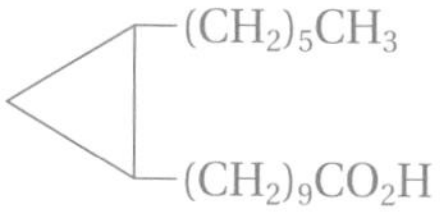

$$\text{(CH}_2)_5\text{CH}_3$$
$$\text{(CH}_2)_9\text{CO}_2\text{H}$$

"first fatty acid." The four-carbon acid, butyric acid, can be obtained from butterfat.

The low-molar-mass members of the aliphatic carboxylic acid series are colorless, volatile liquids. They have sharp odors. The smells of rancid butter and dirty feet are due in part to butyric acid. The smell of goats is caused by the 6-, 8-, and 10-carbon (C-6, C-8, C-10) straight-chain acids. The longer members of the series are nonvolatile, waxy solids with low melting points. Stearic acid (C-18) is obtained from beef fat. It is used to make wax candles. Stearic acid and other long-chain fatty acids have very little odor. Table 14.1 lists the names and formulas for some common saturated aliphatic carboxylic acids.

Many fatty acids contain carbon-carbon double bonds and are referred to as *unsaturated fatty acids.* Table 14.2 lists some names and formulas for some unsaturated aliphatic carboxylic acids. Recall from the Case in Point in the introduction to this chapter that Lana was afflicted with chronic acne. In our follow-up we will see how she was helped by a rather amazing unsaturated carboxylic acid.

Dicarboxylic acids

Dicarboxylic acids *are compounds that have two carboxyl groups.* Here are the first five members of the series; if you want to memorize their names, the expression "**O**h **M**y, **S**uch **G**ood **A**pples" is helpful.

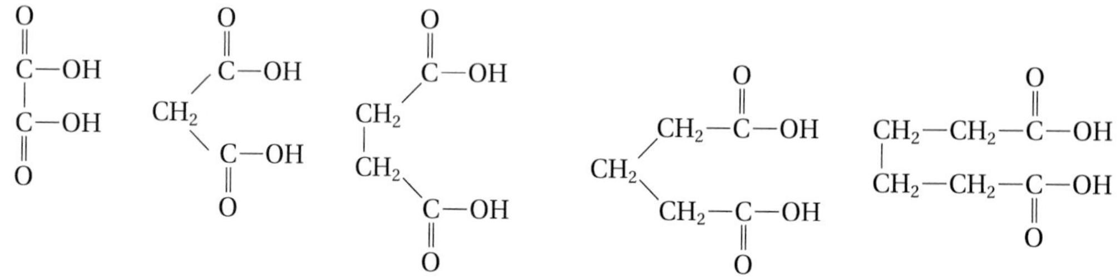

| Oxalic acid | Malonic acid | Succinic acid | Glutaric acid | Adipic acid |

Table 14.1 Saturated Aliphatic Carboxylic Acids

Formula	Carbon atoms	Common name	Melting point (°C)	Derivation of name
HCOOH	1	formic acid	8	Lat. *formica,* ant
CH_3COOH	2	acetic acid	17	Lat. *acetum,* vinegar
CH_3CH_2COOH	3	propionic acid	-22	Gk. *protos,* first; *pion,* fat
$CH_3(CH_2)_2COOH$	4	butyric acid	-6	Lat. *butyrum,* butter
$CH_3(CH_2)_4COOH$	6	caproic acid	-3	Lat. *caper,* goat
$CH_3(CH_2)_6COOH$	8	caprylic acid	16	Lat. *caper,* goat
$CH_3(CH_2)_8COOH$	10	capric acid	31	Lat. *caper,* goat
$CH_3(CH_2)_{10}COOH$	12	lauric acid	44	laurel seed oil
$CH_3(CH_2)_{12}COOH$	14	myristic acid	58	nutmeg (*Myristica fragrans*)
$CH_3(CH_2)_{14}COOH$	16	palmitic acid	63	palm oil
$CH_3(CH_2)_{16}COOH$	18	stearic acid	70	Gk. *stear,* tallow

FOLLOW-UP TO THE CASE IN POINT: Acne and tretinoin

Lana was suffering from the pimples and pustules of common acne, which have long been an embarrassment to millions of adolescents. Many adults are also afflicted by this skin condition, which may affect the shoulders and back as well as the face. Severe acne can be a threat to health, since the lesions can become infected. Even after acne has subsided, it can leave its mark in the form of disfiguring pockmarks. Acne occurs when dead cells and natural oils plug the hair follicles and ducts of oil glands in the skin. The resulting irritation causes pimples that may become infected. The surface of trapped oils may be oxidized by contact with air, forming dark spots known as *blackheads*. People with acne are often advised to keep their skin clean, not to pick at pimples or squeeze blackheads, and to avoid certain foods such as chocolate.

People such as Lana, with severe acne, may require medical intervention. Lana and many other acne sufferers have achieved good results with tretinoin (see figure), a carboxylic acid derived from vitamin A (see A Closer Look: The Chemistry of Vision, on pages 412–413).

area. Two other drugs related to tretinoin, Accutane and Tegison, are administered orally. Tretinoin appears to stimulate the reproduction of the cells of the hair follicles. This activity promotes the extrusion of dead cells and oils that could plug the follicle. The skin of some patients being treated with tretinoin may become very sensitive to sunlight. The use of sunscreens and avoidance of sunlight during treatment usually are recommended. Tretinoin also presents an extremely high risk to a developing fetus. Tretinoin products require a prescription from a physician for these reasons. Since its introduction as an acne medication, tretinoin has been advanced as a kind of miracle drug for the skin. Some studies indicate that tretinoin can reverse sun damage, fade freckles and age spots, reduce the number of fine wrinkles in the skin, and even promote hair growth.

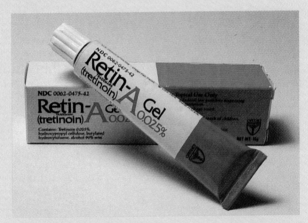

The discomfort and embarrassment of acne often can be relieved with application of tretinoin, a carboxylic acid.

Tretinoin
(Retin-A, retinoic acid, vitamin A acid)

Tretinoin is sold in a cream formulation as Retin-A. This formulation is applied topically to the affected

Table 14.2 Unsaturated Aliphatic Carboxylic Acids

Formula	Carbon atoms	Common name	Melting point ($^\circ$C)
$CH_3(CH_2)_5CH=CH(CH_2)_7COOH$	16	palmitoleic acid	−1
$CH_3(CH_2)_7CH=CH(CH_2)_7COOH$	18	oleic acid	14
$CH_3(CH_2)_4CH=CHCH_2CH=CH(CH_2)_7COOH$	18	linoleic acid	−5
$CH_3CH_2(CH=CHCH_2)_3(CH_2)_6COOH$	18	linolenic acid	−11

The unsaturated dicarboxylic acids maleic acid and fumaric acid are geometric isomers; phthalic acid is an aromatic dicarboxylic acid.

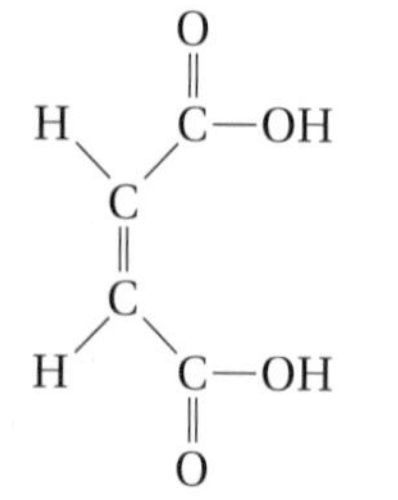

Maleic acid
(*cis* isomer; mp 130.5 °C)

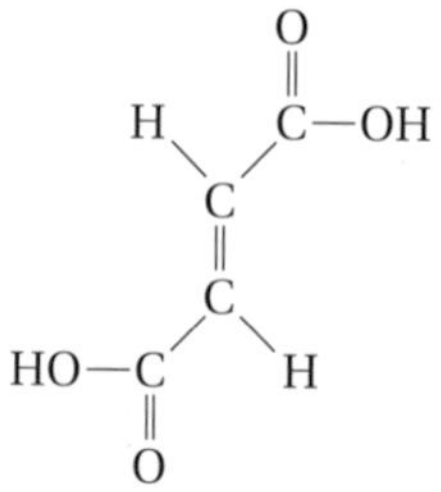

Fumaric acid
(*trans* isomer; mp 302 °C)

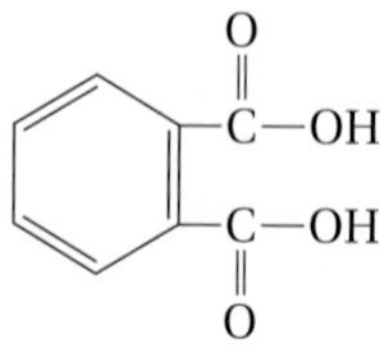

Phthalic acid
(mp 210 °C)

Polyfunctional carboxylic acids

Polyfunctional carboxylic acids are carboxylic acids that contain other functional groups as well. Many important acids are polyfunctional. Lactic acid is a **hydroxy acid**—*that is, it contains a hydroxyl and a carboxy group:*

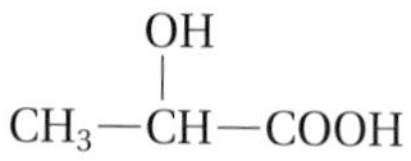

$$CH_3-\overset{\displaystyle OH}{\underset{\displaystyle |}{CH}}-COOH$$

Lactic acid

When lactose (milk sugar) is fermented by the bacterium *Lactobacillus,* lactic acid is the only product. It is responsible for the taste of sour milk. Other hydroxy acids are citric acid, tartaric acid, and salicylic acid.

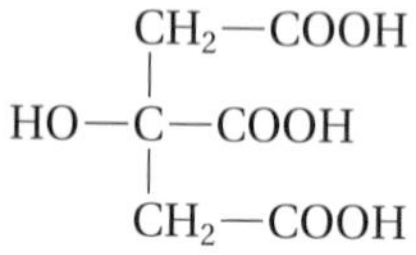

Citric acid
(citrus juice)

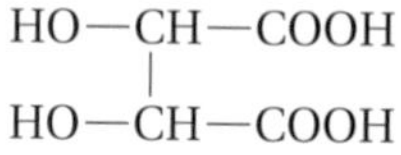

Tartaric acid
(grape juice)

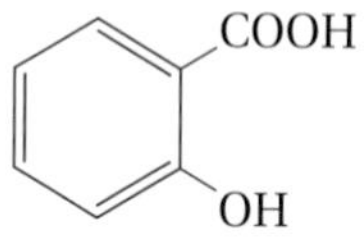

Salicylic acid
(willow bark)

Here are some examples of carboxylic acids with functional groups other than hydroxyl:

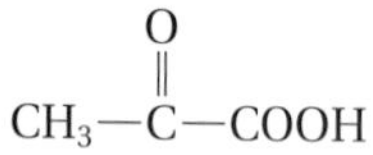

Pyruvic acid
(a keto acid
important in
the energy
production of
living organisms)

Chloroacetic acid
(a halo acid)

Glycine
(an amino acid
found in proteins)

EXAMPLE 14.2

Writing the structure of a carboxylic acid

Write the structure of 4,4-dibromohexanoic acid.

SOLUTION

Hexanoic acid is the six-carbon straight-chain acid with the carboxyl carbon as carbon-1:

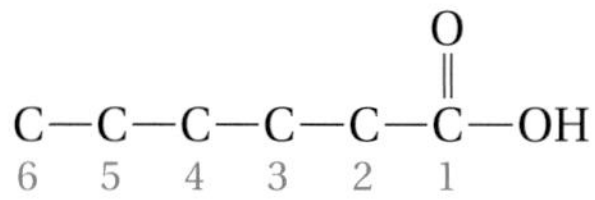

$$\underset{6}{C}-\underset{5}{C}-\underset{4}{C}-\underset{3}{C}-\underset{2}{C}-\underset{1}{\overset{\displaystyle O}{\overset{\|}{C}}}-OH$$

Add two bromine atoms at carbon-4. Then add sufficient hydrogen atoms for each carbon to have four bonds:

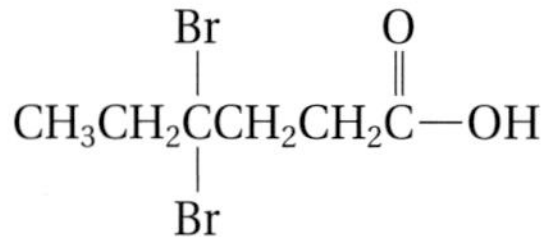

$$CH_3CH_2\underset{\underset{Br}{|}}{\overset{\overset{Br}{|}}{C}}CH_2CH_2\overset{\overset{\displaystyle O}{\|}}{C}-OH$$

PRACTICE EXERCISE 14.1

Give the name (IUPAC or common) of the following carboxylic acids.

(a) CH_3CH_2-COOH

(b) 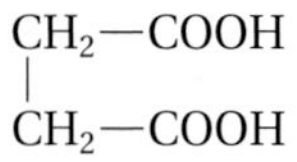 CH_2-COOH
    $|$
    CH_2-COOH

(c) 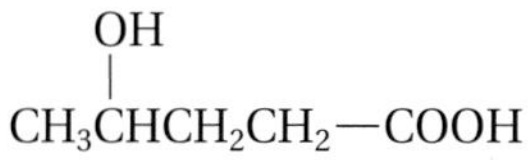 $\quad OH$
    $\quad\;|$
    $CH_3CHCH_2CH_2-COOH$

(d)

COOH
COOH

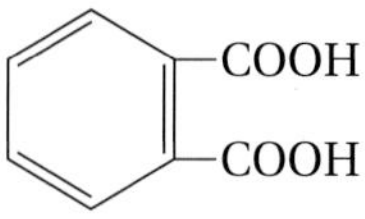

14.2 The carboxyl group

AIM: To relate the structure of the carboxyl group to the relatively high melting and boiling points, as well as the water solubility, of carboxylic acids.

Focus

The carboxyl group avidly forms hydrogen bonds.

Like alcohols, carboxylic acids can form intermolecular hydrogen bonds. **Dimers**—*hydrogen-bonded carboxylic acid pairs*—are found even in the vapor state of the acids with low molar masses.

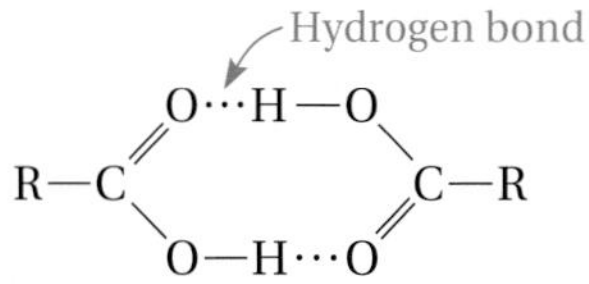

A carboxylic acid dimer

Because of their hydrogen bonding, carboxylic acids have higher boiling points and higher melting points than other compounds of similar molar

Table 14.3 Boiling Points for Different Compounds with Similar Molar Mass

Class	Compound	Molar mass	Boiling point (°C)
alkane	pentane	74	35
aldehyde	butanal	72	75.7
alcohol	butanol	74	118
acid	propanoic acid	74	141

mass. Table 14.3 illustrates this point. All aromatic carboxylic acids and dicarboxylic acids are crystalline solids at room temperature.

The carboxyl group in carboxylic acids is polar and readily hydrogen bonds to water molecules. Formic, acetic, propionic, and butyric acids are completely miscible with water, but the solubility of carboxylic acids of higher molar mass drops off sharply. Because of their long nonpolar hydrocarbon tails and polar carboxyl group, the fatty acids are interesting. When a small quantity of a fatty acid is dropped onto the surface of water, it spreads out to form a **monomolecular layer**—*a film of the acid one molecule thick*. Figure 14.1 shows how the polar carboxyl groups of acid molecules in a monomolecular layer hydrogen bond to water molecules; the nonpolar tails project out of the water to form their own hydrophobic environment. Most carboxylic acids dissolve in organic solvents such as ethanol, acetone, or ether.

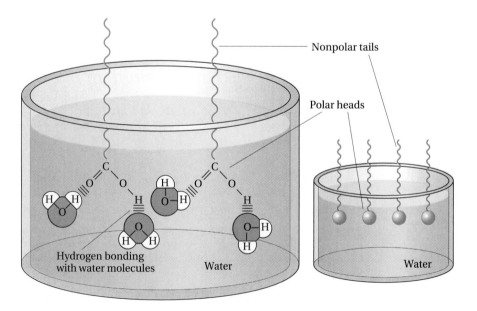

Figure 14.1
The organization of long-chain fatty acids at the surface of water.

14.3 Acidity of carboxylic acids

AIMS: *To explain the acidity of carboxylic acids in terms of* K_a *values, resonance stabilization, and R groups with different electronegativities. To describe, on a molecular scale, the action of soaps and detergents and explain the effects of hard water on each.*

Focus

Carboxylic acids are weak acids.

You may recall from Section 14.1 that carboxylic acids are weak acids because they dissociate to only a slight extent in aqueous solution. As we described in Section 9.2, the ionization constant K_a is a quantitative measure of the extent to which an acid ionizes. The smaller the K_a value, the weaker is the acid, and the larger the K_a value, the stronger is the acid. Table 14.4 gives the K_a values of some carboxylic acids. As the values show, ionization of carboxylic acids is much greater than that of alcohols and phenols, two classes of compounds that also contain O—H bonds. Moreover, acetic acid ($K_a = 1.8 \times 10^{-5}$) is a much weaker acid than trichloroacetic acid ($K_a = 2.2 \times 10^{-1}$). The acidity of a carboxylic acid heavily depends on its molecular structure. Next, we will investigate the reasons for this dependence.

Acidity and structure of the R group

Why do carboxylic acids lose protons more readily than alcohols? As shown in Figure 14.2, the electronegative carbonyl oxygen pulls electrons toward itself and away from the O—H bonds. This polarizes the O—H bond; the oxygen of the O—H bond draws some of the electrons of the bond to itself

Table 14.4 K_a **Values for Some Carboxylic Acids**

Name	Structure	K_a(25 °C)
formic acid	HCOOH	1.8×10^{-4}
acetic acid	CH_3COOH	1.8×10^{-5}
chloroacetic acid	$CH_2ClCOOH$	1.5×10^{-3}
dichloroacetic acid	$CHCl_2COOH$	5.0×10^{-2}
trichloroacetic acid	CCl_3COOH	2.2×10^{-1}
benzoic acid	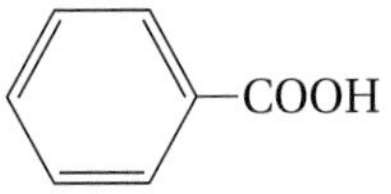 ⟨benzene ring⟩—COOH	6.8×10^{-5}
phenol*	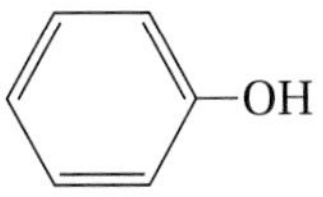 ⟨benzene ring⟩—OH	1.0×10^{-10}
ethanol*	CH_3CH_2OH	1.0×10^{-16}
water*	HOH	1.0×10^{-14}

*Presented for comparison.

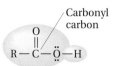

(a) Carboxylic acid

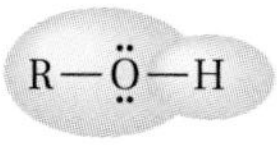

R—Ö—H

(b) Phenol or alcohol

Figure 14.2
The hydroxyl group in a carboxylic acid (a) loses a proton more readily than the hydroxyl group in a phenol or alcohol (b) because the electronegative carbonyl group pulls the bonding electron pair between oxygen and hydrogen away from hydrogen.

and becomes more negative. The hydrogen of the O—H bond loses some of its share of the bonding electrons and becomes more positive. This partial gain of electrons by oxygen and partial loss of electrons by hydrogen makes the O—H bond more ionic and therefore more readily dissociable into a carboxylate ion and a proton. In addition, the carboxylate ion produced by the dissociation of a carboxylic acid is stabilized by resonance—we can draw more than one form of the ion (see Sec. 11.9). Alkoxide ions (R—O$^-$) are not stabilized by resonance:

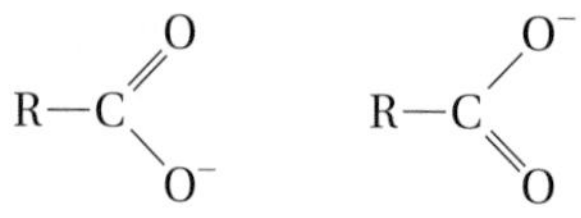

Resonance forms of carboxylate anion (two forms can be drawn)

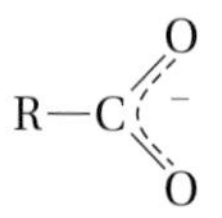

Resonance structures can be drawn as a hybrid structure

The R group attached to the carboxyl group may increase or decrease the acidity of a carboxylic acid. Like the carbonyl oxygen of the carboxyl group itself, an R group that is strongly electron attracting polarizes the O—H bond. The proton is released more easily, and the acid is stronger. Chlorine is electronegative. If the hydrogens on the carbon in acetic acid are replaced by one, two, or three chlorines, the resulting acids are successively stronger. Trichloroacetic acid is a much stronger acid than acetic acid.

EXAMPLE 14.3   **Identifying the stronger acid**

Identify the stronger acid of the following pair, and justify your choice.

$$CCl_3COOH \quad or \quad CBr_3COOH$$

SOLUTION

The more electronegative the groups attached to carbon-2, the weaker is the O—H bond and the stronger is the acid. Since chlorine is more electronegative than bromine (see Sec. 5.8), trichloroacetic acid can more easily lose the carboxyl hydrogen, and it is a stronger acid than tribromoacetic acid.

PRACTICE EXERCISE 14.2

Arrange the following carboxylic acids in order of increasing acidity (you may use Table 14.4):

(a) acetic acid   (b) dichloroacetic acid

(c) benzoic acid   (d) formic acid

Salt formation

Carboxylic acids can be neutralized by bases to produce salts. Organic acid salts are named like inorganic acid salts (see Sec. 5.5). Name the cation first. Name the anion by dropping the *-oic acid* or *-ic acid* ending and adding

A number of salts of carboxylic acids are important food additives. Sodium propionate inhibits mold growth in baked goods. Sodium benzoate is used in soft drinks and other acidic foods to inhibit bacterial growth. These additives both preserve and increase the shelf life of many foods.

-*ate*. For example, the salt of ammonia and formic acid is ammonium formate.

$$\underset{\text{Formic acid}}{H-\overset{\displaystyle O}{\overset{\|}{C}}-OH(aq)} + \underset{\text{Ammonia}}{NH_3(aq)} \longrightarrow \underset{\substack{\text{Ammonium}\\\text{formate}}}{H-\overset{\displaystyle O}{\overset{\|}{C}}-O^-NH_4^+(aq)}$$

When sodium carbonate or sodium bicarbonate neutralizes carboxylic acids, the solution froths as carbon dioxide is liberated.

$$\underset{\text{Acetic acid}}{CH_3-\overset{\displaystyle O}{\overset{\|}{C}}-OH(aq)} + \underset{\substack{\text{Sodium}\\\text{bicarbonate}}}{NaHCO_3(aq)} \longrightarrow \underset{\substack{\text{Sodium}\\\text{acetate}}}{CH_3-\overset{\displaystyle O}{\overset{\|}{C}}-O^-Na^+(aq)} + \underset{\text{Water}}{H_2O(l)} + \underset{\substack{\text{Carbon}\\\text{dioxide}}}{CO_2(g)}$$

The frothing helps distinguish carboxylic acids from neutral or basic organic compounds.

EXAMPLE 14.4

Neutralization of a carboxylic acid

Write an equation and name the products of the neutralization of benzoic acid with sodium hydroxide.

SOLUTION

The reaction of benzoic acid, the simplest aromatic acid, and sodium hydroxide gives water and a salt. The name of the salt is *sodium* (the cation from the base) *benzoate* (the anion from the acid).

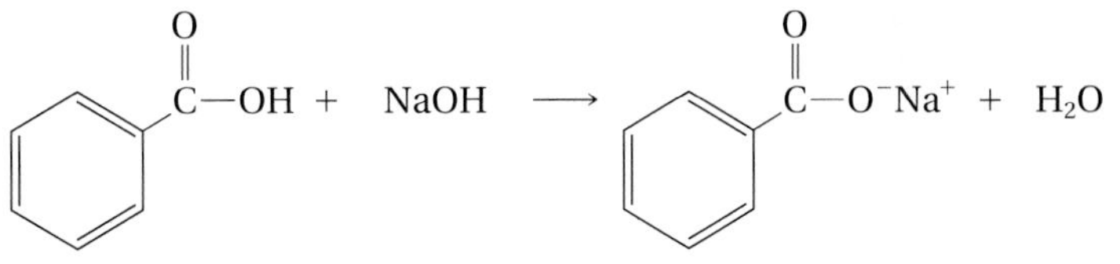

Benzoic acid + Sodium hydroxide → Sodium benzoate (a salt) + Water

PRACTICE EXERCISE 14.3

Write the structure and give the name of the products in each of the following neutralization reactions.

(a) $CH_3CH_2COOH + NaOH \longrightarrow$

(b) $CH_3COOH + Ca(OH)_2 \longrightarrow$

Soaps and detergents

Soaps *are the alkali metal (Li, Na, or K) salts of long-chain fatty acids.* Since most dirt is held to surfaces by an oily film, dirt is difficult to remove unless the oily coating is first emulsified with soaps.

Hydrophobic
hudor (Greek): water
phobos (Greek): fear

Hydrophilic
hudor (Greek): water
philos (Greek): loving

Sodium stearate is a typical soap. Like all soaps, the sodium stearate molecule has two distinct parts—an ionic head (the carboxylate ion) and a long, nonpolar hydrocarbon tail. The charged head is *hydrophilic* ("water-loving"), and the nonpolar tail is *hydrophobic* ("water-hating").

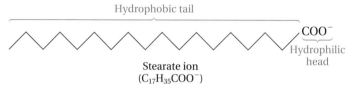

Hydrophobic tail

COO^-
Hydrophilic head

Stearate ion
$(C_{17}H_{35}COO^-)$

Because of their hydrophobic tails, soaps added to water go into colloidal dispersion instead of dissolving. Figure 14.3 shows the dispersed soap molecules arranged in spherical clusters called *micelles*. *In **micelles,** the hydrophobic tails of the soap molecules aggregate to create an oily nonpolar environment protected from water at the center of the cluster.* The charged carboxylate groups form a highly charged, highly solvated shell at the surface of the micelle. The micelles remain dispersed in water because they are like-charged and repel each other. When soapy water contacts an oily surface, the tails of the soap molecules become embedded in the oily film. Micelles then form with small oil droplets at their centers (Fig. 14.4). We might say that the oil dissolves in the soapy water, but it is actually not in true solution. The soap has acted as an emulsifying agent.

Figure 14.3
A micelle. The ionized groups of the soap molecules are in contact with water, and the nonpolar parts of the molecules are protected from water.

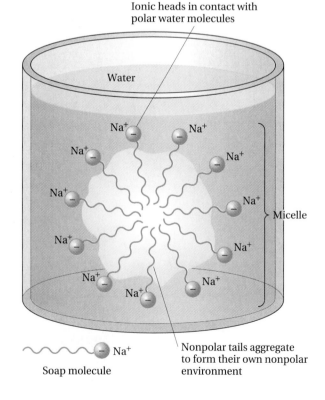

Ionic heads in contact with polar water molecules

Water

Micelle

Na^+

Soap molecule

Nonpolar tails aggregate to form their own nonpolar environment

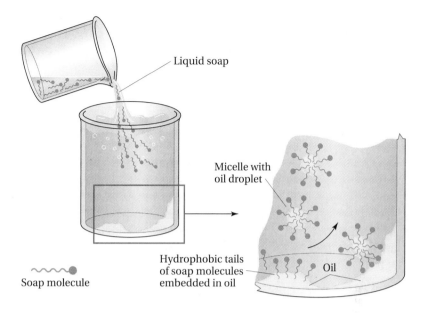

Figure 14.4
Soap "dissolves" oil by packaging it in micelles.

Soaps are essentially useless in acidic water, because in acidic water the carboxylate ions of soap molecules pick up hydrogen ions to form the undissociated fatty acid:

$$RCOO^- \;+\; H^+ \;\rightleftharpoons\; RCOOH$$

$$\begin{array}{ccc} \text{Carboxylate} & \text{Hydrogen} & \text{Fatty acid} \\ \text{ion} & \text{ion} & \text{(undissociated)} \end{array}$$

The fatty acid precipitates as a scum because it cannot form micelles. Soaps are also essentially useless in **hard water**—*water containing calcium, magnesium, or iron ions.* Carboxylate ions form insoluble salts with these metal ions in the water. With a stearate soap, a typical reaction is

$$2CH_3(CH_2)_{16}COO^- \;+\; Ca^{2+} \;\longrightarrow\; [CH_3(CH_2)_{16}COO^-]_2Ca^{2+}$$

$$\begin{array}{ccc} \text{Stearate} & \text{Calcium} & \text{Calcium stearate} \\ \text{ion} & \text{ion} & \text{(precipitate)} \end{array}$$

Such precipitates form unsightly scums that stick to clothes and give whites a gray look. A Closer Look: Hard Water and Water Softening describes the treatment of hard water in more detail.

Detergents *are cleaning agents with the desirable properties of soap without soap's scum-forming property.* Detergent molecules are like soap molecules because they have an ionic or polar head and a nonpolar tail. They also act as emulsifying agents for oils and greases, just as soap molecules do. However, they do not form precipitates with the ions present in hard water, so clothes being washed do not get coated with scum. One of

Hard Water and Water Softening

Hard water can be problematic. It forms precipitates in boilers and hot water systems. When combined with soap, it forms an insoluble scum that coats skin and clings to the bathtub. It also can have a disagreeable taste.

Because natural waters often contain high levels of metal ions, they deliver supplies of hard water to our water systems. Ions are formed when rainwater on its way to these reservoirs leaches calcium from limestone and other metal ions such as magnesium and iron from various mineral rocks.

Hard water exists in two types. *Temporary hardness,* an indication of the presence of hydrogen carbonates (bicarbonates) of calcium and/or magnesium, is produced when rainwater tinged with carbonic acid filters through limestone or chalk rocks. *Permanent hardness* is due to dissolved sulfates and chlorides of calcium and magnesium, whose source is also mineral rock.

Hard water can be softened—the offending ions removed—by several procedures. Temporary hardness can be removed by boiling the water. However, this process produces a precipitate of calcium carbonate that sticks as hard "fur," or scale, on kettles, boilers, and the inside of hot water pipes (see figure). Permanent hardness cannot be removed by boiling, but it can be removed by adding slaked lime, $Ca(OH)_2$, or washing soda (sodium carbonate, Na_2CO_3) to the water. In this process, calcium and magnesium ions, precipitated as insoluble $CaCO_3$ and $Mg(OH)_2$, are replaced in solution by the sodium ions. Another common method of water softening is the addition of sodium triphosphate (Calgon), which combines with the offending ions to form soluble complexes.

Large quantities of hard water are often softened in commercial water softeners by ion exchange. In this process, an undesirable ion in solution displaces an ion, usually sodium, from a solid, insoluble material called a *matrix.* The matrix used in ion-exchange water softeners is usually a natural mineral such as permutite or a synthetic resin. Most of these minerals and resins can be regenerated by pouring strong sodium chloride solution through them to displace the calcium and magnesium ions and replace them with a new stock of sodium ions.

Softening water is not a method of purification, since the ions causing water hardness are made only more soluble or are replaced by other ions. The high sodium ion content in softened water is of concern, especially to people on restricted sodium diets. A solution to the problem is to use soft water for washing and naturally hard water for drinking.

Pipe scale (right) forms when calcium carbonate sticks to the inside of hot water pipes.

the first detergents was the sodium salt of lauryl hydrogen sulfate:

$$CH_3(CH_2)_{10}CH_2-O-\overset{\displaystyle O}{\underset{\displaystyle O}{\overset{\|}{\underset{\|}{S}}}}-O^-Na^+$$

Sodium lauryl sulfate

Today's detergents are mostly salts of sulfonic acids:

$$R-\overset{\displaystyle O}{\underset{\displaystyle O}{\overset{\|}{\underset{\|}{S}}}}-OH \quad \text{or} \quad RSO_3H$$

Sulfonic acid

Probably the most common detergent is sodium dodecylbenzene sulfonate:

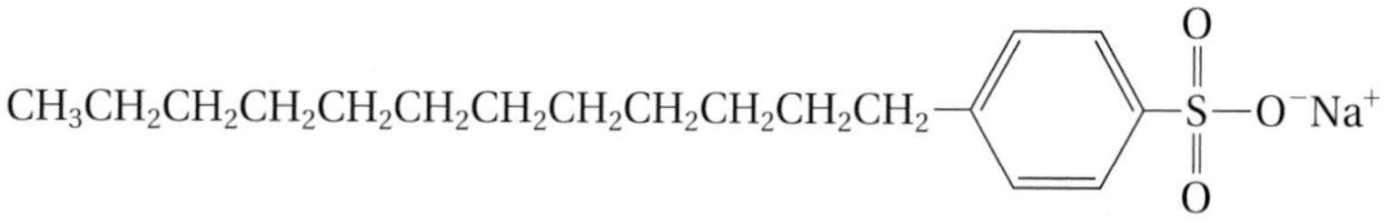

$$CH_3CH_2CH_2CH_2CH_2CH_2CH_2CH_2CH_2CH_2CH_2CH_2 - \text{(benzene ring)} - \overset{\displaystyle O}{\underset{\displaystyle O}{\overset{\|}{\underset{\|}{S}}}} - O^-Na^+$$

Sodium dodecylbenzene sulfonate

This molecule has an obvious physical resemblance to a soap molecule— an ionic head and a nonpolar tail. And like a soap, it is **biodegradable**—*it can be broken down by bacteria in the environment into harmless products and does not become a pollutant.* When sudsing must be kept to a minimum, as in automatic washers for dishes and clothing, nonionic detergents are useful. *A **nonionic detergent** molecule has a nonpolar hydrocarbon tail and a polar but uncharged head.* One such compound is an ester of palmitic acid, pentaerythrityl palmitate:

From palmitic acid | From pentaerythritol

$$CH_3CH_2CH_2CH_2CH_2CH_2CH_2CH_2CH_2CH_2CH_2CH_2CH_2CH_2CH_2 - \overset{\displaystyle O}{\overset{\|}{C}} - O - CH_2\overset{\displaystyle CH_2OH}{\underset{\displaystyle CH_2OH}{C}}CH_2OH$$

Pentaerythrityl palmitate

PRACTICE EXERCISE 14.4

When stearate soap is used in hard water containing iron(III) ions (Fe^{3+}), the scum that forms is iron(III) stearate. Write the formula for this compound.

14.4 Synthesis of carboxylic acids

AIM: To write equations for the preparation of carboxylic acids from alcohols and aldehydes.

Focus

Carboxylic acids may be prepared by oxidation of alcohols or aldehydes.

Sometimes a chemist needs to make a carboxylic acid with a particular structure, say, in the synthesis of a new drug. Two common methods for the preparation of carboxylic acids involve the oxidation of primary alcohols or aldehydes (see Sec. 13.4). The starting alcohol or aldehyde must have the same carbon skeleton as the desired acid.

Oxidation of alcohols

The general reaction for oxidation of primary alcohols to carboxylic acids is

Primary alcohol — Aldehyde — Carboxylic acid

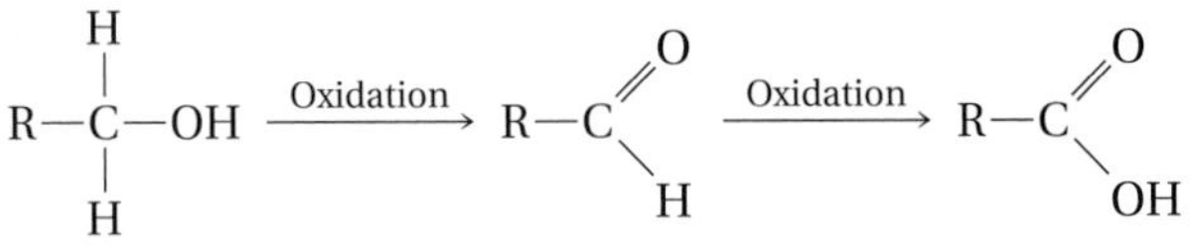

Potassium dichromate ($K_2Cr_2O_7$) and potassium permanganate ($KMnO_4$) are commonly used as oxidizing agents:

$$CH_3CH_2{-}CH_2OH \xrightarrow{K_2Cr_2O_7} CH_3CH_2{-}COOH$$

1-Propanol — Propionic acid

Benzyl alcohol — Benzoic acid

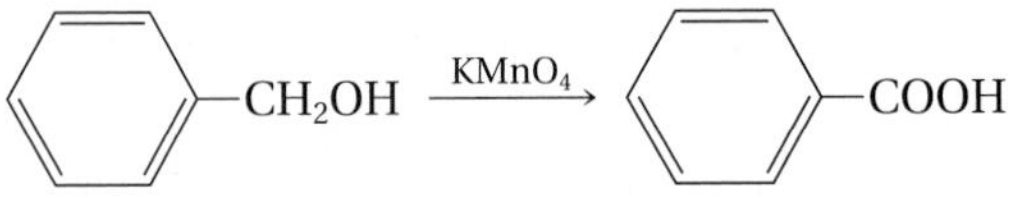

Breaking a C—C bond by oxidation is much more difficult than breaking a C—H bond. The usual oxidizing agents oxidize secondary alcohols to ketones. Tertiary alcohols are not oxidized.

Secondary alcohol — Ketone

2-Propanol — Acetone

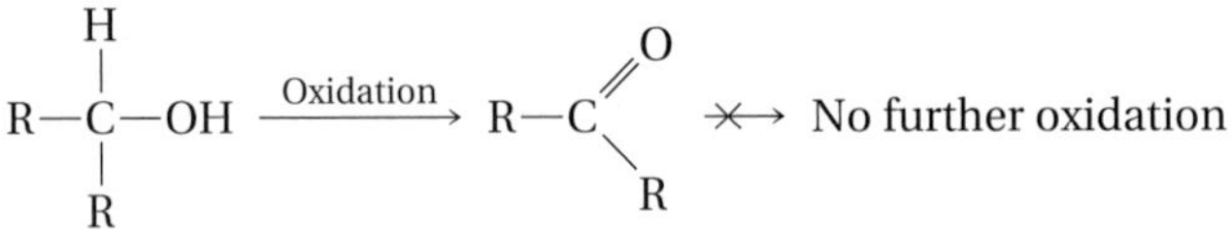

Oxidation of aldehydes

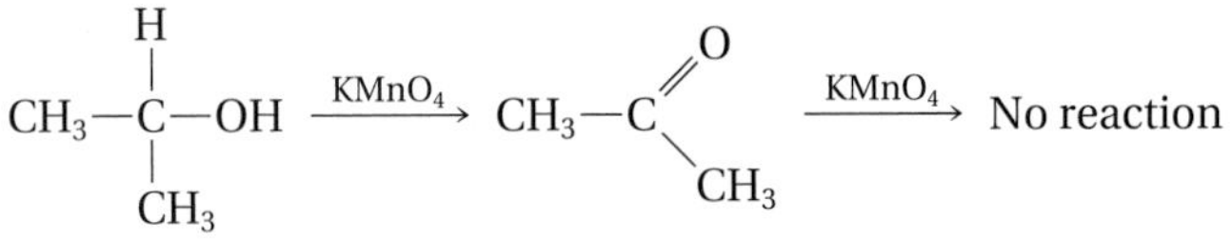

Since an aldehyde is an intermediate in the oxidation of a primary alcohol to a carboxylic acid, it follows that aldehydes can be oxidized to carboxylic acids.

$$CH_3{-}CHO \xrightarrow{K_2Cr_2O_7} CH_3COOH$$

Acetaldehyde — Acetic acid

Benzaldehyde — Benzoic acid

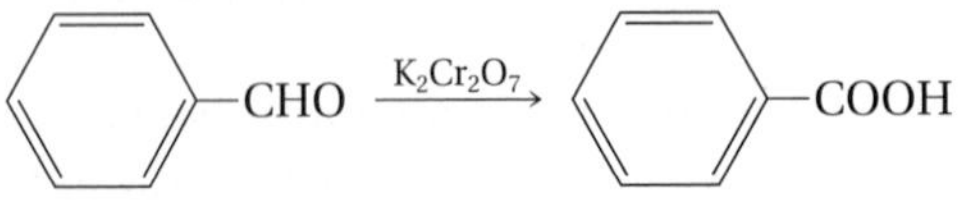

PRACTICE EXERCISE 14.5

The following compounds are oxidized with acidified potassium dichromate. Write the structure for the expected product. If no product is expected, write "no reaction."

(a) CH_3CH_2OH (b) $CH_3\overset{\displaystyle |}{\underset{\displaystyle CH_3}{CHOH}}$

(c) CH_3CH_2CHO (d) $CH_3\overset{\displaystyle O}{\overset{\displaystyle \|}{C}}CH_3$

14.5 Carboxylic acid anhydrides

AIMS: To name and write structures of common acid anhydrides. To write equations for the formation of inter- and intramolecular acid anhydrides. To distinguish between an acyl and an acetyl group and write equations for typical acylation reactions.

Focus

Carboxylic acids lose water to form anhydrides.

Carboxylic acid anhydrides *come from two carboxylic acid molecules, or a single dicarboxylic acid molecule, through the loss of a molecule of water.* The name *anhydride* means "without water."

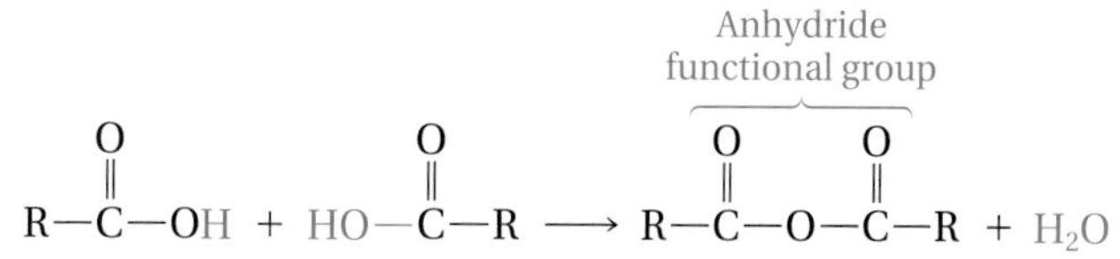

$$\underset{\text{Two carboxylic acid molecules}}{R-\overset{O}{\overset{\|}{C}}-OH \ + \ HO-\overset{O}{\overset{\|}{C}}-R} \longrightarrow \underset{\text{Acid anhydride}}{R-\overset{O}{\overset{\|}{C}}-\overset{\overbrace{\qquad}^{\text{Anhydride functional group}}}{O}-\overset{O}{\overset{\|}{C}}-R} \ + \ H_2O$$

Names of anhydrides

Anhydride
anudros (Greek): without water

Anhydrides are named by adding the word *anhydride* after the parent name of the acid. One common acid anhydride is acetic anhydride:

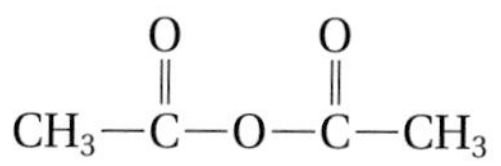

$$CH_3-\overset{O}{\overset{\|}{C}}-O-\overset{O}{\overset{\|}{C}}-CH_3$$
Acetic anhydride
(bp 139.5 °C)

Acetic anhydride, a corrosive liquid, is immensely important as both an industrial chemical and a laboratory reagent.

Certain dicarboxylic acids readily produce anhydrides upon heating if five- or six-membered rings are formed. (Recall that five- and six-membered rings are especially stable.) For example:

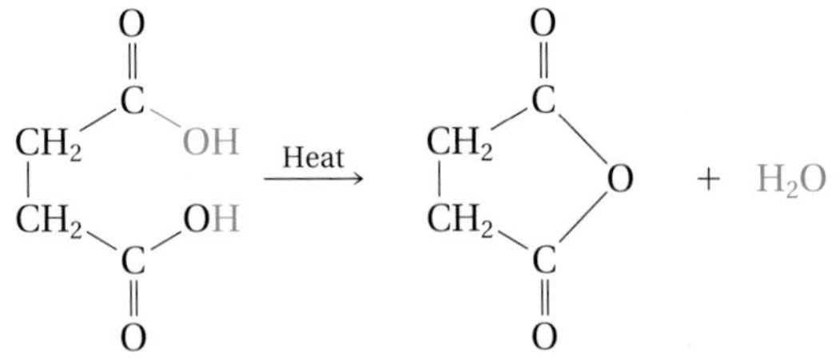

Succinic acid   Succinic anhydride   Water

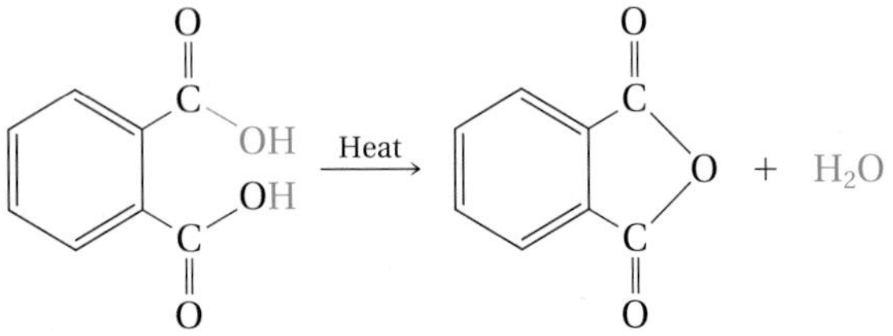

Phthalic acid   Phthalic anhydride   Water

The anhydrides of monocarboxylic acids are liquids. Those of dicarboxylic acids and aromatic carboxylic acids are solids.

Hydrolysis of anhydrides

Processes in which **acyl groups**

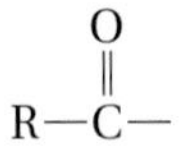

are transferred to another molecule are called **acylation reactions.** Acid anhydrides are very reactive acylating agents. When the acyl group is transferred to water, the reaction is a *hydrolysis.* In organic chemistry, **hydrolysis** *is the splitting of a molecule by water.* When the molecule is split, a hydrogen from water ends up in one product, and the —OH ends up in the other. When acetic anhydride is hydrolyzed, the acetyl group

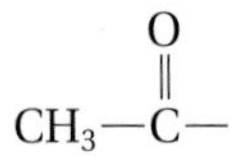

is transferred to water. The products of hydrolysis of an anhydride are two molecules of the acid.

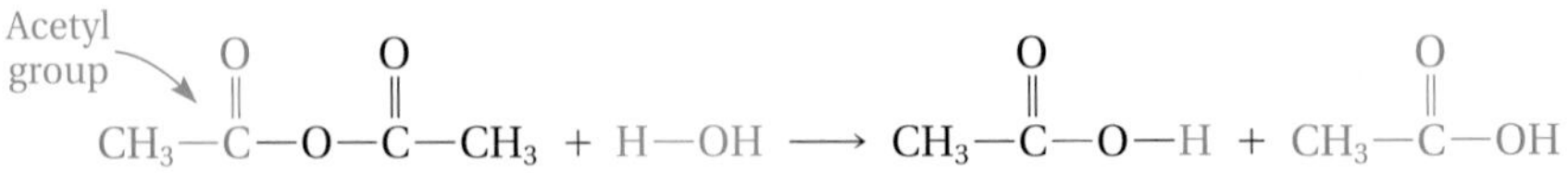

Acetic anhydride   Water   Two molecules of acetic acid

PRACTICE EXERCISE 14.6

Name the following compound, and give the product of its hydrolysis.

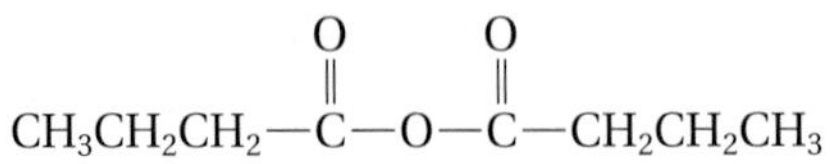

$$CH_3CH_2CH_2 \overset{\overset{\displaystyle O}{\|}}{-C} -O- \overset{\overset{\displaystyle O}{\|}}{C} -CH_2CH_2CH_3$$

14.6 Carboxylic esters

AIMS: *To name and write structures of common esters. To write equations for the formation of an ester from an alcohol and a carboxylic acid or an acid anhydride.*

Carboxylic esters *are derivatives of carboxylic acids in which the hydrogen of the* —OH *of the carboxyl group has been replaced by* —R *of an alkyl or aryl group.* They contain a carbonyl group and an ether link to the carbonyl carbon. The general formula is

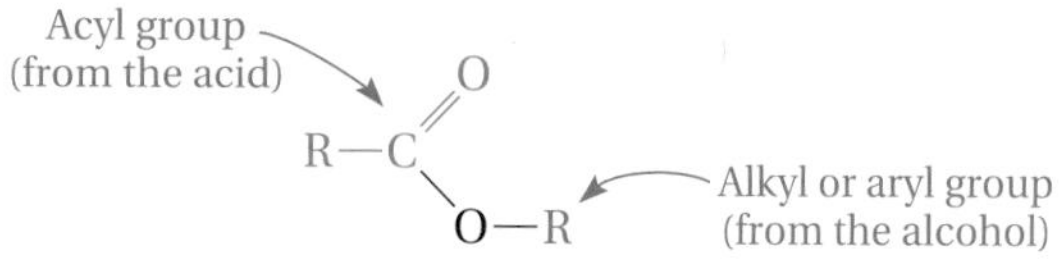

Acyl group (from the acid)

$$R-C \overset{\displaystyle O}{\underset{\displaystyle O-R}{\Big\langle}}$$

Alkyl or aryl group (from the alcohol)

The abbreviated formula for a carboxylic ester is RCOOR. The R groups can be short chains or long chains, aliphatic (alkyl) or aromatic (aryl), saturated or unsaturated.

Naming carboxylic esters

Carboxylic esters are named as derivatives of carboxylic acids. The names contain two words, the name of the acid and the name of the —R group of the ether linkage. The *-ic* ending of the acid name is changed to *-ate* and is preceded by the name of the alkyl or aryl group of the ether linkage. This method is illustrated with some derivatives of acetic acid.

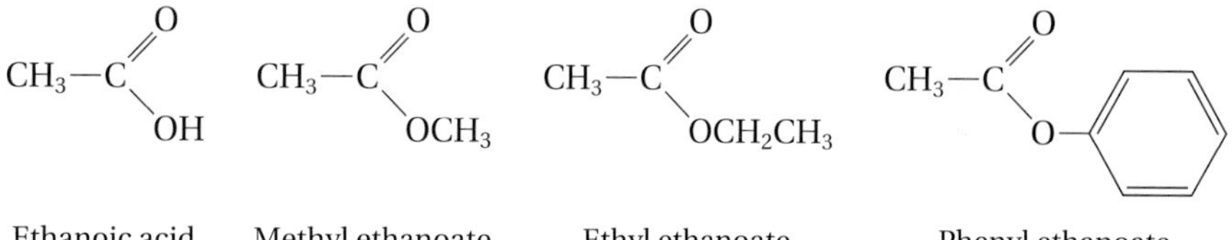

$$CH_3-C \overset{\displaystyle O}{\underset{\displaystyle OH}{\Big\langle}} \qquad CH_3-C \overset{\displaystyle O}{\underset{\displaystyle OCH_3}{\Big\langle}} \qquad CH_3-C \overset{\displaystyle O}{\underset{\displaystyle OCH_2CH_3}{\Big\langle}} \qquad CH_3-C \overset{\displaystyle O}{\underset{\displaystyle O-}{\Big\langle}}$$

Ethanoic acid (acetic acid)   Methyl ethanoate (methyl acetate)   Ethyl ethanoate (ethyl acetate)   Phenyl ethanoate (phenyl acetate)

EXAMPLE 14.5

Naming an ester

Give the IUPAC and common names of the following ester.

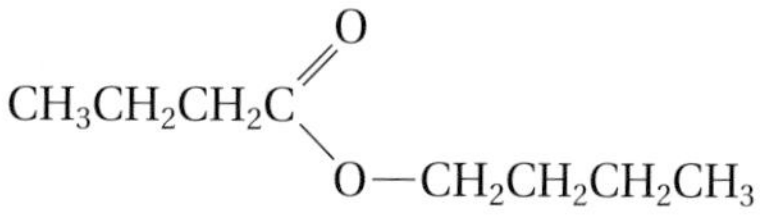

$$CH_3CH_2CH_2C \overset{\displaystyle O}{\underset{\displaystyle O-CH_2CH_2CH_2CH_3}{\Big\langle}}$$

SOLUTION

In order to find the IUPAC name of the ester, we first inspect the portion of the ester that comes from the carboxylic acid.

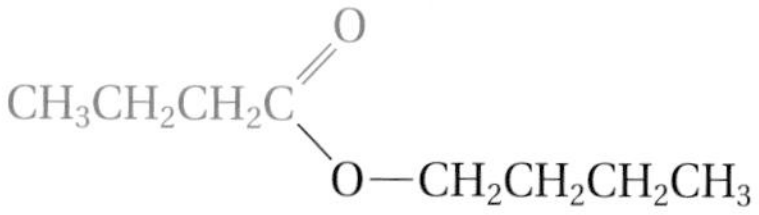

$$CH_3CH_2CH_2C \overset{\displaystyle O}{\underset{\displaystyle O-CH_2CH_2CH_2CH_3}{\Big\langle}}$$

The acid portion is four carbons long and is derived from butanoic acid. By removing the *-ic acid* ending and adding an *-ate* ending, we see that the ester is a *butanoate*. We next name the alkyl group bonded to the ether oxygen of the ester.

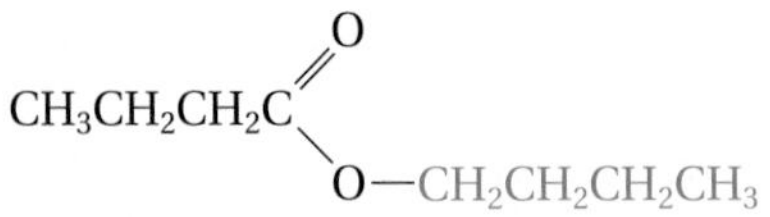

$$CH_3CH_2CH_2C\overset{\displaystyle O}{\underset{\displaystyle O-CH_2CH_2CH_2CH_3}{\big\backslash}}$$

This alkyl group, consisting of four carbons in a straight chain, is called *butyl*. We combine both parts of the name in two words to obtain the complete IUPAC name of the ester: *butyl butanoate*.

The common name is derived similarly, except the common name for butanoic acid is *butyric acid*. The common name of the ester is *butyl butyrate*.

PRACTICE EXERCISE 14.7

Name these esters:

(a)

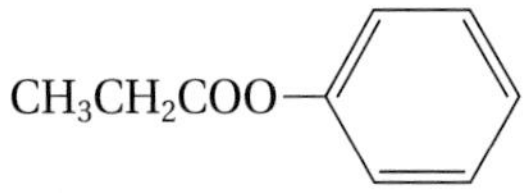

CH_3CH_2COO- ⬡

(b)

⬡ $-COOCH_2CH_3$

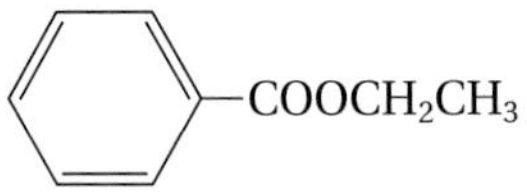

Properties of carboxylic esters

The triesters formed by a molecule of glycerol and three molecules of fatty acids are called *triglycerides*. As we will see in Chapter 25, triglycerides are important energy storage molecules.

Simple carboxylic esters are neutral substances. The molecules are polar but cannot form intermolecular hydrogen bonds. They are much less soluble in water and have much lower boiling points than the carboxylic acids from which they are derived. Esters can hydrogen bond with water. Esters of low molar mass are somewhat soluble, but esters containing more than four or five carbons have very limited solubility in water.

Esters derived from acids and alcohols with low molar masses have delightful odors. They are volatile compounds that occur naturally in fruits and flowers, to which they impart a characteristic taste or fragrance. Table 14.5 lists some of the familiar esters and their characteristic odors. Many of

Table 14.5 Flavor Esters

Name	Formula	Odor/flavor
ethyl formate	$HCOOCH_2CH_3$	rum
octyl acetate	$CH_3COO(CH_2)_7CH_3$	oranges
pentyl acetate	$CH_3COO(CH_2)_4CH_3$	bananas
ethyl butyrate	$CH_3(CH_2)_2COOCH_2CH_3$	pineapples
pentyl butyrate	$CH_3(CH_2)_2COO(CH_2)_4CH_3$	apricots
methyl butyrate	$CH_3(CH_2)_2COOCH_3$	apples
methyl salicylate	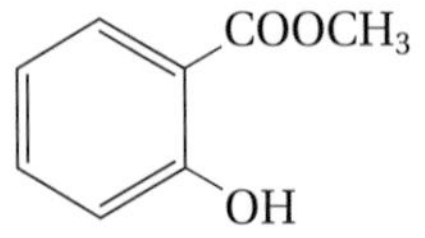 ⬡ with $COOCH_3$ and OH	oil of wintergreen

these esters can be synthesized in the laboratory and are used to flavor foods and drinks artificially. Natural flavors usually consist of more complex mixtures of esters.

Preparation of esters

Esters can be prepared from an acid and an alcohol or from an acid anhydride and an alcohol. **Esterification**—*the formation of esters*—takes place when carboxylic acids are heated with primary or secondary alcohols in the presence of a trace of mineral acid as a catalyst. The reaction is reversible.

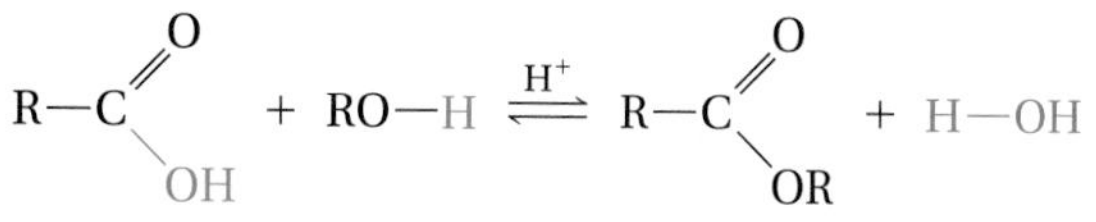

| Carboxylic acid | Alcohol | Carboxylate ester | Water |

Consider this specific example. When salicylic acid and methanol react, the product is methyl salicylate.

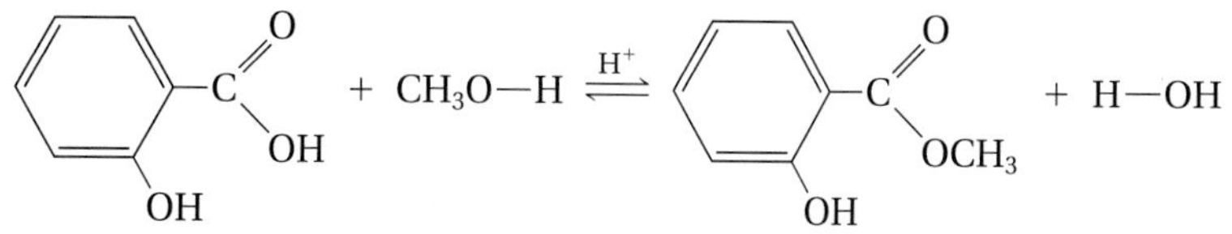

Salicylic acid   Methanol   Methyl salicylate (oil of wintergreen)   Water

Methyl salicylate, also called *oil of wintergreen,* is used as a flavoring and in liniments for soothing sore muscles. Many local anesthetics also contain the ester function. See A Closer Look: Ester Local Anesthetics.

Recall from Chapter 9 that there are ways in which the equilibrium in a reversible reaction can be disturbed to improve the product yield. Esterification reactions can be pushed to completion if we use an excess of one of the reactants (acid or alcohol) or if we remove the water from the reaction mixture as it is produced.

The industrial production of esters often involves reacting an acid anhydride with an alcohol. The ester of greatest commercial importance prepared in this way is acetylsalicylic acid, commonly called *aspirin.* Acetylsalicylic acid is obtained from acetic anhydride and salicylic acid:

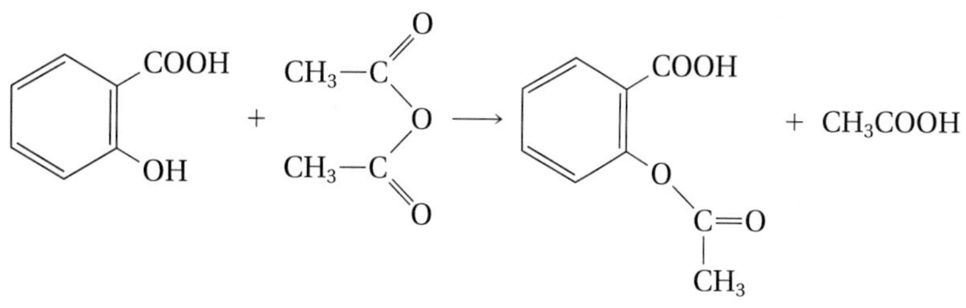

Salicylic acid   Acetic anhydride   Acetylsalicylic acid (aspirin)   Acetic acid

Ester Local Anesthetics

Local anesthetics applied to body surfaces (topical application) or injected around nerves are used primarily to prevent pain during surgical procedures in which the patient stays conscious. Several reliable local anesthetics are available. Some of the most common contain the ester functional group. For example, ethyl-*p*-aminobenzoate (Benzocaine) is a common ingredient in lotions and creams meant to relieve the pain of sunburn.

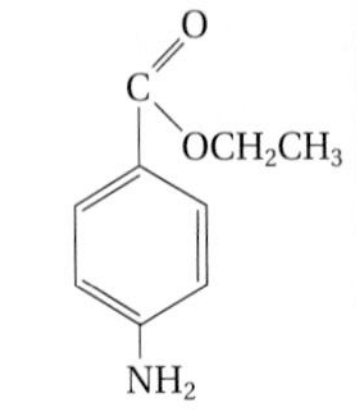

Cocaine, an ester obtained from the coca plant (not to be confused with the cocoa plant), is used in the form of its hydrochloride salt for local anesthesia of the nose, pharynx, and upper respiratory passages. The ability of cocaine to cause constriction of blood vessels is unequaled by other local anesthetics. Cocaine therefore is useful in decreasing

Ethyl-*p*-aminobenzoate
(Benzocaine)

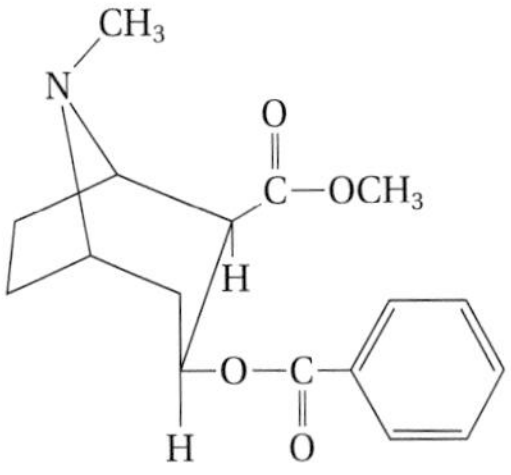

Cocaine

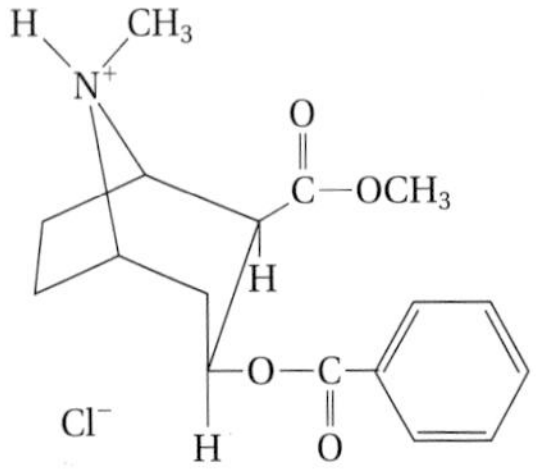

Cocaine (hydrochloride salt)

bleeding and shrinking mucous membranes. Cocaine was once used as a local anesthetic in dentistry. Because cocaine is toxic and habit-forming, its use has diminished as other anesthetics have become available.

Procaine (Novocain) is an ester of *N,N*-diethylaminoethanol and ethyl-*p*-aminobenzoic acid. Procaine must be injected to produce local anesthesia. It has minimal systemic toxicity and produces no local irritation, and its effects rapidly disappear. These features make it popular for injection into the spinal column to produce a spinal block (see figure). An ester similar to procaine, chloroprocaine (or Nesacaine), also must be injected but is more potent and wears off faster than procaine. Chloroprocaine is rapidly hydrolyzed in plasma and is much less toxic than procaine. It is probably the safest local anesthetic.

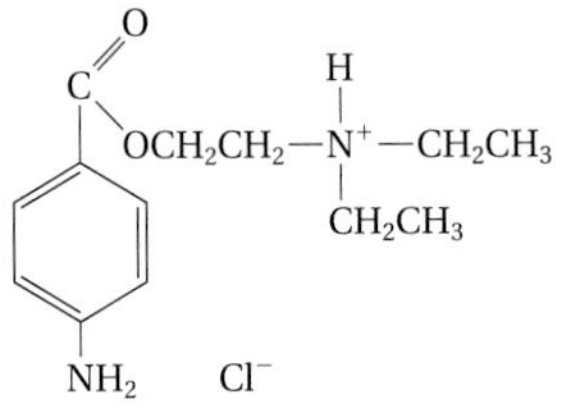

Procaine
(hydrochloride salt)
(Novocaine)

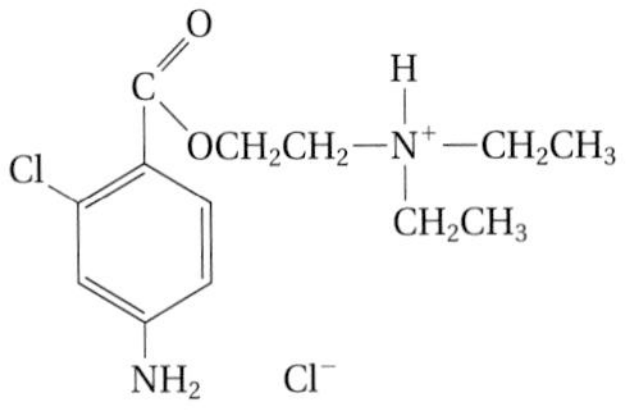

Chloroprocaine
(hydrochloride salt)
(Nesacaine)

Like the hydrolysis of acetic anhydride, this is an acylation reaction—but in this case the acceptor of the acetyl group is an alcohol instead of water. Salicylic acid is a polyfunctional carboxylic acid because it contains both a carboxyl group and a hydroxyl group. This means that it can react as an acid or as an alcohol depending on the other reactants. In the formation of methyl salicylate, salicylic acid behaves as an acid, and reaction takes place at the carboxyl group. In the production of aspirin, however, salicylic acid acts as an alcohol, and reaction with acetic anhydride occurs at the hydroxyl group. Aspirin is an **antipyretic** (*fever killer*) and an **analgesic** (*pain killer*). It is usually dispensed as its sodium salt, sodium acetyl salicylate.

Another drug in this ester series is tetracaine (Pantocaine, Pontocaine). The potency and duration of action of this compound are higher than those of the other esters. The resulting toxicity, however, is also greater. Tetracaine is the most commonly used anesthetic for spinal anesthesia; it is combined with an equal volume of 10% glucose to increase the specific gravity and thus control the spread of the solution. In 1% to 2% concentrations it is used as a topical anesthetic in the pharynx and the upper respiratory passages.

Tetracaine
(hydrochloride salt)
(Pantocaine, Pontocaine)

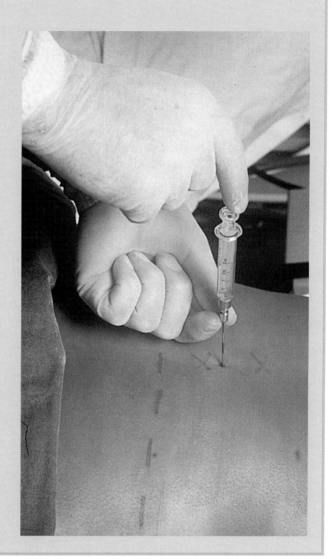

Procaine is injected to produce the local anesthesia of a spinal block.

Sodium acetyl salicylate

See A Closer Look: Aspirin for more details about this medication.

A Closer Look

Aspirin

How would you like to discover a drug that could relieve pain from headaches and arthritis, prevent heart attacks, and reduce fever? Moreover, your drug would be cheap and available without a prescription. Such a miracle drug already exists. It is called *aspirin.*

In 1763, Edward Stone, an English clergyman, reported that the extract of willow bark would reduce fever. The active ingredient, salicylic acid, was isolated from willow bark in 1860. Because salicylic acid is sour and irritating when taken orally, chemists have modified the structure to remove the undesirable properties while retaining or enhancing its desirable properties. A number of derivatives of the acid have been made and used with success as analgesics and antipyretics. The most common derivative is aspirin (acetylsalicylic acid). A typical aspirin tablet contains about 325 mg acetylsalicylic acid held together with starch, an inert binder. Over 50 billion aspirin tablets are produced annually in the United States. Aspirin, which was introduced in 1899, is still the largest-selling and most widely used medical drug in the world (see figure).

Aspirin seems to work as a pain killer by blocking the synthesis of certain substances called *prostaglandins* (see Sec. 17.11). Some prostaglandins are involved in the inflammation of tissues. At elevated levels, prostaglandins activate pain receptors in tissues, making the tissues more sensitive to pain. This is why aspirin is frequently the drug of choice for the treatment of simple headaches and for arthritis, a condition in which the joints and connective tissues become inflamed.

Fever reduction by aspirin is not completely understood but appears to be due to an action on the source of the body's temperature control, the hypothalamus gland in the brain. In addition, blood vessels expand. The resulting increase in the circulation of the blood helps to dissipate the heat of the fever. Blockage of prostaglandin synthesis also may play a role in the fever-reducing effects of aspirin.

Several recent studies have shown that small daily amounts of aspirin decrease the incidence of heart attack in middle-aged or older individuals. Many times heart attacks result from blood clots that block the coronary arteries. The aggregation of blood cells called *platelets* (see Sec. 22.2) is an important component in the formation of these blood clots. Aspirin helps to prevent these clots by reducing the "stickiness" of blood platelet cells. The clots do not form, thereby reducing the likelihood of a heart attack. The use of aspirin for its "blood thinning" effects is often recommended for patients who have undergone heart bypass surgery.

Like any medication, aspirin use can be abused. Taking an aspirin usually causes a small amount of internal bleeding. This is not a big problem for most people, but those who have ulcers or other bleeding problems are advised to avoid aspirin. Aspirin and salicylates should not be given to children to treat the symptoms of flu or chicken pox. There is a strong correlation between aspirin use and the onset of Reye's syndrome, a disease characterized by weakness and paralysis. Reye's syndrome can be fatal unless treatment is begun immediately.

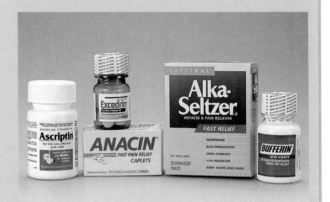

Many medicines contain aspirin or other salicylates.

PRACTICE EXERCISE 14.8

Complete the following reactions by giving the structure and name of the product.

(a) $CH_3CH_2COOH + CH_3OH \xrightarrow{H^+}$

(b)

$\langle\text{benzene ring}\rangle{-}COOH + CH_3CH_2CH_2OH \xrightarrow{H^+}$

(c)

$$\underset{\displaystyle CH_3\overset{\displaystyle O}{\overset{\|}{C}}-O-\overset{\displaystyle O}{\overset{\|}{C}}CH_3 + CH_3CH_2OH \longrightarrow}{}$$

Polyesters

What material can you wear, use to play music, and ride to classes on? The answer is *polyesters.* Chemically speaking, **polyesters** *are polymers consisting of many repeating units of dicarboxylic acids and dihydroxy alcohols joined by ester bonds.* The formation of a polyester can be represented by a block diagram:

$$x\text{HO}-\overset{O}{\overset{\|}{C}}-\boxed{\quad}-\overset{O}{\overset{\|}{C}}-\text{OH} + x\text{HO}-\bigcirc-\text{OH} \longrightarrow \left(\overset{O}{\overset{\|}{C}}-\boxed{\quad}-\overset{O}{\overset{\|}{C}}-O-\bigcirc-O\right)_x + 2x\text{H}_2\text{O}$$

Dicarboxylic acid   Dihydroxy alcohol   Representative polymer unit

Several polyesters are commercially successful. Polyethylene terephthalate (PET) is one of the most important. This polyester is formed from terephthalic acid and ethylene glycol:

$$x\text{HO}-\overset{O}{\overset{\|}{C}}-\langle\text{ring}\rangle-\overset{O}{\overset{\|}{C}}-\text{OH} + x\text{HO}-CH_2CH_2-\text{OH} \longrightarrow \left(\overset{O}{\overset{\|}{C}}-\langle\text{ring}\rangle-\overset{O}{\overset{\|}{C}}-O-CH_2CH_2-O\right)_x + 2x\text{H}_2\text{O}$$

Terephthalic acid   Ethylene glycol   Representative polymer unit of polyethylene terephthalate (PET)

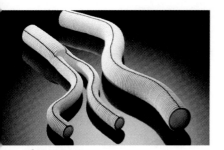

Figure 14.5
Arterial grafts, fabricated from woven Dacron fibers, are used to repair or replace blood vessels. ©1995 Meadox Medicals, Inc.

The melting point of the polymer is about 270 °C. PET fibers are formed when it is melted and forced through tiny holes in devices called *spinnerettes.* The fibers, marketed as Dacron, Fortrel, or Terylene, depending on the manufacturer, are used for tire cord and permanent-press clothing. Woven Dacron tubing has been used to replace major blood vessels (Fig. 14.5). PET fibers are often the polyesters that are blended with cotton to make permanent-press clothing. Garments made from these blends are more comfortable on hot, humid days than those containing 100% polyester, but they retain the latter's wrinkle resistance. Instead of being forced through spinnerettes, PET melts may be forced through a narrow slit to produce sheets of Mylar, which is used extensively as magnetic tapes for tape recorders and computers.

14.7 Thioesters

AIM: To write an equation for the formation of a thioester.

Focus

Thioesters are biologically important.

The carboxylic esters we have seen so far are *oxyesters* because their carboxyl groups contain two oxygen atoms. *In* **thioesters,** *the oxygen in the C—O—C link of an oxyester is replaced by sulfur:*

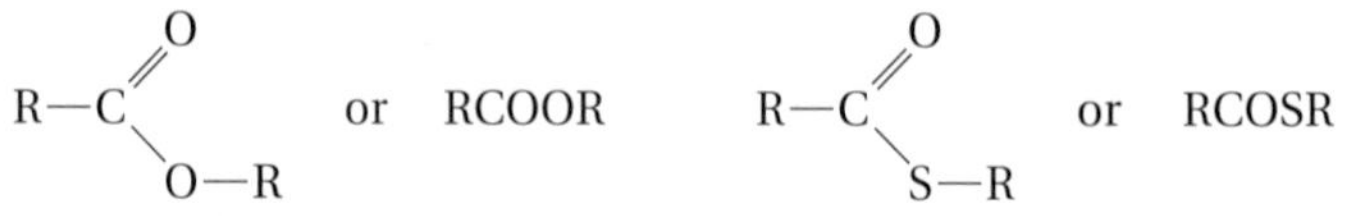

Oxyester Thioester

Thioesters can be prepared in the same way as oxyesters. A thiol is used instead of an alcohol:

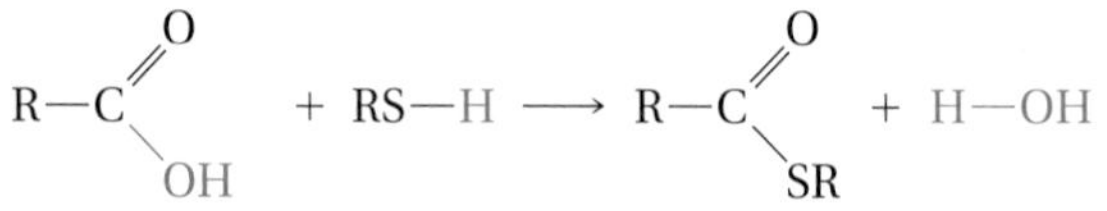

Carboxylic acid Thiol Thioester Water

If an acid anhydride is used in the esterification of a thiol, the process is an acylation:

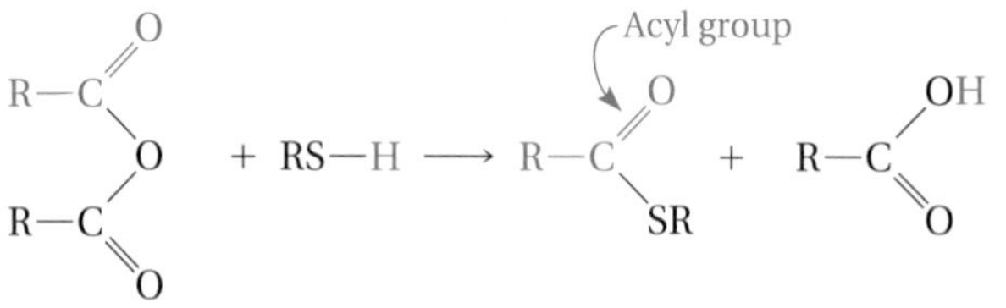

Acid anhydride Thiol Thioester Carboxylic acid

Certain thioesters have biological significance. They transfer acyl groups during the synthesis and degradation of such substances as carbohydrates, amino acids, and fatty acids. Acetyl coenzyme A is the most important acyl transfer agent in living organisms. This compound is the thioester of acetic acid and a thiol, coenzyme A:

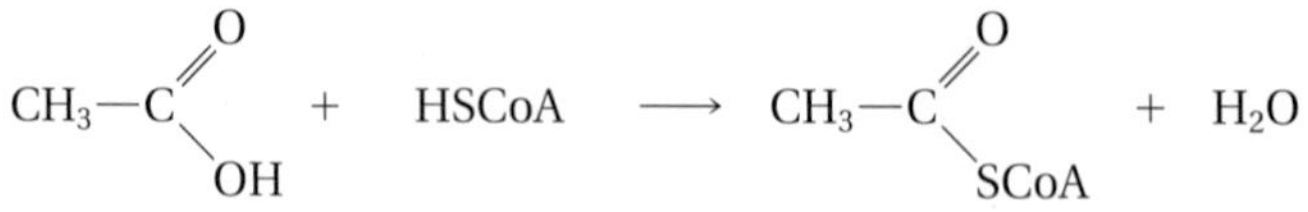

Acetic acid Coenzyme A Acetyl coenzyme A Water

14.8 Ester hydrolysis

AIM: To write an equation for the acid- or base-catalyzed hydrolysis of an ester.

Focus

Esters may be hydrolyzed to their component acids and alcohols.

If an ester is heated with water for several hours, usually very little happens. In strong acid solutions, however, ester hydrolysis is rapid because it is catalyzed by hydrogen ions:

$$CH_3-\overset{\overset{\displaystyle O}{\|}}{C}-OCH_2CH_3 + H-OH \overset{H^+}{\rightleftharpoons} CH_3-\overset{\overset{\displaystyle O}{\|}}{C}-OH + HOCH_2CH_3$$

Ethyl acetate   Acetic acid   Ethanol

Since this hydrolysis process is reversible, a large excess of water pushes the reaction to completion.

An aqueous solution of sodium hydroxide or potassium hydroxide also can be used to hydrolyze esters. Since many esters do not dissolve in water, a solvent such as ethanol is added to make the solution homogeneous. The reaction mixture is usually heated. All the ester is converted to products. At the end of the reaction, the carboxylic acid is in solution as its sodium or potassium salt:

$$CH_3-\overset{\overset{\displaystyle O}{\|}}{C}-OCH_2CH_3 + NaOH \overset{NaOH}{\longrightarrow} CH_3-\overset{\overset{\displaystyle O}{\|}}{C}-O^-Na^+ + HOCH_2CH_3$$

Ethyl acetate   Sodium hydroxide   Sodium acetate   Ethanol

If the reaction mixture is acidified, the carboxylic acid is formed:

$$CH_3-\overset{\overset{\displaystyle O}{\|}}{C}-O^-Na^+ + HCl \longrightarrow CH_3-\overset{\overset{\displaystyle O}{\|}}{C}-OH + NaCl$$

Sodium acetate   Acetic acid

Soap is made by the alkaline hydrolysis of naturally occurring triglycerides (triesters of glycerol and fatty acids). The process of making soap is called *saponification* after *sapon* (Latin): soap. Once hydrolyzed, the fatty acid salts are purified, dried, and shaped into bars. Additional ingredients such as dyes for color, perfume for odor, and antiseptics may be added to the soap.

EXAMPLE 14.6   **Hydrolysis of an ester**

Write the structure and name the products of the acid-catalyzed hydrolysis of isobutyl propanoate.

SOLUTION

The ester is hydrolyzed to an alcohol and an acid:

$$CH_3CH_2COOCH_2\underset{\underset{\displaystyle CH_3}{|}}{C}HCH_3 + H_2O \overset{HCl}{\longrightarrow} CH_3CH_2COOH + HOCH_2\underset{\underset{\displaystyle CH_3}{|}}{C}HCH_3$$

The alkyl part of the name, *isobutyl,* comes from the alcohol, isobutyl alcohol. The acid is *propanoic acid.*

PRACTICE EXERCISE 14.9

Complete the following reactions by writing the names and structures of the expected products.

(a) $CH_3CH_2COOCH_2CH_3$ + NaOH

(b) CH_3COO-⬡$ + KOH$

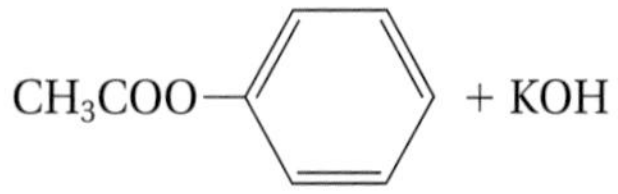

14.9 Phosphoric acids, anhydrides, and esters

AIMS: To name and draw structures of the various forms of phosphoric acid and phosphate esters. To write equations for the hydrolysis of phosphoric acid anhydrides. To describe a phosphoryl group and write an equation showing a phosphorylation reaction.

Recall that phosphorus is one of the important elements found in the human body. This phosphorus is present as phosphate ions, as phosphoric acid anhydrides, and as phosphate esters.

Phosphoric acid is a moderately strong acid. It differs from carboxylic acids because it contains three ionizable protons; a carboxyl function contains only one. The various degrees of ionization are

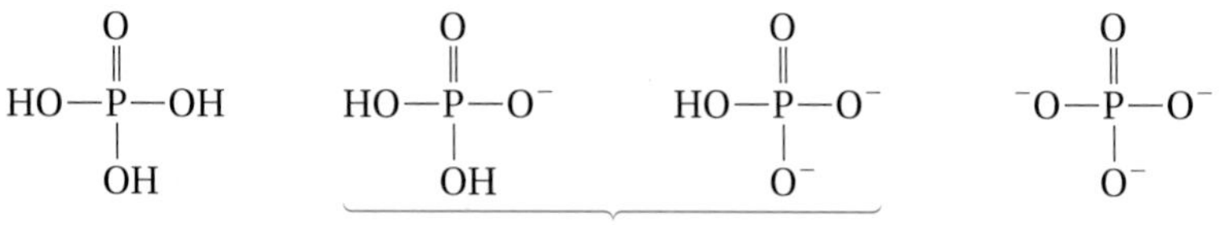

$$
\begin{array}{cccc}
\underset{\substack{\text{O} \\ \| \\ \text{HO}-\text{P}-\text{OH} \\ | \\ \text{OH}}}{} & \underset{\substack{\text{O} \\ \| \\ \text{HO}-\text{P}-\text{O}^- \\ | \\ \text{OH}}}{} & \underset{\substack{\text{O} \\ \| \\ \text{HO}-\text{P}-\text{O}^- \\ | \\ \text{O}^-}}{} & \underset{\substack{\text{O} \\ \| \\ {}^-\text{O}-\text{P}-\text{O}^- \\ | \\ \text{O}^-}}{}
\end{array}
$$

Predominant species at pH values less than 2	About equal amounts of these two forms exist at pH 7.0	This form predominates at pH values greater than 11

As shown, the ionization state varies with the pH of the solution. At pH 7.0, the singly and doubly charged phosphate ions are present in about equal amounts. In biochemistry, the symbol P_i (inorganic phosphate) is often used to represent all possible ionization states of phosphoric acid in solution at pH 7.0.

Anhydrides of phosphoric acid

Like carboxylic acids, phosphoric acid can be dehydrated to form the anhydride. Unlike carboxylic acids, however, phosphoric acid can form more than one anhydride bond due to its multiple hydroxyl groups. The simplest anhydride of phosphoric acid is pyrophosphoric acid. The various ionized forms of pyrophosphoric acid (diphosphate ions) at pH 7.0 are often abbreviated PP_i (inorganic pyrophosphate):

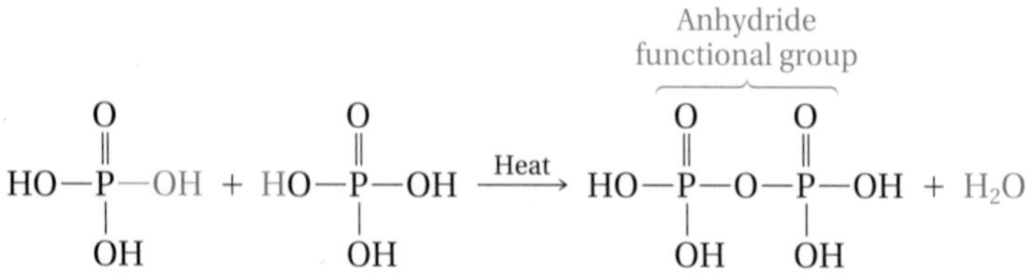

$$
\underset{\substack{\text{O} \\ \| \\ \text{HO}-\text{P}-\text{OH} \\ | \\ \text{OH}}}{} + \underset{\substack{\text{O} \\ \| \\ \text{HO}-\text{P}-\text{OH} \\ | \\ \text{OH}}}{} \xrightarrow{\text{Heat}} \underset{\substack{\text{O} \quad\quad \text{O} \\ \| \quad\quad \| \\ \text{HO}-\text{P}-\text{O}-\text{P}-\text{OH} \\ | \quad\quad\quad | \\ \text{OH} \quad\quad \text{OH}}}{} + \text{H}_2\text{O}
$$

Anhydride functional group

Two molecules of phosphoric acid	Pyrophosphoric acid

Pyrophosphoric acid can react with yet another molecule of phosphoric acid by elimination of water to form triphosphoric acid. Triphosphoric acid is sometimes represented in its various ionized forms (triphosphate ions) at pH 7.0 as PPP$_i$ (inorganic triphosphate):

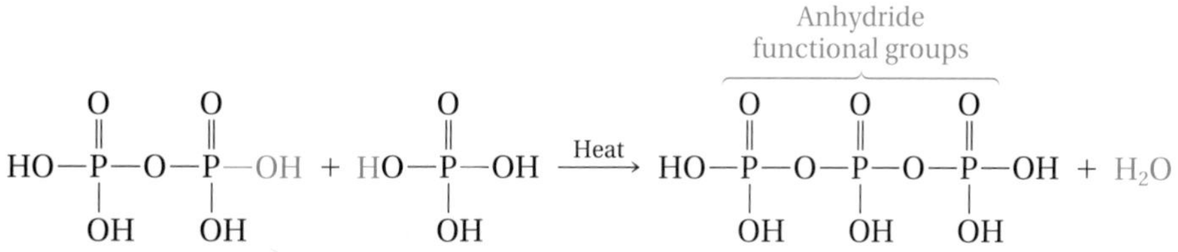

Anhydride functional groups

$$HO-\underset{\underset{OH}{|}}{\overset{\overset{O}{\|}}{P}}-O-\underset{\underset{OH}{|}}{\overset{\overset{O}{\|}}{P}}-OH \ + \ HO-\underset{\underset{OH}{|}}{\overset{\overset{O}{\|}}{P}}-OH \ \xrightarrow{\text{Heat}} \ HO-\underset{\underset{OH}{|}}{\overset{\overset{O}{\|}}{P}}-O-\underset{\underset{OH}{|}}{\overset{\overset{O}{\|}}{P}}-O-\underset{\underset{OH}{|}}{\overset{\overset{O}{\|}}{P}}-OH \ + \ H_2O$$

| Pyrophosphoric acid | Phosphoric acid | | Triphosphoric acid |

Pyrophosphoric acid and triphosphoric acid are the major anhydrides of phosphoric acid. Remember that phosphoric acid anhydrides contain two functional groups: the anhydride and hydroxyl groups. Both these functional groups are important in the reactions we will be discussing.

Esters of phosphoric acid

Phosphate esters are important intermediates in many metabolic pathways. In Chapter 24, for example, we will see that the ester glyceraldehyde-3-phosphate plays an important role in the energy-producing breakdown of glucose.

Just as carboxylic acids and alcohols react to form carboxylic acid esters, phosphoric acid can react with alcohols to form phosphate esters. Because phosphoric acid has three hydroxyl groups, one, two, or all three of these groups can be esterified to form monoesters, diesters, or triesters. This is shown for the reaction of phosphoric acid with methanol.

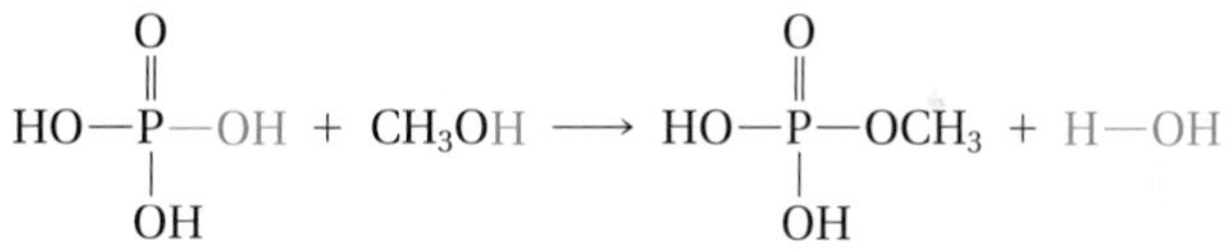

$$HO-\underset{\underset{OH}{|}}{\overset{\overset{O}{\|}}{P}}-OH \ + \ CH_3OH \ \longrightarrow \ HO-\underset{\underset{OH}{|}}{\overset{\overset{O}{\|}}{P}}-OCH_3 \ + \ H-OH$$

Monomethyl phosphate

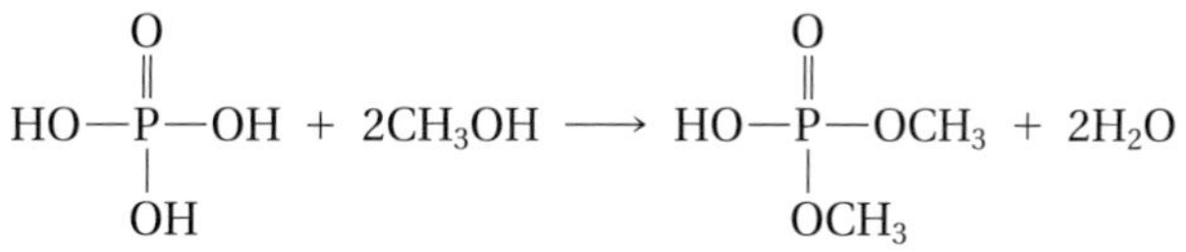

$$HO-\underset{\underset{OH}{|}}{\overset{\overset{O}{\|}}{P}}-OH \ + \ 2CH_3OH \ \longrightarrow \ HO-\underset{\underset{OCH_3}{|}}{\overset{\overset{O}{\|}}{P}}-OCH_3 \ + \ 2H_2O$$

Dimethyl phosphate

$$HO-\underset{\underset{OH}{|}}{\overset{\overset{O}{\|}}{P}}-OH \ + \ 3CH_3OH \ \longrightarrow \ CH_3O-\underset{\underset{OCH_3}{|}}{\overset{\overset{O}{\|}}{P}}-OCH_3 \ + \ 3H_2O$$

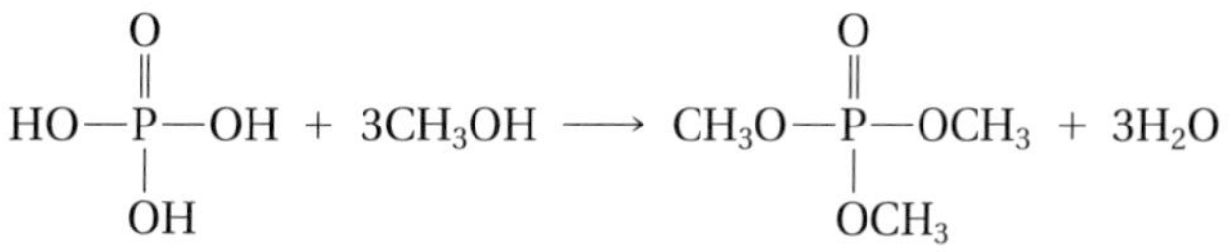

Trimethyl phosphate

The hydroxyl groups of phosphoric acid anhydrides may form esters with alcohols without breaking the anhydride bond. The following example shows a monoester, although more than one hydroxyl group could be esterified. Again, the methyl esters are used as typical examples:

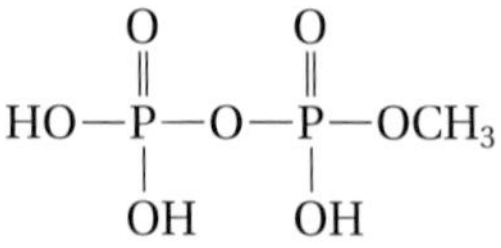

$$
\begin{array}{cc}
\underset{\substack{|\\ \text{OH}}}{\overset{\overset{\displaystyle O}{\|}}{\text{HO}-\text{P}}}-\text{O}-\underset{\substack{|\\ \text{OH}}}{\overset{\overset{\displaystyle O}{\|}}{\text{P}}}-\text{OCH}_3 &
\underset{\substack{|\\ \text{OH}}}{\overset{\overset{\displaystyle O}{\|}}{\text{HO}-\text{P}}}-\text{O}-\underset{\substack{|\\ \text{OH}}}{\overset{\overset{\displaystyle O}{\|}}{\text{P}}}-\text{O}-\underset{\substack{|\\ \text{OH}}}{\overset{\overset{\displaystyle O}{\|}}{\text{P}}}-\text{OCH}_3
\end{array}
$$

Monomethyl ester of pyrophosphoric acid 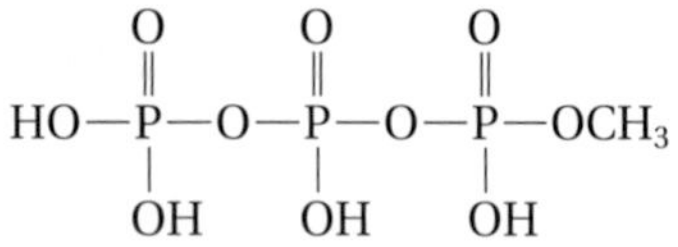   Monomethyl ester of triphosphoric acid

Hydrolysis and phosphorylation

We have seen that acetic anhydride can undergo cleavage by water in hydrolysis to form acetic acid (see Sec. 14.5) or react with alcohols to form esters in acylation (see Sec. 14.6). Phosphoric acid anhydrides can do the same. Upon hydrolysis, phosphoric acid anhydrides are converted back to phosphoric acid:

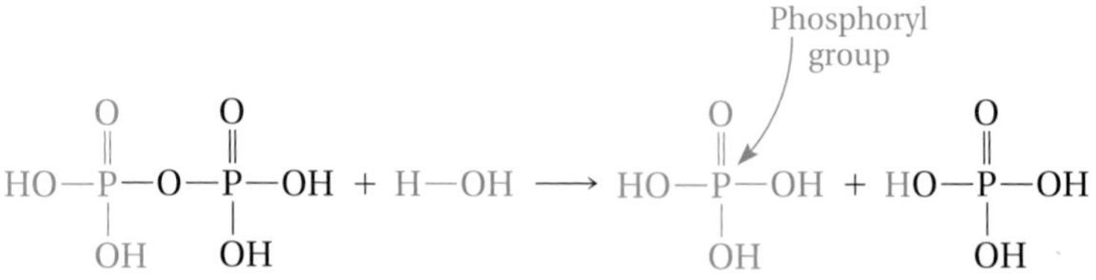

Pyrophosphoric acid   Two molecules of phosphoric acid

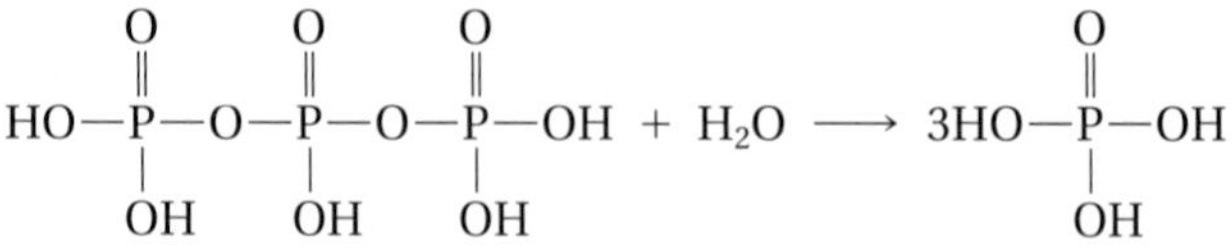

Triphosphoric acid   Three molecules of phosphoric acid

We can consider the hydrolysis of phosphoric acid anhydrides as the transfer of a phosphoryl group ($-\text{PO}_3\text{H}_2$) to water, just as the hydrolysis of acetic anhydride is a transfer of an acetyl group to water. And just as the transfer of an acetyl group to an alcohol or other functional group is called *acetylation, transfer of a* **phosphoryl group** *to an alcohol or other functional group is called* **phosphorylation.** Here are two typical phosphorylation reactions:

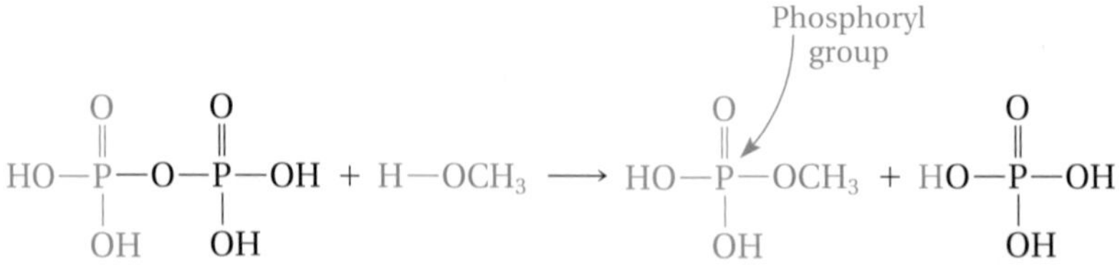

Pyrophosphoric acid   Methanol   Monomethyl phosphate   Phosphoric acid

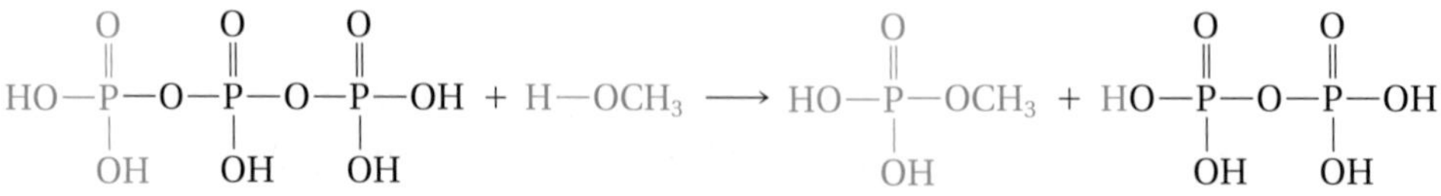

Triphosphoric acid   Methanol   Monomethyl phosphate   Pyrophosphoric acid

Phosphorylation reactions of this type are enormously important in biochemistry. Adenosine triphosphate (ATP), a monoester of triphosphoric acid, is nature's universal phosphorylating agent. The ATP in living systems is used to phosphorylate water, sugars, proteins, and nucleic acids. Transfers of energy in living organisms come from these phosphorylation reactions.

PRACTICE EXERCISE 14.10

Explain what is meant by the term *phosphorylation*.

14.10 Esters of nitric and nitrous acids

AIM: To state the names and uses of important esters of nitrous and nitric acids.

Focus

Nitrous and nitric acids form important esters.

Alcohols react with nitric acid (HNO_3) and nitrous acid (HNO_2) to produce the alkyl nitrates and alkyl nitrites, respectively. Glycerol and nitric acid produce the ester glyceryl trinitrate, more commonly called *nitroglycerin*.

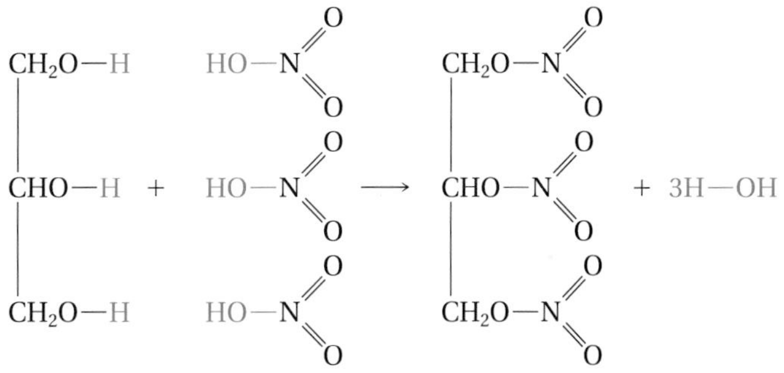

| Glycerol | Three molecules of nitric acid | Glyceryl trinitrate (nitroglycerin) | Water |

Alfred Nobel (1833–1896) conducted experiments to stabilize nitroglycerin on a lake barge. This came after the factory in which he manufactured nitroglycerin was destroyed in an explosion and he was forbidden by the Swedish government to rebuild. Upon his death, he left the bulk of his fortune to establish the Nobel prizes.

Nitroglycerin is an unstable, shock-sensitive, explosive, pale yellow liquid. It was first made in 1846 by the Italian chemist Ascanio Sobrero (1812–1888). The Swedish chemist Alfred Nobel (1833–1896) perfected its synthesis in the mid-19th century and devised a safe method for handling it. Nobel mixed nitroglycerin with *kieselghur,* a claylike absorbent material. The result was dynamite, an explosive that is not sensitive to shock. Since its development, dynamite has been important in the construction of canals, dams, and roads. It also has played a major role in warfare.

Nitroglycerin and isoamyl nitrite, an ester of nitrous acid, have been used as drugs for over a hundred years.

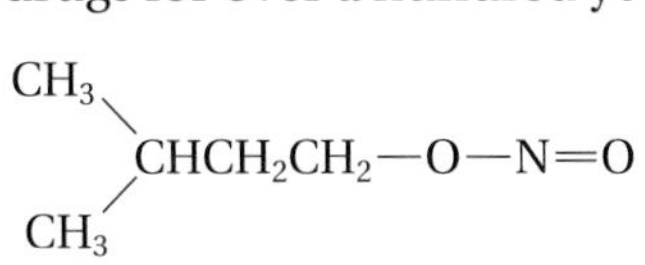

Isoamyl nitrite

Both these esters, when inhaled or taken as tablets, produce immediate relaxation of the smooth muscles of the body and expansion of the blood vessels. They are used to treat people with *angina pectoris*—chest pains caused by an insufficient supply of oxygen to the heart muscle. Isoamyl nitrite is also the active ingredient in the widely abused drug called "poppers."

SUMMARY

Carboxylic acids (RCO_2H) contain the carboxyl functional group ($—CO_2H$). Fatty acids (straight-chain saturated and unsaturated aliphatic carboxylic acids) are widely distributed in nature, as are many other carboxylic acids. Benzoic acid is the parent aromatic carboxylic acid. Fatty acids containing up to four carbons are completely soluble in water. Carboxylic acids of higher molar mass are less soluble, but they tend to form monomolecular layers at water surfaces and micelles within water. Carboxylic acids are weak acids and ionize only to a slight extent in water. The acidity increases if the acid molecules contain electron-withdrawing substituents. Like other acids, carboxylic acids can be neutralized by bases to give salts. Soaps are the alkali metal salts of long-chain carboxylic acids. Detergents are often alkali metal salts of sulfonic acids (RSO_3H). Carboxylic acids may be prepared by either the oxidation of primary alcohols or the oxidation of aldehydes.

Anhydrides (RCO_2COR) and esters (RCO_2R) are important derivatives of carboxylic acids. Anhydrides, formed by the dehydration of carboxylic acids, form esters when they react with an alcohol or a phenol, as in the preparation of aspirin. Esters also can be prepared by the reaction of a carboxylic acid and an alcohol. Many naturally occurring esters impart pleasant odors and flavors to fruits. Polyesters (polymeric esters) can be made into synthetic fibers. Esters can be converted to the component carboxylic acid and alcohol by hydrolysis in acidic or basic solution. Certain thioesters, especially esters of the thiol coenzyme A, are biologically important. Esters and anhydrides of phosphoric acid are also biologically important as agents that can transfer phosphoryl groups ($—PO_3H_2$). ATP, an ester of the phosphoric anhydride triphosphoric acid, is the major phosphorylating agent of living cells. Esters of inorganic acids such as nitric and nitrous acids have such varied applications as explosives (nitroglycerin) and medications for the relief of angina pectoris (isoamyl nitrite).

SUMMARY OF REACTIONS

Here are the reactions of carboxylic acids and inorganic acids and their derivatives described in this chapter.

Carboxylic Acids

1. Preparation of carboxylic acids:

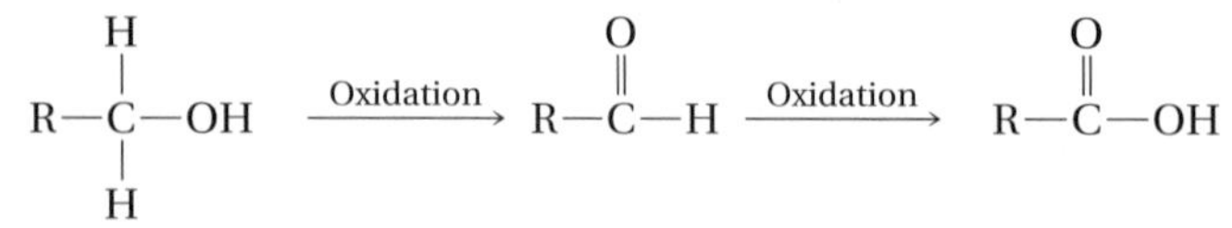

Primary alcohol → Aldehyde → Carboxylic acid

2. Neutralization of carboxylic acids:

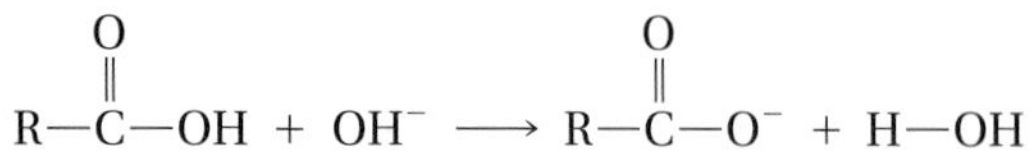

Carboxylic acid + Base → Carboxylate ion + Water

3. Preparation of acid anhydrides:

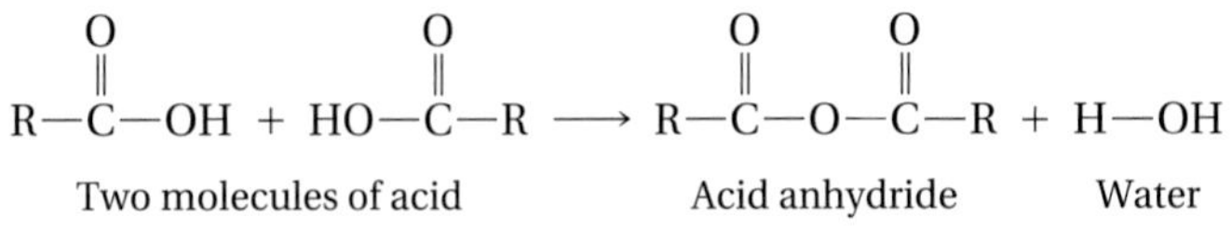

$$\underset{\text{Two molecules of acid}}{\overset{\overset{\text{O}}{\|}}{R-C-OH} + \overset{\overset{\text{O}}{\|}}{HO-C-R}} \longrightarrow \underset{\text{Acid anhydride}}{\overset{\overset{\text{O}}{\|} \quad \overset{\text{O}}{\|}}{R-C-O-C-R}} + \underset{\text{Water}}{H-OH}$$

4. Preparation of esters
   (a) Oxyesters:

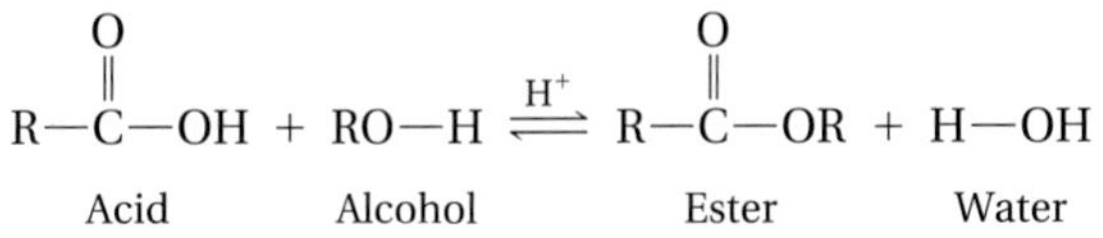

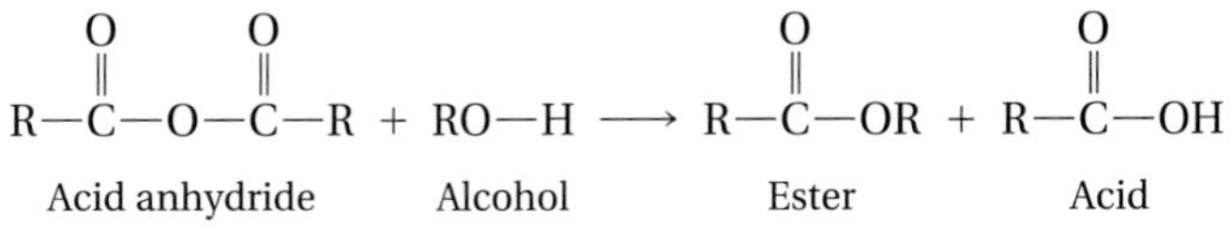

$$\underset{\text{Acid}}{\overset{\overset{\text{O}}{\|}}{R-C-OH}} + \underset{\text{Alcohol}}{RO-H} \overset{H^+}{\rightleftharpoons} \underset{\text{Ester}}{\overset{\overset{\text{O}}{\|}}{R-C-OR}} + \underset{\text{Water}}{H-OH}$$

$$\underset{\text{Acid anhydride}}{\overset{\overset{\text{O}}{\|} \quad \overset{\text{O}}{\|}}{R-C-O-C-R}} + \underset{\text{Alcohol}}{RO-H} \longrightarrow \underset{\text{Ester}}{\overset{\overset{\text{O}}{\|}}{R-C-OR}} + \underset{\text{Acid}}{\overset{\overset{\text{O}}{\|}}{R-C-OH}}$$

   (b) Thioesters:

$$\underset{\text{Acid}}{\overset{\overset{\text{O}}{\|}}{R-C-OH}} + \underset{\text{Thiol}}{RS-H} \longrightarrow \underset{\text{Thioester}}{\overset{\overset{\text{O}}{\|}}{R-C-SR}} + \underset{\text{Water}}{H-OH}$$

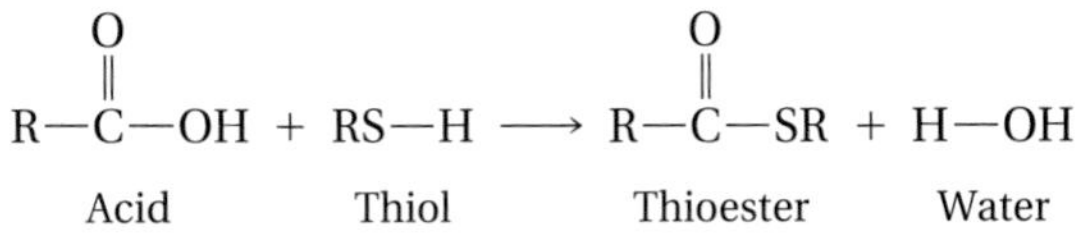

Inorganic Acids

1. Preparation of phosphoric acid anhydrides:

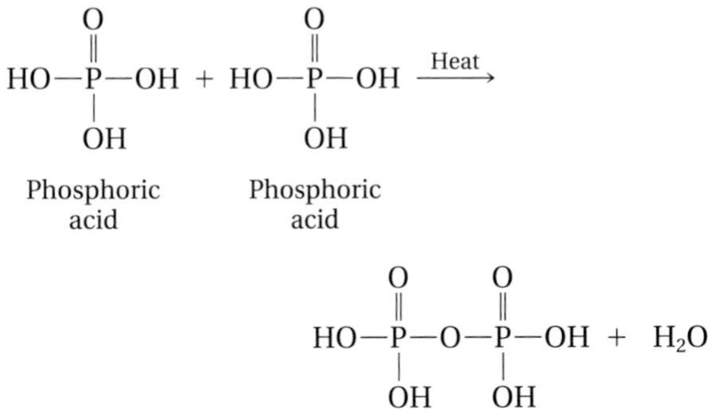

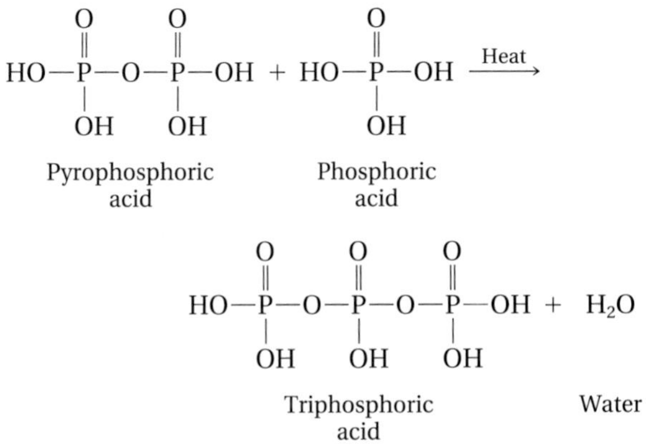

$$\underset{\substack{\text{Phosphoric} \\ \text{acid}}}{\overset{\overset{\text{O}}{\|}}{\underset{\text{OH}}{HO-P-OH}}} + \underset{\substack{\text{Phosphoric} \\ \text{acid}}}{\overset{\overset{\text{O}}{\|}}{\underset{\text{OH}}{HO-P-OH}}} \xrightarrow{\text{Heat}}$$

$$\overset{\overset{\text{O}}{\|} \quad\quad \overset{\text{O}}{\|}}{\underset{\text{OH} \quad\;\; \text{OH}}{HO-P-O-P-OH}} + H_2O$$

$$\underset{\substack{\text{Pyrophosphoric} \\ \text{acid}}}{\overset{\overset{\text{O}}{\|} \quad\quad \overset{\text{O}}{\|}}{\underset{\text{OH} \quad\;\; \text{OH}}{HO-P-O-P-OH}}} + \underset{\substack{\text{Phosphoric} \\ \text{acid}}}{\overset{\overset{\text{O}}{\|}}{\underset{\text{OH}}{HO-P-OH}}} \xrightarrow{\text{Heat}}$$

$$\underset{\substack{\text{Triphosphoric} \\ \text{acid}}}{\overset{\overset{\text{O}}{\|} \quad\quad \overset{\text{O}}{\|} \quad\quad \overset{\text{O}}{\|}}{\underset{\text{OH} \quad\;\; \text{OH} \quad\;\; \text{OH}}{HO-P-O-P-O-P-OH}}} + \underset{\text{Water}}{H_2O}$$

2. Preparation of phosphate esters:

$$\text{HO}-\overset{\overset{\displaystyle O}{\|}}{\underset{\underset{\displaystyle OH}{|}}{P}}-\text{OH} + \text{ROH} \longrightarrow \text{HO}-\overset{\overset{\displaystyle O}{\|}}{\underset{\underset{\displaystyle OH}{|}}{P}}-\text{OR} + \text{H}-\text{OH}$$

Anhydrides

1. Hydrolysis of acid anhydrides
   (a) Carboxylic acid anhydrides:

$$\text{R}-\overset{\overset{\displaystyle O}{\|}}{C}-\text{O}-\overset{\overset{\displaystyle O}{\|}}{C}-\text{R} + \text{H}-\text{OH} \longrightarrow$$

$$\text{R}-\overset{\overset{\displaystyle O}{\|}}{C}-\text{OH} + \text{R}-\overset{\overset{\displaystyle O}{\|}}{C}-\text{OH}$$

   (b) Phosphoric acid anhydrides:

$$\text{HO}-\overset{\overset{\displaystyle O}{\|}}{\underset{\underset{\displaystyle OH}{|}}{P}}-\text{O}-\overset{\overset{\displaystyle O}{\|}}{\underset{\underset{\displaystyle OH}{|}}{P}}-\text{OH} + \text{H}-\text{OH} \longrightarrow 2\text{HO}-\overset{\overset{\displaystyle O}{\|}}{\underset{\underset{\displaystyle OH}{|}}{P}}-\text{OH}$$

2. Acylation of alcohols:

$$\text{R}-\overset{\overset{\displaystyle O}{\|}}{C}-\text{O}-\overset{\overset{\displaystyle O}{\|}}{C}-\text{R} + \text{R}-\text{OH} \longrightarrow$$

$$\text{R}-\overset{\overset{\displaystyle O}{\|}}{C}-\text{OR} + \text{R}-\overset{\overset{\displaystyle O}{\|}}{C}-\text{OH}$$

3. Phosphorylation of alcohols:

$$\text{HO}-\overset{\overset{\displaystyle O}{\|}}{\underset{\underset{\displaystyle OH}{|}}{P}}-\text{O}-\overset{\overset{\displaystyle O}{\|}}{\underset{\underset{\displaystyle OH}{|}}{P}}-\text{OH} + \text{H}-\text{OR} \longrightarrow$$

$$\text{HO}-\overset{\overset{\displaystyle O}{\|}}{\underset{\underset{\displaystyle OH}{|}}{P}}-\text{OR} + \text{HO}-\overset{\overset{\displaystyle O}{\|}}{\underset{\underset{\displaystyle OH}{|}}{P}}-\text{OH}$$

Esters

1. Hydrolysis of carboxylate esters
   (a) In acid:

$$\text{R}-\overset{\overset{\displaystyle O}{\|}}{C}-\text{OR} + \text{H}-\text{OH} \xrightarrow[\text{H}^+]{\text{Heat}} \text{R}-\overset{\overset{\displaystyle O}{\|}}{C}-\text{OH} + \text{RO}-\text{H}$$

   (b) In base:

$$\text{R}-\overset{\overset{\displaystyle O}{\|}}{C}-\text{OR} + \text{OH}^- \xrightarrow{\text{Heat}} \text{R}-\overset{\overset{\displaystyle O}{\|}}{C}-\text{O}^- + \text{RO}-\text{H}$$

KEY TERMS

Acid anhydride (14.5)
Acylation reaction (14.5)
Acyl group (14.5)
Analgesic (14.6)
Antipyretic (14.6)
Biodegradable (14.3)
Carboxyl group (14.1)

Carboxylic acid (14.1)
Detergent (14.3)
Dicarboxylic acid (14.1)
Dimer (14.2)
Ester (14.6)
Esterification (14.6)
Fatty acid (14.1)

Hard water (14.3)
Hydrolysis (14.5)
Hydroxy acid (14.1)
Micelle (14.3)
Monomolecular layer (14.2)
Nonionic detergent (14.3)

Phosphorylation (14.9)
Phosphoryl group (14.9)
Polyester (14.6)
Soap (14.3)
Thioester (14.7)

EXERCISES

Carboxylic Acids and Their Salts
(Sections 14.1, 14.2, 14.3)

14.11 Give the name (IUPAC or common) or write the structure of each of the following carboxylic acids.
(a) H—COOH

(b)
$$CH_3-\underset{\underset{OH}{|}}{CH}-COOH$$

(c)
a benzene ring with —COOH and —OH substituents

(d)
$$CH_3\underset{\underset{O}{\|}}{C}-COOH$$

(e) stearic acid   (f) lactic acid
(g) *p*-chlorobenzoic acid   (h) malonic acid

14.12 Write the structure or give the name of each of the following compounds.
(a) acetic acid
(b) 2-hydroxypropanoic acid
(c) citric acid
(d) benzoic acid

(e)
$$\underset{\underset{COOH}{|}}{COOH}$$

(f)
$$CH_3CH_2\underset{\underset{Br}{|}}{CH}CH_2CH_2COOH$$

(g) $CH_3(CH_2)_5COOH$

(h)
$$\underset{\underset{HO-CH-COOH}{|}}{HO-CH-COOH}$$

14.13 Which of the acids in Exercise 14.11, if any, are dicarboxylic, polyfunctional, or fatty acids?

14.14 Designate each acid in Exercise 14.12 as a dicarboxylic, polyfunctional, or fatty acid.

14.15 Write the structural formulas for the compounds in Table 14.3, and explain why propanoic acid has the highest boiling point.

14.16 Predict which of the following compounds has the highest boiling point. Molar masses are given in parentheses.
(a) CH_3CHO (44)   (b) $HCOOH$ (46)
(c) $CH_3CH_2CH_3$ (44)   (d) CH_3CH_2OH (46)

14.17 Why are carboxylic acids weak acids?

14.18 Why is trichloroacetic acid a much stronger acid than acetic acid?

14.19 Draw the structure and name the products for each of the following reactions.
(a) $HCOOH + KOH \longrightarrow$
(b)
$$\underset{\underset{COOH}{|}}{COOH} + Ca(OH)_2 \longrightarrow$$

(c) Myristic acid + sodium hydroxide $\longrightarrow$

14.20 Write the structure and name the salt formed in each of the following reactions.
(a) Propanoic acid + sodium bicarbonate $\longrightarrow$
(b)
a benzene ring with —COOH substituent + NaOH $\longrightarrow$

(c) $CH_3CH_2CH_2COOH + NH_3 \longrightarrow$

14.21 What is a soap? Describe the two distinct parts of a soap molecule, and explain how soap acts as an emulsifying agent.

14.22 Distinguish between a detergent molecule and a soap molecule, and explain why detergents are better than soaps for washing clothes in hard water.

Preparation of Carboxylic Acids (Section 14.4)

14.23 Complete the following reactions by writing the structure of the expected products.
(a) $CH_3CH_2OH \xrightarrow{K_2Cr_2O_7}$
(b) $CH_3CH_2CHO \xrightarrow{K_2Cr_2O_7}$
(c) $(CH_3)_3COH \xrightarrow{KMnO_4}$

14.24 Write the structure of the acid formed (if any) by the oxidation of the following compounds.
(a)
a benzene ring with —CH_2CHO substituent $\xrightarrow{K_2Cr_2O_7}$

(b)
$$CH_3CH_2\underset{\underset{O}{\|}}{C}CH_3 \xrightarrow{K_2Cr_2O_7}$$

(c) $CH_3CH_2\underset{\underset{CH_3}{|}}{CH}OH \xrightarrow{K_2Cr_2O_7}$

14.25 Name the aldehyde you would oxidize, with acidified $K_2Cr_2O_7$, to produce (a) butanoic acid and (b) formic acid.

14.26 What compound could be oxidized to produce (a) benzoic acid and (b) 3-methylpentanoic acid?

Anhydrides and Esters (Sections 14.5–14.8)

14.27 Draw the general formula for (a) an acid anhydride and (b) an ester.

14.28 What are the general products of (a) hydrolysis of an acid anhydride and (b) hydrolysis of an ester?

14.29 Name the given ester or write the structure and name of the ester that could be produced from the given reactants.
(a) CH_3COOCH_3 (b) $CH_3COOCH_2CH_3$
(c)

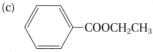

(d) formic acid + methanol $\longrightarrow$
(e) propionic acid + propanol $\longrightarrow$

14.30 Write the structure and name of the ester that could be produced from the given reactants or name the given ester.
(a) butyric acid + ethanol $\longrightarrow$
(b) acetic acid + butanol $\longrightarrow$
(c) $HCOOCH_3$
(d)

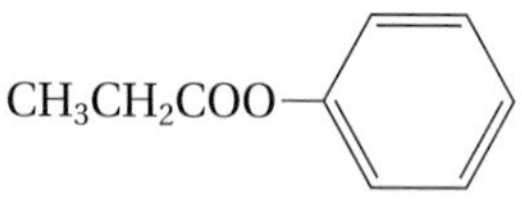

CH_3CH_2COO—⬡

(e) $CH_3COOCH_2CH_2CH_3$

14.31 Show the difference in structure between an oxyester and a thioester.

14.32 Write the structures of the hydrolysis products of the following thioester:

$$CH_3CH_2\overset{\overset{\displaystyle O}{\|}}{C}-S-CH_3$$

14.33 Complete the following reactions by writing the structures of the expected products and by naming each of the reactants and products.
(a) $CH_3COOCH_3 + H_2O \xrightarrow{HCl}$
(b)

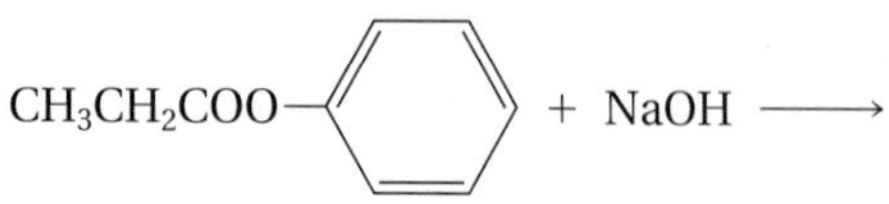

CH_3CH_2COO—⬡ + NaOH $\longrightarrow$

(c)

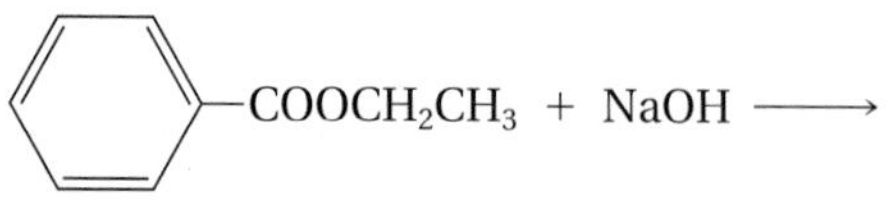

⬡—$COOCH_2CH_3$ + NaOH $\longrightarrow$

(d) $CH_3CH_2CH_2COOCH_2CH_2CH_3 + NaOH \longrightarrow$

(e) $HCOOCH_2CH_3 + KOH \longrightarrow$
(f)

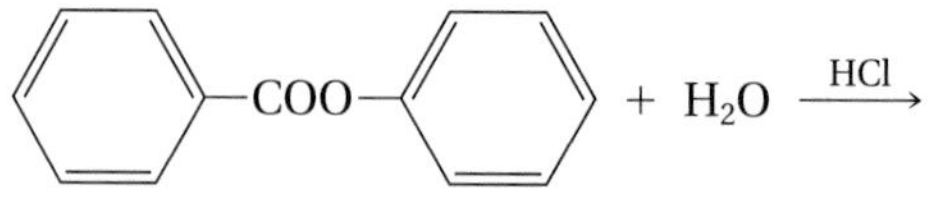

⬡—COO—⬡ + $H_2O \xrightarrow{HCl}$

14.34 Write the structure and name of the expected products for each of the following reactions.
(a) $CH_3COOH + CH_3OH \xrightarrow{H^+}$
(b) $CH_3CH_2CH_2COOCH_2CH_3 + NaOH \longrightarrow$
(c) $CH_3CH_2OH \xrightarrow{K_2Cr_2O_7}$
(d) $CH_3CH_2CH_2COOH + NaOH \longrightarrow$
(e)

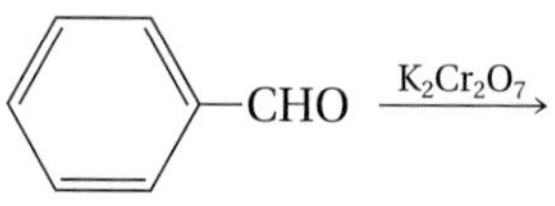

⬡—$CHO \xrightarrow{K_2Cr_2O_7}$

(f)

$$CH_3\overset{\overset{\displaystyle O}{\|}}{C}-O-\overset{\overset{\displaystyle O}{\|}}{C}CH_3 + CH_3OH \longrightarrow$$

Acids of Phosphorus and Nitrogen (Sections 14.9, 14.10)

14.35 Write the structures of phosphoric acid to show the various degrees of ionization. Which forms predominate at pH 7?

14.36 Write the structures of (a) pyrophosphoric acid (PP_i) and (b) triphosphoric acid (PPP_i). Why are these compounds called *anhydrides?*

14.37 One molecule of ethanol reacts with one molecule of pyrophosphoric acid.
(a) Give the name and structure of the organic product.
(b) What is the process called?

14.38 Isoamyl nitrite is an inorganic ester whose formula is

$$\begin{array}{c} CH_3\diagdown \\ \qquad CHCH_2CH_2-O-N=O \\ CH_3\diagup \end{array}$$

What will the products of hydrolysis be?

Additional Exercises

14.39 Arrange the following substances in order of decreasing acidity.
(a)

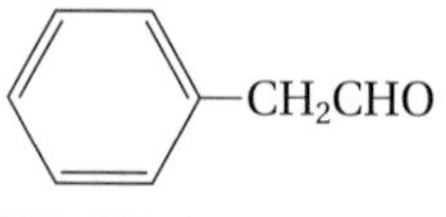

⬡—CH_2CHO

(b) CH_3CH_2OH
(c) CH_3COOH (d) H_3O^+

14.40 Write the equilibrium expression for the following reaction:

$$CH_3COOH + H_2O \rightleftharpoons CH_3COO^- + H_3O^+$$

14.41 In the following reaction, which is the stronger base, water or acetate ion? (The K_a of acetic acid is 1.8×10^{-5}.)

$$CH_3COOH + H_2O \rightleftharpoons CH_3COO^- + H_3O^+$$

14.42 List the following compounds in order of increasing acid strength.

(a) CH$_3$CHCO$_2$H
    |
    Cl

(b) ClCH$_2$CH$_2$CO$_2$H

(c) CH$_3$CH$_2$CO$_2$H

(d) CH$_3$CHCO$_2$H
    |
    F

14.43 Write the structural formula of each of the following compounds:
(a) decane
(b) hexanoic acid
(c) dimethyl ether
(d) octanal
(e) ethyl butyrate
(f) 1-pentanol
(g) 2-chlorobutanoic acid
(h) 3-bromo-4-methyl-5-phenyl-1-pentanol

14.44 Write equations for each of the following reactions, naming the organic products formed.
(a) the neutralization of butyric acid
(b) the acylation of methanol with acetic anhydride
(c) the oxidation of 3,3-dimethylbutanal
(d) the hydrolysis of acetic anhydride
(e) the complete ionization of oxalic acid
(f) the esterification of butyric acid with isobutyl alcohol
(g) the hydrolysis of triphosphoric acid
(h) the hydrolysis of isopropyl propionate with base

SELF-TEST (REVIEW)

True/False

1. A secondary alcohol can be oxidized to form a carboxylic acid.
2. A carboxyl group contains both a hydroxyl group and a carbonyl group.
3. Carboxylic acids of low molar mass are soluble in water.
4. The first word in the name of an ester comes from the alcohol part of the molecule.
5. Pyrophosphoric acid is an anhydride of phosphoric acid.
6. Ester hydrolysis is catalyzed by both acids and bases.
7. The ester formed between formic acid and butyl alcohol is called *butyl methanoate*.
8. The functional group of a thioester is

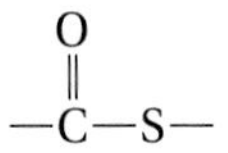

9. Both acylation and phosphorylation yield an ester as a product.
10. Carboxylic acids are able to form intermolecular hydrogen bonds.

Multiple Choice

11. The best name for the compound

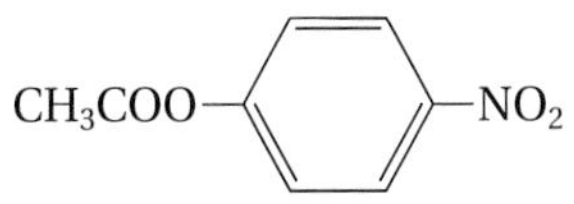

$$CH_3COO-\!\!\!\left\langle\bigcirc\right\rangle\!\!\!-NO_2$$

is
(a) *p*-nitroacetylbenzene.
(b) acetylbenzene nitrite.
(c) *p*-nitrophenyl acetate.
(d) nitrobenzene acetate.

12. The reaction of a carboxylic acid and a thiol results in the formation of a
(a) thioether.　　(b) thioester.
(c) thioacetal.　　(d) thioanhydride.

13. Which of the following statements about soap molecules is *false?*
(a) The hydrophobic ends of the molecules tend to interact favorably with water.
(b) They can be precipitated by high acid concentration and by calcium ions.
(c) They have both a hydrophilic head and a hydrophobic tail.
(d) They are salts of fatty acids.

14. Carboxylic acids have relatively high melting and boiling points because
    (a) they are generally long-chain compounds.
    (b) they have a high density.
    (c) they have a nonpolar end.
    (d) they form intermolecular hydrogen bonds.

15. Which of the following compounds could be oxidized by potassium permanganate to form propionic acid?
    (a) isopropyl alcohol   (b) propanone
    (c) propane   (d) 1-propanol

16. In acetylsalicylic acid, the acetyl group is
    (a) $CH_3—C—O—$.   (b) $—COOH$.
    (c) $—COO^-$.   (d) $CH_3—CO—$.

17. All the following are dicarboxylic acids except
    (a) lactic acid.   (b) succinic acid.
    (c) oxalic acid.   (d) phthalic acid.
    (e) glutaric acid.

18. In the monomethyl ester of pyrophosphoric acid,

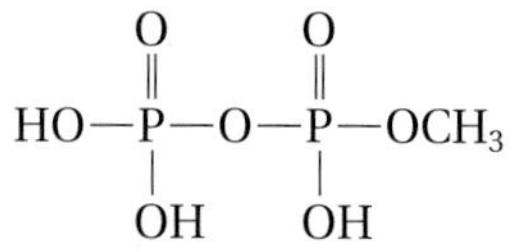

$$HO—\overset{\overset{\displaystyle O}{\|}}{\underset{\underset{\displaystyle OH}{|}}{P}}—O—\overset{\overset{\displaystyle O}{\|}}{\underset{\underset{\displaystyle OH}{|}}{P}}—OCH_3$$

    which of the following functional groups is *not* present?
    (a) hydroxy   (b) anhydride   (c) acid
    (d) ester   (e) all are present

19. Carboxylic acids are weak acids because
    (a) they are always found in low concentration in nature.
    (b) they have large K_a values.
    (c) they do not completely ionize in water.
    (d) they are found in food.

20. Which of the following acids would be least likely to lose a proton?
    (a) acetic acid   (b) chloroacetic acid
    (c) dichloroacetic acid   (d) trichloroacetic acid

21. A carboxylate ester is formed by reacting an alcohol with
    (a) an ether.   (b) a carboxylic acid.
    (c) an anhydride.   (d) another alcohol.
    (e) More than one are correct.

22. A carboxylic acid with six carbons in a straight chain would be named
    (a) succinic acid.   (b) hexanalic acid.
    (c) dimethylbutanoic acid.   (d) hexanoic acid.

23. Which of the following compounds is a product of the hydrolysis of ethyl propionate catalyzed by sodium hydroxide?
    (a) sodium acetate   (b) propanol
    (c) acetic acid   (d) sodium propionate

Amines and Amides

Organic Nitrogen Compounds

Many natural and synthetic nitrogen-containing compounds are used as medicinal drugs.

CHAPTER OUTLINE

CASE IN POINT: Antidepressants

15.1 Amines

15.2 Amine basicity

15.3 Preparation of amines

15.4 Amides

A CLOSER LOOK: The Sulfonamide
Antibiotics

A CLOSER LOOK: Amide Local
Anesthetics

15.5 Preparation of amides

15.6 Amide hydrolysis

15.7 Important amines

FOLLOW-UP TO THE CASE IN POINT:
Antidepressants

15.8 Alkaloids

15.9 Barbiturates

This chapter is about two different classes of organic compounds containing nitrogen: amines and amides. The amines and amides are very significant for the health sciences because they include among their members many important molecules in our bodies. Moreover, many natural and synthetic drugs are organic nitrogen compounds. We will first examine the structures, names, and chemistry of the amines and amides. Then we will learn about some of the major compounds that belong to these categories. Among these compounds are some important drugs that help combat depression, a condition that affected a woman named Maude. Maude's condition and its resolution are described in the following Case in Point.

CASE IN POINT: Antidepressants

Maude always considered herself a person with a sunny disposition. However, over a period of several weeks, and for no discernible reason, Maude was agitated and had difficulty sleeping. After an examination by Maude's physician revealed no physical cause for her symptoms, the doctor suspected that Maude might be suffering from depression and referred her to a psychiatrist. After probing Maude's mood, the psychiatrist suggested that Maude might benefit from an antidepressant drug in addition to taking part in counseling sessions with a group of people similarly affected (see figure). What are the chemical aspects of Maude's depression? We will learn about the use of drugs in the treatment of depression in Section 15.7.

Group sessions along with drug administration are often used in the treatment of psychiatric disorders.

15.1 Amines

AIMS: *To name and write structures of simple aliphatic and aromatic amines. To classify an amine as primary, secondary, or tertiary. To name and write structures of common aliphatic and aromatic heterocyclic amines.*

Amines *are derivatives of ammonia (NH_3) in which alkyl or aryl groups (R—) replace hydrogen.* Amines are called *primary, secondary,* or *tertiary* depending on the number of R groups attached to the nitrogen.

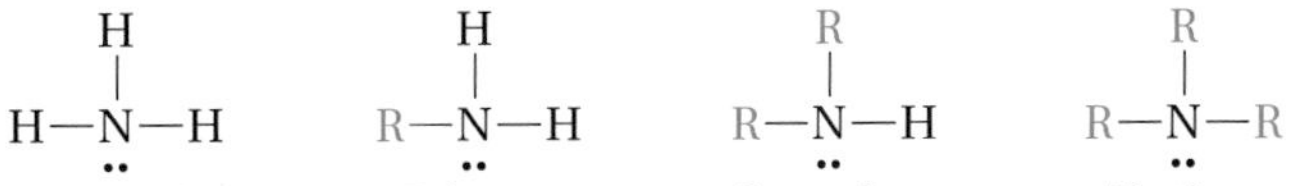

$$
\begin{array}{cccc}
& H & H & R & R \\
& | & | & | & | \\
H-N-H & R-N-H & R-N-H & R-N-R \\
\text{••} & \text{••} & \text{••} & \text{••} \\
\text{Ammonia} & \text{Primary} & \text{Secondary} & \text{Tertiary} \\
& (1°)\ \text{amine} & (2°)\ \text{amine} & (3°)\ \text{amine}
\end{array}
$$

The terms *primary, secondary,* and *tertiary* are abbreviated 1°, 2°, and 3°, respectively. Here primary, secondary, and tertiary mean something quite different from when they are applied to alcohols. Recall that in alcohols these terms refer to the number of carbon groups attached to the *carbon* that bears the hydroxyl function; in amines they refer to the number of carbon groups attached to the amine *nitrogen.* The carbon groups attached to the nitrogen may be aliphatic, aromatic, or both.

Names of aliphatic amines

Aliphatic amines are named by first listing the alkyl groups attached to the amine nitrogen in alphabetical order. The names of the alkyl groups are followed by the word *-amine;* the complete name is written as one word. We use the prefixes *di-* and *tri-* to indicate more than one alkyl group of the same kind. Here are some examples of the structures and names of aliphatic amines:

$$
\begin{array}{cc}
H & H \\
| & | \\
CH_3-N-H & CH_3-N-CH_3 \\
\text{••} & \text{••} \\
\text{Methylamine} & \text{Dimethylamine} \\
(1°) & (2°)
\end{array}
$$

$$
\begin{array}{cc}
CH_3\ \ H & CH_3 \\
|\ \ \ \ | & | \\
CH_3-CH-N-CH_2CH_3 & CH_3CH_2-N-CH_3 \\
\text{••} & \text{••} \\
\text{Ethylisopropylamine} & \text{Ethyldimethylamine} \\
(2°) & (3°)
\end{array}
$$

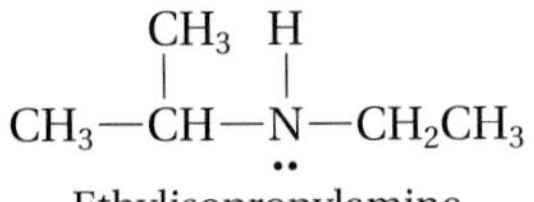

EXAMPLE 15.1   **Naming an aliphatic amine**

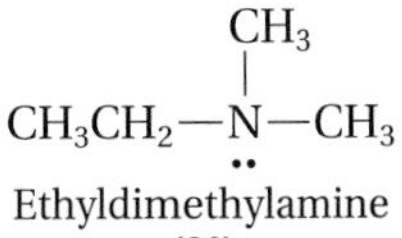

Name the following aliphatic amine, and classify it as primary, secondary, or tertiary.

$$
\begin{array}{c}
H \\
| \\
CH_3CH_2-N-CH_2CH_2CH_3
\end{array}
$$

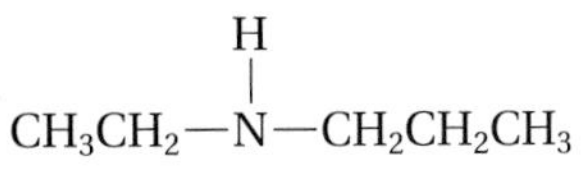

SOLUTION

There are two alkyl groups attached to the amine nitrogen, so the structure is a secondary amine. In alphabetical order, the alkyl groups are ethyl (CH_3CH_2-) and propyl ($CH_3CH_2CH_2-$). The name of the amine is *ethylpropylamine*.

PRACTICE EXERCISE 15.1

Name the following amines, and classify them as primary, secondary, or tertiary.

$$\text{H}$$
(a) $CH_3\overset{|}{N}CH_2CH_3$

(b) $CH_3CH_2CH_2NH_2$

(c) $(CH_3)_2NCH_2CH_3$

(d) $(CH_3CH_2)_3N$

Names of aromatic amines

The parent aromatic amine is called *aniline*. Anilines substituted with alkyl groups on the amine nitrogen are named as derivatives of aniline. The prefix *N-* indicates that the substituent is on the amine nitrogen, not on the aromatic ring. For example:

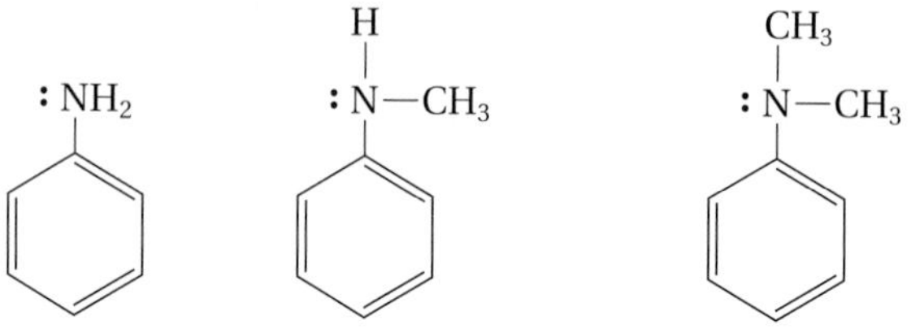

Aniline
(1°)

N-Methylaniline
(2°)

N,N-Dimethylaniline
(3°)

There are exceptions to the aniline naming system. Sometimes, for example, the benzene rings are named as phenyl substituents attached to the amine nitrogen. Two examples are diphenylamine and triphenylamine:

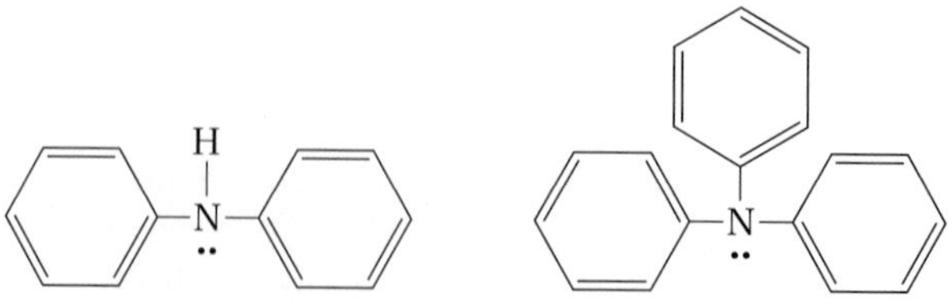

Diphenylamine
(2°)

Triphenylamine
(3°)

Anilines with a single methyl group on the benzene ring are sometimes called *toluidines*. The relative positions of the amino and methyl groups on the aniline rings of the toluidines are indicated by the prefixes *ortho-* (*o-*), *meta-* (*m-*), and *para-* (*p-*). For example:

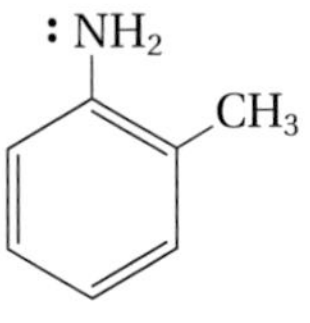

o-Toluidine 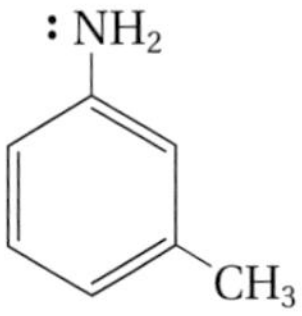   m-Toluidine    p-Toluidine

---

EXAMPLE 15.2   **Naming an aromatic amine**

Name and classify this aromatic amine:

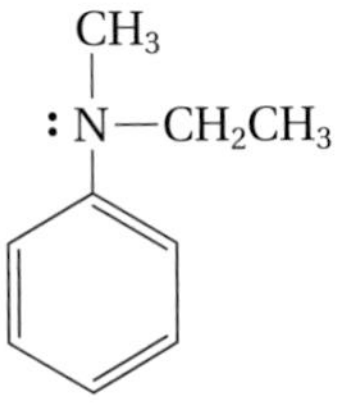

$$CH_3$$
$$: N—CH_2CH_3$$

SOLUTION

This compound has an ethyl group and a methyl group on the amine nitrogen of aniline. It is a tertiary amine named N-*ethyl*-N-*methylaniline*.

PRACTICE EXERCISE 15.2

Name and classify these aromatic amines:

(a) $NHCH_2CH_3$ (b) $N(CH_2CH_3)_2$

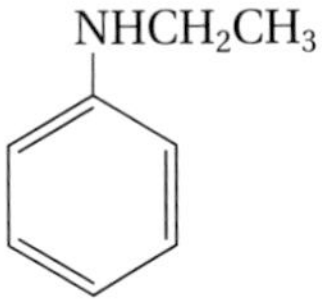

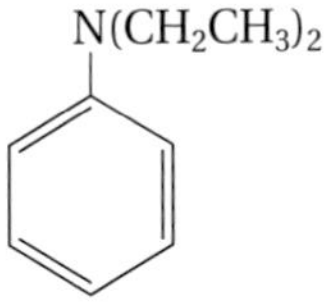

EXAMPLE 15.3   **Drawing an amine structure when the name is given**

Write the structure for each of the following:
(a) triethylamine   (b) *N*-ethylaniline   (c) *p*-chloroaniline

SOLUTION

(a) Substitute an ethyl group for each of the three ammonia hydrogens:

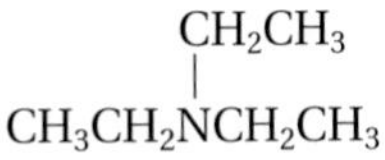

$$CH_2CH_3$$
$$CH_3CH_2NCH_2CH_3$$

(b) Substitute an ethyl group for one of the hydrogens of the amino group of aniline:

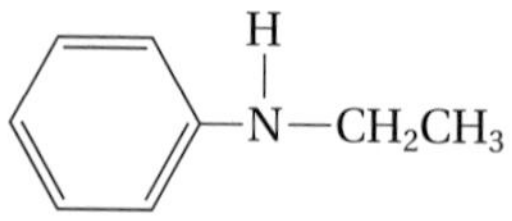

(c) Substitute a chlorine atom for a hydrogen on the phenyl ring of aniline in the *para-* position:

NH$_2$

Cl

Heterocyclic amines

In a *heterocyclic amine*, the amine nitrogen is incorporated into a ring. Many aliphatic heterocyclic amines exist. Among the most commonly encountered are pyrrolidine and piperidine.

Pyrrolidine (2°)	Piperidine (2°)

A number of the water-soluble B-complex vitamins contain heterocyclic amine rings. Deficiencies in either thiamine (vitamin B$_1$), niacin, or pyridoxine (vitamin B$_6$) can lead to nervous system damage.

Pyridine, a liquid amine with a nauseating odor, is an example of an aromatic heterocyclic amine; it is a tertiary amine. Several heterocyclic aromatic amines are present in natural products. The ring system of imidazole occurs in the amino acid histidine. The indole ring system is found in the amino acid tryptophan and in many alkaloids, a group of physiologically active amines obtained from plants.

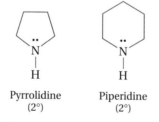

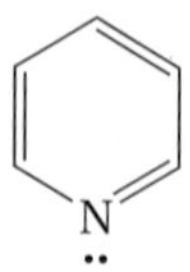

Pyridine (3°) 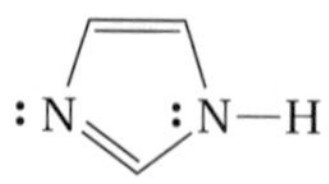   Imidazole 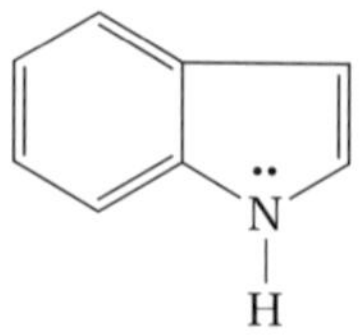   Indole

Other heterocyclic aromatic ring systems in nature are those of purine, pyrimidine, and pyrrole. Derivatives of purine and pyrimidine are impor-

Chlorophyll is the green pigment found in the chloroplast of plants. The structure of the pyrrole ring system in chlorophyll is very similar to that found in hemoglobin. In photosynthesis, chlorophyll absorbs light energy, which is eventually stored as chemical potential energy in carbohydrate and other nutrient molecules.

tant components of the nucleic acids, the molecules of heredity. Caffeine, the stimulant in tea, coffee, cocoa, and many soft drinks, is a purine derivative. The hemoglobin of blood is an example of a naturally occurring molecule that contains the pyrrole ring system.

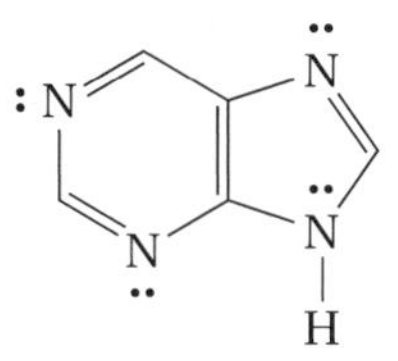

Purine 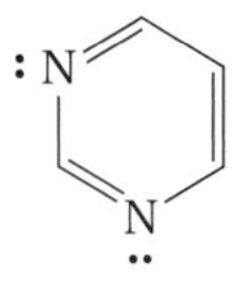   Pyrimidine 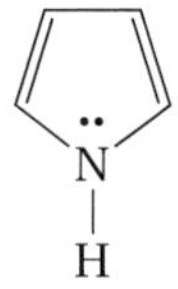   Pyrrole 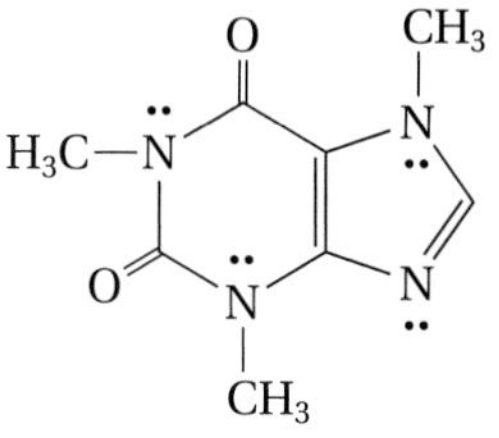   Caffeine

PRACTICE EXERCISE 15.4

Which structure represents an aromatic heterocyclic amine?

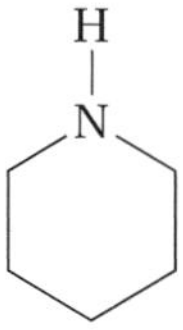

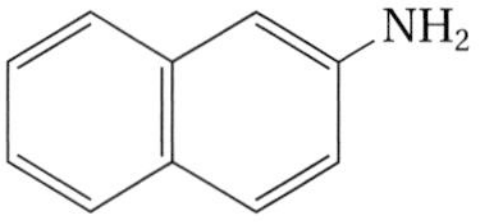

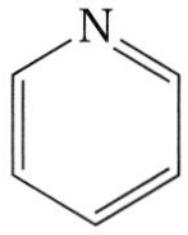

Aliphatic diamines

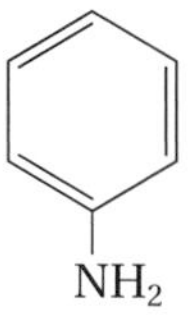

Aliphatic amines that contain two amino groups in the same molecule are called *aliphatic diamines*. Two aliphatic diamines produced by decaying flesh are putrescine and cadaverine. These amines have characteristic odors. You may judge from their names what the odors are like.

$$H_2N-CH_2CH_2CH_2CH_2-NH_2 \qquad H_2N-CH_2CH_2CH_2CH_2CH_2-NH_2$$

Putrescine   Cadaverine

15.2 Amine basicity

AIM: To show, with equations, how amines act as weak bases.

As we discussed in Chapter 9 for the weak base ammonia, the unshared electron pair on the nitrogen of primary, secondary, or tertiary amines can accept a hydrogen ion, or proton, from acids. Amines are weak bases of base strength similar to that of ammonia. The general reaction for the protonation of an amine is

Free amine   Proton   Ammonium ion
(unprotonated amine)   (protonated amine)

An amine that has accepted a hydrogen ion and acquired a positive charge to

become a cation is called a **protonated amine.** *The cations produced by protonation of aliphatic and aromatic amines are called* **alkylammonium ions** *and* **arylammonium ions,** *respectively. An electrically neutral, unprotonated amine is called a* **free amine.**

Amines are bases, so they can react with acids to form salts. Since amines are derivatives of ammonia, *the salts formed from the reactions of amines and acids are called* **ammonium salts.** Here are three specific examples of the formation of ammonium salts from amines and hydrochloric acid:

$$CH_3{-}\overset{\overset{\displaystyle H}{|}}{\underset{\underset{\displaystyle H}{|}}{N}}{:} \quad + \quad HCl \quad \rightleftharpoons \quad CH_3{-}\overset{\overset{\displaystyle H}{|}}{\underset{\underset{\displaystyle H}{|}}{\overset{+}{N}}}{-}H \quad + \quad Cl^-$$

| Methylamine (free amine) | Hydrochloric acid | | Methylammonium ion | Chloride ion |

Methylammonium chloride

$$(CH_3CH_2)_2\overset{\overset{\displaystyle H}{|}}{N}{:} \; + \; HCl \; \rightleftharpoons \; (CH_3CH_2)_2\overset{\overset{\displaystyle H}{|}}{\overset{+}{N}}{-}H \quad Cl^-$$

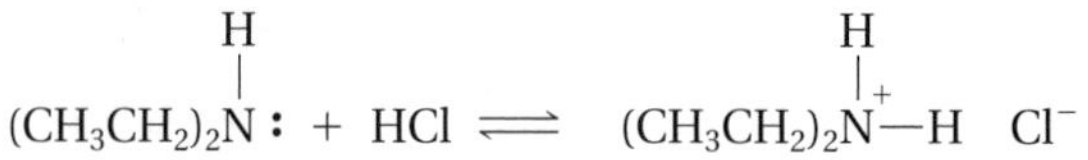

Diethylamine   Diethylammonium chloride

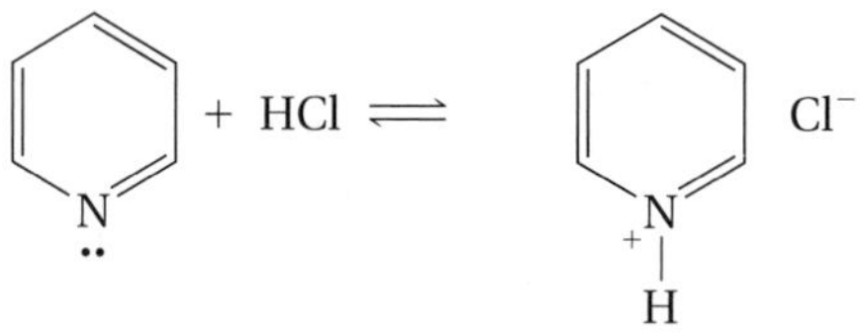

Pyridine   Pyridinium hydrochloride

The methylammonium chloride formed in the reaction of methylamine and hydrochloric acid is the salt of a weak base and a strong acid just as ammonium chloride is. The methylammonium ion undergoes slight dissociation in water in the same way as we discussed for ammonium chloride in Section 9.9:

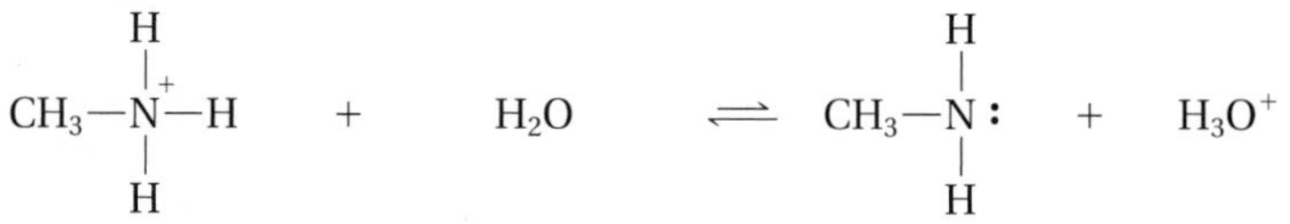

$$CH_3{-}\overset{\overset{\displaystyle H}{|}}{\underset{\underset{\displaystyle H}{|}}{\overset{+}{N}}}{-}H \quad + \quad H_2O \quad \rightleftharpoons \quad CH_3{-}\overset{\overset{\displaystyle H}{|}}{\underset{\underset{\displaystyle H}{|}}{N}}{:} \quad + \quad H_3O^+$$

| Methylammonium ion (proton donor) | Water (proton acceptor) | Methylamine (free base) | Hydronium ion (makes the solution acidic) |

The ammonium salts formed between amines and strong acids make slightly acidic aqueous solutions. Because of their ability to accept a proton and retain it most of the time, amines can act as buffers, or proton sponges, to protect against drastic changes in pH (see Sec. 9.10). Many naturally occurring amines serve as pH buffers in the fluids of living organisms, such as blood.

Like alcohols, the solubilities of free amines in water depend on their molecular structure. Amines with large hydrocarbon groups on the amine nitrogen tend to be insoluble. Those with small hydrocarbon substituents, such as methylamine, are very soluble in water. Protonated amines are almost always soluble in water because they are ionic. Many medicinal drugs that contain the amino group are supplied as the hydrochloride salts. Because of its greater solubility in water, a hydrochloride salt is usually easier to administer and more readily absorbed into the bloodstream than the free amine.

PRACTICE EXERCISE 15.5

Complete the following reactions, and name the products.

(a) $(CH_3)_2NH + HCl$

(b) $CH_3NH_2 + HNO_3$

(c)

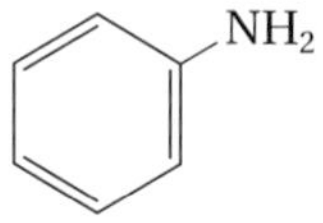

+ HCl

(d) $CH_3CH_2NH_2 + H_3PO_4$

15.3 Preparation of amines

AIMS: **To write an equation for the preparation of an amine from a halocarbon and ammonia or an amine nitrogen. To name and write the structure for a quaternary ammonium salt.**

Focus

Simple aliphatic amines can be prepared by displacement reactions.

There are several ways to prepare amines. One common way to prepare aliphatic amines is to displace the halogen in a halocarbon with ammonia or an amine nitrogen. This reaction produces an ammonium salt.

$$H-\overset{\overset{\displaystyle H}{|}}{\underset{\underset{\displaystyle H}{|}}{N}}: \quad + \quad R-I \quad \longrightarrow \quad R-\overset{+}{N}H_3 \ I^-$$

| Ammonia | Alkyliodide (other halides may be used) | Alkylammonium iodide |

The solution containing the ammonium salt can be treated with a base such as sodium hydroxide to produce the free amine.

$$R-\overset{+}{N}H_3 \ I^- + \quad NaOH \quad \longrightarrow \ R-NH_2 + \quad NaI \quad + \ H_2O$$

| Ammonium salt | Sodium hydroxide | Free amine | Sodium iodide | Water |

It may be difficult to control the number of alkyl groups attaching to the amine nitrogen. Treatment of ammonia with an alkyl halide usually gives a mixture of primary, secondary, and tertiary amines.

$$NH_3 + R-X \quad \longrightarrow R-NH_2 \qquad \text{Primary (1°) amine}$$

$$R-NH_2 + R-X \longrightarrow R-\overset{\overset{\displaystyle R}{|}}{NH} \qquad \text{Secondary (2°) amine}$$

$$R-\overset{\overset{\displaystyle R}{|}}{N}-H + R-X \longrightarrow R-\overset{\overset{\displaystyle R}{|}}{N}-R \qquad \text{Tertiary (3°) amine}$$

Because of its basicity, the nitrogen of a tertiary amine can displace a halogen from a fourth alkyl halide molecule. A nitrogen with *four* R groups attached has a positive charge and is called a *quaternary ammonium ion.* Choline, an important component of some constituents of cell membranes (see Sec. 17.3), is a quaternary ammonium ion.

$$CH_3-\overset{\overset{\displaystyle CH_3}{|}}{\underset{\underset{\displaystyle CH_3}{|}}{N^+}}-CH_2CH_2OH$$

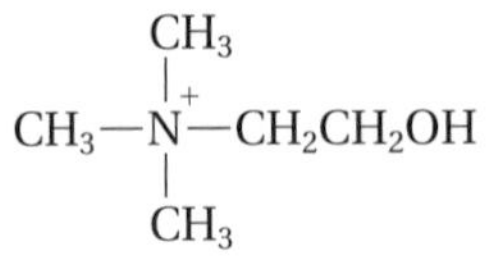

Choline

Quaternary
quator (Latin):   four times

Quaternary ammonium salts having one long alkyl group function as soaps. One quaternary ammonium salt, cetyltrimethyl-ammonium chloride, is found in mouthwashes and is used as a germicide for sterilizing medical instruments.

Salts containing a quaternary ammonium ion are called **quaternary ammonium salts.** The formation of a quaternary ammonium salt can be illustrated by the reaction of trimethylamine with bromomethane.

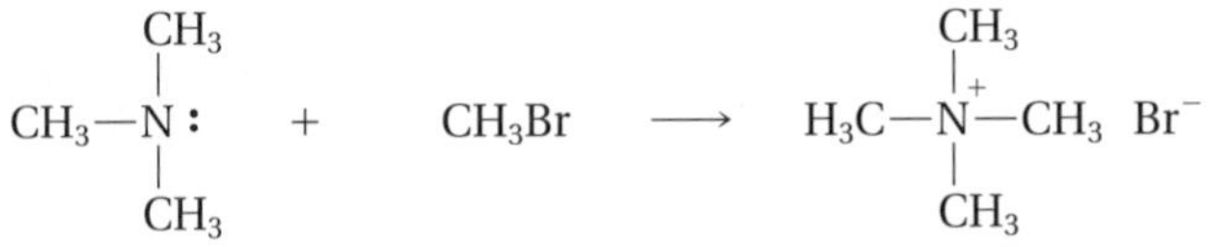

$$CH_3-\overset{\overset{\displaystyle CH_3}{|}}{\underset{\underset{\displaystyle CH_3}{|}}{N}}\!:\quad +\quad CH_3Br\quad \longrightarrow\quad H_3C-\overset{\overset{\displaystyle CH_3}{|}}{\underset{\underset{\displaystyle CH_3}{|}}{N^+}}-CH_3\ \ Br^-$$

| Trimethylamine | Bromomethane (methyl bromide) | Tetramethyl ammonium bromide (a quaternary ammonium salt) |

PRACTICE EXERCISE 15.6

An aqueous solution of ammonia is heated with 1-chlorobutane. (a) What is the organic product? (b) Give the structure and name of the free amine that is liberated when the product in part (a) is treated with sodium hydroxide.

15.4 Amides

AIM: To name and write structures of simple amides.

Focus

Amides are ammonia and amine derivatives of carboxylic acids.

Amides *are ammonia or amine derivatives of organic acids.* Amides may be simple, monosubstituted, or disubstituted. For example:

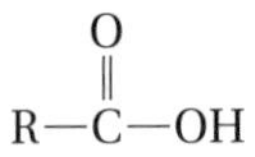

$$R-\overset{\overset{\displaystyle O}{\|}}{C}-OH \qquad \text{Carboxylic acid}$$

$$R-\overset{\overset{\displaystyle O}{\|}}{C}-\overset{\displaystyle}{\underset{\displaystyle \cdot\cdot}{N}}H_2 \qquad \text{Simple amide}$$

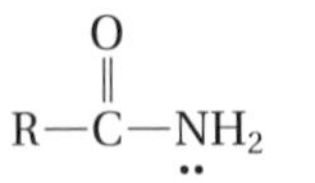

$$R-\overset{\overset{\displaystyle O}{\|}}{C}-\overset{\overset{\displaystyle H}{|}}{\underset{\displaystyle \cdot\cdot}{N}}-R \qquad \text{Monosubstituted amide}$$

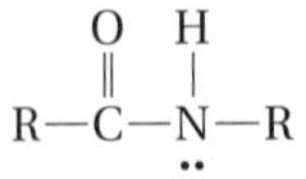

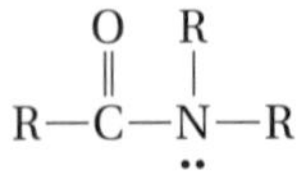

$$R-\overset{\overset{\displaystyle O}{\|}}{C}-\overset{\overset{\displaystyle R}{|}}{\underset{\displaystyle \cdot\cdot}{N}}-R \qquad \text{Disubstituted amide}$$

Although the amides of carboxylic acids are the focus of this chapter, the acid part of the amide need not be a carboxylic acid. The amides of

A Closer Look

The Sulfonamide Antibiotics

Certain bacterial infections, especially bladder infections, are often treated with a class of medications called *sulfa drugs.* All of the sulfa drugs are amides of sulfonic acids (RSO_3H). Benzenesulfonamide is an example of an amide of a sulfonic acid; it is the amide of benzenesulfonic acid. The figure above shows the structures of other sulfa drugs used as antibiotics.

Sulfa drugs are effective antibiotics against bacteria that need folic acid to sustain life. These bacteria make folic acid by assembling it from molecular components that include *p*-aminobenzoic acid (see figure below, right).

Sulfanilamide, a compound similar in structure to *p*-aminobenzoic acid, sticks to the bacterial enzyme at a place on the molecule usually reserved for *p*-aminobenzoic acid. The bacterial enzyme cannot use the sulfanilamide to make folic acid, so the bacteria die from folic acid deficiency. Unlike the bacteria, our body cells are relatively unaffected by sulfa drugs. We need folic acid (a B-complex vitamin) in order to stay alive and well, but our cells lack the enzyme needed to make the compound. We obtain our folic acid from the food we eat and from beneficial bacteria that live in our intestines.

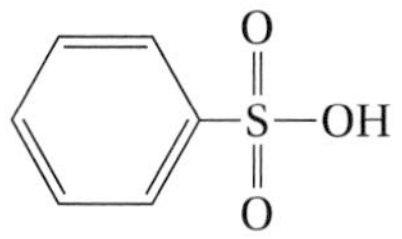

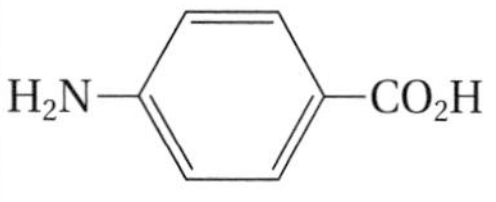

Benzenesulfonic acid   Benzenesulfonamide   Sulfanilamide

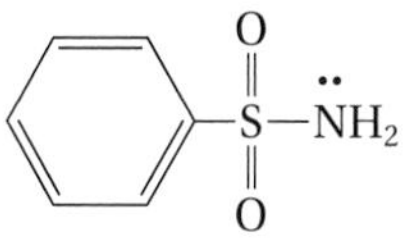

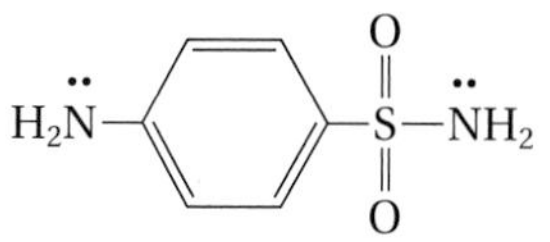

The structures of some sulfa antibiotics.

p-Aminobenzoic acid 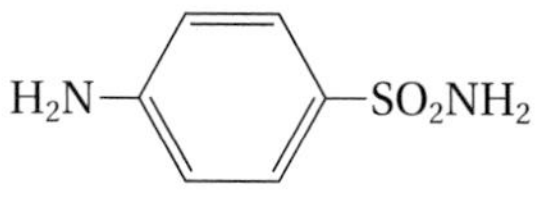   Sulfanilamide

some sulfonic acids are important antibiotics, as discussed above in A Closer Look: The Sulfonamide Antibiotics.

Names of amides

Amides are named by dropping the *-ic* or *-oic* ending from the name of the parent acid and adding the ending *-amide.* Any substituents on the amine nitrogen are named as prefixes preceded by *N-* or *N,N-.* Here are some examples:

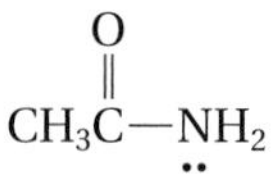

Acetamide 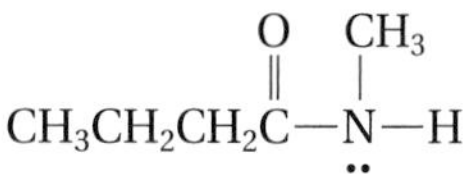   *N*-Methylbutyramide 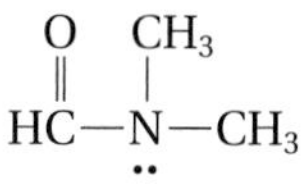   *N,N*-Dimethylformamide

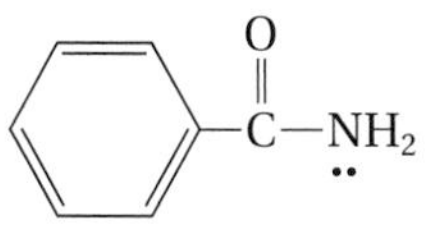

Benzamide

Amides formed from carboxylic acids and aniline are called **anilides.** Acetanilide, formed from acetic acid and aniline, is a simple example. Phenacetin, an analgesic and antipyretic, is also an anilide. APC tablets contain a mixture of aspirin, phenacetin, and caffeine.

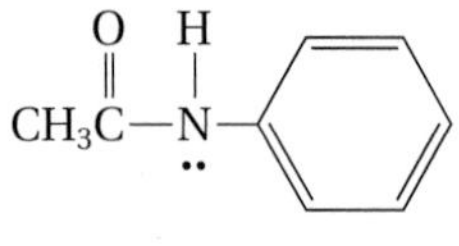

Acetanilide 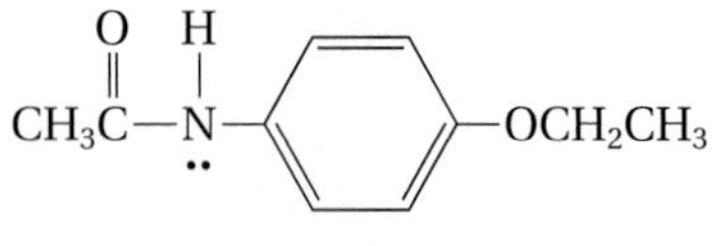   Phenacetin

Some amides are used as local anesthetics. See A Closer Look: Amide Local Anesthetics.

EXAMPLE 15.4	**Naming amides**

Name these amides: (a) $CH_3CH_2CH_2\overset{\displaystyle O}{\overset{\displaystyle \|}{C}}NH_2$ (b) $H\overset{\displaystyle O}{\overset{\displaystyle \|}{C}}NHCH_3$

SOLUTION

(a) This amide is derived from butyric acid:

$CH_3CH_2CH_2\overset{\displaystyle O}{\overset{\displaystyle \|}{C}}OH$

It is butyramide.

(b) This amide is derived from formic acid:

$H\overset{\displaystyle O}{\overset{\displaystyle \|}{C}}OH$

It is a formamide. There is also a methyl group attached to the amide N. The name is N-*methylformamide.*

PRACTICE EXERCISE 15.7

Name the following amides.

(a) $CH_3CH_2\overset{\displaystyle O}{\overset{\displaystyle \|}{C}}NH_2$ (b) $CH_3\overset{\displaystyle CH_3}{\overset{\displaystyle |}{C}}HCH_2\overset{\displaystyle O}{\overset{\displaystyle \|}{C}}NH_2$ (c) $CH_3CH_2\overset{\displaystyle O}{\overset{\displaystyle \|}{C}}NHCH_3$

Chemistry of the amide functional group

The electronegativity of oxygen in the amide bond pulls the unshared pair of electrons in amide nitrogen toward the oxygen. Since these electrons are unavailable to accept a proton, *an amide nitrogen is much less basic*

A Closer Look

Amide Local Anesthetics

Several important local anesthetics contain amide functional groups. Their structures are essentially the same as those of the ester local anesthetics (A Closer Look: Ester Local Anesthetics, page 442) except that nitrogen replaces the ester oxygen. These compounds are less readily metabolized than the esters and tend to accumulate in the plasma.

Lidocaine hydrochloride (Xylocaine) (see figure) has received widespread acceptance since its introduction more than 35 years ago. The major advantages are that it produces rapid anesthesia and is nonirritating and hypoallergenic. Topically it is not as effective as cocaine (A Closer Look: Ester Local Anesthetics, pages 442–443) but is better than procaine. Mepivacaine hydrochloride (Carbocaine) is also an amide. It takes effect faster than lidocaine, and the duration of the anesthesia is somewhat longer. Adverse reactions are few, and tissue irritation is minimal. Bupivacaine hydrochloride (Marcaine) has the same structure as mepivacaine except that a methyl group is replaced by a butyl group. It is more potent and has a longer-lasting action than lidocaine or mepivacaine. This drug has been used occasionally to produce spinal anesthesia.

The most recently introduced local anesthetic with an amide structure is etidocaine (Duranest). It is structurally similar to lidocaine but has greater potency and is longer lasting. Its general toxicologic and pharmacologic actions differ little from the other anesthetics in the group.

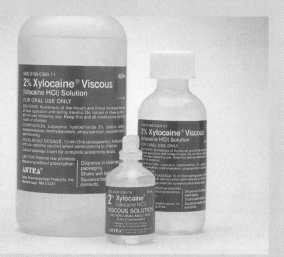

Xylocaine Viscous, a fast-acting, local anesthetic, is sometimes prescribed as a gargle for a severe sore throat.

Lidocaine (hydrochloride salt)
(Xylocaine)

Bupivacaine (hydrochloride salt)
(Marcaine)

Mepivacaine (hydrochloride salt)
(Carbocaine)

Etidocaine (hydrochloride salt)
(Duranest)

than an amine nitrogen. It does not usually accept a proton in an acidic solution. However, amides form hydrogen bonds with other amides and with water.

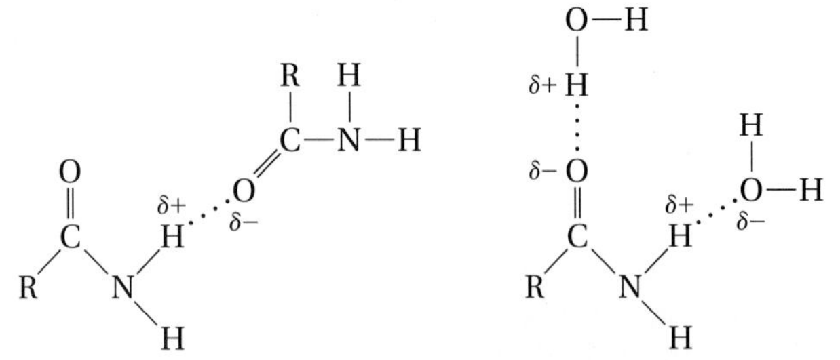

Hydrogen bonding
in amides

Hydrogen bonding
with water

Such hydrogen bonding is very important in holding protein structures in unique shapes (see Chap. 18).

15.5 Preparation of amides

AIMS: To write equations for the preparation of amides from ammonium salts and carboxylic acid derivatives. To describe the preparation and use of some common polyamides.

Focus

Amides can be prepared from amines and carboxylic acids or their derivatives.

There are many ways to prepare amides. One method is the dehydration of ammonium salts of carboxylic acids; another is the reaction of ammonia or an amine with either an ester or a carboxylic acid anhydride.

Dehydration of ammonium salts

If we mix a carboxylic acid with an amine, we will get an ammonium salt, since the acid is a proton donor and the amine is a proton acceptor.

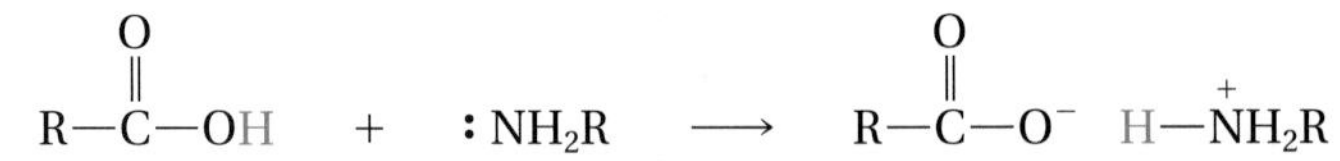

$$R-\underset{\underset{O}{\|}}{C}-OH \ + \ :NH_2R \ \longrightarrow \ R-\underset{\underset{O}{\|}}{C}-O^- \ H-\overset{+}{N}H_2R$$

Carboxylic acid   Amine   Ammonium salt
(proton donor)   (proton acceptor)

We can remove a molecule of water from the dry ammonium salt formed from an amine and a carboxylic acid by heating it. This is a dehydration reaction, and the organic product is an amide.

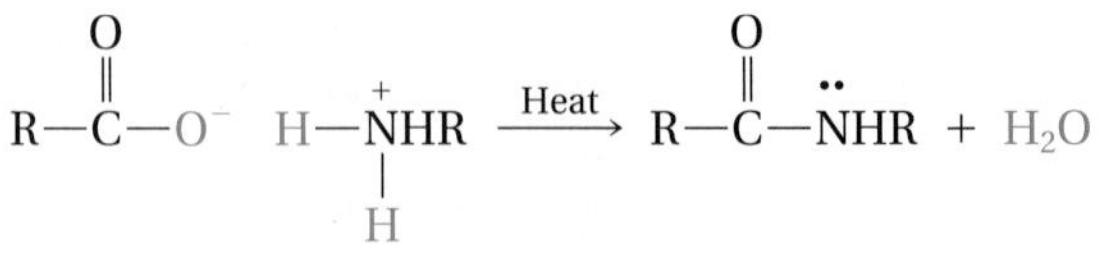

$$R-\underset{\underset{O}{\|}}{C}-O^- \ H-\overset{+}{\underset{\underset{H}{|}}{N}}HR \ \xrightarrow{\text{Heat}} \ R-\underset{\underset{O}{\|}}{C}-\overset{\cdot\cdot}{N}HR \ + \ H_2O$$

Ammonium salt   Amide

Here are two examples of the preparation of amides by dehydration of ammonium salts of carboxylic acids:

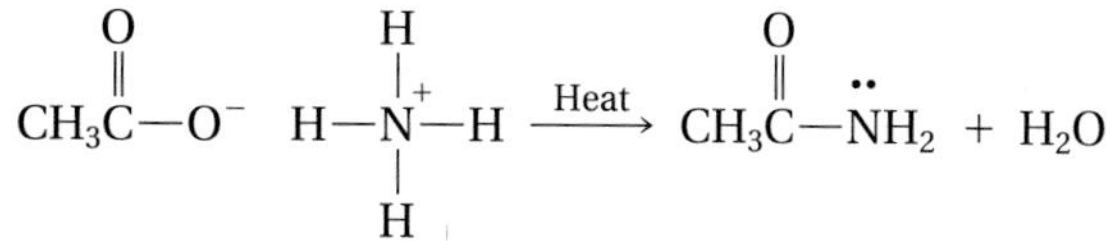

$$\underset{\text{Ammonium acetate}}{CH_3\overset{\displaystyle O}{\overset{\|}{C}}-O^-\quad H-\overset{\displaystyle H}{\underset{\displaystyle H}{\overset{|}{\underset{|}{\overset{+}{N}}}}}-H} \xrightarrow{\text{Heat}} \underset{\text{Acetamide}}{CH_3\overset{\displaystyle O}{\overset{\|}{C}}-\overset{\displaystyle ..}{N}H_2} + H_2O$$

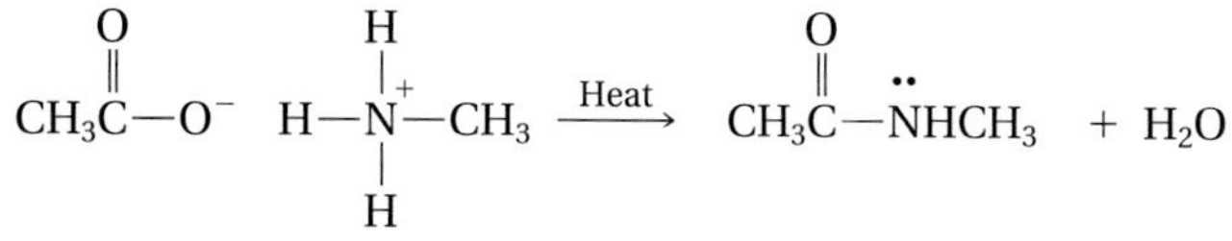

$$\underset{\text{\textit{N}-Methylammonium acetate}}{CH_3\overset{\displaystyle O}{\overset{\|}{C}}-O^-\quad H-\overset{\displaystyle H}{\underset{\displaystyle H}{\overset{|}{\underset{|}{\overset{+}{N}}}}}-CH_3} \xrightarrow{\text{Heat}} \underset{\text{\textit{N}-Methylacetamide}}{CH_3\overset{\displaystyle O}{\overset{\|}{C}}-\overset{\displaystyle ..}{N}HCH_3} + H_2O$$

PRACTICE EXERCISE 15.8

The salt of propanoic acid with ethylamine is heated. Give the structure and name of the compound that is formed.

Reaction with carboxylic acid derivatives

Amides also can be prepared by the reaction of ammonia or amines with derivatives of carboxylic acids. Esters, especially methyl esters, and acid anhydrides are often the carboxylic acid derivatives.

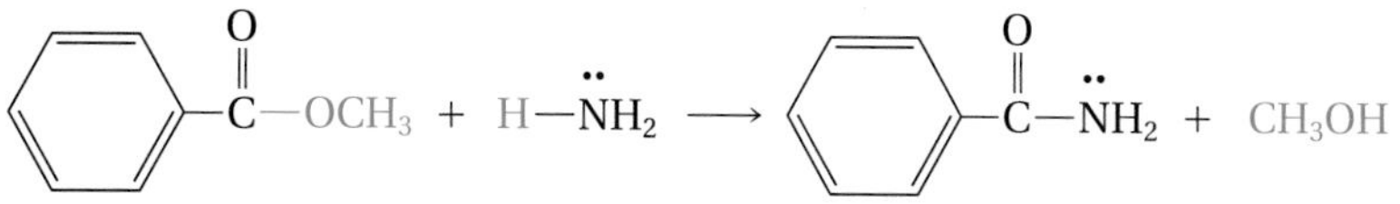

$$\underset{\text{Methyl benzoate}}{C_6H_5-\overset{\displaystyle O}{\overset{\|}{C}}-OCH_3} + \underset{\text{Ammonia}}{H-\overset{\displaystyle ..}{N}H_2} \longrightarrow \underset{\text{Benzamide}}{C_6H_5-\overset{\displaystyle O}{\overset{\|}{C}}-\overset{\displaystyle ..}{N}H_2} + \underset{\text{Methanol}}{CH_3OH}$$

With an ester as the starting material, an alcohol is formed as a by-product of the reaction. With an anhydride, a carboxylic acid is formed as a by-product.

EXAMPLE 15.5

Preparing an amide

Write an equation for the preparation of acetamide from ammonia and an acid anhydride.

SOLUTION

The product, acetamide, has two carbons. We let acetic anhydride react with ammonia. The other product is acetic acid.

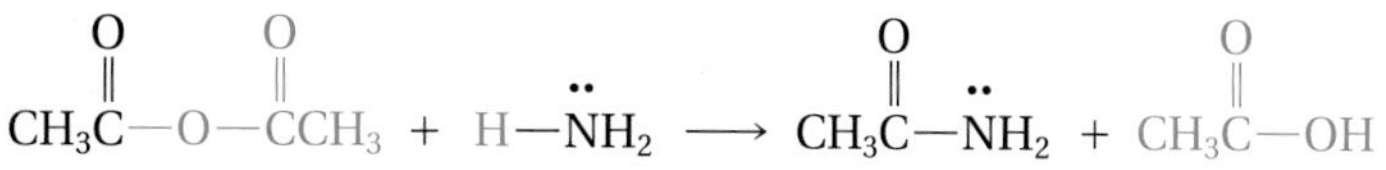

$$\underset{\text{Acetic anhydride}}{CH_3\overset{\displaystyle O}{\overset{\|}{C}}-O-\overset{\displaystyle O}{\overset{\|}{C}}CH_3} + \underset{\text{Ammonia}}{H-\overset{\displaystyle ..}{N}H_2} \longrightarrow \underset{\text{Acetamide}}{CH_3\overset{\displaystyle O}{\overset{\|}{C}}-\overset{\displaystyle ..}{N}H_2} + \underset{\text{Acetic acid}}{CH_3\overset{\displaystyle O}{\overset{\|}{C}}-OH}$$

PRACTICE EXERCISE 15.9

Predict the products of the reaction of ethyl acetate with methylamine.

Polyamides

Polyamides *are polymers in which the monomer units are linked by amide bonds.* Various types of nylon, familiar materials to nearly everyone, are polyamides. Of these, nylon 66 was the first polyamide to find wide commercial application. Nylon 66 gets its name from its repeating units of adipic acid, a six-carbon dicarboxylic acid, and 1,6-diaminohexane, a six-carbon diamine. The polymer is prepared by heating the dicarboxylic acid and the diamine; amide bonds are formed as water is removed.

$$x\text{HO}-\overset{\overset{\text{O}}{\|}}{\text{C}}\text{+CH}_2\text{+}_4\overset{\overset{\text{O}}{\|}}{\text{C}}-\text{OH} + x\text{H}_2\text{N}-\text{CH}_2\text{+CH}_2\text{+}_4\text{CH}_2-\text{NH}_2 \longrightarrow$$

Adipic acid   1,6-Diaminohexane

$$\left(\overset{\overset{\text{O}}{\|}}{-\text{C}}\text{+CH}_2\text{+}_4\overset{\overset{\text{O}}{\|}}{\text{C}}-\overset{\overset{\text{H}}{|}}{\text{N}}-\text{CH}_2\text{+CH}_2\text{+}_4\text{CH}_2-\overset{\overset{\text{H}}{|}}{\text{N}}\right)_x + x\text{H}_2\text{O}$$

Representative unit of nylon 66

The finished polymer has a formula weight of about 10,000 and a melting point of 250 °C. The melted polymer can be spun into very fine, strong fibers. Because of this property, nylon 66 was rapidly substituted for the scarcer and more expensive natural silk in women's sheer hosiery.

Today, nylon 66 has been largely replaced by nylon 6, which has similar properties but is even cheaper to produce. The representative polymer unit of nylon 6 is derived from 6-aminohexanoic acid; the long polymer chain is formed by the successive attachment of the carboxyl group of one molecule of the acid to the amino group of the next by formation of an amide bond.

$$x\text{H}_2\text{N}-\text{CH}_2\text{+CH}_2\text{+}_4\overset{\overset{\text{O}}{\|}}{\text{C}}-\text{OH} \xrightarrow{\text{Heat}} \left(\text{CH}_2(\text{CH}_2\text{+}_4\overset{\overset{\text{O}}{\|}}{\text{C}}-\overset{\overset{\text{H}}{|}}{\text{N}}\right)_x + x\text{H}_2\text{O}$$

6-Aminohexanoic acid   Representative polymer unit of nylon 6

In medicine, nylon was used to make the first synthetic sutures, and it is still used for sutures today. It is also used to make specialized tubing for medical applications (Fig. 15.1). Nylon is spun into fibers for making carpets, tire cord, and textiles and molded into solid objects such as gears, bearings, and zippers.

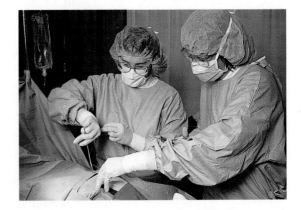

Figure 15.1
Nylon's many uses include surgical applications as well as many other items used by medical personnel.

Aramides, polyamides that contain aromatic rings, have been found to have special properties—bullet resistance and flame resistance. The aromatic rings make the resulting fiber stiffer and tougher. Nomex is a polyamide whose carbon skeleton consists of aromatic rings derived from isophthalic acid and *m*-phenylenediamine.

Isophthalic acid   *m*-Phenylenediamine

Representative unit of Nomex

Figure 15.2
Firefighters and racing car drivers wear fire-resistant clothing made of Nomex

Like nylon, Nomex is a poor conductor of electricity, and since it is more rigid than nylon, it is used to make parts for electrical fixtures. Nomex is also used in the manufacture of flame-resistant clothing for racing car drivers and firefighters (Fig. 15.2). It is also used in the fabrication of flame-resistant building materials. Kevlar has a structure that is similar to that of Nomex. It is a polyamide made from terephthalic acid and *p*-phenylene-diamine.

Terephthalic acid   *p*-Phenylenediamine

Representative unit of Kevlar

Kevlar is used as a replacement for steel in bullet-resistant vests. Since the vests are more flexible and much lighter than their metal counterparts, they can be worn under normal clothing.

Proteins, which are polyamides of naturally occurring amino acids, rank among the most important of all biological molecules. Their structures will be discussed in detail in Chapter 18.

15.6 Amide hydrolysis

AIM: To predict the products of hydrolysis of an amide.

Compared with esters, amides hydrolyze slowly in water to acid and amine.

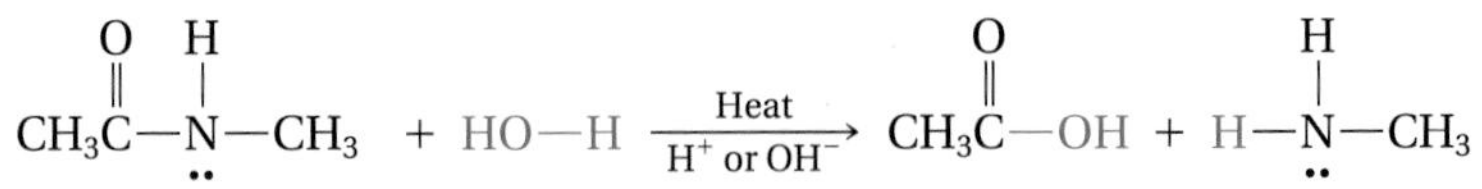

$$\underset{\text{N-Methylacetamide}}{CH_3\overset{\displaystyle O}{\overset{\|}{C}}-\overset{\displaystyle H}{\underset{\displaystyle \cdot\cdot}{N}}-CH_3} + \underset{\text{Water}}{HO-H} \xrightarrow[\text{H}^+ \text{ or OH}^-]{\text{Heat}} \underset{\text{Acetic acid}}{CH_3\overset{\displaystyle O}{\overset{\|}{C}}-OH} + \underset{\text{Methylamine}}{H-\overset{\displaystyle H}{\underset{\displaystyle \cdot\cdot}{N}}-CH_3}$$

The hydrolysis can be speeded up by heating the amide in a strongly acidic or basic solution. The hydrolysis of amide bonds is a very important reaction in the digestion of proteins; the hydrolysis is aided by enzyme catalysts.

EXAMPLE 15.6

Predicting the products of amide hydrolysis

What are the products when *N*-methylpropionamide is hydrolyzed?

SOLUTION

The structure of *N*-methylpropionamide is

$$CH_3CH_2\overset{\displaystyle O}{\overset{\|}{C}}NHCH_3$$

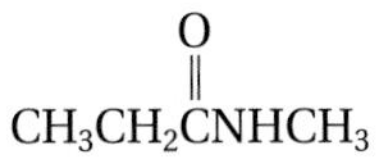

When hydrolyzed with acid, it will give propionic acid and methylamine.

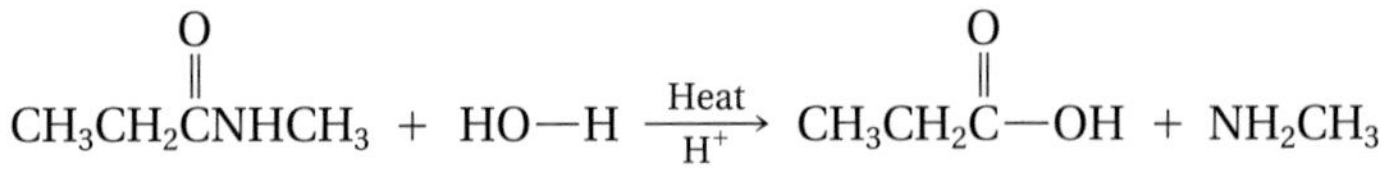

$$CH_3CH_2\overset{\displaystyle O}{\overset{\|}{C}}NHCH_3 + HO-H \xrightarrow[\text{H}^+]{\text{Heat}} CH_3CH_2\overset{\displaystyle O}{\overset{\|}{C}}-OH + NH_2CH_3$$

PRACTICE EXERCISE 15.10

What are the products when *N*-ethylbenzamide is hydrolyzed?

15.7 Important amines

AIM: To name and describe the amines used as neurotransmitters, hallucinogens, decongestants, and antihistamines.

In the central nervous system—the part of the nervous system consisting of the brain, the spinal cord, and the nerves that radiate from the cord—*the most important chemical messengers or **neurotransmitters** between nerve cells are three amines: norepinephrine (also called noradrenaline), dopamine, and serotonin.*

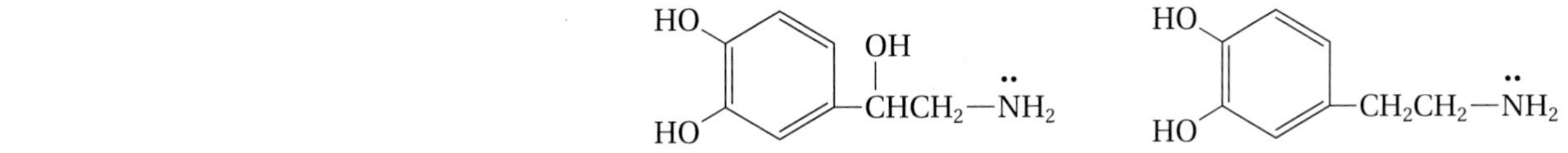

Norepinephrine
(noradrenaline)

Dopamine

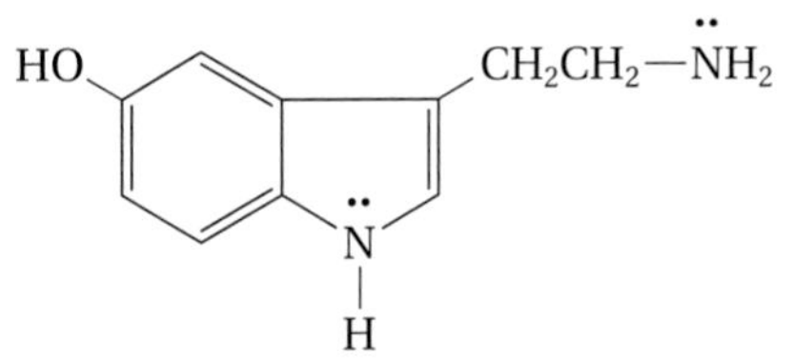

Serotonin

Norepinephrine and dopamine are called *catecholamines* because their molecular structures are derived from the dihydroxy phenol known as *catechol*. The catecholamines also have the carbon skeleton 2-phenylethylamine.

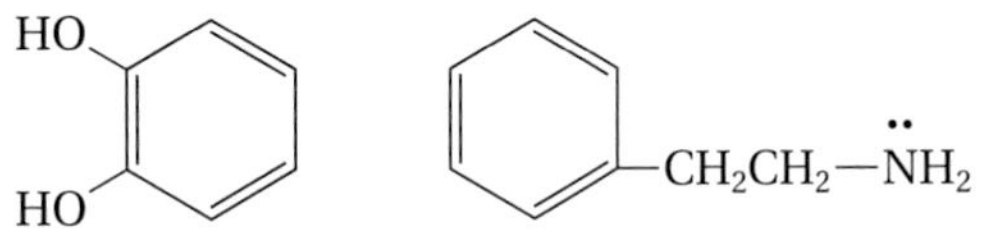

Catechol

2-Phenylethylamine

Dopamine deficiency results in Parkinson's disease, which can cause tremors of the head and extremities, usually in middle-aged and older people. Brain cells of Parkinson's patients have only 5% to 15% of the normal concentration of dopamine. Administration of dopamine does not stem the symptoms of the disease, however, because the amine cannot breach the blood-brain barrier—a natural filter that prevents certain molecules from reaching the brain through the circulation. (How the blood-brain barrier operates is still a puzzle.) Dopa, however, a related carboxylic acid, can pass through the blood-brain barrier. Inside the brain cells, enzymes catalyze the loss of carbon dioxide from the prodrug dopa to produce dopamine.

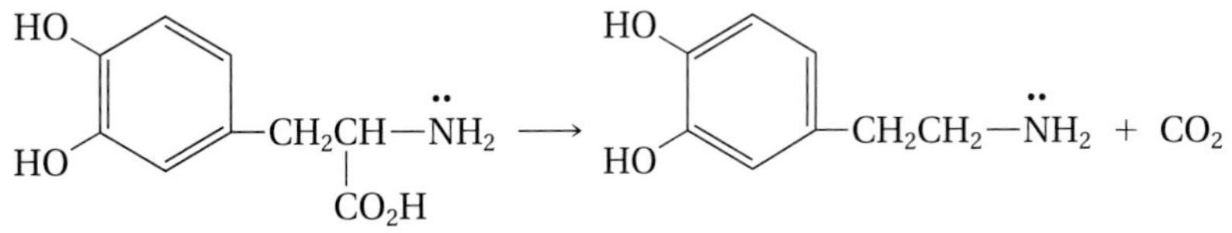

Dopa

Dopamine

The use of dopa to relieve the symptoms of parkinsonism has helped many people resume quite normal, active lives.

Serotonin deficiency has been implicated in mental depression, a disease that strikes millions of people each year. Indeed, depression is the world's most pervasive mental health problem. Our moods appear to rest fundamentally with nerve cells (neurons) in the brain. Signals between neurons are sent by neurotransmitters. Serotonin and norepinephrine are two examples of neurotransmitters that are amines. The neurotransmitters are produced and released by the neurons and then taken up and destroyed

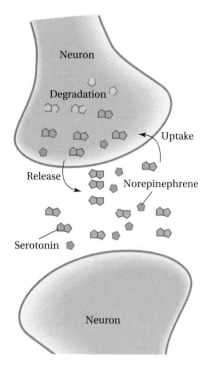

Figure 15.3
Depression can result when the concentration of the neurotransmitter serotonin between neurons drops to a low level.

by neurons. In a normal individual, the ends of the neurons are bathed by fluid containing the released neurotransmitters (Fig. 15.3). It appears that the concentration of the neurotransmitters between the neurons is important in controlling people's moods. Clinical depression is essentially the result of a chemical imbalance in which certain neurotransmitters drop below normal levels. Over time, three generations of drugs have been developed that help to maintain the levels of neurotransmitters. The earliest class of antidepression drugs to be discovered were the m*onoamine* o*xidase* i*nhibitors* (MAO inhibitors). MAO is an enzyme system responsible for the destruction of neurotransmitters; the MAO inhibitors block the destruction. The mood-elevating effect of iproniazid, the first of the MAO inhibitors, was accidentally discovered while it was being used to treat patients with tuberculosis. Physicians noticed that these patients were exhilarated by the drug. Subsequent studies showed that iproniazid keeps norepinephrine and serotonin at normal levels by inhibiting, or blocking, the degradation of these neurotransmitters by MAO. More than a dozen MAO inhibitors are now available.

Iproniazid
(Euphozid, Marsilid)

A second generation of antidepressants, the *tricyclics,* soon followed the MAO inhibitors. The tricyclics are so called because their chemical structures contain three carbon rings. Amitriptyline is one of several possible examples of tricyclic antidepressive drugs.

Phencyclidine (PCP) is also known by its street name, "angel dust." PCP is a hallucinogenic drug that contains a piperdine ring. It is a drug of abuse that causes confusion, paranoia, and delirium.

Phencyclidine
(PCP)

Amitriptyline (hydrochloride salt)
(Domenical, Elavil, Larozyl, Saroten)

The tricyclics' antidepressant action is not completely understood, but they are thought to block the uptake of norepinephrine and serotonin by neurons. Because studies indicated that low serotonin is more important than low norepinephrine in depression, scientists began a search for a third generation of drugs that would prevent the uptake of only serotonin. This search for such medications, called *zero-serotonin-uptake drugs,* has been quite successful. You may recall from the Case in Point to this chapter that a

FOLLOW-UP TO THE CASE IN POINT: Antidepressants

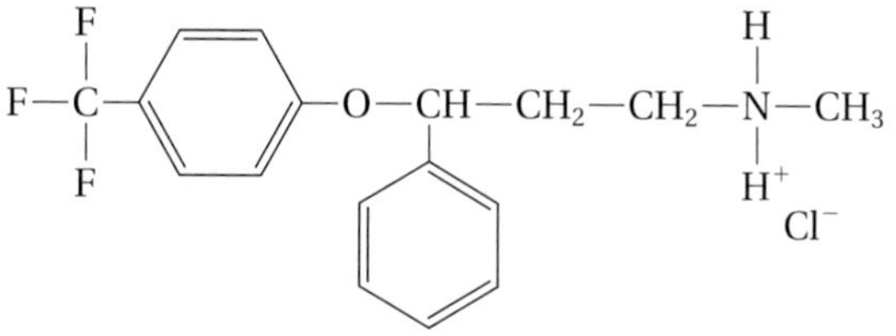

Maude is one of the more than 12 million people in the United States afflicted by depression. Children and adults alike are affected by this malady, which may include weight loss, insomnia, and general feelings of hopelessness or worthlessness. After a careful analysis of symptoms, Maude's psychiatrist thought that she might be suffering from a chemical imbalance in which the concentration of the neurotransmitter serotonin is depleted. For this reason, the psychiatrist suggested that Maude might benefit from a zero-serotonin-uptake drug.

Maude's psychiatrist prescribed the drug fluoxetine (Prozac), but other zero-serotonin-uptake drugs such as paroxetine (Paxil) and sertraline (Zoloft) are also available. These drugs are the most widely used antidepressant drugs on the market today.

Within a few weeks, Maude began to feel more like her old self. After a few months, her depression had disappeared. She is still attending group counseling sessions but is trying life without her antidepressant medication.

The study of depression is far from over. With the recent explosion of knowledge about the complex chemistry of the brain, the next few years hold the promise of new understanding and treatments for this widespread illness.

Fluoxetine (hydrochloride salt)
(Prozac)

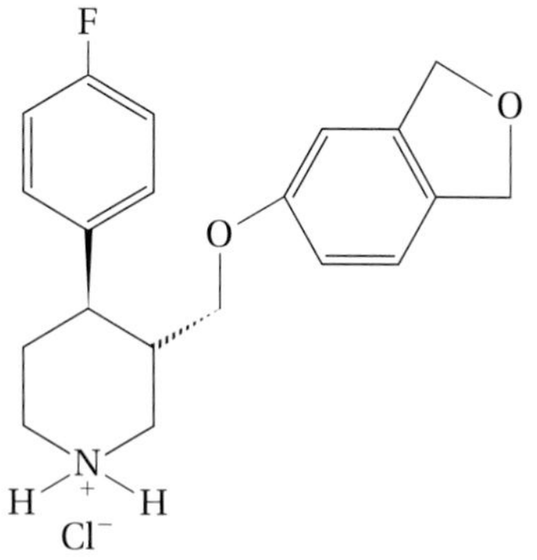

Paroxetine (hydrochloride salt)
(Paxil)

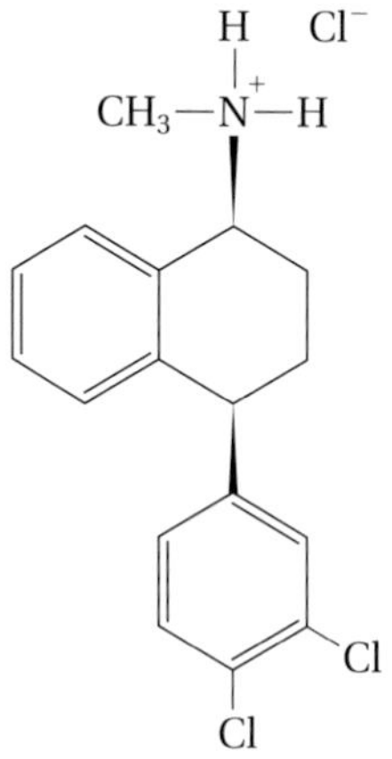

Sertraline (hydrochloride salt)
(Zoloft)

woman named Maude suffered from depression. Above, we learn how she was helped by the third generation of antidepressants.

Epinephrine, commonly called *adrenaline,* is another catecholamine. Its structure is the same as norepinephrine except that epinephrine has an *N*-methyl group.

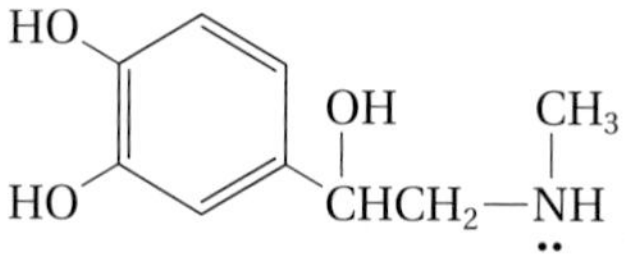

Epinephrine
(adrenaline)

Epinephrine appears to be more important as a hormone than as a neurotransmitter; its effects will be considered in more detail in Chapter 24. Since the catecholamines help determine what stimuli your brain receives, you might very well imagine that blocking these stimuli could approximate the effects of the hallucinogenic drugs and certain narcotics. **Hallucinogens** *are mind-altering drugs that produce visions and distorted views of reality.* Mescaline and DOM (also called STP) are two hallucinogenic drugs whose structures resemble those of the catecholamines.

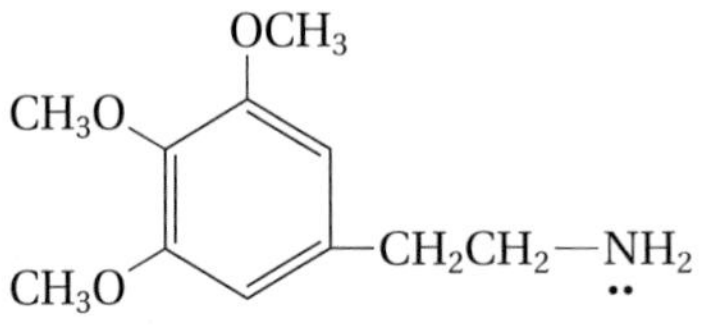

Mescaline
(hallucinogen)

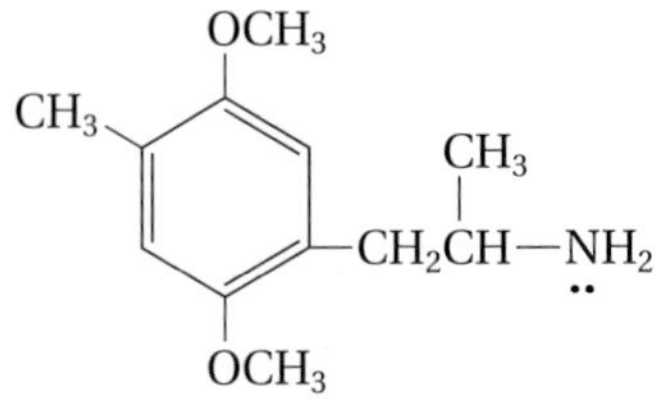

DOM, STP
(hallucinogen)

Mescaline, extracted from the resin pods of the peyote cactus, is used by members of the Native American Church of the Southwest in its rituals.

Scientists have developed several medications using compounds with structures similar to those of the catecholamines. Ephedrine is widely used in cough syrups, while neosynephrine is often an ingredient in nose drops. Both these compounds are **decongestants,** *compounds that cause shrinkage of the membranes lining the nasal passages.*

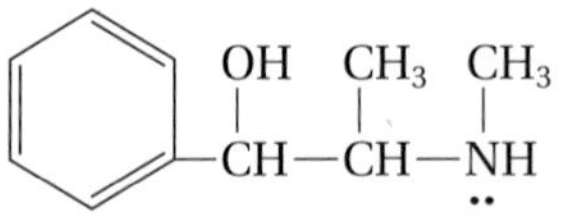

Ephedrine

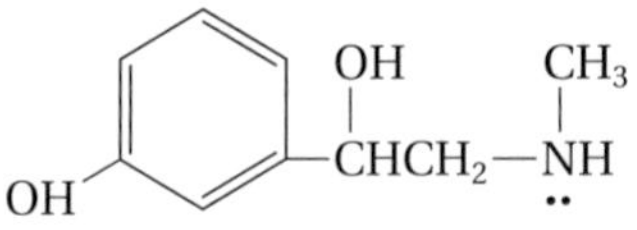

Neosynephrine

Amphetamine is an appetite depressant and nervous system stimulant. Often prescribed for dieters, amphetamines can be abused when the drug is habitually taken as an "upper" for its mood-elevating effects—lessened fatigue, increased confidence, alertness, and increased desire to work.

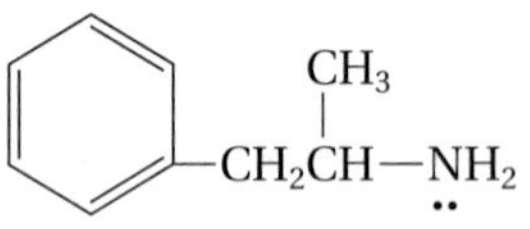

Amphetamine

Do you get hay fever on days when the pollen count is high? If you do, you are allergic to pollen. Your body cells are responding to the invasion of these foreign substances by producing histamine. Histamine is responsible for watery eyes, sniffles, and other hay fever miseries. *The effects of histamine can be relieved to some extent by a class of drugs called* **antihistamines;** two of the most common are pyribenzamine and diphenylhydramine, also called *benadryl.*

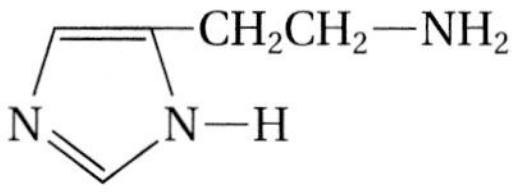

Histamine

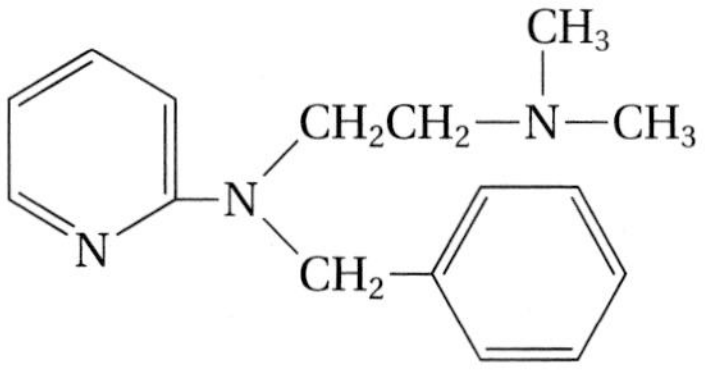

Pyribenzamine

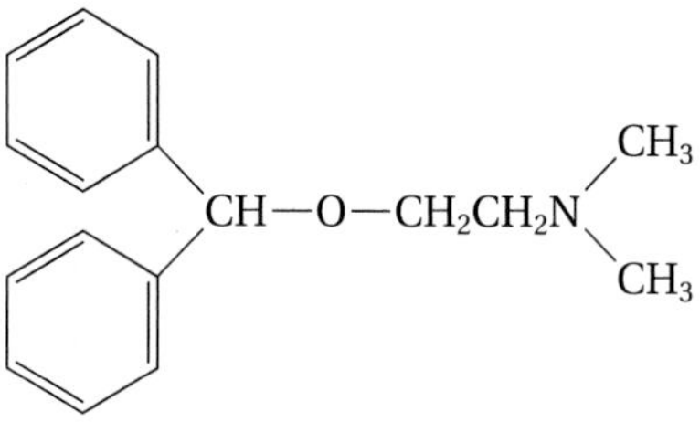

Diphenylhydramine
(benadryl)

Benadryl is also a component of Dramamine, a drug commonly used to prevent motion sickness or nausea.

> **PRACTICE EXERCISE 15.11**
>
> Identify the functional groups in the neosynephrine molecule. What is the physiologic action of this drug?

15.8 Alkaloids

AIM: To name and give examples of uses of medicinal alkaloids.

Focus

Alkaloids are physiologically active amines produced by plants.

Some of the most powerful drugs known are derived from plants. Some have been used for thousands of years. Among the oldest of these drugs are the **alkaloids**—*a group of over 2500 amines obtained from plants.* The name *alkaloid,* meaning "alkali-like," bears testimony to the weakly basic properties of these compounds.

The molecular structures of the alkaloids vary from simple to complex. Nicotine is one of the simplest.

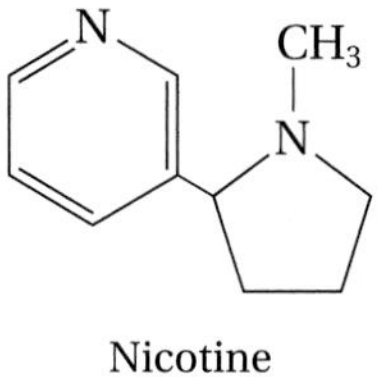

Nicotine

Nicotine is one of several alkaloids present in tobacco. Small doses, such as those obtained by smokers, stimulate the involuntary nervous system. Large doses are toxic and result in nicotine poisoning. Nicotine is a *habituating drug*—one on which people acquire a dependence. An *addicting drug* is one that causes physiologic changes when it is not used for a time.

Several important alkaloids contain the indole ring system. Lysergic acid and reserpine are indole alkaloids. Lysergic acid is produced by ergot, a fungus that grows as brown bodies on infected rye. The diethylamide of lysergic acid is called LSD (lysergic acid diethylamide), a powerful and widely abused hallucinogenic drug. As little as 1 μg of LSD, scarcely enough to see, is sufficient to cause hallucinations. The danger of overdose is high.

Reserpine, an indole alkaloid of the Indian snakeroot (Fig. 15.4a), reduces hypertension (high blood pressure), which, if left unchecked,

Figure 15.4
(a) Indian snakeroot is the source of reserpine. (b) The Cinchona tree yields quinine.

(a) **(b)**

could lead to a stroke or cardiac arrest. Strychnine, a bitter-tasting compound from the plant *Strychnos nux vomica,* has a long and sometimes sinister history as a poison.

Lysergic acid diethylamide (LSD)

Strychnine

Reserpine

The antimalarial drug quinine is an alkaloid that occurs in the Cinchona (pronounced "sinkona") tree (Fig. 15.4b) of the Andes Mountains of South America. Quinine has the aromatic quinoline ring system as part of its molecular structure. Other drugs that act against malaria have been synthesized, but quinine is still used.

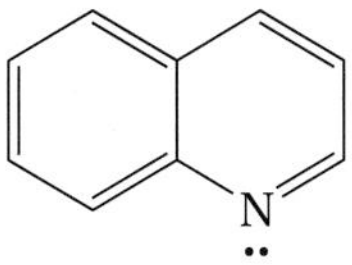

Quinoline

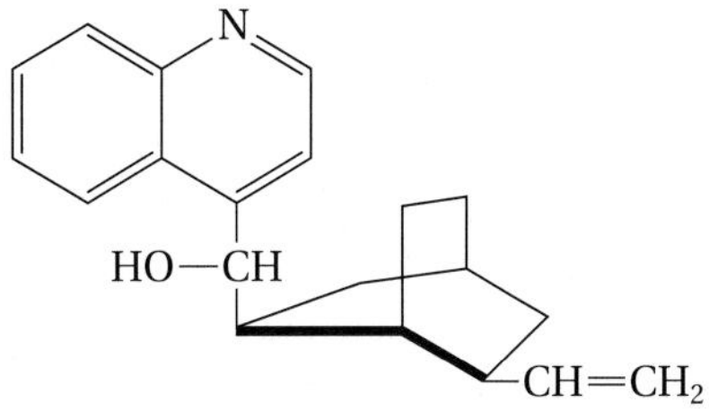

Quinine

The tropane ring system is found in cocaine and atropine. Cocaine stimulates the central nervous system. It is now a widely abused habituating drug. Atropine is an alkaloid obtained from belladonna, hemlane, and deadly nightshade. It has several medicinal purposes, including treatment of certain eye conditions. A tincture of atropine causes dilation of the pupils. Roman women used belladonna to achieve this dilating effect because they thought it attractive. Indeed, belladonna means "beautiful lady."

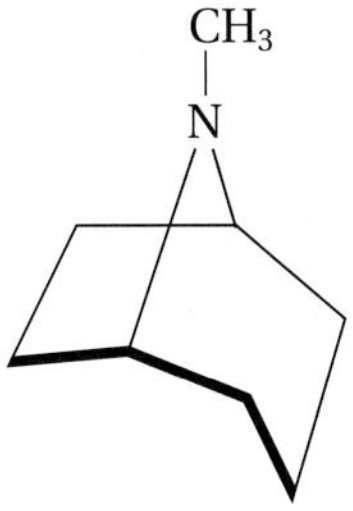

Tropane

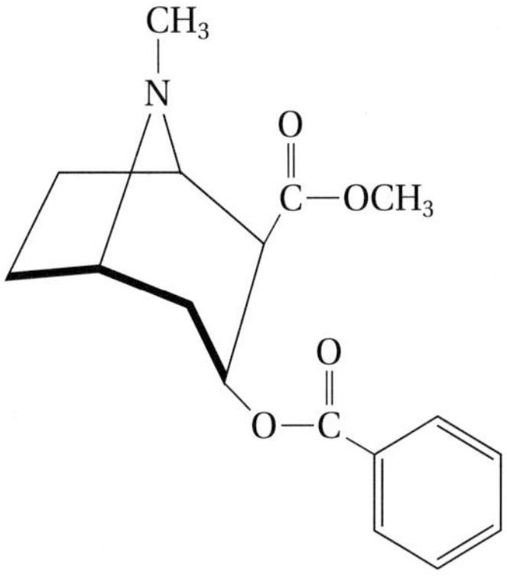

Cocaine

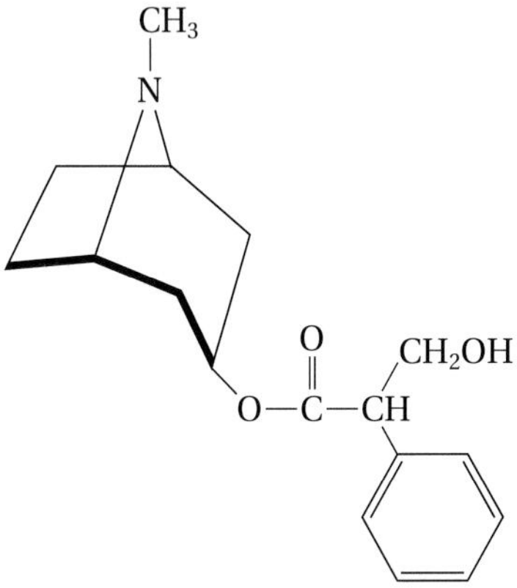

Atropine

Illegally produced "designer" drugs chemically mimic the effects of specific drugs of abuse. Some derivatives of fentanyl are over 1000 times more powerful than morphine. Designer drugs have a high risk of addiction and can cause brain damage and even death.

Opium is the raw resin extracted from the seedpods of the opium poppy (*Papaver somniferum*), not the poppy grown in gardens. When the resin is refined, two important alkaloids, morphine and codeine, can be isolated in pure form. Morphine is the most effective pain-killing drug known. Codeine is a powerful analgesic and cough suppressant. It was an ingredient in cough medicines for many years, but it has been replaced by dextromethorphan, a synthetic alkaloid that is equally effective. Morphine and codeine are addictive drugs.

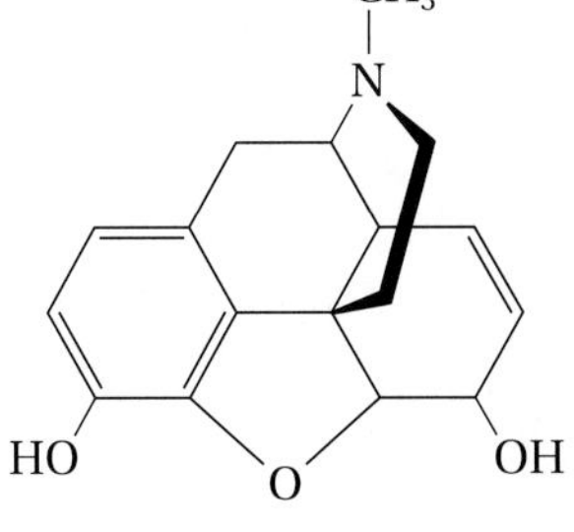

Morphine

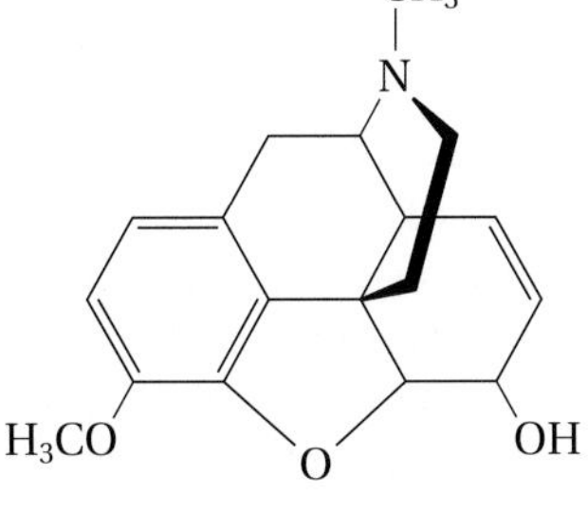

Codeine

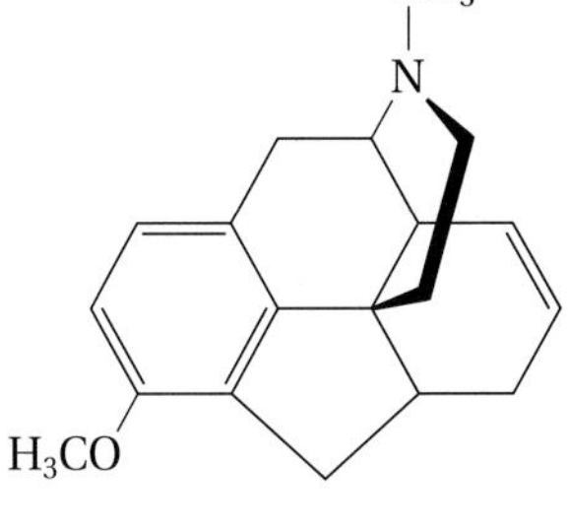

Dextromethorphan

Ironically, heroin, an acetic acid ester derivative of morphine, was introduced because it was thought to be a better analgesic and to lack morphine's addictive properties. It was soon found, however, that heroin is no better than morphine as an analgesic and is even more addictive than other **opiates**—*drugs that produce the psychological and physiologic effects of opium.* Today, heroin is a destructive hard drug. Methadone, a drug that blocks heroin action, has been introduced to wean heroin addicts from their habit. Methadone does not produce the euphoria associated with heroin use. Its use is controversial, since it too is an addictive drug.

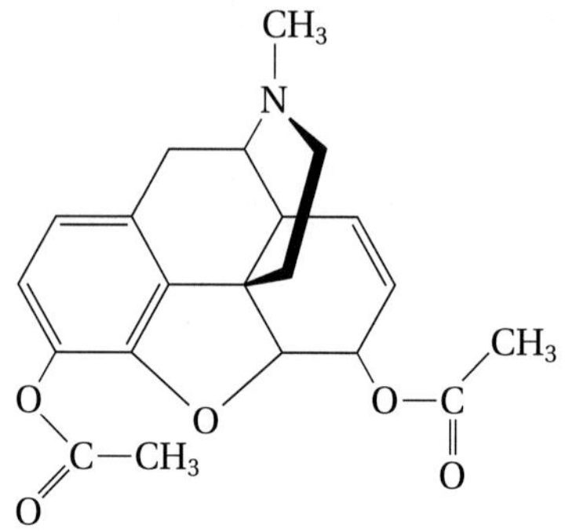

Heroin 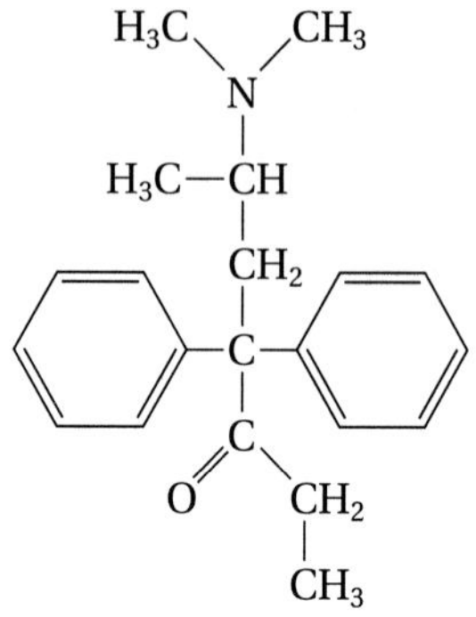   Methadone

PRACTICE EXERCISE 15.12

List some of the physiologic effects and uses of (a) morphine, (b) nicotine, (c) atropine, and (d) quinine.

15.9 Barbiturates

AIM: To name and give examples of uses of barbiturate drugs.

Focus

Barbiturate drugs are derivatives of urea.

Urea has a simple molecular structure. In structural terms, we can consider it the diamide of carbonic acid (H_2CO_3):

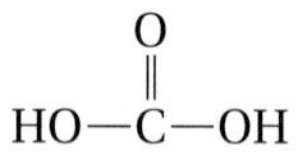

Carbonic acid 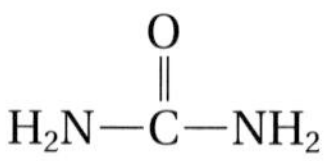   Urea

Urea is important in its own right, since it is the form in which our bodies dispose of excess nitrogen in the urine. Its high water solubility (125 g/100 mL of water) and low toxicity make it ideal for this purpose.

Condensation of malonic acid with urea produces barbituric acid, the parent compound of a number of drugs.

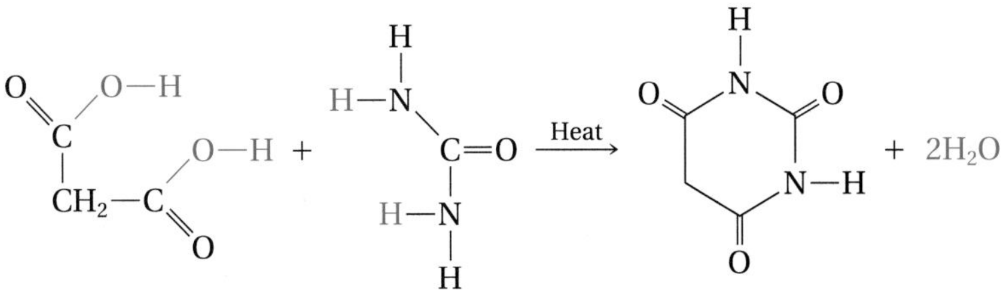

Malonic acid   Urea   Barbituric acid

Note that barbituric acid contains the pyrimidine ring system. It is an acid because the hydrogens on the ring nitrogens readily dissociate in basic solution to form barbiturate salts.

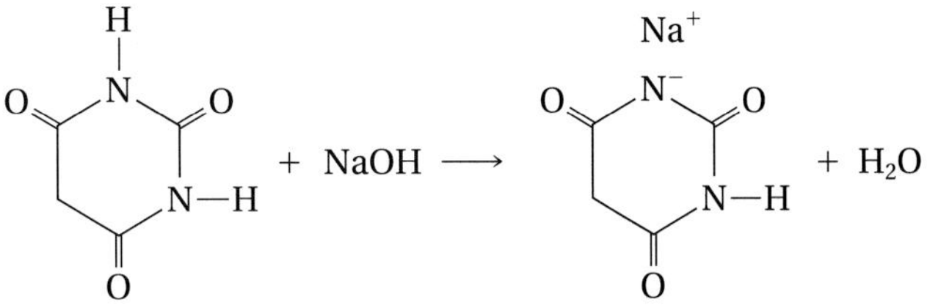

Barbituric acid   Sodium barbiturate

Hypnotic
hupnos (Greek):   sleep

Sedative
sedare (Latin):   to calm

Salts of barbituric acid and its derivatives are called **barbiturates.** The sodium salts are often administered because they ionize in solution and are more water soluble than the acid forms. Barbituric acid is not physiologically active, but some of its substituted derivatives are among the most potent **hypnotics** (*sleep-inducers*) and **sedatives** (*tranquilizers*) known. Barbital was introduced in Germany in 1903 under the name Veronal, and it is still used as a hypnotic. Since the introduction of barbital, medicinal chemists have synthesized thousands of variations on the barbituric acid structure, but only about a dozen are clinically useful.

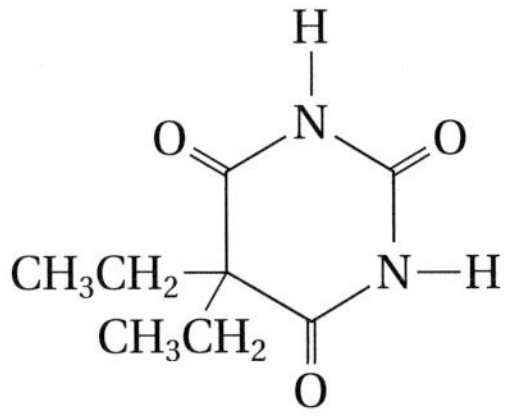

Barbital
(Veronal)

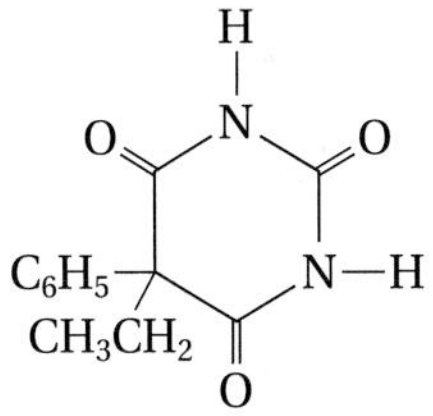

Phenobarbital
(Luminal)

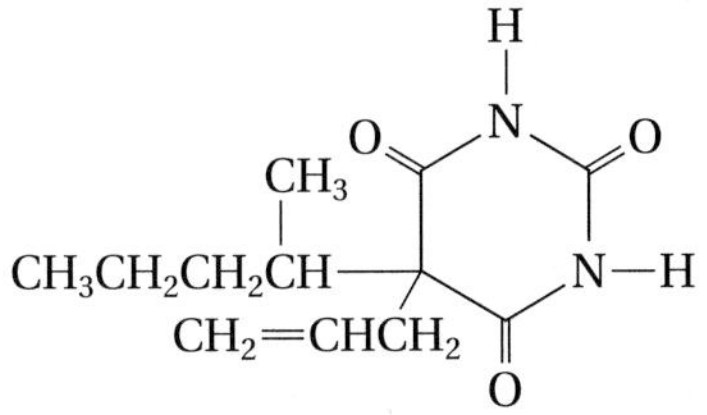

Secobarbital
(Seconal)

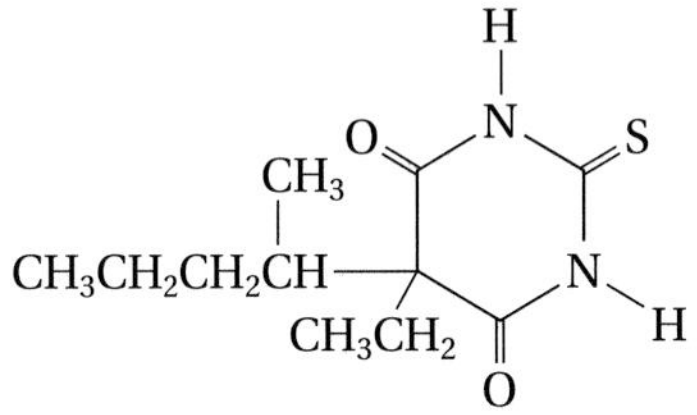

Thiopental

As hypnotics, barbiturates are often classified as long-acting or short-acting. Barbital is long-acting. A 0.3-g dose produces six or more hours of sleep in an adult. Thiopental sodium salt (sodium pentothal), on the other hand, acts only briefly; it is used in surgery to put patients to sleep before a general anesthetic is administered.

PRACTICE EXERCISE 15.13

Write the structure of sodium pentothal.

PRACTICE EXERCISE 15.14

Explain why the hydrogens on the ring nitrogens of barbituric acid and its derivatives are readily lost as protons.

SUMMARY

Amines and amides are major nitrogen-containing classes of organic compounds. Amines are organic derivatives of ammonia. They are classified as primary (1°; RNH_2), secondary (2°; RNHR), and tertiary (3°; RNRR). A quaternary ammonium salt has four carbon groups attached to the amine nitrogen; the nitrogen is positively charged. Carbon rings that contain amine nitrogen are called heterocyclic amines. The acid-base properties of amines are similar to those of ammonia. That is, the ability of the unshared electron pair of amine nitrogen to accept a proton in acidic solutions makes amines weak bases. Because of their weak basicity, certain amines act as pH buffers in biological fluids.

Amides are derivatives of acids and amines. When formed from carboxylic acids, they have the general formulas $RCONH_2$ (simple amides),

RCONHR (monosubstituted amides), and RCONRR (disubstituted amides). Amides are neutral compounds; the unshared electron pair of the amide nitrogen is pulled toward carbonyl oxygen, rendering it less available for protonation than it is in amines.

Many amines are physiologically important. The catecholamines are neurotransmitters of the central nervous system. Amphetamine and several hallucinogenic drugs are similar in structure to catecholamines. Alkaloids are a class of over 2500 amines isolated from plants. Many have useful medicinal properties. Lysergic acid diethylamide (LSD), reserpine, quinine, atropine, and cocaine are a few alkaloids. Morphine, another alkaloid, is the most effective pain killer known. Barbiturates are derivatives of urea (NH_2CONH_2). A dozen or so barbiturates are used in medicine, mainly as hypnotics and sedatives.

SUMMARY OF REACTIONS

Here are examples of the reactions covered in this chapter.

1. Preparation of amines. (The amine is 1° or 2°.)

$$\underset{\underset{H}{|}}{\overset{\overset{H}{|}}{R-N}}:\ +\ R-Cl \longrightarrow \underset{\underset{H}{|}}{\overset{\overset{H}{|}}{R-\overset{+}{N}-R}}\ \ Cl^-$$

Treatment with base liberates the free amine.

2. Preparation of quaternary ammonium salts. (The amine is 3°.)

$$\underset{\underset{R}{|}}{\overset{\overset{R}{|}}{R-N}}:\ +\ R-Cl \longrightarrow \underset{\underset{R}{|}}{\overset{\overset{R}{|}}{R-\overset{+}{N}-R}}\ \ Cl^-$$

3. Preparation of amides.

(a) By dehydration of ammonium salts of acids. (The nitrogen compound is ammonia or a 1° or 2° amine.)

$$\overset{O}{\overset{\|}{R-C}}-O^-\ NH_4^+\ \xrightarrow[\text{Heat}]{P_2O_5}\ \overset{O}{\overset{\|}{R-C}}-NH_2\ +\ H_2O$$

(b) By reactions of ammonia or amines with esters or anhydrides. (The amine is 1° or 2°.)

$$\overset{O}{\overset{\|}{R-C}}-OR\ +\ H-\overset{..}{N}HR \longrightarrow \overset{O}{\overset{\|}{R-C}}-\overset{..}{N}HR\ +\ ROH$$

$$\overset{O}{\overset{\|}{R-C}}-O-\overset{O}{\overset{\|}{C}}-R\ +\ H-\overset{..}{N}HR \longrightarrow \overset{O}{\overset{\|}{R-C}}-\overset{..}{N}HR\ +\ \overset{O}{\overset{\|}{R-C}}-OH$$

4. Hydrolysis of amides. (The amide may be simple, monosubstituted, or disubstituted.)

$$\underset{R-\overset{\overset{\displaystyle O}{\|}}{C}-\overset{\cdot\cdot}{N}H_2}{} + HO-H \xrightarrow[\text{H}^+ \text{ or OH}^-]{\text{Heat}} R-\overset{\overset{\displaystyle O}{\|}}{C}-OH + H-\overset{\cdot\cdot}{N}H_2$$

KEY TERMS

Alkaloid (15.8)
Amide (15.4)
Amine (15.1)
Alkylammonium ion (15.2)
Ammonium salt (15.2)

Anilide (15.4)
Antihistamine (15.7)
Arylammonium ion (15.2)
Barbiturate (15.9)
Decongestant (15.7)

Free amine (15.2)
Hallucinogen (15.7)
Hypnotic (15.9)
Neurotransmitter (15.7)
Opiate (15.8)

Polyamide (15.5)
Protonated amine (15.2)
Quaternary ammonium
  salt (15.3)
Sedative (15.9)

EXERCISES

Amines (Sections 15.1, 15.2, 15.3)

15.15 Name or write a structural formula for the following amines. Classify them as primary, secondary, or tertiary amines.
(a) $(CH_3)_2NH$ (b) *p*-chloro-*N*-methylaniline
(c) NH_2
(d) diethylmethylamine

15.16 Write structural formulas and name the following amines. Classify each amine as primary, secondary, or tertiary.
(a) diethylamine   (b) $(CH_3CH_2)_2NCH_3$
(c) butylamine   (d)

15.17 What is the meaning of the term *heterocyclic amine?*

15.18 Draw the structure and give the name of (a) an aromatic heterocyclic amine and (b) an aliphatic heterocyclic amine.

15.19 Draw structural formulas for (a) pyrimidine and (b) purine. Derivatives of these two compounds are found in what biologically important molecules?

15.20 Draw the structure of pyrrole. List some of the naturally occurring molecules that contain the pyrrole ring system.

15.21 Why are amines weak bases?

15.22 Draw the general formulas for (a) an unprotonated (free) amine and (b) a protonated amine.

15.23 Write the structure and name the expected products of each of the following reactions.
(a) $CH_3CH_2NH_2 + HCl$ (b) $(CH_3)_2NH + HNO_3$
(c) $CH_3NH_2 + H_2SO_4$ (d) $(CH_3)_3CNH_2 + HCl$

15.24 Draw the structure and name the organic product for each of the following reactions.
(a) $CH_3NH_3{}^+I^- + NaOH$
(b) $(CH_3)_2NH + CH_3Cl$
(c) $CH_3I + NH_3$
(d) $NH_3{}^+Cl^-$ + NaOH

(e) $(CH_3CH_2)_3N + CH_3CH_2Cl$

15.25 Write an equation for the dissociation of the dimethylammonium ion. Why does an aqueous solution of dimethylammonium chloride test acidic?

15.26 Draw the structure of tetramethylammonium iodide. What happens if this compound is treated with sodium hydroxide?

Amides (Sections 15.4, 15.5, 15.6)

15.27 Name or write structural formulas for the following amides.

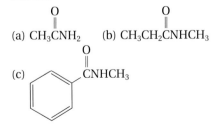

(a) $CH_3\overset{\overset{\displaystyle O}{\|}}{C}NH_2$ (b) $CH_3CH_2\overset{\overset{\displaystyle O}{\|}}{C}NHCH_3$

(c)

(d) *N*-ethyl-*N*-methylpropanamide
(e) acetanilide

15.28 Write structural formulas or name the following.
(a) acetamide   (b) benzamide
(c) *N*-ethylbenzamide

$$\text{(d) } CH_3CH_2\overset{\displaystyle O}{\overset{\displaystyle \|}{C}}N\underset{\displaystyle CH_3}{CH_3} \qquad \text{(e) } HC\overset{\displaystyle O}{\overset{\displaystyle \|}{}}N\underset{\displaystyle CH_3}{CH_3}$$

15.29 Write the structure for the product of each of the following reactions.

$$\text{(a) } CH_3CH_2\overset{\displaystyle O}{\overset{\displaystyle \|}{C}}O^- \ NH_4{}^+ \xrightarrow{\text{Heat}}$$

$$\text{(b) } CH_3\overset{\displaystyle O}{\overset{\displaystyle \|}{C}}OCH_3 + NH_3 \longrightarrow$$

$$\text{(c) } HC\overset{\displaystyle O}{\overset{\displaystyle \|}{}}OCH_3 + NH_2CH_2CH_3 \longrightarrow$$

15.30 The following amides are hydrolyzed at the conditions shown in the equations. What are the products of each reaction?

$$\text{(a) } CH_3CH_2\overset{\displaystyle O}{\overset{\displaystyle \|}{C}}NHCH_3 + H_2O \xrightarrow[H^+]{\text{Heat}}$$

(b) ⬡$-\overset{\displaystyle O}{\overset{\displaystyle \|}{C}}N\underset{\displaystyle CH_3}{CH_3}$ + NaOH $\xrightarrow{\text{Heat}}$

$$\text{(c) } CH_3\overset{\displaystyle O}{\overset{\displaystyle \|}{C}}NHCH_2CH_3 + NaOH \xrightarrow{\text{Heat}}$$

Important amines (Sections 15.7, 15.8, 15.9)

15.31 Parkinson's disease results when the concentration of dopamine in the brain cells is very low. Explain why dopa is an effective drug for parkinsonism and why dopamine is ineffective.

15.32 What effect does amphetamine have on the body?

15.33 What are the alkaloids? How did these compounds get their name?

15.34 Name the nitrogen-containing functional groups in the LSD molecule.

15.35 Name two opiates that occur naturally.

15.36 What is the structural difference between morphine and codeine? What is codeine used for?

15.37 Is barbituric acid physiologically active? What are some of the physiologic effects of barbiturates?

Additional Exercises

15.38 Give the products of the following reactions. Name the products and reactants whenever possible. If there is no reaction, write "no reaction."

$$\text{(a) } CH_3CH_2-\underset{\displaystyle H}{\overset{\displaystyle H}{N}}: + HCl \longrightarrow$$

$$\text{(b) } CH_3-\underset{\displaystyle H}{\overset{\displaystyle H}{\overset{+}{N}}}-H \ Cl^- + NaOH \longrightarrow$$

$$\text{(c) } H-\underset{\displaystyle H}{\overset{\displaystyle H}{N}}: + CH_3I \longrightarrow$$

$$\text{(d) } CH_3-\overset{\displaystyle O}{\overset{\displaystyle \|}{C}}-N\overset{\displaystyle H}{\underset{\displaystyle H}{}} + H_2O \xrightarrow{\text{HCl}}$$

$$\text{(e) } CH_3-C\overset{\displaystyle O}{\underset{\displaystyle OH}{}} + H-\overset{\displaystyle H}{\overset{\displaystyle |}{N}}-⬡ \xrightarrow{\text{Heat}}$$

$$\text{(f) } CH_3-\underset{\displaystyle CH_3}{\overset{\displaystyle H}{\overset{\displaystyle |}{C}}}-\overset{\displaystyle O}{\overset{\displaystyle \|}{C}}-\underset{\displaystyle H}{\overset{}{N}}-CH_2CH_3 \xrightarrow[\text{Heat}]{\text{Acid}}$$

$$\text{(g) } CH_3CH_2C\overset{\displaystyle O}{\underset{\displaystyle NH_2}{}} + P_2O_5 \xrightarrow{\text{Heat}}$$

15.39 Identify each of these heterocyclic compounds.

(a) ⬡ pyrimidine ring (b) ⬡ pyridine ring

(c) ⬠ pyrrole ring (d) fused ring (purine)

(e) ⬠ imidazole ring (f) fused ring (indole)

15.40 Many nitrogen-containing compounds of biochemical significance have been mentioned in this chapter. Match the following compounds with the expression that best fits.

(1) Dopa
(2) Serotonin
(3) Mescaline
(4) Amphetamine
(5) Histamine
(6) Benadryl
(7) Nicotine
(8) Quinine
(9) Atropine
(10) Codeine
(11) Morphine
(12) Barbital
(13) Sodium pentothal
(14) Caffeine

(a) responsible for the symptoms of hay fever
(b) a long-acting hypnotic
(c) used as an antimalarial drug
(d) low levels implicated in mental depression
(e) antihistamine drug
(f) drug administered for Parkinson's disease
(g) hallucinogen of similar structure to a neurotransmitter
(h) a short-acting barbiturate
(i) causes dilation of pupils; used to treat certain eye conditions
(j) a common drug of abuse that stimulates the nervous system and depresses the appetite
(k) most effective pain killer known
(l) an alkaloid present in tobacco
(m) a stimulant in tea, coffee, and cocoa
(n) an alkaloid derived from poppies; used as a cough suppressant

15.41 Write structural formulas for the following compounds: (a) triethylmethylammonium bromide, (b) *N*-propylaniline, (c) *N*-methylacetamide, and (d) *N,N*-dimethylbenzamide.

15.42 Name these compounds and classify each as a primary, secondary, or tertiary amine.

(a) —N(CH₃)₂   (b) CH₃CH₂NH₂

(c) CH₃—CH—NHCH₃   (d)
         |
         CH₃

(e) (CH₃)₃C—NH₂   (f) CH₃CH₂NHCH₂CH₃

15.43 Draw structural formulas for the following compounds.
(a) triethylamine   (b) propionamide
(c) *N*-ethylaniline   (d) *N*-phenylbenzamide
(e) 3-ethylaniline   (f) *N,N*-diethylacetamide

15.44 Name the following amides.

(a) CH₃C—N(CH₃)₂   (b) CH₃CH₂C—NH₂
        ‖   ‖
        O   O

(c)

(d)

15.45 Write general reactions to show the formation of a simple, monosubstituted, and disubstituted amide.

15.46 Name the following compounds. Predict their solubility in water. If soluble, will the solution be acidic, basic, or neutral? Explain.

(a) CH₃CH₂NH₂   (b) CH₃CH₂C—OH
                            ‖
                            O

(c) CH₃CH₂C—H   (d) CH₃—C—CH₃
        ‖   ‖
        O   O

(e) CH₃CH₂C—NH₂   (f) CH₃—C—OCH₃
        ‖   ‖
        O   O

(g)   (h)

15.47 Using ethene and propanol as your only organic starting material, show how you might prepare each of the following compounds. You may use any inorganic reagents that you need. Name the intermediate products of each reaction.
(a) propylamine   (b) *N*-ethylpropanamide

15.48 Draw the structure and name the compound described by each of the following statements.
(a) a six-membered aliphatic heterocyclic amine
(b) the simplest aromatic amine
(c) the amine formed by the reaction of dimethylamine with ethyl chloride
(d) an aromatic amine with an ethyl group that is not on the benzene ring

SELF-TEST (REVIEW)

True/False

1. Pyridine is a six-membered aromatic heterocyclic amine.
2. Amines are relatively weak bases.
3. An amide contains a carbon-nitrogen double bond.
4. A free amine has an unshared electron pair on the amine nitrogen.
5. A tertiary amine is the product of a reaction between a secondary amine and a halocarbon.
6. The reaction between a carboxylic acid and an amine gives an amide.
7. A secondary amine that contains two carbon atoms is dimethylamine.
8. In a quaternary ammonium salt the nitrogen atom is bonded to four carbon atoms.
9. Another name for *o*-toluidine is 2-methylaniline.
10. Trimethylamine is a tertiary amine.

Multiple Choice

11. What is the correct name for the following compound?

$$CH_3CH_2\overset{\overset{\displaystyle O}{\|}}{C}-NHCH_3$$

   (a) *N*-methylpropanamide
   (b) ethylmethylamide   (c) propylmethylamide
   (d) methylaminoproprionic acid

12. Which of these is an antimalarial drug?
   (a) dopa   (b) quinine   (c) benadryl
   (d) atropine

13. The reaction of two moles of ethyl chloride with one mole of ammonia would produce
   (a) ethylamine.   (b) diethylamine.
   (c) triethylamine.   (d) all of the above.

14. Which of the following is an antihistamine?
   (a) mescaline   (b) codeine   (c) caffeine
   (d) benadryl

15. Amines are
   (a) weak bases.   (b) weak acids.
   (c) all physiologically important.
   (d) insoluble in water.

16. The acid hydrolysis of *N*-ethylacetamide gives
   (a) ethanol and ethylamine.
   (b) ethanol and acetic acid.
   (c) acetic acid and ethylamine.
   (d) ethylamine and acetaldehyde.

17. Which of these is not an aromatic amine?
   (a) aniline   (b) diphenylamine
   (c) piperidine   (d) *o*-toluidine

18. An amide can be prepared by
   (a) heating an ammonium salt.
   (b) reacting ammonia with an acid anhydride.
   (c) reacting an ester with an amine.
   (d) all of the above.

19. Which of these is a tertiary amine?
   (a) aniline   (b) pyrrolidine
   (c) triphenylamine   (d) *N*-methyldiethylamide

20. An aqueous solution of ethyl ammonium chloride would be
   (a) strongly acidic.   (b) weakly acidic.
   (c) strongly basic.   (d) neutral.

21. Which of the following terms would *not* describe codeine?
   (a) cough suppressant   (b) barbiturate
   (c) addictive   (d) analgesic

22. The name of the following compound is

   (a) *N*-ethyl-*p*-toluidine.   (b) ethylaminotoluene.
   (c) ethylbenzylamide.   (d) ethylphenylaniline.

23. Which of the following terms does not describe the structure of purine?

   (a) fused ring   (b) heterocyclic
   (c) quaternary   (d) aromatic

24. An anilide would generally have all the following except
   (a) a carboxylic acid group.   (b) an amide group.
   (c) an aromatic ring.   (d) a carbonyl group.

Carbohydrates

The Structure and Chemistry of Sugars

Honey contains fructose,
an important carbohydrate.

CHAPTER OUTLINE

CASE IN POINT: Lactose intolerance

16.1 Carbohydrate structure and stereochemistry

16.2 Monosaccharides

16.3 Cyclic structures

16.4 Haworth projections

16.5 Glycosides

16.6 Polysaccharides

A CLOSER LOOK: Dietary Fiber

16.7 Disaccharides

16.8 Sucrose and lactose

FOLLOW-UP TO THE CASE IN POINT: Lactose intolerance

16.9 Reducing and nonreducing sugars

A CLOSER LOOK: Tests for Blood Glucose in Diabetes

16.10 Chitin, heparin, and acid mucopolysaccharides

The class of compounds called *carbohydrates,* or *sugars,* is widespread throughout nature. These versatile molecules provide food for all living creatures. They are also very important structural components of the cell walls of plants and the shells of crustaceans such as crabs and shrimp. In this chapter we will learn what carbohydrates are and then learn to recognize differences in the molecular structures of simple sugars. We also will learn how simple sugars are linked together to form some important polymer molecules. And finally, we will learn a few tests for the identification of sugars.

Despite the importance of sugars as food, not all people can digest a sugar that most people find edible, as we will learn in the following Case in Point.

CASE IN POINT: Lactose intolerance

Alberto is healthy, but he has always been rather thin. In his sophomore year in college, he tried to gain some weight by lifting weights and increasing his caloric intake. After his first weightlifting session, Alberto and his friends stopped at the campus cafeteria, where Alberto drank two large milkshakes. He awoke in the middle of the night with stomach cramps, gas, and severe diarrhea. Fearing that he had food poisoning, he immediately reported to the college infirmary. After the physician examined Alberto and heard about the suspicious milkshakes, she dismissed the possibility of food poisoning and suggested that Alberto avoid all dairy products until he felt well again. What is the link between ingesting dairy products and the symptoms that Alberto experienced? We will learn more about Alberto's illness and how it is related to the digestion of sugars in Section 16.8.

Lactose, or milk sugar, is present in dairy products.

16.1 *Carbohydrate structure and stereochemistry*

AIMS: *To name and classify a carbohydrate as a monosaccharide, disaccharide, or polysaccharide; as a triose, tetrose, pentose, or hexose; as an aldose or a ketose. To explain what is meant by the handedness of a molecule using the terms* asymmetric carbon *and* stereoisomer. *To interpret two-dimensional Fischer projection formulas of sugars as three-dimensional structures.*

Focus

Glyceraldehyde is the reference substance for the study of carbohydrate structure.

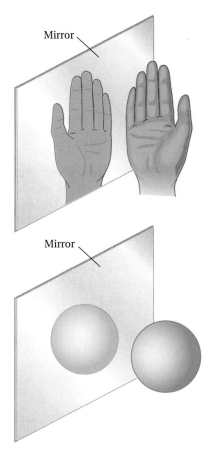

Carbohydrates abound in nature; among the many forms are starch, cotton, table sugar, and wood. **Carbohydrates** *are polyhydroxy aldehydes or ketones.* The name *carbohydrate* means "hydrate of carbon." The name was coined in the early days of carbohydrate chemistry because many of these compounds have molecular formulas that are multiples of CH_2O, such as $C_6H_{12}O_6$ and $C_5H_{10}O_5$. *Carbohydrates are also called* **sugars.**

Glyceraldehyde is the simplest sugar found in nature. If you have a good understanding of the structure of glyceraldehyde, you will be able to understand more complicated sugars. Here is the structure of glyceraldehyde:

$$\overset{1}{C}HO$$
$$H-\overset{2}{C}-OH$$
$$\overset{3}{C}H_2-OH$$

Glyceraldehyde

The ending *-ose* is a characteristic of the names of many sugars and sugar derivatives. *Sugars that contain an aldehyde functional group (—CHO) are called* **aldoses.** The number of carbons in a sugar gives a more detailed description of the sugar. For example, a sugar containing three carbons and an aldehyde functional group is an *aldotriose*. Glyceraldehyde is an aldotriose. The carbon chains of aldoses are numbered starting with the aldehyde group as carbon 1.

Stereoisomerism of glyceraldehyde

Have you ever noticed that placing an object in front of a mirror can give two different results? If the object is symmetrical—a sphere, say—then its mirror image is indistinguishable from the object. That is, the appearance of the sphere and its reflection are superimposable mirror images— the images can be placed on top of each other to obtain a match. However, if you look at your hands in a mirror, your right hand reflects as a left hand and your left hand reflects as a right hand (Fig. 16.1). Your hands are examples of **nonsuperimposable mirror images**—*mirror images that cannot be placed on top of each other to obtain a match.* Many pairs of familiar objects, such as ears, feet, shoes, and bird wings, are related similarly.

Figure 16.1
The reflected image of a right hand in a mirror appears as a left hand. Note the relative positions of the thumbs. A sphere and its reflected image are identical.

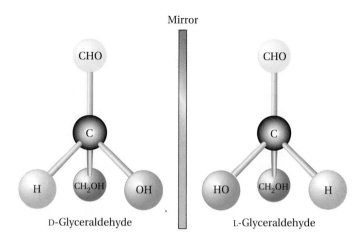

Figure 16.2
Ball-and-stick models of D- and L-glyceraldehyde. These two structures are nonsuperimposable mirror images of one another.

PRACTICE EXERCISE 16.1

Which of the following objects would have a nonsuperimposable mirror image? (Ignore designs or other markings.)

(a) a clam shell   (b) a cup   (c) a ball   (d) a car

(e) a wood screw   (f) a fingerprint   (g) a baseball bat

Looking closely at the glyceraldehyde structure, we see that carbon 2 has four different groups attached: —CHO, —H, —OH, and —CH$_2$OH. *A carbon with four different groups attached is called an* **asymmetric carbon.** Compounds whose molecules contain asymmetric carbons have handedness—that is, there are two kinds of molecules with the same structural formula that are related to one another in much the same way as your hands are related to one another. Glyceraldehyde, with its four different groups attached to the asymmetric carbon in one possible arrangement, is called D-*glyceraldehyde.* Its mirror image is L-glyceraldehyde. Figure 16.2 shows D- and L-glyceraldehyde. Like your hands, D- and L-glyceraldehyde are nonsuperimposable mirror images. No matter how hard you try, there is no way short of breaking bonds to superimpose D- and L-glyceraldehyde so that the four groups attached to the asymmetric carbon of D-glyceraldehyde coincide with the same four groups of L-glyceraldehyde. The D- and L-glyceraldehydes are examples of **stereoisomers**—*molecules whose molecular structure differs only in the arrangement of groups in space.*

Stereochemistry is concerned with the spatial arrangement of the atoms in a molecule. The three-dimensional shapes of molecules are very important in biological systems. In Chapter 19 we will see that the activity of many biological catalysts called *enzymes* are stereospecific—they will work only on a particular molecule that has a particular three-dimensional shape.

EXAMPLE 16.1

Identifying asymmetric carbons

Which of the following compounds have an asymmetric carbon?

(a) CH$_3$CHCH$_3$ (b) CH$_3$CHCH$_2$CH$_3$ (c) CH$_3$CHCHO
    | | |
    OH   OH   OH

SOLUTION

An asymmetric carbon has four different groups attached. It may help to draw structures in a more complete form.

(a) The central carbon has one —H, one —OH, but two —CH₃ groups attached. It is not asymmetric.

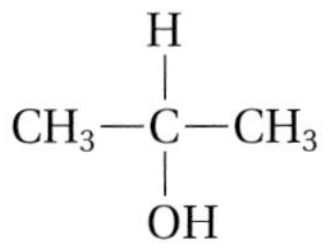

$$CH_3 - \overset{\displaystyle H}{\underset{\displaystyle OH}{C}} - CH_3$$

(b) The central carbon has one —H, one —OH, one —CH₃, and one —CH₂CH₃ group attached. These four groups are different, so the central carbon is asymmetric. It is marked with an asterisk. None of the other carbons in this molecule is asymmetric because no other carbon is attached to four different groups (check to be sure).

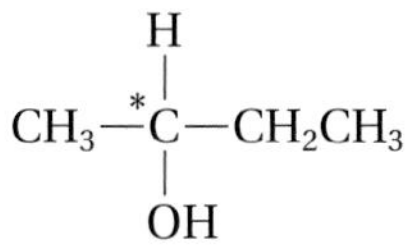

$$CH_3 - \overset{\displaystyle H}{\underset{\displaystyle OH}{\overset{*}{C}}} - CH_2CH_3$$

(c) This molecule also has an asymmetric carbon. It is marked with an asterisk.

$$CH_3 - \overset{\displaystyle H}{\underset{\displaystyle OH}{\overset{*}{C}}} - CHO$$

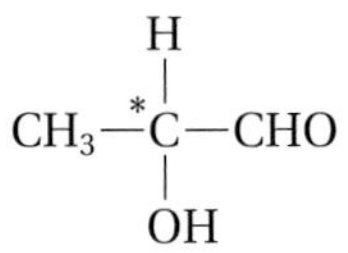

PRACTICE EXERCISE 16.2

Identify the asymmetric carbon, if there is one, in each of the following structures.

(a) CH₃CH₂CHO   (b) CH₃CHCHO   (c) CH₃CHOH
                          | |
                         CH₃   Cl

(d) CH₃CHCHO
         |
       CH₂CH₃

EXAMPLE 16.2   **Finding nonsuperimposable mirror images**

Decide which of the following has a nonsuperimposable mirror image, and draw the stereoisomers.

(a) CH₃CHCH₃   (b) CH₃CHCH₂CH₃
         | |
        OH   OH

SOLUTION

(a) The structure has no asymmetric carbon and exists as only one form—its mirror image is superimposable.

(b) The structure has an asymmetric carbon—its mirror image is non-superimposable. It will exist as a pair of stereoisomers.

   One way of writing the structural formulas of stereoisomers is to consider the central carbon—usually the asymmetric carbon if the molecule has one—to be in the plane of the paper. The two groups attached to the central atom by *tapered bonds* are understood to be above the plane of the paper; the group attached to the central carbon by a dashed bond is understood to be below the plane. The remaining group is in the plane of the paper. The stereorepresentations of (a) and (b) are as follows:

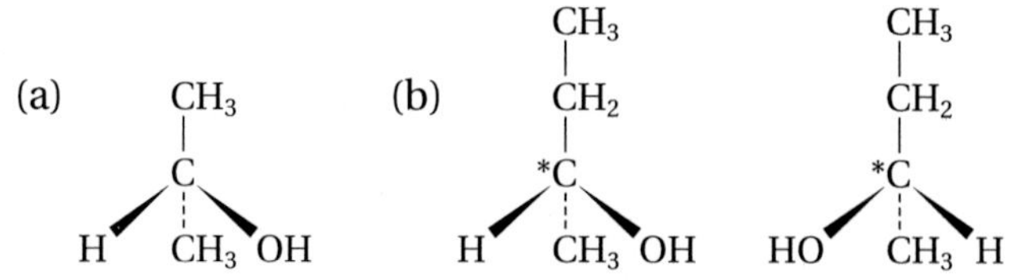

(a)

$$CH_3$$
$$|$$
$$C$$
H CH_3 OH

(b)

$$CH_3$$
$$|$$
$$CH_2$$
$$|$$
$$*C$$
H CH_3 OH

$$CH_3$$
$$|$$
$$CH_2$$
$$|$$
$$*C$$
HO CH_3 H

Stereoisomers

PRACTICE EXERCISE 16.3

Draw stereorepresentations for each of the following compounds. Include stereoisomers if they exist.

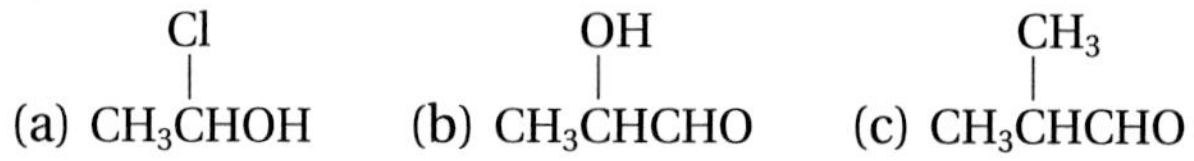

(a) $CH_3\overset{\overset{\displaystyle Cl}{|}}{C}HOH$ (b) $CH_3\overset{\overset{\displaystyle OH}{|}}{C}HCHO$ (c) $CH_3\overset{\overset{\displaystyle CH_3}{|}}{C}HCHO$

Fischer projections

Fischer projections *are a way of representing three-dimensional organic molecular structures in two dimensions.* Devised by the German chemist Emil Fischer (1852–1915), they are a convenient means of differentiating stereoisomers. Here are the Fischer projections for D- and L-glyceraldehyde:

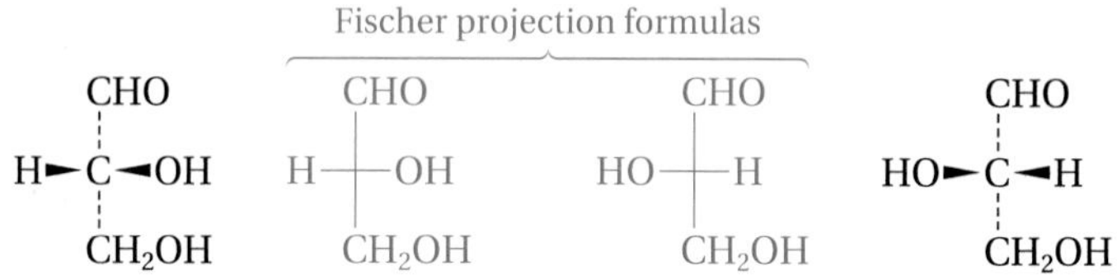

Fischer projection formulas

$$CHO$$
$$H{\blacktriangleright}C{\blacktriangleleft}OH$$
$$CH_2OH$$
D-Glyceraldehyde

$$CHO$$
$$H{-}{|}{-}OH$$
$$CH_2OH$$

$$CHO$$
$$HO{-}{|}{-}H$$
$$CH_2OH$$
L-Glyceraldehyde

$$CHO$$
$$HO{\blacktriangleright}C{\blacktriangleleft}H$$
$$CH_2OH$$

Fischer projections place the principal functional group (in this case aldehyde) at the top of the carbon chain. The carbon chain is written vertically, with substituent groups to the left and right. The asymmetric carbon is in the plane of the paper; the chain carbons attached to the asymmetric carbon are below the plane. The groups to the right and left of the asymmetric carbon are above the plane. The Fischer projection formula of glyceraldehyde that shows the hydroxyl group to the right of the carbon chain is D-glyceraldehyde; the Fischer projection formula that shows the hydroxyl group to the left of the carbon chain is L-glyceraldehyde. The naturally occurring form of glyceraldehyde is D-glyceraldehyde. Virtually all sugars found in nature are of the D family.

Optical activity

The D and L forms of stereoisomers are identical in their physical and chemical properties except that they rotate plane-polarized light in opposite

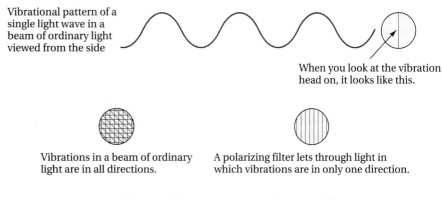

Vibrational pattern of a single light wave in a beam of ordinary light viewed from the side

When you look at the vibration head on, it looks like this.

Vibrations in a beam of ordinary light are in all directions.

A polarizing filter lets through light in which vibrations are in only one direction.

Polarizing filter   Sample tube   Polarizing filter

D-(+) glyceraldehyde   Rotation to right

L-(−) glyceraldehyde   Rotation to left

Figure 16.3
One of a pair of optical isomers rotates plane-polarized light to the right. Its mirror image rotates the light to the left. The instrument used to measure optical rotation is called a *polarimeter.*

directions. **Plane-polarized light** *is light in which vibrations are in only one direction.* Stereoisomers are often called *optical isomers.* **Optical isomers** *are stereoisomers that rotate plane-polarized light in opposite directions.* For example, D-glyceraldehyde rotates plane-polarized light to the right, and L-glyceraldehyde rotates plane-polarized light to the left. Sometimes you may see the name D-(+)-glyceraldehyde. The D refers to the handedness of the particular glyceraldehyde; the plus sign refers to the direction of rotation of plane-polarized light (Fig. 16.3). There is no connection between the D and the (+); other D sugars may rotate plane-polarized light to the left. Then the sugar would be designated D-(−).

16.2 Monosaccharides

AIMS: *To draw open-chain Fischer projections for the common simple sugars. To identify a sugar as D or L by looking at its Fischer projection formula.*

Focus

Most monosaccharides contain four, five, or six carbons.

The simplest carbohydrate molecules, not bonded to any other carbohydrate, are called **simple sugars** *or* **monosaccharides.** Many of the monosaccharides in nature contain four, five, or six carbons. Sugars containing an aldehyde functional group and consisting of four, five, and six carbons are *aldotetroses, aldopentoses,* and *aldohexoses,* respectively. In this section we will examine some of them.

The four-carbon sugars

D-Threose, with a chain of four carbons, is a naturally occurring tetrose; the aldehyde functional group makes threose an aldotetrose. Two of the carbons of threose, carbon 2 and carbon 3, have four different groups attached, so threose has two asymmetric carbons:

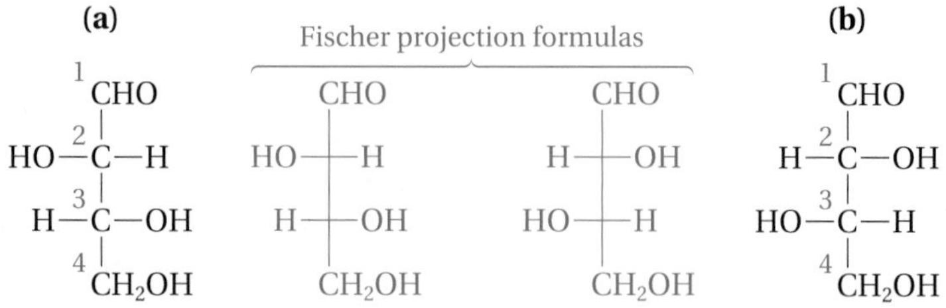

(a)

$$\overset{1}{C}HO$$
$$HO-\overset{2}{C}-H$$
$$H-\overset{3}{C}-OH$$
$$\overset{4}{C}H_2OH$$

D-Threose

Fischer projection formulas

$$CHO$$
$$HO-\!\!\!-\!\!\!-H$$
$$H-\!\!\!-\!\!\!-OH$$
$$CH_2OH$$

$$CHO$$
$$H-\!\!\!-\!\!\!-OH$$
$$HO-\!\!\!-\!\!\!-H$$
$$CH_2OH$$

L-Threose

(b)

$$\overset{1}{C}HO$$
$$H-\overset{2}{C}-OH$$
$$HO-\overset{3}{C}-H$$
$$\overset{4}{C}H_2OH$$

The Fischer projections of the stereoisomer have an —OH group on each of their two asymmetric centers, one pointing to the left and one to the right. If they were not labeled for handedness, how could you tell which is the D and which is the L isomer? First, draw a Fischer projection of the sugar being considered. Then look at the hydroxyl group attached to the last asymmetric carbon in the chain. If the hydroxyl group points to the right, the sugar belongs to the D family; if it points to the left, the sugar belongs to the L family. In the case of threose, the last asymmetric carbon in the chain is at carbon 3, so structure (a) belongs to the D family of sugars. Structure (b) belongs to the L family.

The structures of D- and L-threose are mirror images. However, since there are two asymmetric carbons in a four-carbon sugar molecule, another pair of stereoisomers can exist. These stereoisomers are D- and L-erythrose:

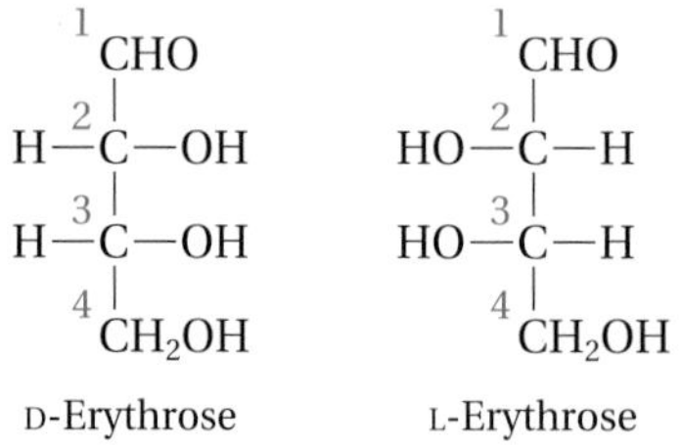

$$\overset{1}{C}HO$$
$$H-\overset{2}{C}-OH$$
$$H-\overset{3}{C}-OH$$
$$\overset{4}{C}H_2OH$$

D-Erythrose

$$\overset{1}{C}HO$$
$$HO-\overset{2}{C}-H$$
$$HO-\overset{3}{C}-H$$
$$\overset{4}{C}H_2OH$$

L-Erythrose

It is easy to calculate the number of possible stereoisomers of a sugar that contains multiple asymmetric carbons. This number is 2^n, where n is the number of asymmetric carbons. Thus glyceraldehyde, with one asymmetric carbon, has 2^1, or 2, stereoisomers; these are the D and L isomers. Four-carbon sugars with two asymmetric carbons can exist as 2^2, or 4, stereoisomers. We can calculate the number of mirror-image pairs by dividing the number of stereoisomers by 2. The four aldotetroses form two pairs of mirror images.

PRACTICE EXERCISE 16.4

Draw Fischer projections for the two aldotetroses that belong to the D family.

The five-carbon sugars

Several aldopentoses are found in nature. D-Arabinose and D-xylose are five-carbon sugars produced by plants. D-Arabinose is sometimes called *pectin sugar*. Pectin, the polysaccharide from which it is obtained, forms gels that are useful in making jelly. Because it is isolated from wood, D-xylose is sometimes called *wood sugar.*

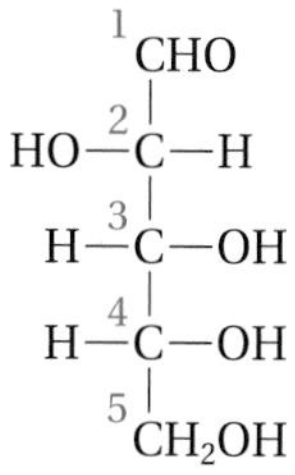

$$
\begin{array}{cc}
\overset{1}{C}HO & \overset{1}{C}HO \\
HO-\overset{2}{C}-H & H-\overset{2}{C}-OH \\
H-\overset{3}{C}-OH & HO-\overset{3}{C}-H \\
H-\overset{4}{C}-OH & H-\overset{4}{C}-OH \\
\overset{5}{C}H_2OH & \overset{5}{C}H_2OH \\
\text{D-Arabinose} & \text{D-Xylose}
\end{array}
$$

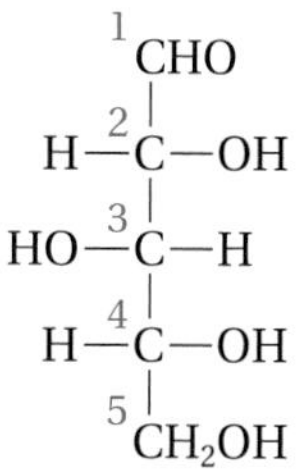

Other important aldopentoses are D-ribose and D-2-deoxyribose, a related compound that lacks an —OH group at carbon 2:

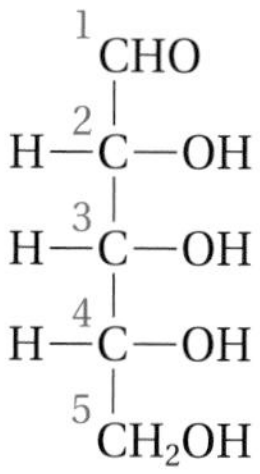

$$
\begin{array}{cc}
\overset{1}{C}HO & \overset{1}{C}HO \\
H-\overset{2}{C}-OH & H-\overset{2}{C}-H \\
H-\overset{3}{C}-OH & H-\overset{3}{C}-OH \\
H-\overset{4}{C}-OH & H-\overset{4}{C}-OH \\
\overset{5}{C}H_2OH & \overset{5}{C}H_2OH \\
\text{D-Ribose} & \text{D-2-Deoxyribose}
\end{array}
$$

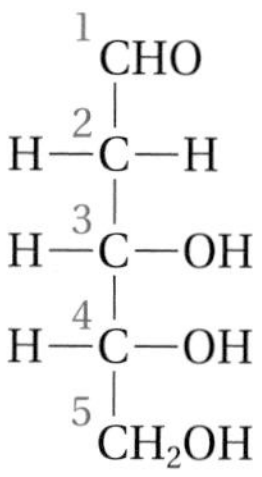

These two sugars are an integral part of the hereditary materials ribonucleic acid (RNA) and deoxyribonucleic acid (DNA). The structures of RNA and DNA are discussed in Chapter 20.

The six-carbon sugars

Only three aldohexoses appear in nature: D-glucose, D-galactose, and D-mannose:

The relative sweetness of sugars and sugar substitutes varies over a wide range. Lactose is about one-sixth as sweet and glucose about three-quarters as sweet as sucrose (table sugar). Fructose is not quite twice as sweet as sucrose. The artificial sweeteners aspartame (NutraSweet) and saccharin are about 150 and 500 times as sweet as sucrose.

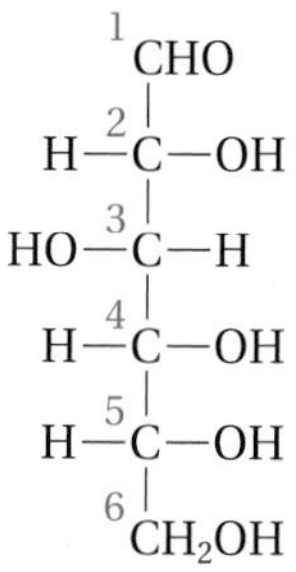

$$
\begin{array}{ccc}
\overset{1}{C}HO & \overset{1}{C}HO & \overset{1}{C}HO \\
H-\overset{2}{C}-OH & H-\overset{2}{C}-OH & HO-\overset{2}{C}-H \\
HO-\overset{3}{C}-H & HO-\overset{3}{C}-H & HO-\overset{3}{C}-H \\
H-\overset{4}{C}-OH & HO-\overset{4}{C}-H & H-\overset{4}{C}-OH \\
H-\overset{5}{C}-OH & H-\overset{5}{C}-OH & H-\overset{5}{C}-OH \\
\overset{6}{C}H_2OH & \overset{6}{C}H_2OH & \overset{6}{C}H_2OH \\
\text{D-Glucose} & \text{D-Galactose} & \text{D-Mannose}
\end{array}
$$

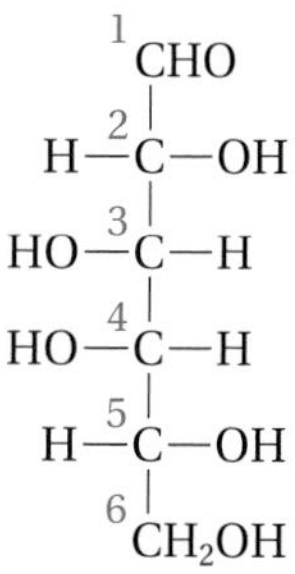

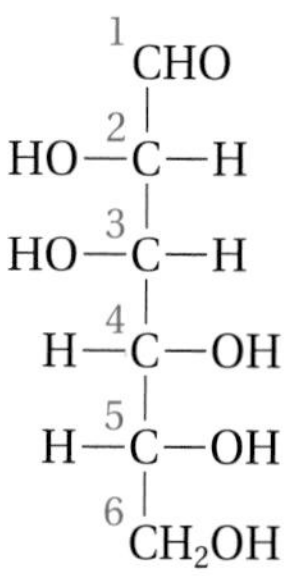

The D form of glucose has a central role in the nutrition of virtually all species, including plants and humans. The biochemistry of glucose is so important that Chapter 24 is devoted to it. D-Glucose is abundant in all life forms. Depending on the source, it has been called *grape sugar, corn sugar,* and *blood sugar.* Urine usually contains a trace of D-glucose, but the concentration is greatly increased in the urine of patients with untreated diabetes mellitus. D-Galactose is a constituent of lactose, also called *milk sugar* (see Sec. 16.7). D-Mannose is a major constituent of polymeric molecules called *mannans,* which are found in several plants.

Monosaccharides that contain a ketone functional group $-\overset{\overset{\text{O}}{\parallel}}{\text{C}}-$ *are called* **ketoses.** Ketoses containing three, four, five, and six carbons are *ketotrioses, ketotetroses, ketopentoses,* and *ketohexoses,* respectively. No discussion of hexoses would be complete without including D-fructose, a ketohexose because of the presence of a ketone carbonyl group in the molecule at carbon 2. D-Fructose and D-glucose differ in structure only at carbons 1 and 2. The identical stereochemistry at carbons 3, 4, and 5 exists because the breakdown of D-glucose in living systems involves conversion of D-glucose to D-fructose.

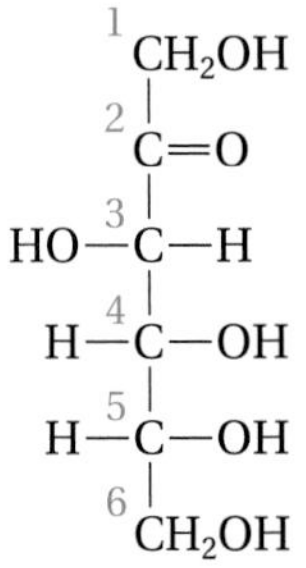

$$
\begin{array}{c}
\overset{1}{\text{C}}\text{H}_2\text{OH} \\
|\\
\overset{2}{\text{C}}=\text{O} \\
|\\
\text{HO}-\overset{3}{\text{C}}-\text{H} \\
|\\
\text{H}-\overset{4}{\text{C}}-\text{OH} \\
|\\
\text{H}-\overset{5}{\text{C}}-\text{OH} \\
|\\
\overset{6}{\text{C}}\text{H}_2\text{OH}
\end{array}
$$

D-Fructose

D-Fructose occurs in a large number of fruits and in honey. It is also the only sugar found in human semen. D-Fructose is one of those sugars which belongs to the D family but rotates plane-polarized light in a left-handed direction.

PRACTICE EXERCISE 16.5

Identify each structure as D or L.

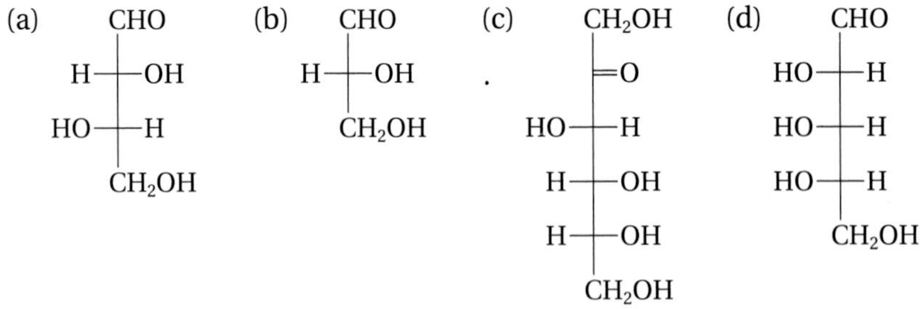

(a)
$$
\begin{array}{c}
\text{CHO} \\
|\\
\text{H}-\!\!\!-\text{OH} \\
|\\
\text{HO}-\!\!\!-\text{H} \\
|\\
\text{CH}_2\text{OH}
\end{array}
$$

(b)
$$
\begin{array}{c}
\text{CHO} \\
|\\
\text{H}-\!\!\!-\text{OH} \\
|\\
\text{CH}_2\text{OH}
\end{array}
$$

(c)
$$
\begin{array}{c}
\text{CH}_2\text{OH} \\
|\\
=\!\!\!\text{O} \\
|\\
\text{HO}-\!\!\!-\text{H} \\
|\\
\text{H}-\!\!\!-\text{OH} \\
|\\
\text{H}-\!\!\!-\text{OH} \\
|\\
\text{CH}_2\text{OH}
\end{array}
$$

(d)
$$
\begin{array}{c}
\text{CHO} \\
|\\
\text{HO}-\!\!\!-\text{H} \\
|\\
\text{HO}-\!\!\!-\text{H} \\
|\\
\text{HO}-\!\!\!-\text{H} \\
|\\
\text{CH}_2\text{OH}
\end{array}
$$

PRACTICE EXERCISE 16.6

Draw the Fischer projection formula for D-glucose. Number the carbons, and identify each asymmetric carbon with an asterisk.

16.3 Cyclic structures

AIMS: To describe the bonding that results in cyclic forms of sugars. To classify simple sugars as a pyranose or a furanose and as a hemiacetal or a hemiketal.

Focus

Pentoses and hexoses exist mainly as ring structures.

Until now, saccharides have been depicted as straight-chain compounds. The reality is somewhat different, however, for pentoses and hexoses exist primarily in five- and six-membered rings or cyclic forms. Examples of two such cyclic forms, those of D-glucose and D-fructose, are shown in Figure 16.4.

The five-membered sugar ring system is given the general name **furanose** *after the parent cyclic ether furan; the six-membered sugar ring system is considered a derivative of pyran and is called a* **pyranose.**

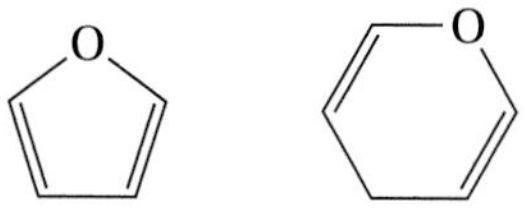

Furan   Pyran

To name a sugar in its cyclic form precisely, remove the *-se* from the sugar name and add *furanose* or *pyranose* according to whether the cyclic form is a five-membered or a six-membered ring. Thus the cyclic form of D-fructose is properly called D-*fructofuranose* and that of D-glucose is called D-*glucopyranose.*

To understand how these cyclic structures are formed from the straight-chain sugars, recall that alcohols can add to carbonyl groups of aldehydes or ketones to form hemiacetals or hemiketals, and five- and six-membered rings are more stable than smaller rings. Hemiacetal formation, discussed previously (Sec. 13.6), is the addition of an alcohol to a carbonyl group of an aldehyde. Hemiketals are formed in a similar way from ketones and alcohols.

Sugars such as D-glucose and D-fructose contain a carbonyl group and several hydroxyl groups in the same molecule. Figure 16.5 illustrates the

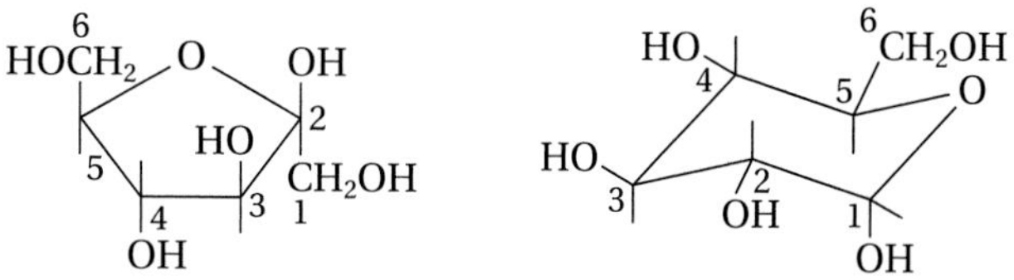

Cyclic form of
D-fructose
(D-fructofuranose)

Cyclic form of
D-glucose
(D-glucopyranose)

Figure 16.4
The cyclic forms of two sugars drawn as in their true shapes. The five- and six-membered rings resemble those of cyclopentane and cyclohexane, respectively.

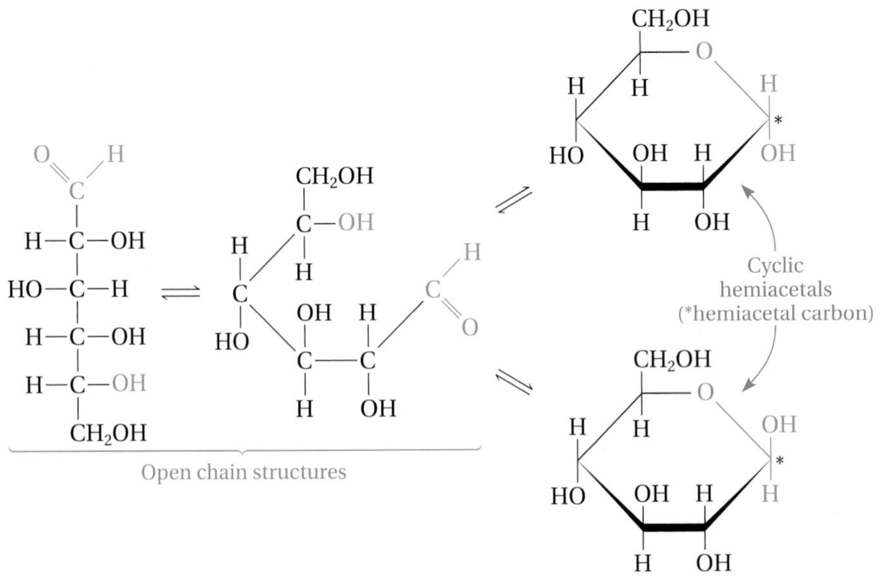

Figure 16.5
An intramolecular reaction between the aldehyde and hydroxyl groups of a sugar, in this case D-glucose, results in formation of a cyclic hemiacetal.

possible internal hemiacetal or hemiketal formation. The internal hydroxyl group that is selected for reaction is one that will give a five- or six-membered ring depending on the saccharide. Smaller rings would be unstable because of ring strain; larger rings are not formed because the more distant ends of the molecule do not often collide in solution.

16.4 Haworth projections

AIMS: To draw Haworth projections for the common simple sugars. To identify a simple sugar as an alpha or beta anomer. To explain the interconversion of closed-chain forms of sugars.

Focus

The stereochemistry of the cyclic forms of sugars is depicted by Haworth projections.

The stereochemistry of the cyclic forms of sugars is often represented by their **Haworth projections**—*standardized ways of depicting the positions of hydroxyl groups in space.* In viewing Haworth projections, envision the plane of the ring as tilted perpendicular to the plane of the paper. The attached groups are above and below the plane of the ring. Figure 16.6 shows Haworth projections for D-glucopyranose and D-fructofuranose. Two forms of each of these sugars are possible; these are designated *alpha* and *beta*. Two cyclic forms are possible because, in going from a straight chain to a ring, a new asymmetric carbon is introduced at carbon 1 (the hemiacetal carbon) of aldoses and carbon 2 (the hemiketal carbon) of

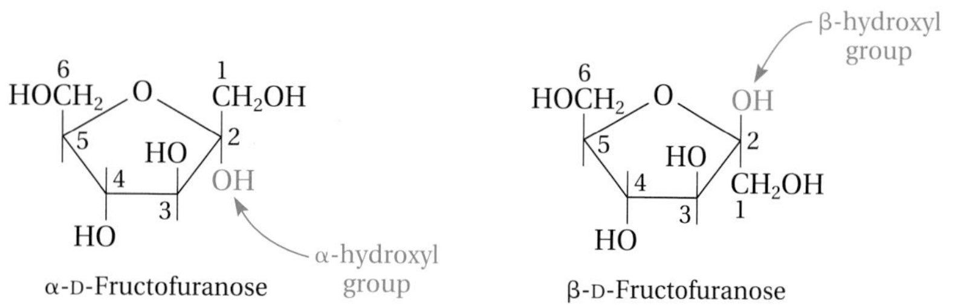

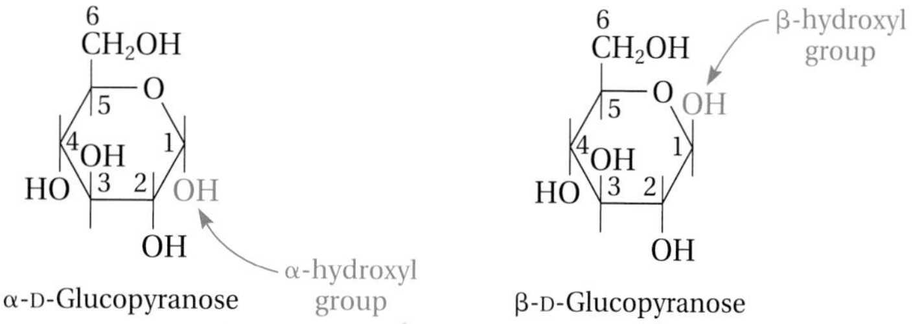

Figure 16.6
Haworth projections of D-glucopyranose and D-fructofuranose. Two different stereoisomers, labeled α and β, may be formed on hemiacetal formation.

ketoses. In the Haworth projections of D-glucopyranose, for example, the hydroxyl group at carbon 1, which is formed from the aldehyde functional group of the straight-chain sugar, may end up below (alpha or Greek α) or above (beta or Greek β) the plane of the pyranose ring.

Anomers *are sugars that differ in stereochemistry only at the hemiacetal or hemiketal carbon.* The alpha and beta anomers of the cyclic forms of sugars have different melting points and different abilities to rotate plane-polarized light. Alpha-D-glucose melts at 146 °C, for example, but beta-D-glucose melts at 150 °C. These differences help to demonstrate again how small changes in molecular shape or structure dramatically affect the physical properties of molecules.

How to draw Haworth projections

Conversion of a straight-chain Fischer projection to a ring in a Haworth projection is easily accomplished for D sugars. Hydroxyl groups that point to the left in a Fischer projection of a D sugar point *up* in the Haworth projection. Hydroxyl groups that point to the right in a Fischer projection point *down* in the Haworth projection.

EXAMPLE 16.3   **Drawing a Haworth projection**

Draw the Haworth projection of α-D-glucopyranose.

SOLUTION

The α-D-glucopyranose molecule is the cyclic form of D-glucose. Draw a projection of D-glucose in the straight-chain form:

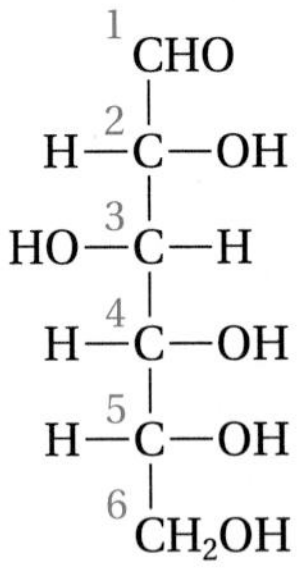

D-Glucose

Draw a six-membered pyranose ring in its abbreviated form as shown:

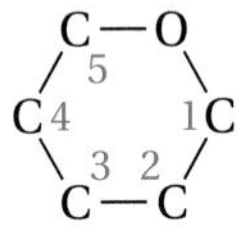

Pyranose ring 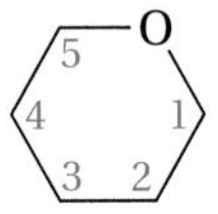   Abbreviated pyranose ring

Put in the —CH$_2$OH group of carbon 6 of the hexose. In a D sugar, the sixth carbon is always above the plane of the ring as shown:

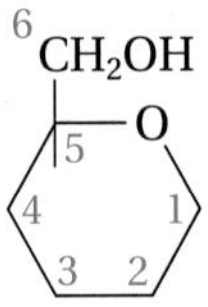

Fill in the —OH groups on carbons 2, 3, and 4. Notice that the oxygen on carbon 5 of the chain form is now in the ring and need not concern us. Hydroxyl groups to the right in the Fischer projection go below the plane of the ring. Those to the left are above the plane. Ring hydrogens are usually omitted for clarity.

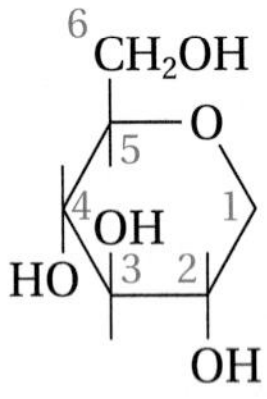

Finally, write the anomeric —OH group at carbon 1—in this case alpha (below the plane of the ring).

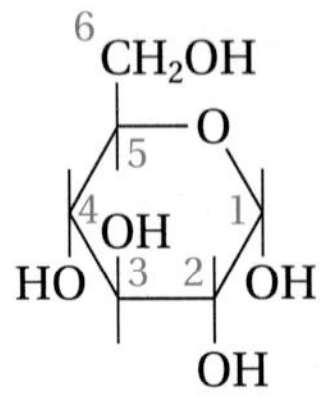

α-D-Glucopyranose

> **PRACTICE EXERCISE 16.7**
>
> Draw the Haworth projection of β-D-glucopyranose. Identify the hemiacetal carbon.
>
> **PRACTICE EXERCISE 16.8**
>
> Draw the Haworth projection of α-D-ribofuranose. Identify the hemiacetal carbon.
>
> **PRACTICE EXERCISE 16.9**
>
> Draw the Haworth projection of α-D-fructofuranose. Identify the hemiketal carbon.

Interconversion of straight-chain and ring forms of sugars

The straight-chain sugar forms are in equilibrium with the ring forms. The ring forms are usually quite predominant. For example, if stereochemically pure α-D-glucopyranose is dissolved in an acidic solution, the ring will open and close repeatedly. In reclosing, some β-D-glucopyranose is formed. The final equilibrium mixture consists of about 63% β-D-glucopyranose, about 37% α-D-glucopyranose, and only a tiny amount of the straight-chain aldehyde. From the percentages of products formed, we can say that β-D-glucopyranose is only slightly more preferred than α-D-glucopyranose and that both D-glucopyranose anomers are much more preferred than the straight-chain aldehyde form of D-glucose.

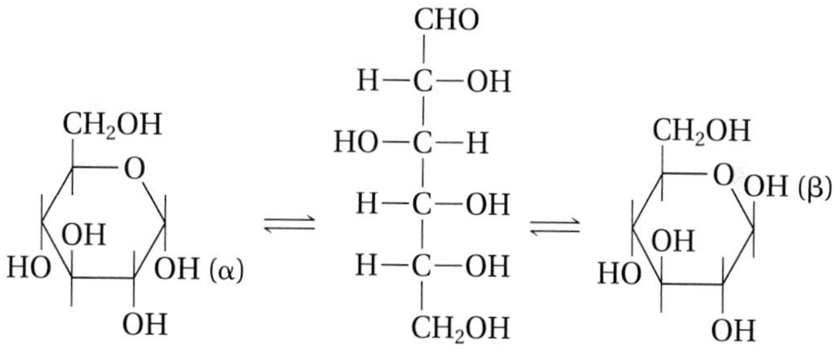

α-D-Glucopyranose (about 37%)   D-Glucose (less than 1%)   β-D-Glucopyranose (about 63%)

16.5 Glycosides

AIM: To describe the formation of glycosidic bonds and the products of their hydrolysis.

Focus

Alcohols react with closed-chain forms of sugars to form glycosidic bonds.

The closed-chain hemiacetal or hemiketal forms of sugars may react with alcohols to form acetals or ketals (see Sec. 13.6). *The acetals or ketals of sugars are called* **glycosides.**

Glycosidic bonds

The covalent ether link between the sugar hydroxyl and the alcohol is a **glycosidic bond.** A simple alcohol such as methanol and a sugar such as α-D-

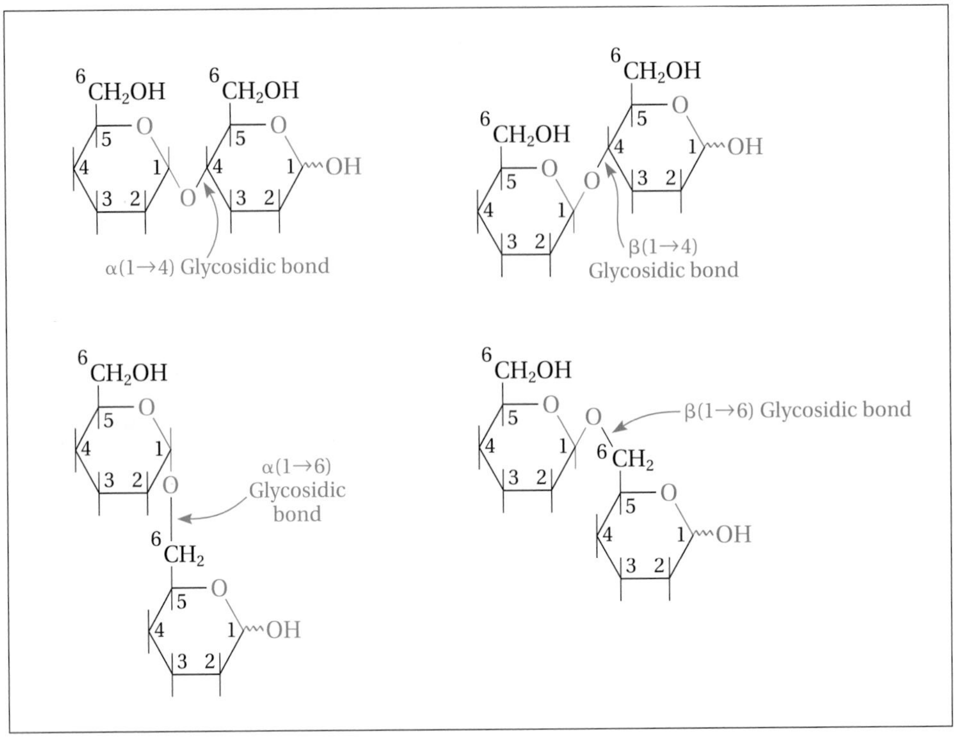

Figure 16.7
Some of the common glycosidic bonds found in polysaccharides. The acetal and hemiacetal portions of the molecules are shown in color. The wavy line connecting the hydroxyl group to carbon 1 indicates that the carbon-oxygen linkage may be either alpha or beta.

glucopyranose produce methyl α-D-glucopyranoside:

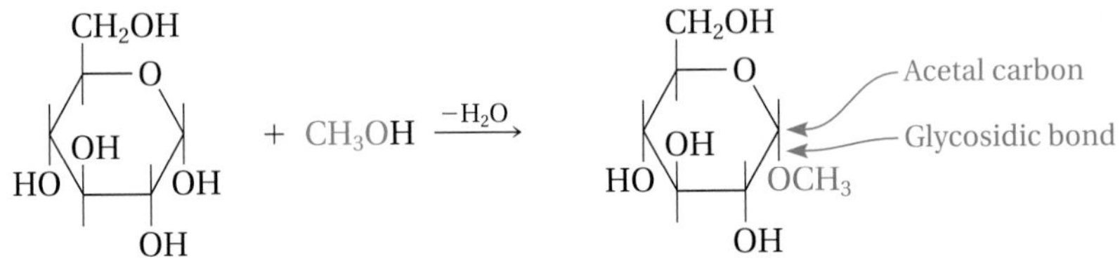

α-D-Glucopyranose Methanol Methyl α-D-glucopyranoside

The alcohol used to make a glycosidic bond is often more complex than methanol—in fact, sugars themselves are alcohols. As shown in Figure 16.7, the individual saccharide units are attached through glycosidic bonds. Glycosidic bonds between sugars are designated according to the position numbers of the carbons of the sugars that are linked and also according to the stereochemistry of the linkage. For example, suppose the beta hydroxyl group at carbon 1 in a hexose is linked by a glycosidic bond to carbon 4 of another hexose. This linkage is called a $\beta(1\rightarrow4)$ glycosidic bond. Other common linkages are $\alpha(1\rightarrow4)$, $\alpha(1\rightarrow6)$, and $\beta(1\rightarrow6)$. Once the anomeric —OH group of a sugar is tied up as an acetal, it is no longer free to go from the ring form to the straight-chain form.

PRACTICE EXERCISE 16.10

The hydroxyl group of carbon 1 in α-D-glucopyranose is linked by a glycosidic bond to carbon 4 of another D-glucopyranose molecule. Draw the structure of the glycoside that is formed. Identify the acetal carbon.

Hydrolysis of glycosidic bonds

Glycosidic bonds may be cleaved by hydrolysis reactions. We can take the hydrolysis of an $\alpha(1\rightarrow4)$ glycosidic bond between two hexoses as an example. For simplicity, only the carbon skeleton and the glycosidic bond are shown:

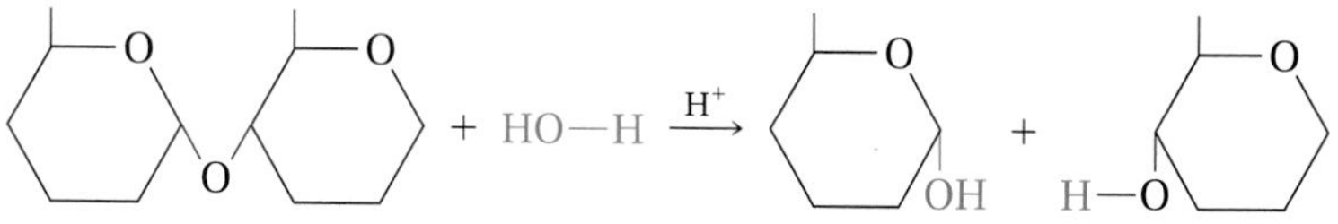

The chemical hydrolysis of most complex sugars can be done by heating an aqueous solution of the carbohydrate. A trace of acid is added as a catalyst. Enzymes act as the catalyst in biological systems. Hydrolysis reactions will be important as we proceed into biochemistry, since they are the means by which sugars, fats, and proteins are broken down to simple materials by digestion.

PRACTICE EXERCISE 16.11

The glycosidic bond in the following compound is hydrolyzed. What are the structures of the products?

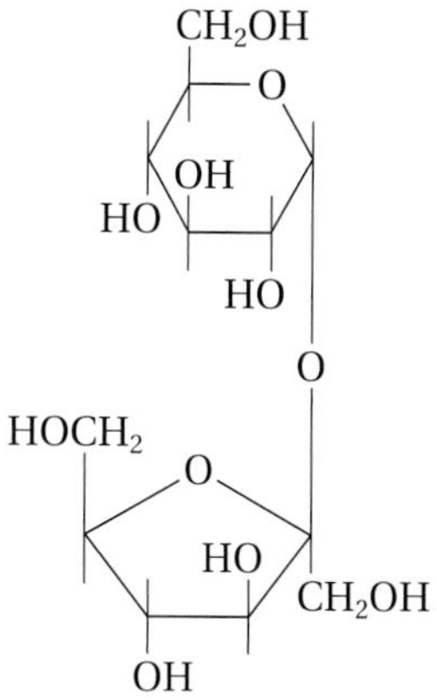

16.6 Polysaccharides

AIM: To list the structures, sources, and uses of the following polysaccharides: starch, amylose, amylopectin, glycogen, and cellulose.

Focus

Polysaccharides are composed of many monosaccharide units connected by glycosidic linkages.

Individual sugar units may be connected to one another to form linear, branched, or circular polymers, as shown in Figure 16.8. **Polysaccharides** *have many monosaccharides bonded together to form a long polymer chain.* The bonds connecting the sugar units are glycosidic. The $1\rightarrow4$ and $1\rightarrow6$ linkages are the ones most commonly found in natural polysaccharides consisting of hexoses.

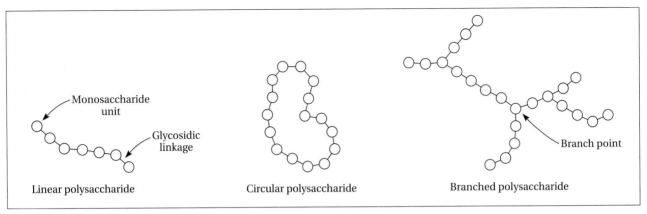

Figure 16.8
Sugar units in a polysaccharide can form linear, circular, or branched structures.

There are four human blood types, A, B, AB, and O. These blood types occur because of differences in composition of relatively small polysaccharide chains on the surface of red blood cells.

Starches *are polymers consisting entirely of* D-*glucose units.* They are the major storage form of D-glucose in plants. Two starches, amylose and amylopectin, are the most common. Amylose is composed of a linear chain of D-glucopyranose molecules linked $\alpha(1\rightarrow4)$ as shown in Figure 16.9. Complete amylose molecules may contain anywhere from a few to about 3000 D-glucopyranose units. Amylopectin consists of chains of D-glucopyranose molecules held together by $\alpha(1\rightarrow4)$ glycosidic linkages, but as shown in Figure 16.10, amylopectin also contains $\alpha(1\rightarrow6)$ cross-linking. These cross-links give amylopectin the appearance of a branched, bushy molecule. There are usually 24 to 30 D-glucopyranose units between the branch points of amylopectin.

Glycogen—*the main storage form of* D-*glucose in animal cells*—has a

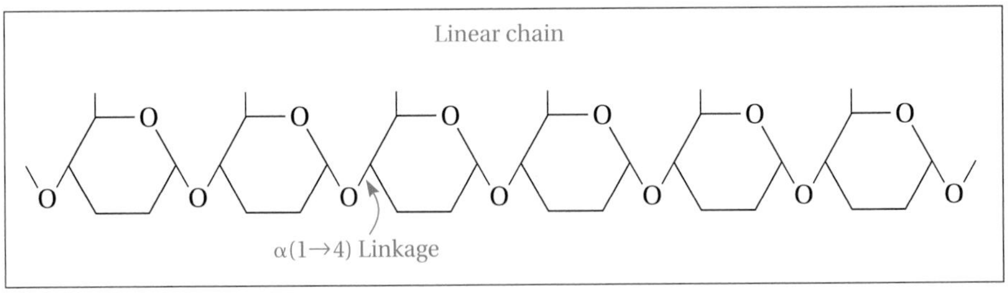

Figure 16.9
Partial structure of the starch molecule amylose.

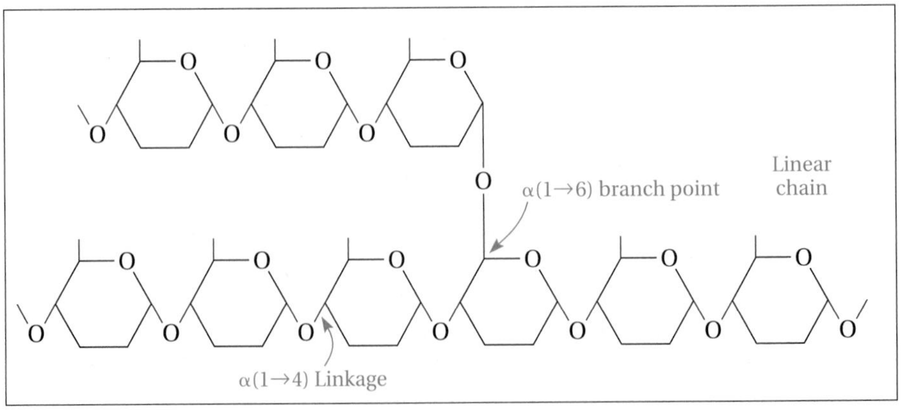

Figure 16.10
Partial structure of an amylopectin molecule.

structure similar to amylopectin. The main difference between amylopectin and glycogen is that glycogen is even more highly branched—there are only 8 to 12 D-glucopyranose units between branch points. Glycogen is especially abundant in the liver, where it may amount to as much as 8% of the dry mass of the organ.

Figure 16.11 shows **cellulose,** *the most important structural polysaccharide.* Cellulose is the material of plant cell walls, and it is the major component in wood. Cotton is about 80% pure cellulose. Like amylose, cellulose is a linear polymer consisting of D-glucose units. These D-glucose units, usually from 300 to 15,000 of them, are connected by 1→4 linkages that are beta rather than alpha. The shapes of amylose and cellulose molecules are quite different because of it—$\alpha(1{\rightarrow}4)$-linked amylose tends to form loose spiral structures (Fig. 16.12) and $\beta(1{\rightarrow}4)$-linked cellulose tends to form straight chains.

The difference in the shapes of amylose and cellulose has a tremendous

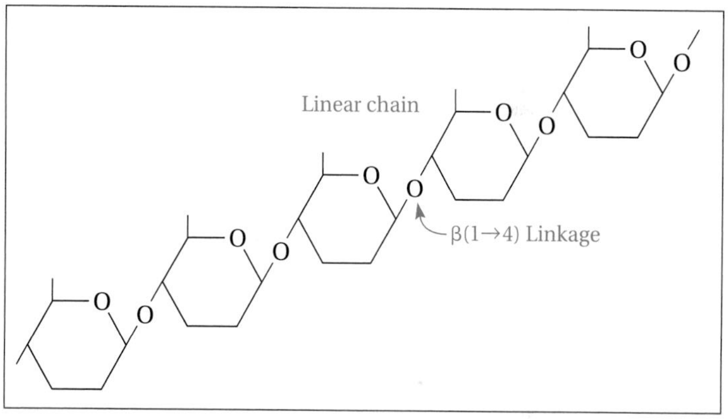

Cellulose is a very important raw material. The wood used in construction, the cotton used in textile production, and the manufacture of paper and paper products all have a large economic impact.

Figure 16.11
Partial structure of a cellulose molecule.

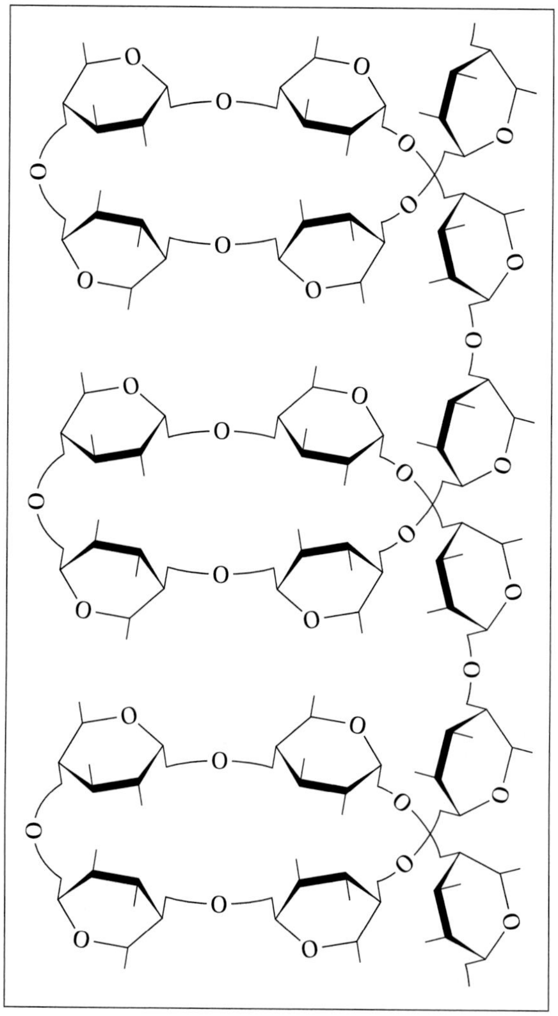

Figure 16.12
Amylose forms loose spiral structures.

biological effect. The linear chains of cellulose present a uniform surface of hydroxyl groups. These hydroxyl groups are involved in hydrogen bonding between adjacent cellulose molecules; the large number of these weak interactions give wood and cotton fibers their strength. A second effect of the two different shapes is that enzymes capable of catalyzing the hydrolysis of starch are not capable of hydrolyzing cellulose. Human beings can convert starch to its fuel form, D-glucose, but people lack enzymes to catalyze the hydrolysis of cellulose to glucose. Sawdust has little potential as snack food for people. On the other hand, undigestible dietary carbohydrates do seem to be important for human health as discussed in A Closer Look: Dietary Fiber. The digestive systems of the cud-chewing animals (ruminants) such as cows, sheep, and goats, as well as those of termites, contain microorganisms whose enzymes catalyze the production of glucose from cellulose. These animals can therefore use cellulose as a nutritional source.

PRACTICE EXERCISE 16.12

How does the structure of amylose differ from the structure of amylopectin?

PRACTICE EXERCISE 16.13

What is the product of the complete hydrolysis of each of the following polysaccharides?

(a) amylose   (b) amylopectin   (c) glycogen   (d) cellulose

A Closer Look

Dietary Fiber

Have you had your fiber today? Many experts recommend that people eat high-fiber diets in order to prevent colon cancer. The nondigestible carbohydrates that we eat constitute dietary fibers. Dietary fibers consist of polysaccharides that cannot be hydrolyzed to monosaccharides and therefore cannot be absorbed into the bloodstream. Cellulose is insoluble fiber. Cellulose from vegetable leaves and stalks provides you with dietary bulk and helps to prevent constipation. There are also soluble fibers. The soluble dietary fibers are noncellulosic. Pectins from fruits, which are used to thicken jams and jellies, are examples of soluble fibers. The pectins are polyhydroxy compounds that have a carboxylic acid group at one end of the molecule and an aldehyde group at the other. Vegetable "gums" are also soluble fibers.

Fruits and vegetables are high in dietary fiber.

Which foods can we eat to ensure that we get enough dietary fiber? Wheat, brown rice, and bran cereals are high in insoluble fiber. Oats, barley, carrots, and fruits are high in soluble fiber. Peas and beans are a good source of both soluble and insoluble fiber.

16.7 Disaccharides

AIM: To write the structures and list the sources and uses of the disaccharides maltose and cellobiose.

Focus

Maltose and cellobiose are derived from the hydrolysis of starch and cellulose.

Disaccharides *are compounds in which two monosaccharides are bonded together.* Disaccharides are glycosides in which the alcohol is a second monosaccharide molecule. Maltose, or malt sugar, is a disaccharide derived from the partial hydrolysis of starch. Diastase, an enzyme found in germinating barley, catalyzes the hydrolysis of starch to maltose. Since starch is composed of many D-glucopyranose units connected by $\alpha(1\rightarrow4)$ glycosidic bonds and maltose is a disaccharide produced from starch, it is not surprising that maltose is composed of two D-glucopyranose molecules connected by an $\alpha(1\rightarrow4)$ glycosidic linkage. Maltose is hydrolyzed to D-glucose by the enzyme maltase.

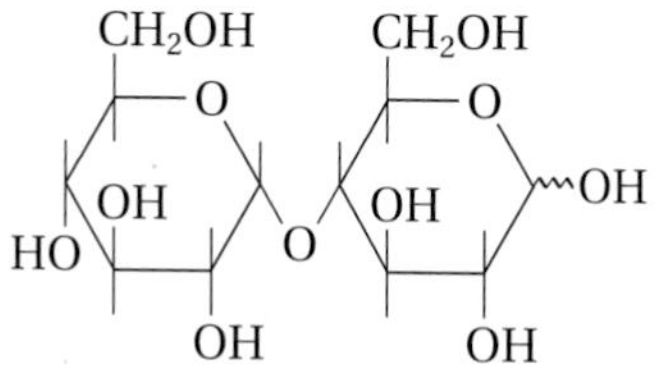

Maltose

Likewise, cellobiose is a disaccharide produced from the partial hydrolysis of cellulose by dilute acid. This disaccharide consists of two D-glucose molecules like maltose, but the $1\rightarrow4$ glycosidic bond that links the two monosaccharides is beta. Cellobiose is not hydrolyzed by the enzyme maltase.

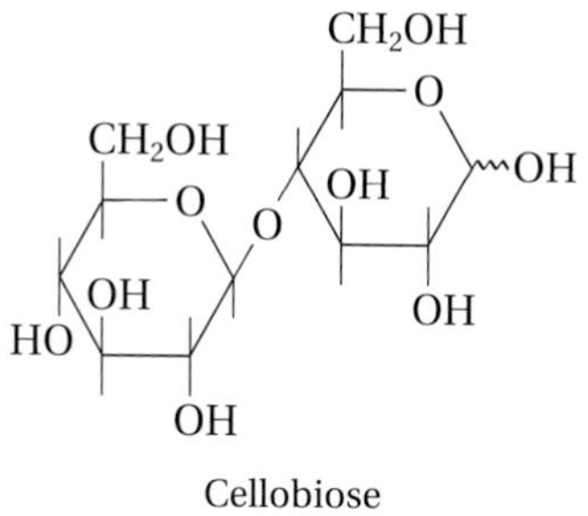

Cellobiose

PRACTICE EXERCISE 16.14

What is the relationship of maltose to cellobiose?

16.8 Sucrose and lactose

AIM: To write the structures and list the sources and uses of the disaccharides sucrose and lactose.

Focus

Sucrose and lactose are the sugars of sap and milk.

The most familiar sugar is sucrose—ordinary table sugar. Sucrose is isolated from the juice or sap of several plants, including sugarcane, sugar beets, and maple trees. The world's production of sucrose from these sources exceeds 7 billion kilograms per year. The sucrose molecule is a disaccharide composed of α-D-glucose and β-D-fructose:

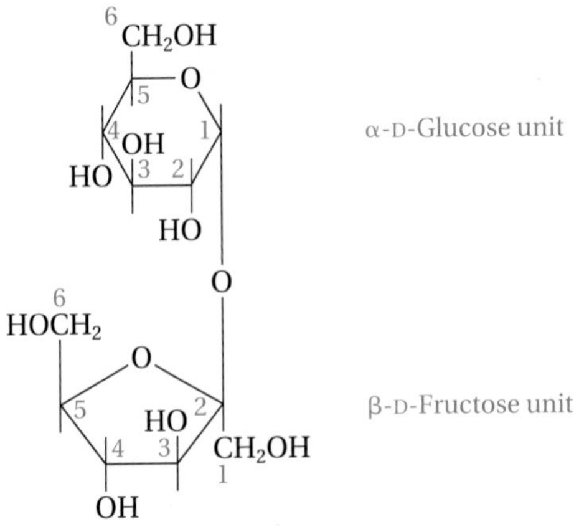

Sucrose

Structurally, the linkage between glucose and fructose in sucrose is unusual among sugars, since both sugars are linked together at the anomeric carbons (carbon 1 of each sugar). Hydrolysis of sucrose by acid or by enzymes gives **invert sugar**—*a mixture of equal molar quantities of glucose and fructose.*

Honey is a rich natural source of invert sugar. Invert sugar is also produced commercially and used when a noncrystalline sweetener is desired. The gooey syrup that traditionally bathes chocolate-covered cherries is one use of invert sugar.

As a food, sucrose has a high caloric value. Many people rely on artificial sweeteners such as saccharin, which is much sweeter than sugar. At one time calcium cyclamate, a chemical 30 times sweeter than sucrose, competed with saccharin as an artificial sweetener; then it was removed from the market as a possible cancer-causing agent. The use of saccharin has been reduced for the same reason.

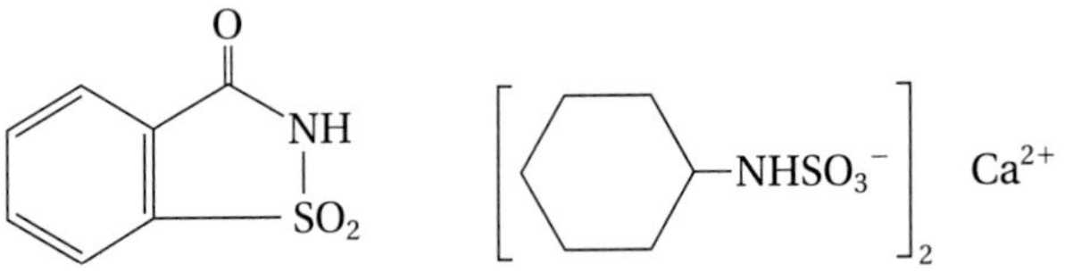

Saccharin   Calcium cyclamate

Lactose, or milk sugar, constitutes 5% of cow's milk and 7% of human milk. Pure lactose is obtained from whey, the watery by-product of cheese production. Lactose is composed of one molecule of D-galactose and one of D-glucose. The linkage between the two sugar units is $\beta(1\rightarrow4)$.

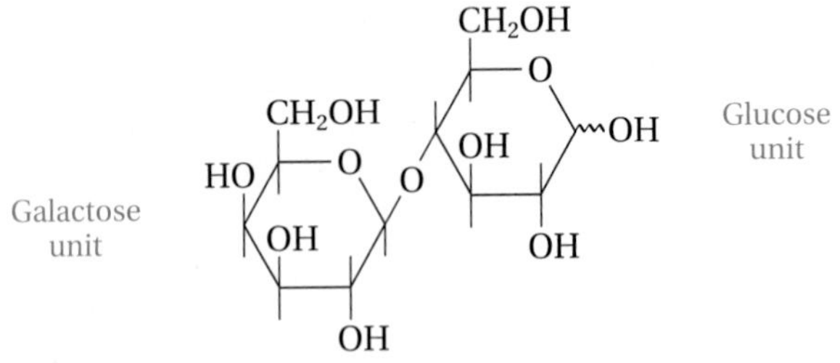

Lactose

Galactose metabolism is impaired in the genetic disease galactosemia. In this disease, galactose builds up to toxic levels, resulting in irreversible nerve damage and early cataract formation. Symptoms do not develop if the intake of galactose and lactose is restricted.

The glycosidic bond between the D-galactose and D-glucose portions of lactose may be cleaved by the enzyme *lactase*. Lactase deficiency is fairly common in people. Lactase deficiency is the reason for the discomfort experienced by Alberto, the student described in the Case in Point earlier in this chapter. In the follow-up, below, we learn more of this deficiency and how it can be treated.

PRACTICE EXERCISE 16.15

Give the names and structures of the compounds produced by the hydrolysis of the glycosidic bond in lactose.

FOLLOW-UP TO THE CASE IN POINT: Lactose intolerance

The physician correctly diagnosed Alberto's stomach cramps, gas, and diarrhea to be the result of lactose intolerance. Like Alberto, many adults and some children lack the enzyme *lactase*. As a result, these people are unable to hydrolyze lactose to its simpler sugar components, D-galactose and D-glucose. The symptoms of lactose intolerance are caused by undigested lactose that sits in the intestinal tract, where it causes cramps and diarrhea. The diarrhea is caused by an imbalance in the osmotic pressure on the walls of the intestine. Lactose draws water through the intestinal wall into the intestines, where the excessive water causes diarrhea, which can lead to dehydration. Some of the lactose can be oxidized by intestinal bacteria, releasing CO_2 into the intestines. The bubbles of gas can cause great discomfort. The effects of lactose intolerance can be avoided by excluding milk and milk products from the diet. Some people with lactose intolerance can eat yogurt, which is a fermented milk product. The fermentation of yogurt by bacteria breaks down the lactose present. In many instances, lactase production diminishes with age. A reduced level of lactase, combined with his consumption of a large amount of milk at one sitting, may have triggered Alberto's illness. Alberto later found that he was able to consume small amounts of dairy products without a repeat of the symptoms of lactose intolerance. Some other sufferers of lactose intolerance—estimated to be as many as one-third of all adult Americans—can avoid the unpleasant symptoms by adding readily available lactase tablets to their milk (see figure). Lactose-free milk is also available at some groceries.

Lactose intolerance is fairly common in infants, especially among babies of Middle Eastern, Asian, and African descent. Europeans are statistically least susceptible. Dehydration from diarrhea from lactose intolerance or any other cause is always a hazard for babies because, compared with adults, their bodies contain a relatively small mass of water. Switching a lactase-deficient baby from natural milk sources to a commercial milk substitute alleviates the diarrhea and other symptoms.

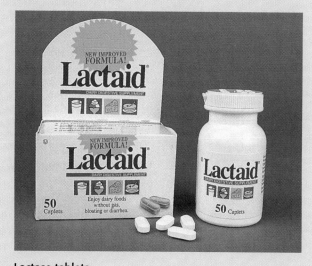

Lactase tablets.

16.9 Reducing and nonreducing sugars

AIM: To predict, on the basis of molecular structure, whether a carbohydrate is reducing or nonreducing.

You may recall from Section 13.5 that Benedict's and Tollens' reagents oxidize aldehydes and alpha-hydroxy ketones. When Benedict's reagent, an alkaline solution of copper(II) sulfate, is the oxidizing agent, the copper(II) ions are reduced to copper(I), and insoluble red copper(I) oxide (Cu_2O) is formed. With Tollens' reagent, silver ions (Ag^+) are the oxidizing agent, and these ions are reduced to silver metal (Ag). The silver deposits as a mirror on the sides of the reaction vessel.

Many sugars behave the same way toward Benedict's and Tollens' reagents as simple aldehydes and alpha-hydroxy ketones. Sugars such as D-glucose and D-fructose give a brick-red color with Benedict's reagent; with Tollens' reagent the silver mirror test is a positive indicator. **Reducing sugars** *give positive tests with Benedict's and Tollens' reagents.* **Nonreducing sugars** *do not give positive tests.* Positive tests are obtained with sugars in which the cyclic hemiacetal or hemiketal forms are in equilibrium with their straight-chain form. For example, even though the equilibrium may greatly favor the hemiacetal, oxidation of all the sugar occurs. As soon as the aldehyde is oxidized by the test reagent, more aldehyde is produced as the equilibrium between hemiacetal and aldehyde is reestablished.

Sugars do not give positive tests with Benedict's and Tollens' reagents if the cyclic and aldehyde forms do not exist in equilibrium with the aldehyde form. Any sugar that is an acetal or a ketal—either because of ether formation of the hemiacetal hydroxyl group with a simple alcohol or by formation of a glycosidic linkage—is nonreducing. Of the simple sugars you have seen in this chapter, only the methyl glycosides and sucrose are nonreducing.

The Benedict's and Tollens' tests for carbohydrates work best for simple sugars. Polysaccharides such as amylose should be reducing sugars, since the hemiacetal form of the terminal sugar unit is in equilibrium with the aldehyde form. If the polysaccharide chain is a long one, however, the number of end groups in a fairly large sample of the material may be so small that a positive Benedict's or Tollens' test is not observed. Thus large polysaccharides such as starch and cellulose are generally not reducing sugars. Chemical and biochemical tests for reducing sugars are very important to patients being treated for diabetes, as discussed in A Closer Look: Tests for Blood Glucose in Diabetes.

EXAMPLE 16.4

Analyzing why a sugar is a reducing sugar

The disaccharide maltose is a reducing sugar. Why?

SOLUTION

Maltose has the structure

The right-hand ring is a cyclic hemiacetal. (The hemiacetal carbon is marked with an asterisk.) This ring can open and close, as you saw earlier for D-glucose (Fig. 16.5), to give an equilibrium between hemiacetal and aldehyde. The aldehyde group can react with Benedict's reagent, reducing copper(II) to copper(I). Therefore, maltose is a reducing sugar.

EXAMPLE 16.5

Deducing why a sugar is nonreducing

Why is sucrose a nonreducing sugar?

SOLUTION

Sucrose (see the structure on page 512) is a disaccharide of α-D-glucose and β-D-fructose. The glucose unit is a cyclic acetal, and the fructose unit is a cyclic ketal. Neither of these two units can ring open, and neither exists in equilibrium with the aldehyde form. Sucrose cannot react with Benedict's reagent and is therefore a nonreducing sugar.

A Closer Look

Tests for Blood Glucose in Diabetes

Patients with diabetes must take steps to keep levels of glucose in the blood as close as possible to the middle of the normal range—4.5 to 8.0 mmol/L. Recent research shows that when patients with diabetes measure their levels of blood sugar at least four times a day and fine-tune their medication, they can avoid the disease's worst complications: kidney disease, blindness, and poor blood circulation that can require limb amputations.

Portable electronic devices called *glucose meters* are available for home testing of blood glucose levels (see figure). The patient must prick his or her finger to get blood for a reading, and the blood glucose level is obtained in less than a minute. These instruments are expensive, however, and some patients may use less expensive methods to check their urine for the presence of glucose. Glucose appears in the urine when the blood glucose concentration rises to 11.0 mmol/L. One kind of tablet called Clinitest uses a reducing sugar test to give the glucose level in the urine. When dissolved in urine, Clinitest tablets produce Benedict's reagent: they contain copper(II) sulfate ($CuSO_4$), sodium carbonate (Na_2CO_3), sodium hydroxide (NaOH), and citric acid ($C_6H_8O_7$). The quantity of glucose in the urine is determined by comparing the color of the test solution against test blocks supplied with the tablets. The Clinitest method is not specific for glucose, and the presence of interfering substances must be considered when interpreting the results. Interferences can be caused by drugs such as aspirin, lactose in nursing mothers, fructose after excessive fruit juice intake, and certain inherited errors of sugar biochemistry. Reagent strips such as Clinistix, BMstix, and Diastix use a color reaction involving enzymes to detect glucose. These products are very specific for glucose in the urine even in the presence of other reducing substances.

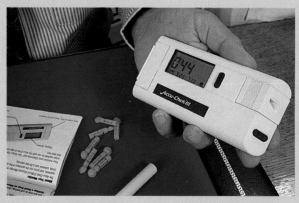

A blood glucose meter can give a result in 20 seconds.

16.10 Chitin, heparin, and acid mucopolysaccharides

AIM: To name and characterize the source of saccharides containing functional groups other than hydroxyl and carbonyl.

Focus

Saccharides may contain functional groups other than carbonyl and hydroxyl.

Two monosaccharides that contain functional groups other than carbonyl and hydroxyl are glucuronic acid and glucosamine. They are related structurally and stereochemically to D-glucose. *N*-Acetylglucosamine is also found in many polysaccharides and is formed from glucosamine.

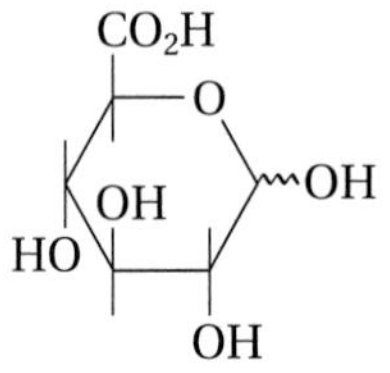

Glucuronic acid

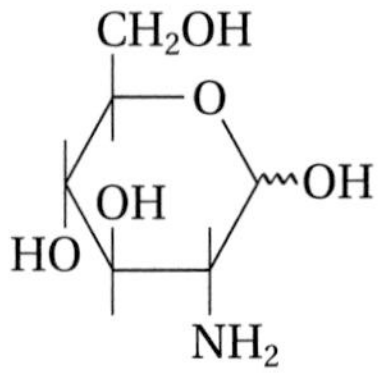

Glucosamine

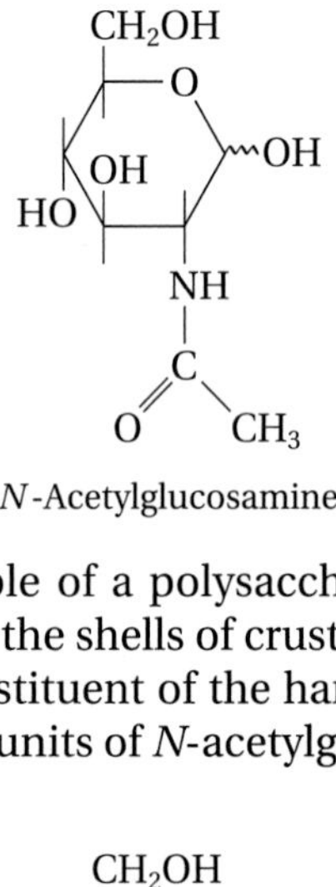

N-Acetylglucosamine

Chitin is an example of a polysaccharide that contains *N*-acetylglucosamine. Chitin forms the shells of crustaceans: crabs, lobsters, and shrimp. It is also a major constituent of the hard skeletons of insects. Chitin is composed of repeating units of *N*-acetylglucosamine:

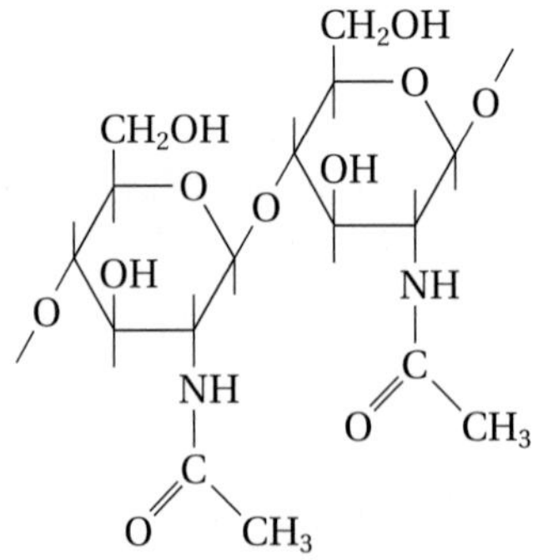

Chitin

Chitin is insoluble in water and is very resistant to hydrolysis of its component saccharides.

Acid mucopolysaccharides *are viscous polysaccharides that contain N-acetylglucosamine and glucuronic acid.* Hyaluronic acid is an acid mucopolysaccharide. It consists of repeating units of *N*-acetylglucosamine linked to glucuronic acid. The bonding pattern is

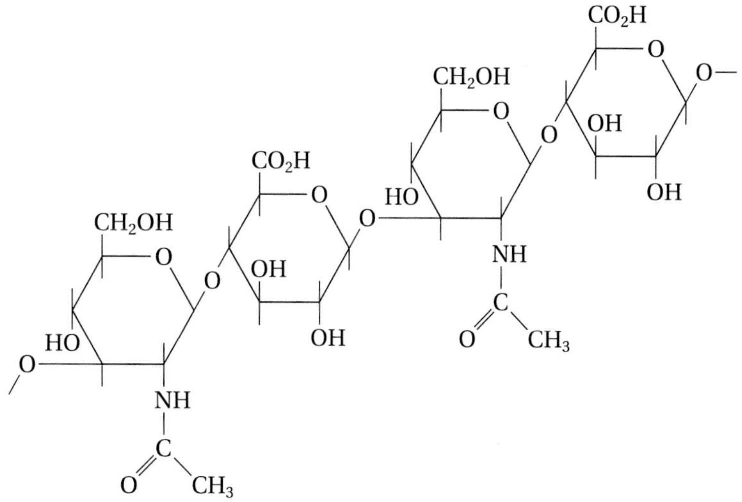

Hyaluronic acid

Hyaluronic acid is found in the connective tissue of animals, where it acts as a glue that helps to hold cells together. Another acid mucopolysaccharide is heparin, an important natural blood anticoagulant:

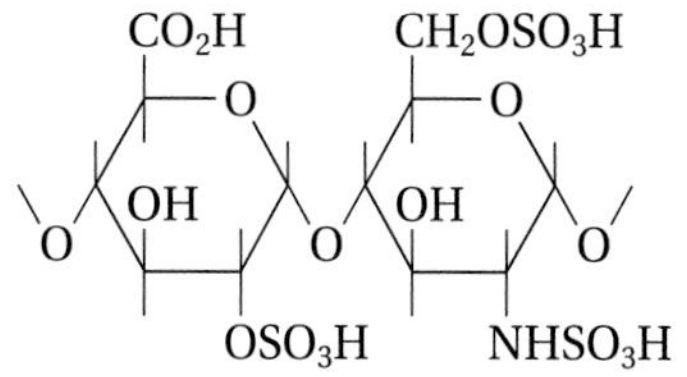

Heparin

Heparin is sometimes used to reduce the possibility of the formation of blood clots in the arteries of people who are at risk of heart attack or who have undergone heart bypass surgery.

The use of anticoagulant drugs such as heparin may cause abnormal bleeding in different parts of the body. Treatment is monitored through the use of periodic blood clotting tests. Using aspirin or alcohol when taking anticoagulant drugs can increase the chance of abnormal bleeding.

PRACTICE EXERCISE 16.16

How many functional groups can you identify in the following structure?

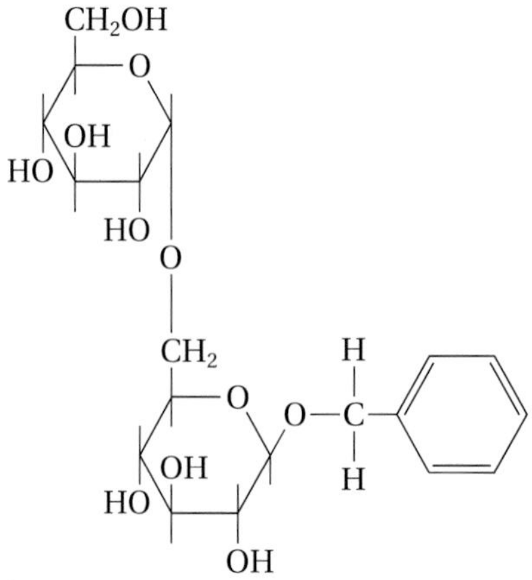

SUMMARY

Simple sugars, called monosaccharides, are poly-hydroxy aldehydes (aldoses) or ketones (ketoses). All sugars contain at least one asymmetric carbon. An important aspect of sugar chemistry is stereochemistry—relating the positions of groups of a compound in space. There is one pair of stereoisomers for compounds containing one asymmetric carbon. Stereoisomers have the same molecular formula, but they are nonsuperimposable mirror images. Stereoisomers may be called optical isomers because they rotate plane-polarized light.

Straight-chain aldoses and ketoses form hemiacetal and hemiketal rings, respectively. The heterocyclic ether rings are five-membered (furanoses) and six-membered (pyranoses). Fischer projections represent straight-chain sugars; Haworth projections represent sugars in ring forms. Cyclization of a straight-chain sugar to a pyranose or furanose ring creates a new asymmetric carbon in the molecule. Molecules whose stereochemistry differs only at the newly created asymmetric carbon are called anomers.

Simple sugars are often linked to other sugars through glycosidic bonds to create disaccharides. Polysaccharides are sugars containing many monosaccharide units.

Starches are polymers of glucose from plants. One form, which is a linear polymer, is amylose. Another, which is a branched polymer, is amylopectin. Glycogen is the animal form of starch; it is more highly branched than amylopectin. Cellulose is another linear polymer of glucose. Cellulose is more resistant to hydrolysis than starch. Many other polymers of monosaccharides are found in nature, where they have many diverse functions—from forming the hard shells of crabs, like chitin, to acting as a blood anticoagulant, like heparin.

Reducing sugars give positive results in the form of a red precipitate with Benedict's reagent or a silver mirror with Tollens' reagent. Nonreducing sugars do not. Most monosaccharides and disaccharides except sucrose are reducing sugars. Most polysaccharides are nonreducing sugars.

KEY TERMS

Acid mucopolysaccharide (16.10)	Fischer projection (16.1)	Ketose (16.2)	Polysaccharide (16.6)
Aldose (16.1)	Furanose ring system (16.3)	Monosaccharide (16.2)	Pyranose ring system (16.3)
Anomer (16.4)	Glycogen (16.6)	Nonreducing sugar (16.9)	Reducing sugar (16.9)
Asymmetric carbon (16.1)	Glycoside (16.5)	Nonsuperimposable mirror image (16.1)	Simple sugar (16.2)
Carbohydrate (16.1)	Glycosidic bond (16.5)	Optical isomer (16.1)	Starch (16.6)
Cellulose (16.6)	Haworth projection (16.4)	Plane-polarized light (16.1)	Stereoisomer (16.1)
Disaccharide (16.7)	Invert sugar (16.8)		Sugar (16.1)

EXERCISES

Carbohydrate Structure and Stereochemistry (Section 16.1)

16.17 Identify the asymmetric carbon (if there is one) in the following compounds.

(a) CH_3CHOH
    $|$
    CH_2CH_3

(b) CH_3CH_2CHOH
    $|$
    CH_3

(c) CH_3CHCH_2OH
    $|$
    OH

(d) CH_3CHCH_2OH
    $|$
    Cl

16.18 Draw stereorepresentations, and stereoisomers if they exist, for each of the following.

(a) $CH_3\overset{\overset{\displaystyle OH}{|}}{C}HCHO$

(b) CH_3CH_2CHO

16.19 Draw stereorepresentations and Fischer projection formulas of D-glyceraldehyde and L-glyceraldehyde. Explain the meaning of D and L.

16.20 Draw the structural formula for 3-hydroxybutanal, and identify the asymmetrical carbon.

Monosaccharides (Section 16.2)

16.21 Name each monosaccharide, and characterize it as to chain length and functional group.

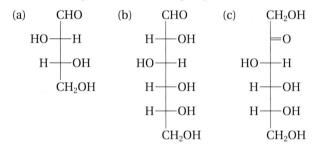

16.22 Draw Fischer projection formulas for D-ribose and D-2-deoxyribose. Number the carbons, and asterisk those which are asymmetrical.

16.23 Identify each of the following monosaccharides as D or L.

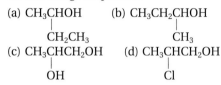

(c)
```
    CH₂OH
     |
     =O
     |
HO──H
     |
 H──OH
     |
 H──OH
     |
    CH₂OH
```

(d)
```
    CHO
     |
 H──OH
     |
 H──OH
     |
 H──OH
     |
    CH₂OH
```

Closed-Chain Structures and Haworth Projections (Sections 16.3, 16.4)

16.24 Draw (a) a structure of the cyclic ether furan and (b) the general structure of a hemiketal.

16.25 Draw (a) the general structure of a hemiacetal and (b) a structure of the cyclic ether pyran.

16.26 Draw the Haworth projections of α-D-glucopyranose and β-D-glucopyranose, the cyclic forms of D-glucose.

16.27 Draw the Haworth projections of α-D-fructofuranose and β-D-fructofuranose, the cyclic forms of D-fructose.

16.28 Draw structures to show the relationship that exists between the two cyclic hemiacetal forms of D-glucose and the open-chain aldehyde form.

Polysaccharides (Sections 16.5, 16.6)

16.29 Write the general structure of (a) an acetal and (b) a ketal.

16.30 Explain the difference between a $\beta(1\rightarrow4)$ and an $\alpha(1\rightarrow6)$ glycosidic bond.

16.31 Give one use for (a) starch, (b) cellulose, and (c) glycogen.

16.32 Name a source of (a) cellulose, (b) glycogen, and (c) starch.

16.33 Describe the structural and stereochemical differences between starch (amylose) and cellulose. What is the biological significance of these differences?

16.34 What is glycogen? How does it differ from amylopectin?

Disaccharides (Sections 16.7, 16.8)

16.35 Give sources, uses, and the products of hydrolysis of (a) sucrose and (b) maltose.

16.36 Give sources, uses, and the products of hydrolysis of (a) lactose and (b) cellobiose.

16.37 Draw a Haworth projection formula of the disaccharide maltose. Label the hemiketal, hemiacetal, ketal, or acetal carbons.

16.38 Draw a Haworth projection formula of the disaccharide sucrose. Label the hemiketal, hemiacetal, ketal, or acetal carbons.

Reducing and Nonreducing Sugars (Section 16.9)

16.39 Explain why lactose is a reducing sugar but sucrose is a nonreducing sugar.

16.40 Why are polysaccharides generally not regarded as reducing sugars?

16.41 Would Benedict's reagent be useful for distinguishing between
(a) glucose and starch
(b) fructose and sucrose
(c) fructose and glucose?

16.42 A solution of invert sugar is treated with (a) Tollens' reagent and (b) Benedict's reagent. What would you expect to happen?

Additional Exercises

16.43 Which of the following compounds exist as optical isomers?

(a)

Br
|
Cl—C—H
|
Cl

(b)

Cl

(c)

CH₃CH₂CHC=O
|
OH

(d)

Cl
|
CH₂=CHCH₂C—
|
H

16.44 Draw the Fischer projection formula for D-galactose.

16.45 Draw the Haworth structures for α-D-galactose and β-D-galactose. Identify the anomeric carbon.

16.46 Draw the structure for the disaccharide formed when two D-galactose molecules are joined β(1→4). Would you expect this to be a reducing sugar? Explain.

16.47 Tell the number of asymmetrical carbons in the open-chain structure of (a) L-ribose, (b) D-2-deoxyribose, (c) D-glucose, (d) D-fructose, (e) L-galactose, and (f) D-mannose.

16.48 How would you explain the water solubility of monosaccharides?

16.49 Identify these monosaccharides as D or L.

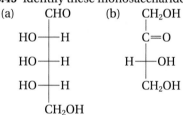

(a)

CHO
|
HO—H
|
HO—H
|
HO—H
|
CH₂OH

(b)

CH₂OH
|
C=O
|
H—OH
|
CH₂OH

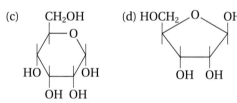

(c)

CH₂OH
O
HO   OH
OH OH

(d) HOCH₂ O OH

OH OH

16.50 Match the description on the right with the appropriate polysaccharides. A number may be used more than once.
(a) starch (amylose)
(b) starch (amylopectin)
(c) glycogen
(d) cellulose

(1) found in plants
(2) α(1→4) linkages
(3) β(1→6) linkages
(4) α(1→6) linkages
(5) found in animals
(6) β(1→4) linkages

SELF-TEST (REVIEW)

True/False

1. A compound with optical isomers must have at least one asymmetrical carbon atom.
2. Anomers are stereoisomers that result from ring closure.
3. The cyclic form of an aldose is an intramolecular hemiacetal.
4. In polysaccharides, glycosidic bonds are formed between the alcohol of one sugar unit and the carbonyl group of a different sugar unit.
5. The disaccharide lactose is composed of the monosaccharides D-glucose and D-galactose.
6. A carbohydrate with a free aldehyde group will give a positive Tollens' test and be classified as a reducing sugar.
7. Molecules of D-threose and L-threose are mirror images.
8. A pyranose is a six-membered sugar ring system; a five-membered sugar ring system is a furanose.
9. A molecule of L(−)-glyceraldehyde is identical to a molecule of D(+)-glyceraldehyde.
10. Starch, cellulose, and glycogen are composed of the monosaccharide glucose.

Multiple Choice

11. Which of the following statements is *false* for β-D-fructofuranose?
    (a) It is a reducing sugar.   (b) It is a hemiacetal.
    (c) It has a five-membered ring.
    (d) It is a structural isomer of D-glucose.

12. When pure β-D-galactopyranose is dissolved in an acidic solution
    (a) some of it is changed into β-L-galactopyranose.
    (b) all of it changes to the open-chain form.
    (c) some α-D-galactopyranose is formed.
    (d) no change takes place.

13. All the following are aldohexoses except
    (a) galactose.   (b) fructose.   (c) mannose.
    (d) glucose.

14. α-D-Glucopyranose
    (a) has the —OH group on carbon 1 pointing up.
    (b) has a five-membered ring.
    (c) is an intramolecular hemiacetal.
    (d) is an anomer of α-L-glucopyranose.

15. The disaccharide lactose
    (a) is found in milk.
    (b) is nonreducing.
    (c) is made up of the simple sugars D-galactose and D-fructose.
    (d) all of the above are true.

16. All the following are nonreducing sugars except
    (a) starch.   (b) sucrose.   (c) methyl glucoside.
    (d) none of the above.

17. The reaction of ethanol with α-D-ribofuranose would give
    (a) a glycoside.   (b) ethyl α-D-ribofuranoside.
    (c) an acetal.   (d) all of the above.

18. Complete hydrolysis of starch would give
    (a) D-glucose.   (b) amylopectin.   (c) maltose.
    (d) D-fructose.

19. The Fischer projection formula for a stereoisomer of xylose is

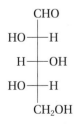

From this formula we know that this stereoisomer is
    (a) D-xylose.   (b) (+)-xylose.   (c) (−)-xylose.
    (d) L-xylose.

20. Which of the following statements is *not* true about an aldotetrose?
    (a) It has four asymmetrical carbons.
    (b) It gives a positive Benedict's test.
    (c) It has an aldehyde group.
    (d) It has three hydroxyl groups.

21. When starch is partially hydrolyzed, the disaccharide obtained is
    (a) maltose.   (b) amylose.   (c) lactose.
    (d) xylose.

22. For a carbohydrate molecule to have a mirror image, it must
    (a) be very complex.
    (b) be an organic compound.
    (c) not have a stereoisomer.
    (d) have an asymmetrical carbon.

23. One thing we can say for certain about the pentose L-ribose is that
    (a) it rotates plane-polarized light to the left.
    (b) it is a mirror image of D-ribose.
    (c) it rotates plane-polarized light to the right.
    (d) it has no asymmetrical carbons.

24. The most branched natural polysaccharide is
    (a) cellulose.   (b) amylose.   (c) glycogen.
    (d) amylopectin.

25. Human beings cannot use cellulose as a nutritional source because
    (a) it is found only in inedible materials.
    (b) it contains β(1→4) glycosidic linkages.
    (c) its structure is too highly branched.
    (d) all of the above are true.

26. Sucrose is unique among the common disaccharides because
    (a) it is not a reducing sugar.
    (b) it has a 1→2 linkage.
    (c) it has no hemiacetal group.
    (d) all of the above are true.

17

Lipids

A Potpourri of Fatty Molecules

Safflower is an Old World herb related to daisies. Its seeds are rich in an edible oil and its large orange-red flowers yield a dye.

CHAPTER OUTLINE

CASE IN POINT: Obesity

17.1 Waxes

17.2 Triglycerides

FOLLOW-UP TO THE CASE IN POINT: Obesity

A CLOSER LOOK: Cosmetic Creams

17.3 Lipid composition of cell membranes

A CLOSER LOOK: Cells

17.4 Structure of liposomes and cell membranes

17.5 Steroids

17.6 Steroid hormones

17.7 Male sex hormones

17.8 Female sex hormones

17.9 The pill

17.10 Plant steroids

17.11 Prostaglandins and leukotrienes

Lipids are a class of naturally occurring compounds that dissolve in organic solvents such as ether and chloroform. This property sets them apart from carbohydrates, proteins, and nucleic acids, the other great classes of biological molecules. Lipids encompass such a wide variety of molecular structures that their study will allow us to review many of the functional groups we have seen before while we learn about their biological function.

Obesity, or being overweight, occurs when the body stores excessive amounts of one kind of lipid that we discuss in this chapter. Because many people are concerned about controlling their weight, we explore the measurement of obesity and its treatment in this chapter's Case in Point.

CASE IN POINT: Obesity

Fat, fat, fat! Our culture is obsessed with fat. Almost every day we can expect to hear somebody say, "I'm too fat!" While people often want to reduce their weight in order to look better, obesity is a health problem as well as a cosmetic problem. Medical results of being overweight include increases in blood pressure, gallstones, and diabetes and an increased risk of certain cancers. Obese people are also increasing their risk of having a heart attack. How is obesity measured? What are the possible treatments for obesity? We will learn the answers to these questions in Section 17.2.

Besides a reduced-calorie diet, regular aerobic exercise assists in weight loss by increasing the metabolic rate and by burning calories from excess stored fat.

17.1 Waxes

AIM: To describe the general structure, source, and use of waxes.

Lipids *are a large class of relatively water-insoluble compounds found in nature.* Waxes are among the simplest members of the lipid family. **Waxes** *are esters of long-chain fatty acids and long-chain alcohols.*

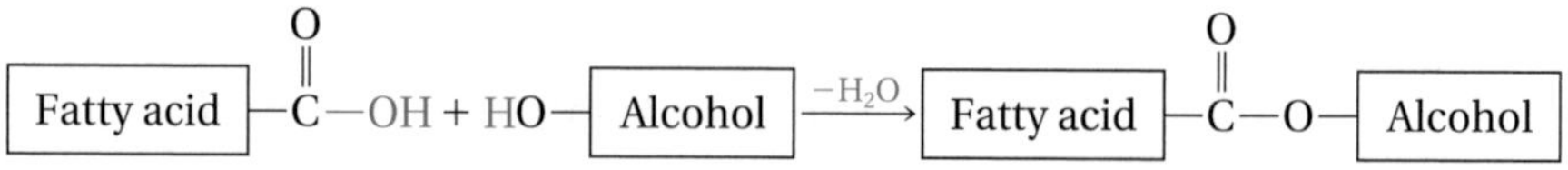

$$\boxed{\text{Fatty acid}} - \overset{\displaystyle O}{\overset{\|}{C}} - OH + HO - \boxed{\text{Alcohol}} \xrightarrow{-H_2O} \boxed{\text{Fatty acid}} - \overset{\displaystyle O}{\overset{\|}{C}} - O - \boxed{\text{Alcohol}}$$

You may recall from Section 14.1 that the fatty acids are straight-chain aliphatic carboxylic acids (RCO_2H). (Throughout the first three sections of this chapter you may wish to refer to Tables 14.1 and 14.2 for fatty acid structures.) In waxes, the hydrocarbon chains for both the acid and the alcohol (R—OH) usually contain from 10 to 30 carbons. Waxes are low-melting, stable solids that appear in nature in both plants and animals. A wax coat protects surfaces of many plant leaves from water loss and attack by microorganisms. Carnauba wax, a major ingredient in car wax and floor polish, comes from the leaves of a South American palm tree. Beeswax is largely myricyl palmitate, the ester of myricyl alcohol and palmitic acid.

$$CH_3(CH_2)_{14} - \overset{\displaystyle O}{\overset{\|}{C}} - O - (CH_2)_{29}CH_3$$

Myricyl palmitate

Waxes also coat skin, hair, and feathers and help keep them pliable and waterproof.

17.2 Triglycerides

AIMS: To characterize triglycerides, fats, and oils by source, structure, and use. To distinguish between hydrolytic rancidity and oxidative rancidity and explain how each can be prevented. To describe the production of soap by saponification.

Natural **triglycerides**—*triesters of long-chain fatty acids (C_{12} through C_{24}) and glycerol*—are the major components of animal lipids. The following equation shows the general reaction for the formation of triglycerides:

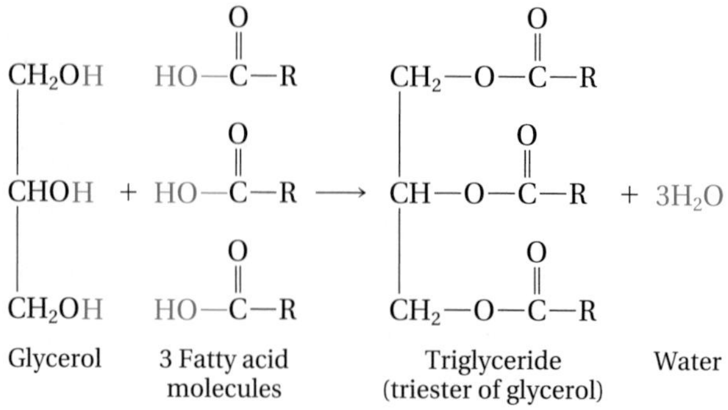

$$\begin{array}{l}
CH_2OH \\
| \\
CHOH \\
| \\
CH_2OH
\end{array}
+
\begin{array}{l}
HO - \overset{\displaystyle O}{\overset{\|}{C}} - R \\
HO - \overset{\displaystyle O}{\overset{\|}{C}} - R \\
HO - \overset{\displaystyle O}{\overset{\|}{C}} - R
\end{array}
\longrightarrow
\begin{array}{l}
CH_2 - O - \overset{\displaystyle O}{\overset{\|}{C}} - R \\
| \\
CH - O - \overset{\displaystyle O}{\overset{\|}{C}} - R \\
| \\
CH_2 - O - \overset{\displaystyle O}{\overset{\|}{C}} - R
\end{array}
+ \; 3H_2O$$

Glycerol   3 Fatty acid molecules   Triglyceride (triester of glycerol)   Water

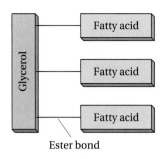

Figure 17.1
Schematic diagram of a triglyceride.

The bonding of the triglyceride building blocks may be represented by a diagram, as shown in Figure 17.1.

The many different types of triglycerides vary with the positions and identities of the fatty acids. **Simple triglycerides** *are triesters made from glycerol and three molecules of one kind of fatty acid.* For example, glycerol and three molecules of stearic acid give a simple triglyceride called *glyceryl tristearate,* or *tristearin:*

$$CH_2-O-\overset{\overset{\displaystyle O}{\|}}{C}-(CH_2)_{16}CH_3$$
$$CH-O-\overset{\overset{\displaystyle O}{\|}}{C}-(CH_2)_{16}CH_3$$
$$CH_2-O-\overset{\overset{\displaystyle O}{\|}}{C}-(CH_2)_{16}CH_3$$

Glyceryl tristearate
(tristearin)

Simple triglycerides are rare. More often, natural triglycerides are **mixed triglycerides**—*triesters of glycerol with different fatty acid components.* The mixed triglycerides in butterfat, for example, contain at least 14 different fatty acids. Triglycerides are the form in which fats are stored in the human body. In the Case in Point earlier in this chapter, we questioned how body fat is measured and how obesity is treated. Answers to these questions can be found in the Follow-up to the Case in Point below.

Sometimes reactions of glycerol with fatty acids may produce mono- or diesters. For example, see glycerol monostearate, a commonly used emulsifier, in A Closer Look: Cosmetic Creams.

FOLLOW-UP TO THE CASE IN POINT: Obesity

The degree of obesity is measured by the excess of body weight relative to height. Because everyone is different, an *ideal body weight (IBW)* for different heights has been established statistically. The IBW is the weight that gives the lowest incidence of sickness and death in the statistical sample of people of a certain height. The terms *overweight* and *obesity* are based on the IBW. People with a relative weight up to 20% above the IBW are defined as being overweight. Those with a relative weight greater than 20% above the IBW are defined as being obese. Another measure of overweight and obesity is the *body mass index (BMI).* This measure is defined as body mass in kilograms divided by height squared (height is in meters). Overweight is defined as a BMI of 25 to 30 kg/m^2. Obesity is a BMI of >30 kg/m^2.

The causes of most obesity are not very well understood. A few diseases, such as hypothyroidism, tumors of the hypothalamus, and Cushing's disease, can cause obesity, but these diseases are very rare. Inherited and environmental factors may be involved, since about 80% of the offspring of obese parents will be obese. Only about 15% of the offspring of normal-weight parents are obese.

The treatment of obesity usually involves restricting dietary caloric intake and increasing exercise. However, many people go on "crash" diets and lose weight only to gain it back. Permanent reductions in weight often require that, besides diet and exercise, eating habits be changed. No medications are currently available that are effective in the long term. Surgery in which the size of the stomach is restricted can be part of a reducing program for patients who weigh double their IBW.

A Closer Look

Cosmetic Creams

Skin care products such as cold creams, moisturizers, lotions, sunscreens, and cleansing creams are basically emulsions of oil and water (see figure). They are used to protect the skin from the drying effects of detergents and exposure to wind and sun. Emulsifiers are used to stabilize oil-water mixtures. One commonly used emulsifier is the oil glycerol monostearate.

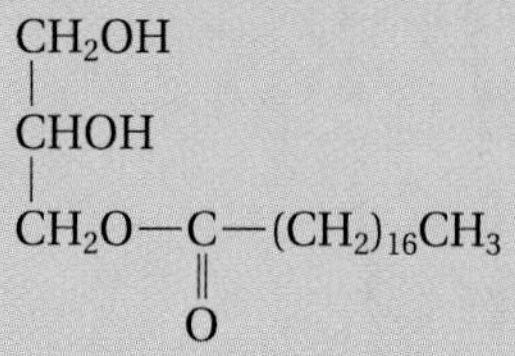

CH$_2$OH
|
CHOH
|
CH$_2$O—C—(CH$_2$)$_{16}$CH$_3$
‖
O

Glycerol monostearate

Glycerol monostearate is an emulsifier commonly used in cosmetic products.

Cold creams are oil-in-water emulsions. That is, there is more water than oil in the emulsion. As the water evaporates, it produces a cooling effect on the skin. The thin film of oil that remains is quickly absorbed by the outer skin layers. Water-in-oil emulsions, which contain more oil than water, are also used in cosmetics. Water-in-oil emulsions are sometimes called *warm emulsions* because there is little cooling effect due to water evaporation. These emulsions permit the oil to contact the skin almost immediately. Cleansing creams, which dissolve oily materials on the surface of the skin, and moisturizing lotions, which soften the skin and help replace lost water, are water-in-oil emulsions.

PRACTICE EXERCISE 17.1

Write one structure for the mixed triglyceride that is made from glycerol, stearic acid, palmitic acid, and oleic acid.

PRACTICE EXERCISE 17.2

Name the following triglyceride:

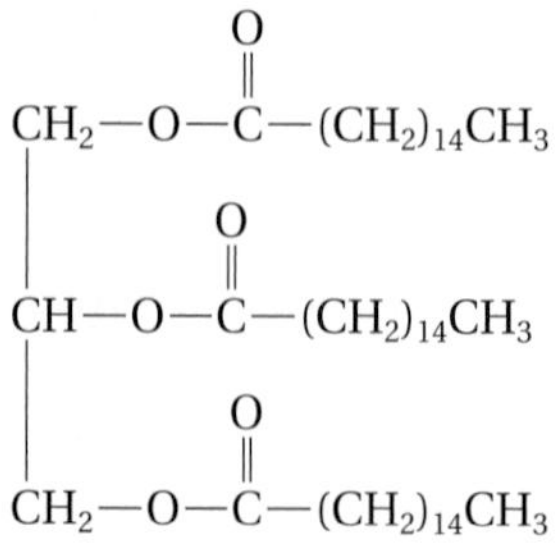

$$
\begin{array}{l}
\quad\quad\quad\quad O \\
\quad\quad\quad\quad \| \\
CH_2\text{—}O\text{—}C\text{—}(CH_2)_{14}CH_3 \\
\quad\quad\quad\quad O \\
\quad\quad\quad\quad \| \\
CH\text{—}O\text{—}C\text{—}(CH_2)_{14}CH_3 \\
\quad\quad\quad\quad O \\
\quad\quad\quad\quad \| \\
CH_2\text{—}O\text{—}C\text{—}(CH_2)_{14}CH_3
\end{array}
$$

Hydrogenation of fats and oils

Animal fats tend to be rich in saturated fatty acids; vegetable oils are rich in unsaturated fatty acids.

The distinction between fats and oils is usually based on melting point. *At room temperature,* **fats** *are solid triglycerides and* **oils** *are liquid triglycerides.* The melting point of a fat or oil depends on its structure, usually increasing with the number of carbons. The number of carbon-carbon double bonds in the fatty acid component also has an effect: Triglycerides rich in unsaturated fatty acids, such as oleic and linoleic, are generally oils; triglycerides rich in saturated fatty acids, such as stearic and palmitic, are generally fats. The triglycerides in liquid olive oil contain mostly unsaturated oleic acid, for example, but those of solid beef fat, or tallow, contain mostly saturated stearic acid. Hydrogenation (see Sec. 12.2) converts vegetable oils into fats. It is commercially important in the manufacture of margarine and lard substitutes. Powdered metallic nickel (which is removed later) is dispersed in the hot oil as a catalyst. Hydrogen adds to some of the double bonds in the unsaturated fatty acid carbon chains, saturating them and thereby converting the oil to a fat. For example, the hydrogenation of triolein (melting point $-17\ °C$) produces tristearin (melting point $55\ °C$):

Two polyunsaturated fatty acids with 18 carbon atoms, linolenic and linoleic acid, are essential fatty acids that cannot be synthesized by the body. These fatty acids are found in fish oils and in many plants. Some studies suggest that the consumption of linolenic acid may reduce the risk of heart disease.

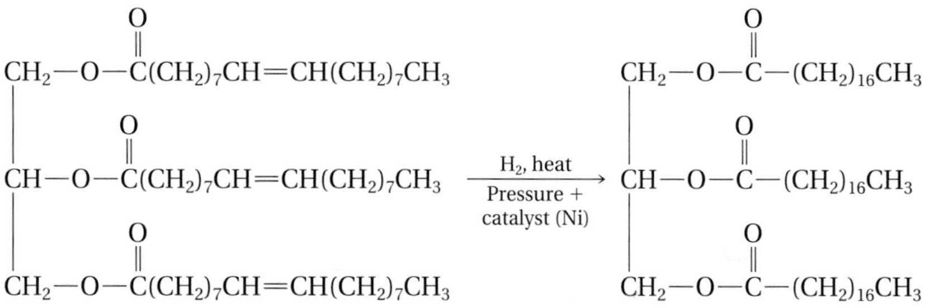

$$
\begin{array}{l}
CH_2-O-\overset{\displaystyle O}{\overset{\|}{C}}(CH_2)_7CH=CH(CH_2)_7CH_3 \\
\ \ | \\
CH-O-\overset{\displaystyle O}{\overset{\|}{C}}(CH_2)_7CH=CH(CH_2)_7CH_3 \quad \xrightarrow[\substack{\text{Pressure +}\\ \text{catalyst (Ni)}}]{H_2,\ \text{heat}} \\
\ \ | \\
CH_2-O-\overset{\displaystyle O}{\overset{\|}{C}}(CH_2)_7CH=CH(CH_2)_7CH_3
\end{array}
$$

$$
\begin{array}{l}
CH_2-O-\overset{\displaystyle O}{\overset{\|}{C}}-(CH_2)_{16}CH_3 \\
\ \ | \\
CH-O-\overset{\displaystyle O}{\overset{\|}{C}}-(CH_2)_{16}CH_3 \\
\ \ | \\
CH_2-O-\overset{\displaystyle O}{\overset{\|}{C}}-(CH_2)_{16}CH_3
\end{array}
$$

Triolein (an oil)   Tristearin (a fat)

The consistency of the fat is controlled by the degree of hydrogenation; the higher the degree of saturation of the fat by hydrogenation, the harder is the product at room temperature.

PRACTICE EXERCISE 17.3

A mixed triglyceride with the following structure is hydrogenated. What is the hydrogenation product?

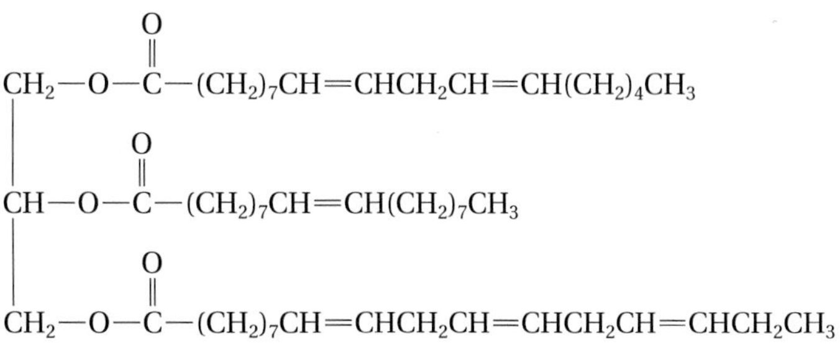

$$
\begin{array}{l}
CH_2-O-\overset{\displaystyle O}{\overset{\|}{C}}-(CH_2)_7CH=CHCH_2CH=CH(CH_2)_4CH_3 \\
\ \ | \\
CH-O-\overset{\displaystyle O}{\overset{\|}{C}}-(CH_2)_7CH=CH(CH_2)_7CH_3 \\
\ \ | \\
CH_2-O-\overset{\displaystyle O}{\overset{\|}{C}}-(CH_2)_7CH=CHCH_2CH=CHCH_2CH=CHCH_2CH_3
\end{array}
$$

Rancidity

Triglycerides soon become **rancid,** *developing an unpleasant odor and flavor on exposure to moist air at room temperature.* The release of volatile fatty acids (particularly butyric acid) from butterfat causes the disagreeable odor of rancid butter. These acids are formed either by the hydrolysis of ester bonds or by the oxidation of double bonds. The hydrolysis of a fat or oil is often catalyzed by enzymes, called *lipases,* present in airborne bacteria. *Hydrolytic rancidity* is prevented or delayed by storing food in a refrigerator. The unwelcome odor of sweat results when bacterial lipases catalyze the hydrolysis of oils and fats on the skin.

Oxidative processes, however, not hydrolysis, are the major cause of rancidity in food. Warmth and exposure to air induce oxidative rancidity. In *oxidative rancidity,* double bonds in the unsaturated fatty acid components of triglycerides rupture, forming low-formula-weight aldehydes with objectionable odors. The aldehydes then oxidize to the equally offensive low-formula-weight fatty acids. Oxidative rancidity shortens the shelf life of cookies and similar foods. Antioxidants are compounds that delay the onset of oxidative rancidity. Two naturally occurring substances often used as antioxidants are ascorbic acid (vitamin C) and α-tocopherol (vitamin E).

Saponification of fats and oils

Saponification
sapon (Latin):   soap

Fats and oils are simply esters. Like other esters, they are easily hydrolyzed in the presence of acids and bases (Sec. 14.8). *Hydrolysis of oils or fats by boiling them with aqueous sodium hydroxide is called* **saponification.** This process is used to make soap. Soaps are the alkali metal (Na, K, or Li) salts of fatty acids.

Soap is made by heating beef tallow or coconut oil in large kettles with an excess of sodium hydroxide. When sodium chloride is added to the saponified mixture, the sodium salts of the fatty acids separate as a thick curd of crude soap. If the fat is tristearin, the soap is sodium stearate:

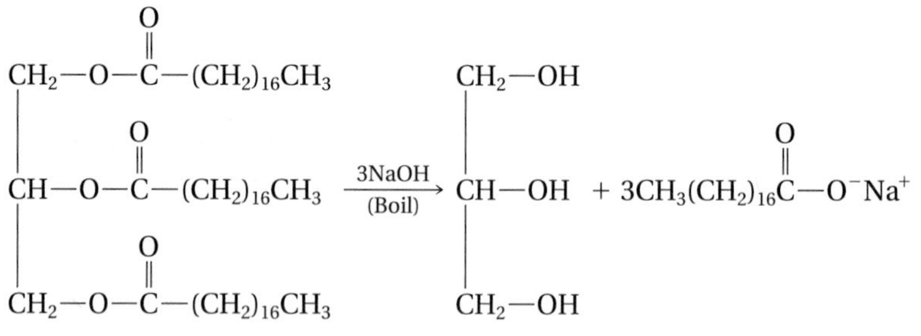

$$
\begin{array}{l}
CH_2-O-\overset{\displaystyle O}{\overset{\|}{C}}-(CH_2)_{16}CH_3 \\[2mm]
CH-O-\overset{\displaystyle O}{\overset{\|}{C}}-(CH_2)_{16}CH_3 \\[2mm]
CH_2-O-\overset{\displaystyle O}{\overset{\|}{C}}-(CH_2)_{16}CH_3
\end{array}
\xrightarrow[\text{(Boil)}]{3NaOH}
\begin{array}{l}
CH_2-OH \\[2mm]
CH-OH \\[2mm]
CH_2-OH
\end{array}
+ 3CH_3(CH_2)_{16}\overset{\displaystyle O}{\overset{\|}{C}}-O^-Na^+
$$

Tristearin (a triester)   Glycerol   Sodium stearate (soap)

Glycerol is an important by-product of the reaction. It is recovered by evaporating the water layer. The crude soap is then purified, and coloring agents and perfumes are added according to market demands.

PRACTICE EXERCISE 17.4

What are the products when a mixed triglyceride with the following structure is saponified?

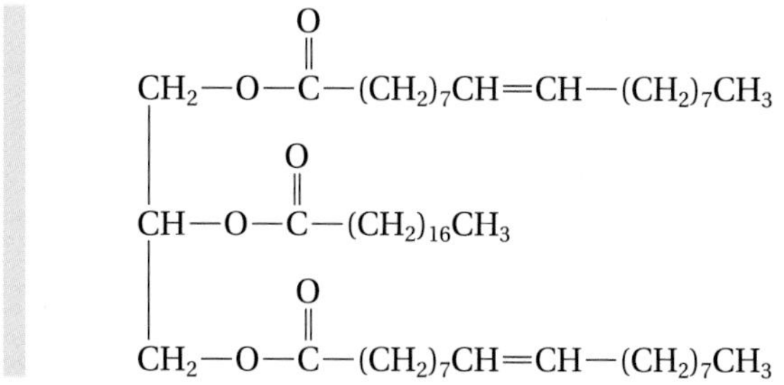

$$CH_2-O-\overset{\overset{\displaystyle O}{\|}}{C}-(CH_2)_7CH=CH-(CH_2)_7CH_3$$

$$CH-O-\overset{\overset{\displaystyle O}{\|}}{C}-(CH_2)_{16}CH_3$$

$$CH_2-O-\overset{\overset{\displaystyle O}{\|}}{C}-(CH_2)_7CH=CH-(CH_2)_7CH_3$$

17.3 Lipid composition of cell membranes

AIM: To recognize the general structures of the following three types of lipid molecules: phosphoglycerides, sphingomyelins, and glycolipids.

Focus

Cellular membranes are made primarily of complex lipids.

It is possible to break cells, empty them of their contents, and isolate the *cell membranes. The* **cell membrane** *is the "sack" that holds the contents of cells and acts as a selective barrier for the passage of certain substances in and out of the cell.* The interior of cells also contains membrane structures, as described in A Closer Look: Cells. Chemical analysis of the isolated membranes shows that lipids are the major components. These lipids are not triglycerides, but another group of compounds called *complex lipids.* **Complex lipids** *contain parts made from substances besides fatty acids and glycerol; some contain no glycerol.* The complex lipids fall into two categories: *phospholipids* and *glycolipids.*

Phospholipids *are lipids that are esters of phosphoric acid.* There are two main types of phospholipid molecules in cell membranes: phosphoglycerides and sphingomyelins.

Phosphoglyceride *molecules are built from long-chain fatty acids (14 to 24 carbons), glycerol, and phosphoric acid.* Two fatty acids are covalently bonded to adjacent hydroxyl groups of glycerol by ester linkages. The phosphoric acid is bonded through phosphate ester linkages to the remaining hydroxyl function of glycerol. The resulting molecule is called a *phosphatidic acid,* which is a phosphoglyceride.

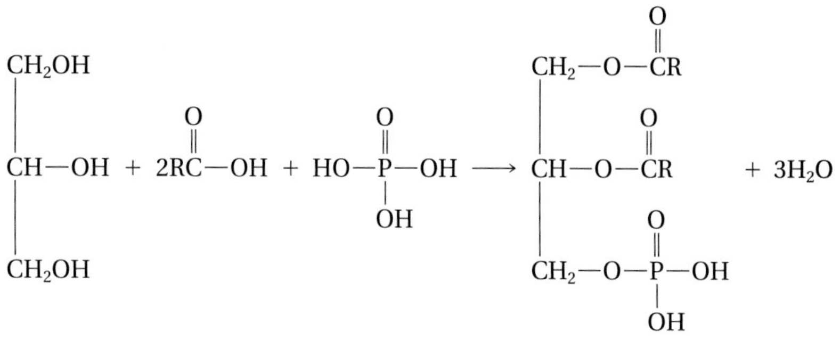

| Glycerol | Fatty acids | Phosphoric acid | A phosphatidic acid |

A Closer Look

Cells

The two major cell designs are *prokaryotic* and *eukaryotic*. The former is the more ancient of the two. Microscopic examination of fossilized remains shows that prokaryotes were present on Earth at least 3 billion years ago, whereas eukaryotes did not appear until 2 billion years later. In the modern world, the prokaryotic cell design is limited to bacteria and blue-green algae. The cells of other cellular organisms, including green plants and people, are eukaryotic.

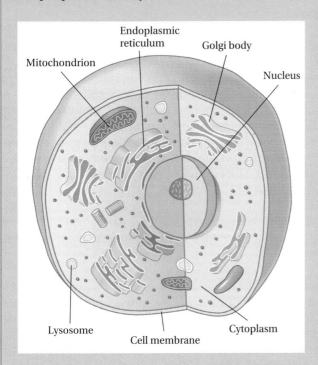

The eukaryotic cell

Both types of cells are essentially packages of chemicals necessary for life encased in a cell membrane. Eukaryotic cells are considerably larger and somewhat more complicated than prokaryotes, but the chemical processes carried out by both types of cells are very similar, and both are exceedingly efficient chemical factories. The major feature that distinguishes prokaryotes from eukaryotes is the latter's *organelles* ("little organs")—small membrane-enclosed bodies suspended in the interior cellular fluid or cytoplasm (see figure). The organelles are the sites of many specialized functions in eukaryotes. The most prominent membrane-encased organelles and their functions are as follows:

Organelle	*Function*
Nucleus	Cell reproduction
Mitochondrion	Production of most cellular energy in cells using oxygen for respiration
Golgi body	Processing of proteins into glycoproteins
Lysosome	Digestion of cellular wastes and substances taken into cells

Yet another membrane structure in eukaryotes is the highly folded, netlike endoplasmic reticulum (ER). Among its various functions, the ER serves as an attachment site for ribosomes—small organelles that are not membrane-encased but are the sites where proteins are made. Endoplasmic reticula with and without attached ribosomes are called, respectively, *rough ER* and *smooth ER* because of their appearance under a microscope.

Living cells contain little or no free phosphatidic acid. Usually, the phosphorus of the phosphatidic acid is linked to the hydroxyl group of a second alcohol. Choline, a common amino alcohol constituent of cell membrane phospholipids, is an example. In Figure 17.2, the hydroxyl group of choline is attached to the phosphorus of the phosphatidic acid through a phosphate ester bond. This phospholipid is a phosphoglyceride

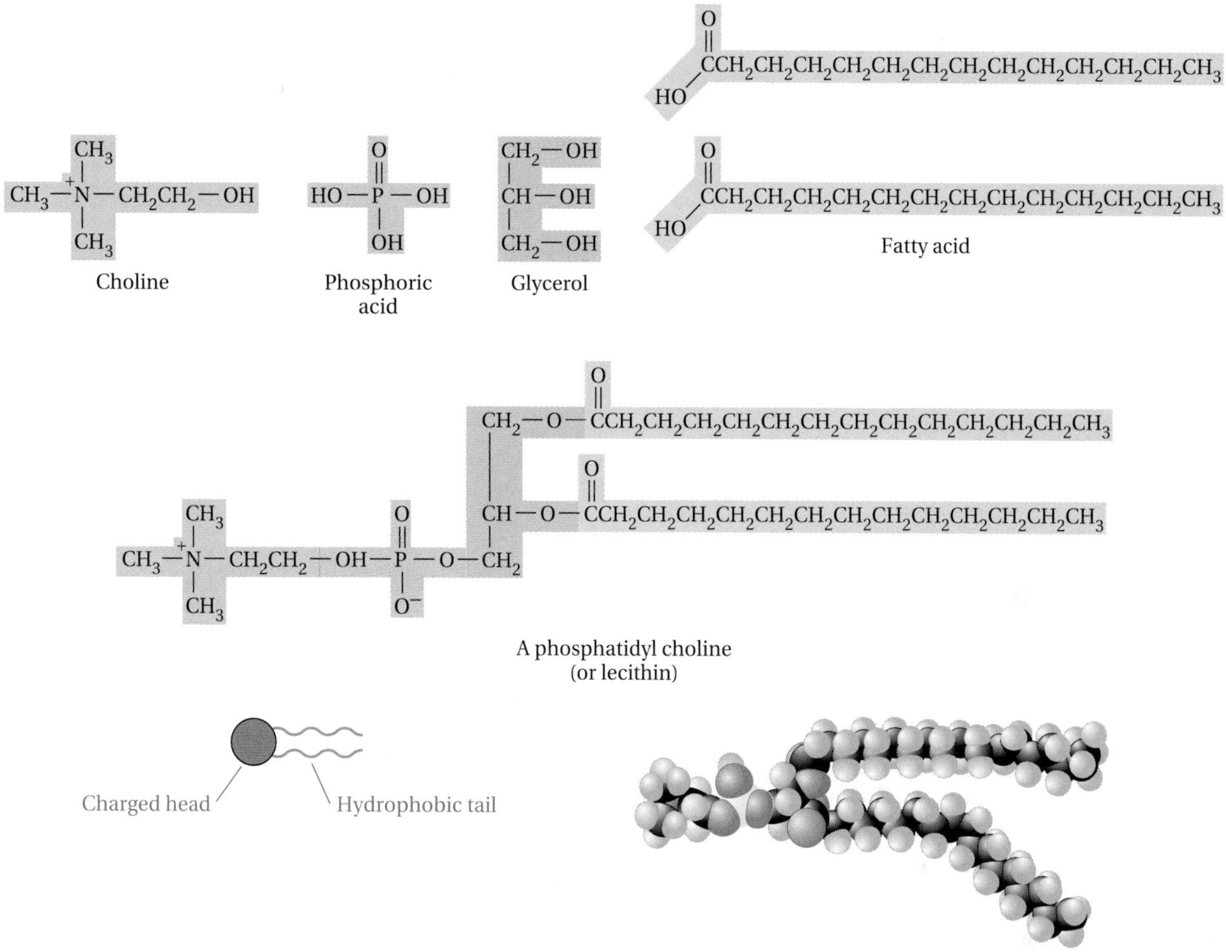

Figure 17.2
A phosphatidyl choline (phosphoglyceride) or lecithin molecule is constructed from the amino alcohol choline, phosphoric acid, glycerol, and two fatty acid molecules. A simplified representation of the molecule (lower left) has been adopted for all membrane lipids. The hydrophilic (charged) head is shown as a purple sphere, and the hydrophobic tails as red wavy lines. A space-filling model of a phosphatidyl choline molecule is shown at the lower right of the figure.

because it contains phosphorus and has a backbone of glycerol. It also may be called a *phosphatidyl choline* because it is an ester of phosphatidic acid and choline. *An older name for phosphatidyl choline is* **lecithin.** It is important to recognize that the names *phosphoglyceride, phosphatidic acid,* and *phosphatidyl choline* are only general names for classes of compounds. The lengths of the hydrocarbon chains of the fatty acids may vary and these chains may be saturated or contain one or more double bonds.

PRACTICE EXERCISE 17.5

When the phosphorus of a phosphatidic acid is linked to the hydroxyl group of the amino alcohol ethanolamine, $NH_2CH_2CH_2OH$, the compound formed is phosphatidyl ethanolamine. Phosphatidyl ethanolamine, also called *cephalin,* is found in brain tissue and is important in blood clotting. Draw the general structure for cephalin.

The second type of phospholipid molecules encountered in cell membranes is **sphingomyelins.** Sphingomyelins do not contain glycerol. Instead, they contain sphingosine, a long-chain unsaturated amino alcohol. Only one fatty acid is attached to sphingosine, as shown in Figure 17.3, through an amide linkage. The structure of the nonpolar end of

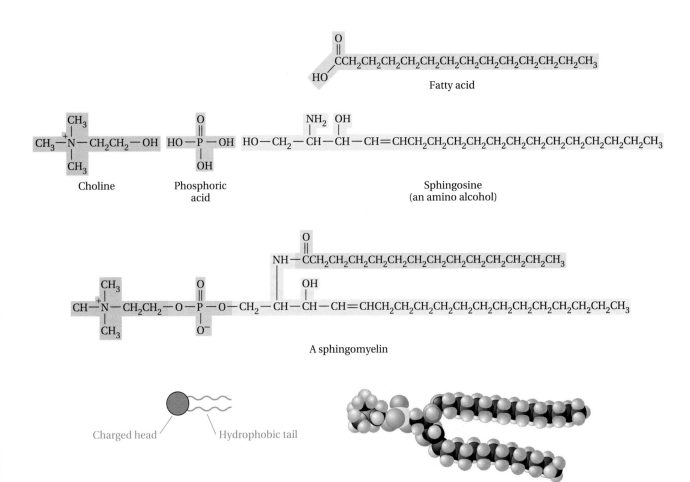

Figure 17.3
A sphingomyelin molecule consists of one molecule each of choline, phosphoric acid, sphingosine, and a fatty acid. A space-filling model of a sphingomyelin molecule is shown at the bottom of the figure.

The abnormal metabolism and accumulation of certain types of lipid molecules occur in a number of genetic diseases. For example, a glycolipid accumulates and damages the brain in Tay-Sachs disease.

sphingomyelins may differ somewhat, depending on the length and degree of saturation of the fatty acid attached to the sphingosine amino group. Sphingomyelins are the only phospholipids that are not built on glycerol. Large amounts of sphingomyelins are found in brain and nervous tissue and in the myelin sheath, the protective coat of nerves.

Glycolipids *are lipid molecules that contain carbohydrates, usually simple sugars such as glucose or galactose.* Figure 17.4 shows a glycolipid that consists of sphingosine, a fatty acid, and a sugar. These are called *cerebrosides* because of their abundance in the brain. The cerebrosides are not phospholipids, because they do not contain phosphorus.

The classification of lipids is summarized in Figure 17.5.

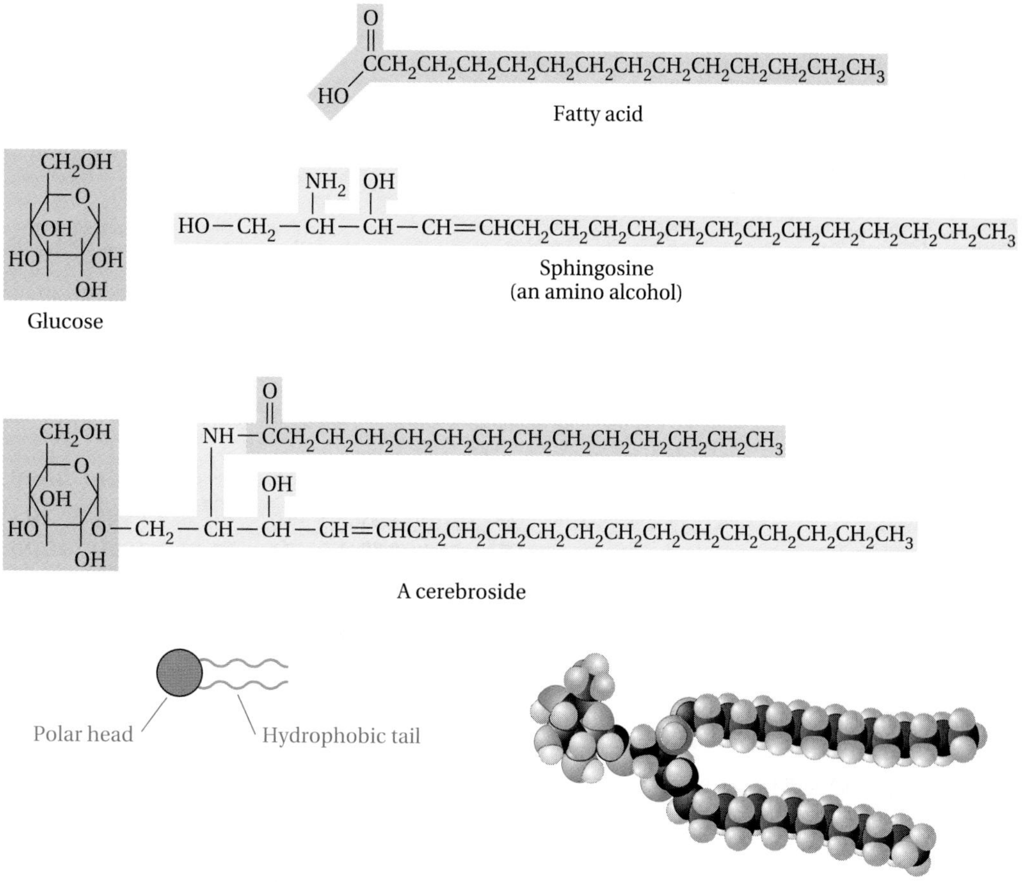

Figure 17.4
A cerebroside (glycolipid) consists of a sugar, a sphingosine, and one fatty acid molecule. The sugar unit shown is glucose, but it may be galactose. A space-filling model of a cerebroside is shown at the bottom of the figure.

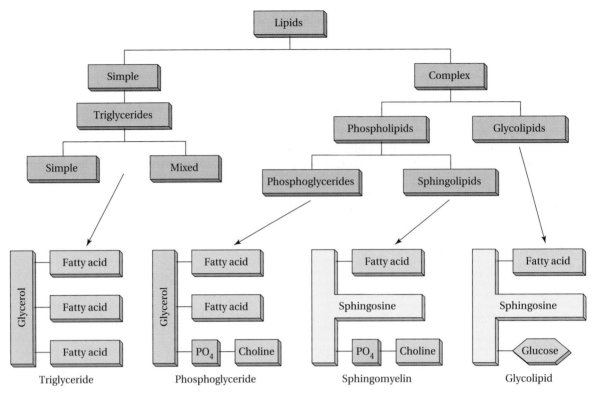

Figure 17.5
Lipid classification diagram.

17.4 Structure of liposomes and cell membranes

AIMS: *To sketch sections of the liposomal bilayer in water, labeling the polar end of the lipid molecules. To explain the relationship between the degree of unsaturation in phospholipid molecules and membrane flexibility. To use the fluid mosaic model to describe the movement of lipid molecules in membranes.*

Focus

Complex lipids form the lipid bilayers of liposomes and cell membranes.

Phosphatidyl choline is a typical membrane phospholipid. It contains a charged head consisting of negatively charged phosphate and positively charged choline attached through glycerol to two hydrophobic fatty acid tails. If we vigorously shake a mixture of phosphatidyl choline and water, the lipid molecules form microscopic spheres rather than dispersing evenly in water. *These lipid spheres, or* **liposomes,** *are packages of water surrounded by a* **lipid bilayer**—*a two-layer-thick wall of phosphatidyl choline.* Figure 17.6 shows a cross section of a liposome. The lipid molecules of the liposomal bilayer are more ordered than the sulfonic acid molecules in detergent micelles (Sec. 14.3).

All the hydrophobic hydrocarbon tails of the lipids are protected from water, and all the hydrophilic phospholipid heads interact with water. Lipo-

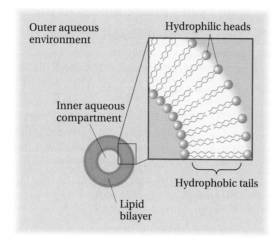

Figure 17.6
Cross-section of a liposome.
The whole liposome is spherical.

somes are stable structures. When broken to expose the lipid hydrocarbons to water, liposomes spontaneously reseal. Lipid bilayers are tight, so any leakage of the contents of a liposome to the outside is quite slow.

PRACTICE EXERCISE 17.6

Predict the result if the liposome experiment (shaking phosphatidyl choline in a solvent) is repeated using a nonpolar solvent (such as carbon tetrachloride) instead of water.

Structure of cell membranes

Figure 17.7
The space-filling models show that the molecule of the saturated fatty acid palmitic acid (a) is straighter than that of the unsaturated fatty acid oleic acid (b). The highly ordered packing of three molecules of palmitic acid (c) is disrupted by the presence of a molecule of oleic acid between two molecules of palmitic acid (d). For this reason, the phospholipids rich in saturated fatty acids make more tightly packed, less flexible membranes (e) than those rich in unsaturated fatty acids (f).

Cell membranes also consist of lipid bilayers similar to those described for liposomes. Some cell membranes are more flexible than others, due to differing properties of the different fatty acids of the lipids that make the lipid bilayers. Lipid bilayers made of phospholipids having a high percentage of unsaturated fatty acids are more flexible than those having a high percentage of saturated fatty acids. As shown in Figure 17.7, unsaturated fatty acid chains are bent and fit into bilayers more loosely than saturated fatty acid chains. The looser the packing in the bilayer, the more flexible is the membrane. A high percentage of glycolipids also tends to increase membrane flexibility.

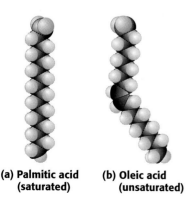

(a) Palmitic acid (saturated)   **(b) Oleic acid (unsaturated)**

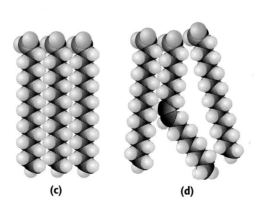

(c)   **(d)**

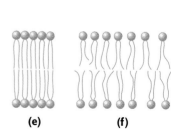

(e)   **(f)**

The fluid mosaic model of membrane structure

The **fluid mosaic model** *of membrane structure proposes that lipids of the bilayer are in constant motion, gliding from one part of their bilayer to another at high speed.* Although lipids move freely within their own layer of the bilayer, they cannot easily cross or flip-flop to the other lipid layer (see Fig. 17.8).

Stuck in the mass of moving lipids are protein molecules, some moving and some apparently anchored in place. Figure 17.9 shows two kinds of membrane proteins: peripheral and integral. **Peripheral proteins** *perch on either side of the lipid bilayer.* They can usually be removed from the membrane by high salt concentrations. **Integral proteins** *may be partially embedded in one side of the bilayer or jut all the way through.*

Many integral proteins are *glycoproteins*—protein molecules with attached carbohydrates. The carbohydrate portion of the embedded glycoproteins is found at the surface of the bilayer, where its hydroxyl groups can hydrogen bond with water. The protein portion of membrane glycoproteins is usually hydrophobic. This ensures a favorable interaction of the protein with the hydrophobic lipid tails of the bilayer. Many membrane proteins facilitate the transport of ions and molecules into and out of the cell.

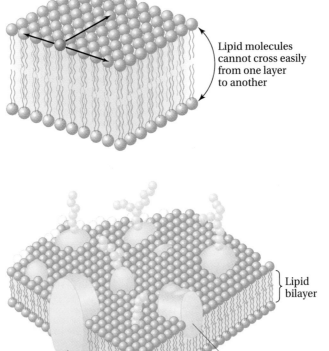

Figure 17.8
Lipid molecules move easily within their own layer of the bilayer but do not readily flip-flop to the other layer.

Lipid molecules cannot cross easily from one layer to another

Figure 17.9
Fluid mosaic model. Most membrane proteins are integral; some membrane proteins are peripheral. The beadlike structures attached to the proteins represent the carbohydrate portions of glycoproteins.

Lipid bilayer

Integral protein

Peripheral protein

17.5 Steroids

AIMS: To draw the fundamental chemical structure of all steroid molecules. To describe the function of cholesterol.

Steroid
stereos (Greek):   solid

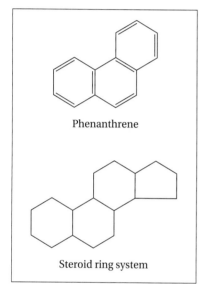

Phenanthrene

Steroid ring system

Figure 17.10
The phenanthrene and steroid ring system.

The amount of artery-plugging cholesterol in the blood is closely related to the dietary intake of saturated fats. Cooking oils, which contain less saturated triglyceride, may be a healthier alternative to the hydrogenated products.

The next class of lipids we will examine is **steroids**—*a family of lipids found in plants and animals.* The connection between steroids and other classes of lipids is mainly historical. Scientists first isolated steroids along with fats by extracting plant and animal tissues with chloroform or ether. Like other lipids, steroids are insoluble in water but soluble in organic solvents. As we will soon see, however, the structures of steroids are quite different from those of triglycerides, phospholipids, and glycolipids. Medical research has revealed that certain steroids, the *steroid hormones,* are enormously important to the proper functioning of the human body and, when misused, to the dysfunctioning of the human body. Hence it will be worth our while to examine the chemical structure and biological function of steroids.

All steroid molecules are saturated derivatives of phenanthrene, a tricyclic aromatic hydrocarbon. The fundamental structure of every steroid molecule also contains a fused cyclopentane ring. Figure 17.10 compares the structures of phenanthrene and the steroid ring system.

All steroid molecules are formed from the steroid ring system by replacing ring hydrogens with hydrocarbon chains or functional groups or by introducing carbon-carbon double bonds into one or more of the cyclohexane rings. *Steroid molecules containing a hydroxy function and no carboxyl or aldehyde groups are called* **sterols.** Cholesterol, a waxy solid, is an example of a sterol:

Cholesterol

Cholesterol is found only in animal cells; plants do not make it. A typical animal cell membrane contains about 60% phospholipid and 25% cholesterol. Cholesterol appears to impart rigidity to cell membranes. The higher the percentage of cholesterol, the more rigid is the membrane.

PRACTICE EXERCISE 17.7

What structural features of cholesterol permit cholesterol molecules to become part of cell membrane structures?

17.6 Steroid hormones

AIM: To state the source and at least one function of cortisone, prednisone, and aldosterone.

Hormones *are chemical messengers produced by the endocrine glands in the human body.* Figure 17.11 shows the locations and names of the endocrine glands. Hormones produced by these glands are secreted directly into the bloodstream. Some of the most important hormones in the human body are steroids. Cholesterol is not a steroid hormone, but many steroid hormones are probably made from it.

    Steroid hormones *are produced and secreted at two major places in the human body: the adrenal glands and the gonads (the testes in males and the ovaries in females).* Adrenal glands are small mounds of tissue at the top of each kidney. An outer layer of the gland, the adrenal cortex, produces a number of potent steroid hormones. *The adrenal cortex steroids are often called adrenal corticoids or simply* **corticoids.** The chemical structures of more than 30 corticoids are known, but only a few seem to function as hormones. Active corticoids fall into three categories according to function: *glucocorticoids, mineralocorticoids,* and *secondary sex hormones.* We will cover the glucocorticoids and mineralocorticoids in this section. Secondary sex hormones will be discussed with other male and female hormones in Sections 17.7 and 17.8, respectively.

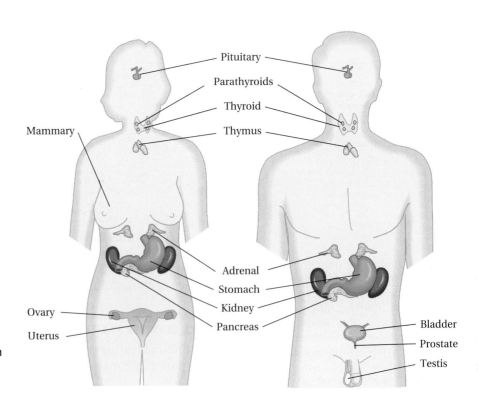

Figure 17.11
The human endocrine system. Steroid hormones are produced in the adrenal cortex, the outer layer of the adrenal glands.

Glucocorticoids

Glucocorticoids *regulate the body's use of glucose—how much is burned as fuel to provide energy and how much is stored as glycogen for future energy needs.* Glucocorticoids tend to prevent the uptake of glucose by tissues and therefore promote its storage as glycogen (see Sec. 16.6). Other nonsteroid hormones, especially insulin, promote the uptake of glucose by tissues. Together, the glucocorticoids and insulin maintain the body's delicate balance between too much and too little glucose. Glucocorticoids also affect body protein and fat utilization in a similar way. Cortisone, cortisol, and corticosterone are the major glucocorticoids:

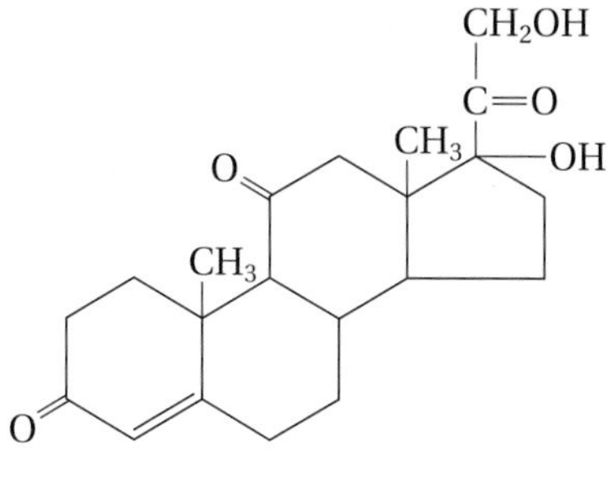

Cortisone

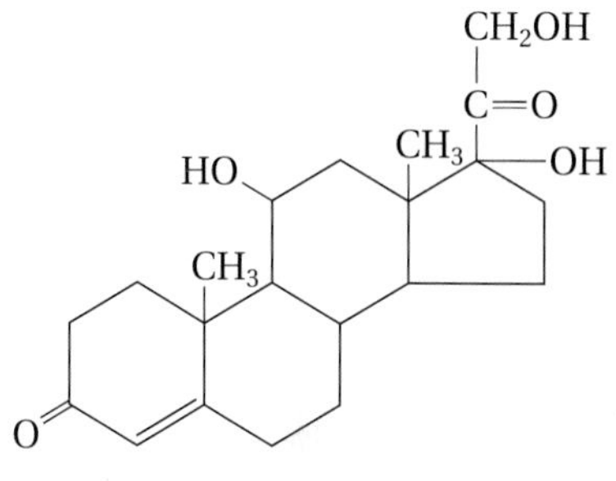

Cortisol

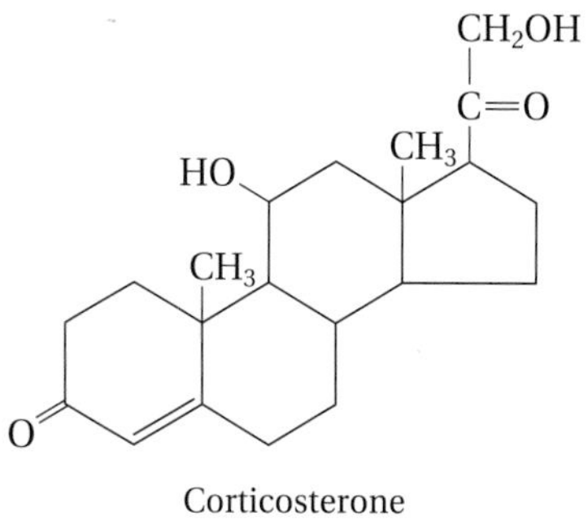

Corticosterone

Quite a few years ago scientists learned that cortisone reduces the symptoms of rheumatoid arthritis, allergic diseases, and inflammation. Many people hoped it would become a new wonder drug. However, since the side effects of cortisone injections are often severe, the drug is rarely used for these purposes. Fortunately, medicinal chemists have been able to synthesize new steroids that give many of the advantages of cortisone therapy with fewer undesirable side effects. Two of the main cortisone substitutes in use today are prednisone and prednisolone:

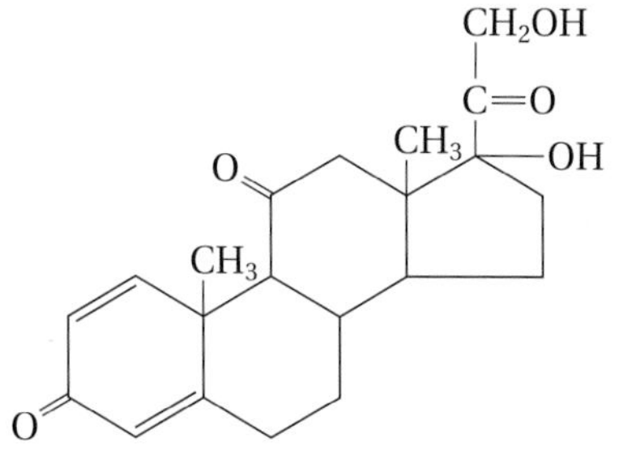

Prednisone

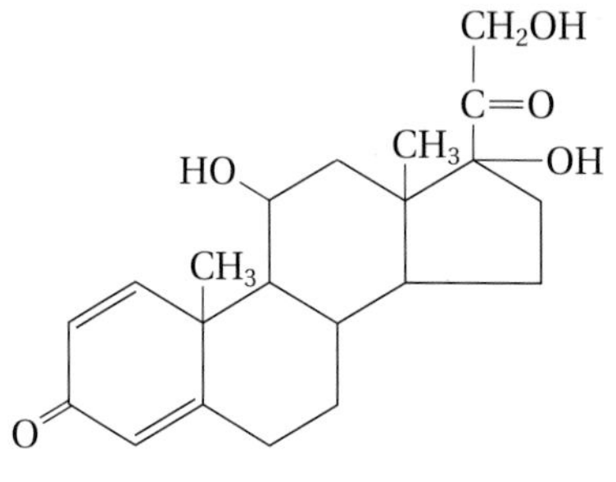

Prednisolone

Glucocorticoids reduce the body's immunity to the introduction of foreign materials. (This may be why these compounds are effective in the treatment of allergic diseases, which are a severe response to foreign materials.) For this reason, physicians sometimes give injections of cortisone or related drugs to organ transplant patients to help suppress their immune system's response to the new tissue or organ, thereby preventing rejection.

Mineralocorticoids

Mineralocorticoids *help retain sodium chloride and maintain fluid balance in the body.* All corticoids affect salt and water balance to some extent, but aldosterone is by far the most important mineralocorticoid. Aldosterone's sodium chloride–retaining activity is about 1000 times the activity of cortisol, for example. Aldosterone is the principal hormone in maintaining life in an animal whose adrenals have been removed.

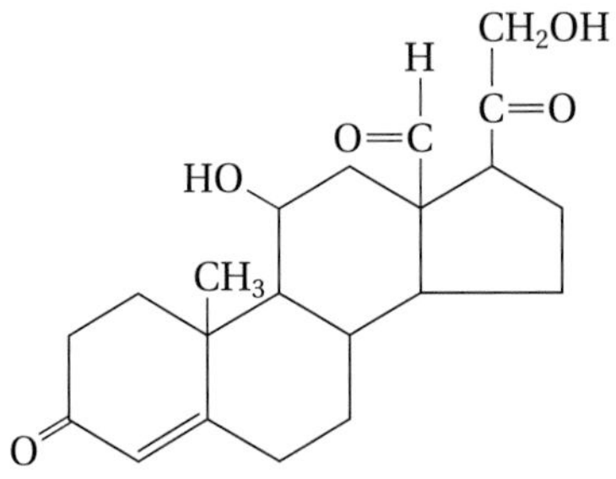

Aldosterone

PRACTICE EXERCISE 17.8

Name the functional groups in (a) aldosterone and (b) cortisone.

17.7 Male sex hormones

AIM: To state the source and at least one function of testosterone and androgens.

Focus

The primary male sex hormone is produced in the testes.

Testosterone *is the principal sex hormone in males and is produced in the testes (testicles).*

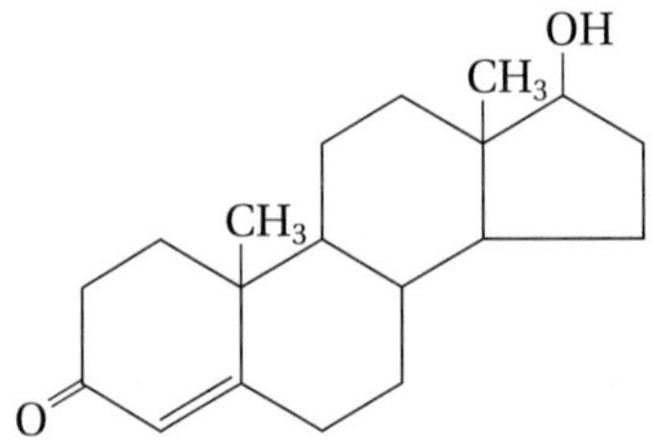

Testosterone

The testes perform two functions: They produce sperm, and they produce testosterone. At puberty, testosterone promotes the maturation and growth of the male sex organs. It also aids in the development of male secondary sex characteristics such as deep voices and beards, and it contributes to the greater muscular development and bone growth of males compared with females.

Women also produce testosterone. Masculinization does not occur in normal women, however, because chemical reactions in a woman's body rapidly convert testosterone into female hormones. The adrenal cortex also produces sex hormones—*androgens* in males and *estrogens* in females. Beginning at puberty, androgens and estrogens produce the secondary sex characteristics considered masculine or feminine: deep voices and beards in males, higher voices, breast development, and lack of facial hair in females. The major androgens are dehydroepiandrosterone and androstenedione.

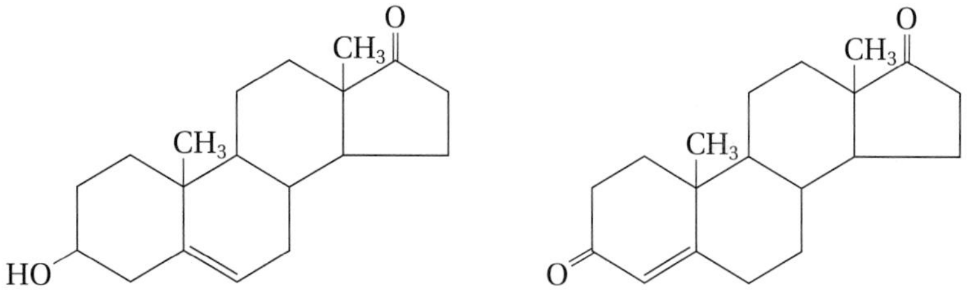

Dihydrotestosterone is a hormone in which the double bond of testosterone has been reduced by enzyme reactions in the body. Dihydrotestosterone appears to be involved in two conditions experienced by men: male-pattern baldness and growth of the prostate, an exclusively male gland that controls the flow of urine from the bladder.

Dehydroepiandrosterone   Androstenedione

Overproduction of androgens can cause masculinization, also called *virilization*, in females. The "bearded ladies" sometimes seen in circus sideshows usually suffer from excessive androgen production.

Certain world-class athletes have used synthetic steroids structurally related to testosterone. These drugs, the anabolic steroids, promote muscular development. Norancholane is one of the anabolic steroids.

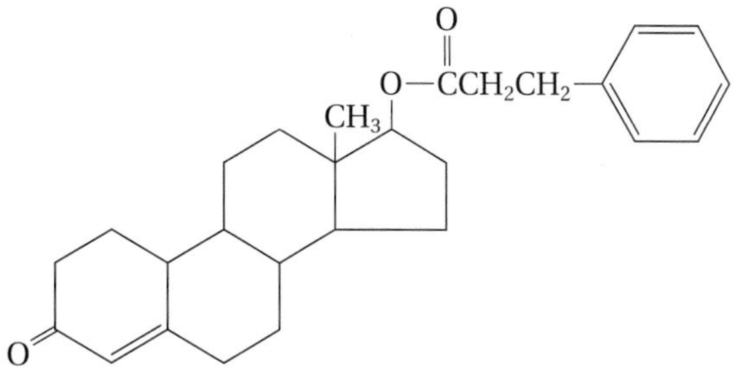

Norancholane

The use of any drug for this purpose is officially banned by the world's amateur athletic unions. Anabolic steroids can be detected in the urine for only a few days after their use is discontinued, however; so this rule is hard to enforce. In the short term, there are reports of testicular atrophy in some users of anabolic steroids. The long-term effects of synthetic bodybuilding steroid drugs are not well understood but may include cancer.

17.8 Female sex hormones

AIM: To state the source and at least one function of estrogens and progesterone.

The primary female sex hormones are produced by the ovaries. *Two estrogens—estrone and estradiol—and* **progesterone** *are the principal female sex hormones.* An apparently small change in the molecular structure of a steroid causes a profound change in its effect on the body. Testosterone and progesterone, two steroid hormones that help divide all humanity into male and female, differ by only two atoms of carbon and two of hydrogen.

Estrone   Estradiol   Progesterone

Estrogens are produced in the adrenal cortex and contribute to the secondary sex characteristics of women—breast development, high voices, and lack of facial hair. They are also extremely important as primary sex hormones. Estrogens are important to the development of the egg, the ovum, in the ovaries. Progesterone causes changes in the wall of the uterus that prepare it for pregnancy and that maintain pregnancy.

Figure 17.12 shows how the levels of estrogens and progesterone affect egg and uterine wall development in the human female's ovarian cycle. The cycle usually takes 28 days; the first day of the menstrual period is considered the beginning of the cycle. Near the end of menstruation, development of a new egg begins in the follicle, a small body inside the ovary. Estrogens promote growth of the follicle and also stimulate thickening of the uterine lining. By about the fourteenth day of the cycle, the follicle, now called the *graafian follicle,* has matured. At ovulation, the matured follicle ruptures, releasing the egg, which enters the oviduct (fallopian tube), a channel between the ovary and the uterus.

Once the egg is released, the ruptured follicle, now called the *corpus luteum,* begins to secrete progesterone. The action of progesterone maintains the thickness of the uterine lining and stimulates the lining's final development. If the egg is not fertilized, the corpus luteum begins to shrink and secrete less progesterone. Eventually, the concentration of progesterone is too small to maintain the uterine lining. The lining begins to slough off, and hemorrhagia takes place. This marks the onset of menstruation and a new ovarian cycle. A woman loses an average of 50 to 150 mL of blood during menstruation. If the egg is fertilized, the corpus

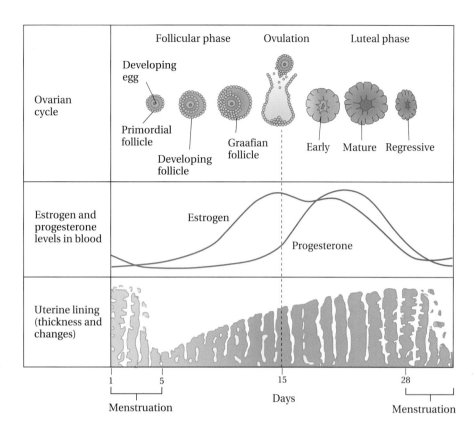

Figure 17.12

Figure 17.12
Effects of blood levels of estrogens and progesterone on the development of the egg and the uterine lining during the ovarian cycle.

luteum continues to produce progesterone throughout the course of pregnancy. The progesterone makes implantation of the fertilized egg on the uterine wall possible, promotes the development of the mammary glands, and prevents new follicles and eggs from maturing.

17.9 The pill

AIM: To explain how birth control pills prevent conception.

Focus

The original oral contraceptive pill prevented ovulation.

Since progesterone prevents maturation and release of the egg, this hormone would seem to be a good candidate for a birth control drug—a *contraceptive*—that could be taken by mouth. However, progesterone itself is an ineffective oral contraceptive, because a woman's body breaks it down before it reaches the ovaries. In the late 1940s, scientists began a search for a synthetic oral drug that mimicked the hormonal effect of progesterone. Several such drugs are now on the market; they are collectively called "the pill." Many early preparations of the pill contained rather large doses (up to 1.5 mg) of norethindrone or norethynodrel. A synthetic estrogen, mestranol (usually about 0.06 mg), is also included in this class of drugs.

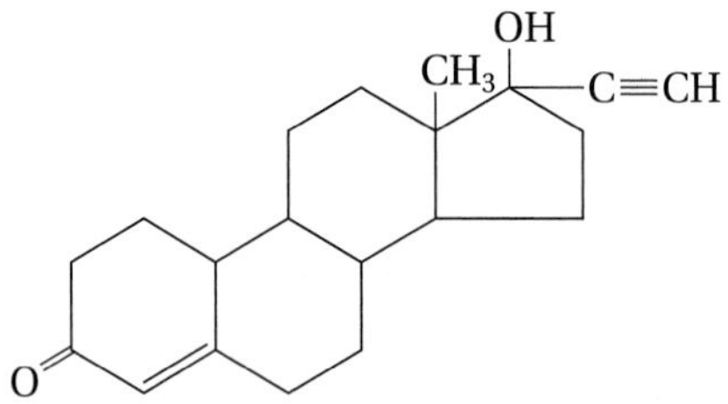

Norethindrone
(Norlutin)

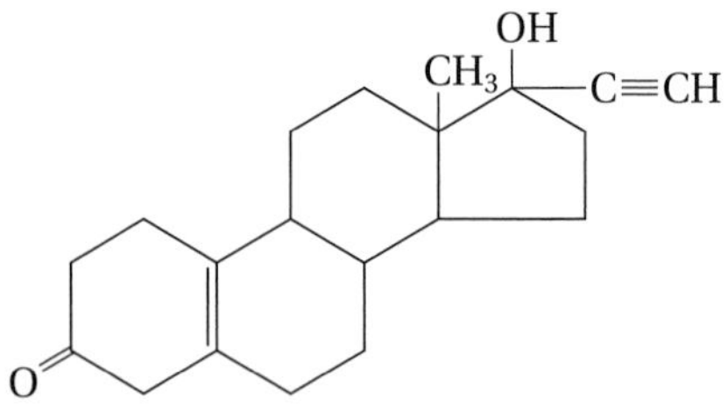

Norethynodrel
(Enovid ketone)

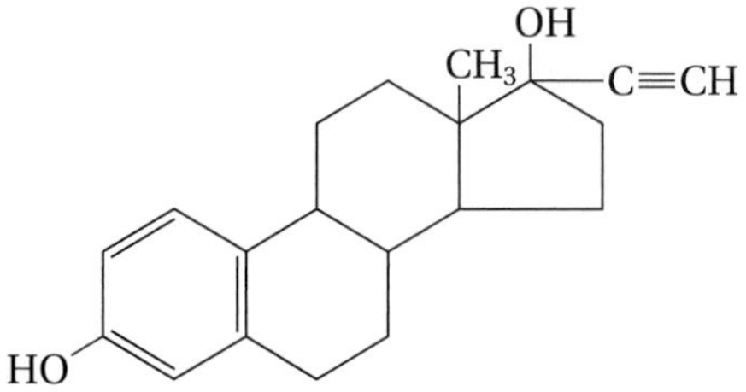

Mestranol
(synthetic estrogen)

Women who take the pill often experience the symptoms of pregnancy—dizziness, headaches, vomiting, and increases in weight and breast sensitivity—because of the high levels of progesterone-like drug in their bodies. Put another way, women's bodies have no way of "knowing" that they are not pregnant.

The "minipill" was introduced around 1971. Minipills contain much smaller amounts (0.1 to 0.2 mg) of the synthetic progesterones and no synthetic estrogen. The minipill was introduced in the hope that its users would experience fewer side effects. The action of the newer generations of the pill appears to be somewhat different from that of the ancestors. Minipills seem to stop conception by preventing sperm from entering the oviduct, by preventing release of the egg, and by making the uterus less receptive to any fertilized egg that reaches it.

Scientists have had somewhat less success in developing a "morning-after pill." Diethylstilbestrol (DES), a nonsteroid compound that mimics the effects of the estrogens, is effective in preventing pregnancy if taken in large doses for 3 days after intercourse; DES prevents implantation of a fertilized egg by causing excessive thickening of the uterine lining.

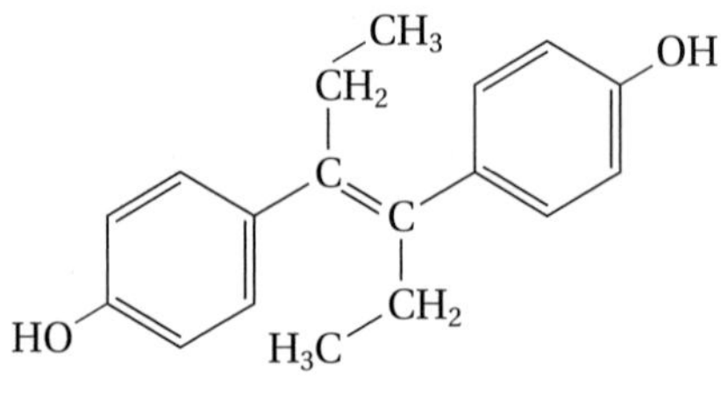

A steroid-like drug called RU-486 is an *abortifacient*, a drug used to terminate pregnancy.

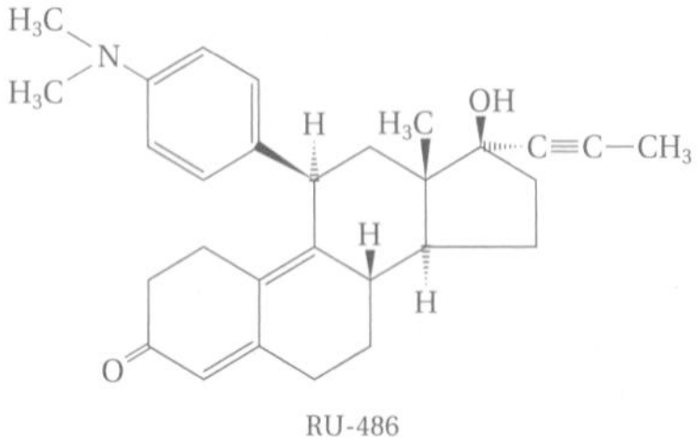

RU-486

Diethylstilbestrol
(DES)

The undesirable side effects of DES include vomiting, nausea, and excessive vaginal bleeding. In the 1940s and early 1950s, DES was routinely used to prevent spontaneous abortions (miscarriages). This practice was stopped when several studies showed that daughters born to women who took DES have a higher-than-average incidence of vaginal cancer after reaching puberty.

17.10 Plant steroids

AIM: To state the source and at least one function of digitoxin.

Focus

Digitalis is a mixture of plant steroids used to treat heart disease.

Many plants produce steroids also. Nobody knows, though, what function many of these compounds have in plants. Many plant steroids are toxic to humans. Nevertheless, one of the oldest drugs, in use since 1785, is in fact a mixture of toxic steroids. The drug is *digitalis,* obtained from the common foxglove plant. Digitalis is potentially deadly, but in very small doses it improves the tone of heart muscles and is widely used to treat congestive heart failure. Digitoxin is one of the major components of digitalis. Digitoxin has a steroid ring system, but it is unlike animal steroids in that the steroid part of the molecule is attached to a carbohydrate part.

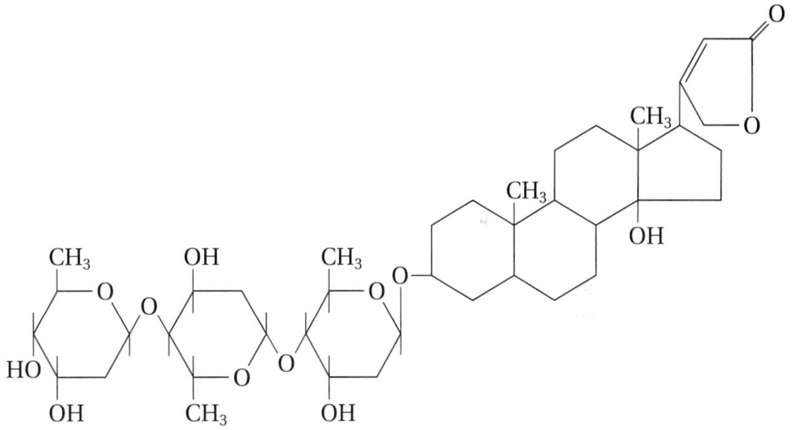

Digitoxin

17.11 Prostaglandins and leukotrienes

AIM: To recognize prostaglandins and leukotrienes and state several of their biological effects.

Focus

Prostaglandins and leukotrienes influence many body functions.

Prostaglandins *are a class of fatty acid derivatives that have a wide range of physiologic activity.* These substances have hormone-like effects. Unlike hormones, however, they are not transported to their site of action in the bloodstream but are synthesized in the same environment in which they act. Prostaglandins are involved in the control of pain and fever, acid secretion into the stomach, relaxation and contraction of smooth muscle, blood pressure, and the wake/sleep cycle. They also stimulate many inflammatory responses, particularly of the joints (rheumatoid arthritis), skin (psoriasis), and eyes. It seems likely that prostaglandins play a role in almost every stage of reproduction.

Since their initial discovery in the 1930s, more than a dozen different prostaglandins have been identified. Slight structural differences account for their distinct biological effects. Prostaglandins are synthesized in the body by oxidation and cyclization of unsaturated continuous-chain fatty acids containing 20 carbons. Arachidonic acid is one such acid.

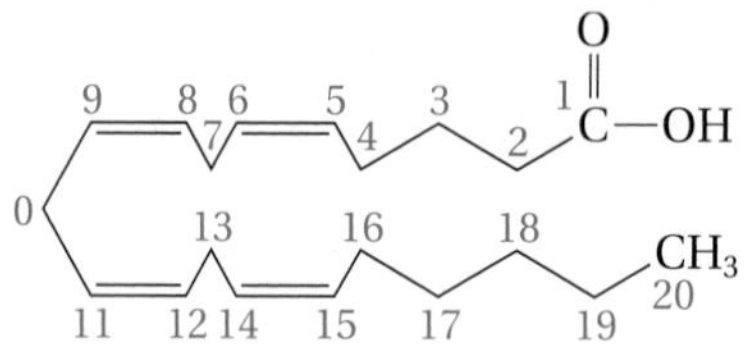

Arachidonic acid

This acid is converted into a prostaglandin structure when the eighth and twelfth carbons fuse to make a cyclopentane ring. Every prostaglandin molecule contains 20 carbons arranged in the basic structure of prostanoic acid.

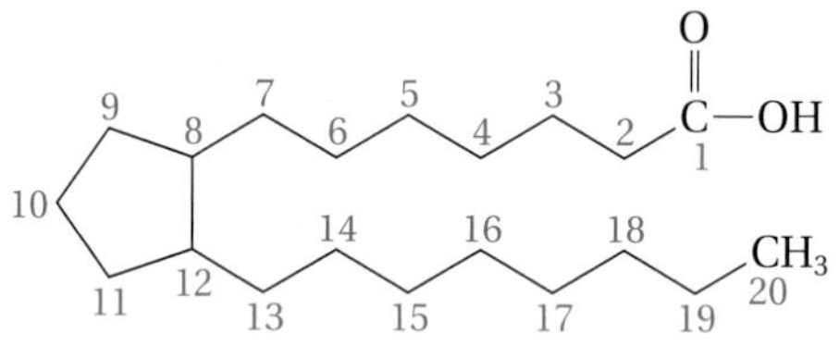

Prostanoic acid

Prostanoic acid is so named because prostaglandins were first isolated from human semen, and scientists thought that they were produced in the prostate gland. It is now known that they are present in minute amounts in nearly all animal tissues and fluids.

Prostaglandin E_2 demonstrates the widespread effects of prostaglandins on a variety of body functions. Prostaglandin E_2 does not itself cause pain, but it enhances the intensity of the pain induced by other pain-producing chemicals in the body. This prostaglandin also induces the signs of inflammation such as swelling, redness, and heat in the inflamed body part. It also lowers arterial blood pressure and promotes blood clotting. Prostaglandin E_2 and prostaglandin $F_{2\alpha}$ have been used extensively as drugs in reproductive medicine, where they can be employed to induce labor and terminate pregnancy.

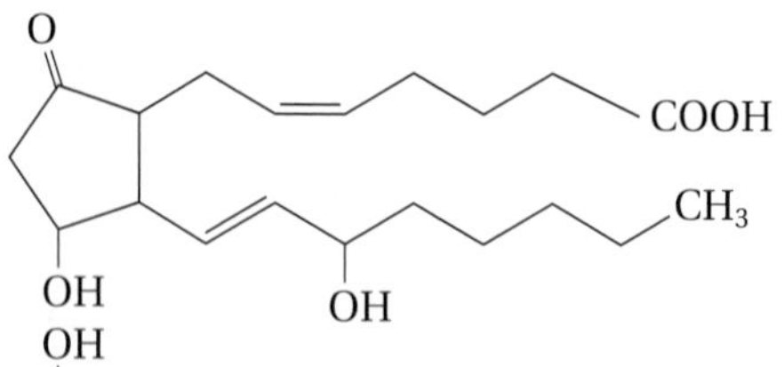

Prostaglandin E_2

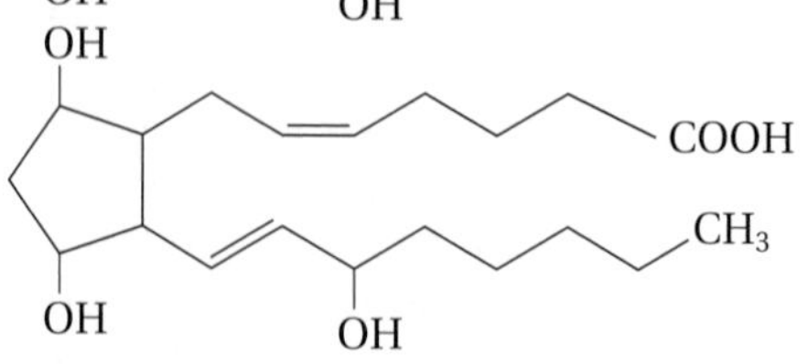

Prostaglandin $F_{2\alpha}$

Aspirin blocks the synthesis of prostaglandins that induce pain and fever. It also has been reported that low doses of aspirin reduce the danger of heart attack and stroke by blocking the synthesis of prostaglandins that promote blood clotting. Aspirin's inhibition of the formation of these prostaglandins accounts for its effectiveness as an analgesic (pain killer), antipyretic (fever killer), and anticlotting agent.

Leukotrienes *are another class of hormone-like substances synthesized from arachidonic acid.* Leukotriene B_4 and leukotriene E_4 are examples of leukotrienes. Unlike the prostaglandins, the leukotrienes contain three carbon-carbon double bonds in a row in their molecular structures.

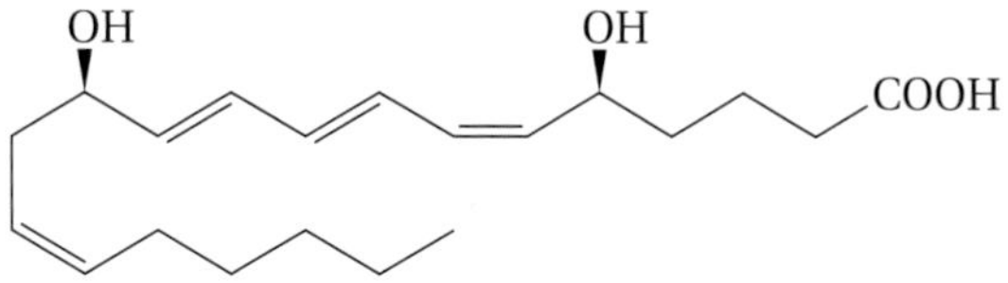

Leukotriene B_4

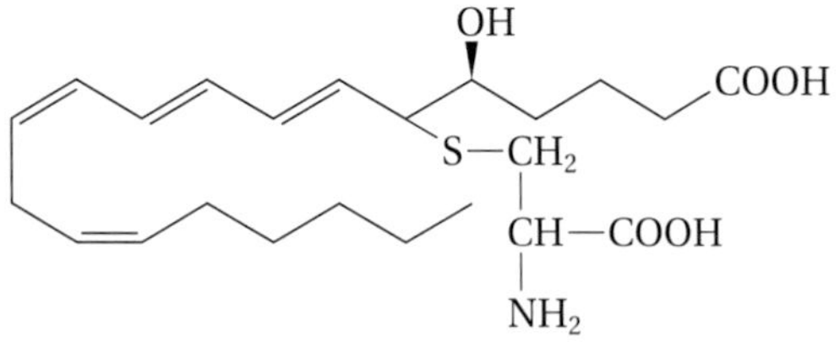

Leukotriene E_4

Certain leukotrienes constitute what is referred to as *the slow-acting substances of anaphylactic shock.* Anaphylactic shock is a drastic allergic response of the body that can be fatal. Severe allergies to foods such as shellfish or peanuts can trigger anaphylactic shock in susceptible people. The symptoms of anaphylactic shock include closing of the throat, blocking the airways. Heart stoppage also can occur, leading to death. Leukotrienes are also implicated in less severe allergic reactions, asthma, inflammations, and heart attacks. Recent studies suggest that dietary fish oils can reduce levels of leukotrienes and reduce the risk of heart attack.

SUMMARY

Lipids are a broad class of naturally occurring, relatively water-insoluble molecules. Triglycerides—triesters of glycerol and fatty acids—are the most abundant lipids in animal tissue.

The membranes of all cells are composed of lipids. The major lipids of cell membranes are the phospholipids, the glycolipids, and in animal cells, cholesterol. Phospholipids have a backbone of glycerol (phosphoglycerides) or of sphingosine (sphingomyelins). Cerebrosides, containing glucose or galactose, with a backbone of sphingosine are common glycolipids. Phospholipid and glycolipid molecules have polar heads and hydrophobic tails. The lipid bilayer is the fundamental structure of liposomes and the membranes of cells and organelles. Its behavior is best described by a fluid mosaic model. Proteins, which may be peripheral or integral, are associated with cell and organelle membranes. Membrane proteins are often glycoproteins.

Steroid molecules have no structural features in common with other lipids. They are often classified as lipids, however, because, like fatty acid esters, they are extractable from plant and animal tissues with organic solvents. Aldehyde and ketone functions as well as hydroxyl groups appear frequently in steroid molecules, which sometimes contain a double bond.

Cholesterol is the most abundant steroid in animals. Many steroids present in lesser amounts in the animal body act as hormones. The adrenal cortex, the outer layer of the adrenal gland, produces three groups of steroid hormones: the glucocorticoids, the mineralocorticoids, and the secondary sex hormones. Glucocorticoids (cortisone, cortisol, corticosterone) regulate the body's use of glucose and affect salt and water balance. Mineralocorticoids

also affect the body's salt and water balance; aldosterone is the most important mineralocorticoid. Secondary sex hormones produce secondary sex characteristics of the male and female. They are the androgens in the male and the estrogens in the female. Primary sex hormones produced by the gonads are responsible for the maturation of the reproductive system.

Prostaglandins and leukotrienes, two groups of fatty acid–related molecules, exert many hormone-like effects on the body. Prostaglandins are used for inducing labor and terminating pregnancy and also have been used for controlling blood pressure and reducing inflammation.

KEY TERMS

Cell membrane (17.3)	Integral protein (17.4)	Peripheral protein (17.4)	Sphingomyelin (17.3)
Complex lipid (17.3)	Lecithin (17.3)	Phosphoglyceride (17.3)	Steroid (17.5)
Corticoid (17.6)	Leukotriene (17.11)	Phospholipid (17.3)	Steroid hormone (17.6)
Estrogen (17.8)	Lipid (17.1)	Progesterone (17.8)	Sterol (17.5)
Fat (17.2)	Lipid bilayer (17.4)	Prostaglandin (17.11)	Testosterone (17.7)
Fluid mosaic model (17.4)	Liposome (17.4)	Rancid (17.2)	Triglyceride (17.2)
Glucocorticoid (17.6)	Mineralocorticoid (17.6)	Saponification (17.2)	Wax (17.1)
Glycolipid (17.3)	Mixed triglyceride (17.2)	Sex hormone (17.5)	
Hormone (17.6)	Oil (17.2)	Simple triglyceride (17.2)	

EXERCISES

Waxes and Simple Lipids (Sections 17.1, 17.2)

17.9 Write the general formula for an ester.

17.10 Name the two classes of organic compounds that are produced when waxes are hydrolyzed.

17.11 (a) Name the following fatty acids, and state whether they are saturated or unsaturated.
(b) Which of them are solids and which are liquids at room temperature?

(a) $CH_3(CH_2)_{10}COOH$
(b) $CH_3(CH_2)_4CH=CHCH_2CH=(CH_2)_7COOH$
(c) $CH_3(CH_2)_{14}COOH$
(d) $CH_3(CH_2)_7CH=CH(CH_2)_7COOH$

17.12 Decide whether the following fatty acids are most abundant in animal fat or vegetable oil: (a) stearic acid, (b) oleic acid, and (c) linoleic acid.

17.13 What is a triglyceride?

17.14 Write the structure for glyceryl trioleate, a simple triglyceride. Would you expect this compound to be an oil or a fat? Why?

17.15 (a) What happens, chemically, when an oil is hydrogenated? (b) Why is this process of great commercial importance?

17.16 Why do animal fats and vegetable oils become rancid when exposed to moist warm air?

17.17 Palmolive soap is mostly sodium palmitate. Write the structure of this compound.

17.18 The following compound is hydrolyzed by boiling with sodium hydroxide. What are the saponification products?

$$CH_2-O-\overset{\overset{O}{\|}}{C}-(CH_2)_{14}CH_3$$
$$CH-O-\overset{\overset{O}{\|}}{C}-(CH_2)_{10}CH_3$$
$$CH_2-O-\overset{\overset{O}{\|}}{C}-(CH_2)_7CH=CH(CH_2)_7CH_3$$

Complex Lipids and Cell Membranes (Sections 17.3, 17.4)

17.19 Name the two categories of complex lipids.

17.20 Write a general structure for the following complex lipids: (a) a phosphoglyceride and (b) a sphingomyelin.

17.21 What is the difference, structurally, between sphingomyelins and phosphoglycerides?

17.22 Draw structural formulas for the products of complete hydrolysis of phosphatidyl choline.

17.23 Use diagrams to show the difference between a liposome and a micelle.

17.24 What is the role of phospholipids in cell structure?

17.25 Explain how the degree of unsaturation in membrane lipids affects the flexibility of cell membranes.

17.26 Describe the major features of the fluid mosaic model of membrane structure.

17.27 What are the functions of membrane proteins?

17.28 Describe the two types of membrane proteins.

Steroids (Sections 17.5–17.10)

17.29 Draw the fundamental chemical structure that applies to all steroid molecules.

17.30 What steroid forms a large part of animal cell membranes but is not found in plants?

17.31 What is a hormone, and where are steroid hormones produced?

17.32 Cortisone is a hormone. Where is it produced in the body, and what is its biological function?

17.33 Explain why a physician might prescribe cortisone shots before a patient has a kidney transplant operation.

17.34 What is (a) an androgen and (b) an estrogen?

17.35 Name (a) the primary male sex hormone and (b) the three primary female sex hormones.

17.36 Compare the structural formulas of testosterone and progesterone. How are these two molecules different?

17.37 Comment on the use and action of anabolic steroids.

17.38 What are the functions of the estrogens and progesterone in the female ovarian cycle?

17.39 Why is progesterone ineffective as an oral contraceptive?

17.40 What is the action of each of the following as an oral contraceptive?
 (a) the pill   (b) the minipill   (c) DES

17.41 How does the structure of digitoxin (the major component of the drug digitalis) differ from that of all the other steroids discussed?

17.42 What is the medicinal use of digitalis?

Prostaglandins and Leukotrienes (Sections 17.10, 17.11)

17.43 Draw the basic structure for a prostaglandin molecule.

17.44 How are prostaglandins synthesized in the body?

17.45 Prostaglandins are among the most potent biological compounds known. What are some of their physiologic effects?

Additional Exercises

17.46 Match the following.
 (a) phosphoglyceride   (1) long-chain esters
 (b) lipid bilayer   (2) absent in plants
 (c) testosterone   (3) fluid mosaic model
 (d) lecithin   (4) body's salt balance
 (e) saponification   (5) congestive heart failure
 (f) fats   (6) phosphate ester
 (g) wax   (7) fat hydrolysis
 (h) digitalis   (8) highly saturated
 (i) cholesterol   (9) male sex hormone
 (j) aldosterone   (10) phosphatidyl choline

17.47 Which type of lipids are found in cell membranes?

17.48 Explain the relationship between the nature of the fatty acids of complex lipids (that is, whether they are saturated or unsaturated) and the fluidity of cell membranes.

17.49 Name all the functional groups in the following compounds:
 (a) cortisone
 (b) phosphatidyl choline
 (c) testosterone
 (d) diethylstilbestrol
 (e) prostaglandin E_2
 (f) sphingosine
 (g) digitoxin
 (h) choline
 (i) triolein.

17.50 Distinguish between a fat and an oil.

17.51 Explain why saturated fatty acids have higher melting points than unsaturated fatty acids.

SELF-TEST (REVIEW)

True/False

1. A triglyceride contains three ester bonds.
2. Integral proteins of membranes help move small molecules through membranes.
3. The major component of most membranes is carbohydrate.
4. Cholesterol is a sterol found in virtually all plants and animals.
5. Increasing the percentage of unsaturated fatty acids in a membrane increases its flexibility.
6. Because the lipid molecules in one layer of a membrane can move about freely, membranes are very permeable.
7. The sex hormone testosterone is produced only by males.
8. Triglycerides containing only saturated fatty acids are often spoiled by oxidative rancidity.
9. In a liposome, the hydrophobic hydrocarbon tails form the juncture of the lipid bilayer.
10. Aldosterone is a mineralocorticoid hormone that helps maintain the body's salt and water balance.

Multiple Choice

11. The polar end of a lipid molecule
    (a) is characterized by long hydrocarbon chains.
    (b) is hydrophobic.
    (c) interacts favorably with water.
    (d) Both (a) and (b) are correct.
12. The chemical structure common to all steroid molecules is

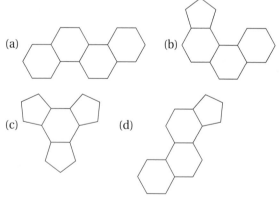

13. Hydrolysis of a mixed triglyceride may
    (a) give three different alcohols.
    (b) give ethylene glycol.
    (c) give three different fatty acids.
    (d) give phosphoric acid.
14. A lipid molecule composed of glycerol, phosphoric acid, and two fatty acids is called a
    (a) glycolipid.   (b) phosphoglyceride.
    (c) cholesterol.   (d) sphingomyelin.

15. Prostaglandins are involved in all the following regulatory functions *except*
    (a) timing of menstrual cycle.
    (b) body temperature regulation.
    (c) muscle contraction.
    (d) stomach acid secretion.
16. Which of the following statements is *not* true about a sphingomyelin?
    (a) It has an amide bond.
    (b) It has a phosphate ester bond.
    (c) It has a glycerol backbone.
    (d) Choline is found at the polar head of the molecule.
17. Sources of hormones in the body include
    (a) the adrenal glands.   (b) the ovaries.
    (c) the testes.   (d) all of the above.
18. The adrenal corticoids called *glucocorticoids*
    (a) produce the body's sex characteristics.
    (b) are used to reduce the risks of organ transplant operations.
    (c) include the primary sex hormones.
    (d) all contain at least one carbohydrate molecule.
19. Liquid oils are changed into solid fats by
    (a) hydrogenation.   (b) solidification.
    (c) hydration.   (d) rancidization.
20. Waxes are formed from long-chain
    (a) acids and amines.   (b) amines and alcohols.
    (c) alcohols and aldehydes.   (d) acids and alcohols.
21. Which of the following is *not* primarily a female sex hormone?
    (a) progesterone   (b) estradiol   (c) mestranol
    (d) estrone
22. In hydrolytic rancidity of triglycerides, which part of the molecule is attacked?
    (a) double bond   (b) alcohol group
    (c) acid group   (d) ester group
23. Which of the following drugs is an important steroid produced by a plant?
    (a) androgen (b) digitalis   (c) progestin
    (d) estrogen
24. Alkaline hydrolysis of a fat to make soap is called
    (a) glycolysis.   (b) saponification.
    (c) hydration.   (d) anabolication.
25. The "power stations" of a eukaryotic cell are the
    (a) nucleolus.   (b) Golgi bodies.
    (c) mitochondria.   (d) peroxisomes.
26. Lysosomes function in
    (a) breaking down complex molecules.
    (b) cellular reproduction.
    (c) transporting proteins across cell membranes.
    (d) protein synthesis.

Amino Acids, Peptides, and Proteins

18

Molecular Structures and Biological Roles

Proteins are the source of amino acids. Egg whites are rich in protein.

CHAPTER OUTLINE

CASE IN POINT: Runner's high

18.1 The amino acids

18.2 Stereoisomers of amino acids

18.3 Zwitterions

18.4 Peptides

A CLOSER LOOK: Aspartame

FOLLOW-UP TO THE CASE IN POINT: Runner's high

18.5 Primary structure of proteins

18.6 Secondary structure of proteins

18.7 Tertiary structure of proteins

A CLOSER LOOK: X-ray Crystallography

18.8 Quaternary structure of proteins

18.9 Hemoglobin function

18.10 Sickle cell anemia

A CLOSER LOOK: Electrophoresis

18.11 Glycoproteins

A CLOSER LOOK: Glycoproteins: Control of Blood Glucose in Diabetes

18.12 Denaturation

This chapter begins with a discussion of the structures and chemistry of the most common naturally occurring amino acids. Then we will see how combinations of the natural amino acids form large molecules called *peptides* and *proteins* (see chapter opener figure). Along the way we will encounter a few important peptides and proteins that highlight the relationship between molecular structure and biological function. The Case in Point for this chapter involves a class of molecules called the *enkephalins,* which promote feelings of comfort and sometimes exhilaration.

CASE IN POINT: Runner's high

Karla, a college student, does not consider herself to be athletic. A few months ago, however, she decided to try running for exercise. Her first few times out she was able to run for only a few minutes. Gradually, she began to cover more and more distance, until she could run about 5 miles in 35 minutes, three times a week. In addition to her increased endurance, Karla has lost a few pounds and feels mentally more alert. To her surprise, Karla has become such a dedicated runner that she becomes edgy if her schedule forces her to miss a running session. Part of Karla's attraction to running is that during her runs she sometimes experiences a feeling that can only be described as exhilaration. Karla's exhilaration is "runner's high." What causes Karla's runner's high? We will find out in the discussion of certain peptide hormones in Section 18.4.

Physical exercise can be exhilarating.

18.1 The amino acids

AIM: To classify the 20 common amino acids according to their side-chain structures.

Focus

There are 20 common natural amino acids.

An **amino acid** *is any carboxylic acid (RCO_2H) that contains an amino group (—NH_2) in the same molecule.* To most chemists and biochemists, however, the term is usually reserved for the 20 amino acids that are found and used in living organisms. All 20 amino acids common in nature have a skeleton consisting of a carboxylic acid group and an amino group covalently bonded to a central atom. *The central atom of amino acids is called the* **alpha carbon.** The remaining two groups on the alpha carbon are hydrogen and an *R group—the* **amino acid side chain.**

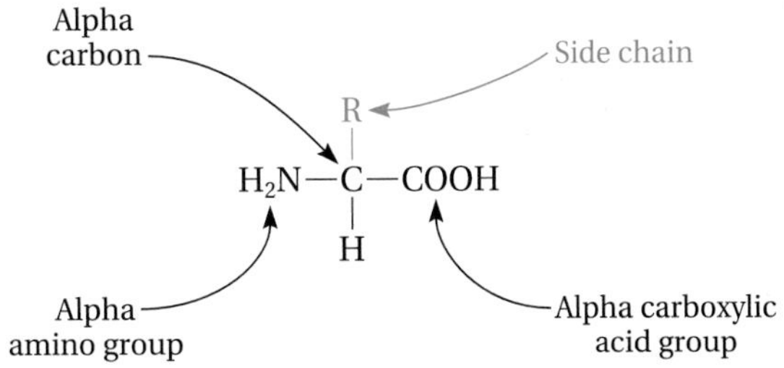

The chemical nature of the side chains accounts for differences in amino acid properties. The 20 common naturally occurring alpha amino acids are divided into seven categories according to their side-chain structures.

Glycine
glycos (Greek): sweet
(The amino acid glycine has a sweet taste.)

Aliphatic side chains This group consists of amino acids having hydrocarbon (aliphatic) side chains. Glycine, the simplest amino acid, has a hydrogen rather than an aliphatic side chain, but it is still placed in this category. Alanine, with a methyl side chain, is the smallest true member of this group. Other aliphatic R groups are those of the amino acids valine (isopropyl), leucine (isobutyl), and isoleucine (*sec*-butyl). Leucine and isoleucine have the same molecular formulas and are structural isomers. Proline is the only amino acid with its alpha amino group incorporated into a ring. The amino group of proline is a secondary amine.

Glycine

Alanine

Valine

Leucine

Isoleucine

Proline

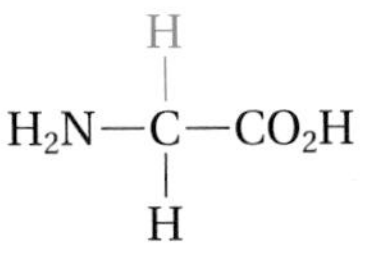

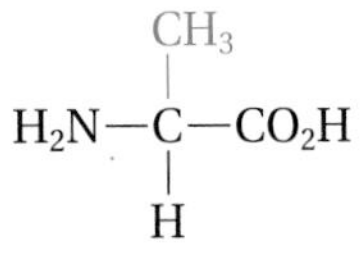

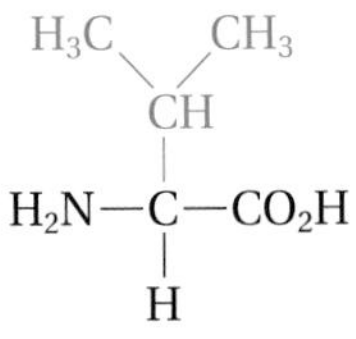

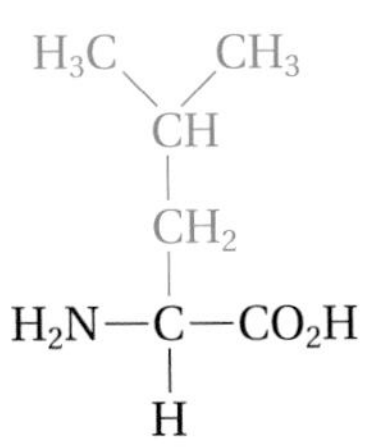

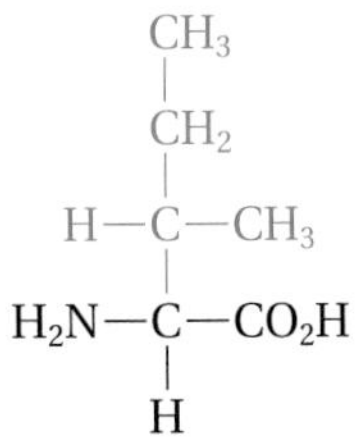

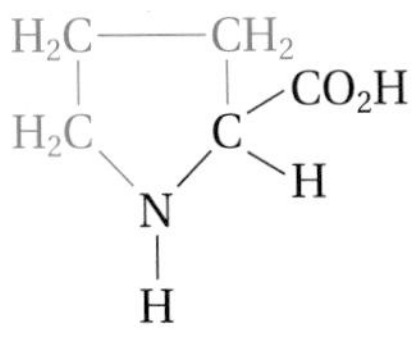

Hydroxylic side chains   The amino acids in this group are serine and threonine. Both have aliphatic side chains containing hydroxy functional groups.

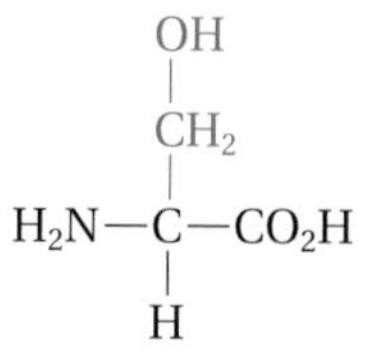

$$\underset{\text{Serine}}{\underset{|}{\overset{\text{OH}}{\underset{\overset{|}{\text{H}}}{\overset{|}{\underset{\text{H}_2\text{N}-\overset{|}{\text{C}}-\text{CO}_2\text{H}}{\text{CH}_2}}}}}} \qquad \underset{\text{Threonine}}{\underset{\overset{|}{\text{H}}}{\text{H}_2\text{N}-\overset{|}{\text{C}}-\text{CO}_2\text{H}}}$$

Serine 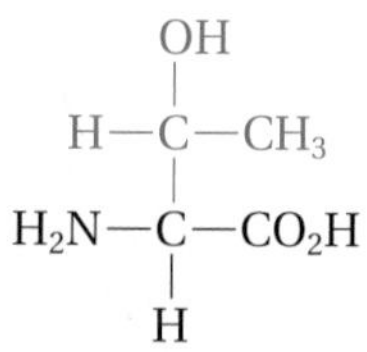   Threonine

Eight of the amino acids, isoleucine, leucine, lysine, methionine, phenylalanine, threonine, tryptophan, and valine, are essential amino acids. These amino acids cannot be synthesized by the body but are essential to its proper functioning. They must be supplied in our diet.

Aromatic side chains   There are three amino acids that have aromatic rings in their side chains. Phenylalanine is an amino acid in which one of the methyl hydrogens of alanine is replaced by a phenyl group. In tyrosine, the phenyl ring of phenylalanine is substituted in the *para* position with a phenolic hydroxyl group. The aromatic ring system of tryptophan is derived from indole (see Sec. 15.1).

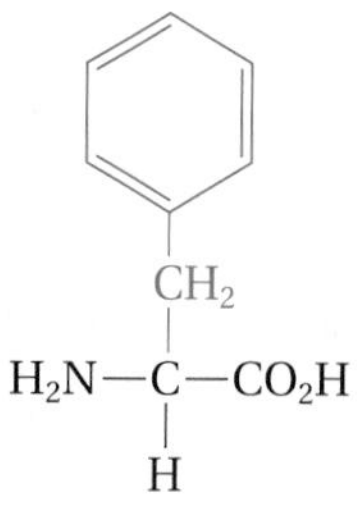

Phenylalanine 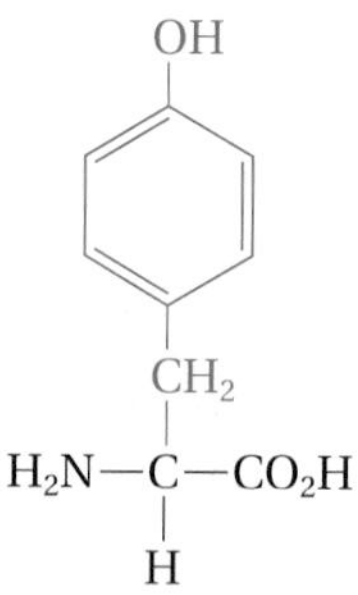   Tyrosine 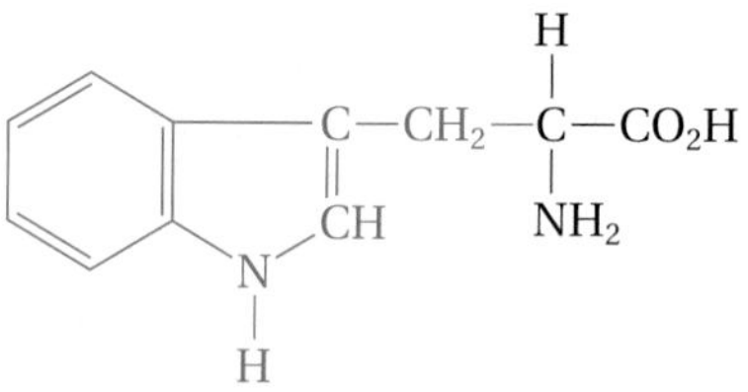   Tryptophan

Acidic side chains   Aspartic and glutamic acids have side chains that are terminated by carboxylic acid groups. At the usual biological pH, slightly above pH 7, these carboxylic acid groups are ionized. For this reason, aspartic acid and glutamic acid are often referred to as their carboxylate ions, aspartate and glutamate.

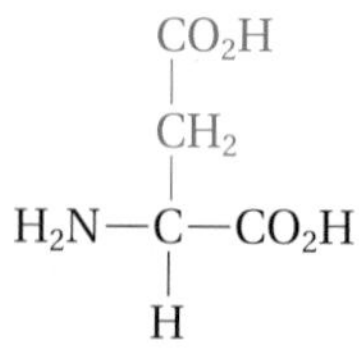

Aspartic acid 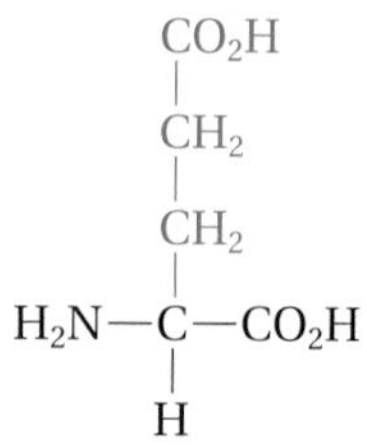   Glutamic acid

Amide side chains   Asparagine and glutamine are the amides of aspartic and glutamic acids, respectively. Their side chains are electrically neutral at pH 7.0.

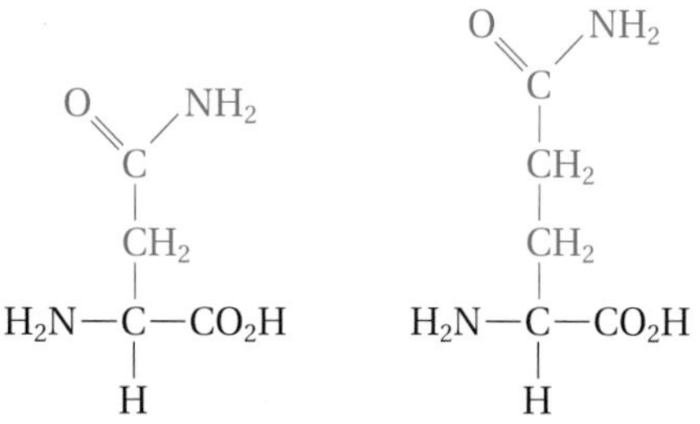

Asparagine   Glutamine

Basic side chains   This group consists of three amino acids containing weakly basic nitrogens. The nitrogens on the side chains of lysine and arginine are such sufficiently strong bases that they remove a proton from water at neutral pH. At pH 7.0, therefore, the side chains of arginine and lysine are positively charged. The nitrogens of the side chains of histidine are more weakly basic than those of lysine and arginine. In solution at pH 7.0, the side chains of only about half of the histidine molecules are positively charged at any moment.

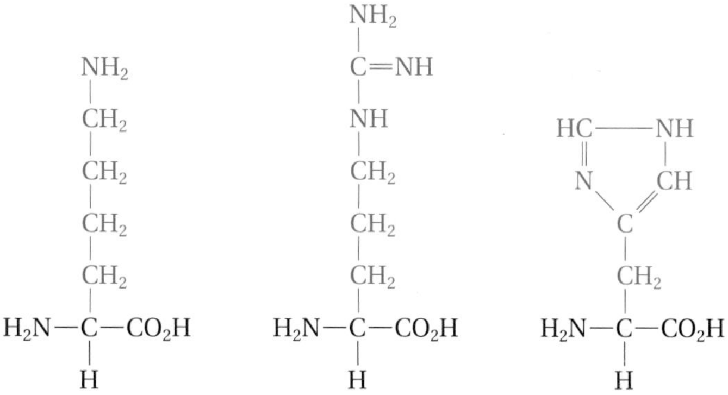

Lysine   Arginine   Histidine

Sulfur-containing side chains   Methionine and cysteine are two common sulfur-containing amino acids. Cysteine is often found linked to another cysteine through formation of a disulfide bond (—S—S—) to form the amino acid *cystine*. Cystine is not considered one of the 20 common amino acids, since it is formed from two molecules of cysteine.

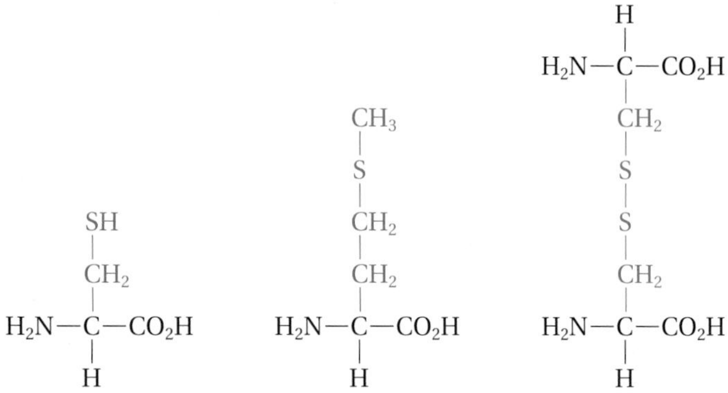

Cysteine   Methionine   Cystine

PRACTICE EXERCISE 18.1

Categorize the following amino acids according to their side-chain groups:

(a) leucine   (b) lysine   (c) serine   (d) tyrosine

18.2 Stereoisomers of amino acids

AIMS: To state the handedness of amino acids found in nature. To draw Fischer projection formulas for alpha amino acids.

Focus

All the amino acids but glycine have handedness.

A glance at the structures of common amino acids will show that, except for glycine, there are four different groups attached to the alpha carbons. This is a criterion for handedness in organic molecules (see Sec. 16.1). Amino acids fit this rule and may exist as D or L forms that are nonsuperimposable mirror images of each other. The handedness of amino acids is related to that of D- and L-glyceraldehyde as follows:

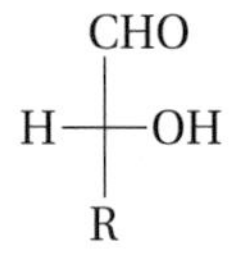

D-Glyceraldehyde 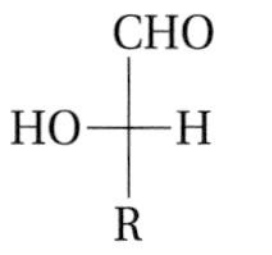   L-Glyceraldehyde

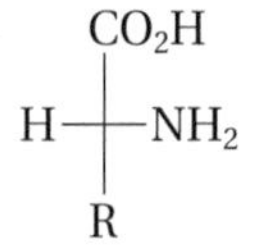

D-Amino acid 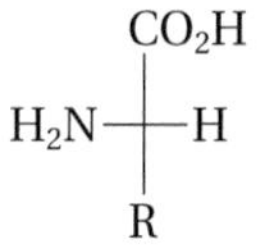   L-Amino acid 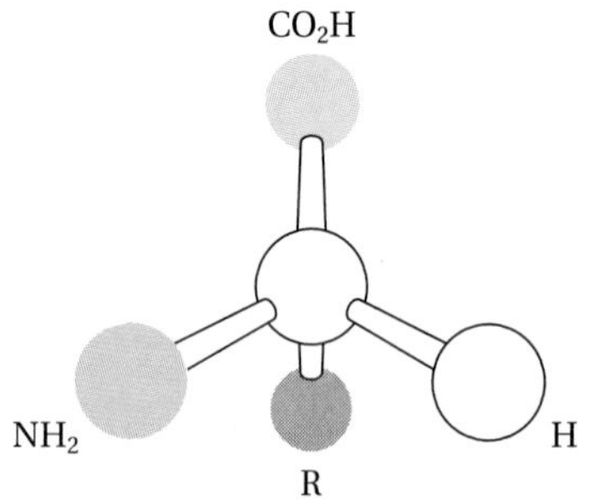   Configuration of L-amino acids

Although the amino acids found in nature are mostly the L forms, the D forms of a few amino acids are found in the cell walls of certain bacteria.

Nature shows a marked preference for the L forms of amino acids. With few exceptions, the amino acids found in living systems belong to the L family.

EXAMPLE 18.1

Drawing Fischer projection formulas

Draw the Fischer projection formula of L-alanine.

SOLUTION

Place the carboxylic acid group at the top and the methyl group (the R group) at the bottom. For the L isomer, the amino group is on the left.

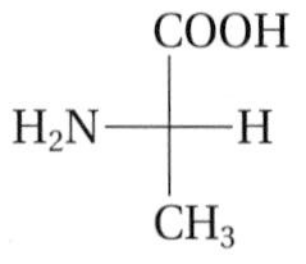

PRACTICE EXERCISE 18.2

Give (a) the Fischer projection formula for L-cysteine and (b) the stereorepresentation of D-cysteine. (c) Which isomer predominates in nature?

18.3 Zwitterions

AIM: To describe the formation of zwitterions and their effect on the properties of amino acids.

Focus

Amino acids form internal salts called *zwitterions.*

Consider the proximity of the carboxylic acid and amino groups in the glycine molecule. The weakly acidic proton of the carboxylic acid group easily transfers to the weakly basic amino group to form an internal salt.

Amino acid　　　Internal salt (zwitterion)

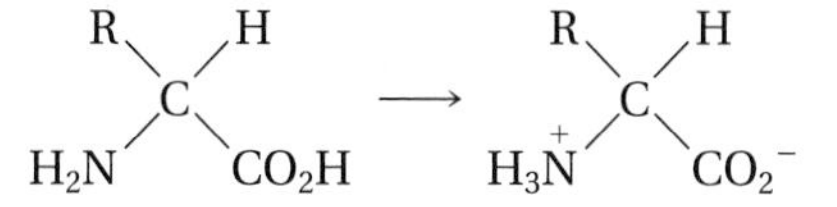

Amino acids are really ionic compounds. And as you may recall, ionic compounds generally have much higher melting points than molecular compounds. Glycine is a solid with a melting point of 233 °C—much greater than the temperature required to boil water. To see how the internal salt formation affects such properties as melting point, we can compare the melting point of glycine with those of structurally similar amines and carboxylic acids. Glycine is a two-carbon compound with amino and carboxylic acid groups, so we might expect glycine to be a liquid with a melting point between those of ethylamine ($CH_3CH_2NH_2$), −80.6 °C, and acetic acid (CH_3CO_2H), 16.6 °C. The much higher melting point of glycine is the result of internal salt formation.

$H_2NCH_2CH_3$	CH_3CO_2H	$H_2NCH_2CO_2H$
Ethylamine (mp −80.6 °C)	Acetic acid (mp 16.6 °C)	Glycine (mp 233 °C)

Zwitterion
zwitter (German): hybrid

The internal salts of such compounds as amino acids are called **zwitterions.** In the pure solid state and in aqueous solution near neutral pH, the amino acids exist almost completely as zwitterions.

Isoelectric points

The isoelectric point of the milk protein casein is 4.6. Casein is least soluble and will precipitate at this acidic pH. This acidic condition is met in the cheese-making process by the action of certain bacteria in producing lactic acid.

In zwitterions of amino acids with uncharged side chains, the positive and negative charges cancel one another. There is a net zero charge on the molecule. *Any amino acid in which the positive and negative charges are balanced is at its* **isoelectric point.** *For amino acids in aqueous solution, the pH at which this balancing of charges occurs is the* **isoelectric pH.** The isoelectric point of amino acids with uncharged side chains occurs near pH 7 in aqueous solution. An amino acid tends to be least soluble at its isoelectric pH because of the net zero charge. At lower pH values (where more protons are available), the solubility of the amino acid increases because the carboxylate ion of the zwitterion picks up a proton from the solution; the amino acid acquires a net positive charge.

The formation of zwitterions is not limited to amino acids. For example, *p*-aminosulfonic acid, the parent acid of the antibiotic sulfanilamide, exists mainly in the zwitterion form.

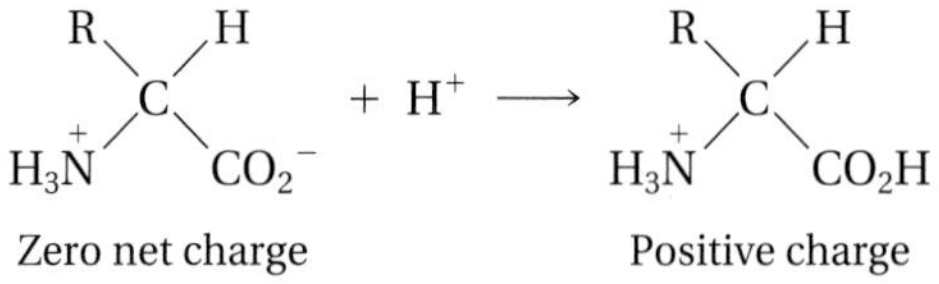

$$\begin{array}{c} \text{R} \quad \text{H} \\ \diagdown \quad \diagup \\ \text{C} \\ \diagup \quad \diagdown \\ \overset{+}{\text{H}_3\text{N}} \quad \text{CO}_2^{-} \end{array} \quad + \text{H}^+ \longrightarrow \quad \begin{array}{c} \text{R} \quad \text{H} \\ \diagdown \quad \diagup \\ \text{C} \\ \diagup \quad \diagdown \\ \overset{+}{\text{H}_3\text{N}} \quad \text{CO}_2\text{H} \end{array}$$

Zero net charge   Positive charge

At higher pH values, solubility of the amino acid also increases because a proton is removed from the positively charged ammonium ion of the zwitterion and the net charge of the amino acid is negative.

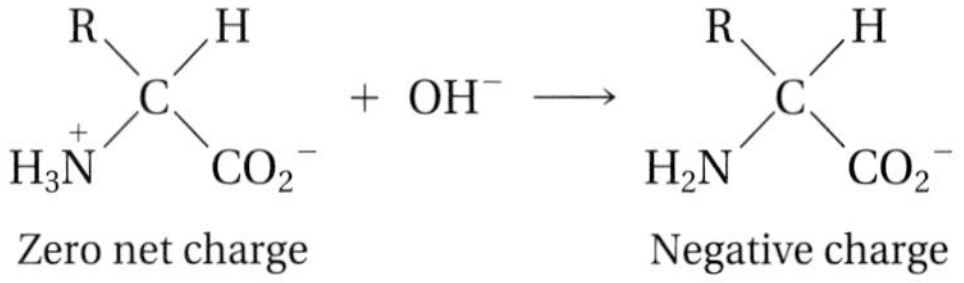

$$\begin{array}{c} \text{R} \quad \text{H} \\ \diagdown \quad \diagup \\ \text{C} \\ \diagup \quad \diagdown \\ \overset{+}{\text{H}_3\text{N}} \quad \text{CO}_2^{-} \end{array} \quad + \text{OH}^- \longrightarrow \quad \begin{array}{c} \text{R} \quad \text{H} \\ \diagdown \quad \diagup \\ \text{C} \\ \diagup \quad \diagdown \\ \text{H}_2\text{N} \quad \text{CO}_2^{-} \end{array}$$

Zero net charge   Negative charge

EXAMPLE 18.2   **Drawing formulas of zwitterions**

Draw the structural formula for the amino acid alanine at (a) pH 2 and (b) pH 11.

SOLUTION

(a) At pH 2 the amino group and the carboxylate ion will be protonated.

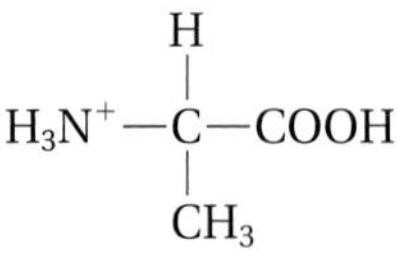

$$\begin{array}{c} \text{H} \\ | \\ \text{H}_3\text{N}^+\!-\!\text{C}\!-\!\text{COOH} \\ | \\ \text{CH}_3 \end{array}$$

At this pH the amino acid has a net positive charge.

(b) At pH 11 the functional groups are the unprotonated amino group and the carboxylate ion.

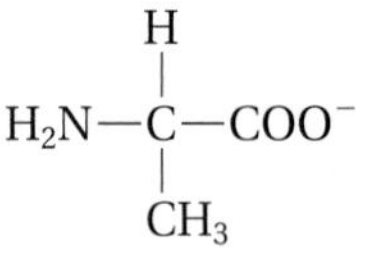

$$\begin{array}{c} \text{H} \\ | \\ \text{H}_2\text{N}\!-\!\text{C}\!-\!\text{COO}^- \\ | \\ \text{CH}_3 \end{array}$$

At this pH the amino acid has a net negative charge.

PRACTICE EXERCISE 18.3

Consider the amino acids glycine and aspartic acid. Draw structural formulas for the species that predominate in aqueous solution at

(a) pH 2   (b) pH 10

18.4 Peptides

AIMS: To name and describe the bond that links amino acids together. To draw complete structural formulas for simple peptides. To contrast the biological functions of some peptide hormones.

*A **peptide** is any combination of amino acids in which the alpha amino group ($-NH_2$) of one acid is united with the alpha carboxylic group ($-CO_2H$) of another through an amide bond.*

$$\begin{array}{c} R \quad O \\ | \quad \parallel \\ H_2N-C-C-OH \\ | \\ H \end{array} + \begin{array}{c} R \quad O \\ | \quad \parallel \\ H-N-C-C-OH \\ | \quad | \\ H \quad H \end{array} \longrightarrow$$

Amino acid   Amino acid

$$\begin{array}{c} R \quad O \quad\quad R \quad O \\ | \quad \parallel \quad\quad | \quad \parallel \\ H_2N-C-C-N-C-C-OH \\ | \quad\quad | \quad | \\ H \quad\quad H \quad H \end{array} + H_2O$$

Peptide

The amide bonds formed in peptides always involve the alpha amino and alpha carboxylic acid groups and never those of side chains. More amino acids may be added in the same fashion to form chains such as those in Figure 18.1. *The amide bond between the carbonyl group of one amino acid and the nitrogen of the next amino acid in the peptide chain is called a* **peptide bond** *or peptide link. Amino acids that have been incorporated into peptides are called* **amino acid residues.** As more amino acid residues are added, a backbone common to all peptide molecules is formed. The amino acid residue with a free amino group at one end of the chain is the *N-terminal residue;* the residue with a free carboxylic acid at the other end of the chain is the *C-terminal residue.* The number of amino acid residues in a peptide is often indicated by a set of prefixes for peptides of up to 10

Figure 18.1
Parts of a peptide—in this case, a tetrapeptide. The peptide bonds of the zigzag backbone are shown in color. Note that the C-terminal is at the right and the N-terminal is at the left.

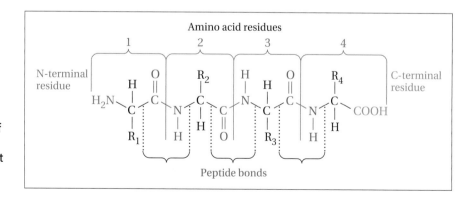

Table 18.1 Prefixes for Short Peptides

Residues	Prefix
2	*di-*
3	*tri-*
4	*tetra-*
5	*penta-*
6	*hexa-*
7	*hepta-*
8	*octa-*
9	*nona-*
10	*deca-*

Table 18.2 Three-Letter Abbreviations for Amino Acids

Amino acid	Abbreviation
alanine	Ala
arginine	Arg
asparagine	Asn
aspartic acid	Asp
cysteine	Cys
glutamine	Gln
glutamic acid	Glu
glycine	Gly
histidine	His
isoleucine	Ile
leucine	Leu
lysine	Lys
methionine	Met
phenylalanine	Phe
proline	Pro
serine	Ser
threonine	Thr
tryptophan	Trp
tyrosine	Tyr
valine	Val

residues, as shown in Table 18.1. *We call any peptide with more than 20 amino acid residues a* **polypeptide.** In theory, the process of adding amino acids to a peptide chain may be continued indefinitely. Names of peptides are derived from names of amino acid residues. By convention, names of peptides are always written from left to right starting with the N-terminal end; a peptide that contains N-terminal glycine, followed by a histidine, followed by C-terminal phenylalanine is named *glycyl-histidyl-phenylalanine.* The sequence is extremely important; glycyl-histidyl-phenylalanine is a different molecule from phenylalanyl-histidyl-glycine. The methyl ester of the dipeptide aspartyl phenylalanine is an artificial sweetener (see A Closer Look: Aspartame).

Structural formulas or even full word names for large peptides become very unwieldy and time-consuming to write. To simplify matters, chemists write peptide and protein structures by using three-letter abbreviations for the amino acid residues with dashes to show peptide bonds. Table 18.2 lists the abbreviations. Peptide structures written with these abbreviations always start with the N-terminal group to the left and end with the C-terminal group to the right. For example, we can write glycyl-histidyl-phenylalanine as Gly-His-Phe and phenylalanyl-histidyl-glycine as Phe-His-Gly.

EXAMPLE 18.3   Naming a peptide

Write the (a) the three-letter amino acid name and (b) the name using full abbreviations for the peptide in Figure 18.1 assuming that R_1 is methyl, R_2 is isopropyl, R_3 is methyl, and R_4 is hydrogen.

SOLUTION

(a) The full name, starting at the N-terminal end, is alanyl-valyl-alanyl-glycine.
(b) Using abbreviations: Ala-Val-Ala-Gly.

PRACTICE EXERCISE 18.4

Glutathione, a tripeptide that is widely distributed in all living tissues, is named glutamyl-cysteinyl-glycine. (a) Draw the complete structural formula for this peptide. (b) Write the amino acid sequence of the peptide using the three-letter abbreviations.

PRACTICE EXERCISE 18.5

Write three-letter amino acid sequences for all possible tripeptide structures that contain one residue each of glutamic acid, cysteine, and glycine.

Chemists have isolated over 200 peptides important to the smooth functioning of the human body. The peptide hormones oxytocin and vasopressin are two examples showing that apparently minor differences in the order of amino acid residues can result in profoundly different biological actions.

Aspartame

Aspartame, an artificial sweetener with the brand name NutraSweet®, is the methyl ester of the dipeptide aspartylphenylalanine.

Aspartylphenylalanine methyl ester (aspartame)

Aspartame is a substitute for sucrose and saccharin.

Aspartame is more than 50 times sweeter than sucrose. Now approved for use in more than 30 countries, this product of amino acid chemistry has found wide acceptance in the food industry as a substitute for both sucrose and saccharin (see figure). The main advantage of aspartame over saccharin is its taste, which is very similar to that of cane sugar. Its chief disadvantage is its instability. Aspartame is not recommended for the preparation of foods in which cooking temperatures exceed 150 °C. High temperatures and extremes of pH can cause aspartame in solution to hydrolyze to the unesterified dipeptide, aspartylphenylalanine and methanol, with a simultaneous loss of sweetness. Methanol is a toxic alcohol, and there is disagreement among scientists whether the breakdown of the small amounts of aspartame used for sweetening produces sufficient amounts of methanol to cause harm. Food scientists have found that aspartame's stability is improved by using it in combination with saccharin. This combination is now used in a number of soft drinks. Products that contain aspartame, however, must carry a warning label for people who suffer from the hereditary disease phenylketonuria. People who have phenylketonuria are unable to break down phenylalanine and must therefore limit its intake, as we will see in Section 26.8.

Oxytocin and vasopressin are formed in the hypothalamus (pituitary gland) and enter the bloodstream. Each hormone is a nonapeptide (contains nine amino acid residues) with six of the amino acid residues drawn into a loop by a disulfide bond. The disulfide bond is formed by the coupling of cysteine residues in the first and sixth positions of their peptide chains, as shown in Figure 18.2. *In peptides and proteins, disulfide bonds formed between two cysteine —SH groups that draw a single peptide chain into a loop or hold two peptide chains together are called* **disulfide bridges.**

Figure 18.2
Oxytocin and vasopressin. These peptide hormones differ by two amino acid residues (shown in color). The C-terminal residues have amide functional groups rather than carboxylic acid groups.

Although the amino acid composition differs at only the third and eighth positions of their peptide chains (counting from the N-terminal end), the biological roles of these two peptides are different. Oxytocin stimulates milk ejection in females and contraction of the smooth muscle of the uterus in labor. Oxytocin has been called the "cuddle drug," because in females it stimulates sensations during lovemaking and produces feelings of relaxed satisfaction and attachment. Vasopressin is an **antidiuretic**—*it helps to maintain a proper water balance in both sexes by helping to retain water.* Defective production of vasopressin results in diabetes insipidus, characterized by the production of massive volumes of urine. Injections of the hormone control the volume of urine produced.

Another example of how the sequence of amino acid residues affects biological function can be seen when we compare the blood pressure–controlling activities of two peptides, bradykinin and boguskinin. Bradykinin is a nonapeptide formed directly in the bloodstream when a fragment is chopped from a large protein, α-2-globulin. Boguskinin is a synthetic octapeptide that lacks only the proline residue at position 7 of bradykinin.

$$\underset{1}{\text{Arg}}-\underset{2}{\text{Pro}}-\underset{3}{\text{Pro}}-\underset{4}{\text{Gly}}-\underset{5}{\text{Phe}}-\underset{6}{\text{Ser}}-\underset{7}{\text{Pro}}-\underset{8}{\text{Phe}}-\underset{9}{\text{Arg}}$$

Bradykinin

$$\underset{1}{\text{Arg}}-\underset{2}{\text{Pro}}-\underset{3}{\text{Pro}}-\underset{4}{\text{Gly}}-\underset{5}{\text{Phe}}-\underset{6}{\text{Ser}}-\underset{7}{\text{Phe}}-\underset{8}{\text{Arg}}$$

Boguskinin

Bradykinin is partially responsible for triggering pain, welt formation (as in scratches), movement of smooth muscle, and lowering of blood pressure. Blood pressure is lowered when, in response to a signal, bradykinin and related peptides relax muscles of blood vessel walls. Blood vessels dilate, or expand, and blood flows into the expanded volume, lowering blood pressure. Less than 1 μg of bradykinin lowers blood pressure in an average-sized adult. Boguskinin, on the other hand, is completely inactive—hence the name *bogus,* meaning "false."

Parts of the brain contain **enkephalins**—*peptides involved with feelings of emotion and sensation of pain.* Two major enkephalins are methionine enkephalin and leucine enkephalin, which differ in structure by only one amino acid residue:

$$\text{Tyr}-\text{Gly}-\text{Gly}-\text{Phe}-\text{Met}\qquad\text{Tyr}-\text{Gly}-\text{Gly}-\text{Phe}-\text{Leu}$$

Methionine enkephalin   Leucine enkephalin

These two pentapeptides are messengers in brain processes associated with emotional euphoria and relief of pain—the same processes affected by morphine, heroin, and other opiate drugs. Researchers hope that administration of these peptides, or similar synthetics, will bring relief to people with chronic pain without the danger of addiction. Karla, the jogger in the Case in Point earlier in this chapter, experiences the effects of these peptide hormones during her workouts.

FOLLOW-UP TO THE CASE IN POINT: Runner's high

Karla sometimes experiences runner's high during her runs. What is causing this effect? Many sports physiologists believe that runner's high comes from increases in the levels of enkephalins in the brain. These increases, which can cause remarkable feelings of well-being, appear to be stimulated by endurance exercises such as distance running. The effects of enkephalins appear to last beyond the period of exercise to increase all-around mental alertness and a sense of comfort. Exercise is a good thing, but like most good things, it can be overdone, since the body may suffer from exhaustion and other breakdowns when overtaxed without sufficient time to recover. Enkephalins may be responsible in part for the potentially harmful addiction of some people to excessive exercise.

18.5 *Primary structure of proteins*

AIM: To characterize the primary structure of a protein.

Focus

The primary structure is the first level of organization of a protein.

When the number of amino acid residues becomes greater than about 40, a naturally occurring peptide is called a **protein.** On average, a peptide molecule containing 100 amino acid residues has a molar mass of about 10,000 g. Proteins are so vital to living organisms that many of the remaining topics in this book concern them. The structures of proteins are usually studied at four levels of organization: *primary structure, secondary structure, tertiary structure,* and *quaternary structure. The* **primary structure** *of a peptide or protein is the order in which the amino acid residues of a peptide or protein molecule are linked by peptide bonds.* The primary structure of a protein molecule also includes any disulfide bridges that the molecule contains. Just as we saw for peptides in the preceding section, differences in the chemical and biological properties of proteins result from differences in the order of amino acids in the polypeptide chain (primary structure). The secondary, tertiary, and quaternary structures of proteins are discussed in the following three sections.

Protein
proteios (Greek):   of first
                      importance

18.6 *Secondary structure of proteins*

AIMS: To describe the bonding and structure in the following secondary structures of proteins: alpha helix, beta-pleated sheet, and collagen helix. To distinguish among the following fibrous proteins: alpha keratin, beta keratin, and collagen.

Focus

The alpha helix, beta-pleated sheet, and collagen helix are three types of secondary structures found in proteins.

The polypeptide chains of many proteins contain specific, repeating patterns of folding of the peptide backbone. *These specific, repeating patterns of folding of the peptide backbone constitute the protein's* **secondary structure.** Three commonly found patterns are the *alpha helix,* the *beta-pleated sheet,* and the *collagen helix.*

Figure 18.3
A corkscrew must be turned in a right-handed, or clockwise, direction to penetrate a cork.

Alpha helix

In some proteins, regions of the backbone of the peptide chain are coiled into a spiral shape called an **alpha helix,** *similar to a corkscrew.* As Figure 18.3 shows, a corkscrew must be turned in a right-handed, or clockwise, direction to penetrate a cork. The alpha helixes of proteins are always right-handed. The helixes are held together by hydrogen bonds, shown in Figure 18.4, formed between the hydrogen of an N—H of a peptide bond and the carbonyl oxygen of another peptide bond group four residues away in the same peptide chain.

The tightness of coiling is such that 3.6 amino acid residues of the peptide backbone make each full turn of the alpha helix. There are no amino acid residue side chains inside the alpha helix; they are located on the outside. The cyclic amino acid proline does not fit well into the peptide backbone of alpha helixes. Alpha helixes in long protein chains often end at a place where proline residues occur in the primary structure. Proline is

Figure 18.4
A peptide chain twisted into a right-handed alpha helix constitutes a protein's secondary structure. The N-terminal to C-terminal direction is from top to bottom. The dotted lines show the hydrogen bonds between the carbonyl oxygen of one amino acid residue and the N—H hydrogen of another, four amino acid residues further down the chain.

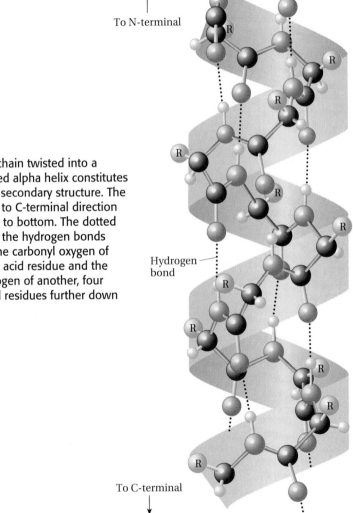

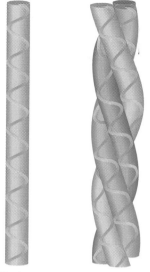

(a) Helical peptide chain   **(b) Alpha keratin**

Figure 18.5
Three helical peptide chains, like the one shown in (a), are twisted or supercoiled to form a rope in alpha keratin (b).

sometimes called an *alpha helix disrupter* for this reason. *The inability of proline to fit into the alpha helix is one of many pieces of evidence that the secondary structure of a protein is determined by its primary structure.*

Much of what is known about protein alpha helixes comes from studies of fibrous proteins. **Fibrous proteins** *tend to be long, rod-shaped molecules with great mechanical strength.* Such proteins are usually insoluble in water, dilute salt solutions, and other solvents. *The polypeptide chains of a class of fibrous proteins called* **alpha keratins** *consist mainly of alpha helixes.* Alpha keratins form the hard tissue of hooves, horns, outer skin layer (epidermis), hair, wool, and nails of mammals. We see in Figure 18.5 complex protein molecules consisting of long alpha helixes that are twisted together—*supercoiled*—in ropes of three or seven strands. It takes many supercoiled ropes to make a strong but elastic wool fiber. If you have ever washed a wool sweater, you know that warm, wet wool fibers can be stretched, but they eventually return to their original length. This is because the alpha helixes of the damp fibers are easily pulled into an extended form. The extended form is less stable than the alpha helix. It will, in time, return to the original alpha helix. Disulfide bridges (see Sec. 18.1) between alpha helixes help to make alpha keratins rigid; the alpha keratins of a hard hoof have more disulfide linkages than relatively elastic epidermis.

Beta-pleated sheet

Figure 18.6 shows the beta-pleated sheet, another kind of secondary structure commonly found in proteins. *The* **beta-pleated sheet** *consists of peptide chains arranged side by side to form a structure that resembles a piece of paper folded into many pleats.* The carbonyl groups of one zigzag peptide backbone are hydrogen bonded to peptide N—H hydrogens of adjacent peptide chains that run in opposite directions in the N-terminal to C-ter-

Figure 18.6
A beta-pleated sheet is another kind of secondary structure. Hydrogen bonds (shown as dotted lines) hold adjacent strands of the sheet together.

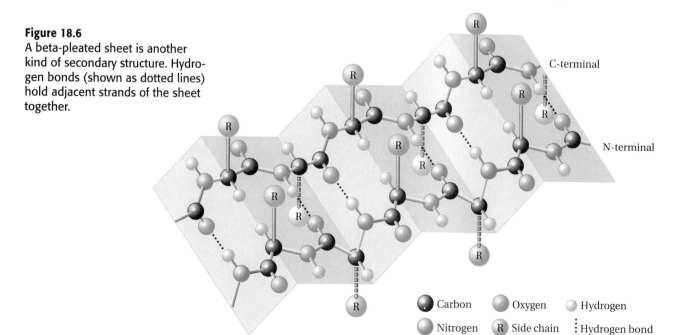

C-terminal

N-terminal

● Carbon   ● Oxygen   ○ Hydrogen
○ Nitrogen   (R) Side chain   ⋮ Hydrogen bond

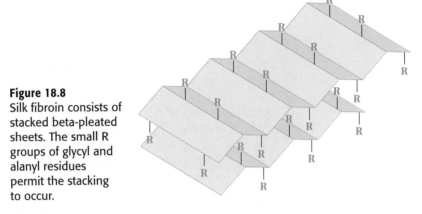

Figure 18.8
Silk fibroin consists of stacked beta-pleated sheets. The small R groups of glycyl and alanyl residues permit the stacking to occur.

Figure 18.7
Spider webs are made of fibroin, a protein that exists mainly as beta-pleated sheets.

minal sense. Beta-pleated sheets are formed by separate strands of protein or by a single chain looping back on itself.

Beta keratins *are a class of fibrous proteins that consist mainly of beta-pleated sheets.* The long, thin fibers secreted by silkworms and spiders (Fig. 18.7) are composed of fibroin, a well-studied beta keratin. Fibroin consists mainly of layers of beta-pleated sheets, as shown in Figure 18.8. Again, the primary structure of a protein determines its secondary structure. Fibroin peptide chains are particularly rich in glycyl and alanyl residues. The small side chains of these residues are important to the organization of fibroin, since larger side chains would interfere with the packing of the sheets in layers.

Collagen helix

Collagen is yet another kind of fibrous protein. **Collagen** *is the most abundant protein in human beings and many other animals with spinal columns (vertebrates).* About one-third of the protein in the human body is present as the collagen of bones, teeth, inner skin layer (dermis), tendons, and cartilage. The inner material of the eye lens is almost pure collagen. Collagen occurs in all organs, where it imparts strength and stiffness.

Collagen is formed from three peptide chains, each a helix, wound into a rope. There are important differences between alpha helixes and collagen helixes, however. Collagen is rich in proline (40%), which does not fit into regular alpha helixes, and collagen helixes are not as tightly coiled as alpha helixes. Unlike alpha helixes, which are right-handed, collagen helixes are left-handed. Collagen contains large quantities of glycine (35%) and the unusual amino acids 4-hydroxyproline (5%) and 5-hydroxylysine (1%).

4-Hydroxyproline

5-Hydroxylysine

In addition, collagen is a glycoprotein (see Sec. 18.11). Sugar units (usually disaccharides of glucose and galactose) are covalently attached to the peptide chains.

18.7 Tertiary structure of proteins

AIMS: To describe the forces that help determine the tertiary structure of proteins. To define a conjugated protein. To describe the tertiary structure of myoglobin.

Focus

Folding of polypeptide chains produces tertiary structure.

The polypeptide chains of proteins fold to form three-dimensional structures. *The folding of a polypeptide chain of a protein molecule into a relatively stable three-dimensional shape constitutes the protein's* **tertiary structure.** Like the formation of a secondary structure, the formation of the tertiary structure of a protein is determined by its primary structure. The tertiary structure may contain within it regions of secondary structure, such as alpha helix and beta-pleated sheet. The overall folding of a polypeptide into its tertiary structure results from interactions between the side chains of the amino acid residues within the primary structure of the protein. For example, disulfide bridges can hold two regions of the same polypeptide chain together (Fig. 18.9). Weak attractions between amino acid side chains are also important. A polypeptide chain folds in the way that maximizes energetically favorable hydrogen bonds. *Ionic bonds, also called* **salt bridges,** *are usually formed between the negatively charged carboxylate ion side-chain groups of aspartate or glutamate and the positively charged amino side-chain groups of lysine or arginine.* One major factor in how a protein folds is the presence of aliphatic or aromatic amino acid residues in the peptide chain. The hydrophobic side chains of these residues tend to face the interior of protein molecules, much as the hydrophobic tails of soap molecules exclude water by forming micelles. Protein molecules that have different primary structures experience different side-chain interactions and therefore fold into different tertiary structures. Conversely, protein molecules that have the same primary structure experience the same side-chain interactions and therefore fold into the same tertiary structure.

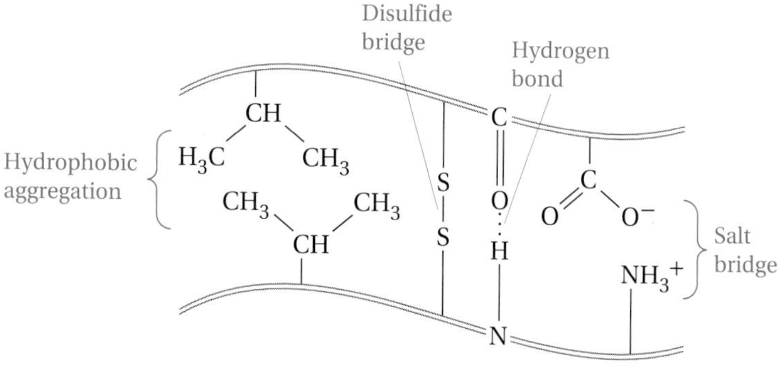

Figure 18.9
Many forces stabilize the tertiary structure of proteins.

The unusually large amount of myoglobin in whale muscle enables whales to remain submerged in water for long periods of time without rising to the surface for oxygen.

Myoglobin—*the oxygen-storage protein of mammalian muscle*—was the first protein to have its tertiary structure determined. Myoglobin is a *globular protein.* In contrast to rod-shaped fibrous proteins, **globular proteins** *have more or less spherical shapes.* Like many other globular proteins, myoglobin is soluble in water and dilute salt solutions. Myoglobin consists of a single peptide chain of 153 amino acids. This relatively small protein contains 1260 atoms, not counting hydrogens. Myoglobin is an example of a *conjugated protein.* **Conjugated proteins** *have structures that incorporate nonprotein portions called* **prosthetic groups.** Prosthetic groups are necessary for the protein to perform its biological function. They are usually organic molecules that are permanently attached to the protein by covalent bonds. The myoglobin molecule contains a prosthetic group called a *heme group.* The heme group contains an iron atom in the iron(II) or ferrous (Fe^{2+}) state (Fig. 18.10). Oxygen is bound to the heme iron to form oxymyoglobin.

In 1957, John C. Kendrew of the Medical Research Council laboratories in Cambridge, England, and his colleagues determined the three-dimensional structure of myoglobin by a method called *X-ray crystallography.* (See A Closer Look: X-ray Crystallography, for more about the X-ray method for determining protein structures.) The X-ray crystallographic picture of myoglobin in Figure 18.11 shows a globular structure in which almost three-fourths of the peptide chain is alpha helix. The tertiary structure of myoglobin's single chain consists of segments of alpha helix between turns of the polypeptide chain. Because of the turns, the molecule takes on a

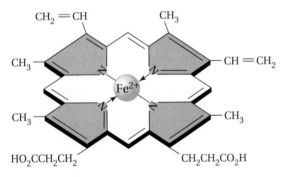

Figure 18.10
Heme. Notice that the structure contains four linked pyrrole rings (color) with an iron(II) ion (Fe^{2+}) at the center.

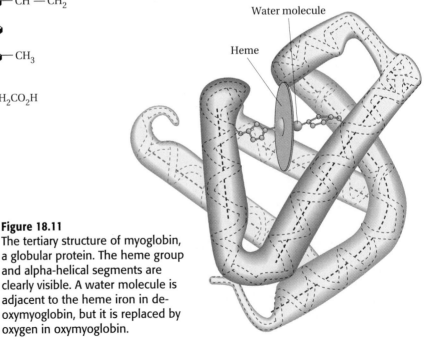

Figure 18.11
The tertiary structure of myoglobin, a globular protein. The heme group and alpha-helical segments are clearly visible. A water molecule is adjacent to the heme iron in deoxymyoglobin, but it is replaced by oxygen in oxymyoglobin.

X-ray Crystallography

The determination of the three-dimensional structure of a biologically interesting molecule is fascinating and practical. The fascination comes from learning how the atoms in the molecule are arranged in space and trying to figure out how this arrangement enables the molecule to perform its biological function. The practical aspect is that this knowledge, once obtained, can sometimes be used to design compounds that will either mimic or block the action of the biological molecule.

X-ray crystallography is a very powerful tool for determination of the three-dimensional structures of molecules, including large molecules such as those of proteins and deoxyribonucleic acids (DNA). Exposing a crystal of a substance to a narrow beam of X rays causes the X rays to be scattered, or diffracted, by their interactions with the electrons of the atoms in the crystal. Hundreds or thousands of diffraction patterns are recorded from different angles all around the crystal.

The spacing and intensity of the spots on diffraction patterns contain information about the electron densities in molecules. Computers are used to translate the diffraction information into three-dimensional maps of the electron densities all over the molecule (see figure). The electron-density maps give very accurate, complete molecular structures that pinpoint the positions of the atoms in space. For example, a region in the molecule that has a high electron density must be a

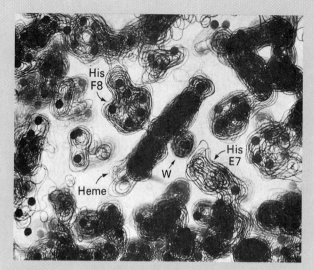

Electron-density map of myoglobin. The heme group is seen edge-on with its two associated histidine (His) residues and a water molecule (W).

covalent bond, because electrons are concentrated in covalent bonds. Conversely, a region that contains essentially no electron density must be a nucleus of an atom, because electrons are not found in the nucleus. The identity of a chemical bond, say, a carbon-carbon bond, is confirmed if the distance between two nuclei connected by a region of high electron density is 0.15 nm, the length of an ordinary C—C bond. Hydrogen atoms are too small to be directly detected by X-ray crystallography. However, the positions of hydrogens can be inferred once the positions of the heavier atoms in a molecule are known.

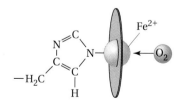

Figure 18.12
Heme is attached to myoglobin through a bond from the iron of the heme group to the nitrogen of a histidyl residue.

globular shape with the heme group resting in a "basket" formed by segments of alpha helix. There are other interesting features. The entire molecule is very compact—no more than two water molecules will fit on the inside. All the hydrophobic amino acid residues, those with aliphatic and aromatic side chains, are turned toward the interior of the molecule so that they are not exposed to water. All the charged side chains are exposed to water on the exterior of the molecule. Finally, as Figure 18.12 shows, the heme group is attached to the protein by a bond between the iron(II) ion of the heme group and a nitrogen of a histidyl residue of the polypeptide chain.

18.8 Quaternary structure of proteins

AIMS: To define the terms subunit *and* quaternary structure. *To describe the quaternary structure of hemoglobin. To distinguish among oxyhemoglobin, deoxyhemoglobin, and methemoglobin.*

Focus

In some proteins, polypeptide chains aggregate to form quaternary structures.

Some proteins consist of more than one polypeptide chain. *These individual chains are called* **subunits** *of the protein. Proteins composed of subunits are said to have* **quaternary structure.** Many proteins have structures that contain subunits. Proteins consisting of dimers (two subunits), tetramers (four subunits), and hexamers (six subunits) are fairly common. The proteins that comprise the individual subunits may be identical, or they may be different. Like the secondary and tertiary structures, the quaternary structure of a protein is determined by its primary structure. The polypeptide chains of subunits are held in place by the same forces that determine tertiary structure—hydrogen bonds, salt bridges, and sometimes disulfide bridges—except the forces are *between* the polypeptide chains of the subunits instead of *within* them. Hydrophobic aliphatic and aromatic side chains of subunits can aggregate to exclude water.

Hemoglobin—*the globular oxygen-transport protein of blood*—is an example of a protein that has a quaternary structure. Max Perutz, also of the Medical Research Council laboratories, determined the structure of horse blood hemoglobin in 1959. Hemoglobin is a larger molecule than myoglobin. The hemoglobin molecule has a molar mass of 64,500. It contains about 5000 individual atoms, excluding hydrogens, in 574 amino acid residues.

The quaternary structure of hemoglobin consists of four peptide subunits. Two of the subunits are identical and are called the *alpha subunits*. The remaining two subunits, called the *beta subunits*, are identical to each other but different from the alpha subunits. Figure 18.13 shows the four subunits of hemoglobin interlocked in a compact globular structure held together by ionic and hydrogen bonds between the amino acid side chains of the polypeptide subunits.

A chromoprotein has a color because of a colored prosthetic group. Hemoglobin is a chromoprotein that gets its red color from the heme group. Plants have their green color because of chlorophyll, a porphyrin ring with a structure similar to the heme group.

A heme prosthetic group containing an iron(II) ion capable of carrying one oxygen molecule is associated with each of the four protein subunits of hemoglobin. **Globin** *is the protein from which the heme groups have been removed. The hemoglobin complex with oxygen is* **oxyhemoglobin;** *that without oxygen is* **deoxyhemoglobin.** Oxyhemoglobin, the major hemoglobin of arterial blood, is bright red; deoxyhemoglobin of venous blood is purplish. *Long exposure to oxygen will convert the heme iron of hemoglobin to the iron(III), or ferric state (Fe^{3+}), to give* **methemoglobin.** The brown color of dried blood results from the conversion of hemoglobin to methemoglobin. Oxygen does not form complexes with methemoglobin. *In* **methemoglobinemia,** *a hereditary blood disease, the heme iron of hemoglobin, in either the alpha or the beta chains, exists in the iron(III) state.* People with iron(III) ions in both the alpha- and the beta-chain heme groups would be unable to live because their blood could not transport oxygen.

The folding of the polypeptide chains of the alpha and beta chains of the hemoglobin subunits is very similar. Moreover, the folding of

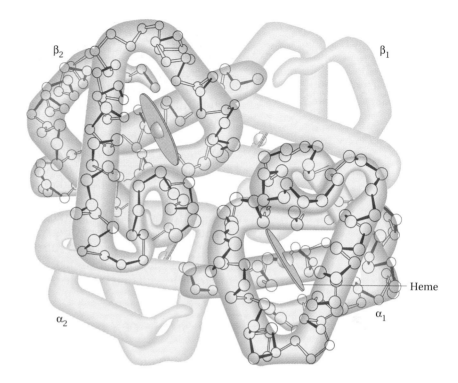

Figure 18.13
The quaternary structure of hemo-globin, showing the arrangement of alpha and beta chains.

myoglobin is very similar to the folding of the subunits of hemoglobin. This surprised biochemists, since the primary structures of sperm whale myo-globin and horse hemoglobin are very dissimilar. It is now clear that the ter-tiary structures devised by nature for myoglobin and the hemoglobin sub-units are crucial to oxygen transport by these proteins.

18.9 Hemoglobin function

AIM: To describe the mechanism of oxygen transport by hemoglobin.

Focus

The environment of the heme groups gives myoglobin and hemoglobin the ability to carry oxygen.

Nature has gone to considerable trouble to construct complicated proteins such as myoglobin and hemoglobin for the apparently simple task of bind-ing oxygen. In oxygen transport, however, as in many biological processes, a delicate balance must be maintained. The binding between iron(II) ions and oxygen must be strong enough that the oxygen can be stored or trans-ported yet weak enough to provide a means for releasing the oxygen where it is needed.

Free iron(II) ions in water form very unstable complexes with molecu-lar oxygen. Complexes of oxygen with iron(II) ions contained in heme groups are only slightly more stable. In myoglobin and hemoglobin, how-ever, the heme groups are tucked into hydrophobic folds in the globin pep-tide chains. This hydrophobic environment fosters formation of relatively stable complexes between molecular oxygen and heme iron. It is the hydrophobic environment of the heme group that makes myoglobin and

More than 100 different types of protein molecules are found in the blood. The normal concentration of protein in blood plasma ranges from 7.0 to 8.0 g/mL. Blood proteins help maintain osmotic pressure, play a part in blood coagulation, transport lipids, and help fight disease.

hemoglobin useful oxygen carriers. Some molecules and ions form more stable complexes with the heme iron of hemoglobin than oxygen. *Carbon monoxide (CO) reacts with the heme iron of hemoglobin to form a complex called* **carboxyhemoglobin.** This complex is about 200 times more stable than oxyhemoglobin. Breathing even low levels of carbon monoxide raises the amount of carboxyhemoglobin and decreases the amount of oxyhemoglobin in the blood. The result is carbon monoxide poisoning (discussed in Sec. 11.11). *Cyanide ions (CN⁻) bind more tightly than either oxygen or carbon monoxide to the heme iron of hemoglobin to form* **cyanohemoglobin.** Compounds that produce cyanide ions are deadly poisons.

The fact that the hemoglobin molecule consists of four subunits increases the efficiency of oxygen transport. Once the first oxygen of the hemoglobin molecule is complexed, the second, third, and fourth follow with increasing ease. This property guarantees that hemoglobin that is oxygenated will always carry a full load of oxygen. Myoglobin, which is involved only in oxygen storage and not transport, performs its function satisfactorily with only a single peptide chain.

18.10 Sickle cell anemia

AIM: To explain the molecular basis for sickle cell anemia.

Focus

Sickle cell anemia stems from a single substitution in the primary structure of hemoglobin.

Sickle cell anemia affects about 3 in every 1000 African-Americans. You may recall from the introduction to Chapter 1 that Linus Pauling discovered that the hemoglobin of people who suffer from sickle cell anemia (HbS) is different from the hemoglobin of normal adults (HbA). Now that we are familiar with protein structure, we can understand the molecular difference between HbS and HbA, and the reason this difference causes the symptoms of sickle cell anemia. HbA and HbS differ by only one amino acid residue. A negatively charged glutamic acid residue at the sixth position of the beta chain of normal HbA is replaced by an uncharged valine residue in HbS. The difference in electric charge causes HbS, especially in its deoxygenated form, to be much less soluble than HbA. Precipitation of HbS in the red blood cell causes these cells to sickle and sometimes burst. (For a comparison of normal and sickled red blood cells, see the figure on page 2.) The sickled cells as well as clumped HbS and cellular debris from burst cells block the flow of blood to capillaries, triggering the painful clinical episodes of sickle cell anemia.

The sickle cell trait is exhibited by about 1 in every 10 African-Americans. In people who carry the sickle cell trait, half the hemoglobin is HbA and half is HbS. These people often lead normal lives. Exposure to low oxygen levels, however, as in the mountains or on an airplane flight, can be dangerous. There is no cure for sickle cell anemia; treatment of its symptoms may involve hyperbaric oxygenation—administration of oxygen at higher than atmospheric pressure. The idea behind hyperbaric oxygenation is to keep the number of deoxygenated red blood cells as low as possible to prevent sickling and subsequent clumping. Hydroxyurea ($H_2NCONHOH$), formerly used as an anticancer drug, has been found to reduce the painful episodes associated with sickle cell anemia by 50 percent.

A hereditary disease giving symptoms similar to sickle cell anemia is thalassemia, which is often found in people whose family origins are from Mediterranean countries. In thalassemia, however, either the alpha or beta chains of hemoglobin are not made (which is fatal) or are made in reduced amounts (thalassemia trait). Like sickle cell trait, thalassemia trait confers resistance to malaria.

Why is sickle cell anemia found almost exclusively in people of African descent? Although patients with sickle cell anemia often die very young, persons with sickle cell trait have a high resistance to malaria, a disease prevalent in certain parts of Africa. Apparently, ancestral Africans with sickle cell trait had a distinct survival advantage over those who lacked it. Sickle cell trait and sickle cell anemia are passed from generation to generation, accounting for the high incidence of the two conditions today.

The electrophoresis of hemoglobin obtained from red blood cells is a simple and effective technique for screening individuals for sickle cell trait or sickle cell anemia. **Electrophoresis** *is a method for the separation of ions according to their charge,* as described in A Closer Look: Electrophoresis. At pH 8.4, HbA migrates more rapidly than HbS. Since both HbA and HbS are red, they are readily visible without staining.

A Closer Look

Electrophoresis

Since they are charged, ions in solution move in electrical fields. Anions migrate to the anode (the positively charged electrode), and cations migrate to the cathode (the negatively charged electrode). The more highly charged an ion, the more rapidly it migrates. The difference in migratory rates of differently charged ions is the basis for electrophoresis, a powerful tool for the separation of mixtures of proteins.

Electrophoresis succeeds as a protein-separation method because different kinds of protein molecules behave as complex ions with slightly different charges. The size of the charge on a given protein molecule depends not only on the number and kind of acidic and basic side chains on the amino acid residues of the protein but also on the pH of the solution containing the protein. In acidic solutions, most proteins are positively charged, because the amino groups are present as positively charged ions; carboxylic acid groups are also protonated and are therefore electrically neutral. In basic solutions, most proteins are negatively charged, because the amino groups are unprotonated and are electrically neutral; carboxylic acid groups are unprotonated and negatively charged. At the isoelectric pH of a protein, the positive and negative charges balance, and the protein behaves as if it had no charge—it will not migrate in an electrical field.

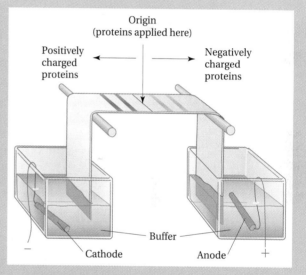

A procedure called *electrophoresis* is used to separate a mixture of proteins.

Biochemists perform electrophoresis experiments by applying a sample of solution containing a mixture of proteins to the center of a strip of porous material such as thick filter paper. Both ends of the paper are dipped like wicks into troughs containing a buffer solution of appropriate pH, and an electric current is passed through the system (see figure). Charged proteins separate by migrating at different rates toward the electrode of opposite charge. The separated proteins can be seen directly if they are colored, but often it is necessary to make them visible by staining with a suitable dye.

PRACTICE EXERCISE 18.6

Draw the structure of a glutamic acid residue, and compare it with a valine residue. How do their side chains differ?

18.11 Glycoproteins

AIM: To characterize glycoproteins.

Glycoproteins *are protein molecules that have sugar molecules (glycosides) covalently bonded to them.* Mucous secretions consist in part of mucoproteins, a special class of glycoproteins. Some of the glycoproteins we mention are many cell membrane proteins (see Chap. 17), collagen (see Sec. 18.9), fibrinogen of blood clotting (see Chap. 19), and immunoglobulin G (see Chap. 22).

The measurement of certain glycoproteins in the blood of patients with diabetes can be an important indicator of how well the level of blood sugar is being controlled over a period of time, as described in A Closer Look: Glycoproteins: Control of Blood Glucose in Diabetes.

18.12 Denaturation

AIM: To state three ways to denature proteins.

A protein that is folded into its normal, biologically active structure is in its **native state.** **Denaturation** *occurs when a native protein unfolds owing to the disruption of weak attractive forces or cleavage of disulfide bridges,* as shown in Figure 18.14. Denaturation can disrupt the secondary, tertiary,

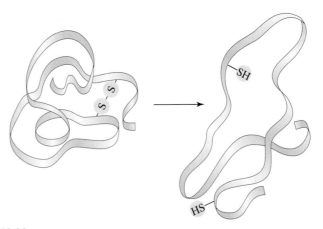

Figure 18.14
Protein denaturation results from disruption of weak forces and, in some instances, cleavage of disulfide bridges.

A Closer Look

Glycoproteins: Control of Blood Glucose in Diabetes

Glycoproteins play an important role in control of the level of blood glucose in patients with diabetes. The goal of those patients is usually to maintain blood glucose as near as possible to the midpoint of the nondiabetic range, about 6 to 7 mmol of glucose per liter of blood. The absence of glucose in urine tests (A Closer Look: Test for Glucose in Diabetes) or a normal amount of blood glucose from blood tests done four times a day does not mean that the level of blood glucose is constant at all times. The most accurate way to find out how the blood glucose concentration changes over time would be to monitor it constantly. This is usually impractical, but a test for a variant of normal adult hemoglobin (HbA) can indicate the degree of control of blood glucose levels. Normally, between 5% and 8% of HbA has a glucose molecule attached to it. This glycoprotein is called *hemoglobin A₁ (HbA₁)*. People with untreated diabetes have much higher levels of HbA₁ than healthy people. The blood levels of HbA₁ reflect the average blood glucose level during the previous 6 to 8 weeks. Therefore, the percentage of HbA₁ in the blood can be used to estimate the degree of control of blood sugar during this period. Blood levels of HbA₁ below 7% repre-

sent good control, levels of 7% to 11% represent moderate control, and levels greater than 11% represent poor control. This information is useful for adjusting the diet or the amount and frequency of insulin injections.

Evidence suggests that glycoproteins are responsible for circulation problems sometimes encountered in diabetes. A test used to monitor the amount of glycoproteins and glucose control involves albumin, the major protein of the fluid part of the blood. Similar to hemoglobin's reaction with glucose, albumin and other proteins react with a sugar in the blood to make glycoproteins. The sugar is fructosamine, an amine derivative of fructose. Fructosamine reacts with albumin and other proteins in direct proportion to its concentration in the blood. The amount of fructosamine in the blood therefore provides a measure of the proportion of glycoproteins being made in the body. Since the body makes fructosamine from glucose, the amount of circulating fructosamine also reflects the amount of glucose in the blood over a period of time. Fructosamine lasts a shorter time in the body than hemoglobin, so the level of the sugar reflects the control of blood glucose for the previous period of only 3 to 6 weeks instead of the 6 to 8 weeks for HbA₁. The normal fructosamine level is less than 2.8 mmol/L. Higher levels in a patient with diabetes indicate poor control of blood glucose.

and quaternary structures of a protein, but it does not break the peptide chains or in any other way alter the primary structure.

Protein denaturation may or may not be reversible. The proteins of egg white unfold and congeal into a rubbery mass when we boil an egg. This denaturation is irreversible because the protein can never return to its original state. Not all proteins are so heat-sensitive. Thermolysin, a protein-cutting enzyme of *Bacillus thermoproteolyticus,* a microorganism that lives in hot springs, resists unfolding even in boiling water. Heat, extremes of pH, and many chemicals, especially organic solvents, cause the irreversible

Heat sterilization of medical instruments kills bacteria because the heat irreversibly denatures their proteins.

denaturation of proteins by disrupting weak attractive forces. Figure 18.15 shows disulfide bonds cleaved in the presence of oxidants and reductants.

In reversible denaturation, the protein unfolds in the presence of a denaturing agent, such as a concentrated urea solution, but is restored to its native state on removal of the agent. Whether a protein undergoes reversible or irreversible denaturation varies with the kind of protein reacting and the conditions of reaction.

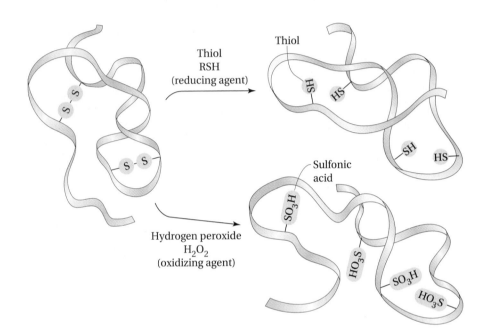

Figure 18.15
Disulfide bridges are cleaved by oxidation or reduction

SUMMARY

Twenty alpha amino acids commonly occur in nature. These compounds are grouped according to the chemical properties of their side chains. All the amino acids but glycine belong to the L family of stereoisomers. At biological pH, most amino acids exist as electrically neutral zwitterions, in which the positive charge of the protonated alpha amino group is compensated by the negatively charged alpha carboxylate ion.

The formation of an amide bond (peptide bond) between the alpha amino group of one amino acid and the alpha carboxylic acid group of another produces a peptide. Peptides have a common backbone and N-terminal and C-terminal ends. The order in which amino acids are linked in a peptide is the peptide's amino acid sequence. The peptide's amino acid sequence determines its biological role.

Proteins are peptides of greater than about 100 amino acid residues. Structural proteins are often fibrous, water-insoluble molecules; other proteins are globular and water-soluble. Conjugated proteins have nonprotein portions called prosthetic groups, which are often relatively small organic molecules.

Protein molecules fold into unique shapes. The chemical nature of the amino acid side chains and the order of their occurrence within the peptide chain (primary structure) govern the type of folding. Two patterns of chain folding (secondary structure) are found in many proteins; these are the alpha helix and the beta-pleated sheet. Collagen forms a secondary structure of collagen helixes. Each molecule of a particular kind of protein has the same overall three-dimensional folding (tertiary structure). Forces that hold proteins in their respective foldings include hydrophobic aggregation, hydrogen bonds, ionic bonds, and disulfide bridges.

Myoglobin is a conjugated globular protein that stores oxygen in muscle. The folding of its single-peptide chain is mostly alpha helix. The heme prosthetic group is tucked into folds in the chain. Hemoglobin, the oxygen carrier of blood, consists of four single-chain subunits comprising its quaternary structure—two alpha chains and two beta chains. Sickle cell anemia is an example of how a minor modification of the primary structure of a protein affects the protein's biological function.

Peptide chains of proteins are unfolded (denatured) by chemical agents, heat, or extremes of pH. Denaturation may be reversible or irreversible.

KEY TERMS

Alpha carbon (18.1)
Alpha helix (18.6)
Alpha keratin (18.6)
Amino acid (18.1)
Amino acid residue (18.4)
Amino acid side chain (18.1)
Antidiuretic (18.4)
Beta keratin (18.6)
Beta-pleated sheet (18.6)
Carboxyhemoglobin (18.9)

Collagen (18.6)
Conjugated protein (18.7)
Cyanohemoglobin (18.9)
Denaturation (18.12)
Deoxyhemoglobin (18.8)
Disulfide bridge (18.4)
Electrophoresis (18.10)
Enkephalin (18.4)
Fibrous protein (18.6)
Globin (18.8)
Globular protein (18.7)

Glycoprotein (18.11)
Hemoglobin (18.8)
Isoelectric pH (18.3)
Isoelectric point (18.3)
Methemoglobin (18.8)
Methemoglobinemia (18.8)
Myoglobin (18.7)
Native state (18.12)
Oxyhemoglobin (18.8)
Peptide (18.4)
Peptide bond (18.4)

Polypeptide (18.4)
Primary structure (18.5)
Prosthetic group (18.7)
Protein (18.5)
Quaternary structure (18.8)
Salt bridge (18.7)
Secondary structure (18.6)
Subunit (18.8)
Tertiary structure (18.7)
Zwitterion (18.3)

EXERCISES

Amino Acids (Sections 18.1, 18.2, 18.3)

18.7 Write the structure and identify the alpha carbon and R group for the following amino acids: (a) alanine, (b) serine, and (c) glutamic acid.

18.8 Categorize the following amino acids according to their side-chain groups.
(a) valine   (b) glutamine   (c) cysteine
(d) phenylalanine

18.9 Draw a stereorepresentation of (a) D-serine and (b) L-alanine.

18.10 Identify the following compounds represented by their Fischer projection formulas.

(a) CO_2H (b) CO_2H (c) CO_2H

H——NH_2 H——NH_2 H_2N——H

CH_2OH H CH_2SH

18.11 Define the term *zwitterion*. Draw the amino acid leucine as a zwitterion.

18.12 Consider the amino acids valine, glutamic acid, and lysine. Draw structural formulas for the species that predominate in solution at (a) pH 2 and (b) pH 10.

18.13 Define *isoelectric pH*.

18.14 At which pH will alanine be least soluble, (a) pH 2, (b) pH 7, or (c) pH 10? Why?

Peptides (Section 18.4)

18.15 In peptide chemistry, what is the meaning of the word *residue*?

18.16 What is the name given to the amide bond in a peptide chain?

18.17 Consider the tripeptide seryl-glycyl-phenylalanine. (a) Draw the complete structural formula, and (b) write the three-letter abbreviation. (c) How many peptide bonds does this molecule have?

18.18 Translate the following three-letter abbreviations: (a) Ala-Ser-Gly and (b) Gly-Ser-Ala. Are the structures of these tripeptides the same? Explain your answer.

18.19 Explain the biological functions of the peptide hormones (a) oxytocin and (b) vasopressin.

18.20 Name two pentapeptides that act as the body's own opiates.

Primary and Secondary Structures of Proteins (Sections 18.5, 18.6)

18.21 What is meant by the *primary structure* of a protein?

18.22 Define the *secondary structure* of a protein.

18.23 Describe three common repeating patterns that are found in the secondary structure of proteins.

18.24 Consider the structure of the alpha helix. Are the amino acid residue side chains all inside the helix, part of the helix, or all outside the helix?

18.25 What is the function of collagen in the body?

18.26 Compare the molecular structures of collagen and alpha keratin with regard to secondary structure and amino acid composition.

Tertiary and Quaternary Structures of Proteins (Sections 18.7, 18.8)

18.27 Define what is meant by a *conjugated protein*.

18.28 (a) In what ways are the molecular structures of myoglobin and hemoglobin similar? (b) How are they different?

18.29 With the aid of diagrams, describe the factors that contribute to the folding of peptide chains in the tertiary structures of proteins.

18.30 Describe the heme group and explain its function.

18.31 What is the oxidation state of iron in (a) oxyhemoglobin and (b) methemoglobin?

Hemoglobin Function (Section 18.9)

18.32 Discuss the role of hemoglobin in the transport of oxygen in the body.

18.33 Why are carbon monoxide and cyanide ions poisons?

Sickle Cell Anemia (Section 18.10)

18.34 What is the basic structural difference between normal hemoglobin (HbA) and sickle cell hemoglobin (HbS)?

18.35 Why is exposure to low oxygen levels potentially dangerous for a person who has sickle cell trait?

18.36 What technique is used for screening individuals for sickle cell trait or sickle cell anemia?

18.37 The symptoms of sickle cell anemia can be treated by hyperbaric oxygenation. Why does this procedure offer relief for the patient?

Glycoproteins and Denaturation (Sections 18.11, 18.12)

18.38 Name and describe the biological functions of at least two glycoproteins.

18.39 What happens when a protein is denatured?

18.40 Describe one way a protein can be denatured irreversibly and one way a protein can be denatured reversibly.

Additional Exercises

18.41 Match the following.
(a) lysine
(b) vasopressin
(c) denaturation
(d) peptide link
(e) globular protein
(f) beta keratin
(g) zwitterion
(h) collagen

(1) fibrous protein
(2) antidiuretic hormone
(3) basic amino acid
(4) protein of tendon
(5) unfolding
(6) net charge equals zero
(7) spherical shape
(8) amide bond

18.42 (a) Write the three-letter abbreviation and draw the structural formula of the tripeptide glutamyl-cysteinyl-glycine. (b) Draw the different ionic species of this tripeptide at low (pH = 2), neutral (pH = 7), and high pH (pH = 11). (c) A disulfide bridge is formed when this tripeptide is treated with a mild oxidizing agent. Draw the structural formula of the resulting hexapeptide.

18.43 Write the three-letter abbreviations of all the different tripeptides that can be made if the peptides contain one amino acid residue each of histidine, methionine, and glutamic acid.

18.44 For each of the following fibrous proteins, match all the items at the right that apply. Some answers may be used more than once, some not at all.
(a) beta keratin
(b) collagen
(c) alpha keratin

(1) spider webs
(2) cartilage
(3) predominantly alpha helixes
(4) rich in proteins
(5) a glycoprotein
(6) found in hair
(7) rich in glycine and alanine

18.45 Draw the structural formulas of the expected products of acid hydrolysis of (a) Cys-Gly-Ala, (b) Glu-Leu-Val-Pro, and (c) Ser-His-Phe-Tyr-Trp.

18.46 What unique structural features do each of the following amino acids have that help to distinguish them from all the others?
(a) serine
(b) phenylalanine
(c) glycine
(d) cysteine
(e) tyrosine
(f) proline

18.47 Define (a) salt bridge, (b) prosthetic group, (c) tertiary structure, (d) globular protein, (e) beta-pleated sheet, (f) secondary structure, (g) alpha-helix, and (h) primary structure.

18.48 Distinguish among alpha keratin, beta keratin, and collagen by biological function and molecular structure.

18.49 Match the following. Each answer will be used once only.
(a) fibrous protein
(b) native state
(c) disulfide bridge
(d) peptide bond
(e) globular protein
(f) denaturation
(g) prosthetic group
(h) salt bridge
(i) conjugated protein

(1) water-soluble protein, easily denatured
(2) can be reversible or irreversible
(3) the metal ion of a conjugated protein
(4) a protein in its normal, biologically active form
(5) ionic bond between carboxylate ion and protonated amine group
(6) important structural protein
(7) covalent bond that can hold two peptide chains together
(8) protein that has a prosthetic group
(9) covalent bond that holds two amino acids together

SELF-TEST (REVIEW)

True/False

1. Amino acids found in nature are generally the L forms.
2. At the isoelectric pH of an amino acid, its solubility in water is maximum.
3. A basic amino acid such as asparagine would have a net negative charge at high pH.
4. The amino acid glycine is the C-terminal amino acid of the pentapeptide Gln-Asp-Pro-Val-Gly.
5. In aqueous solutions, the hydrocarbon side chains of a protein tend to point inward, avoiding interaction with the water.
6. Protein denaturation is always irreversible.
7. The amino acid proline disrupts the alpha-helical secondary structure of a protein.
8. Myoglobin is used for oxygen storage in the muscle.
9. Proteins are composed of amino acids.
10. All protein molecules have a quaternary structure.

Multiple Choice

11. All the following amino acids have aliphatic side chains *except*
(a) valine.   (b) alanine.   (c) serine.
(d) leucine.
12. Disulfide bridges can be broken by
(a) dissolving the protein in water.
(b) cooling the protein.
(c) oxidizing agents.
(d) dehydrating agents.
13. Proteins that tend to be water soluble and generally have spherical shapes are
(a) conjugated proteins.   (b) globular proteins.
(c) beta keratins.   (d) fibrous proteins.
14. Which of the following contributes most importantly to the secondary structure of a protein?
(a) hydrogen bonds   (b) disulfide bridges
(c) salt bridges   (d) none of the above

15. A disruption of the secondary and tertiary structures of a protein
(a) is never reversible.
(b) is caused only by heat.
(c) results in a conjugated form.
(d) is called *denaturation*.

16. The peptide Phe-Glu-Ala-Val
(a) has valine at the N-terminal end.
(b) contains four peptide bonds.
(c) is the same as the peptide Val-Ala-Glu-Phe.
(d) has four amino acid residues.

17. Two commonly occurring peptide secondary structures in proteins are
(a) alpha helix and beta-pleated sheet.
(b) protofibril and globular.
(c) globular and beta-pleated sheet.
(d) protofibril and alpha helix.

18. The most abundant protein in higher vertebrates is
(a) collagen.   (b) oxytocin.
(c) alpha keratins.   (d) hemoglobin.

19. Amino acids found in biological systems are usually
(a) in the left-handed, or L, form.
(b) in the D form.
(c) found in an uncombined state.
(d) beta amino acids.

20. The heme group in a hemoglobin molecule
(a) is absent in sickle cell anemia.
(b) is called a *prosthetic group*.
(c) is held jointly by the four peptide subunits.
(d) can never contain Fe(III) ions.

21. The complex present in carbon monoxide poisoning is
(a) cyanohemoglobin.   (b) deoxyhemoglobin.
(c) carboxyhemoglobin.   (d) methemoglobin.

22. Myoglobin, a globular, conjugated protein,
(a) has most of its aliphatic side chains on the exterior surface of the molecule.
(b) would be expected to be water soluble.
(c) carries oxygen from the lungs to the tissues.
(d) has no prosthetic group.

23. A structural protein found in mammalian hair is
(a) alpha keratin.   (b) collagen.
(c) beta keratin.   (d) fibroin.

24. A molecule of the amino acid valine in a strongly acid solution would be a
(a) zwitterion.
(b) negatively charged species.
(c) positively charged species.
(d) neutral molecule.

25. Which of these is *not* usually disrupted in the denaturation of a protein?
(a) ionic bonds   (b) hydrogen bonds
(c) peptide bonds   (d) hydrophobic aggregation

26. The major type of hemoglobin of arterial blood is
(a) cyanohemoglobin.   (b) oxyhemoglobin.
(c) methemoglobin.   (d) deoxyhemoglobin.

27. The correct general formula for a dipeptide at neutral pH is

(a)

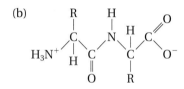

(b)

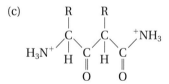

(c)

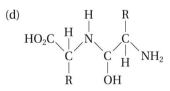

(d)

HO$_2$C ... NH$_2$ structure

28. Which type of bonding holds proteins in the beta-pleated sheet?
(a) ionic bonds   (b) hydrogen bonds
(c) hydrophobic aggregation   (d) covalent bonds

19

Enzymes

Catalysis of the Reactions of Life

Enzymes of clinical significance are derived from many unusual sources, including the venom of snakes.

CHAPTER OUTLINE

CASE IN POINT: Enzyme therapy for heart attacks

19.1 Enzymes

19.2 Names of enzymes

19.3 Enzyme specificity

19.4 Enzyme-substrate complexes

19.5 Active sites

19.6 Cofactors

19.7 Enzyme assay

19.8 Effects of pH, temperature, and heavy metals

A CLOSER LOOK: Lead Poisoning

19.9 Induction and degradation of enzymes

19.10 Control of enzyme activity

A CLOSER LOOK: HIV Protease and Its Inhibition

19.11 Zymogens

A CLOSER LOOK: Trypsin: Anatomy of an Enzyme

19.12 Blood clotting

FOLLOW-UP TO THE CASE IN POINT: Enzyme therapy for heart attacks

19.13 Antibiotics and other therapeutic drugs

A CLOSER LOOK: The Toxicity of Pesticides

The operation of living cells may be compared with the operation of highly efficient chemical factories. *Enzymes,* substances that catalyze biological reactions, are the workers in these cell factories, busily disassembling raw materials (nutrients taken into the cell) and reassembling them into products used to help the cell survive and grow. Depending on the tasks they perform, particular enzymes may be distributed throughout the cell or limited to a specific location.

In this chapter we will learn about the structure of enzymes and see how they function. Enzymes are proving useful as medical drugs, as we will learn in the Case in Point.

CASE IN POINT: Enzyme therapy for heart attacks

Art, a 45-year-old stockbroker, awoke early Monday morning. The markets were not performing well, and he was feeling a great deal of stress over his clients' and his own financial losses. By 11 o'clock in the morning, the markets had dropped further. Art began to feel a tightness in his chest and pains running down his left arm. Fifteen minutes later a coworker found Art slumped over his desk. After a quick examination, the emergency room physician was certain that Art had suffered a heart attack. Among other steps, the physician ordered an intravenous drip of a solution containing an enzyme. What enzyme did the doctor order? What effect did the enzyme have on Art's recovery? We will learn the answers to these questions in Section 19.12.

Clinical laboratory technicians use automated equipment to measure enzyme concentrations in blood serum. This procedure helps in the diagnosis of medical problems such as heart attacks.

19.1 Enzymes

AIM: To state three properties that show that enzymes are catalysts.

Enzymes *are biological catalysts.* In 1926, American chemist James B. Sumner reported the first purification of an enzyme. The enzyme he isolated was urease, which is able to hydrolyze urea, a constituent of urine, into ammonia and carbon dioxide. The reaction is this:

$$NH_2-\overset{\overset{O}{\|}}{C}-NH_2 \ + \ H_2O \ \xrightarrow{\text{Urease}} \ 2NH_3 \ + \ CO_2$$

Urea   Water   Ammonia   Carbon
                                                    dioxide

Sumner demonstrated that urease is a protein. Since Sumner's pioneering work, hundreds of proteins that serve as enzymes have been isolated and structurally characterized. Until the 1980s, it was believed all enzymes were proteins. An exciting recent finding is that certain ribonucleic acids (RNAs) are capable of cleaving out sections of other long, polymeric RNA molecules. These catalytic RNA molecules, called *ribozymes,* also can splice the ends of two RNA molecules together. The existence of ribozymes means that some enzymes are not proteins.

Enzyme
enzymos (Greek):   leavened

Enzymes constitute a substantial portion of the total protein of the cell. A typical cell contains about 3000 different enzymes and many molecules of each kind. Enzymes can speed up chemical reactions, while other proteins cannot. Besides being able to speed up reactions, enzymes have two other properties of true catalysts. First, they are unchanged by the reaction they catalyze. Second, and very important, even though they speed up reactions, enzymes cannot change the normal position of a chemical equilibrium. In other words, an enzyme can help a reaction-product to be formedfaster, but the same amount of product is eventually formed whether or not an enzyme is present. (Few reactions in cells ever reach equilibrium because the products are rapidly converted to another substance in a further enzyme-catalyzed reaction. You may recall from Le Châtelier's principle, discussed in Chapter 6, that the removal of a reaction product pulls the reaction toward completion.)

People have used the catalytic power of enzymes to suit their desires since prehistoric times. The fermentation of fruit sugars to alcohol by yeast enzymes was a very early discovery. Herdsmen who made canteens from the stomachs of goats and sheep found that when they filled them with milk instead of water the milk soon clumped into cheese. The agent for this transformation is *rennin,* an enzyme produced in the stomachs of cud-chewing animals (ruminants). Yogurt, an ancient food with modern popularity, is prepared by the action of enzymes produced by several bacteria. Brewing beer from grain, leavening bread with yeast, and fermenting apple cider to vinegar are other practical applications of the catalytic power of enzymes.

19.2 Names of enzymes

AIM: To identify the function of an enzyme from its name.

Focus

Many enzymes are named for the reactions they catalyze.

A class of enzymes responsible for energy-producing reactions consists of the oxidoreductases. These enzymes catalyze oxidation-reduction reactions—reactions that involve the transfer of electrons. Many of these enzymes are found in the mitochondria of a cell.

Proteases are used as meat tenderizers, in laundry detergents as stain removers, in beer production to remove cloudiness, and in leather tanning.

Biochemists often name enzymes by taking the name of the compound undergoing change and adding the ending *-ase*. The ending *-ase* serves as a signpost that says that the substance in question is an enzyme. In the hydrolysis of urea, the substance undergoing change is urea, and the name given to the enzyme that catalyzes this change is *urease*. Sometimes the enzyme name also reflects the kind of chemical transformation that occurs. For example, an enzyme catalyzes the removal of hydrogen from ethanol (CH_3CH_2OH) to give acetaldehyde (CH_3CHO). The enzyme that catalyzes this dehydrogenation reaction is called *alcohol dehydrogenase*. In another example, *enzymes that catalyze the hydrolysis of one or more peptide bonds of proteins are given the general name* **protease** *or* **peptidase.** Some proteases are digestive enzymes that have names established long ago: *pepsin, trypsin, chymotrypsin*. *Thrombin* and *plasmin* are peptidases involved in blood clotting.

EXAMPLE 19.1 Discerning the function of an enzyme

What is the role of the enzyme sucrase?

SOLUTION

Based on the name of the enzyme, the substance undergoing a change is sucrose. Sucrose is a disaccharide. Most likely this enzyme catalyzes the breakdown (hydrolysis) of the disaccharide into two monosaccharides.

PRACTICE EXERCISE 19.1

Predict the function of the following enzymes: (a) lipase and (b) cellulase.

PRACTICE EXERCISE 19.2

Suggest general names for enzymes that catalyze (a) an oxidation reaction involving a gain of oxygen and (b) a hydrolysis reaction.

19.3 Enzyme specificity

AIM: To explain what is meant by the specificity of an enzyme.

Focus

Enzymes are specific catalysts.

Most of the chemical changes that occur in the cell are catalyzed by enzymes. **Substrates** *are reactants that are transformed to products by the catalytic action of enzymes.* As in nonenzymatic reactions, substrates are transformed into products by bond-making and bond-breaking processes. In an enzymatic reaction, these processes occur through interactions of the enzyme with its substrate. One property of enzymes is **specificity**—*catalyzing one chemical reaction with only one substrate.*

Enzymes exhibit stereospecificity. They will act on only one stereo-isomer of a compound, for example, the L form of an amino acid but not the D form.

To get some idea of the power and specificity of enzymes, consider carbonic anhydrase. This enzyme catalyzes only one reversible reaction, the breakdown of carbonic acid to water and carbon dioxide.

$$H_2CO_3 \underset{\text{anhydrase}}{\overset{\text{Carbonic}}{\rightleftharpoons}} CO_2 + H_2O$$

At ideal conditions, a single molecule of carbonic anhydrase is capable of catalyzing the breakdown of about 36 million molecules of carbonic acid in 1 minute. The next section examines how enzymes do their work.

19.4 Enzyme-substrate complexes

AIM: To use the lock-and-key model and the induced-fit model to explain binding and specificity in enzyme action.

Focus

Lock-and-key and induced-fit models of binding of substrates to enzymes help explain enzyme specificity.

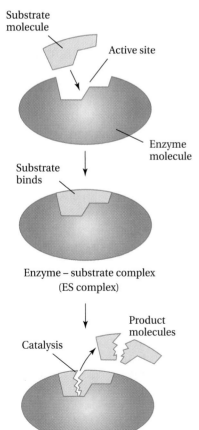

Enzyme – substrate complex
(ES complex)

A substrate molecule must contact an enzyme molecule before it can be transformed into products. Once the substrate has made contact, it must bind to the enzyme at the **active site**—*a region of the enzyme where the processes that convert substrates to products can take place.* The active site is usually a dimple, pocket, or crevice formed by folds in the tertiary structure of the protein. One model that explains the binding of substrates to enzymes is the **lock-and-key model**—*the active site of a given enzyme is shaped so that only the substrate molecule for that enzyme will fit into it, much as only one key will fit into a certain lock.* The interaction of a substrate molecule with an enzyme molecule is illustrated in Figure 19.1. *When the substrate key fits the enzyme lock, an* **enzyme-substrate complex** *is formed.*

$$E + S \rightleftharpoons ES$$

Enzyme　　Substrate　　Enzyme-substrate complex

The formation of an enzyme-substrate complex is an example of *complementarity.* The **complementarity,** *or fit,* between enzyme and substrate in binding helps to produce enzyme specificity. Without complementarity between the enzyme and its substrate, the enzyme would be unable to exert its catalytic power and transform substrate to products. Complementarity between enzyme and substrate is governed by factors such as the overall shapes of the active site and the substrate. Electric charge also can be important. The substrate may be positively charged and the enzyme active site negatively charged, or vice versa. Hydrogen bonds and other weak forces also may hold the enzyme-substrate complex together. Moreover, hydrophobic regions of the substrate associate with similar regions on the

Figure 19.1 (left)
Interaction of a substrate molecule with an enzyme molecule. The substrate fits the active site of the enzyme as a key fits a lock. The result is an enzyme-substrate complex. The bond-breaking and bond-making processes that transform substrates to products occur while the substrate is bound to the active site.

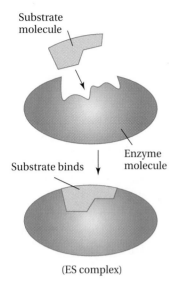

Substrate molecule

Substrate binds

Enzyme molecule

(ES complex)

Figure 19.2
The shape of the enzyme molecule changes to accommodate the substrate in the induced-fit model of enzyme action.

active site of the enzyme. For most enzymes, hydrophobic association and contributions from shape, charge, and a large number of weak forces are responsible for the formation and strength of the enzyme-substrate complex.

The lock-and-key analogy for the formation of enzyme-substrate complexes suffers from an important fact that we have not considered yet: Locks and keys are rigid objects, but substrates and enzymes are flexible. The chains and functional groups of substrate molecules are vibrating and rotating. Enzymes in solution "breathe"—undergo small changes in tertiary structure. Because of this flexibility, the *induced-fit model* is a better model for the formation of enzyme-substrate complexes than the lock-and-key model. *In the* **induced-fit model** *of enzyme-substrate binding, the substrate and enzyme make continuous structural adjustments as the substrate approaches the active site of the enzyme.* The complementarity between enzyme and substrate achieves perfection only at the moment of contact. Figure 19.2 illustrates the induced-fit model of enzyme-substrate complex formation. We should recognize that the lock-and-key and induced-fit models for the formation of enzyme-substrate complexes are just that—models. In reality, the detailed processes of enzyme-substrate complex formation are as numerous as the number of substrates and enzymes themselves.

19.5 Active sites

AIM: To describe what happens at the active site of an enzyme.

We have seen that enzyme binding specificity is analogous to a lock-and-key relationship between enzyme and substrate. But what accounts for the marvelous efficiency of enzymes in transforming substrates to products? The beauty of the precise binding of substrates to enzymes is that it achieves two important ends. First, binding brings specific substrates and enzymes together. This increases the effective concentration of the substrate at the enzyme's active site. You may recall that the speed of chemical reactions depends on the concentrations of the reactants (see Sec. 6.7). Even in dilute solution, the substrate and enzyme are in a one-to-one correspondence at the active site. Second, when the substrate is bound to the active site, it is in precisely the right position for the substrate to undergo the bond making and bond breaking catalyzed by the enzyme. Both these features contribute to the speed of enzyme-catalyzed reactions, which may be millions of times faster than the same reaction in the absence of the enzyme.

Active sites are usually only a small portion of the enzyme protein. Most proteins have molar masses of at least 25,000 g/mol, but only a small part, the active site, is involved in catalysis. Perhaps proteins are large because a firm foundation is needed for the active site to ensure precise alignment of substrate and enzyme. Many mechanical devices are based on this principle. A precision lathe, for example, may be used to turn out small metal spindles. If close tolerances are required, the framework of the lathe is apt to be large and made of cast iron to prevent errors due to vibration.

19.6 Cofactors

AIM: To describe the functions of cofactors.

> **Focus**
>
> Cofactors are essential to the catalytic activity of some enzymes.

Some enzymes consisting only of polypeptide chains can catalyze transformations of substrates, but others need nonprotein *cofactors,* also called *coenzymes,* to assist the transformation. *A* **cofactor** *(coenzyme) is a nonprotein portion of an enzyme necessary for the enzyme's function.* Cofactors may be metal ions of elements such as magnesium, potassium, iron, or zinc, but they also may be small organic molecules. Usually, a cofactor is easily removed from an enzyme molecule, whereas a prosthetic group (see Sec. 18.7) is covalently attached. *An enzyme that contains its bound cofactor is called a* **holoenzyme** *("whole enzyme"); if the cofactor is missing, the enzyme is called an* **apoenzyme.** As Figure 19.3 shows, the job of the enzyme superstructure is to bring the substrate into proximity with the cofactor. The cofactor participates in the chemical transformation catalyzed by the enzyme. Many cofactors are members, or are synthesized from members, of the vitamin B complex group (see Chap. 21).

19.7 Enzyme assay

AIM: To discuss the importance of assaying enzyme activity in the diagnosis of disease.

> **Focus**
>
> Assays of enzyme activity are useful in the diagnosis of disease.

Enzyme activity *is expressed as the rate at which the enzyme catalyzes the conversion of a substrate to products. Experiments in which enzyme activity is measured are called* **enzyme assays.** Many clinical testing procedures consist of assays for enzymes in blood serum—the fluid that remains when blood has clotted. The concept that makes these testing procedures useful is simple: Slight leakage of enzymes from tissues and organs to the circulatory system is normal; additional leakage is abnormal and indicates damage or disease. For instance, damage to the heart muscle in heart attacks

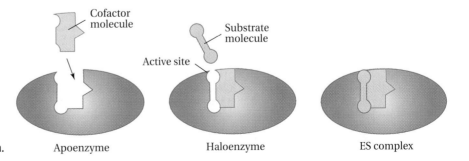

Figure 19.3
Many enzymes contain binding sites for cofactor and substrate molecules. The closeness of these sites in the protein superstructure of the enzyme brings the cofactor and substrate together for reaction.

Some enzymes, such as lactate dehydrogenase (LDH), have a number of different forms that differ slightly structurally. The relative amounts of these different forms in the blood have diagnostic value. A large elevation of LDH_5 is linked to acute hepatitis, while a large elevation of LDH_1 typically occurs after a heart attack.

results in the release of the enzyme *creatine phosphokinase* (CPK) into the bloodstream. Comparison of the range of CPK levels found in normal patients with that of the victim of a possible heart attack can indicate whether an attack has occurred and, if it has, the extent of the damage.

Over 40 other blood serum enzymes can be subjected to clinical testing. Information about a few selected groups of serum enzymes is given in Table 19.1. Enzyme activities are often measured by following the appearance of colored products or the disappearance of colored substrates with time. Alternatively, reactions that involve the uptake or release of protons can be followed by measuring changes in the pH of the assay solution with time.

Because of an increased demand for clinical tests, the assay procedures for many serum enzymes have been automated and computerized. The enzyme activities are recorded on a form that also shows ranges for normal patients.

The activity levels of enzymes are almost always reported in *international units*. The international unit defines a standard level of enzyme activity that will convert a specified amount of substrate to product within a certain time. *The internationally agreed on value of* **1 international unit (1 IU) is that quantity of enzyme that catalyzes the conversion of 1 micromole ($1 \mu mol$) of substrate per minute at a specified set of reaction conditions.** For example, an enzyme preparation with a value of 40 IU presumably contains an amount of the enzyme 40 times greater than the standard. It is the level of enzyme activity with respect to normal activity that is significant in the diagnosis of disease.

Table 19.1 Some Serum Enzymes Used to Diagnose Disease

Enzyme	Site	Physiologic function	Elevated serum levels in
amylases	salivary glands, pancreas	starch digestion	mumps, pancreatic obstruction or inflammation
peptidases	digestive tract, tissue cells	protein digestion	tissue injury, shock, fever, anemia
lipases	pancreas	fat digestion	pancreatic disorders
alkaline phosphatases	bone marrow, liver	cleavage of phosphate ester bonds at alkaline pH	bone inflammation (Paget's disease), bone softening (osteomalacia), hepatitis, obstructive jaundice
acid phosphatases	prostate	cleavage of phosphate ester bonds at acidic pH	prostate cancer
transaminases	heart, liver	control of nitrogen balance	hepatitis, myocardial infarction
dehydrogenases	heart muscle, skeletal muscle	oxidation-reduction reactions	myocardial infarction, hepatitis, acute and chronic leukemia
creatine phosphokinase (CPK)	heart muscle, skeletal muscle	formation of creatine phosphate in muscle	myocardial infarction

EXAMPLE 19.2   **Interpreting enzyme assays**

What disease is indicated in a patient with elevated blood serum levels of both transaminase and alkaline phosphatase enzymes?

SOLUTION

Although other factors would enter into a diagnosis, elevation of these two enzymes occurs in the disease hepatitis.

PRACTICE EXERCISE 19.4

A patient was vomiting, complained of stomach pains, and had a fever. An enzyme assay of the patient's blood revealed abnormally high levels of serum amylase and serum lipase. Suggest a possible diagnosis of the patient's illness.

19.8 Effects of pH, temperature, and heavy metals

AIM: To interpret, on a molecular scale, changes in enzyme activity that occur as a result of changing pH and changing temperature.

Focus

Enzyme activity is sensitive to environmental conditions.

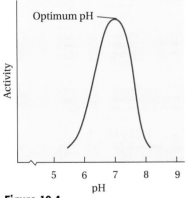

Figure 19.4
Typical pH-activity profile for an enzyme. The pH value at which the maximum activity is observed is called the enzyme's pH optimum. Here, the enzyme has its optimum pH near 7.

Enzyme assays must be carried out at identical conditions for comparisons of activity to be valid. Particular attention must be paid to pH and temperature, since small variations in these two conditions often cause large changes in the activity values obtained. Suppose, for example, that the anionic form of a single carboxylic acid side chain of aspartic acid is involved in binding or catalysis by a certain enzyme. At pH values near or above neutrality, this group is completely dissociated into a carboxylate anion and a proton, and the enzyme has the greatest possible activity. If the pH of the solution is gradually lowered, the fraction of carboxylate anion decreases, and the fraction of undissociated carboxylic acid increases. At low pH, where all the carboxylic acid groups are in the undissociated form, the enzyme would become completely inactive.

The variation of an enzyme's activity with changes in pH is shown in a **pH-activity profile.** Figure 19.4 gives an example. Most enzymes exhibit maximum activity at a pH near 7, the pH of most body fluids. *The pH where maximum activity occurs is called the* **pH optimum** *of the enzyme.* The pH-activity profiles of different enzymes may be very complex if ionization of more than one group on the enzyme is important to binding or catalysis or if substrate ionizations are also important.

The pH-activity profiles of enzymes generally reflect reversible changes in activity of enzymes. That is, a decline in the activity of an enzyme, say, with a decrease in pH, is reversed if the pH is increased to its original value. Like other proteins, however, most enzymes are stable only within a narrow range of pH. Extremes of pH will irreversibly denature most enzymes, and a

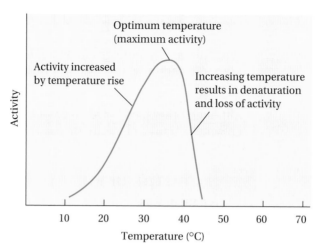

Figure 19.5
The activity of an enzyme depends on the temperature. If the temperature is raised past a certain point, heat denaturation causes a sharp decrease in enzyme activity.

Denaturation leaves peptide bonds intact but disrupts the secondary, tertiary, and quaternary structures of a protein molecule. If the protein is an enzyme, denaturation will damage the active site and destroy the activity of the enzyme.

return to the pH optimum will not restore the activity. What constitutes an extreme of pH depends on the particular enzyme being studied.

Enzyme activity usually varies with temperature. At temperatures near 20 °C, most enzymes are active and relatively stable. As with nonbiological catalysts, an increase in temperature increases the reaction rate. Figure 19.5 illustrates that activity can rapidly decrease beyond the optimal temperature. The decrease in activity is usually due to heat denaturation of the enzyme.

Although some enzymes contain metal ion cofactors, ions of heavy metal ions such as those of lead and cadmium are often damaging to enzymes, as discussed in A Closer Look: Lead Poisoning.

PRACTICE EXERCISE 19.5

Enzyme assays must be done at a specified pH and specified temperature. Why?

PRACTICE EXERCISE 19.6

Pepsin, an enzyme found in gastric juice, has an optimal pH of about 2. Draw its pH-activity profile.

19.9 Induction and degradation of enzymes

AIM: To explain how induction and degradation can control enzyme concentrations in a cell.

Focus

Induction and degradation help control enzyme concentrations in cells.

The concentrations of various body chemicals must be strictly controlled. To assert this control, cells need to slow down or speed up the production of cellular products. Cells use several means to control levels of cellular products. Two of these means involve enzyme **induction**—*the synthesis of enzymes on demand*—and **degradation**—*the hydrolysis of enzymes to their constituent amino acids.*

Each enzyme molecule, given a sufficient supply of raw materials, will catalyze the formation of products at a rapid, steady pace. One way to

A Closer Look

Lead Poisoning

Lead is a heavy metal poison that affects the functioning of the brain, blood, liver, and kidneys. Lead poisoning often leads to mental retardation and neurologic disorders through damage to the brain and central nervous system. Some of the harmful effects of lead poisoning result from the effects of lead(II) ions (Pb^{2+}) on the action of enzymes. Many enzymes contain thiol groups (—SH groups) from the amino acid cysteine. Lead ions, as well as other heavy metal ions, bind tightly to —SH groups and denature the enzymes. This renders the enzymes inactive. Lead ions also form strong bonds with the side-chain carboxylate ions of acidic amino acid residues with similar results.

Compounds of lead are widespread in the environment. Leaded gasoline was once a major contributor to environmental lead. This problem source has been eliminated by the introduction of lead-free gasoline. However, another more insidious source of environmental lead pollution—lead paint—still exists.

Lead poisoning is a current major problem with children. The interior woodwork of most homes built before 1950 was coated with paint that contained up to 50% lead. Lead paint poses little hazard unless it flakes and powders. Children may eat the sweet-tasting chips of lead-based paint. These same children also may breathe lead-containing paint dust, which further increases their body burden of lead.

It is estimated that more than 3 million children in the United States have blood levels of 10 μg of lead or more per deciliter. Such blood levels are considered a cause for concern by the U.S. Centers for Disease Control and Prevention. Children with these blood levels of lead are at risk of

The presence of lead paint in many older buildings makes its abatement essential in preventing lead poisoning, especially in children.

underdeveloped nervous systems and lowered intelligence.

A child with lead poisoning is often constipated and vomits. A radiograph of the child's abdomen will reveal the presence of radiopaque ingested paint fragments. A blood sample can be taken and measured for lead content. Blood tests also reveal a low red blood cell count and low hemoglobin levels.

Chronic lead poisoning can be treated by giving the patient calcium EDTA (ethylenediaminetetraacetic acid) intravenously. The EDTA complexes (binds tightly) with the lead ions in the body and allows them to be excreted as the Pb-EDTA complex in the urine. During treatment, the lead content of the urine is monitored. A rise in the lead content of the patient's urine indicates that lead is being eliminated from the body.

increase production is for the cell to increase the number of enzyme molecules. *Enzymes that are synthesized in response to a temporary need of the cell are called* **inducible enzymes.** Cells grown in a medium that is deficient in a certain nutrient may not produce the enzymes required to transform that nutrient to useful products. If the cells are transferred to a medium that contains the nutrient, they begin to produce the necessary enzymes. When the cells are returned to the original medium, the induced enzymes soon disappear, degraded by the digestive machinery of the organism.

19.10 Control of enzyme activity

AIMS: To compare the control mechanisms of competitive inhibition and enzyme modulation. To define an allosteric enzyme and a pacemaker enzyme.

Focus

Inhibitors and modulators help control enzyme activity in cells.

Induction and degradation of enzymes are important ways for the cell to economize on the variety of enzymes it produces. Used exclusively, however, they would be a rather coarse way of controlling production of substances in the cell. A finer control for the slowdown of cellular processes is provided by *inhibitors*.

Competitive inhibitors *are molecules that are similar in shape or charge to substrate molecules and capable of binding to enzyme active sites.* However, competitive inhibitors are not transformed into products. If the inhibitor in Figure 19.6 binds at the active site, for example, it will block further use of the enzyme by substrates for as long as it remains bound. Certain enzymes normally present in the body are complexed with natural competitive inhibitors. When a need for such an enzyme arises, events occur that cause the complex to dissociate. The enzyme is then free to go about its appointed task. A Closer Look: HIV Protease and Its Inhibition, describes one way that scientists are trying to enlist enzyme inhibitors in the fight against HIV.

Cells have other ways of using competitive inhibitors to control the activity of enzymes. A fine slowdown control of enzyme activity built into some enzymes relies on **feedback inhibition**—*the concentration of a product at the end of a series of steps builds up and then "feeds back" to inhibit the enzyme in a preceding step.* Assume that substrate A is converted by a team of three enzymes, E_1, E_2, and E_3, to a product D as shown in this equation:

$$A \xrightarrow{E_1} B \xrightarrow{E_2} C \xrightarrow{E_3} D$$

with Feedback from D back to the first step.

Each of the enzymes works at transforming the product made by the previous enzyme into a new product, finally turning out the end product D. Now suppose that enough D has been produced to satisfy the needs of the cell. A slowdown of the production line will occur if the product D, now present in a relatively high concentration, is a competitive inhibitor of the enzyme E_1.

More often, an enzyme such as E_1 is *not* competitively inhibited by a product such as D. Instead, enzymes that catalyze the early stages of a long

Figure 19.6
Competitive inhibitors block the active site of an enzyme so that the substrate molecule cannot enter. Naturally, a substrate molecule that cannot enter the active site cannot be transformed to products.

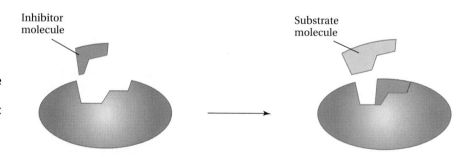

Inhibitor molecule

Substrate molecule

HIV Protease and Its Inhibition

The human immunodeficiency virus (HIV) is the causative agent in acquired immunodeficiency syndrome (AIDS). The destruction of cells of the immune system by HIV heightens the susceptibility of patients with AIDS to infections and tumors that inevitably lead to death. Responsible for one of only four enzyme activities produced by the virus, *HIV protease* is a target for the development of therapeutic drugs for AIDS. We can describe the enzyme and the path that research on the HIV protease is taking, but it is too early to say whether the work will lead to effective ways of slowing or stopping the progress of AIDS.

The three-dimensional structure of HIV protease has been determined by X-ray crystallography (see figure). The enzyme is a protein dimer consisting of two identical subunits. Each subunit contains only 99 amino acid residues. HIV protease binds to large proteins produced during reproduction of the virus and splits them into smaller polypeptides. These smaller polypeptides are needed for the assembly of new, infectious viruses. The biological function of HIV protease provides an approach to the design of medicines to treat AIDS. If HIV protease could be inhibited in an HIV-infected person, only incomplete, noninfectious viruses would be formed. Very good inhibitors of HIV protease have been discovered. Some of the best can be considered "decoy" molecules. These decoy molecules consist of peptides similar to the peptide substrates cleaved by HIV protease, except the peptide bond that would ordinarily be cleaved by the protease is replaced by a carbon-carbon or other hydrolysis-resistant bond.

Like ducks drawn to decoys, the protease is attracted to the peptide parts of the inhibitor molecules and binds them tightly to its active site. However, the enzyme cannot cleave the hydrolysis-resistant bond of the decoy molecule. The protease is rendered unable to process viral proteins as long as the decoy inhibitor remains bound to the active site. Other compounds are also being tried in order to block the action of HIV protease. Many of these molecules have no obvious structural resemblance to the natural protein substrate of HIV protease. For example, a chemically modified version of buckyball (A Closer Look: Buckyball) has been found to block the active site of the enzyme.

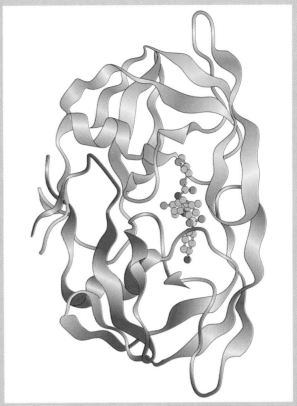

The dimer of HIV protease with bound inhibitor. The identical subunits of the dimer are colored in blue and red. Each subunit contains a short section of alpha helix (shown as a curly ribbon) and several regions of beta-pleated sheet structure (shown as sections of parallel ribbons).

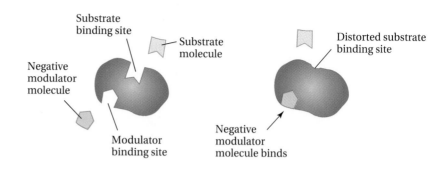

Figure 19.7
An allosteric enzyme has a site for a modulator and a site for a substrate molecule. Negative modulators distort the peptide chain conformation of the active site so that substrate molecules cannot bind.

Allosteric
allo (Greek):   other
stereos (Greek):   site or space

Allosteric enzymes are used as control points in many metabolic pathways. For example, a high concentration of a metabolic intermediate can serve as a negative modulator of an enzyme that catalyzes a reaction producing more of the intermediate.

sequence of reactions often contain another binding site in addition to the active site. *This second site is called the* **modulator binding site.** *Enzymes containing both an active site and a modulator binding site are called* **allosteric enzymes.** *A molecule that binds to the modulator binding site and slows down a reaction is a* **negative modulator.** In our example, if E_1 is an allosteric enzyme, product D will often serve as a negative modulator.

Binding of a negative modulator to the modulator binding site of an allosteric enzyme usually distorts the enzyme's active site (Fig. 19.7). This distortion hampers the formation of the enzyme-substrate complex. The enzyme's activity is decreased when the negative modulator is bound to the modulator binding site. *Some allosteric enzymes have* **positive modulators**—*modulators that increase the enzyme activity.* In these enzymes, the active site may be distorted in the *absence* of the modulator. Figure 19.8 shows a positive modulator binding to the modulator site and inducing refolding of an enzyme to the active form.

Pacemaker enzymes *control the rates of cellular processes.* Often allosteric enzymes, pacemakers are usually found near the start of a sequence of enzyme-catalyzed reactions. Negative modulation of allosteric pacemaker enzymes by end products of the sequence is important to the economy of the cell. After all, there is not much sense in expending the material and energy resources of the cell by going through a series of complex reactions to make a final product that already is present in sufficient quantities to meet cellular needs. On the other hand, positive modulation provides a means to speed up a process in a way that is more rapid, more

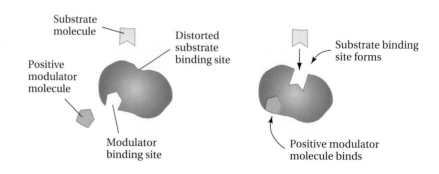

Figure 19.8
Positive modulators change the peptide chain conformation of the active site so that substrate molecules can bind.

direct, and more economical than the synthesis of whole new enzyme molecules.

PRACTICE EXERCISE 19.7

Explain the difference between a positive modulator and a negative modulator.

PRACTICE EXERCISE 19.8

Distinguish between a cofactor and a positive modulator.

19.11 Zymogens

AIMS: *To describe the action of a peptidase on a zymogen to form an active enzyme. To describe at least two physiologic processes that rely on activation of zymogens.*

Focus

Active enzymes are sometimes produced from zymogens.

Some enzymes are synthesized in biologically inactive forms that are transformed into active enzymes as the organism needs them. *These inactive precursors are called* **zymogens** *or* **proenzymes.** Along with induction and positive modulation, the activation of zymogens represents a way to control enzyme activity in an organism. A peptidase is usually required for the conversion of a zymogen to an active enzyme. The peptidase catalyzes the hydrolysis of one or more specific peptide bonds in the zymogen. Once these peptide bonds are broken, the remaining protein chains are free to fold into a catalytically active conformation.

Table 19.2 lists many peptidases that are formed from zymogen precursors. *Pepsin,* for example, a protein-digesting enzyme of the stomach, is produced by the action of stomach acid on the zymogen *pepsinogen.* Pepsin itself is also capable of converting pepsinogen into the active enzyme. *The process by which an enzyme molecule catalyzes the activation of its own zymogen is called* **autoactivation.**

Zymogens of some other peptidases are formed in the pancreas; these include *trypsinogen, chymotrypsinogen,* and *procarboxypeptidase.* As the zymogens are formed, they are packaged as **zymogen granules**—*zymogens in coats of lipid and protein.* The zymogen granules are then stored in the

Table 19.2 Peptidases Formed from Zymogen Precursors

Zymogen ⟶ active enzyme	Site of zymogen formation	Site of zymogen activation	Activating agent
pepsinogen ⟶ pepsin	stomach cell walls	stomach	stomach acid or autoactivation
trypsinogen ⟶ trypsin	pancreatic cells	small intestine	enterokinase or autoactivation
chymotrypsinogen ⟶ chymotrypsin	pancreatic cells	small intestine	trypsin
procarboxypeptidase ⟶ carboxypeptidase	pancreatic cells	small intestine	trypsin

pancreatic cells and, when required, are released into the small intestine. Trypsinogen is converted to trypsin by chopping small peptides from the zymogen. (See A Closer Look: Trypsin: Anatomy of an Enzyme.) These peptide cleavages are catalyzed by *enteropeptidase.* Autoactivation of trypsinogen also may occur. Trypsin then activates chymotrypsinogen and procarboxypeptidase to their active enzymes by peptide bond cleavages.

Once activated, pepsin, trypsin, and chymotrypsin help to digest food proteins. All three enzymes are **endopeptidases**—*peptidases that catalyze the hydrolysis of peptide bonds in the interior of peptide chains but not those at the C-terminal or N-terminal end. Carboxypeptidase* is also a digestive enzyme, but it is an **exopeptidase**—*a peptidase that catalyzes the removal of amino acid residues, one at a time, from the C-terminal or N-terminal end of peptide chains.* Carboxypeptidase starts from the C-terminal end. Working in concert, the four digestive peptidases are capable of completely degrading a protein molecule to its constituent amino acids.

EXAMPLE 19.3 **Determining the C-terminal group of a peptide**

You may recall from Chapter 18 that methionine enkephalin, Tyr-Gly-Gly-Phe-Met, is one of nature's pain killers. What products will be produced if a solution of methionine enkephalin is treated for a short time with carboxypeptidase?

SOLUTION

Carboxypeptidase is an exopeptidase that catalyzes the hydrolysis of peptide bonds, one at a time, starting at the C-terminal end of the peptide. After a short exposure of the methionine enkephalin to carboxypeptidase, the products would be the C-terminal amino acid methionine (Met) and the tetrapeptide Tyr-Gly-Gly-Phe.

EXAMPLE 19.4 **Determining the N-terminal group of a peptide**

Aminopeptidase is an exopeptidase that catalyzes the hydrolysis of peptide bonds, one at a time, starting at the N-terminal end of the peptide. When we treat a tripeptide with aminopeptidase and isolate the products, we obtain the amino acid leucine (Leu) and the dipeptide Asp-Gly. What is the sequence of amino acids in the original tripeptide?

SOLUTION

Aminopeptidase cleaves peptide bonds from the N-terminal end of peptides, so the leucine obtained from the hydrolysis must have been the N-terminal residue of the tripeptide. The original tripeptide must be Leu-Asp-Gly.

EXAMPLE 19.5 **Enzyme hydrolysis of peptides**

Chymotrypsin preferentially catalyzes the hydrolysis of peptide bonds in which the amino acid residue on the C-terminal side of the cleaved peptide bond has an aromatic side chain. What are the products if a solution of the following hexapeptide is treated with chymotrypsin?

Arg-Gly-Phe-Gly-Gly-Phe

SOLUTION

The only hydrophobic side chains in the hexapeptide are two phenylalanine (Phe) residues: at the third position from the N-terminal end and at the C-terminal end. Since chymotrypsin is an endopeptidase, it will not remove the C-terminal Phe residue. It will, however, cleave the Phe-Gly peptide bond. The products will be two tripeptides: Arg-Gly-Phe and Gly-Gly-Phe.

PRACTICE EXERCISE 19.9

Consider the general structure for a hexapeptide with its five peptide bonds:

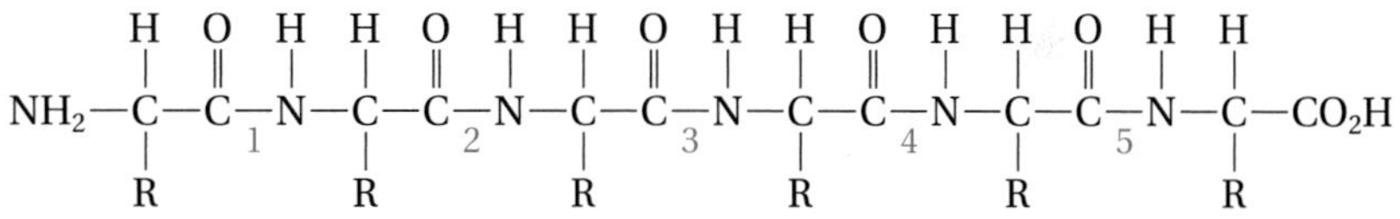

$$NH_2-\underset{R}{\overset{H}{C}}-\underset{1}{\overset{O}{C}}-\underset{R}{\overset{H}{N}}-\underset{}{\overset{H}{C}}-\underset{2}{\overset{O}{C}}-\underset{R}{\overset{H}{N}}-\underset{}{\overset{H}{C}}-\underset{3}{\overset{O}{C}}-\underset{R}{\overset{H}{N}}-\underset{}{\overset{H}{C}}-\underset{4}{\overset{O}{C}}-\underset{R}{\overset{H}{N}}-\underset{}{\overset{H}{C}}-\underset{5}{\overset{O}{C}}-\underset{R}{\overset{H}{N}}-\overset{H}{C}-CO_2H$$

Which peptide bonds could be hydrolyzed by (a) endopeptidases and (b) exopeptidases?

Activation of digestive peptidases from their proenzymes is a way to control levels of peptidase activity in digestion, but it is also a necessary protective device for the pancreas. Premature activation of trypsinogen and chymotrypsinogen occurs in one form of acute pancreatitis. In this disease, the patient's pancreas is destroyed by its own protein-digesting enzymes.

19.12 Blood clotting

AIM: To describe the function of zymogens and enzymes in the blood-clotting and clot-dissolution mechanisms.

Focus

Blood clot formation and dissolution are controlled by activation of zymogens.

Along with protein digestion, blood clotting is another important process that involves the activation of zymogens. Blood clotting must be precisely controlled, because blood must clot when it is shed, but it must not clot in the blood vessels. The early steps of the process leading to blood clotting are extremely complex. At the end of the process, the cleavage of peptides from the zymogen *prothrombin* to give *thrombin* sets the stage for actual clot formation.

Prothrombin $\longrightarrow$ Thrombin

(Zymogen)   (Active enzyme)

Thrombin is a peptidase that accelerates clot formation by catalyzing the

Trypsin: Anatomy of an Enzyme

Most of our knowledge about how enzymes work comes from detailed studies of a relatively few enzymes, one of which is trypsin. Over the years, biochemists have probed nearly every facet of trypsin's action and structure, ranging from its pH dependence to its tertiary structure (see figure, part a). Trypsin is a globular protein containing 245 amino acid residues with five disulfide bridges. Trypsin aids digestion by catalyzing the

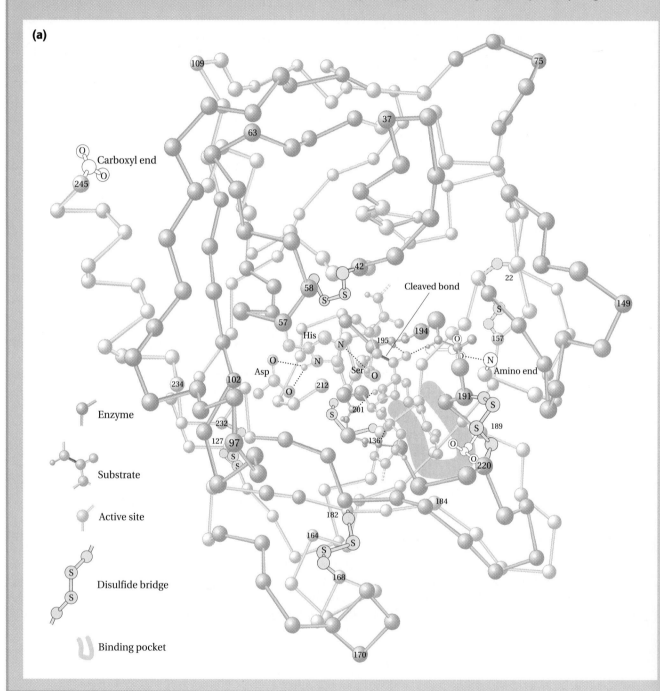

(a)

cleavage of peptide bonds of dietary proteins and peptides. However, the specificity of trypsin is such that it cleaves only peptide bonds in which the carboxyl group of the peptide bond to be cleaved comes from lysine and arginine, two residues whose side chains are positively charged at the enzyme's optimum pH.

As shown in part (b) of the figure, a lysine side chain of the polypeptide binds selectively to the active site pocket. It is drawn there by the negatively charged carboxylate ion of an aspartate residue at position 189 of the amino acid sequence of the enzyme's peptide chain. Except for lysine and arginine, the side chains of other amino acid residues do not bind because of a lack of complementarity of size or charge.

Once the substrate is bound in the proper orientation, the catalytic groups of the enzyme go to

work. Three side-chain groups on the enzyme seem to be necessary, but the one involved most directly is the hydroxyl group of serine 195. The cleavage of the substrate's peptide bond occurs in two stages.

First, the hydroxyl group of serine 195 displaces the amide nitrogen of the peptide bond of the substrate. As a result of the displacement, the substrate's peptide bond is broken, and an ester bond forms between enzyme and substrate. This intermediate is called an *acyl enzyme*. The piece of substrate protein just past the bond cleavage diffuses away.

Second, a water molecule at the enzyme active site hydrolyzes the ester bond of the acyl enzyme. This reaction is also catalyzed by the enzyme. The released peptide fragment floats away, and the free enzyme is able to do its catalytic work on yet another peptide bond.

Although the process is complex, one trypsin molecule can cleave more than 100 peptide bonds per second under ideal conditions. Several other enzymes that catalyze hydrolysis processes also employ a crucial serine hydroxyl group at the active site, although the binding sites of these enzymes are quite different from trypsin's. Included in this family of trypsin-like enzymes are chymotrypsin, thrombin, plasmin, and acetylcholinesterase.

(a) (left) The chain conformation of trypsin as deduced by X-ray crystallography. Only the positions of the alpha carbons of the amino acid residues of the enzyme's peptide chain are shown. The numbers refer to the positions of some of these residues in the enzyme's peptide chain. The shaded area at the lower right of the molecule is the active site pocket; part of a bound peptide substrate is shown in dark blue. The arrow marks the position of the catalytically important serine 195. Residues of histidine 57 and aspartate 102 also contribute to the speed of trypsin's catalysis of certain peptide bonds.

(b) (below) The mechanism of cleavage of a peptide bond at the active site of trypsin.

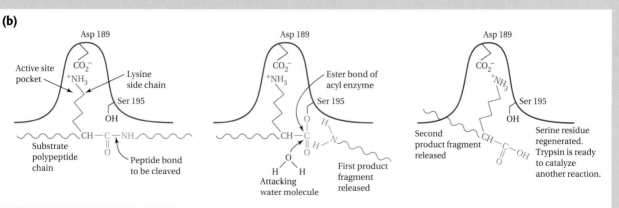

(b)

599

Fibrinogen is not a zymogen. Here the -*ogen* ending simply indicates that fibrinogen is the precursor molecule to fibrin, which is a fibrous protein and *not* an enzyme.

conversion of *fibrinogen* to *fibrin*. Clot formation begins when thrombin cleaves the peptide bonds of four Arg-Gly pairs in fibrinogen, a large (molar mass 340,000 g/mol), soluble blood protein. Cleavage of these Arg-Gly bonds releases four highly charged peptides that are responsible for maintaining the solubility of fibrinogen. Following the release of these peptides, the resultant fibrin molecules form long, insoluble fibers that precipitate as a mass held together by hydrogen bonds. The eventual formation of covalent bonds between fibrin molecules makes a stable clot.

After blood clots serve their function, they are dissolved and reabsorbed into the bloodstream. *Plasmin*, a peptidase activated from its zymogen *plasminogen*, is responsible for clot breakdown. A clot that prematurely breaks loose from the point of formation and enters a blood vessel may cause an *embolism*, a blockage of the flow of blood to tissues and organs. Embolisms also may be caused by air bubbles, clumps of bacteria, or other foreign matter, including bullets.

Especially severe damage or death results when the blockage causes a decreased oxygen supply to the brain (stroke), lung (pulmonary embolism), or heart (myocardial infarction). Potentially damaging blood clots are prevented from forming in the bloodstream of a normal individual by the presence of inhibitors that prevent completion of the clotting activation sequence. Among these inhibitors is heparin, a complex polysaccharide concentrated in the lungs and arterial cell walls (see Sec. 16.10). You may recall from the Case in Point earlier in this chapter that a stockbroker named Art suffered a heart attack. The administration of a solution containing an enzyme was among his treatments, which are discussed below in the Follow-up to the Case in Point.

PRACTICE EXERCISE 19.10

How is blood clot formation controlled by zymogen activation?

FOLLOW-UP TO THE CASE IN POINT: Enzyme therapy for heart attacks

Many heart attacks involve the formation of blood clots that clog (infarct) the arteries of the heart. These clots block the flow of blood into the heart, starving the heart tissues for oxygen. Failure to rapidly restore the flow of blood to the heart muscle results in permanent damage to the heart or even death. What can be done to dissolve the clots? In Art's case, the doctor ordered intravenous administration of a solution containing 100 mg of an enzyme called *tissue plasminogen activator* (TPA). The TPA solution was administered over a 2-hour period. TPA is a peptidase that activates plasminogen, the zymogen of plasmin. The plasmin then goes about

its natural work of dissolving the fibrin in blood clots that may be blocking the coronary arteries. Studies show that the survival of victims of coronary infarctions is significantly increased if steps are taken to dissolve the clots within about 2.5 hours of the onset of the heart attack. Thanks to the early administration of TPA and other excellent medical care, Art is now recuperating from his heart attack; his prospects for a full recovery are good. TPA is not the only enzyme used for the dissolution of clots in heart attacks. *Streptokinase*, a peptidase obtained from certain bacteria, also can be used with good results.

19.13 Antibiotics and other therapeutic drugs

AIM: To discuss the action of antibiotics and other therapeutic drugs as enzyme inhibitors.

Knowledge of how enzymes and inhibitors work has been of enormous benefit to medical scientists in their search for ways to cure or control diseases. **Antibiotics**—*drugs used to control harmful bacteria*—show how enzyme inhibitors can be put to therapeutic use. You may recall that sulfanilamide and other sulfa drugs work as antibiotics because they inhibit the synthesis of folic acid from *p*-aminobenzoic acid (A Closer Look: The Sulfonamide Antibiotics). Bacteria require folic acid to survive but cannot use the inhibitor to make it, so the microorganism dies from folic acid deficiency.

In good antibiotic design, the drug must interfere with the metabolism of the infecting organism but not with that of the infected individual. In 1928, Alexander Fleming, working at St. Mary's Hospital in London, was searching for a universal antibiotic. He discovered that a growth culture of a pus-producing bacterium disappeared in an area of the culture where green mold was growing. Figure 19.9 shows the mold component responsible—*penicillin*. In 1941, the drug was first applied to severe bacterial infections in humans. The results were dramatic. The patients who received penicillin made rapid, complete recoveries. It is now believed that one action of peni-

Figure 19.9
Below, penicillin mold and, right, the structure of penicillin and some of its commonly used derivatives.

cillin is to inhibit an enzyme that catalyzes the synthesis of cell walls of reproducing bacteria. The new cells have defective cell walls, the cell contents leak out, and the cell dies. Since the cell wall material is not present in humans, penicillin can be used to treat bacterial infections in people.

As we will learn in A Closer Look: The Toxicity of Pesticides, some enzyme inhibitors are very effective poisons.

Since the discoveries of sulfanilamide and penicillin, scores of new antibiotics have been developed. Indeed, the search for new antibiotics has stepped up in recent years because of the evolution of the so-called superbugs. Because of the widespread use of antibiotics, these highly antibiotic-resistant bacteria have evolved. Whatever antibiotics are used now and in the future, probably no antibiotic will ever be capable of winning the war against infecting bacteria. Instead, antibiotics fight a holding action that gives the infected organism's natural defenses time to mobilize against the invader.

Many therapeutic drugs besides antibiotics are also enzyme inhibitors. Several drugs introduced to lower blood pressure, including captopril and linisopril, are inhibitors of a peptidase, the *angiotensin-converting enzyme* (ACE).

$$HS-\underset{\underset{CH_3}{|}}{\overset{\overset{H}{|}}{C}}-CH_2-\overset{\overset{O}{||}}{C}-N$$

Captopril (Capoten)

$$\text{C}_6\text{H}_5-CH_2-CH_2-\underset{\underset{CO_2H}{|}}{\overset{\overset{H}{|}}{C}}-\overset{\overset{H}{|}}{N}-\underset{\underset{(CH_2)_4}{|} \; \underset{NH_2}{|}}{\overset{\overset{H}{|}}{C}}-\overset{\overset{O}{||}}{C}-N$$

Linisopril (Prinivil)

ACE catalyzes the release of the octapeptide angiotensin II from the decapeptide angiotensin I.

$$\text{Asp}-\text{Arg}-\text{Val}-\text{Tyr}-\text{Ile}-\text{His}-\text{Pro}-\text{Phe}-\text{His}-\text{Leu} \xrightarrow{\text{ACE}}$$

Angiotensin I

$$\text{Asp}-\text{Arg}-\text{Val}-\text{Tyr}-\text{Ile}-\text{His}-\text{Pro}-\text{Phe} + \text{His}-\text{Leu}$$

Angiotensin II

An assay for elevated levels of angiotensin-converting enzyme in the blood can be used to detect *sarcoidosis*—an accumulation of certain tissue called *granuloma*. Sarcoidosis usually affects the lungs, where it causes coughing and shortness of breath, but it may appear in other parts of the body such as the joints.

Angiotensin II causes the constriction of arteries. Since the heart still tries to pump the same amount of blood through the narrowed arteries, blood pressure increases. The ACE inhibitors help lower blood pressure by preventing ACE from converting angiotensin I to angiotensin II. You may recall from A Closer Look: HIV Protease and Its Inhibition that scientists are also working to find therapeutically useful inhibitors of enzymes produced by the human immunodeficiency virus (HIV), the causative agent for the acquired immunodeficiency syndrome (AIDS). Whether the quest will eventually lead to new drugs that delay the onset of AIDS or control the disease is still open to question.

The Toxicity of Pesticides

Substances that inhibit bacterial enzymes while leaving human enzymes unaffected have become important antibiotics, such as sulfa drugs. However, several powerful insecticides, including parathion and malathion, are inhibitors of human *and* insect enzymes.

$$CH_3CH_2O-\overset{\overset{\displaystyle S}{\|}}{P}-O-\!\!\!\left\langle\!\!\bigcirc\!\!\right\rangle\!\!-NO_2$$
$$\underset{CH_3CH_2O}{}$$

Parathion

$$CH_3O-\overset{\overset{\displaystyle S}{\|}}{P}-S-\underset{\underset{CH_2\overset{\overset{\displaystyle O}{\|}}{C}-OCH_2CH_3}{|}}{CH}-\overset{\overset{\displaystyle O}{\|}}{C}-OCH_2CH_3$$
$$\underset{CH_3O}{}$$

Malathion

Poisons, such as these insecticides, inhibit the enzyme acetylcholinesterase, which is responsible for a crucial event in the process of muscle contraction. Its inhibition results in paralysis of the body's muscles and eventually death. Inhibitors of this type were originally called "nerve gases," although nowadays many liquids and solids that have a similar effect are known. Muscle contraction begins with signals from the nervous system that are communicated to muscles by acetylcholine (the ester of acetic acid and the nitrogen-containing alcohol choline).

$$CH_3\overset{\overset{\displaystyle O}{\|}}{C}-OH + H-OCH_2CH_2-\overset{\overset{\displaystyle CH_3}{|}}{\underset{\underset{CH_3}{|}}{N^+}}-CH_3 \longrightarrow$$

Acetic acid   Choline

$$CH_3\overset{\overset{\displaystyle O}{\|}}{C}-OCH_2CH_2-\overset{\overset{\displaystyle CH_3}{|}}{\underset{\underset{CH_3}{|}}{N^+}}-CH_3 + H_2O$$

Acetylcholine   Water

Acetylcholine is normally released by nerve cells at their junction with muscle. It flows into the muscle cells and triggers a chain of events leading to muscle contraction. In order for muscles to alternately contract and relax, such as in the functioning of the

Crops are often sprayed with insecticides.

cardiorespiratory system, acetylcholine must be released and destroyed alternately. Acetylcholinesterase is the destroyer, promoting the hydrolysis of acetylcholine back to acetic acid and choline.

$$CH_3-\overset{\overset{\displaystyle CH_3}{|}}{\underset{\underset{CH_3}{|}}{N^+}}-CH_2CH_2O-\overset{\overset{\displaystyle O}{\|}}{C}CH_3 + H-OH$$

Acetylcholine   Water

Acetylcholinesterase ↓

$$CH_3-\overset{\overset{\displaystyle CH_3}{|}}{\underset{\underset{CH_3}{|}}{N^+}}-CH_2CH_2OH + CH_3\overset{\overset{\displaystyle O}{\|}}{C}-OH$$

Choline   Acetic acid

Without acetylcholinesterase, muscles cannot relax, and paralysis sets in. The action of parathion is similar to that of many pesticides. Parathion blocks the action of acetylcholinesterase by forming a stable phosphate ester with the hydroxyl group of a serine residue of the enzyme. This hydroxyl group is crucial for the enzyme to do its catalytic work, just as it is in trypsin.

$$CH_3CH_2O-\overset{\overset{\displaystyle S}{\|}}{P}-O-\!\!\!\left\langle\!\!\bigcirc\!\!\right\rangle\!\!-NO_2 + HO-\boxed{\text{Active enzyme}}$$
$$\underset{CH_3CH_2O}{}$$

↓

$$CH_3CH_2O-\overset{\overset{\displaystyle S}{\|}}{P}-O-\boxed{\text{Inactive enzyme}} + HO-\!\!\!\left\langle\!\!\bigcirc\!\!\right\rangle\!\!-NO_2$$
$$\underset{CH_3CH_2O}{}$$

A dose of parathion sufficient to paralyze the muscles of the respiratory system would be lethal.

SUMMARY

Enzymes, which except for ribozymes are proteins, serve as catalysts of chemical reactions that take place in biological systems. Enzyme-catalyzed reactions begin when a substrate molecule binds to an enzyme. In binding, the substrate fits a region of the folded peptide chains of the enzyme called the active site. The fit of substrate to enzyme occurs through complementarity of shape, charge, or weaker forces.

Conversion of bound substrates to products often results from chemical reactions involving the substrate and amino acid side chains of the enzyme. In many instances, enzymatic reactions involve a cofactor, a small organic molecule, or a metal ion contained within the enzyme active site.

Compared with their nonenzymatic counterparts, enzyme-catalyzed reactions are more rapid and occur at milder conditions of temperature and pH. They are also very specific; only one kind of product is formed by interaction of one kind of substrate with a particular kind of enzyme. Enzyme activity levels in body fluids may be used to diagnose

diseases. In certain illnesses, the patient's levels of blood serum enzymes are higher than those of a normal individual.

Enzymatic reactions in living systems may be subject to sensitive control. Such reactions may be speeded up by induction, activation of zymogens or proenzymes, and positive modulation. Enzymatic reactions may be slowed down by enzyme degradation, inhibition by a substrate mimic, or negative modulation. Enzymes whose activities are controlled by positive or negative modulation, or both, are usually allosteric, or second-site, enzymes. Allosteric enzymes are often pacemakers, appearing at the beginning of a sequence of several reactions and controlling the manufacture of products for the entire sequence. Protein digestion and blood clotting are examples of biological events in which zymogens must be converted to active enzymes.

Antibiotics permit people to fight off bacterial infections until their natural defenses can take effect. Many antibiotics are designed to inhibit the normal enzymatic processes of bacteria.

KEY TERMS

Active site (19.4)
Allosteric enzyme (19.10)
Antibiotics (19.13)
Apoenzyme (19.6)
Autoactivation (19.11)
Cofactor (19.6)
Competitive inhibitor (19.10)
Complementarity (19.4)
Degradation (19.9)

Endopeptidase (19.11)
Enzyme (19.1)
Enzyme activity (19.7)
Enzyme assay (19.7)
Enzyme-substrate complex (19.4)
Exopeptidase (19.11)
Feedback inhibition (19.10)
Holoenzyme (19.6)
Induced-fit model (19.4)

Inducible enzyme (19.9)
Induction (19.9)
International unit (19.7)
Lock-and-key model (19.4)
Modulator binding site (19.10)
Negative modulator (19.10)
Pacemaker enzyme (19.10)
Peptidase (19.2)
pH-activity profile (19.8)

pH optimum (19.8)
Positive modulator (19.10)
Proenzyme (19.11)
Protease (19.2)
Specificity (19.3)
Substrate (19.3)
Zymogen (19.11)
Zymogen granule (19.11)

EXERCISES

Enzymes (Sections 19.1–19.6)

19.11 Describe what an enzyme is and what it does.

19.12 Define the terms (a) *substrate* and (b) *specificity*.

19.13 What type of reaction is catalyzed by each of the following enzymes? (a) esterase (b) lactase (c) dehydrogenase (d) amylase

19.14 Suggest a name for the enzyme that catalyzes (a) the hydrolysis of maltose and (b) the hydrolysis of a lipid.

19.15 Explain how the lock-and-key model describes the action of an enzyme.

19.16 What is an *enzyme-substrate complex*?

19.17 What is the *active site* of an enzyme?

19.18 Explain why enzyme molecules are very large structures but their active sites are small.

19.19 Why do some enzymes need cofactors (coenzymes)?

19.20 List some typical cofactors.

Enzyme Assay (Section 19.7)

19.21 Define enzyme activity.

19.22 Why is an enzyme assay a valuable tool in clinical testing?

Effects of pH, Temperature, and Heavy Metals (Section 19.8)

19.23 What effect does an enzyme have on the energy of activation of a reaction?

19.24 Explain the term *optimal pH* of an enzyme.

19.25 The optimal temperature for a given enzyme is 37 °C. What will probably happen to the enzyme and its activity when (a) the temperature is lowered to 0 °C and (b) the temperature is raised to 100 °C?

19.26 Trypsin, an enzyme present in pancreatic juice, has an optimal pH of 8.2. Draw a pH-activity profile for trypsin.

Control of Enzyme Activity (Sections 19.9–19.13)

19.27 What is an *inducible* enzyme?

19.28 An enzyme that has its maximum activity at 37 °C is denatured by raising the temperature to 100 °C. The temperature is then lowered to 37 °C. Comment on the activity of the enzyme.

19.29 What is a competitive inhibitor, and how does it work?

19.30 How does feedback inhibition control enzyme activity?

19.31 What type of enzyme contains both a substrate binding site and a modulator binding site?

19.32 Distinguish between a competitive inhibitor and a negative modulator.

19.33 What is a *zymogen*?

19.34 Why is a peptidase usually required to convert a zymogen into an active enzyme?

19.35 Explain how soluble fibrinogen is converted to insoluble fibrin—the blood clot.

19.36 Describe how blood clots are dissolved.

19.37 How does penicillin kill bacteria?

19.38 Inhibitors of the enzyme acetylcholinesterase are extremely effective poisons. Briefly describe how they work.

Additional Exercises

19.39 Describe one way in which poisons work.

19.40 List at least three ways in which the activity of an enzyme is regulated.

19.41 Describe what is meant by the *specificity* of an enzyme.

19.42 What is the difference between (a) urea and urease, (b) lactose and lactase, and (c) pepsin and pepsinogen?

19.43 Explain the relationship of a zymogen to the corresponding enzyme.

19.44 Discuss how the concentration of enzymes in blood serum can give useful information about specific disease conditions.

19.45 Define (a) competitive inhibitor, (b) positive modulator, (c) allosteric enzyme, and (d) feedback inhibition.

SELF-TEST (REVIEW)

True/False

1. The strength of the enzyme-substrate complex is determined entirely by the fit of the substrate into the active site.

2. The shape of a substrate molecule does not affect the rate of an enzyme-catalyzed reaction.

3. In the clinical assay for blood serum enzymes, disease or tissue injury is indicated by increased enzyme activity as compared with normal.

4. An increase in the temperature of an enzyme always increases its activity.

5. Both competitive inhibitors and negative modulators affect the active sites of the enzymes they regulate.

6. Trypsinogen is the zymogen of the endopeptidase trypsin.

7. An exopeptidase hydrolyzes the terminal peptide bonds of a peptide chain.

8. Antibiotics are produced by the body to combat bacterial infections.

9. The product of a series of consecutive enzyme-catalyzed reactions may function in an allosteric manner in feedback inhibition.

10. Plasmin is responsible for the formation of blood clots.

Multiple Choice

11. A sample of blood serum to be assayed for enzyme Z is mixed with a known amount of substrate for this enzyme. Which of the following conditions must be controlled?
    (a) pH   (b) amount of blood serum
    (c) temperature   (d) all of these

12. The demand for a particular substance in the body varies widely with time. The production of this substance is most likely controlled by a(n)
    (a) changeable enzyme.   (b) allosteric enzyme.
    (c) proenzyme.   (d) accelerin enzyme.

13. A patient's lactate dehydrogenase (LDH) and glutamic-oxaloacetic transaminase (SGOT) levels are both elevated. The patient's creatine phosphokinase (CPK) level is normal. Which of the following is a probable diagnosis? (Use Table 19.1.)
    (a) myocardial infarction   (b) hepatitis
    (c) pancreatic disorder   (d) leukemia

14. Which of the following substances is most directly involved in the formation of a blood clot?
    (a) fibrinogen   (b) prothrombin
    (c) heparin   (d) thrombin

15. In an enzyme-catalyzed reaction, a substrate is changed to products
    (a) at a number of different locations on the enzyme.
    (b) at the active site of the enzyme.
    (c) after the enzyme-substrate complex is formed.
    (d) More than one are correct.

16. The enzyme maltase would show specificity for
    (a) all carbohydrates.
    (b) all disaccharides.
    (c) maltose.
    (d) more than one are correct.

17. Which of the following statements is *not* generally true of enzymes?
    (a) They are not changed by the reaction they catalyze.
    (b) They are proteins.
    (c) They shift the equilibrium toward the product side of the equation.
    (d) They speed up chemical reactions.

18. The activity of some enzymes is dependent on the availability of necessary
    (a) cofactors.   (b) inhibitors.
    (c) antigens.   (d) inductors.

19. Peptidase proenzyme can be activated by the action of
    (a) cofactors.   (b) a peptidase.
    (c) an inductor.   (d) an acinar cell.

20. In the lock-and-key model of enzyme action, a competitive inhibitor
    (a) is like a key that will not fit the lock.
    (b) is like the wrong lock for the available key.
    (c) is like a key that will not work but still fits the lock.
    (d) more than one are correct.

21. The function of an antibiotic such as penicillin is best described as
    (a) phagocytosis.   (b) positive modulation.
    (c) enzyme inhibition.   (d) immune response.

22. Enzyme-catalyzed reactions in the cell can be slowed down by
    (a) negative modulation.   (b) enzyme degradation.
    (c) inhibition.   (d) all of these.

Nucleic Acids

The Molecular Basis of Heredity

These twins are
genetically identical.

CHAPTER OUTLINE

CASE IN POINT: Treatment for an inherited disease

20.1 Nucleic acids, nucleotides, and nucleosides

20.2 The DNA double helix

20.3 The central dogma

A CLOSER LOOK: HIV Reverse Transcriptase

20.4 Replication

20.5 Genes

A CLOSER LOOK: The Human Genome Project

A CLOSER LOOK: DNA Fingerprinting

20.6 Classes of RNA

20.7 Transcription

20.8 Translation

20.9 The genetic code

20.10 Gene mutations and molecular diseases

A CLOSER LOOK: Oncogenes and Tumor Suppressor Genes

20.11 Recombinant DNA and gene therapy

FOLLOW-UP TO THE CASE IN POINT: Treatment for an inherited disease

Mice beget mice, cats beget cats, and people beget people. The observation that organisms reproduce their own species is widely known and self-evident. Yet the mystery of how this occurs is one of the toughest cases that scientific sleuths have had to crack. The study of the chemical composition and biological functions of the molecules of heredity—*nucleic acids*—has already led to enormous advances in our understanding of human heredity and the origins and treatment of inherited diseases. Our Case in Point, a true story, illustrates one of these advances.

CASE IN POINT: Treatment for an inherited disease

In September 1990, a 4-year-old girl named Ashanthi was treated by a revolutionary new method for the cure of inherited diseases. When she was a baby, Ashanthi's parents and physician were concerned about her repeated infections. Eventually, they learned that the infections resulted from a lack of the enzyme adenosine deaminase (ADA). ADA-deficient patients are subject to all kinds of infections. Because even a simple infection can be fatal, ADA-deficient patients traditionally have been kept in isolation. What revolutionary treatment was Ashanthi given, and what was its outcome? We will find out in Section 20.11.

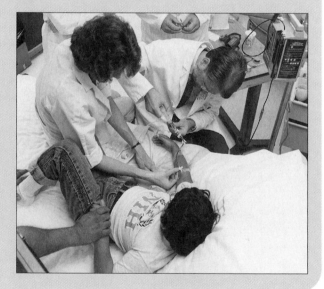

Ashanthi received a newly developed treatment for ADA deficiency when she was very young.

20.1 Nucleic acids, nucleotides, and nucleosides

AIMS: To give the names and one-letter symbols for the five major nitrogen bases found in nucleic acids. To state two differences between the molecular composition of DNA and RNA. To name and draw structures of nucleosides and nucleotides. To describe the bonding that joins nucleotides together in nucleic acids.

Focus

Nucleotides are the fundamental building blocks of nucleic acids.

Cells contain two types of **nucleic acids: deoxyribonucleic acid (DNA)** *and* **ribonucleic acid (RNA).** *DNA and RNA molecules are* **polynucleotides—** *polymers composed of many repeating units of nucleotides. Each* **nucleotide** *consists of a nitrogen base, a sugar unit, and a phosphate group attached to the sugar unit. The combination of a base and a sugar unit makes a* **nucleoside.** Adding a phosphate group to the sugar unit of a nucleoside makes a nucleotide.

$$\boxed{\text{Base}} + \boxed{\text{Sugar}} \longrightarrow \boxed{\text{Base}} - \boxed{\text{Sugar}}$$

A nucleoside

$$\boxed{\text{Base}} - \boxed{\text{Sugar}} + \boxed{\text{Phosphate}} \longrightarrow \boxed{\text{Base}} - \boxed{\text{Sugar}} - \boxed{\text{Phosphate}}$$

A nucleoside   A nucleotide

Structure of the sugar units

The sugar unit of the nucleotides strung together to make RNA is β-D-ribose—hence the name *ribonucleic acid.* The sugar unit of DNA is β-D-2-deoxyribose—hence the name *deoxyribonucleic acid.*

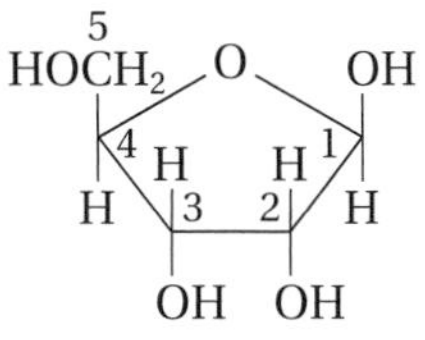

β-D-Ribose 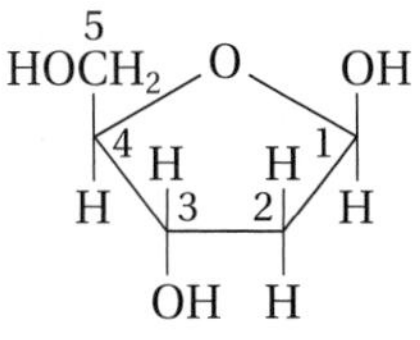   β-D-2-Deoxyribose

Structure of the base units

The molecular mass of a DNA molecule can be as high as several billion. RNA molecules are much smaller, often falling in the range of 20,000 to 40,000.

Four different nitrogen bases (heterocyclic amines) are found in DNA. (Recall from Section 15.2 that amines are weak bases.) Two of these bases, adenine (A) and guanine (G), are derivatives of purine; the other two, thymine (T) and cytosine (C), are derivatives of pyrimidine (Fig. 20.1). *Adenine and guanine are the* **purine bases** *of DNA; thymine and cytosine are the* **pyrimidine bases** *of DNA.* Except for thymine, these same bases are found in RNA. The pyrimidine base uracil (U) is found instead of thymine in RNA.

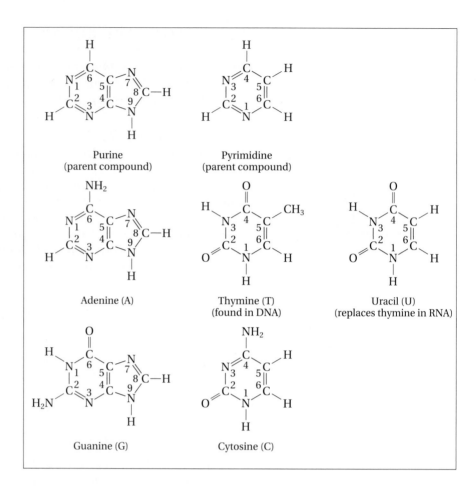

Figure 20.1
Structures of the nitrogen bases found in DNA and RNA.

Structure of nucleosides

The base and sugar units of nucleosides are held together by a covalent bond between a nitrogen of the purine or pyrimidine base and a ring carbon of the sugar unit, as shown for the deoxyribonucleosides of DNA in Figure 20.2. Ribonucleosides (the nucleosides found in RNA) are similar in structure to the deoxyribonucleosides, except that ribose rather than deoxyribose is the sugar, and uracil replaces thymine. The name for these compounds—*nucleosides*—reflects their role as constituents of nucleic acids (*nucleo-*) and their carbohydrate character (*-oside,* as in *glycoside*).

Since both the ring atoms of the sugar unit and the nitrogen bases of nucleosides are numbered, the conventional way to name nucleosides is to number the atoms of the base unit as 1, 2, 3, 4, and so forth. Carbons of the sugar ring are numbered $1'$, $2'$, $3'$, $4'$, and $5'$ (read as "one prime," "two prime," and so forth). In Figure 20.2 we see that purine nucleosides are formed by a covalent bond between nitrogen 9 of the base and carbon $1'$ of the sugar unit. The base and sugar of pyrimidine nucleosides are joined together by a covalent bond between nitrogen 1 of the base and carbon $1'$ of the sugar.

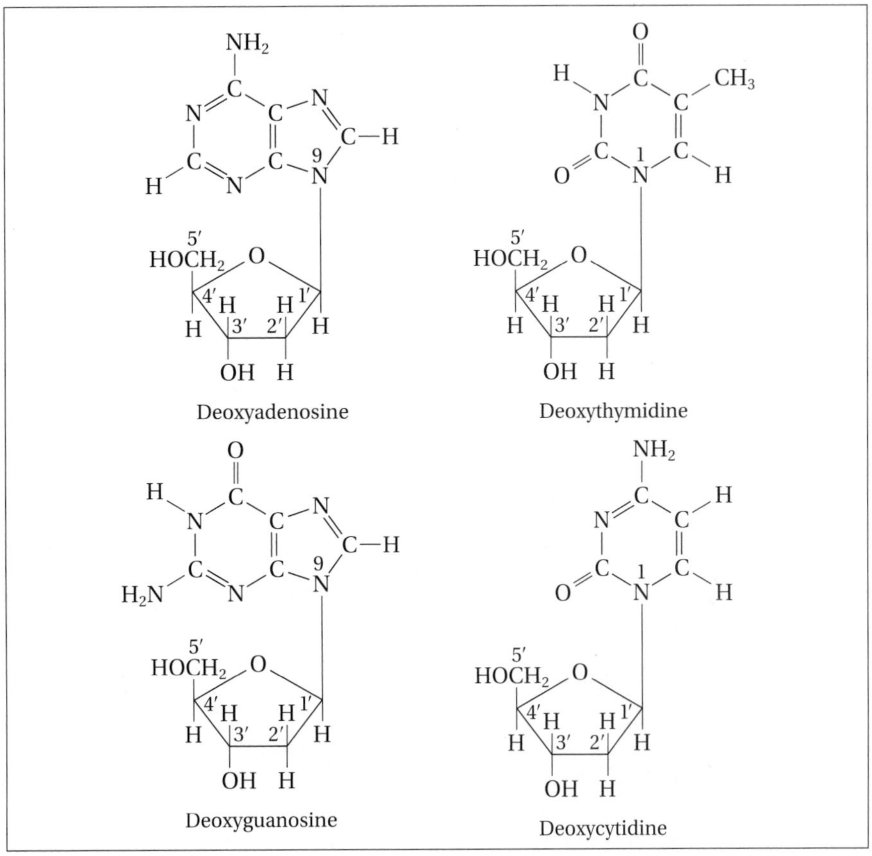

Figure 20.2
Structures of the nucleosides found in DNA.

PRACTICE EXERCISE 20.1

Draw the structure of the nucleoside formed by the combination of the nitrogen base cytosine and the sugar β-D-ribose. Name this nucleoside.

Structure of nucleotides

Addition of a phosphate group to a sugar hydroxyl group of a nucleoside forms a *nucleotide.* In other words, nucleotides are phosphoric acid esters of nucleosides. (See Section 14.9 to review phosphate esters.) In nucleotides of DNA and RNA, the phosphate ester is formed at the 5′-hydroxyl group of the nucleoside. Table 20.1 summarizes the names of the nucleosides and nucleotides of DNA and RNA. In the following structure of the nucleotide, the base is A, T, G, or C in DNA and A, U, G, or C in RNA.

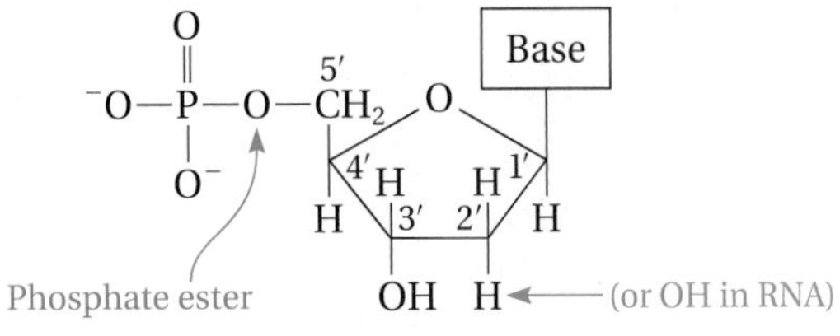

A nucleotide

Table 20.1 Names of Bases, Nucleosides, and Nucleotides Found in DNA and RNA

Base	Nucleoside	Nucleotide
DNA		
adenine (A)	deoxyadenosine	deoxyadenosine 5′-monophosphate (dAMP)
guanine (G)	deoxyguanosine	deoxyguanosine 5′-monophosphate (dGMP)
thymine (T)	deoxythymidine	deoxythymidine 5′-monophosphate (dTMP)
cytosine (C)	deoxycytidine	deoxycytidine 5′-monophosphate (dCMP)
RNA		
adenine (A)	adenosine	adenosine 5′-monophosphate (AMP)
guanine (G)	guanosine	guanosine 5′-monophosphate (GMP)
uracil (U)	uridine	uridine 5′-monophosphate (UMP)
cytosine (C)	cytidine	cytidine 5′-monophosphate (CMP)

Figure 20.3 shows how DNA and RNA molecules consist of nucleotides linked together through phosphate ester bridges between the 3′-hydroxyl group of one nucleotide and the 5′-hydroxyl group of the next nucleotide in the chain. *Since the phosphate ester bridges holding the nucleotides together each contain two phosphate ester linkages, these bridges are called* **phosphodiesters.** Notice that even though two of the four oxygens attached to the phosphorus of each bridge are tied up as phosphate esters and one is present in a phosphorus-oxygen double bond, one oxygen is free to lose a proton as in the ionization of other phosphoric acid esters (see Sec. 14.9).

$$\underset{\underset{\text{OH}}{\overset{\text{O}}{\|}}}{\text{RO}-\text{P}-\text{OR}} \rightleftharpoons \underset{\underset{\text{O}^-}{\overset{\text{O}}{\|}}}{\text{RO}-\text{P}-\text{OR}} + \text{H}^+$$

It is the presence of many such dissociating groups that gives DNA and RNA their highly acidic character.

PRACTICE EXERCISE 20.2

A phosphate group is added to the 5′-hydroxyl group of the nucleoside in Practice Exercise 20.1 to form a nucleotide. What is the structure and name of this nucleotide?

Shorthand structures for RNA and DNA nucleotide sequences

Most DNA and RNA molecules are too long to write out in full, so biochemists often use a shorthand form. The shorthand shows the sequence of their bases. The sugar units and phosphate groups are identical in all the nucleotides. Thus we can describe the structure of an RNA molecule by ignoring the sugars and phosphodiester bridges and writing only the sequence of the nitrogen bases. By convention, we start at the left with the end of the molecule that has a free 5′-hydroxyl group (not attached to another nucleotide) and work toward the end that has the free 3′-hydroxyl

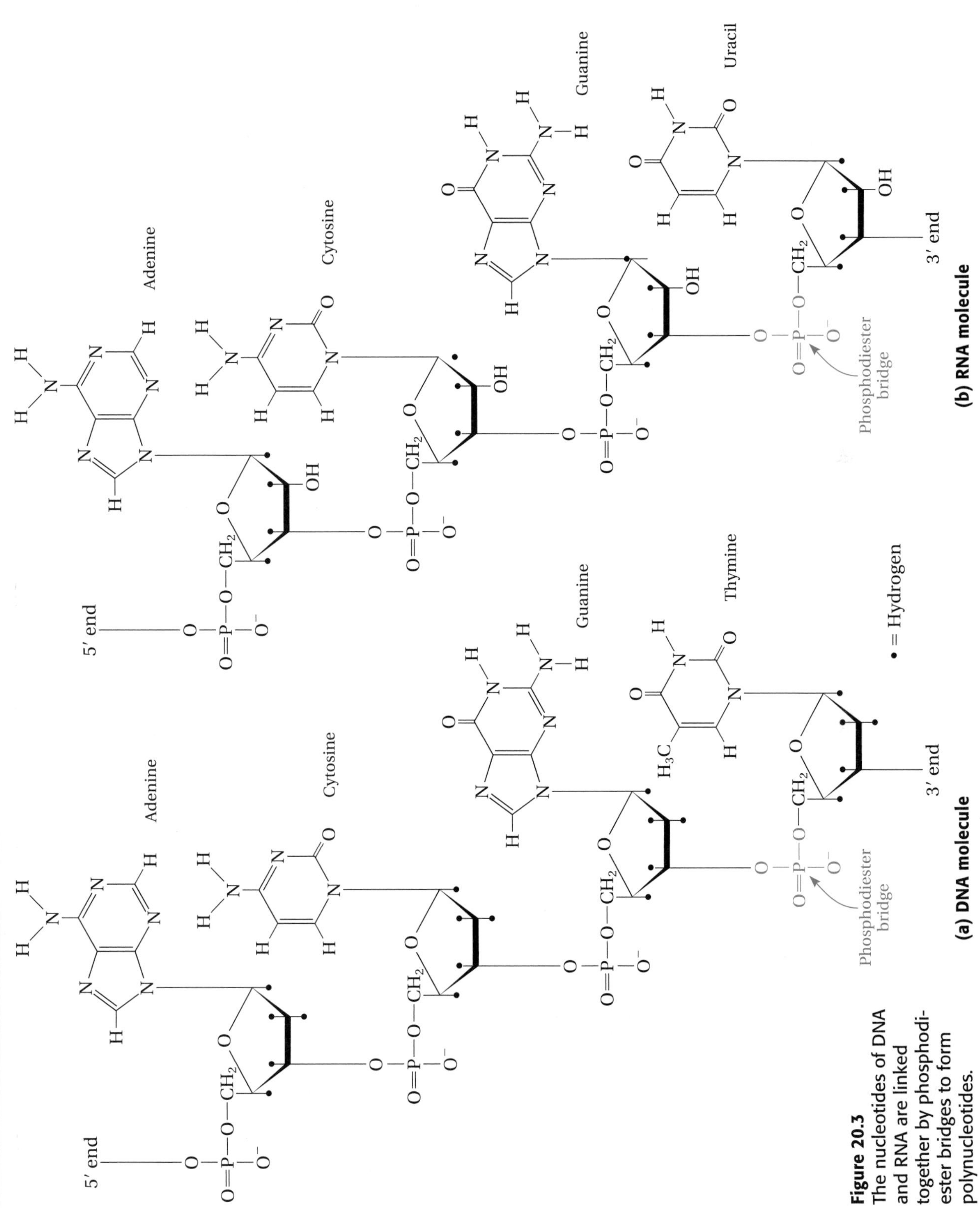

Figure 20.3
The nucleotides of DNA and RNA are linked together by phosphodiester bridges to form polynucleotides.

group (not attached to another nucleotide). Another way of saying this is that we work in the 5′ to 3′ direction.

U—A—G—C—U—G—C—C
5′ ————————————→3′

We can write abbreviations for the base sequence in DNA by putting a lowercase *d* in front of the base sequence. This indicates that all the sugar units of the sugar-phosphodiester backbone of the molecule are deoxyribose in DNA.

dA—A—T—G—T—C—A—C
5′ ————————————→3′

EXAMPLE 20.1   **Writing base sequence abbreviations**

Write the base sequence abbreviation of a strand of DNA composed of the bases adenine, thymine, guanine, guanine, and cytosine. Assume adenine is not bonded at the 5′ position.

SOLUTION

Use the one-letter abbreviation of each base starting with adenine that has the free 5′ end. Since this is DNA, begin the sequence with a lowercase *d*.

dA—T—G—G—C

PRACTICE EXERCISE 20.3

Translate the following base sequence abbreviations:

(a) A—C—G—U   (b) dT—A—C—G

PRACTICE EXERCISE 20.4

What is a phosphodiester bridge, and where is it found?

20.2 The DNA double helix

AIM: To discuss the significance of A = T and G = C as it relates to the formation of the double-helical structure of DNA.

Focus

Strands of DNA form a double helix.

The three-dimensional structure of DNA was unknown until 1953, when James D. Watson and Francis Crick proposed that these polymeric molecules form a **DNA double helix**—*a double-stranded DNA molecule in which the deoxyribose-phosphate backbones of the DNA strands are wound in a helix.* Figure 20.4 depicts the structure of the now-famous DNA double helix. In the double helix, adenine can form two hydrogen bonds to a

Figure 20.4 (below)
The DNA double helix.

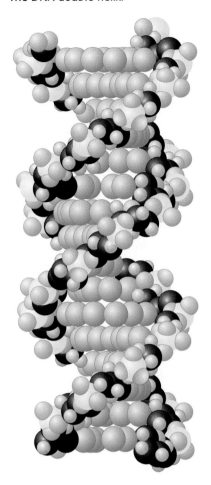

Thymine   Adenine   Cytosine   H   Guanine

(a) AT pair   **(b) GC pair**

Figure 20.5
Complementary base pairing:
(a) adenine (A) forms two hydrogen bonds with thymine (T);
(b) guanine (G) forms three hydrogen bonds with cytosine (C).

In 1962, James D. Watson and Francis Crick shared the Nobel prize in medicine and physiology with Maurice Wilkins for their discovery of the structure of DNA.

thymine on the opposing DNA chain (Fig. 20.5). Guanine can form three hydrogen bonds to cytosine on an opposing chain. *This specific matching of opposing base pairs by hydrogen bonding is termed* **complementary base pairing.** In any DNA molecule, A = T and G = C.

The strands of DNA constituting the DNA double helix run in opposite directions, much as traffic on a two-way street moves in opposite directions. In other words, *the two DNA strands of the double helix are* **antiparallel.** The sugar-phosphate units of the nucleotide are on the outside of the DNA molecule, with the complementary base pairs stacked neatly inside the molecule. The hydrogen bonds between the base pairs hold the double helix together. A and T each constitute about 30% of the bases of human DNA. The bases C and G each constitute about 20%. (Since there are only four different bases in DNA, these numbers always add up to 100%.)

PRACTICE EXERCISE 20.5

Describe the DNA double helix in your own words.

20.3 The central dogma

***AIMS:** To name the three processes that constitute the central dogma of molecular biology. To describe a unique characteristic of all retroviruses.*

DNA is the bearer of genetic information in all living organisms. The entire basis of information storage and transmission in cells is embodied in three steps:

1. *DNA is replicated.* **Replication** *means copying;* it is the way DNA is supplied to new cells formed by cell division and to a new organism formed by reproduction.

2. *The information contained in DNA is passed to a form of RNA called* **messenger RNA** *(mRNA) by* **transcription** *("rewriting").*

3. The mRNA directs **translation,** *the synthesis of proteins.* In everyday parlance, the word *translate* means "to change from one language to another." The meaning in molecular biology is quite similar: The "language" of DNA that is transcribed to mRNA is translated to a completely different language—the primary structure of a protein.

Replication, transcription, and translation can be summarized as

$$\text{DNA} \xrightarrow{\text{Replication}} \text{DNA} \xrightarrow{\text{Transcription}} \text{mRNA} \xrightarrow{\text{Translation}} \text{Protein}$$

Replication, transcription, and translation are so important to an understanding of the relationship between heredity and protein synthesis that they have been called the **central dogma** *of molecular biology.* The enormous diversity of living creatures ultimately resides in the central dogma. The information needed to make enzymes and other proteins is stored in DNA. The base composition of the DNA of various species is different, but the base composition of all the DNA from an individual is the same. Differences among species result from differences in the proteins constructed with this information. The cells of trees contain enzymes to make cellulose for cell walls, but humans lack these enzymes. At least some of the proteins in cats—hemoglobin, for example—are not found in bacteria. Moreover, the hemoglobin of cats differs slightly in structure from the hemoglobin of mice and humans. Differences in proteins are also found among individuals of the same species. Your eyes may be brown and your friend's eyes blue because your friend lacks enzymes needed to manufacture the pigment that makes eyes brown. It is the immense number of possible variations in our protein makeup that turns each of us into a unique individual.

Retrovirus
retro (Latin): backward
virus (Latin): poison, slime

The steps of replication, transcription, and translation are followed in every kind of organism except the *retroviruses.* In **retroviruses,** *the genetic material is RNA.* In order for the retroviruses to reproduce, the message contained in the viral RNA must be transcribed—or more accurately, *reverse transcribed*—into DNA.

$$\text{RNA} \xrightarrow{\text{Reverse transcription}} \text{DNA}$$

Once formed, the retroviral DNA is transcribed and translated much as it is for other organisms. The viruses responsible for many human diseases, such as the common cold, poliomyelitis, measles, influenza, and some tumors, are retroviruses. The human immunodeficiency virus (HIV), which causes AIDS, is also a retrovirus. Because of its centrality in the reproduction of HIV, *reverse transcriptase*, the enzyme that catalyzes the transcription of the virus's RNA into DNA, is the target of researchers seeking treatments for AIDS. This work is described in A Closer Look: HIV Reverse Transcriptase.

20.4 Replication

AIM: To outline the process of replication.

Focus

DNA is replicated on a DNA template.

Watson and Crick also proposed a model for DNA replication. They suggested that each strand of DNA in the double helix acts as a **template**—*a pattern*—for the synthesis of its complement, the other strand. Since DNA is double-stranded, complementary replication would produce two DNA molecules, each a double helix containing a strand of the original DNA and a new strand complementary to it. In other words, suppose the structure of the double helix is

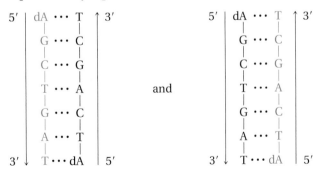

Then complementary replication of each strand produces

This type of replication is called *semiconservative replication* because the double-stranded DNA in each round of replication contains one strand from the previous round (the conserved strand) and one new strand. Figure 20.6 illustrates semiconservative replication in a slightly different way. The growth of the DNA chain requires **DNA polymerase**—*a large enzyme that catalyzes the formation of DNA.* The synthesis of DNA uses as starting mate-

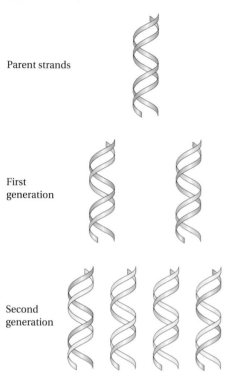

Parent strands

First
generation

Second
generation

Figure 20.6 (above)
Semiconservative replication.
In semiconservative replication,
the DNA of each succeeding
generation of cells contains one
strand from the generation before
it and one new strand that is
complementary to the old strand.

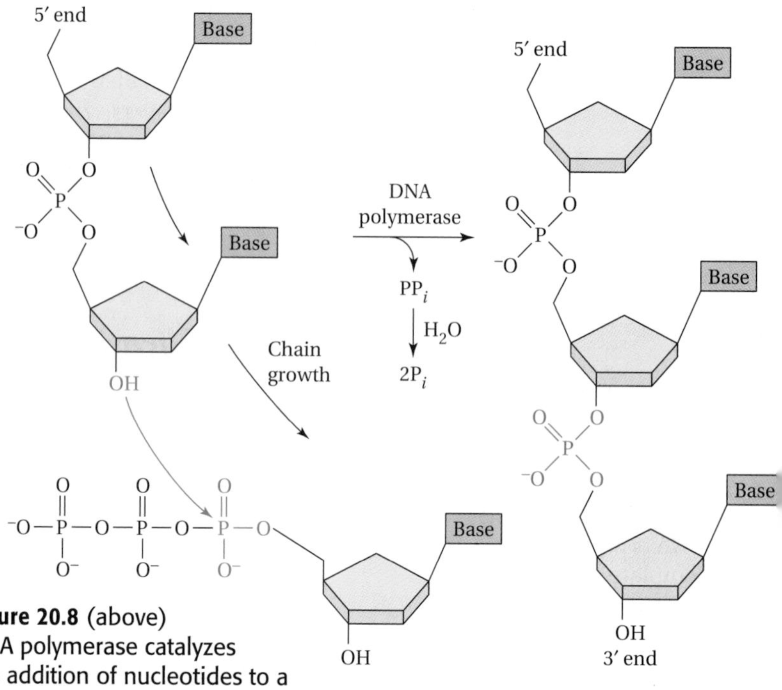

Figure 20.7 (right)
General structure of
nucleoside 5'-triphosphates.
In the nucleoside triphosphates
required for the synthesis of DNA,
the sugar units are 2'-deoxyribose
and the bases are A, T, G, and C.
The nucleoside 5'-triphosphates are
named after the parent nucleoside
(Table 20.1) by adding the word *5'-
triphosphate*—as in deoxyguanosine
5'-triphosphate (dGTP) or adenosine 5'-triphosphate (ATP).

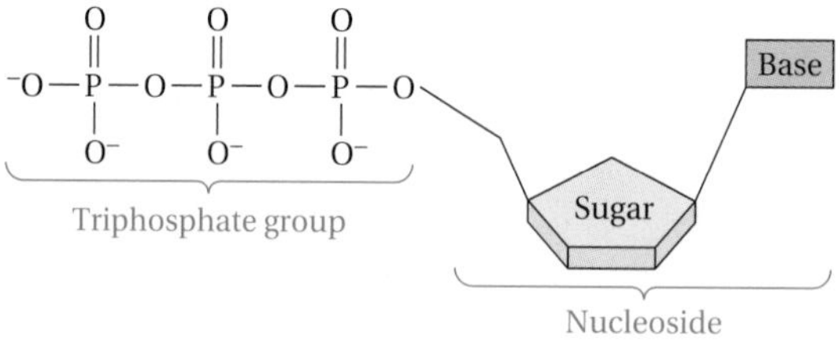

Figure 20.8 (above)
DNA polymerase catalyzes
the addition of nucleotides to a
growing DNA chain. A "high energy"
phosphate bond is broken in the formation of the phosphodiester linkage,
releasing a pyrophosphate ion (PP_i). Notice that the direction of chain growth
is in the 5' to 3' direction. Recall from Section 14.9 that triphosphates are
anhydrides of phosphoric acid and therefore phosphorylate alcohols to make
phosphate esters. Thus the esterification reaction is spontaneous and proceeds with a large release of free energy. The hydrolysis of the PP_i to $2P_i$ also
releases free energy, which further aids in driving esterification to completion.

rials the nucleoside 5'-triphosphates of the bases found in DNA (Fig. 20.7).
DNA must be present for new DNA to be formed, which agrees with Watson
and Crick's idea that DNA serves as a template for DNA synthesis. DNA
strands elongate when an incoming nucleoside triphosphate phosphory-
lates the 3'-hydroxyl group of ribose at the end of a growing DNA chain (Fig.
20.8). Stated another way, the direction of chain growth is always from the
5' end to the 3' end of the elongating DNA molecule.

A Closer Look

HIV Reverse Transcriptase

HIV is a retrovirus—it contains RNA as the instructions for its reproduction. In addition to the protease discussed in A Closer Look: HIV Protease and Its Inhibition, on page 593, the HIV virus produces a reverse transcriptase (see figure). HIV reverse transcriptase is needed to make DNA from the RNA. The information in the DNA is then transcribed into mRNA. These instructions are in turn translated into proteins needed to assemble a mature, infectious virus. Like HIV protease, HIV reverse transcriptase is an important target for medical drugs to slow down or prevent the reproduction of HIV. Zidovudine (AZT) and DDI are two drugs used in the treatment of AIDS. These drugs are designed to impede HIV reverse transcriptase from synthesizing DNA from viral RNA. As the chemical structures of these drugs show, they are chemical modifications of the deoxynucleosides normally found in DNA.

Zidovudine and DDI bind to the active site of HIV reverse transcriptase. They are added to a growing DNA copy of the viral RNA. However, they lack a 3′-hydroxyl group. This group is needed to lengthen the DNA chain, so chain growth stops before the DNA has reached its full length. Incomplete, defective proteins are made, and infectious viruses cannot be assembled. The deoxynucleoside analogues are helpful to patients with symptoms of AIDS, but none cures the disease. High toxicity is one reason for the limited effectiveness of these antiviral drugs; toxicity limits the size of the doses that can be administered, so not all viral reproduction can be stopped. More important, rapid changes (mutations) in HIV produce versions of the reverse transcriptase that are insensitive to the drugs.

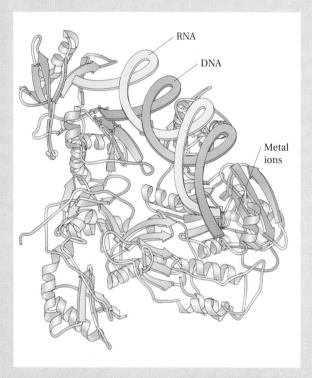

The structure of HIV reverse transcriptase. The structure is a dimer consisting of a subunit of molar mass 66,000 g/mol (blue) and a subunit of molar mass 51,000 g/mol (green). The molecule is shown with viral RNA (gold) and a strand of DNA (brown) bound to the enzyme. Two metal ions (blue spheres) are thought to be important in the reverse transcription of RNA to DNA.

3′-Azido-3′-deoxythymidine
(Zidovudine, AZT)

2′, 3′-Dideoxyinosine (DDI)

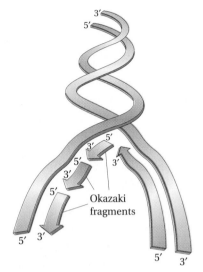

Figure 20.9
DNA is replicated in the 5′ to 3′ direction. Since the strands of DNA run in opposite directions, one complementary strand is synthesized in a continuous chain in the 5′ to 3′ direction. The other is also synthesized in the 5′ to 3′ direction, but in sections of about 1000 nucleotides that are connected by an enzyme. The short sections are called *Okazaki fragments.*

Although replication seems to proceed essentially by the method described by Watson and Crick, there are other considerations. The bases in the double helix are inside the DNA molecule. How are the bases of the strands of DNA exposed so that replication can occur? It appears that the DNA strands unwind as replication proceeds. *Special proteins, termed* **unwinding proteins,** *catalyze the unwinding of double-stranded DNA.*

Another problem concerns the direction of growth of the new DNA chains. As we have seen, the strands of DNA are antiparallel, yet when double-stranded DNA unwinds during replication, only one new strand can grow continuously in the 5′ to 3′ direction. How is the other new strand formed? In 1968, R. Okazaki showed that the other new strand is formed in the 5′ to 3′ direction in pieces that are subsequently connected by the action of an enzyme (Fig. 20.9). *These pieces of DNA, which have only a fleeting existence independent of each other, are called* **Okazaki fragments.**

EXAMPLE 20.2   Writing a complementary DNA strand

Write the abbreviated base sequence (written 5′ to 3′) for the strand of DNA complementary to the following strand:

dA—G—G—A—C—T—C—C

SOLUTION

In DNA replication, A pairs with T, G pairs with C. Using the given strand as a template, and remembering that the complementary strand is antiparallel, gives

3′ dT—C—C—T—G—A—G—G 5′

Written 5′ to 3′, the complementary strand is

5′ dG—G—A—G—T—C—C—T 3′

PRACTICE EXERCISE 20.6

Complete the following segment of a DNA double helix. Write symbols for the missing bases, and draw the hydrogen bonds.

20.5 Genes

AIMS: To characterize a gene. To describe the function of exons and introns.

Focus

A gene is a DNA segment that directs the synthesis of a single kind of protein.

Table 20.2 Number of Nucleotide Base Pairs in the DNA of Various Organisms

Organism	Nucleotide base pairs (millions)
Homo sapiens	5500*
birds	2000
higher plants	2300
sponges	100
fungi	20
bacteria	2†
complex virus	0.17
simple virus	0.05

*1 million genes.
†100 genes.

Early geneticists defined a *gene* as the smallest unit of heredity that would lead to an observable trait. In 1943, George Beadle and Edward Tatum made the connection between heredity and protein synthesis when they showed that hereditary differences are the result of differing abilities to synthesize proteins. They then suggested a refined definition of a gene as the unit of heredity that directs the synthesis of an enzyme. This definition is known as the *one-gene, one-enzyme hypothesis.*

Two later discoveries somewhat changed this definition. First, scientists discovered that the synthesis of all proteins, not just enzymes, is governed by genes. Second, they found that information for protein synthesis resides in the number and sequence of bases in the DNA molecule. *We can regard a* **gene** *as a segment of the DNA molecule that directs the synthesis of one kind of protein molecule.* The proteins expressed (produced) in protein synthesis are enzymes, hemoglobin, collagen, and so forth.

DNA molecules are among the largest molecules known. For example, the DNA found in one human cell contains about 5.5 billion nucleotide base pairs that make up perhaps 1 million genes (Table 20.2). If we could make each nucleotide base pair of human DNA one book page and make the pages into books of 1000 pages each, our library would contain 5.5 million volumes. The number of nucleotide base pairs of a bacterial DNA is much smaller, about 2 million, but they still would occupy 2000 volumes. Despite the enormous number of base pairs, scientists are determined to learn the base sequence of human DNA, as discussed in A Closer Look: The Human Genome Project.

Introns and exons

The genes of eukaryotic cells differ from those of prokaryotic cells in an important respect. Essentially 100% of the bases of DNA in bacteria (prokaryotes) are used as a source of information for making proteins or for regulating the making of proteins. In contrast, only about 5% of the DNA of human cells (eukaryotes) is used in protein synthesis or its regulation. *The base sequences of the genes that contain the information (code) for protein synthesis in eukaryotic cells are called* **exons.** The remaining 95% of DNA of eukaryotic cells interrupts the coding sequences of genes (Fig. 20.10). *The noncoding base sequences that interrupt eukaryotic genes are called intervening sequences or* **introns.** A large human gene, such as the gene for the synthesis of collagen, may have as many as 50 introns interspersed between coding segments of DNA. Introns often contain much more DNA than the genes they interrupt. Scientists do not understand why human genes contain so much "junk" DNA. The base sequences of introns are somewhat similar among families and ethnic

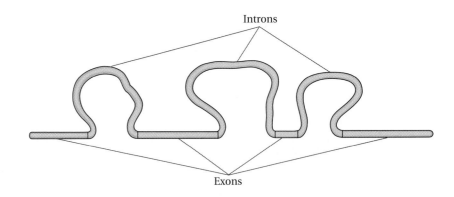

Figure 20.10
In eukaryotic cells, the base sequences of the DNA that code for protein synthesis are called *exons.* The intervening base sequences, known as *introns,* are noncoding.

groups. Since each individual inherits portions of these base sequences from both parents, a person's DNA structure is characteristic of that person. The genetic individuality of people makes it possible to identify a

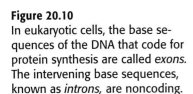

A Closer Look

The Human Genome Project

The long-term goal of scientists involved in a venture called the *Human Genome Project* (HGP) is the determination of the complete base sequence of human DNA (the human genome). Long before this goal is accomplished, these scientists expect to map signposts at close intervals within the genome. They also will have sequenced short regions of interest in the human genome as well as in such organisms as bacteria, yeast, fruit flies, and mice.

Why is it important to know about human DNA, down to the last base pair? Scientific curiosity is one perfectly valid reason. Like a mountain to be climbed, the sequence of human DNA is of interest because it exists. Supporters of the HGP are also quick to point out that much fundamental information about human heredity is still lacking. For example, the number of human genes is not known with certainty. Furthermore, the biological function of many genes is still unknown. Undoubtedly, the HGP will produce evidence of previously unknown genes with previously unknown functions. Supporters of the HGP correctly point out that many diseases are of genetic origin. Knowledge of the human genome will help to identify genes that are linked to about 4000 hereditary diseases. In time, the knowledge gained from the HGP might be applied to the cure of these diseases.

Despite the many excellent justifications advanced by its supporters, the HGP remains somewhat controversial. When resources are limited, critics say, satisfying scientific curiosity is not a particularly good reason to start a grand project like the HGP. Critics of the project are also concerned that its cost will divert scientists and dollars from other equally worthy research projects. Originally, it was estimated that a DNA base pair could be sequenced at an average cost of about $1. This estimate, however, assumed great improvements in efficiency relative to today's sequencing technology. With today's technology, it costs about $5 to determine the sequence of a base pair. Some critics also say that it is unclear that advancements toward gene therapies could not be made by simpler, less expensive methods for studying genes. An alternative plan might be better. For example, it might be more efficient to first identify a gene of interest before going to the trouble of sequencing the DNA of the gene. This criticism is supported by evidence suggesting that only about 5% of the base pairs in DNA play a functional role in human cells.

person with great certainty by examination of the base sequences of his or her DNA, as described in A Closer Look: DNA Fingerprinting.

A Closer Look

DNA Fingerprinting

Violent criminals often leave evidence in the form of blood and hair at the scene of their crimes. Perpetrators of sexual assaults may leave evidence as seminal fluid. A method called *DNA fingerprinting*, or *DNA profiling*, is helping to solve crimes. DNA fingerprinting also can be used to establish family relationships. For example, the paternity of a child may be at issue.

Except for identical twins, the base sequences of introns and, to a lesser extent, exons are slightly different for different individuals. These small variations among individuals form the basis for DNA fingerprinting. To make a DNA fingerprint, enzymes called *restriction enzymes* are used to cleave DNA at specific base sequences (see Sec. 20.11). Because of the differences in the base sequences among individuals, the action of a restriction enzyme may cut a fragment of 1000 bases from the DNA of one person and 3000 bases from the DNA of another person. The DNA fragments produced by the restriction enzymes can be separated by electrophoresis. The pattern of separated DNA fragments is the DNA fingerprint (see figure). The variations in DNA sequences among individuals make it very improbable that two people will have the same DNA fingerprint. In forensic work, DNA fingerprints from evidence taken at the scene of a crime are compared with the "fingerprint" obtained from DNA provided by a suspect. In principle, the chance that a DNA fingerprint is unique to one individual can be as high as 10^{19} to 1. This is a very high degree of certainty, since the world contains fewer than 6×10^9 people. In practice, the chance of identifying an individual from a DNA fingerprint is usually lower. This has caused controversy in the use of DNA finger-

A scientist compares DNA fingerprints.

printing for criminal proceedings. Is a 100 to 1 chance, a 1000 to 1 chance, or a 10,000 to 1 chance of having identified a murderer adequate to convict an accused person? Despite the controversy, more than half the states permit the use of evidence based on DNA fingerprints.

20.6 Classes of RNA

AIM: To describe the function of the three major classes of RNA: tRNA, rRNA, and mRNA.

In addition to DNA, the nucleic acids of cells include three major classes of RNA molecules: *ribosomal RNA, messenger RNA,* and *transfer RNA.*

Ribosomal RNA

Ribosomal RNA *(rRNA) molecules are an integral part of* **ribosomes,** *the organelles where proteins are synthesized* (refer to A Closer Look: Cells, on page 530). Ribosomes are composed of two subunits, one light subunit and one heavy subunit (Fig. 20.11). Each subunit is composed of about 65% rRNA and 35% protein. The rRNA maintains the structure of the ribosome and provides sites for the binding of messenger RNA. Ribosomes are self-assembling units. When the proteins and rRNA needed to make a complete ribosome are mixed together in solution, the ribosome forms spontaneously.

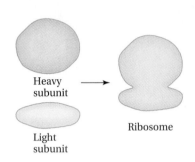

Heavy subunit

Light subunit

Ribosome

Ribosomal subunits

Figure 20.11
Heavy and light subunits combine to form a complete ribosome.

Messenger RNA

Messenger RNA *(mRNA) molecules carry the information for protein synthesis to ribosomes in prokaryotic and eukaryotic cells.* In prokaryotes, which have no nucleus, mRNA is transcribed from DNA, and proteins are synthesized on ribosomes in the cytoplasm. In eukaryotes, the mRNA is transcribed from DNA in the nucleus. The mRNA then carries the information for protein synthesis to ribosomes in the cytoplasm. In contrast to DNA, which remains intact and unchanged throughout the life of the cell, mRNA has a short lifetime—usually less than an hour. Then it is rapidly degraded back to the constituent nucleotides.

Transfer RNA

Molecules of **transfer RNA** *(tRNA) deliver amino acids to mRNA during translation.* Cells have at least one type of tRNA for each of the 20 common amino acids found in proteins. These tRNA molecules are relatively small; most contain fewer than 100 nucleotides. Moreover, scientists have found some bases other than the usual A, U, G, and C in the nucleotides that make up the tRNA molecules. Unlike DNA molecules, most RNA molecules are single-stranded. However, a single strand of RNA may sometimes loop back on itself to make "hairpin turns" held in place by hydrogen bonds between complementary bases on the same strand. The maximum number of complementary base pairs of this type are found in tRNA molecules. Figure 20.12(a) shows in a schematic way how the alanine-carrying tRNA of yeast is shaped like a cloverleaf. The tRNAs for all amino acids are quite similar in appearance. The cloverleaf shape is present but twisted in the three-dimensional structures (see Fig. 20.12b).

Figure 20.12
Typical shapes of tRNA molecules:
(a) two-dimensional cloverleaf structure;
(b) three-dimensional structure.

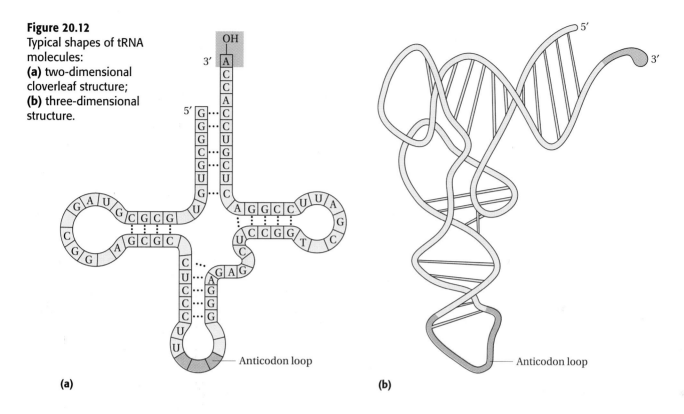

(a)   **(b)**

Anticodon loop

20.7 Transcription

AIM: To describe the process of transcription.

Focus
RNA is made on a DNA template.

Now we will turn our attention to *transcription*—the means by which information for protein synthesis passes from DNA to mRNA. Molecules of tRNA and rRNA are made by a similar process. The mechanics of transcription are quite similar to the mechanics of replication because DNA is the template upon which RNA is formed. The major differences are that *transcription requires an enzyme called* **RNA polymerase** rather than DNA polymerase, that the sugar units of the nucleoside triphosphates used to make RNA are ribose rather than deoxyribose, and that U substitutes for T in RNA. Suppose, for example, the order of bases in a segment of DNA is this:

$$5' \quad \text{dT} - \text{A} - \text{A} - \text{G} - \text{C} - \text{T} \quad 3'$$

Then the DNA molecule and its complementary RNA strand formed in transcription are

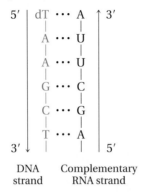

In prokaryotic cells, each mRNA molecule is transcribed from start to finish in a continuous chain. Because of introns, eukaryotic cells handle the transcription of DNA into mRNA somewhat differently than prokaryotic cells. You may recall that the genes of eukaryotic DNA are interrupted by stretches of introns that do not code for proteins, and only the exons contain the information needed to synthesize proteins. The transcription of DNA in eukaryotic cells is a three-step process (Fig. 20.13). First, the entire

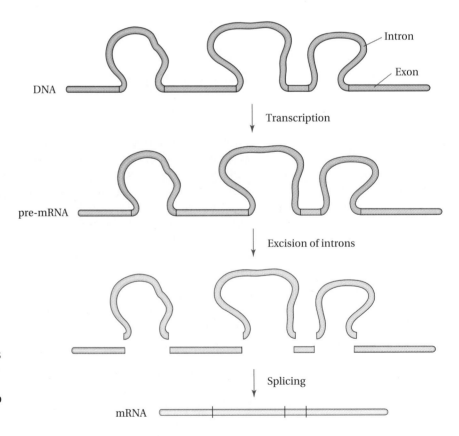

Figure 20.13
Formation of mRNA of eukaryotic cells involves transcription of genes containing exons and introns, excision of the introns of the RNA transcript, and splicing the exons to make the final mRNA.

gene containing both the introns and the exons is transcribed into an RNA molecule called a *pre-mRNA,* much as we just described. Second, the segments of the pre-mRNA molecules that correspond to the introns are cut out by *ribonuclease* (RNase) enzymes. Third, the portions of pre-mRNA that correspond to the exons of the DNA are spliced together by *ligase* enzymes to make a finished mRNA. The finished mRNA contains the information needed for translation into a protein.

EXAMPLE 20.3 **Writing a complementary RNA strand**

Consider the following segment of DNA:

  5′ dACCGTCCAAGATAAT 3′

Write the sequence of bases in the complementary segment of RNA.

SOLUTION

In forming RNA, G pairs with C, C with G, T with A, and A with U. The lowercase *d* is dropped because this is RNA.

  3′ UGGCAGGUUCUAUUA 5′

PRACTICE EXERCISE 20.7

The sequence on a DNA strand is

5′ dA—T—C—G—A—A 3′

Write the sequence of the complementary mRNA strand.

20.8 Translation

AIMS: To outline, with the aid of sketches, the steps of protein synthesis, indicating the function of tRNA, mRNA, ribosomes, and enzymes. To write the anticodon for a given codon.

Focus

Translation involves tRNA, mRNA, ribosomes, and enzymes.

The translation of the information contained in mRNA into protein requires a supply of amino acids, tRNA molecules, mRNA, ribosomes, and a number of enzymes. As we shall see, translation involves four steps: *activation* of tRNA, *initiation, elongation,* and *termination* of a polypeptide chain.

Activation of tRNA

In translation, tRNA molecules carry amino acids to mRNA bound to ribosomes. *A tRNA molecule carrying its specified amino acid is termed an* **activated tRNA.** Activation of tRNAs occurs in two steps. In the first step, the amino acid reacts with adenosine triphosphate (ATP) to form a mixed car-boxylic-phosphoric acid anhydride.

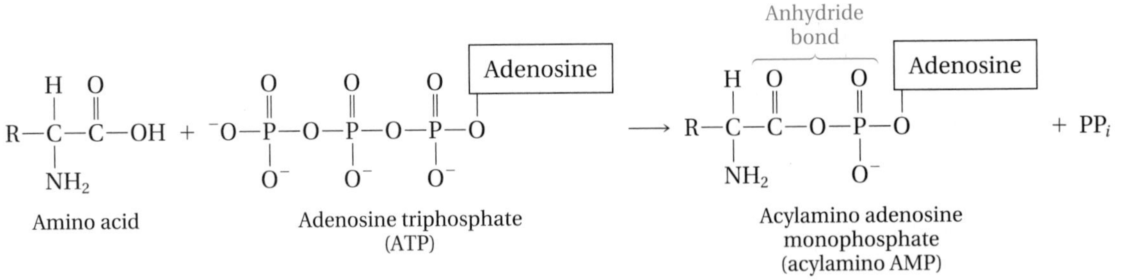

The mixed anhydride bond is extremely energetic and reacts readily with the 3′-hydroxyl group on the ribose ring at the 3′ end of the tRNA to form an ester.

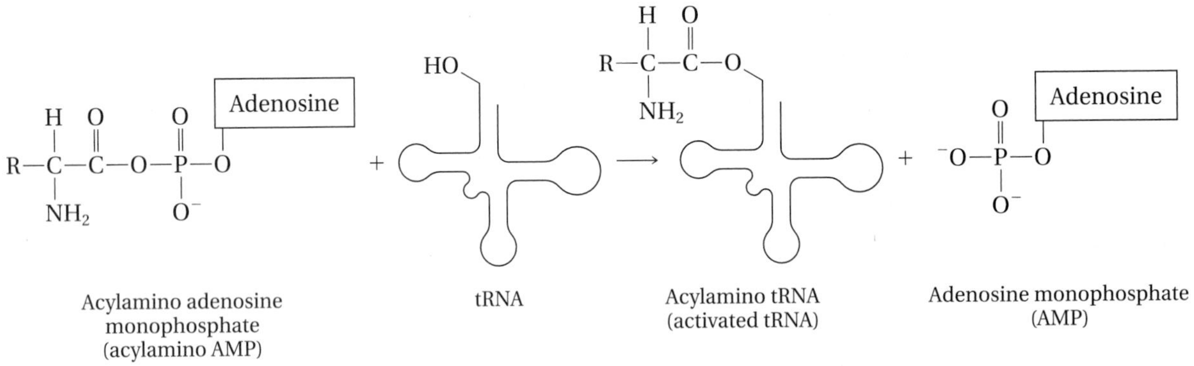

The structures of all tRNA molecules appear quite similar. How then does a given tRNA molecule recognize the amino acid it is to carry to mRNA on the ribosome? Cells contain an entire set of enzymes—one for each amino acid tRNA combination—that match tRNA molecules to their proper amino acids. These enzymes are very specific for both the structure of the amino acid and the tRNA. Almost never does an amino acid attach to the wrong tRNA.

Initiation

The beginning step in the growth of a polypeptide chain of a protein is called **initiation.** For initiation to begin, molecules of mRNA bind at a ribosomal site. A tRNA molecule carrying the N-terminal amino acid (AA 1) of the protein to be synthesized meets the mRNA at the first of two **tRNA binding sites**—*specific locations where tRNA molecules bind to ribosomes* (Fig. 20.14). The first amino acid incorporated into proteins in eukaryotic cells is always methionine. The activated tRNA carrying methionine must land on the correct spot on mRNA or else the protein that is synthesized will have the wrong sequence of amino acids. The mRNA landing point for the activated tRNA consists of three adjacent bases at the 5′ end of the mRNA molecule. *These three bases or* **base triplets** *on mRNA are called a* **codon.** *A loop in each tRNA contains three complementary bases called* **anticodons.** The anticodon of the tRNA molecule forms three complementary base pairs with the first triplet or codon of the mRNA molecule (Fig. 20.15). As in the

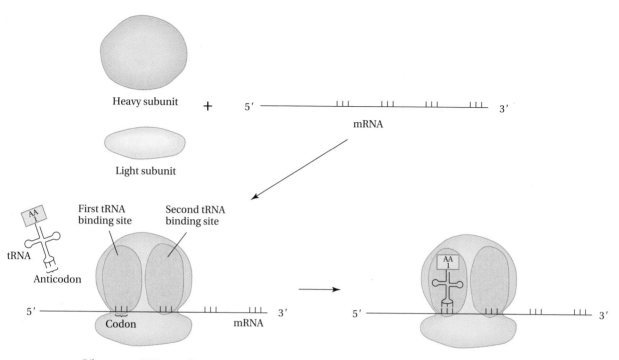

Figure 20.14
Initiation of translation.
Subunits of ribosome combine
with mRNA to form a ribosome-
mRNA complex. Activated tRNA
carrying the first amino acid
(methionine) to be incorporated
into the polypeptide chain enters
the first tRNA binding site. The
anticodon of the tRNA forms
three complementary base
pairs with the codon of mRNA.

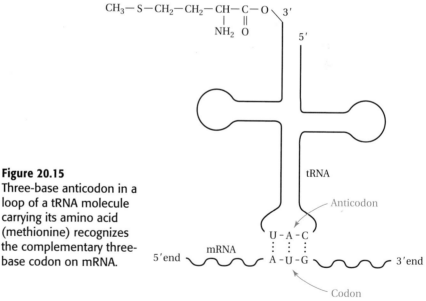

Figure 20.15
Three-base anticodon in a
loop of a tRNA molecule
carrying its amino acid
(methionine) recognizes
the complementary three-
base codon on mRNA.

DNA double helix, a stable complex between tRNA and mRNA forms only
when the bases of the codon and anticodon are complementary. In the initiation of translation, a codon for methionine on mRNA signals the proper
starting place for the initiation of translation. Later in this chapter we will
learn about the codons for each amino acid.

Elongation of the polypeptide chain

A number of antibiotics interfere with protein synthesis in bacteria. The tetracyclines, for example, block binding at the second tRNA binding site. Erythromycin blocks the shift to expose a new base triplet once a peptide bond has been formed.

Once the growth of a polypeptide chain has been initiated, the next step is *elongation* of the peptide chain. *In* **elongation,** *the polypeptide chain grows an amino acid at a time.* The tRNA molecule bearing the second amino acid (AA 2) to be incorporated into the protein enters the second tRNA binding site on the ribosome (Fig. 20.16). The anticodon of this tRNA molecule forms complementary base pairs with the second base triplet of the mRNA molecule. An enzyme promotes the formation of a peptide bond between the first (N-terminal) and the second amino acids. The first tRNA molecule leaves its binding site. The mRNA and the remaining tRNA, which now bears the dipeptide, shift to the first tRNA binding site (to the left in the figure). The shift exposes a new base triplet of mRNA to the second tRNA binding site. A tRNA molecule carrying the third amino acid (AA 3) to be incorporated into the protein being synthesized enters the second tRNA binding site. These steps are repeated with appropriately activated tRNA molecules until a complete protein is synthesized.

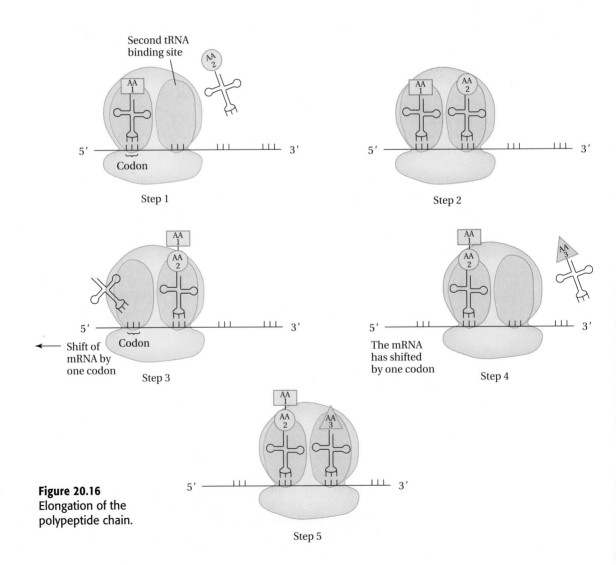

Figure 20.16
Elongation of the polypeptide chain.

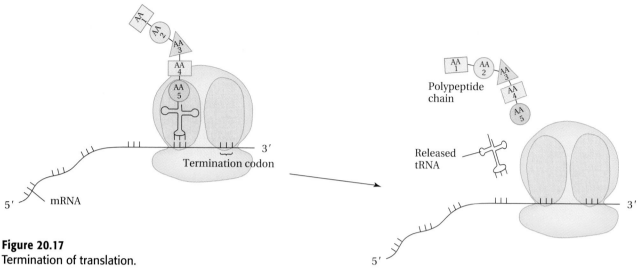

Figure 20.17
Termination of translation. A growing polypeptide chain terminates when it encounters a noncoding base triplet (termination codon) on mRNA. The tRNA is cleaved from the completed polypeptide chain, which diffuses away from the ribosome.

Termination

Termination *is the cessation of growth of the polypeptide chain.* The protein chain undergoes termination when incoming tRNA molecules encounter a **termination codon**—*a base triplet of mRNA that does not code for an amino acid.* On termination of a protein chain, an enzyme cleaves the bond between the protein and tRNA. The protein diffuses away from the ribosome to complete translation (Fig. 20.17).

Posttranslation processing

Although the translation of proteins in eukaryotic cells is initiated with methionine as the N-terminal amino acid, few finished proteins have methionine as the N-terminal amino acid residue. The reason is that most new proteins undergo **posttranslation processing**—*modification subsequent to translation.* Cleavage of N-terminal methionine from proteins is part of posttranslation processing. Other modifications may occur as well. Necessary disulfide (—S—S—) bonds are formed, for example. If the protein is to be a glycoprotein, the carbohydrate molecules are attached. Other posttranslation processes include methylations, hydroxylations, and attachment of coenzymes or prosthetic groups. With the completion of posttranslation processing, the new protein molecule is ready to make its contribution to the life of the cell.

20.9 The genetic code

AIMS: To write the amino acid sequence of a peptide given the DNA or RNA base sequence. To explain why the genetic code is termed degenerate.

Focus

The genetic code is a dictionary of three-letter words.

The proper amino acid sequence of a protein molecule depends on the formation of a complex between a base triplet, or codon, on mRNA and a complementary triplet, or anticodon, on tRNA. What are the codons for the 20 common amino acids? We will find out in this section.

It is sometimes helpful to compare mRNA to a dictionary. Actually, DNA is the master dictionary, and mRNA is a copy, but nature protects the DNA of eukaryotic cells by locking it in the nucleus. We, with our copy of the dictionary, or the cell with its mRNA, can work just as well from the copy as from the original. Whereas all the words in an English dictionary are composed from 26 letters, the words in the mRNA dictionary are composed from only 4 letters: the bases A, U, G, and C. The words in the mRNA dictionary have 20 meanings, the 20 common amino acids. *The collection of words or codons in the mRNA dictionary that is translated from one language to another—nucleic acid language to protein language—is called the* **genetic code.**

A sequence of three bases is required to code for an amino acid. Table 20.3 lists the amino acids and their mRNA codons. Three codons do not code for an amino acid. These codons—*UAA, UAG, and UGA*—are the *termination codons.* No anticodon on a tRNA molecule is complementary to them, so they signal "stop" at the end of a newly synthesized chain.

Over the years, scientists have examined the spellings of genetic code words in many different organisms, and so far the code has been the same in almost all of them. The genetic code appears to be nearly *universal.* As the genetic code was being deciphered, it became apparent that the code is

Table 20.3 Three-Letter Code Words for the Amino Acids*

	U		C		A		G	
U	UUU	Phe	UCU	Ser	UAU	Tyr	UGU	Cys
	UUC	Phe	UCC	Ser	UAC	Tyr	UGC	Cys
	UUA	Leu	UCA	Ser	UAA	End	UGA	End
	UUG	Leu	UCG	Ser	UAG	End	UGG	Trp
C	CUU	Leu	CCU	Pro	CAU	His	CGU	Arg
	CUC	Leu	CCC	Pro	CAC	His	CGC	Arg
	CUA	Leu	CCA	Pro	CAA	Gln	CGA	Arg
	CUG	Leu	CCG	Pro	CAG	Gln	CGG	Arg
A	AUU	Ile	ACU	Thr	AAU	Asn	AGU	Ser
	AUC	Ile	ACC	Thr	AAC	Asn	AGC	Ser
	AUA	Ile	ACA	Thr	AAA	Lys	AGA	Arg
	AUG	Met	ACG	Thr	AAG	Lys	AGG	Arg
G	GUU	Val	GCU	Ala	GAU	Asp	GGU	Gly
	GUC	Val	GCC	Ala	GAC	Asp	GGC	Gly
	GUA	Val	GCA	Ala	GAA	Glu	GGA	Gly
	GUG	Val	GCG	Ala	GAG	Glu	GGG	Gly

* Codes are read in the 5′ to 3′ direction. "End" denotes a termination codon.

degenerate—*that is, there is more than one code word for most of the amino acids.* The term *degenerate* does not imply that the code is defective, but only that different code words can have the same meaning. On the other hand, no codon specifies more than one amino acid. Of the 64 possible combinations of triplets formed by 4 bases, 61 code for an amino acid, and 3 are termination codons.

EXAMPLE 20.4	**Writing an amino acid sequence**

Write the amino acid sequence formed by the strand of DNA in Example 20.3.

SOLUTION

The mRNA rewritten in the 5′ → 3′ direction from the solution to Example 20.3 is

5′ AUUAUCUUGGACGGU 3′

Divide the mRNA into 3-base codons, and use Table 20.3 to find the corresponding amino acids.

5′ AUU   AUC   UUG   GAC   GGU 3′
 Ile——Ile——Leu——Asp——Gly

PRACTICE EXERCISE 20.8

Use Table 20.3 to write a base sequence for mRNA that codes for the tripeptide Ala-Gly-Ser.

PRACTICE EXERCISE 20.9

Determine the amino acid sequence of a tetrapeptide if the corresponding base sequence on DNA is

5′ dA—G—T—G—T—T—T—C—T—C—C—T 3′

20.10 Gene mutations and molecular diseases

AIMS: To show how the addition, deletion, or substitution of a nucleotide can result in a gene mutation. To explain the relationship between a gene mutation and a molecular disease using sickle cell anemia as an example.

Focus

Gene mutations are changes in the base sequence of DNA.

Substitutions, additions, or deletions of one or more nucleotides in the DNA molecule are called **gene mutations.** If mutated DNA is faithfully transcribed, the mutation is reflected in a change in the nucleotide sequence of mRNA. As a result, the primary structure of a protein may be changed, or a given protein may not be synthesized at all. The functioning of genes involved in the multiplication and growth of cells must be kept under strict cellular controls. When gene mutations cause the failure of these controls, the result can be cancer, as described in A Closer Look: Oncogenes and Tumor Suppressor Genes.

Oncogenes and Tumor Suppressor Genes

Researchers have made significant advances in understanding the molecular basis of cancer. Essentially a family of diseases, cancer is characterized by uncontrolled cell division and cell growth. Because DNA is involved in the regulation of cell multiplication, cancer may be thought of as a genetic disease. The formation of tumors is related to mutations in DNA that cause unregulated growth of cells. If the DNA in a cell undergoes mutations that cause inappropriate multiplication, the result may be formation of a tumor. The cells in an entire tumor are often the descendants of a single cell that has started unregulated and unrestrained multiplication. Not every tumor is cancerous. Cells of malignant tumors invade surrounding normal tissue and metastasize, or spread to new sites, to start new tumors. Benign tumors stay where they begin and do not invade surrounding tissues.

How do mutations in DNA lead to cancers? The answer to this simple question is complex and still incomplete. Certainly part of the answer lies in the activities of two types of genes normally found in cells: *oncogenes* and *tumor suppressor genes.*

Oncogenes are genes whose normal activity is necessary for cells to divide and grow. In normal cells of adults, the activity of these genes is strictly controlled, since most cells in organs and other tissues need to divide and grow rather slowly, if at all. If the controls on these genes are turned off, cell division and growth begins and continues unabated. Many of the known oncogenes have been shown to encode the information for a group of polypeptides called *growth factors.* The controls for the production of growth factors are encoded in the same or a nearby gene. Mutation of the DNA in the control portion of the gene inactivates the control, and unrestrained cell growth results.

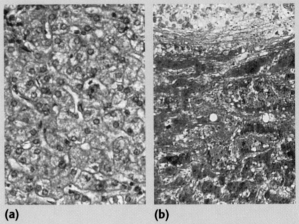

(a)   (b)

Oncogenes appear to trigger the processes that convert normal cells (a), to cancerous cells (b).

Another family of oncogenes consists of those which encode certain *protein kinase enzymes.* The protein kinases stimulate many activities of cells, including growth. One or more protein kinases important for cellular growth are normally active only if they are bound to growth factors. However, mutations in the normal DNA that encodes the kinases can cause the enzymes to be active even in the absence of growth factors. The result is unrestrained cell growth.

The normal job of tumor suppressor genes is to prevent unrestrained cellular growth. In an inherited disease called *retinoblastoma,* children lack both copies of a gene called *RB.* As a result of the missing tumor-suppressing *RB* gene, the children develop tumors in both eyes.

In addition to the effects of both oncogenes and tumor suppressor genes, additional steps are probably required to transform abnormal, rapidly growing cells into cancer cells. However, the nature of the transformation is still murky. Knowledge of the role of DNA mutation and cellular transformation may or may not lead to better treatments for diseases, but improved treatments will probably not be devised without it.

PRACTICE EXERCISE 20.10

Explain how gene mutations occur.

Mutations may be harmful or beneficial to the survival of a species. Sickle cell trait and sickle cell anemia illustrate both the harm and the benefit. As mentioned in Section 18.10, the difference between normal adult hemoglobin (HbA) and sickle cell hemoglobin (HbS) is the replacement of a glutamic acid residue by a valine residue in the sixth position of the beta chain of the protein.

HbA: Val—His—Leu—Thr—Pro—Glu—Glu—Lys—————
HbS: Val—His—Leu—Thr—Pro—Val—Glu—Lys—————
         1   2   3   4   5   6   7   8

The code words for glutamic acid and valine are

Glutamic acid	Valine
GAA	GUA
GAG	GUG
	GUU
	GUC

Substitution of the middle base, A, by U changes the meaning of the code words for glutamic acid to two code words for valine. People with sickle cell trait have immunity to malaria, so in this sense the mutation is beneficial. People with sickle cell anemia are very sick and often die young, which certainly is a harmful effect of the mutation. The benefits and the harm of hemoglobin S come from the substitution of a single nucleotide base in the gene that directs the synthesis of the beta chain of hemoglobin.

Among the hardest mutations to recognize are those that cause no detectable change in the activity of enzymes. At the other extreme, a mutation allowing no active protein at all is among the easiest to detect. A cell may be unable to synthesize a certain protein if a mutation changes a codon for an amino acid into one of the termination codons: UAA, UGA, or UAG. For example, one of the codons for tyrosine is UAU. A change of the last base from U to G produces the termination codon UAG. Translation of the message for a polypeptide chain stops when the termination codon is reached. A complete protein chain is usually necessary to produce an active protein. The single base change prevents an active protein from being synthesized.

The idea that certain diseases stem from faulty genes originated long before anyone knew about DNA's role in heredity. About 1900, Archibald Garrod, an English physician, proposed that certain lifelong diseases arise because an enzyme necessary to good health is defective or missing. Garrod's concept has been extended to include other proteins as well. Over 4000 **molecular diseases**—*inborn errors*—are known. Although they have not been labeled previously as such, we have already met two molecular diseases: methemoglobinemia (Sec. 18.9) and sickle cell anemia (Sec. 18.10).

When a human cell divides and replicates its DNA, approximately 4×10^9 bases are copied. It has been estimated that there are perhaps 2000 errors made during each replication.

Albinism is a genetic disease in which the enzyme tyrosinase is not produced. Tyrosinase is necessary for the production of melanin, the pigment of the skin, hair, and eyes. Persons affected by albinism are very sensitive to sun exposure.

EXAMPLE 20.5   **Finding peptide sequences from mRNA base sequences**

Consider the following segment of mRNA:

    5′ CGGGGUUGCAAU 3′

(a) What is the amino acid sequence formed by translation? (b) What amino acid sequence would result from the substitution of adenine for the second uracil?

SOLUTION

(a) Divide the mRNA sequence into 3-base codons, using Table 20.3.

$$5' \text{ CGG} \quad \text{GGU} \quad \text{UGC} \quad \text{AAU } 3'$$
$$\text{Arg}——\text{Gly}——\text{Cys}——\text{Asn}$$

(b) The substitution gives a different amino acid sequence:

$$5' \text{ CGG} \quad \text{GGU} \quad \text{AGC} \quad \text{AAU } 3'$$
$$\text{Arg}——\text{Gly}——\text{Ser}——\text{Asn}$$

PRACTICE EXERCISE 20.11

The following base sequence is a portion of an mRNA:

$$5' \quad \text{A—C—G—G—U—C—A—G—C—G—A—G—C—C—C} \quad 3'$$

(a) Translate this genetic message to give the amino acid sequence of a pentapeptide. (b) Uracil is replaced by adenine in a mutation. What is the new amino acid sequence? (c) Suppose uracil is deleted. What is the peptide composition?

20.11 Recombinant DNA and gene therapy

AIM: Describe how recombinant DNA technology can be used to treat a molecular disease.

Focus

Scientists have learned to transplant genes.

Since the late 1970s, scientists have taken giant steps in **recombinant DNA technology**—*transplantation of genes from one organism into another. Recombination* consists of cleaving DNA chains, inserting a new piece of DNA into the gap created by the cleavage, and resealing the chains. When the recombination is successful, the transplanted gene will express (synthesize) its normal protein product. For example, bacteria that receive human genes can be induced to express human proteins of value for research or for the treatment of disease.

Therapeutic drugs from recombinant research

Recombinant DNA research has produced several notable proteins for human therapy. Human insulin produced by bacteria is on the market for use by diabetics. This product is said to be preferable to insulin from hogs, which is very similar in amino acid composition to human insulin but which sometimes causes an adverse allergic response. A protein called *beta-interferon* (see Sec. 22.2), made by incorporation of the human interferon gene into bacteria, has been approved for treatment of multiple sclerosis patients. The protein has been shown to reduce the number and severity of attacks that strike patients. Another human protein produced in large quantities by recombinant techniques and of therapeutic value is tissue plasminogen activator (TPA; see Case in Point in Chapter 19). You may recall that this enzyme converts the zymogen plasminogen to plasmin, the enzyme that dissolves blood clots (see Sec. 19.12), and is useful in the treatment of heart attacks. The number of protein products of recombinant DNA entering testing for therapeutic use seems to increase almost daily, and hundreds more are the subject of intense research. Undoubtedly, the surface of the potential for new pharmaceuticals from recombinant DNA has been barely scratched.

Gene therapy

The explosion of knowledge about the structure, function, and manipulation of the human genome has led scientists to the dawn of a new era—the era of *gene therapy*—in the treatment of disease. **Gene therapy** *is the transfer of new genetic material to the cells of an individual with resulting therapeutic benefit to the individual.* Ashanthi, the 4-year-old girl described in the Case in Point earlier in this chapter, was the first human recipient of gene therapy. In the Follow-up to the Case in Point, below, we will learn the outcome of this therapy.

Procedures for gene therapy are now being worked out for a number of hereditary diseases such as hemophilia, in which clotting factors are defective or missing. In 1994, researchers successfully applied gene therapy to a woman to reduce the symptoms of familial hypercholesteremia,

FOLLOW-UP TO THE CASE IN POINT: Treatment for an inherited disease

In September of 1990, the first gene therapy was done on Ashanthi, a 4-year-old girl who suffered from an inherited disease called *adenosine deaminase (ADA) deficiency.* Adenosine deaminase (ADA) plays an important role in the functioning of a type of cells called *T cells.* One role of T cells is to destroy virus-infected cells, parasites, and cancer cells. Another function is to activate other cells that are responsible for the destruction of infecting bacteria. In the pioneering gene therapy treatment, Ashanthi received some of her own T cells, which had been removed and infected with a virus. The viral gene was cleverly modified; a gene for the production of ADA had been inserted into the viral gene. The virus was of a type that attaches its own genetic material to the DNA of the cells it infects. It was hoped that when the virus infected the isolated T cells, the DNA necessary to make ADA also would be inserted into the T cell gene, along with the usual viral gene.

The gene therapy for ADA deficiency seems to have worked. The T cells Ashanthi received produced ADA and strengthened her ability to ward off

Recovered from her ADA deficiency, Ashanthi now attends public school.

infections. The procedure also was effective for a second girl who started treatment in January 1991. Both girls are enrolled in public schools and have had no more than the average number of infections.

an inherited disease whose symptoms are characterized by high blood cholesterol and clogging of the arteries. The researchers removed some of the woman's liver cells, which contained a flawed gene for a key enzyme that processes cholesterol. They added a correct form of the gene to the DNA in the liver cells and returned these cells to the liver. The woman now has cleaner arteries and a lower blood cholesterol level.

Gene therapy also could prove useful for treating a number of non-hereditary diseases. For example, new research might make it possible to insert into cells certain genes that would produce factors to shrink tumors or stop the progression of AIDS. In the AIDS work, researchers are using an approach that allows them to insert a gene for a ribozyme (see Sec. 19.1) into cells infected with HIV. The scientists hope that this catalytic RNA will disable HIV by cleaving its RNA.

The demonstrated feasibility of gene transplantation in humans has aroused the attention of scientists and other concerned citizens, for it opens the door to profound ethical problems. Human control of the human genome offers unlimited opportunities for moral or immoral choices. Who will benefit and who will not? And for what reasons? Biomedical ethicists generally agree that the cure of diseases of existing human beings is an ethical use of gene therapy. However, the techniques of gene therapy are potentially applicable to the creation of humans to suit various purposes: superintelligent people to rule the world, tall people for basketball, workers who would not be bored by repetitive tasks or harmed by dirty environments. For this reason, the genetic manipulation of sperm or egg (germ line cells) is more ethically problematic than the use of gene therapy to cure diseases.

PRACTICE EXERCISE 20.12

Why do you think recombinant DNA research is controversial? What are your personal thoughts on the matter?

SUMMARY

Ribonucleic and deoxyribonucleic acids (RNA and DNA) consist of giant molecules consisting of nucleotides bound into a polymer through phosphodiester linkages. The sugar units of the nucleotide units in RNA are ribose; in DNA they are deoxyribose. The nucleotide base groups of DNA are two purines, adenine (A) and guanine (G), and two pyrimidines, thymine (T) and cytosine (C). Uracil (U) substitutes for T in RNA.

The complete DNA molecule exists as a double-stranded helix—two single strands of DNA running in opposite directions. The bases of one DNA strand in the helix are paired through hydrogen bonding with complementary bases from the other; A is always paired with T, and G is always paired with C.

DNA acts as an instruction manual for the synthesis of all the proteins made by the cell. The process by which DNA is copied is called replication. The information contained in DNA can be copied (replicated) and passed to a kind of RNA, messenger RNA (mRNA), in transcription. Information contained in mRNA can be used to direct synthesis of specific proteins (translation).

Protein synthesis requires molecules of transfer RNA (tRNA). The tRNA molecules, at least one kind for each amino acid, carry amino acids to the ribosomes for protein synthesis. Molecules of tRNA recognize their appropriate positions on mRNA by an anticodon—a base triplet that is complementary to a triplet on mRNA. The triplet code words for the amino acids have been worked out; the genetic code is degenerate and nearly universal.

The fidelity of replication, transcription, and translation is remarkable, but sometimes mutations can occur. Mutations can result in the synthesis of a fully active protein, a less active protein, or no protein at all. Some mutations are harmful and lead to death or illnesses called hereditary diseases, molecular diseases, or inborn errors. Others help increase the organism's ability to survive. Recombinant techniques in the past few years have increased the scientist's ability to manipulate genes. In gene therapy, a faulty gene responsible for a disease is replaced by a normal gene.

KEY TERMS

Activated tRNA (20.8)
Anticodon (20.8)
Antiparallel strands (20.2)
Base triplet (20.8)
Central dogma (20.3)
Codon (20.8)
Complementary base pair (20.2)
Degenerate code (20.9)
Deoxyribonucleic acid, DNA (20.1)
DNA double helix (20.2)
DNA polymerase (20.4)

Elongation (20.8)
Exon (20.5)
Gene (20.5)
Gene mutation (20.10)
Gene therapy (20.11)
Genetic code (20.9)
Initiation (20.8)
Intron (20.5)
Messenger RNA (20.3)
Molecular disease (20.10)
Nucleic acid (20.1)
Nucleoside (20.1)
Nucleotide (20.1)

Okazaki fragments (20.4)
Phosphodiester (20.1)
Polynucleotide (20.1)
Posttranslation processing (20.8)
Purine bases (20.1)
Pyrimidine bases (20.1)
Recombinant DNA technology (20.11)
Replication (20.3)
Retrovirus (20.3)
Ribonucleic acid, RNA (20.1)

Ribosomal RNA (20.6)
Ribosome (20.6)
RNA polymerase (20.7)
Template (20.4)
Termination (20.8)
Termination codons (20.8)
Transcription (20.3)
Transfer RNA (20.6)
Translation (20.3)
tRNA binding site (20.8)
Unwinding proteins (20.4)

EXERCISES

Nucleic Acids and the Double Helix (Sections 20.1, 20.2)

20.13 Cells contain two types of nucleic acids. What are they called?

20.14 What is the structural difference between the sugar unit in RNA and the sugar unit in DNA?

20.15 Explain the difference between a nucleoside and a nucleotide.

20.16 Explain the difference between a nucleotide and a nucleic acid.

20.17 (a) Name the nitrogen bases found in DNA. (b) Identify them as derivatives of purine or pyrimidine.

20.18 Are the nitrogen bases in RNA the same as those found in DNA? Explain.

20.19 (a) Draw the structure of the nucleoside formed by the combination of the nitrogen base adenine and the sugar β-D-2-deoxyribose. (b) What is its name?

20.20 How are nucleotides linked together in RNA and DNA molecules?

20.21 What gives DNA and RNA their highly acidic character?

20.22 (a) What does the following abbreviation mean?

U—G—C—A—G

(b) Which end is 3′ and which end is 5′?

20.23 What type of bonding holds a DNA double helix together?

20.24 Which of the following base pairs are found in a DNA molecule?
(a) A—A   (b) A—T
(c) C—G   (d) G—C
(e) G—A   (f) A—U

20.25 The two strands of DNA in the DNA double helix are *antiparallel*. What does this mean?

20.26 Consider the double helix. Are the base pairs
(a) part of the backbone structure
(b) inside the helix
(c) outside the helix?

The Central Dogma (Section 20.3)

20.27 Define the three steps that make up the central dogma of molecular biology.

20.28 What is a common characteristic of all retroviruses?

Replication and Genes (Sections 20.4, 20.5)

20.29 Describe the process of DNA replication.

20.30 A segment of a DNA strand has the following base sequence: dC—G—A—T—C—C—A. Write the base sequence that appears on its complementary strand in the Double Helix. Label the 3′ and 5′ ends of the sequence.

20.31 What is the function of DNA polymerase?

20.32 Why does the replication of DNA require a DNA template?

20.33 Distinguish between a gene and a polypeptide.

20.34 Contrast the function of exons and introns in a gene.

Classes of RNA and Transcription (Sections 20.6, 20.7)

20.35 Describe the function of (a) rRNA and (b) mRNA.

20.36 What is the function of tRNA? Sketch the shape of a typical tRNA molecule.

20.37 What components are needed for the synthesis of RNA?

20.38 Write the base sequence for the complementary RNA strand when the DNA base sequence is as follows:

5′ dA—T—C—G—C—T—A 3′

Label the ends of the RNA strand.

Translation and the Genetic Code (Sections 20.8, 20.9)

20.39 Give the location and describe the general structure of (a) a codon and (b) an anticodon.

20.40 Does every base triplet of mRNA code for an amino acid in protein synthesis? Explain.

20.41 The codon for methionine is AUG. What is the anticodon?

20.42 (a) Name the first amino acid that is always incorporated into proteins in eukaryotic cells. (b) Predict what would happen if this amino acid were missing from the diet.

20.43 The base sequence of a DNA fragment is a gene:

5′ dA—G—C—T—G—G—G—A—C 3′

(a) Write the base sequence of the mRNA strand.
(b) Write the amino acid sequence of the tripeptide. (Remember to start reading codons at the 5′ end of mRNA.)

20.44 Describe the changes that may occur in posttranslation processing.

Gene Mutations and Molecular Diseases (Section 20.10)

20.45 What causes a gene mutation?

20.46 (a) Name some molecular diseases. (b) Can these diseases be cured? Explain.

Recombinant DNA and Gene Therapy (Sections 20.11)

20.47 What are some outcomes of recombinant DNA research?

20.48 Describe a possible use of gene therapy.

Additional Exercises

20.49 Match the following.
(a) anticodon
(b) A, T, G, C
(c) tRNA
(d) transcription
(e) ribosome
(f) phosphodiester bridge
(g) codon
(h) degenerate
(i) molecular disease
(J) translation

(1) RNA → protein
(2) links nucleotides
(3) inborn error
(4) triplet on mRNA
(5) cloverleaf shape
(6) bases of DNA
(7) DNA → RNA
(8) site of protein synthesis
(9) genetic code
(10) triplet on tRNA

20.50 Answer the questions below based on the following strand of DNA:

5′ T T A G A T T C G C A T C C G 3′

(a) What is the structure of the mRNA formed from this DNA template?
(b) What polypeptide would be formed from this strand of mRNA?
(c) What polypeptide would be formed
(1) if the first base at the 5′ end of mRNA is deleted?
(2) if the first G from the 5′ end in the mRNA sequence is replaced by an A?
(3) if the terminal C at the 5′ end of the mRNA is replaced by a G?

20.51 To make the tetrapeptide Phe-Ser-Ala-His, (a) what is the sequence of bases on the mRNA strand? (b) What is the sequence of bases on the complementary segment of the DNA molecule?

20.52 The genetic code is *degenerate*. What does this mean?

20.53 What are the different types of RNA in the cell? Describe the functions of each in protein synthesis.

20.54 Why are nucleic acids *acidic*?

20.55 One of the following triplets cannot be a codon: UCG, ATC, GCU. Explain.

20.56 In which direction is the DNA molecule (a) synthesized continuously? (b) synthesized discontinuously (using Okazaki fragments)?

20.57 What are the possible consequences if a DNA molecule fails to replicate itself *every time* without error?

20.58 Match one of the processes at the right to each of the phrases on the left.
(a) utilizes codon-anticodon interaction
(b) *not* carried out in the nucleus
(c) produces mRNA
(d) uses tRNA and mRNA
(e) produces a DNA copy
(f) results in single-stranded RNA
(g) produces a specific protein
(h) semiconservative in nature

(1) replication
(2) transcription
(3) translation

SELF-TEST (REVIEW)

True/False

1. RNA and DNA are structurally the same except that uracil is found in place of thymine in RNA.
2. The backbone of a DNA molecule is held together by phosphodiester bridges.
3. RNA and DNA are termed *nucleic acids* because their nitrogen bases can donate protons.
4. Gene mutations are always harmful to the organism.
5. In any organism the A + G to T + C ratio in DNA equals 1.
6. In the double helix of DNA, the largest number of stabilizing double bonds is achieved when purines oppose purines and pyrimidines oppose pyrimidines.
7. Protein synthesis takes place in the nucleus, where the genetic information is found.
8. The tRNA molecules are responsible for the sequence of amino acids in a protein.
9. An anticodon on a tRNA molecule must be complementary to a codon on the mRNA for a stable tRNA-mRNA structure to form.
10. The replacement of one letter in the genetic code by another letter always leads to a gene mutation.

Multiple Choice

11. Human beings are different from monkeys because of
(a) differences in types of nucleotides.
(b) a different A + G/T + C ratio.
(c) differences in proteins.
(d) a different genetic code.

12. The *ordered* sequence of steps in the central dogma of genetic information transmission is
(a) translation, replication, transcription.
(b) replication, translation, transcription.
(c) translation, transcription, replication.
(d) replication, transcription, translation.

13. The RNA strand transcribed from the strand of DNA molecule (-G-A-T-T-C-G-) would be
(a) -C-U-A-A-G-C-. (b) -C-T-U-U-G-C-.
(c) -T-U-G-G-A-T-. (d) -C-T-A-A-G-C-.

14. The type of RNA molecule that brings amino acids to the site of protein synthesis is
(a) rRNA. (b) aRNA. (c) mRNA.
(d) tRNA.

15. A segment of mRNA is composed of the alternating bases adenine and guanine, A-G-A-G-A-G-A-G. The protein formed from this molecule
(a) would probably contain only one amino acid.
(b) would probably contain two different amino acids.
(c) would contain three or more different amino acids.
(d) would be the same as one from one with an mRNA sequence of G-A-G-A-G.

16. Which of the following statements is *incorrect* concerning the structure of DNA?
(a) The structure is a double helix.
(b) The bases are arranged on the outside of the molecule.
(c) The two DNA strands are antiparallel.
(d) The two strands are held together by hydrogen bonds.

17. A nucleic acid is made up of
    (a) a sugar, a nucleoside, and a nitrogen base.
    (b) a nitrogen base, a phosphate group, and a sugar.
    (c) a protein, a sugar, and a phosphate group.
    (d) a nitrogen base, an amino acid, and a sugar.

18. When the DNA molecules of two different species are compared, it is found that
    (a) the percent of guanine in each is always equal.
    (b) the lengths of the molecules are the same.
    (c) the A-to-T ratios are the same.
    (d) the age of the subjects compared must be the same, or results can vary.

19. A nitrogen base bonded to a sugar gives
    (a) a nucleoside.   (b) a nucleotide.
    (c) a nucleic acid.   (d) all of these.

20. The following units are found in RNA:
    (a) β-D-2-deoxyribose, A, T, G, C.
    (b) β-D-2-deoxyribose, C, U, A, G.
    (c) β-D-ribose, A, T, G, C.
    (d) β-D-ribose, A, G, U, C.

21. A segment of a DNA molecule that directs the synthesis of a particular protein molecule is a(n)
    (a) gene.   (b) deoxyribozyme.
    (c) codon.   (d) operon.

Digestion and Nutrition

Materials for Living

Eating a selection from the recommended food groups provides a well-balanced diet.

CHAPTER OUTLINE

CASE IN POINT: Osteoporosis

21.1 Essential needs

21.2 Digestion

21.3 The stomach

21.4 The small intestine

A CLOSER LOOK: Blood Lipoproteins and Heart Disease

A CLOSER LOOK: Therapies for Cystic Fibrosis

21.5 Balanced diet

A CLOSER LOOK: The Nutrition Pyramid

21.6 Proteins and amino acids in the diet

21.7 Fat-soluble vitamins

21.8 Water-soluble vitamins

21.9 Minerals

FOLLOW-UP TO THE CASE IN POINT: Osteoporosis

21.10 Trace elements

In previous chapters we have introduced the four major classes of large biological molecules: carbohydrates, lipids, proteins, and nucleic acids. We have learned that enzymes are vital to our bodies' proper functioning. The final chapters of this book explore the ways in which the human body maintains its internal environment, produces energy, and makes molecular building blocks for growth. Our bodies must be supplied with substances through dietary intake to perform these tasks, and these substances must be processed by digestion so that they can be used by our bodies. In this pivotal chapter we will examine nutrition and digestion.

Everyone knows that the building of strong bones in children is important. However, the maintenance of bone strength is often a problem for older adults, particularly women, as we will learn in the Case in Point.

CASE IN POINT: Osteoporosis

Bessie lived on a farm her entire life where she was admired by her family and neighbors for her capacity for hard work. In her eighty-sixth year she planted, hoed, weeded, and carried water to a large garden beside the farmhouse. One evening as she stood on the porch surveying her day's work, she fell suddenly. Like many elderly women, Bessie had broken a hip. What was the cause of Bessie's fracture, and how was it related to the special nutritional needs of older women? We will learn the answers to these questions in Section 21.9.

Many women suffer from osteoporosis in later life.

21.1 Essential needs

AIM: To list five essential needs of the human body.

Focus

The human body has five essential needs.

Our bodies depend on external sources for five essential requirements:

1. *Oxygen. Cells that require oxygen to live are called* **aerobic cells.** Human cells are aerobic. Oxygen is not a nutrient, but human cells require it for energy production. The length of time human cells can survive without oxygen depends on the type of cell. The cells of heart muscle survive only about 30 seconds without oxygen, and brain cells die within about 5 minutes without it. Skeletal muscle cells live much longer than either heart muscle cells or brain cells in the absence of oxygen.
2. *Water.* Like oxygen, water is not a nutrient. But it is indispensable as the medium in which all body functions take place.
3. *Carbon compounds.* Carbon compounds supply energy and raw materials for growth and maintenance through the diet—mostly in the form of carbohydrates, lipids, proteins, and nucleic acids.
4. *Vitamins.* **Vitamins** *are compounds of carbon needed for growth and health that must be supplied in the diet.* Vitamins are divided into two major categories: fat-soluble and water-soluble (or B-complex) vitamins. Vitamin C is also water-soluble.
5. *Salts or minerals.* The various salts or minerals needed by the body for the maintenance of the acid-base balance, the electrolyte balance, and the growth of teeth and bones must be supplied through the diet.

Vitamin
vita (Latin): life

The focus of this chapter is carbon compounds, vitamins, and salts or minerals. We begin this chapter by discussing the intake and digestion of dietary carbon compounds.

21.2 Digestion

AIM: To name the type of chemical reaction that is common to the digestion of carbohydrates, proteins, lipids, and nucleic acids.

Focus

Digestion begins in the mouth.

The bulk of the food we eat is composed of complex carbohydrates (such as starch and glycogen), complex lipids, proteins, and nucleic acids, most of which are molecules too large to be absorbed by body tissues without further processing. Organs of the digestive system (Fig. 21.1) are sites of hydrolysis of these substances into simpler molecules that are absorbed into the bloodstream and transported to body cells.

When food is chewed, it mixes thoroughly with **saliva,** *the fluid secreted by the salivary glands.* An average adult secretes almost 1.5 L of saliva per day. Saliva is about 99% water and has a pH slightly above 7. It also contains small amounts of salts and two proteins: *mucin* and the enzyme *amylase* (ptyalin). Mucin is a glycoprotein that gives saliva its viscosity and stringiness. Amylase promotes the hydrolysis of starch and

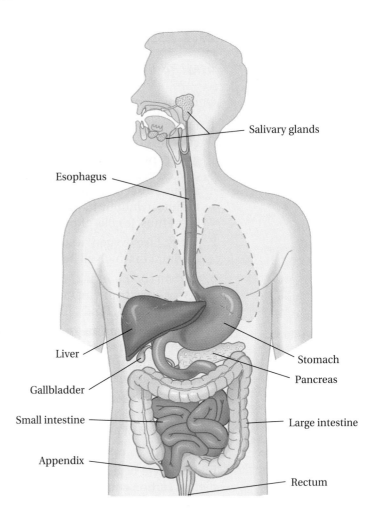

Figure 21.1
Organs of the human digestive system.

glycogen to the disaccharide maltose. No digestion of fats, proteins, or nucleic acids takes place in the mouth.

21.3 The stomach

AIM: To name the enzyme that hydrolyzes protein molecules in the stomach and characterize the environment in which it works.

Focus

Protein digestion begins in the stomach.

The mass of food mixed with saliva is swallowed and enters the stomach. The environment of the stomach is quite different from that of the mouth. The fluid of the stomach is very acidic owing to the presence of **gastric juice,** *a fluid secreted by cells that line the stomach.* The gastric juice is about 0.1 *M* hydrochloric acid. (A 0.1 *M* solution of a strong acid has a pH of 1.)

Most enzymes have a pH optimum in solutions near pH 7. Pepsin, however, the protease of the stomach, is an exception; it is most active at the pH of gastric juice. Pepsin, formed from pepsinogen by autoactivation or by the action of hydrochloric acid, begins to catalyze the hydrolysis of large proteins in the acid environment of the stomach, breaking them down to smaller peptides.

Considering that it contains strong acid and a voracious protease, the stomach usually holds up well. Pepsin and hydrochloric acid do not digest the cells that line the stomach because the outer membranes of these cells are mainly lipids, not proteins. Moreover, cells of the stomach lining rapidly slough off and are completely replaced every 3 days or so. Bacterial infection, nervous tension, and excessive alcohol or aspirin consumption can cause the stomach lining to break down in places. The breakdown may result in peptic ulcers, a form of self-digestion of the stomach.

PRACTICE EXERCISE 21.1

Explain why the frequent ingestion of large doses of an antacid, such as milk of magnesia, can affect protein digestion in the stomach.

21.4 The small intestine

AIM: To outline the digestive process for carbohydrates, proteins, lipids, and nucleic acids, citing locations and enzymes.

Focus

Digestion is completed in the small intestine.

The contents of the stomach, now thick and creamy, enter the upper part of the small intestine. Secretions from the pancreas and gallbladder flow into the upper part of the small intestine. These secretions are alkaline, and they neutralize stomach acid entering the small intestine.

Digestion and absorption of proteins and sugars

Pancreatic juice—*an aqueous fluid with a high bicarbonate ion concentration and a slightly basic pH (about pH 8.0)*—contains several proteases, including chymotrypsin, trypsin, and carboxypeptidase. These proteases, formed from their zymogens, catalyze the hydrolysis of proteins into a mixture of small peptides and individual amino acids. Digestion of proteins is completed within a few hours, but without proteases, the process under the same conditions would take about 7 years.

Digestion of sugars also continues in the small intestine. Maltose is hydrolyzed to glucose by maltase, sucrose is hydrolyzed to fructose and glucose by sucrase, and lactose is hydrolyzed to glucose and galactose by lactase (unless a person has a lactase deficiency). The absorption of monosaccharides, amino acids, and a few small peptides into the bloodstream starts as the products of digestion begin their long journey through the small intestine.

The concentrations of sugars and amino acids in the small intestine are lower than in the cells of the intestinal wall. Since the normal flow of substances through membranes is from higher to lower concentration, these compounds must be transported into intestinal cells by means of energy-requiring cellular pumps. From the cells of the small intestine, the sugars and amino acids make their way into the bloodstream. The liver extracts from the blood the sugars and amino acids it needs for its cells. The rest remain in the bloodstream for delivery to other tissues.

Digestion and absorption of lipids

Hydrolysis of lipids also begins in the small intestine. The lipids we consume as part of our diet are mainly the triglycerides, cholesterol, and the complex phospholipids—the phosphatidyl cholines, sphingomyelins, and cerebrosides of cell membranes. You may recall that cholesterol comes entirely from animal products.

Fats form small, insoluble globules in water rather than dissolving in it. The fats in these globules are relatively inaccessible to enzymes. The globules are broken up by being mixed with bile when they enter the duodenum. **Bile** *is a soaplike fluid produced in the liver and stored in the gallbladder.* The digestion of fats by enzymes requires bile to form water-soluble micelles with dietary fats just as the cleansing action of soaps depends on micelle formation with greasy dirt. Bile consists of cholesterol, *bile salts,* and **bile pigments**—*a group of colored compounds.* **Bile salts** *are derivatives of cholesterol;* Figure 21.2 shows the major ones.

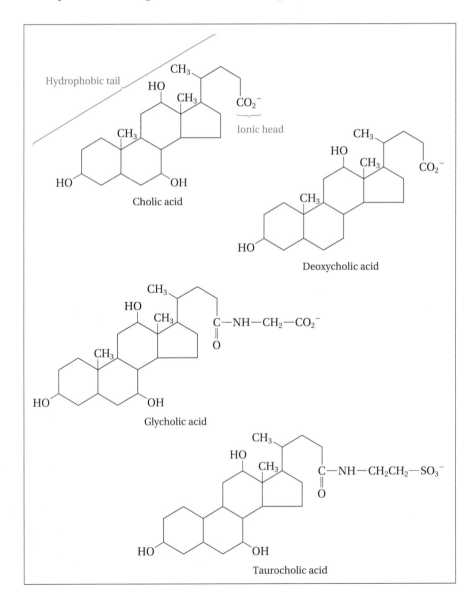

Figure 21.2
The bile salts are soaplike molecules derived from cholesterol that help dissolve dietary lipids. Notice that the bile salts, like soaps, have ionic heads and hydrophobic tails.

Dietary lipids that enter the small intestine are hydrolyzed into their simpler component molecules. *Lipase,* a pancreatic enzyme, promotes the hydrolysis of lipids. Lipase hydrolyzes only lipids that are solubilized by forming micelles with bile salts. The breakdown of tripalmitin, the triester of palmitic acid and glycerol, is a typical example of the hydrolysis of a triglyceride. The hydrolysis products are glycerol and the 16-carbon fatty acid palmitic acid.

$$CH_3 \!\!-\!\!(CH_2)_{\overline{14}}\!\!-\!\!\overset{\overset{\textstyle O}{\|}}{C}\!\!-\!\!O\!\!-\!\!CH_2$$

$$CH_3 \!\!-\!\!(CH_2)_{\overline{14}}\!\!-\!\!\overset{\overset{\textstyle O}{\|}}{C}\!\!-\!\!O\!\!-\!\!CH + 3H_2O \xrightarrow{\text{Pancreatic lipase}} \begin{matrix} HO\!\!-\!\!CH_2 \\ | \\ HO\!\!-\!\!CH \\ | \\ HO\!\!-\!\!CH_2 \end{matrix} + 3CH_3 \!\!-\!\!(CH_2)_{\overline{14}}\!\!-\!\!\overset{\overset{\textstyle O}{\|}}{C}\!\!-\!\!OH$$

$$CH_3 \!\!-\!\!(CH_2)_{\overline{14}}\!\!-\!\!\overset{\overset{\textstyle O}{\|}}{C}\!\!-\!\!O\!\!-\!\!CH_2 \qquad\qquad\qquad \text{Glycerol} \qquad\qquad \text{Palmitic acid}$$

Tripalmitin

Likewise, phospholipids such as the phosphatidyl cholines are hydrolyzed to their component parts.

$$R\!\!-\!\!\overset{\overset{\textstyle O}{\|}}{C}\!\!-\!\!O\!\!-\!\!CH_2$$

$$R\!\!-\!\!\overset{\overset{\textstyle O}{\|}}{C}\!\!-\!\!O\!\!-\!\!CH + 4H_2O \xrightarrow{\text{Pancreatic lipase}}$$

$$\begin{matrix} & CH_3 & & O \\ & | & & \| \\ CH_2\!\!-\!\!\overset{+}{N}\!\!-\!\!CH_2\!\!-\!\!CH_2\!\!-\!\!O\!\!-\!\!P\!\!-\!\!O\!\!-\!\!CH_2 \\ & | & & | \\ & CH_3 & & O^- \end{matrix}$$

A phosphatidyl choline
(lecithin)

$$2R\!\!-\!\!\overset{\overset{\textstyle O}{\|}}{C}\!\!-\!\!OH + \begin{matrix} HO\!\!-\!\!CH_2 \\ | \\ HO\!\!-\!\!CH \\ | \\ HO\!\!-\!\!CH_2 \end{matrix} + \begin{matrix} CH_3 \\ | \\ CH_3\!\!-\!\!\overset{+}{N}\!\!-\!\!CH_2\!\!-\!\!CH_2\!\!-\!\!OH \\ | \\ CH_3 \end{matrix} + P_i$$

Fatty acid $\qquad$ Glycerol $\qquad\qquad\qquad$ Choline $\qquad\qquad$ Phosphate ion

The breakdown products of dietary lipids are absorbed into intestinal cells as is cholesterol. The intestinal cells contain another lipase—one that catalyzes the synthesis of lipids. Resynthesized lipids form as a result of the action of this lipase (Fig. 21.3). Because of their relative nonpolarity, cholesterol and other lipids are too insoluble in body fluids to be transported by the blood to other body tissues. However, insoluble cholesterol and other lipids are solubilized by interacting with **lipoproteins**—*proteins designed to bind to lipids*—in the intestinal lining. The lipoproteins with their bound lipids form bodies called *chylomicrons.* The structures of chylomicrons are similar to those of micelles. *In* **chylomicrons** *the nonpolar portions of lipids are surrounded by a film of the polar lipoprotein, a combination that is soluble in body fluids.*

Figure 21.3
The path taken by dietary lipids in going from the small intestine to the bloodstream. Triglycerides and complex lipids are hydrolyzed to their simpler components. These components pass into cells of the intestinal mucosa, where they are resynthesized into lipids. The newly synthesized lipids complex with lipoproteins and cholesterol to make soluble bodies called *chylomicrons.* The chylomicrons enter the bloodstream through the lymphatics, a network of vessels that carries substances from body tissues to the blood.

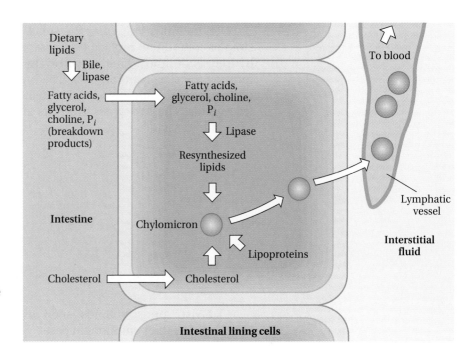

The incorporation of triglycerides into chylomicrons prepares the lipids for transport through the bloodstream, much as a ship transports passengers who cannot swim. Chylomicrons cannot enter the blood through capillaries, probably because they are too large, but they can enter lymphatic vessels. After a meal heavy in fats, the lymph may acquire a milky white appearance because of the high concentration of triglycerides. The fat-laden lymph enters the bloodstream through the thoracic duct (see A Closer Look: Blood Lipoproteins and Heart Disease).

Digestion of nucleic acids

A group of enzymes known as **nucleotidases** *catalyze the hydrolysis of nucleic acids to their component nucleotides.* Nucleotidases are produced in the pancreas and transported to the small intestine, where they perform their catalytic work. The nucleotides produced by the action of nucleotidases are then absorbed into intestinal cells. The nucleotidase that catalyzes the hydrolysis of DNA is called *deoxyribonuclease (DNase); ribonuclease (RNase)* catalyzes the hydrolysis of RNA. An interesting therapeutic use of DNase is in cystic fibrosis, as described in A Closer Look: Therapies for Cystic Fibrosis, on page 652.

Food is almost completely absorbed by the time it reaches the end of the small intestine. A watery mixture of undigested materials, the feces, passes into the large intestine (the colon). There, much of the water is reabsorbed into the body. The reabsorption gives the feces their characteristic semisolid texture. The color of feces comes from bile pigments that have escaped the small intestine.

Blood Lipoproteins and Heart Disease

Chylomicrons are the largest lipid-carrying particles of blood, but they are usually found in the blood only after a fatty meal. Blood also contains smaller particles consisting of lipoproteins and their associated lipids. The role of these particles is to transport lipids between tissues independent of the chylomicrons.

One general method of classifying lipoproteins is by density. Since lipids are less dense than proteins, particles containing lipoproteins associated with large amounts of lipid are less dense than those associated with smaller amounts. The lipoproteins of chylomicrons haul more lipids than any other lipoproteins. The protein content of chylomicrons is often as little as 1% of the total mass of the particle, and chylomicrons have densities of less than 0.96 g/cm^3. On the other hand, the mass of proteins in the so-called high-density lipoproteins (HDLs) may be as much as 60% of the total mass of the particles. The HDL particles have densities from 1.06 to 1.21 g/cm^3. Two other types of lipoproteins, the very-low-density lipoproteins (VLDLs) and low-density lipoproteins (LDLs), have densities between these two extremes.

Biomedical scientists studying the link between blood cholesterol levels and the risk of heart disease have become interested in the cholesterol content of HDLs and LDLs. So far they have learned that the risk of heart disease increases as the amount of LDL cholesterol (see figure) rises in proportion to the amount of HDL cholesterol. As the following table shows, the LDL cholesterol/HDL cholesterol ratio is presently the best method for assessing the risk of heart disease.

Risk	LDL/HDL ratio
Men	
$\frac{1}{2}$ average	1.00
average	3.55
2 times average	6.25
3 times average	7.99
Women	
$\frac{1}{2}$ average	1.47
average	3.22
2 times average	5.03
3 times average	6.14

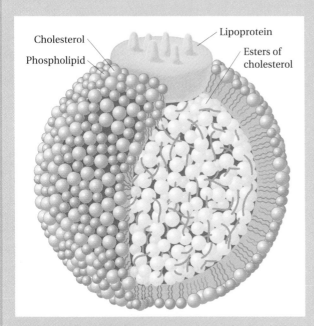

Cholesterol
Phospholipid
Lipoprotein
Esters of cholesterol

Schematic diagram of an LDL particle. As the level of LDL increases in the blood, so does the deposition of atherosclerotic plaque. For this reason, LDL is often called "bad cholesterol."

Several inherited diseases are connected with defects in lipoprotein synthesis or function. Depending on the defect, these diseases may result in either hypolipoproteinemia or hyperlipoproteinemia (low or high concentrations of lipoprotein in the blood, respectively). Many individuals with such diseases as diabetes mellitus and atherosclerosis (hardening of the arteries) also show one or the other condition. The treatment depends on the cause of the disease and nature of the condition, but it often consists of adjusting the amount of fat in the diet.

Therapies for Cystic Fibrosis

A treatment for cystic fibrosis uses deoxyribonuclease (DNase), the enzyme normally used in the body to hydrolyze DNA to its constituent nucleotides. Cystic fibrosis is a genetic disease that affects 20,000 Americans annually. The disease, which historically has killed 50% of its victims by age 20, is characterized by the thick, infected mucous secretions that plug air passages in the lungs. These secretions cause inflammation and damage to lung tissue and ultimately death. Traditional treatment for cystic fibrosis has involved sessions of thumping the patient on the back and chest to clear the mucus from the passages and administering antibiotics to control infections. The high viscosity of the mucus of cystic fibrosis results from large amounts of DNA in the secretions. The DNA comes from the nuclei of white blood cells that have gathered in the lungs to fight infections and have died there. This unwanted DNA is the target of the DNase used in the treatment of cystic fibrosis. A solution of DNase is inhaled into the lungs through an atomizer, which produces a fine spray (see figure). In the lungs, the DNase catalyzes the hydrolysis of the DNA to thin the mucous secretions. The thinned secretions are much more easily expelled from the lungs, improving lung function and reducing infections. DNase treatment is not a cure for cystic fibrosis, but it can improve life for many patients with the disease. The DNase used for cystic fibrosis treatment is a human enzyme obtained by recombinant DNA technology and marketed as Pulmozyme. Since it is identical to the enzyme in our bodies, the recombinant DNase does not cause an

An inhalant spray of deoxyribonuclease is effective in treating the symptoms of cystic fibrosis.

allergic response. Its major side effect appears to be hoarseness in some patients.

In addition to the DNase treatment, there has been progress toward the treatment or cure for cystic fibrosis by gene therapy. The secretions of cystic fibrosis are caused by mutations in the gene for cystic fibrosis transmembrane conductance regulator (CFTR). These mutations lead to a marked decrease in chloride transport across membranes of airway cells. Researchers have used a virus carrier to deliver the normal CFTR gene to the nasal cells of patients with cystic fibrosis. They found that normal chloride transport was achieved in treated cells for a few days, with no adverse effects. Many questions remain to be answered before this promising gene therapy can be used on a large scale.

Undigested materials are subjected to the action of bacteria in the large intestine. One action of these bacteria is to degrade the amino acid tryptophan into skatole and indole.

Skatole

Indole

These two heterocyclic amines are largely responsible for the odor of feces. Feces are stored in the rectum, where more water reabsorption occurs, until they are excreted.

> **PRACTICE EXERCISE 21.2**
> Describe the digestion and absorption of carbohydrates in the body.

> **PRACTICE EXERCISE 21.3**
> Outline the digestion and absorption of dietary lipids.

21.5 Balanced diet

AIM: To explain why basal metabolism varies among individuals.

Focus

A balanced diet provides for basal metabolism, activity, and growth.

Heats of combustion of foods are determined using a bomb calorimeter. A food sample is burned inside a well-insulated container surrounded by a known quantity of water. The energy released by the combustion of the food is calculated from the in-crease in temperature of the water.

Recall that the amount of energy that can be obtained from a molecule is related to the energy contained in the molecule's chemical bonds. When an organic compound oxidizes, the amount of energy released is independent of the method of oxidation. Both combustion and oxidation in aerobic cells ultimately oxidize carbon compounds to carbon dioxide and water. The same amount of energy is released by burning a food substance in air as is released when the food is oxidized in steps by a living organism. It is possible to burn various foods and measure the heat evolved. *The quantity of heat evolved per gram of food burned is that food's* **caloric value.** Each gram of carbohydrate or protein burned produces about 4 kcal of heat energy. Fats produce more energy; just over 9 kcal is produced for each gram burned. Nutritionists usually use the dietary Calorie (note the capital C) in place of the kilocalorie; 1 Calorie equals 1 kcal or 1000 cal.

EXAMPLE 21.1

Calculating the caloric value of food

A college student eats a snack that contains 12 g of carbohydrate, 7 g of protein, and 3 g of fat. What is the caloric value of the snack in dietary Calories?

SOLUTION

Each of the 19 total grams of carbohydrates and proteins supplies 4 kcal of energy. Each of the 3 grams of fat supplies 9 kcal. The total caloric value of this snack is:

$$(19 \ g \times \frac{4 \ \text{kcal}}{1 \ g}) + (3 \ g \times \frac{9 \ \text{kcal}}{1 \ g}) = 103 \ \text{kcal}$$

When we are at rest, our body cells are using nutrients to produce energy at a level near the minimum to sustain life. *This minimum level of energy production is called the* **basal metabolism.** (Chapters 23 through 26 present a more detailed picture of metabolism.)

The basal metabolism of healthy people of the same age, size, and physical condition is remarkably constant. It varies, however, according to age, size, and sex: Children have a higher basal metabolism than adults; obese people have a higher basal metabolism than thin people; males have a higher basal metabolism than females. A typical healthy 35-year-old male of average size and weight has a basal metabolism requiring about 1700 Calories per day. Such a man needs a daily intake of food with a caloric value of 1700 Calories even if he never wiggles a finger. The caloric requirements of women are slightly less than those of men of comparable size, age, and weight.

Most of us do eat and wiggle our fingers, or even lift a leg or two, in a typical day. These activities require the expenditure of energy beyond that required for basal metabolism. Adults in sedentary occupations require a food intake equivalent to 2000 to 2500 Calories per day. Very active people may need as many as 4000 to 6000 Calories per day to meet their energy needs. Children need more food than comparably active adults to provide for new tissue growth.

When we eat less than our daily minimum caloric requirement, our bodies call on storage reserves of fat to supply energy, and we lose weight. When we eat too much, energy reserves are stored as fat, and we gain weight. Only when our caloric intake equals our caloric output do we stay the same weight. However, the diet of every human being must be well balanced with regard to intake of carbohydrates, fats, proteins, minerals, and vitamins, since the body needs some of each to efficiently produce substances needed to stay healthy. In other words, getting enough calories is not necessarily good nutrition. Guidelines for a balanced diet have been issued, as described in A Closer Look: The Nutrition Pyramid.

Table 21.1 Essential and Nonessential Amino Acids

Essential	Nonessential
isoleucine	alanine
leucine	arginine*
threonine	asparagine
lysine	aspartic acid
methionine	cysteine
phenylalanine	glutamic acid
tryptophan	histidine*
valine	glutamine
	glycine
	proline
	serine
	tyrosine

*Needed in the diet of growing children.

Focus

Dietary protein must contain all the essential amino acids.

Recombinant DNA research in plants holds promise for producing corn and wheat that contain all essential amino acids. Success would greatly help to solve the world's food problems.

21.6 Proteins and amino acids in the diet

AIMS: *To distinguish between essential and nonessential amino acids. To distinguish between complete and incomplete proteins. To distinguish between nutritional marasmus and kwashiorkor.*

The amino acids synthesized by the body are called the **nonessential amino acids.** *Amino acids not synthesized by the body must be obtained in the diet and are called the* **essential** *(or indispensable)* **amino acids.** Table 21.1 lists 8 essential and 12 nonessential amino acids for humans. Two of the amino acids listed as nonessential, arginine and histidine, are nonessential for adults but are not synthesized in sufficient amounts to satisfy the needs of growing children.

Not all foods contain all the amino acids. This means that we must pay attention to the sources of protein eaten. We must make certain that the proteins we consume contain enough of the essential amino acids to satisfy our bodies' needs. **Complete proteins** *are those which contain all the essential amino acids.* Eggs, dairy products, kidneys, and liver are sources of complete proteins. Zein, the principal corn protein, is incomplete because it contains no lysine or tryptophan. Wheat protein is lacking in lysine also.

A Closer Look

The Nutrition Pyramid

In order to promote public health, the U.S. Department of Agriculture issues guidelines on the content and amount of food in a well-balanced diet. The latest guidelines, issued in 1992, replace an earlier set nearly 50 years old. The new guidelines are summarized in a pyramid, with the number of daily servings of the various food groups decreasing from the bottom to the top of the pyramid (see figure). At the base of the pyramid are grains—bread, cereal, rice, and pasta. At the top are sweets and fats, which should be eaten only sparingly. Many people are confused by what constitutes a daily serving, since this is not given on the chart. The following list may be helpful.

Breads, cereals, rice, and pasta (6 to 11 servings): A serving could be one slice of bread, half a bun or bagel, 1 ounce of dry cereal, or a half-cup of cooked cereal, rice, or pasta.

Vegetables (3 to 5 servings): A serving is 1 cup of raw, leafy greens or a half-cup of any other vegetable.

Fruit (2 to 4 servings): A serving could be one medium apple, banana, or orange; a half-cup of fresh,

cooked, or canned fruit; or three-quarters of a cup of fruit juice.

Dairy products (2 to 3 servings): A serving could be 1 cup of milk, 8 ounces of yogurt, 1.5 ounces of natural cheese, or 2 ounces of processed cheese.

Meat, poultry, dry beans, eggs, and nuts (2 to 3 servings): A total of 5 to 7 ounces of lean cooked meat, poultry, or fish a day. Count a half-cup of cooked beans, one egg, or 2 tablespoons of peanut butter as 1 ounce of meat.

The food guide pyramid.

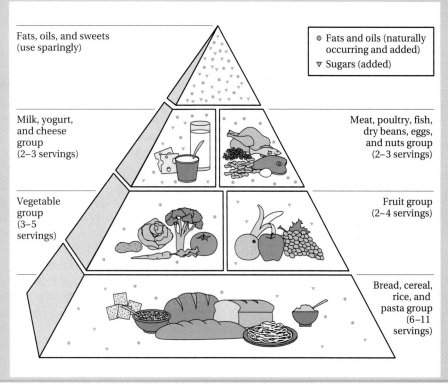

Fats, oils, and sweets (use sparingly)

- Fats and oils (naturally occurring and added)
- ▽ Sugars (added)

Milk, yogurt, and cheese group (2–3 servings)

Meat, poultry, fish, dry beans, eggs, and nuts group (2–3 servings)

Vegetable group (3–5 servings)

Fruit group (2–4 servings)

Bread, cereal, rice, and pasta group (6–11 servings)

Marasmus
marasmos (Greek):   a wasting away.

Gelatin, another incomplete protein, contains no tryptophan. Incomplete corn, wheat, and gelatin protein are easily converted to complete protein by adding the missing amino acid. Soybeans, peanuts, and potatoes are sources of complete protein, as are poultry, fish, and red meats. A combination of several foods containing protein is more likely to contain sufficient amounts of all essential amino acids than a limited selection.

A lack of total dietary calories and a lack of sufficient protein are the two most important nutritional problems that an individual may

Figure 21.4
The disease kwashiorkor is caused by a protein deficiency. The symptoms include bloating of the belly and stunted growth.

Focus

Vitamins A, D, E, and K are the fat-soluble vitamins.

The term *vitamine* was coined in 1912 by a Polish biochemist, Casimir Funk, who believed that all vitamins were amines. When it was discovered that not all vitamins were amines, the "e" was dropped.

Retinoic acid (13-*cis*-retinoic acid), a derivative of vitamin A, appears promising as a drug for treating severe acne and other skin conditions.

encounter. *Nutritional* **marasmus**—*starvation—is the name given to malnutrition caused by an insufficient total food intake. The most common protein-deficiency disease is* **kwashiorkor.** The name of this disease is an African word that means "weaned child." Kwashiorkor is most prevalent in certain parts of Africa and Asia. As the name implies, kwashiorkor often occurs in infants who have just left their mother's breast. These children are often fed a diet consisting solely of thin, starchy gruel that contains virtually no protein. Their plump faces often look healthy enough, but their bodies have underdeveloped muscles owing to a lack of protein. Moreover, such children have swollen bellies (Fig. 21.4). The blood protein serum albumin is important in maintaining a balance between blood fluid and tissue fluid volumes. The swelling of kwashiorkor is caused by the escape of water from the blood to the tissues owing to a lack of sufficient serum albumin to maintain the osmotic pressure of the blood.

21.7 Fat-soluble vitamins

AIMS: *To name the fat-soluble vitamins, indicating their dietary sources and biologic functions. To identify fat-soluble vitamins responsible for particular vitamin deficiency diseases.*

Even an intake of food of adequate caloric value and protein quality does not guarantee good nutrition. We also need vitamins and minerals. The definition of a *vitamin* has two aspects. First, vitamins are organic compounds absolutely necessary, usually in small amounts, for an animal's growth and health. And second, they are substances the animal cannot synthesize and must therefore be supplied in the diet.

Fat-soluble vitamins essential to humans are our chief interest. These are vitamins A, D, E, and K. Their chemical structures have little in common, but they are all insoluble in water, dissolve in fat, and are stored in body fat, especially liver fat. A person who has a problem absorbing dietary fats will probably also have a problem absorbing enough fat-soluble vitamins. On the other hand, taking excessive amounts of fat-soluble vitamins can cause **hypervitaminosis**—*a dangerous overaccumulation of vitamins in the tissues.*

Vitamin A Vitamin A (retinol) occurs only in animal tissues. It is an unsaturated primary alcohol and was first isolated from fish oils. The double bonds of the unsaturated carbon chain are all of the *trans* configuration.

Vitamin A

Polar bear liver is toxic to humans because of the high concentration of vitamin A it contains.

Cod liver oil is one of the richest sources of vitamin A, but dairy products usually provide enough in a normal diet. Certain vegetables are another excellent source of vitamin A, but only indirectly. Carrots and spinach contain a yellow compound, called β-*carotene,* which the body can transform to vitamin A.

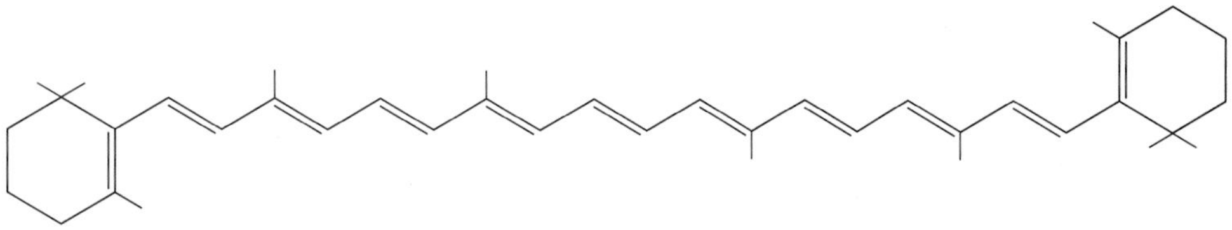

β-Carotene

One of the earliest signs of vitamin A deficiency in humans is a loss of the ability to see in dim light called *night blindness,* or *nyctalopia.* Vitamin A keeps the mucous membranes of the eye and the respiratory, digestive, and urinary tracts in healthy condition. When vitamin A is deficient, *mucous membranes become hard and dry—a process called* **keratinization.** When tear ducts become keratinized, tear secretions stop, and the outer surface of the eye becomes dry and scaly. Serious eye infections are then likely because tears normally remove bacteria from the eye. If left untreated, keratinization can cause blindness.

Vitamin D   The substances called *vitamin D* are actually a number of different compounds. Of these compounds, vitamin D_2 is most critical to human health. This vitamin promotes the uptake of calcium and phosphorus, both of which help form strong teeth and bones in growing children. Vitamin D is also involved in the maintenance of bone mass in adults. Ergosterol, a steroid abundant in yeast and many fungi, is the starting material, or precursor, for the formation of vitamin D_2. Irradiation by light transforms ergosterol into the vitamin by cleavage of one carbon-carbon bond in the ergosterol steroid nucleus. You may have noticed that the labels on some cartons of vitamin D–fortified milk say "contains irradiated ergosterol."

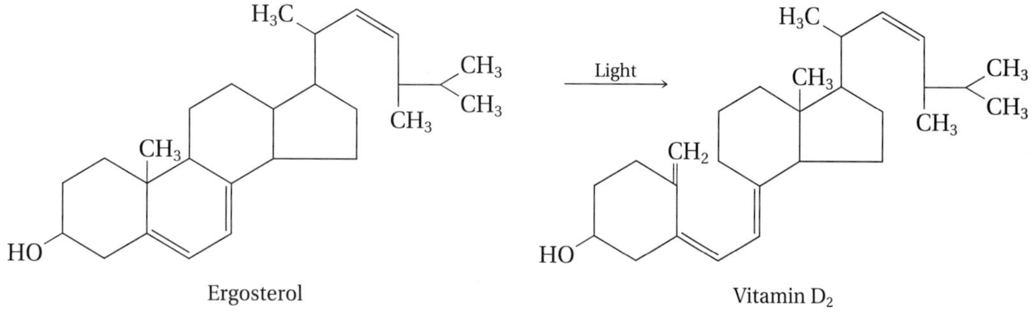

Ergosterol   Vitamin D_2

Steroids in the skin can be converted to vitamin D by sunlight, and children who play in the sun usually get enough vitamin D this way. This is why vitamin D is called the "sunshine vitamin."

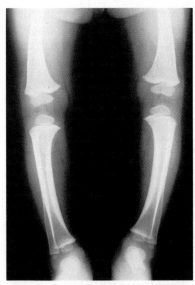

(a)

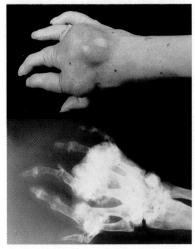

(b)

Figure 21.5
The distortion of the leg bones in rickets (a) is due to a deficiency of vitamin D; the deposition of additional bonelike material in joints (b) is caused by too much vitamin D.

Rickets, *a disease of children, is caused by a vitamin D deficiency.* Some symptoms of severe rickets are bowed legs (Fig. 21.5), an underdeveloped and abnormal formation of the ribs called "pigeon breast," and poor tooth development. These symptoms can be completely overcome by adding vitamin D to the diet. The best natural sources of vitamin D are cod, tuna, and other fish liver oils, as well as sardines and herring.

Vitamin E   Several naturally occurring compounds have vitamin E activity. Of these, α-tocopherol is the most potent. α-Tocopherol is abundant in fish oils and vegetable oils; wheat germ oil is the richest source.

α-Tocopherol

Vitamin E is used as a food additive to prevent oxidation—an antioxidant—and thereby prevent food spoilage. It extends the shelf life of foods containing unsaturated fatty acids that otherwise quickly become rancid. Vitamin E also functions as an antioxidant in the body by inhibiting the oxidation of unsaturated fats in much the same way it prevents food spoilage. An important use for vitamin E in the body appears to be as an antioxidant for vitamin A. In the absence of vitamin E, symptoms of vitamin A deficiency appear. Vitamin E is also necessary for the proper development and operation of the membranes of muscle cells and red blood cells.

Vitamin K   Vitamin K is known as the coagulation or antihemorrhagic vitamin because it is essential for the production of prothrombin, the precursor of the blood-clotting enzyme thrombin.

Vitamin K

Spinach and leafy green vegetables are good sources of vitamin K. Deficiencies are not usually due to a lack of the vitamin in the diet, because bacteria in the large intestine synthesize this vitamin, and people absorb it from the bacteria. Certain drugs such as warfarin and dicoumarin are vitamin K *antagonists*—drugs that interfere with the action of the vitamin. They are used in anticoagulation therapy to prevent potentially harmful blood clots. Both these drugs have structures similar to vitamin K and probably interfere with the synthesis of prothrombin. Warfarin is also used as a rat poison; the poisoned rodents die of severe internal bleeding.

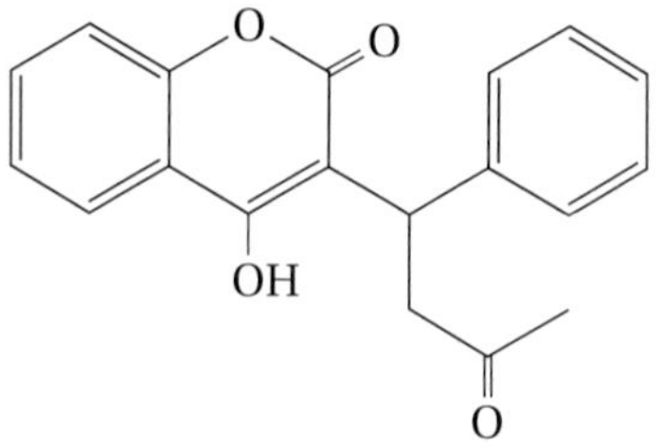

Warfarin

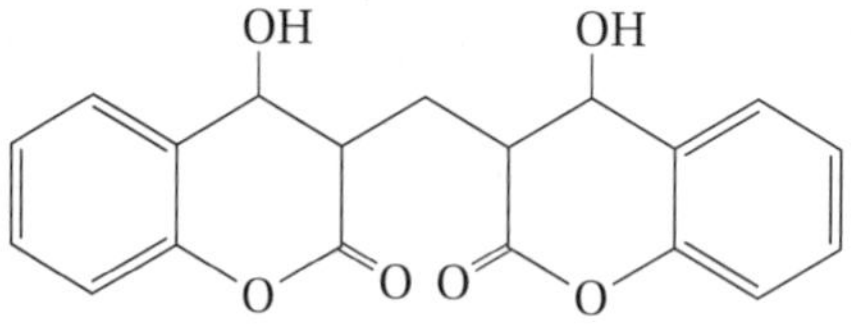

Dicoumarin

21.8 Water-soluble vitamins

AIMS: To name the water-soluble vitamins, indicating their dietary sources and biologic functions. To identify water-soluble vitamins responsible for particular vitamin deficiency diseases.

Focus

The B-complex vitamins are water-soluble.

The B-complex vitamins have only one property in common: They are soluble in water. As we will see, their biological functions are as diverse as their number. From Chapter 19 you may recall that many of the B-complex vitamins are important as enzyme cofactors. Vitamin C is also water-soluble, and we will consider it along with the B vitamins.

Because of their water solubility, the B vitamins and vitamin C are rapidly eliminated by our bodies. Hypervitaminosis is not usually a problem with these substances, as it may be with the fat-soluble vitamins. Because they are not stored in body tissue like the fat-soluble vitamins, however, the intake of water-soluble vitamins in the diet must be more frequent to maintain adequate levels.

Ascorbic acid (vitamin C)   Vitamin C is essential for the formation of the collagen in tendons, ligaments, teeth, bones, and skin.

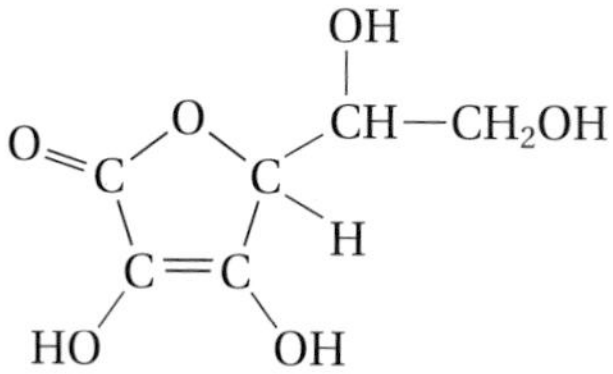

Ascorbic acid
(vitamin C)

Although vitamin C is abundant in citrus fruits and fresh vegetables, it is slowly oxidized by air and destroyed in appreciable amounts during cooking.

A deficiency of vitamin C results in **scurvy**—*symptoms include skin lesions, loose teeth, and rotting gums.* Scurvy is easily prevented by including sources of vitamin C, especially citrus fruits, in the diet. Once the bane of seafarers who had no fresh fruits and vegetables on long voyages, scurvy is still reported in the United States. Today it exists mostly as the "bachelor's scurvy" sometimes found among older widowers and bachelors, a group of individuals whose nutritional needs are often neglected. Vitamin C also promotes the absorption of dietary iron in the body.

In 1970, Linus Pauling claimed that large doses of vitamin C (1 to 1.5 g per day) could prevent the common cold. A fierce controversy ensued that is still going on.

Thiamine (vitamin B$_1$)   Meat, yeast, and unpolished grain are the main sources of thiamine in the diet. A modified form of thiamine is a cofactor in the breakdown of fatty acids.

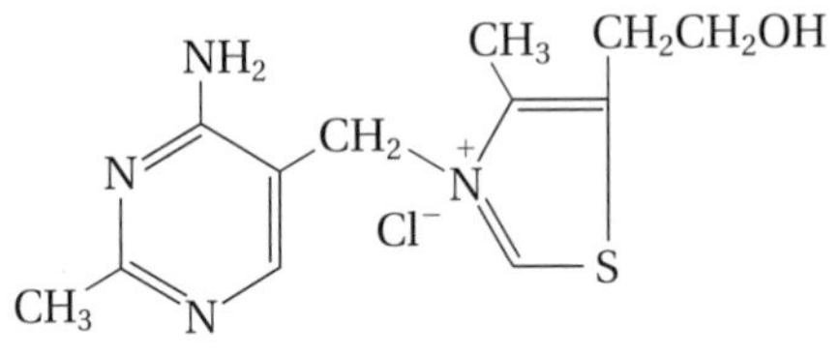

Thiamine chloride
(vitamin B$_1$)

A thiamine deficiency causes **beriberi**—*a disease that affects the nervous system and the heart.* Natives of Java coined the name *beriberi*, which means "sheep" in Javanese, because they thought that its victims walked like sheep. Typical symptoms of beriberi include pain in the arms and legs, weak muscles, and distorted skin sensations. Since most foods are low in thiamine, good nutrition is necessary to prevent beriberi. Even someone who eats sensibly may have a slight thiamine deficiency.

Addition of thiamine to some processed foods, such as white bread, has helped to lower the incidence of beriberi in the general population of the United States. It is sometimes seen in people who abuse alcohol and other malnourished people. Beriberi is still a problem in parts of Asia, where the chief food staple is rice. Rice is low in thiamine, and polished rice has none at all, since all the thiamine is in the outer layer.

Nicotinic acid (niacin)   Milk, yeast, and meat are important dietary sources of nicotinic acid. Many enzymes that catalyze oxidation-reduction reactions use a modified form of this vitamin (nicotinamide) as a cofactor.

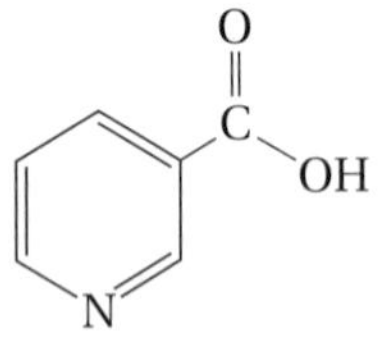

Nicotinic acid
(niacin)

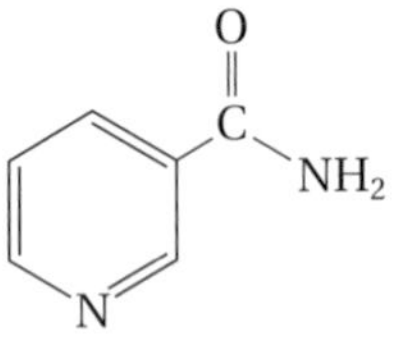

Nicotinamide
(niacinamide)

A niacin deficiency results in **pellagra,** *a disease whose symptoms include damage to the nerves as well as to the linings of the skin and intestines.* The vitamin is sometimes administered as nicotinamide, the amide of nicotinic acid. Despite the similarity in name and structure, nicotine is not converted to nicotinic acid in the body. Indeed, nicotine may interfere with the proper utilization of nicotinic acid.

Riboflavin (vitamin B$_2$) Cheese, eggs, liver, and milk are good sources of riboflavin. A modified form of riboflavin is an enzyme cofactor in certain biological oxidation-reduction reactions.

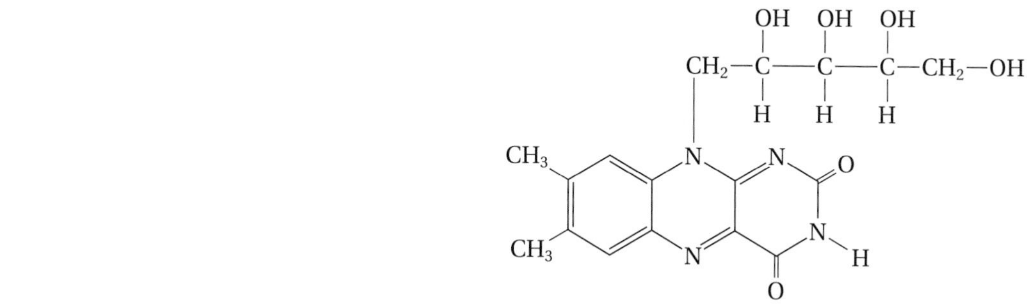

Riboflavin
(vitamin B$_2$)

The symptoms caused by a deficiency of riboflavin are rather vague, but they include general weakness, eye damage, and a reddening of the tongue.

Pyridoxine (vitamin B$_6$) Pyridoxine is the recognized form of vitamin B$_6$. In our bodies, however, the related aldehyde (pyridoxal) and the related amine (pyridoxamine) are the active forms. Modified forms of pyridoxal and pyridoxamine are cofactors for several enzymes needed in the breakdown of amino acids. Wheat germ, yeast, peanuts, corn, and meat are good sources of this vitamin.

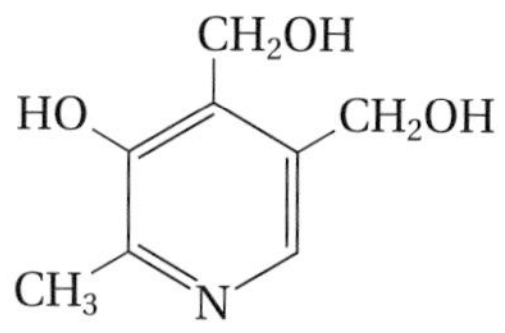

Pyridoxine 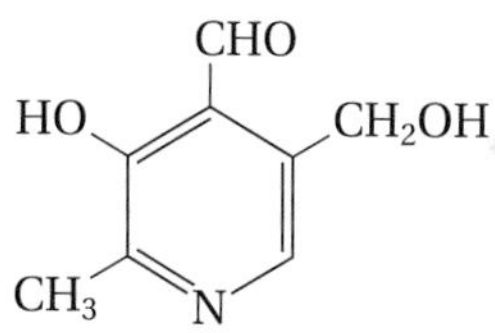   Pyridoxal 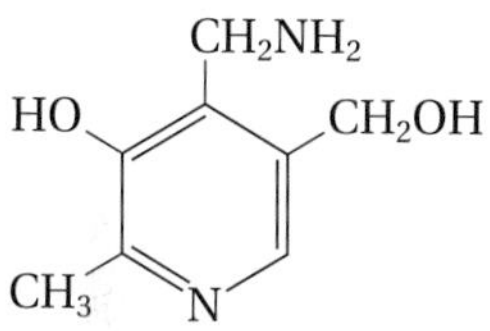   Pyridoxamine

Deficiencies of pyridoxine result in nervous system damage.

Folic acid Folic acid is a yellow compound that is only slightly soluble in water. One good dietary source is leafy green vegetables. Bacteria also manufacture the vitamin, and it is possible for humans to absorb it from intestinal bacteria. Folic acid is the cofactor for a group of enzymes that catalyze the transfer of methyl groups (CH$_3$—) between biological molecules. In this role, it is particularly important for the synthesis of the purine and pyrimidine bases of nucleic acids.

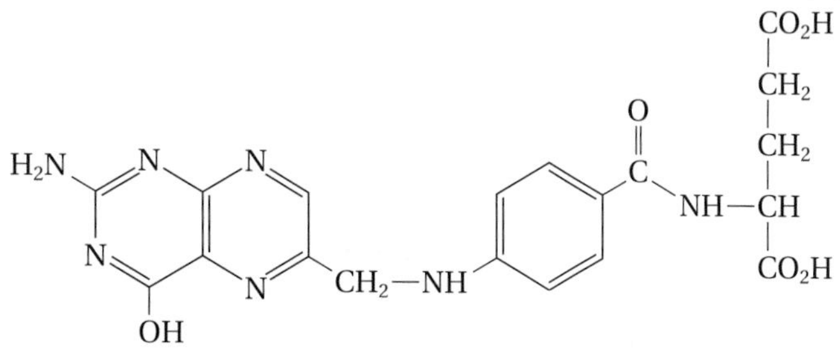

Folic acid

Anemia, a deficiency of red blood cells, results from a severe folic acid deficiency.

Pantothenic acid Pantothenic acid is a component of a cofactor that plays an important role in carbohydrate and lipid biochemistry. Pantothenic acid is abundant in such foods as liver, egg yolk, cabbage, fruits, and sweet potatoes.

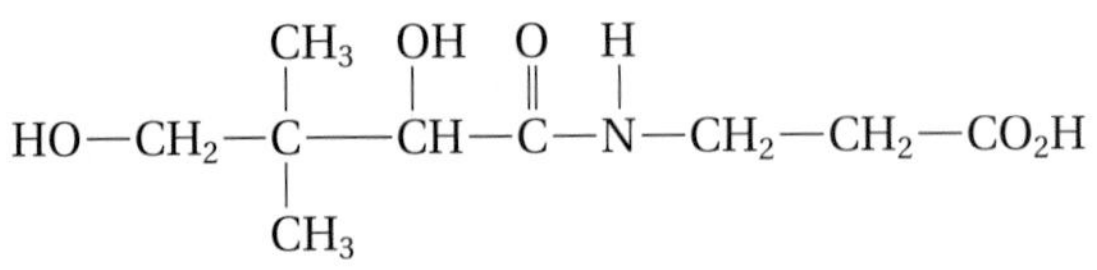

$$HO-CH_2-\underset{\underset{CH_3}{|}}{\overset{\overset{CH_3}{|}}{C}}-\underset{\overset{|}{OH}}{CH}-\overset{\overset{O}{\|}}{C}-\underset{\overset{|}{H}}{N}-CH_2-CH_2-CO_2H$$

Pantothenic acid

A deficiency results in fatigue, muscle spasms, and intestinal disturbances.

Biotin Biotin is a human vitamin, yet deficiencies hardly ever occur—apparently, enough biotin from intestinal bacteria is absorbed through the intestinal walls. Liver, kidney, milk, and molasses are excellent sources. Biotin serves as a cofactor for enzymes that use carbon dioxide to synthesize certain biological molecules.

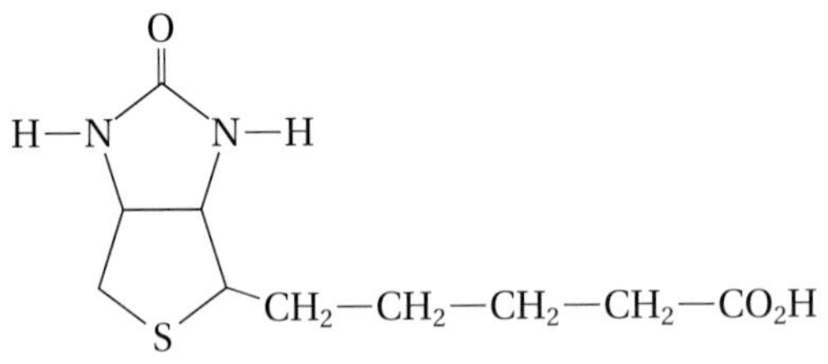

$$H-N\overset{\overset{O}{\|}}{\underset{}{C}}N-H$$

$$CH_2-CH_2-CH_2-CH_2-CO_2H$$

Biotin

Cobalamin (vitamin B$_{12}$) This vitamin is unusual because it is the only cobalt-containing organic compound found in nature (Fig. 21.6). The cobalt is present as a cobaltous or Co(I) ion bound to a corrin ring—a ring that is similar to the heme ring of hemoglobin.

Vitamin B$_{12}$ is essential for the synthesis of red blood cells. Liver is the chief source of vitamin B$_{12}$, although the vitamin is also found in eggs, milk, meat, and seafood. We need only about 2×10^{-5} g of vitamin B$_{12}$ daily. Normally, there is no difficulty in obtaining this minute amount in the diet. **Pernicious anemia** *is most often caused by an inability to absorb vitamin B$_{12}$.* Victims of pernicious anemia suffer from general fatigue and weakness. Before 1926, older people often died from pernicious anemia. Their red blood cells are immature, very large, and low in number. In 1926 it was discovered that people with pernicious anemia could benefit from eating half a pound of lightly cooked liver every day. Some 20 years later it was shown that the active ingredient in the liver is vitamin B$_{12}$. Daily injections of 1×10^{-6} g (1 μg) of pure vitamin B$_{12}$ control pernicious anemia. Since vitamin B$_{12}$ is formed principally by bacteria, there is little or

Pernicious anemia
pernecare (Latin): to kill
anaimia (Greek): without blood

We now know that pernicious anemia results from immunologic damage to the stomach lining, preventing the secretion of a substance called intrinsic factor, which aids the absorption of vitamin B$_{12}$.

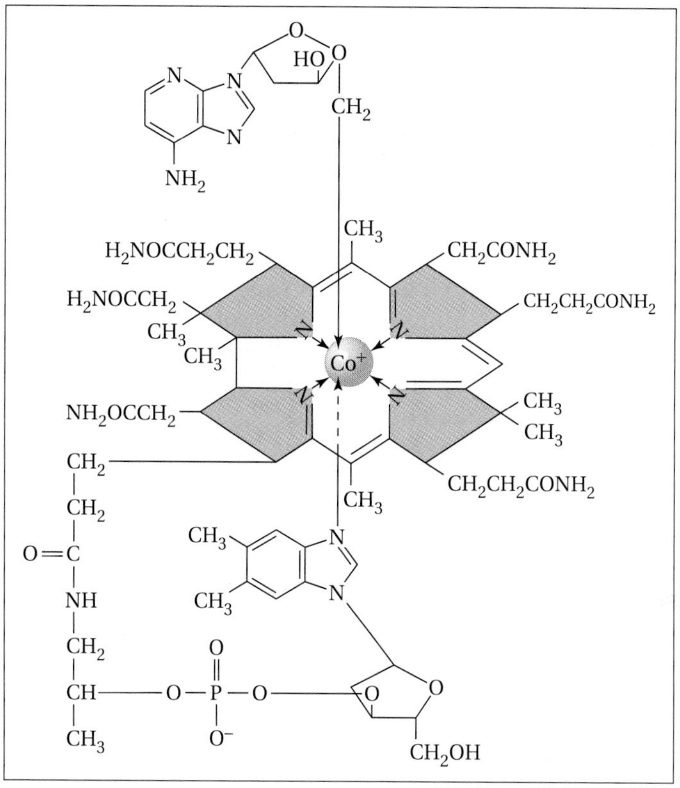

Figure 21.6
The structure of vitamin B$_{12}$. The corrin ring system (brown) contains a cobalt(I) ion (Co$^+$) (blue/gray).

none in most plants, and strict vegetarians may develop symptoms of pernicious anemia.

EXAMPLE 21.2

Determining the number of molecules of vitamin B$_{12}$ needed daily

The 2×10^{-5} g of vitamin B$_{12}$ we require daily is a small mass, but how many molecules of the vitamin does this mass represent? (Use a molar mass for vitamin B$_{12}$ of 1400 g/mol.)

SOLUTION

Remember that 1 mole of any substance is 6.02×10^{23} molecules. Use the molar mass to find the number of moles in the given mass. Then calculate the number of molecules.

$$2 \times 10^{-5}\,g \times \frac{1\ mol}{1400\ g} \times \frac{6.02 \times 10^{23}\ \text{molecules}}{1\ mol} = 9 \times 10^{15}\ \text{molecules}$$

PRACTICE EXERCISE 21.4

The recommended daily allowance (RDA) of niacin (molar mass = 122 g/mol) is 18 mg. How many molecules is this?

21.9 Minerals

AIM: To list the macronutrients and micronutrients by name and chemical formula, indicating dietary sources, functions in the body, and consequences of deficient or excessive intake.

Hydrogen, oxygen, carbon, and nitrogen are the most abundant elements in living organisms. Together they account for 99.4% of all the atoms in the human body. Most of the hydrogen and oxygen is combined as water; the remainder, together with carbon, nitrogen, sulfur, and phosphorus, makes up the molecular building blocks of life: sugars, fatty acids, amino acids, and nucleotides.

Experiments have demonstrated which **mineral elements**—*elements other than H, O, C, and N*—are essential for life. One procedure involves incineration. Technicians burn samples of plants and dead animals and analyze the elemental composition of the ashes. Scientists then vary the diets fed to test animals, excluding these elements one at a time. A detrimental effect on health, growth rate, or life span establishes the excluded element as essential. The essential elements found in relatively large amounts include metals and nonmetals. *The metals calcium, potassium, sodium, and magnesium and the nonmetals phosphorus, sulfur, and chlorine are called* **macronutrients.** *Trace amounts of many other elements are also found; those essential for life are called* **micronutrients.**

Table 21.2 lists the approximate abundance of the elements in the human body. Although these essential ingredients are described as elements, no free element, either metal or nonmetal, is present in the body. Rather, they exist as simple ions, polyatomic ions, or in covalent molecules. Besides the elements listed in Table 21.2, lead, mercury, silver, cadmium, barium, and antimony are often found in the body in trace amounts. All these metals are highly toxic. None of them seems to have any beneficial role in the human body.

Macronutrients

Calcium and phosphorus Dairy products such as milk and cheese are good sources of calcium and phosphorus. Nuts, beans, egg yolk, and shellfish also contain calcium, but the calcium in milk is more readily absorbed from the digestive tract than the calcium in vegetables. Meats and whole-wheat flour are other sources of phosphorus. Nutrition experts believe that the most desirable ratio of calcium to phosphorus in the diet is 1:1—the ratio present in milk and cheese.

Human infants need extra calcium in the first weeks of life. The demand is met by drinking mother's milk, which has a calcium/phosphorus ratio of 2:1. (A newborn infant fed only cow's milk can develop a calcium deficiency.) The recommended daily intake of calcium for young adults ages 11 to 24 and nursing and pregnant women is 1.2 g every day, the amount obtained by drinking a liter of milk. People under age 11 and over age 25 require about 0.8 g of calcium daily. Some experts suggest that women over 50 should consume 1.2 g of calcium each day.

Table 21.2 The Elemental Composition of the Body

Element	Percentage of total mass of body	Percentage of total number of atoms in body	Kilograms per 70-kg person
most abundant			
oxygen	65.0	25.5	45.500
carbon	18.0	9.5	12.600
hydrogen	10.0	63.0	7.000
nitrogen	3.0	1.4	2.100
macronutrients			
calcium	1.5	0.31	1.050
phosphorus	1.0	0.22	0.700
potassium	0.35	0.06	0.245
sulfur	0.25	0.05	0.175
chlorine	0.15	0.03	0.105
sodium	0.15	0.03	0.105
magnesium	0.05	0.01	0.035
micronutrients			
iron	0.006	0.05	0.004
zinc	0.003	0.01	0.002
copper	0.0001	< 0.01	0.0001
manganese cobalt chromium selenium iodine molybdenum	< 0.0001	< 0.01	< 0.0001
*trace amounts**			
tin vanadium nickel fluorine silicon arsenic	< 0.0001	< 0.01	< 0.0001

*Need has not been established.

About 90% of the calcium and 80% of the phosphorus in the body are present in the bones and teeth. Bone is a combination of inorganic salts and collagen. *A network of collagen fibers forms the basic structure of bone: the* **bone matrix.** A complex salt of calcium phosphate with a composition similar to hydroxyapatite $[Ca_{10}(PO_4)_6(OH)_2]$ deposits as crystals around the collagen matrix. The fibers lend flexibility and toughness; the salt lends hardness and rigidity. The dentine and enamel of teeth are similar to bone. Dentine is moderately hard and has a mineral content of 70%; enamel is harder, with a mineral content of 98%.

A calcium deficiency results in bones and teeth that do not form properly. In children, a calcium deficiency causes rickets. In the elderly, calcium is sometimes released from bone to keep blood calcium levels constant. *As a result of the loss of calcium, bones can become brittle and porous, a condition called* **osteoporosis.** In the Case in Point earlier in this chapter we learned that Bessie, an elderly farm woman, had broken her hip. In the Follow-up to the Case in Point, below, we will learn more about these fractures.

Calcium and phosphorus are found in the body not only in bones and teeth. They are present, for example, in the blood as calcium and phosphate ions. Calcium ions are required for clotting blood, maintaining heartbeat rhythm, and forming milk. Phosphorus is a component of DNA and RNA. Another phosphorus-containing molecule, adenosine triphosphate (ATP), transmits energy for nearly all body functions.

Sodium, potassium, and chlorine   Sodium ions (Na^+), potassium ions (K^+), and chloride ions (Cl^-) are three of the principal electrolytes (conduc-

FOLLOW-UP TO THE CASE IN POINT: Osteoporosis

It is common to have a great-aunt or grandmother who, like Bessie, has broken a hip. Most elderly victims of broken hips suffer from osteoporosis, a crippling disease characterized by the loss of calcium from bones. The bones of the spine, hips, and wrists are in general affected mostly in older people, postmenopausal women in particular. Initially, health scientists thought that raising the dietary intake of calcium would cure the disease or at least prevent it from occurring. However, increased calcium intake does not always slow the loss of calcium from bone. Instead, recent research shows that the cause of osteoporosis is closely related to regulation of the release and uptake of calcium in bones. Several hormones are involved in regulation of the calcium levels in the blood and bones. Of these, the principal one is *parathyroid hormone* (PTH). When the concentration of calcium in the blood is low, PTH is released. PTH stimulates the release of calcium from bone into the bloodstream. PTH also stimulates calcium retention in the kidneys and the adsorption of dietary calcium in the intestines. When the blood level of calcium is high, the thyroid gland secretes the hormone calcitonin. The effects of calcitonin are the opposite of the effects of PTH. Consequently, in response to the calcium level in the blood, bones are constantly being broken down and built up. Osteoporosis results if more calcium from bone is lost than is deposited (see figure, part a). Postmenopausal women may lose bone at a rate of more than 1% per year.

Load-bearing exercises such as walking and running help prevent the bone depletion of osteoporosis. Eating foods that are rich in calcium and vitamin D also is helpful, since vitamin D aids in the absorption of dietary calcium. The steroid hormone estrogen, which declines in postmenopausal women, also plays a role in calcium deposition in bone. Many postmenopausal women are aided by estrogen replacement therapy, which helps prevent bone depletion. The current belief is that healthful habits begun in the teens and continued throughout adulthood will develop dense bones. As a result, the chance of developing osteoporosis in later life will be reduced.

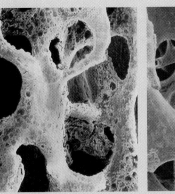

(a) **(b)**

The bone of a person with osteoporosis (a) has a lower density than normal bone (b).

tors of electricity in aqueous solutions) in the body. Electrolytes help keep the volume of body fluids constant. Sharp changes in body fluid volume or electrolyte concentration often indicate illness. Abnormal loss of water (*dehydration*) and abnormal retention of water (*edema*) are examples of the loss of control of body fluid volumes. Although sodium and potassium ions are similar in their physical and chemical properties—both are Group 1 metals—they cannot replace each other in the body. Sodium ions are the principal cations of blood, and potassium ions are the principal cations inside cells. Both sodium and potassium ions help maintain nerve responses at normal levels; potassium ions exert a relaxing effect on heart muscle between heartbeats.

Sodium chloride is the main source of sodium in the diet. Bread, cheese, carrots, celery, eggs, milk, oatmeal, and clams all are high in sodium. The recommended daily intake for people who do not have a history of high blood pressure is about 5 g of sodium chloride per day, which is about one-half the amount many people consume. For people with high blood pressure, a daily intake of less than 1 g of sodium chloride is usually recommended. Most of the sodium entering the body is absorbed in the small intestine; about 95% of the sodium ions excreted are in the urine.

Potassium occurs in all food, so a dietary deficiency is uncommon. The daily requirement for an adult is about 1 g. Foods high in both sodium and potassium are not always suitable, however, because people who need potassium are often on low-sodium diets. Foods high in potassium but low in sodium include beef and beef liver, chicken, pork, bananas, dried apricots, orange juice, broccoli, potatoes, and pineapples.

The main source of chlorine in the diet is table salt (sodium chloride). In both ingestion and excretion, sodium ions and chloride ions are inseparable; a low-salt diet produces decreased excretion of both sodium ions and chloride ions in the urine. Excessive sweating and diarrhea produce a simultaneous loss of both sodium ions and chloride ions from the body. A very important use of chloride ions by the body is in the production of hydrochloric acid in the stomach.

Magnesium Most of the magnesium (Mg^{2+}) in the body is combined with calcium and phosphorus in the bone. High concentrations are found in muscles and red blood cells, while the remainder is distributed throughout body tissues and fluids. A normal diet provides an adequate supply of magnesium. Foods rich in this element include green vegetables, soybeans, nuts, dried peas and beans, and whole-grain cereals. The recommended dose is 300 mg per day.

Magnesium is essential in nerve impulse transmission and muscle contraction. A magnesium deficiency causes muscle tremors, twitches, and convulsions, sometimes accompanied by behavioral disturbances. Alcohol consumption leads to an increased loss of magnesium from the body. Since magnesium is a depressant, intravenous injection of magnesium salts produces anesthesia and even paralysis, but it also can reduce convulsions. Magnesium ions are not readily absorbed through the intestinal wall, so they pull water into the intestine from adjacent tissues by osmosis. Compounds such as Epsom salts ($MgSO_4 \cdot 7H_2O$) are used as laxatives for this reason.

Sulfur   Sulfur is present in every cell of the body and in most food proteins. A normal diet provides an adequate supply. In proteins, sulfur is a component of certain amino acids. The proteins of hair, fingernails, and feathers are rich in sulfur-containing amino acids, which is why these materials smell so offensive when burned. Sulfur is also present as sulfate ions in the blood and other body fluids. In the breakdown of food in the body, sulfur is eventually oxidized to sulfate and excreted in the urine. Low concentrations of sulfuric and phosphoric acids give normal urine a slightly acidic pH.

Micronutrients

Iron ions are cofactors of the cytochromes, proteins that transport electrons in cellular respiration.

Iron   The human body contains less than 5 g of iron. The greatest dietary need is during the first 2 years of life and for women during childbearing years. The recommended allowance is 10 mg per day for an adult male and 20 mg per day for an adult female. The best sources of iron are liver, kidney, and green vegetables. Other sources are egg yolk, brewer's yeast, fish, whole wheat, nuts, oatmeal, molasses, and beans. Absorption of iron as iron(II) ions (Fe^{2+}) takes place in the stomach and upper part of the small intestine. The presence of vitamin C, which promotes the reduction of dietary iron(III) ions (Fe^{3+}) to iron(II) ions (Fe^{2+}) assists the absorption. Nearly all the iron in the body is continually reabsorbed and reused.

A healthy adult loses about 1 mg of iron every day in the urine, sweat, and feces. Ordinarily, this loss is easily made up by the diet. However, women who have excessive loss of menstrual blood or who are in the late stages of pregnancy or persons suffering from iron-deficient anemia need to take 100 mg of iron daily, as iron(II) salts, to correct the deficiency. Surgical removal of the stomach or intestines results in *diminished iron absorption leading to* **iron-deficient anemia,** *which is characterized by fatigue.* High concentrations of iron in the body are harmful and may lead to congestive heart failure or cirrhosis of the liver.

Carbonic anhydrase, carboxypeptidase, DNA polymerase, and RNA polymerase are a few of the enzymes that require Zn^{2+} ions for catalytic activity.

Zinc, copper, manganese, molybdenum, and cobalt   Zinc is essential for normal growth, life span, and reproduction in plants and animals. About 2 g of zinc is present in the body of an adult. Most of it is concentrated in the skin; bones and teeth contain lesser amounts. Zinc is also present as a cofactor in many enzymes. The recommended daily allowance for healthy adults is 15 mg of zinc per day, which is easily obtained in a normal diet. Liver, eggs, meat, milk, whole-grain products, and shellfish are rich sources of zinc. Zinc deficiency results in an impaired sense of taste and a poor appetite; it also causes slow healing of wounds and in severe cases produces dwarfism.

The human body contains about 100 mg of copper, distributed mainly in the muscles, bone, liver, and blood. The copper concentration in the blood of an adult is normally in the range 100 to 200 $\mu g/100$ mL of blood. The amount of copper in the human body, though minute, is an essential component of several vital enzymes. One such enzyme is involved in the formation of melanin, a skin pigment that helps protect us from harmful rays of the sun. The proper formation and maintenance of the protective coat of nerves also depend on copper. In instances of severe copper deficiency, this coat becomes defective, and degeneration of the nervous system occurs.

The formation of hemoglobin in the body depends on traces of both copper and iron. A decrease in the formation of hemoglobin occurs when there is a normal intake of iron but a deficiency of copper. *People with an inherited disorder called* **Wilson's disease** *accumulate copper in the liver and brain.* These people excrete copper in their urine, but copper is often undetectable in the blood. The large deposits of copper in the liver can cause damage, and cirrhosis often develops. A balanced diet supplies the recommended 2 mg of copper per day for an adult. Foods that are a good source of copper include liver, kidney, raisins, nuts, dried peas and beans, and shellfish.

The functioning of the central nervous system and the thyroid gland, as well as the formation of normal bone and cartilage, depends on a small amount of manganese in the body. The total manganese content of the body is only about 15 mg, most of which is stored in the liver and kidney. An average daily intake of about 5 mg is satisfactory.

Molybdenum is probably absorbed into living systems as the molybdate ion (MoO_4^{2-}) from such foods as yeast, liver, kidney, beans, and peas. If the levels of molybdenum or zinc in the body are high, the absorption of copper decreases, and symptoms of copper deficiency appear.

Cobalt, as we have seen, is an essential component of vitamin B_{12}.

Selenium and chromium   Selenium is a nonmetal of the same chemical group as oxygen and sulfur. We learned in Chapter 4 that selenium is extremely toxic; in minute amounts it is an essential human nutrient. Chromium is an essential trace element in all plant and animal tissues. The estimated amount of chromium in the body of an adult is only about 6 mg. Chromium deficiency in animals has been shown to impair growth and reduce life span. Some nutrition experts believe that the impaired ability of many middle-aged people to use glucose might be reversed by chromium supplements.

Iodine   Nearly all the iodine in the body is concentrated in the thyroid gland. Iodine is needed only for the synthesis of iodine-containing thyroid hormones. One such hormone, thyroxine, is involved in controlling body growth and regulating basal metabolism.

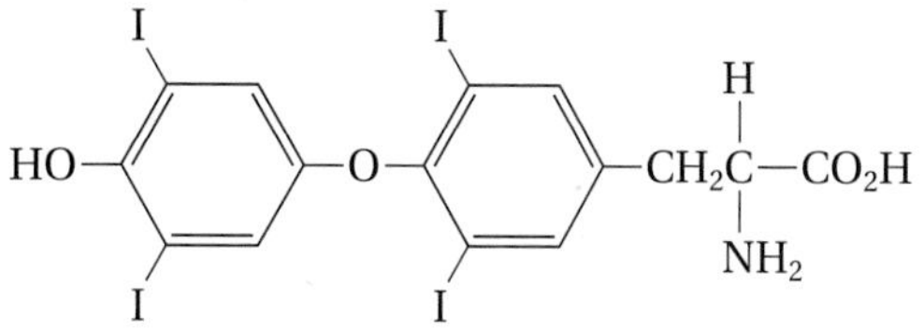

Thyroxine

A normal diet usually provides the daily requirement for iodine, about 100 μg. The need for iodine is highest during adolescence and in pregnancy. People who have an iodine deficiency may develop **goiter**—*an enlarged thyroid gland.* Simple goiter occurs as the gland grows larger in an attempt to compensate for the low iodine content in the diet. The widespread use of *iodized salt*—table salt to which a small amount of sodium iodide has been added—has virtually eliminated simple goiter.

21.10 Trace elements

AIM: To name the trace elements found in the body, and state some of their possible functions.

Focus

Fluorine, nickel, tin, vanadium, silicon, and arsenic may be essential trace elements.

Many other elements are present in trace quantities in living systems, but it is hard to establish whether they are essential or merely passing through. Fluorine, nickel, tin, vanadium, silicon, and arsenic are essential trace elements in animals and also may be required by humans, though their exact functions have not been established. Deficiencies of these elements in test animals result in reduced growth rates. Since nickel deficiency also results in damage to liver cells, it has been suggested that nickel might play a role in the structure of liver cell membranes. Tin and vanadium appear to have an effect on fat biochemistry, silicon seems to be involved in the structure of skin and connective tissue, and arsenic appears to influence the ability to reproduce.

Fluorine accumulates in bones and teeth as fluoride ion. The concentration there depends on a person's age and fluoride ion intake. Fluoride ions are rapidly absorbed by the body and are distributed mostly in the fluid surrounding cells, though some are retained in the bones and teeth. The fluoride ions that are not retained are rapidly excreted in the urine.

Years of research have established that fluorine, as the fluoride ion, is effective in protecting teeth against dental caries (cavities). Today, many municipalities add fluoride to drinking water at a level of about 1 ppm. This concentration corresponds to a daily intake of about 1 or 2 mg. The fluoride ion replaces a small fraction of the hydroxide ions (OH^-) in the hydroxyapatite mineral of the bones and teeth; the modified hydroxyapatite has much greater mechanical strength than the non-fluorine-containing mineral.

Although fluoride ion intake of about 1 to 2 mg per day is beneficial for the maintenance of strong, healthy teeth, higher or lower amounts are detrimental. A regular daily intake of about 2 to 4 mg per day of fluoride causes teeth to become discolored or mottled; the teeth are hard but brittle. When the daily intake is about 10 to 20 mg per day, fluorosis occurs. **Fluorosis** *is characterized by severe mottling of the teeth (dental fluorosis) and abnormal changes in bone growth (skeletal fluorosis).* If fluoride ion is absent from drinking water or present below 0.5 ppm, there is a noticeable increase in the incidence of dental caries, but there is no discoloration or mottling.

PRACTICE EXERCISE 21.5

Match the following:

Medical condition

(a) simple goiter

(b) pancreatitis

(c) pellagra

(d) kwashiorkor

(e) poor coagulation of blood

(f) night blindness and keratinization of mucous membranes

(g) Wilson's disease

(h) fluorosis

(i) osteoporosis

(j) scurvy

(k) pernicious anemia

Cause or result of condition

(1) niacin deficiency

(2) vitamin K deficiency

(3) copper accumulation in liver and brain, but no copper in blood

(4) iodine deficiency

(5) vitamin C deficiency

(6) premature activation of pancreatic enzymes

(7) vitamin B_{12} deficiency

(8) protein deficiency

(9) excessive fluoride ion in diet

(10) vitamin A deficiency

(11) excessive loss of bone calcium

SUMMARY

Sugars, lipids, proteins, and nucleic acids, the sources of energy and materials for growth in living organisms, are absorbed by the human body in digestion. All digestion processes involve enzyme-catalyzed hydrolysis reactions: hydrolysis of complex carbohydrates to monosaccharides, hydrolysis of lipids to fatty acids and other products, hydrolysis of proteins to amino acids, and hydrolysis of nucleic acids to nucleotides. Bile salts aid the digestion of fats and lipids by emulsifying them so that they are susceptible to attack by hydrolyzing enzymes called lipases.

A balanced diet requires proper amounts of the carbon compounds mentioned above, as well as vitamins and minerals. The protein consumed must contain sufficient amounts of the eight essential amino acids, since these acids are not synthesized by humans. Moreover, human beings do not make vitamins, the organic molecules needed for health and growth. The vitamins are divided into two categories: fat-soluble and water-soluble. The fat-soluble vitamins required by humans are A, D, E, and K. The water-soluble vitamins include the B-complex vitamins and vitamin C. The water-soluble vitamins are precursors of cofactors of enzymes that catalyze many important chemical reactions in the body.

Four elements—hydrogen, oxygen, carbon, and nitrogen—account for more than 99% of all the atoms in the human body. Twenty-two other elements, known as the mineral elements, make up the remainder. Although the mineral elements constitute a very small portion of the total number of atoms in the body, they are nevertheless essential—life cannot exist without them. The mineral elements are classified as either macronutrients or micronutrients. The seven macronutrients are the metals calcium, magnesium, sodium, and potassium and the nonmetals phosphorus, sulfur, and chlorine. Ten elements (iron, copper, zinc, manganese, cobalt, molybdenum, chromium, selenium, iodine, and fluorine) are all essential but are present only in trace amounts in the body—these are the micronutrients. Other trace elements that are present include nickel, tin, vanadium, arsenic, and silicon. These elements are essential in test animals and may be essential to humans.

KEY TERMS

Aerobic cell (21.1)
Basal metabolism (21.5)
Beriberi (21.8)
Bile (21.4)
Bile pigment (21.4)
Bile salt (21.4)
Bone matrix (21.9)
Caloric value (21.5)
Chylomicron (21.4)

Complete protein (21.6)
Essential amino acid (21.6)
Fluorosis (21.10)
Gastric juice (21.3)
Goiter (21.9)
Hypervitaminosis (21.7)
Iron-deficient anemia (21.9)
Keratinization (21.7)

Kwashiorkor (21.6)
Lipoprotein (21.4)
Macronutrient (21.9)
Marasmus (21.6)
Micronutrient (21.9)
Mineral element (21.9)
Nonessential amino acid (21.6)
Nucleotidase (21.4)

Osteoporosis (21.9)
Pancreatic juice (21.4)
Pellagra (21.8)
Pernicious anemia (21.8)
Rickets (21.7)
Saliva (21.2)
Scurvy (21.8)
Vitamin (21.1)
Wilson's disease (21.9)

EXERCISES

Digestion (Sections 21.1–21.4)

21.6 What is digestion, and where does it begin?

21.7 Name the enzyme contained in saliva, and describe its function. What is mucin?

21.8 What makes gastric juice acidic?

21.9 What is the principal zymogen in gastric juice?

21.10 How is pepsinogen converted to pepsin?

21.11 Why is pepsin called a *peptidase?* At what pH is pepsin most active?

21.12 What classes of compounds are digested in the small intestine?

21.13 Name the enzymes secreted in pancreatic juice.

21.14 Why must fats be emulsified before they can be digested?

21.15 (a) What is bile? (b) Where is it produced? (c) What is its function in lipid digestion?

Dietary Needs (Sections 21.5, 21.6)

21.16 Explain basal metabolism.

21.17 Which nutrients have the highest caloric value?

21.18 What other items do we need in our diet apart from sufficient calories?

21.19 Distinguish between essential and nonessential amino acids.

21.20 What are complete proteins?

21.21 Which foods are a source of (a) complete protein and (b) incomplete protein?

21.22 Describe the difference between nutritional marasmus and kwashiorkor.

Vitamins (Sections 21.7, 21.8)

21.23 List the fat-soluble vitamins, and describe an important function of each.

21.24 Why is it dangerous to take excessive amounts of fat-soluble vitamins?

21.25 Why must we have a frequent intake of water-soluble vitamins?

21.26 The following diseases are all caused by a vitamin deficiency. Name the vitamin that is responsible, and identify it as fat-soluble or water-soluble.
(a) rickets   (b) pellagra   (c) scurvy
(d) beriberi   (e) nyctalopia
(f) pernicious anemia

Macronutrients, Micronutrients, and Trace Elements (Sections 21.9, 21.10)

21.27 List the seven macronutrients, and write the formula for each.

21.28 Identify the macronutrients as metallic elements or nonmetallic elements.

21.29 Name three principal electrolytes in the body.

21.30 Respond to this statement: "A potassium deficiency in the diet can be compensated by an increased sodium intake."

21.31 Explain why the calcium/phosphorus ratio in the diet is an important factor in calcium absorption.

21.32 What is the result of (a) an excessive intake of calcium and (b) an excessive intake of phosphorus?

21.33 In what way is growth affected if there is an inadequate supply of both calcium and phosphorus in the diet?

21.34 Organic compounds, such as oxalic acid, form insoluble substances with calcium. Why should this reaction lead to a calcium deficiency?

21.35 What are the effects of rickets?

21.36 What is the major difference between sodium ions and potassium ions with respect to their occurrence in the body?

21.37 List the nine micronutrients, and write the chemical symbol for each element.

21.38 Identify the micronutrients as metallic elements, nonmetallic elements, or metalloids.

21.39 What foods are rich in iron?

21.40 Iron exists as Fe^{2+} and Fe^{3+}. (a) Which of these forms is most readily absorbed by the body? (b) How does the presence of vitamin C help the absorption?

21.41 Why do we need iron? What happens if our diet is deficient in iron?

21.42 Comment on this statement: "The majority of iron in the diet is absorbed by the body and used in the synthesis of hemoglobin."

21.43 What metallic element, besides iron, is responsible for hemoglobin formation? Give the name of the inherited disease that causes an abnormal distribution of this element in the body.

21.44 What is the function of iodine in the body?

21.45 Explain why a lack of iodine in the diet causes simple goiter.

21.46 What is the major source of fluorine in the diet? Is fluorine an essential element?

21.47 In what respect is fluorine beneficial? What is the cause of mottled teeth?

21.48 Describe one function of each of the following trace elements.
(a) molybdenum   (b) zinc   (c) manganese
(d) cobalt   (e) chromium   (f) selenium
(g) nickel   (h) tin   (i) vanadium
(j) silicon   (k) arsenic

Additional Exercises

21.49 Name at least two sources of folic acid.

21.50 What are the end products of the complete digestion of
(a) proteins?   (b) carbohydrates?
(c) triglycerides?   (d) nucleic acids?

21.51 Name the enzymes or zymogens in (a) saliva, (b) gastric juice, and (c) pancreatic juice.

21.52 Describe, briefly, how proteins and sugars are digested and absorbed in the intestinal tract.

21.53 Discuss the digestion and absorption of lipids.

21.54 Give a definition and example for the following.
(a) macronutrient   (b) micronutrient
(c) mineral element

21.55 Name the water-soluble vitamins.

21.56 Describe the disease condition that results in a deficiency of each of the following:
(a) thiamine   (b) vitamin C   (c) vitamin B_{12}
(d) vitamin D   (e) vitamin A

21.57 What is the major function in the body of each of the following mineral elements?
(a) cobalt   (b) chromium   (c) copper
(d) zinc   (e) manganese   (f) magnesium

SELF-TEST (REVIEW)

True/False

1. Potassium ions can substitute for sodium ions in the body.
2. Some micronutrients can be toxic to the body.
3. A common function of vitamins is to act as an enzyme cofactor.
4. Most grains are sources of high-quality protein.
5. A person can never take too much of a vitamin.
6. Marasmus and kwashiorkor describe the same condition.
7. Essential amino acids cannot be synthesized in the body.
8. The four most abundant elements in the body are carbon, hydrogen, oxygen, and calcium.
9. Some trace elements function as enzyme cofactors.
10. A lipase catalyzes the breakdown of lipid molecules.

Multiple Choice

11. The enzyme responsible for the hydrolysis of proteins in the stomach is
    (a) pepsin.   (b) chymotrypsin.
    (c) amylase.   (d) hydrolase.
12. Compounds of the following macronutrient are used as laxatives.
    (a) iron   (b) sulfur
    (c) magnesium   (d) calcium
13. An iodine deficiency causes
    (a) simple goiter.   (b) Wilson's disease.
    (c) pellagra.   (d) osteoporosis.
14. An incomplete protein is one that
    (a) has a low amino acid concentration.
    (b) is a result of a defect in amino acid metabolism.
    (c) lacks one or more essential amino acids.
    (d) is produced in the urea cycle.

15. The minimum energy output for a person to survive
    (a) is the same regardless of the person's age or sex.
    (b) is the basal metabolism.
    (c) is greater the older the person gets.
    (d) more than one are correct.

16. Which of the following is *not* a fat-soluble vitamin?
    (a) vitamin A   (b) vitamin K
    (c) vitamin C   (d) vitamin D

17. Most nutrients are absorbed into the bloodstream in the
    (a) stomach.   (b) large intestine.
    (c) small intestine.   (d) colon.

18. Which of the following is *not* a trace element?
    (a) tin   (b) arsenic
    (c) vanadium   (d) sulfur

19. A patient who has her gallbladder removed should control the dietary intake of
    (a) carbohydrates.   (b) fats.
    (c) proteins.   (d) all of these.

20. Carboxypeptidase is associated with the digestion of
    (a) protein.   (b) fats.
    (c) carbohydrates.   (d) nucleic acids.

21. Lipids are transported to body tissues
    (a) through the lymphatic system only.
    (b) in a combined state with lipoprotein.
    (c) in bodies called *chylomicrons*.
    (d) more than one are correct.

22. The following fat-soluble vitamin is important in the blood-clotting process.
    (a) vitamin K   (b) vitamin B_1
    (c) vitamin D   (d) folic acid

23. Vitamin B_{12}
    (a) contains a cobalt(II) ion.
    (b) aids in the synthesis of red blood cells.
    (c) deficiency causes pernicious anemia.
    (d) all of the above are correct.

Body Fluids

Maintaining the Body's Internal Environment

Replenishing liquids to maintain proper fluid balance is important for everyone.

CHAPTER OUTLINE

CASE IN POINT: A diabetic imbalance

22.1 Body water

A CLOSER LOOK: Artificial Skin

22.2 Blood

A CLOSER LOOK: Blood: Risks and Replacements

22.3 Antibodies and interferons

22.4 Blood buffers

22.5 Oxygen and carbon dioxide transport

22.6 The urinary system

22.7 Acid-base balance

22.8 Acidosis and alkalosis

FOLLOW-UP TO THE CASE IN POINT: A diabetic imbalance

22.9 Water and salt balance

The maintenance of a consistent internal environment in the human body is an interactive and dynamic process. Organs and organ systems working together continuously adjust imbalances that occur. A breakdown in any part of the collaborating systems of the body can cause serious illness or even death. What role does chemistry play in these systems? Consider Larry's plight in the following Case in Point.

CASE IN POINT: A diabetic imbalance

Larry, a 32-year-old teacher, collapses on the steps of the public library. The paramedics discover that he wears a medical alert tag identifying him as a diabetic. Upon his arrival at the hospital, the attending physician immediately orders several lab tests. One of these tests is measurement of the pH of Larry's blood. What imbalance does Larry's doctor suspect? And what does the chemical concept of pH have to do with Larry's illness? We will find out in Section 22.8.

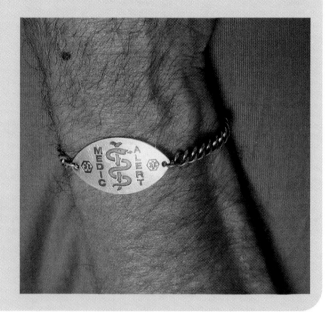

In case of an emergency, many people display their medical condition on a Medic Alert tag.

22.1 Body water

AIMS: To describe the distribution of body water and compare the electrolyte compositions of blood plasma, interstitial fluid, and intracellular fluid. To define homeostasis and cite some examples of homeostatic control in the body.

Focus

Tissue cells are bathed in fluid inside and out.

The average adult body contains 42 L (42 kg) of water. (Remember that the mass of 1 L of water is 1 kg.) This 42 L of water is compartmentalized in three regions of the body. The majority, about 28 L, is located in the **intracellular fluid**—*the fluid inside cells.* About 11 L of water is in the **interstitial fluid**—*fluid that fills the space between tissue cells.* *Lymph,* a fluid discussed later, may be considered part of the interstitial fluid. Most of the remaining 3 L of water is present in the fluid portion of the blood. Most body fluids consist of water and its dissolved substances. The retention of these fluids is important. See A Closer Look: Artificial Skin.

Figure 22.1 shows the composition of the major body fluids. As the figure shows, the salt composition of *blood plasma*—the fluid portion of the

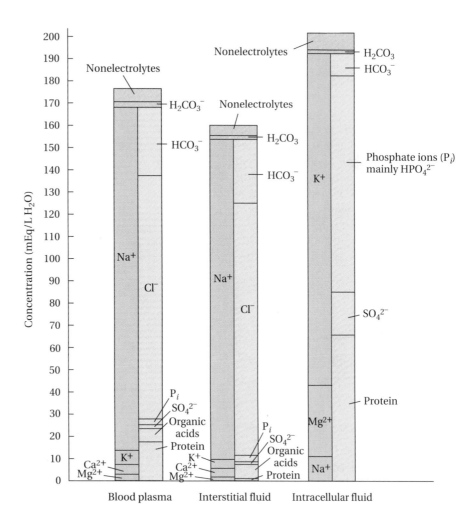

Figure 22.1
Composition of the major body fluids.

Artificial Skin

Many burn victims lose large areas of skin. These people often die because vital fluids ooze out of their bodies, and they are too weakened to fight off bacterial infection. The essential requirement in treating burns is covering the damaged area quickly. When there is too little unburned skin on the patient's body for grafting, surgeons use skin from human cadavers or pigs. These grafts are usually rejected by the body after a few days and must be replaced.

An artificial skin that shows encouraging results was first used in the 1980s. The body does not recognize the artificial skin as foreign, and drugs that prevent rejection are not required. The wounds heal with little scarring.

Like natural skin, artificial skin has two layers (see figure). The bottom layer is a blend of a complex carbohydrate obtained from shark cartilage and collagen (a protein) extracted from cowhide. These components are added to water and acidified to produce short white fibers. The mixture is poured into a shallow pan and freeze-dried to remove the water. The fibers form a thin white sheet of material that is light and highly porous. The sheet is then baked in an oven to preserve its shape. A top layer is added. This layer, made from a thin sheet of a rubber-like plastic, is bonded with an adhesive to the fibrous sheet. The completed sheet of synthetic skin is soft and pliable as natural skin and about as thick as a paper towel. It is freeze-dried again and stored in a sterile container at room temperature until required. Ten square feet of material, enough to cover wounds over 50% of an adult body, can be made in a few days.

Artificial skin reacts with human flesh as does natural skin. The fibrous bottom layer, which is placed next to the burned area, is porous and provides scaffolding into which body cells migrate to make more collagen. Over a period of months, collagen in the artificial skin breaks down in the same way as collagen in healthy skin and sloughs off. Nerve fibers, which are still alive, and blood vessels in the flesh grow up into the new material. The upper plastic layer does not become part of the body but serves only as a protective flexible covering. Within a week or so after grafting, small areas of this layer are removed and replaced with thin patches of the patient's own skin. The need to cover the artificial graft with natural skin, although inconvenient, is not a serious drawback because it can be done later when the patient's condition is improved.

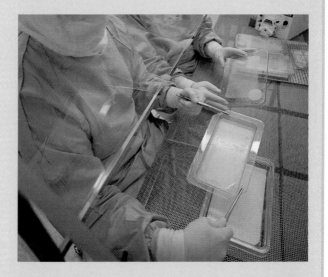

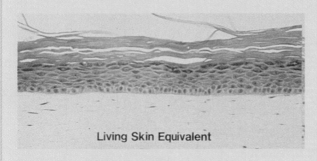

Living Skin Equivalent

A skin graft using artificial skin can be a lifesaving procedure.

blood—and interstitial fluid is nearly identical. Sodium ions and chloride ions are the major electrolytes of these two fluids. In contrast, the electrolyte composition of intracellular fluid is quite different from that of the extracellular fluids. Potassium ions and HPO_4^{2-} ions are the major inorganic electrolytes of the intracellular fluid—a reflection of the tendency of cells to transport potassium ions and HPO_4^{2-} ions from low concentrations in the extracellular fluid to higher concentrations in the intracellular fluid (cytoplasm). Cells also transport sodium ions from a low concentration in the cell to higher concentrations in the extracellular fluid. The higher protein concentration of intracellular fluid compared with the extracellular fluids results from the high concentration of water-soluble enzymes in the cytoplasm of cells.

Temperature, energy production, pH of fluids, and the levels of salts, water, and nutrients all must be carefully controlled to maintain an environment in which body cells can operate at peak efficiency. **Homeostasis** *is the process that tends to keep all these crucial factors in balance.*

The composition of seawater and interstitial fluid are very similar. Does this give credence to the theory that the first living cells originated in the sea?

Homeostasis
homo (Greek):   the same
stasis (Greek):   condition

22.2 Blood

AIM: *To list the functions of blood, the formed elements of blood, and the plasma proteins and distinguish between blood serum and blood plasma.*

Focus

Blood has many important functions.

Supplies of food and oxygen are carried to tissues and wastes are carried away from tissues by circulating blood. Fluid received from the blood bathes tissues and maintains their pH at slightly alkaline values. Blood helps to equalize body temperature by evenly distributing heat generated by active cells. Blood carries hormones that help coordinate the activities of the different body organs. Cells and proteins of the blood help defend the body against infecting organisms.

Formed elements of blood

On average, 1 mL of blood from an adult contains about 5×10^6 red cells, 7×10^3 white cells, and 250×10^3 platelets.

Within the blood are red blood cells, white blood cells, and platelets. **Erythrocytes** *(red blood cells) carry oxygen to tissues and remove carbon dioxide.* Hemoglobin, the chief carrier of oxygen and carbon dioxide in the blood, is packed into the erythrocytes. Each erythrocyte contains about 70 million molecules of hemoglobin. Blood also contains *leukocytes* and *thrombocytes.* **Leukocytes** *(white blood cells) perform many functions in cells, including the destruction of foreign organisms.* **Thrombocytes** *(platelets) are cells that play an important role in blood clotting. Erythrocytes, leukocytes, and thrombocytes are called the* **formed elements** *of the blood.*

Soluble elements of blood

Plasma *is unclotted blood from which the formed elements have been removed.* The plasma contains clotting agents and other proteins, as well as various salts and glucose. When whole blood is permitted to stand in a test

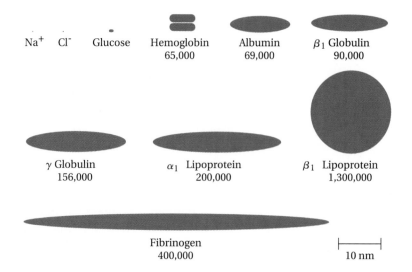

Figure 22.2
Relative shapes and sizes of blood proteins. The molar masses are given below the names of the proteins. Relative sizes of a sodium ion, a chloride ion, and a glucose molecule are shown for comparison.

Blood drawn for some medical tests is prevented from clotting by the addition of the natural polysaccharide heparin, citrate ion, or oxalate ion. The anticlotting agent that is added depends on the test to be performed.

tube, it clots. *The straw-colored liquid that separates from the clot is* **serum.** Serum lacks the formed elements and clotting agents.

The major plasma proteins are *fibrinogen,* the *albumins,* the *globulins,* and the *lipoproteins* (Fig. 22.2). Recall that fibrinogen is essential for clot formation (see Sec. 19.12). *Albumins* help to maintain plasma volume by increasing the osmotic pressure of the blood. When the concentration of serum albumin is too low, water flows from the blood to the tissues. Owing to the force of gravity, the excess tissue fluid often collects in the lower extremities, leading to swelling (edema). *Globulins* perform many functions, such as defense against infecting organisms and the transport of metal ions such as iron and copper. *Lipoproteins* carry lipids in the bloodstream, and most plasma lipids are found associated with them (refer to A Closer Look: Blood Lipoproteins and Heart Disease, page 651). Recent concerns about the safety of blood supplies have led to new research to find blood replacements and to prevent the loss of blood during medical procedures: See A Closer Look: Blood: Risks and Replacements.

The circulatory system

The heart, the blood vessels, and the lymphatic network constitute the major organs of the circulatory system (Fig. 22.3). The heart pumps oxygen-carrying blood from the lungs through the large blood vessels, the arteries. Oxygen-laden hemoglobin is red, and arterial blood has a bright scarlet hue. Like the trunk of a tree, arteries separate into thinner vessels, the branchlike arterioles and the twiglike capillaries.

A vast network of capillaries enmeshes every tissue, and no tissue cell is very far from a capillary. Scientists estimate that the length of capillaries in an average person's body is in the neighborhood of 100,000 km (over 60,000 miles). The microscopic capillaries are so thin that erythrocytes must squeeze through single file. Hemoglobin molecules of erythrocytes exchange their loads of oxygen for loads of carbon dioxide in the capillaries.

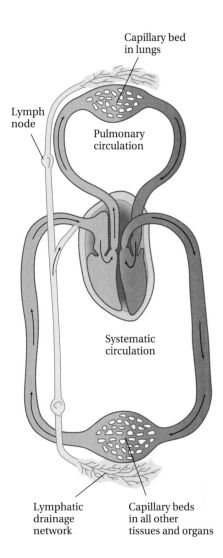

Figure 22.3
The circulatory and lymphatic systems.

Blood leaves the capillaries by entering small veins that lead in succession to larger veins. The large veins carry the blood back to the heart. Venous blood—which has a bluish cast—is pumped into the capillaries of the lungs, where carbon dioxide is lost. A new load of oxygen is picked up, and the circulatory cycle begins again.

The lymphatic system consists of the lymphatics, the thymus, the spleen, and the lymph nodes. The lymphatics are tubes through which interstitial fluid drains into the bloodstream. **Lymph**—*the fluid that enters small capillary-sized lymphatics*—flows to progressively larger lymphatics and drains into the bloodstream at the thoracic duct or the right lymphatic duct.

PRACTICE EXERCISE 22.1

Describe the functions of the blood and the lymph, and explain how these two fluids function in the circulatory system.

Blood: Risks and Replacements

Some loss of blood in surgical procedures must be expected. However, the loss of too much blood can be fatal. Transfusions of blood obtained from blood bank programs have saved many lives, but there are disadvantages to the use of donated blood. The blood type of donors and recipients must be carefully matched to avoid a potentially fatal allergic response caused by transfusing mismatched blood. There is also a possibility of catching diseases from blood transfusions. In the early 1980s, many hemophiliacs who were treated with whole blood or blood products became infected with HIV, the virus that causes AIDS. Blood is now heat-treated to kill HIV. Today, the chances of being infected with HIV by a transfusion of blood pooled from many donors are small, but they are finite—from 1 in 45,000 to 1 in 225,000. The risk of catching hepatitis is about 1 in 3500. One way that patients are avoiding these risks is to bank their own blood before surgery. Hospitals are trying to use less blood and transfuse only as absolutely necessary. Blood lost during surgery is also recovered and returned to the patient.

Because of real risks and people's fears about blood transfusions, scientists are trying to find synthetics to replace natural blood. There will likely never be a substitute to perform all of blood's many functions, so current research is geared to synthetics that can temporarily perform the oxygen-carrying function of red blood cells. One promising line of research employs perfluo-

rocarbons as the oxygen-carrying materials. *Perfluorocarbons* are organic compounds in which all the hydrogens are replaced by fluorines. A unique property of perfluorocarbons is their ability to dissolve oxygen. One perfluorocarbon product, Fluosol DA, is at present the only product to win approval in the United States for clinical use as a temporary blood replacement. Since 1989, Fluosol DA has been approved for the single purpose of oxygenating heart muscles during artery-widening balloon angioplasty. Fluosol DA is a milky aqueous emulsion containing 14% perfluorodecalin with 6% perfluorotripropyl amine as the emulsifier. The product carries oxygen to tissue cells and transports carbon dioxide from tissues cells like blood. It appears to be safe except for flulike symptoms in some patients. It can be transfused without delay for blood typing because it contains no antigens. One disadvantage is that the concentration of perfluorocarbons in the emulsion is low, and thus has a low oxygen-carrying capacity. However, emulsions containing more of the perfluorocarbon tend to separate into aqueous and organic layers and are not stable enough to use. In order to improve the stability of the emulsions, scientists are trying new fluorocarbons and new emulsifiers such as lecithin, egg yolk phospholipids, and triglycerides.

Another line of blood-replacement research involves the use of hemoglobin, the natural oxygen carrier of blood that has been removed from its red blood cells. The hemoglobin research is in relative infancy, and we do not know what its future holds.

22.3 Antibodies and interferons

AIM: To compare and contrast phagocytosis and the immune response.

Focus

Antibodies and interferons form an important natural line of defense against infection.

The body uses blood components in two major defenses against infectious organisms or other foreign matter. The first of these mechanisms, *phagocytosis,* is a process by which certain leukocytes called *macrophages* engulf and subsequently destroy foreign objects. The pus that forms around a cut on an infected finger or at the root of an abscessed tooth is composed of macrophages that have fought and died in the battle with the foreign invader. Phagocytosis forms an immediate first line of defense in infec-

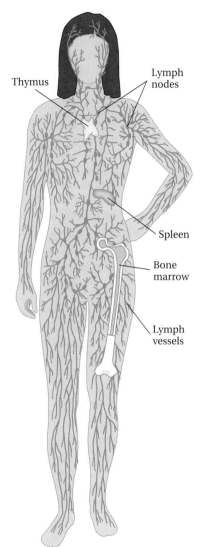

Figure 22.4
Antibody-producing cells, the lymphocytes, are produced in the thymus, the spleen, the lymph nodes, and the bone marrow.

tions. The second line of defense takes about a week to achieve its full power following an infection. This second defense is the *immune response.* In the immune response, special cells of the body are called on to produce proteins called *antibodies.* Another class of proteins, the *interferons,* is produced specifically to defend against infections by viruses.

Antibodies

Antibodies *are large protein molecules whose job is to react with infecting agents.* The antibody-producing cells are *lymphocytes.* Lymphocytes are produced in bone marrow and in the other lymphatic tissues illustrated in Figure 22.4: the thymus, the spleen, and the lymph nodes. The lymph nodes act as filters of lymph, trapping bacteria and other foreign substances. When infecting agents are trapped in a lymph node, lymphocytes begin to produce antibodies against them. The antibodies then enter the bloodstream by means of the lymphatic system. *The infecting agents that are acted on by antibodies are called* **antigens.** In a normal individual, antigens may be bacteria, viruses, proteins, carbohydrates, or just about anything the body recognizes as foreign.

The antibody proteins released into the bloodstream make up a part of blood serum called **immunoglobulins.** *There are five classes of immunoglobulins, of which the most common are immunoglobulins G (IgG or gamma globulins).* The number of antibodies produced by the lymphocytes is enormous. Since you began reading this section, your body has produced over a million billion antibody molecules. Even more astonishing, the different antibodies you have produced probably number in the millions. Despite the great numbers of different antibodies being continually produced, all antibody molecules have certain structural features in common. They all consist of two identical large-protein portions (heavy chains) and two identical smaller-protein portions (light chains). Figure 22.5(a) shows the Y shape of IgG molecules.

The peptide chains of all antibody molecules have very similar primary structures. However, small regions of the light and heavy chains differ in primary structure from antibody to antibody. These small regions are called the *variable regions* of the light and heavy chains. The interaction of the antibody molecule with the antigen occurs in these variable regions, as

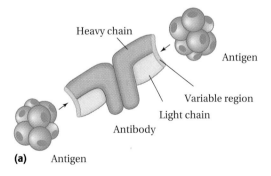

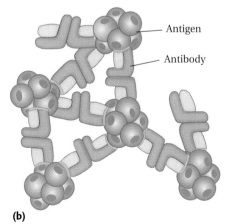

Figure 22.5
(a) Representation of an immunoglobulin G antibody molecule (IgG) with antigens. Each light chain of IgG has a molar mass of about 25,000, and each heavy chain has a molar mass of about 50,000. (b) An antibody-antigen complex is formed when antibody molecules interact with antigens.

Multiple sclerosis is an example of an *autoimmune disease*—the body mistakenly recognizes a normal component as foreign and generates an immune response to destroy the component. Other autoimmune diseases include insulin-dependent diabetes mellitus, rheumatoid arthritis, myasthenia gravis, and systemic lupus erythematosis.

illustrated in Figure 22.5(b). This interaction occurs by a lock-and-key or "sticky patch" mechanism similar to that which binds substrates to enzymes. It is likely that only a small part of the total antigen is recognized by the antibody. Once antibody-antigen complexes have been formed, they are destroyed by phagocytosis. A system called the *complement system* also contributes to the destruction of foreign cells and viruses. *The* **complement system** *consists of at least 11 plasma proteins that are responsible for the lysis (breakage) of the cell membrane and the death of foreign cells.* The attachment of antibodies to a foreign cell triggers the complement system.

Why are millions of different antibodies needed? Obviously, the organism manufacturing the antibodies does not know what antigens it will encounter, so apparently it synthesizes many light and heavy chains to provide enough different antibody "locks" to accept almost any antigen "key." Each lymphocyte produces only one kind of antibody molecule. When a particular antigen has been detected, an order is sent to the proper lymphocytes to produce many more copies of that antibody.

Sometimes defects occur in the combination of light and heavy chains to form complete antibodies. In 1845, English physician Henry Bence-Jones observed that the urine of people with multiple myeloma, a cancer of antibody-producing cells, was of abnormally high specific gravity. A component of the urine was later found to be a protein substance. The appearance of these proteins in the urine called for a diagnosis of multiple myeloma, but the source of Bence-Jones proteins remained unknown for over a hundred years. In 1962, Gerald M. Edelman and Joseph A. Gaily of Rockefeller University showed that **Bence-Jones proteins** *are antibody light chains made by the myeloma tumor but not incorporated into complete antibodies.*

Because each kind of lymphocyte produces only one kind of antibody, we might expect to obtain homogeneous antibodies by growing a single

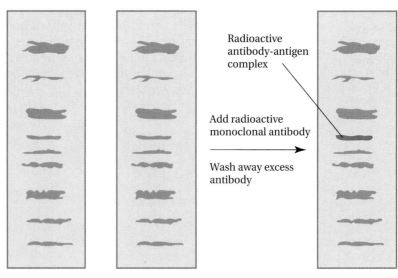

Figure 22.6
Method for detection of an antigen using a radioactive monoclonal antibody.

Electrophoresis of protein mixture containing antigen

Proteins transferred to nitrocellulose paper by blotting

Radioactive antibody-antigen complex

Add radioactive monoclonal antibody

Wash away excess antibody

kind of lymphocyte and harvesting the antibody that the cells produce. This is not possible because lymphocytes cannot be grown in cell cultures. However, crossing (cloning) lymphocytes with myeloma cells produces hybrid cells (hybridomas) that make homogeneous antibodies and can be grown in culture. **Monoclonal antibodies**— *homogeneous antibodies produced by a single kind of hybridoma*—can be obtained in large quantities. Monoclonal antibodies have become indispensable tools in the detection of infecting organisms. Consider, for example, a blood sample from a patient suspected of having a bacterial or viral infection. Blood that is infected by a foreign organism contains protein antigens characteristic of the organism. To test for these antigens, a mixture of proteins from the blood can be separated by electrophoresis and then blotted on to nitrocellulose paper (Fig. 22.6). The proteins stick tightly to the nitrocellulose paper and are not removed by washing. A solution containing a radioactively labeled monoclonal antibody of the antigen is then placed on the paper, where the antibody and any antigen make a tight antibody-antigen complex. The paper is washed, which removes any excess antibody; washing does not remove the proteins or the antibody-antigen complexes. The presence of the antigen can be detected as a radioactive spot of antibody-antigen complex.

Interferons

People infected with one virus are resistant to infection by another virus at the same time. This suggests that besides producing antibodies, the body is capable of mounting special molecular defenses against viral infections. In 1957, virologists Alick Isaacs and Jean Lindenmann discovered that cells infected by viruses produce minute quantities of glycoproteins that travel to nearby cells, where they stimulate these cells to produce protective antiviral proteins. *These antiviral proteins are called* **interferons.** There are three families of interferons. Each interferon molecule contains about 150 amino acid residues. Interferons are among the most powerful known biological agents. They can be effective antiviral agents in concentrations as low as 3×10^{-14} *M.* Because of the low concentrations of interferons in the body, it is impossible to isolate from human sources the quantity needed for research and therapeutic use. Although early interferon research was hampered because so little of the substances were available, the techniques of recombinant DNA (see Sec. 20.13) now produce sufficient amounts for research and for the treatment of disease.

Recent clinical tests of interferons against several viral diseases give encouraging results. Since interferons also slow cell division, they are now used against some types of cancer, a family of diseases characterized by uncontrolled, explosive rates of cell multiplication. One type of interferon, beta-interferon (Fig. 22.7), is the only available drug for the treatment of multiple sclerosis. This neuromuscular disease affects 350,000 Americans and causes vision problems, partial paralysis, and memory loss. Multiple sclerosis is not cured by beta-interferon, but administration of the protein slows the progress of the disease for those who suffer periodic attacks.

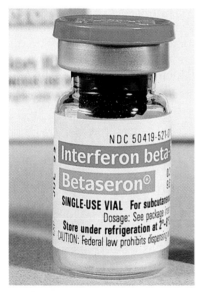

Figure 22.7
Beta-interferon, a genetically engineered human protein, is used for the treatment of multiple sclerosis.

22.4 Blood buffers

AIM: To name the major blood buffer systems and describe how each helps maintain a constant blood pH.

Focus

The pH of the blood should be maintained at about 7.4.

Our bodies function properly only when the pH of various fluids lies within certain narrow limits. To keep these fluids in a proper **acid-base balance**—*the control of pH of body fluids within a narrow range necessary for life and health*—our bodies contain buffer systems. Three independent buffer systems keep the normal pH of blood between 7.35 and 7.45, with an average normal value of 7.40. **Acidosis** *is any shift in the blood pH to below the normal value;* **alkalosis** *is a pH shift to above the normal value.* A drop in the blood pH to below 6.8 or an increase to above 7.8 is life-threatening.

You may recall from A Closer Look: pH of Body Fluids, on page 262, that the most important buffer system of the blood consists of carbonic acid (H_2CO_3) and the hydrogen carbonate (bicarbonate) ion (HCO_3^-). The HCO_3^-/H_2CO_3 ratio is normally 20:1. When protons (hydrogen ions) enter the blood, they are picked up by the bicarbonate ion to form carbonic acid.

$$HCO_3^-(aq) + H^+(aq) \longrightarrow H_2CO_3(aq)$$

As long as bicarbonate ions are present, excess protons are removed, and the pH of the blood changes very little. When hydroxide ions (bases) enter the blood, carbonic acid molecules lose a proton by reacting with hydroxide ions to form more bicarbonate ions.

$$H_2CO_3(aq) + OH^-(aq) \longrightarrow HCO_3^-(aq) + H_2O(l)$$

As long as carbonic acid molecules are present, excess hydroxide ions are removed, and the pH of the blood changes very little.

The second important buffer system in the blood consists of monohydrogen phosphate and dihydrogen phosphate ($HPO_4^{2-}/H_2PO_4^-$). This buffer system works much like bicarbonate and carbonic acid. Protons react with monohydrogen phosphate to produce the dihydrogen form, and the pH changes very little.

$$HPO_4^{2-}(aq) + H^+(aq) \longrightarrow H_2PO_4^-(aq)$$

Bases in the blood remove a proton from dihydrogen phosphate to produce the monohydrogen compound, and the pH changes very little.

$$H_2PO_4^-(aq) + OH^-(aq) \longrightarrow HPO_4^{2-}(aq) + H_2O(l)$$

The third important buffer system in the blood consists of the large protein molecules that are present in colloidal dispersion. These complex molecules contain many acidic (—COOH) and basic (—NH_2) groups that donate or accept protons.

$$\begin{aligned} &\text{—COOH} + OH^- &&\longrightarrow \text{—COO}^- + H_2O \\ &\text{—COO}^- + H^+ &&\longrightarrow \text{—COOH} \\ &\text{—NH}_2 + H^+ &&\longrightarrow \text{—NH}_3^+ \\ &\text{—NH}_3^+ + OH^- &&\longrightarrow \text{—NH}_2 + H_2O \end{aligned}$$

22.5 Oxygen and carbon dioxide transport

AIM: *To describe with chemical equations the exchange of oxygen and carbon dioxide between a red blood cell and lung alveolus and between a red blood cell and a tissue cell.*

Focus

Hemoglobin facilitates the transport of oxygen and carbon dioxide.

Without the hemoglobin of red blood cells, 1 L of arterial blood at room temperature could dissolve and carry only about 3 mL of oxygen—not nearly enough to supply the oxygen needs of the body. With hemoglobin, the amount of oxygen that can be dissolved and carried by a liter of blood increases 70 times.

Transport of oxygen

Transport of oxygen to tissues begins when air inhaled into the lungs enters the *alveoli*—grapelike clusters of sacs responsible for the absorption of oxygen. Figure 22.8 shows how the exchange of oxygen and carbon dioxide takes place between the alveoli and red blood cells in the alveolar capillaries. The five basic steps shown in the figure are as follows:

1. Oxygen pressure is higher in the alveoli than in the red blood cell. Therefore, oxygen diffuses through the capillary wall and into the red blood cell from the alveoli. (Recall that diffusion of gases is always from higher partial pressure to lower partial pressure.)

Oxygen transported by hemoglobin is bound to the Fe^{2+} of the heme group. If Fe^{2+} is oxidized to Fe^{3+}, it loses its ability to carry oxygen molecules.

2. The concentration of protons in the red blood cell is low. This low concentration of protons means that the reaction of protonated hemoglobin (HHb) with oxygen, a reaction that releases protons, is favored. (Review Le Châtelier's principle in Section 6.8, page 162.) Oxyhemoglobin (HbO_2) is formed. Each molecule of oxyhemoglobin carries four molecules of oxygen.

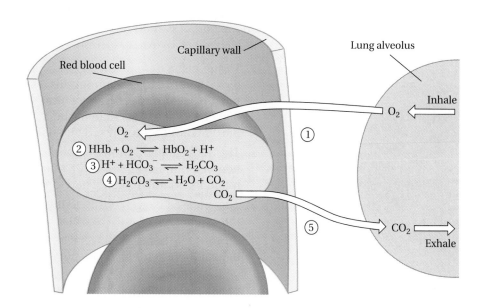

Figure 22.8
In five basic steps, red blood cells exchange carbon dioxide for oxygen at the lungs.

3. Protons released by hemoglobin as it is oxygenated react with bicarbonate ions to form carbonic acid. (The bicarbonate ions come from active tissue or the kidneys.)

4. An enzyme in the red blood cell, carbonic anhydrase, promotes the breakdown of carbonic acid to water and carbon dioxide. The reaction is reversible, but the low pressure of carbon dioxide in the lungs pulls the reaction in the direction of formation of carbon dioxide and water (Le Châtelier's principle again).

5. Since the carbon dioxide pressure is high in the red blood cell but low in the alveoli, carbon dioxide molecules diffuse out of the red blood cell and into the lung. The lungs expire the carbon dioxide and some of the water formed in step 4.

What would happen if the bicarbonate ions lost in expiration were not replaced? First of all, the body would soon deplete its store of bicarbonate ions, since a molecule of carbon dioxide is expired for each proton neutralized. Buffering of the blood by formation of carbonic acid from excess protons and bicarbonate ions would become impossible. Thus, with its load of excess protons, the blood would become acidic. Hemoglobin does not pick up oxygen in an acidic environment because the equilibrium position in the reaction shifts far to the left in the presence of a high concentration of protons.

$$HHb + O_2 \rightleftharpoons HbO_2 + H^+$$

Consequently, in severe acidosis, oxygen loses the competition with protons for hemoglobin, and body cells die for lack of oxygen. To prevent this, the necessary bicarbonate ions are partly replaced by carbon dioxide from active tissue. And as we will see later in this chapter, bicarbonate ions are also replaced by the kidneys.

Transport of carbon dioxide

Oxygenated red blood cells are carried by the bloodstream to the capillaries of active tissues. Figure 22.9 shows how oxygen is delivered to the tissues and a load of carbon dioxide is picked up by the deoxyhemoglobin of the red blood cells. We can see from the reactions in the figure that

1. Tissue cells are using oxygen to produce acids (protons) and carbon dioxide from carbon compounds. These substances are toxic to cells and must be removed. The concentration of protons and carbon dioxide is higher in the tissues than in the red blood cells, so protons and carbon dioxide diffuse into the red blood cells.

2. The high concentration of protons promotes addition of protons to oxyhemoglobin. Recall, however, that protonated hemoglobin tends not to retain its bound oxygen. Therefore, oxygen is released.

3. The released oxygen diffuses through the red blood cell membrane and the capillary wall, and oxygen is delivered to the tissue cells.

4. Protonated hemoglobin can react with carbon dioxide to form **carbaminohemoglobin** (abbreviated HHb-CO$_2$), a form of hemoglobin to which carbon dioxide is attached, and now does so.

5. The rest of the carbon dioxide that enters the red blood cell is converted to carbonic acid by the enzyme carbonic anhydrase. This reaction is the

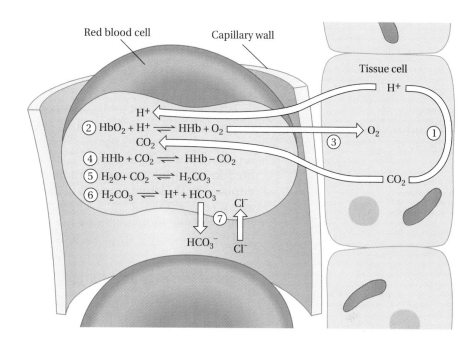

Figure 22.9
In seven steps, red blood cells exchange oxygen for carbon dioxide at the tissues.

reverse of the one carried out by carbonic anhydrase in the lungs. It is the high concentration of carbon dioxide that drives the reaction in the direction of formation of carbonic acid.

6. Some of the carbonic acid dissociates into protons and bicarbonate ions. *The released protons can be sponged up by oxyhemoglobin as in step 2 and carried to the lungs for neutralization in a step called the* **isohydric shift.**

7. Some bicarbonate ions leave the red blood cell and enter the plasma to serve as blood buffers. *The bicarbonate ions are replaced by chloride ions in a process called the* **chloride shift.** The chloride shift maintains a charge balance so that osmotic pressure relationships between the electrolytes in the blood cells and the plasma are not upset.

About 25% of the total carbon dioxide output of cells leaves tissues as carbaminohemoglobin. Some 70% is converted to bicarbonate ions. The remaining 5% is transported to the lungs as carbon dioxide dissolved in the plasma.

> **PRACTICE EXERCISE 22.2**
>
> Write the equation for the breakdown of carbonic acid in the lungs, and name the enzyme responsible for catalyzing the process. How would the pH of the blood be affected if this process could not occur?

> **PRACTICE EXERCISE 22.3**
>
> In hypochloremia, a deficiency of chloride ions in the blood caused by vomiting or upper intestinal blockage, bicarbonate ions replace the missing chloride ions. How would you expect hypochloremia to affect blood pH?

22.6 The urinary system

AIM: To state two important functions of the kidneys.

As we have just seen, carbon dioxide molecules produced by body cells are expelled from the blood through the lungs. A selective filtration process in the urinary system removes other waste products of cells from the blood. For example, ammonia is a by-product of amino acid breakdown in the cell. Ammonia is toxic to human cells in high concentrations and is converted to urea in the liver. The urea enters the bloodstream, circulates until it is removed by the kidneys, and is then excreted in the urine. Apart from acting as a selective filter of wastes in the bloodstream, the kidneys also play an important role in maintaining the pH of the blood by eliminating protons and replenishing bicarbonate ions lost in expiration.

Parts of the urinary system

A pair of kidneys, two ureters, a bladder, and a urethra constitute the *urinary system,* illustrated in Figure 22.10(a). Each kidney contains about a million *nephrons—narrow tubes twisted into a shape rather like the letter U* (Fig. 22.10b). Each nephron has its own miniature circulatory system. With every heartbeat, blood is forced from the renal artery into a progressively smaller system of capillaries or arterioles that enter the *renal capsule* of each nephron. Inside the capsule, the arteriole branches into even smaller vessels that are gathered into a tuft, the *glomerulus.* The vessels of the glomerulus enlarge again to form the outgoing arteriole that branches and intertwines with the nephron. The large number of very small blood vessels and their close association with the nephrons mean that all the blood that passes into the kidney can be cleansed.

Figure 22.10
(a) The human urinary system.
(b) A kidney nephron and its
circulatory system.

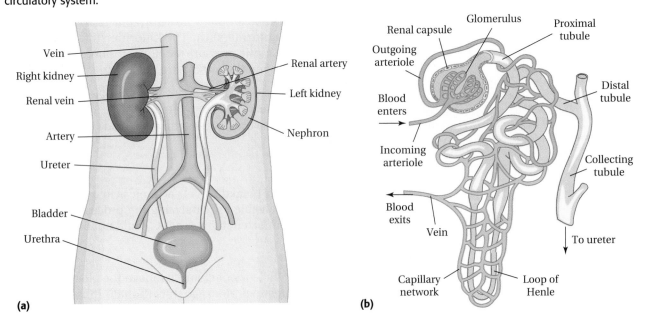

(a)

(b)

Filtration and reabsorption

The blood that enters the glomerulus is under considerable pressure. Since the walls of the capillaries of the glomerulus are very thin, water containing almost all the salts, sugars, amino acids, and other small molecules that are in the blood plasma is forced through small pores in the capillary walls. Cells and large molecules, such as proteins, do not pass through. The walls of the renal capsule are also porous, and the fluid passing through the capillaries—the *glomular filtrate*—enters the nephron at the renal capsule. The filtration of the blood by the renal capsule is rather indiscriminate, so a process of selective reabsorption must now begin.

Water and other substances needed by the body are reabsorbed into the capillaries that surround the nephron. Wastes are not reabsorbed to any great extent. As the filtered fluid passes into the *proximal tubule* of the nephron, about 80% of the water originally removed is returned to the bloodstream. All but a trace of glucose is reabsorbed in normal individuals. In untreated diabetes, where there is an excessively high concentration of blood glucose, some of the excess glucose passes into the urine. When glucose is found in the urine, the renal threshold of glucose is said to have been exceeded. In a normal adult, the renal threshold for glucose is usually exceeded when the blood glucose level rises above 140 to 170 mg of glucose per 100 mL of blood. Salt levels of blood are also controlled by the kidneys. About 75 g of sodium chloride is reabsorbed by the kidneys every day, but some is lost in the urine. About 500 g of sodium bicarbonate enters the kidneys daily, but only a few grams are excreted.

22.7 Acid-base balance

AIMS: *To relate the bicarbonate ion-producing capabilities of the kidneys to the maintenance of the acid-base balance in the blood. To describe how the pH of the urine is controlled.*

Focus

The respiratory and urinary systems help maintain the body's acid-base balance.

The control of acidity in the blood is carried out jointly by the respiratory and urinary systems. To begin this discussion, recall what happens to protons in the circulatory system. In the blood, the most important buffer system is HCO_3^-/H_2CO_3. The carbonic acid was formed by the combination of bicarbonate ions in the blood with protons expelled by active tissues.

$$H^+(aq) + HCO_3^-(aq) \longrightarrow H_2CO_3(aq)$$

We have seen that when blood arrives at the lungs, carbonic acid is degraded to water and carbon dioxide by the enzyme carbonic anhydrase.

$$H_2CO_3 \underset{\text{anhydrase}}{\overset{\text{Carbonic}}{\rightleftharpoons}} H_2O + CO_2$$

In other words, bicarbonate ions act as "proton sponges" that protect the blood against changes in pH that would be caused by production of protons if no buffer ions were present. The water produced by the decomposi-

One aspirin (acetylsalicylic acid) tablet daily may be recommended for people with a history or risk of heart attack. Aspirin interferes with the aggregation of platelets in the blood and prevents heart attacks by inhibiting thrombosis. Could aspirin affect the blood's acid-base balance?

tion of carbonic acid is expired as vapor. The carbon dioxide formed by the breakdown of carbonic acid is also expired. For every proton produced by tissues and trapped by a bicarbonate ion, a bicarbonate ion is lost as carbon dioxide. A continuation of this process without replacement of bicarbonate ions would soon exhaust the blood's most important buffer system. This is where the kidneys make their contribution.

The kidneys are the source of new bicarbonate ions. Carbon dioxide being transported away from the tissues by the blood enters the kidney nephrons as part of the filtrate. The carbon dioxide in the fluid of the *distal tubule* enters the cells of the distal tubule walls. Figure 22.11 traces what happens to the carbon dioxide in the distal tubule cells.

1. Carbon dioxide enters the cells of the distal tubule wall.
2. Carbonic anhydrase promotes the formation of carbonic acid.
3. Carbonic acid dissociates to protons and bicarbonate ions.
4. An exchange of protons and sodium ions now occurs. For every sodium ion pumped into the tubule wall cells, a proton is pumped into the developing urine.
5. The sodium ions and bicarbonate ions enter the bloodstream through capillaries near the distal tubule.

The process that has been described results in sodium bicarbonate being added to the bloodstream and protons being added to the developing urine. The protons that enter the developing urine would make the urine very acidic if they did not encounter buffer ions in the filtrate. However, HPO_4^{2-} ions are present in the developing urine. These ions act as proton sponges.

$$H^+(aq) + HPO_4^{2-}(aq) \longrightarrow H_2PO_4^-(aq)$$

The presence of the phosphate buffer system $HPO_4^{2-}/H_2PO_4^-$ usually keeps the urine from going below a pH of 6, but too great an excess of protons can overload the system. In severe acidosis, the pH of urine may drop as low as 4.5.

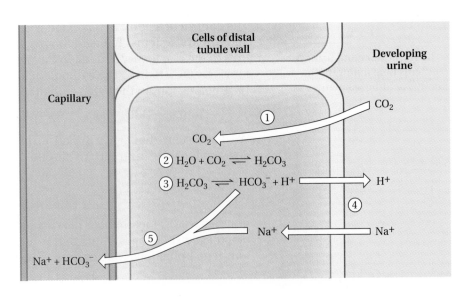

Figure 22.11
Bicarbonate ions are restored to the blood by the kidneys in five steps.

22.8 Acidosis and alkalosis

AIM: To explain alkalosis and acidosis of both of the respiratory and the metabolic types using Le Châtelier's principle.

As you may recall from Section 22.2, the ratio of the concentration of bicarbonate ions to the concentration of carbonic acid in blood at the normal pH of 7.4 is 20:1. The pH of blood does not, however, depend on the concentrations of these two species. Rather, as Table 22.1 shows, blood pH depends on the HCO_3^-/H_2CO_3 concentration ratio. As the ratio increases, the blood pH increases (blood becomes more basic); as the ratio decreases, the blood pH decreases. Notice that the HCO_3^-/H_2CO_3 ratio can vary only over a tenfold range—from 50:1 to 5:1—for human life to exist.

The HCO_3^-/H_2CO_3 ratio, and therefore the blood pH, may increase in two ways: by an increase in the concentration of HCO_3^- or by a decrease in the concentration of H_2CO_3. Similarly, the ratio, and therefore the blood pH, may be lowered by a decrease in the concentration of HCO_3^- or by an increase in the concentration of H_2CO_3. When the respiratory and urinary systems work together to maintain the blood at pH 7.4, they do so primarily by maintaining the HCO_3^-/H_2CO_3 ratio at 20:1. Events that tend to make the ratio larger than 20:1 can lead to alkalosis; events that make the ratio less can lead to acidosis. Alkalosis and acidosis occur only when the body's control systems cannot respond fast enough, or cannot respond at all, to maintain the 20:1 ratio.

Respiratory acidosis *and* **respiratory alkalosis** *are acid-base imbalances that result from abnormal breathing.* **Metabolic acidosis** *and* **metabolic alkalosis** *are acid-base imbalances that result from causes other than abnormal breathing.* In all types of acid-base imbalances, the major system of concern is the carbon dioxide–carbonic acid–bicarbonate equilibrium.

$$H_2O + CO_2 \rightleftharpoons H_2CO_3 \rightleftharpoons H^+ + HCO_3^-$$

Table 22.1 Relationship of Blood pH to the HCO_3^-/H_2CO_3 Concentration Ratio

HCO_3^-/H_2CO_3	pH	Remarks
50:1	7.8	highest pH compatible with life
40:1	7.7	
32:1	7.6	
25:1	7.5	slight alkalosis
20:1	7.4	pH of normal blood
16:1	7.3	slight acidosis
12.5:1	7.2	
10:1	7.1	
8:1	7.0	
6.25:1	6.9	
5:1	6.8	lowest pH compatible with life

Respiratory alkalosis

The most common way to acquire respiratory alkalosis is by **hyperventilation**—*breathing too deeply and too rapidly*. Sometimes respiratory alkalosis is called *hyperventilation alkalosis*. Hysteria, anxiety, or heavy physical exercise (such as giving birth) may result in hyperventilation, in which too much carbon dioxide is "blown off" by exhalation. The partial pressure of carbon dioxide in the lungs becomes lower than it is in the tissues. More than the normal amount of carbon dioxide therefore diffuses from the blood into the lungs. This disturbs the blood's normal carbon dioxide–carbonic acid–bicarbonate equilibrium. According to Le Châtelier's principle, the system shifts to the left—in favor of reactants—as it attempts to restore the equilibrium.

$$\text{Loss of CO}_2 \nwarrow$$
$$H_2O + CO_2 \rightleftharpoons H_2CO_3 \rightleftharpoons H^+ + HCO_3^-$$

Direction of shift
to restore equilibrium

The result of this shift is a major loss of carbonic acid. The HCO_3^-/H_2CO_3 ratio rapidly increases to greater than 20:1, and the blood pH rises to as high as 7.7 within a few minutes. There is also a corresponding loss of bicarbonate ions, but this is not great enough to maintain the HCO_3^-/H_2CO_3 ratio at 20:1.

To understand why a similar loss of both carbonic acid and bicarbonate ion causes the HCO_3^-/H_2CO_3 ratio to *increase* instead of staying the same, we can do a simple calculation. Normal venous blood of pH 7.4 has a bicarbonate ion concentration of about 27 mEq/L and a carbonic acid concentration of about 1.35 mEq/L. The ratio is 27.0 mEq ÷ 1.35 mEq, or 20:1. If the carbonic acid concentration is reduced by half, to 0.68 mEq, the *ratio* becomes 27.0 mEq ÷ 0.68 mEq, or 40:1, corresponding to a pH of 7.7 (see Table 22.1). On the other hand, if the bicarbonate ion concentration is reduced by 0.68 mEq to 26.32 mEq, the HCO_3^-/H_2CO_3 ratio is 26.32 mEq ÷ 1.35 mEq, or 19.5:1, and the pH change is negligible. If we assume that the loss of carbonic acid equals the loss of bicarbonate ion, 0.68 mEq, we get the ratio 26.32 ÷ 0.68, or 38.7:1, corresponding to a pH very near 7.7.

Respiratory acidosis

Respiratory acidosis is sometimes the result of **hypoventilation**—*too-shallow breathing*. Lung diseases such as emphysema, an object lodged in the windpipe, or other causes of impaired breathing may result in hypoventilation. Anesthetists need to be particularly aware of respiratory acidosis because most inhalation anesthetics depress respiration. In respiratory acidosis, not enough carbon dioxide is exhaled, and its partial pressure in the lungs therefore increases. The high partial pressure of carbon dioxide in the lungs is higher than it is in the blood. Carbon dioxide therefore diffuses into the blood, pushing the carbon dioxide–carbonic acid–bicarbonate equilibrium system to the right to restore the equilibrium.

$$\overset{\text{Increase of } CO_2}{H_2O \ + \ CO_2} \rightleftharpoons H_2CO_3 \rightleftharpoons H^+ \ + \ HCO_3^-$$

Direction of shift
to restore equilibrium

As a result of the shift, the concentration of carbonic acid in the blood increases. The ratio rapidly decreases to less than 20:1, and respiratory acidosis ensues.

Metabolic acidosis

Body tissues produce protons as waste products of various body processes. These processes are collectively called *metabolism*. In metabolic acidosis, the diffusion of protons from the tissues into the bloodstream pushes the carbon dioxide–carbonic acid–bicarbonate system to the left to restore the equilibrium.

$$H_2O \ + \ CO_2 \rightleftharpoons H_2CO_3 \rightleftharpoons \overset{\text{Increase of } H^+}{H^+ \ + \ HCO_3^-}$$

Direction of shift
to restore equilibrium

The shift decreases the concentration of bicarbonate ions but increases the concentration of carbonic acid. Since such changes affect the carbonic acid concentration much more than the bicarbonate ion concentration, the HCO_3^-/H_2CO_3 ratio decreases, and so does the blood pH. In the Follow-up to the Case in Point on the following page, we will learn more about how metabolic acidosis affects people.

Metabolic alkalosis

Metabolic alkalosis is less common than the acid-base imbalances we have seen so far. It may be caused by several conditions—including a loss of anions (primarily chloride ions), a deficiency of potassium ions, or the administration of alkaline salts in drug therapy. Excessive use of sodium bicarbonate, for example, a common home remedy for gastric acidity, increases the concentration of bicarbonate ions in the blood.

In response to a small influx of bicarbonate ions, the carbon dioxide–carbonic acid–bicarbonate equilibrium shifts to the left, decreasing the hydrogen ion concentration.

$$H_2O \ + \ CO_2 \rightleftharpoons H_2CO_3 \rightleftharpoons H^+ \ + \overset{\text{Influx of } HCO_3^-}{HCO_3^-}$$

Direction of shift
to restore equilibrium

We might expect the pH to increase, but another result of the shift is a corresponding increase in the carbonic acid concentration. The

FOLLOW-UP TO THE CASE IN POINT: A diabetic imbalance

Metabolic acidosis is often a serious problem in uncontrolled diabetes, and it also occurs on a temporary basis after heavy exercise. Both conditions result in large influxes of protons from active tissues into the bloodstream. Larry, the teacher in the Case in Point earlier in this chapter, failed to keep his schedule of insulin injections. The acid his body produced as a result of a deficiency of insulin caused the pH of Larry's blood to drop to life-threatening levels. Larry's doctor suspected that diabetes-induced metabolic acidosis might be one of Larry's immediate problems. When the acidosis was confirmed by lab tests (see figure), Larry was given an intravenous drip of a sodium bicarbonate solution to restore his blood pH to normal. The sodium bicarbonate neutralized the excess acids in his blood, and his blood was literally titrated back to normal. With a proper schedule of insulin injections, Larry was able to leave the hospital in a few days.

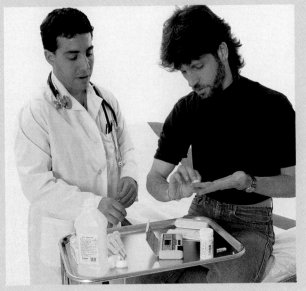

Testing for diabetes.

HCO_3^-/H_2CO_3 ratio does not change very much, even though carbonic acid and bicarbonate ions are at higher than normal concentrations. When the bicarbonate ion influx is large, however, the HCO_3^-/H_2CO_3 ratio may become greater than 20:1, and measurable alkalosis is observed.

PRACTICE EXERCISE 22.4

A first aid treatment for respiratory alkalosis is having the victim breathe into a paper bag. Using the carbon dioxide–carbonic acid–bicarbonate equilibrium and Le Châtelier's principle, explain why this treatment is effective in restoring the pH of blood to normal.

22.9 Water and salt balance

AIM: To describe the action of vasopressin and aldosterone in controlling body water and salt levels.

Focus

The endocrine system controls body water volume and salt concentration.

The volume of blood passed through the kidneys is phenomenal. In an average adult, about 100 L of blood passes through the kidneys every day. Of the water in this 100 L, only 1 or 2 L is excreted as urine. Water reabsorption and excretion by the kidneys are carefully controlled to maintain a constant volume of water in the body.

By the time the glomular filtrate reaches the loop of Henle in the kidney (Fig. 22.12), most of the solutes that are not wastes have been reabsorbed into the bloodstream. Most of the water also has been reabsorbed. However, if the dietary intake of water has been high—during a meal, for instance—a large volume of dilute urine is excreted soon after. If the intake of water has been low, a lower volume of more concentrated urine is excreted.

Vasopressin, a peptide hormone of the pituitary gland, normally controls the volume of urine. Vasopressin, also called *antidiuretic hormone* (ADH), exerts its action by affecting the permeability of the distal tubule and collecting duct to water. In the absence of vasopressin, these parts of the nephron are impermeable to water (see Fig. 22.12). Water cannot pass from the inside of the tubule or the duct to the surrounding tissues. The result is a large volume of light yellow, dilute urine. Patients with the rare disease diabetes insipidus do not produce vasopressin; their urine is extremely voluminous and dilute. In the presence of vasopressin, the distal tubule and collecting duct membranes become permeable to water. Water therefore leaves these parts of the nephron and is drawn into the surrounding tissues by osmosis, since the fluid of the tissues is saltier than the developing urine. The reabsorption of water by the tissues produces a low volume of dark yellow, concentrated urine.

The action of vasopressin is usually sufficient to maintain the proper level of body water. Sometimes, however, the water level dips dangerously low because of insufficient water intake or diarrhea. Heavy exercise or a fever also may cause dehydration due to excessive sweating. Under such circumstances, the adrenal gland secretes the steroid hormone *aldosterone.*

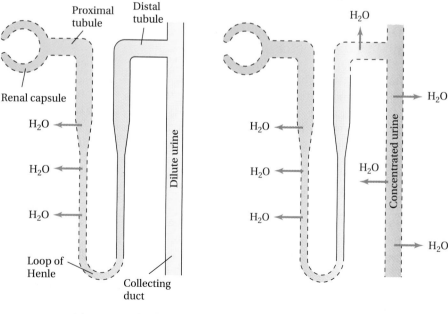

Figure 22.12
Production of dilute and concentrated urine by the kidneys. Dashed lines indicate sections of the nephron that are permeable to water in the (a) absence and (b) presence of vasopressin.

(a) Vasopressin absent

(b) Vasopressin present

Aldosterone activates cellular pumps that transport sodium ions from the nephrons to the interstitial fluid. Chloride ions follow the sodium ions to maintain electrical neutrality, and water moves to this more concentrated salt solution. The action of aldosterone conserves both salt and water in the body. When the water level of the body returns to normal, release of aldosterone drops.

Body water can be lost by sweating. Evaporation of sweat helps to cool the body. An active person not engaged in heavy exercise loses 600 to 700 mL of water per day through perspiration. Heavy exercise, especially in a hot climate, causes more profuse sweating—up to 4 L a day. Sweat is almost pure water, but it also contains small, varying concentrations of sodium chloride. Excessive loss of water through sweating can cause dehydration unless water intake balances water loss. Excessive loss of sodium ions by sweating causes muscle cramps.

SUMMARY

The functions performed by complex living things complement one another, contributing to health and well-being. Healthy cells, tissues, and organs work in harmony to maintain homeostasis—balancing the body's needs with respect to temperature, pH of fluids, energy production, and the supply of oxygen, nutrients, water, and salts.

Body cells live in a fluid environment. The major body fluids are intracellular fluid (the fluid inside of cells), interstitial fluid (the fluid outside of cells), and plasma (the fluid portion of the blood). Cells do most of their chemical work in the intracellular fluid. The interstitial fluid provides the proper amount of salt and water for cells. Among other functions, blood carries oxygen and nutrients to cells and takes away wastes, mostly carbon dioxide and protons. Antibodies are large blood proteins produced by the body in response to a foreign substance or organism (antigen). Interferons help the body ward off viral infections.

The respiratory system and kidneys cleanse the blood of wastes and control blood pH at a value near 7.40 to avoid alkalosis or acidosis. Excess protons are neutralized by combining with bicarbonate ions (HCO_3^-) to form carbonic acid (H_2CO_3). The enzyme carbonic anhydrase breaks down carbonic acid to carbon dioxide (which is exhaled) and to water, a neutral nontoxic waste easily removed by the kidneys. The kidneys also help restore bicarbonate ions to the bloodstream by exchanging one proton of carbonic acid in the blood for one sodium ion in developing urine. If the supply of bicarbonate ions runs low either as a result of excess proton production or by retention of carbonic acid through shallow breathing, acidosis ensues. Hyperventilation causes respiratory alkalosis; too much carbon dioxide is lost through heavy breathing, the concentration of carbonic acid decreases, and the blood pH rises.

Vasopressin, a peptide hormone, controls urine concentration by affecting the permeability of parts of the kidney nephrons to water. When the hormone is present, the nephron is permeable to water, water flows from the nephron back to the saltier interstitial fluid, and the urine is scant and concentrated. When the hormone is not present, the nephron is impermeable to water, water cannot be reabsorbed, and the urine is voluminous and dilute. Aldosterone controls the body's salt retention–water volume balance. Aldosterone activates pumps that carry sodium ions from the nephron to the interstitial fluid. Chloride ions follow sodium ions, and water follows the sodium chloride.

KEY TERMS

Acid-base balance (22.4)
Acidosis (22.4)
Alkalosis (22.4)
Antibody (22.3)
Antigen (22.3)
Bence-Jones proteins (22.3)
Carbaminohemoglobin (22.5)

Chloride shift (22.5)
Complement system (22.3)
Erythrocyte (22.2)
Formed element (22.2)
Homeostasis (22.1)
Hyperventilation (22.8)
Hypoventilation (22.8)
Immunoglobulins (22.3)

Interferon (22.3)
Interstitial fluid (22.1)
Intracellular fluid (22.1)
Isohydric shift (22.5)
Leukocyte (22.2)
Lymph (22.2)
Metabolic acidosis (22.8)
Metabolic alkalosis (22.8)

Monoclonal antibody (22.3)
Plasma (22.2)
Respiratory acidosis (22.8)
Respiratory alkalosis (22.8)
Serum (22.2)
Thrombocyte (22.2)

EXERCISES

Body Water (Section 22.1)

22.5 Describe the distribution of body water in an average adult.

22.6 Define homeostasis.

22.7 Compare the electrolyte composition of intracellular fluid and interstitial fluid.

22.8 Compare the electrolyte composition of interstitial fluid and blood plasma.

Blood, Antibodies, and Blood Buffers (Sections 22.2, 22.3, 22.4)

22.9 Name five functions of the blood.

22.10 Distinguish among whole blood, plasma, and serum.

22.11 Name the formed elements of the blood, and describe their functions in the body.

22.12 List the plasma proteins, and describe their roles.

22.13 Describe the relationship between an antigen and an antibody.

22.14 Where are antibodies produced?

22.15 Describe a use of monoclonal antibodies.

22.16 What is the function of interferons?

22.17 What is the normal pH of the blood?

22.18 Distinguish between acidosis and alkalosis.

22.19 Name the three buffer systems of the blood.

22.20 Explain how the HCO_3^-/H_2CO_3 buffer helps to maintain a constant pH when small amounts of acid or base are added to the blood.

Oxygen and Carbon Dioxide Transport (Section 22.5)

22.21 How is oxygen transported from the lungs to active tissue?

22.22 Explain how partial-pressure differences favor uptake of oxygen by red blood cells from the lungs and release of oxygen from HbO_2 to active tissues.

22.23 Predict what would happen to the oxygen level in the tissue cells of a person who inhaled pure oxygen.

22.24 How does the pH of the blood affect the formation of oxyhemoglobin?

22.25 Describe the ways in which CO_2 is transported from the cells to the lungs.

22.26 What is the function of carbonic anhydrase?

22.27 What happens in the chloride shift?

22.28 Why does the inhalation of pure CO_2 cause asphyxiation?

Urinary System and Acid-Base Balance (Sections 22.6, 22.7, 22.8)

22.29 What essential functions do the kidneys perform?

22.30 Amino acids are broken down to ammonia in the cell. How is ammonia removed from the body?

22.31 Name three substances that are reabsorbed by the kidneys.

22.32 What has happened if glucose is present in the urine?

22.33 Explain how the kidneys supply bicarbonate ions to the bloodstream.

22.34 Why is normal urine not very acidic?

22.35 What is the cause of respiratory alkalosis?

22.36 There is a change in the blood pH when a person hyperventilates. What happens to the pH and what leads to the change?

Water and Salt Balance (Section 22.9)

22.37 What hormones control loss and retention of water by the body?

22.38 Describe the action of antidiuretic hormone (ADH).

22.39 Dehydration or low water intake causes the adrenal gland to secrete a hormone. Name this hormone, and describe its action.

22.40 What are the symptoms and cause of diabetes insipidus?

Additional Exercises

22.41 Would you expect the blood's oxygen-carrying capacity to increase or decrease in acidosis? Explain your answer.

22.42 Discuss the oxygen-carrying efficiency of hemoglobin when the pH of the blood falls below 7.35.

22.43 Explain how oxyhemoglobin releases oxygen to active tissue cells.

22.44 In what ways are the following conditions the same and in what ways are they different?
(a) respiratory acidosis and metabolic acidosis
(b) respiratory alkalosis and metabolic alkalosis

22.45 Clearly distinguish between *hypoventilation* and *hyperventilation*. Explain how each condition disturbs the HCO_3^-/H_2CO_3 ratio in the blood.

22.46 What waste products are eliminated in the urine?

22.47 How do the kidneys help to reduce acidosis?

22.48 The circulatory and urinary systems work together to maintain the fluid environment of the body. Two important reversible reactions that take place in both these systems are

Reaction A $\quad H_2CO_3(aq) \rightleftharpoons H_2O(l) + CO_2(g)$
Reaction B $\quad H_2CO_3(aq) \rightleftharpoons H^+(aq) + HCO_3^-(aq)$

For each of the three locations in the body listed below, write these two equations in the correct sequence and with the favored product on the right.
(a) at the juncture of capillary and tissue cell
(b) at the juncture of capillary and lung alveolus
(c) in the cells of the distal tubules of the kidneys

SELF-TEST (REVIEW)

True/False

1. Monoclonal antibodies are homogeneous.
2. The hormone aldosterone helps control the pH of the blood.
3. Most of the water found in the body is in the blood.
4. Removing the formed elements from unclotted blood leaves blood plasma.
5. Lymphocytes are responsible for antibody production in the body.
6. Respiratory acidosis is usually associated with high carbon dioxide concentration in the blood.
7. A high concentration of H^+ in red blood cells causes HbO_2 to dissociate.
8. Most of the carbon dioxide produced by the cells of the body leaves the tissues as $HHb\text{-}CO_2$.
9. The pH of the urine in the body is controlled by the $HPO_4^{2-}/H_2PO_4^-$ buffer system.
10. Perspiration is the major process by which sodium chloride is lost from the body.

Multiple Choice

11. The hormone that controls the volume of urine in the body under normal conditions
(a) works by causing changes in the osmotic pressure in the nephrons.
(b) is aldosterone.
(c) is vasopressin.
(d) both (a) and (b) are correct.

12. The enzyme carbonic anhydrase may
(a) speed the breakdown of carbonic acid to water and carbon dioxide.
(b) be found in the red blood cells.
(c) speed the formation of carbonic acid from carbon dioxide and water.
(d) all of the above are true.

13. The process of selective reabsorption by the kidneys
(a) applies equally to all substances in the blood.
(b) should let only a trace of glucose pass into the urine of a normal person.
(c) leads to a net increase of the concentration of CO_3^{2-} in the blood.
(d) allows more than half the water in the blood to pass into the urine.

14. The kidneys play an important part in maintaining the pH of the blood by
    (a) eliminating protons in the urine.
    (b) producing OH^- ions.
    (c) producing bicarbonate ions.
    (d) More than one are correct.

15. Hyperventilation can lead to
    (a) too much carbon dioxide being retained in the blood.
    (b) respiratory alkalosis.
    (c) respiratory acidosis.
    (d) a high blood level of carbonic acid.

16. When conditions in the body are such that body cells can operate at peak efficiency, then
    (a) equilibrium has been achieved.
    (b) interactions between the body's systems are minimal.
    (c) homeostasis has been achieved.
    (d) Both (a) and (b) are correct.

17. In the body, the majority of oxygen is carried to the cells as
    (a) dissolved O_2.   (b) oxyhemoglobin.
    (c) H^+-O_2.   (d) bicarbonate ions.

18. The kidneys are important in all the following functions *except*
    (a) conversion of ammonia to urea.
    (b) selective filtration of waste products from the blood.
    (c) control of body water volume.
    (d) regulation of blood pH.

19. The most important buffering system of the blood is the
    (a) HCO_3^-/H_2CO_3 system.
    (b) H_3PO_4/$H_2PO_4^-$ system.
    (c) HPO_4^{2-}/$H_2PO_4^-$ system.
    (d) SO_4^{2-}/HSO_4^- system.

20. At the junction between tissue cell and capillary, both CO_2 and H^+ move into the red blood cell. Then
    (a) some of the CO_2 reacts with HHb to form carbaminohemoglobin (HHb-CO_2).
    (b) all the carbon dioxide reacts with water to form carbonic acid.
    (c) the H^+ ions react with Cl^- to form HCl.
    (d) None of the above is true.

21. Which of the following is *not* a function of the blood?
    (a) carrying hormones   (b) maintenance of pH
    (c) infection fighting   (d) none of the above

22. Blood serum contains
    (a) hemoglobin.   (b) albumin.
    (c) leukocytes.   (d) clotting agents.

23. In comparing the electrolyte composition of various body fluids, we find that sodium and chloride ions predominate in interstitial fluid and that in intracellular fluid the major inorganic electrolytes are
    (a) Na^+ and CO_3^{2-}.   (b) K^+ and HPO_4^{2-}.
    (c) Na^+ and PO_4^{3-}.   (d) K^+ and SO_4^{2-}.

24. Antibodies are made of which of the following plasma proteins?
    (a) lipoprotein   (b) hemoglobin
    (c) albumin   (d) globulin

23

Energy and Life

Sources and Uses of Energy in Living Organisms

Life requires energy.

CHAPTER OUTLINE

CASE IN POINT: A mysterious
  fatigue

23.1 Metabolism

23.2 Photosynthesis

A CLOSER LOOK: Phototherapies

23.3 The energy and carbon cycle

23.4 Adenosine triphosphate

23.5 Cellular energetics

23.6 Oxidative phosphorylation

A CLOSER LOOK: Mitochondria

A CLOSER LOOK: Oxygen, Disease,
  and Aging

FOLLOW-UP TO THE CASE IN POINT:
  A mysterious fatigue

23.7 Cellular work

E*nergy* and *life*—the two terms are practically synonymous. We may speak of someone who has "boundless energy" or who is "glowing with energy." And we are not far wrong, because all living organisms need ample energy to maintain their vital functions. This chapter discusses how the cells of living creatures produce and use energy. Sometimes we can better understand and appreciate what we have by seeing someone who is lacking our possessions or qualities. Our Case in Point concerns a person who lacked energy.

CASE IN POINT: A mysterious fatigue

Emily had been plagued by fatigue since birth. Numerous tests eliminated the possibility of anemia or other causes of fatigue. Puzzled, her doctor referred her to a nearby medical research center for further evaluation. After extensive testing, the medical researchers pinpointed the source of Emily's fatigue as a defect in her ability to generate the central molecule in the transmission of energy. What is this molecule, and what defect interfered with its formation? We will learn the answers to these questions in Section 23.6.

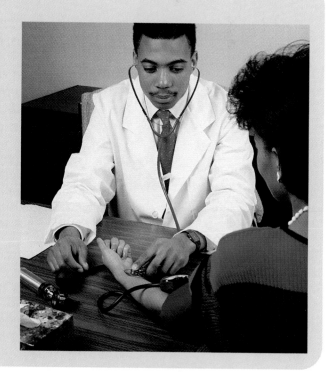

Extensive examination and testing allowed doctors to diagnose the cause of Emily's fatigue.

23.1 Metabolism

AIM: To differentiate among metabolism, catabolism, and anabolism.

Metabolism
meta (Greek): beyond
ballein (Greek): to cast or throw

Catabolism
cata (Greek): down

Anabolism
ana (Greek): up

There are three important terms that describe the chemical reactions in living organisms: *metabolism, catabolism,* and *anabolism.* **Metabolism** *is all the chemical reactions that occur in living organisms.* Virtually all metabolic reactions are catalyzed by enzymes. **Catabolism** *is a part of metabolism—the breakdown of molecules by an organism.* **Anabolism** *is another part of metabolism—the synthesis of molecules for cell growth and multiplication.* Nutrients are also converted to their storage forms by anabolic processes. Conversion of fatty acids to triglycerides for storage in fatty tissue is one example. Conversion of glucose to glycogen for storage in liver and muscle cells is another.

Anabolism and catabolism are quite distinct from each other. Cells usually employ different chemical reactions for the breakdown and synthesis of the same molecule. The reactions used to synthesize glucose, for example, are not the reverse of the reactions used to degrade it. Apart from being chemically separated, catabolic and anabolic reactions are frequently separated physically. Many important catabolic reactions occur in the mitochondria, whereas many anabolic reactions occur in the cytoplasm. The chemical and physical separation of anabolism from catabolism enables cells to regulate metabolism to make it responsive to current needs.

23.2 Photosynthesis

AIMS: To write a chemical equation for photosynthesis indicating the energy-rich and energy-poor carbon compounds. To distinguish among chloroplasts, thyalkoids, and chlorophyll.

Photosynthesis
photos (Greek): light
synthesis (Greek): to place
              together

Energy production by cells involves the catabolism of carbon compounds that serve as nutrients—mainly sugars, fats, and amino acids. Oxidation reactions are generally energy-producing. Oxidation reactions that are part of cellular catabolism release the energy stored in the chemical bonds of nutrient molecules, making it available to perform the work that cells must do to stay alive.

Where do sugars, fats, and amino acids originate? Carbon dioxide in the Earth's atmosphere is the ultimate source of all carbon compounds. Carbon dioxide is an energy-poor compound because it cannot be oxidized further. Animal cells discard it as a waste product. Green plants, blue-green algae, and certain bacteria, however, conduct **photosynthesis**—*harness the energy of sunlight, convert it to useful chemical energy, and use that energy to synthesize glucose, a more reduced molecule, from carbon dioxide. The*

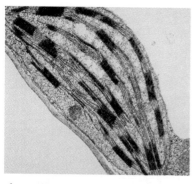

Figure 23.1
Transmission electron micrograph
of a chloroplast.

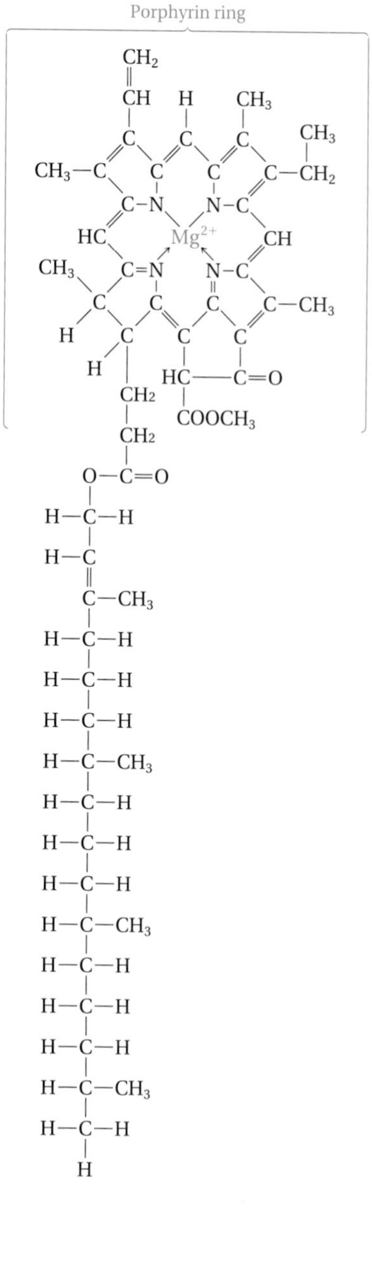

Porphyrin ring

processes that occur in photosynthesis are often summarized by a single
equation.

$$6CO_2 \;+\; 6H_2O \xrightarrow{\text{Light energy}} C_6H_{12}O_6 \;+\; 6O_2$$

Carbon   Water   Glucose   Oxygen
dioxide   (more reduced)
(more oxidized)

Reoxidation of glucose back to carbon dioxide and water releases the
energy expended in forming the chemical bonds of the glucose molecule.

$$C_6H_{12}O_6 \;+\; 6O_2 \longrightarrow 6CO_2 \;+\; 6H_2O \;+\; \text{Energy}$$

Glucose   Oxygen   Carbon   Water
(more reduced)   dioxide
(more oxidized)

Glucose is an energy-rich molecule and a potential source of chemical
energy for any organism that obtains it.

The equation for photosynthesis is deceptively simple. Molecules of
carbon dioxide and water do not simply bang together in sunlight and pop
out molecules of glucose and oxygen. To carry out photosynthesis, cells
must be able to absorb light energy and convert it to chemical energy. *For
this purpose, photosynthetic eukaryotic cells—those of green plants are the
most familiar example—contain special organelles, the* **chloroplasts,**
shown in Figure 23.1. Chloroplasts can carry out this conversion. *Located
within the chloroplasts are a large number of connected disks called*
thyalkoids. *The thyalkoids contain the molecules that constitute the* **light
system** *of photosynthesis.* Responsibility for trapping light energy and converting it to useful chemical energy rests with the light system. **Chlorophyll**—*the pigment that gives green plants their color*—is a molecule of
chief importance to the light system. Its molecular structure is depicted in
Figure 23.2. Chlorophyll initially absorbs light energy and passes it on to
other molecules of the light system. The light systems of prokaryotic cells
that are capable of conducting photosynthesis—certain bacteria and the
blue-green algae—are located on the interior of the cell membranes.

The importance of light to life on Earth through photosynthesis in
plants cannot be overemphasized. However, researchers are beginning to
find that healthy human beings respond to light in ways unimaginable only
a few years ago. The cure rate of certain cancers may be improved by a treatment involving light. See A Closer Look: Phototherapies, for a discussion of
some connections between light and health.

PRACTICE EXERCISE 23.1

(a) Identify the structural feature that is common to vitamin B_{12} (see
Figure 21.3 on page 650), hemoglobin, and chlorophyll. (b) What metals
are essential to each of these molecules?

Figure 23.2 (left)
The molecular structure of chlorophyll. The magnesium ion is essential for chlorophyll's light-trapping function. Notice the similarity between the structure of heme
(see Sec. 18.11) and the porphyrin ring of chlorophyll.

A Closer Look

Phototherapies

You may recall that the transformation of steroids in the skin by sunlight is important in the formation of vitamin D_2, but medical specialists and researchers are just beginning to recognize many of the human body's responses to light. Several conditions, including some forms of disturbed sleep, "jet lag," and *seasonal affective disorder* (SAD), appear to be related to the biological cycles of wakefulness and sleep called the *circadian rhythms*. A small portion of the hypothalamus, a region at the base of the brain, supervises our bodies' clocks, telling us when it is time to eat, sleep, and wake up. This small portion of the hypothalamus, called the *suprachiasmatic nucleus*, responds to light and darkness to affect body temperature fluctuations, hormone release, blood pressure, heart rate, and the sleep-wake cycle. Light treatment is proving useful in the treatment of several human conditions related to the circadian rhythm. More than half of all Americans over age 65 suffer from disturbed sleep, waking too early in the morning and becoming sleepy early in the evening. Experts suggest that this sleep disorder is caused by a speeding up of the body's circadian rhythms in older people. To combat this troublesome shift, researchers are exposing older patients to bright light in the early evening to delay their sleep-wake cycles. Jet lag similarly affects people who fly by jet across international time zones and shift workers, such as nurses, who rotate shifts. The body's sleep-wake cycle is set to have night during the day in the new location or on the new shift. A person responds by feeling exhausted during the day and wide awake at night until the cycle is reset. This usually happens after a few days in the new location, but researchers suggest it can be speeded up by spending the first 2 days after a trip in sunlight. Exposure to bright light is also said to exert a positive effect on SAD (see figure), the name given to a form of depression experienced by some people. SAD people may feel fine in summer but become depressed during the winter months, when the days are short and sunshine may be scarce. SAD is an emotional illness more severe

Phototherapy often relieves the distressing symptoms of SAD, seasonal affective disorder.

than the tired, run-down feeling many people have during the winter months. Preliminary results indicate that light treatment also may be effective against some nonseasonal forms of depression. Light treatments are also important in the treatment of jaundice, or yellowing, in newborns, as we will see in A Closer Look: Hyperbilirubinemia, in Chapter 25.

Recent studies indicate that synchronizing cancer treatment with the body's internal rhythms gives chemotherapy drugs a stronger effect, perhaps as much as doubling their power to fight tumors. Some chemotherapy drugs work better at night; others seem more potent when given during the day. Light is also being pressed into service against certain blood cell cancers by a technology called *photopheresis*. In photopheresis, mixtures of normal and cancerous blood cells are removed from the patient and treated with a drug such as psoralen.

Psoralen

The psoralen is selectively absorbed by the cancerous cells. The interaction of light with the psoralen cross-links the DNA in the cancerous cells, making it impossible for the cells to reproduce. The treated blood is then returned to the patient.

23.3 The energy and carbon cycle

AIM: To describe the energy and carbon cycle.

Environmentalists are extremely troubled over the rapid destruction of vast areas of the world's rain forests. What do you think are some of their concerns? Do you share these concerns?

Life depends on the *energy and carbon cycle* (Fig. 23.3). *In the* **energy and carbon cycle,** *the photosynthetic organisms in the Earth's forests and oceans produce glucose and use it as a source of chemical energy and to build the carbon skeletons of carbohydrates, fats, amino acids, and other biological molecules.* Animals obtain these substances by eating plants, by eating animals that eat plants, or by a combination of both.

Animals and plants in need of energy unleash the energy stored in the chemical bonds of these carbon nutrients by oxidizing them back to carbon dioxide and water. The energy is used for the work that plant and animal cells must do, and the carbon dioxide and water can be recycled in photosynthesis. Plant life on Earth could probably survive without animals. However, animal life could never survive without plants, since without photosynthesis there would be no new supply of the crucial carbon compounds needed by animals for energy production.

PRACTICE EXERCISE 23.2

Explain why our existence depends on photosynthesis.

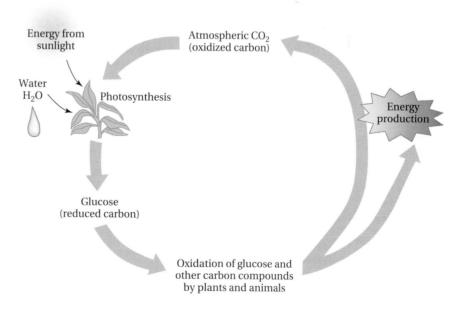

Figure 23.3
The energy and carbon cycle.

23.4 Adenosine triphosphate

AIMS: To name the energy-transmitting molecule in nature and list its hydrolysis products. To explain why ATP could not be a very high-energy compound and still fulfill its role in cellular energetics.

Focus

Adenosine triphosphate ties energy production to energy use.

The chemical energy released by the oxidation reactions of cellular catabolism is divisible into two parts. One part of the energy is lost as heat and is unavailable for biological use. The other part, the free energy (see Sec. 6.6), is available to do useful work in the cell. The free energy is transmitted to a cellular carrier molecule, which then transmits this energy to cellular processes that need it.

Adenosine triphosphate: The energy carrier

Adenosine triphosphate (ATP) *is the molecule that carries and transmits the energy needed by cells of all living things.* We could compare the function of ATP in cellular metabolism to a belt that connects a motor to a pump. The motor generates energy that is capable of operating the pump. If the motor and the pump are not connected by a belt, however, the energy produced by the motor is useless. ATP is the belt that couples the motor (catabolism) and the pump (anabolism) (Fig. 23.4). To understand how cells couple energy production and energy use, we will need to know more about the chemistry of ATP.

Structure and hydrolysis of ATP

The ATP molecule (Fig. 23.5) consists of the nucleoside adenosine covalently bonded to a triphosphate group. The triphosphate portion of the ATP molecule is an anhydride of phosphoric acid. Like other phosphoric acid anhydrides, the anhydride group of ATP is readily hydrolyzed. The

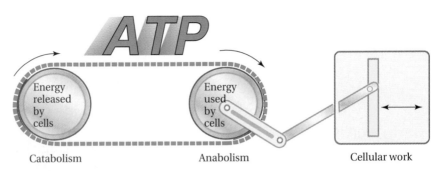

Figure 23.4
ATP is the energy carrier between the metabolic reactions that release energy and those that use energy.

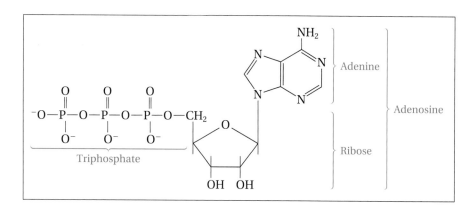

Figure 23.5
The structure of ATP.

hydrolysis products are adenosine diphosphate (ADP) and phosphate (P_i).

$$Adenosine\!-\!\overset{\overset{\displaystyle O}{\|}}{\underset{\underset{\displaystyle OH}{|}}{P}}\!-\!O\!-\!\overset{\overset{\displaystyle O}{\|}}{\underset{\underset{\displaystyle OH}{|}}{P}}\!-\!O\!-\!\overset{\overset{\displaystyle O}{\|}}{\underset{\underset{\displaystyle OH}{|}}{P}}\!-\!OH \;+\; H_2O \;\longrightarrow\; Adenosine\!-\!\overset{\overset{\displaystyle O}{\|}}{\underset{\underset{\displaystyle OH}{|}}{P}}\!-\!O\!-\!\overset{\overset{\displaystyle O}{\|}}{\underset{\underset{\displaystyle OH}{|}}{P}}\!-\!OH \;+\; HO\!-\!\overset{\overset{\displaystyle O}{\|}}{\underset{\underset{\displaystyle OH}{|}}{P}}\!-\!OH$$

Adenosine triphosphate   Water   Adenosine diphosphate   Inorganic phosphate
(ATP)   (ADP)   (P_i)

Or

$$ATP + H_2O \rightarrow ADP + P_i$$

The other anhydride bond also can be broken by hydrolysis. The products are adenosine monophosphate (AMP) and pyrophosphate (PP_i).

$$Adenosine\!-\!\overset{\overset{\displaystyle O}{\|}}{\underset{\underset{\displaystyle OH}{|}}{P}}\!-\!O\!-\!\overset{\overset{\displaystyle O}{\|}}{\underset{\underset{\displaystyle OH}{|}}{P}}\!-\!O\!-\!\overset{\overset{\displaystyle O}{\|}}{\underset{\underset{\displaystyle OH}{|}}{P}}\!-\!OH \;+\; H_2O \;\longrightarrow\; Adenosine\!-\!\overset{\overset{\displaystyle O}{\|}}{\underset{\underset{\displaystyle OH}{|}}{P}}\!-\!OH \;+\; HO\!-\!\overset{\overset{\displaystyle O}{\|}}{\underset{\underset{\displaystyle OH}{|}}{P}}\!-\!O\!-\!\overset{\overset{\displaystyle O}{\|}}{\underset{\underset{\displaystyle OH}{|}}{P}}\!-\!OH$$

ATP   Water   Adenosine   Inorganic
                                     monophosphate   pyrophosphate
                                         (AMP)   (PP_i)

Or

$$ATP + H_2O \longrightarrow AMP + PP_i$$

The hydrolysis of ATP to ADP and P_i or to AMP and PP_i releases a good deal of free energy—7 kcal for each mole of ATP hydrolyzed.

$$ATP \longrightarrow ADP + P_i + 7 \text{ kcal free energy}$$
$$ATP \longrightarrow AMP + PP_i + 7 \text{ kcal free energy}$$

There is no guarantee, of course, that cells will harness this free energy to do work. Indeed, if ATP were simply hydrolyzed in cells, all the energy would be lost as heat, and no useful work would be done. But cells have evolved ways of harnessing this energy. We will learn how later in this chapter, but first we will see where ATP stands in the energetics of cells and how it is made.

PRACTICE EXERCISE 23.3

How much free energy would be released for each mole of ATP hydrolyzed according to the following reaction?

$$ATP + 2H_2O \longrightarrow AMP + 2P_i$$

PRACTICE EXERCISE 23.4

Calculate the free energy released when 750 g of ATP is hydrolyzed according to the reaction in Practice Exercise 23.3.

23.5 Cellular energetics

AIM: To state the general relationships between ATP, anabolism, and catabolism.

Focus

ATP occupies an intermediate position in the energetics of the cell.

Newborn babies and hibernating animals have brown fat tissue that contains an enzyme that prevents ATP production. The energy from brown fat tissue is released as heat used to maintain body temperature.

ATP is often described as a high-energy compound. The energy released by the breakdown of ATP to ADP is not particularly high, however, compared with the energy released in some cellular oxidation reactions. To see how this works to the advantage of the cell, consider the hydrolysis of ATP, a process that releases a total of 7 kcal of free energy for each mole of ATP hydrolyzed. The reverse of hydrolysis, formation of ATP from ADP and P_i, requires at least the same amount of energy. This energy can be supplied by the energy produced by a few higher-energy oxidation reactions of catabolism. ATP is important not because it is a high-energy compound but because it occupies an intermediate position in the energetics of the cell. The energy conserved in the anhydride bonds of ATP is passed to other cellular processes.

Cells maintain a careful balance between ATP production and use. When a cell has a good supply of nutrients to catabolize, ATP production is high, and there is enough ATP to maintain the cell with some left over. Under these conditions, the excess of ATP is used in anabolic reactions to synthesize substances needed for cell growth and multiplication. In an environment poor in nutrients, the ATP produced is used to maintain cell vitality. Under these conditions, synthesis (anabolism) is deemphasized or reversed. There is an important principle here: *Catabolism generally produces ATP; anabolism generally uses it.*

PRACTICE EXERCISE 23.5

Explain why a lack of nutrients stops cell growth.

23.6 Oxidative phosphorylation

AIMS: To outline the production of ATP in aerobic cells, showing the relationship between oxidation reactions, formation of reducing power, cellular respiration, and oxidative phosphorylation. To classify members of the NAD$^+$-NADH and FAD-FADH$_2$ pairs as oxidizing or reducing agents. To write net equations for the oxidation of NADH and FADH$_2$ in cellular respiration. To explain what happens to carriers of the electron transport chain as electrons move down the chain. To indicate the number of ATP molecules formed for every NADH and FADH$_2$ molecule that enters the electron transport chain.

Focus

The ATP requirements of aerobic cells are met by oxidative phosphorylation.

Aerobic
aer (Greek): air
bios (Greek): life

Cellular respiration is concerned with oxygen use by aerobic cells and is *not* to be confused with pulmonary respiration, or breathing, in creatures with lungs.

Anaerobic
an (Greek): without
aer (Greek): air
bios (Greek): life

We now know in a general way that cells trap free energy released by the oxidation of sugars, fats, and amino acids. But how do cells extract this energy and use it to make ATP? This extraction of energy in **aerobic cells**—*oxygen-requiring cells*—can be described as three basic processes:

1. Some of the energy released by the oxidation of carbon compounds is conserved as **reducing power**—*the ability of the reduced forms of certain enzyme cofactors to reduce other biological molecules.*

2. Much of the energy conserved as reducing power is used for **cellular respiration**—*the energy-releasing reduction of molecular oxygen (O$_2$) to water.*

3. The energy released in cellular respiration is used in **oxidative phosphorylation**—*the process by which most of the cellular ATP is produced from ADP and P$_i$.*

These three points explain why aerobic cells need nutrients and oxygen to live. Without nutrients, there is no energy released by oxidation reactions; without energy, there is no formation of reducing power; without reducing power and oxygen, there is no cellular respiration; without cellular respiration, there is no ATP production by oxidative phosphorylation; and without ATP, there can be no life.

The need for oxygen is not universal. Certain bacteria—for example, the bacterium that causes tetanus—do not require oxygen to sustain life. *Organisms that do not need oxygen to live are called* **anaerobes.** For some strict anaerobes, oxygen is a poison, and they die in its presence.

Reducing power

Nicotinamide adenine dinucleotide (NAD$^+$) (Fig. 23.6a) and flavin adenine dinucleotide (FAD) (Fig. 23.6b) are compounds synthesized by cells from B-complex vitamins. NAD$^+$ is structurally related to nicotinic acid; FAD is structurally related to riboflavin. NAD$^+$ and FAD are oxidizing agents—that is, electron acceptors. Several different oxidation reactions of catabolism are catalyzed by enzymes that use either NAD$^+$ or FAD as their cofactor.

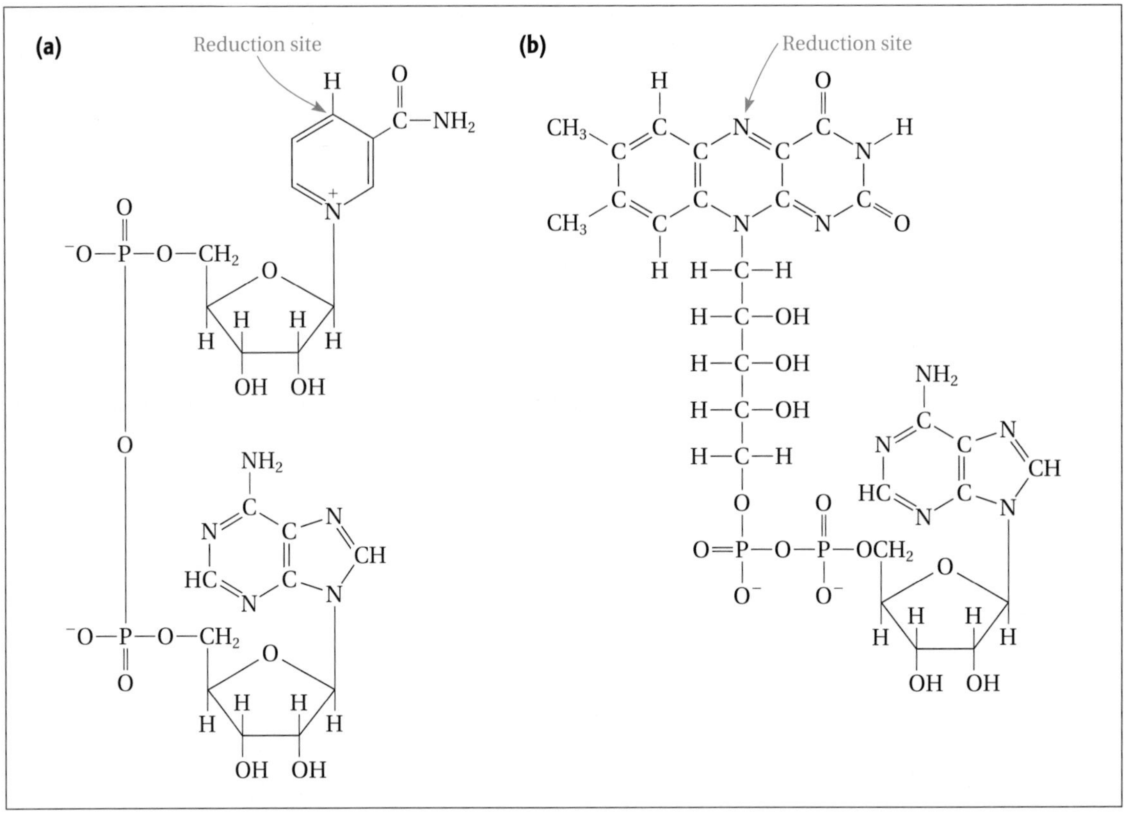

Figure 23.6
(a) The oxidized form of nicotinamide adenine dinucleotide. Notice that the oxidized form of the coenzyme has a positive charge. Hence its abbreviation is NAD^+. (b) The oxidized form of flavin adenine dinucleotide (FAD) does not carry a charge.

During these oxidation reactions, the substrates are oxidized, and the cofactors are reduced. NADH is the reduced form of NAD^+, and $FADH_2$ is the reduced form of FAD. NADH and $FADH_2$ produced by the oxidation of certain substrates are then used to reduce a variety of other molecules. *This capability of NADH and FADH₂ to act as reducing agents is called their reducing power.*

Let's see how the reduction of NAD^+ and FAD works. During oxidation reactions catalyzed by enzymes that use NAD^+ as the coenzyme, an NAD^+ molecule accepts a pair of electrons and a proton from a substrate molecule.

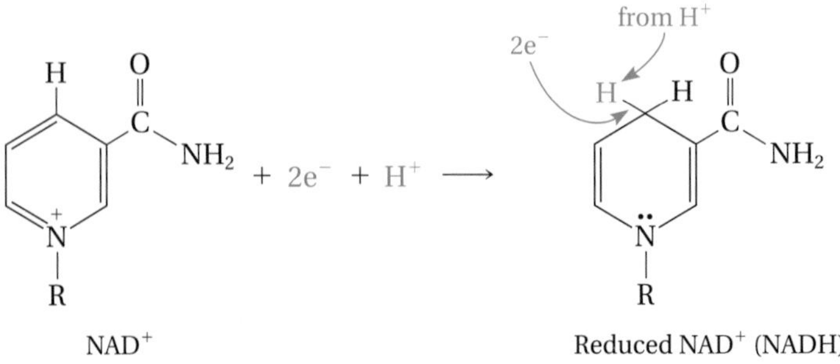

Because it has gained electrons, the NAD^+ molecule is reduced to NADH. A substrate molecule has lost two electrons and a proton. The substrate has therefore been oxidized. A typical cellular reaction that results in the reduction of NAD^+ and the oxidation of a substrate is the oxidation of an alcohol to an aldehyde.

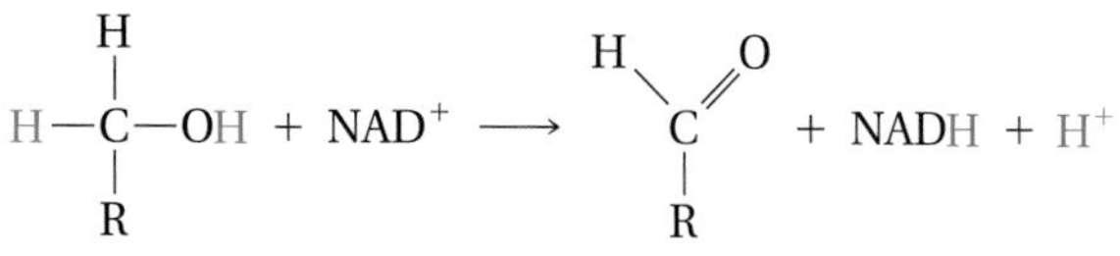

$$H-\overset{\overset{\displaystyle H}{|}}{\underset{\underset{\displaystyle R}{|}}{C}}-OH \; + \; NAD^+ \; \longrightarrow \; \underset{\underset{\displaystyle R}{|}}{\overset{H}{C}}\!\!\diagup\!\!\overset{O}{} \; + \; NADH \; + \; H^+$$

Alcohol   Aldehyde

NAD^+ is also the cofactor usually involved in the oxidation of aldehydes to carboxylic acids.

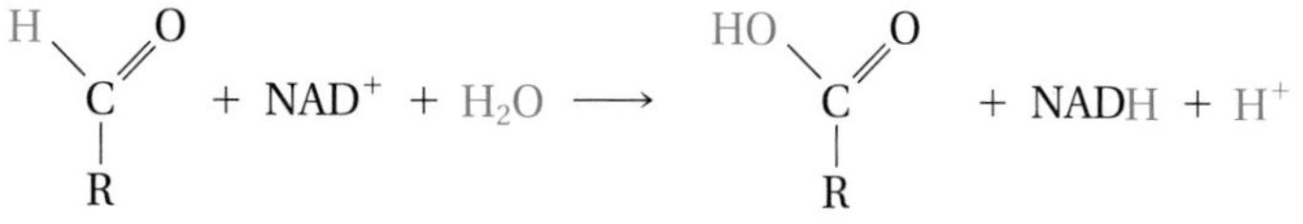

$$\underset{\underset{\displaystyle R}{|}}{\overset{H}{C}}\!\!\diagup\!\!\overset{O}{} \; + \; NAD^+ \; + \; H_2O \; \longrightarrow \; \underset{\underset{\displaystyle R}{|}}{\overset{HO}{C}}\!\!\diagup\!\!\overset{O}{} \; + \; NADH \; + \; H^+$$

Aldehyde   Carboxylic acid

In each of these reactions, the substrate is oxidized and loses energy. The NAD^+ is reduced and gains energy. This means that *some of the energy produced by the oxidation of the substrate has been conserved as reducing power by the formation of NADH.* As we will see, the reducing power of NADH will be used in two important ways: It is one of two sources of electrons that can be used to reduce oxygen to water in cellular respiration, and it can be used to reduce carbon compounds in anabolic reactions.

The reduced form of FAD ($FADH_2$) is the second source of cellular reducing power. As in the reduction of NAD^+, substrates of enzymes that use FAD as their cofactor give up two electrons to the cofactor. Upon reduction, however, FAD picks up one more proton than NAD^+. Thus, the abbreviation for the reduced form of FAD is $FADH_2$.

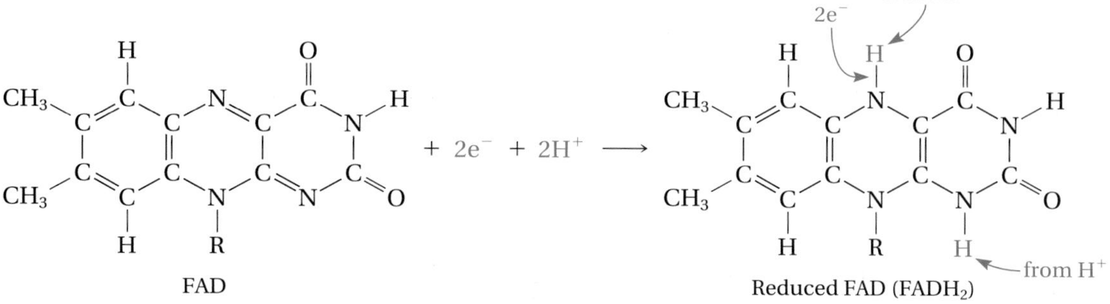

FAD   + 2e⁻ + 2H⁺ ⟶   Reduced FAD ($FADH_2$)

The substrates for oxidation reactions in which FAD is the cofactor are different from those involving NAD^+. FAD is involved in oxidation reactions in which a $-CH_2-CH_2-$ linkage of the substrate is oxidized to a double bond.

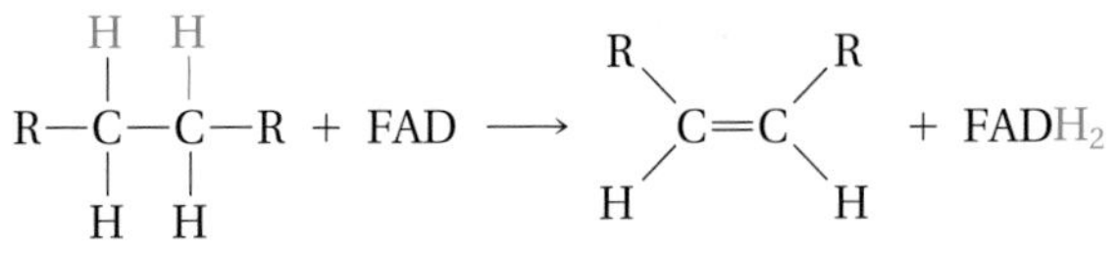

$$R-\underset{\underset{H}{|}}{\overset{\overset{H}{|}}{C}}-\underset{\underset{H}{|}}{\overset{\overset{H}{|}}{C}}-R + FAD \longrightarrow \underset{H}{\overset{R}{\diagdown}}C=C\underset{H}{\overset{R}{\diagup}} + FADH_2$$

Saturated   Unsaturated
(less oxidized)   (more oxidized)

PRACTICE EXERCISE 23.6

Predict the products of the following enzyme-catalyzed oxidation reactions. The molecule on the left in the reaction is the substrate.

(a) $CH_3CHO + NAD^+ + H_2O \longrightarrow$

(b) $HOOCCH_2CH_2COOH + FAD \longrightarrow$

(c) $CH_3CH_2OH + NAD^+ \longrightarrow$

PRACTICE EXERCISE 23.7

How many moles of NADH are formed in the enzyme-catalyzed oxidation of 1 mol of ethanol (CH_3CH_2OH) to 1 mol of acetic acid (CH_3CO_2H)? Give the reaction.

Cellular respiration

Of all the metabolic processes of aerobic cells, by far the most ATP is produced by events that depend on cellular respiration. *Molecular oxygen (O_2) is reduced to water (H_2O) by NADH and FADH$_2$ in cellular respiration.* The following equation summarizes cellular respiration for NADH:

$$2NADH + 2H^+ + O_2 \longrightarrow 2NAD^+ + 2H_2O$$

Four electrons are transferred from NADH to molecular oxygen as this reduction occurs. Each molecule of NADH supplies two electrons. Two of the four protons needed come from NADH, and two come from the surroundings. FADH$_2$ is also important in cellular respiration, since it reduces oxygen in a similar way.

$$2FADH_2 + O_2 \longrightarrow 2FAD + 2H_2O$$

In the preceding equation, each FADH$_2$ molecule supplies two electrons and two protons.

Although these equations can be used to represent cellular respiration, the reduction of oxygen by NADH or FADH$_2$ does not occur in a single step. Instead, the two electrons stripped from the reduced cofactors (NADH and FADH$_2$) are passed through an *electron transport chain* (also called the *respiratory chain*). The **electron transport chain (respiratory chain)** *consists of a group of molecules that accept electrons from NADH and FADH$_2$ and transfer these electrons to molecular oxygen in a sequence of energy-releasing steps.* The components of the electron transport chain are assembled on the inner membrane of the mitochondria of eukaryotic cells (see A Closer Look:

Approximately 90% of all ATP molecules are produced in the mitochondria. A liver cell contains hundreds of mitochondria.

Mitochondria). The components of the respiratory chain of aerobic prokaryotes are located on the interior of the cell membrane. The location of the respiratory assembly in the mitochondria and the importance of respiration in ATP production give mitochondria a well-deserved reputation as the "power plants" of the eukaryotic cell.

Six intermediate electron carriers separate NADH from the ultimate electron acceptor (oxygen) in the respiratory chain. The electron carriers of the respiratory chain are lined up in order of increasing affinity for elec-

A Closer Look

Mitochondria

All eukaryotic cells have mitochondria (singular: mitochondrion). Mitochondria (see figures) are often called the "power plants" of the cell, since most of the cell's energy is generated there by oxidation-reduction reactions. To understand the membrane structure of mitochondria, imagine a soft purse. A mitochondrion has a double membrane that corresponds to the outside of the purse and its lining. The interior membrane of the mitochondrion is highly folded, much like pleats in the purse lining. These folds, which extend into the interior of the mitochondrion, are *cristae*. Many of the mitochondrion's enzymes are attached to the inside surface of the interior membrane.

The chemical composition of the inner membrane of mitochondria closely resembles that of the membranes of prokaryotic cells. Moreover, mitochondria contain DNA that is different from nuclear DNA and reproduces independently of the reproduction processes of the cell. Because of these similarities, scientists believe that mitochondria are highly specialized remains of prokaryotic cells. They speculate that a mutually beneficial (symbiotic) relationship resulted when an anaerobic eukaryotic cell encapsulated an aerobic prokaryotic cell. This arrangement allowed the prokaryotic cell to use the waste products of the eukaryotic cell for ATP production; in return, the eukaryotic cell used some of the ATP produced by the prokaryotic cell. These specialized functions of the eukaryotic host and its prokaryotic guest eventually developed to the point that today neither host nor guest can survive without the other.

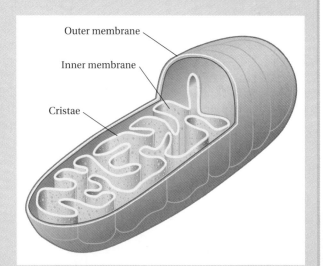

Outer membrane

Inner membrane

Cristae

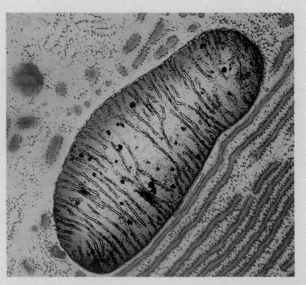

A mitochondrion.

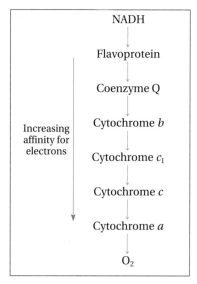

NADH

↓

Flavoprotein

↓

Coenzyme Q

↓

Cytochrome *b*

↓

Cytochrome *c*₁

↓

Cytochrome *c*

↓

Cytochrome *a*

↓

O₂

Increasing affinity for electrons

Figure 23.7
Six electron carriers move electrons from NADH to oxygen in the electron transport chain.

trons. Figure 23.7 shows these electron carriers. The first electron carrier is an enzyme that contains a tightly bound cofactor. This cofactor is similar in structure to riboflavin and FAD. Hence *the enzyme is called a* **flavoprotein.** In the first step of the respiratory chain, the cofactor portion of the flavoprotein is changed from the oxidized (ox) to the reduced (red) state by accepting an electron pair from NADH.

$$NADH + Flavoprotein_{ox} \longrightarrow NAD^+ + Flavoprotein_{red} + H^+$$

Since oxidation is a loss of electrons and reduction is a gain of electrons, NADH is oxidized and the flavoprotein is reduced. NADH has now exercised its reducing power.

Coenzyme Q (CoQ) is the second electron carrier of the respiratory chain. It accepts two electrons from the reduced flavoprotein.

$$Flavoprotein_{red} + CoQ_{ox} \longrightarrow Flavoprotein_{ox} + CoQ_{red}$$

The CoQ is reduced and the flavoprotein is reoxidized. Figure 23.8 shows the structures of the oxidized and reduced forms of CoQ.

The electron carriers beyond CoQ are the cytochromes. **Cytochromes** *are proteins that contain an iron-heme complex very similar to that of hemoglobin* (Fig. 23.9). However, there is a major chemical difference between the iron of hemoglobin and that of the cytochromes. Whereas the iron of normal hemoglobin remains in the ferrous (Fe^{2+}) state, the iron of

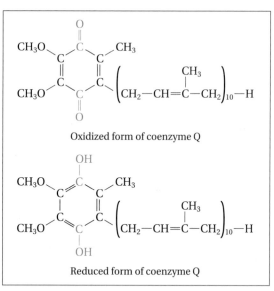

Oxidized form of coenzyme Q

Reduced form of coenzyme Q

Figure 23.8
The oxidized and reduced forms of coenzyme Q.

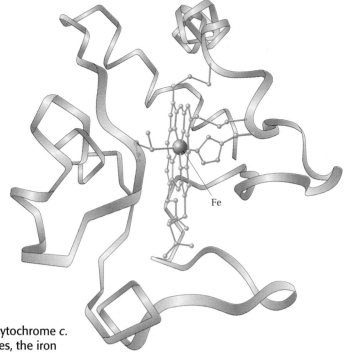

Fe

Figure 23.9
The structure of cytochrome *c*. In the cytochromes, the iron changes from 2+ in the reduced state to 3+ in the oxidized state.

the cytochromes flip-flops between the reduced ferrous (Fe^{2+}) and oxidized ferric (Fe^{3+}) states by the addition or loss of one electron.

$$Fe^{3+} + e^{-} \xrightleftharpoons[\text{Oxidation}]{\text{Reduction}} Fe^{2+}$$

Cytochrome
cyto (Greek):   cell
chroma (Greek):   color

Thus the first cytochrome of the respiratory chain readily accepts electrons from reduced CoQ and just as readily loses them to the next cytochrome in the chain. The pattern of oxidations and reductions continues as the electrons pass through the chain single file.

$$CoQ_{red} + Cyt\ b_{ox} \longrightarrow CoQ_{ox} + Cyt\ b_{red}$$
$$Cyt\ b_{red} + Cyt\ c_{1_{ox}} \longrightarrow Cyt\ b_{ox} + Cyt\ c_{1_{red}}$$
$$Cyt\ c_{1_{red}} + Cyt\ c_{ox} \longrightarrow Cyt\ c_{1_{ox}} + Cyt\ c_{red}$$
$$Cyt\ c_{red} + Cyt\ a_{ox} \longrightarrow Cyt\ c_{ox} + Cyt\ a_{red}$$

Finally, at reduced cytochrome *a*, an oxygen molecule picks up four electrons that have traveled through the chain. With the addition of four protons from the surroundings, the reduction of oxygen to water has occurred, and the passage of an electron pair from NADH through the electron chain is complete.

$$2e^{-} + \cdot \ddot{O} \cdot + 2H^{+} \longrightarrow H{-}\ddot{O}{-}H$$

From   Oxygen   Water
respiratory   atom
chain

A Closer Look: Oxygen, Disease, and Aging discusses in more detail how this water is formed and its ramifications for health.

As each carrier is reduced by the addition of electrons, the carrier from which it received the electrons becomes reoxidized and ready to receive more electrons. This feature permits electrons to flow through the respiratory chain as long as there is sufficient NADH available to donate them and sufficient oxygen available to accept them. Figure 23.10 summarizes the oxidation-reduction reactions of the respiratory chain.

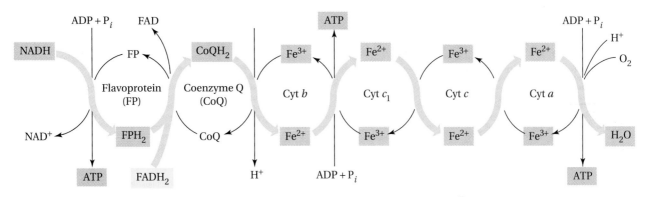

Figure 23.10
Oxidation-reduction reactions of the respiratory chain. The brown arrows show the path of the electrons as they pass from one carrier to the next. The reduced species are shown in yellow, blue, and purple. Electrons from NADH and $FADH_2$ pass through the chain in pairs until they reach the cytochromes. They pass through the cytochromes one at a time. The final electron acceptor is oxygen.

Cyanide ions are a powerful poison of the respiratory chain. These ions form an irreversible complex with the iron of cytochrome *a*, the last cytochrome in the chain. Formation of the cytochrome *a*–cyanide complex prevents electrons from being drained away by oxygen. Under these conditions, the electron carriers preceding cytochrome *a* become clogged with electrons—that is, all the carriers are soon reduced, and further electron

Lou Gehrig, a well-known baseball player, was a victim of ALS, amyotrophic lateral sclerosis.

A Closer Look

Oxygen, Disease, and Aging

The reduction of molecular oxygen (O_2) to water (H_2O) in the electron transport chain involves the formation of several *reactive oxygen species (ROS)*. The major ROS are the *hydroxyl radical* ($\cdot OH$) (see Sec. 10.8), the *superoxide anion* ($\cdot O_2^-$), and *hydrogen peroxide* (H_2O_2). Hydroxyl radicals and superoxide anions are free radicals (contain an unpaired electron). The rapid decomposition of ROS is vital because ROS are strong oxidizing agents that are capable of causing extensive cell damage and even death. Normal cells contain enzymes that destroy superoxide anions and hydrogen peroxide. The enzyme *superoxide dismutase* catalyzes the conversion of superoxide anions to hydrogen peroxide. Superoxide dismutase contains zinc, copper, and manganese cofactors that are important in this reaction. *Catalase* (recall A Closer Look: Catalase, on page 166), an iron-containing enzyme, catalyzes the decomposition of the hydrogen peroxide to molecular oxygen and water. *Peroxidase*, another iron-containing enzyme, catalyzes the same reaction. Vitamin C (ascorbic acid) and vitamin E destroy ROS, especially hydroxyl radicals, by reacting with them.

The effects of faulty destruction of ROS can be seen in several human diseases and in aging. Amyotrophic lateral sclerosis (ALS), also known as Lou Gehrig's disease, is an invariably fatal degenerative disease of the nervous system (see figure). ALS results from the mutation of the single gene that codes for superoxide dismutase. The defective gene produces defective superoxide dismutase, and the damage to the nervous system by high levels of superoxide anion causes the symptoms of ALS. Oxidative damage by ROS is thought to be a factor in such diseases as arthritis, emphysema, and some cancers. A fairly extensive body of evidence indicates that normal aging is in part the result of oxidative damage by ROS and other free radicals in body cells. As we age, free radicals damage our proteins, lipids, and nucleic acids by oxidative reactions. Oxidized proteins such as enzymes may not function properly. Lipid oxidation can produce fragile cell membranes. Nucleic acid oxidation can lead to gene mutations that result in the expression of defective or nonfunctional proteins. Experimental animals fed diets high in vitamins C and E live longer and exhibit fewer of the debilitating effects of aging than control animals fed regular diets. Scientists believe that these beneficial effects of vitamins C and E are the result of their reactions with ROS and other free radicals.

transport is blocked. The blockage of electron transport has the same effect on the cell as a lack of oxygen. Respiration stops; no ATP is produced by oxidative phosphorylation; the cell, and soon the organism, dies because it cannot pass electrons to oxygen.

Oxidative phosphorylation

Oxidative phosphorylation ties cellular respiration to ATP production. Cellular respiration is an energy-yielding process. For every pair of electrons that flow through the electron transport chain—this is the number of electrons given up by one molecule of NADH—enough energy to phosphorylate a molecule of ADP is released at three points in the chain (Fig. 23.11). *For every NADH oxidized to NAD$^+$ in cellular respiration, three ATP molecules may be formed by oxidative phosphorylation.* The electron pairs from FADH$_2$ enter the respiratory chain further down the line than electrons from NADH. *For every FADH$_2$ oxidized to FAD, only two ATP molecules are formed by oxidative phosphorylation.* How is the energy of oxidation of NADH and FADH$_2$ used to make ATP in oxidative phosphorylation? The energy is used to activate an enzyme called an *ATP synthase.* ATP synthase catalyzes the coupling of ADP with P$_i$ to make ATP.

Recall from the Case in Point earlier in this chapter that Emily was suffering from a mysterious fatigue. In the Follow-up to the Case in Point, on the next page, we will learn the nature of her affliction.

Figure 23.11
Energy is released as electrons pass through the electron transport chain. When two electrons from the oxidation of NADH to NAD$^+$ pass through the chain, enough energy is released to phosphorylate one molecule of ADP to ATP by the process of oxidative phosphorylation at each of the three sites shown. The pair of electrons from the oxidation of FADH$_2$ to FAD enters the chain at a later point. Only two molecules of ATP are formed for every two electrons from FADH$_2$ that pass through the chain.

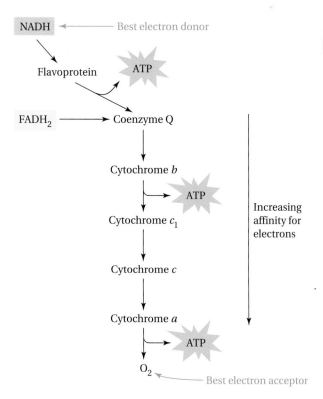

FOLLOW-UP TO THE CASE IN POINT: A mysterious fatigue

After eliminating usual causes of fatigue, the medical researchers who investigated Emily's condition sought the source of her problem in the biochemical processes of energy production: oxygen delivery, formation of NADH and $FADH_2$, the respiratory chain, and oxidative phosphorylation. They found that oxygen uptake by her tissues was normal. This eliminated the possibility of a flawed delivery of oxygen to the tissues. Her body cells' levels of NADH, NAD^+, $FADH_2$, and FAD also were normal, suggesting that enough reducing power was available to the respiratory chain and that the respiratory chain was operative. Her cellular ATP levels were considerably lower than normal.

The researchers concluded that Emily was suffering from an extremely rare defect of oxidative phosphorylation: The respiratory chain and the production of ATP were partially uncoupled. In such a situation, oxygen uptake, cellular respiration, and levels of NADH, NAD^+, $FADH_2$, and FAD would be normal. However, ATP levels would be low, leading to chronic fatigue. The molecular basis of the defect was not determined, but a mutant gene could be responsible for the defective expression of the ATP synthase that couples the respiratory chain to the production of ATP in oxidative phosphorylation. A flawed ATP synthase would result in impaired ATP production.

23.7 Cellular work

AIMS: To describe the three major types of work done by cells. To show how coupled reactions enable a cell to carry out chemical work.

Focus

Energy conserved in ATP drives cellular processes.

We have seen that cleavage of the phosphoric acid anhydride bonds in the ATP molecule releases free energy that can do cellular work. Cells do three major kinds of work: chemical, osmotic, and mechanical. In this section we will examine each kind of work and see how ATP is involved.

Chemical work

ATP often supplies energy to chemical reactions that need energy to make a product. Most of these reactions involve ATP as a phosphorylating agent. That is, because of its very reactive phosphoric acid anhydride functional groups, ATP readily transfers phosphoryl

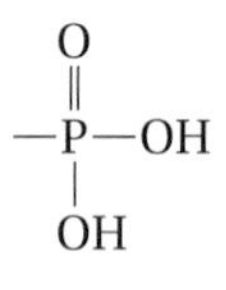

$$\begin{array}{c} O \\ \parallel \\ -P-OH \\ | \\ OH \end{array}$$

or, less commonly, pyrophosphoryl

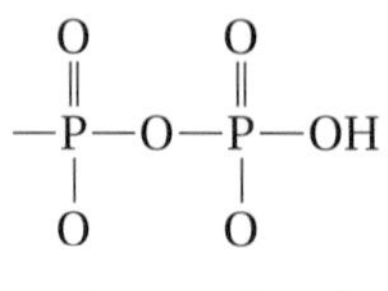

$$\begin{array}{c} O \quad\;\; O \\ \parallel \qquad \parallel \\ -P-O-P-OH \\ | \qquad\;\; | \\ O \qquad\;\; O \end{array}$$

groups to other molecules. Just as reducing power is the ability of NADH and $FADH_2$ to transfer electrons in reduction reactions, **phosphorylating power** *is the ability of ATP to transfer phosphoryl groups in phosphorylation reactions.*

Some phosphorylated enzyme substrates are activated for subsequent reactions they would not ordinarily undergo. The process of activation often involves a **coupled reaction**—*an energetically unfavorable reaction is made to occur by being linked to a reaction that is energetically very favorable (very exergonic)*. In a coupled reaction, the net energy change for the linked reactions is favorable (exergonic). For example, liver cells need to use ammonia and carbon dioxide to make citrulline from the amino acid ornithine. (This reaction is important in the urea cycle, and we will return to it in Chapter 26.) A hypothetical equation for this reaction can be written:

$$
\underset{\text{Ornithine}}{
\begin{array}{c}
\text{H} \\
| \\
\text{NH} \\
| \\
\text{CH}_2 \\
| \\
\text{CH}_2 \\
| \\
\text{CH}_2 \\
| \\
\text{H}_2\text{N}-\text{CH} \\
| \\
\text{CO}_2\text{H}
\end{array}}
\;+\;
\underset{\substack{\text{Carbon} \\ \text{dioxide}}}{
\begin{array}{c}
\text{O} \\
\| \\
\text{C} \\
\| \\
\text{O}
\end{array}}
\;+\;
\underset{\text{Ammonia}}{\text{NH}_3}
\;\longrightarrow\;
\underset{\text{Citrulline}}{
\begin{array}{c}
\quad\;\;\text{NH}_2 \\
\diagup \\
\text{O}{=}\text{C} \\
\diagdown \\
\text{NH} \\
| \\
\text{CH}_2 \\
| \\
\text{CH}_2 \\
| \\
\text{CH}_2 \\
| \\
\text{H}_2\text{N}-\text{CH} \\
| \\
\text{CO}_2\text{H}
\end{array}}
\;+\;
\underset{\text{Water}}{\text{H}_2\text{O}}
$$

This hypothetical reaction is energetically unfavorable (Fig. 23.12a). The equilibrium position greatly favors the reactants over the product. What about using an enzyme to catalyze the reaction? This would not make the reaction exergonic. As catalysts, enzymes only speed up reactions that are already exergonic; they do not alter the position of equilibrium. Cells are not stymied by this state of affairs, however, because they have ATP on their side. They carry out an exergonic reaction instead, forming a molecule of carbamoyl phosphate by using one molecule of ammonia, one of carbon

Figure 23.12
The direct conversion of ammonia to citrulline is not a favorable process (a). When the reaction is coupled with the breakdown of ATP, however, it is exergonic (favorable) (b).

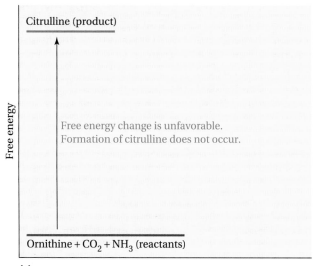

(a)

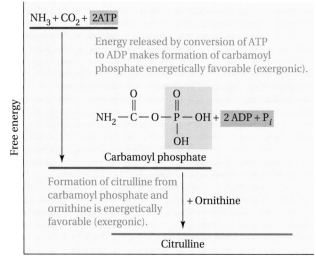

(b)

dioxide, and two of ATP (Fig. 23.12b). An enzyme speeds up the reaction.

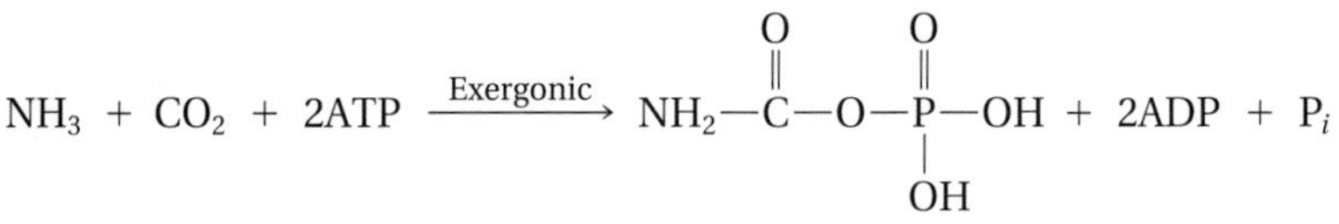

$$NH_3 + CO_2 + 2ATP \xrightarrow{\text{Exergonic}} NH_2-\overset{\overset{O}{\|}}{C}-O-\overset{\overset{O}{\|}}{\underset{\underset{OH}{|}}{P}}-OH + 2ADP + P_i$$

Carbamoyl
phosphate

The structure of carbamoyl phosphate contains a mixed anhydride linkage, one formed from a carboxylic acid and phosphoric acid. This linkage is very energetic. Carbamoyl phosphate readily reacts with ornithine to produce citrulline in another exergonic enzyme-catalyzed reaction (Fig. 23.12b).

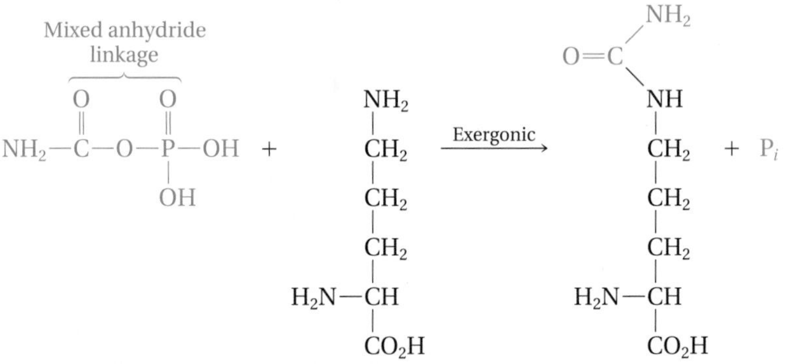

Carbamoyl   Ornithine   Citrulline
phosphate

ATP appears in the equation for the overall reaction for the formation of citrulline, and it looks as if it were used up, as it is in hydrolysis.

$$\text{Ornithine} + CO_2 + NH_3 + 2ATP \xrightarrow{\text{Exergonic}} \text{Citrulline} + 2ADP + 2P_i + H_2O$$

The reality is quite different. The free energy stored in the energetic anhydride bonds of two ATP molecules was unleashed by breaking these bonds, but some was conserved in the anhydride bond of carbamoyl phosphate. By expending the phosphorylating power of ATP to make carbamoyl phosphate, liver cells boosted the reactivity of ammonia and carbon dioxide and achieved the synthesis of citrulline. In our example, an energetically unfavorable reaction—the formation of citrulline from ammonia, carbon dioxide, and ornithine—could not occur without coupling the reaction to the favorable breakdown of ATP. Sometimes the route to energetically activated molecules requires several chemical steps, but the principle is the same. The inevitable result of substrate phosphorylation is the production of a more reactive molecule.

Like substrates, some enzymes are also phosphorylated by ATP. Phosphorylation often makes the difference between an active and inactive enzyme. A good example of enzyme activation caused by phosphorylation comes from catabolism and involves phosphorylase, the enzyme that catalyzes the breakdown of stored glycogen. The inactive form of phosphorylase (phosphorylase *b*) is a dimer consisting of identical protein subunits.

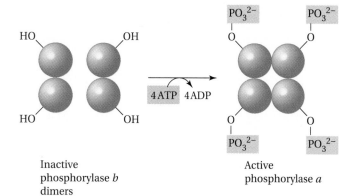

Figure 23.13
Phosphorylase is converted from an inactive to an active form upon phosphorylation.

Inactive phosphorylase *b* dimers

Active phosphorylase *a*

The active form of phosphorylase (phosphorylase *a*) is a tetramer of these identical subunits.

Phosphorylation of the hydroxyl groups of a single serine residue in the amino acid sequence of each subunit causes two inactive phosphorylase *b* dimers to form the active phosphorylase *a* tetramer (Fig. 23.13). The phosphorylase *a* can now go about the work nature intended: the breakdown of glycogen. It may appear that the phosphorylation of phosphorylase, a catabolic enzyme, violates the rule that catabolism produces ATP. However, one molecule of phosphorylase *a* catalyzes the cleavage of glycogen into many molecules of glucose. And as we will see in Chapter 24, the complete oxidation of only one glucose molecule freed by the action of phosphorylase *a* on glycogen gives a return of ATP many times greater than the four molecules of ATP it takes to phosphorylate the enzyme.

Osmotic work

Cells contain molecular pumps that transport ions and molecules through cell membranes. These pumps work against the normal concentration gradient—the transport is from a lower to a higher concentration of the substance being transported—and require the expenditure of energy in the form of ATP. Such pumps often require the phosphorylation of transport proteins embedded in the cell membrane.

Mechanical work

At this point we understand that ATP is a phosphorylating agent in the chemical and osmotic work done by cells. But in mechanical work—that is, cellular movements including muscle contraction—ATP is hydrolyzed rather than acting as a phosphorylating agent. The sliding-filament model is the generally accepted depiction of muscle contraction. In this model, muscles contract when thick and thin filaments slide past each other, with ATP providing the energy of propulsion (Fig. 23.14).

Muscle contraction requires so much ATP that muscle cells cannot keep enough on hand. Nature uses ATP to *transmit* energy, not to store it. In

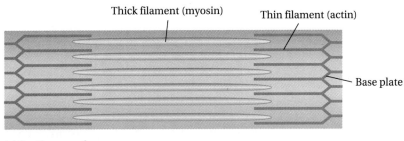

Thick filament (myosin)   Thin filament (actin)

Base plate

(a) Resting muscle

ATP → ADP + P$_i$

Figure 23.14
Muscle fibers contract when thick filaments made of myosin and thin filaments made of actin slide past each other, propelled by energy released in the hydrolysis of ATP.

(b) Contracted muscle

resting muscle, the phosphorylating power of ATP is stored in another high-energy compound called *creatine phosphate*, or *phosphocreatine*. On demand, the phosphoryl groups of creatine phosphate are transferred back to ADP.

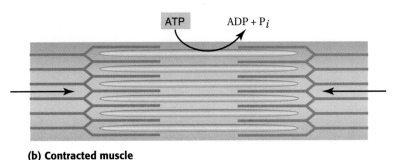

The enzyme involved in this reversible reaction is *creatine phosphokinase*. (Enzymes that catalyze the transfer of a phosphoryl group to or from ATP are called *kinases*.) As soon as the demand for ATP exceeds the supply, as in heavy exercise, muscles become tired and weak.

PRACTICE EXERCISE 23.8

Outline the steps to show why the body's need for oxygen increases as physical activity increases. What happens to muscles when the demand for ATP exceeds the supply?

SUMMARY

The Sun supplies the energy needed for life through the carbon and energy cycle. Photosynthesis by plants is perhaps the most important event in the carbon and energy cycle, since it is by photosynthesis that the Sun's light energy is converted to the chemical energy stored in the bonds of sugars, fats, and proteins. Chemical energy stored in these bonds is released in catabolism—the chemical reactions that cells use to oxidize the carbons of sugars, fats, and proteins back to carbon dioxide.

Cells conserve some of the energy of oxidation as reducing power in the form of NADH and $FADH_2$. Reduction of molecular oxygen by NADH and $FADH_2$ releases enough energy to drive oxidative phosphorylation—the phosphorylation of ADP to ATP. Cells use ATP for chemical work, osmotic work, and mechanical work, as in muscle contraction.

KEY TERMS

Adenosine triphosphate (ATP) (23.4)
Anabolism (23.1)
Anaerobe (23.6)
Catabolism (23.1)
Cellular respiration (23.6)
Chlorophyll (23.2)

Chloroplast (23.2)
Coupled reaction (23.7)
Cytochrome (23.6)
Electron transport chain (23.6)
Energy and carbon cycle (23.3)

Flavoprotein (23.6)
Light system (23.2)
Metabolism (23.1)
Oxidative phosphorylation (23.6)
Phosphorylating power (23.7)

Photosynthesis (23.2)
Reducing power (23.6)
Respiratory chain (23.6)
Thyalkoids (23.2)

EXERCISES

Metabolism (Section 23.1)

23.9 Define *metabolism*.

23.10 How do *anabolism* and *catabolism* differ from each other?

The Energy and Carbon Cycle (Sections 23.2, 23.3)

23.11 Write an equation that summarizes the process of photosynthesis.

23.12 Give a balanced equation for the oxidation of glucose.

23.13 Explain what happens in photosynthesis.

23.14 Where do sugars, fats, and amino acids originate?

23.15 How do cells produce energy?

23.16 What is the ultimate source of all carbon compounds? Explain.

23.17 Interpret this statement: "Carbon dioxide is an energy-poor molecule, but glucose is an energy-rich molecule."

23.18 Arrange the following compounds starting with the most reduced (energy-rich) and ending with the most oxidized (energy-poor).

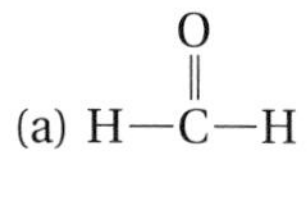

(a) Formaldehyde

(b) Formic acid

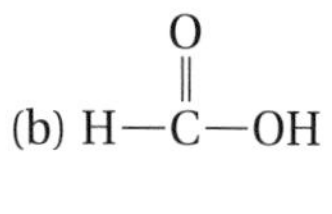

(c) Methanol

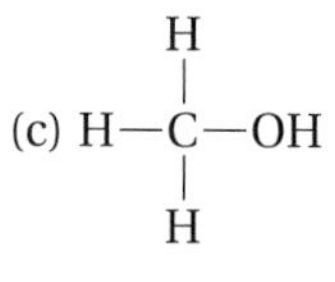

(d) Carbon dioxide

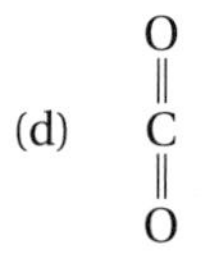

(e) Methane

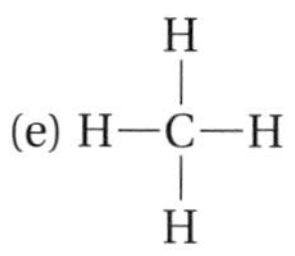

ATP (Sections 23.4, 23.5)

23.19 Draw the structure of ATP. Label the characteristic features of the molecule.

23.20 Discuss the function of ATP in cellular metabolism.

23.21 What is the charge on a fully unprotonated ATP molecule? What is the gram formula weight of the unprotonated structure?

23.22 Calculate the free energy released when (a) 2 mol of ATP is hydrolyzed to ADP and P_i and when (b) 50 g of ATP is hydrolyzed to ADP and P_i.

23.23 ATP occupies an intermediate position in the energetics of the cell. Explain what this means and why it is important.

23.24 What processes are responsible for controlling ATP concentrations in cells?

Oxidative Phosphorylation (Section 23.6)

23.25 What are aerobic cells? Outline the processes that occur when aerobic cells use energy released by the oxidation of nutrient molecules to phosphorylate ADP to ATP.

23.26 Write the abbreviations for the reduced and oxidized forms of nicotinamide adenine dinucleotide and flavin adenine dinucleotide.

23.27 Complete the following reactions.

(a) $NAD^+ + H^+ +$ _____ $\longrightarrow NADH$

(b) $FAD +$ _____ $+$ _____ $\longrightarrow FADH_2$

23.28 The coenzymes nicotinamide adenine dinucleotide (NAD^+) and flavin adenine dinucleotide (FAD) are oxidizing agents. What is their role in the enzyme-catalyzed oxidation of a substrate molecule?

23.29 Write a balanced equation for the oxidation of methanol to formaldehyde by NAD^+. What happens to NAD^+ in the course of this reaction?

23.30 NADH and $FADH_2$ are sources of cellular reducing power. What does this mean, and why is it important?

23.31 What happens during (a) cellular respiration and (b) oxidative phosphorylation?

23.32 What happens in the respiratory (or electron transport) chain?

23.33 State where cellular respiration takes place in eukaryotic cells and where ATP is produced.

23.34 What are the first and last steps in the respiratory chain?

23.35 Give the general name for the heme-containing proteins that are electron carriers in the respiratory chain. What is the major chemical difference between these molecules and hemoglobin?

23.36 Compare the effects of oxygen deficiency and cyanide poisoning on ATP production by oxidative phosphorylation.

23.37 How many molecules of ATP may be formed by oxidative phosphorylation when one molecule of NADH is oxidized to NAD^+ in cellular respiration?

23.38 If two molecules of $FADH_2$ are oxidized to FAD in cellular respiration, how many molecules of ADP may be phosphorylated?

Cellular Work (Section 23.7)

23.39 Some of the chemical energy released when nutrients are oxidized is conserved in ATP. What are the three major types of cellular work that require ATP energy?

23.40 ATP is a good phosphorylating agent. What takes place when a molecule is phosphorylated?

23.41 Why must some enzyme substrates be phosphorylated before they can undergo reaction?

23.42 Explain the importance of ATP in coupled reactions.

23.43 Why is ATP required in the transport of certain ions and molecules across cell membranes?

23.44 Creatine phosphate contains a high-energy phosphate bond. Explain its function as a store of the phosphorylating power of ATP in muscle cells.

Additional Exercises

23.45 Give the name of an important biological molecule that contains each of the following:
(a) heme ring   (b) corrin ring   (c) porphyrin ring

23.46 What do the abbreviations NAD^+ and FAD stand for?

23.47 Where do NADH and $FADH_2$ enter the electron transport chain?

23.48 Identify each of the following changes as a reduction or oxidation.
(a) $Fe^{3+} \longrightarrow Fe^{2+}$ (b) $NADH \longrightarrow NAD^+$
(c) $FAD \longrightarrow FADH_2$ (d) $RCH_2OH \longrightarrow RCHO$
(e) $RCHO \longrightarrow RCO_2H$

23.49 In the electron transport chain, how does the reduction of coenzyme Q differ from the reduction of cytochromes?

23.50 The ATP molecule has *phosphorylating power.* What does this term mean?

23.51 Draw the structure of the phosphoryl group. Why is this group important in certain metabolic reactions?

SELF-TEST (REVIEW)

True/False

1. Anabolic and catabolic reactions are reversible, the direction of the reaction depending on the needs of the cell.

2. In photosynthesis, an oxidized, energy-poor carbon-containing compound is changed into an energy-rich carbon-containing compound.

3. Photosynthesis takes place in the chloroplasts of the cells of green plants.

4. If either plant life or animal life had to exist without the other, animals would have the better chance of survival.

5. In the respiratory chain, a mole of $FADH_2$ produces twice as many molecules of ATP as does a mole of NADH.

6. At the end of a marathon race, the supply of creatine phosphate in the muscle cells of a runner would probably be low.

7. Typically, when ATP is hydrolyzed, it loses two of its phosphate groups.

8. If a cell has an excess of ATP, anabolic processes will generally take place.

9. The energy needed to drive the process of oxidative phosphorylation comes from the process of cellular respiration.

10. NADH is a reducing agent.

11. Cellular respiration takes place in the mitochondria.

12. The cytochromes in the respiratory chain are almost always in the oxidized (Fe^{3+}) state.

13. Cyanide poisons kill an organism by interfering with the transport of oxygen to the cells.

Multiple Choice

14. In an enzyme-catalyzed reaction in which the cofactor FAD is changed to $FADH_2$,
    (a) the substrate is reduced.
    (b) the FAD is oxidized.
    (c) the substrate loses two electrons.
    (d) both (a) and (b) are correct.

15. In the overall process of photosynthesis,
    (a) energy is given off.
    (b) an energy-poor compound is used to make an energy-rich compound.
    (c) water is produced.
    (d) carbon dioxide is further oxidized.

16. Cells often accomplish a desired reaction by pairing an energetically unfavorable reaction with an ener-getically favorable one. These reactions are called
    (a) paired reactions.
    (b) assisting reactions.
    (c) coupled reactions.
    (d) enzyme-catalyzed reactions.

17. Catabolic oxidation reactions in the cell
    (a) usually release energy.
    (b) produce compounds with higher reducing power.
    (c) consume nutrients supplied to the cell.
    (d) all of the above are true.

18. ATP is used as a phosphorylating agent by a cell in doing
    (a) chemical work.
    (b) osmotic work.
    (c) mechanical work.
    (d) both (a) and (b) are true.

19. The major energy-transmitting molecule of the cell is
    (a) nicotinamide adenine dinucleotide (NAD^+).
    (b) adenosine triphosphate (ATP).
    (c) flavin adenine dinucleotide (FAD).
    (d) adenosine monophosphate (AMP).

20. A secondary alcohol is changed to a ketone in an enzymatically catalyzed reaction using nicotinamide adenine dinucleotide as a coenzyme. In this reaction,
    (a) the NAD^+ acts as a reducing agent.
    (b) NADH is produced.
    (c) the alcohol has been reduced.
    (d) the NAD^+ has been oxidized.

21. A cell converts amino acids into a protein molecule. This is an example of
    (a) catabolism.
    (b) oxidative phosphorylation.
    (c) photosynthesis.
    (d) anabolism.

22. In which of the following pairs of substances is the member with the greater reducing power listed first?
    (a) $FAD/FADH_2$ (b) C_2H_5OH/C_2H_6
    (c) CH_3CO_2H/CH_3CHO
    (d) $NADH/NAD^+$

23. The overall equation for the cellular respiration of $FADH_2$ shows
    (a) oxygen as a reactant.
    (b) the $FADH_2$ is reduced to FAD.
    (c) hydroxide ion as a product.
    (d) that oxygen is oxidized to water.

24. The formation of ATP from ADP and P_i
    (a) requires no more than 7 kcal of energy per mole.
    (b) usually occurs in an anabolic process.
    (c) allows energy in a cell to be conserved.
    (d) is a hydrolysis reaction.

25. During cellular catabolism, the energy released
   (a) is all used to do work.
   (b) is immediately used to do work.
   (c) is immediately lost as heat.
   (d) is partially lost as heat.

26. The passage of electrons through the electron transport chain results in all the following *except*
   (a) the oxidative phosphorylation of ADP to ATP.
   (b) a series of oxidation-reduction reactions within the chain.
   (c) the formation of water molecules at the end of the chain.
   (d) the use of the oxidizing power of NADH.

27. The hydrolysis of a mole of ATP by a mole of water molecules gives
   (a) a mole of ADP.
   (b) a mole of 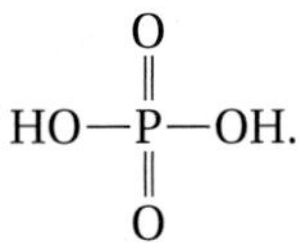
$$HO-\overset{\displaystyle O}{\underset{\displaystyle O}{\overset{\|}{\underset{\|}{P}}}}-OH.$$
   (c) about 7 kcal of free energy.
   (d) all of the above.

Carbohydrates in Living Organisms

At the Core of Metabolism

Bread is rich in carbohydrates; it is also a good source of vitamins and minerals.

CHAPTER OUTLINE

CASE IN POINT: Carbohydrate loading

24.1 Catabolism

24.2 Glucose oxidation

24.3 Glycolysis

24.4 Acetyl coenzyme A

24.5 The citric acid cycle

24.6 ATP yield

24.7 Fermentation

A CLOSER LOOK: Monitoring Heart Attacks with Serum LDH Tests

24.8 Oxygen debt

24.9 Glucose storage

24.10 Glycogen breakdown

FOLLOW-UP TO THE CASE IN POINT: Carbohydrate loading

24.11 Metabolic regulation

A CLOSER LOOK: Nitric Oxide and Carbon Monoxide as Second Messengers

24.12 Control of glycogenolysis

24.13 Glucose absorption

In this chapter we will explore the relationship between carbohydrate metabolism and energy production in the cell. Why is carbohydrate metabolism important? It is because the carbohydrates occupy center stage in the metabolic drama of the cell. The star of carbohydrate metabolism is glucose. In this chapter we will discover how glucose is broken down and synthesized. Carbohydrate metabolism is important for everyone, but it can be of special concern to athletes who wish to improve their performance. This is illustrated in the following Case in Point.

CASE IN POINT: Carbohydrate loading

Barbara, the coach of a college cross-country team, has been trying to improve her runners' performance. Because Barbara studied biochemistry in her college days, she encourages her runners to eat a lot of pasta in the 2 days before a meet. What biochemical result is Barbara trying to achieve in her runners? We will see in Section 24.10.

24.1 Catabolism

AIMS: To write an equation for the catabolism of one mole of glucose. To name the three stages of aerobic catabolism of glucose.

One tomato and one potato each contain about 3.5 g of sugar; an orange contains about 12.5 g, and a banana, about 18 g.

Glycolysis
glykos (Greek):   sugar or sweet
lysis (Greek):   splitting

The glucose produced by digestion of the carbohydrates a person eats enters the bloodstream. In a normal individual, this glucose is extracted from the bloodstream by body cells. Once inside the cells, the glucose is completely oxidized to carbon dioxide and water.

$$C_6H_{12}O_6 \ + \ O_2 \ \longrightarrow \ 6CO_2 \ + \ 6H_2O$$

Glucose   Oxygen   Carbon   Water
dioxide

The paths that cells use to oxidize glucose completely to carbon dioxide involve many individual chemical reactions. But all aerobic cells have adopted essentially the same three-stage master plan to do the job. Briefly these three stages, shown in Figure 24.1, consist of initial breakdown in *glycolysis,* further degradation to *acetyl coenzyme A,* and, finally, complete oxidation in the *citric acid cycle.*

Glycolysis

Glycolysis *is the sequence of chemical reactions by which glucose, a six-carbon sugar, is cleaved to two molecules of pyruvate, a three-carbon acid.* Biochemists usually call organic acids produced in metabolism by the names of their dissociated forms, since these are the forms that exist at pH 7.0. Pyruvate is simply the anion of pyruvic acid.

Pyruvic acid   Pyruvate

Glycolysis is sometimes called the *glycolytic pathway* or the *Embden-Meyerhof pathway* after two German biochemists, Gustav Embden and Otto Meyerhof, who proposed it in the latter part of the nineteenth century.

Acetyl coenzyme A

The second stage of glucose catabolism occurs when pyruvate ions lose carbon dioxide. The remaining two-carbon fragments end up attached to coenzyme A to form acetyl coenzyme A (Fig. 24.2). **Coenzyme A,** *usually abbreviated CoA, is a thiol;* **acetyl CoA** *is the thioester of acetic acid and CoA.* The structure of CoA contains pantothenic acid, one of the B-complex vitamins.

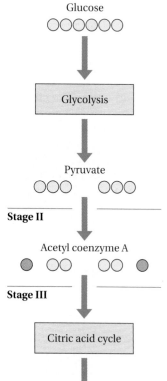

Figure 24.1 (left)
The oxidation of glucose, a six-carbon sugar (the carbon atoms are represented by yellow circles), to six molecules of carbon dioxide occurs in three stages. Two molecules of carbon dioxide (purple circles) are produced in the second stage, and four are produced in the third stage.

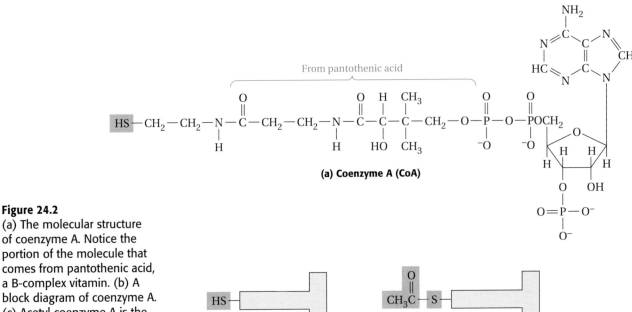

Figure 24.2
(a) The molecular structure of coenzyme A. Notice the portion of the molecule that comes from pantothenic acid, a B-complex vitamin. (b) A block diagram of coenzyme A. (c) Acetyl coenzyme A is the thioester of acetic acid and coenzyme A.

Like ATP, acetyl CoA is often considered an energy-rich compound. Just as the transfer of a phosphoryl group of ATP to some other molecule is made easier by its attachment to ADP, the transfer of an acetyl group is made easier by its attachment to CoA.

Citric acid cycle

The acetyl group of acetyl CoA enters the citric acid cycle. *The **citric acid cycle** is the pathway used by most organisms to oxidize completely to carbon dioxide the acetyl carbons of acetyl CoA formed in the breakdown of sugars, fats, and amino acids.* Production of two molecules of carbon dioxide and two of water for each molecule of acetyl CoA entering the cycle completes the third stage of glucose catabolism. The citric acid cycle is also called the *Krebs cycle* after Sir Hans Krebs, the English biochemist who proposed it in 1937.

24.2 Glucose oxidation

AIM: To state the entry points of sugars, fatty acids, and amino acids into the three stages of the aerobic catabolism of glucose.

Most biochemists consider the oxidation of glucose to be the core of catabolism because the cells of virtually all organisms rely on glucose as their main source of energy. Human brain cells, for example, will accept no other nutrient to obtain energy except under dire circumstances such as starva-

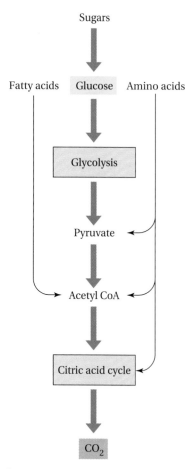

Figure 24.3
Proteins are hydrolyzed to amino acids; lipids are hydrolyzed to fatty acids and glycerol. Above, we see that the amino acids and fatty acids are converted to simpler compounds that enter the main pathways of glucose metabolism.

Focus

One glucose molecule produces two ATP and two NADH molecules in glycolysis.

The phosphorylation of glucose to form glucose 6-phosphate is so energetically favorable that essentially all the glucose that enters the cell is immediately phosphorylated.

tion. Many other cells do oxidize other sugars, fatty acids, and amino acids to obtain energy, however. Certain enzymes contained in such cells degrade these substances to compounds that eventually enter catabolism through the central core of glucose metabolism. Figure 24.3 shows the entry points.

Many organisms have the enzymes necessary to convert galactose, fructose, and other hexoses into glucose. These sugars therefore enter glycolysis as glucose. Fatty acids are oxidized and enter the central core of glucose catabolism as acetyl CoA. Because of the variety of amino acid structures, their degradation products enter the central core at several different points: at the tail end of glycolysis as pyruvate, as acetyl CoA, or as chemical intermediates of the citric acid cycle.

24.3 Glycolysis

AIM: To list the steps in the aerobic glycolysis of one molecule of glucose.

The enzymes that catalyze the steps of glycolysis are found in the cytoplasm of cells. This is where glycolysis occurs. Glycolysis begins with the phosphorylation of glucose to glucose 6-phosphate.

The names and structures of some of the intermediate compounds in metabolism are complex. You do not need to memorize them, but they will be used in the text to make it easier to follow what is happening. Remember also that all the steps of these reactions are catalyzed by enzymes.

The phosphoryl group of glucose 6-phosphate comes from ATP. This may seem a little surprising. Since glycolysis is a pathway of catabolism, we might expect it to *produce* ATP, not to *use* it! The important point here is that the cell is investing ATP, just as you might invest money in the stock market. Your investment—if you are lucky—will be returned with dividends of more money. The cell's investment will be repaid with dividends of more ATP. The glucose 6-phosphate is converted to fructose 6-phosphate.

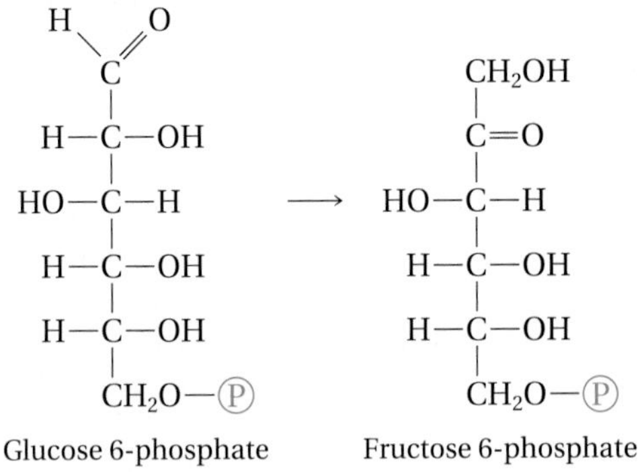

$$
\begin{array}{ll}
\underset{\displaystyle \text{H}}{}\overset{\displaystyle \text{O}}{\underset{\displaystyle}{\text{C}}} & \text{CH}_2\text{OH} \\
\text{H--C--OH} & \text{C=O} \\
\text{HO--C--H} & \text{HO--C--H} \\
\text{H--C--OH} & \text{H--C--OH} \\
\text{H--C--OH} & \text{H--C--OH} \\
\text{CH}_2\text{O--}\textcircled{P} & \text{CH}_2\text{O--}\textcircled{P} \\
\text{Glucose 6-phosphate} & \text{Fructose 6-phosphate}
\end{array}
$$

Fructose 6-phosphate undergoes phosphorylation to fructose 1,6-bisphosphate at the expense of another molecule of ATP invested.

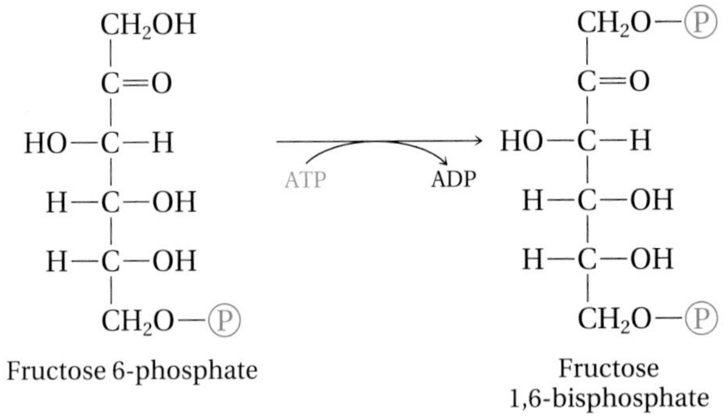

$$
\begin{array}{ll}
\text{CH}_2\text{OH} & \text{CH}_2\text{O--}\textcircled{P} \\
\text{C=O} & \text{C=O} \\
\text{HO--C--H} & \text{HO--C--H} \\
\text{H--C--OH} & \text{H--C--OH} \\
\text{H--C--OH} & \text{H--C--OH} \\
\text{CH}_2\text{O--}\textcircled{P} & \text{CH}_2\text{O--}\textcircled{P} \\
\text{Fructose 6-phosphate} & \text{Fructose} \\
& \text{1,6-bisphosphate}
\end{array}
$$

(ATP → ADP)

The cell has now invested two molecules of ATP for every molecule of glucose to be degraded. The conversion of fructose 6-phosphate to fructose 1,6-bisphosphate is an important control step in glycolysis. Once fructose 1,6-bisphosphate is formed, it cannot escape the glycolytic pathway. The phosphorylation of fructose 6-phosphate to fructose 1,6-bisphosphate is called the *committed step* of glycolysis.

Fructose 1,6-bisphosphate is now cleaved to give a pair of three-carbon compounds, dihydroxyacetone phosphate and glyceraldehyde 3-phosphate.

$$
\begin{array}{lll}
\text{CH}_2\text{O--}\textcircled{P} & & \\
\text{C=O} & \text{CH}_2\text{O--}\textcircled{P} & \underset{\displaystyle \text{H}}{}\overset{\displaystyle \text{O}}{\underset{\displaystyle}{\text{C}}} \\
\text{HO--C--H} & \text{C=O} & \text{H--C--OH} \\
\text{H--C--OH} & \text{CH}_2\text{OH} & \text{CH}_2\text{O--}\textcircled{P} \\
\text{H--C--OH} & & \\
\text{CH}_2\text{O--}\textcircled{P} & & \\
\text{Fructose} & \text{Dihydroxyacetone} & \text{Glyceraldehyde} \\
\text{1,6-bisphosphate} & \text{phosphate} & \text{3-phosphate}
\end{array}
$$

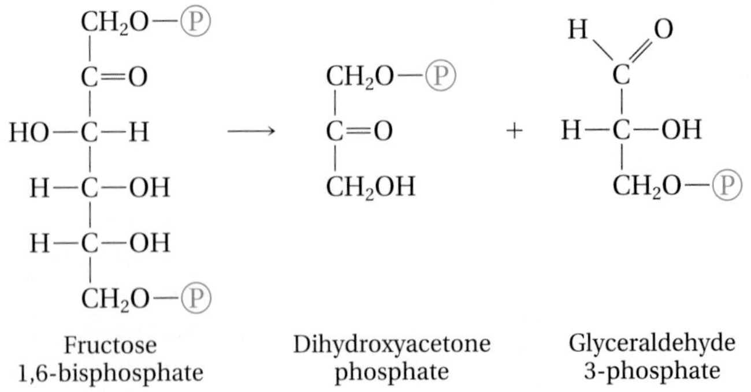

Only glyceraldehyde 3-phosphate will be used in further steps of glycolysis.

The dihydroxyacetone is not wasted. Nature is economical, and cells have an enzyme that promotes the conversion of dihydroxyacetone phosphate to glyceraldehyde 3-phosphate.

$$\begin{array}{ccc}
\text{CH}_2\text{O}-\circled{P} & & \text{CH}_2\text{O}-\circled{P} \\
| & & | \\
\text{C}=\text{O} & \rightleftharpoons & \text{H}-\text{C}-\text{OH} \\
| & & | \\
\text{CH}_2\text{OH} & & \text{C} \\
& & \diagup\ \diagdown \\
& & \text{H}\quad\text{O}
\end{array}$$

Dihydroxyacetone phosphate   Glyceraldehyde 3-phosphate

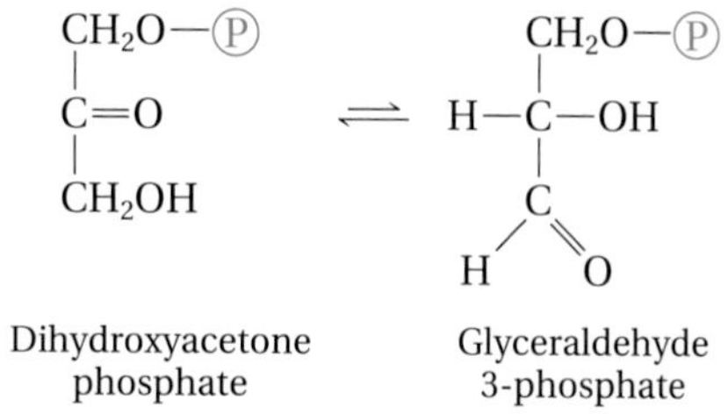

Since one molecule of glucose has provided two molecules of glyceraldehyde 3-phosphate, we will have to take this into account in our future bookkeeping. From now on, we will have to multiply the reactants and products of our reactions by 2.

An enzyme next converts glyceraldehyde 3-phosphate to 1,3-bisphosphoglycerate in the first energy-yielding oxidation reaction of glucose catabolism.

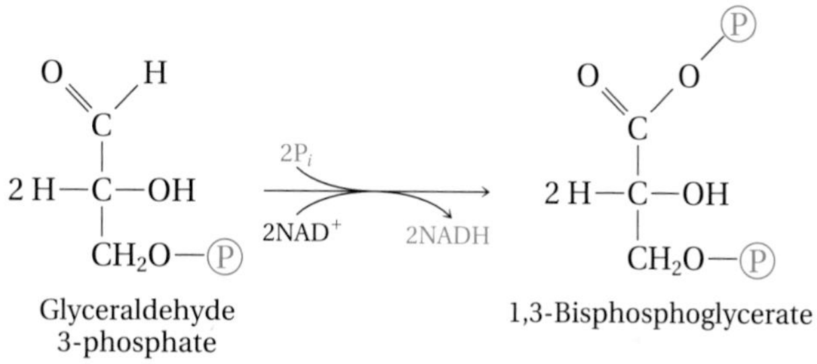

Glyceraldehyde 3-phosphate   1,3-Bisphosphoglycerate

The enzyme uses NAD^+ as a cofactor. The NAD^+ is reduced to NADH—it receives two electrons and a proton from the aldehyde substrate—in the course of the reaction. The new phosphoryl group of the organic product comes from inorganic phosphate ions present in the cytoplasm, so no ATP is expended here. In fact, 1,3-bisphosphoglycerate is itself a high-energy compound—a mixed anhydride of a carboxylic acid and phosphoric acid (see Sec. 23.7) that can transfer its new phosphoryl group to ADP. (The phosphorylation of ADP to ATP outside oxidative phosphorylation is called *substrate-level phosphorylation.*) This transfer occurs in the next step for glycolysis. Two ATP molecules are gained.

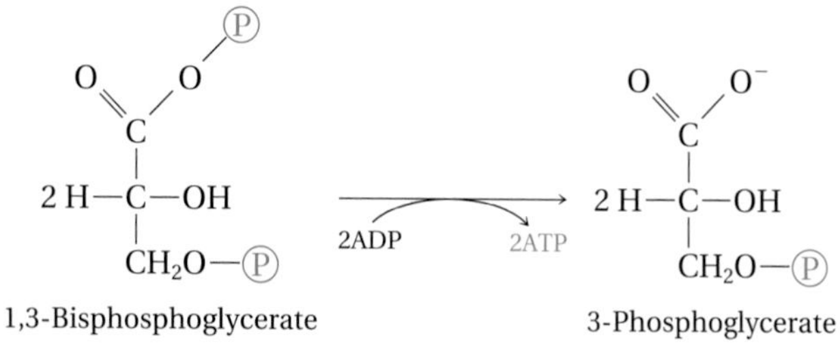

1,3-Bisphosphoglycerate   3-Phosphoglycerate

Since the cell invested two ATP molecules and now has two back, it is even in the ATP stock market. Any ATP produced from this point on is profit.

The next step in glycolysis is a shift of the phosphoryl group of 3-phosphoglycerate.

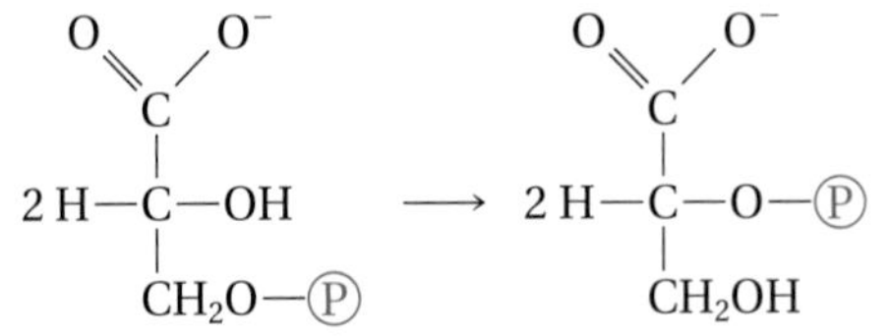

3-Phosphoglycerate   2-Phosphoglycerate

The product of this reaction, 2-phosphoglycerate, loses a molecule of water to give phosphoenolpyruvate.

2-Phosphoglycerate   2-Phosphoenolpyruvate

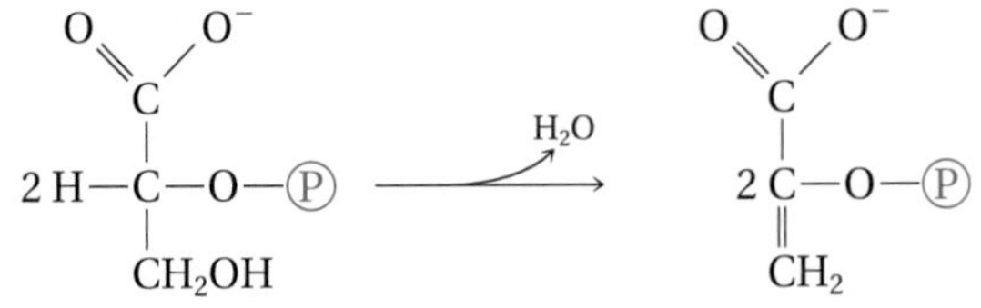

Phosphoenolpyruvate is another energy-rich phosphate molecule capable of passing its phosphoryl group to ADP in another substrate-level phosphorylation. Two ATP molecules are gained.

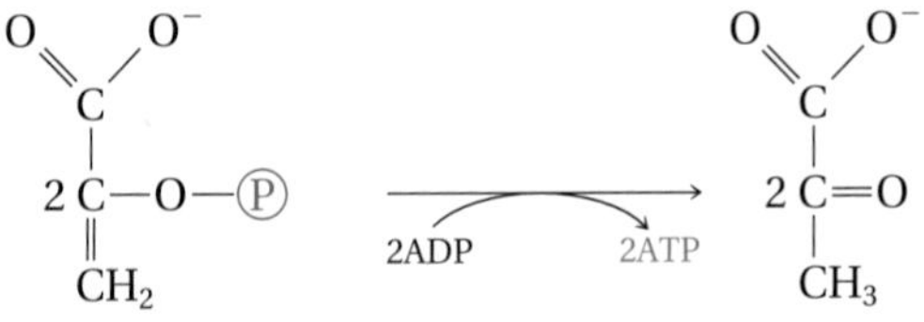

2-Phosphoenolpyruvate   Pyruvate

Since the degradation of one glucose molecule eventually produces two molecules of phosphoenolpyruvate, two molecules of ADP can be phosphorylated to ATP when the phosphoenolpyruvate from one molecule of glucose is converted to pyruvate. These two molecules of ATP are the ATP dividends earned in glycolysis.

The formation of pyruvate is the final step of aerobic glycolysis. Here is what has happened in the oxidation of one molecule of glucose:

1. Two molecules of pyruvate have been formed.
2. Two molecules of NAD^+ have been reduced to NADH.
3. A net total of two ADP molecules have been phosphorylated to ATP (four ATP molecules gained minus two invested).

Table 24.1 summarizes the reactions of glycolysis.

PRACTICE EXERCISE 24.1

Write a net equation that summarizes glycolysis.

Table 24.1 The Reactions of Glycolysis

1. Glucose $\xrightarrow[\text{ATP} \quad \text{ADP}]{}$ Glucose 6-phosphate

2. Glucose 6-phosphate $\rightleftharpoons$ Fructose 6-phosphate

3. Fructose 6-phosphate $\xrightarrow[\text{ATP} \quad \text{ADP}]{}$ Fructose 1,6-bisphosphate

4. Fructose 1,6-bisphosphate

Dihydroxyacetone phosphate $\rightleftharpoons$ Glyceraldehyde 3-phosphate

5. Glyceraldehyde 3-phosphate $+ \; P_i \xrightarrow[\text{NAD}^+ \quad \text{NADH}]{}$ 1,3-Bisphosphoglycerate

6. 1,3-Bisphosphoglycerate $\xrightarrow[\text{2ADP} \quad \text{2ATP}]{}$ 3-Phosphoglycerate

7. 3-Phosphoglycerate $\rightleftharpoons$ 2-Phosphoglycerate

8. 2-Phosphoglycerate $\rightleftharpoons$ Phosphoenolpyruvate

9. Phosphoenolpyruvate $\xrightarrow[\text{2ADP} \quad \text{2ATP}]{}$ Pyruvate

PRACTICE EXERCISE 24.2

Calculate the moles of ATP produced when 90 g of glucose is broken down in glycolysis. The molar mass of glucose is 180 g.

24.4 Acetyl coenzyme A

AIM: To describe the formation and function of acetyl CoA.

When an aerobic cell is operating with a good supply of oxygen, pyruvate molecules flow into the mitochondria. Two of the carbons of each pyruvate ion end up as acetyl CoA, and one molecule of NAD^+ is reduced to NADH. Carbon dioxide is formed as a waste product. This process can be summarized by a single equation:

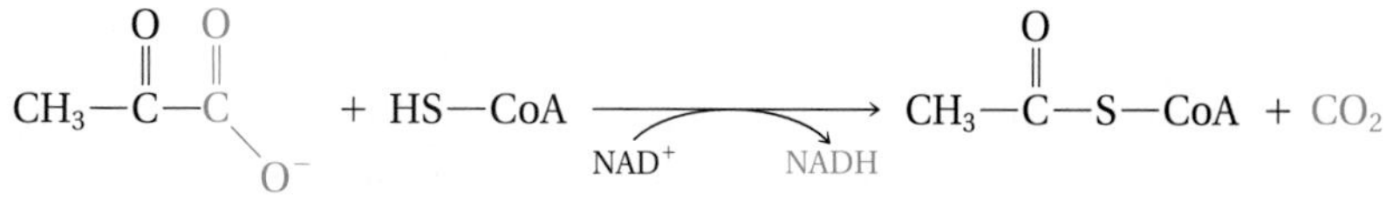

$$CH_3-\overset{\overset{O}{\|}}{C}-\overset{\overset{O}{\|}}{C}\diagdown_{O^-} + \; HS-CoA \xrightarrow[\text{NAD}^+ \quad \text{NADH}]{} CH_3-\overset{\overset{O}{\|}}{C}-S-CoA \; + \; CO_2$$

Pyruvate$\qquad\qquad\qquad\qquad\qquad\qquad\qquad\qquad$ Acetyl CoA

An organized assembly of three different kinds of enzyme molecules called a **multienzyme complex** *is responsible for the formation of acetyl CoA from pyruvate.* This multienzyme complex is named as if it were one enzyme—*pyruvate dehydrogenase.* Several copies of each type of enzyme are present in the pyruvate dehydrogenase multienzyme complex. Five coenzymes—

thiamine pyrophosphate, lipoic acid, FAD, NAD^+, and coenzyme A—are also present. Thiamine was mentioned previously (Sec. 21.8) as vitamin B_1. Now we see that there is a need for it, as its pyrophosphate, in converting pyruvate to acetyl CoA.

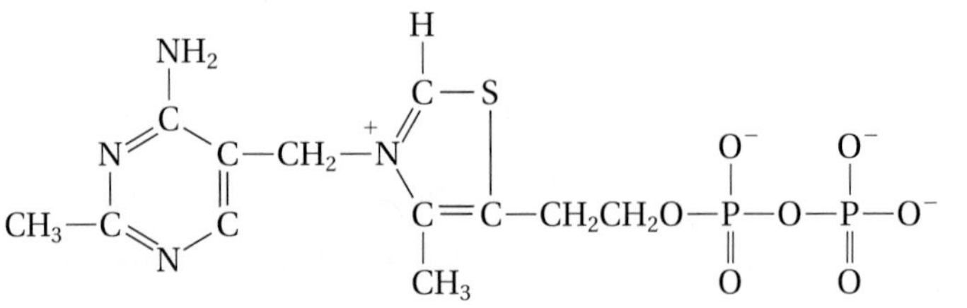

Thiamine pyrophosphate

Lipoic acid is not classified as a vitamin. Evidently humans can make their own lipoic acid. No case of lipoic acid deficiency in a human being has ever been reported.

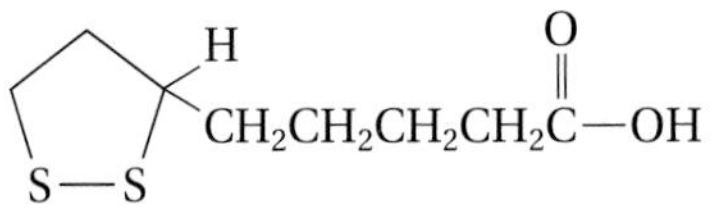

Lipoic acid

PRACTICE EXERCISE 24.3

Acetyl CoA is often considered an energy-rich compound like ATP. Explain why.

24.5 The citric acid cycle

AIM: To list the steps for the degradation of one acetyl group in the citric acid cycle.

Focus

The oxidation of glucose carbons is completed in the citric acid cycle.

The two molecules of acetyl CoA from one molecule of glucose now pass into the citric acid cycle. Figure 24.4 shows the complete cycle, which takes place in the mitochondria of eukaryotic cells. As in other metabolic pathways, all the reactions of the citric acid cycle are catalyzed by enzymes. Some of the necessary enzymes are located in the fluid contained inside the mitochondrial inner membrane; others are attached to the inner surface of the interior membrane. As we go through the steps of the cycle, be especially alert to the fates of the carbons of the reacting molecules, the various types of transformations that are occurring, and the production of NADH, $FADH_2$, and ATP.

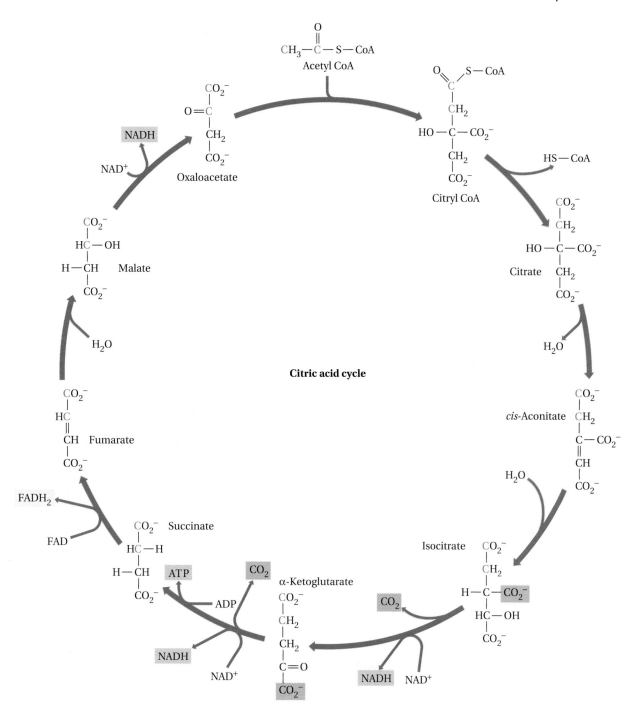

Figure 24.4
The citric acid cycle begins at the 12 o'clock position. As we follow the fate of the carbon atoms of the acetyl group of acetyl CoA, we see that they are not the ones lost as carbon dioxide in one turn of the cycle. Will either of these carbon atoms be oxidized to carbon dioxide during their second pass through the cycle?

The citric acid cycle begins when an acetyl group of acetyl CoA condenses with a molecule of oxaloacetate to give a molecule of citryl CoA.

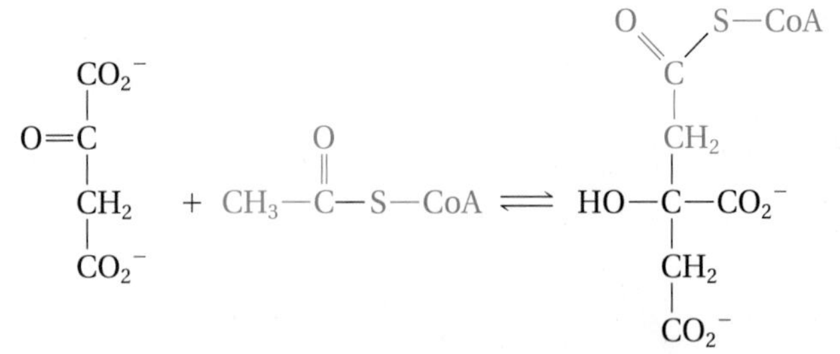

$$
\begin{array}{c}
\underset{\text{Oxaloacetate}}{
\begin{array}{c}
CO_2^- \\
| \\
O{=}C \\
| \\
CH_2 \\
| \\
CO_2^-
\end{array}}
\;+\;
\underset{\text{Acetyl CoA}}{
\begin{array}{c}
O \\
\| \\
CH_3{-}C{-}S{-}CoA
\end{array}}
\;\rightleftharpoons\;
\underset{\substack{\text{Citryl CoA}\\ \text{(CoA ester of citrate)}}}{
\begin{array}{c}
O\;\diagdown\;S{-}CoA \\
C \\
| \\
CH_2 \\
| \\
HO{-}C{-}CO_2^- \\
| \\
CH_2 \\
| \\
CO_2^-
\end{array}}
\end{array}
$$

The thioester bond of citryl CoA is rapidly hydrolyzed to give citrate and CoA.

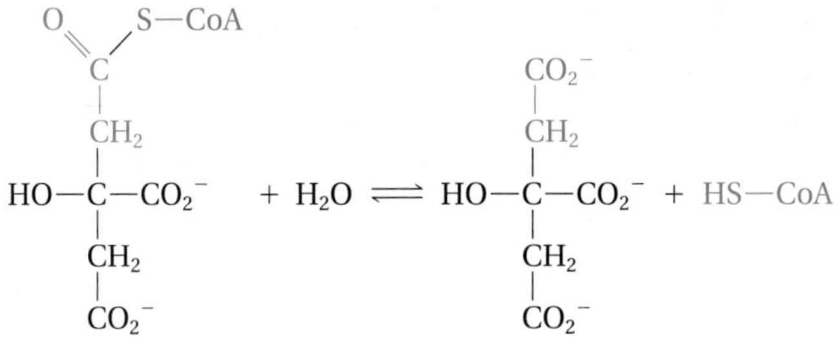

$$
\begin{array}{c}
\underset{\text{Citryl CoA}}{
\begin{array}{c}
O\;\diagdown\;S{-}CoA \\
C \\
| \\
CH_2 \\
| \\
HO{-}C{-}CO_2^- \\
| \\
CH_2 \\
| \\
CO_2^-
\end{array}}
\;+\;H_2O\;\rightleftharpoons\;
\underset{\text{Citrate}}{
\begin{array}{c}
CO_2^- \\
| \\
CH_2 \\
| \\
HO{-}C{-}CO_2^- \\
| \\
CH_2 \\
| \\
CO_2^-
\end{array}}
\;+\;HS{-}CoA
\end{array}
$$

The early discovery of citrate is the reason the pathway is called the *citric acid cycle.*

The next step of the cycle involves the dehydration of citrate to *cis-*aconitate.

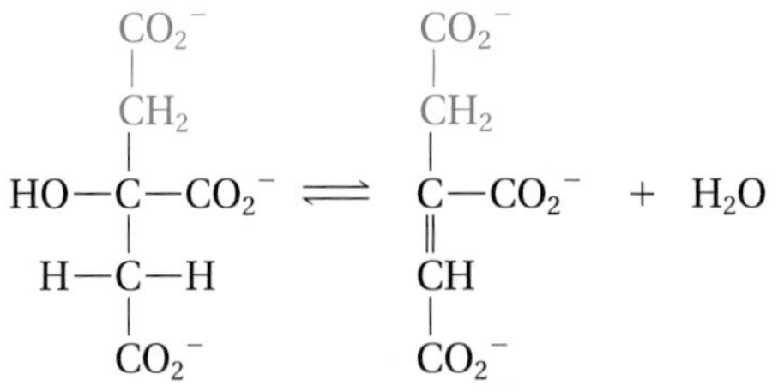

$$
\begin{array}{c}
\underset{\text{Citrate}}{
\begin{array}{c}
CO_2^- \\
| \\
CH_2 \\
| \\
HO{-}C{-}CO_2^- \\
| \\
H{-}C{-}H \\
| \\
CO_2^-
\end{array}}
\;\rightleftharpoons\;
\underset{\textit{cis-}\text{Aconitate}}{
\begin{array}{c}
CO_2^- \\
| \\
CH_2 \\
| \\
C{-}CO_2^- \\
\| \\
CH \\
| \\
CO_2^-
\end{array}}
\;+\;\underset{\text{Water}}{H_2O}
\end{array}
$$

Now a hydration reaction occurs in which water is added to the double bond. Isocitrate is the product of the rehydration.

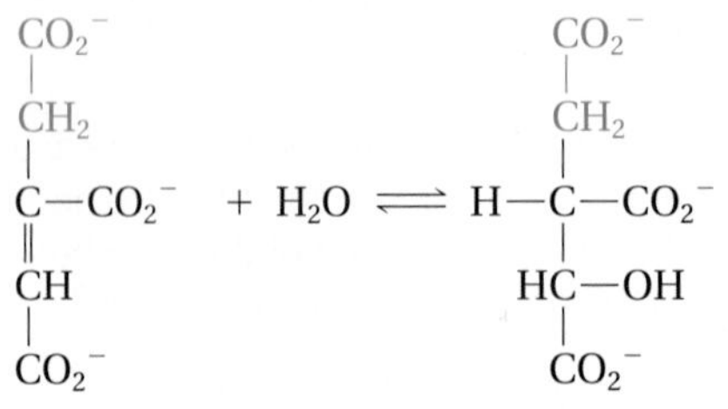

$$
\begin{array}{c}
\underset{\textit{cis-}\text{Aconitate}}{
\begin{array}{c}
CO_2^- \\
| \\
CH_2 \\
| \\
C{-}CO_2^- \\
\| \\
CH \\
| \\
CO_2^-
\end{array}}
\;+\;H_2O\;\rightleftharpoons\;
\underset{\text{Isocitrate}}{
\begin{array}{c}
CO_2^- \\
| \\
CH_2 \\
| \\
H{-}C{-}CO_2^- \\
| \\
HC{-}OH \\
| \\
CO_2^-
\end{array}}
\end{array}
$$

At this point the production of reducing power and ATP is about to begin as isocitrate is oxidized to α-ketoglutarate.

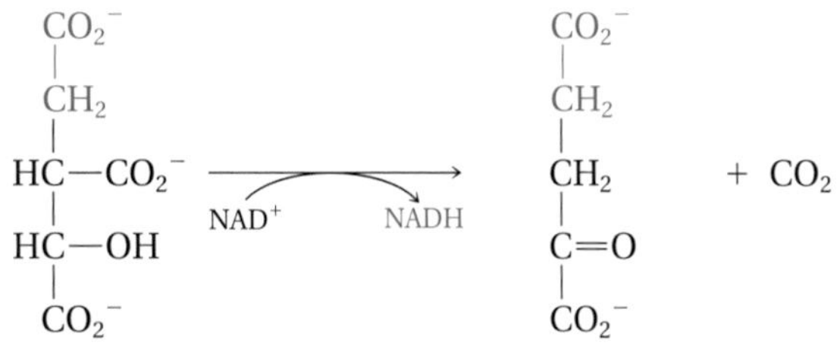

$$\begin{array}{ccc}
\text{CO}_2^- & & \text{CO}_2^- \\
| & & | \\
\text{CH}_2 & & \text{CH}_2 \\
| & & | \\
\text{HC}-\text{CO}_2^- & \xrightarrow[\text{NAD}^+ \quad \text{NADH}]{} & \text{CH}_2 \qquad + \text{ CO}_2 \\
| & & | \\
\text{HC}-\text{OH} & & \text{C}=\text{O} \\
| & & | \\
\text{CO}_2^- & & \text{CO}_2^-
\end{array}$$

Isocitrate   α-Ketoglutarate

NAD^+ is the coenzyme in the reaction, and for every molecule of isocitrate oxidized, one molecule of NAD^+ is reduced to NADH. The oxidation also involves the loss of a molecule of carbon dioxide, but notice that the carbon dioxide comes from the oxaloacetate and not from the acetyl group that entered the cycle.

The next step in the cycle results in the loss of carbon dioxide, reduction of a second NAD^+, and phosphorylation of ADP to ATP (a substrate-level phosphorylation). Succinate is the product. Note that once again the carbon dioxide comes from oxaloacetate instead of acetyl CoA. A second NADH and one ATP are produced.

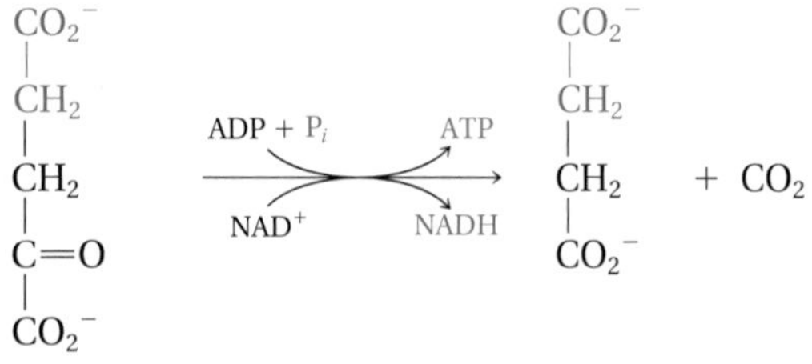

$$\begin{array}{ccc}
\text{CO}_2^- & & \text{CO}_2^- \\
| & & | \\
\text{CH}_2 & \text{ADP} + \text{P}_i \quad \text{ATP} & \text{CH}_2 \\
| & \xrightarrow{\quad\quad\quad} & | \\
\text{CH}_2 & & \text{CH}_2 \qquad + \text{ CO}_2 \\
| & \text{NAD}^+ \quad \text{NADH} & | \\
\text{C}=\text{O} & & \text{CO}_2^- \\
| & & \\
\text{CO}_2^- & &
\end{array}$$

α-Ketoglutarate   Succinate

Succinate is next oxidized to fumarate. Recall that oxidation-reduction enzymes that catalyze the formation of carbon-carbon double bonds usually use FAD as the coenzyme. That is the case here, where one $FADH_2$ is produced.

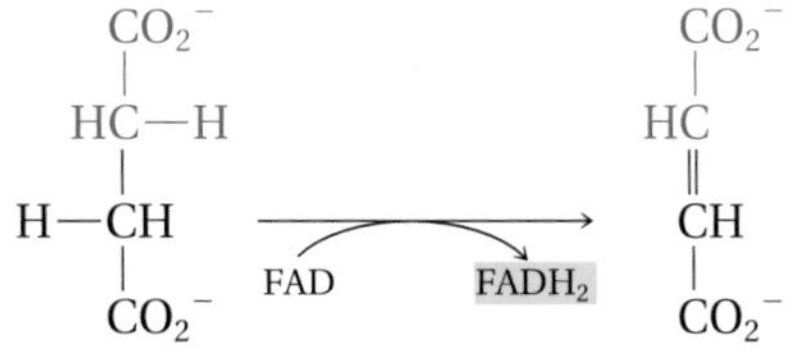

$$\begin{array}{ccc}
\text{CO}_2^- & & \text{CO}_2^- \\
| & & | \\
\text{HC}-\text{H} & & \text{HC} \\
| & \xrightarrow[\text{FAD} \quad \text{FADH}_2]{} & \| \\
\text{H}-\text{CH} & & \text{CH} \\
| & & | \\
\text{CO}_2^- & & \text{CO}_2^-
\end{array}$$

Succinate   Fumarate

A second hydration reaction now occurs as fumarate is converted to malate.

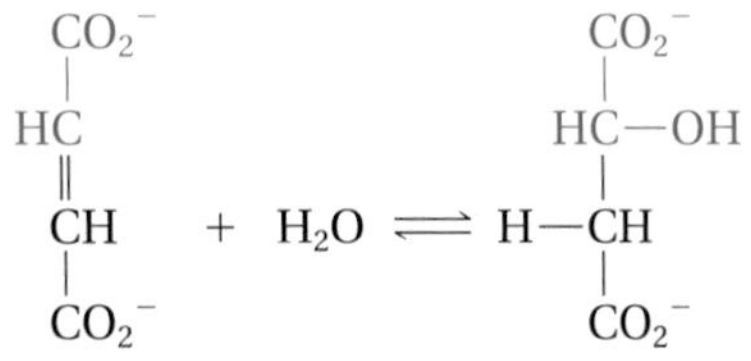

$$\begin{array}{ccc}
\text{CO}_2^- & & \text{CO}_2^- \\
| & & | \\
\text{HC} & & \text{HC}-\text{OH} \\
\| & + \text{ H}_2\text{O} \rightleftharpoons & | \\
\text{CH} & & \text{H}-\text{CH} \\
| & & | \\
\text{CO}_2^- & & \text{CO}_2^-
\end{array}$$

Fumarate   Malate

Malate is oxidized back to oxaloacetate in the final step of the cycle, and NAD^+ is simultaneously reduced. A third NADH is produced.

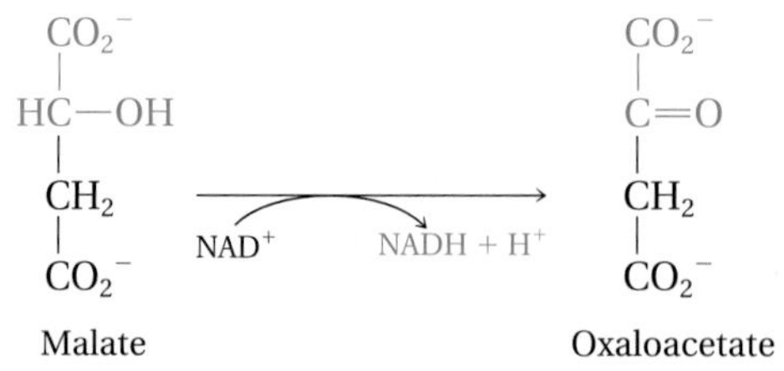

The oxaloacetate molecule is now available to start another turn of the cycle by reacting with another molecule of acetyl CoA.

This is what happens in the citric acid cycle:

1. Acetyl CoA and oxaloacetate combine to form citrate.

2. Citric acid eventually loses two carbon atoms as carbon dioxide. The carbons in the two molecules of carbon dioxide are not the same carbons that entered the citric acid cycle as acetyl groups of acetyl CoA. Nevertheless, the *net effect* of the cycle is the same as if these carbons were oxidized.

3. At the end of the pathway, a molecule of oxaloacetate remains, which is why the pathway is called a *cycle*. (The original oxaloacetate molecule could have come from several places in metabolism, but we need not worry about that here.)

4. Each turn of the citric acid cycle yields three molecules of NADH, one of $FADH_2$, and one of ATP.

PRACTICE EXERCISE 24.4

Write a net equation that summarizes one turn of the citric acid cycle.

24.6 ATP yield

AIM: To give an accounting of the ATP produced by the complete oxidation of a molecule of glucose.

Focus

The complete oxidation of 1 molecule of glucose yields 38 ATP molecules.

Two turns of the citric acid cycle are necessary to oxidize completely the equivalent of two molecules of acetyl CoA (obtained from one molecule of glucose) to four molecules of carbon dioxide. Using this information, along with the yields of ATP and NADH obtained in glycolysis, we can calculate the total amount of ATP generated by the aerobic catabolism of one molecule of glucose. Remember that each NADH molecule can generate three ATP molecules and each $FADH_2$ molecule can generate two ATP molecules by oxidative phosphorylation. Table 24.2 summarizes the

Table 24.2 ATP Production from Complete Aerobic Catabolism of One Glucose Molecule

Pathway	ATP yield
glycolysis (4ATP generated minus 2 invested)	2
citric acid cycle (2 acetyl groups of acetyl CoA oxidized to carbon dioxide)	2
oxidative phosphorylation:	
2NADH from glycolysis	6
2NADH from acetyl CoA formation	6
6NADH from the citric acid cycle	18
2FADH$_2$ from the citric acid cycle	4
	38

Burning a mole of glucose to form CO_2 and H_2O produces 686 kcal of energy. When we oxidize a mole of glucose, the 38 mol of ATP represents 277 kcal of stored energy. Our bodies, then, are about 277 kcal/686 kcal $\times$ 100 = 40% efficient.

results. The data show that enough of the energy released in the complete oxidation of 1 molecule of glucose is trapped by aerobic cells to generate 38 ATP molecules.

24.7 Lactic Fermentation

AIM: To explain why cells sometimes use lactic fermentation for energy production.

Focus

Aerobic cells switch to lactic or alcoholic fermentation in the absence of oxygen.

So far our discussion of glucose metabolism has assumed that the cell doing the metabolism has plenty of oxygen available. What if it does not? We know that the mitochondria of aerobic cells need oxygen so that the electron transport chain can operate. When there is no oxygen available to drain electrons from NADH and FADH$_2$ in respiration, the electron carriers of the electron transport chain become completely reduced. More electrons cannot be passed down the chain, and oxidative phosphorylation stops. However, the levels of NADH and FADH$_2$ in the mitochondrion increase as the citric acid cycle continues to operate. Soon, not enough NAD$^+$ and FAD are regenerated by respiration to sustain the operation of the citric acid cycle, and the mitochondrial power plant shuts down.

Now the only place that ATP is being produced is in the cytoplasm. Here, two molecules of ATP are produced for every glucose molecule converted to two molecules of pyruvate in glycolysis. Cytoplasmic NAD$^+$ is also being reduced to NADH, and NAD$^+$ is needed for glycolysis to continue. (It is needed to change glyceraldehyde 3-phosphate to 1,3-bisphosphoglycerate.) If there were no way to regenerate NAD$^+$, glycolysis too would stop. With no energy production, the cell would die.

In such an emergency, the cells of many aerobic organisms regenerate NAD$^+$ from the NADH formed in glycolysis by using the NADH to reduce pyruvate to lactate.

$$ \text{Pyruvate} \xrightarrow[\text{NADH + H}^+ \quad \text{NAD}^+]{} \text{Lactate} $$

Pyruvate   Lactate

The reduction of pyruvate to lactate is called **lactic fermentation.** (For more information, see A Closer Look: Monitoring Heart Attacks with Serum LDH Tests.) Lactic fermentation keeps glycolysis going. Since aerobic catabolism produces 38 ATP molecules from 1 molecule of glucose and lactic fermentation produces only 2, the aerobic catabolism of glucose is *19 times* more efficient than lactic fermentation. Nevertheless, considering the choices between death and life at a lower level of ATP production, lactic fermentation is not a bad bargain for the cell.

A Closer Look

Monitoring Heart Attacks with Serum LDH Tests

Many clinical laboratory tests in hospitals measure the amount of a critical enzyme in the bloodstream. One such enzyme, *lactate dehydrogenase* (LDH), is an important catalyst, promoting the reduction of pyruvate to lactate in your body. Consisting of four polypeptide chains or tetramer subunits, LDH exhibits multiple molecular forms called *isozymes*. LDH is composed of two principal kinds of subunits, H and M, that differ slightly in their primary structures. The five isozymes of LDH, H_4, H_3M, H_2M_2, HM_3, and M_4, are various combinations of these subunits. Heart and liver LDH is rich in H subunits; muscle LDH is rich in M subunits.

Because damaged tissues often release LDH into the bloodstream, certain diseases can be detected by a clinical laboratory test that measures *serum LDH* levels. In this procedure, a patient's blood sample is analyzed, and the rate at which the serum converts pyruvate to lactate is obtained. This rate, in turn, determines the level of LDH that may indicate a general abnormality. Often it is useful to track the course of disease in specific tissues. A diagnosis can be refined by using another technique called *electrophoresis* to separate the total LDH in a blood sample into its various isozymes (see figure). Abnormal amounts

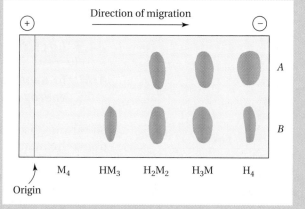

Electrophoresis of serum LDH isozymes at pH 8.6: pattern *A* belongs to a heart attack patient; pattern *B* is that of a normal individual. Leakage of cell contents of damaged heart muscle in the heart attack patient has substantially increased the serum level of the H_4 isozyme.

of a given isozyme can, in turn, narrow the search for disease to specific organs or help monitor the magnitude and course of disease in the body.

During a heart attack, for example, damaged heart muscle releases LDH, mainly the H_4 isozyme. Within 24 hours of the heart attack episode, serum LDH reaches a peak; it then returns to a normal level within 5 or 6 days. A physician often can get a good idea of the extent of a patient's heart damage by carefully monitoring the amount of H_4 isozyme over this period of time.

The product of fermentation processes is not always lactate. Some organisms—brewer's yeast is a particularly well-known example—oxidize the pyruvate formed in glycolysis to acetaldehyde.

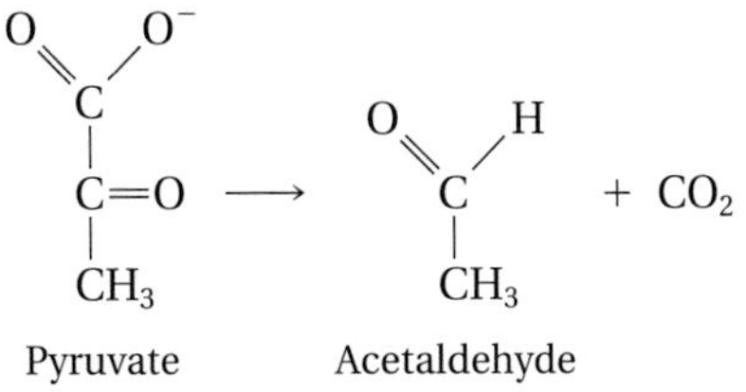

Pyruvate   Acetaldehyde

The reduction of acetaldehyde regenerates NAD^+ to keep glycolysis going. The reduced product of acetaldehyde reduction is ethanol.

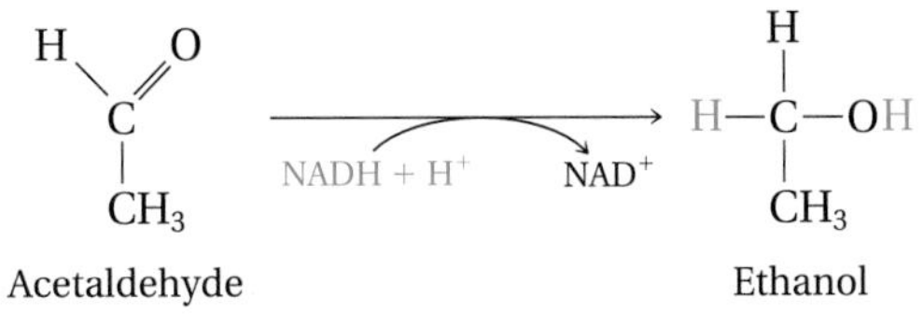

Acetaldehyde   Ethanol

The process by which glucose is degraded to ethanol is called **alcoholic fermentation.**

PRACTICE EXERCISE 24.5

Aerobic catabolism of glucose is much more efficient than fermentation. What does this statement mean?

24.8 Oxygen debt

AIMS: To describe the conditions that could cause a state of oxygen debt. To outline two possible fates of lactate once cellular respiration is returned to normal.

Focus

Lactate is oxidized to carbon dioxide or made into glucose.

Your respiratory system supplies the mitochondria of your body cells with enough oxygen to operate their respiratory chains efficiently at normal levels of activity. If you undertake vigorous exercise, such as sprinting or long-distance running, your cells step up respiration to make ATP to meet their increased energy needs. You breathe more rapidly and more deeply, but eventually you may develop an **oxygen debt**—*that is, not enough oxygen is available for cellular respiration.* At this point, your muscle cells begin to use the much less efficient lactic fermentation route to make ATP.

Lactate buildup

Lactate builds up as your cells regenerate NAD^+ from NADH by lactic fermentation. If you have an oxygen debt and do not stop to rest, the lactic acid concentration in your muscles continues to build. Lactic acid is toxic to muscle cells in high concentrations and results in muscle cramps and soreness. (Remember how you felt the day after a long-distance run or swim?) If you push on much further, you will soon collapse, because lactic fermentation cannot supply enough ATP to take care of your body's heavy demands.

If you do stop to rest, you will pant as your respiratory system supplies the oxygen necessary to pay the oxygen debt. Your cellular respiration is soon restored to normal, but body cells must still deal with the lactate built up during lactic fermentation.

Lactate disposal

Some of the lactate is oxidized back to pyruvate by a reversal of the lactic dehydrogenase–catalyzed reactions.

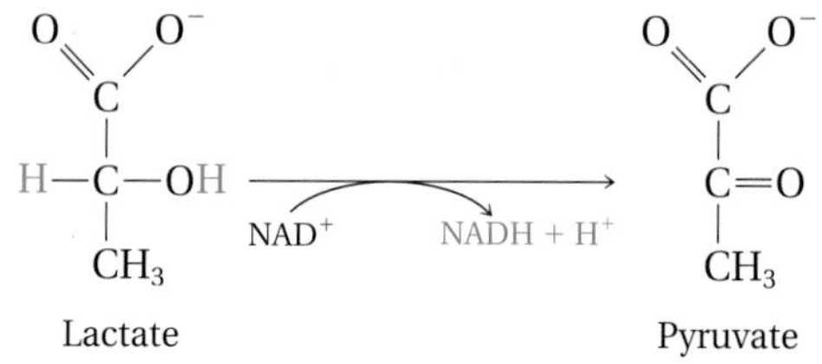

The NADH that is formed enters the respiratory chain, where ATP is produced by oxidative phosphorylation. The pyruvate is converted to acetyl CoA, which enters the citric acid cycle. There the acetyl group of acetyl CoA undergoes complete oxidation to carbon dioxide.

Some of the lactic acid formed in your muscles is not oxidized. It drains from the muscle cells and enters the bloodstream, where it could give you a severe case of acidosis. As we learned in Chapter 22, however, blood is buffered. Your blood buffers, mainly the HCO_3^-/H_2CO_3 system, will absorb the extra protons added to your bloodstream by the lactic acid. And if need be, your kidneys will generate new bicarbonate buffer ions that are lost as carbon dioxide through breathing. Your kidneys also will expel excess protons in your urine.

Gluconeogenesis

The lactic acid in your bloodstream is absorbed by your liver. Liver cells, but not muscle cells, can convert lactate back to glucose in several steps.

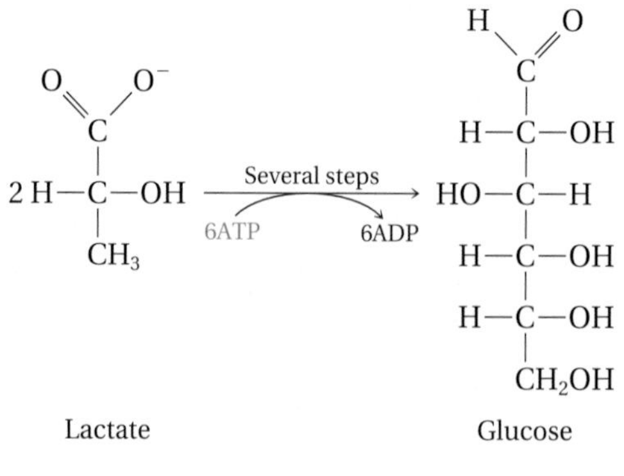

Gluconeogenesis
gluco (Greek): sweet, glucose
neo (Greek): new
genesis (Greek): creation

The pathway by which lactate is converted to glucose is called **gluconeogenesis.** Gluconeogenesis, which is the synthesis of glucose from starting materials that are not carbohydrates, is an example of an anabolic (synthetic) pathway. Like most anabolic pathways, it requires the expenditure of ATP. Six molecules of ATP are required to convert two molecules of lactate to one molecule of glucose. However, only two molecules of ATP were gained by converting one molecule of glucose to two molecules of lactate in lactic fermentation. Your liver cells must pay for this deficit of ATP. They do so with the abundant ATP produced in oxidative phosphorylation.

We will not be studying gluconeogenesis in detail. Nevertheless, the first step of the pathway is particularly interesting because it is also relevant to the citric acid cycle. It involves the chemical combination of pyruvate and carbon dioxide, a *carboxylation reaction,* to give oxaloacetate.

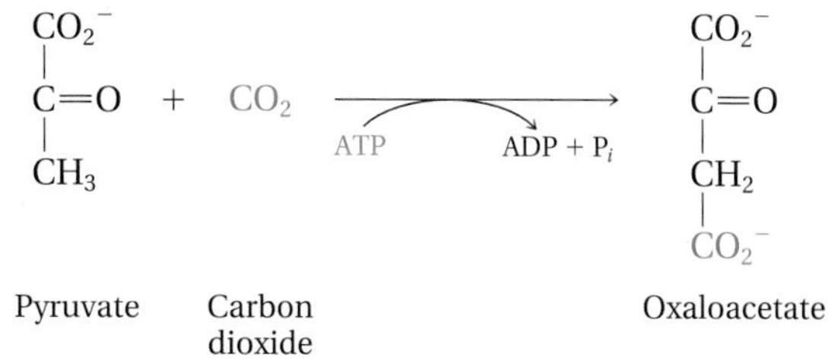

Pyruvate   Carbon dioxide   Oxaloacetate

The carboxylation of pyruvate is energetically unfavorable and requires the expenditure of one of the six ATP molecules invested in gluconeogenesis. Pyruvate carboxylase, the enzyme that catalyzes the reaction, requires the B-complex vitamin biotin as the coenzyme. The biotin serves as a carrier of carbon dioxide in this and other biological carboxylation reactions in the form of carboxybiotin. It is attached to the side-chain terminal amino group (not the alpha amino group) of a lysine residue of the protein through an amide bond.

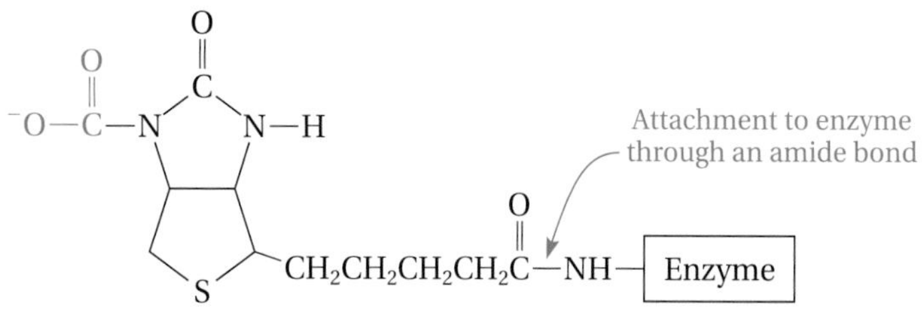

Carboxybiotin

The oxaloacetate produced in the reaction meets one of two fates: It is converted to glucose in the remainder of the reactions of gluconeogenesis, or alternatively, it may enter the citric acid cycle. Indeed, the carboxylation of pyruvate is the major source of oxaloacetate for the citric acid cycle.

24.9 Glucose storage

AIM: To describe how excess glucose is temporarily stored in the body.

Foods containing carbohydrates are one source of the glucose used for energy production in the cell. As you have seen, another source of glucose is gluconeogenesis. *Glucose that is not required to meet the immediate energy needs of the body is assembled into glycogen, the animal form of starch; this anabolic process is called* **glycogenesis.** Glycogenesis requires two enzymes: *glycogen synthetase* and *branching enzyme.* Glycogen synthetase promotes the formation of the $\alpha(1\rightarrow4)$ glycosidic bonds that form the straight-chain portions of glycogen. As its name implies, the branching enzyme promotes the formation of $\alpha(1\rightarrow6)$ glycosidic linkages at branch points of the bushy glycogen molecule.

Glucose adds to an existing glycogen chain only if it is first activated to uridine diphosphate glucose (UDP-glucose). The starting materials for the activation are glucose 1-phosphate and uridine triphosphate (UTP), one of the nucleotide triphosphates used in the synthesis of RNA. The glucose 1-phosphate comes from glucose 6-phosphate by way of the following enzyme-catalyzed reaction:

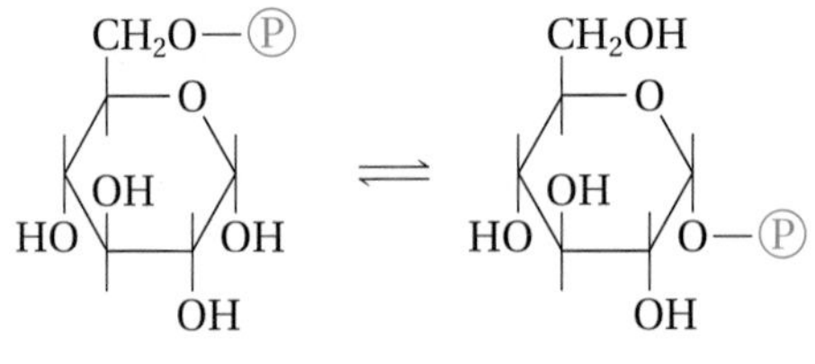

Glucose 6-phosphate   Glucose 1-phosphate

The glucose 1-phosphate now reacts with the UTP to form UDP-glucose and inorganic pyrophosphate (PP_i).

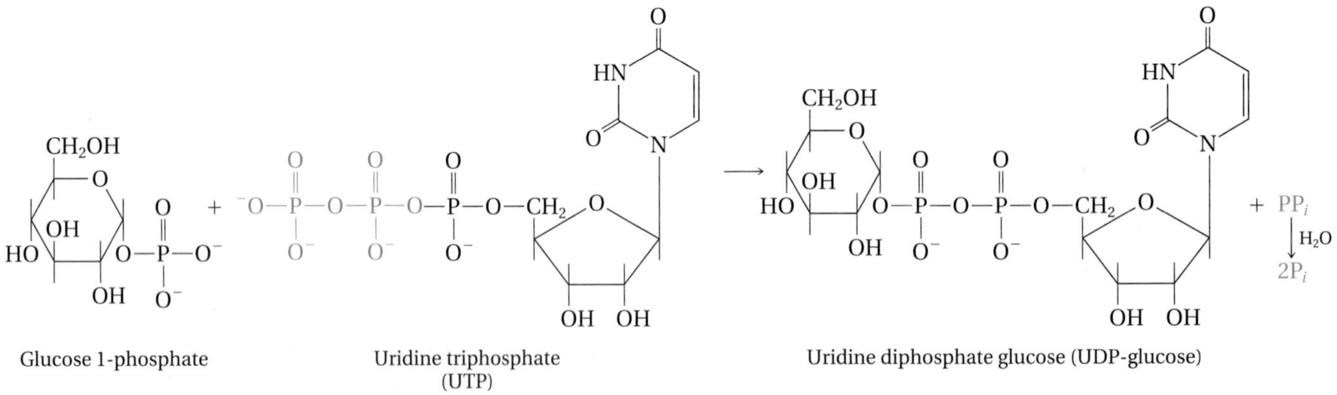

Glucose 1-phosphate   Uridine triphosphate (UTP)   Uridine diphosphate glucose (UDP-glucose)

Note that the hydrolysis of the high-energy anhydride bond of the PP_i

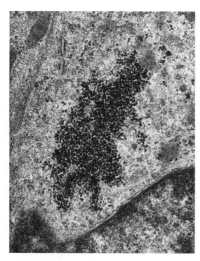

Figure 24.5
Liver cell of a rat fed a high-carbo-hydrate diet. The dark areas are masses of glycogen.

releases substantial energy and therefore helps drive the formation of the UDP-glucose. Once the UDP-glucose is formed, its glucose unit attaches to the end of an existing glycogen chain by an $\alpha(1 \longrightarrow 4)$ linkage.

UDP-glucose   Glycogen chain

Uridine diphosphate (UDP)   Glycogen chain plus 1 new glucose unit

No high-energy phosphoric acid anhydride bonds are broken in this reaction, so the cell's total investment for each molecule of glucose incorporated into glycogen is two high-energy phosphate bonds—one to make glucose 6-phosphate from glucose and one in the hydrolysis of PP_i to $2P_i$.

Glycogen is temporarily stored in muscle cells and liver cells until glucose is needed (Fig. 24.5). The amount of liver glycogen is seldom more than 5% of the weight of fresh liver tissue. Muscle glycogen seldom exceeds 1% of the weight of fresh muscle.

A typical runner can call on his or her glycogen reserves to provide glucose for energy for 2 to 3 hours. Once the glycogen reserves are gone, a runner "hits the wall" and is unable to maintain the previous level of activity.

PRACTICE EXERCISE 24.6

In glycogenesis, the glucose 1-phosphate comes from glucose 6-phosphate by way of an enzyme-catalyzed reaction. Where does the glucose 6-phosphate come from?

24.10 Glycogen breakdown

AIM: To distinguish between the body's use of liver glycogen and muscle glycogen.

Focus

Stored glycogen is degraded to meet glucose needs.

When your body's glucose supply gets low, glycogen is broken down to meet energy needs. *The process by which glycogen is broken down is called* **glycogenolysis** ("splitting of glycogen"). Glycogenolysis starts with the conversion of glycogen into glucose 1-phosphate. The enzyme *phosphorylase* catalyzes this conversion.

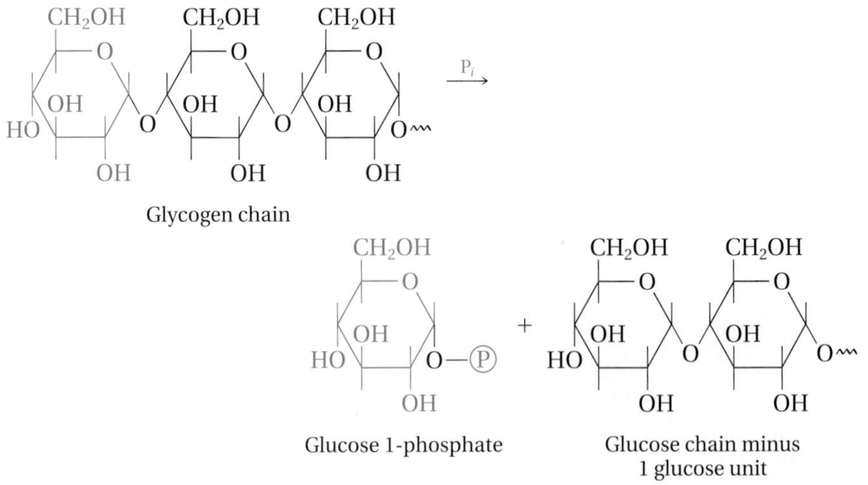

Glycogen chain

Glucose 1-phosphate

Glucose chain minus
1 glucose unit

In the process of cleaving glycogen, a phosphoryl group of inorganic phosphate (P_i) is transferred to glucose. To enter glycolysis, the glucose 1-phosphate must be converted to glucose 6-phosphate. The catalysis of this reaction is accomplished by a reversal of the reaction used to make glucose 1-phosphate.

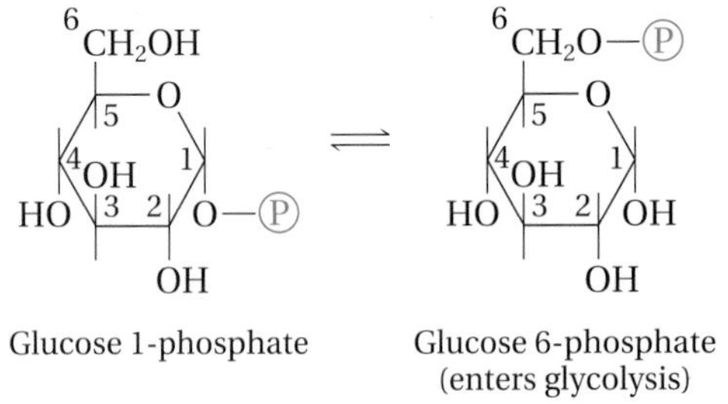

Glucose 1-phosphate

Glucose 6-phosphate
(enters glycolysis)

Negatively charged glucose 6-phosphate cannot readily pass through the membranes of cells. In muscle cells, all the glucose 6-phosphate produced by glycogenolysis is used for glycolysis. Liver cells have a broader responsibility: They must distribute glucose to other tissues via the blood. Therefore, liver cells, but not muscle cells, contain an enzyme that catalyzes the hydrolysis of glucose 6-phosphate to glucose and P_i.

$$\text{Glucose 6-phosphate} + H_2O \longrightarrow \text{Glucose} + P_i$$

The glucose leaves the liver cells and enters the blood, which delivers it to tissues where it is urgently needed. Stored glycogen does not last long. After a 24-hour fast, there is virtually no glycogen left in either the liver or the muscle.

In the preceding section we noted that the synthesis of glycogen requires the expenditure of two high-energy phosphate bonds for each molecule of glucose incorporated into glycogen (the equivalent of converting two molecules of ATP to ADP). Since the aerobic catabolism of each glucose molecule produces 38 ATP molecules, a net total of 36 ATP molecules is produced from each glucose molecule stored as muscle glycogen. Thus glycogen is nearly as good a source of cellular energy as glucose itself. Ath-

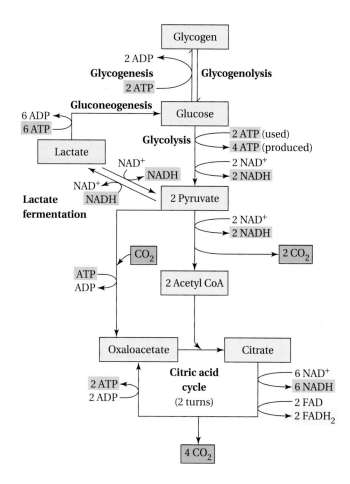

Figure 24.6
The many pathways of glucose metabolism.

letic endeavors require energy. Barbara, the biochemically savvy cross-country coach from the Case in Point earlier in this chapter, would like her athletes to store as much glycogen as possible before a meet. In the Follow-up to the Case in Point, below, we will learn more about this strategy.

An overview of the metabolic pathways we have discussed is presented in Figure 24.6.

FOLLOW-UP TO THE CASE IN POINT: Carbohydrate loading

Athletes, especially those involved in long-distance running, often try to improve their performances by maximizing the amount of stored muscle glycogen. The practice of eating large amounts of carbohydrate-rich foods such as pasta before strenuous activities is known as *carbohydrate loading*. By prescribing a pasta diet to her team, Barbara, the college cross-country coach, is attempting to maximize her runners' stores of muscle glycogen so that the glyco-gen will be available to meet the glucose needs of their muscle cells during the meet. Because these glycogen stores are fairly small and rapidly depleted, they must be built up within a day or two of the athletic event. The maximum amount of glycogen that can be stored is different among individuals and is determined by genetics, but people can approach their personal maximum by physical conditioning.

PRACTICE EXERCISE 24.7

Explain why liver glycogen, but not muscle glycogen, is a source of glucose to other body tissues.

24.11 Metabolic regulation

AIM: To describe how hormones regulate metabolism through the use of second messengers.

Focus

Many metabolic processes are subject to hormonal control.

One of the most fascinating aspects of metabolism is its regulation—how cellular processes are turned on when they are needed and turned off when they are not. So that you can clearly grasp the regulation of glucose metabolism, you will first need to know more about how hormones control cellular processes in general.

Examples of hormonal control

Since metabolic reactions are enzyme-catalyzed, you might reasonably expect the control of cellular metabolism to involve the regulation of enzyme activity, and it often does. In complex organisms such as human beings, many processes are controlled by hormones. In fact, over 30 effects of a single hormone, insulin, are known. You have already seen that hormones control basal metabolism (thyroxine), permeability of cellular membranes (vasopressin), and the activity of ion pumps (aldosterone). The prostaglandins also exhibit hormone-like effects. The study of prostaglandins is complicated, though, by the fact that they seem to have opposite effects on different kinds of tissue cells, whereas hormones influence all their target cells in the same way.

Second messengers

Hormones are powerful regulators of processes that occur inside cells, but few hormones are ever found inside the cells they act on. The discovery that hormones that reach target cells bind to specific protein receptors on the outer surface of the cell membrane was an important contribution to the study of hormones and cells. This discovery did not, however, explain how hormones influence processes *inside* cells. Scientists had few leads to the solution of this mystery until Earl Sutherland and his colleagues discovered *cyclic AMP* (cAMP) in 1957 (Fig. 24.7). Their work, and the work of those who followed them, has shown that formation of a hormone-receptor complex on the exterior of a cell membrane activates an enzyme, *adenylate cyclase,* that is attached to the interior of the cell membrane (Fig. 24.8).

Adenylate cyclase catalyzes the conversion of ATP to cAMP. In most instances, cAMP then binds to the allosteric sites of a group of enzymes called *protein kinases.* In other words, cAMP is a positive modulator (see Sec. 19.10) of protein kinases. Activated protein kinases catalyze the transfer of phosphoryl groups from ATP to various proteins. These proteins are

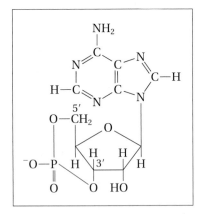

Figure 24.7
Cyclic adenosine monophosphate (cAMP). Note that in cAMP a cyclic phosphodiester linkage is formed between the hydroxyl groups on the 3′ and 5′ positions of the ribose ring.

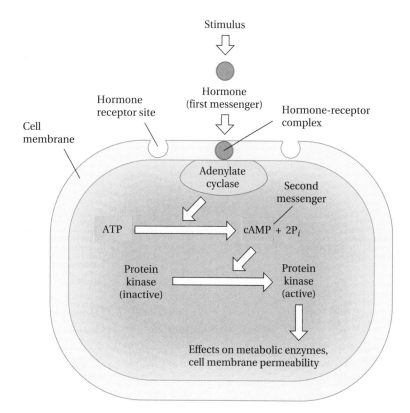

Figure 24.8
Many hormones exert their influence on cells by stimulating the production of cAMP, the second messenger.

mainly enzymes that catalyze important cellular processes. Phosphorylation activates many of these enzymes, but it deactivates some; ultimately, the presence of cAMP starts some cellular processes and stops others. Cyclic AMP also may play roles other than that of activating protein kinases. In at least one instance, cAMP activates an enzyme by releasing it from an enzyme-inhibitor complex (Fig. 24.9).

Hormone action is often compared with a messenger service in which the hormone is the **first messenger,** *delivering a regulatory message to the cell. Cyclic AMP is often the* **second messenger,** *taking the message to its final destination: the enzymes within the cell.* There are other second messengers besides cAMP. For example, scientists have found that the action of some hormones is impaired in the absence of calcium ions. They see this effect at both low and high levels of cAMP in the cellular cytoplasm.

Figure 24.9
Cyclic AMP activates some enzymes by releasing them from inactive complexes.

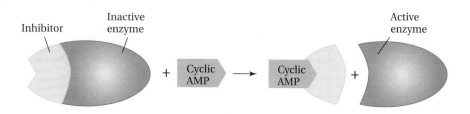

This indicates that calcium ions rather than cAMP are the second messengers. Sometimes calcium ions and cAMP work together to cause a certain effect on cellular metabolism. In addition to cAMP and calcium ions, two very simple molecules have been found to serve as second messengers in many important physiologic processes. These molecules are discussed in A Closer Look: Nitric Oxide and Carbon Monoxide as Second Messengers.

Nitric Oxide and Carbon Monoxide as Second Messengers

In a world of complex biological molecules such as cyclic AMP, simple nitric oxide (NO) seems an unlikely candidate for an important role in the body. Its simplicity aside, nitric oxide (NO) is one of the most important messengers in the body. A specific enzyme, nitric oxide synthase, makes NO from nitrogen in the amino acid arginine. Nitric oxide is a gas under atmospheric conditions. In the body it lasts only 6 to 10 seconds before it is oxidized by oxygen and water to nitrite ions (NO_2^-) and nitrate ions (NO_3^-). Despite its fleeting existence, NO is responsible for a host of effects in the body. The molecule is toxic to many kinds of cells, but certain white blood cells called *macrophages* are resistant to its toxicity. Macrophages contain NO and use it to kill tumor cells, fungi, and bacteria. NO also relaxes muscles of blood vessels, which permits the vessels to expand and reduce blood pressure. Many patients who suffer from *angina pectoralis*, chest pain that is caused by spasms of the arteries of the heart, are treated with nitroglycerin. The relaxation of these arteries relieves the symptoms of angina. Nitroglycerin is degraded to NO in the body, the spasms stop, and angina is relieved. NO simultaneously behaves much like a neurotransmitter in the brain and other parts of the body. Learning and memory are thought to involve the passage of neurotransmitter molecules between the nerve cells of the brain (neurons). NO may enable us to learn and to remember what we have learned.

Perhaps even more surprising than NO as a messenger in the body is the recent implication of carbon monoxide (CO). Most of us know that carbon monoxide is a toxic component of automobile exhaust and other incomplete combustion of hydrocarbons (Case in Point: Carbon monoxide poisoning, page 316). Scientists have found that our bodies naturally make a small amount of carbon monoxide by the oxidation of heme, the oxygen-carrying part of the hemoglobin molecule in red blood cells. An enzyme called *heme oxygenase* catalyzes the formation of carbon monoxide from heme. The details of the action of carbon monoxide are sketchy at present, but it appears that this tiny molecule is the main regulator of the amount of the second messenger cGMP (see Sec. 24.12) in at least some brain cells.

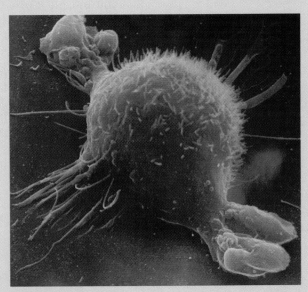

Macrophages are resistant to the toxicity of nitric oxide (NO).

24.12 *Control of glycogenolysis*

AIM: To describe the hormonal control of glycogenolysis.

Recall from Section 23.7 that phosphorylase, an important enzyme catalyst of glycogenolysis, is converted from an inactive form (*phosphorylase b*) to an active form (*phosphorylase a*) upon phosphorylation by ATP. This conversion is subject to hormonal control by glucagon, a peptide hormone. When levels of blood glucose are low, glucagon produced by the pancreas enters the bloodstream and is carried to its target cells—the primary target of glucagon is liver cells. Glucagon stimulates the production of cAMP (Fig. 24.10), which activates a protein kinase by binding to it. The protein kinase catalyzes the phosphorylation of phosphorylase *b*, producing phosphorylase *a*. Phosphorylase *a* then begins to catalyze the breakdown of glycogen. The end result of the action of glucagon is to raise the level of glucose in the blood.

Another hormone that stimulates the breakdown of glycogen is epinephrine (adrenaline). Epinephrine is often called the "flight or fight" hormone because it is secreted by the adrenal glands in response to a threat of bodily harm. The main effect of epinephrine is on muscle cells, which may need energy to meet the threat. The nervous system is also mobilized

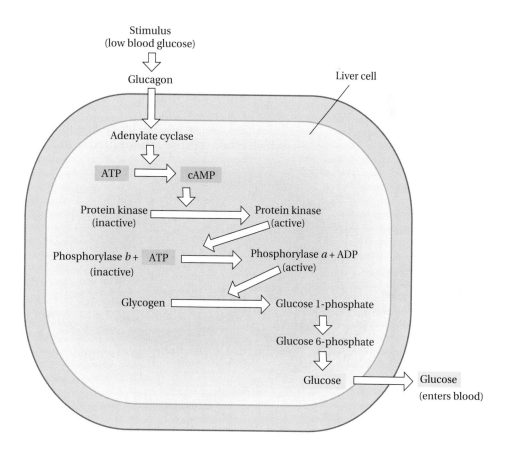

Figure 24.10
Glucagon promotes the breakdown of glycogen stored in liver cells to glucose. Some of the glucose produced by liver cells is carried by the blood to other tissues.

simultaneously. Within moments of the release of epinephrine, all systems are *go*. You may have experienced the unique feeling that results—perhaps at a crucial point in an athletic contest, in an automobile accident or near-accident, or by stepping on a snake on a woodland trail.

The path leading from epinephrine production to glycogenolysis in muscle is similar to the path used in the liver. There is, however, one major difference. Because muscle cells lack the enzyme necessary to hydrolyze glucose 6-phosphate to glucose and glucose 6-phosphate cannot easily pass through cell membranes, all the glucose 6-phosphate produced by glycogenolysis in muscle cells is used for glycolysis within the muscle.

24.13 Glucose absorption

AIMS: To describe the production and function of insulin. To name the terms for low and high sugar levels in the blood. To describe the physiologic effects of low and high sugar levels in the blood. To explain the phrase "starvation in the midst of plenty" as applied to diabetes.

Focus

Insulin stimulates the absorption of glucose by tissue cells.

Although all cells degrade glucose, most carbohydrate metabolism occurs in three places in the body: the muscle, the fatty tissue, and the liver. Liver cells present no barrier to the entry of glucose from the blood. The passage of blood glucose through the membranes of muscle and fat cells must be stimulated by the peptide hormone insulin (Fig. 24.11).

Proinsulin, an inactive form of insulin, is synthesized as a single peptide chain in the pancreas. Proinsulin is packaged in granules and converted to active insulin within the granules by the action of peptidases. The activation of proinsulin to insulin is very similar to the activation of zymogens to active enzymes.

Upon release into the bloodstream, insulin initiates a chain of events by which blood glucose is permitted to enter muscle and fat cells (Fig. 24.12).

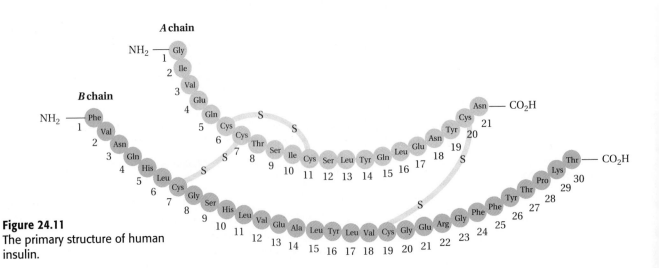

Figure 24.11
The primary structure of human insulin.

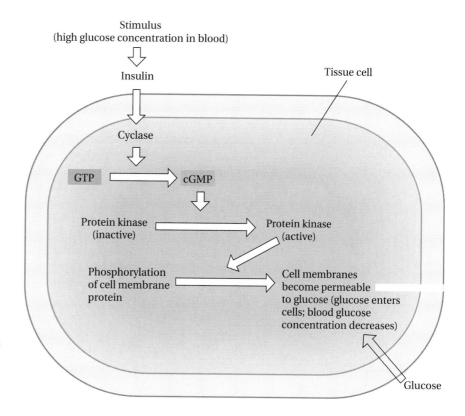

Stimulus
(high glucose concentration in blood)

Insulin

Tissue cell

Cyclase

GTP → cGMP

Protein kinase
(inactive) → Protein kinase
(active)

Phosphorylation
of cell membrane
protein

Cell membranes
become permeable
to glucose (glucose enters
cells; blood glucose
concentration decreases)

Glucose

Figure 24.12
Insulin stimulates the uptake of glucose by tissue cells by starting a process that eventually renders the cell membranes permeable to glucose. The second messenger for the process is probably cGMP rather than cAMP.

Scientists believe that the second messenger for insulin is cyclic guanosine monophosphate (cGMP), a compound similar in structure to cAMP. When glucose enters tissue cells, the blood glucose concentration decreases. The drop in blood sugar is the *hypoglycemic effect* of insulin. A flaw in any link of the chain—from the production of insulin to the uptake of glucose by tissues—is a serious health problem.

Hypoglycemia and hyperglycemia

A blood glucose level between 70 and 90 mg/100 mL of blood is normal. And it is important that the blood sugar concentration be kept in this range. People who experience mild **hypoglycemia**—*low blood sugar*—may become pale, weak, and nervous. Severe hypoglycemia, a blood sugar level of less than 20 mg of glucose per 100 mL of blood, leads to shock, convulsions, coma, and finally death. There also can be too much of a good thing. **Hyperglycemia**—*high blood sugar*—for a short time is not usually harmful, but prolonged hyperglycemia is a symptom of diabetes mellitus.

Diabetes mellitus

Diabetes mellitus *is a family of diseases in which the uptake of blood glucose by tissues is impaired.* Blood glucose levels may become so high that **glycosuria**—*the appearance of glucose in the urine*—begins to occur. Loss of body water that is needed to eliminate excess glucose can severely

Hypoglycemia
hypo (Greek):   below
glyc: glucose
haima (Greek):   blood

Hyperglycemia
hyper (Greek):   above
glyc: glucose
haima (Greek):   blood

Glycosuria
glyco (Greek):   sweet
uria (Greek):   urine

dehydrate a patient with untreated diabetes. The name *diabetes mellitus* literally means "sweet siphon." Although there is plenty of glucose in the blood, glucose cannot reach tissues where it is urgently needed. Thus these untreated patients experience weight loss and constant hunger. This is nothing more or less than starvation in the midst of plenty.

Glycosuria occasionally occurs during pregnancy because of hormonal changes. As long as blood glucose levels are normal, this is usually not a serious condition.

There are many causes of diabetes, not all of which are well understood. In type I (insulin-dependent or juvenile-onset) diabetes, which usually appears in children before age 10, practically no insulin is produced by the victim's pancreas. Type I diabetes is usually treated by daily injections of insulin. The insulin dose must be carefully regulated, though; too much insulin produces *insulin shock. Caused by severe hypoglycemia,* **insulin shock** *bears the potential for coma and death.* Diabetics often carry candy bars to eat if they feel the onset of insulin shock. Type II (non-insulin-dependent or maturity-onset) diabetics, those in whom symptoms appear during young adulthood or middle age, appear to produce insulin but the release is delayed. About 20% of type II diabetics require insulin. In many others, release of their own insulin can be stimulated by drugs such as tolbutamide.

$$CH_3-CH_2-CH_2-CH_2-\underset{H}{N}-\underset{\underset{O}{\parallel}}{C}-\underset{H}{N}-\underset{\underset{O}{\parallel}}{\overset{\overset{O}{\parallel}}{S}}-\!-CH_3$$

Tolbutamide

Diabetes has many more effects on metabolism and health than the few mentioned here. More of these effects are considered in the next chapter.

The insulin used to treat human diabetics formerly came from rabbits and cattle. The amino acid sequences of the insulin of these species and humans are similar enough that no severe allergic response usually occurred. Today, human insulin is being produced on a commercial scale by bacteria. A functional gene for the synthesis of human insulin was inserted into the bacteria by the gene-splicing method described in Section 20.13. The genetically engineered bacteria are a convenient source of human insulin for treatment of diabetes.

PRACTICE EXERCISE 24.8

Insulin is a peptide of 51 amino acid residues. Explain why this hormone cannot be administered orally to patients with diabetes but must be given intravenously.

SUMMARY

Glucose is a major source of energy for living creatures, and some of the reactions by which it is oxidized to carbon dioxide provide a common pathway for the oxidation of fats and amino acids. Glucose is oxidized in three stages: (1) oxidation to pyruvate in glycolysis, (2) formation of acetyl CoA from pyruvate, and (3) oxidation of the acetyl group of acetyl CoA to carbon dioxide in the citric acid cycle.

For each molecule of glucose oxidized, the citric acid cycle also produces six molecules of NADH, two of FADH$_2$, and two of ATP. When oxygen is available, the reducing power (NADH and FADH$_2$) formed in the oxidation of glucose is swept into cellular respiration and used to generate 36 additional ATP molecules by oxidative phosphorylation. When oxygen is not available, the citric acid cycle shuts down, and cells revert to fermentation to reoxidize NADH to NAD and thereby keep glycolysis going for a low level of ATP production.

Glycogen, the storage form of glucose, is stored mainly in the liver and muscle. Glycogen is converted on demand to glucose 1-phosphate and then to glucose 6-phosphate. Muscle cells keep glucose 6-phosphate for their own glycolysis. Liver contains an enzyme that converts glucose 6-phosphate to glucose, which passes from the liver to other tissues that need it.

Many metabolic reactions are under hormonal control. When a particular need is sensed, a hormone is sent through the bloodstream from the site of production to target cells. The hormone initiates a sequence of processes inside cells by binding to protein receptors in the target cell membrane. This triggers the formation of a second messenger, usually cyclic AMP (cAMP), which in turn triggers a series of events that ends with an effect on metabolism. Hormones also affect cell membrane permeability. The action of insulin in promoting the uptake of glucose by muscle and fat cells is one example.

Diabetes mellitus is a family of diseases that result when insulin is not produced, not released, or released at the wrong time. In all cases, the uptake of glucose by tissue cells is impaired.

KEY TERMS

Acetyl coenzyme A (24.1)
Alcoholic fermentation (24.7)
Citric acid cycle (24.1)
Coenzyme A (24.1)
Diabetes mellitus (24.13)

First messenger (24.11)
Gluconeogenesis (24.8)
Glycogenesis (24.9)
Glycogenolysis (24.10)
Glycolysis (24.1)
Glycosuria (24.13)

Hyperglycemia (24.13)
Hypoglycemia (24.13)
Insulin shock (24.13)
Lactic fermentation (24.7)
Multienzyme complex (24.4)

Oxygen debt (24.8)
Second messenger (24.11)

EXERCISES

Glycolysis (Sections 24.1, 24.2, 24.3)

24.9 Write a balanced equation for the complete oxidation of glucose.

24.10 Explain the difference between oxidation of glucose inside body cells and oxidation of glucose by burning in air.

24.11 What are the three stages by which glucose is broken down in aerobic cells?

24.12 What is the end product of glycolysis?

24.13 Where does glycolysis take place?

24.14 Write an equation for the first step in glycolysis.

24.15 Is the first step in glycolysis energetically favorable? Explain your answer.

24.16 How many molecules of ATP are required for the conversion of one molecule of glucose to two molecules of pyruvate? What is the net gain in ATP?

24.17 Write equations for the two reactions in glycolysis that lead to ATP production.

24.18 Write the equation for the reaction in glycolysis that requires NAD$^+$ as a coenzyme.

24.19 What is the fate of NAD$^+$ in Exercise 24.18?

24.20 The first oxidation reaction of glucose catabolism occurs in glycolysis. Write the equation for the reaction, and identify the oxidizing agent.

Acetyl Coenzyme A (Section 24.4)

24.21 Where does the formation of acetyl CoA take place? What is its function?

24.22 Name the B-complex vitamin that is part of the structure of CoA. What foods are a good source of this vitamin?

Citric Acid Cycle (Sections 24.5, 24.6)

24.23 Where does the citric acid cycle take place in eukaryotic cells?

24.24. How many molecules of carbon dioxide are produced by each turn of the citric acid cycle?

24.25 Write the equation for the only step in the citric acid cycle that leads directly to phosphorylation of ADP.

24.26 Draw the structure of (a) the organic molecule that ends the citric acid cycle and (b) the organic molecule that starts the cycle.

24.27 How many molecules of ATP are generated in oxidative phosphorylation by (a) $FADH_2$ and (b) NADH?

24.28 What is the total amount of ATP generated by the aerobic catabolism of one molecule of glucose?

Fermentation (Sections 24.7, 24.8)

24.29 What happens in aerobic cells when no oxygen is available to accept electrons in the electron transport chain?

24.30 Is the citric acid cycle able to operate in the absence of oxygen?

24.31 Why is lactic fermentation important?

24.32 What is produced in alcoholic fermentation?

24.33 (a) How do you get an oxygen debt? (b) What are the consequences of this condition?

24.34 What happens to the lactate that builds up in muscle cells during anaerobic glycolysis when you stop to rest and cellular respiration returns to normal?

24.35 What is gluconeogenesis?

24.36 Where does gluconeogenesis take place?

Glycogen Synthesis and Degradation (Sections 24.9, 24.10)

24.37 Describe the structure, function, and storage of glycogen.

24.38 Describe the process of glycogenesis. Why are two enzymes required?

24.39 What is glycogenolysis, and when does it occur?

24.40 Explain why muscle glycogen is not a source of glucose to other body tissues.

Hormonal Control of Glucose Metabolism (Sections 24.11, 24.12, 24.13)

24.41 Explain how hormones regulate processes that occur inside cells even though they are never found inside the cells they act on.

24.42 Where is glucagon produced, and what is its function?

24.43 (a) What stimulates the secretion of epinephrine? (b) Where is it produced, and what is its function?

24.44 Where does the majority of carbohydrate metabolism occur in the body?

24.45 Explain how insulin controls blood glucose levels.

24.46 How do blood glucose levels in hyperglycemia and hypoglycemia compare with normal blood glucose levels?

24.47 Describe the difference between juvenile-onset diabetes and maturity-onset diabetes.

24.48 What is meant by *insulin shock?*

24.49 A screening test for diabetes mellitus is done by analyzing urine for the presence of sugar. Explain why diabetes mellitus causes glycosuria.

Additional Exercises

24.50 What is the fate of the hydrogen atoms removed in oxidation reactions in glycolysis and the Krebs cycle?

24.51 Explain the role of cAMP in glycogenolysis.

24.52 Match the following:

(a) glycogenesis
(b) cAMP
(c) citric acid cycle
(d) gluconeogenesis
(e) diabetes mellitus
(f) coenzyme A
(g) lactic fermentation
(h) oxygen debt
(i) insulin
(j) glycolysis

(1) reduction of pyruvate
(2) oxidation of acetyl CoA
(3) uptake of glucose by cells
(4) insufficient oxygen for oxidative phosphorylation
(5) synthesis of glycogen
(6) oxidation of glucose to pyruvate
(7) glycosuria
(8) synthesis of glucose
(9) thiol
(10) second messenger

24.53 Why do we breathe heavily when engaged in vigorous exercise?

24.54 Describe the process by which glucose is made from noncarbohydrate substances.

24.55 Discuss the major cause of hyperglycemia, and explain how it may be treated.

24.56 Explain what happens to pyruvate in the body under (a) anaerobic conditions and under (b) aerobic conditions.

SELF-TEST (REVIEW)

True/False

1. Oral administration of insulin is an ineffective treatment of diabetes.
2. Gluconeogenesis is an anabolic process.
3. Acetyl CoA is a product of the citric acid cycle.
4. To be useful as a source of energy, all nutrients must first be changed into either glucose or pyruvate.
5. Every step in the glycolysis process is an energy-yielding one.
6. When eukaryotic cells are deprived of oxygen, oxidative phosphorylation stops, and the cell dies.
7. Lactic fermentation is a more efficient source of ATP than respiration.
8. A person's energy requirements could be met from stored glycogen for approximately 1 week.
9. Insulin decreases the blood sugar level by increasing the rate of metabolism of sugar in the liver.
10. A diabetic with a consistently high blood sugar level may feel hungry all the time and actually experience weight loss.

Multiple Choice

11. A chemical messenger produced in the endocrine glands is called
    (a) a hormone.   (b) cAMP.
    (c) a target cell.   (d) a nerve impulse.
12. The major energy product of the citric acid cycle is
    (a) ATP.   (b) cAMP.
    (c) NADH.   (d) $FADH_2$.
13. A person whose blood sugar level drops below 60 mg/100 mL of blood
    (a) will probably feel weak.
    (b) is hyperglycemic.
    (c) probably has glucose in their urine.
    (d) more than one are correct.
14. All the following occur in the mitochondria of a eukaryotic cell *except*
    (a) the citric acid cycle.
    (b) formation of acetyl CoA from pyruvate.
    (c) glycolysis.
    (d) cellular respiration.
15. cAMP is best classed as a(n)
    (a) enzyme-hormone receptor.
    (b) enzyme modulator.
    (c) first messenger.
    (d) calcium ion inhibitor.

16. The glycolysis of one molecule of glucose results in the formation of
    (a) a net total of two ATP molecules.
    (b) two molecules of NADH.
    (c) two molecules of pyruvate.
    (d) all of the above.
17. A low blood concentration of glucose
    (a) is called *hyperglycemia.*
    (b) can be countered by glycogenolysis of glycogen in the liver.
    (c) can be countered by injection of insulin.
    (d) both (a) and (c) are correct.
18. Pyruvate is the end product of
    (a) glycogenesis.
    (b) the citric acid cycle.
    (c) oxidative phosphorylation.
    (d) glycolysis.
19. Glycogenolysis can be stimulated by
    (a) epinephrine.   (b) amylase.
    (c) glucagon.   (d) Both (a) and (c) are correct.
20. Alcoholic fermentation
    (a) allows glycolysis to keep going.
    (b) results in the production of lactate.
    (c) is an anaerobic process.
    (d) answers (a) and (c) are correct.
21. Which of the following processes requires a net expenditure of ATP?
    (a) glycolysis
    (b) citric acid cycle
    (c) gluconeogenesis
    (d) lactic fermentation
22. Glucose is temporarily stored in the body as
    (a) fat.   (b) glycogen.
    (c) lipids.   (d) glucose 6-phosphate.
23. A cell in the state of oxygen debt can produce energy by
    (a) cellular respiration.
    (b) glycolysis.
    (c) glycogenesis.
    (d) gluconeogenesis.
24. You are climbing a mountain, and your partner who has diabetes is slipping into insulin shock. You should
    (a) not give her anything to eat or drink.
    (b) let her eat the last chocolate bar you own.
    (c) let her drink water but nothing else.
    (d) head for the nearest phone and call a doctor.

25

Lipid Metabolism

Fat Chemistry in Cells

Animals that live in cold climates, such as these crabeater seals, metabolize lipids for body heat. They also rely on lipids for insulation against the extremes of temperature.

CHAPTER OUTLINE

CASE IN POINT: Carnitine deficiency

25.1   Body lipids

A CLOSER LOOK: Lipid Storage
Diseases

25.2   Fat mobilization

25.3   Fatty acid oxidation

FOLLOW-UP TO THE CASE IN POINT:
Carnitine deficiency

25.4   ATP yield

25.5   Glycerol metabolism

25.6   The key intermediate—
acetyl CoA

25.7   Fatty acid synthesis

25.8   Ketone bodies

25.9   Ketosis

25.10 Cholesterol synthesis

25.11 Blood cholesterol

A CLOSER LOOK: Drug Strategies for
Reducing Serum Cholesterol

Lipids are essential to cellular life. As the material of membranes, this class of oily biological molecules is a wrapper around cells and organelles of eukaryotic cells. As sources of energy, fatty acids are preferred by liver cells and resting muscle cells, among others. Body cells also synthesize lipids and other fatty substances such as cholesterol.

This chapter is about how lipids are stored, degraded, and synthesized. Carbohydrate metabolism and lipid metabolism are intimately related, so we will examine the ways in which carbohydrate metabolism is linked to fatty acid metabolism and how diabetes affects lipid metabolism as well as glucose metabolism. Our Case in Point concerns a disorder of lipid metabolism.

CASE IN POINT: Carnitine deficiency

 Baby Clarence began to suffer from vomiting, slow heart rate, and breathing problems about 6 months after birth. One morning Clarence's mother found him, blue and not breathing, in his crib. Paramedics managed to restore Clarence's breathing, but the doctors were uncertain of the cause of his illness. Clarence's prognosis was not good. Within a few weeks, however, doctors were able to determine that the cause of Clarence's condition was a lipid metabolism defect. What was Clarence's disorder? Could it be treated? We will find out in Section 25.3.

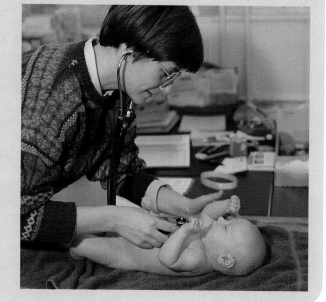

An infant suspected of having a defect in lipid metabolism is examined by a specialist.

25.1 Body lipids

AIM: To differentiate between the lipids of adipose tissue and those of working tissue on the basis of structure and function.

About 12% of the body weight of the average male is depot fat; about 20% of the body weight of the average female is depot fat.

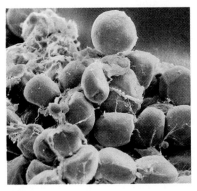

Figure 25.1
Scanning electron micrograph of fat cells (yellow) in adipose tissue (magnified 1575×).

Lipids that enter the bloodstream are distributed in the body tissues in two ways: as **storage fat (depot fat)** *or as* **working lipid (tissue lipid).** *Storage fat is collected in the cells of fat tissue, also called* **adipose tissue.** Most of the lipids stored in fat cells (Fig. 25.1) are triglycerides. Adipose tissue has a tendency to collect around the waist, hips, and buttocks.

Layers of adipose tissue also surround many body organs. These layers of body fat protect the body and its organs against bumps and blows. Fat is also a good insulator against cold. The bodies of whales and seals, mammals exposed to severe cold, are covered by thick layers of fat. At one time many biologists and physiologists thought of adipose tissue as a metabolically inactive insulating blanket. However, it now appears that metabolism in fatty tissue cells is an important source of body heat.

Complex lipids such as the phospholipids, sphingomyelins, and cerebrosides (see Chap. 17) become part of the working lipids of the body. The working lipids are mainly incorporated into cell membranes and the membranes of cell organelles such as the mitochondria, endoplasmic reticula, nuclei, and lysosomes.

Normally, both triglycerides and complex lipids are constantly being broken down and synthesized in the body. However, there are several diseases in which the enzymes required to catalyze the breakdown of complex lipids are lacking (see A Closer Look: Lipid Storage Diseases, on pages 766–767).

25.2 Fat mobilization

AIM: To describe the function of epinephrine and serum albumin in lipid metabolism.

Brown adipose tissue is found in newborn infants and hibernating animals. It is loaded with mitochondria and generates *heat* rather than ATP, as does the normal (*white*) adipose tissue.

Carbohydrates—through glucose catabolism—are the first choice of some tissues for energy production. The brain and active skeletal muscles are examples of such tissues. As we have seen, however, the body's stores of glycogen are used up after only a few hours of fasting. Well before the glycogen supply is used up, many body cells, such as those of resting muscle and liver, begin to break down fatty acids to produce energy. By using fatty acids for energy production, these cells help to conserve the rest of the body's glycogen store for brain cells and other cells of the nervous system. Brain and nerve cells demand a constant supply of glucose. In fact, except in emergencies such as starvation, they produce their energy only from glucose. When other body cells need fatty acids, signaled by a low blood glucose level, the endocrine system sends several hormones, including epinephrine, to the adipose tissue storehouse. In fat cells, epinephrine, through a second messenger system, stimulates **mobilization**—*the hydrolysis of triglycerides to fatty acids and glycerol, which enter the blood-*

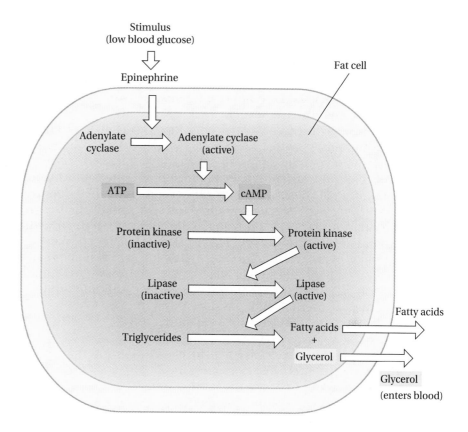

Figure 25.2
Mobilization of fatty acids from
adipose tissue.

stream. Fat mobilization is shown in Figure 25.2. The mobilized fatty acids
become tightly attached to a plasma protein, serum albumin. The serum
albumin transports the fatty acids to tissues where they are taken up by
cells that need them. The glycerol produced by the hydrolysis of the
triglyceride also enters these cells.

25.3 Fatty acid oxidation

AIMS: *To name and describe the process of the catabolism of
fatty acids, indicating the end products. To predict the
hydrolysis products of typical lipid molecules. To determine
the number of molecules of acetyl CoA, NADH, and FADH$_2$
that result from the beta oxidation of a given fatty acid.*

Focus

Fatty acids are degraded to
acetyl coenzyme A.

Fatty acids enter tissue cells that need energy. Before a fatty acid molecule
can be oxidized to produce energy, however, it must be converted to a **fatty
acyl CoA**—*a thiol ester formed from a fatty acid and CoA.*

$$\underset{\text{Fatty acid}}{R-\overset{\overset{\displaystyle O}{\|}}{C}-OH} + HS-CoA \longrightarrow \underset{\text{Fatty acyl CoA}}{R-\overset{\overset{\displaystyle O}{\|}}{C}-S-CoA} + H_2O$$

A Closer Look

Lipid Storage Diseases

Inborn errors associated with lipid metabolism include Tay-Sachs disease, Gaucher's disease, and Niemann-Pick disease. All three conditions are lipid storage diseases. Normally, both triglycerides and complex lipids are constantly being broken down and synthesized in the body. The lipid storage diseases seem to occur because the body lacks enzymes required to catalyze the breakdown of certain complex lipids. This enzyme deficiency leads to accumulation of abnormally high levels of complex lipids in specific tissues, particularly the brain, spleen, and liver. In all lipid storage diseases, there is retardation of mental and physical development for reasons that are not well understood.

Tay-Sachs disease, in which lipids primarily accumulate in the brain, is characterized by paralysis, a deteriorated mental state, and blindness. In Tay-Sachs, the stored lipids are primarily gangliosides, the most complex members of the lipids containing sphingosine (see figure, part a). As shown in the figure, the gangliosides contain four different sugar units.

Gaucher's disease has many variants with certain clinical features, including elevated concentrations of serum acid phosphatase. The structure of a typical cerebroside that accumulates in this disease is shown in part (b) of the figure. The spleen becomes enlarged because spleen cells accumulate cerebrosides (see Sec. 16.3). Gaucher's disease may progress rapidly and be fatal in infancy, but some patients live fairly normal life spans. This suggests that Gaucher's disease is a group of inheritable diseases. The inherited mutation may be different in each form of the disease, but all these mutations affect cerebroside metabolism.

The chemical abnormality found in Niemann-Pick disease is an increased content of sphingolipids (see Sec. 16.3) in the organs, but especially the spleen (see figure, part c). As in Gaucher's disease, there are many variants of Niemann-Pick disease, indicating a variety of mutations affecting sphingolipid metabolism.

At right: The three lipids whose excessive storage causes Tay-Sachs disease, Gaucher's disease, and Niemann-Pick disease. In Tay-Sachs disease, the stored lipids are gangliosides, the most complex members of the lipids containing sphingosine (a). The sugar units of the gangliosides contain glucose (Glu), galactose (Gal), and two unusual sugars: *N*-acetylgalactosamine (Naga) and *N*-acetylneurominic acid (Nana). In Gaucher's and Niemann-Pick diseases, the lipids that cannot be broken down are the cerebrosides (b) and the sphingomyelins (c), respectively.

The conversion of fatty acids to fatty acyl CoA occurs in the cytoplasm of cells, but the subsequent oxidation of fatty acyl CoA occurs in the mitochondria. This presents a problem, since fatty acyl CoA molecules cannot pass through the mitochondrial membrane. The passage of fatty acyl CoA from the cytoplasm into the mitochondria is done indirectly. The fatty acyl CoA is first converted into an ester of the amino alcohol carnitine.

In cytoplasm

$$
\underset{\text{Fatty acyl CoA}}{R-\overset{O}{\overset{\|}{C}}-SCoA} + \underset{\text{Carnitine}}{(H_3C)_3\overset{+}{N}-CH_2-\overset{OH}{\underset{|}{C}H}-CO_2H} \longrightarrow \underset{\text{Fatty acyl carnitine}}{(H_3C)_3\overset{+}{N}-CH_2-\overset{\overset{O}{\overset{\|}{R-C-O}}}{\underset{|}{C}H}-CO_2H} + H-SCoA
$$

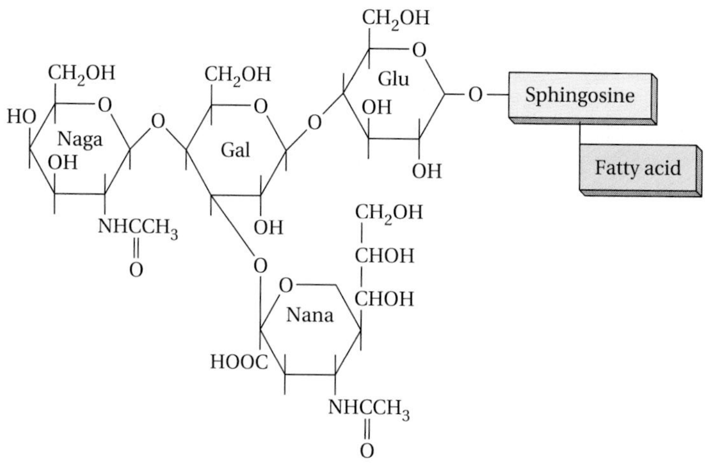

(a) A ganglioside (accumulates in Tay–Sachs disease)

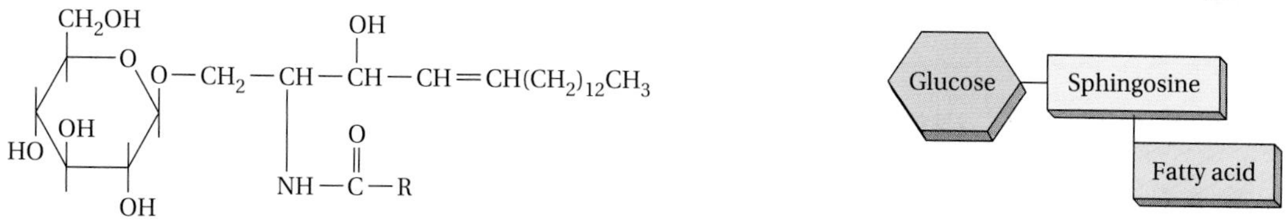

(b) A cerebroside (accumulates in Gaucher's disease)

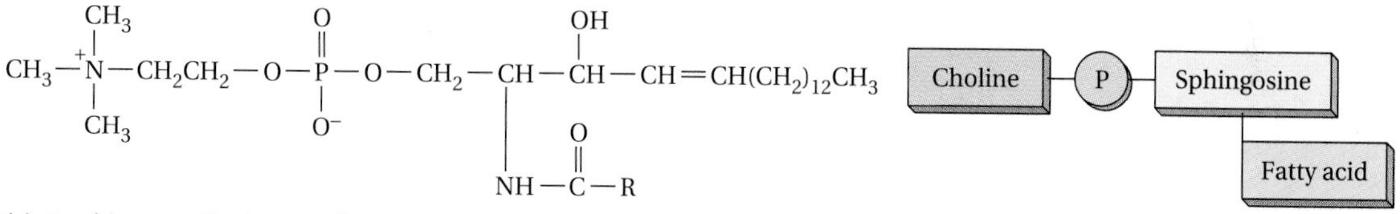

(c) A sphingomyelin (accumulates in Niemann–Pick disease)

The fatty acyl carnitine passes through the mitochondrial membrane, where reaction with CoA converts it back to fatty acyl CoA.

In mitochondrion

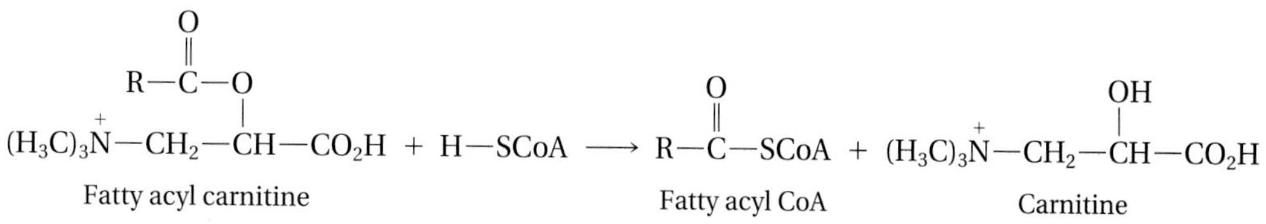

$$\underset{\text{Fatty acyl carnitine}}{(H_3C)_3\overset{+}{N}-CH_2-\underset{\underset{\displaystyle O-\underset{\displaystyle \,}{\overset{\displaystyle R-\overset{\displaystyle O}{\overset{\|}{C}}-O}{|}}}{CH}-CO_2H} + H-SCoA \longrightarrow \underset{\text{Fatty acyl CoA}}{R-\overset{\displaystyle O}{\overset{\|}{C}}-SCoA} + \underset{\text{Carnitine}}{(H_3C)_3\overset{+}{N}-CH_2-\underset{\overset{\displaystyle |}{\displaystyle OH}}{CH}-CO_2H}$$

The fatty acyl carnitine is so important in the metabolism of fatty acids that a carnitine deficiency can be life-threatening. Clarence, the sick baby in the Case in Point earlier in this chapter, is an example.

FOLLOW-UP TO THE CASE IN POINT: Carnitine deficiency

Clarence was found to be suffering from a potentially fatal carnitine deficiency. This deficiency was the result of a defect in the genes responsible for fatty acid metabolism. Clarence's symptoms—vomiting, slow heart rate, and breathing problems—were general, and the condition is rare. He was fortunate that his carnitine deficiency was diagnosed before it took its fatal course. Most of the symptoms of his disease disappeared when carnitine was added to his formula three times a day. Since his carnitine deficiency was accompanied by other defects in his fatty acid metabolism, Clarence also was placed on a low-fat diet. Clarence, whose prognosis was poor, seems to be thriving on his carnitine supplement and low-fat diet.

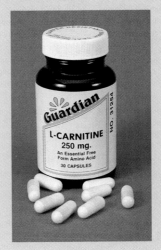

Carnitine supplements help in the oxidation of long-chain fatty acids.

Fatty acid activation

Formation of fatty acyl CoA molecules shows how the energy stored in the phosphate bonds of ATP can be used to drive chemical processes that normally could not occur. Neither the fatty acid nor the CoA is sufficiently energetic to form the fatty acyl CoA directly, so the cell boosts the chemical reactivity of the fatty acid carboxyl group by investing ATP. The triphosphate group of the ATP molecule is cleaved to give pyrophosphate (PP_i), and the remaining adenosine monophosphate (AMP) forms a high-energy anhydride bond with the fatty acid.

$$\text{Fatty acid} + \text{ATP} \longrightarrow \text{Mixed anhydride} + PP_i$$

The anhydride linkage formed between the fatty acid and AMP in this reaction is energetic enough to be broken by reaction with the thiol group of CoA.

$$\text{Mixed anhydride} + \text{HS—CoA} \longrightarrow \text{Fatty acyl CoA} + \text{Adenosine monophosphate (AMP)}$$

The thioester bond of the fatty acyl CoA is a high-energy bond, just as it is in acetyl CoA. Therefore, some of the energy invested in the breakdown of ATP to AMP has been conserved in the thioester bond of the fatty acyl CoA. The cost to the cell of activating one molecule of fatty acid is the loss of two high-energy phosphoric acid anhydride bonds: one in the formation of the anhydride between the fatty acid and AMP and one when PP_i is hydrolyzed to $2P_i$.

Beta oxidation

The formation of fatty acyl CoA molecules prepares fatty acids for entry into the mitochondria. *In the mitochondria, fatty acyl CoA molecules are degraded to acetyl CoA in the catabolic process called* **beta oxidation.** During beta oxidation, the third (beta) carbon of the saturated fatty acid chain of the fatty acyl CoA molecule is oxidized to a ketone.

$$C-C-C-\overset{\overset{H}{|}}{\underset{\underset{H}{|}}{\underset{3}{C}}}-\overset{}{\underset{2}{C}}-\overset{\overset{O}{\|}}{\underset{1}{C}}-S-CoA \xrightarrow{\text{Oxidation}} C-C-C-\overset{\overset{O}{\|}}{\underset{\beta}{C}}-C-\overset{\overset{O}{\|}}{C}-S-CoA$$

Fatty acyl CoA

There are four reactions involved in beta oxidation. An example is the fatty acyl CoA formed from a 16-carbon acid, palmitic acid. Any NADH or $FADH_2$ produced in these reactions must be accounted for, since the reducing power of these molecules may be used to produce ATP in respiration. As in all the metabolic reactions we have discussed, every step of beta oxidation is catalyzed by enzymes.

Step 1: Dehydrogenation. A carbon-carbon double bond is formed between the second and third carbons of the palmitic acid chain. One molecule of $FADH_2$ is produced.

$$CH_3\!\!-\!\!(CH_2\,)_{\overline{12}}\,\overset{\overset{H}{|}}{\underset{\underset{H}{|}}{\underset{\beta}{C}}}\!-\!\overset{\overset{H}{|}}{\underset{\underset{H}{|}}{\underset{\alpha}{C}}}\!-\!\overset{\overset{O}{\|}}{C}\!-\!S\!-\!CoA \xrightarrow[\underset{FAD \quad FADH_2}{}]{} CH_3\!\!-\!\!(CH_2\,)_{\overline{12}}\,\underset{\beta}{CH}\!\!=\!\!\underset{\alpha}{CH}\!-\!\overset{\overset{O}{\|}}{C}\!-\!S\!-\!CoA$$

As we have seen, FAD is often the coenzyme for enzymes that catalyze the removal of hydrogen from saturated carbon compounds. This is the case here, and one $FADH_2$ molecule is produced.

Step 2: Hydration. A molecule of water is added to the double bond of the fatty acyl CoA. This step is the hydration reaction of beta oxidation.

$$CH_3\!\!-\!\!(CH_2\,)_{\overline{12}}\,\underset{\beta}{CH}\!\!=\!\!\underset{\alpha}{CH}\!-\!\overset{\overset{O}{\|}}{C}\!-\!S\!-\!CoA \;+\; HO\!-\!H \longrightarrow CH_3\!\!-\!\!(CH_2\,)_{\overline{12}}\,\overset{\overset{OH}{|}}{\underset{\underset{H}{|}}{\underset{\beta}{C}}}\!-\!\overset{\overset{H}{|}}{\underset{\underset{H}{|}}{\underset{\alpha}{C}}}\!-\!\overset{\overset{O}{\|}}{C}\!-\!S\!-\!CoA$$

Step 3: Oxidation. The hydroxyl group is now oxidized to a ketone. NAD^+ is the coenzyme that acts as the oxidizing agent in this second oxidation reaction of beta oxidation. Thus beta oxidation has now produced one molecule of $FADH_2$ and one molecule of NADH.

$$CH_3\text{-}(CH_2)_{12}\text{-}\underset{\beta}{\underset{|}{\overset{OH}{\overset{|}{C}}}}\text{-}\underset{\alpha}{\underset{|}{\overset{H}{\overset{|}{C}}}}\text{-}\overset{O}{\overset{\parallel}{C}}\text{-}S\text{-}CoA \xrightarrow[\quad NAD^+ \qquad NADH + H^+ \quad]{} CH_3\text{-}(CH_2)_{12}\text{-}\overset{O}{\overset{\parallel}{C}}\text{-}\underset{|}{\overset{H}{\overset{|}{C}}}\text{-}\overset{O}{\overset{\parallel}{C}}\text{-}S\text{-}CoA$$

Step 4: Carbon-carbon bond cleavage. The carbon-carbon bond between the second and third carbons is cleaved to give acetyl CoA and a new fatty acyl CoA that is two carbons shorter than the starting thioester. This reaction involves another molecule of CoA, but no ATP is expended.

$$\underset{\text{16 carbons}}{CH_3\text{-}(CH_2)_{12}\text{-}\overset{O}{\overset{\parallel}{C}}\text{-}\underset{|}{\overset{H}{\overset{|}{C}}}\text{-}\overset{O}{\overset{\parallel}{C}}\text{-}S\text{-}CoA} + HS\text{-}CoA \longrightarrow \underset{\text{14 carbons}}{CH_3\text{-}(CH_2)_{12}\text{-}\overset{O}{\overset{\parallel}{C}}\text{-}S\text{-}CoA} + \underset{\text{2 carbons}}{CH_3\overset{O}{\overset{\parallel}{C}}\text{-}S\text{-}CoA}$$

The fragmentation of fatty acyl CoA molecules into molecules of acetyl CoA is similar to other metabolic cycles. The steps of beta oxidation repeat over and over again until the fatty acyl CoA is completely degraded to acetyl CoA. Look at Figure 25.3. The fatty acyl CoA that starts each round of the

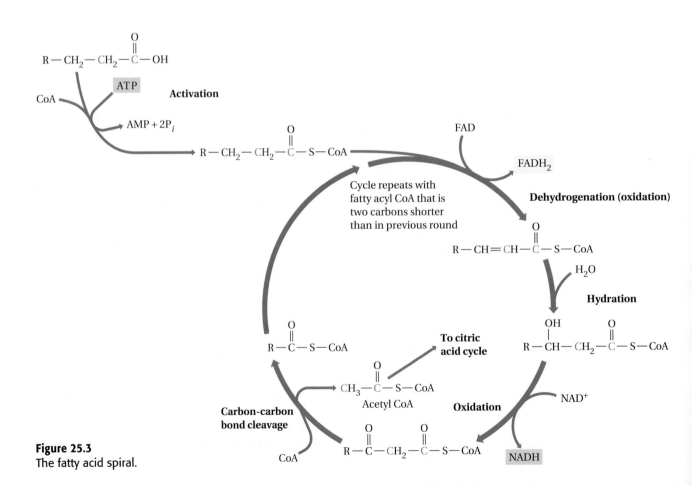

Figure 25.3
The fatty acid spiral.

beta-oxidation cycle is two carbons shorter than in the previous round. *For this reason, the pathway for the degradation of fatty acids to acetyl CoA is often called the* **fatty acid spiral.**

Every round of the spiral produces one molecule each of acetyl CoA, NADH, and $FADH_2$ until the fatty acyl CoA molecule is only four carbons long. At this point, the first three steps of the final round of beta oxidation produce the compound acetoacetyl CoA. The fourth step, the reaction of acetoacetyl CoA with CoA, produces an extra molecule of acetyl CoA from the tail end of the fatty acid without formation of NADH and $FADH_2$.

$$CH_3-\overset{\overset{O}{\|}}{C}-CH_2-\overset{\overset{O}{\|}}{C}-S-CoA \ + \ HS-CoA \ \longrightarrow$$

Acetoacetyl CoA

$$CH_3-\overset{\overset{O}{\|}}{C}-S-CoA \ + \ CH_3-\overset{\overset{O}{\|}}{C}-S-CoA$$

In other words, the complete conversion of a fatty acyl CoA to two-carbon fragments of acetyl CoA always produces one more molecule of acetyl CoA than of NADH or $FADH_2$. To summarize, the breakdown of palmitic acid gives eight molecules of acetyl CoA, but only seven molecules of NADH and seven molecules of $FADH_2$ are produced.

PRACTICE EXERCISE 25.1

Lauric acid is converted to acetyl CoA in beta oxidation. Determine the yields of (a) acetyl CoA, (b) NADH, and (c) $FADH_2$.

25.4 ATP yield

AIM: To calculate the number of ATP molecules formed by the oxidation of a fatty acid molecule.

In Chapter 24 we saw that the carbons of the acetyl CoA produced by the catabolism of glucose can be completely oxidized to carbon dioxide in the citric acid cycle. Each molecule of acetyl CoA oxidized in this fashion yields enough energy to make one molecule of ATP, one molecule of $FADH_2$, and three molecules of NADH. The reducing power of each molecule of NADH can make three molecules of ATP by cellular respiration; $FADH_2$ produces two molecules of ATP in the same way. It was shown that 38 ATP molecules is the total useful energy yield of aerobic glucose catabolism.

Molecules of acetyl CoA are the same, regardless of their source. Like acetyl CoA molecules produced from glucose, the acetyl CoA molecules formed in the fatty acid spiral can be oxidized in the citric acid cycle. Since we can find the yield of NADH, $FADH_2$, and ATP from the beta-oxidation reactions of the fatty acid spiral and from the citric acid cycle, we can calculate how many molecules of ATP are produced by the total oxidation of one molecule of any fatty acid to carbon dioxide and water. Table 25.1 shows a calculation of this kind for palmitic acid.

In calculating the total ATP yield obtained from the complete oxidation of the fatty acid, we can count the investment of two high-energy phos-

Studies have shown that exercising less frequently but for a longer duration is a good approach to burning body fat. After 40 minutes of exercising, the percentage of energy supplied by fat is greater than that supplied by carbohydrates. Still longer exercise periods lead to an even higher percentage of energy being supplied by body fat.

Focus

The complete oxidation of 1 molecule of palmitic acid yields 129 molecules of ATP.

Table 25.1 ATP Production from Complete Aerobic Catabolism of One Molecule of Palmitic Acid

Pathway	ATP yield
fatty acid spiral (2 ATP invested to make palmitoyl CoA to start spiral)	-2
citric acid cycle (8 molecules of acetyl CoA degraded)	8
oxidative phosphorylation	
7NADH from fatty acid spiral	21
7FADH$_2$ from fatty acid spiral	14
24NADH from citric acid cycle	72
8FADH$_2$ from citric acid cycle	16
	129

The oxidation of 1 g of a fatty acid yields about 0.5 mol ATP. The oxidation of 1 g of glucose yields about 0.2 mol ATP. This ratio, 0.5:0.2, is approximately the same as the ratio of the energy yield for the complete oxidation of fats and carbohydrates, 9 kcal:4 kcal.

phate bonds required to activate the fatty acid as two ATP molecules. We can do this because hydrolysis of one molecule of ATP to AMP and 2P$_i$ is equivalent to the hydrolysis of 2ATP to 2ADP and 2P$_i$. Table 25.1 shows that for every molecule of palmitic acid completely oxidized to carbon dioxide and water, 129 molecules of ATP are formed. No wonder fats are an important source of energy for cellular work.

Energy production is not the only useful function of beta oxidation. Cells using NADH and FADH$_2$ to reduce oxygen also produce a good deal of water as a by-product of cellular respiration. Certain animals have become physiologically adapted to take advantage of this fact. A camel's hump, for example, consists of fat that has been stored in times of plenty. Aerobic catabolism of this fat supplies enough energy and water to permit camels to survive during long periods of famine and drought.

PRACTICE EXERCISE 25.2

Calculate how many molecules of ATP are produced by the total oxidation of lauric acid to carbon dioxide and water.

25.5 Glycerol metabolism

AIM: To describe how glycerol is used as an energy source.

Focus

Glycerol produced by hydrolysis of triglycerides enters glycolysis as dihydroxyacetone phosphate.

The hydrolysis of triglycerides produces glycerol as well as fatty acids. Glycerol is converted by cells to dihydroxyacetone phosphate in two steps.

$$
\begin{array}{ccccc}
\text{H}_2\text{C}-\text{OH} & & \text{H}_2\text{C}-\text{OH} & & \text{H}_2\text{C}-\text{OH} \\
| & & | & & | \\
\text{H}-\text{C}-\text{OH} & \xrightarrow[\text{ATP} \quad \text{ADP}]{} & \text{H}-\text{C}-\text{OH} & \xrightarrow[\text{NAD}^+ \quad \text{NADH}+\text{H}^+]{} & \text{C}=\text{O} \\
| & & | & & | \\
\text{H}_2\text{C}-\text{OH} & & \text{H}_2\text{C}-\text{O}-\textcircled{P} & & \text{H}_2\text{C}-\text{O}-\textcircled{P} \\
\text{Glycerol} & & \text{Glycerol} & & \text{Dihydroxy-} \\
& & \text{3-phosphate} & & \text{acetone} \\
& & & & \text{phosphate}
\end{array}
$$

Dihydroxyacetone phosphate is one of the chemical intermediates of glycolysis. Thus the glycerol produced by hydrolysis of triglycerides contributes to energy production by entering glycolysis as dihydroxyacetone phosphate.

25.6 The key intermediate–acetyl CoA

AIMS: To name the shared intermediate of both carbohydrate and fatty acid metabolism. To list four fates of acetyl CoA in the liver.

Now that we have seen how the body oxidizes fatty acids, we can form an overall picture of the various parts of fatty acid metabolism. We can examine the relationships between carbohydrate metabolism and fatty acid metabolism at the same time. Since the liver conducts more carbohydrate metabolism and fatty acid metabolism than any other organ, this discussion will focus on it.

Figure 25.4 shows the relationships we will be examining in the remainder of this chapter. Refer to it often as you read on. The figure shows that

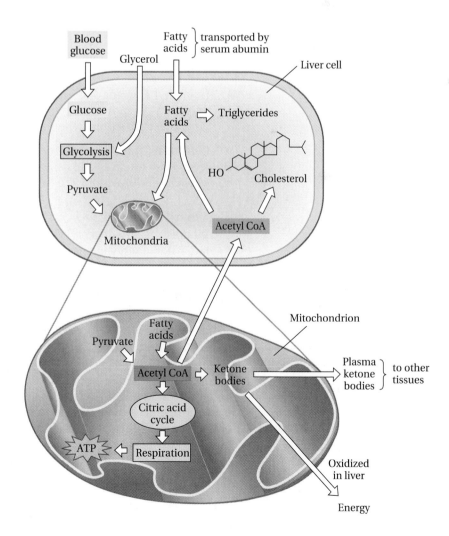

Figure 25.4
The major pathways of lipid metabolism in the liver and their relationship to carbohydrate metabolism.

fatty acids entering the liver from the blood may be resynthesized into triglycerides and stored in the adipose tissue there. Alternatively, fatty acids may be broken down to acetyl CoA. Glucose is also broken down to acetyl CoA. If you are beginning to suspect that acetyl CoA must be a key compound in the metabolic interplay between carbohydrate and fatty acid metabolism, you are certainly correct. Four possible fates await the acetyl CoA produced from fatty acids or glucose in the liver:

1. The acetyl CoA in the mitochondria may be oxidized to carbon dioxide and water in the citric acid cycle and respiration. This pathway, which is used if the liver cells need to generate energy through respiration, makes it clear that the citric acid cycle is shared by both glucose metabolism and fatty acid metabolism.

2. The acetyl CoA in mitochondria may be used to synthesize the substances called *ketone bodies*. Some ketone bodies are oxidized for energy production by the liver. The remainder are transported to other tissues that can use them for energy production. Extensive production of ketone bodies from acetyl CoA occurs only when no glucose is available, as in starvation or diabetes. When no glucose is available to starving brain cells, for example, they adapt to the use of ketone bodies as an energy source in about 48 hours.

3. The cytoplasmic acetyl CoA may be used to resynthesize fatty acids. Since some of the acetyl CoA may have come from glucose, this means that carbohydrates may be converted to fatty acids.

4. The acetyl CoA may be used to synthesize cholesterol in the cellular cytoplasm. As with other synthetic processes, the synthesis of fatty acids and cholesterol occurs when the body is well supplied with energy.

Cells other than liver cells can carry out most of these four functions. However, only the liver adds significant quantities of ketone bodies to the bloodstream for transport to other tissues.

> **PRACTICE EXERCISE 25.3**
>
> What is the fate of acetyl CoA when the body needs energy? What is its fate when energy needs are met?

25.7 Fatty acid synthesis

AIM: To explain the advantages of lipid energy storage over carbohydrate energy storage.

Focus

Converting carbohydrates to fatty acids is an efficient way to store energy.

Carbohydrates can be converted to fats. The key compound that links carbohydrate metabolism to fatty acid synthesis is acetyl CoA. When body cells have more glucose than they require to meet their energy needs, they can divert some of the acetyl CoA produced by glucose catabolism to the synthesis of fatty acids.

Fatty acid synthesis and fatty acid oxidation are not the reverse of one another, though both processes do occur in cycles. In fatty acid synthesis, a two-carbon fragment is added to the growing fatty acid chain in each turn of the cycle rather than being removed, as it is in fatty acid oxidation. Moreover, the synthesis and degradation of fatty acids involve different sets of enzymes. The cellular locations of the two processes are different as well: Fatty acid synthesis occurs in the cytoplasm; fatty acid oxidation takes place in the mitochondria. Like most synthetic or anabolic pathways, the synthesis of fatty acids requires the expenditure of cellular reducing power and ATP. This requirement is clearly shown by the overall equation for the synthesis of palmitic acid.

$$8CH_3\overset{O}{\overset{\|}{C}}-S-CoA \ + \ 7ATP \ + \ 14NADH \ \longrightarrow \ CH_3\!\!\left(CH_2\right)_{\!14}\!\!\overset{O}{\overset{\|}{C}}-OH \ + \ 7ADP \ + \ 14NAD^+ \ + \ 8CoA \ + \ 6H_2O \ + \ 7P_i$$

Acetyl CoA   Palmitic acid

The amount of energy that is released upon complete oxidation of a fuel can be measured in a *bomb calorimeter*. A weighed sample of a fuel is burned in a closed chamber situated in a water-filled, insulated container. The heat given off is calculated from the measured increase in the temperature of the known mass of water.

The newly synthesized fatty acids are incorporated into triglycerides and are stored as depot fat in adipose tissues.

Lipids are a concentrated storage form of energy compared with carbohydrates. There are two reasons for this. First, fatty acids are in a more reduced state than carbohydrates; the carbon chains of fatty acids are generally saturated hydrocarbons, whereas the carbon chains of most carbohydrates are already partially oxidized to alcohols. Thus the bonds of fatty acids have more energy to release upon complete oxidation to carbon dioxide and water than the bonds of carbohydrates. This is why 1 g of fat releases 9 kcal of energy but 1 g of carbohydrate releases only 4 kcal. Second, fats are stored in a nearly anhydrous, or dry, state. In contrast, for each gram of carbohydrate stored as glycogen, nearly 2 g of water is carried along.

A reasonable amount of body fat is necessary and beneficial, but maintaining proper body weight is a problem for many people. From the metabolic point of view, dieting consists of reducing caloric intake so that stored fats must be degraded to meet energy needs. Some people try to reduce by eliminating all fatty foods from their diets. By eliminating foods that contain fats, however, one runs the risk of also eliminating essential vitamins and minerals contained in these foods. Moreover, if the decreased fat intake is more than compensated by an increased intake of carbohydrates—in starchy foods, sweets, soda, and so forth—a person not only fails to lose weight but actually gains.

Human beings convert carbohydrates to fats, but people do not convert fats to carbohydrates. Human cells have no enzyme that can catalyze the conversion of acetyl CoA to pyruvate, a compound required for gluconeogenesis. Some bacteria, however, do have such an enzyme; they convert fats to carbohydrates as part of their normal metabolism.

PRACTICE EXERCISE 25.4

A person has a basal metabolism of 1800 Calories per day. Calculate (a) the weight of fat that must be oxidized to CO_2 and H_2O to provide this amount of energy and (b) the weight of glucose that must be metabolized to release this amount of energy.

25.8 Ketone bodies

AIM: To list the three ketone bodies and the conditions that cause their production.

Focus

Liver cells produce ketone bodies when glucose is in short supply.

Under certain circumstances, body cells do not have enough glucose even for brain cells to use as an energy source. This happens most often in starvation or in untreated diabetes. In starvation, no supply of glucose is available; in diabetes, glucose is present in the blood, but it cannot penetrate cell membranes.

A lack of glucose causes the cells of many organs to step up the beta oxidation of fatty acids. However, when glucose levels are low, there is not enough oxaloacetate available to condense with acetyl CoA in the first step of the citric acid cycle. This is so because oxaloacetate comes from the carboxylation of pyruvate, and pyruvate comes from the breakdown of glucose in glycolysis. At low glucose levels, therefore, the concentration of acetyl CoA produced by the beta oxidation of fatty acids builds up. Under these conditions, the liver manufactures three special compounds from the excess acetyl CoA—the ketone bodies. Ketone bodies may be oxidized by many tissues to meet energy needs.

The **ketone bodies** *are acetoacetic acid, β-hydroxybutyric acid, and acetone.* We can see from their structural formulas that one of these compounds, β-hydroxybutyric acid, is inaccurately named as a ketone body, since it does not contain a ketone group.

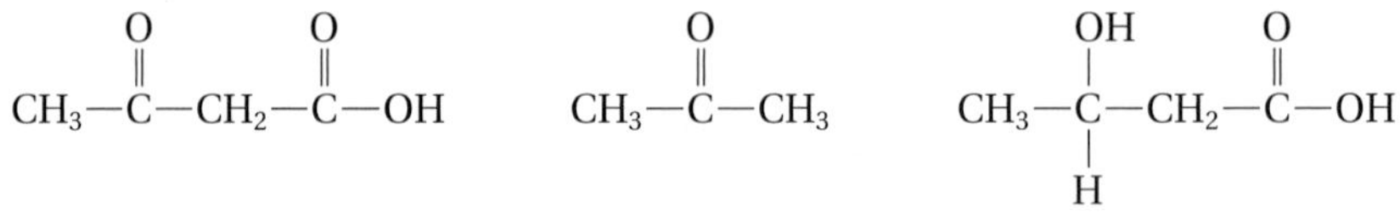

Acetoacetic acid | Acetone | *β*-Hydroxybutyric acid (not a ketone)

The liver does not use ketone bodies for energy production but releases them into the bloodstream. From the bloodstream, the ketone bodies reach other tissues—mainly the brain, the heart, and skeletal muscle.

The only ketone body that is in a form that can be used directly to produce energy is acetoacetic acid. The acetoacetic acid is converted to its thioester with CoA.

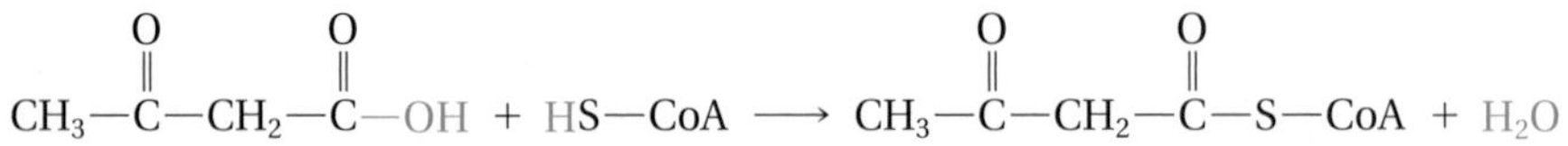

Acetoacetic acid | Acetoacetyl CoA

The thioester that is formed, acetoacetyl CoA, may look familiar. If you recall our discussion of beta oxidation, you will see that acetoacetyl CoA is the same compound that is formed at the end of the fatty acid spiral. Tissue cells can cleave the acetoacetyl CoA back to two molecules of acetyl CoA.

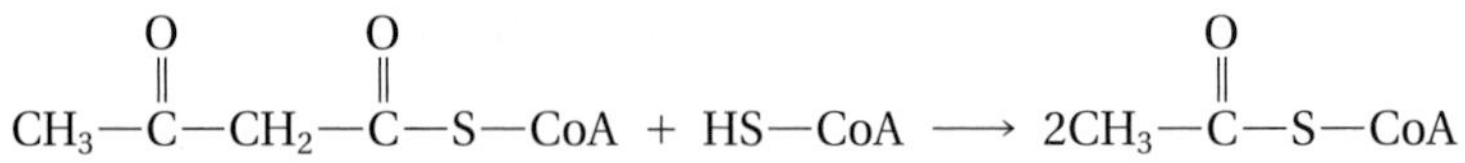

Acetoacetyl CoA

The acetyl CoA is then oxidized to carbon dioxide in the citric acid cycle, thereby providing NADH and $FADH_2$ for ATP production by cellular respiration. The thioester of the ketone body β-hydroxybutyric acid is also formed in cells, but the hydroxyl group of the acid portion of the ester must be oxidized to a ketone. Acetoacetyl CoA, useful for energy production, is formed as a result of this oxidation.

$$CH_3-\underset{\underset{H}{|}}{\overset{\overset{OH}{|}}{C}}-CH_2-\overset{\overset{O}{\|}}{C}-S-CoA \xrightarrow[\quad NAD^+ \qquad NADH + H^+ \quad]{} CH_3-\overset{\overset{O}{\|}}{C}-CH_2-\overset{\overset{O}{\|}}{C}-S-CoA$$

β-Hydroxybutyryl CoA   Acetoacetyl CoA

Acetone, the third ketone body, is not used as an energy source.

25.9 Ketosis

AIMS: To characterize the following aspects of ketosis: ketonemia, ketonuria, acetone breath, and ketoacidosis. To describe how the effects of ketosis are counteracted by mechanisms within the body and by the administration of external agents.

Focus

Prolonged ketosis starves cells for oxygen.

In normal metabolism, some ketone bodies are continuously produced and broken down in energy production. The normal blood level of ketone bodies seldom exceeds 3 mg/100 mL of blood. In diabetes, however, the liver produces large quantities of ketone bodies, releasing them into the bloodstream for delivery to other tissues. This causes a substantial increase in the level of ketone bodies in the blood of untreated diabetics. *A level of ketone bodies greater than about 20 mg/100 mL of blood is called* **ketonemia** *("ketones in the blood").*

Tissue cells cannot use all the ketone bodies produced. But the liver does not stop production, and eventually, a surplus builds up. *At a level of about 70 mg/100 mL of blood, ketone bodies are excreted in the urine. This condition is* **ketonuria** *("ketones in the urine").* At high levels of ketone bodies in the blood, acetone is excreted by the lungs. The sweet, minty smell of *acetone breath* becomes apparent.

The conditions of ketonuria, ketonemia, and acetone breath together are symptoms of **ketosis** (*also called* **ketoacidosis**)—*blood acidosis caused by an excess of the ketone body acids, acetoacetic acid, and β-hydroxybutyric acid.* Diabetic ketosis involves the same problem as respiratory acidosis and lactic acidosis. This is the problem: The ketone body acids in the blood will lower the blood pH unless enough bicarbonate ions are present to act as buffers (proton sponges).

$$\xrightarrow{\hspace{1cm}} H^+ \ + \ HCO_3^- \ \rightleftharpoons \ H_2CO_3$$

from ketone   Proton   Bicarbonate   Carbonic
body acid   ion   acid

The pH of the blood is maintained at 7.40 as long as the kidneys can regenerate new bicarbonate ions. In severe cases of diabetic ketosis, however, the kidneys cannot supply enough bicarbonate ions to keep up with the production of ketone bodies. More ketone bodies are produced, insufficient bicarbonate ions are available, and the blood pH drops. This sequence has a disastrous effect. Hemoglobin can pick up oxygen only in an environment with a low concentration of protons. The lower the pH of the blood, the higher is the concentration of protons, and the less oxygen can be transported by hemoglobin. Brain cells become starved for oxygen. If this lack of oxygen continues, coma and death will follow.

The first step in the treatment of patients with diabetes who are exhibiting ketosis is usually the administration of insulin. This should restore normal glucose metabolism and reduce the formation of ketone bodies. Like glycosuria, ketonuria results in the loss of a large volume of body water, often causing dehydration. In diabetic patients with severe dehydration and ketosis, fluids and buffering power are restored by intravenous administration of solutions containing sodium bicarbonate.

25.10 Cholesterol synthesis

AIM: To show how the synthesis of cholesterol and ketone bodies illustrates the compartmentalization of cellular processes.

Focus

The carbon skeleton of cholesterol is formed from acetyl CoA.

Cholesterol is formed in the cytoplasm of liver cells. Biochemists have shown that the 27 carbons of the carbon skeleton of this important steroid come entirely from acetyl CoA (Fig. 25.5). The total biosynthesis of 1 molecule of cholesterol uses up 15 molecules of acetyl CoA and involves at least 13 separate chemical reactions. A few of the main steps in cholesterol synthesis are shown in Figure 25.6.

The synthesis of cholesterol in the cytoplasm and the synthesis of ketone bodies in mitochondria are excellent examples of the compartmentalization of cellular processes. The cytoplasm lacks the enzymes necessary to synthesize ketone bodies from acetyl CoA; the mitochondrion lacks the enzymes necessary to synthesize cholesterol from acetyl CoA. Compartmentalization is often important in balancing the synthetic and energy needs of cells. When a cell needs energy, most of the acetyl CoA produced in its mitochondria is oxidized in the citric acid cycle. When large amounts of acetyl CoA are being oxidized, lesser amounts can be furnished to the cytoplasm. The decreased level of acetyl CoA in the cytoplasm slows down or

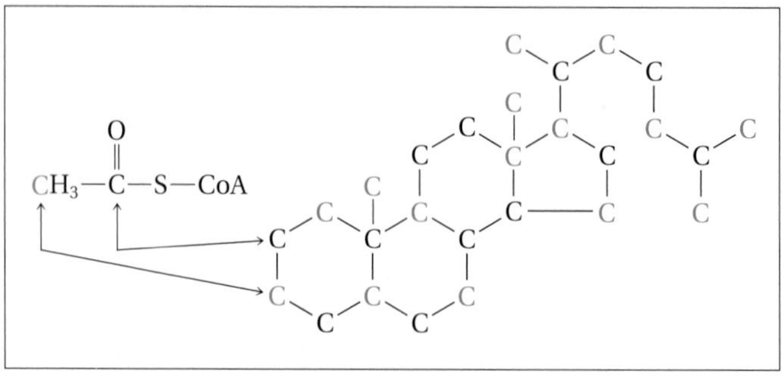

Figure 25.5
The incorporation of the acetyl carbons of acetyl CoA into the carbon skeleton of cholesterol. The carbons of cholesterol shown in color come from the methyl group of acetyl CoA. The remainder of the carbons come from the carbonyl group (−C=O).

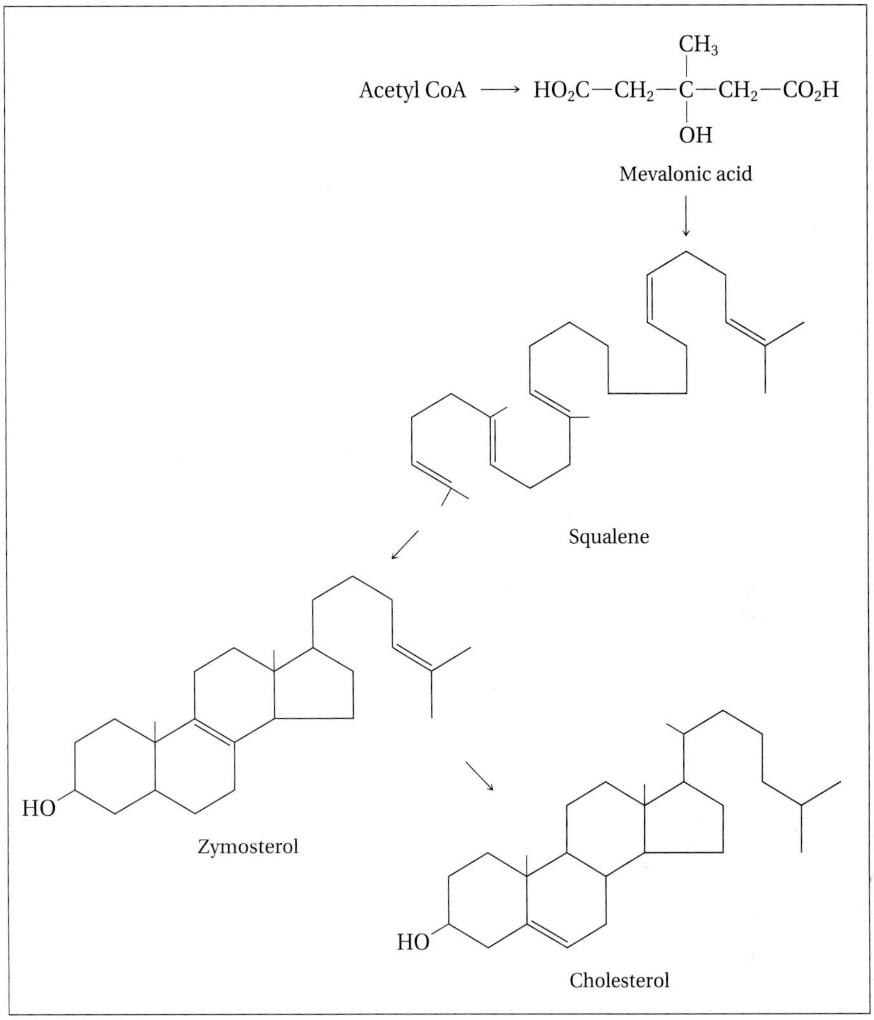

Figure 25.6
Some intermediates in the synthesis of cholesterol.

The level of cholesterol in the blood is influenced by a combination of heredity, diet, and metabolic diseases.

stops cholesterol synthesis. The synthesis of cholesterol is an anabolic (synthetic) pathway. As such, it requires the expenditure of the reducing power of NADH and the energy of ATP. The slowdown of cholesterol synthesis has the effect of conserving reducing power and energy in the cell.

25.11 Blood cholesterol

AIM: To explain how an increase in the percentage of unsaturated fatty acids in the diet could lead to lower plasma cholesterol levels.

Focus

Blood cholesterol levels are related to the incidence of atherosclerosis.

Cholesterol is the most abundant steroid in the animal body. About 90% of the cholesterol synthesized by mammals is formed in the liver and distributed to blood serum and other tissues. The serum cholesterol is bound to lipoproteins (see A Closer Look: Blood Lipoproteins and Heart Disease, page 651). The amount of cholesterol bound to a lipoprotein increases in proportion to the density of the lipoprotein. *Chylomicrons,*

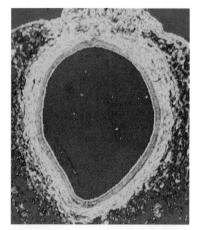

(a)

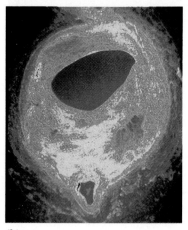

(b)

Figure 25.7
(a) A cross-section of a normal artery. (b) A cross-section of an artery constricted by the formation of atherosclerotic plaque.

Atherosclerosis
athera (Greek): gruel
skleros (Greek): hard

According to a study by the Center for Science in the Public Interest, the American Cancer Society, and the American Public Health Association, if Americans reduced their daily intake of saturated fats by just 8 grams, health care costs would go down as much as $17 billion per year.

which have the lowest density of all lipoproteins, mainly contain triacyl glycerols and a relatively small amount of cholesterol. *Very-low-density lipoproteins* (VLDLs) contain more cholesterol than is present in chylomicrons. *Low-density lipoproteins* (LDLs) contain more cholesterol than VLDLs, and *high-density lipoproteins* (HDLs) are very rich in cholesterol. Normal blood contains 120 to 200 mg of cholesterol per 100 mL of serum. Elevated levels of 200 to 300 mg of cholesterol per 100 mL of serum are usually considered a threat to health, having been associated with *atherosclerosis.*

Atherosclerosis *is a disease in which cholesterol and arterial tissue cells build up in layers, called* **plaques** (Fig. 25.7). The buildup of arterial plaque can block arteries and cause impaired blood circulation. Heart failure from blocked coronary (heart) arteries is the leading cause of death in industrialized nations. Doctors advise many people to take steps to reduce their serum cholesterol because high levels stimulate the onset of atherosclerosis.

Three major strategies for the reduction of serum cholesterol are increasing the level of HDLs, a low-saturated-fat diet, and cholesterol-lowering medications. HDLs can extract cholesterol from the blood, which tends to lower serum cholesterol levels. This explains the protection from heart attack conferred by high levels of serum HDLs mentioned in A Closer Look: Blood Lipoproteins and Heart Disease (page 651). Because of this protection, the cholesterol in HDLs is sometimes called "good cholesterol." Physically fit people tend to have higher levels of HDLs than unfit people, so exercise is often an important component of a program to reduce serum cholesterol.

A low-saturated-fat diet—one in which fat intake is mainly restricted to lipids containing unsaturated fatty acids—is sometimes helpful in reducing serum cholesterol. The biochemical basis for the low-saturated-fat diet is the relationship between the lipid composition of cell membranes and the cholesterol concentration of plasma. Cell membranes need just the right degree of rigidity to function normally, and body cells control membrane rigidity by continuously adjusting the mix of lipids they use to construct their membranes. As we learned in Chapter 17, saturated lipids and cholesterol tend to make membranes rigid; unsaturated lipids tend to make them flexible. A person on a low-saturated-fat diet consumes enough unsaturated lipids to make very flexible membranes, but the dietary supply of saturated lipids and cholesterol needed to rigidize the membranes is limited, so the dieter's body cells use plasma cholesterol instead. Consequently, the low-saturated-fat diet results in increased tissue cholesterol and a corresponding decrease in plasma cholesterol. For information on the control of serum cholesterol by medications, see A Closer Look: Drug Strategies for Reducing Serum Cholesterol.

A Closer Look

Drug Strategies for Reducing Serum Cholesterol

Exercise and a low-saturated-fat diet may not bring blood cholesterol under control for two reasons: (1) As dietary cholesterol is reduced, the body increases its synthesis to compensate for the decrease and there is no net change, and (2) the human body lacks a way to excrete cholesterol. When exercise and a low-saturated-fat diet fail to reduce serum cholesterol, a physician may prescribe medication. Several drugs, called *hypolipidemic* drugs, are available for the reduction of serum cholesterol.

The action of hypolipidemic drugs is based on two different strategies. The first strategy is to promote the excretion of cholesterol. The patient takes oral doses of a cholestyramine or similar resin. The resin is an edible but indigestible polymer that contains hydrophobic benzene rings and hydrophilic quaternary amine groups.

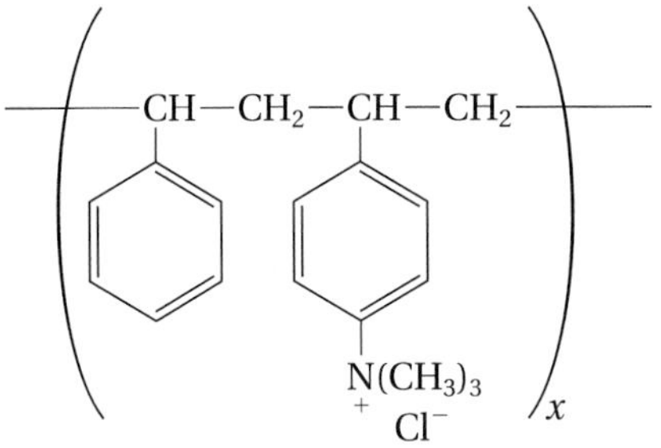

Cholestyramine resin (Questran)

Nutrition Facts	Amount/Serving	%Daily Value*	Amount/Serving	%Daily Value*
	Total Fat 13g	20%	Total Carbohydrate 37g	12%
Serving Size 1 cupcake (66g)	Saturated Fat 3g	16%	Dietary Fiber 0g	0%
Servings Per Container 6	Cholesterol 30mg	10%	Sugars 31g	
	Sodium 180mg	8%	Protein 2g	
Calories 270				
Calories from Fat 120	Vitamin A 0% • Vitamin C 0% • Calcium 2% • Iron 0%			

00378 ASSORTED CUPCAKES R3281

Cholesterol content and fat content are given on the labels of most consumable items. A reduced intake of cholesterol does not necessarily result in lowered cholesterol levels in the blood.

Ordinarily, cholesterol in the intestines is reabsorbed into the bloodstream and is not excreted. However, cholesterol binds tightly to the resin present in the intestines. The resin and its bound cholesterol are indigestible, and the resin-cholesterol complex is excreted in the feces. This loss results in lowered serum cholesterol. The second strategy aims to inhibit cholesterol synthesis. An oral drug called *lovastatin* exemplifies this strategy.

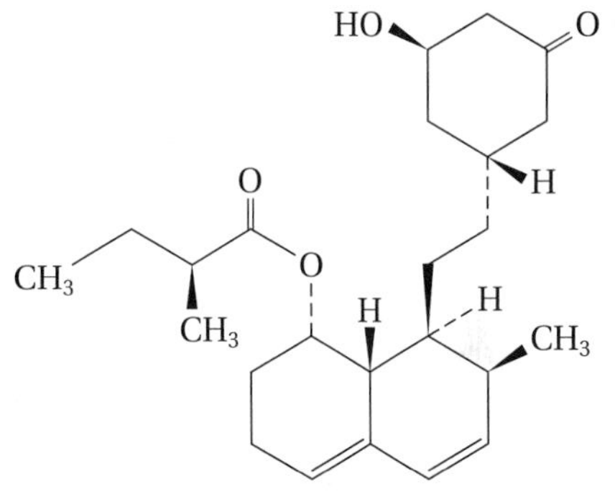

Lovastatin (Mevacor)

Lovastatin is an inhibitor of the enzyme *hydroxymethylglutaryl-CoA reductase* (HMG-CoA reductase). HMG-CoA reductase catalyzes the reduction of its substrate, hydroxymethylglutaryl-CoA (HMG-CoA), to the product mevalonate. This reaction is a necessary early step in the synthesis of cholesterol from acetyl CoA.

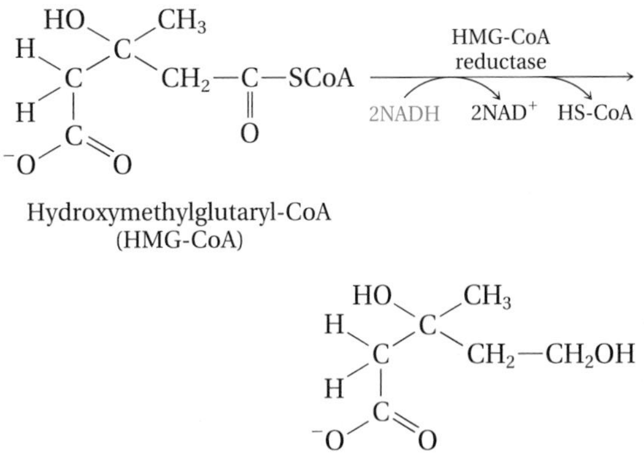

Hydroxymethylglutaryl-CoA (HMG-CoA)

Mevalonate

The inhibition of the HMG-CoA reductase reduces the amount of cholesterol synthesized, so the level of serum cholesterol decreases.

SUMMARY

Lipids are extremely important as nutrients and in cell membranes. Reserves of stored triglycerides are mobilized as needed for energy production. Fat mobilization is stimulated by epinephrine (adrenalin). The triglycerides are hydrolyzed into fatty acids and glycerol, which enter the bloodstream. In tissues, free fatty acids are converted to fatty acyl CoA molecules, which are broken down to acetyl CoA by beta oxidation in the mitochondrion. The acetyl CoA may be used for energy production by way of the citric acid cycle and the electron transport chain.

Beta oxidation of fatty acids is a good source of energy. One molecule of palmitic acid produces enough energy to phosphorylate 129 molecules of ADP to ATP. When ATP is in abundant supply, acetyl CoA is resynthesized back to fatty acids or to cholesterol. When ATP is in short supply and glucose is not available, as in starvation or diabetes, the liver uses acetyl CoA to make the ketone bodies acetone, β-hydroxybutyric acid, and acetoacetic acid; the latter

two compounds are used for energy production in the citric acid cycle.

The brain usually needs glucose, but it adapts to ketone body oxidation in about 2 days. Excess production of ketone body acids leads to ketonemia (ketone bodies in the blood). The ketone bodies eventually appear in the urine (ketonuria). Chronic overproduction of ketone body acids leads to ketone body acidosis (ketosis) of the blood because the kidneys cannot supply enough bicarbonate ions to neutralize the acids.

The liver synthesizes cholesterol from acetyl CoA. The cellular cytoplasm is the site of cholesterol synthesis. Too much cholesterol in the blood may cause the formation of atherosclerotic plaque—a buildup of layers of cholesterol and cells on arterial walls. The resulting constriction of the blood vessels causes harmful effects on circulation; a blockage in the arteries of the heart can cause heart failure. Blood cholesterol is sometimes helped by exercise, a low-saturated-fat diet, or medication.

KEY TERMS

Adipose tissue (25.1)
Atherosclerosis (25.11)
Beta oxidation (25.3)
Depot fat (25.1)

Fatty acid spiral (25.3)
Fatty acyl CoA (25.3)
Ketoacidosis (25.9)
Ketone bodies (25.8)

Ketonemia (25.9)
Ketonuria (25.9)
Ketosis (25.9)
Mobilization (25.2)

Plaque (25.11)
Storage fat (25.1)
Tissue lipid (25.1)
Working lipid (25.1)

EXERCISES

Body Lipids (Sections 25.1, 25.2)

25.5 Draw the structure of a triglyceride that contains stearic, oleic, and linoleic acids.

25.6 What are the products of hydrolysis of the triglyceride in Exercise 25.5? Name the enzyme that catalyzes the process.

25.7 Describe how lipids are distributed in body tissues.

25.8 Discuss the functions of adipose tissue.

25.9 What stimulates the hydrolysis of triglycerides?

25.10 Explain the function of serum albumin in fatty acid transport.

25.11 Which tissues prefer glucose to fatty acids for energy production?

25.12 What is the cause of lipid storage diseases?

Fatty Acid Metabolism and Storage (Sections 25.3, 25.4, 25.5, 25.6, 25.7)

25.13 Where in the cell does fatty acid catabolism occur?

25.14 How is a fatty acid prepared for entry into a mitochondrion?

25.15 What is meant by the term *beta oxidation*?

25.16 Why is the degradation pathway of fatty acids called the fatty acid *spiral* rather than the fatty acid cycle?

25.17 The fatty acyl CoA formed from the 10-carbon acid capric acid is degraded by beta oxidation to acetyl CoA. What is the yield of (a) acetyl CoA, (b) NADH, and (c) $FADH_2$?

25.18 What is the maximum number of ATP molecules that can be produced in the complete oxidation of capric acid to carbon dioxide and water? (See Table 25.1.)

25.19 Why is acetyl CoA a key compound in the interplay between fatty acid and carbohydrate metabolism?

25.20 Describe the four fates that await acetyl CoA in the liver.

25.21 How does the camel's hump permit it to survive famine and drought?

25.22 Explain why fatty acids are more efficient energy stores than glycogen.

Ketone Bodies and Diabetes (Sections 25.8, 25.9)

25.23 What is the difference between starvation and diabetes?

25.24 What are the ketone bodies?

25.25 How do ketone bodies form?

25.26 Explain why untreated diabetics accumulate large amounts of ketone bodies.

25.27 Define (a) *ketonemia,* (b) *ketonuria,* (c) *acetone breath,* and (d) *ketosis.*

25.28 Describe how untreated diabetes mellitus leads to coma and death.

Cholesterol (Sections 25.10, 25.11)

25.29 Where does cholesterol synthesis take place? What is used as "starting material"?

25.30 Explain why cholesterol synthesis slows down when a cell needs energy.

25.31 What is atherosclerosis?

25.32 Explain why an increased intake of unsaturated fats is more helpful in reducing plasma cholesterol than decreasing dietary cholesterol.

Additional Exercises

25.33 Match the following:
- (a) beta oxidation
- (b) ketonemia
- (c) fatty acid synthesis
- (d) ketonuria
- (e) mitochondria
- (f) cholesterol
- (g) epinephrine
- (h) glycerol
- (i) acetone
- (j) ketosis

- (1) ketone body
- (2) blood acidosis caused by ketone bodies
- (3) synthesized from acetyl CoA
- (4) degradation of fatty acids
- (5) ketone bodies in blood
- (6) enters glycolysis
- (7) site of beta oxidation
- (8) fatty acid mobilization
- (9) utilizes ATP
- (10) ketone bodies in urine

25.34 What is the total ATP yield from the complete oxidation of 1 mol of myristic acid, $CH_3(CH_2)_{12}CO_2H$, to carbon dioxide and water?

25.35 How are fatty acid molecules prepared for transport across the mitochondrial membrane?

25.36 Name and describe the several conditions that are symptoms of ketosis.

25.37 Acetoacetic acid and β-hydroxybutyric acid, two of the ketone bodies, are used by the brain, the heart, and skeletal muscle as a source of energy. Why do these compounds present a serious problem in diabetes or starvation?

25.38 Where in the cell does (a) fatty acid catabolism take place? (b) fatty acid activation occur?

25.39 How do low blood glucose levels trigger the mobilization of fatty acids from adipose tissue?

25.40 Explain why the mobilization of fatty acids from adipose tissue stops when blood glucose level rises.

25.41 For each statement, write the number of the correct name and the number for the correct structure.

- (a) A ketone body that is not used as an energy source
- (b) Produced by reaction of a fatty acid with CoA
- (c) A product of the hydrolysis in adipose tissue
- (d) A ketone body that is not really a ketone
- (e) Reacts with CoA to produce two molecules of substance (14)
- (f) One of the ketone bodies
- (g) Intermediate of both carbohydrate and lipid metabolism

- (1) acetoacetyl CoA
- (2) fatty acyl CoA
- (3) β-hydroxybutyric acid
- (4) acetyl CoA
- (5) acetone
- (6) acetoacetic acid
- (7) a fatty acid

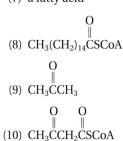

(8) $CH_3(CH_2)_{14}\overset{\overset{\displaystyle O}{\|}}{C}SCoA$

(9) $CH_3\overset{\overset{\displaystyle O}{\|}}{C}CH_3$

(10) $CH_3\overset{\overset{\displaystyle O}{\|}}{C}CH_2\overset{\overset{\displaystyle O}{\|}}{C}SCoA$

(11) $CH_3\overset{\overset{\displaystyle OH}{|}}{C}HCH_2CO_2H$

(12) $CH_3\overset{\overset{\displaystyle O}{\|}}{C}CH_2CO_2H$

(13) $CH_3(CH_2)_{14}CO_2H$

(14) $CH_3\overset{\overset{\displaystyle O}{\|}}{C}SCoA$

SELF-TEST (REVIEW)

True/False

1. Working lipids are generally localized in muscle tissue.

2. For equal weights, fats are better sources of energy than carbohydrates.

3. Epinephrine stimulates the hydrolysis of glycogen and triglycerides.

4. Once acetyl CoA is in the mitochondria, it is always used for energy production in the citric acid cycle.

5. Fatty acids can be converted to carbohydrates through an acetyl CoA intermediate.

6. The principal site of fatty acid metabolism is the liver.

7. A body with low ATP demand and excess glucose responds by making and storing fatty acid molecules.

8. The beta oxidation of a molecule of

$$CH_3(CH_2)_{16}CO_2H$$

would yield nine molecules each of acetyl CoA, NADH, and FADH$_2$.

9. Glucose metabolism has no effect on the formation of ketone bodies.

10. The rate of synthesis of cholesterol increases as the rate of acetyl CoA oxidation in the citric acid cycle increases.

Multiple Choice

11. An untreated severe diabetic would probably experience
    (a) acetone breath. (b) glucosuria.
    (c) ketonemia. (d) all of these.

12. Which of the following does not occur in the mitochondria of liver cells?
    (a) synthesis of ketones
    (b) beta oxidation
    (c) citric acid cycle
    (d) synthesis of cholesterol

13. The glycerol produced by the hydrolysis of triglycerides enters the metabolic pathway in
    (a) the citric acid cycle.
    (b) oxidative phosphorylation.
    (c) glycolysis.
    (d) the fatty acid spiral.

14. Fat tissue in the body is
    (a) an important source of body heat.
    (b) also called *adipose tissue.*
    (c) a good insulator.
    (d) all of the above.

15. In the aerobic catabolism of a molecule of a fatty acid, a majority of the energy is produced
    (a) directly as ATP in the citric acid cycle.
    (b) from NADH from the fatty acid spiral.
    (c) from NADH from the citric acid cycle.
    (d) from FADH$_2$ from the fatty acid cycle.

16. In the actual oxidation step of beta oxidation,
    (a) an alcohol is oxidized to an acid.
    (b) an alcohol is oxidized to a ketone.
    (c) an alcohol is oxidized to an aldehyde.
    (d) an aldehyde is oxidized to an acid.

17. Acetyl CoA in the liver may be converted to all the following *except*
    (a) cholesterol. (b) glycogen.
    (c) fatty acid. (d) ketone bodies.

18. Efficient energy storage is achieved by
    (a) converting fats to glycogen.
    (b) eliminating all fatty foods from the diet.
    (c) converting excess acetyl CoA into stored triglycerides.
    (d) avoiding the use of ATP in forming storage compounds.

19. Tay-Sachs disease is associated with
    (a) a defect in lipid anabolism.
    (b) the unavailability of complex lipids for use in the brain and liver.
    (c) a defect in lipid catabolism.
    (d) a defect in the process by which cholesterol is synthesized.

20. Suppose that you haven't eaten for 5 hours and you begin jogging. Which of the following statements about your system is *not* true?
    (a) Triglycerides are being hydrolyzed.
    (b) Glucose is being changed to glycogen.
    (c) Fatty acids are being broken down to acetyl CoA.
    (d) Cholesterol synthesis is minimal.

21. A deficiency of glucose in cells
    (a) increases the rate of beta oxidation of fatty acid.
    (b) can be a result of severe diabetes.
    (c) results in the formation of ketone bodies.
    (d) all of the above are true.

22. In severe diabetic ketosis,
    (a) the concentration of H$^+$ eventually drops.
    (b) oxygen transport by hemoglobin is impaired.
    (c) the pH of the blood remains unchanged.
    (d) the bicarbonate ion concentration increases.

23. Ketone bodies are not utilized as an energy source in the
    (a) brain. (b) liver.
    (c) heart. (d) skeletal muscles.

Metabolism of Nitrogen Compounds

Nitrogen and Life

Because of nitrogen fixation, legumes, such as these soybeans, can incorporate nitrogen into amino acids and eventually into important dietary proteins.

CHAPTER OUTLINE

CASE IN POINT: A painful episode

26.1 Nitrogen fixation

26.2 Protein turnover

26.3 Transamination reactions

26.4 The urea cycle

A CLOSER LOOK: Hyperammonemia

26.5 Catabolism of amino acids

26.6 Synthesis of amino acids

26.7 Defects of amino acid metabolism

A CLOSER LOOK: The Amino Acidurias

26.8 Hemoglobin and bile pigments

A CLOSER LOOK: Hyperbilirubinemia

26.9 Purines and pyrimidines

FOLLOW-UP TO THE CASE IN POINT: A painful episode

This chapter is about the metabolism of biological molecules containing nitrogen. Nitrogen compounds are so diverse—amino acids, proteins, vitamins, and nucleic acids, to name a few—that we will cover only the principal aspects of nitrogen metabolism. In this chapter we will learn how nitrogen compounds are degraded and synthesized and how the metabolism of nitrogen compounds is related to the metabolism of carbohydrates and lipids. We also will learn about the diseases that can result from defects in the metabolism of nitrogen compounds. There are many such defects because there are so many nitrogen compounds and their metabolism is so complex. The following Case in Point describes the beginning of symptoms caused by one of these defects.

CASE IN POINT: A painful episode

Ralph, a 45-year-old accountant for a large soda bottler, was not very concerned about maintaining a healthy diet. One recent night Ralph awoke from a sound sleep with an excruciating pain in the big toe of his right foot. In the morning his toe was still hurting and had become inflamed at the joint. What disease has caused Ralph's symptoms? What does his condition have to do with nitrogen metabolism, and how can it be treated? These questions will be answered in Section 26.9.

When uric acid crystals deposit in joints they cause pain, inflammation, and swelling of the affected region (magnified 150×).

26.1 Nitrogen fixation

AIM: *To name and describe the process by which atmospheric nitrogen is made available to plants and animals.*

Few plants can form nitrogen-containing compounds from nitrogen in the air, no animals can, but certain bacteria can. It is through nitrogen-fixing bacteria that atmospheric nitrogen enters the *biosphere*—the domain of living things. **Nitrogen-fixing bacteria** *are organisms that reduce atmospheric nitrogen to ammonia, a water-soluble form of nitrogen that can be used by plants and animals.*

$$N_2 + 3H_2 \longrightarrow 2NH_3$$

In soil and biological fluids, most ammonia is present as ammonium ions.

$$NH_3 \quad + \quad H^+ \quad \rightleftharpoons \quad NH_4^+$$

Ammonia   Proton   Ammonium ion

Plants and animals incorporate ammonia into nitrogen compounds such as proteins and nucleic acids. The plants and animals die and decay, aided by other bacteria. Decaying matter returns nitrogen to the soil as ammonia, nitrite ions (NO_2^-), or nitrate ions (NO_3^-). Moreover, some nitrogen gas is returned to the atmosphere. *The flow of nitrogen between the atmosphere and the Earth and its living creatures is the* **nitrogen cycle,** shown in Figure 26.1.

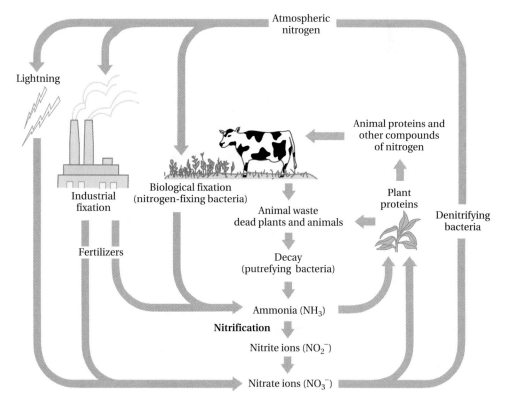

Figure 26.1
The nitrogen cycle.

It has been estimated that each year upwards of 10 million tons of nitrogen is fixed by natural and industrial processes. Some of this fixed nitrogen accumulates in soil and lakes and promotes algae growth. There are bacteria that convert fixed nitrogen back to N_2.

Modern agriculture intervenes to a great extent in the nitrogen cycle. For the past several years, the daily amount of atmospheric nitrogen fixed by industrial processes for the production of fertilizers has actually exceeded the amount fixed by living organisms in the Earth's forests and oceans. Besides bacterial and industrial nitrogen fixation, a smaller amount of atmospheric nitrogen is fixed by lightning discharges, which produce the soluble nitrogen oxides (NO, NO_2, N_2O_4, N_2O_5).

26.2 Protein turnover

AIM: To explain the function of cathepsins and state their cellular location.

Focus

Proteins are continuously hydrolyzed and synthesized by the body.

The rate of turnover varies by protein function. Plasma proteins have a half-life of about 10 days; 50% are hydrolyzed in a 10-day period. Muscle protein has a half-life of 180 days, some connective tissue proteins, as high as 1000 days.

In the human body, dietary proteins are hydrolyzed to their constituent amino acids in the small intestine as part of digestion. Many of the body's proteins are continuously hydrolyzed and synthesized within body cells. *The continuous hydrolysis and synthesis of proteins in the body is called* **protein turnover.** The enzymes responsible for the intracellular hydrolysis of proteins are a class of proteases called the *cathepsins.* In cells, the cathepsins are confined to the lysosomes, where protein degradation occurs. We discussed translation, the process of protein synthesis from genetic information, in Section 20.9.

PRACTICE EXERCISE 26.1

Why are cathepsins confined to the lysosomes in cells?

26.3 Transamination reactions

AIMS: To describe some uses of the amino acids in the amino acid pool. To use words and equations to describe transamination reactions, indicating the function of pyridoxal phosphate. To interpret the significance of increased levels of transaminases in a person's blood serum.

Focus

Amino acids part with their amino groups by transamination reactions.

Amino acids produced by digestion of dietary protein and during protein turnover in body cells become part of the body's amino acid pool. *The* **amino acid pool** *is the total quantity of free amino acids present in tissue cells, plasma, and other body fluids.* The amino acids of the amino acid pool are available for cellular needs, as shown in Figure 26.2.

Many of the reactions of amino acid metabolism require that amino acids first lose their alpha amino groups. The most common way for this to occur is by **transamination**—*the transfer of an amino group from one molecule to another.* An intermediate of the citric acid cycle, α-ketoglutaric

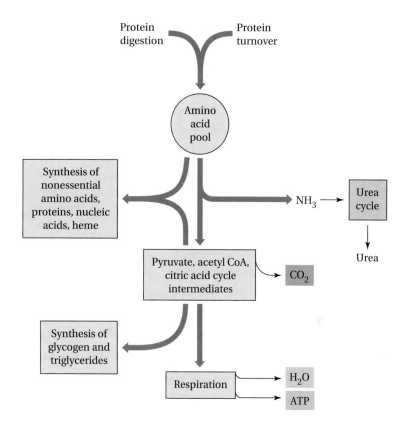

Figure 26.2
The major pathways of amino acid metabolism.

acid, is the usual acceptor of the amino group. The products of the reaction are an α-ketoacid and glutamic acid.

$$R-\underset{\underset{NH_2}{|}}{\overset{\overset{H}{|}}{C}}-\overset{\overset{O}{\|}}{C}-OH \ + \ HO-\overset{\overset{O}{\|}}{C}-CH_2CH_2-\overset{\overset{O}{\|}}{C}-\overset{\overset{O}{\|}}{C}-OH \ \rightleftharpoons \ R-\overset{\overset{O}{\|}}{C}-\overset{\overset{O}{\|}}{C}-OH \ + \ HO-\overset{\overset{O}{\|}}{C}-CH_2CH_2\underset{\underset{NH_2}{|}}{CH}-\overset{\overset{O}{\|}}{C}-OH$$

An α-amino acid   α-Ketoglutaric acid   An α-keto acid   Glutamic acid

Transamination reactions are catalyzed by enzymes called *transaminases*. The transaminases were mentioned in Table 19.1 as indicators of disease or trauma that affects tissues. The principal transaminase of the liver is *glutamic-pyruvic transaminase* (GPT), an enzyme that catalyzes the formation of pyruvate from alanine.

$$CH_3-\underset{\underset{NH_2}{|}}{\overset{\overset{H}{|}}{C}}-\overset{\overset{O}{\|}}{C}-OH \ + \ \alpha\text{-Ketoglutaric acid} \ \underset{}{\overset{GPT}{\rightleftharpoons}} \ CH_3\overset{\overset{O}{\|}}{C}-\overset{\overset{O}{\|}}{C}-OH \ + \ \text{Glutamic acid}$$

Alanine   Pyruvic acid

Liver disease causes an increase of blood serum GPT (SGPT). Elevated levels of the enzyme in serum can be used as a diagnostic test for liver disease. The principal transaminase of the heart muscle is *glutamic-oxaloacetic transaminase* (GOT), which catalyzes the formation of oxaloacetate, one of the intermediates of the citric acid cycle, from aspartic acid.

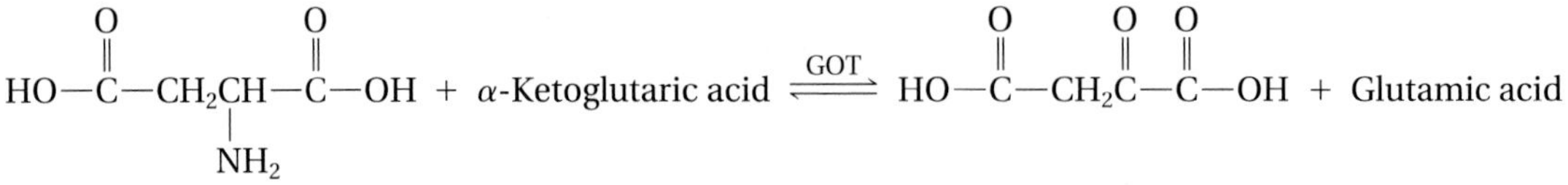

$$\underset{\text{Aspartic acid}}{\underset{\quad\;\;NH_2}{HO-\overset{O}{\overset{\|}{C}}-CH_2\overset{\qquad}{\underset{|}{CH}}-\overset{O}{\overset{\|}{C}}-OH}} + \alpha\text{-Ketoglutaric acid} \rightleftharpoons[\;]{GOT} \underset{\text{Oxaloacetic acid}}{HO-\overset{O}{\overset{\|}{C}}-CH_2\overset{O}{\overset{\|}{C}}-\overset{O}{\overset{\|}{C}}-OH} + \text{Glutamic acid}$$

Damaged heart cells die and split open, but some of the GOT molecules that spill into the blood are still active. The levels of serum GOT (SGOT) activity are a measure of the extent of damage to heart muscle caused by a heart attack.

All known transaminases require pyridoxal phosphate, a derivative of pyridoxol (vitamin B_6), as the cofactor.

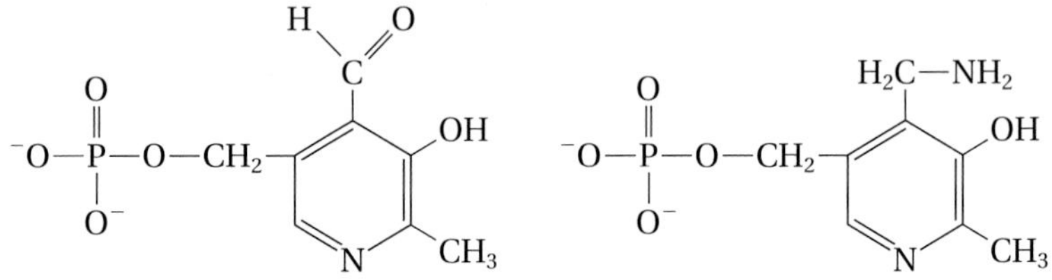

Pyridoxal phosphate   Pyridoxamine phosphate

The aldehyde carbon of pyridoxal phosphate is the acceptor of the amino group of amino acids in the first stage of reactions catalyzed by transaminases. An α-ketoacid and pyridoxamine phosphate are the products of the reaction.

Amino acid + Pyridoxal phosphate $\longrightarrow$ α-Ketoacid + Pyridoxamine phosphate

The second stage of transamination consists of the transfer of the amino group of the pyridoxamine phosphate to α-ketoglutarate with regeneration of the pyridoxal phosphate.

α-Ketoglutaric acid + Pyridoxamine phosphate $\longrightarrow$ Glutamic acid + Pyridoxal phosphate

PRACTICE EXERCISE 26.2

Write an equation for the transamination of tyrosine with α-ketoglutarate as the acceptor.

PRACTICE EXERCISE 26.3

How does an amino acid lose its amino group in transamination?

26.4 The urea cycle

AIMS: To explain the net results of the urea cycle on a molecular and a physiologic scale. To list two causes of a negative nitrogen balance.

Focus

Ammonia produced from the oxidation of glutamic acid is excreted as urea.

Glutamic acid serves as the depot for receiving amino groups removed from amino acids by transamination reactions. Since there is a limited quantity of α-ketoglutarate in cells, it must be regenerated so that transamination reactions and the citric acid cycle can continue. The α-

ketoglutarate is regenerated from glutamic acid by an oxidation reaction catalyzed by glutamate dehydrogenase. This enzyme uses NAD^+ as the coenzyme. *The removal of the amino group by oxidation is called* **oxidative deamination.**

$$HO-\overset{\overset{O}{\|}}{C}-CH_2CH_2\overset{\overset{H}{\overset{|}{}}}{\underset{NH_2}{C}}-\overset{\overset{O}{\|}}{C}-OH + H_2O \xrightarrow[NAD^+ \qquad NADH]{\text{Glutamate dehydrogenase}} NH_4^+ + HO-\overset{\overset{O}{\|}}{C}-CH_2CH_2\overset{\overset{O}{\|}}{C}-\overset{\overset{O}{\|}}{C}-OH$$

Glutamic acid   Ammonium   α-Ketoglutarate
                                              ion

With the regeneration of α-ketoglutarate, the goal of the cell has been met—the cellular concentration of α-ketoglutarate is restored. Oxidative deamination also has produced ammonia, however. And ammonia, even in low concentrations, is especially toxic to brain cells and can result in coma and death. A Closer Look: Hyperammonemia examines some reasons for this toxicity. It is important that ammonia be removed from the body. This is accomplished by the reactions of the *urea cycle,* shown in Figure 26.3.

The **urea cycle** *is a metabolic pathway by which land-dwelling animals prepare waste ammonia and ammonium ions for excretion.* The excretion of nitrogen as urea helps to conserve body water and prevent dehydration. If ammonia were excreted directly in the urine, the ammonia would have to be kept very dilute. The resulting loss of large volumes of body water would make dehydration an ever-present danger. Urea, on the other hand, is very soluble in water and is nontoxic except at high concentrations. Urea formation permits land-dwellers to eliminate large quantities of nitrogen while losing relatively little water as urine. Other animals, primarily the fishes,

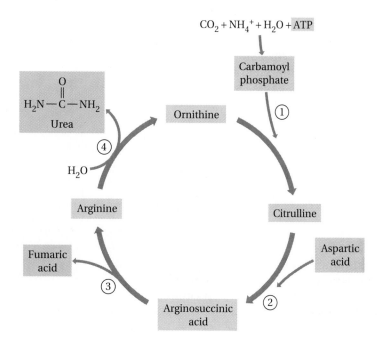

Figure 26.3
The urea cycle.

excrete ammonia as a waste product. Because they live in a water environment, fishes are not in danger of dehydration.

> **PRACTICE EXERCISE 26.4**
> What toxic substance is produced when α-ketoglutarate is regenerated from glutamic acid? How is this substance removed from the body?

Steps of the urea cycle

Let's examine the chemical reactions of the urea cycle in more detail. Before ammonia can enter the urea cycle, it is converted to a more energetic form: the compound carbamoyl phosphate. The formation of carbamoyl phos-

---

A Closer Look

Hyperammonemia

Hyperammonemia is an increase in ammonia in the blood to higher concentrations than the normal 30 to 60 mM. This condition is usually caused by the inability of patients to form urea fast enough to keep ammonia in the blood at normal levels. Several conditions can lead to hyperammonemia. Some newborns exhibit a delay in the development of urea cycle enzymes. The hyperammonemia in these infants clears up when the enzymes begin to function. In adults, causes of hyperammonemia include hereditary defects in the enzymes of the urea cycle and blockage of the blood flow through the liver, as might occur in cirrhosis (see figure).

Coma (an unconscious state) may occur in cases of hyperammonemia that are caused by liver (hepatic) disease. The reason for hepatic coma is uncertain, but it may be due to the depletion of ATP in the brain. To help us understand how ammonia may contribute to ATP depletion, let's review some aspects of metabolism. You may recall that the citric acid cycle is an important source of reducing power in the form of NADH and FADH$_2$ for oxidative phosphorylation and that oxidative phosphorylation is responsible for the production of most of the ATP needed to provide the energy to keep cells alive and healthy. In hyperammonemia, ammonia may cross the blood-brain barrier and enter brain cells. In the brain cells, the excess ammonia could drive the reaction between glutamate and α-ketoglutarate in the direction of the formation of glutamate. The shift to glutamate would reduce the amount of α-ketoglutarate available for the citric acid cycle. As a result, the concentrations of NADH and FADH$_2$ also would decrease, less ATP would be formed in oxidative phosphorylation, and hepatic coma could result from the depletion of ATP. Some cases of hyperammonemia can be treated by the addition of α-ketoacids to the diet of the hyperammonemic patient. This diverts some of the excess ammonia into amino acids through transamination of the α-ketoacids. The amino acids are used by body cells to make proteins instead of having the ammonia contribute to the patient's hyperammonemia.

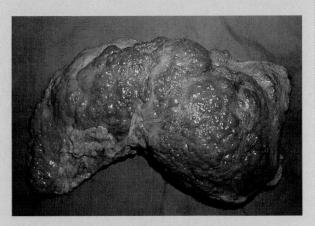

Cirrhosis of the liver impairs the urea cycle and can cause hyperammonemia.

phate requires, in addition to ammonia, one bicarbonate ion and the expenditure of two molecules of ATP.

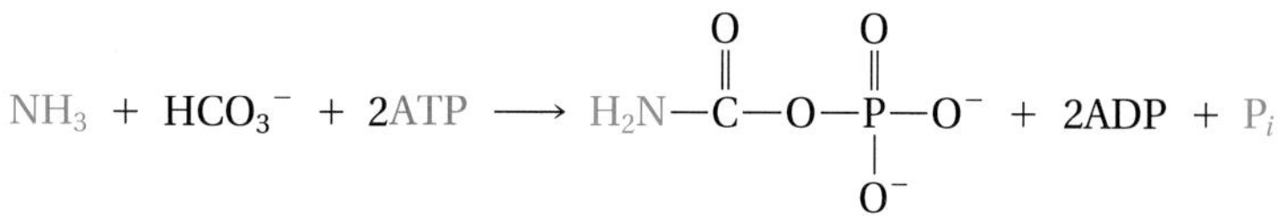

$$NH_3 + HCO_3^- + 2ATP \longrightarrow H_2N-\overset{\overset{\displaystyle O}{\|}}{C}-O-\overset{\overset{\displaystyle O}{\|}}{\underset{\underset{\displaystyle O^-}{|}}{P}}-O^- + 2ADP + P_i$$

Carbamoyl
phosphate

With the formation of carbamoyl phosphate, ammonia has been energetically primed for entry into the urea cycle. The cycle can be summarized in four steps:

Step 1: Carbamoyl phosphate reacts with ornithine (see Sec. 23.7). Ornithine, an amino acid not found in proteins, is already present in the cell. The product of this reaction is citrulline.

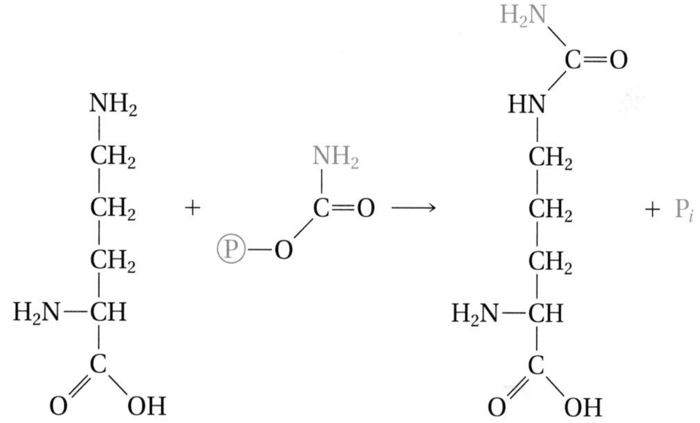

Ornithine   Carbamoyl   Citrulline
                   phosphate

Step 2: Citrulline reacts with aspartic acid to give a complex molecule called *arginosuccinic acid.* This reaction requires the expenditure of two more high-energy phosphate bonds of ATP. You need not memorize such complex molecular structures, but you should be aware of the path that nitrogen follows through the cycle.

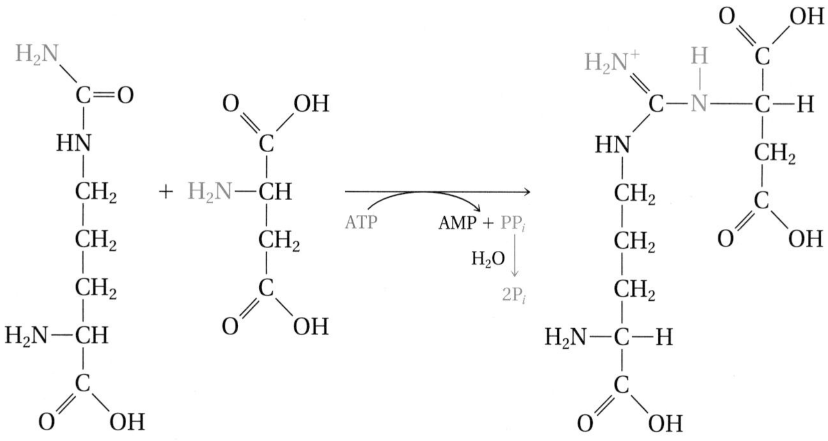

Citrulline   Aspartic   Arginosuccinic acid
                    acid

Keep in mind that both the new nitrogens that have been added to ornithine can originate as amino groups of amino acids: The first nitrogen can come from ammonia produced by the deamination of an amino acid and used to form carbamoyl phosphate; the second comes from succinic acid.

Step 3: An enzyme now cleaves arginosuccinic acid into arginine and fumaric acid. Fumaric acid is one of the intermediates of the citric acid cycle.

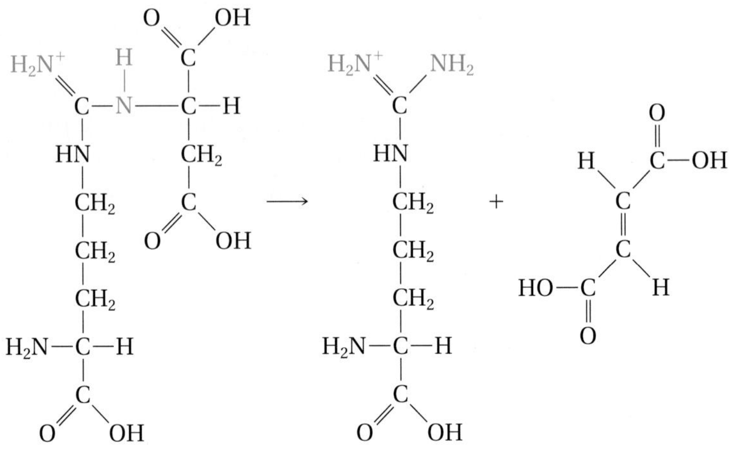

Arginosuccinic acid   Arginine   Fumaric acid

We have followed the synthesis of the nonessential amino acid arginine.

Step 4: The production of urea occurs when arginine is hydrolyzed to ornithine and urea.

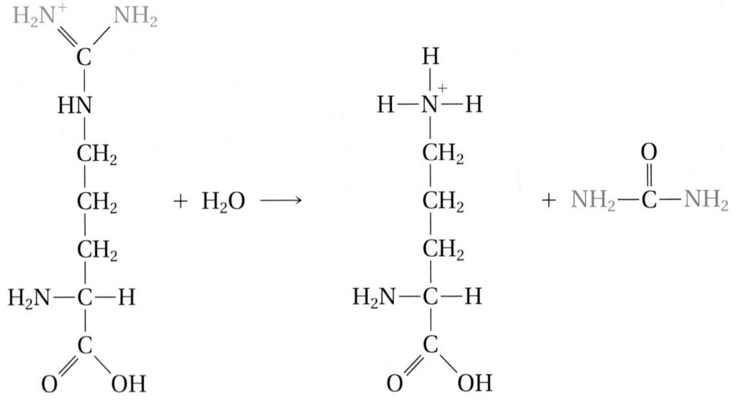

Arginine   Ornithine (as its ammonium ion)   Urea

The ornithine reenters the urea cycle by reacting with another molecule of carbamoyl phosphate. After entering the bloodstream, urea is filtered by the kidneys and disposed of in the urine.

Nitrogen balance

Normal adults generally maintain a **nitrogen balance**—*the quantity of nitrogen excreted daily equals the intake*. About 80% of the nitrogen excreted in the urine is in the form of urea, and almost all the body's waste nitrogen of metabolism is excreted in the urine; the nitrogen compounds in the feces come mainly from indigestible materials. Children have a **positive nitrogen balance**—*the excretion of less nitrogen than is consumed.* The nitrogen balance is positive because children are growing and their cells are making new proteins and other nitrogen compounds. Several conditions result in a **negative nitrogen balance**—*the excretion of more nitrogen than is consumed*. During starvation and certain diseases, the carbon skeleton of amino acids derived from the breakdown of muscle proteins must be catabolized as an energy source. Since no new protein is available to eat, starving people excrete more nitrogen than they consume. The lack of even one essential amino acid in the diet results in a negative nitrogen balance. With an essential amino acid missing, the other amino acids cannot be used to make complete proteins. These other amino acids are deaminated, and the nitrogen is excreted as urea.

26.5 Catabolism of amino acids

AIMS: To distinguish between glucogenic and ketogenic amino acids. To discuss how amino acids can be used for energy production, gluconeogenesis, and the synthesis of fats.

Focus

The carbon skeletons of amino acids are converted to intermediates of glucose metabolism and fatty acid metabolism.

Once α-ketoacids have been formed from amino acid by transamination reactions, their carbon skeletons are subjected to further chemical changes. One set of amino acids is converted to pyruvate, oxaloacetate, or α-ketoglutarate (Fig. 26.4). *Amino acids that are converted to these intermediates are called* **glucogenic,** *since these compounds are also important to glucose metabolism.* Pyruvate, formed at the end of glycolysis, and oxaloacetate are intermediates of the citric acid cycle. The remainder of the amino acids are converted to acetyl CoA, which is also a product of fatty acid metabolism. *The amino acids that are converted to acetyl CoA are called* **ketogenic.**

The conversion of all the amino acids to intermediates of glucose or fatty acid metabolism demonstrates the highly organized character of metabolism and the economy of nature. By using a single, central pathway for the metabolism of sugars, fats, and amino acids, the cell greatly decreases the number of enzymes and chemical steps that otherwise might be required to accomplish the same task.

Cells have priorities for the utilization of amino acids present in the amino acid pool. In a normal individual, well nourished with carbohydrates, fats, and proteins, the synthesis of nonessential amino acids, proteins, and other nitrogen-containing compounds is at the top of the prior-

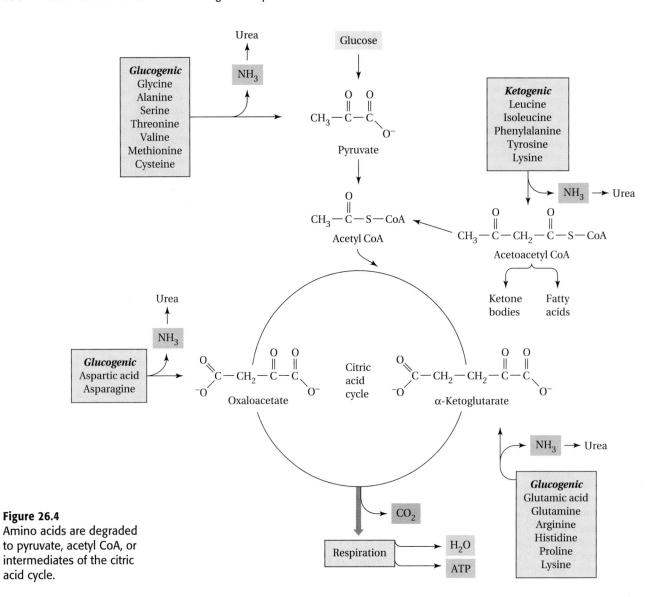

Figure 26.4
Amino acids are degraded to pyruvate, acetyl CoA, or intermediates of the citric acid cycle.

ity list. Energy production comes second. The body does not store amino acids as such. Any amino acids that remain after synthetic and energy needs are met are converted to glucose and fatty acids.

PRACTICE EXERCISE 26.5

Explain how transamination provides the link between amino acid metabolism and glucose metabolism or fat metabolism.

Energy production

The carbons of amino acids are used for energy production when needed. Upon demand, pyruvate and acetyl CoA derived from amino acids are oxidized to carbon dioxide in the citric acid cycle. Moreover, by forming oxaloacetate and α-ketoglutarate from glucogenic amino acids, cells can replenish or increase the concentrations of intermediates of the citric acid cycle. An increase in these intermediates enables cells to step up energy production. You may recall that oxaloacetate for the citric acid cycle can come from several places in metabolism. We have seen that the carboxylation of pyruvate in gluconeogenesis is one of those places. Now we see that amino acid metabolism is another.

Certain emergencies such as diabetes or starvation result in a reduction in the amount of acetyl CoA in the liver. Liver cells respond by using acetyl CoA produced in amino acid metabolism to make ketone bodies. The ketone bodies are transported to other tissues, where they are oxidized for energy production.

Synthesis of glycogen and triglycerides

From our study of glucose metabolism, we know that glucose is formed from pyruvate by gluconeogenesis, in which oxaloacetate is an intermediate (see Sec. 24.8). The cell does not recognize whether the pyruvate has come from glucose or from amino acid metabolism. Once glucose has been synthesized, it can be assembled into glycogen and stored in muscle or liver cells. Oxaloacetate from amino acid metabolism also can be converted to glucose in gluconeogenesis.

Fatty acids, as we have seen, are synthesized from acetyl CoA. The acetyl CoA can come from glucose metabolism, from fatty acid metabolism, or from amino acid metabolism. Newly synthesized fatty acids are either used immediately for energy production or converted to triglycerides or membrane lipids. The triglycerides are stored in adipose tissue as an energy reserve. Humans cannot synthesize glucose from acetyl CoA, since people lack the enzyme that converts acetyl CoA to pyruvate.

26.6 Synthesis of amino acids

AIM: To show the relationship between the citric acid cycle and the synthesis of nonessential amino acids.

Most animal proteins have a higher nutritional value than vegetable proteins because they have more essential amino acids.

Our bodies need to synthesize nonessential amino acids (see Sec. 21.6) because their proportions in our diet seldom match our bodies' needs. The main starting materials for this synthesis are pyruvate and two intermediates of the citric acid cycle: α-ketoglutarate and oxaloacetate. As we have seen, α-ketoglutarate accepts amino groups from other amino acids in transamination to give *glutamic acid,* and *arginine* is formed in the urea cycle. Two other nonessential amino acids—*aspartic acid* and *alanine*—may be synthesized directly from α-ketoacids because the reactions catalyzed by the transaminases are reversible. Reversals of transamination reactions form alanine from pyruvic acid and aspartic acid from oxaloacetic acid.

$$CH_3-\overset{\displaystyle O}{\underset{\displaystyle \|}{C}}-\overset{\displaystyle O}{\underset{\displaystyle \|}{C}}-OH + \text{Glutamic acid} \rightleftharpoons CH_3\underset{\displaystyle \underset{\displaystyle NH_2}{|}}{CH}-\overset{\displaystyle O}{\underset{\displaystyle \|}{C}}-OH + \alpha\text{-Ketoglutarate}$$

Pyruvic acid   Alanine

$$HO-\overset{\displaystyle O}{\underset{\displaystyle \|}{C}}-CH_2-\overset{\displaystyle O}{\underset{\displaystyle \|}{C}}-\overset{\displaystyle O}{\underset{\displaystyle \|}{C}}-OH + \text{Glutamic acid} \rightleftharpoons HO-\overset{\displaystyle O}{\underset{\displaystyle \|}{C}}-CH_2\underset{\displaystyle \underset{\displaystyle NH_2}{|}}{CH}-\overset{\displaystyle O}{\underset{\displaystyle \|}{C}}-OH + \alpha\text{-Ketoglutarate}$$

Oxaloacetic acid   Aspartic acid

Glutamine and *asparagine* are formed from glutamic acid and aspartic acid by reaction of the side-chain carboxyl groups with ammonia.

$$HO-\overset{\displaystyle O}{\underset{\displaystyle \|}{C}}-CH_2\underset{\displaystyle \underset{\displaystyle NH_2}{|}}{CH}-\overset{\displaystyle O}{\underset{\displaystyle \|}{C}}-OH + NH_3 \longrightarrow H_2N-\overset{\displaystyle O}{\underset{\displaystyle \|}{C}}-CH_2\underset{\displaystyle \underset{\displaystyle NH_2}{|}}{CH}-\overset{\displaystyle O}{\underset{\displaystyle \|}{C}}-OH + H_2O$$

Aspartic acid   Asparagine

$$HO-\overset{\displaystyle O}{\underset{\displaystyle \|}{C}}-CH_2CH_2\underset{\displaystyle \underset{\displaystyle NH_2}{|}}{CH}-\overset{\displaystyle O}{\underset{\displaystyle \|}{C}}-OH + NH_3 \longrightarrow H_2N-\overset{\displaystyle O}{\underset{\displaystyle \|}{C}}-CH_2CH_2\underset{\displaystyle \underset{\displaystyle NH_2}{|}}{CH}-\overset{\displaystyle O}{\underset{\displaystyle \|}{C}}-OH + H_2O$$

Glutamic acid   Glutamine

Tyrosine, the only nonessential amino acid with an aromatic side chain, is produced from the essential amino acid phenylalanine. The conversion requires a single oxidation step catalyzed by the enzyme phenylalanine hydroxylase.

Phenylalanine   Tyrosine

So far we have seen how 7 of the 12 nonessential amino acids are synthesized. The syntheses of the remaining 5—cysteine, histidine, glycine, proline, and serine—are more complex. Their synthesis will not be considered here. The syntheses of 4 of the nonessential amino acids from oxaloacetate and α-ketoglutarate, two intermediates in the citric acid cycle, demonstrate an important metabolic principle: Besides being a pathway of catabolism, the citric acid cycle is a pathway of anabolism—a metabolic switch-hitter.

26.7 Defects of amino acid metabolism

AIM: To describe the cause, effects, and treatment of phenylketonuria.

Focus

There are many known defects of amino acid degradation.

Since there are a large number of amino acids, the possibilities for diseases related to amino acid metabolism are also great. Many of the diseases caused by defects of amino acid metabolism are rare. The most common are the **amino acidurias**—*conditions in which amino acids or related compounds are excreted in large quantities in the urine.* One example of an amino aciduria is phenylketonuria, which occurs about once for every 10,000 births.

Phenylketonuria (PKU) *is the result of an inborn error of metabolism in which phenylalanine hydroxylase, the enzyme responsible for the conversion of phenylalanine to tyrosine, is inactive.* This conversion is necessary for the complete catabolism of the benzene ring of phenylalanine. Since phenylalanine cannot be degraded further without being converted to tyrosine, the levels of phenylalanine and its deamination product, phenylpyruvic acid, build up until they are excreted in the urine in large quantities.

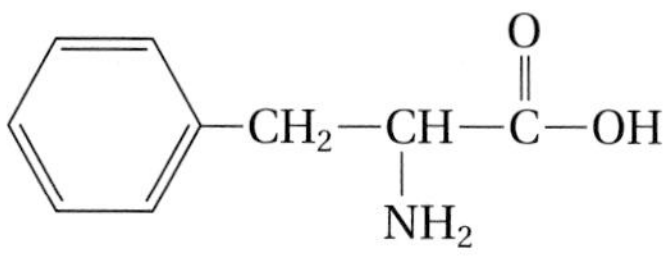

Phenylalanine

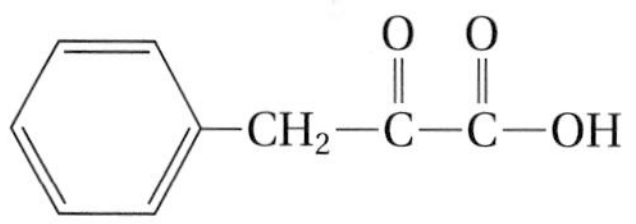

Phenylpyruvic acid

In the inherited disease iminoglycinuria, the urine contains excessive amounts of proline, hydroxyproline, and glycine.

Left untreated, phenylketonuria results in severe mental retardation by age 6 months. Why retardation occurs is not known, but it can be prevented or greatly alleviated by feeding the newborn infant a diet low in phenylalanine. (Phenylalanine cannot be eliminated from the diet entirely because it is an essential amino acid.) Since the retardation due to phenylketonuria can be prevented but not reversed, many states require the routine screening of urine samples of all newborn infants for signs of the disease. One screening test is very simple. Addition of a few drops of a dilute solution of ferric chloride to a small urine sample gives an olive-green color if the baby has phenylketonuria. The color results from a complex formed by ferric ions and phenylpyruvic acid. A Closer Look: The Amino Acidurias examines other defects of amino acid metabolism.

26.8 Hemoglobin and bile pigments

AIMS: To trace the degradation of hemoglobin, indicating two important results of the process. To list some possible causes of jaundice.

Focus

Heme groups of hemoglobin are degraded to the bile pigment bilirubin.

Hemoglobin is synthesized by means of many complex reactions that take place in immature red blood cells. In the mature red blood cells, the hemoglobin does its work by transporting oxygen to body tissues. The life span of

The Amino Acidurias

Phenylketonuria is only one of many amino acidurias that are recorded in the medical literature (see figure). All are the result of inborn errors of metabolism. *Alkaptonuria* is a disease related to the inability to break down phenylalanine, although the error in the catabolism of the phenylalanine is different from that in phenylketonuria. In phenylketonuria, as we have mentioned, the body lacks the enzyme necessary to put a phenolic hydroxyl group on the benzene ring of phenylalanine in order to make tyrosine. Alkaptonuria is a disease of tyrosine metabolism. In alkaptonuria, the conversion of phenylalanine to tyrosine occurs, but an enzyme necessary to degrade the benzene ring of tyrosine is lacking or defective. Ingested tyrosine in the body is incompletely converted into its normal degradation products, and the partially degraded products are excreted in the urine. These products are colorless when excreted but soon form a dark red pigment when exposed to air. The urine of people who have alkaptonuria may be colored from wine red to black depending on the concentration of the pigment. The pigment also forms in the bones, connective tissue, and organs of alkaptonuric patients. This deposition is thought to be the cause of arthritis that develops in many individuals with alkaptonuria. Except for the discomfort of arthritis, many alkaptonuric individuals have lived long and reasonably healthy lives.

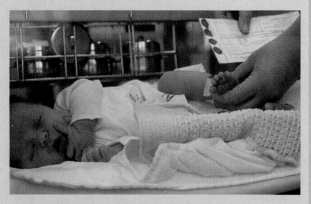

The blood of newborn infants is routinely screened for phenylketonuria (PKU).

An interesting group of amino acidurias is the family of diseases grouped as the *maple syrup urine diseases.* Many genetic diseases are found mainly in families and small groups that are highly intermarried. Maple syrup urine disease is a thousand times more common among the Old Order Mennonites of Pennsylvania than in the general population. The genetic flaws in the maple syrup urine diseases are in the catabolism of the branched-chain amino acids such as leucine and valine. The excretion of the products of incomplete breakdown of branched-chain amino acids imparts the odor of maple syrup to the urine. The product or products that give the urine this characteristic odor are unknown. Many individuals with maple syrup urine disease are mentally retarded and generally have short life spans.

the average red blood cell is about 120 days. The spleen—an organ of the lymphatic system—filters out cells that have reached the end of their useful lives. In the spleen, red blood cell membranes are broken down and hemoglobin spills out. Globin, the protein portion of hemoglobin, is hydrolyzed to its individual amino acids.

Degradation of heme begins with the removal of a single carbon from the heme ring by an oxidation reaction (Fig. 26.5). *The product of oxidation of heme is* **biliverdin,** *a green pigment. The iron released when heme is oxidized to biliverdin is retained in the body as a complex with the protein* **ferritin.** *Reduction converts biliverdin to* **bilirubin,** *an orange-red pigment.* Bilirubin enters the circulation and is transported to the liver as a complex with serum albumin. From the liver it moves to the gallbladder, where it is

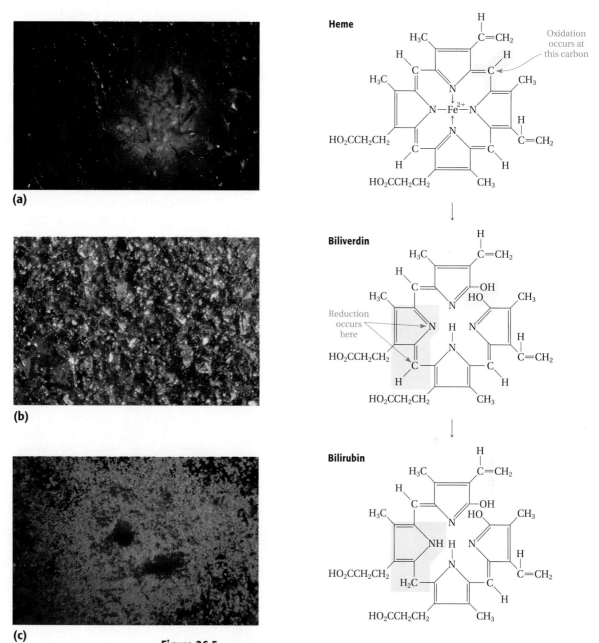

Figure 26.5
Hemoglobin (a) releases heme which is degraded to the bile pigments biliverdin (b) and bilirubin (c).

stored as part of the bile and eventually excreted into the small intestine along with the bile salts.

An excess of bilirubin in circulating blood is responsible for **jaundice**—*a yellow color of the skin and the whites of the eyes.* Jaundice is caused by any of a number of malfunctions of the bile production and storage system (see A Closer Look: Hyperbilirubinemia). If the bile duct is obstructed, the bile enters the circulation rather than the small intestine. The obstruction may be due to gallstones, which often consist of nearly pure cholesterol. Why the cholesterol forms these hard, insoluble lumps in the gallbladders of some people and not in others is still not understood. *In certain diseases such as* **infectious hepatitis,** *the liver cannot remove bile pigments as they are formed, and they enter the circulation.* **Hemolytic jaundice** *occurs when breakdown of heme groups by the spleen is faster than the liver can remove the bile that is produced.*

Bilirubin that passes from the small intestine into the large intestine is oxidized to colorless **urobilinogen** *by bacteria residing there. The urobilinogen oxidizes in air to orange-yellow* **urobilin.** The excreted urobilin gives feces their color. The yellow color of urine is the result of a small amount of urobilin filtered from the bloodstream by the kidneys.

A Closer Look

Hyperbilirubinemia

Before birth, biliverdin (from the breakdown of fetal hemoglobin) is converted to bilirubin and crosses the placenta into the mother's liver. It is then secreted in her bile. At birth an infant's liver must take over this vital function; to do this, the liver cells must be mature. Usually the liver becomes fully functional within the first week after birth. But if it does not, high serum bilirubin levels accumulated in the blood (hyperbilirubinemia) and skin (jaundice) cause the infant's skin to turn yellow. If the maturation period of the liver is prolonged, bilirubin may start to accumulate in brain tissue. Left untreated, this condition can lead to cerebral palsy, brain damage, and death.

Phototherapy (see A Closer Look: Phototherapies, on page 706) is used to treat hyperbilirubinemia. When bilirubin is exposed to white fluorescent light or sunlight, it is converted to *photobilirubin.* Therefore, if an infant suffering from hyperbilirubinemia is exposed to fluorescent light, some of the bilirubin in blood flowing near the skin is converted to photobilirubin. Photobilirubin is more water-soluble than bilirubin because of a slight difference in molecular structure. Because this photoproduct is more soluble in water than in fatty tissues, it leaves the skin and enters the blood circulation. From the blood it passes to the liver and is readily secreted in bile and excreted in the urine and feces.

During phototherapy, infants may be exposed to florescent light (see figure) for periods of 8 to 10 hours daily for a week or until the liver cells reach maturity. The infant's eyes are covered to prevent possible damage. He or she is fed intravenously to minimize dehydration and turned frequently; only the skin exposed to light loses its yellow color as the bilirubin is changed to photobilirubin and excreted.

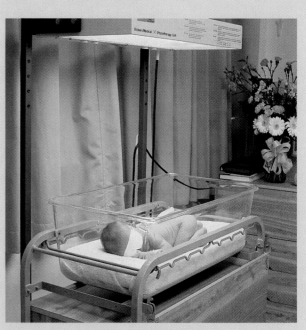

Infant undergoing phototherapy for hyperbilirubinemia.

26.9 Purines and pyrimidines

***AIMS:** To show the link between amino acids and nucleic acids in the body. To name the product of purine catabolism and name the disease that results from excessive concentrations of this substance in the blood.*

Focus

Many of the ring atoms of pyrimidines and purines come from amino acids.

Patients with high serum uric acid concentrations may have to take drugs for life to prevent development of hypertension or kidney disease. The drugs act either by preventing the formation of uric acid or by increasing its rate of excretion by the kidneys.

Organisms need to synthesize purine and pyrimidine bases for incorporation into the nucleic acids RNA and DNA. Moreover, nucleosides such as adenosine are found as part of ATP, cyclic AMP, CoA (coenzyme A), NAD^+ (nicotinamide adenine dinucleotide), and FAD (flavin adenine dinucleotide). The atoms that constitute both pyrimidine and purine ring systems come from amino acids, ammonia, and carbon dioxide, as shown for uracil and adenine in Figure 26.6.

The pathway of purine degradation is shown in Figure 26.7. The end product of this pathway is uric acid. Uric acid does not have a carboxyl group, but it is an acidic compound because one of the hydrogens on its five-membered ring readily dissociates. Human beings normally excrete uric acid as a minor waste product in the urine. For reptiles and birds, uric acid is the major form of excreted waste nitrogen. The white part of the deposits that pigeons leave on statues in the park is almost pure uric acid.

Some people, almost always male, make too much uric acid or fail to excrete it. Uric acid is quite insoluble, and it also readily forms an insoluble sodium salt, sodium urate. *In **gout,** uric acid or sodium urate exceeds its solubility in plasma and forms needle-like crystals that are deposited in joints, especially in the big toe.* Gout attacks, which result in swelling and inflammation of the affected joint, are agonizingly painful. Ralph, the accountant in the Case in Point earlier in this chapter, has gout.

Lesch-Nyhan syndrome, *a genetic disease, is caused by the lack of an enzyme needed to catalyze the synthesis of purines.* At about 2 or 3 years of

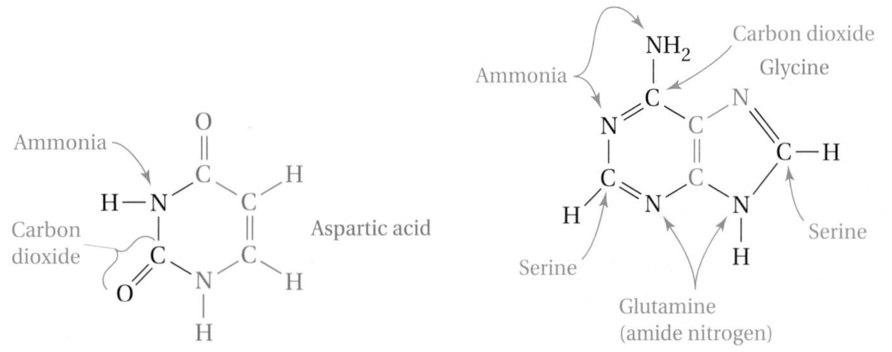

(a) Uracil   (b) Adenine

Figure 26.6
The atoms that comprise pyrimidines and purines come from amino acids. The figure shows the pyrimidine uracil (a) and the purine adenine (b).

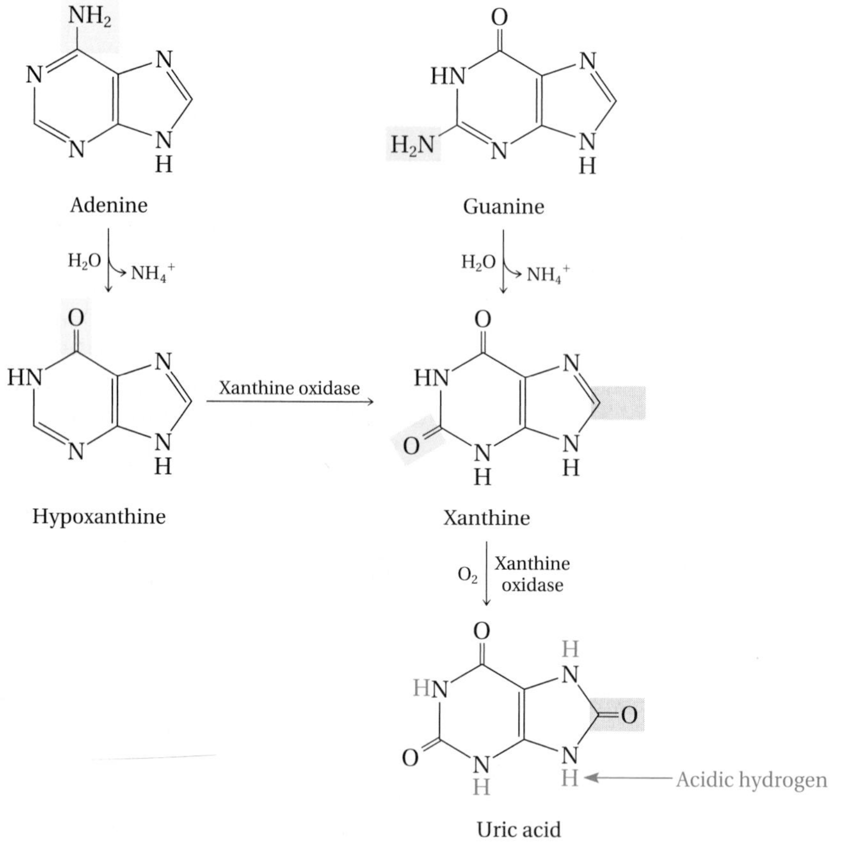

Figure 26.7
Purines are degraded to uric acid.

age, Lesch-Nyhan victims begin to show an uncontrollable urge to bite themselves. Unless they are forcibly restrained, some Lesch-Nyhan patients will literally bite off their fingers and lips. This disease is rare; about 60 cases are reported in the medical literature. Victims of the Lesch-Nyhan syndrome often show symptoms of gout. These symptoms are alleviated by allopurinol, but there is no alleviation of the tendency for self-destruction.

FOLLOW-UP TO THE CASE IN POINT: A painful episode

Ralph's painful symptoms indicated gout, and a blood test showed that he had elevated levels of uric acid in his blood. The symptoms of gout may often be relieved by a diet that restricts the intake of foods high in purines, such as shellfish, bacon, beef, and turkey. In severe cases, the drug allopurinol is sometimes effective.

Allopurinol prevents the synthesis of uric acid by inhibiting *xanthine oxidase*, the enzyme that converts xanthine to uric acid. A low-purine diet and allopurinol have helped put Ralph back on his feet.

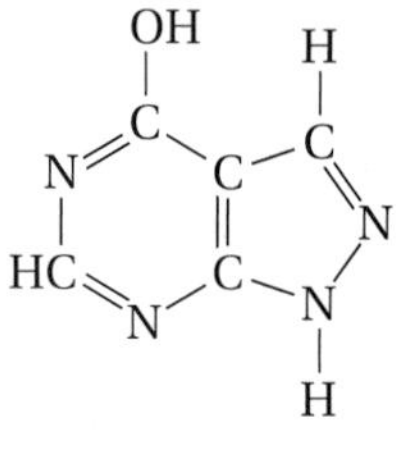

Allopurinol

PRACTICE EXERCISE 26.6

Match the following:

Medical condition	*Cause or result of condition*
(a) bile duct obstruction	(1) gout
(b) high serum GOT	(2) absence of methionine from diet
(c) high serum GPT	(3) jaundice
(d) phenylpyruvic acid in urine	(4) liver disease
(e) negative nitrogen balance	(5) damage to heart muscle
(f) deposit of uric acid in joints	(6) PKU

SUMMARY

Long before nitrogen was fixed (reduced to ammonia) by industrial processes, nitrogen-fixing bacteria were the link between atmospheric nitrogen and living organisms. Fixed nitrogen is used by living creatures to synthesize important classes of biological molecules: amino acids, nucleic acids, and heme, to name but three. Upon dying, organisms decay and return nitrogen compounds to the soil, from which molecular nitrogen eventually returns to the atmosphere to complete the nitrogen cycle.

The chief use of amino acids in the body is for making proteins. Proteins are continuously synthesized and hydrolyzed in the body. The lack of even one essential amino acid causes death because the amino acid is not available for protein synthesis. Amino groups are liberated from amino acids by transamination reactions. The amino groups are transferred to α-ketoglutarate to form glutamic acid, which serves as a nitrogen storage pool. The release of ammonia from glutamic acid by oxidative deamination restores the supply of α-ketoglutarate. Ammonia is very toxic to humans. It is converted to relatively nontoxic urea by the reactions of the urea cycle and excreted in the urine.

The carbon skeletons of the lipogenic amino acids are degraded to acetyl CoA. Those of the glucogenic amino acids are degraded to pyruvate or intermediates of the citric acid cycle. When ATP is abundant, acetyl CoA is used to make fatty acids for storage as triglycerides. Pyruvate and intermediates of the citric acid cycle are used to make glycogen. When ATP supplies are low, acetyl CoA, pyruvate, and intermediates of the citric acid cycle are degraded by the cycle to produce ATP by way of respiration and oxidative phosphorylation. Many malfunctions of amino acid metabolism are known.

Hemoglobin is synthesized in immature red blood cells and degraded in the spleen. Globin, the protein part of hemoglobin, is hydrolyzed to its individual amino acids. The heme is degraded to green biliverdin and orange-red bilirubin. Bilirubin is the coloring agent in jaundice.

The carbons and amine nitrogens of the purines and pyrimidines come from amino acids. The ultimate breakdown product of purine metabolism is uric acid.

KEY TERMS

Amino acid pool (26.3)
Amino aciduria (26.7)
Bilirubin (26.8)
Biliverdin (26.8)
Ferritin (26.8)
Glucogenic amino acid (26.5)
Gout (26.9)

Hemolytic jaundice (26.8)
Infectious hepatitis (26.8)
Jaundice (26.8)
Ketogenic amino acid (26.5)
Lesch-Nyhan syndrome (26.9)
Negative nitrogen balance (26.4)

Nitrogen balance (26.4)
Nitrogen cycle (26.1)
Nitrogen-fixing bacteria (26.1)
Oxidative deamination (26.4)
Phenylketonuria (PKU) (26.7)

Positive nitrogen balance (26.4)
Protein turnover (26.2)
Transamination (26.3)
Urobilin (26.8)
Urobilinogen (26.8)

EXERCISES

Nitrogen Fixation (Section 26.1)

26.7 How does atmospheric nitrogen enter the biosphere?

26.8 Describe the nitrogen cycle in your own words.

Protein and Amino Acid Catabolism (Sections 26.2, 26.3, 26.4, 26.5)

26.9 What is meant by the term *protein turnover?*

26.10 What are cathepsins?

26.11 What is the amino acid pool, and where is it in the body?

26.12 Discuss two metabolic fates of amino acids in the body.

26.13 What is a transamination reaction?

26.14 Explain the function of pyridoxal phosphate in transamination reactions.

26.15 Write an equation for the transamination of valine with α-ketoglutarate as the amino group acceptor.

26.16 Describe how an assay for GPT and GOT in blood serum can give information about liver or heart damage.

26.17 What is the function of the urea cycle?

26.18 How does the excretion of nitrogen as urea conserve body water in land-dwelling animals?

26.19 Define nitrogen balance. Why do growing children have a positive nitrogen balance?

26.20 Describe one medical condition that produces a negative nitrogen balance.

26.21 Amino acids are glucogenic or ketogenic. What do these terms mean?

26.22 Identify the locations at which the carbon skeletons derived from amino acids enter the citric acid cycle.

Amino Acid Synthesis (Section 26.6)

26.23 Describe the difference between an essential amino acid and a nonessential amino acid.

26.24 Write a reaction for the synthesis of a nonessential amino acid.

Defects of Amino Acid Metabolism (Section 26.7)

26.25 Draw the structural formula of phenylalanine. What compound is phenylalanine converted to early in its catabolism?

26.26 Explain why newborn infants are often given PKU tests.

Degradation of Hemoglobin (Section 26.8)

26.27 What is the average life span of a red blood cell?

26.28 Outline the degradation of a hemoglobin molecule.

26.29 What are the bile pigments?

26.30 Name two malfunctions of bile production and storage that give rise to a jaundiced appearance.

Degradation of Purines (Section 26.9)

26.31 Draw the structural formula of uric acid. Why is this compound acidic?

26.32 Relate the cause of gout to its symptoms. How does the drug allopurinol relieve the symptoms of gout?

Additional Exercises

26.33 How many high-energy phosphate bonds are expended in the urea cycle? In which steps?

26.34 What α-ketoacid is the transamination product of alanine? Of glutamic acid?

26.35 In what form is most of the nitrogen in amino acids and proteins excreted from the body? Give the name of the compound and draw its structure.

26.36 What are the two kinds of compounds that react in transamination reactions?

26.37 Describe the structure of the compounds that are catabolized in the synthesis of uric acid.

26.38 The purine and pyrimidine ring systems are synthesized in the body. Where do the atoms that make up these ring systems come from?

26.39 The body synthesizes urea. What is the source of each amino group and the carbonyl group in this compound?

26.40 Write equations to show how a molecule of glycine is deaminated in the body.

SELF-TEST (REVIEW)

True/False

1. The following compounds are both α-ketoacids:

$$CH_3-\overset{\overset{\displaystyle O}{\|}}{C}-\overset{\overset{\displaystyle O}{\|}}{C}-OH$$

pyruvate

$$HO-\overset{\overset{\displaystyle O}{\|}}{C}-CH_2\overset{\overset{\displaystyle O}{\|}}{C}-\overset{\overset{\displaystyle O}{\|}}{C}-OH$$

oxaloacetate

2. In periods of starvation, amino acids are used for energy production.

3. Symbiotic nitrogen-fixing bacteria always work in conjunction with plants.

4. A protein that contains 14 of the 20 common amino acids would have to be an incomplete protein.

5. The only dietary source of nitrogen is protein.

6. Damage to liver tissue is usually indicated by an increased level of serum glutamic-pyruvic transaminase (SGPT).

7. All animals eliminate nitrogen from their system primarily in the form of urea.

8. Overall, the urea cycle is energy producing.

9. A normal, healthy adult should, on average, take in more nitrogen than he or she excretes.

10. The supply of iron in the body is conserved by complexing with the protein ferritin.

Multiple Choice

11. The carbon skeletons of glucogenic amino acids are used
(a) for energy production.
(b) to synthesize triglycerides.
(c) for gluconeogenesis.
(d) all of the above are true.

12. A nitrogen atom from an ingested protein could eventually be found in which of the following types of molecules?
(a) heme   (b) urea
(c) nonessential amino acid   (d) all of the above

13. Inorganic nitrogen is usually found in the body in the form of
(a) N_2 (b) NO_3^-
(c) NH_3 (d) NH_4^+

14. An elevation of serum glutamic-oxaloacetic transaminase (SGOT) activity is indicative of
(a) heart muscle damage.   (b) starvation.
(c) liver damage.   (d) Lesch-Nyhan syndrome.

15. The orange-red bile pigment produced by the degradation of heme is
(a) ferritin.   (b) bilirubin.
(c) citrulline.   (d) biliverdin.

16. Energy produced from the metabolism of ketogenic amino acids would be stored as
(a) triglycerides.   (b) glycogen.
(c) uric acid.   (d) More than one are correct.

17. Phenylketonuria is the result of
(a) the faulty metabolism of phenylalanine.
(b) the lack of the essential amino acid phenylalanine in the diet.
(c) an inborn error of metabolism.
(d) more than one are correct.

18. A transamination reaction results in
(a) the formation of ammonium ion.
(b) an intermolecular amino group transfer.
(c) the formation of urea.
(d) amino aciduria.

19. The process by which bacteria change atmospheric nitrogen to ammonia is called
(a) ammoniation.   (b) transamination.
(c) nitrogen cycle.   (d) nitrogen fixation.

20. Which of the following pairs of compounds is used both as intermediates in the citric acid cycle and as starting materials for the synthesis of nonessential amino acids?
(a) pyruvate and oxaloacetate
(b) α-ketoglutarate and glutamic acid
(c) oxaloacetate and α-ketoglutarate
(d) aspartic acid and pyruvate

21. The chemical equation below represents a reaction called

$$HO-\overset{\overset{\displaystyle O}{\|}}{C}-CH_2CH_2\overset{\overset{\displaystyle H}{|}}{\underset{\underset{\displaystyle NH_2}{|}}{C}}-\overset{\overset{\displaystyle O}{\|}}{C}-OH + H_2O \xrightarrow[\text{NAD}]{\quad\text{NADH}\quad}$$

$$NH_4^+ + HO-\overset{\overset{\displaystyle O}{\|}}{C}-CH_2\overset{\overset{\displaystyle O}{\|}}{C}-\overset{\overset{\displaystyle O}{\|}}{C}-OH$$

(a) hydrolysis.   (b) transamination.
(c) oxidative deamination.   (d) α-ketogenesis.

22. Gout is characterized by high blood levels of
    (a) uric acid.
    (b) ornithine.
    (c) urea.
    (d) phenylalanine.

23. The enzymes found in the lysosomes of cells that hydrolyze proteins are called
    (a) cathepsins.
    (b) peptidases.
    (c) transaminases.
    (d) hydrolases.

24. On a daily basis, a child consumes more nitrogen than he excretes. This child has
    (a) a positive nitrogen balance.
    (b) a serious medical problem.
    (c) a negative nitrogen balance.
    (d) more than one are correct.

25. The majority of the nitrogen eliminated from the body is in the form of
    (a) urea.   (b) ammonium ions.
    (c) uric acid.   (d) bilirubin.

Glossary

Absolute zero The temperature at which all molecular motions theoretically stop; this occurs at −273 °C. (1.3)

Accuracy The closeness of a measurement to the true value of what is being measured. (2.1)

Acetals Compounds produced by the reaction of two molecules of an alcohol with an aldehyde. The general structure is

$$R—CH—OR$$
$$\overset{|}{OR} \qquad (13.6)$$

Acetyl coenzyme A (acetyl CoA) A thioester of acetic acid and coenzyme A, which is important for the transfer of acetyl groups in many processes of metabolism. (24.1)

Acid A substance that releases hydrogen ions (H^+) when dissolved in water (Arrhenius definition); a substance that donates a proton (H^+) to another substance (Brønsted-Lowry definition). (9.2)

Acid anhydride Substances formed when a total of one molecule of water is removed from two acid molecules. (14.5)

Acid-base balance A state in which the pH of body fluids is controlled so that no dramatic changes in acidity or basicity occur; blood has a normal pH of 7.40. (22.4)

Acid dissociation constant, K_a Equilibrium constant for the ionization of an acid, taking the form $K_a = [H^+][A^-]/[HA]$, where HA is a monoprotic acid and A^- is the anion of the dissociated acid. (9.2)

Acid mucopolysaccharide A class of carbohydrates in which some carbohydrate units contain acidic ($—CO_2H$ or $—SO_3H$) groups; the fluid that lubricates joints contains these substances. (16.10)

Acidic solution A solution in which the hydrogen-ion concentration exceeds the hydroxide-ion concentration. (9.1)

Acidosis A condition in which excessive acid production or faulty acid elimination lowers the body's alkali reserves. (22.4)

Activated transfer RNA (activated tRNA) A tRNA molecule attached to an amino acid through an ester bond; the amino acid is transferred to a growing peptide chain in protein synthesis. (20.8)

Activation energy The amount of energy colliding molecules must have to produce a chemical reaction. (6.7)

Active site A groove or pocket in an enzyme molecule into which the substrate (reactant molecule) fits; the substrate is converted to products in a subsequent reaction. (19.4)

Acyl group A group with the structure 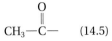; the acetyl group has the structure

$$CH_3—\overset{\overset{\textstyle O}{\|}}{C}— \qquad (14.5)$$

Acylation reaction A reaction in which an acyl group ($R—\overset{\overset{\textstyle O}{\|}}{C}—$) is transferred from one molecule to another; the accepting group is usually an alcohol (to form an ester) or an amine (to form an amide). (14.5)

Addition reaction A reaction in which one reactant divides into two parts, each part joining to a carbon of a multiple carbon-carbon bond to form the product. (12.2)

Adenosine triphosphate (ATP) The major source of usable energy in all cells. When an ATP molecule loses a phosphate group to form ADP, free energy is released. (23.4)

Adipose tissue Tissue that stores large amounts of fats, mainly as triglycerides. (25.1)

Aerobic cell Cell that requires the presence of free oxygen to carry out oxidation reactions vital to cellular life. (21.1)

Alcohol A compound with the general structure R—OH. (12.3)

Alcoholic fermentation The process by which yeast converts glucose to ethyl alcohol (CH_3CH_2OH) in the absence of oxygen. (24.7)

Aldehyde Compound containing an aldehyde functional group, $—\overset{\overset{\textstyle O}{\|}}{C}—H$; the general formula is RCHO. (13.1)

Aldose A carbohydrate containing an aldehyde group; glucose, galactose, and ribose are all simple aldoses. (16.1)

Aliphatic compound An open-chain compound of carbon that contains no aromatic rings. (11.9)

Alkali metal Any metal of Group 1A of the periodic table; lithium, sodium, and potassium are alkali metals. (4.6)

Alkaline earth metal Any element of Group 2A of the periodic table; magnesium, calcium, and barium are among the alkaline earth metals. (4.6)

Alkaloids A class of nitrogen-containing organic compounds obtained from plants. Many have striking biological effects on animals and humans. The alkaloids are weak bases (hence their name). (15.8)

Alkalosis   A condition in which the pH of body fluids such as blood increases when the buffering capacity of the body is exceeded. (22.4)

Alkane   A compound composed only of hydrogen and carbon; all the bonds are single covalent bonds. Alkanes are saturated hydrocarbons. (11.2)

Alkene   A hydrocarbon containing one or more double bonds, such as 1-butene ($CH_3CH_2CH{=}CH_2$). Alkenes are unsaturated hydrocarbons. (11.8)

Alkoxide   A salt produced by the reaction of a reactive metal with an alcohol. Sodium methoxide ($CH_3O{-}Na^+$) is the alkoxide produced by the reaction of methanol and sodium. (12.6)

Alkyl group   A hydrocarbon group such as methyl ($CH_3{-}$), ethyl ($CH_3CH_2{-}$), or *n*-propyl ($CH_3CH_2CH_2{-}$). (11.5)

Alkyl halide   A derivative of a hydrocarbon in which one or more hydrogens are replaced by halogens. Alkyl halides are also called halocarbons. (12.1)

Alkylammonium ion   The cation produced by protonation of an aliphatic amine. (15.2)

Alkyne   A hydrocarbon containing a triple bond, such as ethyne ($CH{\equiv}CH$). Alkynes are unsaturated hydrocarbons. (11.8)

Allosteric enzyme   An enzyme whose activity is changed by the binding of a small molecule (a modulator) to a site other than the active site. (19.10)

Alloy   Usually a homogeneous mixture (solution) of metals, although a nonmetal such as carbon may be included; steel, bronze, and brass are alloys. (4.6)

Alpha carbon   The carbon to which the carboxyl and amino groups are attached in the twenty common amino acids. (18.1)

Alpha emission   Penetrating radiation consisting of high-speed alpha particles (helium nuclei). (10.1)

Alpha helix   A winding chain conformation in proteins that is formed and maintained by hydrogen bonds. (18.6)

Alpha-hydroxy ketone   A compound of the general structure

$$R-\overset{\overset{\textstyle O}{\|}}{C}-\overset{\overset{\textstyle H}{|}}{\underset{\underset{\textstyle OH}{|}}{C}}-R \quad (13.5)$$

Alpha keratin   A type of protein found in hair, wool, nails, and hooves. The protein chain conformation is mostly alpha helix. (18.6)

Alpha particle   A positively charged particle emitted from a nucleus. The nucleus of the helium atom and an alpha particle are identical, since both are composed of two protons and two neutrons. (10.1)

Amide   An organic compound in which nitrogen is linked to a carbonyl carbon; general structures are

$$R-\overset{\overset{\textstyle O}{\|}}{C}-NH_2 \qquad R-\overset{\overset{\textstyle O}{\|}}{C}-\overset{\overset{\textstyle H}{|}}{N}-R \quad \text{and} \quad R-\overset{\overset{\textstyle O}{\|}}{C}-\overset{\overset{\textstyle R}{|}}{N}-R$$
$$(15.4)$$

Amine   An organic compound of general structure

$$R-\overset{\overset{\textstyle H}{|}}{N}-H \qquad R-\overset{\overset{\textstyle R}{|}}{N}-H \quad \text{and} \quad R-\overset{\overset{\textstyle R}{|}}{N}-R \quad (15.1)$$

Amino acid   Any carboxylic acid that contains an amino ($-NH_2$) group; the name is usually reserved for the twenty naturally occurring alpha amino acids. (18.1)

Amino acid pool   The total supply of amino acids present in the body. (26.3)

Amino acid residue   The portion of an amino acid that remains after it has been incorporated into a protein molecule; the general formula is

$$-\overset{\overset{\textstyle H}{|}}{N}-\overset{\overset{\textstyle R}{|}}{\underset{\underset{\textstyle H}{|}}{C}}-\overset{\overset{\textstyle O}{\|}}{C}- \quad (18.4)$$

Amino acid side chain   The substituent group attached to the alpha carbon of an alpha amino acid in addition to the carboxyl and amino groups. (18.1)

Amino aciduria   A condition in which amino acids are excreted in excessive amounts in the urine. (26.7)

Ammonium salt   Salt formed by the neutralization of aqueous ammonia or an amine with an acid. (15.2)

Anabolism   The synthetic processes in the metabolism of cells; it usually requires the expenditure of ATP and reducing power in the form of NADH and FADH$_2$. (23.1)

Anaerobe   A cell that can live in the absence of oxygen. (23.6)

Analgesic   Any painkilling drug. (14.6)

Anilide   An amide in which the amine portion comes from the parent amine aniline or a derivative of aniline. (15.4)

Anion   A negatively charged ion. (5.4)

Anomers   Two stereoisomers of the closed-chain form of a sugar that differ only in the position of the hydroxyl group at carbon 1. (16.4)

Antibiotic   A drug that kills bacteria, usually by interfering with some vital process. (19.13)

Antibody   Protein synthesized in response to a foreign substance (antigen) in the body; the antigen is destroyed after the formation of an antibody-antigen complex. (22.3)

Anticodon   A three-base group on a tRNA molecule that pairs with a three-base codon of mRNA, leading to the correct incorporation of an amino acid into a growing protein chain in protein synthesis. (20.8)

Antidiuretic A substance that blocks the loss of body water through urination. (18.4)

Antigen A substance that stimulates production of antibodies and reacts with them. (22.3)

Antihistamine Any drug that relieves the symptoms of colds and allergies by blocking the release of histamine. (15.7)

Antiparallel strands In nucleic acids, the structural feature in which the two strands of a double helix run in opposite directions. (20.2)

Antipyretic Any drug used to reduce a fever. (14.6)

Apoenzyme The protein portion of an enzyme that requires a cofactor for activity. (19.6)

Aqueous solution A solution in which the solvent is water. (8.3)

Arene Any compound that contains a benzene ring. (11.9)

Aromatic compound An organic compound that contains a benzene ring or closely related ring. (11.9)

Aromatic substitution A chemical reaction in which a substituent replaces hydrogen on an aromatic ring. (12.2)

Aryl halide A benzene ring with one or more hydrogens replaced by halogen. (12.1)

Arylammonium ion The cation produced by protonation of an aromatic amine. (15.2)

Asymmetric carbon atom A carbon atom covalently bonded to four different groups; two different tetrahedral arrangements of these groups in space are possible, leading to stereoisomerism. (16.1)

Atherosclerosis Hardening of the arteries; caused by the buildup of a high-cholesterol substance called plaque; also called arteriosclerosis. (25.11)

Atmosphere The pressure exerted by a column of air extending from the earth's surface to the top of the atmosphere. Atmospheric pressure is measured in units called atmospheres; 1 atmosphere (atm) = 760 mm Hg at STP. (7.3)

Atom The smallest unit of an element that retains the chemical properties of that element. It consists of protons and neutrons in a nucleus surrounded with electrons. (3.1)

Atomic mass The relative mass of an atom compared to one-twelfth the weight of carbon-12; approximately the sum of the masses of protons and neutrons in the nucleus. (3.4)

Atomic mass unit (amu) A unit of mass equaling one-twelfth the mass of a carbon-12 atom; approximately the mass of a proton or neutron. (3.3)

Atomic number The number of protons in the nucleus of an atom; since atoms are electrically neutral, the atomic number also equals the number of electrons in a complete atom. (3.3)

Atomic orbital A region of space around the nucleus of an atom in which an electron is most likely to be found. (3.5)

Atomic radius Distance from the center of the nucleus of an atom to the outermost electron(s). (4.4)

Atomic theory Matter is composed of small particles called atoms that combine in whole-number ratios to form compounds. (3.1)

Aufbau principle The arrangement of electrons in an atom is the one requiring the least amount of energy to attain. (3.6)

Autoactivation The process by which active proteases (protein-cutting enzymes) produce more of the same protease by catalyzing cleavage of peptide bonds in the zymogen. (19.11)

Average kinetic energy Total kinetic energy of a substance divided by the number of particles in the substance; temperature is a measure of average kinetic energy. (7.2)

Avogadro's hypothesis Equal volumes of different gases at the same conditions of temperature and pressure contain the same numbers of molecules. (7.4)

Avogadro's number The number of particles (atoms, formula units, molecules, ions) that constitutes 1 mol of the particle; equal to 6.02×10^{23} particles. (6.1)

Background radiation The amount of radiation produced by natural sources in the environment. (10.11)

Balanced chemical equation An equation in which mass is conserved; the number of atoms in the reactants equals the number of atoms in the products. (6.2)

Barbiturates Salts of barbituric acid and its derivatives used (and abused) as sedatives and hypnotics. (15.9)

Basal metabolism The minimum metabolic activity required to sustain life in a resting but awake human being. (21.5)

Base (Arrhenius definition) A substance that produces hydroxide ions (OH^-) when dissolved in water; (Brønsted-Lowry definition) A substance capable of accepting a proton or hydrogen ion (H^+) from some other substance. (9.2)

Base triplet A sequence of three nitrogen bases that codes for an amino acid in DNA or mRNA. (20.8)

Basic (alkaline) solution An aqueous solution in which the hydroxide-ion concentration is greater than the hydrogen-ion concentration. (9.1)

Bence-Jones protein Light chains of single antibody species produced by myeloma cells; detected in the urine of multiple myeloma patients. (22.3)

Benedict's reagent An alkaline solution of copper(II) sulfate ($CuSO_4$) used to test for the presence of reducing sugars. (13.5)

Beriberi A disease affecting the nervous system, caused by a thiamine deficiency. (21.8)

Beta emission Fast-moving electrons formed by the decomposition of a nucleus of an atom into a proton and an electron. (10.1)

Beta keratin The major protein of spiderwebs and silk; the protein chain conformation is primarily the beta pleated sheet. (18.6)

Beta oxidation The metabolic pathway in which fatty acids are broken down in steps into two-carbon fragments of acetyl coenzyme A. (25.3)

Beta particles High-speed electrons emitted from the nuclei of radioactive substances; a stream of these electrons is beta radiation. (10.1)

Beta pleated sheet A protein chain conformation in which two protein strands running in opposite directions are held together by hydrogen bonds. (18.6)

Bile A soap-like substance produced by the gallbladder. Primarily consisting of carboxylic acid derivatives of cholesterol, bile is used in the dissolution of fats in digestion. (21.4)

Bile pigment A group of substances produced by the breakdown of hemoglobin; many, such as biliverdin and bilirubin, are highly colored. (21.4)

Bile salt Salts of one of several carboxylic acids derived from cholesterol. (21.4)

Bilirubin An orange-red bile pigment. (26.8)

Biliverdin A green bile pigment. (26.8)

Biodegradable Substances capable of being broken down to simple nontoxic components, usually by the action of microorganisms, but sometimes by water, sunlight, and other means. (14.3)

Boiling point The temperature at which the vapor pressure of a liquid equals atmospheric pressure. (7.9)

Bone matrix The molecular framework of collagen upon which calcium salts precipitate to form hard bone. (21.9)

Boyle's law The volume of a gas is inversely proportional to the pressure exerted upon it if the temperature remains constant. (7.7)

Branched-chain alkane An alkane containing carbon groups attached to the longest continuous carbon chain. (11.5)

Brønsted-Lowry theory Acids are hydrogen-ion (proton) donors; bases are hydrogen-ion acceptors. (9.2)

Buffer A solution containing a weak acid and the salt of the weak acid and a strong base, or a weak base and the salt of the weak base and a strong acid; buffers resist changes in pH since their components release or bind hydrogen ions (H^+), depending on the pH of the solution. (9.10)

Caloric value The quantity of heat evolved (in calories) per gram of food burned. (21.5)

Calorie The amount of heat needed to raise the temperature of 1 g of water 1 °C. (2.10)

Carbaminohemoglobin The product of the reaction of hemoglobin with carbon dioxide; it is a carrier of carbon dioxide away from tissue cells that produce the carbon dioxide as a waste product. (22.5)

Carbohydrates Polyhydroxyaldehydes, polyhydroxyketones, or substances that yield these substances upon hydrolysis. (16.1)

Carbonyl group The functional group $-\overset{\overset{\displaystyle O}{\|}}{C}-$, found in aldehydes, ketones, esters, and amides. (13.1)

Carboxyhemoglobin A hemoglobin molecule in which carbon monoxide has reacted with the heme iron; it does not react with oxygen. (18.9)

Carboxyl group The functional group $-\overset{\overset{\displaystyle O}{\|}}{C}-OH$ found in carboxylic acids. (14.1)

Carboxylic acid One of a class of organic molecules containing the carboxyl group ($-\overset{\overset{\displaystyle O}{\|}}{C}-OH$); the general formula is $R-CO_2H$. (14.1)

Catabolism The reactions in living cells in which substances are broken down and energy is produced. (23.1)

Catalyst A substance that speeds up a chemical reaction without being used up; catalysts do not change the position of equilibrium. (6.7)

Cation A positively charged ion. (5.4)

Cell membrane The partition made of phospholipid and protein that forms the outermost edge of the cell; it is the barrier between the cell and its environment. (17.3)

Cellular respiration The process in living cells in which oxygen is reduced to water and NADH and $FADH_2$ are oxidized to NAD^+ and FAD, respectively. It is an energy-releasing process that is tightly coupled to the production of ATP by oxidative phosphorylation. (23.6)

Cellulose A polysaccharide produced by plants as a structural material. Humans and most other animals cannot digest it, since they lack the enzyme to hydrolyze it. A typical cellulose molecule contains from 1000 to 3000 glucose units. (16.6)

Celsius temperature The temperature scale on which the freezing point of pure water is 0 ° and the boiling point is 100 ° at standard pressure. (1.3)

Central dogma Hypothesis that hereditary information flows from DNA to RNA to protein. (20.3)

Chain reaction A self-sustaining reaction usually involving neutrons reacting with atoms to produce new atoms and additional neutrons. (10.5)

Charles's law The volume of a gas is directly proportional to the Kelvin temperature if the pressure remains constant. (7.7)

Chemical equilibrium A state of dynamic balance between two competing processes taking place at equal rates. (6.8)

Chemical formula A symbolic representation of the kinds and numbers of atoms present in a compound; the formula of water, with two hydrogens and one oxygen, is H_2O. (5.3)

Chemical properties The characteristics of a substance with respect to the manner in which it reacts with other substances. (1.7)

Chemical reaction The breaking of old chemical bonds and the formation of new bonds resulting in the rearrangement of atoms. (1.7)

Chemistry The science concerned with the composition of the substances of which the universe is composed, the properties of these substances, and the changes they undergo. It is also concerned with the energy relationships involved in the changes. (1.1)

Chloride shift A process in which chloride ions (Cl^-) replace bicarbonate ions (HCO_3^-); it is important in maintaining a constant osmotic pressure in red blood cells. (22.5)

Chlorophyll The green pigment found in plants that is a major light-trapping molecule of photosynthesis. (23.2)

Chloroplast An organelle found in photosynthetic eukaryotic cells; chloroplasts contain the chlorophyll necessary for the conversion of light energy into chemical energy in photosynthesis. (23.2)

Chylomicron A large particle consisting of lipoproteins and their associated lipids found in the bloodstream after a fatty meal. (21.4)

***Cis* configuration** Alkenes in which groups attached to each carbon of the double bond are on the same side of the bond. (11.8)

Citric acid cycle The central metabolic pathway, in which the acetyl group of acetyl coenzyme A is oxidized to two molecules of carbon dioxide. (24.1)

Codon A sequence of three adjacent nucleotides in mRNA that codes for a specific amino acid. (20.8)

Coenzyme A Coenzyme that functions as a carrier in the acetyl group and other acyl groups. (24.1)

Cofactor A small organic molecule or metal ion necessary for an enzyme's biological activity; one of the B-complex vitamins is often part of coenzyme molecules. (19.6)

Collagen The structural protein of tendons, ligaments, bones, and teeth. (18.6)

Colligative property A property of solutions that depends on the number (concentration) of solute particles but not on their structures. (8.10)

Colloid A liquid mixture in which particles larger than those of a dissolved solute but smaller than particles of a suspension are evenly distributed throughout the liquid. They do not settle out when the colloid is allowed to stand. (8.9)

Colloidal dispersion See *Colloid*. (8.9)

Combined gas law States that
$$P_1 \times V_1/T_1 = P_2 \times V_2/T_2.$$ (7.7)

Competitive inhibitor A molecule resembling the normal substrate molecule of an enzyme; it blocks the catalytic site of the enzyme by binding to it so that the normal enzymatic reaction cannot occur. (19.10)

Complement system A complex series of interacting plasma proteins that "complements" the function of antibodies in destroying antigens. (22.3)

Complementarity The fit between substrate and enzyme; also refers to the pairing of purine and pyrimidine bases in DNA. (19.4)

Complementary base pair Interacting pair of nitrogen bases on opposing strands of DNA; adenine (A) pairs with thymine (T), and cytosine (C) pairs with guanine (G). (20.2)

Complete protein A protein containing the essential amino acids. (21.6)

Complex lipid A lipid molecule that is not a triglyceride; the major complex lipids are the phospholipids, the sphingomyelins, and the glycolipids. (17.3)

Compound Substance formed by a chemical combination of one or more elements in which component elements are in a fixed whole number ratio. (1.8)

Concentration Amount of solute dissolved in a solvent, usually expressed in chemistry as moles per liter. (8.7)

Condensed structural formula Formulas of organic molecules that leave out part of the structure of a molecule. (11.4)

Conjugated protein A protein molecule that requires tight binding of a nonprotein ion or small organic molecule for biological activity. (18.7)

Conversion factor A ratio that relates a quantity in one set of units to the same quantity in another set of units. (2.7)

Coordinate covalent bond A covalent bond between two atoms in which the shared electron pair comes from only one of the atoms; one of the bonds in the ammonium ion (NH_4^+) is coordinate covalent since both electrons come from nitrogen. (5.7)

Corticoids A group of steroid hormones produced by the adrenal cortex. (17.6)

Coupled reactions Two chemical reactions, one spontaneous and the other nonspontaneous, that have an intermediate in common; because of the common intermediate, the free energy released from the spontaneous reaction may be used to drive the nonspontaneous reaction. (23.7)

Covalent bond A chemical bond in which two atoms share a pair of electrons. (5.6)

Covalent compound   Chemical compound in which the constituent atoms are joined by covalent bonds. (5.6)

Cracking   Decomposing large hydrocarbons to smaller ones by heat or pressure with or without a catalyst; this process is used to increase the supply of low-formula mass hydrocarbons from petroleum for gasoline. (11.10)

Crystal   A substance in which the component particles are arranged in an orderly, three-dimensional repeating pattern. (5.5)

Curie (Ci)   The mass of a radioactive isotope that gives 3.7×10^{10} disintegrations per second. (10.9)

Cyanohemoglobin   The binding of carbon monoxide to the heme iron of hemoglobin. (18.9)

Cycloalkane   A saturated hydrocarbon in which the ends are joined to form a ring. (11.7)

Cytochromes   A group of heme proteins that transport electrons in the respiratory chain. (23.6)

Dalton's law of partial pressures   Each gas in a mixture of gases behaves as if it were the only gas present in the same volume. The total pressure of the mixture is therefore the sum of the partial pressures of all the gases present. (7.6)

Decongestant   Medical drug that causes shrinkage of membranes lining the nasal passages, relieving stuffed-up feeling. (15.7)

Degenerate code   Property of the genetic code; more than one three-letter code word exists for each amino acid. (20.9)

Degradation   Hydrolysis of enzymes to their constituent amino acids as part of the protein turnover in cells. (19.9)

Dehalogenation   Elimination of a halogen (Cl_2, Br_2) from an organic molecule containing halogen constituents to produce a carbon-carbon double bond. (12.5)

Dehydration   Any loss of water; in organic chemistry, loss of water from a primary or secondary alcohol to form an alkene. (12.5)

Dehydrogenation reaction   A reaction in which two hydrogens are lost from a compound; all dehydrogenations are oxidations. (13.3)

Dehydrohalogenation   Loss of a hydrohalogen (HF, HCl, HBr, HI) from a halocarbon to produce an alkene. (12.5)

Denaturation   The partial or complete unfolding of the biologically active chain conformation of a protein; the process may be reversible or irreversible. (18.12)

Denatured alcohol   Ethanol (CH_3CH_2OH) to which a poisonous substance has been added to make it unfit to drink. (12.3)

Density   The mass of a substance in a given unit of volume; usually expressed in grams per cubic centimeter. (2.9)

Deoxyhemoglobin   A hemoglobin molecule to which no oxygen is bound. (18.8)

Deoxyribonucleic acid (DNA)   A giant polymer of repeating deoxyribonucleotide units joined by phosphodiester linkages; the storage form of genetic information in living organisms. (20.1)

Depot fat   Stored fat, mainly in the form of triglycerides. (25.1)

Detergent   A cleaning agent, often the sodium salt of a long-chain sulfonic acid. (14.3)

Diabetes mellitus   A disease characterized by the inability of glucose to enter tissue cells; caused by a deficiency of the peptide hormone insulin, which is needed to make cell membranes permeable to the sugar. (24.13)

Dialysis   The separation of colloidal particles from those in true solution by using a semipermeable membrane; the dissolved particles flow from higher to lower concentration. (8.11)

Dicarboxylic acid   Any carboxylic acid molecule containing two carboxylic acid groups ($-CO_2H$). (14.1)

Diffusion   The tendency for molecules to move to lower concentration until the concentration is uniform throughout the system. (7.5)

Dimer   A hydrogen-bonded carboxylic acid pair. (14.2)

Dipole interaction   Attractions between the positive and negative poles of polar molecules. (5.9)

Disaccharide   Carbohydrate made from two monosaccharide units joined by a glycosidic linkage. (16.7)

Dispersion forces   Weak attractive forces between molecules associated with the number of electrons in the molecule; the greater the number of electrons, the stronger are the dispersion forces. (5.9)

Displacement reaction   Chemical reaction in which a substituent on carbon is replaced by a different substituent. (12.4)

Disulfide   An organic compound containing the $-S-S-$ functional group; the general formula is R$-$S$-$S$-$R. (12.9)

Disulfide bridge   A covalent bond between sulfurs of two cysteine residues in a protein; important in holding proteins in their native chain conformations. (18.4)

DNA double helix   A form of DNA in which two antiparallel single polynucleotide strands are twisted into a helix in which opposing nitrogen bases are paired (A = T and G = C). (20.2)

DNA polymerase   An enzyme that catalyzes the synthesis of DNA from its deoxyribonucleoside 5′-triphosphate precursors. (20.4)

Double covalent bond   A covalent bond involving two pairs of electrons; each atom donates one pair of electrons. (5.6)

Electrolyte   A substance that conducts electricity by itself or in aqueous solution; all ionic compounds are electrolytes, but most covalent compounds are not. (8.4)

Electron   A subatomic particle with 1840 times the mass of the proton and a charge of -1. (3.2)

Electron configuration   The arrangement of electrons around the nucleus of an atom. (3.6)

Electron dot structure   Notation for writing the valence-shell electron configurations of atoms; valence electrons are depicted as dots surrounding the inner electrons, and the nucleus is represented by the symbol of the atom being considered. Also called a Lewis dot structure. (5.2)

Electron transport chain   A sequence of molecules that pass electrons from one molecule to the next in line; the original electron donors are NADH and $FADH_2$, and the final receptor is oxygen. Energy released in the process may be used to phosphorylate ADP in oxidative phosphorylation. Also called the respiratory chain. (23.6)

Electronegativity   A measure of the ability of an atomic nucleus in a covalent bond to attract electrons. (5.8)

Electrophoresis   A method of separating substances of different charges by placing them in an electrical field; positively charged ions (cations) migrate to the negatively charged electrode (cathode), and negatively charged ions (anions) migrate to the positively charged electrode (anode). (18.10)

Element   A substance that cannot be broken down into simpler substances by chemical means; one of the fundamental forms of matter. There are over 100 known chemical elements. (1.8)

Elimination reaction   Reaction of organic molecule in which substituents on adjacent carbons leave and an alkene is formed. (12.5)

Elongation   Steps in protein synthesis in which amino acid residues are added to a growing peptide chain. (20.8)

Emulsion   A colloidal dispersion in which the particles are very small droplets of a liquid. (8.9)

Endopeptidase   A protein-cutting enzyme that specifically cleaves peptide bonds in a protein chain. (19.11)

Endothermic   A heat-absorbing reaction. (1.11)

Energy   A concept that includes both heat and work. Every substance contains some energy, and substances with high energies are less stable and therefore more reactive than substances with low energies. (1.10)

Energy and carbon cycle   The vast and complex process in which atmospheric carbon dioxide is reduced to organic molecules in photosynthesis, animals eat the organic molecules, and then they oxidize these molecules to carbon dioxide that reenters the atmosphere. (23.3)

Enkephalins   A group of hormonelike peptides that transmit sensations of pain and pleasure in the brain; opiate drugs appear to act by binding to brain receptors normally reserved for enkephalins. (18.4)

Entropy   A measure of the disorder or chaos of a system; systems tend to go from a state of order (low entropy) to a state of maximum disorder (high entropy). (6.5)

Enzyme   A biological catalyst, always a protein; enzymes are very specific with respect to the reactions they catalyze. (19.1)

Enzyme activity   The speed at which an enzyme catalyzes a biological reaction; a high activity may be due to a high concentration of enzyme or a favorable set of reaction conditions. (19.7)

Enzyme assay   An experiment done to find the concentration of an enzyme, usually in a biological fluid such as blood; important in medical diagnosis. (19.7)

Enzyme-substrate complex   The weak association of an enzyme and its substrate; necessary for the enzyme to exert its catalytic activity. (19.4)

Equilibrium constant   A constant that relates the concentrations of all species in a chemical reaction at a given temperature. Every chemical reaction has its characteristic equilibrium constant. (6.8)

Equilibrium position   The composition of a chemical reaction or physical process at equilibrium. (6.8)

Equivalence point   The point in a chemical reaction, usually a titration, when equivalent amounts of the reactants have reacted. (9.8)

Equivalent (equiv) of acid or base   The amount of an acid that will react with 1 mol of hydroxide ions or the amount of base that will react with 1 mol of hydrogen ions; the mass of an acid or a base that will furnish 1 mol of hydrogen ions or hydroxide ions, respectively. (9.6)

Erythrocyte   A red blood cell. (22.2)

Essential amino acids   The 8 common amino acids, out of a total of 20, that must be included in the diet because they cannot be synthesized by humans. (21.6)

Ester   The product of an acid and an alcohol. Organic esters contain the functional group $-\overset{\overset{\displaystyle O}{\|}}{C}-O-$ and have the general structure $R-\overset{\overset{\displaystyle O}{\|}}{C}-OR$; esters of phosphoric acid are important in biological systems. (14.6)

Esterification (reaction)   A chemical reaction in which an alcohol reacts with an acid to produce an ester. (14.6)

Estrogen   A female sex (steroid) hormone that produces secondary sex characteristics; it also prepares the lining of the uterus for pregnancy. (17.8)

Ether   An organic compound that contains the functional group $-\overset{|}{\underset{|}{C}}-O-\overset{|}{\underset{|}{C}}-$; the general formula is $R-O-R$. (12.7)

Evaporation   The escape of the most energetic molecules from the surface of a liquid below the boiling point. (7.9)

Exergonic   A reaction that releases free energy; in principle, the free energy released can be harnessed to do useful work. (6.6)

Exon   Portion of the DNA of eukaryotic cells that contains genetic information; about 5% of the DNA of eukaryotes consists of exons. (20.5)

Exopeptidase   A protein-cutting enzyme that cleaves consecutive amino acid residues from a protein, starting from the N-terminal or C-terminal end. (19.11)

Exothermic   A heat-releasing reaction. (1.11)

Experiment   Observation or measurement of physical phenomena made under controlled conditions. (1.2)

Fahrenheit temperature   The temperature scale on which the freezing point of pure water is 32 ° and the boiling point is 212 ° at standard pressure. (1.3)

Fats   Lipids (especially triglycerides) that exist as solids at room temperature. (17.2)

Fatty acid   Any continuous-chain carboxylic acid, but especially those with chains containing 12 or more carbon atoms; the natural fatty acids contain an even number of carbons. (14.1)

Fatty acid spiral   Consecutive rounds of beta oxidation in which a long-chain fatty acid molecule is degraded to acetyl coenzyme A with a loss of one molecule of acetyl coenzyme A per round. (25.3)

Fatty acyl coenzyme A   The thioester between a fatty acid and coenzyme A. (25.3)

Feedback inhibition   Process in which the end product of a sequence of enzyme-catalyzed reactions "feeds back" to inhibit an earlier step in the sequence; many feedback inhibitors are allosteric modulators; also called end-product inhibition. (19.10)

Fehling's reagent   An alkaline solution of copper(II) sulfate used to detect the presence of reducing sugars. (13.5)

Ferritin   The protein used to store iron in the body. (26.8)

Fibrous proteins   A class of long, rod-shaped protein molecules that play mainly structural roles in cells; usually insoluble in water and dilute salt solutions. (18.6)

First messenger   Hormones; act to carry signals for cellular activity to receptors on the external surfaces of eukaryotic cells. (24.11)

Fischer projection   A method for depicting the three-dimensional shapes of molecules, especially carbohydrates. (16.1)

Fission   The splitting of an atomic nucleus into two nuclei, accompanied by the release of two or three neutrons and a large amount of energy. (10.5)

Flavoprotein   An oxidation-reduction protein having a molecule related to riboflavin as its prosthetic group. (23.6)

Fluid mosaic model   A theory describing cell membranes as dynamic, with lipids and some proteins moving about within the lipid bilayer. (17.4)

Fluorosis   A condition in which excessive fluoride ion (F^-) is deposited in bones and teeth. (21.10)

Formed element   The insoluble substances of blood, consisting mainly of erythrocytes, leukocytes, and thrombocytes. (22.2)

Formula mass   The mass of an atom, ion, ionic compound, or molecular compound expressed in atomic mass units. (6.1)

Free amine   An amine in which the nitrogen is not protonated to an ammonium ion; free amines are electrically neutral. (15.2)

Free energy   The energy available to do work. (6.6)

Free radical   A species containing one or more unpaired electrons; destructive free radicals are implicated in diseases and aging. (10.8)

Functional group   A specific arrangement of atoms in an organic compound that is capable of characteristic chemical reactions; the chemistry of an organic compound is determined by its functional groups. (12.2)

Furanose ring system   The fundamental closed-chain form of several sugars; contains four carbons and one oxygen. (16.3)

Fused-ring aromatic compound   An organic compound containing two or more benzene rings; each ring shares two ring carbons with one or more other rings. (11.9)

Fusion   In nuclear chemistry, a reaction in which two light nuclei combine to form a heavier nucleus accompanied by the release of a large amount of energy; also refers to the melting of a solid to a liquid. (10.6)

Gamma radiation   A type of radiation, similar to X radiation but more penetrating, emitted by certain radioactive substances. (10.1)

Gas   A substance that has no definite shape or volume; the particles in a gas are far apart because there is little or no attraction between them. (1.4)

Gas pressure   The force exerted by a given mass of gas on an object; caused by particles in the gas striking the object. (7.3)

Gastric juice   The fluid of the stomach; it consists of 0.01 M hydrochloric acid and enzymes that aid digestion. (21.3)

Geiger counter   A device that detects the presence of all types of penetrating radiation. (10.7)

Gene   The segment of a DNA molecule coding for a single polypeptide chain. (20.5)

Gene mutation Any change in the base sequence of DNA; a protein synthesized from a mutated gene may be fully functional, partially functional, or nonfunctional. (20.10)

Gene therapy Cure or alleviation of a hereditary disease by replacement of a missing or defective gene. (20.11)

Genetic code The set of 61 triplet code words used to specify amino acids in protein synthesis plus three triplet code words that signal "stop" when a peptide chain has been synthesized. (20.9)

Genetic damage Damage from radiation that affects the body's reproductive system and genetic material. (10.8)

Geometric isomer An organic structure that differs from another only in the geometry of the molecule and not in the order of linkage of atoms. (11.8)

Globin The protein portion of hemoglobin; that is, hemoglobin from which the heme prosthetic group has been removed. (18.8)

Globular protein Any protein that has its peptide chain folded into a roughly spherical shape; most globular proteins are soluble in water and dilute salt solutions. (18.7)

Glucocorticoids A group of steroid hormones produced in the adrenal cortex; they regulate several body functions, including glucose breakdown. (17.6)

Glucogenic amino acids Amino acids that are broken down to pyruvic acid in cellular metabolism. (26.5)

Gluconeogenesis The metabolic pathway leading to the synthesis of glucose. (24.8)

Glycogen Animal starch consisting of polymers of D-glucose; storage form of glucose in animals. (16.6)

Glycogenesis The series of chemical reactions by which living organisms convert glucose to glycogen. (24.9)

Glycogenolysis The cleavage of stored glycogen to glucose-1-phosphate in living organisms. (24.10)

Glycolipids Lipids that have polar, hydrophilic carbohydrate "heads" attached to nonpolar, hydrophobic "tails." (17.3)

Glycolysis The sequence of chemical reactions that living organisms use to oxidize glucose to pyruvic acid; also called the glycolytic pathway or Embden-Meyerhof pathway. (24.1)

Glycoprotein Protein molecules that have carbohydrate units attached to side chains of certain amino acid residues; many membrane proteins are glycoproteins, as are the protein blood-clotting factors and collagen. (18.11)

Glycoside A carbohydrate in which two or more sugar units are connected by glycosidic bonds; also any acetal or ketal of a saccharide. (16.5)

Glycosidic bond The ether linkage produced by the reaction between the anomeric hydroxyl group of one sugar unit and a hydroxy group of another sugar or an alcohol. (16.5)

Glycosuria A condition in which reducing sugars, mainly glucose, are found in the urine; a symptom of diabetes mellitus. (24.13)

Goiter A swelling of the neck due to the enlargement of the thyroid gland; the enlargement is sometimes caused by an iodine deficiency. (21.9)

Gout A painful hereditary disorder of purine metabolism, mainly in males, in which crystals of uric acid precipitate in the joints. (26.9)

Gram A unit of mass in the metric system; the mass of 1 cm^3 of water is 1 g. (2.8)

Group A vertical column of elements in the periodic table; the constituent elements have similar physical and chemical properties. (4.1)

Half-life ($t_{1/2}$) The time required for one-half of a given mass of radioactive material to disintegrate. (10.2)

Hallucinogen A substance that produces hallucinations and other bizarre mental effects. (15.7)

Halocarbon A hydrocarbon in which one or more halogens replace hydrogen. (12.1)

Halogenation Addition of a halogen (Cl_2, Br_2) to the carbon-carbon double bond of an alkene. (12.2)

Halogens The elements in Group 7A of the periodic table, including fluorine, chlorine, bromine, and iodine. (4.6)

Hard water Water that contains dissolved ions of calcium, magnesium, or iron. (14.3)

Haworth projection A method of depicting the structures of carbohydrates in three dimensions. (16.4)

Heat Thermal energy that is being transferred from one body to another. (1.10)

Heat capacity The thermal energy content of a body at a given temperature. (2.10)

Heat of condensation The heat, in calories, released by 1 g of a substance as it changes from a vapor to a liquid at the liquid's boiling point. (7.9)

Heat of fusion The heat, in calories, absorbed by 1 g of a substance as it changes from a solid to a liquid at the solid's melting point. (7.10)

Heat of solidification The heat, in calories, released by 1 g of a substance as it changes from a liquid to a solid at the solid's melting point. (7.10)

Heat of vaporization The heat, in calories, absorbed by 1 g of a substance as it changes from a liquid to a vapor at the liquid's boiling point. (7.9)

Hemiacetal The product of the reaction between an aldehyde molecule and one molecule of an alcohol; the general formula is

$$\text{R}-\overset{\displaystyle \text{OH}}{\underset{\displaystyle \text{H}}{\overset{|}{\underset{|}{\text{C}}}}}-\text{OR} \quad (13.6)$$

Hemiketal   The product of the reaction between a ketone molecule and one molecule of an alcohol; the general formula is

$$R—\underset{\underset{R}{|}}{\overset{\overset{OR}{|}}{C}}—OR \qquad (13.6)$$

Hemodialysis   Process used to purify the blood of individuals who have suffered kidney failure. (8.11)

Hemoglobin   The oxygen-carrying heme protein of blood; it consists of two pairs of identical polypeptide chains and an iron-heme prosthetic group. (18.8)

Hemolysis   The rupture of red blood cells that occurs as a result of an osmotic pressure imbalance when they are placed in a hypotonic solution. (8.11)

Hemolytic jaundice   A yellow appearance of the skin due to the excessive breakdown of red blood cells, with the resultant formation and circulation of excessive amounts of bile pigments. (26.8)

Henry's law   At constant temperature the solubility of a gas in a liquid is directly proportional to the partial pressure of the gas above the liquid. (8.6)

Heterocyclic ring   A ring in which one or more atoms is an element other than carbon. (12.7)

Heterogeneous mixture   Aggregation of two or more substances that is not uniform throughout. (1.5)

Heteronuclear molecule   A molecule containing more than one kind of atom. (5.3)

Holoenzyme   An enzyme that contains both the protein portion and a cofactor required for activity. (19.6)

Homeostasis   The maintenance of constant conditions within the body, even when external conditions change. (22.1)

Homogeneous mixture   Aggregation of two or more substances that is uniform throughout. (1.5)

Homonuclear molecule   A molecule consisting of atoms of the same element; chlorine (Cl_2) is a homonuclear diatomic molecule. (5.3)

Hormone   A substance produced by the body in minute amounts that causes profound physiologic effects; many, but not all, are either peptides or steroids. (17.6)

Hund's rule   In orbitals of equal energy, each orbital gets an electron until all of the orbitals contain one electron; additional electrons add to the orbitals until they all contain two electrons. (3.6)

Hydrate   Any substance that contains water in a tightly bound form, such as sodium sulfate decahydrate ($Na_2SO_4 \cdot 10H_2O$). (13.6)

Hydration reaction   Addition of water to an alkene to form an alcohol. (12.4)

Hydrocarbon   An organic compound containing only carbon and hydrogen. (11.2)

Hydrogen bond   A weak attractive force; very important in determining the properties of water, proteins, and other biological substances. (8.1)

Hydrogen ion acceptor   A base in Brønsted-Lowry acid-base theory. (9.2)

Hydrogen ion donor   An acid in Brønsted-Lowry acid-base theory. (9.2)

Hydrogenation   The addition of hydrogen to a multiple covalent bond; the hydrogenation product of an alkene is an alkane. (12.2)

Hydrohalogenation   Addition of a hydrohalogen (HCl, HBr, HI) to the carbon-carbon double bond of an alkene to produce a halocarbon. (12.2)

Hydrolysis   Reaction that involves the splitting of a covalent O—H bond of water. (14.5)

Hydrometer   A device used to measure the specific gravity of liquids. (2.9)

Hydronium ion   A hydrated proton (H_3O^+); all protons in acidic aqueous solution are present as hydronium ions that impart to these solutions their acidic character. (9.1)

Hydrophilic   Water-loving; polar and ionic species are hydrophilic owing to the favorability of their interactions with water. (12.8)

Hydrophobic   Water-hating; nonpolar molecules are hydrophobic owing to the lack of favorable interactions with water. (12.8)

Hydroxide ion   The ionic species OH^-; these ions impart an alkaline character to aqueous solutions that contain them. (9.1)

Hydroxy acid   A carboxylic acid (RCO_2H) that contains a hydroxyl group (HO—). (14.1)

Hydroxy function   The —OH functional group in alcohols; also called a hydroxyl group. (12.3)

Hydroxyl group   See *Hydroxy function*. (12.3)

Hyperglycemia   A condition in which the level of blood sugar is abnormally high; a symptom of diabetes mellitus. (24.13)

Hypertonic solution   A solution that exerts an osmotic pressure higher than that of the fluid in a body cell; body cells shrivel in hypertonic solutions. (8.11)

Hyperventilation   Respiratory condition characterized by deep, rapid breathing that can lead to respiratory alkalosis. (22.8)

Hypervitaminosis   A condition resulting from excessive intake and storage of a vitamin; a fat-soluble vitamin is usually involved. (21.7)

Hypnotic   A sleep-producing drug. (15.9)

Hypoglycemia   A condition in which the level of blood sugar is abnormally low. (24.13)

Hypothesis   A reasonable explanation of an observation or experimental result supported by additional observations or experiments. (1.2)

Hypotonic solution   A solution that exerts an osmotic pressure lower than that of the fluid in a body cell; body cells swell and split in hypotonic solutions. (8.11)

Hypoventilation   Respiratory condition caused by too-shallow breathing that can lead to respiratory acidosis. (22.8)

Ideal gas constant   The quantity 0.0821 L $\times$ atm/K $\times$ mol, used in the solution of gas law problems involving the ideal gas law. (7.8)

Ideal gas law   States that $P \times V = n \times R \times T$, where P, V, and T are respectively pressure, volume, and Kelvin temperature, n is the number of moles of gas, and R is the ideal gas constant. (7.8)

Immunoglobulin   Y-shaped antibody protein molecule that binds to antigens and thereby neutralizes them. (22.3)

Induced-fit model   A description of the binding of substrates to enzymes: The enzyme active site continuously adjusts to fit the substrate as it approaches so that the fit is perfectly complementary at contact. (19.4)

Inducible enzyme   An enzyme that is synthesized in response to a cellular need. (19.9)

Induction   The synthesis of enzymes by cells as the enzymes are needed. (19.9)

Infectious hepatitis   A disease of the liver in which the liver cannot remove bile pigments from the body, allowing the pigments to circulate. (26.8)

Initiation   The first step in protein synthesis; in eukaryotic cells initiation involves the binding of tRNA for methionine to mRNA. (20.8)

Inner transition metal   A metal whose outermost s sublevel and nearby f sublevel contain electrons. (4.3)

Insulin shock   A serious condition of hypoglycemia caused by an overdose of insulin; insulin lowers blood sugar by increasing the permeability of cell membranes to glucose. (24.13)

Integral protein   A protein molecule that is embedded in the lipid bilayer of a cell or organelle membrane. (17.4)

Interferon   Protein of the immune system that has evolved specifically for defense against viral infections. (22.3)

International System of Units (SI)   Units of measurement accepted worldwide, consisting of seven base units from which all other units can be derived. (2.2)

International unit   A measure of enzyme activity; 1 international unit (1 IU) is the amount of enzyme required to convert 1 micromole of substrate to product in 1 minute at specified conditions. (19.7)

Interstitial fluid   The water and dissolved and suspended substances that surround tissue cells. (22.1)

Intracellular fluid   The water and dissolved and suspended substances contained inside cells. (22.1)

Intron   The noncoding portion of the DNA of eukaryotic cells; about 95% of human DNA consists of introns. (20.5)

Invert sugar   The syrupy mixture of equal amounts of D-glucose and D-fructose produced by the hydrolysis of sucrose. (16.8)

Ion   Atoms or group of atoms bonded together that have acquired a charge by gaining or losing electrons. (5.3)

Ion-product constant for water (K_w)   The product of the hydrogen ion (H^+) and hydroxide ion (OH^-) concentrations of water; it is 1×10^{-14} at 25 °C. (9.1)

Ionic bond   The force of attraction between positively and negatively charged ions. (5.5)

Ionic compounds   Electrically neutral substances consisting of particles of positive and negative charge (ions). (5.3)

Ionization energy   The quantity of energy needed to remove an electron from a gaseous atom. (4.5)

Ionizing radiation   Penetrating radiation of sufficient energy to knock electrons from the bombarded substance. (10.7)

Iron-deficient anemia   A defect in red blood cell production owing to a lack of dietary iron with which to make hemoglobin. (21.9)

Isoelectric pH   The pH at which there is no net charge on a protein or amino acid; a substance at its isoelectric point will not migrate in an electric field. (18.3)

Isoelectric point   The condition of having no net charge; a protein is at its isoelectric point when the positive and negative charges on the side chains of its amino acid residues balance. (18.3)

Isohydric shift   Transport of protons from red blood cells to the lungs. (22.5)

Isotonic solution   A solution that exerts an osmotic pressure identical to that exerted by the intracellular fluid of body cells; body cells do not shrink or expand in an isotonic solution. (8.11)

Isotopes   Atoms of the same element that have the same atomic number but different atomic weights because they have different numbers of neutrons in their nuclei. (3.4)

IUPAC system   A method for naming organic compounds proposed by the International Union of Pure and Applied Chemistry. (11.4)

Jaundice   Yellow coloration of skin and, sometimes, whites of eyes due to accumulation of bilirubin or other bile pigments in the bloodstream. Caused by a number of diseases involving the liver and gallbladder. (26.8)

Joule Basic unit of energy in the SI system of measurement; 4.184 J(oule) = 1 calorie. (2.10)

Kelvin temperature scale Scale on which absolute zero is taken as 0 K; also called the absolute temperature scale. (1.3)

Keratinization The hardening of soft tissues, especially those of the eyes, due to hypervitaminosis A. (21.7)

Ketal The organic product of the reaction between one molecule of a ketone and two molecules of an alcohol; the general formula is

$$R{-}\underset{\displaystyle R}{\overset{\displaystyle OR}{C}}{-}OR \quad (13.6)$$

Ketoacidosis Lowering of the blood pH due to excessive production of ketone body acids. (25.9)

Ketogenic amino acids Amino acids that are broken down to acetyl coenzyme A in metabolism. (26.5)

Ketone An organic compound with the general formula $R{-}\overset{\displaystyle O}{\overset{\displaystyle \|}{C}}{-}R$. (13.1)

Ketone bodies Substances produced by the liver as an alternative energy source in a carbohydrate-deficient diet: acetoacetic acid, β-hydroxybutyric acid, and acetone. (25.8)

Ketonemia A condition characterized by an excess of ketone bodies in the blood; often found in untreated diabetics. (25.9)

Ketonuria A condition in which ketone bodies are excreted in the urine; often found in diabetics. (25.9)

Ketose A carbohydrate that contains a ketone functional group. (16.2)

Ketosis Another name for ketoacidosis. (25.9)

Kilocalorie A unit of heat or energy equal to 1000 cal; in nutrition, the Calorie. (2.10)

Kilogram The fundamental unit of mass in the metric system; equal to 1000 g. (2.8)

Kinetic energy The energy a body possesses by virtue of its motion. (7.2)

Kinetic-molecular theory Theory explaining the states of matter as well as other chemical and physical properties on the basis of the motions of the submicroscopic particles that constitute matter. (7.1)

Kwashiorkor A protein deficiency disease, especially prevalent in certain parts of Africa. (21.6)

Lactic fermentation The process by which glucose is degraded to lactic acid in the absence of oxygen. (24.7)

Law A general statement about a natural phenomenon whose validity is supported by substantial experimentation. (1.2)

Law of conservation of energy Energy is neither created nor destroyed in an ordinary chemical or physical process. (1.10)

Law of conservation of mass Mass can be neither created nor destroyed in an ordinary chemical or physical process. (1.9)

Le Châtelier's principle When stress is placed on a system in equilibrium, the system shifts to relieve the stress. (6.8)

Lecithin Another name for phosphatidyl choline. (17.3)

Lesch-Nyhan syndrome A hereditary disorder of purine metabolism characterized by the self-destructive tendencies of its victims. (26.11)

Leukocyte A white blood cell. (22.2)

Leukotriene A group of C_{20} compounds related to the prostaglandins; leukotrienes are involved in triggering the physiological effects of such processes as inflammation and allergic reactions. (17.11)

Light system A highly organized group of molecules in the chloroplasts of photosynthetic cells; responsible for trapping light energy and converting it to chemical energy. (23.2)

Lipid bilayer A model of the structure of cell membranes; polar head groups of the constituent lipid molecules face the solvent, and nonpolar tail groups cluster toward the interior. (17.4)

Lipids A broad class of organic compounds from natural sources; soluble in organic solvents and insoluble in water. (17.1)

Lipoprotein A protein that associates with and transports lipids; interaction with a lipoprotein is necessary to make most lipids soluble in body fluids. (21.4)

Liposome A spherical body consisting of a lipid bilayer that surrounds a droplet of aqueous solvent. (17.4)

Liquid A state of matter in which the molecules are weakly associated; it has a definite volume but an indefinite shape. (1.4)

Liter The fundamental unit of volume in the metric system; the volume occupied by 1 kg of water, equal in volume to 1000 mL or 1000 cm^3. (2.6)

Lock-and-key model A description of the binding of a substrate to an enzyme active site; the fit between the substrate and enzyme active site is analogous to the fit of a key into a lock. (19.4)

Lymph A clear fluid generated by filtration of blood into tissues from the capillaries; it contains no blood cells but is otherwise similar in composition to blood plasma. (22.2)

Macronutrient Essential mineral element present in relatively large amounts in the body. (21.9)

Marasmus Another name for starvation.

Markovnikov's rule   In additions of hydrohalogens (H—Cl, H—Br) or water (H—OH) to a double bond of an alkene, the hydrogen of the adding reagent ends up on the hydrogen of the double bond that already has the most hydrogens. (12.2)

Mass   A measure of the quantity of matter a body contains; mass is independent of the force of gravity and therefore does not change when the location of the body changes. Mass is not the same as *weight*. (1.9)

Mass number   The total number of protons and neutrons in the nucleus of an atom. (3.3)

Matter   Anything that occupies space and has mass. (1.1)

Melting point   The temperature at which a solid changes to a liquid. (7.10)

Mercaptan   An older name for a thiol or thioalcohol (general formula R—SH). (12.9)

Messenger RNA (mRNA)   A class of RNA molecules, complementary to one strand of cellular DNA, that carry hereditary information from the nucleus to the ribosomes for protein synthesis. (20.3)

Metabolic acidosis   A decrease of blood pH to lower than normal because of excessive influx of protons into the bloodstream; metabolic acidosis is a common complication of diabetes mellitus. (22.8)

Metabolic alkalosis   An increase in blood pH to higher than normal values, usually because of excessive influx of alkaline substances, such as the bicarbonate ions of antacids, into the bloodstream. (22.8)

Metabolism   The sum of the enzyme-catalyzed chemical and energy changes that occur in cells. (23.1)

Metal   A group of elements that shine, conduct electricity, and are malleable (can be hammered into shapes) and ductile (can be drawn into wire). (4.2)

Metalloid   A group of elements that possess certain properties of both metals and nonmetals; silicon, a metalloid, is a semiconductor. (4.2)

Meter   The fundamental unit of length in the metric system and in the SI. (2.6)

Methemoglobin   A form of hemoglobin in which the heme iron is Fe^{3+} rather than Fe^{2+}; methemoglobin does not bind oxygen. (18.8)

Methemoglobinemia   A hereditary disease in which the heme iron of either the alpha or beta chains of hemoglobin is Fe^{3+} rather than Fe^{2+}. (18.8)

Metric system   A system of measurement based on units of 10; the meter is the base unit of length, the kilogram is the base unit of mass, and the liter is the base unit of volume. (2.2)

Micelle   An association in water of molecules possessing polar heads and nonpolar tails; the polar heads face the solvent, and the nonpolar tails cluster together in the interior. (14.3)

Micronutrient   Essential mineral element present in trace amounts in the body. (21.9)

Mineral element   In a living system, any element other than carbon, oxygen, hydrogen, nitrogen, phosphorus, or sulfur. (21.9)

Mineralocorticoids   A class of steroid hormones produced in the adrenal cortex; they regulate mineral and water metabolism in the body. (17.6)

Mixed triglycerides   A triglyceride in which more than one kind of long-chain fatty acid is esterified to the glycerol hydroxyl groups. (17.2)

Mixture   A combination of two or more substances that are not chemically combined. (1.5)

Mobilization   Hydrolysis of stored fats in preparation for the degradation of the product fatty acids in energy production within cells. (25.2)

Modulator binding site   A location in an allosteric enzyme, not the active site, to which a regulatory substance (modulator) binds. (19.10)

Molar mass   Mass of 1 mole of a substance in grams. (6.1)

Molar volume of a gas   Volume occupied by 1 mole of gas at STP; the molar volume of any ideal gas is 22.4 liters per mole. (7.4)

Molarity (*M*)   The concentration of solute in solution expressed as the number of moles per liter of solution. (8.8)

Mole (mol)   Avogadro's number (6.02×10^{23}) of particles of any substance; a molar mass of any substance. (6.1)

Molecular compound   A compound in which the constituent atoms are joined together by covalent bonds. (5.3)

Molecular disease   Any disease caused by an absent or defective enzyme or other protein as a result of heredity; also known as inborn errors of metabolism. (20.10)

Molecular formula   The lowest ratio of atoms that correctly describes the composition of a molecule. (5.6)

Molecule   Two or more atoms of the same or different elements joined by chemical bonds. (5.3)

Monoclonal antibody   Homogeneous antibody; antibody made by one kind of lymphocyte. (22.3)

Monomer   The individual building block unit from which a polymer is constructed. (12.2)

Monomolecular layer   One-molecule-thick layer of a fatty acid or detergent on the surface of water. (14.2)

Monosaccharide   A carbohydrate consisting of one sugar unit; also called a simple sugar. (16.2)

Multienzyme complex   A highly organized assembly of enzyme molecules that together catalyze one or more steps in a metabolic pathway. (24.4)

Myoglobin   A protein used to store oxygen in muscle; it consists of a heme-iron prosthetic group associated with a single polypeptide chain. (18.7)

Native state   The normal, physiologically active chain conformation of a protein. (18.12)

Negative modulator   A molecule that can bind at the modulator binding site of an allosteric enzyme and inhibit enzyme action by distorting the peptide chain conformation of the enzyme active site. (19.10)

Negative nitrogen balance   The excretion of more nitrogen than is consumed, a situation that occurs in starvation. (26.4)

Neurotransmitter   A chemical substance released from an activated neuron that diffuses across a synapse to an adjacent neuron or muscle cell; acts as a chemical bridge in nerve impulse transmission. (15.7)

Neutral solution   An aqueous solution in which there is an equal concentration of hydrogen ions and hydroxide ions; pH = 7. (9.1)

Neutralization reaction   The reaction of an acid and a base to form a salt and water; a proton transfer reaction. (9.5)

Neutron   A neutral particle found in the nucleus of an atom; its mass is about the same as a proton's. (3.2)

Nitrogen balance   Equilibrium reached when the quantity of nitrogen a person excretes daily equals the intake. (26.4)

Nitrogen cycle   A chemical cycle in which nitrogen, as an element or in compounds, circulates between the atmosphere and the earth and its living creatures. (26.1)

Nitrogen-fixing bacteria   Bacteria that convert atmospheric nitrogen to ammonia, a water-soluble compound of nitrogen that can be used by growing plants. (26.1)

Noble gas   An element of Group O; an unreactive gas with a filled outer energy level. (4.3)

Nonelectrolyte   A substance whose aqueous solution does not conduct an electric current. (8.4)

Nonessential amino acid   An amino acid synthesized by the body. (21.6)

Nonionic detergent   A detergent whose molecules have a nonpolar hydrocarbon tail and a polar, but uncharged, head. (14.3)

Nonmetal   Any element that does not conduct electricity; a nonmetal may be a solid or a liquid or a gas at STP. (4.2)

Nonpolar covalent bond   The equal sharing of bonding electrons by two atoms. (5.8)

Nonreducing sugar   A sugar that fails to give a positive test with Benedict's or Tollens' reagents. (16.9)

Nonsuperimposable mirror image   A molecule or other object that exists in right-handed and left-handed forms. (16.1)

Normal boiling point   The temperature at which the vapor pressure of a liquid equals 1 atm (760 mm Hg). (7.9)

Normality (N)   The concentration of a solution expressed as the number of equivalents of solute per liter of solution. (9.7)

Nuclear transformation reaction   The transformation of one element to another by radioactive decay or by bombardment with high-energy particles. (10.4)

Nucleic acid   A polymer of ribonucleotides (RNA) or deoxyribonucleotides (DNA); found primarily in the nucleus of eukaryotic cells, they play a role in the transmission of hereditary characteristics, the control of cellular activities, and protein synthesis. (20.1)

Nucleoside   A compound that on hydrolysis yields a purine or pyrimidine base and a pentose. (20.1)

Nucleotidase   An enzyme that catalyzes the hydrolysis of nucleotides to give nucleosides and phosphate. (21.4)

Nucleotide   A compound consisting of a nitrogen-containing base (a purine or pyrimidine), a sugar (ribose or deoxyribose), and a phosphate; a monomer of a nucleic acid. (20.1)

Nucleus   The dense, positively charged center of an atom containing protons and neutrons; also refers to the organelle found in all eukaryotic cells that contains the chromatin. (3.2)

Observation   An act of seeing or some fact or occurrence; observation in science often involves a measurement. (1.2)

Oil   In biochemistry, a triglyceride that is a liquid at room temperature. (17.2)

Okazaki fragments   Pieces of DNA that are connected to form a new DNA strand, by action of an enzyme, during replication. (20.4)

Opiate   A substance that acts as an analgesic and narcotic; for example, morphine. (15.8)

Optical isomers   Mirror-image isomers that contain at least one asymmetric carbon. They are identical in their chemical and physical properties with one exception—they rotate plane-polarized light in different directions; also called stereoisomers. (16.1)

Orbital hybridization   The mixing of atomic orbitals to produce the same number of orbitals of lower energy than the original orbitals. (11.2)

Osmosis   The net flow of water through a semipermeable membrane from a solution of low solute concentration to a solution of higher solute concentration. (8.11)

Osmotic membrane   A selective semipermeable membrane that permits only water molecules to pass through. (8.11)

Osmotic pressure   The external pressure required to stop osmosis. (8.11)

Osteoporosis   A condition of bone fragility caused by the desorption of calcium from the bones. (21.9)

Oxidation The complete or partial loss of electrons by a substance. (6.4)

Oxidation reaction A chemical reaction in which a reactant gains oxygen or loses electrons. (4.6)

Oxidative deamination Loss of the amino group of glutamic acid to produce ammonium ions and α-ketoglutarate; oxidative deamination is catalyzed by the enzyme glutamate dehydrogenase. (26.4)

Oxidative phosphorylation The phosphorylation of ADP to ATP in cellular respiration. (23.6)

Oxide The product of a reaction in which a reactant gains oxygen. (4.6)

Oxidizing agent The substance in an oxidation-reduction reaction that accepts the shifted electrons. (6.4)

Oxygen debt The condition that exists when not enough oxygen is available for cellular respiration; it occurs during vigorous exercise. (24.8)

Oxyhemoglobin The compound formed when oxygen unites with hemoglobin; it is the red pigment of red blood cells. (18.8)

Pacemaker enzyme An enzyme that helps control the rate of cellular processes; often an allosteric enzyme. (19.10)

Pancreatic juice A digestive secretion that flows from the pancreas into the upper region of the small intestine; enzymes in pancreatic juice help hydrolyze proteins, carbohydrates, and fats. (21.4)

Partial pressure The pressure that each gas in a mixture contributes to the total pressure. (7.6)

Pascal The SI unit of pressure; 1 atm = 101.3 kPa. (7.3)

Pauli exclusion principle An atomic orbital can contain at most two electrons and two electrons must have opposite spins to occupy the same orbital. (3.6)

Pellagra A disease caused by a deficiency of niacin (nicotinic acid); it is characterized by diarrhea, dementia, and dermatitis. (21.8)

Peptidase An enzyme that catalyzes the hydrolysis of peptide bonds. (19.2)

Peptide A compound formed when two or more amino acids are joined by peptide bonds. (18.4)

Peptide bond The amide bond joining amino acid residues in a peptide; it has the structure

$$\begin{array}{cc} O & H \\ \| & | \\ -C & -N- \end{array} \quad (18.4)$$

Period A horizontal row of elements in the periodic table. (4.1)

Periodic law When the elements are arranged in order of increasing atomic number, there is a periodic repetition of their physical and chemical properties. (4.3)

Periodic table An arrangement of the chemical elements in tabular form to show the periodic recurrence of properties. (4.1)

Peripheral protein A cell membrane protein that perches on either side of the lipid bilayer. (17.4)

Pernicious anemia A disease caused by an inability to absorb vitamin B_{12} (cobalamin); it is characterized by general fatigue. (21.8)

Petroleum refining The distillation of crude oil to divide it into fractions according to boiling point. (11.10)

pH A number used to denote the hydrogen-ion concentration, or the acidity, of a solution; a pH of 7 is neutral. Mathematically: $pH = -\log [H^+]$. (9.3)

pH activity profile A measure showing how the activity of an enzyme varies with changes in pH. (19.8)

pH optimum The pH at which an enzyme has maximum catalytic activity. (19.8)

Phenol A class of organic compounds having the hydroxyl group (—OH) attached directly to a carbon of a benzene ring. (12.3)

Phenylketonuria (PKU) A disease caused by an inborn error of metabolism in which the enzyme responsible for the conversion of phenylalanine to tyrosine is inactive. If untreated, PKU results in severe mental retardation by the age of 6 months. (26.7)

Phosphodiester Any molecule that has the general formula

$$R-O-\underset{\underset{OH}{|}}{\overset{\overset{O}{\|}}{P}}-O-R \quad (20.1)$$

Phosphoglyceride A type of phospholipid molecule that is built from long-chain fatty acids, glycerol, and phosphoric acid. (17.3)

Phospholipid A complex lipid that contains a phosphate group and is a major component of most cell membranes. There are two main types of phospholipids: phosphoglycerides and sphingomyelins. (17.3)

Phosphoryl group A functional group with the structure

$$-\underset{\underset{OH}{|}}{\overset{\overset{O}{\|}}{P}}-OH \quad (14.9)$$

Phosphorylating power The ability of ATP to transfer phosphoryl groups to other molecules. (23.7)

Phosphorylation The transfer of a phosphoryl group to a molecule. (14.9)

Photosynthesis The process in which green plants and algae convert radiant energy from the sun into useful chemical energy in order to synthesize glucose from carbon dioxide and water. (23.2)

Physical change   A change in the state of a substance that does not involve a change in its composition; for example, the change of ice to water. (1.6)

Physical property   The property of a substance that can be measured without changing the composition of the substance; for example, boiling point, color, density. (1.6)

Pi bond   Chemical bond formed by the side-by-side overlap of two atomic *p* orbitals, each containing one electron. (11.8)

Plane-polarized light   Light that has only one plane of vibration. (16.1)

Plaque   The layers of cholesterol and arterial tissue cells that build up in the arteries. The disease in which plaque formation occurs—arteriosclerosis—often leads to a heart attack due to impaired blood circulation. (25.11)

Plasma   The amber fluid that remains after red blood cells, white blood cells, and platelets are removed from whole blood. (22.2)

Plasmolysis   The rupture of any kind of body cells that occurs as a result of an osmotic pressure imbalance when they are placed in a hypotonic solution. (8.11)

Polar bond   A covalent bond in which the bonding electrons are shared unequally by the bonding atoms. (5.8)

Polar covalent bond   A covalent bond between two atoms of different electronegativities in which the bonding electron pairs are not shared equally by the bonded atoms; also called a polar bond. (5.8)

Polar molecule   A molecule that has a dipole; molecules of water and ammonia are polar. (5.8)

Polyamide   A polymer in which the constituent units are joined by amide bonds; nylon and proteins are polyamides. (15.5)

Polyatomic ion   A tightly bound group of atoms that behaves as a unit and has a negative or positive charge. (5.4)

Polycyclic aromatic compound   A derivative of benzene in which carbons are shared between benzene rings; also called a fused-ring compound. (11.9)

Polyester   A polymer that consists of many repeating units of dicarboxylic acids and dihydroxy alcohols joined by ester bonds. (14.6)

Polyfunctional molecule   An organic compound that contains two or more functional groups. (12.10)

Polymer   A very large molecule formed when large numbers of small molecules, known as monomers, are joined by covalent bonds. (12.2)

Polynucleotide   A polymer in which the repeating units are nucleotides; DNA and RNA molecules are polynucleotides. (20.1)

Polypeptide   Any peptide with more than ten amino acid residues. (18.4)

Polysaccharide   A polymer of monosaccharide units; examples include starch, cellulose, and glycogen. (16.6)

Positive modulator   A molecule that binds to the modulator site of an allosteric enzyme and induces formation of an active site. (19.10)

Positive nitrogen balance   The excretion of less nitrogen than is consumed. Children have a positive nitrogen balance; since they are growing, their cells are making new proteins and other nitrogen compounds. (26.4)

Posttranslational processing   The modification of proteins after the proteins are synthesized at the ribosomes. (20.8)

Precision   The deviation of a set of estimates from their average; precise measurements may be accurate or inaccurate. (2.1)

Primary structure   The sequence of amino acids in a protein molecule. (18.5)

Product   A substance formed in a chemical reaction. (1.7)

Proenzyme   A physiologically inactive form of an enzyme; also called a zymogen. (19.11)

Progesterone   A female sex hormone that helps ready the uterus for pregnancy, prevents the release of further ova (eggs), and prepares the breasts for lactation. (17.8)

Prostaglandin   A potent hormonelike substance; prostaglandins are present in nearly all tissues and organs of the body and are derived from unsaturated fatty acids. (17.11)

Prosthetic group   The nonprotein coenzyme attached to a conjugated protein. Prosthetic groups are often metal ions or small organic molecules. (18.7)

Protease   Any enzyme that catalyzes the hydrolysis of one or more peptide bonds in a protein. (19.2)

Protein   Any peptide with more than 100 amino acid residues. (18.5)

Protein turnover   The dynamic process by which many of the body's proteins are continuously hydrolyzed and synthesized within body cells. (26.2)

Proton   A positively charged subatomic particle with a mass of 1 amu or 1.67×10^{-24} g found in the nucleus of an atom; also called a hydrogen ion (H^+). (3.2)

Protonated amine   A cation produced by the combination of an amine with a hydrogen ion. (15.2)

Purine base   A component of nucleic acids related to purine. Adenine and guanine are purine bases. (20.1)

Pyranose ring system   A six-membered sugar ring system that is considered a derivative of pyran. (16.3)

Pyrimidine base   A component of nucleic acids related to pyrimidine. Cytosine, thymine, and uracil are pyrimidine bases. (20.1)

Qualitative   An approximate evaluation. (2.1)

Quantitative   An exact quantity usually involving a measurement with a number and a unit. (2.1)

Quantum   A small package of energy. (3.5)

Quaternary ammonium salt   An ammonium salt with four organic groups (the same or different); the general formula is $R_4N^+Cl^-$. (15.3)

Quaternary structure   The organization of protein subunits into a biologically active assembly. (18.8)

Rad   A measure of the energy absorbed by an object exposed to a radiation source; 1 rad is the quantity of ionizing radiation that delivers 100 erg of energy to 1 g of a substance. (10.9)

Radiation   The penetrating rays emitted by a disintegrating radioactive isotope or an X-ray machine. (10.1)

Radiation therapy   The use of ionizing radiation to kill cancer cells. (10.10)

Radioactive decay   The loss of energy, as radiation, by a radioisotope and the conversion of an atom of one element into an atom of a different element. (10.1)

Radioactivity   The name used to describe the spontaneous disintegration of an atomic nucleus; material that undergoes spontaneous disintegration is radioactive. (10.1)

Radioisotope   An isotope that is radioactive; a radioisotope spontaneously emits ionizing radiation. (10.1)

Rancid   Having an unpleasant smell or taste because of chemical decomposition. (17.2)

Rate of decay   The number of radioactive atoms that decay in a given period of time; the intensity of radiation is proportional to the rate of decay. (10.9)

Rate of reaction   The amount of reactant consumed, or product formed, in a chemical reaction in a specified time. (6.7)

Reactant   A starting substance in a chemical reaction. (1.7)

Recombinant DNA   New DNA chains produced in the laboratory by breaking apart and recombining the DNA chains of different organisms. (20.11)

Redox reaction   Another name for an oxidation-reduction reaction. (6.4)

Reducing agent   The substance in an oxidation-reduction reaction that donates the shifted electrons. (6.4)

Reducing power   The ability of NADH to supply electrons that can be used to reduce oxygen to water in cellular respiration and to reduce carbon compounds in synthetic or anabolic reactions. (23.6)

Reducing sugar   A sugar with an aldehyde group or potential aldehyde group that reduces cupric ions (Benedict's reagent) or silver ions (Tollens' reagent) in alkaline solution. (16.9)

Reduction   The complete or partial gain of electrons by a substance. (6.4)

Rem   The quantity of any type of radiation that produces the same biological effect in humans as those resulting from the absorption of 1 R of X rays or gamma rays. (10.8)

Replication   The process by which a single DNA molecule produces two exact copies of itself; replication occurs during mitosis (cell division). (20.3)

Representative element   Any of the Group A elements of the periodic table, so-called because they represent examples of many chemical and physical properties. (4.3)

Resonance hybrid   An average of two or more resonance structures of a molecule or ion; the true molecule is a resonance hybrid of all of these structures. (11.9)

Resonance structure   Structural formulas for a molecule in which the nuclei have the same arrangement, but the arrangement of the electrons is different. (11.9)

Respiratory acidosis   A decrease in blood pH caused by hypoventilation (very shallow breathing). (22.8)

Respiratory alkalosis   A rise in blood pH caused by hyperventilation (rapid, deep breathing). (22.8)

Respiratory chain   See *Electron transport chain.* (23.6)

Retrovirus   A virus that contains RNA as its genetic material. (20.3)

Reversible reaction   A process in which the conversion of reactants into products (the forward reaction) and the conversion of products into reactants (the backward reaction) occur simultaneously; for example, $A + B \rightleftharpoons C + D$. (6.8)

Ribonucleic acid (RNA)   A nucleic acid important in the synthesis of proteins; a polymer of ribonucleotides. See also *Messenger RNA (mRNA), Ribosomal RNA (rRNA),* and *Transfer RNA (tRNA).* (20.1)

Ribosomal RNA (rRNA)   The nucleic acid component of ribosomes. (20.6)

Ribosome   A small spherical body or organelle found in cells; composed largely of ribosomal RNA and protein. Ribosomes are the site of protein synthesis in a cell. (20.6)

Rickets   A disease of children caused by a vitamin D deficiency; the characteristic symptoms of rickets are bowed legs and malformed ribs. (21.7)

RNA polymerase   An enzyme that catalyzes the formation of RNA from ribonucleotides, using a strand of DNA as a template. (20.7)

Roentgen (R)   A measure of the output from a radiation source; 1 R is the amount of X rays or gamma radiation that produces ions carrying a total of 2.1×10^9 units of electrical charge in 1 cm^3 of dry air at 0 °C and 1 atm pressure. (10.9)

Saliva   The digestive juice secreted in the mouth by the salivary glands; it contains the enzyme amylase (ptyalin) that catalyzes starch hydrolysis. (21.2)

Salt   An ionic compound; the product of the reaction of an acid with a base. (9.5)

Salt bridge   An ionic bond that contributes to peptide chain conformations in proteins. (18.7)

Saponification   The alkaline hydrolysis of fats or oils (triglycerides) yields glycerol and salts of fatty acids; soaps are the alkali metal salts of fatty acids. (17.2)

Saturated compound   An organic compound in which the carbons are joined to each other by single covalent bonds. (11.8)

Saturated solution   A solution that can dissolve no more solute at the existing conditions of temperature and pressure; the dissolved solute and the undissolved solute are in dynamic equilibrium. (8.6)

Saytzeff's rule   The major product in an elimination reaction is the alkene with the largest number of carbon groups on the double bond. (12.5)

Scientific method   Method of inquiry involving experimentation, observation, and explanation. (1.2)

Scientific notation   Expression of number is the form $N \times 10n$, where N is equal to a number greater than 1 and less than 10 and n is an integer. (2.3)

Scintillation counter   An instrument used to detect ionizing radiation. (10.7)

Scurvy   A disease caused by lack of vitamin C (ascorbic acid) in the diet. (21.8)

Second messenger   A molecule used to transmit signals within cells; cyclic AMP (cAMP), calcium ions, and nitric oxide (NO) are second messengers. (24.11)

Secondary structure   Certain regular arrangements of protein chains, typically the beta pleated sheet and the alpha helix. (18.6)

Sedative   A substance that tends to tranquilize the human mind. (15.9)

Self-dissociation of water   The dissociation of H_2O into H^+ ions and OH^- ions due to collisions between water molecules; pure water contains very low concentrations of H^+ ions and OH^- ions in dynamic equilibrium with unionized water molecules. (9.1)

Semipermeable membrane   A natural material such as a cell membrane or a synthetic such as cellophane; semipermeable membranes allow small molecules and ions to pass through but exclude larger particles. (8.11)

Serum   The straw-colored liquid that separates from blood that has been allowed to clot; serum lacks the formed elements and clotting agents. (22.2)

Sex hormones   Steroid hormones that produce the primary and secondary sex characteristics in males and females. (17.5)

SI   See *International System of Units.* (2.2)

Sigma bond   Chemical bond formed by overlap of two spherical half-filled *s* orbitals, overlap of half-filled *s* and *p* orbitals, or end-to-end overlap of two half-filled *p* orbitals. (11.2)

Significant figure   All the digits in a measurement that can be known with certainty plus a last digit that is estimated. (2.4)

Simple sugar   Another name for a monosaccharide. (16.2)

Simple triglyceride   A triester made from glycerol and three molecules of one kind of fatty acid. (17.2)

Single covalent bond   A covalent bond in which only one pair of electrons is shared by two bonded atoms. (5.6)

Soap   The alkali metal salts of long-chain fatty acids; sodium stearate ($C_{17}H_{35}COO^-Na^+$) is a typical soap. (14.3)

Solid   A state of matter in which the component particles are in contact with each other; a solid has a definite volume and definite shape. (1.4)

Solubility   The quantity of a substance that dissolves in a given amount of solvent at specified conditions of temperature and pressure. (8.6)

Solute   A dissolved substance. (8.3)

Solution   Another name for a homogeneous mixture, but more usually referring to a liquid (solvent) in which another liquid or solid (solute) is dissolved. (1.5)

Solvation   Interactions of solute particles with a solvent; solvation by water tends to stabilize ions in aqueous solutions, hence many ionic compounds dissociate to ions in water. (8.4)

Solvent   The dissolving medium in a solution; water is the most common solvent. (8.3)

Somatic damage   Radiation damage that affects the irradiated organism during its own lifetime. (10.8)

Specific gravity   The ratio of the density of a substance to the density of a standard substance; water at 4 °C is the usual standard. Specific gravity has no units. (2.9)

Specific heat   The quantity of heat, in calories, required to raise the temperature of 1 g of a substance 1 °C. (2.10)

Specific heat capacity   See *Specific heat.* (2.10)

Specificity   The ability of many enzymes to catalyze only one chemical reaction using only one substrate. (19.3)

Sphingomyelin   A phospholipid that contains sphingosine, a long-chain unsaturated amino alcohol; large amounts of sphingomyelins are found in brain and nerve tissue. (17.3)

Standard temperature and pressure (STP)   A temperature of 0 °C and a pressure of 760 mm Hg or 1 atm. (7.3)

Starch A polysaccharide found in plants that yields only glucose upon complete hydrolysis; amylose starch is a linear chain, whereas amylopectin is a branched molecule. (16.6)

Stereoisomers Isomers that are nonsuperimposable mirror images of each other. (16.1)

Steroid A class of organic compounds that contain four fused rings; many biologically important compounds are steroids (cholesterol, the sex hormones, vitamin D). (17.5)

Steroid hormone An organic compound that acts as a chemical messenger and has a steroid nucleus; progesterone and testosterone are important steroid hormones. (17.6)

Sterol A steroid molecule containing the hydroxy function; cholesterol is a sterol. (17.5)

Stoichiometry The branch of chemistry concerned with the amounts of substances involved in chemical reactions; stoichiometry is a form of chemical bookkeeping. (6.3)

Storage fat Fat collected in the cells of adipose tissue (fat tissue); also called depot fat. Triglycerides are the major component of storage fat. (25.1)

Straight-chain alkane A saturated open-chain hydrocarbon in which all carbons are arranged consecutively; that is, there is no branching. (11.4)

Strong acid A substance that is capable of donating hydrogen ions and is completely (or almost completely) ionized in aqueous solution; HCl, H_2SO_4, and HNO_3 are strong acids. (9.2)

Strong base A substance that is capable of accepting protons and is completely (or almost completely) ionized in aqueous solution; NaOH and KOH are strong bases. (9.2)

Structural formula A formula that shows the arrangement of atoms in a molecule; each covalent bond is drawn as a dash. (11.2)

Structural isomers Compounds that have the same molecular formula but different molecular structures. (11.6)

Sublimation The conversion of a solid to vapor without passing through the liquid state. (1.6)

Substituent Any group attached to a parent hydrocarbon chain. (11.5)

Substrate The molecule on which an enzyme acts. (19.3)

Subunit One of the individual protein molecules in the organization of a protein's quaternary structure. (18.8)

Sugar A general name for any carbohydrate, but most often applied to monosaccharides and disaccharides. (16.1)

Supersaturated solution A solution that has a higher solute concentration than in a saturated solution at the same temperature and pressure; the solution is not at equilibrium and is unstable. (8.6)

Surface tension A phenomenon in which the surface of a liquid behaves as if it were covered by a thin skin; this property results from unbalanced attractive forces (hydrogen bonds in the case of water) acting on the molecules at the surface of the liquid. (8.1)

Surfactant A surface active agent; any substance that will lower the surface tension of water; soap and detergents are surfactants. (8.1)

Suspension A mixture from which some of the particles slowly settle upon standing, for example, muddy water. (8.9)

Temperature-pressure relationship The pressure of a gas is directly proportional to the Kelvin temperature if the volume is kept constant. (7.7)

Template In DNA replication: A DNA strand used as a pattern for the formation of a complementary strand of DNA. In transcription: A DNA strand used as a pattern for the formation of a complementary strand of mRNA. (20.4)

Termination The end of peptide chain growth in protein synthesis (translation). (20.8)

Termination codons Three noncoding base triplets of mRNA that block further peptide chain growth in translation: UAA, UAG, and UGA. (20.8)

Tertiary structure The overall folding in space of the peptide chain of a protein; the tertiary structure includes the secondary structure of the protein and is determined by the primary structure. (18.7)

Testosterone The major male sex hormone. (17.7)

Tetrahedron A solid geometric figure formed by four equilateral triangles; a term used to describe the regular arrangement of the four covalent bonds to carbon. (5.10)

Theory A physical or mathematical model designed to explain some observation or phenomenon. (1.2)

Thioalcohol A group of organic compounds with the general formula R—SH; also called a thiol. (12.9)

Thioester An ester that contains sulfur and has the general structure

$$R-\overset{\overset{\displaystyle O}{\|}}{C}-S-R \quad (14.7)$$

Thioether An organic derivative of hydrogen sulfide; a thioether has the general formula R—S—R. (12.9)

Thiol Another name for a thioalcohol. (12.9)

Thrombocyte A formed element of the blood that functions in blood-clotting; also known as a platelet. (22.2)

Thyalkoid Structure located in the chloroplasts of green plants; thyalkoids contain the molecules that comprise the light system. (23.2)

Tissue lipid   Those lipids that comprise the membrane structures of cells; also called working lipids. (25.1)

Titration   A method of reacting a solution of unknown concentration with one of known concentration; this procedure is often used to determine the concentrations of acids and bases. (9.8)

Tollens' reagent   A mild oxidizing agent used to test for the presence of an aldehyde; the reagent is an alkaline solution of silver nitrate. (13.5)

Torr   A unit of pressure: 1 torr = 1/760 atm = 1 mm Hg. (7.3)

***Trans* configuration**   Geometric isomer of an alkene in which groups attached to each carbon of the carbon-carbon double bond are on different sides of the bond. (11.8)

Transamination   The enzyme-catalyzed transfer of an amino group from an amino acid to a keto acid; some nonessential amino acids are synthesized by this reaction. (26.3)

Transcription   A process involving base-pairing in which genetic information contained in a DNA molecule is used to assemble a complementary sequence of bases in an RNA chain. (20.3)

Transfer RNA (tRNA)   A class of short-stranded RNA molecules (molecular weight about 25,000) that bind with amino acids and carry them to the ribosomes in protein synthesis; the nucleotide triplet on tRNA is known as an anticodon. (20.6)

Transition metal   Any of the Group B elements of the periodic table; the transition metals include iron, tin, copper, and many other familiar metals. (4.3)

Translation   The process by which the genetic information in an mRNA strand directs the sequence of amino acids during protein synthesis. (20.3)

Transmutation   The conversion of an atom of one element into an atom of a different element by the emission of radiation. (10.1)

Triglyceride   An ester in which all three hydroxyl groups on glycerol have been esterified by saturated or unsaturated fatty acids; natural oils and fats are triglycerides. (17.2)

Triple covalent bond   A covalent bond in which three pairs of electrons are shared by two bonded atoms. (5.6)

tRNA binding site   Positions on ribosomes at which activated tRNA molecules bind to complementary mRNA codons during translation; there are two tRNA binding sites on a ribosome. (20.8)

Tyndall effect   The visible path produced by a beam of light passing through a colloidal dispersion or suspension; the phenomenon is caused by the scattering of light by the colloidal and suspended particles. (8.9)

Unsaturated compound   An organic compound with one or more double or triple bonds. (11.8)

Unsaturated solution   A solution that can dissolve more solute at a given temperature and pressure. (8.6)

Unshared pair   Pair of valence electrons not involved in covalent bonding in molecules. (5.6)

Unwinding protein   A polypeptide that tends to unwind the double helix because it can bind to and stabilize single-stranded DNA. (20.4)

Urobilin   A colorless product of the degradation of heme, the prosthetic group of hemoglobin. (26.8)

Urobilinogen   An orange-yellow product of the degradation of heme, the prosthetic group of hemoglobin, by oxidation of bilirubin. (26.8)

Valence electron   An electron in the highest occupied energy level of an element's atoms; valence electrons largely determine the chemistry of an element. (5.1)

van der Waals forces   Weakest of the attractive forces between molecules, they include dispersion forces and dipole interactions. (5.9)

Vapor   The gaseous state of a substance that normally exists as a solid or liquid at room temperature. (1.4)

Vapor pressure   The pressure exerted by a vapor that is in dynamic equilibrium with its liquid or solid in a closed system. (7.9)

Vaporization   The conversion of a liquid to a gas or vapor below its boiling point. (7.9)

Vitamin   An organic compound essential in the diet in small amounts; vitamins are fat-soluble or water-soluble. (21.1)

Volume   The space occupied by matter. (2.6)

Water of crystallization   See *Water of hydration.* (8.4)

Water of hydration   Water held in chemical combination in a chemical substance; also called water of crystallization. (8.4)

Wax   An ester of a long-chain fatty acid and a long-chain monohydric alcohol. (17.1)

Weak acid   A substance that is capable of donating hydrogen ions but is only weakly ionized in aqueous solution; acetic acid is a weak acid. (9.2)

Weak base   A substance that is capable of accepting protons but is only weakly ionized in aqueous solution; ammonia is a weak base. (9.2)

Weight   A measure of the force of attraction between the earth and the mass of an object; weight is often used when the word *mass* is meant. (2.8)

Wilson's disease   An inherited disease in which large amounts of copper accumulate in the liver and brain. (21.9)

Working lipid   The lipid in cell membranes and the membranes of cell organelles such as mitochondria. (25.1)

X ray High-energy electromagnetic radiation, extremely penetrating and potentially very dangerous, produced by specially built machines. X rays are essentially the same as gamma radiation except for their origin. (10.1)

Zwitterion The internal salt of an amino acid; a dipolar ion. (18.3)

Zymogen The inactive precursor of an enzyme; for example, pepsinogen is the zymogen of pepsin. (19.11)

Zymogen granule Package of enzyme precursors (zymogens) encased in coats of lipids and proteins. (19.11)

Answers to Selected Exercises

Chapter 1

Practice Exercises
1.1 (a) 338 °F (b) 443 K
1.2 (a) 180 °C (b) 356 °F
1.3 (a) gas (b) liquid (c) solid (d) liquid
1.4 (a) homogeneous (b) homogeneous
(c) heterogeneous (d) heterogeneous
1.5 (a) physical change (b) physical change
(c) chemical change (d) physical change
1.6 (a) sulfur (b) chlorine (c) magnesium
(d) cobalt

Exercises
1.7 A law is a statement of a natural phenomenon to which no exceptions are known. A theory is a tested (by experiments) hypothesis.
1.9 (a) False (b) False (c) True
1.11 (a) 176 °F (b) 353 K
1.13 (a) 78.5 °C (b) 173.3 °F
1.15 (a) 134 °F (b) 329.7 K
1.17 (a) −182.8 °C (b) −297 °F
1.19 The physical states of matter are solid (gold, glass, and wood), liquid (gasoline, wine, and mercury), and gas (oxygen, helium, and carbon monoxide).
1.21 The term *gas* describes only those substances such as nitrogen and oxygen that are in the gaseous state at room temperature. The term *vapor* applies to substances in the gaseous form such as rubbing alcohol and gasoline that are solids or liquids at room temperature.
1.23 The components of a heterogeneous mixture can be seen with either the naked eye or a microscope; a heterogeneous mixture has more than one phase. A homogeneous mixture is a single phase; it looks like a pure substance.
1.25 Mixture (a) (c) (d) (e) (g) (h)
1.27 In a chemical change a new substance with new properties is formed. In a physical change, the composition of the substance does not change even though the substance may undergo a change of state or form.
1.29 (a) Use a magnet to pull out the iron filings (magnetism), or add water to dissolve the sugar (solubility). (b) Distill the water from the salt (boiling point).
1.31 Physical properties: (a), (b), (d), and (e).
1.33 Chemical properties are observed only when a substance undergoes a chemical reaction; iron rusts and iron "dissolves" in some acids. No change in composition is necessary to observe a physical property; iron is shiny, is hard, conducts electricity, and melts at 1535 °C.
1.35 (a) change in color and odor; change not reversible (b) gas produced; change not reversible (c) color change; change in properties; not easily reversed (d) energy given off; not easily reversed
1.37 An element is a pure substance, is homogeneous, and cannot be broken down by chemical means. A compound is homogeneous, has a definite composition, and can be broken down into simpler substances by a chemical change. A mixture has a variable composition; its components can be separated by physical means.
1.39 (a) The red powder is a compound because it was broken down chemically by heating; a new substance with new properties was formed. (b) The shiny liquid is probably an element, but it is not yet proven.
1.41 (a) Ca (b) O (c) K (d) Au (e) Fe (f) H
1.43 The law of conservation of mass states that mass is neither created nor destroyed in ordinary physical and chemical changes.
1.45 Energy is neither created nor destroyed in ordinary physical and chemical changes; energy can be transformed.
1.47 From coldest to hottest: 340 K, 70 °C, 170 °F.
1.49 (a) −153 °C (b) 149 °C (c) 20 °C (d) 16 °C
1.51 (a) 233 K (b) 233 K (c) 298 K (d) 373 K
1.53 Possible answers include: (a) sunlight falling on a solar cell (b) burning gasoline in a lawnmower (c) recharging a calculator battery (d) striking a match
1.55 Murphy's Law has not been tested by experiment, and the generalizations that it makes would be difficult to prove.

Self-Test
1. False 2. True 3. False 4. True 5. True
6. False 7. True 8. False 9. False 10. True
11. b 12. b 13. d 14. d 15. b 16. a
17. b 18. d 19. b 20. b 21. a 22. d
23. b 24. b 25. d 26. d 27. b 28. c
29. c 30. d

Chapter 2

Practice Exercises

2.1 (a) 3.688×10^2 (b) 6.07×10^{-4}
(c) 6.7×10^{10} (d) 1.02×10^{-2}

2.2 (a) 1.39×10^5 (b) 2.34×10^{-2}
(c) 9.089×10^{-5}

2.3 (a) 6.5×10^{-6} (b) 8.5×10^8 (c) 1.71×10^2

2.4 The significant digits in each measurement are underlined.
(a) 16.030 m (5) (b) 0.00450 m (3)
(c) 4.5000 m (5) (d) 9.80×10^{-4} m (3)
(e) 7864 m (4) (f) 0.543 m (3)
(g) 4.00×10^3 m (3) (h) 0.0502 m (3)

2.5 (a) 9.4×10^{-4} m (b) 0.046 m (c) 8600 m
(d) 0.032 m (e) 7.0 m (f) 3.9×10^7 m

2.6 (a) area = 9.8 m^2; perimeter = 13.1 m (b) area = 78.2 m^2; perimeter = 37.0 m (c) area = 2×10^1 m^2; perimeter = 21 m (d) area = 21 m^2; perimeter = 21.2 m

2.7 (a) 5.4×10^{-1} m (b) 5.4×10^2 mm
(c) 5.4×10^5 μm

2.8 (a) 1.5×10^1 mL (b) 1.5×10^4 μL

2.9 (a) 4.3×10^2 mg (b) 4.3×10^{-1} g
(c) 4.3×10^5 μg

2.10 (a) 8.4×10^{-3} kg (b) 8.4×10^3 mg
(c) 8.4×10^6 μg

2.11 (b) has the largest mass.

2.12 (a) 9.6 g (b) $115.20

2.13 (a) 0.859 g/cm^3 (b) Yes

2.14 2.1×10^4 cal

2.15 $0.21 \dfrac{\text{cal}}{\text{g} \times {}^\circ\text{C}}$

Exercises

2.17 (a) mass (b) volume (c) length
(d) volume

2.19 (a) micro- (b) milli- (c) kilo- (d) centi-

2.21 (a) 8.64×10^4 s = 86,400 s (b) 7.5×10^{-6} m = 0.0000075 m (c) 5×10^8 red blood cells = 500,000,000 red blood cells

2.23 (a) 1×10^{-5} (b) 5.5×10^{-8} (c) 1.0×10^7

2.25 (a) 0.064 μL (b) 950 K (c) 0.68 g
(d) 3.0×10^3 km (e) 530×10^7 m (f) 3.8 mm

2.27 (a) 6.4×10^{-2} μL (b) 9.5×10^2 K
(c) 6.8×10^{-1} g (d) 3.0×10^3 km
(e) 5.3×10^9 m (f) 3.8×10^0 mm

2.29 The statement is incorrect, because a significant figure is never dropped. The number to the right of the last significant figure is dropped in rounding if it is less than 5.

2.31 (a) 196 K (b) 18 cm (c) 32.4 m^3
(d) 104 cm^3

2.33 (a) 1.96×10^2 K (b) 1.8×10^1 cm
(c) 3.24×10^1 m^3 (d) 1.04×10^2 cm^3

2.35 (a) 4.56×10^{-5} m (b) 23.4 g (c) 1.7×10^{-2} L
(d) 9.88×10^4 cg

2.37 (a) 4.26 cm (b) 4.26×10^{-2} m
(c) 4.26×10^4 μm

2.39 (d) has the largest mass.

2.41 68.0 L

2.43 (a) centigram (b) liter (c) centisecond
(d) kilocalorie (e) milliliter (f) microgram

2.45 The weight of an astronaut depends on the pull of gravity. The mass of the astronaut is constant, regardless of location.

2.47 1.20 g/mL

2.49 4.44 cm^3

2.51 (a) 1.3×10^4 g (b) 13 kg (c) 1.3×10^7 mg

2.53 8.8 g/cm^3. No, the density of silver is 10.5 g/cm^3.

2.55 1.0×10^1 L

2.57 23.6 kg

2.59 11 cm tall

2.61 A joule is a unit of energy. There are 4.18 J in 1 calorie.

2.63 (a) 3.89×10^4 J (b) 38.9 kJ

2.65 68.3 °C

2.67 (a) 1.5 kcal (b) 6.3 kJ

2.69 12 g

2.71 (a) 25 °C (b) 5.0×10^1 °C

2.73 The significant figures in each measurement are underlined: (b), (c), (d), and (f) all have more than two significant figures.
(a) 0.00072 (b) 8.07×10^4 (c) 155
(d) 6.00×10^{-3} (e) 8,700 (f) 1.0003

2.75 28 cal/sec

2.77 4.6×10^3 gal/day

2.79 (b) is the longest.

2.81 (a) second (b) kelvin (c) mole
(d) meter (e) kilogram

2.83 0.804

2.85 47.6 mL

2.87 Lissa's measurements are accurate but they are not precise. Manuel's measurements are both accurate and precise. Yukari's measurements are precise, but they are not accurate. Leigh Anne's measurements are neither accurate nor precise.

2.89 4.60×10^{22} atoms

2.91 2.2×10^3 g of water

2.93 (a) 0.0170 g/L (b) 1.70×10^{-5} g/cm^3

Self-Test

1. False 2. True 3. False 4. True 5. False
6. False 7. True 8. False 9. True 10. True
11. c 12. c 13. b 14. c 15. b 16. a
17. c 18. b 19. b 20. d 21. c 22. c
23. c 24. c 25. c 26. d 27. a 28. a
29. d 30. b

Chapter 3

Practice Exercises

3.1 Eight protons and eight electrons.

3.2 Sodium (Na)

3.3

Atomic number	Mass number	Number of protons	Number of neutrons	Number of electrons	Symbol of element
4	9	4	5	4	Be
11	23	11	12	11	Na
24	52	24	28	24	Cr
18	40	18	22	18	Ar

3.4 potassium-39, $^{39}_{19}$K; potassium-41, $^{41}_{19}$K.

3.5 (a) 24   (b) 48   (c) 117

3.6 69.8 amu

3.7 (a) 3   (b) 1   (c) 7   (d) 3   (e) 5   (f) 9

3.8 $3d$ $4s$ $3p$ $3s$ $2p$.

3.9 (a) $1s^2$ $2s^2$ $2p^1$; one unpaired electron.
     (b) $1s^2$ $2s^2$ $2p^5$; one unpaired electron.

3.10 (a) 5   (b) 6   (c) 5   (d) 7

Exercises

3.11 Each element is composed of atoms that are different from the atoms of any other element. Compounds are a chemical combination of the atoms of two or more different elements. In a chemical reaction atoms are rearranged to form new substances with new properties.

3.13 In a chemical reaction atoms are not created or destroyed, rather, the way in which the atoms are joined together is changed.

3.15 (a) A +1 charge and a mass of about 1 amu.
      (b) A charge of 0 and a mass of about 1 amu.
      (c) A charge of −1 and essentially "massless."

3.17 Moseley showed that atoms of different elements have different numbers of protons. Before Moseley's discovery, elements were arranged on the periodic table in order of increasing atomic mass, rather than increasing atomic number, as is done today.

3.19 (a) 9   (b) 7   (c) 6   (d) 38

3.21 An atomic mass unit is defined as one-twelfth of the mass of a carbon atom that contains six protons and six neutrons.

3.23 Atomic number, 18; mass number, 40; symbol, Ar.

3.25 Isotopes of an element have different (1) mass numbers, (2) numbers of neutrons and (3) atomic masses.

3.27 (a) oxygen-17   (b) $^{17}_8$O   (c) 8 protons and 9 neutrons

3.29 $^{60}_{27}$Co

3.31 (a) 13 protons; 14 neutrons; 13 electrons
      (b) 29 protons; 36 neutrons; 29 electrons

(c) 17 protons; 20 neutrons; 17 electrons
(d) 3 protons; 4 neutrons; 3 electrons

3.33 107.9 amu; the atomic mass of silver on the periodic table is 107.868 amu.

3.35 Quantized energies of electrons mean that electrons can exist only at certain energy levels.

3.37 (1) Aufbau principle, electrons enter orbitals of lowest energy first; (2) Pauli exclusion principle, an atomic orbital may describe at most two electrons; (3) Hund's rule, when electrons occupy orbitals of equal energy they do so with spins parallel as much as possible.

3.39 (a) 10   (b) 6   (c) 2   (d) 14

3.41 The symbol $4d^3$ represents the three electrons in the $4d$ sublevel. The 4 indicates the principle energy level and the d shows the energy type of orbital.

3.43 (a) $1s^2$ $2s^2$ $2p^2$; carbon   (b) two unpaired electrons

3.45 (a) 7   (b) 0   (c) 8

3.47 (a) Na, sodium   (b) N, nitrogen   (c) Si, silicon
      (d) K, potassium

3.49 The first energy level ($1s$) is a single orbital that is complete with two electrons. The third electron must be in the $2s$ orbital of the second energy level.

3.51 (a) $1s^2$ $2s^2$ $2p^6$ $3s^2$ $3p^3$ (b) $1s^2$ $2s^2$ $2p^5$
      (c) $1s^2$ $2s^2$ $2p^6$ $3s^2$ $3p^6$

3.53 See Figure 3.4 in the text. The three p orbitals are oriented at right angles, each directed along an axis of a three-dimensional coordinate system.

3.55 Hund's rule states that you don't pair up electrons in an energy sublevel until you have to. Silicon, $1s^2$ $2s^2$ $2p^6$ $3s^2$ $3p^2$, has 2 unpaired electrons. Phosphorus, $1s^2$ $2s^2$ $2p^6$ $3s^2$ $3p^3$, has 3 unpaired electrons. Sulfur, $1s^2$ $2s^2$ $2p^6$ $3s^2$ $3p^4$, has 2 paired and 2 unpaired electrons in the $3p$ sublevel.

3.57 (a) 2   (b) 2   (c) 2

3.59 (a) $1s^2$ $2s^2$ $2p^6$ $3s^1$; sodium   (b) $1s^2$ $2s^2$ $2p^6$ $3s^2$ $3p^3$; phosphorus   (c) $1s^2$ $2s^2$ $2p^4$; oxygen   (d) $1s^2$ $2s^2$ $2p^6$ $3s^2$ $3p^5$; chlorine   (e) $1s^2$ $2s^2$ $2p^6$ $3s^2$ $3p^1$; aluminum

3.61 (a) $^{34}_{16}$Q and $^{36}_{16}$Q; $^{17}_8$Q and $^{18}_8$Q   (b) $^{36}_{18}$Q and $^{36}_{16}$Q
      (c) $^{10}_5$Q and $^9_4$Q; $^{34}_{16}$Q and $^{36}_{18}$Q   (d) $^{34}_{16}$Q and $^{36}_{16}$Q; $^{17}_8$Q and $^{18}_8$Q

Self-Test

1. True   2. False   3. True   4. True   5. False
6. True   7. True   8. True   9. False   10. False
11. a   12. b   13. d   14. b   15. c   16. d
17. c   18. d   19. c   20. b   21. c   22. b
23. b   24. a   25. a   26. c   27. b   28. a
29. c   30. b

Chapter 4

Practice Exercises

4.1 (a) $1s^2 2s^2 2p^3$ (b) $1s^2 2s^2 2p^6 3s^1$ (c) $1s^2 2s^2 2p^6 3s^2 3p^6 4s^2 3d^8$ (d) $1s^2 2s^2 2p^6 3s^2 3p^6 4s^2 3d^{10} 4p^6 5s^2$

4.2 (a) $1s^2 2s^2 2p^6 3s^2 3p^6 4s^2 3d^{10} 4p^6$ (b) $1s^2 2s^2 2p^6 3s^2 3p^6 4s^2 3d^{10} 4p^1$ (c) $1s^2 2s^2 2p^6 3s^2 3p^6 4s^2 3d^{10} 4p^6 5s^1$

4.3 Compared with oxygen, fluorine has one additional proton and electron. Since the shielding effect is constant in moving across a period, the increase in nuclear charge pulls the outermost electrons closer to the nucleus. Chlorine has electrons in the next higher energy level, resulting in a larger atom.

4.4 (a) C, carbon (b) Cl, chlorine (c) Mo, molybdenum (d) V, vanadium

Exercises

4.5 Mendeleev arranged the columns so that the elements with the most similar properties were side-by-side.

4.7 (a) A group is a vertical column on the periodic table. (b) A group is identified by a number-letter combination, for example, 1A and 4B.

4.9 (a) 7A (b) 5A (c) 3A (d) 1A

4.11 All metals are solids at room temperature except mercury, which is a liquid. When freshly cut, metals have a high luster. Metals are malleable, are ductile, and have high electrical conductivity.

4.13 (a) C and P (b) Hg (c) Ar (d) Si (e) C (as graphite)

4.15 (a) P, (c) S, and (e) Ar are nonmetals; (b) K, (d) Fe, and (f) Ca are metals.

4.17 The periodic law states: When elements are arranged in order of increasing atomic number, there is a periodic repetition of their physical and chemical properties.

4.19 The representative elements are the seven group A elements and the group 0 elements. The inner transition elements are found at the bottom of the periodic table with atomic numbers in the ranges of 58 to 71 and 90 to 103.

4.21 The noble gases, the Group 0 elements, have outer s and p sublevels that are filled. The representative elements have outer s and p sublevels that are only partially filled. In the transition metals the outermost s sublevel and neighboring d sublevel contain electrons. In the inner transition metals the outermost s sublevel and the neighboring f sublevel contain electrons.

4.23 (a) $1s^2 2s^2 2p^5$ (b) $1s^2 2s^2 2p^6 3s^2 3p^1$

(c) $1s^2 2s^2 2p^6 3s^2 3p^6 4s^2 3d^{10}$ (d) $1s^2 2s^2 2p^6 3s^2 3p^6 4s^2 3d^{10} 4p^6 5s^2 4d^{10} 5p^2$

4.25 (a) The outermost energy level is filled. (b) The outermost energy level has a number of electrons equal to the group number.

4.27 (a) $1s^2 2s^2 2p^6 3s^2 3p^6 4s^2$ (b) $1s^2 2s^2 2p^6 3s^2 3p^6 4s^2 3d^{10} 4p^5$ (c) $1s^2 2s^2 2p^6 3s^1$ (d) $1s^2 2s^2 2p^6 3s^2 3p^6 4s^2 3d^5$

4.29 (a) fluorine, oxygen, boron, lithium (b) Atomic size tends to decrease moving from left to right across a period in the periodic table so this is a periodic trend.

4.31 Ionization energy is energy required to remove an electron from an atom in the gaseous state.

4.33 (a) boron (b) magnesium (c) aluminum

4.35 (a) potassium, K (b) argon, Ar (c) fluorine, F (d) barium, Ba

4.37 (a) representative elements, transition metals, and noble gases (b) representative elements (40%); transition metals (55%); noble gases (5%)

4.39 helium, He

4.41 sodium, Na, and lithium, Li

4.43 Alkaline earth metals react more slowly with water and are harder than alkali metals. When comparing elements in the same period, alkali metals tend to have lower ionization energies and larger atomic size than the corresponding alkaline earth metal.

4.45 Both gallium (mp = 29.8 °C, bp = 2403 °C) and mercury (mp = −38.87 °C, bp = 356.58 °C) are liquids at or near room temperature. Both metals expand upon heating, making them suitable for use in thermometers.

4.47 increase

4.49 high luster, malleable, ductile, and good conductors of heat and electricity.

4.51 Refer to Figure 4.6 on page 90 in the text. (a) The inner transition metals are at the bottom of the periodic table. (b) The representative elements are on each side of the periodic table; groups 1A through 7A and group 0. (c) The transition metals are in the middle of the periodic table; the group B metals.

4.53 (a) calcium, Ca (b) helium, He (c) carbon, C (d) chlorine, Cl (e) sodium, Na (f) neon, Ne (g) calcium, Ca (h) lithium, Li

Self-Test

1. False 2. False 3. False 4. True 5. False
6. False 7. True 8. False 9. False
10. False 11. d 12. c 13. d 14. c 15. c
16. c 17. d 18. d 19. c 20. c 21. b
22. b 23. c 24. a 25. c 26. c 27. d
28. b 29. b 30. c

Chapter 5

Practice Exercises

5.1 (a) 1   (b) 4   (c) 2   (d) 6

5.2 (a) $\cdot \overset{\cdot}{\underset{\cdot}{C}} \cdot$ (b) $:\overset{\cdot}{\underset{\cdot}{S}}:$ (c) $:\overset{\cdot}{P}\cdot$ (d) $\cdot Mg \cdot$

5.3 (a) Ca^{2+} (b) K^+ (c) Al^{3+}

5.4 (a) S^{2-} (b) N^{3-} (c) F^-

5.5 (a) $MgBr_2$ (b) $AlCl_3$ (c) $Ca(HCO_3)_2$
(d) $Mg_3(PO_4)_2$

5.6 (a) magnesium oxide   (b) sodium sulfide
(c) magnesium hydroxide   (d) ammonium chloride

5.7 (a) KI   (b) MgF_2 (c) $BaSO_4$ (d) $Cu_3(PO_4)_2$

5.8 (a) $:\overset{\cdot\cdot}{\underset{\cdot\cdot}{Cl}}:\overset{\cdot\cdot}{\underset{\cdot\cdot}{Cl}}:$ (b) $:\overset{\cdot\cdot}{\underset{\cdot\cdot}{Br}}:\overset{\cdot\cdot}{\underset{\cdot\cdot}{Br}}:$ (c) $:\overset{\cdot\cdot}{\underset{\cdot\cdot}{I}}:\overset{\cdot\cdot}{\underset{\cdot\cdot}{I}}:$

5.9 (a) $H:\overset{\cdot\cdot}{\underset{\cdot\cdot}{S}}:H$ (b) $H:\overset{\cdot\cdot}{\underset{H}{P}}:H$

(c) $:\overset{\cdot\cdot}{\underset{\cdot\cdot}{Cl}}:\overset{\cdot\cdot}{\underset{\cdot\cdot}{F}}:$ (d) $H:\overset{H}{\underset{H}{C}}:H$

5.10 (a) polar covalent   (b) ionic
(c) polar covalent   (d) polar covalent

5.11 The most polar bond is (a), H—Cl.

5.12 The single covalent bond in the HF molecule is polar because of an electronegativity difference of 1.9. Hydrogen fluoride is a polar molecule because this polar bond creates unbalanced electrical poles.

5.13 $\overset{\cdot\cdot}{\underset{}{Cl}}$
$:\overset{\cdot\cdot}{\underset{\cdot\cdot}{Cl}}:\overset{}{\underset{}{C}}:\overset{\cdot\cdot}{\underset{\cdot\cdot}{Cl}}:$
$:\overset{\cdot\cdot}{\underset{\cdot\cdot}{Cl}}:$ Since there are four bonding pairs of electrons, this molecule has the same shape as methane, a regular tetrahedron.

5.14 Phosphorus, like nitrogen, has 5 valence electrons. The phosphorus atom forms three single covalent bonds with hydrogen and has one unshared pair of electrons. A molecule of PH_3 would not be planar. PH_3 has the same pyramidal shape as ammonia, NH_3.

Exercises

5.15 Valence electrons are those in the highest energy level. For the group A elements, the group number equals the number of valence electrons.

5.17 (a) 5   (b) 6   (c) 7   (d) 1

5.19 The elements in set (c) are all in Group 2A and all have 2 valence electrons. The elements in set (d) are all in Group 7A and all have 7 valence electrons.

5.21 (a) Cations are positively charged ions.
(b) Cations are formed when atoms of metallic elements lose one or more valence electrons.

5.23 (a) 2   (b) 2   (c) 1   (d) 3   (e) 1

5.25 (a) Ca^{2+} (b) Na^+ (c) K^+ (d) Sr^{2+}

5.27 (b)

5.29 An ionic bond is a force of attraction between oppositely charged ions.

5.31 (a) K_2S (b) CaO   (c) Na_2SO_4 (d) $AlPO_4$

5.33 (a) potassium ion, K^+; chloride ion, Cl^-
(b) potassium ion, K^+; permanganate ion, MnO_4^-
(c) calcium ion, Ca^{2+}; sulfate ion, SO_4^{2-}
(d) magnesium ion, Mg^{2+}; bromide ion, Br^-

5.35 (a) sodium bromide   (b) silver oxide
(c) calcium hydrogen sulfate   (d) iron(II) oxide or ferrous oxide   (e) copper(I) cyanide or cuprous cyanide   (f) sodium hydrogen carbonate or sodium bicarbonate

5.37 (b) and (d)

5.39 (a) +1   (b) +2   (c) +2   (d) +3

5.41 (a) $:\overset{\cdot}{\underset{\cdot}{S}}:$

(b) Sulfur and oxygen are in the same group (6A) on the periodic table. Atoms in the same group have the same number of valence electrons.

5.43 Both a sulfide ion, S^{2-}, and a chloride ion, Cl^-, have 8 valence electrons.

$$:\overset{\cdot\cdot}{\underset{\cdot\cdot}{S}}:^{2-} \quad :\overset{\cdot\cdot}{\underset{\cdot\cdot}{Cl}}:^-$$

5.45 In forming chemical bonds, atoms achieve a state of lower energy. Atoms transfer electrons to form an ionic bond; atoms share electrons in a covalent bond.

5.47 Helium atoms have a stable electron configuration; hydrogen atoms attain a noble gas configuration by sharing electrons.

5.49 (a) $H:\overset{\cdot\cdot}{\underset{H}{O}}:$ (b) $:\overset{\cdot\cdot}{\underset{\cdot\cdot}{Cl}}:\overset{\cdot\cdot}{\underset{\underset{\cdot\cdot}{:\overset{\cdot\cdot}{Cl}:}}{P}}:\overset{\cdot\cdot}{\underset{\cdot\cdot}{Cl}}:$

(c) $H:\overset{\cdot\cdot}{\underset{\cdot\cdot}{O}}:\overset{\cdot\cdot}{\underset{\cdot\cdot}{O}}:H$ (d) $H:\overset{H}{\underset{H}{C}}:\overset{\cdot\cdot}{\underset{\cdot\cdot}{Cl}}:$

5.51 (a) covalent   (b) covalent   (c) ionic
(d) covalent   (e) ionic   (f) covalent

5.53 Electronegativity is a measure of the electron attracting power of an atom in a covalent bond. The greater the electronegativity, the greater the pull on the electrons.

5.55 Electronegativity increases moving across a period from left to right. Moving down a group, electronegativity decreases. The exception is the noble gas group; the electronegativity of each element in Group 0 is 0.0.

5.57 (a) ionic   (b) nonpolar covalent   (c) ionic
(d) polar covalent (shift toward oxygen)

(e) polar covalent (shift toward oxygen)

(f) polar covalent (shift toward oxygen)

5.59 Valence electrons orient themselves so that they are as far apart as possible. In methane the four pairs of bonding electrons meet this criterion when they are 109.5 degrees apart; this gives methane its tetrahedral shape. In water the two unshared pairs of electrons force the bonding pairs closer together giving water its characteristic bent shape with an angle of about 105 degrees.

5.61 (a) 1 Na, 2 H, 1 P, and 4 O   (b) 2 N, 8 H, 1 S, and 4 O   (c) 1 K, 1 Mn, and 4 O   (d) 1 Al, 3 O, and 3 H

5.63 (a) 6A   (b) 4A   (c) 1A   (d) 0

5.65 (a) $:\overset{..}{\underset{..}{Br}}:\overset{..}{\underset{..}{Cl}}:$ (b) $H:C:::N:$

(c) $:\overset{..}{O}::\overset{..}{\underset{..}{S}}:\overset{..}{\underset{..}{O}}:$ (d) $H:C:::C:H$

5.67 If a hydrogen atom gains an electron it forms an anion with a 1− charge. The name of this ion is derived like any other monatomic anion: the suffix of the name is changed to -ide, the hydride ion.

5.69 The electron dot structure of an element shows the number of valence electrons in an atom's outer shell; the location of the element on the periodic table is related to the number of valence electrons, because the group A number is the number of valence electrons.

5.71

	Oxide	Nitrate	Iodide	Phosphate	Hydroxide
Potassium	K_2O potassium oxide	KNO_3 potassium nitrate	KI potassium iodide	K_3PO_4 potassium phosphate	KOH potassium hydroxide
Iron(III)	Fe_2O_3 iron(III) oxide	$Fe(NO_3)_3$ iron(III) nitrate	FeI_3 iron(III) iodide	$FePO_4$ iron(III) phosphate	$Fe(OH)_3$ iron(III) hydroxide
Magnesium	MgO magnesium oxide	$Mg(NO_3)_2$ magnesium nitrate	MgI_2 magnesium iodide	$Mg_3(PO_4)_2$ magnesium phosphate	$Mg(OH)_2$ magnesium hydroxide
Copper(I)	Cu_2O copper(I) oxide	$CuNO_3$ copper(I) nitrate	CuI copper(I) iodide	Cu_3PO_4 copper(I) phosphate	$CuOH$ copper(I) hydroxide

Self-Test

1. False   2. True   3. False   4. True   5. False

6. False   7. True   8. True   9. True   10. True

11. b   12. d   13. c   14. a   15. a   16. b

17. d   18. d   19. d   20. c   21. b   22. b

23. b   24. b   25. c   26. b   27. b   28. c

29. a   30. d

Chapter 6

Practice Exercises

6.1 (a) 119.0 amu   (b) 136.2 amu   (c) 84.0 amu   (d) 180.0 amu

6.2 (a) 16.0 g   (b) 342.0 g   (c) 164.1 g   (d) 71.0 g

6.3 (a) 34.0 g H_2O_2 (b) 14.2 g Cl_2 (c) 43.6 g K_2SO_4 (d) 3.00 g $CaCO_3$

6.4 (a) 0.0500 mol NaOH   (b) 5.00 mol Na   (c) 0.0499 mol H_2SO_4 (d) 0.0250 mol Na_3PO_4

6.5 (a) $2C + O_2 \longrightarrow 2CO$ (b) $2Al + 3S \longrightarrow Al_2S_3$ (c) $BaCl_2 + Na_2SO_4 \longrightarrow BaSO_4 + 2NaCl$ (d) $2Na + Br_2 \longrightarrow 2NaBr$

6.6 (a) $N_2 + 3H_2 \longrightarrow 2NH_3$ (b) $4Na + O_2 \longrightarrow 2Na_2O$ (c) $H_2 + S \longrightarrow H_2S$

6.7 (a) 1.0 mol O_2 (b) 0.40 mol H_2O (c) 6.40 g CH_4

6.8 (a) Sodium is oxidized and acts as a reducing agent. Sulfur is reduced and acts as an oxidizing agent.   (b) Potassium is oxidized and acts as a reducing agent. Chlorine is reduced and acts as an oxidizing agent.   (c) Aluminum is oxidized and acts as a reducing agent. Oxygen is reduced and acts as an oxidizing agent.   (d) Oxygen "gains" electrons by an electron shift; it is reduced and acts as an oxidizing agent. Nitrogen "loses" electrons; it is oxidized and acts as a reducing agent.

6.9 Because the products of (a)–(c) are covalently bonded, we are looking for electron shifts.

(a) Chlorine is reduced and is an oxidizing agent. Hydrogen is oxidized and is a reducing agent.

(b) Nitrogen is reduced and is an oxidizing agent. Hydrogen is oxidized and is a reducing agent.

(c) Chlorine is reduced and is the oxidizing agent. Sulfur is oxidized and is the reducing agent.

(d) Lithium is oxidized and is itself a reducing agent. Fluorine is reduced and acts as an oxidizing agent.

6.10

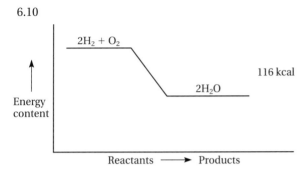

6.11 (a) playing cards in use   (b) salt dissolved in water   (c) cup of water at 75 °C   (d) 1 g of powdered sugar

6.12 At the higher temperature outside the refrigerator, the chemical reactions that take place when food spoils proceed at a much faster rate.

6.13 (1) increase temperature; increases both the frequency and energy of collisions (2) use of catalyst; lowers activation energy (3) increase reactant concentration; increases number of collisions (4) decrease particle size; increases number of collisions

6.14 Cooling removes a "reactant" and the equilibrium position shifts to the left.

6.15 $K_{eq} = \dfrac{[NH_3]^2}{[H_2]^3[N_2]}$

6.16 $12 \, (mol/L)^{-2}$

6.17 $[I_2] = 0.500$ mol; $[HI] = 3.54$ mol

6.18 Products are favored in (a) and (b).

Exercises

6.19 (a) 28.0 amu   (b) 85.0 amu   (c) 16.0 amu
      (d) 24.3 amu   (e) 14.0 amu   (f) 132.1 amu

6.21 (a) $32.0 \, g \, O_2$ (b) 146 g NaCl   (c) 59.3 g $Mg(NO_3)_2$ (d) 147 g Al   (e) $35.5 \, g \, Na_2SO_4$ (f) $25.5 \, g \, NH_3$

6.23 (a) $1.27 \, mol \, CaCl_2$ (b) $0.100 \, mol \, H_2O$
      (c) 0.0250 mol NaOH   (d) 0.296 mol C
      (e) $0.191 \, mol \, Ba(NO_3)_2$ (f) $0.0500 \, mol \, CO_2$

6.25 In any physical or chemical change, matter is neither created nor destroyed.

6.27 (a) $H_2 + Cl_2 \longrightarrow 2HCl$
      (b) $2Al + 6HCl \longrightarrow 2AlCl_3 + 3H_2$
      (c) $2C_2H_6 + 7O_2 \longrightarrow 4CO_2 + 6H_2O$
      (d) $CaCO_3 + 2HCl \longrightarrow CaCl_2 + H_2O + CO_2$
      (e) $2C_3H_6 + 9O_2 \longrightarrow 6CO_2 + 6H_2O$
      (f) $Mg + Cl_2 \longrightarrow MgCl_2$

6.29 (a) $3.3 \, mol \, NH_3$ (b) $43.2 \, mol \, H_2$ (c) $2.8 \, g \, N_2$
      (d) $14 \, g \, NH_3$

6.31 Redox is short for oxidation-reduction.

6.33 (a) Magnesium is oxidized and acts as a reducing agent. Bromine is reduced and acts as an oxidizing

agent.   (b) Iron is oxidized and acts as a reducing agent. Chlorine is reduced and acts as an oxidizing agent.   (c) Copper is oxidized and acts as a reducing agent. Sulfur is reduced and acts as an oxidizing agent.   (d) Carbon gains oxygen and is oxidized. Carbon acts as a reducing agent. Oxygen acts as an oxidizing agent and is reduced. (e) Phosphorus gains hydrogen and is reduced. Phosphorus acts as an oxidizing agent. Hydrogen acts as a reducing agent and is oxidized.

6.35 A reaction in which heat is released and the energy of the products is less than the energy of the reactants is an exothermic reaction. An example is burning.

6.37

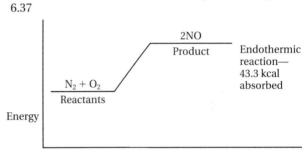

6.39 Unfavorable, because reactions are favored in which disorder increases.

6.41 (a) 100 g of water   (b) a teaspoon of table salt dissolved in water   (c) hot tea

6.43 The rate of reaction expresses how fast reactants form products.

6.45 In order for a reaction to take place the reacting particles must collide. In order for bond-breaking or bond-making to occur the colliding particles must have a certain amount of energy (the activation energy). The reaction rate depends upon the number of collisions and the energy of the collisions.

6.47 (a) An increase in temperature increases the rate of the reaction because it increases the energy of the particles.   (b) A decrease in particle size increases the rate of the reaction because it increases the surface area of the reactants and increases the number of particles.

6.49 A reversible reaction is one in which reactants form products (the forward reaction   from left to right) and products form reactants (the reverse reaction   from right to left).

6.51 A catalyst increases the speed at which chemical equilibrium is attained; a catalyst does not affect the position of equilibrium.

6.53 (a) shift left   (b) no change   (c) shift right

6.55 1.51×10^{24} molecules H_2O_2

6.57 (a) 2,2,1   (b) 1,3,1,1   (c) 1,1,4,1
      (d) 1,4,1,2,1   (e) 3,1,2,3   (f) 1,1,1,2
      (g) 4,1,2

6.59 $[Br_2]$ = 4.00 mol; $[BrCl]$ = 1.20 mol
6.61 (a) 5.93 mol $Ca(C_2H_3O_2)_2$ (b) 1.12 mol Cl_2
      (c) 0.106 mol KOH   (d) 0.397 mol C_6H_6O
6.63 (a) 1.26 mol CS_2 (b) 18.1 mol C
      (c) 364 mol SO_2
6.65 (a) Adding *C* causes the equilibrium to shift to the
      right, converting reactants to products. This de-
      creases the amount of *Y*, increases the amount of
      B, and causes a color change from green to blue.
      (b) Since the reaction is exothermic, cooling the
      system has the same effect as removing a product
      of the reaction. It causes the reaction to shift to the
      right. As more product is formed, the color changes
      from green to blue.   (c) A catalyst speeds up
      both the forward and the reverse reaction, favoring
      neither. The solution remains green.
      (d) Removing *B* shifts the equilibrium to the right.
      As more *B* is formed, *Y* is used up. Once a new
      equilibrium is established, the amount of *B* is
      greater than the amount of *Y*, and the color is
      changed from green to blue.

Self-Test

1. False   2. False   3. True   4. True   5. True
6. True   7. True   8. False   9. True   10. True
11. b   12. b   13. d   14. d   15. b   16. b
17. c   18. c   19. d   20. c   21. a   22. a
23. c   24. c   25. a   26. b   27. a   28. d

Chapter 7

Practice Exercises

7.1   Since the temperatures of the two samples of mat-
      ter are identical, the average kinetic energies of the
      particles in each sample are identical.
7.2   increase
7.3   They are the same.
7.4   It is tripled.
7.5   0.428 atm
7.6   253 mm Hg = 0.333 atm. The pressure is greater
      than 0.30 atm.
7.7   (a) 0.145 mol O_2 (b) 8.73 × 10^{22} molecules O_2
7.8   53.8 L
7.9   1.4 × 10^2 L H_2
7.10 5.23 L CO
7.11 (a) hydrogen   (b) hydrogen
7.12 300 mL
7.13 A balloon shrinks. The pressure inside the balloon
      decreases because the average kinetic energy of the
      gas particles inside the balloon is lower at the lower
      temperature. The reduced kinetic energy results in
      fewer and less forceful collisions with the inside
      walls of the balloon.

7.14 820 mm Hg = 1.1 atm
7.15 7.5 L
7.16 10.2 L
7.17 8.5 mm Hg
7.18 0.93 L
7.19 1.5 × 10^2 mol He
7.20 2.5 g air

Exercises

7.21 (a) Compared with liquids and solids, the mole-
      cules of substances in the gaseous state are very far
      apart.   (b) Molecules of a gas move rapidly and
      randomly, distributing themselves throughout a
      container.
7.23 Decrease the temperature.
7.25 Absolute zero (0 K) is the temperature at which all
      molecular motion would cease.
7.27 Decreases by a factor of 2.
7.29 Gas pressure is a force that acts on a given area. It
      is caused by gas particles colliding with an object,
      and it is measured in the units of atmospheres
      (atm), kilopascals (kPa), millimeters of mercury
      (mm Hg), or torr.
7.31 (a) 0.25 atm   (b) 25 kPa
7.33 Convert each pressure in the ratio to the new unit.
      (a) 0.150/0.11 (atm)   (b) 15.2/11 (kPa)
7.35 STP is standard temperature and pressure. STP for
      any gas is 0 °C and 1 atm (760 mm Hg or 101.3 kPa).
7.37 (a) 0 °C   (b) 273 K
7.39 0.125 mol N_2
7.41 5.6 L H_2
7.43 10.4 L O_2
7.45 The pressure must increase by a factor of 3.
7.47 217 mm Hg
7.49 110 mm Hg
7.51 34 L
7.53 24 L
7.55 1.31 × 10^3 K = 1.04 × 10^3 °C
7.57 13 atm
7.59 1.1 × 10^3 K
7.61 (a) decreases   (b) increases   (c) decrease
      (d) increase   (e) decrease
7.63 17.6 L
7.65 4.40 atm
7.67 (a) When a gas is heated the average kinetic energy
      of the gas particles increases. The particles are
      moving faster and hitting the walls of the container
      with more force. The only way to maintain a con-
      stant pressure is for the volume of the container to
      increase, decreasing the number of collisions be-
      tween the particles and the container walls.
      (b) Gas particles are in constant, rapid, random
      motion; therefore, they can easily mix together by
      diffusion.   (c) When a gas at constant volume is

heated, the increased average kinetic energy of the particles hitting the walls of the container is reflected in an increase in pressure.   (d) At constant temperature there is no change in the average kinetic energy of the gas particles. A pressure increase indicates that more particles are hitting the walls of the container. This occurs when the volume of the container decreases.

7.69 Vaporization is the conversion of particles from the liquid state to the gaseous state. When a substance vaporizes in a sealed container, the vapor formed above the liquid has a vapor pressure that will vary with the temperature. The boiling point is the temperature at which the vapor pressure of a liquid equals the external air pressure.

7.71 The forces of attraction between ions in ionic solids are much stronger than forces of attraction between molecules in molecular solids. Therefore, ionic compounds tend to have higher melting points than molecular compounds.

7.73 (a) 41.0 L O_2 (b) 22.0 L CO_2

7.75

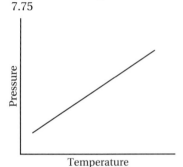

7.77 1.04 g/L

7.79 They are identical.

7.81 For any quantity of ice that melts to water at 0 °C, a corresponding amount of water freezes to ice at 0 °C.

7.83

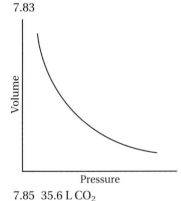

7.85 35.6 L CO_2

Self-Test

1. True   2. False   3. True   4. True   5. True
6. True   7. True   8. True   9. False   10. True

11. b   12. c   13. d   14. b   15. a   16. d
17. d   18. c   19. a   20. d   21. b   22. d
23. d   24. b   25. c   26. a   27. a   28. b
29. d   30. c   31. b

Chapter 8

Practice Exercises

8.1 Water is a polar molecule and there is hydrogen bonding between water molecules. Methane is a nonpolar molecule with weak polar bonds and has no intermolecular hydrogen bonding. Substances whose molecules form intermolecular hydrogen bonds tend to have higher boiling points.

8.2 Water is a polar solvent. Solutes that are ionic or polar generally dissolve in water.   (a) HCl is polar and will dissolve.   (b) NaI is ionic and will dissolve.   (c) NH_3 is polar and will dissolve.   (d) $MgSO_4$ is ionic and will dissolve.   (e) CH_4 is nonpolar and will not dissolve.   (f) $CaCO_3$ is ionic but will not dissolve; some ionic compounds are insoluble in water because of the strong attractive forces within the solid crystal.   (g) Gasoline is nonpolar and will not dissolve.

8.3 36.1%

8.4 272 g NaCl

8.5 2.6 atm

8.6 5.0 % (v/v)

8.7 4.0 g $MgSO_4$

8.8 3.6% $CuSO_4$ (w/v)

8.9 3×10^1 ppb

8.10 2.8 M

8.11 0.10 M

8.12 0.50 mol $CaCl_2$; 56 g $CaCl_2$

8.13 Use a pipet to transfer 5.0×10^1 mL of the 1.0 M solution to a 250-mL volumetric flask. Then add the distilled water up to the mark.

8.14 101.6 °C

8.15 (a) −5.58 °C   (b) −3.72 °C

8.16 (a) The volume of solution A will increase.   (b) Solution A will become more dilute.

Exercises

8.17 The bonds between the oxygen atom and hydrogen atoms in water are polar. Since the water molecule has a nonlinear (bent) shape, the bond dipoles do not cancel, and the water molecule as a whole is polar.

8.19 Hydrogen bonding is a weak force of attraction between the hydrogen atom of one molecule and an oxygen or some other atom with a partial negative charge of another molecule.

8.21 Surface water molecules are strongly attracted to the body of water through hydrogen bonding.

Water molecules do not hydrogen bond with molecules in the air.

8.23 Generally, the higher the vapor pressure of a liquid, the lower its boiling point, and vice versa.

8.25 Ice is a very regular open framework of water molecules; water does not have a regular structure and the molecules are packed closer together than in ice. The density of ice is lower than the density of water at 0 °C (because of its greater volume for a given mass); therefore, ice floats on water.

8.27 Sweat is mostly water; for each gram of water that evaporates from the surface of your skin, 540 cal of heat is absorbed from the body.

8.29 The general rule is that "like dissolves like." Water molecules can solvate ions and polar covalent molecules but not nonpolar molecules.

8.31 Solutions are homogeneous mixtures in which a solute is dissolved in a solvent. In aqueous solutions the solvent is water.

8.33 20.9%

8.35 A molten sample or an aqueous solution of an electrolyte conducts an electric current; a molten sample or an aqueous solution of a nonelectrolyte does not conduct an electric current.

8.37 Sodium ions, Na^+; potassium ions, K^+; calcium ions, Ca^{2+}; magnesium ions, Mg^{2+}

8.39 Agitation (stirring), temperature of the solvent, and particle size of the solute.

8.41 (a) 44.2 g KCl   (b) 5.8 g KCl

8.43 As the pressure of a gas above a liquid increases, the solubility of a gas in a liquid increases.

8.45 3 g/L

8.47 11%

8.49 (a) 23 g NaCl   (b) 3.0 g $MgCl_2$
(c) 7.0 g glucose   (d) 0.25 g $MgSO_4$

8.51 (a) 0.4 ppm   (b) 4×10^2 ppb

8.53 (a) 1.3 M KCl   (b) 0.33 M $MgCl_2$
(c) 0.050 M NaCl   (d) 0.397 M C_2H_5OH

8.55 (a) 0.066 mol K_2SO_4 (b) 0.0055 mol K_2SO_4
(c) 0.17 mol K_2SO_4

8.57 63 mL

8.59 A solution and a colloidal dispersion are similar in that they are both mixtures, their particles cannot be separated by filtration, and their particles do not separate on standing. A colloidal dispersion gives the Tyndall effect; a solution does not. The particles in a colloidal dispersion are larger (1–100 nm) than in a solution (< 1 nm).

8.61 The Tyndall effect is the scattering of visible light in all directions by particles in a colloid or a suspension.

8.63 When a solute is dissolved in a pure solvent, the solution formed has a higher boiling point and a lower freezing point than the pure solvent.

8.65 Sea water contains more dissolved solutes than fresh water; it therefore has a lower vapor pressure. A low vapor pressure means a slower rate of evaporation.

8.67 (a) 100.26 °C   (b) 101.6 °C

8.69 Osmosis is the flow of water molecules across semipermeable membranes from a more dilute solution to a more concentrated solution.

8.71 (a) osmosis   (b) There is a net flow of water from the 2% solution to the 4% solution.   (c) The 4% solution level rises and the 2% level drops.

8.73 (a) Water will flow by osmosis from the cell to the solution causing the cell to shrink and become crenated.   (b) Water flows into the cell, causing the cell to swell.   (c) There is no net change in the size of the cell.

8.75 0.1 M KCl

8.77 (a) Sodium and chloride ions pass across the membrane until their concentrations on both sides of the membrane are the same.   (b) Sodium and chloride ions pass across the membrane. At equilibrium the sodium chloride concentration outside the bag is higher than inside the bag due to the presence of the protein.   (c) Sodium and chloride ions would pass through the membrane into the protein dispersion until the system was like (b).

8.79 The amount the freezing point of a solution is depressed depends on the number of solute particles. An electrolyte will dissociate into ions (that is why it is an electrolyte); a nonelectrolyte does not dissociate. Therefore, the electrolyte solution contains more particles than a nonelectrolyte solution when both solutes are present in equal molar quantities.

8.81 (a) For a suspension: (1) particles can be filtered, (2) heterogeneous mixture, (3) gives Tyndall effect. (b) For a solution: (4) particle size less than 1.0 nm, (5) particles are invisible to the naked eye, (6) does not settle out on standing.   (c) For a colloidal dispersion: (3) gives Tyndall effect, (5) particles are invisible to the naked eye, (6) does not settle out on standing.

8.83	$C_6H_{12}O_6$	NH_4NO_3	LiF
Mass (g)	56.8	22.4 (c)	14.6
Volume (mL)	450	8.00×10^2	376 (e)
Percent (w/v)	13% (a)	2.80%	3.88% (f)
Molarity (M)	0.70 (b)	0.350 (d)	1.50

Self-Test

1. True   2. False   3. True   4. True   5. True
6. True   7. False   8. True   9. True   10. True
11. b   12. a   13. c   14. c   15. b   16. a
17. a   18. b   19. d   20. d   21. b   22. c
23. b   24. a   25. b   26. a   27. b   28. c
29. b   30. b

Chapter 9

Practice Exercises
9.1 (a) $3.8 \times 10^{-12} M$ (b) $1.4 \times 10^{-8} M$
    (c) $1.1 \times 10^{-7} M$
9.2 $[H^+] = 1 \times 10^{-11} M$, basic
9.3 (a) diprotic   (b) triprotic   (c) monoprotic
9.4 (a) $Ca(s) + 2H_2O(l) \longrightarrow Ca(OH)_2(aq) + H_2(g)$
    (b) $Mg(s) + 2H_2O(l) \longrightarrow Mg(OH)_2(aq) + H_2(g)$
9.5 $Mg(OH)_2(aq) \longrightarrow Mg^{2+}(aq) + 2OH^-(aq)$
9.6

$$HNO_3(aq) + H_2O(l) \longrightarrow NO_3^-(aq) + H_3O^+(aq)$$
H$^+$ donor   H$^+$ acceptor

$$H_2CO_3(aq) + H_2O(l) \rightleftharpoons HCO_3^-(aq) + H_3O^+(aq)$$
H$^+$ donor   H$^+$ acceptor

9.7 (a) Hydrochloric acid (HCl)
    (b) Carbonic acid (H_2CO_3)
9.8 (a) basic   (b) basic   (c) acidic   (d) basic
    Solution (c) is the most acidic; solution (d) is the most basic.
9.9 (a) 9.7   (b) 2.8   (c) 2.0   (d) 10.1
9.10 4.7
9.11 (a) $1.0 \times 10^{-5} M$ (b) $1.6 \times 10^{-6} M$ (c) $6.3 \times 10^{-13} M$ (d) $2.5 \times 10^{-3} M$
9.12 (a) $1 \times 10^{-8} M$ (b) $2.0 \times 10^{-6} M$ (c) $5.0 \times 10^{-2} M$
9.13 (a) 5.3   (b) 9.1   (c) 9.3   (d) 3.7
9.14 (a) $[H^+] = 1.6 \times 10^{-4} M$, $[OH^-] 6.3 \times 10^{-11} M$
    (b) $[H^+] = 2.0 \times 10^{-6} M$, $[OH^-] 5.0 \times 10^{-9} M$
    (c) $[H^+] = 5.0 \times 10^{-11} M$, $[OH^-] 2.0 \times 10^{-4} M$
9.15 (a) $HNO_3(aq) + KOH(aq) \longrightarrow KNO_3(aq) + H_2O(l)$
    (b) $2HCl(aq) + Mg(OH)_2(aq) \longrightarrow$
$$MgCl_2(aq) + 2H_2O(l)$$
    (c) $3NH_4^+(aq) + 3OH^-(aq) + H_3PO_4(aq) \longrightarrow$
$$(NH_4)_3PO_4(aq) + 3H_2O(l)$$
9.16 0.60 mol KOH
9.17 (a) 56.1 g KOH   (b) 36.5 g HCl   (c) 80.9 g HBr
9.18 (a) 0.10 equiv $Ca(OH)_2$ (b) 2.0 equiv H_2SO_4
    (c) 0.30 equiv H_3PO_4 (d) 0.25 equiv NaOH
9.19 (a) $2 N$ HCl   (b) $1 N$ NaOH
    (c) $0.1 N$ CH_3COOH (d) $0.4 N$ H_2SO_4
    (e) $0.9 N$ H_3PO_4
9.20 (a) 0.5 equiv NaOH   (b) 0.5 equiv HCl
    (c) 0.02 equiv H_2SO_4
9.21 (a) 2.5 mEq HCl   (b) 1×10^1 mEq H_3PO_4
    (c) 5 mEq HNO_3
9.22 Add 25 mL of $4.0 N$ H_2SO_4 to some distilled water in a 500-mL flask; add distilled water up to the mark.
9.23 18.8 mL
9.24 $1.20 N$

9.25 (a) $HPO_4^{2-}(aq) + H^+(aq) \rightleftharpoons H_2PO_4^-(aq)$
    (b) $H_2PO_4^-(aq) + OH^-(aq) \rightleftharpoons$
$$HPO_4^{2-}(aq) + H_2O(l)$$
9.26 $NH_4^+(aq) + OH^-(aq) \rightleftharpoons NH_3(aq) + H_2O(l)$

Exercises
9.27 $H_2O(l) + H_2O(l) \rightleftharpoons H_3O^+(aq) + OH^-(aq)$
    $H_2O(l) \rightleftharpoons H^+(aq) + OH^-(aq)$
9.29 1×10^{-14}
9.31 $1 \times 10^{-9} M$, basic
9.33 (a) An Arrhenius acid is a compound that dissociates to give hydrogen ions in aqueous solution.
    (b) An Arrhenius base is a compound that dissociates in aqueous solution to give hydroxide ions.
9.35 $2Li(s) + 2H_2O(l) \longrightarrow 2LiOH(aq) + H_2(g)$
9.37 The Brønsted-Lowry theory is more comprehensive; it includes compounds such as NH_3 that are bases even though they do not dissociate to give hydroxide ions in aqueous solution.
9.39 Strong acids and bases are completely dissociated in aqueous solution; weak acids and bases are only slightly dissociated in aqueous solution.
9.41 (a) strong base   (b) strong acid   (c) strong base   (d) strong acid   (e) weak base
9.43 $pH = -\log[H^+]$
9.45 (a) 5.0; acidic   (b) 3.0; acidic   (c) 9.0; basic
    (d) 11.0; basic
9.47 A neutralization reaction is the reaction of an acid and a base in aqueous solution to produce a salt and water.
9.49 (a) $HCl + KOH \longrightarrow H_2O + KCl$
                    [potassium chloride]
    (b) $H_2SO_4 + 2KOH \longrightarrow 2H_2O + K_2SO_4$
                    [potassium sulfate]
    (c) $2H_3PO_4 + 3Ca(OH)_2 \longrightarrow 6H_2O + Ca_3(PO_4)_2$
                    [calcium phosphate]
    (d) $CH_3COOH + NaOH \longrightarrow$
$$H_2O(l) + CH_3COONa(aq) \text{ [sodium acetate]}$$
9.51 (a) 4.0 mol NaOH   (b) 36 mol NaOH
    (c) 0.56 mol NaOH
9.53 (a) 0.016 equiv KOH   (b) 4.53 equiv H_3PO_4
    (c) 1.25 equiv HCl   (d) 0.644 equiv $Mg(OH)_2$
    (e) 0.024 equiv NaOH   (f) 2.02 equiv $Ca(OH)_2$
9.55 (a) $1.8 N$ NaOH   (b) $6.3 N$ H_3PO_4
    (c) $5.0 N$ KOH   (d) $0.18 N$ HNO_3
9.57 $0.050 N$
9.59 pH 6.3–8.0.
9.61 (a) 5.0×10^1 mL   (b) 69 mL   (c) $0.15 N$
    (d) 44 mL   (e) $1.8 N$
9.63 Salts formed by the reaction of a weak acid and a strong base, or by the reaction of a strong acid and a weak base, hydrolyze water.
9.65 (a) neutral   (b) acidic   (c) basic   (d) acidic

9.67 (d), (c), (e), (b), (a)

9.69 (a) $CH_3COO^-(aq) + H^+(aq) \rightleftharpoons CH_3COOH(aq)$
(b) $HCO_3^-(aq) + H^+(aq) \rightleftharpoons H_2CO_3(aq)$

9.71 (a) Use a pH indicator or pH meter to measure the pH (b) neutralize by addition of a buffering weak base such as sodium bicarbonate.

9.73 An aqueous solution of a strong acid is a better conductor because the strong acid undergoes a greater degree of dissociation into ions.

9.75 0.42 *M.*

9.77 (a) CH_3COOH (acid), H_2O (base)
(b) CN^- (base), H_2O (acid)
(c) $H_2PO_4^-$ (acid), OH^- (base)

Self-Test

1. True   2. False   3. False   4. True   5. True
6. True   7. True   8. False   9. True   10. False
11. d   12. a   13. c   14. a   15. c   16. b
17. c   18. c   19. a   20. c   21. d   22. a
23. c   24. c   25. a   26. c   27. c

Chapter 10

Practice Exercises

10.1 $^{222}_{86}Rn \longrightarrow\, ^{218}_{84}Po\, +\, ^{4}_{2}He$

10.2 $^{131}_{53}I \longrightarrow\, ^{131}_{54}Xe\, +\, ^{0}_{-1}e$

10.3 $^{60}_{27}Co \longrightarrow\, ^{60}_{28}Ni\, +\, ^{0}_{-1}e\, +$ gamma

10.4 (a) 1.25 mg   (b) 8.75 mg

10.5 (a) 0.0625 mg   (b) 6.25%

10.6 1.72×10^4 years old

Exercises

10.7

	Particle	Symbol	Charge
(a)	alpha	α or $^{4}_{2}He$	2+
(b)	beta	β or $^{0}_{-1}e$	1−
(c)	gamma	γ	0

10.9 Isotopes are atoms of the same element that have different numbers of neutrons. Radioisotopes are atoms that have unstable nuclei.

10.11 (a) $^{226}_{88}Ra \longrightarrow\, ^{222}_{86}Rn\, +\, ^{4}_{2}He$
radon-222   alpha particle
(b) $^{230}_{90}Th \longrightarrow\, ^{226}_{88}Ra\, +\, ^{4}_{2}He$
radium-226   alpha particle
(c) $^{233}_{92}U \longrightarrow\, ^{229}_{90}Th\, +\, ^{4}_{2}He$
thorium-229   alpha particle
(d) $^{222}_{86}Rn \longrightarrow\, ^{218}_{84}Po\, +\, ^{4}_{2}He$
polonium-218   alpha particle

10.13 0.03 g

10.15 28.1 years

10.17 Nuclear fission is a process in which nuclei break apart while releasing energy. Nuclear fusion is a process in which small nuclei combine and release energy.

10.19 A nuclear chain reaction is a reaction in which the products cause the reaction to continue or even speed up.

10.21 Ionizing radiation produces ions by knocking electrons off atoms.

10.23 (a) The *curie* measures how fast the radioisotope decays and (b) the *roentgen* measures the amount of ionization in air caused by the radiation source.

10.25 Tracers are useful in medical diagnosis because they can be detected as they travel through, or accumulate in, certain regions of an organism.

10.27 Radioactive iodine-131 given orally to a patient concentrates in the thyroid gland, where radiation therapy takes place.

10.29 (a) $^{234}_{92}U + ^{4}_{2}He$ (b) $^{206}_{81}Tl + ^{4}_{2}He$
(c) $^{170}_{70}Yb + ^{4}_{2}He$ (d) $^{206}_{82}Pb + ^{4}_{2}He$
(e) $^{226}_{88}Ra + ^{4}_{2}He$

10.31 (a) $^{32}_{16}S$ (b) $^{14}_{6}C$ (c) $^{4}_{2}He$ (d) $^{141}_{57}La$

10.33 (a) $^{12}_{6}C$ (b) $^{1}_{1}H$ (c) $^{16}_{8}O$ (d) $^{14}_{7}N$

10.35 1.56×10^4 atoms of xenon-133

10.37 The conversion of an atom of one element into an atom of a different element by emission of radiation. (Section 10.4)

10.39 The Geiger-Müller tube is limited to the detection of beta radiation.

Self-Test

1. False   2. False   3. True   4. False   5. False
6. False   7. True   8. False   9. True   10. False
11. (a)   12. (c)   13. (c)   14. (d)   15. (a)
16. (c)   17. (d)   18. (d)   19. (b)   20. (b)
21. (b)   22. (c)

Chapter 11

Practice Exercises

11.1

11.2 $CH_3—CH_2—CH_2—CH_2—CH_3$
$CH_3CH_2CH_2CH_2CH_3$
$CH_3(CH_2)_3CH_3$
C—C—C—C—C

$CH_3—CH_2—CH_2—CH_2—CH_2—CH_3$
$CH_3CH_2CH_2CH_2CH_2CH_3$
$CH_3(CH_2)_4CH_3$
C—C—C—C—C—C

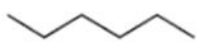

11.3 (a) propane (b) hexane (c) pentane

11.4 (a) 2-methylbutane (b) 3-methylpentane
(c) 3-ethylhexane

11.5

(a)

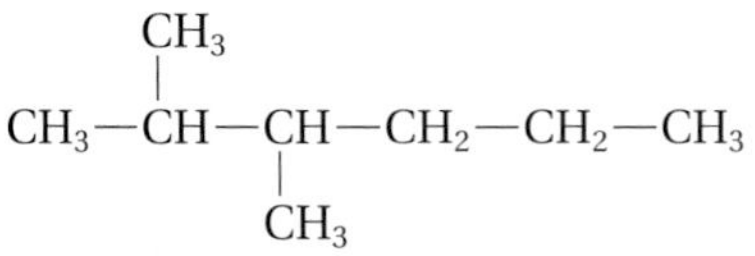

(b)

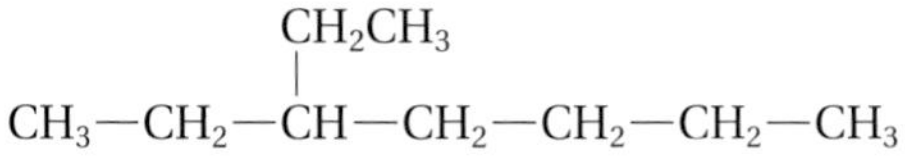

(c)

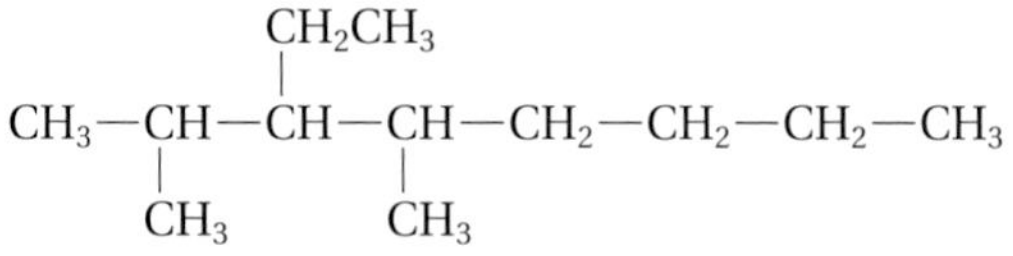

11.6

(a) $CH_3CH_2CH_2CH_2CH_2CH_3$
Hexane

(b) 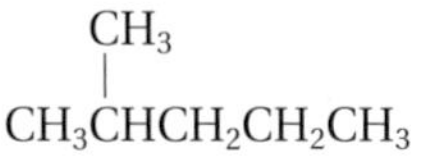
2-Methylpentane

(c) 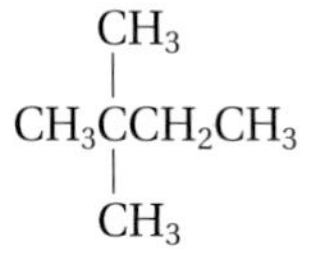
2,2-Dimethylbutane

(d) 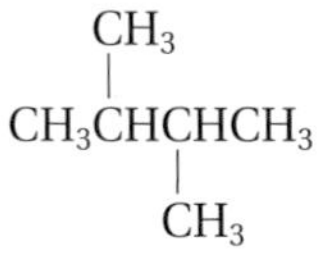
2,3-Dimethylbutane

(e) 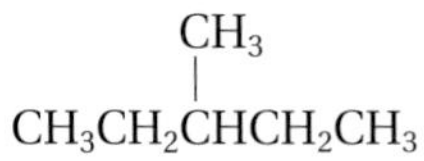
3-Methylpentane

11.7

(a) 1,2,3-trimethylcyclobutane (b) methylcyclohexane

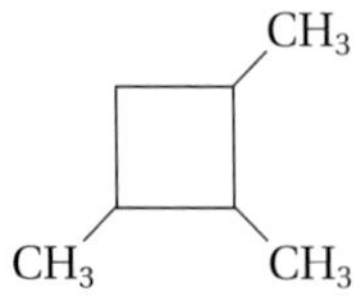

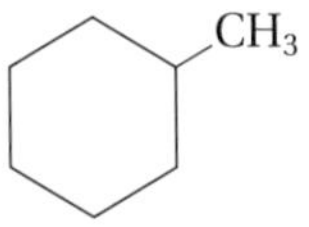

(c) 1,3-dimethylcyclopentane

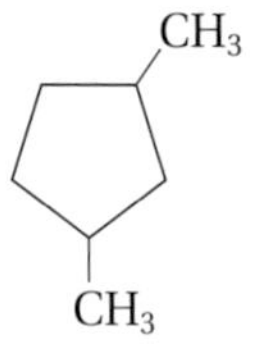

11.8

(a)

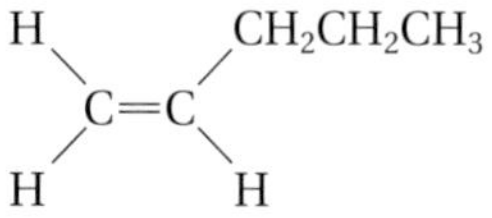

(b) 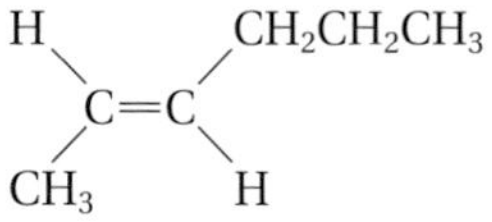
Trans 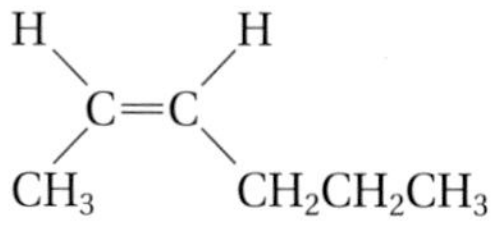   *Cis*

(c)

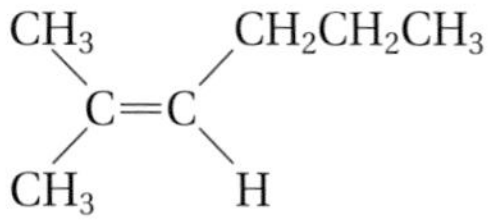

(d)

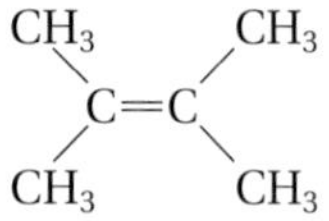

11.9 (a) 1-phenylpropane (propylbenzene)
(b) 2-phenylpropane (isopropylbenzene)
(c) 2,5-dimethyl-3-phenylhexane

Exercises

11.11 (a) ethyl (b) octane (c) propane
(d) pentane (e) pentane (f) propyl

11.13 A group that replaces a hydrogen on a parent alkane; the isopropyl group, $(CH_3)_2CH—$.

11.15 (a) methylpropane (b) 3-ethyl-2,5-dimethyl-hexane (c) 2-methylhexane (d) 2-methyl-hexane (e) 4-ethyl-2,5-dimethylheptane (f) 4-ethylheptane

11.17 (a) $CH_3—CH_2—CH_2—CH_2—CH_3$
(b) 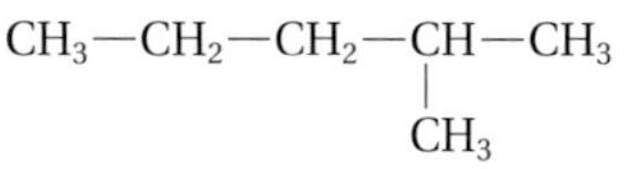
(c) 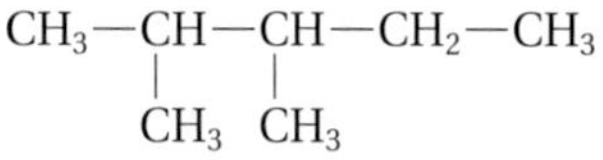
(d) $CH_3—CH_2—CH_2—CH_3$

11.19 $CH_3—CH_2—CH_2—CH_2—CH_3$
Pentane

11.21 (a) 1,2-dimethylcyclohexane (b) methylcy-clopentane (c) 1,1-dimethylcyclopropane

11.23 Number the longest carbon chain containing the carbon-carbon double bond so that the double bond is given the lowest number. The name of this longest chain is the name of the corresponding alkane, but with an *-ene* rather than an *-ane* ending. Substituents on the longest chain are numbered and named according to the rules for naming alkanes.

11.25 No

11.27

(a)

$$CH_3\quad CH_2CH_3$$
$$C=C$$
$$H\quad H$$

Cis-2-pentene

$$CH_3\quad H$$
$$C=C$$
$$H\quad CH_2CH_3$$

Trans-2-pentene

(b)

$$CH_3CH_2\quad CH_2CH_3$$
$$C=C$$
$$H\quad H$$

Cis-3-hexene

$$CH_3CH_2\quad H$$
$$C=C$$
$$H\quad CH_2CH_3$$

Trans-3-hexene

11.29 (a) 1-butene (b) 2-methyl-2-nonene
(c) 2-methyl-2-butene (d) 1-hexene

11.31 The term *aromatic character* refers to hybrid bonding in benzene and related arenes.

11.33 Each structure depicts a different resonance form of 1,2-diethylbenzene.

11.35 (a) cyclohexene (b) 1-3-ethylmethylbenzene
(c) naphthalene (d) methylcyclohexane; b, c, e, and f are aromatic.

(e)
$$CH_3$$
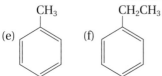

Toluene

(f)
$$CH_2CH_3$$

Ethylbenzene

11.37 The hydrocarbons in crude oil are mostly alkanes and cycloalkanes; the hydrocarbons in coal are mostly aromatic.

11.39 (a) hydrocarbons contain only carbon and hydrogen (b) alkane is a saturated hydrocarbon
(c) alkene contains the carbon-carbon double bond (d) alkyne contains the carbon-carbon triple bond (e) cycloalkane is an alkane with ends of the carbon chain joined together to form a ring (f) arene molecules contain the benzene ring

11.41 (a) 3 (b) 10 (c) 6 (d) 7
(e) 8 (f) 1 (g) 9 (h) 2
(i) 5 (j) 4

11.43 (a) cyclopentane (b) 2-methyl-2-butene
(c) 1,2,4-trimethylcyclohexane
(d) 1,3-dimethylcyclobutane

11.45

$$CH_3\quad CH_3$$

1,1-Dimethylcyclopropane

$$CH_3$$
$$CH_3$$

1,2-Dimethylcyclopropane

$$CH_3$$

Methylcyclobutane

Cyclopentane

Self-Test

1. False 2. False 3. False 4. True 5. False
6. True 7. False 8. True 9. True 10. False
11. c 12. c 13. c 14. c 15. b 16. a
17. b 18. b 19. b 20. a 21. b 22. c
23. a 24. d 25. a 26. b 27. d

Chapter 12

Practice Exercises

12.1 (a) bromobenzene (b) chloroethane
(c) 3-chloro-1-butene

12.2

(a)
$$CH_3$$
$$H-C-Cl$$
$$CH_3$$

(b)
$$CH_3$$
$$CH_3-CH_2-CH_2-C-CH_2-I$$
$$CH_3$$

(c)
$$Br$$
(ring)
$$CH_2CH_3$$

12.3

(a)
$$Br\quad Br$$
$$CH_2-CHCH_2CH_3$$

(b)
(ring with)
$$Cl$$
$$Cl$$

(c)
$$I\quad I$$
$$CH_3CH-CHCH_3$$

12.4

(a)

$$Br-\overset{\overset{\displaystyle CH_3}{|}}{\underset{\underset{\displaystyle CH_3}{|}}{C}}-\overset{\overset{\displaystyle H}{|}}{\underset{\underset{\displaystyle H}{|}}{C}}-H \quad + \quad H-\overset{\overset{\displaystyle CH_3}{|}}{\underset{\underset{\displaystyle CH_3}{|}}{C}}-\overset{\overset{\displaystyle H}{|}}{\underset{\underset{\displaystyle H}{|}}{C}}-Br$$

(major product)   (minor product)

(b)

+ HCl

(c)

12.5   (a) 1-butanol   (b) 2-propanol (isopropyl alcohol)   (c) 2-methyl-1-butanol   (d) cyclohexanol (cyclohexyl alcohol)   (e) 2-methyl-2-propanol (*tert*-butyl alcohol)

12.6   (a) primary alcohol   (b) secondary alcohol
(c) primary alcohol   (d) secondary alcohol
(e) tertiary alcohol

12.7   (a) $CH_2=CH-CH_3$
(b) $CH_3CH=CHCH_3$
(c) $CH_3CH=CHCH_3$ $CH_2=CHCH_2CH_3$
(major product)   (minor product)

12.8   (a) $2Na + 2CH_3OH \longrightarrow 2CH_3O^-Na^+ + H_2$

(b) $2Na + 2$ $-OH \longrightarrow$

2 $-O^-Na^+ + H_2$

12.9   (a) ether   (b) halocarbon   (c) alcohol
(d) aliphatic hydrocarbon (alkane); (c) is the most polar compound.

12.10 (d), (b), (a), and (c)

12.11 Urushiol contains two phenolic hydroxyl groups, two carbon-carbon double bonds, and an aromatic ring.

12.12 Estradiol contains the hydroxyl group of an aliphatic alcohol, a phenolic hydroxyl group, and an aromatic ring.

Exercises

12.13 (a)

(b) 3-chloropropene

(c) 1,2-dichloro-4-methylpentane

(d)

12.15 (a) $CH_3CH_2\overset{\overset{\displaystyle Cl}{|}}{CH}-\overset{\overset{\displaystyle Cl}{|}}{CH_2}$

(b) $CH_3CH_2\overset{\overset{\displaystyle Br}{|}}{CH}-\overset{\overset{\displaystyle H}{|}}{CH_2}$ $CH_3CH_2\overset{\overset{\displaystyle H}{|}}{CH}-\overset{\overset{\displaystyle Br}{|}}{CH_2}$

(major product)   (minor product)

(c) $CH_3CH_2\overset{\overset{\displaystyle CH_3}{|}}{\underset{\underset{\displaystyle H}{|}}{C}}-\overset{\overset{\displaystyle H}{|}}{CH}CH_3$

12.17

$$Cl-\overset{\overset{\displaystyle Cl}{|}}{CH}-CH_2-CH_3$$

1,1-Dichloropropane

$$Cl-CH_2-\overset{\overset{\displaystyle Cl}{|}}{CH}-CH_3$$

1,2-Dichloropropane

$$CH_3-\overset{\overset{\displaystyle Cl}{|}}{\underset{\underset{\displaystyle Cl}{|}}{C}}-CH_3$$

2,2-Dichloropropane

$$Cl-CH_2-CH_2-CH_2-Cl$$

1,3-Dichloropropane

12.19 (a)

$$XCH_3CH_2CH=CH_2 \longrightarrow -(-\overset{\overset{\displaystyle CH_3}{\underset{\displaystyle |}{CH_2}}{|}}{CH}CH_2-)_x-$$

1-Butene   Poly-1-butene

(b)

$$X\overset{\overset{\displaystyle Cl}{|}}{CH}=\overset{\overset{\displaystyle }{}}{\underset{\underset{\displaystyle Cl}{|}}{CH}} \longrightarrow -(-\overset{\overset{\displaystyle Cl}{|}}{CH}\overset{\overset{\displaystyle }{}}{\underset{\underset{\displaystyle Cl}{|}}{CH}}-)_x-$$

1,2-Dichloroethene   Poly-1,2-dichloroethene

12.21 (a) 2-butanol
(b) 2-methyl-1-propanol
(c) $HOCH_2CH_2OH$

(d)   OH

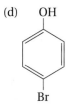

Br

12.23 (a) secondary   (b) secondary   (c) tertiary
(d) primary

12.25 Addition of water to an alkene; displacement of halide ion from an aliphatic halocarbon.

$$CH_2{=}CH_2\ +\ H_2O\ \xrightarrow[100\ °C]{H^+}\ CH_3CH_2OH$$

Ethene   Water   Ethanol

$$CH_3CH_2Br\ +\ OH^-\ \longrightarrow\ CH_3CH_2OH\ +\ Br^-$$

Bromoethane   Hydroxide   Ethanol   Bromide
               ion   ion

12.27 (a) CH_3CH_2OH

Ethanol

OH
|
(b) $CH_3CHCH_2CH_3$

2-Butanol

(c) $\underset{\overset{|}{OH}}{CH_3CH_2\overset{\overset{CH_3}{|}}{C}CH_3}$ $CH_3CH_2\overset{\overset{CH_3}{|}}{\underset{\overset{|}{H}}{C}}CH_2OH$

2-Methyl-2-butanol   2-Methyl-1-butanol
(major product)   (minor product)

12.29 (a)

Cyclohexene

(b) $CH_3CH_2{=}CH_2$

Propene

(c) $\overset{\overset{CH_3}{|}}{CH_3CH{=}CH_2}$

2-Methylpropene

(d) $CH_3CH_2CH{=}CHCH_3$

2-Pentene

12.31 (a) $CH_3CH_2{-}O{-}$⟨phenyl⟩

(b) ⟨furan ring with O⟩

(c) butylethyl ether
(d) dipropyl ether

12.33 (a) ⟨cyclohexyl⟩$-O-CH_3$

(b) ⟨cyclohexyl⟩$-O-CH_3$

(c) $CH_3\underset{\overset{|}{CH_3}}{CH}OCH_2CH_3$

12.35 Polarity of the hydroxyl group of ethanol overcomes hydrophobicity of the short carbon chain; ethanol is very soluble. Hydrophobicity of the long carbon chain of decanol dominates polarity of the hydroxyl group; the decanol is almost insoluble.

12.37 (a) ⟨water hydrogen bonding structure⟩

(b) ⟨water–methanol hydrogen bonding structure⟩

(c) ⟨methanol hydrogen bonding structure⟩

12.39 phenolic hydroxyl group, a primary alcohol, an ether, an alkene, and a chloro group

12.41 (a) phenol, more hydrogen bonding
(b) propanol, more hydrogen bonding
(c) 1-butanol, branched molecules tend to have lower boiling points   (d) methanol, more hydrogen bonding   (e) chlorocyclohexane, higher molar mass and polar molecule

12.43 (a) ⟨cyclohexane with Br, CH₃, H, H⟩   ⟨cyclohexane with H, CH₃, Br, H⟩

(major product)   (minor product)

(b) ⟨cyclohexene with CH₃⟩ + ⟨cyclohexene with CH₃⟩

(major product)   (minor product)

(c) ⟨cyclohexene with CH₃⟩

(d) ⟨cyclohexene with CH₂CH₃⟩   ⟨cyclohexene with CH₂CH₃⟩

(major product)   (minor product)

12.45 (d), (c), (a), (b)

Self-Test
1. True   2. True   3. True   4. False   5. True
6. True   7. True   8. False   9. True   10. True
11. a   12. b   13. b   14. b   15. d   16. a
17. b   18. d   19. b   20. d   21. b   22. b
23. c   24. d   25. a   26. c

Chapter 13

Practice Exercises
13.1 propanal   (b) 3-methylpentanal
(c) pentanal   (d) 3-chlorobutanal
13.2 (a) 3-hexanone   (b) butanone   (c) 4-methyl-2-pentanone

13.3 (a) CH_3CHO (b) $CH_3\overset{\displaystyle O}{\overset{\|}{C}}CH_3$

(c) $CH_3CH_2\overset{\displaystyle O}{\overset{\|}{C}}CH_2CH_3$ (d) $CH_3\overset{\displaystyle CH_3}{\overset{|}{C}}HCH_2CHO$

13.4 (d), (e), (c), (b), (a)
13.5 (a) 1-butyne   (b) propanal   (c) cyclohexanol
(d) 3-pentanone

13.6 (a) $CH_3CH_2CH_2CHO$ (b) $CH_3CH_2\overset{\displaystyle O}{\overset{\|}{C}}CH_3$
        Butanal   Butanone

(c) No reaction   (d)

Cyclobutanone

13.7 (a) $CH_3CH_2CH_2OH$ (b) $CH_3CH_2\overset{\displaystyle OH}{\overset{|}{C}}HCH_3$
        1-Propanol   2-Butanol

(c) $CH_3CH_2\overset{|}{C}HCH_2OH$
           CH_3
        2-Methyl-1-butanol

13.8 Positive Tollens' and Benedict's tests for pentanal; same tests negative for 2-pentanone.
13.9 (b), (c), (e)
13.10 Three ketone carbonyl groups, one carbon-carbon double bond, two aliphatic —OH groups.

Exercises
13.11 The carbonyl group is $-\overset{\displaystyle O}{\overset{\|}{C}}-$; bonding electrons pulled closer to electronegative oxygen, making oxygen partially negative and carbon partially

positive. The electronegativities of carbons in the carbon-carbon double bond are equal; therefore, the carbon-carbon double bond is nonpolar.
13.13 (a) ethanal (acetaldehyde)   (b) benzaldehyde
(c) 3-methylbutanal   (d) diphenyl ketone (benzophenone)   (e) 5-methyl-3-hexanone
13.15 Acetaldehyde molecules attract one another through polar-polar interactions; polar-polar interactions are not possible between propane molecules.
13.17 (a) oxidation   (b) oxidation   (c) reduction
(d) oxidation
13.19 (a) CH_3CH_2CHO (b) No reaction; the alcohol
        Propanal   is tertiary.

(c) $CH_3CH_2\overset{\displaystyle O}{\overset{\|}{C}}CH_2CH_3$ (d)
        3-Pentanone

                    Cyclohexanone

13.21 (a) CH_2O (b) $CH_3\overset{\displaystyle O}{\overset{\|}{C}}CH_3$
        Methanal   Propanone
        (formaldehyde)   (acetone)

(c) $CH_3\overset{\displaystyle CH_3}{\overset{|}{C}}HCHO$
        Methylpropanal

13.23 Of oxidation products from Exercise 13.20, (b) octanal and (c) 2-methylbutanal will give a positive Benedict's test.
13.25

$$CH_3CH_2OH + CH_3CH_2\overset{\displaystyle H}{\overset{|}{C}}=O \longrightarrow CH_3CH_2\overset{\displaystyle H}{\underset{\displaystyle OH}{\overset{|}{\underset{|}{C}}}}-OCH_3$$

13.27 (a) neither   (b) neither   (c) acetal
(d) acetal
13.29

(a) $CH_3\overset{\displaystyle O}{\overset{\|}{C}}CH(CH_3)_2$ (b) $CH_3CH_2\overset{\displaystyle O}{\underset{\displaystyle CH_3}{\overset{|}{\underset{|}{C}}}}HC-OH$

(c) no reaction   (d) $CH_3\overset{|}{C}HCH_2\overset{\displaystyle O}{\overset{\|}{C}}-OH$
                        CH_3

(e) no reaction   (f) CH_3CH_2OH

(g) CH$_3$C—OCH$_3$ (with H above and OH below the central C)   (h) CH$_3$—C—OCH$_2$CH$_3$ (with OH above and CH$_3$ below the central C)

13.31 The structures given are (a) ketal, (b) hemiacetal, (c) acetal, (d) hemiketal, (e) acetal, and (f) acetal.

13.33 Ketone molecules attract each another through polar-polar interactions; propanone boils at a higher temperature than ethyl methyl ether.

13.35 CH$_3$CH$_2$CH$_2$CHO gives positive tests with Benedict's reagent or Tollens' reagent; CH$_3$CH$_2$COCH$_3$ gives negative tests.

13.37 (a) methanal (formaldehyde)   (b) propanone (dimethyl ketone, acetone)   (c) butanal (butyraldehyde)   (d) benzaldehyde   (e) cyclohexanone   (f) methylphenylmethanone (methyl phenyl ketone)   (g) 3-pentanone

13.39 (a) butanol   (b) ethane   (c) formaldehyde   (d) isopropyl alcohol

Self-Test
1. False   2. False   3. False   4. False   5. False
6. True   7. False   8. True   9. False   10. False
11. b   12. c   13. d   14. c   15. b   16. d
17. d   18. c   19. d   20. c   21. b

Chapter 14

Practice Exercises

14.1 (a) propanoic acid (propionic acid)
(b) succinic acid   (c) 4-hydroxypentanoic acid
(d) phthalic acid

14.2 (a), (c), (d), (b)

14.3 (a) CH$_3$CH$_2$COO$^-$Na$^+$ + H$_2$O
   Sodium propanoate   Water
(b) (CH$_3$COO$^-$)$_2$Ca^{2+} + 2H$_2$O
   Calcium acetate   Water

14.4 (CH$_3$(CH$_2$)$_{16}$COO$^-$)$_3$Fe^{3+}

14.5 (a) CH$_3$COOH   (b) CH$_3$CCH$_3$ (with O above the C)
(c) CH$_3$CH$_2$COOH   (d) no reaction

14.6 Butanoic anhydride; hydrolysis gives butanoic acid.

CH$_3$(CH$_2$)$_2$C—O—C(CH$_2$)$_2$CH$_3$ + H$_2$O ⟶ (with O above each C)
   Butanoic anhydride

2CH$_3$(CH$_2$)$_2$C—OH (with O above the C)
   Butanoic acid

14.7 (a) phenyl propanoate (b) ethyl benzoate

14.8 (a) CH$_3$CH$_2$CH$_2$COCH$_3$ (with O above)
   Methyl propanoate (methyl propionate)

(b) ⟨phenyl⟩—COCH$_2$CH$_2$CH$_3$ (with O above)
   Propyl benzoate

(c) CH$_3$COCH$_2$CH$_3$ + CH$_3$COOH (with O above)
   Ethyl ethanoate   Ethanoic acid
   (ethyl acetate)   (acetic acid)

14.9 (a) CH$_3$CH$_2$COO$^-$Na$^+$ + HOCH$_2$CH$_3$
   Sodium propanoate   Ethanol

(b) CH$_3$COO$^-$K$^+$ + HO—⟨phenyl⟩
   Potassium acetate   Phenol

14.10 Transfer of a phosphoryl group (—PO$_3$H$_2$) to an alcohol or other functional group.

Exercises

14.11 (a) methanoic acid (formic acid)   (b) lactic acid
(c) salicylic acid   (d) pyruvic acid
(e) CH$_3$(CH$_2$)$_{16}$COOH

(f) CH$_3$CHCOOH (with OH above)

(g) ⟨benzene ring with COOH above and Cl below⟩

(h) HOOCCH$_2$COOH

14.13 (h) dicarboxylic acid; (b), (c), (d), (f), (g) polyfunctional acids; (e) fatty acid.

14.15 Because propanoic acid has the strongest intermolecular hydrogen bonding, it has the highest boiling point.
(a) CH$_3$CH$_2$CH$_2$CH$_2$CH$_3$ (b) CH$_3$CH$_2$CH$_2$CHO
   Pentane   Butanal
(c) CH$_3$CH$_2$CH$_2$CH$_2$OH   (d) CH$_3$CH$_2$COOH
   Butanol   Propanoic acid

14.17 Carboxylic acids dissociate to $RCOO^-$ and H^+ to a small degree in aqueous solution.

14.19 (a) $HCOO^-K^+ \ + \ H_2O$

Potassium   Water
formate

(b) $\underset{\text{Calcium}\atop\text{oxalate}}{\begin{matrix} COO^- \\ | \\ COO^- \end{matrix}}\ Ca^{2+} \ + \ \underset{\text{Water}}{H_2O}$

(c) $CH_3(CH_2)_{12}COO^-Na^+ \ + \ H_2O$

Sodium   Water
myristate

14.21 A soap is an alkali-metal salt of a long-chain carboxylic acid. The hydrophobic hydrocarbon chains of soap molecules emulsify grease droplets, enabling them to be rinsed away with water.

14.23 (a) CH_3COOH (b) CH_3CH_2COOH
(c) No reaction

14.25 (a) butanal   (b) methanal (formaldehyde)

14.27 (a) $R-\overset{\overset{\displaystyle O}{\|}}{C}-O-\overset{\overset{\displaystyle O}{\|}}{C}-R$

(b) $R-\overset{\overset{\displaystyle O}{\|}}{C}-OR$

14.29 (a) methyl ethanoate (methyl acetate)   (b) ethyl ethanoate (ethyl acetate)   (c) ethyl benzoate

(d) $H\overset{\overset{\displaystyle O}{\|}}{C}OCH_3$

Methyl formate
(methyl methanoate)

(e) $CH_3CH_2\overset{\overset{\displaystyle O}{\|}}{C}OCH_2CH_2CH_3$

Propyl propanoate
(propyl propionate)

14.31 $R-\overset{\overset{\displaystyle O}{\|}}{C}-O-R$ $R-\overset{\overset{\displaystyle O}{\|}}{C}-S-R$

Oxyester   Thioester

14.33
(a) Methyl acetate + Water $\xrightarrow{\text{HCl}}$ CH_3COOH + CH_3OH

Acetic acid   Methanol

(b) Phenyl propanoate + Sodium hydroxide $\longrightarrow$

$CH_3CH_2COO^-Na^+ \ +$ —OH

Sodium propanoate   Phenol

(c) Ethyl benzoate + Sodium hydroxide $\longrightarrow$

—$COO^-Na^+ \ + \ CH_3CH_2OH$

Sodium benzoate   Ethanol

(d) Propyl butanoate + Sodium hydroxide $\longrightarrow$
$CH_3CH_2CH_2COO^-Na^+ \ + \ CH_3CH_2CH_2OH$

Sodium butanoate   1-Propanol

(e) Ethyl formate + Potassium hydroxide $\longrightarrow$
$HCOO^-K^+ \ + \ CH_3CH_2OH$

Potassium   Ethanol
formate

(f) Phenyl benzoate + Water $\xrightarrow{\text{HCl}}$

—COOH + —OH

Benzoic acid   Phenol

14.35 The monoanion and the dianion dominate at pH 7.

$HO-\overset{\overset{\displaystyle O}{\|}}{\underset{\underset{\displaystyle OH}{|}}{P}}-OH$ Completely undissociated

$HO-\overset{\overset{\displaystyle O}{\|}}{\underset{\underset{\displaystyle OH}{|}}{P}}-O^-$ Monoanion

$HO-\overset{\overset{\displaystyle O}{\|}}{\underset{\underset{\displaystyle O^-}{|}}{P}}-O^-$ Dianion

$^-O-\overset{\overset{\displaystyle O}{\|}}{\underset{\underset{\displaystyle O^-}{|}}{P}}-O^-$ Trianion

14.37 (a) $HO-\overset{\overset{\displaystyle O}{\|}}{\underset{\underset{\displaystyle OH}{|}}{P}}-O-\overset{\overset{\displaystyle O}{\|}}{\underset{\underset{\displaystyle OH}{|}}{P}}-OH \ + \ CH_3CH_2OH \ \longrightarrow$

Pyrophosphoric acid

$HO-\overset{\overset{\displaystyle O}{\|}}{\underset{\underset{\displaystyle OH}{|}}{P}}-O-OCH_2CH_3 \ + \ HO-\overset{\overset{\displaystyle O}{\|}}{\underset{\underset{\displaystyle OH}{|}}{P}}-OH$

Ethyl phosphate   Phosphoric acid

(b) phosphorylation

14.39 (d), (c), (b), (a)

14.41 Acetic acid (K_a 1.8×10^{-5}) dissociates only to a small extent in aqueous solution. Water abstracts a hydrogen ion from only a small fraction of acetic acid molecules, so acetate ion must be a stronger base than water.

14.43 (a) $CH_3CH_2CH_2CH_2CH_2CH_2CH_2CH_2CH_2CH_3$
   (b) $CH_3CH_2CH_2CH_2CH_2COOH$
   (c) CH_3OCH_3
   (d) $CH_3CH_2CH_2CH_2CH_2CH_2CH_2CHO$

   (e) $CH_3CH_2CH_2\overset{\overset{\displaystyle O}{\|}}{C}OCH_2CH_3$

   (f) $CH_3CH_2CH_2CH_2CH_2OH$

   (g) $CH_3CH_2\overset{\overset{\displaystyle Cl}{|}}{C}HCOOH$

   (h)

$$-CH_2\overset{\overset{\displaystyle Br}{|}}{C}H CH\ CH_2CH_2OH$$
$$\underset{\displaystyle CH_3}{|}$$

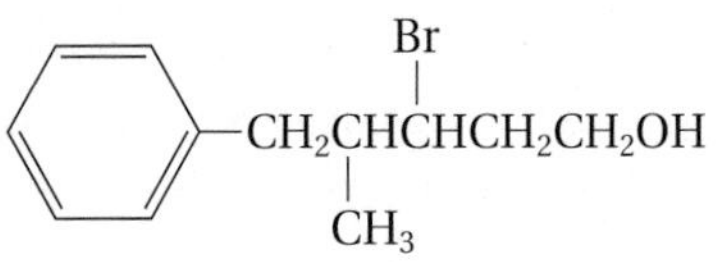

Self-Test

1. False   2. True   3. True   4. True   5. True
6. True   7. True   8. True   9. True   10. True
11. c   12. b   13. a   14. d   15. d   16. d
17. a   18. e   19. c   20. a   21. e   22. d
23. d

Chapter 15

Practice Exercises

15.1 (a) ethylmethylamine (secondary)   (b) propy-lamine (primary)   (c) ethyldimethylamine (tertiary)   (d) triethylamine (tertiary)

15.2 (a) N-ethylaniline (secondary)   (b) N,N-diethylaniline (tertiary)

15.3 (a) $CH_3CH_2NH_2$ (b) 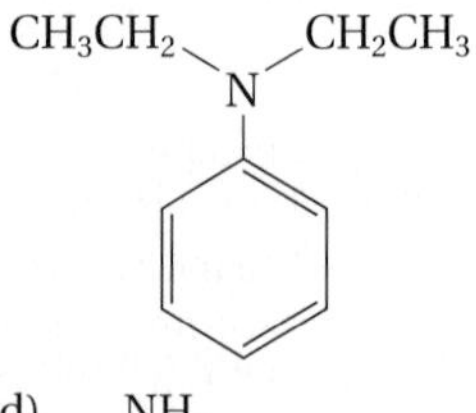 $CH_3CH_2\underset{\displaystyle N}{\diagdown\diagup}CH_2CH_3$

   (c) $(CH_3CH_2CH_2)_3N$ (d)

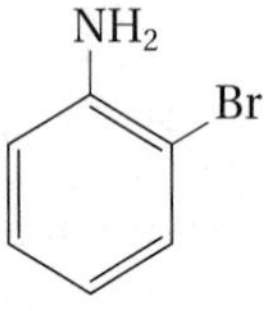

15.4 (d) pyridine

15.5
(a) $(CH_3)_2NH + HCl \longrightarrow (CH_3)_2NH_2^+Cl^-$
Dimethylammonium chloride

(b) $CH_3NH_2 + HNO_3 \longrightarrow CH_3NH_3^+NO_3^-$
Methylammonium nitrate

(c)

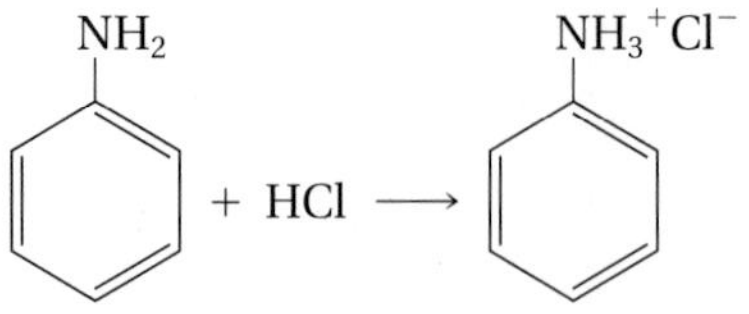

Anilinium chloride

(d) $CH_3CH_2NH_2 + H_3PO_4 \longrightarrow CH_3CH_2NH_3^+H_2PO_4^-$
Ethylammonium dihydrogen phosphate

15.6
(a) $CH_3CH_2CH_2CH_2Cl + NH_3 \longrightarrow$
   1-Chlorobutane   Ammonia

   $CH_3CH_2CH_2CH_2NH_3^+Cl^-$
   Butylammonium chloride

(b) $CH_3CH_2CH_2CH_2NH_3^+Cl^- + NaOH \longrightarrow$
   Butylammonium chloride   Sodium hydroxide

   $CH_3CH_2CH_2CH_2NH_2 + NaCl + H_2O$
   Butylamine   Sodium chloride   Water

15.7 (a) propanamide   (b) 3-methylbutanamide
   (c) N-methylpropanamide

15.8

$$CH_3CH_2\overset{\overset{\displaystyle O}{\|}}{C}-O^-H-\overset{\overset{\displaystyle H}{|}}{\underset{\underset{\displaystyle H}{|}}{N}}^+-CH_2CH_3 \xrightarrow{Heat}$$

$$CH_3CH_2\overset{\overset{\displaystyle O}{\|}}{C}-NHCH_2CH_3 + H_2O$$
N-Ethylpropanamide   Water

15.9

$$CH_3\overset{\overset{\displaystyle O}{\|}}{C}-O-CH_2CH_3 + H-\overset{\overset{\displaystyle H}{|}}{N}-CH_3 \longrightarrow$$
Ethyl acetate   Methylamine

$$CH_3\overset{\overset{\displaystyle O}{\|}}{C}-NHCH_3 + CH_3CH_2OH$$
N-Methylacetamide   Ethanol

15.10

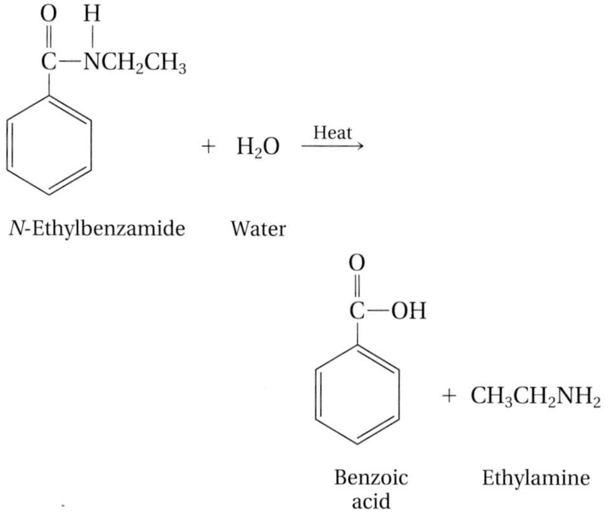

N-Ethylbenzamide   Water

Benzoic acid   Ethylamine

15.11 phenolic hydroxyl group, secondary hydroxyl group, secondary amine group; decongestant

15.12 (a) painkiller   (b) nervous system stimulant   (c) dilates pupils of eyes   (d) antimalarial agent

15.13

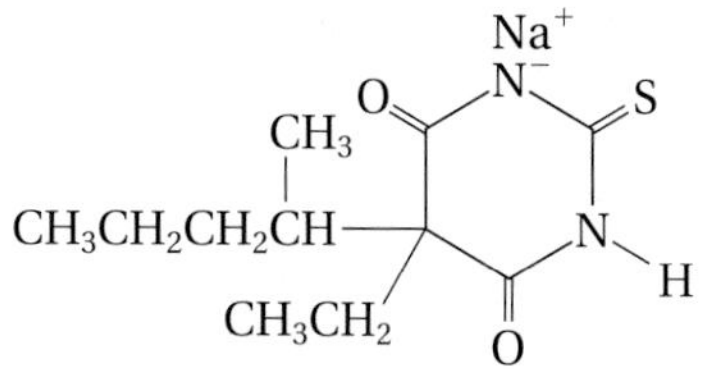

Sodium pentothal

15.14 Electron-withdrawing power of the adjacent carbonyl groups weakens the N—H bond.

Exercises

15.15 (a) Dimethylamine (secondary)

(b)   (c) 3-Chloroaniline (primary)

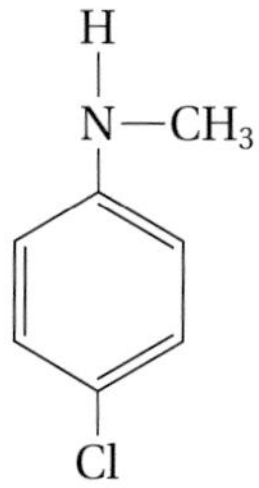

(secondary)

(d) CH_3CH_2—N—CH_2CH_3 with CH_3 (tertiary)

15.17 The amine nitrogen is contained within a ring in a heterocyclic amine.

15.19

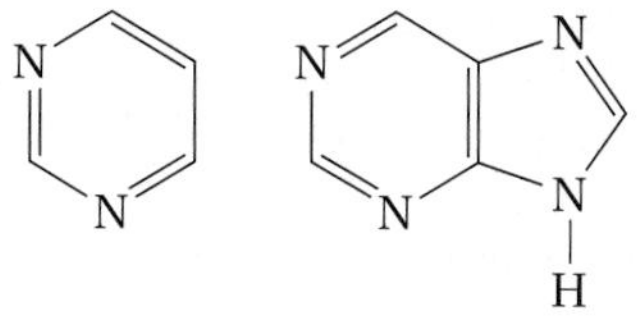

Pyrimidine   Purine

15.21 Amines are weak bases because they abstract a hydrogen ion (proton) from water to a slight extent.

15.23 (a) $CH_3CH_2NH_3{}^+Cl^-$ (b) $(CH_3)_2NH_2{}^+NO_3{}^-$

Ethylammonium chloride   Dimethylammonium nitrate

(c) $CH_3NH_3{}^+HSO_4{}^-$ (d) $(CH_3)_3CNH_3{}^+Cl^-$

Methylammonium hydrogen sulfate   tert-Butylammonium chloride

15.25 The solution tests acidic because a small fraction of the alkylammonium ions dissociate to produce hydrogen ions.

15.27 (a) ethanamide (acetamide)   (b) N-methylpropanamide   (c) N-methylbenzamide

(d) CH_3CH_2C—N—CH_2CH_3 with O above C and CH_3 below N

(e)

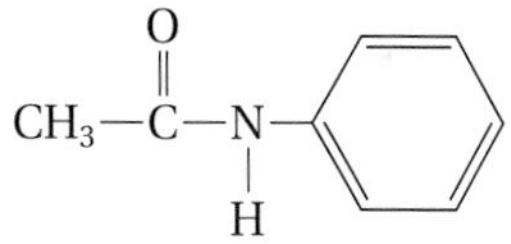

15.29

(a) $CH_3CH_2\overset{O}{\underset{\|}{C}}NH_2$ (b) $CH_3\overset{O}{\underset{\|}{C}}NH_2$

Propanamide   Ethanamide (acetamide)

(c) $H\overset{O}{\underset{\|}{C}}NHCH_2CH_3$

N-Ethylmethanamide (N-ethylformamide)

15.31 Dopamine cannot cross the blood-brain barrier. Dopa penetrates the barrier and is converted to dopamine in the brain.

15.33 Class of amines obtained from plants; gives alkaline (basic) aqueous solutions

15.35 Morphine and codeine

15.37 Barbituric acid is not physiologically active. Barbiturates can be hypnotics (sleep-inducers) and sedatives (tranquilizers).

15.39 (a) pyrimidine; (b) pyridine; (c) pyrrole; (d) purine; (e) imidazole; (f) indole.

15.41

(a) $(CH_3CH_2)_3N^+Br^-$

(b)
$$\underset{\text{(N-phenyl with }CH_2CH_2CH_3)}{N-CH_2CH_2CH_3}$$

(b) phenyl ring attached to N—H and —CH$_2$CH$_2$CH$_3$

(c) $CH_3\overset{\overset{O}{\|}}{C}-\overset{\overset{H}{|}}{N}-CH_3$

(d) $\overset{\overset{O}{\|}}{C}-N\begin{smallmatrix}CH_3\\CH_3\end{smallmatrix}$ attached to phenyl ring

15.43 (a) $(CH_3CH_2)_3N$

(b) $CH_3CH_2\overset{\overset{O}{\|}}{C}NH_2$

(c) phenyl ring attached to $N-CH_2CH_3$ with N—H

(d) $\overset{\overset{O}{\|}}{C}-\overset{\overset{H}{|}}{N}-$ phenyl, attached to phenyl ring

(e) phenyl ring with NH_2 and CH_2CH_3

(f) $CH_3\overset{\overset{O}{\|}}{C}N(CH_2CH_3)_2$

15.45

$$RCOOH + NH_3 \xrightarrow{\text{Heat}} RC\overset{\overset{O}{\|}}{N}H_2 + H_2O$$

$$RCOOH + RNH_2 \xrightarrow{\text{Heat}} RC\overset{\overset{O}{\|}}{N}HR + H_2O$$

$$RCOOH + R_2NH \xrightarrow{\text{Heat}} RC\overset{\overset{O}{\|}}{N}R_2 + H_2O$$

15.47
(a) Preparation of propylamine

$$CH_3CH_2CH_2OH \xrightarrow[\text{Heat}]{H_2SO_4} CH_3CH=CH_2 + H_2O$$

$$CH_3CH=CH_2 + HCl \longrightarrow$$

$$\underset{\text{(major product)}}{CH_3\overset{\overset{Cl}{|}}{C}H-\overset{\overset{H}{|}}{C}H_2} + \underset{\text{(minor product)}}{CH_3\overset{\overset{H}{|}}{C}H-\overset{\overset{Cl}{|}}{C}H_2}$$

$$CH_3CH_2CH_2Cl + NH_3 \longrightarrow CH_3CH_2CH_2NH_3^+Cl^-$$
$$CH_3CH_2CH_2NH_3^+Cl^- + NaOH \longrightarrow$$
$$CH_3CH_2CH_2NH_2 + NaCl + H_2O$$

(b) Preparation of N-ethylpropanamide

$$CH_3CH_2CH_2OH \xrightarrow{K_2Cr_2O_7} \underset{\text{Propanoic acid}}{CH_3CH_2COOH}$$

$$CH_2=CH_2 + HCl \longrightarrow CH_3CH_2Cl$$
$$CH_3CH_2Cl + NH_3 \longrightarrow CH_3CH_2NH_3^+Cl^-$$
$$CH_3CH_2NH_3^+Cl^- + NaOH \longrightarrow$$
$$\underset{\text{Ethylamine}}{CH_3CH_2NH_2 + NaCl + H_2O}$$

$$CH_3CH_2COOH + CH_3CH_2NH_2 \xrightarrow{\text{Heat}}$$

$$\underset{\text{N-Ethylpropanamide}}{CH_3CH_2\overset{\overset{O}{\|}}{C}NHCH_2CH_3 + H_2O}$$

Self-Test

1. True	2. True	3. False	4. True	5. True
6. True	7. True	8. True	9. True	10. True
11. a	12. b	13. d	14. d	15. a	16. c
17. c	18. d	19. c	20. b	21. b	22. a
23. c	24. a				

Chapter 16

Practice Exercises

16.1 (a), (d), (e), (f)

16.2 (asymmetric carbons marked with asterisks)

(a) $CH_3CH_2-\overset{\overset{O}{\|}}{C}-H$

(b) $CH_3-\overset{\overset{H}{|}}{\underset{CH_3}{C}}-\overset{\overset{O}{\|}}{C}-H$

(c) $CH_3-\overset{\overset{H}{|}}{\underset{Cl}{\overset{*}{C}}}-OH$

(d) $CH_3-\overset{\overset{H}{|}}{\underset{CH_2CH_3}{\overset{*}{C}}}-\overset{\overset{O}{\|}}{C}-H$

16.3

(a)
$$\begin{array}{c} CH_3 \\ H\blacktriangleright\!\!-\!\!-\!\!\blacktriangleleft OH \\ Cl \end{array} \qquad \begin{array}{c} CH_3 \\ HO\blacktriangleright\!\!-\!\!-\!\!\blacktriangleleft H \\ Cl \end{array}$$

(b)
$$\begin{array}{c} CHO \\ H\blacktriangleright\!\!-\!\!-\!\!\blacktriangleleft OH \\ CH_3 \end{array} \qquad \begin{array}{c} CHO \\ HO\blacktriangleright\!\!-\!\!-\!\!\blacktriangleleft H \\ CH_3 \end{array}$$

(c)
$$\begin{array}{c} CHO \\ H\blacktriangleright\!\!-\!\!-\!\!\blacktriangleleft CH_3 \\ CH_3 \end{array}$$

16.4

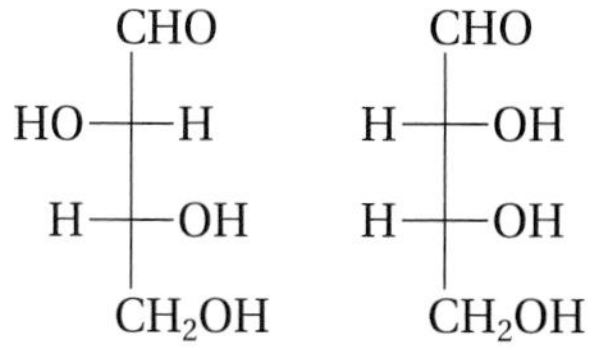

CHO   CHO
HO——H   H——OH
H——OH   H——OH
CH₂OH   CH₂OH

16.5 (a) L, (b) D, (c) D, (d) L isomer.

16.6

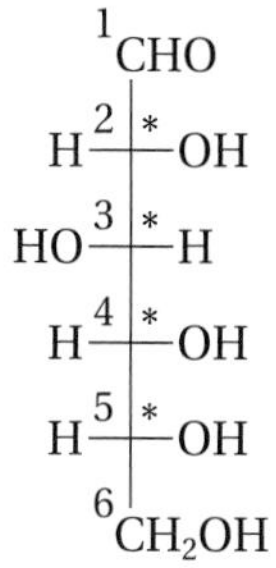

^{1}CHO
H——^{2}*——OH
HO——^{3}*——H
H——^{4}*——OH
H——^{5}*——OH
^{6}CH₂OH

16.7

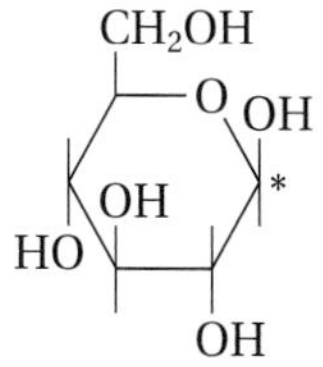

CH₂OH
O OH
OH   *
HO
OH

16.8

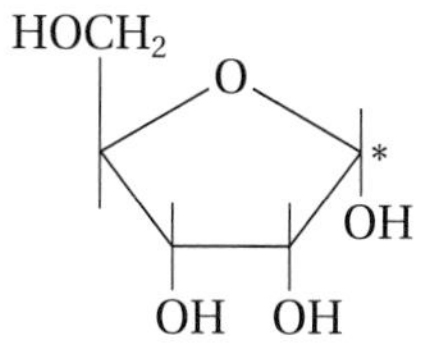

HOCH₂
O
*
OH
OH OH

16.9

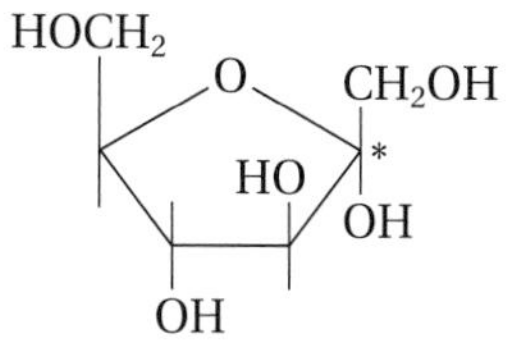

HOCH₂
O   CH₂OH
HO   *
OH
OH

16.10

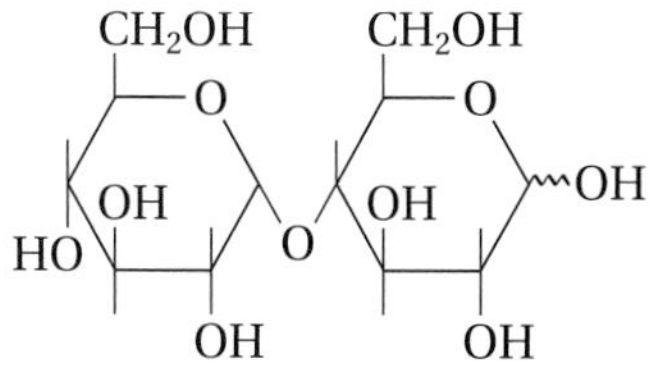

CH₂OH   CH₂OH
O   O
OH   OH   ᵐᵐOH
HO   O
OH   OH

16.11

CH₂OH   HOCH₂
O   O   OH
OH   OH
HO   OH   CH₂OH
OH   OH

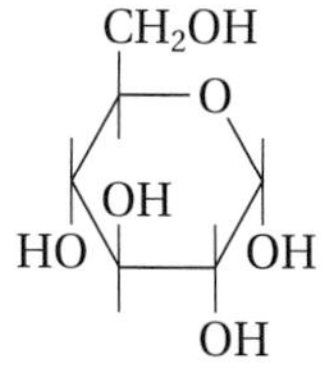

D-Glucopyranose 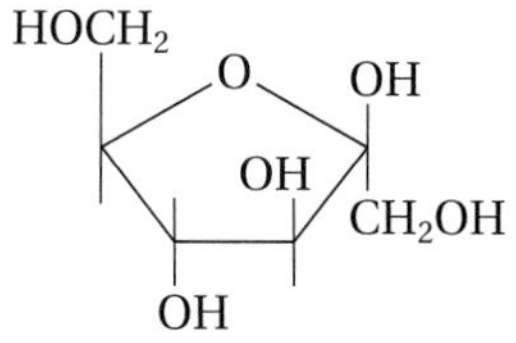   D-Fructofuranose

16.12 Amylose: long polymer chains of D-glucose linked
α(1 ⟶ 4). Amylopectin: polymeric chains with
α(1 ⟶ 4) and α(1 ⟶ 6) cross-links.

16.13 All four polymers yield D-glucose upon hydrolysis.
16.14 Both are disaccharides composed of D-glucose, but
the linkage between the sugars is α(1 ⟶ 4) in
maltose and β(1 ⟶ 4) in cellobiose.

16.15

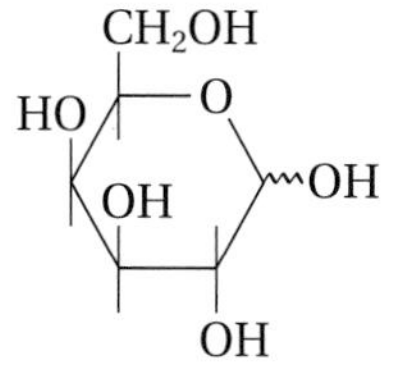

CH₂OH   CH₂OH
HO   O   O
OH   ᵐᵐOH   OH   ᵐᵐOH
HO
OH   OH

D-Galactose 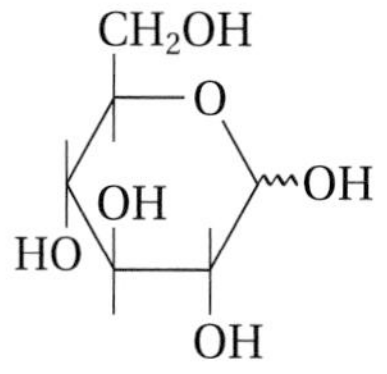   D-Glucose

16.16 1 primary hydroxyl group, 6 secondary hydroxyl
groups, 2 acetal groups, and 1 phenyl group

Exercises

16.17

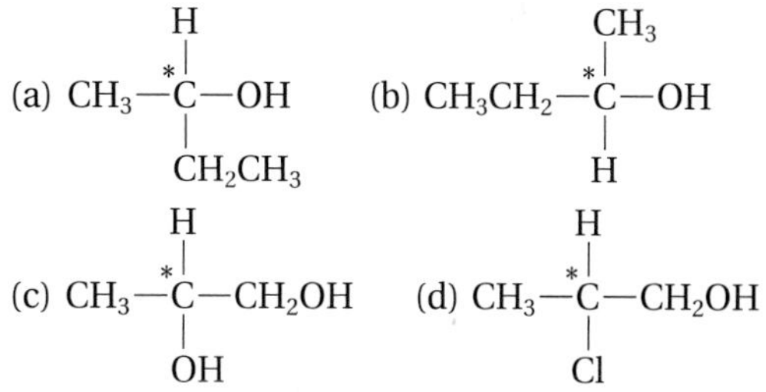

             H   CH₃
(a) CH₃—*C—OH   (b) CH₃CH₂—*C—OH
             CH₂CH₃   H

             H   H
(c) CH₃—*C—CH₂OH   (d) CH₃—*C—CH₂OH
             OH   Cl

16.19

Fischer
projection

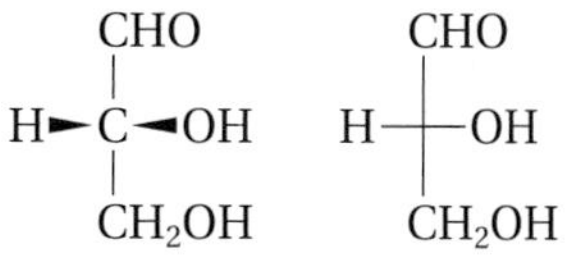

CHO   CHO
H►C◄OH   H——OH
CH₂OH   CH₂OH

D-Glyceraldehyde

Fischer
projection

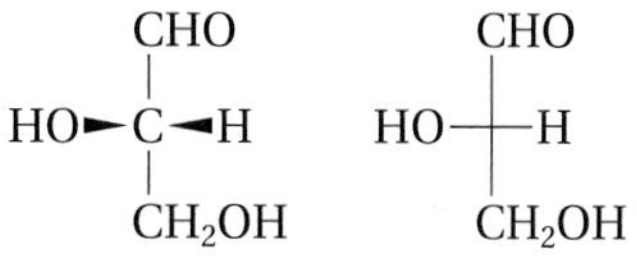

CHO   CHO
HO►C◄H   HO——H
CH₂OH   CH₂OH

L-Glyceraldehyde

16.21 (a) D-threose (an aldotetrose)   (b) D-glucose (an
aldohexose)   (c) D-fructose (a ketohexose)
16.23 (a) L   (b) D   (c) D   (d) D

16.25

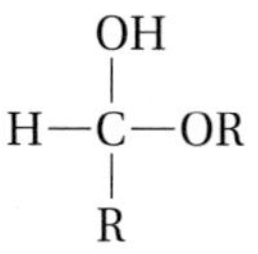

OH   O
H—C—OR
R

Hemiacetal 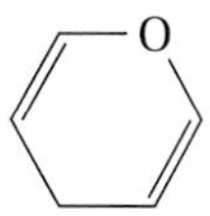   Pyran

16.27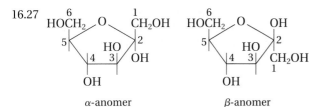

α-anomer   β-anomer

16.29

OR   OR
| |
R—C—OR   R—C—OR
| |
H   R

Acetal   Ketal

16.31 (a) edible   (b) wood and cotton   (c) storage form of glucose

16.33 Starch (amylose): linear chain of glucose linked α(1 ⟶ 4). Cellulose: linear chain of glucose linked β(1 ⟶ 4). Humans can digest (hydrolyze) amylose, but not cellulose.

16.35 (a) Sucrose: from juice or sap of sugar cane, sugar beets, and maple trees; sweetener and constitutes ordinary table sugar; a disaccharide, hydrolysis gives D-glucose and D-fructose.   (b) Maltose (malt sugar): from hydrolysis of starch; a disaccharide; hydrolysis D-glucose.

16.37

Acetal   Hemiacetal
carbon   carbon

CH₂OH   CH₂OH

Maltose

16.39 Open-chain form of lactose is an easily reduced aldehyde group; sucrose has neither a cyclic hemiacetal group that could ring-open to form an easily oxidized aldehyde group nor a cyclic hemiketal group that could ring-open to form an easily reduced alpha-hydroxy ketone.

16.41 (a) glucose positive, starch negative   (b) fructose positive, sucrose negative   (c) fructose and glucose positive

16.43 (a) and (b) no assymetric carbon and not optically active   (c) assymetric carbon and optically active   (d) asymmetric carbon and optically active

16.45

α anomer   β anomer

CH₂OH   CH₂OH

16.47 (a) 3   (b) 2   (c) 4   (d) 3   (e) 4   (f) 4
16.49 (a) L   (b) D

Self-Test

1. True	2. True	3. True	4. False	5. True
6. True	7. True	8. True	9. False	10. True
11. b	12. c	13. b	14. c	15. a	16. d
17. d	18. a	19. d	20. a	21. a	22. d
23. b	24. c	25. b	26. d	

Chapter 17

Practice Exercises

17.1

$$CH_2-O-\overset{O}{\overset{\|}{C}}-(CH_2)_{16}CH_3$$

$$CH-O-\overset{O}{\overset{\|}{C}}-(CH_2)_{14}CH_3$$

$$CH_2-O-\overset{O}{\overset{\|}{C}}-(CH_2)_7CH=CH(CH_2)_7CH_3$$

17.2 Tripalmitin

17.3 Tristearin (glyceryl tristearate)

17.4

CH_2-OH $Na^+\ ^-O-\overset{O}{\overset{\|}{C}}-(CH_2)_7CH=CH(CH_2)_7CH_3$

$CH-OH$ Sodium oleate (2 mols)

CH_2-OH $Na^+\ ^-O-\overset{O}{\overset{\|}{C}}-(CH_2)_{16}CH_3$

Glycerol (1 mol)   Sodium stearate (1 mol)

17.5

$$CH_2-O-\overset{O}{\overset{\|}{C}}-R$$

$$CH-O-\overset{O}{\overset{\|}{C}}-R$$

$$CH_2-O-\overset{O}{\overset{\|}{P}}-OCH_2CH_2NH_2$$
$$\underset{OH}{|}$$

17.6 No micelles formed; hydrophobic tails of phosphatidyl choline molecules would not aggregate to exclude water.

17.7 Polar —OH group on one end and the remainder of the molecule is hydrophobic.

17.8 (a) Aldosterone: 2 ketone carbonyl groups, 1 aldehyde group, 1 C—C double bond, 1 secondary hydroxyl group, and 1 primary hydroxyl group.
(b) Cortisone: 3 ketone carbonyl groups, 1 primary hydroxyl group, 1 tertiary hydroxyl group, and 1 C—C double bond.

Exercises

17.9

$$\underset{RCOR}{\overset{\overset{\displaystyle O}{\|}}{}}$$

17.11 (a) solid (mp 44 °C); (b) liquid (mp −5 °C); (c) solid (mp 63 °C); (d) liquid (mp 14 °C).

17.13 Triester made from 1 molecule of glycerol and 3 molecules of fatty acids.

17.15 (a) Carbon-carbon double bonds of fatty acids become saturated; (b) used to make butter and lard substitutes.

17.17

$$CH_3(CH_2)_{14}\overset{\overset{\displaystyle O}{\|}}{C}O^-Na^+$$

17.19 Phospholipids, glycolipids

17.21 Sphingomyelins: 1 fatty acid, sphingosine, phosphoric acid, and choline. Phosphoglycerides: 2 fatty acids, glycerol, and phosphoric acid.

17.23 Micelle: hydrophobic tails of the lipids point to interior of the aggregate, away from water; polar or ionic heads face water. Liposome: hydrophobic tails of the lipids form a bilayer; polar or ionic heads face solvent water both inside and outside of the liposome.

17.25 Membrane flexibility increases in proportion to the amount of unsaturated fatty acids in membrane lipids.

17.27 Facilitate the transport of ions and molecules into and out of cells.

17.29

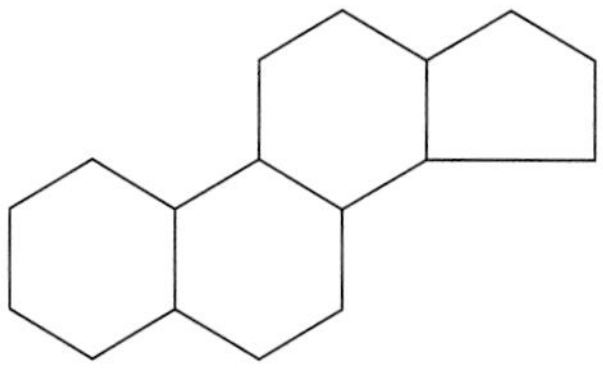

17.31 Chemical messengers produced by endocrine glands; carried through blood to sites where they produce dramatic physiological effects.

17.33 To reduce body's immunity to the introduction of an organ.

17.35 (a) testosterone   (b) estrone, estradiol, and progesterone

17.37 Increase muscle mass; can cause testicular atrophy and cancer.

17.39 Degraded before it reaches the ovaries.

17.41 Contains an attached trisaccharide.

17.43

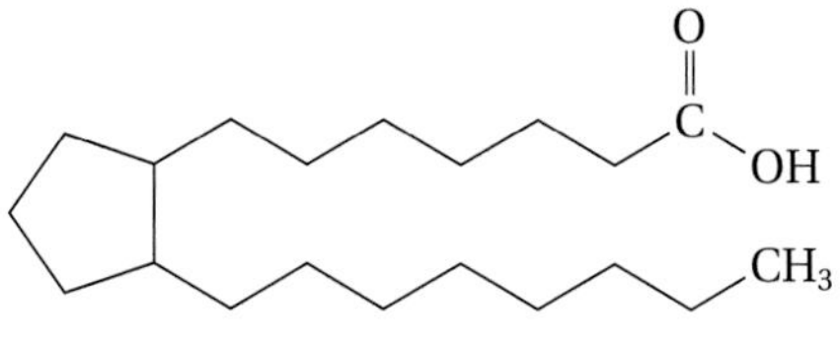

17.45 Control of acid secretions in stomach, relaxation and contraction of smooth muscle, vascular permeability, inflammation, and body temperature.

17.47 Phospholipids and glycolipids

17.49

(a)

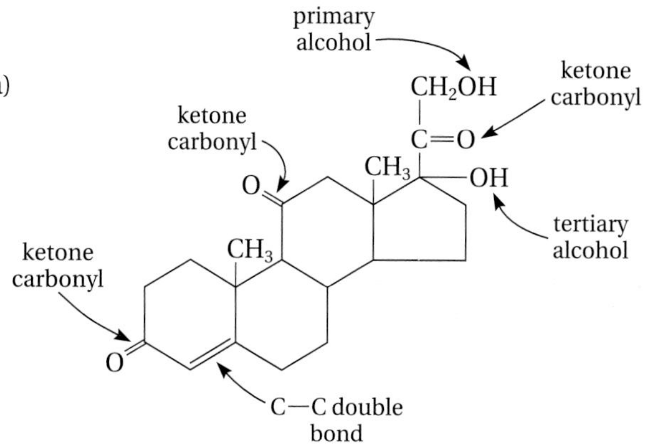

(b)

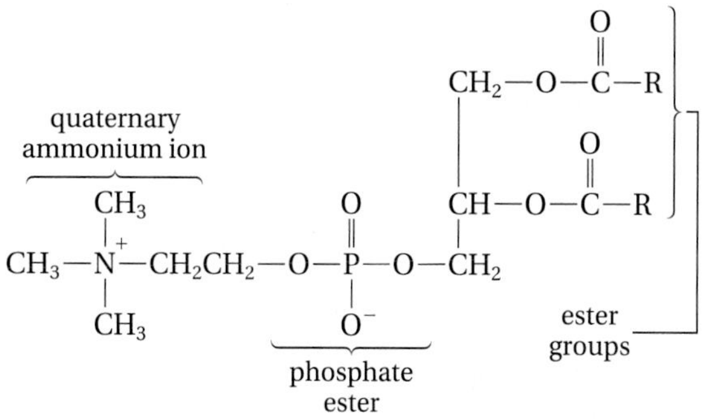

(c)

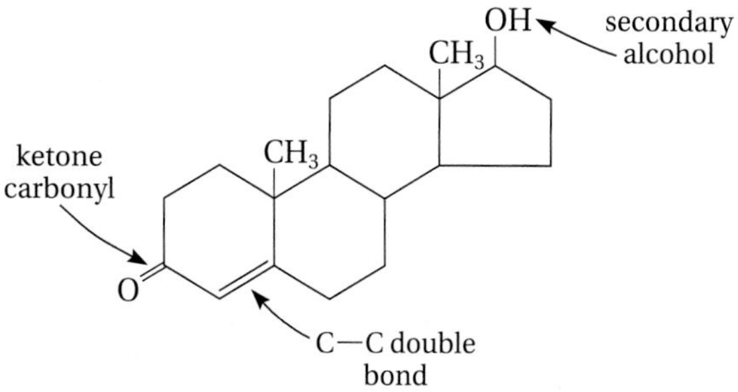

(d)

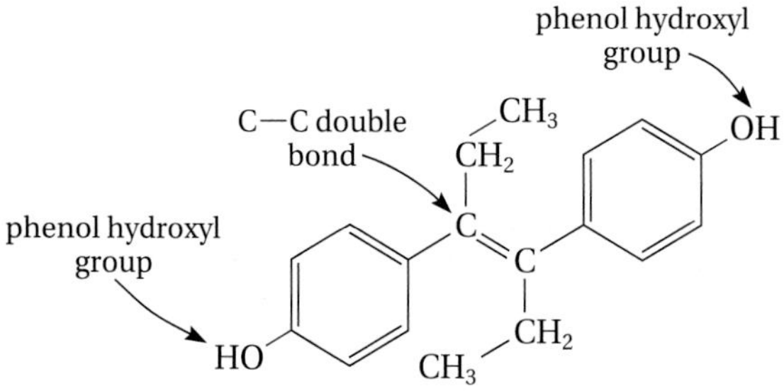

(e)

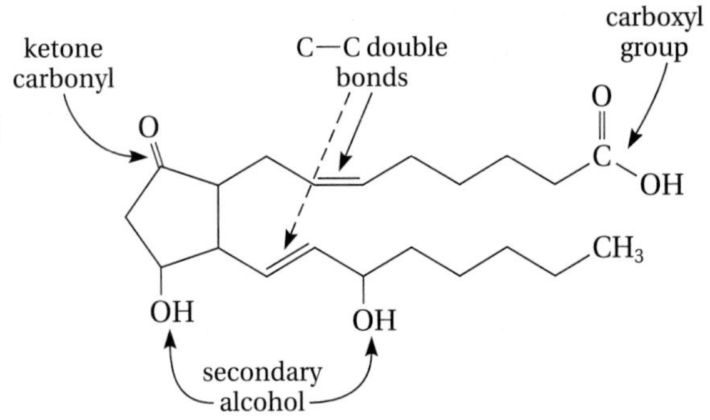

ketone carbonyl

C—C double bonds

carboxyl group

secondary alcohol

(f) primary amine

secondary alcohol

$$NH_2 \quad OH$$

(f) $HO-CH_2-CH-CH-CH=CH(CH_2)_{12}CH_3$

primary alcohol

C—C double bond

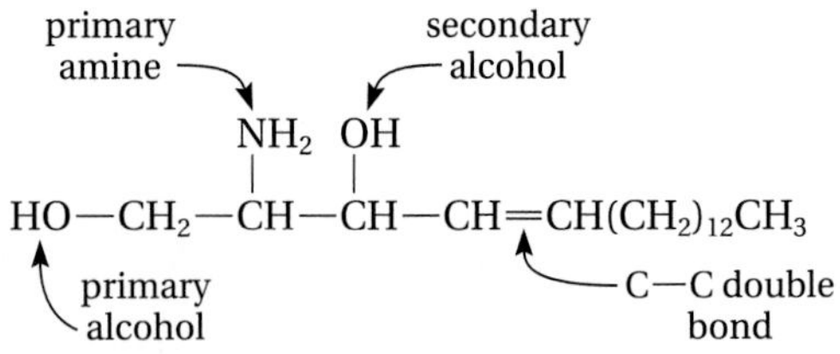

(g)

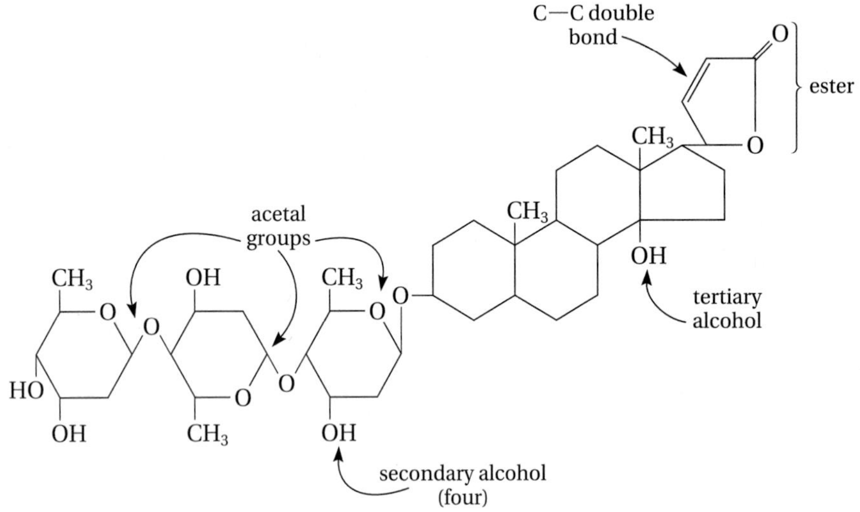

C—C double bond

ester

acetal groups

tertiary alcohol

secondary alcohol (four)

(h) primary alcohol

$$CH_3$$
(h) $CH_3-\overset{+}{\underset{CH_3}{N}}-CH_2CH_2OH$

quaternary ammonium ion

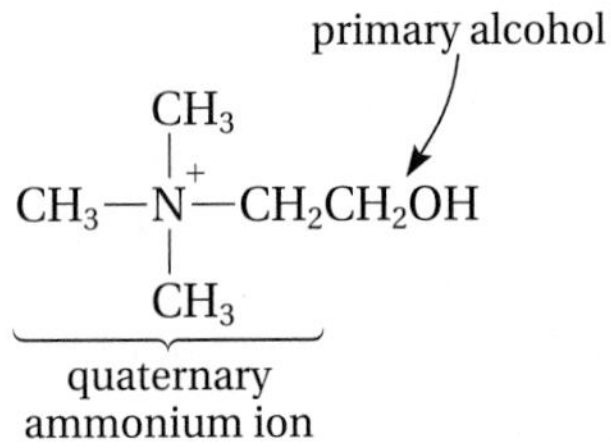

(i) ester group

$$CH_2-O-\overset{O}{\overset{\|}{C}}-(CH_2)_7CH=CH(CH_2)_7CH_3$$

$$CH-O-\overset{O}{\overset{\|}{C}}-(CH_2)_7CH=CH(CH_2)_7CH_3$$

$$CH_2-O-\overset{O}{\overset{\|}{C}}-(CH_2)_7CH=CH(CH_2)_7CH_3$$

C—C double bond

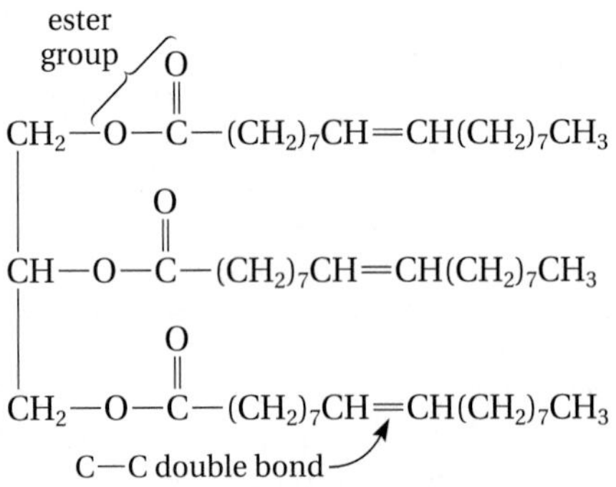

17.51 Hydrocarbon chains of saturated fatty acids fit better into crystals than hydrocarbon chains of unsaturated fatty acids.

Self-Test

1. True 2. True 3. False 4. False 5. True
6. False 7. False 8. False 9. True 10. True
11. c 12. d 13. c 14. b 15. a 16. c
17. d 18. b 19. a 20. d 21. c 22. d
23. b 24. b 25. c 26. a

Chapter 18

Practice Exercises

18.1 (a) aliphatic (b) basic (c) hydroxylic
(d) aromatic

18.2 (a) L-Cysteine (b) D-Cysteine

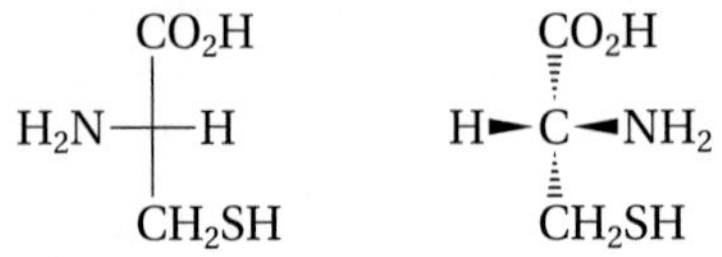

$$CO_2H$$
$$H_2N\!-\!\!\!\!-\!\!\!\!-\!H$$
$$CH_2SH$$

$$CO_2H$$
$$H\!-\!C\!-\!NH_2$$
$$CH_2SH$$

(c) L isomer

18.3 (a)

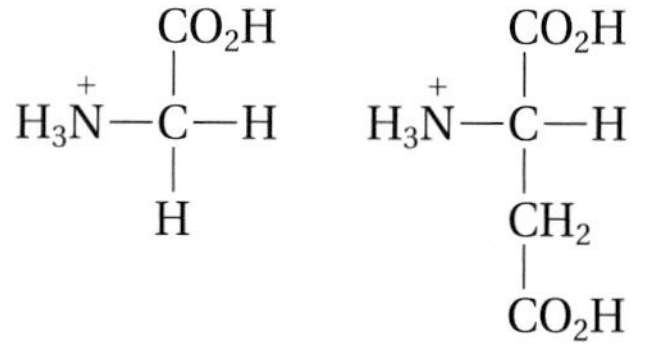

(b)

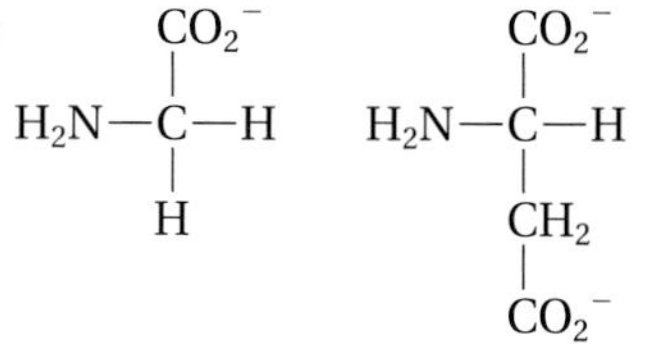

18.4 (a)

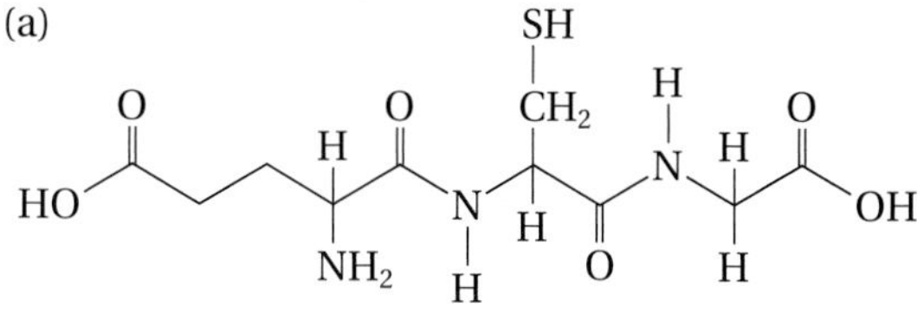

(b) Glu-Cys-Gly

18.5 Glu-Cys-Gly; Glu-Gly-Cys; Cys-Gly-Glu; Cys-Glu-Gly; Gly-Glu-Cys; Gly-Cys-Glu.

18.6

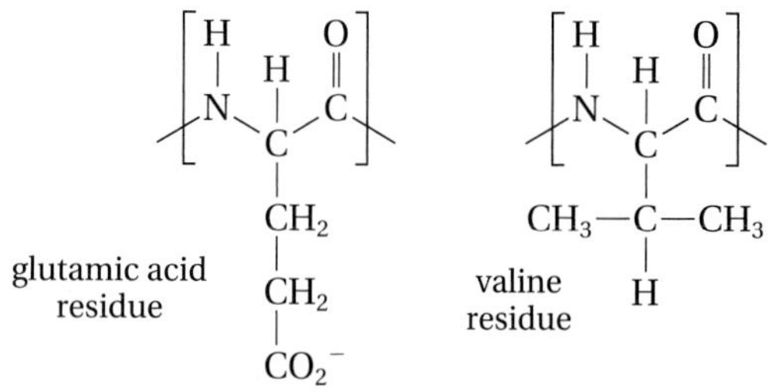

glutamic acid residue   valine residue

Side chain of glutamic acid residue has a carboxyl group dissociated at pH 7. Side chain of valine residue is uncharged at all values of pH and is hydrophobic.

Exercises

18.7 (Alpha carbons circled and side chain R groups boxed)

(a)

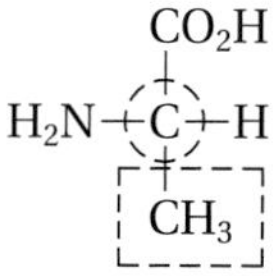

(b)

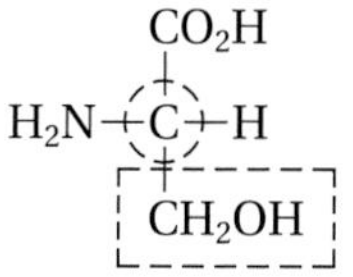

(c)

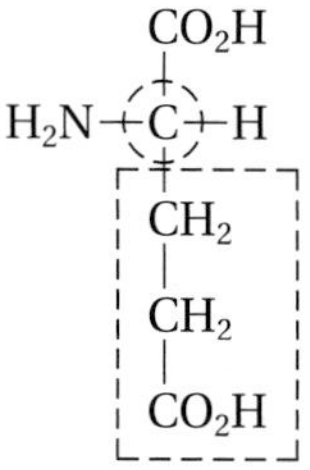

18.9 (a) D-Serine   (b) L-Alanine

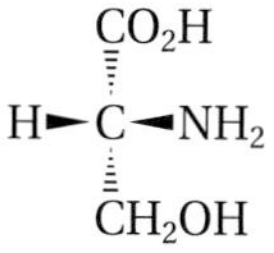

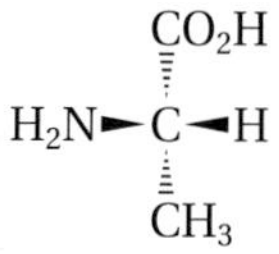

18.11 The internal salt of an amino acid.

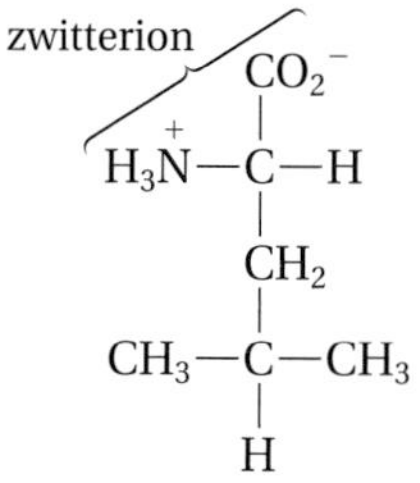

18.13 The pH at which the positive and negative charges on an amino acid balance each other, giving a net charge of zero.

18.15 Portion of an amino acid that is incorporated into a peptide chain.

18.17 (a)

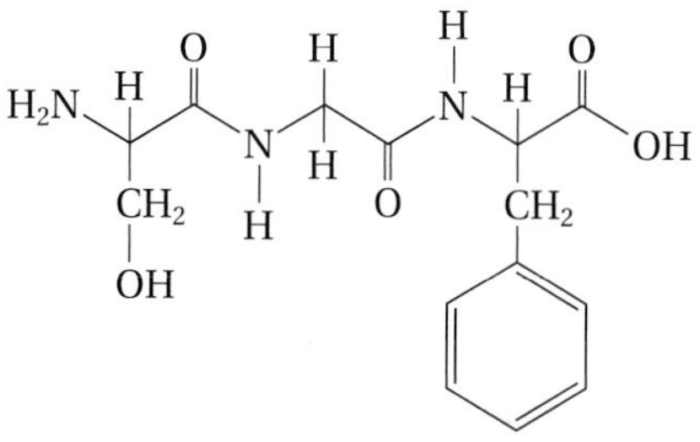

(b) Ser-Gly-Phe   (c) Two peptide bonds

18.19 (a) Stimulates milk ejection in females, contraction of the uterus in labor, and feelings of satisfaction. (b) Antidiuretic in both sexes.

18.21 The *primary structure* of a protein is the sequence of amino acid residues in a protein.

18.23 Alpha helix, beta pleated sheet, and collagen helix.

18.25 Major structural protein of the body, found in skin, bone, teeth, cartilage, and tendon.

18.27 A *conjugated protein* contains an organic non-protein portion called a prosthetic group.

18.29 Salt bridges, hydrogen bonds, hydrophobic aggregation, and disulfide bridges.

18.31 (a) Iron(II) (Fe^{2+})   (b) Iron(III) ion (Fe^{3+})

18.33 They bind to the heme iron of hemoglobin, preventing oxygen from binding.

18.35 Sickle cell hemoglobin (HbS) in its deoxygenated form is less soluble than normal adult hemoglobin (HbA). Precipitation of HbS causes red blood cells to sickle and burst.

18.37 High oxygen pressure used in hyperbaric oxygenation keeps sickle cell hemoglobin (HbS) oxygenated.

18.39 Subunits (quaternary structure) dissociate; secondary and tertiary structure unfolds.

18.41 The correct matches are as follows.
(a) (3) (b) (2) (c) (5) (d) (8) (e) (7)
(f) (1) (g) (6) (h) (4)

18.43 His-Met-Glu; His-Glu-Met; Met-His-Glu; Met-Glu-His; Glu-Met-His; Glu-His-Met.

18.45 (a)

$$
\begin{array}{ccc}
CO_2H & CO_2H & CO_2H \\
| & | & | \\
H_2N-C-H & H_2N-C-H & H_2N-C-H \\
| & | & | \\
CH_2 & H & CH_3 \\
| & & \\
SH & & \\
\text{Cysteine} & \text{Glycine} & \text{Alanine}
\end{array}
$$

(b)

$$
\begin{array}{cc}
CO_2H & CO_2H \\
| & | \\
H_2N-C-H & H_2N-C-H \\
| & | \\
CH_2 & CH_2 \\
| & | \\
CH_2 & CH \\
| & \diagdown \\
CO_2H & CH_3 \quad CH_3 \\
\text{Glu} & \text{Leu}
\end{array}
$$

$$
\begin{array}{cc}
CO_2H & \\
| & \\
H_2N-C-H & \\
| & \\
CH_3-C-CH_3 & \text{(Pro ring)}-CO_2H \\
| & \\
H & \\
\text{Val} & \text{Pro}
\end{array}
$$

(c)

$$
\begin{array}{ccc}
CO_2H & CO_2H & CO_2H \\
| & | & | \\
H_2N-C-H & H_2N-C-H & H_2N-C-H \\
| & | & | \\
CH_2 & CH_2 & CH_2 \\
| & & \\
OH & \text{(His ring)} & \text{(Phe ring)} \\
\text{Ser} & \text{His} & \text{Phe}
\end{array}
$$

$$
\begin{array}{cc}
CO_2H & CO_2H \\
| & | \\
H_2N-C-H & H_2N-C-H \\
| & | \\
CH_2 & CH_2 \\
\text{(Tyr ring)} & \text{(Trp ring)} \\
\text{Tyr} & \text{Trp}
\end{array}
$$

18.47 (a) Ionic bond formed between the negatively charged carboxylate ion of the side chains of aspartate or glutamate and the positively charged amino side chains of lysine or arginine. (b) Nonprotein group attached to a protein by covalent bonds. (c) Folding of a polypeptide chain into a relatively stable three-dimensional shape. (d) A protein with a more or less spherical shape. (e) Peptide chains arranged side by side to form a structure that resembles pleats. (f) Specific, repeating patterns of folding of the peptide backbone of a protein. (g) Coiling of the peptide backbone of a protein into a spiral shape resembling a corkscrew. (h) Order in which the amino acids of a peptide or protein are linked by peptide bonds.

18.49 (a) (6) (b) (4) (c) (7) (d) (9) (e) (1)
(f) (2) (g) (3) (h) (5) (i) (8)

Self-Test
1. True 2. False 3. True 4. True 5. True
6. False 7. True 8. True 9. True 10. False
11. c 12. c 13. b 14. a 15. d 16. d
17. a 18. a 19. a 20. b 21. c 22. b
23. a 24. c 25. c 26. b 27. b 28. b

Chapter 19

Practice Exercises
19.1 (a) lipase (breakdown of lipids) (b) cellulase (breakdown of cellulose)

19.2 (a) oxidase (b) hydrolase

19.3 When a substrate is bound at the active site of an enzyme, (a) there is a high concentration of substrate and enzyme, and (b) the substrate is "locked" in the ideal position for bond-breaking or bond-making processes to occur.

19.4 Pancreatic disorder

19.5 Enzyme assays measure the activity of an enzyme. The activity of an enzyme depends upon the pH and the temperature.

19.6

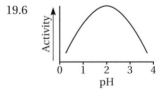

19.7 A positive modulator increases the activity of the enzyme; a negative modulator decreases the activity of an enzyme.

19.8 An enzyme cofactor takes part in the chemical reaction that the enzyme catalyzes; a positive modulator activates the enzyme but does not take part in the reaction.

19.9 (a) bonds 2, 3, 4   (b) bonds 1, 5

19.10 The conversion of fibrinogen to fibrin is catalyzed by thrombin. Active thrombin must be made from its inactive zymogen, prothrombin.

Exercises

19.11 A biological catalyst that speeds up chemical reactions.

19.13 (a) hydrolysis of esters   (b) hydrolysis of lactose (c) removal of hydrogen   (d) hydrolysis of amylose

19.15 The active site of the enzyme is the "lock" that the substrate "key" must fit.

19.17 The location where the action takes place.

19.19 To form an active enzyme.

19.21 Enzyme activity measures the speed at which an enzyme changes substrate to product(s).

19.23 An enzyme lowers the energy of activation.

19.25 (a) activity decreases   (b) denatured irreversibly, all activity lost.

19.27 An enzyme produced in response to a temporary need of the cell.

19.29 Competes with the substrate for the active site of the enzyme.

19.31 allosteric enzyme

19.33 inactive form of an enzyme

19.35 Refer to the drawing in Section 19.12 of your text.

19.37 Penicillin inhibits an enzyme and as a consequence the bacterial cell walls are defective.

19.39 enzyme inhibition

19.41 Ability for an enzyme to catalyze one chemical reaction with only one substrate.

19.43 A zymogen is an inactive precursor of an enzyme (Table 19.2).

19.45 See Section 19.10 of the text.

Self-Test

1. False   2. False   3. True   4. False   5. True
6. True   7. True   8. False   9. True   10. False
11. d   12. b   13. b   14. a   15. d   16. c
17. c   18. a   19. b   20. c   21. c   22. d

Chapter 20

Practice Exercises

20.1

Cytidine

20.2

Cytidine 5′-monophosphate (CMP)

20.3 (a) adenine-cytosine-guanine-uracil (from RNA) (b) thymine-adenine-cytosine-guanine (from DNA)

20.4 A phosphodiester bridge links the 3′-hydroxyl group of one nucleotide to the 5′-hydroxyl group of another nucleotide in DNA and RNA.

20.5 You should include the following ideas: (1) hydrogen bonding, (2) complementary base pairing, and (3) antiparallel strands.

20.6 *d* T=A
       C≡G
       C≡G
       C≡G
       A=T
       A=T
       G≡C *d*

20.7   3′   5′
    RNA: U—A—G—C—U—U

20.8 5′   3′
      G—C—U—G—G—U—U—C—U

20.9 The amino acid sequence of the tetrapeptide is Arg-Arg-Asn-Thr.

20.10 When the sequence of bases on DNA is altered through the deletion, addition, or substitution of a single nucleotide, the DNA is faulty. Faulty DNA leads to gene mutations. Faulty DNA gives faulty mRNA that translates into proteins with the wrong amino acid sequence or no protein at all.

20.11 (a) The pentapeptide is Thr-Val-Ser-Glu-Pro. (b) The mutation leads to the pentapeptide Thr-Asp-Ser-Glu-Pro.   (c) Deletion of uracil gives the tetrapeptide Thr-Ala-Ala-Ser.

20.12 Answers will depend on personal opinions.

Exercises

20.13 Deoxyribonucleic acid, DNA, and ribonucleic acid, RNA.

20.15 A nucleoside is composed of a nitrogen base and a sugar unit. Adding a phosphate group to a nucleoside forms a nucleotide.

20.17 (a) adenine, guanine, thymine, and cytosine (b) Adenine and guanine are derivatives of purine thymine and cytosine are derivatives of pyrimidine

20.19 (a)

Deoxyadenosine

(b) deoxyadenosine.

20.21 Nucleic acids can lose a proton from the phosphate group of their phosphodiester bridges, making them highly acidic.

20.23 Hydrogen bonds between complementary pairs of bases hold the two DNA strands together in the double helix. Adenine and thymine form one base pair and cytosine and guanine form the other base pair.

20.25 Antiparallel strands run in different directions. One DNA strand runs in the 3′ to 5′ direction and the complementary DNA strand runs in the 5′ to 3′ direction.

20.27 (a) *replication*   (b) *transcription* (c) *translation*

20.29 Replication involves the following steps: (1) unwinding the original double helix, (2) synthesis of complementary strands in the 5′ to 3′ direction,

(3) joining together the segments of 3′ to 5′ strand, and (4) assembly of two identical DNA double helixes.

20.31 DNA polymerase catalyzes the joining of nucleotides to form complementary strands of DNA.

20.33 A gene is a segment of a DNA molecule that codes for the synthesis of one kind of protein molecule; a polypeptide is a chain of amino acids that is formed when the genetic message is translated in protein synthesis.

20.35 (a) rRNA is a structural component of a ribosome. (b) mRNA carries the genetic information from DNA to the ribosome, where the message is translated in protein synthesis.

20.37 (1) DNA template, (2) RNA polymerase, (3) nucleoside triphosphates containing the nitrogen bases adenine, guanine, cytosine, and uracil

20.39 (a) A codon is a base triplet on mRNA.   (b) An anticodon is a base triplet on tRNA.

20.41 Written from 5′ to 3′ the anticodon for methionine is CAU.

20.43 (a) 3′ U—C—G—A—C—C—C—U—G 5′ (b) Val-Pro-Ala

20.45 Substitution, addition, or deletion of one or more nucleotides in the DNA molecule.

20.47 Increased ability to manipulate genes and genetic material. Answers will vary.

20.49 (a) 10   (b) 6   (c) 5   (d) 7   (e) 8   (f) 2 (g) 4   (h) 9   (i) 3   (j) 1

20.51 (a) One correct answer is 5′ UUU—UCU—GCU—CAU 3′   (b) 5′ *d*A—T—G—A—G—C—A—G—A—A—A—A 3′

20.53 rRNA: A structural component of ribosomes. mRNA: Carries information from DNA to the ribosome to direct protein synthesis and contains the codons. tRNA: Carries amino acids to the ribosome. The anticodon on tRNA base pairs with a codon of mRNA to ensure that the correct amino acid is incorporated into the peptide chain.

20.55 ATC

20.57 Answers will vary. A change of a single base could result in premature termination of protein synthesis. Insertion or deletion of a base will cause a change in the codons and result in a nonsense protein or no protein.

Self-Test

1. False   2. True   3. False   4. False   5. True
6. False   7. False   8. False   9. True
10. False   11. c   12. d   13. a   14. d
15. b   16. b   17. b   18. c   19. a   20. d
21. a

Chapter 21

Practice Exercises

21.1 Protein breakdown in the stomach is catalyzed by the enzyme pepsin, which works best at very low pH. Antacids neutralize stomach acid, raising the pH and lowering the catalytic activity of pepsin.

21.2 The enzyme amylase in saliva catalyzes the breakdown of carbohydrates to maltose. Maltose, sucrose, and lactose are hydrolyzed to give monosaccharides in the small intestines, where they pass into cells of the intestinal wall and then into the bloodstream.

21.3 Lipase catalyzes the hydrolysis of lipids in the small intestines. The products of lipid hydrolysis pass into the intestinal cells, where they are re-assembled into lipid molecules. Insoluble lipids interact with lipoproteins to form chylomicrons, which then pass into the bloodstream.

21.4 8.9×10^{19} molecules niacin

21.5 (a) 4   (b) 6   (c) 1   (d) 8   (e) 2   (f) 10
(g) 3   (h) 9   (i) 11   (j) 5   (k) 7

Exercises

21.7 amylase; catalyzes the hydrolysis of starch and glycogen. Mucin is a glycoprotein.

21.9 pepsinogen

21.11 It catalyzes the hydrolysis of peptide bonds.

21.13 chymotrypsin, trypsin, carboxypeptidase, maltase, sucrase, lactase, lipases, nucleotidases

21.15 (a) a liquid consisting of cholesterol, bile salts, and bile pigments   (b) in the liver   (c) to emulsify fats so they can be hyrolyzed by lipase action

21.17 fats

21.19 Essential amino acids cannot be synthesized by the body; nonessential amino acids can be synthesized by the body.

21.21 (a) eggs, dairy products, kidney, and liver
(b) corn, wheat, and gelatin

21.23 vitamin A, vision; vitamin D, promotion of calcium and phosphorus uptake; vitamin E, antioxidant; vitamin K, blood clotting.

21.25 They are not stored in the body and are readily excreted.

21.27 calcium (Ca), phosphorus (P), potassium (K), sulfur (S), chlorine (Cl), sodium (Na), magnesium (Mg)

21.29 sodium ion (Na^+), potassium ion (K^+), and chloride ion (Cl^-)

21.31 If phosphorus is too high, the excess is eliminated from the body as calcium phosphate. This leads to a calcium deficiency.

21.33 A deficiency in children causes rickets; in adults a deficiency causes osteoporosis.

21.35 bowed legs, pigeon breast, and poor tooth development

21.37 iron (Fe), zinc (Zn), copper (Cu), manganese (Mn), cobalt (Co), chromium (Cr), selenium (Se), iodine (I), and molybdenum (Mo)

21.39 animal organs, such as liver and kidney, and green vegetables

21.41 for the formation of hemoglobin; deficiency can lead to anemia

21.43 copper; Wilson's disease

21.45 An iodine deficiency causes the thyroid gland to grow large—a goiter—as it tries to meet the body's demand for thyroid hormones.

21.47 Fluoride ions make teeth and bones stronger. Excess of fluoride makes teeth mottled.

21.49 leafy green vegetables; produced by intestinal bacteria

21.51 (a) amylase   (b) pepsinogen or pepsin
(c) trypsinogen or trypsin, chymotrypsinogen or chymotrypsin, procarboxypeptidase or carboxypeptidase, amylase, maltase, sucrase, lactase, lipase, nucleotidases

21.53 See Section 21.4.

21.55 vitamin C and the B vitamins

21.57 (a) part of vitamin B_{12} (b) trace element for growth   (c) hemoglobin formation   (d) enzyme cofactor   (e) functioning of the CNS
(f) nerve impulse transmission, muscle contraction, and bond formation

Self-Test

1. False	2. True	3. True	4. False	5. False
6. False	7. True	8. False	9. True	10. True
11. a	12. c	13. a	14. c	15. b	16. c
17. c	18. d	19. b	20. a	21. c	22. a
23. d					

Chapter 22

Practice Exercises

22.1 Blood: carries food, oxygen, hormones, and anti-bodies and removes wastes. The lymphatic system traps bacteria and other foreign substances, produces antibodies, and transports lipids to the bloodstream.

22.2 Carbonic anhydrase, $H_2CO_3 \rightleftharpoons CO_2 + H_2O$; blood would be more acidic.

22.3 The pH would increase.

22.4 Carbon dioxide/carbonic acid/bicarbonate ion equilibrium shifts to the right, the HCO_3^-/H_2CO_3 ratio is restored to 20:1, and blood pH decreases.

Exercises

22.5 42 L water in 70-kg person: 28 L intracellular, 11 L interstitial, and 3 L in blood plasma.

22.7 Intracellular fluids are high in K^+ and HPO_4^{2-}; interstitial fluids are high in Na^+ and Cl^-.

22.9 carrying food, oxygen, wastes, hormones, antibodies; distributing body heat; maintaining pH

22.11 erythrocytes, carry oxygen; leukocytes, fight bacteria; thrombocytes, role in blood clotting

22.13 An antigen is a foreign particle; an antibody is formed by the immune response to antigen.

22.15 identification of infecting organisms

22.17 7.40

22.19 H_2CO_3/HCO_3^-; $HPO_4^{2-}/H_2PO_4^-$; —COOH/—COO$^-$ and —NH$_2$/—NH$_3^+$

22.21 as oxyhemoglobin, HbO_2

22.23 It would increase.

22.25 carbaminohemoglobin, dissolved in blood, bicarbonate ion

22.27 Chloride ions move from plasma into red blood cells and bicarbonate ions move from red blood cells into the plasma.

22.29 filtration of wastes from bloodstream and maintenance of blood pH

22.31 water, glucose, and electrolytes

22.33 Carbonic acid dissociates into bicarbonate ions and hydrogen ions; hydrogen ions are exchanged for sodium ions in developing urine; sodium ions and bicarbonate ions diffuse into the bloodstream.

22.35 hyperventilation

22.37 vasopressin and aldosterone

22.39 aldosterone, conserves salt and water in the body

22.41 decrease

22.43 Addition of H^+ to oxyhemoglobin releases the oxygen.

22.45 hypoventilation, decreases HCO_3^-/H_2CO_3 ratio; hyperventilation, increases HCO_3^-/H_2CO_3 ratio

22.47 The kidneys are the source of bicarbonate ions.

Self-Test

1. True   2. False   3. False   4. True   5. True
6. True   7. True   8. False   9. True   10. False
11. c   12. d   13. b   14. d   15. b   16. c
17. b   18. a   19. a   20. a   21. d   22. b
23. b   24. d

Chapter 23

Practice Exercises

23.1 (a) porphyrin ring   (b) vitamin B$_{12}$, cobalt; hemoglobin, iron; and chlorophyll, magnesium

23.2 Plants in photosynthesis produce complex molecules that all animals need for survival. Animals cannot make these essential energy-rich compounds.

23.3 14 kcal per mole of ATP

23.4 21 kcal free energy

23.5 A cell needs ATP to grow. Without nutrients the cell cannot make ATP and dies.

23.6 (a) $CH_3CH_2COOH + NADH + H^+$
(b) $HOOCCH=CHCOOH + FADH_2$
(c) $CH_3CHO + NADH + H^+$

23.7 $CH_3CH_2OH + 2NAD^+ + H_2O \longrightarrow$
$CH_3COOH + 2NADH + H^+$

23.8 Energy for muscle contraction comes from the hydrolysis of ATP. With heavy exercise, the demand for ATP exceeds the supply and muscles become tired and weak.

Exercises

23.9 all chemical reactions in living organisms

23.11
$$6CO_2 + 6H_2O \xrightarrow[\text{energy}]{\text{Light}} C_6H_{12}O_6 + 6O_2$$
carbon   water   glucose   oxygen
dioxide

23.13 Energy-poor carbon dioxide is converted into energy-rich glucose.

23.15 catabolism

23.17 Carbon dioxide is an oxidized compound of carbon; it is energy-poor. Glucose is a reduced compound of carbon; it is energy-rich.

23.19 Structure of ATP (adenosine triphosphate)

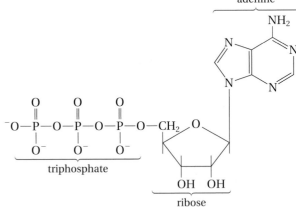

23.21 Unprotonated ATP has a charge of 4−; its molar mass is 504.

23.23 To form ATP requires 7 kcal/mol of free energy. Since some energy is always lost as heat, any reaction used to make ATP must release more than 7 kcal/mol.

23.25 Aerobic cells need oxygen to live. Energy is conserved as "reducing power" by formation of NADH and FADH$_2$; NADH and FADH$_2$ reduce oxygen to water and produce energy; energy is used to convert ADP to ATP.

23.27 (a) $NAD^+ + H^+ + 2e^- \longrightarrow NADH$
(b) $FAD + 2H^+ + 2e^- \longrightarrow FADH_2$

23.29 $CH_3OH + NAD^+ \longrightarrow$

methanol

$HCHO + NADH + H^+$

methanal
(formaldehyde)

NAD^+ is reduced to NADH.

23.31 (a) oxygen is reduced to water and energy is released   (b) energy released is used to convert
ADP to ATP.

23.33 inside mitochondria

23.35 cytochromes; the iron in cytochromes changes
back and forth between Fe^{2+} and Fe^{3+}, the iron in
hemoglobin is Fe^{2+}.

23.37 three

23.39 chemical, osmotic, mechanical

23.41 to make them active

23.43 to supply energy to transport them against the
osmotic gradient

23.45 (a) hemoglobin, cytochromes   (b) vitamin B_{12}
(c) chlorophyll

23.47 NADH enters at the top; $FADH_2$ enters at the second step.

23.49 Reduction of coenzyme Q requires *two* electrons;
reduction of a cytochrome requires *one* electron.

23.51

$$HO-\overset{\overset{\displaystyle O}{\|}}{\underset{\underset{\displaystyle OH}{|}}{P}}-\quad \text{for energy-transfer}$$

Self-Test
1. False   2. True   3. True   4. False   5. False
6. True   7. False   8. True   9. True   10. True
11. True   12. False   13. False   14. c   15. b
16. c   17. d   18. d   19. b   20. b   21. d
22. d   23. a   24. c   25. d   26. d   27. d

Chapter 24

Practice Exercises
24.1

$C_6H_{12}O_6 + 2NAD^+ + 2ADP + 2P_i \longrightarrow$

$$2CH_3\overset{\overset{\displaystyle O}{\|}}{C}-\overset{\overset{\displaystyle O}{\|}}{C}-O^- + 2NADH + 2ATP$$

24.2   1 mol ATP

24.3   An acetyl group attached to CoA is easily transferred to some other molecule, such as water, with
the release of energy.

24.4   oxaloacetate + acetyl CoA + $3NAD^+$ + FAD + ADP
+ $P_i \longrightarrow$ oxaloacetate + $2CO_2$ + 3NADH +
$FADH_2$ + ATP

24.5   Aerobic glycolysis produces 38 ATP molecules per
molecule of glucose; anaerobic glycolysis produces
2 ATP per molecule of glucose.

24.6   phosphorylation of glucose

24.7   Muscle glycogen is converted to glucose 6-phos-
phate but not to glucose. Glucose 6-phosphate cannot escape from muscle cells. In liver cells glucose
6-phosphate is hydrolyzed to glucose that can leave
liver cells to be distributed throughout the body.

24.8   Insulin taken orally would be hydrolyzed in the
stomach and small intestine

Exercises
24.9 $C_6H_{12}O_6 + 6O_2 \longrightarrow 6CO_2 + 6H_2O$

24.11 (1) glycolysis, (2) acetyl CoA, (3) citric acid cycle

24.13 the cytoplasm

24.15 Yes, it involves ATP.

24.17

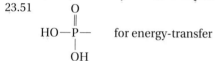

2(1,3-bisphosphoglycerate)   2ADP   2ATP

2(3-phosphoglycerate)

2(2-phosphoenolpyruvate) $\longrightarrow$ 2(pyruvate)
                2ADP   2ATP

24.19 reduced to NADH

24.21 in mitochondria; to transfer an acetyl group

24.23 in mitochondria

24.25 alpha-ketoglutarate + ADP + P_i + NAD^+ $\longrightarrow$
succinate + CO_2 + NADH + ATP

24.27 (a) two molecules of ATP   (b) three molecules of
ATP

24.29 oxidative phosphorylation stops

24.31 Lactic fermentation produces NAD^+ from NADH
when cellular respiration has slowed down or
stopped.

24.33 (a) vigorous exercise results in insufficient oxygen
for cellular respiration   (b) concentration of
lactate increases, causing muscle cramps

24.35 synthesis of glucose from noncarbohydrate
materials

24.37 a branched polysaccharide; source of glucose
formed and stored in muscle and liver cells

24.39 Glycogenolysis is the release of glucose from glycogen. It occurs when glucose is low in the body.

24.41 Hormones act through "second messengers" like
cyclic AMP.

24.43 (a) a scare (b) adrenal gland; stimulates breakdown
of glycogen in muscle cells

24.45 controls the movement of glucose into cells

24.47 Juvenile-onset: pancreas does not produce insulin.
Maturity-onset: delayed insulin release.

24.49 In diabetes mellitus the renal threshold for glucose is
exceeded; glucose appears in the urine (glycosuria).

24.51 Cyclic AMP catalyzes the production of phosphorylase *a*.

24.53 To increase the supply of oxygen for oxidative phosphorylation.

24.55 Chronic hyperglycemia is a symptom of diabetes mellitus. It may be treated by injections of insulin.

Self-Test
1. True   2. True   3. False   4. False   5. False
6. False   7. False   8. False   9. False
10. True   11. a   12. c   13. a   14. c   15. b
16. d   17. b   18. d   19. d   20. d   21. c
22. b   23. b   24. b

Chapter 25

Practice Exercises
25.1 (a) six acetyl CoA molecules   (b) five NADH molecules   (c) five FADH$_2$ molecules
25.2 95 ATP
25.3 Body needs energy; used in citric acid cycle and for production of ketone bodies. Body does not need energy; used to synthesize fatty acids or cholesterol.
25.4 (a) 200 g   (b) 450 g

Exercises
25.5

$$CH_2-O-\overset{\displaystyle O}{\overset{\|}{C}}-(CH_2)_{16}CH_3 \quad \text{(stearic)}$$

$$CH-O-\overset{\displaystyle O}{\overset{\|}{C}}-(CH_2)_7CH=CH(CH_2)_7CH_3 \quad \text{(oleic)}$$

$$CH_2-O-\overset{\displaystyle O}{\overset{\|}{C}}-(CH_2)_7CH=CHCH_2CH=CH(CH_2)_4CH_3$$
$$\text{(linoleic)}$$

25.7 as adipose tissue and working lipids

25.9 the hormone epinephrine
25.11 brain, nervous system, and active skeletal muscle
25.13 mitochondria
25.15 oxidation of the *beta* carbon of a fatty acid to a ketone
25.17 (a) five   (b) four   (c) four
25.19 Both fatty acids and carbohydrates are degraded to acetyl CoA.
25.21 It is a store of triglycerides that produces energy and water when oxidized under aerobic conditions.
25.23 starvation, no glucose in the blood; diabetes, glucose in the blood but none in cells
25.25 When glucose in cells is low the citric acid cycle ceases to function; acetyl CoA therefore produces ketone bodies in the liver.
25.27 (a) ketone bodies in blood greater than 20 mg/100 mL   (b) ketone bodies in the urine   (c) acetone expired by the lungs   (d) a form of blood acidosis
25.29 liver cells; acetyl CoA
25.31 buildup of cholesterol in the arteries
25.33 (a) 4   (b) 5   (c) 9   (d) 10   (e) 7   (f) 3   (g) 8   (h) 6   (i) 1   (j) 2
25.35 converted to fatty acyl CoA
25.37 They contribute to ketosis.
25.39 Epinephrine stimulates hydrolysis of triglycerides to fatty acids and glycerol for use by cells in energy production.
25.41 (a) 5, 9   (b) 2, 8   (c) 7, 13   (d) 3, 11   (e) 1, 10   (f) 3, 5 or 6; 9, 11   (g) 4, 14

Self-Test
1. False   2. True   3. True   4. False   5. False
6. True   7. True   8. False   9. False   10. False
11. d   12. d   13. c   14. d   15. c   16. b
17. b   18. c   19. c   20. b   21. d   22. b
23. b

Chapter 26

Practice Exercises
26.1 Cathepsins are peptidases, and would destroy all proteins found in the cytoplasm if not confined.

26.2

Tyrosine   α-Ketoglutaric acid

Glutamic acid

26.3 initially transferred to the cofactor pyridoxal phosphate

26.4 ammonia; converted to urea, excreted in urine

26.5 Loss of amino group from amino acid gives alpha-keto acid, which enters glucose metabolism (glucogenic) or is converted to acetyl CoA, as are fatty acids (ketogenic).

26.6 (a) 3   (b) 5   (c) 4   (d) 6   (e) 2   (f) 1

Exercises

26.7 nitrogen-fixing bacteria, lightning discharge, industrial processes

26.9 Protein molecules are continually being degraded and synthesized.

26.11 all free amino acids in cells and tissues of the body

26.13 transfer of an amino group from one molecule to another

26.15

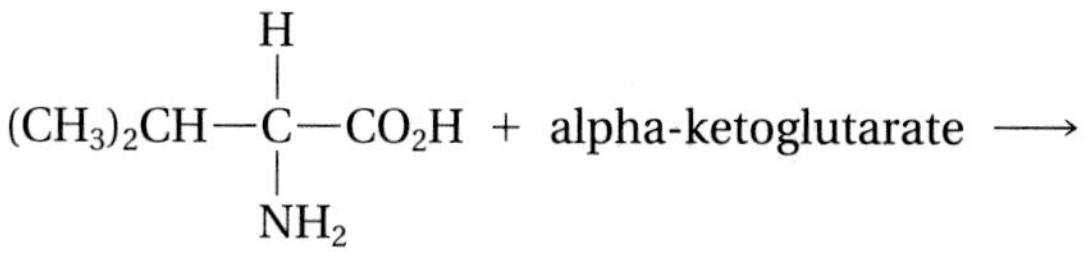

$$(CH_3)_2CH-\underset{\underset{NH_2}{|}}{\overset{\overset{H}{|}}{C}}-CO_2H \;+\; \text{alpha-ketoglutarate} \longrightarrow$$

valine

$$(CH_3)_2CH-\overset{\overset{O}{\|}}{C}-CO_2H \;+\; \text{glutamic acid}$$

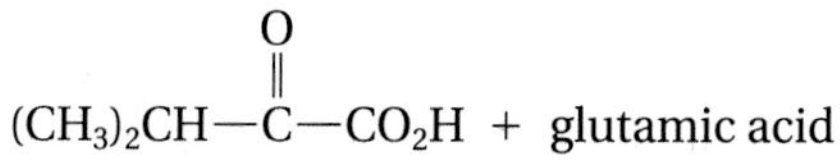

an alpha-keto acid

26.17 converts ammonia to urea

26.19 difference between nitrogen intake and nitrogen excretion; a positive nitrogen balance in children means amino acids are being used for growth

26.21 Glucogenic amino acids form intermediates of glucose metabolism. Ketogenic amino acids are degraded to acetyl CoA.

26.23 Essential amino acids are essential in the diet. Nonessential amino acids are made by the body.

26.25 converted to tyrosine

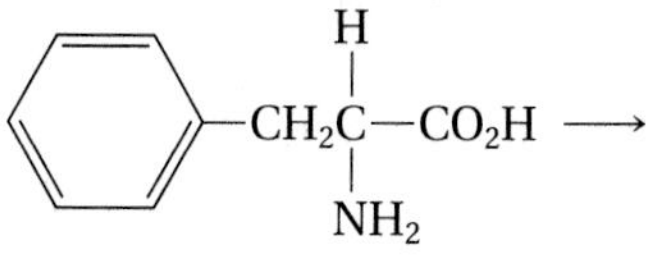

Phenylalanine

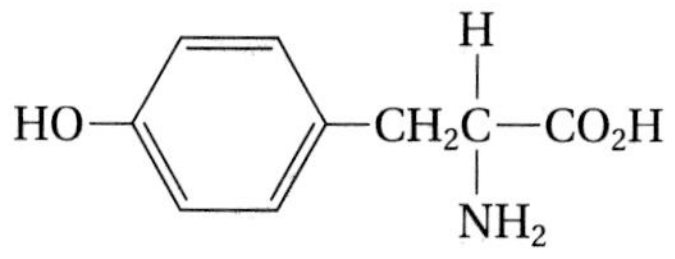

Tyrosine

26.27 120 days

26.29 partially degraded heme

26.31 All of the hydrogens attached to nitrogen in uric acid can be lost as hydrogen ions.

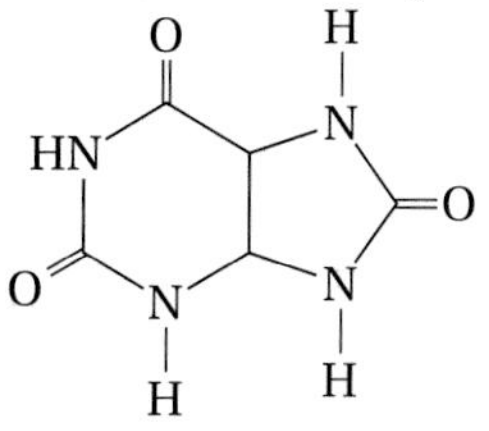

Uric acid

26.33 four; two to make carbamoyl phosphate, two to make arginosuccinic acid

26.35 urea, NH_2CONH_2

26.37 purine ring system

26.39

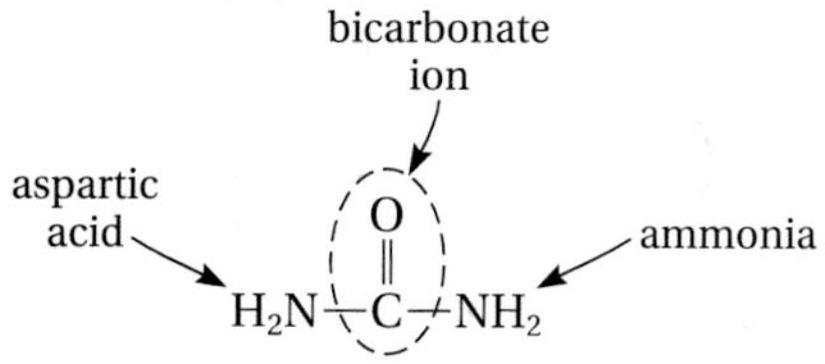

Self-Test

1. True   2. True   3. True   4. False   5. False
6. True   7. False   8. False   9. False   10. True
11. d   12. d   13. d   14. a   15. b   16. a
17. d   18. b   19. d   20. c   21. c   22. a
23. a   24. a   25. a

Photograph Credits

Index

The page on which a term is defined is in **boldface** type. The letters t, f, c, and p following page numbers indicate table, figure, A Closer Look, and Case in Point respectively.

Abortifacient, 544
Abortion, spontaneous (miscarriage), 545
Absolute alcohol, 370
Absolute zero, **8,** 182
Acceptor atoms, 98c, 99f
Accuracy, **30**–31, 31f
ACE (angiotensin-converting enzyme), 602, 602f
Acelation reactions, 446
Acetal, **409**–410
Acetaldehyde, 392, 395t, 396, 414, 414f, 584, 745f
Acetanilide, 470, 470f
Acetic acid, 254, 259, 422, 423, 424t, 557f
Acetic acid-acetate buffer, 280
Acetic anhydride, 437
Acetoacetic acid, 776f
Acetoacetyl CoA, 771f, 776f, 777f
Acetone, 395t, 396, 414, 776f
Acetone breath, 777
Acetyl coenzyme A, **731**
    in cholesterol synthesis, 778, 778f
    in fatty acid synthesis, 774–775, 775f
    in glucose catabolism, 731–732, 732f, 737–738
    ketogenic amino acids, 795, 796f
    as metabolic key, 773–774
    in production of ketone bodies, 776
    starvation and, 774, 797
Acetylation, 450
Acetylcholine, 603c
Acetylcholinesterase, 603c
Acetylene (ethyne), 338, 338f, 339f
Acetylsalicylic acid, 441–443, 444c
Acid(s), **253**–254
    Arrhenius, 253–254
    Brønsted-Lowry, 256–258
    buffers and, 279–281, 280t
    common examples of, 254t
    dental health and, 255c
    equivalents, 272–274, 275
    neutralization of, 269–271
    normality, 274–276
    strong, **258**
    titration, 276–278, 277f
    weak, **259**
Acid anhydrides. *See* Anhydrides
Acid-base balance, 679, **686,** 691–692, 692f, 696p
Acid-base indicators, 267–269, 268f, 269f, 277
Acid dissociation constant, **259,** 259t
Acid mucopolysaccharides, **517,** 517f
Acid phosphatases, 588t

Acidic side chains, 554, 554f
Acidic solutions, **252**–253
Acidosis, **686**
    ketosis, **777**–778
    lactic acid and, 746
    metabolic, 676p, **693,** 695, 696p
    respiratory, **693,** 694–695
Acne, 421p, 425p
cis-Aconitate, 740f
Actin, 724f
Activated tRNA, **627**–628, 628f
Activation energy, **162**–163, 162f
Active oxygen, 166c
Active sites, **585,** 586
Acyl enzyme, 599c
Acyl groups, **438**
Acylation reactions, **438**
ADA deficiency, 608p, 637p
Addicting drug, 481
Addition reactions, **362**–365, 408–410
Adenine, 609, 610f
Adenosine deaminase (ADA) deficiency, 608p, 637p
Adenosine diphosphate (ADP), 710, 735–737, 741
Adenosine monophosphate (AMP), 612t, 768, 769
Adenosine triphosphate (ATP), 451, 628, 666, **708**–710, 720p
    in cellular respiration, 714, 715, 719f
    in citric acid cycle, 741, 742
    as energy carrier, 708, 708f
    in fatty acid activation, 768, 769
    in fatty acid synthesis, 775
    formed by oxidative phosphorylation, 719f
    in gluconeogenesis, 747
    in glycolysis, 732, 733–736, 737t
    hydrolysis of, 723–724, 724f
    in lactic fermentation, 743, 744
    in mitochondria, 715c
    in muscle contraction, 723–724, 724f
    in oxygen debt, 745, 746
    phosphorylating power, 720
    production from glucose catabolism, 742, 743t
    production in cells, 711, 715c, 719, 719f
    structure of, 708–710, 709f
    in urea cycle, 721–722, 793
    yield from fatty acid oxidation, 771–772, 772t
Adenylate cyclase, 752
ADH (antidiuretic hormone), 697
Adipic acid, 424

Adipose tissue, **764,** 764f
ADP. *See* Adenosine diphosphate
Adrenal glands, 538, 538f
Adrenaline. *See* Epinephrine
Aerobic cells, **645, 711**
Aging, and oxygen, 718c
Agitation, and solution formation, 223
AIDS, 593c, 602, 617, 619c, 638, 682c
Alanine, 553f, 797, 798f
Albumins, 680, 765
Alcohol(s), **367**–376
    absolute, 370
    acidity, 429–430
    addition to aldehydes and ketones, 409–410
    boiling points of, 379, 380t
    denatured, **370**
    displacement reactions, 372–373
    elimination reactions, 373–375
    esterification, **441**–443, 446
    fetal alcohol syndrome and, 392p, 405p
    hydration reaction, 372
    nomenclature, 368–369
    oxidation of, 402–405
    oxidation to carboxylic acids, 436
    phenols, 370–371
    physical properties of, 379–381, 380t
    synthesis of, 372–373
    uses of, 369–370
    as very weak acids, 375–376
Alcohol dehydrogenase, 584
Alcoholic fermentation, **745**
Aldehydes, **393**–414
    addition of alcohols, 409–410
    addition of water, 408–409
    aromatic, 411
    Benedict's test for, 407
    boiling points of, 398, 398t
    common examples of, 395t
    detection of, 406–408
    Fehling's test for, 407
    fetal alcohol syndrome and, 392
    flavors and fragrances and, 391f, 399c
    nomenclature, 393–394
    oxidation to carboxylic acids, 436–437
    physical constants, 398t
    redox reactions involving, 402–406
    structure of, 393, 393f, 397–400
    Tollens' test for, 406–407
    uses of, 410–414
    water solubility of, 399–400, 400f
Aldohexoses, 497, 499–500, 499f
Aldopentoses, 497, 499, 499f
Aldoses, **493**

Aldosterone, 540, 540f, 697–698
Aldotetroses, 497, 498, 498f
Aldotriose, 493
Algae, 530c
Aliphatic amines, 461–462
Aliphatic compounds, **338**
Aliphatic diamines, 465, 465f
Aliphatic side chains, 553, 553f
Alkali metals, **95**–96, 96f, 256
Alkaline earth metals, **96**, 96f
Alkaline phosphatases, 588t
Alkaline (basic) solutions, **253**
Alkaloids, **481**–484
Alkalosis, **686**
  metabolic, **693**, 695–696
  respiratory, **693**, 694
Alkanes, **317**
  branched-chain, **325**–330
  cycloalkanes, **331**–334, 331t, 333f
  straight-chain, **321**–324, 322f, 323t, 324f
Alkenes, **334**–338
  addition reactions, 362–365
  double bond shape of, 335–337, 335f, 336f
  geometric isomerism, 337–338, 337f
  halocarbons from, 362–367
  nomenclature, 334–335
Alkoxides, 375–**376**
Alkyl chlorides, 380t
Alkyl groups, **325**, 357t
Alkyl halides, **356**
Alkyl nitrates, 451
Alkyl nitrites, 451
Alkylammonium ions, **466**
Alkynes, **338**, 338f
Allergic diseases, 539–540
Allopurinol, 804p
Allosteric enzymes, **594**, 594f
Alloys, **97**
Alpha carbon, **553**
Alpha emission, 288, **289**
Alpha helix, **564**–565, 564f, 565f
Alpha-hydroxy ketones, 407–**408**
Alpha keratins, **565**
Alpha particles, 288, **289**, 291f
ALS (amyotrophic lateral sclerosis), 718c
Altimeter, 184
Aluminum, 97
Alveoli, 687
Amide(s), **468**–476
  hydrogen bonding of, 472, 472f
  hydrolysis of, 476
  as local anesthetics, 471c
  nomenclature, 469–470
  polyamides, **474**–475
Amide bonds, 474
Amide side chains, 554–555, 555f
Amines, **461**–468
  alkaloids, 481–484
  basicity of, 465–467
  examples of, 476–481
  free, **466**
  nomenclature, 461–464
  protonated, **466**

  solubility of, 466–467
  synthesis of, 467–468
Amino acid(s), **553**–577
  abbreviations for, 560t
  alpha carbon, **553**
  in diet, 654–656
  essential, **654**, 654t
  glucogenic, **795**, 796f
  ketogenic, **795**, 796f
  nonessential, **654**, 654t
  peptides, 552, **559**–562, 560t
  sequence of, 560, 562, 633
  stereoisomers of, 556–557, 556f
  three-letter code words for, 632t
  zwitterions, **557**–558, 557f
Amino acid metabolism, 788–790, 789f, 799
Amino acid pool, **788**
Amino acid residue, **559**–560
Amino acid side chains, **553**–556
Amino acidurias, **799**
p-Aminobenzoic acid, 601
Aminopeptidase, 596
Ammonia, 124, 125, 131f, 135, 135f, 257, 259, 270, 690, 787, 787f, 792c
Ammonium carbonate, 118
Ammonium chloride, 279
Ammonium ion, 114–115
Ammonium salts, **466**–468, 472–473
AMP. See Adenosine monophosphate
Amphetamine, 480f
amu (atomic mass unit), **66**, 143–144, 144t
Amylases, 588t, 645–646
Amylose, 508, 508f, 509–510, 509f
Amyotrophic lateral sclerosis (ALS), 718c
Anabolic steroids, 541
Anabolism, **704**, 710
Anaerobes, **711**
Analgesic, **442**, 443c
Anaphylactic shock, 547
Androgens, 541
Androstenedione, 541f
Anemia, 662
  iron-deficiency, **668**
  pernicious, **662**–663
  sickle cell, 2p, 7p, 572–573
Anesthetics
  general, 354f, 377c
  local, 377c, **442**–443c, 471c
Angina pectoris, 452, 754c
Angiotensin-converting enzyme (ACE), 602, 602f
Anhydrides
  of carboxylic acid, **437**–438, 473
  hydrolysis of, 438, 450
  outline of reactions, 454
  of phosphoric acid, 448–449, 450
  in preparation of amides, 473
Anilides, **470**
Aniline, 462–464
Anions, **113**, 115–117, 115t, 117t, 718c
Anomers, **503**
Antacids, 250p, 271p
Antagonists, 658

Antibiotics, 469c, **601**–602, 601f, 603c, 630
Antibodies, **683**–685, 683f
  monoclonal, **685**
Anticoagulant drugs, 517
Anticodons, **628**, 629f
Antidepressants, 460p, 478–479, 479p
Antidiuretic, **562**
Antidiuretic hormone (ADH), 697
Antigens, **683**, 684f
Antihistamines, **480**–481, 481f
Antimony, 99–100
Antioxidants, 528
Antiparallel strands, **615**
Antipyretic, **442**, 443c
Antiseptics, 371c, 384
Apoenzyme, **587**
Aqueous solutions, **218**–219
Arabinose, 499, 499f
Arenes, **338**–343
Arginine, 555f, 794f, 797
Arginosuccinic acid, 793–794, 793f, 794f
Argon, 94–95, 115f
Aromatic aldehyde, 411
Aromatic amines, 462–464
Aromatic compounds, 338–343, **339**, 343f
Aromatic side chains, 554, 554f
Aromatic substitution, **363**–364
Arrhenius, Svante, 253
Arrhenius acids and bases, 253–256
Arsenic, 98c, 99, 665t, 670
Arteries, 680
Artificial joints, 100c
Artificial skin, 678c
Artificial sweeteners, 499, 561c
Aryl halides, **356**
Arylammonium ions, **466**
Ascorbic acid, 528, 659–660
Asparagine, 555f, 798, 798f
Aspartame, 499, 561c
Aspartic acid, 554f, 793f, 797, 798f
Aspirin, 441–443, 444c, 547, 692
Astatine, 102
Asymmetric carbon, **494**–495
Atherosclerosis, 651c, **780**
Atmosphere, **183**, 348c, 361c
Atmospheric pressure, 183–184
Atom(s), **63**–65
  energy levels in, 72–75, 72f, 75t, 76
  of first ten elements, 66t
  images of, 64c
  nucleus of, **65**
  parts of, 64–65
  size of, 63
Atomic absorption, 73c
Atomic mass, **69**–71, 69t
Atomic mass unit (amu), **66**, 143–144, 144t
Atomic number, **65**, 66t, 67t
Atomic orbitals, **72**–75, 74f, 75t, 76f, 77
Atomic radius, 87f, 91–92, 92f
Atomic structure, 63, 72–75
Atomic theory, 63–65
ATP. See Adenosine triphosphate
Atropine, 483, 483f

Aufbau diagram, 76f
Aufbau principle, **75,** 76
Autoactivation, **595**
Autoclave, 205f
Autoimmune disease, 684
Average kinetic energy, 181–**182**
Avogadro, Amedeo de Quarenga, 144
Avogadro's hypothesis, 184–186, **185**
Avogadro's number, **144**
AZT (zidovudine), 619c

Bacillus thermoproteolyticus, 575
Background radiation, **310**–311, 311t
Bacteria, 530c
    anaerobic, 711
    heat sterilization and, 575–576
    nitrogen-fixing, **787**
    protein synthesis in, 630
Bakelite, 413–414
Balance, 47, 47f
Balanced chemical equations, 147–150,
    149f
Balanced diet, 653–654, 655c. *See also*
    Nutrition
Barbital, 485
Barbiturates, 484–**485**
Barbituric acid, 484–485, 485f
Barium sulfate, 118, 220, 306
Barometer, 183f, 184
Bartlett, Neil, 94
Basal metabolism, **653**–654
Base(s), **253**
    amines as, 465–467
    Arrhenius, 254–256
    balance of acids with, 679, **686,**
        691–692, 692f, 696p
    Brønsted-Lowry, 256–258
    buffers and, 279–281, 280t
    common examples of, 255t
    in DNA and RNA, 612t, 614
    equivalents, 272–274, 275
    neutralization of, 269–271
    normality, 274–276
    strong, **259**
    titration, 276–278, 277f
    weak, **259**
Base sequences, coding vs. noncoding,
    621–623
Base triplets, **628,** 629f
Basic side chains, 555, 555f
Basic (alkaline) solutions, **253**
B-complex vitamins. *See* Vitamin
    B-complex group
Beadle, George, 621
Becquerel, Antoine H., 288
Beeswax, 524
Belladonna, 483
Benadryl, 481f
Bence-Jones, Henry, 684
Bence-Jones proteins, **684**
Bends, 227c
Benedict's reagent, **407,** 408, 514
Benzaldehyde, 395t, 410–411, 411f
Benzene, 339–343
    addition reactions and, 365
    derivatives, 341–343

disubstituted, 342
    fused-ring aromatics, **342**–343, 343f
    leukemia and, 343c
    reduction reactions and, 365
    structural isomers, 342, 342f
    structure of, 340–341, 340f
Benzocaine, 442c
Benzoic acid, 423
Benzpyrene, 343c
Beriberi, **660**
Beta-carotene, 413c, 657, 657f
Beta emission, 288t, **290**–291, 302, 302f
Beta-interferon, 636, 685, 685f
Beta-ionine, 411, 411f
Beta keratin, **566**
Beta oxidation, **769**–771
Beta particles, 288t, **290,** 291f
Beta-pleated sheet, **565**–566, 565f, 566f
BHT (butylated hydroxy toluene), 370f,
    371c
Bicarbonate ions, 115, 117t, 688, 689,
    692, 692f, 693
Big bang theory, 93c
Bile, **648**
Bile pigments, **648,** 799–802
Bile salts, **648,** 648f
Bilirubin, **800**–801, 801f, 802c
Biliverdin, **800,** 801f, 802c
Bioactive materials, 99, 100c
Biodegradable, **435**
Biological effects of radiation, 291–292,
    303–304, 306, 311
Biomedical implants, 89c, 100c
Biotin, 662, 662f
1,3-Biphosphoglycerate, 735f
Birth control, 543–545
Bismuth, 99–100
Bladder, 690, 690f
Blood, 679–689
    carbon dioxide transport in, 688–689,
        689f
    circulation of, 194f, 680–681, 681f
    formed elements of, **679**
    hemoglobin in. *See* Hemoglobin
    nitrogen in, 227c
    oxygen transport in, 687–688, 687f
    pH of, 686, 691–696, 692f, 693f
    soluble elements of, 679–680
Blood buffers, 686, 691–692, 693f
Blood cells
    red, 679, 687–689, 799–800
    white, 679, 683–685
Blood cholesterol, 779–780, 780f
Blood clotting, 444c, 516, 584, 597, 600,
    658, 680
Blood lipoproteins, 651c
Blood plasma, 677, **679**–680
Blood pressure
    high, 189c, 481–482
    measuring, 189c
    sequence of amino acid residues and,
        562
Blood proteins, 680, 680f
Blood serum enzymes, 587–589, 588t
Blood sugar (glucose), 515c, 575c
Blood transfusions, 682c

Body, human
    elemental composition of, 665t
    essential requirements of, 645
    fluids in. *See* Body fluids
    ideal weight of, 525p
    responses to light, 706c
    temperature and, 142p, 164p
Body fluids, 675f, 676–698
    acid-base balance in, 679, 686,
        691–692, 692f, 696p
    blood, 679–689
    composition of, 677t
    diabetes and, 676p, 696p
    pH of, 262c
    water, 677–679, 696–698
Body lipids, 764, 764f
Body mass index (BMI), 525p
Body temperature, 142p, 164p
Body water, 213, 677–679, 696–698
Boguskinin, 562, 562f
Boiling points, **205**
    of alcohols, 379, 380t
    of aldehydes and ketones, 398, 398t
    elevation of, 238
    of low molar mass substances, 216t
    of methane, 323t
    of water, 216, 216t
Bonds
    covalent, **121**–123, 125–126
    electronegativity differences and, 128t
    glycosidic, **505**–507, 506f
    hydrogen, **214**–216, 214f, 317–320,
        397, 397f, 400f, 472, 472f
    ionic, **118,** 567
    nonpolar covalent, **127**
    peptide, **559,** 559f, 595
    pi, **336**–337
    polar, **127**–129, 214, 214f
    predicting polarity of, 128–129
    sigma, **320**
    See also Carbon bonding
Bone, 665–666
    artificial, 100c
    thinning of, 644p, 666, 666p
    vitamin D deficiency and, 658, 658f
Bone matrix, **665**
Boron, 70, 86f
Boyle, Robert, 187, 193
Boyle's law, **193**–195, 195f
Bradykinin, 562, 562f
Brain scan, 286f, 308
Branched-chain alkanes, **325**–330
Breathing, 191c, 193, 194c
Brewer's yeast, 745
Bromine, 86, 86f, 102, 112
Brønsted, Johannes, 256
Brønsted-Lowry acids and bases,
    256–258
Brønsted-Lowry theory, **256**–258
Brown, Robert, 236
Brownian motion, 236
Buckyball, 134c
Buffers, 279–281, **280,** 280t
    amine, 466
    blood, 686, 691–692, 693f
Bupivacaine hydrochloride, 471c

Buret, 42, 42f
Burn victims, 678c
Butane, 323t, 330, 346–347
Butanoate, 440
Butanoic acid, 422
Butyl butyrate, 440
Butylated hydroxy toluene (BHT), 370f, 371c
Butyric acid, 422, 424, 424t, 440

C-terminal residue, 559, 559f
Cadaverine, 465f
Caffeine, 465f
Calcitonin, 666p
Calcium, 19c, 116t, 664–666, 665t, 666p
Calcium carbonate, 255c
Calcium cyclamate, 512f
Calcium ion-channel blockers, 132p
Calgon, 434c
Caloric value, 53t, **653**–654
Calories, **53**, 53t, 55c, 653
cAMP, 752–754, 752f, 753f, 755f
Camphor, 411, 411f
Cancer, 634c, 706c
Cannabis sativa (marijuana), 384
Capillaries, 680–681, 681f
Capric acid, 424t
Caproic acid, 424t
Caprylic acid, 424t
Carbaminohemoglobin, **688**, 689
Carbamoyl phosphate, 721–722, 722f, 793, 793f
Carbocaine, 471c
Carbohydrate(s), 492, **493**–518
  disaccharides, **511**, 511f
  in glycolipids, 533
  glycosides, **505**–507
  monosaccharides, **497**–500, 498f, 499f, 500f
  nonreducing sugars, **514**–515
  polysaccharides, **507**–510, 508f, 509f
  reducing sugars, **514**–515, 514f
  starch, **508**–510, 508f, 509f
  See also Glycogen
Carbohydrate loading, 730p, 751p
Carbolic acid, 371c
Carbon, 68, 89, 124, 125
  alpha, **553**
  asymmetric, **494**–495
  forms of, 86c, 97, 97f, 134c
Carbon bonding
  with carbon, 320
  with halogens, 356–367
  with hydrogen, 317–320, 318f, 319f
  molecular orbitals in. *See* Molecular orbitals
  multiple bonds, 334–338, 335f, 336f, 337f, 338f
  valence electron in, 317–318
Carbon compounds
  body's need for, 645
  boiling points of, 398t
  See also Halocarbons; Hydrocarbons
Carbon dioxide, 112, 124, 135, 135f
  greenhouse effect and, 348c
  molar mass of, 145

in photosynthesis, 704–705
  rain forests and, 169c
Carbon dioxide transport, 688–689, 689f
Carbon-14 dating, 295–296
Carbon monoxide, 112, 316p, 347p, 754c
Carbonic acid, 260, 688, 689, 692, 693
Carbonic anhydrase, 585, 585f, 688, 692
Carbonyl group, **393**, 397–400. *See also*
  Aldehydes; Ketone(s)
Carboxybiotin, 747f
Carboxyhemoglobin, **572**
Carboxyl group, **422**, 427–428
Carboxylation reaction, 747
Carboxylic acid(s), **422**–438
  dicarboxylic acids, **424**, 426
  K_a values for, 429t
  neutralization of, 431
  nomenclature, 422–423
  outline of reactions, 452–453
  polyfunctional, **426**–427
  R group and acidity, 429–430
  salt formation, 430–431
  saturated aliphatic, 424, 424t
  synthesis of, 435–437
  tretinoin, 421p, 425p
  unsaturated aliphatic, 424, 425t
  as weak acids, 429
Carboxylic acid anhydrides, **437**–438, 473
Carboxylic esters, **439**–447
  nomenclature, 439–440
  preparation of, 441–445
  properties of, 440–441
Carboxypeptidase, 596, 647
Carcinogenic compounds, 343c
Carnauba wax, 524
Carnitine, 766–767
Carnitine deficiency, 763p, 768p
β-Carotene, 413c, 657, 657f
CAT scanner, 306, 306f
Catabolism, **704**, 710. *See also* Glucose catabolism
Catalase, 166c, 718c
Catalysts, **164**
  enzymes as, 583, 586
  equilibrium position and, 169
  reaction rate and, 164–165, 165f, 166c
Cataracts, 513
Catechol, 477
Catecholamines, 477
Cathepsins, 788
Cations, **113**–115, 114t
Cell(s), 530c, 530f
Cell membrane, **529**–536, 530c, 535f, 536f
Cell nucleus, 530c, 530f
Cellobiose, 511, 511f
Cellular energetics, 710
Cellular processes, compartmental-ization of, 778–779
Cellular respiration, 115, **711**, 714–719
Cellulose, **509**–510, 509f
Celsius, Anders, 8
Celsius scale, **8**, 9f, 10, 182
Centimeter, 39, 39t, 45
Central dogma, **616**–617

Cerebrosides, 533, 533f
CFCs (chlorofluorocarbons), 102, 361c
cGMP (cyclic guanosine monophos-phate), 757
Chadwick, James, 298
Chain reaction, **298**, 299
Charles, Jacques, 187, 196
Charles's law, **196**–197, 196f
Chemical calculations, 150–153
Chemical change, 15–16, 16f
Chemical equations, 146–150
Chemical equilibrium. *See* Equilibrium
Chemical formulas, **111**, 144t
Chemical pollution, 5c
Chemical properties, **16**
Chemical reactions, 15–**16**, 15f, 16f, 63, 146–147, 160t
  rate of. *See* Reaction rate
  reversible, **167**–173
Chemical symbols, 18–20, 18t, 19t
Chemistry, **3**
Chemophobia, 5c
Chemotherapy, 126c
Chemotrypsin, 584
Chernobyl incident, 301c
Chitin, 516–517, 516f
Chloride ions, 115f, 116t
Chloride shift, **689**
Chlorine, 70, 92, 102, 112, 665t, 666–667
Chloroacetic acid, 426
Chlorofluorocarbons (CFCs), 102, 361c
Chloromethanes, 380t
Chlorophyll, **705**, 705f
Chloroplasts, **705**, 705f
Chloroprocaine, 442c
Choking, 180p, 196p
Cholesterol, 537, 537f, 638
  in blood, 779–780, 780f
  digestion of, 649, 651c
  reducing with drugs, 781c
  synthesis of, 774, 778–779, 779f
Cholestyramine, 781c
Choline, 530–531
Chromium, 665t, 669
Chromophores, 413c
Chylomicrons, **649**–650, 650f, 651c, 779–780
Chymotrypsin, 596–597, 647
Chymotrypsinogen, 595, 595t
Cinnamaldehyde, 411, 411f
Circadian rhythms, 706c
Circulatory system, 194f, 680–681, 681f
Cirrhosis, 404, 792f
Cis configuration, **337**, 337f
Cisplatin, 126c
Citrate, 740f
Citric acid, 420f, 426
Citric acid cycle, 731, **732**, 738–742, 739f, 774
Citrulline, 721–722, 721f, 722f, 793f
Citryl coenzyme A, 740f
Clotting, 444c, 516, 584, 597, 600, 658, 680
Coal, 344–345, 345f
Cobalamin, 662–663, 663f
Cobalt, 78, 103, 662, 665t, 668–669

Cocaine, 442c, 483, 483f
Cod liver oil, 657
Codeine, 483, 483f
Codons, **628,** 629f, 631
Coefficients, 147
Coenzyme A (CoA), **731**–732, 732f. *See also* Acetyl coenzyme A
Cofactors, **587**
Cold, effect on human body, 142p, 164p
Cold creams, 526c
Collagen, **566**–567, 665, 678c
Colligative properties, **237**–238
    boiling point elevation, 238
    dialysis, **242,** 242f, 243f
    freezing point depression, 238
    osmosis, **239**–241, 239f, 241f
    vapor pressure lowering, 237
Colloid(s), **235**–236, 235t, 236t
Colloid dispersion, **235**–236, 235t, 236t
Colon, 650
Combined gas law, **199**–200
Committed step of glycolysis, 734
Compartmentalization of cellular processes, 778–779
Competitive inhibitors, **592,** 593c
Complement system, **684**
Complementarity, **585**
Complementary base pair, **615,** 615f
Complete proteins, **654**
Complex lipids, **529**–533, 534f
Compounds, **17**
    covalent, **121**–125, 130t
    elements vs., 17–18
    ionic, **112.** *See also* Ionic compounds
    molecular, **112, 121**–125, 122f
    organic, 97, 322–323, 400–402, 401f. *See also* Carbon bonding
    polyfunctional, **383**–385
Concentrated, vs. dilute, 228, 260
Concentration, **228**
    equilibrium position and, 170–171
    reaction rate and, 165
Concentration, units of
    molarity, **231**–234
    normality, 274–276
    parts per billion, 230–231
    parts per million, 230–231
    percent (mass/volume), 229–230
    percent (volume/volume), 228–229
Condensation, 13f
    heat of, **205**
Condensed structural formulas, 321–**322**
Conductivity, 98c, 222f
Configurations. *See* Electron configurations
Conjugated proteins, **568**
Conservation
    of energy, 21–**22,** 157
    of mass, **20**
Contraceptive, 543–545
Conversion factors, **44**
Coordinate covalent bonds, **125**–126, 126c
Copper, 103, 665t, 668–669
Core electrons, 92
Corn sugar. *See* Glucose

Corpus luteum, 542–543
Corrin ring, 662, 663f
Corticoids, **538**–545
    glucocorticoids, **539**–540, 539f
    mineralocorticoids, **540,** 540f
    sex hormones, 540–545
Corticosterone, 539, 539f
Cortisol, 539, 539f
Cortisone, 412, 539, 539f, 540
Coupled reaction, **721**
Covalent bonds, **121**–123
    coordinate, **125**–126, 126c
    double, **123**
    nonpolar, **127**
    polar, **127**–129, 214, 214f
    single, 121–123, **122**
    triple, **123**
    See also Carbon bonding
Covalent compounds, **121**–125, 130t
CPK (creatine phosphokinase), 588, 588t, 724
Cracking, **344**
Creatine phosphate, 724, 724f
Creatine phosphokinase (CPK), 588, 588t, 724
Crenated cell, 241, 241f
Cresols, 371c
Crick, Francis, 614, 615, 617, 618, 620
Cristae, 715c
Crystal, 108f, **120,** 120f
    seed, 225, 225f
Crystallization, water of, **221**–222
Crystallography, X-ray, 568–569, 569c, 569f
Curie, Marie, 288, 289f, 305, 306
Curie, Pierre, 288, 306
Curie (unit), **305**
Cyanide, 115, 117t, 718–719
Cyanohemoglobin, **572**
Cyclic AMP (cAMP), 752–754, 752f, 753f, 755f
Cyclic guanosine monophosphate (cGMP), 757
Cyclic sugars, 501–502, 501f, 502f
Cycloalkanes, **331**–334, 331t, 333f
Cyclobutane, 332, 333f
Cyclohexane, 332–334, 333f
Cyclopentane, 332, 333f
Cyclopropane, 332, 333f, 343c
Cysteine, 555f
Cystic fibrosis, 652c
Cystic fibrosis transmembrane conductance regulator (CFTR), 652c
Cystine, 555f
Cytochromes, **716**–717
Cytoplasm, 530f, 774, 778
Cytosine, 609, 610f

Dacron, 445
Dalton, John, 63
Dalton's atomic theory, 63–65
Dalton's law of partial pressures, **192**–193, 194
Decimeter, 39t
Decongestants, **480**
Degenerate code, **633**

Degradation, **590**–591
Dehalogenation, **374**
Dehydration, **374,** 472–473, 667, 698
Dehydroepiandrosterone, 541f
Dehydrogenases, 401, 584, 588t, 737, 744c
Dehydrogenation reaction, **401**
Dehydrohalogenation, **374**
Democritus of Abdera, 63
Denaturation, **574**–576, 574f, 576f
Denatured alcohol, **370**
Density, 49–52, **50**
    of common substances, 50t
    of ice, 217t
    units of, 32t
    volume and, 50f, 51
    of water, 217t
Dental health, 255c, 657, 665–666, 670
Dentine, 255c, 665
Deoxyribonuclease (DNase), 650, 652c
Deoxyribonucleic acid. *See* DNA
Depot fat, **764**
Depression, 460p, 477–479, 479p
DES (diethylstilbestrol), 544–545, 544f
Detergents, **433**–435
    nonionic, **435**
Deuterium, 71
Dextromethorphan, 483, 483f
Diabetes insipidus, 562
Diabetes mellitus, 515c, **757**–758
    causes of, 758
    lipoproteins and, 651c
    maturity-onset, 29p, 49p
    metabolic acidosis and, 676p, 696p
    treatment of, 758
    See also Insulin
Diabetic ketosis, 777–778
Dialysis, **242,** 242f, 243f
Diamagnetic substance, 121c
Diamond, 86c, 97, 97f, 134c
Diamond structure, 133, 133f
Diaphragm, 191c
Diastolic pressure, 189c
Dicarboxylic acids, **424,** 426
Dicoumarin, 658, 659f
Diet
    balanced, 653–654, 655c
    low-saturated fat, 780
    sources of ions in, 116t
    See also Nutrition; Starvation
Dietary fiber, 510c
Diethylstilbestrol (DES), 544–545, 544f
Dieting, 775
Diffusion, 186–**187**
Digestion, 645–653
    of cholesterol, 649, 651c
    of lipids, 648–650, 648f, 650f
    of nucleic acids, 650–653, 652f
    of proteins, 646–647
    of sugars, 647
    of triglycerides, 648–650, 650f
Digestive system, 646f
Digitalis, 169c, 545, 545f
Digitoxin, 545, 545f
Dihydrogen phosphate ion, 115, 117t

Dihydroxyacetone phosphate, 734f, 735f, 772f
Dilute, vs. concentrated, 228, 260
Dimensional analysis, 43–44
Dimers, **427**
Dimethylbenzene, 342
Diphenylhydramine, 481f
Dipolar molecule, 129
Dipole interactions, **130,** 130f
Disaccharides, **511,** 511f
Disease, and oxygen, 718c. *See also specific diseases*
Disinfectants, 371c
Dispersed phase, 235
Dispersion forces, **130**
Dispersion medium, 235–236
Displacement reactions, **372**–373
Distal tubule, 691
Distillation, 12, 13f
Disubstituted benzenes, 342
Disulfide(s), **382**–383
Disulfide bridges, **561,** 565, 574, 574f, 576f
DNA (deoxyribonucleic acid), 499, **609**
  bases in, 612t, 614
  central dogma of, **616**–617
  coding vs. noncoding base sequences of, 621–623
  cystic fibrosis and, 652c
  Human Genome Project (HGP) and, 622c
  length of, 621
  mutations in, 633–636
  nucleosides in, 609, 610, 611f, 612t
  nucleotide base pairs in, 615, 615f, 621t
  nucleotides in, 611–612, 612t, 613f
  recombinant, **636**–637
  replication of, 616, 617–620, 617f, 618f, 620f
DNA double helix, **614**–615
DNA fingerprinting, 623c, 623f
DNA polymerase, **617**–618, 618f
DOM (STP), 480f
Donor atoms, 98c, 99f
Dopa, 477
Dopamine, 477, 477f
Doped silicon, 98c
Double covalent bonds, **123**
Double helix, **614**–615
Dramamine, 481
Drinking water, 213p, 219p
Drugs, medicinal, 109f, 109p, 132f, 132p, 169c. *See also specific drugs*
Duranest, 471c
Dwarfism, 668
Dynamic equilibrium, 204, 224
Dynamite, 451

Edelman, Gerald M., 684
Edema, 667, 680
EDTA (ethylenediaminetetraacetic acid), 591c
Egg (ovum), 542, 544
Electrolytes, **222,** 223

Electron(s), **64**–65
  in atomic structure, 72–75
  core, 92
  energy levels of, 72–75, 72f, 75t, 76
  in fireworks, 73c
  location of, 72–75, 74f, 75t
  unshared pairs of, **123**
  valence, **110,** 317–318
Electron cloud, 340f
Electron configurations, **75**–79
  for first 36 elements, 78t
  for selected elements, 77t
  for stable ions, 113–115, 113f, 115f
  writing, 77–79, 90–91
Electron dot structures, **110**–111, 111t, 126
Electron transport chain, **714**–718, 717f, 719f
Electronegativity, **127,** 127t, 128t
Electronic thermometers, 9, 9f
Electrophoresis, **573,** 573c, 744c
Elements, **17**
  atoms of, 66t
  blocks of, 90, 90f
  categories of, 88–89, 88f
  compounds vs., 17–18
  electron configurations for, 75–79, 77t, 78t
  essential for life, 19c
  Latin names and symbols of, 18–20, 18t, 19t
  natural isotopes of, 69t
  number of protons in atoms of, 65
  origin of, 93c
  origin of names of, 19t
  representative, **88**–89, 94–102
  trace, 670
Elimination reactions, **373**–375
Elongation, **630,** 630f
Embden, Gustav, 731
Embden-Meyerhof pathway, 731
Embolism, 600
Emulsifiers, 526c
Emulsifying agent, 432
Emulsions, **236**–237, 526c
Enamel, tooth, 665
Endocrine system, 538f
Endopeptidases, **596**
Endoplasmic reticulum (ER), 530c, 530f
Endothermic process, **22**–23, 22f
Endothermic reactions, 157–158, 158f
Energy, **21,** 702f, 703–725
  activation, **162**–163, 162f
  in chemical reactions, 156–158, 160t
  free, **160**–161, 161t
  interconvertibility of forms of, 21
  kinetic, **181**–182, 203, 204f
  law of conservation of, 21–**22,** 157
  units of, 32t
Energy and carbon cycle, **707**
Energy diagrams, 157, 157f, 158f
Energy levels of electrons, 72–75, 72f, 76
  principal, 74–75, 75t
  sublevels, 74–75, 75t
Enflurane, 377c
Enkephalins, 552, **562,** 563p

Enteropeptidase, 596
Entropy, **158**–160, 159f, 160t, 161f
Enzyme(s), 142, 165, 166c, 262c, **583**–604
  active sites of, 585, 586
  acyl, 599c
  allosteric, **594,** 594f
  apoenzyme, **587**
  blood serum, 587–589, 588t
  as catalyst, 583, 586
  cofactors, **587**
  degradation, **590**–591
  for heart attacks, 582p, 600p
  historical applications of, 583
  holoenzyme, **587**
  inducible, **591**
  induction, **590**–591
  modulators, 594–595
  nomenclature, 584
  pacemaker, **594**–595
  protein kinase, 634c
  restriction, 623c
  sources of, 581f
  specificity of, **584**–585
  zymogens, **595**–597, 595t
Enzyme activity, **587**
  control of, 592–595, 592f, 594f
  pH and, 589–590, 589f
  temperature and, 590, 590f
Enzyme assays, **587**–589, 588t
Enzyme-substrate complexes, **585**–586, 585f, 586f
Ephedrine, 480f
Epinephrine (adrenaline), 479–480, 479f, 755–756, 764
Epoxides, 378–379
Epoxyethane, 378–379
Epsom salts, 118
Equations, chemical, 146–150
Equilibrium, 167–**168,** 168f
  concentration and, 170–171
  dynamic, 204, 224
  temperature and, 171
Equilibrium constants, **171**–173
  acid dissociation (ionization) constant, **259,** 259t
  ion-product constant for water, **251**–252
Equilibrium position, **168**–171
Equivalence point, **277**
Equivalent (equiv) of acid or base, **272**–274, 275
ER (endoplasmic reticulum), 530c, 530f
Ergosterol, 657
Erythrocytes, **679**
Erythromycin, 630
Erythrose, 498f
Essential amino acids, **654,** 654t
Ester(s)
  carboxylic, **439**–447
  hydrolysis of, 446–447
  hydrolysis of fats and oils, 528
  as local anesthetics, 442–443c
  nitric acid, 451–452
  nomenclature, 439–440
  odor and flavor of, 440–441, 440t

outline of reactions, 454
oxyesters, 446
of phosphoric acid, 449–450
polyesters, **445**
preparation of, 441–445
in preparation of amides, 472, 473
thioesters, **446**
triglycerides, 440, **524**–529, 534f
waxes, **524**
Esterification, **441**–443, 446
Estradiol, 542, 542f
Estrogens, 541, **542**, 543f, 666p
Estrone, 542, 542f
Ethanal, 396
Ethane, 320
Ethanoic acid, 422
Ethanol (ethyl alcohol), 369, 370, 392, 404–405, 745
Ethene (ethylene), 335, 335f, 336f, 366, 397, 397f
Ethers, **376**–379
as anesthetics, 354f, 377c
physical properties of, 379–381
Ethyl alcohol (ethanol), 369, 370, 392, 404–405, 745
Ethylamine, 557f
Ethylene (ethene), 335, 335f, 336f, 366, 397, 397f
Ethylene glycol, 355p, 369–370, 378p
Ethylenediaminetetraacetic acid (EDTA), 591c
Ethylpropylamine, 462
Ethyne (acetylene), 338, 338f, 339f
Etidocaine, 471c
Eugenol, 371c
Eukaryotic cells, 530c, 530f
genes of, 621–623, 622f, 625–626, 626f
mitochondria of, 715c
photosynthetic, 705
Evaporation, 203–204, 204f
Exergonic reactions, **161**, 161t
Exons, **621**–623, 622f
Exopeptidases, **596**
Exothermic process, 22–23, 22f
Exothermic reactions, 156–157, 157f
Experiments, **4**, 5–6
Exponent, 32
Exponential notation, 32
Exposed dose, 305

Factor-label method, 43–44
FAD. See Flavin adenine dinucleotide
Fahrenheit, Daniel, 8
Fahrenheit scale, **8**, 9f, 10
Fats, **527**
hydrogenation of, 527
hydrolysis of, 528
mobilization of, **764**–765, 765f
storage (depot), **764**
See also Lipid(s)
Fat-soluble vitamins, 656–659
Fatty acid(s), **423**–424
activation of, 768–769, 768f
saturated, 424, 424t
unsaturated, 424, 425t
Fatty acid oxidation, 765–771

Fatty acid spiral, 770f, **771**
Fatty acid synthesis, 774–775, 775f
Fatty acyl CoA, **765**–769
Feces, 650–653, 795
Feedback inhibition, **592**
Fehling's reagent, **407**
Female sex hormones, 542–545
Fermentation
alcoholic, **745**
lactic, 743–**744**
Ferric ion, 114
Ferritin, **800**
Ferrous ion, 114
Fetal alcohol effects (FAE), 405p
Fetal alcohol syndrome (FAS), 392p, 405p
Fever reduction, 444c
Fiber, dietary, 510c
Fibrin, 600
Fibrinogen, 597, 600, 680
Fibroin, 566, 566f
Fibrous proteins, **565**
Film badge, 303, 303f
Fingerprinting, DNA, 623c, 623f
Fireworks, 73c
First messenger, **753**
Fischer, Emil, 496
Fischer projections, **496**
Fission, **298**–300, 299f
Flagpole hydrogens, 334
Flavin adenine dinucleotide (FAD)
ATP production and, 742, 743t, 771–772
in cellular respiration, 714, 719f
in citric acid cycle, 741, 742
in fatty acid oxidation, 769–771
formation of acetyl CoA, 738
in lactic fermentation, 743
reduction to $FADH_2$, 711–714, 713–714f
Flavoprotein, **716**
Flavor esters, 440–441, 440t
Fleming, Alexander, 601
Fluid(s). See Body fluids
Fluid mosaic model of membrane structure, **536**, 536f
Fluoride ions, 255c
Fluorine, 91, 92, 102, 112, 122–123, 665t, 670
Fluorosis, **670**
Fluosol DA, 682c
Fluoxetine, 479p
Foam, of soda, 227c
Folic acid, 601, 661–662, 661f
Follicle, 542
Food
caloric values of, 53t, **653**–654
counting calories in, 55c
irradiation of, 292c
oxidation of, 653
Food guide pyramid, 655c
Formaldehyde (methanal), 395t, 396, 397f, 413
Formalin, 413
Formed elements of blood, **679**
Formic acid, 422, 423, 424t

Formula(s)
chemical, **111**, 144t
of ionic compounds, 118–120
molecular, **122**
structural, **122**, **318**, 321–322, 323, 328–330
Formula mass, **143**–146, 144t
Fortrel, 445
Frasch process, 101
Free amine, **466**
Free energy, **160**–161, 161t
Free radical, **303**–304
Freezing point depression, 238
Freons, 361c
Frostbite, 218
Fructosamine, 575c
Fructose, 491f, 499, 500f, 501f, 647
Fructose 1,6-biphosphate, 734f
Fructose 6-phosphate, 734f
Fumarate, 741f
Fumaric acid, 426, 794f
Functional groups, **362**
Furanose ring system, **501**
Fused-ring aromatic compounds, **342**–343, 343f
Fusion, 93c, **300**–301
heat of, **206**

Gaily, Joseph A., 684
Galactose, 499f, 500, 513, 647
Galactosemia, 513
Gallbladder, 646f, 647
Gallium, 71, 86, 97
Gallstones, 225
Gamma radiation, 288t, **291**–292, 291f
Gas(es), **11**, 11f
behavior of, 187–193
contained vs. uncontained, 181
diffusion of, 186–187
ideal vs. real, 187–188
molar volume of, **185**
natural, 344
noble, **88**, 94–95, 95f
solubility in liquids, 225–228
speed of particles in, 181, 181f
Gas laws, 187–188, 193–203
Boyle's law, **193**–195, 195f
Charles's law, **196**–197, 196f
combined, **199**–200
Henry's law, **225**–226
ideal, **201**–203
of temperature-pressure relationship at constant volume, **197**–199, 198f
Gas pressure, **183**, 188–193
container size and, 190
gas added, 188–189, 188f
partial pressures, 192–193, 194
removal of gas and, 190
temperature change and, 190, 190f, 193–199, 195f, 196f, 198f
Gasohol, 370
Gasoline, 343c, 345c
Gastric juice, **646**
Gaucher's disease, 766c, 767f
Geiger (Geiger-Müller) counter, **302**, 302f

Gene(s), **621**–623, 622c, 634c
Gene mutations, **633**–636
Gene therapy, **637**–638, 652c
Genetic code, 631–633, **632**, 632f
Genetic damage, **304**
Geometric isomers, **337**–338, 337f
Germanium, 97, 98c
Glaucoma, 384
Global warming, 348c
Globin, **570**
Globular proteins, **568**, 568f
Globulins, 680
Glomerulus, 690, 691
Glomular filtrate, 691, 697
Glucagon, 755
Glucocorticoids, **539**–540, 539f
Glucogenic amino acids, **795**, 796f
Gluconeogenesis, 746–**747**
Glucosamine, 516f
Glucose, 499–500, 499f
    absorption of, 756–758, 757f
    in blood, 515c, 575c
    cyclic form of, 501f
    in digestion, 647
    oxidation of, 731f, 732–733
    phosphorylation of, 733–736
    in photosynthesis, 705
    renal threshold for, 691
    starvation and, 732–733, 758, 764
    storage of, 748–749, 749f
    water vs., 144t
Glucose catabolism, 731–742
    acetyl coenzyme A in, **731**–732, 732f,
        737–738
    ATP produced from, 742, 743t
    citric acid cycle, 731, **732**, 738–742,
        739f
    glycolysis, **731**, 733–737, 737t
    three stages of, 731–732
Glucose meter, 515f
Glucose 1-phosphate, 748f, 750f
Glucose 6-phosphate, 633f, 734f, 748f,
    750f
Glucose tolerance test, 29p, 49f, 49p
Glucuronic acid, 516f
Glutamic acid, 554f, 790–791, 790f, 797,
    798f
Glutamic-oxaloacetic transaminase
    (GOT), 789–790
Glutamic-pyruvic transaminase (GPT),
    789
Glutamine, 555f, 798, 798f
Glutaric acid, 424f
Glyceraldehyde, 493–494, 493f, 494f,
    496f, 497
Glyceraldehyde 3-phosphate, 734f, 735f
Glycerol, 369, 370, 649, 649f
    in fat mobilization, 764–765
    metabolism of, 772–773
    as soap by-product, 528
    in synthesis of triglycerides, 524–525
Glycerol monostearate, 526c
Glycerol tristearate, 525
Glycine, 426, 553f, 557f
Glycogen, **508**–509
    breakdown of, 722, 749–751, 755f

carbohydrate loading and, 751p
    storage of, 748–749
    synthesis in body, 797
Glycogenesis, **748**–749
Glycogenolysis, **749**–751, 755–756
Glycol(s), 369
Glycolipids, **533**, 533f, 534f
Glycolysis, **731**, 733–737, 737t
Glycolytic pathway, 731
Glycoproteins, 536, **574**, 575c
Glycosides, **505**–507
Glycosidic bonds, **505**–507, 506f
Glycosuria, **757**–758
Goiter, **669**–670
Gold, 86f, 103
Golgi body, 530c, 530f
GOT (glutamic-oxaloacetic transam-
    inase), 789–790
Gout, **803**, 804p
GPT (glutamic-pyruvic transaminase),
    789
Graafian follicle, 542
Grain alcohol, 370
Gram, **46**, 48
Granuloma, 602
Grape sugar. *See* Glucose
Graphite, 97, 97f, 134c
Greenhouse effect, 348c
Group(s), **85**
    0 elements, 94–95, 95f
    1A elements, 95–96, 96f
    2A elements, 96, 96f
    3A elements, 97
    4A elements, 97, 97f
    5A elements, 98–100
    6A elements, 100–101, 101f
    7A elements, 102
Group trends
    in atomic radius, 92
    in ionization energy, 94
Growth factors, 634c
Guanine, 609, 610f
Gypsum, 221

Habituating drug, 481
Half-life, **292**–294, 293f, 293t, 294t,
    305
Halide ions, 115
Hallucinogens, **480**
Halocarbons, **356**–367
    from alkenes, 362–367
    as anesthetics, 354f, 377c
    elimination reactions, 373–375
    monosubstituted, 363
    nomenclature, 356–359, 356t, 357t,
        362
    physical properties of, 379–381
    uses of, 360t
Halogen(s), **102**, 112, 122–123
Halogenation, **362**–363, 363f
Halothane, 377c
Hard water, **433**, 434c
Harrick, James, 2p
Haworth projections, **502**–505, 503f, 504f
HDLs (high-density lipoproteins), 651c,
    780

Health physicists, 302
Heart attacks
    enzyme therapy for, 582p, 600p
    plaque buildup and, 780
    serum LDH levels and, 744c
Heartburn, 250p, 271p
Heat, **21**–22
    specific, **53**–56, 54t, 215–216
Heat capacity, **53**
Heat changes, 156–158, 157f, 158f, 160,
    160t, 161t
Heat of condensation, **205**
Heat of fusion, **206**
Heat of solidification, **206**
Heat of vaporization, **205**, 216
Heat sterilization, 575–576
Heavy water, 71
Heimlich maneuver, 180p, 196p
Helium, 68, 88, 95, 95f
Heme, 800–802, 801f
Heme group, 568, 568f, 569f
Heme oxygenase, 754c
Hemiacetal, **409**–410
Hemiacetal formation, 501–502, 502f
Hemiketal, **410**
Hemiketal formation, 501–502
Hemodialysis, **242**, 242f, 243f
Hemoglobin, 103, 145, 465, **570**–573,
    571f, 679, 680
    in control of blood glucose, 575c
    oxygen transport and, 687–688
    in sickle cell anemia, 7p, 572–573
    synthesis and hydrolysis of, 799–802,
        801f
Hemolysis, 241f
Hemolytic jaundice, **802**
Hemophiliacs, 682c
Henry's law, **225**–226
Heparin, 517f, 600
Hepatitis, 682c, **802**
Heptanal, 394
Heroin, 484, 484f
Heterocyclic amines, 464–465
Heterocyclic rings, **378**
Heterogeneous mixtures, **12**
Heteronuclear molecules, **112**, 123–124
Hexachlorophene, 384
Hexane, 323t, 324f, 325
High blood pressure, 189c, 481–482
High-density lipoproteins (HDLs), 651c,
    780
Histamine, 480–481, 481f
Histidine, 555f
HIV (human immunodeficiency virus),
    602, 617, 619c, 638, 682c
HIV protease, 593c, 593f
HMG-CoA reductase, 781c, 781f
Holoenzyme, **587**
Homeostasis, **679**
Homogeneous mixtures, **12**
Homonuclear molecules, **112**
Honey, 491f, 512
Hormones, **538**
    female, 542–545
    male, 540–541
    metabolism and, 752

peptide, 560–561, 561f, 562
steroid, **538**–541
Human body. *See* Body, human
Human Genome Project (HGP), 622c
Hund's rule, **75,** 77
Hyaluronic acid, 517
Hybrid, resonance, **340**
Hybridization, orbital, **319,** 319f, 338, 339f, 340
Hydrates, **408**
Hydration, water of, **221**–222
Hydration reaction, **372**
Hydrides, 406
Hydrocarbons, **317**–320, 318f, 319f
    burning, 155–156, 346–347, 348c
    fractions, 344, 344t
    greenhouse effect and, 348c
    health and, 343c
    properties of, 346–347, 379–381
    saturated, 318–320, **334**
    sources of, 344–346, 344t, 346f
    structural isomers of, 330, 342, 342f
    unsaturated, **334**–338, 340–341
Hydrochloric acid, 258, 260, 646, 647
Hydrochloride salts, 467
Hydrogen, 102, 112
    chemical formula for, 122
    flagpole, 334
    isotopes of, 68, 71
Hydrogen bonds, 130, **214**–216, 214f, 317–320, 397, 397f, 400f, 472, 472f
Hydrogen chloride, 124, 131f, 254, 258
Hydrogen ion acceptor, **256**
Hydrogen ion donor, **256**
Hydrogen peroxide, 166c, 718c
Hydrogenation, **362,** 365, 527
Hydrohalogenation, **362,** 363–365
Hydrolysis, **438**
    of amides, 476
    of anhydrides, 438, 450
    of esters, 446–447
    of fats and oils, 528
    of glycosidic bonds, 507
    of salts, 279
    of sugars, 507
Hydrolytic rancidity, 528
Hydrometer, **52**
Hydronium ion, **251**
Hydrophilic, **381,** 432
Hydrophobic, **380,** 432
Hydroxide ion, 115, 117t, **251**–253
Hydroxy acid, **426**
Hydroxy function, **367**
Hydroxyapatite, 99, 100c, 255c, 670
β-Hydroxybutyric acid, 776–777
Hydroxyl group, **367**
Hydroxyl radicals, 303–304, 718c
Hydroxylic side chains, 554, 554f
Hyperammonemia, 792c
Hyperbilirubinemia, 802c
Hypercholesteremia, 637–638
Hyperglycemia, **757**
Hyperlipoproteinemia, 651c
Hypertension, 189c, 481–482
Hyperthyroidism, 287p, 307p
Hypertonic solution, **241,** 241f

Hyperventilation, **694**
Hypervitaminosis, **656**
Hypnotics, **485**
Hypoglycemia, **757**
Hypolipidemic drugs, 781c
Hypolipoproteinemia, 651c
Hypothalamus, 561
Hypothermia, 142p, 164p
Hypothesis, **4**–6
Hypotonic solution, **241,** 241f
Hypoventilation, **694**–695

Ice, properties of, 217–218, 217f
Ideal body weight (IBW), 525p
Ideal gas, 187–188
Ideal gas constant, **201**
Ideal gas law, **201**–203
Imidazole, 464f
Immune response, 683
Immunoglobulins, **683,** 683f
Implants, biomedical, 89c, 100c
Incomplete proteins, 654–655
Indicators, acid-base, 267–269, 268f, 269f, 277
Indium, 97
Indole, 464f, 652–653, 652f
Indole ring system, 481–482
Induced-fit model, **586,** 586f
Inducible enzyme, **591**
Induction, **590**–591
Infections, 601–602, 682–683
Infectious hepatitis, **802**
Inflammation, 539
Initiation, **628**–629, 629f
Inner transition metals, **89,** 103
Inorganic acids, reactions, 453–454
Insecticides, 603c
Insulin, 539, 636, 696p, 756–758, 756f, 757f, 778
Insulin shock, **758**
Integral proteins, **536**
Interferons, 683, **685,** 685f
Internal salts, 557–558, 557f
International System of Units (SI), **31**–32, 31t. *See also* Metric system
International Units (IU), **588**
Interstitial fluid, **677,** 677t, 679
Intracellular fluid, **677,** 677t
Introns, **621**–623, 622f
Invert sugar, **512**
Iodine, 102, 112, 287f, 665t, 669
Iodine-123, 307p
Iodine-131, 311
Ion(s), **112**–120
    in diet, 116c, 116t
    formation of, 113–117
    polyatomic, **114**–115, 117t, 126
Ion channels, 132p
Ion exchange, 434c
Ionic bonds, **118,** 567
Ionic compounds, **112**
    anions, **113,** 115–117, 115t, 117t, 718c
    cations, **113**–115, 114t
    covalent compounds vs., 130t
    as electrolytes, 222, 222f
    formulas of, 118–120

molecular compounds vs., 122f
    naming, 118, 119
    solvation and, 219–220, 220f
    structure of, 120, 120f
Ionization constant, **259**
Ionization energy, **93**–94
    group trends in, 94
    periodic trends in, 94
    of representative elements, 94t
Ionized atoms, 112. *See also* Ion(s)
Ionizing radiation, 288t, **302**–303
Ion-product constant for water, **251**–252
Iproniazid, 478f
Iron, 103, 116t
    in diet, 665t, 668, 669
    rusting of, 155, 162–163
Iron-deficiency anemia, **668**
Iron ion, 114
Irradiation, of food, 292c
Isaacs, Alick, 685
Isoamyl nitrite, 451
Isocitrate, 740–741, 741f
Isoelectric pH, **557**–558
Isoelectric point, **557**–558
Isoflurane, 377c
Isohydric shift, **689**
Isoleucine, 553f
Isomers
    geometric, **337**–338, 337f
    optical, **497,** 497f
    stereoisomers, 493–**494,** 496f
    structural, **330,** 342, 342f
Isotonic solution, **241,** 241f
Isotopes, **67**–71. *See also* Radioisotope(s)
Isozymes, 744c
IUPAC system, **322**
    alcohols, 368–369
    aldehydes, 393–394
    aliphatic amines, 461–462
    alkenes, 334–335
    amides, 469–470
    arenes, 341–343
    aromatic amines, 462–464
    branched-chain alkanes, 325–328
    carboxylic acid, 422–423
    carboxylic acid anhydrides, 437–438
    cycloalkanes, 331–332
    esters, 439–440
    ethers, 376
    halocarbons, 356–359, 356t, 357t, 362
    ketones, 394–396
    organic acid salts, 430–431
    straight-chain alkanes, 322–323

Jaundice, **801**–802
Joints, artificial, 100c
Joliot-Curie, Irene, 306
Joule, 32t, **53**

Kelvin, Lord, 8
Kelvin (unit of measurement), 32t
Kelvin scale, **8,** 9f, 10, 182
Kendrew, John C., 568
Keratinization, **657**
Keshan disease, 84p, 101p
Ketal, **410**

Ketoacidosis, **777**–778
Ketogenic amino acids, **795,** 796f
α-Ketoglutarate, 739f, 741, 797, 798f
α-Ketoglutaric acid, 790f, 791f
Ketohexoses, 500
Ketone(s), **393**–414
    addition of alcohols, 409–410
    addition of water, 408–409
    alpha-hydroxy, 407–**408**
    boiling points of, 398, 398t
    common examples of, 395t
    flavors and fragrances and, 391f,
        399c
    nomenclature, 394–396
    physical constants, 398t
    redox reactions involving, 402–406
    structure of, 393, 393f, 396, 397–400
    uses of, 410–414
    water solubility of, 399–400, 400f
Ketone bodies, 774, **776**–777
Ketonemia, **777**
Ketonuria, **777**
Ketopentoses, 500
Ketoses, **500**
Ketosis, **777**–778
Ketotetroses, 500
Ketotrioses, 500
Kevlar, 475
Kidney(s), 690–692, 690f, 692f, 696
Kidney stones, 225, 226c
Kieselghur, 451
Kilocalorie, **53**
Kilogram, 32t, **46,** 48
Kilometer, 39t, 45
Kinases, 724
Kinetic energy, **181**–182
    average, 181–**182**
    of liquids, 203, 204f
Kinetic-molecular theory, **181,** 181f,
    187–193
Krebs, Sir Hans, 732
Krebs cycle, 732, 738–742
Krypton, 94–95
Kwashiorkor, **656**

Lactase, 513p, 515, 647
Lactate, 744f, 745–746
Lactate dehydrogenase (LDH), 744c
Lactic acid, 426, 745–746
Lactic fermentation, 743–**744**
Lactobacillic acid, 424f
Lactose, 499, 512–513, 647
Lactose intolerance, 492p, 513p
Lanthanum, 103
Lauric acid, 424t
Law
    periodic, **87**
    scientific, **6**
Law of conservation of energy, 21–**22,**
    157
Law of conservation of mass, **20**
LDH (lactate dehydrogenase), 744c
LDLs (low-density lipoproteins), 651c,
    780
Le Châtelier's principle, **170,** 583, 687,
    688, 694

Lead, 63, 97
Lead poisoning, 591c
Lecithin, **531,** 531f
Length, units of, 32t, 38–39, 39f, 39t
Lesch-Nyhan syndrome, **803**–804
Leucine, 553f
Leucine enkephalin, 562f
Leukemia, 306, 343c
Leukocytes, **679**
Leukotrienes, **547,** 547f
Levitation, 121f
Lewis, G. N., 110
Lewis dot structures, **110**–111, 111t, 126
Lidocaine, 112
Lidocaine hydrochloride, 471c
Ligase enzymes, 627
Light, human responses to, 706c
Light system of photosynthesis, **705**
Like dissolves like, 380–381
Lindenmann, Jean, 685
Linisopril, 602, 602f
Lipases, 528, 588t, 649
Lipid(s), **524**–548, 763–782
    in cell membranes, 529–536
    complex, **529**–533, 534f
    digestion of, 648–650, 648f, 650f
    glycolipids, **533,** 533f, 534f
    phospholipids, **529**–533, 534f
    stored, 764, 764f
    triglycerides, 440, **524**–529, 534f
    working (tissue), **764**
Lipid bilayer, **534,** 535, 535f
Lipid storage diseases, 766c, 767f
Lipoic acid, 738f
Lipoproteins, **649,** 651c, 680, 780
Liposomes, **534**–535, 535f
Liquid, **11,** 11f
    properties of, 203–205, 204f
    solubility of gases in, 225–228
Lister, Joseph, 371c
Liter, **40,** 41f, 41t, 45–46
Lithium, 89
Lithotripter, 226c
Liver, 721–722
    cholesterol synthesis in, 778
    cirrhosis of, 404, 792f
    in gluconeogenesis, 746
    glycogen storage in, 749f
    in glycogenolysis, 750, 756
    infectious hepatitis and, 802
    metabolic pathways in, 773–774, 773f
    oxidation of alcohols in, 404–405
    in pernicious anemia, 662
    production of ketone bodies, 776
Lock-and-key model, **585**–586, 585f
Loop of Henle, 697, 697f
Lou Gehrig's disease, 718c
Lovastatin, 781c, 781f
Low-density lipoproteins (LDLs), 651c,
    780
Low-saturated fat diet, 780
Lowry, Thomas, 256
LSD (lysergic acid diethylamide), 481,
    482f
Lungs, 194f
Lymph, 677, **681**

Lymph nodes, 683, 683f
Lymphatic system, 681, 681f
Lymphocytes, 683–685, 683f
Lysergic acid diethylamide (LSD), 481,
    482f
Lysine, 555f
Lysosomal enzymes, 262c
Lysosome, 530c, 530f

Macronutrients, **664**–668
Macrophages, 682, 754c, 754f
Magnesium, 19c, 96, 96f, 116t, 665t, 667
Magnesium ion, 114
Magnesium sulfate, 118
Magnet, 121f
Magnetic resonance imaging (MRI), 62p,
    71p
Malaria, 482
Malate, 739f, 741f, 742f
Malathion, 603c
Male sex hormones, 540–541
Maleic acid, 426
Malonic acid, 424, 485f
Maltase, 647
Maltose, 511, 511f, 647
Manganese, 665t, 669
Mannans, 500
Mannose, 499f, 500
MAO (monoamine oxidase) inhibitors,
    478
Marasmus, 655, **656**
Marcaine, 471c
Marijuana, 383–384
Markovnikov, Vladimir, 363
Markovnikov's rule, **363,** 364, 372
Masculinization, 541
Mass, **20,** 46
    law of conservation of, **20**
    units of, 32t, 46–48, 46t
Mass number, **66**–67, 66t, 67t
Matter, **3**
    classification of, 18f
    states of, 10–11, 11f
Maturity-onset diabetes, 29p, 49p
Mean free path, 181
Measurement
    accuracy vs. precision of, 30–31, 31f
    converting units of, 43–46
    of length, 32t, 38–39, 39f, 39t
    liquid volume apparatus for, 42, 42f
    of mass, 46–48, 46t, 47f
    quantitative vs. qualitative, 30
    systems of, 31–32, 31t
    units of, 32t, 38–44, 46–48
    of volume, 32t, 40–44, 41f, 41t, 42, 42f
    See also Metric system
Melanin, 668
Melting points, **206,** 206f, 216t
Mendeleev, Dmitri, 85
Menstruation, 542, 543f
Mental depression, 460p, 477–479, 479p
Mepivacaine hydrochloride, 471c
Mercaptans, **381**–382
Mercury, 86f, 103
Mercury barometer, 183f
Mercury thermometers, 9

Mescaline, 480f
Messenger RNA (mRNA), **616, 624**–631, 626f
Mestranol, 544, 544f
Metabolic acidosis, 676p, **693**, 695, 696p
Metabolic alkalosis, **693**, 695–696
Metabolic regulation, 752–754
Metabolism, **653**–654, 695, **704**
Metal(s), **85**–86, 86f. *See also specific metals*
Metalloids, **86**, 86f
Meter(s), 32t, **38**–39, 39f, 39t, 45
Meter stick, 39f
Methadone, 484, 484f
Methanal (formaldehyde), 395t, 396, 397f, 413
Methane, 317–320
  boiling point of, 323t
  bonding of, 317–318
  burning, 155–156, 346
  shape of, 131f, 132–133, 133f, 318–320, 318f, 319f
Methanoic acid, 422
Methanol (methyl alcohol), 368–369, 370, 404–405
Methemoglobin, **570**
Methemoglobinemia, **570**
Methionine, 555f
Methionine kephalin, 562f, 596
Methyl alcohol (methanol), 368–369, 370, 404–405
Methyl phenols, 371c
Methyl salicylate, 371c, 440t, 441
Methylmalonic acidemia, 378p
Methylpropane, 326
Metric system, **31**–32, 31t
  conversion factors in, **44**
  prefixes in, 39t
  units of length in, 32t, 38–39, 39f, 39t
  units of mass in, 46–48, 46t
  units of volume in, 32t, 40–44, 41f, 41t
Meyerhof, Otto, 732
Micelles, **432**, 432f, 433f
Microcurie, 305
Micrometer, 39t
Micronutrients, **664**, 668–670
Microscope, scanning-tunneling, 64c
Milk sugar, 500
Millicurie, 305
Milliequivalent (mequiv) of acid or base, 275
Milligram, 48
Milliliters, 45–46
Millimeter, 39t
Mineral(s), body's need for, 645
Mineral elements, **664**–670
Mineral oil, 343c
Mineralocorticoids, **540**, 540f
Minipill, 544
Mirror image, 493, 493f, 495–496
Miscarriage (spontaneous abortion), 545
Mitochondria, 715c, 715f
  citric acid cycle, 738
  as part of eucaryotes, 530c, 530f
  in synthesis of ketone bodies, 774

Mixed triglycerides, **525**
Mixtures, **11**–13, 12f
  heterogeneous, **12**
  homogeneous, **12**
Mobilization, of fat, 764–765, 765f
Modulator binding site, **594**
Molar mass, 144–146, 144t, 145f
Molar volume of a gas, **185**
Molarity, **231**–234, 274–275
Mole, 32t, **144**, 145–146
Molecular biology, central dogma of, **616**–617
Molecular compounds, **112, 121**–125, 122f
Molecular diseases, 634c, **635**–636
Molecular formulas, **122**
Molecular modeling, 132p
Molecular orbitals
  in saturated hydrocarbons, 318–320, 319f
  in unsaturated hydrocarbons, 335–336, 336f, 339f, 340–341, 340f
Molecules, **111**–112
  attractions between, 129–130, 130f
  heteronuclear, **112**, 123–124
  homonuclear, **112**
  polar, **129**
  shapes of, 131–136, 131f, 133f, 135f
Molybdenum, 665t, 669
Molybdenum cow, 308, 308f
Monoamine oxidase (MAO) inhibitors, 478
Monoclonal antibodies, **685**
Monohydrogen phosphate ion, 115, 117t
Monomers, **366**–367
Monomolecular layer, **428**, 428f
Monoprotic acid, 254
Monosaccharides, **497**–500, 498f, 499f, 500f
  five-carbon sugars, 499, 499f
  four-carbon sugars, 498, 498f
  six-carbon sugars, 499–500, 499f, 500f
Monosubstituted halocarbon, 363
Morning after pill, 544
Morphine, 169c, 483, 483f
Moseley, Henry, 65, 85
Mouth, digestion in, 645–646
MRI (magnetic resonance imaging), 62p, 71p
mRNA. *See Messenger RNA*
Mucin, 645
Multienzyme complex, **737**
Multiple sclerosis, 636, 684, 685
Muscles
  contraction of, 723–724, 724f
  in control of glycogenolysis, 756
  glycogen storage in, 749
Muscone, 411, 411f
Mutations, gene, **633**–636
Mylar, 445
Myocardial infarction. *See Heart attacks*
Myoglobin, **568**, 568f, 569f, 572
Myristic acid, 424t

N-terminal residue, 559, 559f
n-type semiconductor, 98c, 99f

NAD⁺. *See* Nicotinamide adenine dinucleotide
Nanometer, 39t
Native state, **574**
Natural gas, 344
Negative modulator, **594**
Negative nitrogen balance, **795**
Neon, 94–95, 95f
Neosynephrine, 480f
Nephrons, 690, 690f, 697f
Nesacaine, 442c
Neurotransmitters, **476**–477
Neutral solutions, **252**
Neutralization reactions, 269–271, **276**–278, 277f
Neutrons, **64**
Niacin, 660
Nickel, 665t, 670
Nicotinamide, 660, 660f
Nicotinamide adenine dinucleotide (NAD⁺)
  in alcoholic fermentation, 745
  ATP production and, 742, 743t
  in cellular respiration, 714, 715, 716, 718, 719f
  in citric acid cycle, 739f, 741–742
  in fatty acid oxidation, 769–770
  formation of acetyl CoA and, 737, 738
  in glycolysis, 735
  in lactic fermentation, 743–744
  in oxidative deamination, 791
  in oxygen debt, 745
  reduction to NADH, 711–713, 712f
Nicotine, 481, 481f
Nicotinic acid, 660, 660f, 711
Niemann-Pick disease, 766c, 767f
Night blindness, 657
Nitric acid, esters of, 451–452
Nitric oxide, as second messenger, 754c
Nitrogen, 98–100, 112, 123
  in blood, 227c
  urea cycle, 791–795, 791f
  See also Amino acid(s); Nucleic acids; Protein(s)
Nitrogen balance, **795**
Nitrogen cycle, **787**–788, 787f
Nitrogen fixation, 787–788
Nitrogen-fixing bacteria, **787**
Nitroglycerin, 451, 754c
Nitrous acid, 451
Nobel, Alfred, 451
Noble gases, **88**, 94–95, 95f
Nomenclature. *See* IUPAC system
Nomex, 475
Nonelectrolytes, **222**–223
Nonessential amino acids, **654**, 654t
Nonionic detergent, **435**
Nonmetals, **85**–86, 86f
Nonpolar covalent bonds, **127**
Nonreducing sugars, **514**–515
Nonsuperimposable mirror images, **493**, 493f, 495–496
Norancholane, 541f
Norepinephrine, 477, 477f
Norethindrone, 543, 544f

Norethynodrel, 543, 544f
Normal boiling point, **205**
Normality, **274**–276
Novocain, 442c
Nuclear fission, 298–300, 299f
Nuclear fusion, 93c, **300**–301
Nuclear magnetic resonance, 71p
Nuclear reactor, 299–300, 299f, 301c
Nuclear transformation reactions, **298**
Nucleic acids, **609**–639, 650–653, 652f.
　See also DNA; Nucleosides;
　　Nucleotide(s); RNA
Nucleosides, **609**, 610, 611f, 612t
Nucleotidases, **650**
Nucleotide(s), **609**
　base units, 609
　in DNA and RNA, 611–612, 612t, 613f
　nomenclature, 612t
　structure of, 611–614, 613f
　sugar units, 609
Nucleotide base pairs, 615, 615f, 621t
Nucleus, **65**
　atomic number and, 65
　cell, 530c, 530f
　composition of, 66t
　electron configuration around, 75–79
　isotopes and, 67–71
　mass number and, 66
NutraSweet, 561c
Nutrition, 653–671
　balanced diet and, 653–654, 655c
　essential needs of human body, 645
　fat-soluble vitamins and, 656–659
　food guide pyramid and, 655c
　minerals and, 664–670
　proteins and amino acids and,
　　654–656
　trace elements and, 670
　water-soluble vitamins and, 659–663
Nyctalopia, 657
Nylon, 474–475

Obesity, 523, 525p
Observation, **4**, 4f, 6f
Octane ratings, 345c
Oil(s), **527**
　hydrogenation of, 527
　hydrolysis of, 528
Oil of wintergreen, 441
Okazaki, R., 620
Okazaki fragments, **620**
Oleic acid, 535f
Oncogenes, 634c
One gene, one enzyme hypothesis,
　621
Opiates, **484**
Opium, 483
Opsin, 413c
Optical activity, 496–497
Optical isomers, **497**, 497f
Optimal pH, 589, 589f
Orbital(s), atomic, **72**–75, 74f, 75t, 76f,
　77. *See also* Molecular orbitals
Orbital hybridization, **319**, 319f, 338,
　339f, 340
Organelles, 262c, 530c

Organic compounds, 97, 322–323,
　400–402, 401f. *See also* Carbon
　bonding
Organic molecules, 317
Ornithine, 721–722, 722f, 793f, 794f
Osmosis, **239**–241, 239f, 241f, 723
Osmotic membranes, **239**, 239f
Osmotic pressure, 240–**241**, 240f
Osteoporosis, 644p, **666**, 666p
Oviduct, 542
Ovulation, 542
Ovum (egg), 542, 544
Oxalic acid, 424
Oxaloacetate, 739f, 740f, 742f, 747f, 795,
　796f
Oxaloacetic acid, 790f, 798f
Oxidation, 153, **154**–156
　of alcohols, 402–405
　of fatty acids, 765–771
　of food, 653
　of glucose, 731f, 732–733
　identifying relative degree of, 402
　of organic compounds, 400–402,
　　401f
Oxidative deamination, **791**
Oxidative phosphorylation, 703p,
　711–719, 720p
Oxidative rancidity, 528
Oxide ion, 115
Oxidizing agent, **154**, 155–156
Oxidoreductases, 584
Oxyesters, 446
Oxygen, 100–101, 112, 123, 125
　active, 166c
　body's need for, 645
　in breathing, 193, 194
　disease and aging and, 718c
Oxygen debt, **745**–747
Oxygen transport, 571–572, 687–688,
　687f
Oxyhemoglobin, **570**, 687
Oxytocin, 560–561, 561f, 562
Ozone layer, 361c

p-type semiconductor, 98c, 99f
Pacemaker enzyme, **594**–595
Palmitic acid, 424t, 535f, 649, 649f, 772,
　772t, 775f
Pancreas, 646f, 647
Pancreatic juice, **647**
Pantocaine, 443c
Pantothenic acid, 662, 662f, 731,
　732f
Paraffin wax, 315f
Paraformaldehyde, 413
Paraldehyde, 414, 414f
Parathion, 603c
Parathyroid hormone (PTH), 666p
Parkinson's disease, 477
Paroxetine, 479p
Partial pressure, **192**–193, 194
Particle size
　reaction rate and, 165–166
　solution formation and, 223
Pascal, Blaise, 183
Pascal (unit), 32t, **183**

Pauli exclusion principle, **75**, 77
Pauling, Linus, 2p, 7f, 7p, 660
Paxil, 479p
PCP (phencyclidine), 477
Pectin sugar, 499, 499f
Pellagra, **660**
Penicillin, 601–602, 601f
Pentane, 323t, 330
Pentanoic acid, 423
Pentanones, 395
Pepsin, 584, 595, 646, 647
Pepsinogen, 595, 595t
Peptic ulcers, 647
Peptidases, **584**, 588t, 595
Peptide(s), 552, **559**–562, 560t
Peptide bonds, **559**, 559f, 595
Peptide hormones, 560–561, 561f, 562
Peptide link, 559
Perfluorocarbons, 682c
Period, **85**
Periodic law, **87**
Periodic table, 69, **85**–104
　development of, 85
　modern, 87–91
　organization of, 87–91, 87f, 88f, 90f
　symbolism in, 85f
　writing electron configurations by
　　using, 90–91
Periodic trends, 87
　in atomic radius, 92
　in ionization energy, 94
Peripheral proteins, **536**
Pernicious anemia, **662**–663
Peroxidase, 718c
Perspiration, 698
Pesticides, 603c
PET (polyethylene terephthalate), 445
Petroleum jelly, 343c
Petroleum refining, **344**, 345c
pH, **260**–269, 261t
　of blood, 686, 691–696, 692f, 693t
　of body fluids, 262c
　buffers, 279–281, **280**, 280t
　calculations, 260–267
　measurement of, 267–269, 268f, 269f
pH-activity profile, **589**–590, 589f
pH meter, 269, 269f
pH optimum, **589**, 589f
Phagocytosis, 682–683
Phenacetin, 470, 470f
Phenanthrene, 537, 537f
Phencyclidine (PCP), 477
Phenol, **370**–371
Phenolphthalein, 268, 277, 277f
Phenylalanine, 554f, 597, 798f, 799, 799f
Phenylketonuria (PKU), **799**
Phenylpyruvic acid, 799f
Phosphate esters, 449–450
Phosphatidic acid, 529–530
Phosphatidyl choline, 531, 531f, 534,
　649, 649f
Phosphocreatine, 724, 724f
Phosphodiesters, **612**
2-Phosphoenolpyruvate, 736f
2-Phosphoglycerate, 736f
3-Phosphoglycerate, 735f, 736f

Phosphoglycerides, **529**–532, 529f, 531f
Phospholipids, **529**–533, 534f
  digestion of, 648–650
  phosphoglycerides, **529**–532, 529f, 531f
  sphingomyelins, **532**–533, 532f
Phosphor, 302
Phosphoric acid, 254, 260, 448–451
  anhydrides of, 448–449, 450
  esters of, 449–450
Phosphorus, 19c, 99, 664–666, 665t
Phosphorus-32, 308
Phosphoryl group, **450,** 720–721
Phosphorylase, 722–723, 723f, 755
Phosphorylating power, **720**
Phosphorylation, **450**–451, 720–723, 733–736
Photobilirubin, 802c
Photoconductivity, 101
Photopheresis, 706c
Photosynthesis, 100, **704**–705, 707
Phototherapies, 706c
Phthalic acid, 426
Physical change, 12f, **14**–15, 15f
Physical properties, **14,** 14t
Physical states of matter, 10–11, 11f, 14–15, 15f
Physiologic gases, exchange of, 194c
Pi bond, **336**–337
Pi-bonding molecular orbital, 336f, 339f, 340f
Pill (birth control), 543–545
Pipe scale, 434f
Piperidine, 464f
Pipet, 42, 42f
Pituitary gland, 561
PKU (phenylketonuria), **799**
Plane-polarized light, **497,** 497f, 500
Plant steroids, 545, 545f
Plaques, **780,** 780f
Plasma, blood, 677, **679**–680
Plasmin, 584, 600
Plasminogen, 600
Plasmolysis, **241**
Platelets, 444c
Platinum, 103
Plumber's solder, 97
Plutonium-239, 298
Poisons, 603c, 718–719
Polar covalent bond, **127**–129, 214, 214f
Polar molecule, **129**
Pollution
  chemical, 5c
  thermal, 22
  water, 213p, 219p
Polonium, 100
Polyamides, **474**–475
Polyatomic anions, 115, 117t
Polyatomic ions, **114**–115, 117t, 126
Polycyclic aromatic compounds, **342**–343, 343f
Polyesters, **445**
Polyethylene, 366
Polyethylene terephthalate (PET), 445
Polyfunctional carboxylic acids, **426**–427
Polyfunctional compounds, **383**–385

Polyisoprene, 367
Polymer(s), **366**–367, 367f
Polymerization reactions, 366–367
Polynucleotides, **609,** 613f
Polypeptide, **560**
Polypropylene, 367
Polysaccharides, **507**–510, 508f, 509f
Polystyrene, 367
Polytetrafluoroethene, 366
Polyvinyl chloride (PVC), 366
Pontocaine, 443c
Positive modulator, **594**–595, 594f
Positive nitrogen balance, **795**
Posttranslation processing, **631**
Potassium, 89, 95, 116t, 665t, 666–667
Potassium hydroxide, 255–256
Potassium permanganate, 118
Precision, **30**–31, 31f
Prednisolone, 539f
Prednisone, 539f
Pregnancy, 542–543, 544
Pre-mRNA molecules, 627
Pressure
  atmospheric, 183–184
  blood, 189c, 481–482, 562
  gas, **183,** 188–193
  osmotic, 240–**241,** 240f
  partial, **192**–193, 194
  standard, 184, 185–186
  temperature-pressure relationship at constant volume, **197**–199, 198f
  units of, 32t
  vapor, **204,** 215, 237
Primary structure of proteins, **563**
Procaine, 442c
Procarboxypeptidase, 595, 595t
Products, **16**
Proenzymes, **595**–597, 595t
Progesterone, **542**–543, 542f, 543f
Proinsulin, 756
Prokaryotic cells, 530c, 626–627, 705
Proline, 553f
Propane, 323t, 325, 326, 346–347, 347f
Propanoic acid, 422, 447
Propanol, 369
Propionic acid, 422, 424t
Propyl chloride, 358
Prostaglandins, 444c, **545**–547, 546f
Prostanoic acid, 546, 546f
Prosthetic groups, **568**
Protease, **584,** 593c, 593f, 647
Protein(s), 552, **563**
  alpha helix of, **564**–565, 564f, 565f
  Bence-Jones, **684**
  beta-pleated sheet, **565**–566, 565f, 566f
  blood, 680, 680f
  complete, **654**
  conjugated, **568**
  denaturation of, **574**–576, 574f, 576f
  in diet, 654–656
  digestion of, 646–647
  fibrous, **565**
  globular, **568,** 568f
  glycoproteins, 536, **574,** 575c
  incomplete, 654–655
  integral, **536**

  peripheral, **536**
  primary structure of, **563**
  quaternary structure of, **570**–571, 571f
  secondary structure of, **563**–567, 564f, 565f, 566f
  subunits of, **570**
  synthesis of, 627–631
  tertiary structure of, **567**–569, 567f, 568f, 569f
  unwinding, **620**
  See also Enzyme(s)
Protein kinases, 634c, 752–753
Protein turnover, **788**
Prothrombin, 597, 658
Proton(s), **64**
  atomic number and, 65
  isotopes and, 67–71
Protonated amine, **466**
Prozac, 479p
Psoralen, 706c, 706f
PTH (parathyroid hormone), 666p
Ptyalin, 645–646
Pulmonary embolism, 600
Purine, 465f, 803–804, 803f, 804f
Purine bases, **609**
Putrescine, 465f
Pyranose ring system, **501**
Pyribenzamine, 481f
Pyridine, 464f
Pyridoxal, 661, 661f
Pyridoxal phosphate, 790f
Pyridoxamine, 661, 661f
Pyridoxamine phosphate, 790f
Pyridoxine, 661, 661f
Pyrimidine, 465f, 803, 803f
Pyrimidine bases, **609,** 610f
Pyrophosphate, 448, 748
Pyrophosphoric acid, 448–449
Pyrrole, 465f
Pyrrolidine, 464f
Pyruvate, 731, 731f, 736f, 737f
  in alcoholic fermentation, 745f
  in glucogenic amino acids, 795, 796f
  in gluconeogenesis, 747f
  in lactic fermentation, 744, 744f
Pyruvate carboxylase, 747
Pyruvate dehydrogenase, 737
Pyruvic acid, 426, 731, 731f, 798f

Qualitative measurements, **30**
Quantitative measurements, **30**
Quantum, **72**
Quaternary ammonium salt, **468**
Quaternary structure of proteins, **570**–571, 571f
Quinine, 482, 483f
Quinoline ring system, 482, 483f

R groups
  alcohols, 368
  ethers, 376
Rad, **305**
Radiation, **288**–292
  alpha, 288, **289**
  average doses of, 311t
  background, **310**–311, 311t

Radiation (*continued*)
  benefits of, 292c
  beta, 288t, **290**–291, 302, 302f
  biological effects of, 291–292, 303–304, 306, 311
  detection of, 302–303, 302f, 303f
  gamma, 288t, **291**–292, 291f
  ionizing, 288t, **302**–303
  as medical diagnostic tool, 306–308, 306f, 307t, 308f, 308t, 311, 311t
  units of, 304–305
  X rays, 288t, **291**–292, 291f, 306, 306f
Radiation absorbed dose (rad), **305**
Radiation therapy, **309**–310, 309f
Radioactive decay, **288**
  carbon-14 dating and, 295–296
  half-life and, 292–294, 293f, 293t, 294t
  rate of, 293t, 294, **305**
  series, 296f
  types of radiation in, 288–289, 288t
Radioactive tracers, 306–307
Radioactivity, **288**
Radiographs, 306, 306f
Radioisotope(s), 287, **288**, 292–294
  carbon-14 dating and, 295–296
  diagnostic use of, 307t, 308t
  in radiation therapy, 309–310, 309t
  rate of decay of, 293t, 294, **305**
Radioisotope seeds, 310
Radiolytic products, 292c
Radiopharmaceuticals, 310
Radium-226, 311
Radon, 297c
Rain forests, 169c, 707
Rainmaking, 225
Rancidity, **528**
Random walk, 181
Rate of decay, 293t, 294, **305**
Rate of reaction. *See* Reaction rate
Reactants, **16**
Reaction. *See* Chemical reactions
Reaction rate, **162**–166
  activation energy and, 162–163, 162f
  catalysts and, 164–165, 165f, 166c
  concentration and, 165
  particle size and surface area and, 165–166
  temperature and, 163, 163f
Reactive oxygen species (ROS), 718c
Recombinant DNA technology, **636**–637
Red blood cells, 679, 687–689, 799–800
Redox reactions, **154**–156
  involving aldehydes and ketones, 402–406
  of organic compounds, 400–402, 401f
Reducing agent, **154**, 155–156
Reducing power, **711**–714
Reducing sugars, **514**–515, 514f
Reduction, 153, **154**–156, 365, 400–401
Refractive index, 52c
Refractometer, 52c
Rem, **305**
Renal calculi (kidney stones), 225, 226c
Renal capsule, 690, 690f

Renal threshold for glucose, 691
Rennin, 583
Replication, **616,** 617–620, 617f, 618f, 620f
Representative elements, **88**–89, 94–102
Representative unit, 143
Reserpine, 481–482, 482f
Resonance hybrid, **340**
Resonance structures, **340**
Respiration, cellular, 115, **711,** 714–719
Respirator, 191c
Respiratory acidosis, **693,** 694–695
Respiratory alkalosis, **693,** 694
Respiratory chain, **714**–718, 717f, 719f
Respiratory therapist, 187
Restriction enzymes, 623c
Retinal, 412–413c
Retinoblastoma, 634c
Retinoic acid, 656
Retinol, 656–657
Retroviruses, **616**–617, 619c
Reverse transcriptase, 617, 619c, 619f
Reversible reactions, **167**–173
Reye's syndrome, 444c
Rheumatoid arthritis, 539
Rhodopsin, 413c
Riboflavin, 661, 711
Ribonuclease (RNase) enzymes, 627, 650
Ribonucleic acids. *See* RNA
Ribonucleosides, 610
Ribose, 499, 499f
Ribosomal RNA (rRNA), **624,** 624f, 625
Ribosome, 530c, **624**
Ribozymes, 583
Rickets, **658,** 665
RNA (ribonucleic acid), 499, 583, **609**
  bases in, 612t, 614
  classes of, 624, 625f
  messenger (mRNA), **616, 624**–631, 626f
  nucleosides in, 611–612, 612t
  nucleotides in, 609, 611–614, 612t, 613f
  ribonucleosides in, 610
  ribosomal (rRNA), **624,** 624f, 625
  transfer (tRNA), **624,** 625, 625f, 627–631
  writing complementary RNA strand, 627
RNA polymerase, **625**
Roentgen, Wilhelm, 305
Roentgen (unit), **305**
Roentgen equivalent medical (rem), **305**
Rounding off numbers, 36–38
rRNA, **624,** 624f, 625
Runner's high, 552p, 563p
Rust, 155, 162–163
Rutherford, Ernest, 298

Saccharin, 499, 512f
Safflower, 522f
Salicylic acid, 426
Saliva, **645**
Salivary amylase, 645–646
Salivary glands, 645

Salmonella bacteria, 292c
Salt(s), 102, 112, 120f, 122f, **269,** 270t
  ammonium, **466**–468, 472–473
  body's need for, 645
  concentration in body, 679, 691, 698
  hydrolysis of, 279
  internal, 557–558, 557f
  iodized, 670
  of strong acid and weak base, 279
  of weak acid and strong base, 278–279
Salt bridges, **567**
Saponification, 447, **528**
Sarcoidosis, 602
Saturated fatty acid(s), 424, 424t
Saturated fatty acid chains, 535, 535f
Saturated hydrocarbons, 318–320, **334**
Saturated solution, **224**
Saytzeff's rule, **375**
Scandium, 103
Scanner, 303
Scanning-tunneling microscope (STM), 64c
Scientific law, **6**
Scientific method, **3**–6
Scientific notation, **32**–35
Scintillation counter, **302**–303
Scurvy, **659**–660
Seasonal affective disorder (SAD), 706c, 706f
Second (time), 32t
Second messenger, **753**–754, 754c
Secondary sex characteristics, 541
Secondary structure of proteins, **563**–567, 564f, 565f, 566f
Sedatives, **485**
Seed crystal, 225, 225f
Selenide ion, 117t
Selenium, 84p, 100, 101p, 665t, 669
Self-dissociation of water, **251**–253
Semiconductors, 98c, 99f, 101
Semiconservative replication, 618f
Semipermeable membranes, **239**
Serine, 554f
Serotonin, 477–478, 477f
Sertraline, 479p
Serum, **680**
Serum albumin, 765
Serum LDH levels, 744c
Sex hormones, 540–545
Shock
  anaphylactic, 547
  insulin, **758**
SI (International System of Units), **31**–32, 31t. *See also* Metric system
Sickle cell anemia, 2p, 7p, 572–573
Sigma bond, **320**
Sigma-bonding molecular orbitals, 319f
Silicon, 78, 86f, 89, 97, 98c, 99f, 665t, 670
Silver, 103
Simple sugars, **497**–500
Simple triglycerides, **525**
Single covalent bonds, 121–123, **122**
Skatole, 652–653, 652f
Skin, artificial, 678c
Skin care products, 526c

Skin graft, 678c, 678f
Sklodowska, Marie. *See* Curie, Marie
Small intestine, 646f, 647–658
Soaps, 431–433, 447, 528
Sobrero, Ascanio, 451
Sodium, 89, 92, 95, 96f, 116t, 665t, 666–667
Sodium acetate, 278–279
Sodium barbiturate, 485f
Sodium bicarbonate, 257
Sodium carbonate, 257
Sodium chloride, 112, 120f, 122f, 143. *See also* Salt(s)
Sodium hydroxide, 254–256
Sodium hypochlorite, 102
Sodium ion, 113f
Sodium stearate, 432, 528f
Sodium triphosphate (Calgon), 434c
Solid, **11,** 11f, 206, 206f
Solidification, heat of, **206**
Solubility, 224–228, 224f
    of aldehydes and ketones in water, 399–400, 400f
    of amines, 466–467
    of gases in liquids, 225–228
    Henry's law and, 225–226
Solute, **218**
Solutions, **12**
    acidic, **252**–253
    aqueous, **218**–219
    basic (alkaline), **253**
    colligative properties of, 237–238
    common types of, 12t
    concentration (units) of, 228–231, 274–276
    formation of, 223
    hypertonic, **241,** 241f
    hypotonic, **241,** 241f
    isotonic, **241,** 241f
    neutral, **252**
    properties of, 236t
    saturated and unsaturated, **224**
    supersaturated, 224–**225,** 225f
Solvation, 219–**220,** 220f
Solvent, **218**
Somatic damage, **304**
Specific gravity, **52**
Specific heat, **53**–56, 54t, 215–216
Specificity, **584**–585
Sphingomyelins, **532**–533, 532f
Sphingosine, 532, 532f
Spinnerettes, 445
Spleen, 800
Spontaneous abortion (miscarriage), 545
Standard temperature and pressure (STP), **184,** 185–186
Starch, **508**–510, 508f, 509f
Starvation
    emergency use of acetyl CoA, 774, 797
    glucose demand and, 732–733, 758, 764
    kwashiorkor and, 656
    marasmus and, 655, 656
Stearic acid, 424, 424t
Stereoisomers, 493–**494,** 496f, 556–557, 556f

Sterilization, 575–576
Steroid, **537**–541, 537f
Steroid hormones, **538**–541
Steroid ring system, 537f
Sterols, **537,** 537f
STM (scanning-tunneling microscope), 64c
Stoichiometry, **151**
Stomach, 646–647, 646f
Stone, Edward, 444c
Storage fat, **764**
STP (DOM), 480f
STP (standard temperature and pressure), **184,** 185–186
Straight-chain alkanes, **321**–324, 322f, 323t, 324f
Streptokinase, 600p
Stroke, 600
Strong, vs. weak, 260
Strong acids, **258**
Strong bases, **259**
Strontium ion, 117t
Strontium-90, 311
Structural formulas, **122, 318**
    of branched-chain alkanes, 328–330
    condensed, 321–**322**
    drawing, 321–322, 322f
    of straight-chain alkanes, 323t
Structural isomers, **330,** 342, 342f
Strychnine, 482, 482f
Subatomic particles, 64–65, 66t
Sublimation, **14**–15, 15f
Substituents, **325**
Substitution, aromatic, **363**–364
Substrate, **584,** 585–586, 585f, 586f
Substrate-level phosphorylation, 735
Subunits, **570**
Succinate, 739f, 741f
Succinic acid, 424
Sucrase, 647
Sucrose, 499, 511–513, 512f, 647
Sugars, **493**
    in blood, 515c, 575c
    cyclic, 501–502, 501f, 502f
    digestion of, 647
    hydrolysis of, 507
    invert, **512**
    nonreducing, **514**–515
    pectin, 499, 499f
    reducing, **514**–515, 514f
    simple, **497**–500
    sweetness of, 499
    wood, 499, 499f
    See also Carbohydrate(s); Glycogen
Sulfa drugs, 469c, 601, 603
Sulfanilamide, 469c, 601, 602
Sulfonamide antibiotics, 469c
Sulfur, 86f, 92, 100–101, 101f, 665t, 668
Sulfur compounds, 381–383
Sulfur-containing side chains, 555f
Sulfur dioxide, 101
Sulfuric acid, 101, 254, 258
Sumner, James B., 583
Supercoiled molecules, 565, 565f
Superconductivity, 121c
Supernovas, 93c

Superoxide anion, 718c
Superoxide dismutase, 718c
Supersaturated solutions, 224–**225,** 225f
Suprachiasmatic nucleus, 706c
Surface tension, **215,** 215f
Surfactants, **215**
Suspensions, **235,** 236t
Sutherland, Earl, 752
Sweating, 698
Symbols, chemical, 18–20, 18t, 19t
Syringe, 42, 42f
Système Internationale (SI), **31**–32, 31t. *See also* Metric system
Systolic pressure, 189c

T cells, 637p
Tartaric acid, 426
Tatum, Edward, 621
Taxol, 109f, 132f
Tay-Sachs disease, 766c, 767f
Technetium-99m, 308, 308f
Teeth, 255c, 657, 665–666, 670
Teflon, 102, 366
Teletherapy, 309–310, 309f
Tellurium, 100, 101
Temperature, 8–10
    of body, 142p, 164p
    enzymes and, 590, 590f
    equilibrium position and, 171
    gas pressure and, 190, 190f, 193–199, 195f, 196f, 198f
    kinetic energy and, 182
    reaction rate and, 163, 163f
    solubility and, 224
    solution formation and, 223
    standard, 184, 185–186
    units of, 32t
Temperature scales, 8, 9f, 10, 182
Temperature-pressure relationship at constant volume, **197**–199, 198f
Template, **617**
Termination, **631,** 631f
Termination codons, **631**
Tertiary butyl fluoride, 358
Tertiary structure of proteins, **567**–569, 567f, 568f, 569f
Terylene, 445
Testes, 540–541
Testosterone, **540,** 540f, 541
Tetracaine, 443c
Tetracyclines, 630
Tetraethyl lead, 97
Tetrahedron, **132**–133, 133f, 135, 135f
Tetrahydrocannabinol (marijuana), 383–384
Thalassemia, 573
Thalium, 97
Theory, **6**
Thermal pollution, 22
Thermolysin, 575
Thermometers, 9, 9f
Thermonuclear reaction, 300
Thiamine, 660
Thiamine pyrophosphate, 738f
Thioalcohols, **381**–382

Thioesters, **446**
Thioethers, **382**
Thiols, **381**–382, 446
Thoracic cavity, 191c
Three Mile Island, 301c
Threonine, 554f
Threose, 498, 498f
Thrombin, 584, 597
Thrombocytes, **679**
Thymine, 609, 610f
Thymol, 370f, 371c
Thyroid gland, 287p, 307p, 669
Thyroxine, 287p, 669, 669f
Time, units of, 32t
Tin, 97, 665t, 670
Tissue lipid, **764**
Tissue plasminogen activator (TPA), 600p, 637
Titration, **276**–278, 277f
α-Tocopherol, 528, 658, 658f
Tolbutamide, 758f
Tollen's reagent, **406**–407, 514
Toluene, 343c, 370f, 371c
Toluidines, 462–463, 463f
Torr, **183**
Torricelli, Evangelista, 183
Total-body dose, 305
TPA (tissue plasminogen activator), 600p, 637
Trace elements, 670
Tranquilizers, 485
Trans configuration, **337,** 337f
Transaminases, 588t, 789
Transamination, **788**–790
Transcription, **616,** 625–627
Transfer RNA (tRNA), **624,** 625, 625f, 627–631
    activated, **627**–628, 628f
    binding sites, **628,** 629f
Transition metals, **89,** 103, 103f
Translation, **616,** 627–631
    initiation of, **628**–629, 629f
    termination of, **631,** 631f
Transmutation, **288**
Tretinoin, 421p, 425p
Trichinella spiralis, 292c
Tricyclics, 478
Triglycerides, 440, **524**–529, 534f
    as adipose tissue, 764
    bonding, 524f, 525f
    digestion of, 648–650, 650f
    hydrogenation, 527
    mixed, **525**
    rancidity, 528
    saponification, **528**
    simple, **525**
    synthesis in body, 797
Triolein, 527f
Tripalmitin, 649, 649f
Triphosphate, 618f
Triple covalent bonds, **123**
Tristearin, 527f, 528f
Tritium, 71
tRNA. *See* Transfer RNA
Tropane ring system, 483, 483f

Trypsin, 584, 598–599c, 603c, 647
Trypsinogen, 595, 595t, 596, 597
Tryptophan, 554f
Tumor suppressor genes, 634c
Tungsten, 103
Tyndall effect, **236,** 236f
Tyrosine, 554f, 798, 798f

UDP glucose, 748f, 749f
Ulcers, peptic, 647
Unit-factor method, 43–44
Unsaturated compounds, **334**–338, 340–341
Unsaturated fatty acid(s), 424, 425t
Unsaturated fatty acid chains, 535, 535f
Unsaturated hydrocarbons, **334**–338, 340–341
Unsaturated solution, **224**
Unshared pairs, **123,** 135, 135f
Unwinding proteins, **620**
Uracil, 609, 610f
Uranium-235, 298–299, 299f
Uranium-238, 296, 296f
Urea, 484, 484f, 584, 690
Urea cycle, **791**–795, 791f
Urease, 584
Ureters, 690, 690f
Urethra, 690, 690f
Uric acid, 803, 804f, 804p
Uridine diphosphate (UDP) glucose, 748f, 749f
Uridine triphosphate (UTP), 748f
Urinary system, 690–692, 690f, 692f
Urinometers, 52c
Urobilin, **802**
Urobilinogen, **802**
UTP (uridine triphosphate), 748f

Valence electrons, **110,** 317–318
Valine, 553f
van der Waals, Johannes, 130
van der Waals forces, **130**
Vanadium, 91, 665t, 670
Vanillin, 411, 411f
Vapor, **11**
Vapor pressure, **204,** 215, 237
Vaporization, **203**–204, 204f
    heat of, **205,** 216
Variable regions, 683, 683f
Vasopressin, 560–561, 561f, 697
Veins, 681
Veronal, 485
Very-low-density lipoproteins (VLDLs), 651c, 780
Vinblastine, 169c
Vincristine, 169c
Virilization, 541
Viruses
    interferons and, 685
    retroviruses, **616**–617, 619c
Vision, chemistry of, 412–413c
Vitamin(s), **645**
    fat-soluble, 656–659
    water-soluble, 659–663

Vitamin A, 413c, 656–657
Vitamin B$_1$, 660
Vitamin B$_2$, 661
Vitamin B$_6$, 661, 661f
Vitamin B$_{12}$, 662–663, 663f
Vitamin B-complex group, 660–663
Vitamin C, 528, 659–660
Vitamin D, 657–658
Vitamin E, 528, 658
Vitamin K, 658–659
VLDLs (very-low-density lipoproteins), 651c, 780
Volume, **40**
    devices for measuring, 42, 42f
    units of, 32t, 40–44, 41f, 41t
Volumetric flask, 42, 42f

Warfarin, 658, 659f
Water, 213–219
    in body, 213, 677–679, 696–698
    body's need for, 645
    boiling point of, 216, 216t
    bonds in, 129f, 130
    colloids, 235–236, 235t, 236t
    density of, 217t
    distillation of, 12, 13f
    for drinking, 213p, 219p
    emulsions, 236–237
    equilibrium constants for, 251–252
    formula for, 112
    glucose vs., 144t
    hard, **433,** 434c
    heat of vaporization of, 216
    heavy, 71
    hydrogen bonds in, 214–216, 214f
    hydrogen ions from, 251–253
    ice, 217–218, 217f
    molecular shape of, 131f, 135, 135f
    polar bonds in, 214, 214f
    self-dissociation of, **251**–253
    softening, 434c
    as solvent, 218
    specific heat of, 215–216
    surface tension of, 215, 215f
    suspensions, 235, 236t
    vapor pressure of, 215
Water of hydration (crystallization), **221**–222
Water-soluble vitamins, 659–663
Watson, James D., 614, 615, 617, 618, 620
Watson-Crick double-helix model, 614–615
Waxes, **524**
Weak, vs. strong, 260
Weak acids, **259**
Weak attractive forces, 129–130, 130f
Weak bases, **259**
Weight, **47**
    ideal, 525p
White blood cells, 679, 683–685
Wilkins, Maurice, 615
Wilson's disease, **669**
Wood alcohol (methanol), 368–369, 370, 404–405
Wood sugar, 499, 499f
Working lipid, **764**

X ray, 288t, **291**–292, 291f, 306, 306f
X-ray crystallography, 568–569, 569c, 569f
Xanthine oxidase, 804p
Xenon, 94–95
Xylenes, 342
Xylocaine, 471c
Xylose, 499, 499f

Yttrium, 103

Zero, absolute, **8,** 182
Zero-serotonin-uptake drugs, 478, 479p
Zidovudine (AZT), 619c
Zinc, 103, 665t, 668
Zoloft, 479

Zwitterions, **557**–558, 557f
Zymogen(s), 595t, **595**–597
Zymogen granules, **595**–596